Recent Advances in Material, Manufacturing, and Machine Learning

Proceedings of 1st International Conference (RAMMML-22), Volume 1

About the conference

The international conference on recent advances in material, manufacturing, and machine learning processes-2022 (RAMMML-22), is organized by Yeshwantrao Chavan College of Engineering, Nagpur, Maharashtra, India during April 26–27, 2022. The conference received more than 640 papers. More than 177 papers were accepted and orally presented in the conference.

The role of manufacturing in the country's economy and societal development has long been established through their wealth generating capabilities. To enhance and widen our knowledge of materials and to increase innovation and responsiveness to ever-increasing international needs, more in-depth studies of functionally graded materials/tailor-made materials, recent advancements in manufacturing processes and new design philosophies are needed at present. The objective of this conference is to bring together experts from academic institutions, industries and research organizations and professional engineers for sharing of knowledge, expertise and experience in the emerging trends related to design, advanced materials processing and characterization, advanced manufacturing processes.

The conference is structured with plenary lectures followed by parallel sessions. The plenary lectures introduces the theme of the conference delivered by eminent personalities of international repute. Each parallel session starts with an invited talk on specific topic followed by contributed papers. Papers are invited from the prospective authors from industries, academic institutions and R&D organizations and from professional engineers. This conference brings academicians, industrial experts, researchers, and scholars together from areas of Mechanical Design Engineering, Materials Engineering and Manufacturing Processes. The topics of interest includes Design, Materials and Manufacturing engineering and other related areas such as Mechatronics, Prosthetic design and Bio inspired design and Smart materials. This conference is to provide a platform for learning, exchange of ideas and networking with fellow colleagues and participants across the globe in the field of Mechanical Design, Materials and Manufacture.

This conference RAMMML-2022 have paved the way to understand the latest technological and innovative advancements especially in the fields of manufacturing, design and materials engineering. The conference has developed the solutions to physical problems, questions how things work, make things work better, and create ideas for doing things in new and different ways in the manufacturing, design and materials engineering. The conference focuses on the frontier themes of recent advances in manufacturing, design and materials engineering, as applied to multiple disciplines of engineering. Researchers, Academicians, Industrialist and Students has benefitted with the latest trends and developments in design, manufacturing and materials engineering applied to various disciplines of engineering. The objective of the conference is to have the orientation of research and practice of professionals towards attaining global supremacy in manufacturing, design and materials engineering. Also this conference aims in understanding the recent trends in manufacturing, design and materials engineering including optimization and innovation. This conference RAMMML-2022 also aims in improving Research culture in the minds of faculty in exploring the knowledge base, establishing better insights and maintaining dynamism in the teaching - learning process.

Recent Advances in Material, Manufacturing, and Machine Learning

Proceedings of 1st International Conference (RAMMML-22), Volume 1

Edited by

Dr. Rajiv Gupta
North Carolina State University, Raleigh, NC, United States
ORCID 0000-0003-2684-1994

Dr. Devendra Deshmukh
Indian Institute of Technology, Indore
ORCID 0000-0003-4636-3301

Dr. Awanikumar P. Patil
Visvesvaraya National Institute of Technology, Nagpur
ORCID 0000-0002-4511-7558

Dr. Naveen Kumar Shrivastava
Birla Institute of Technology and Science, Pilani, Dubai (UAE)
ORCID 0000-0002-5773-473X

Dr. Jayant Giri
Associate Professor, Department of Mechanical Engineering, Yeshwantrao Chavan College of Engineering, Nagpur
ORCID 0000-0003-4438-2613

Dr. R.B. Chadge
Assistant Professor, Department of Mechanical Engineering, Yeshwantrao Chavan College of Engineering, Nagpur
ORCID 0000-0001-9072-0607

CRC Press is an imprint of the
Taylor & Francis Group, an **informa** business

First edition published 2023
by CRC Press
4 Park Square, Milton Park, Abingdon, Oxon, OX14 4RN

and by CRC Press
6000 Broken Sound Parkway NW, Suite 300, Boca Raton, FL 33487-2742

British Library Cataloguing-in-Publication Data
A catalogue record for this book is available from the British Library

Library of Congress Cataloging-in-Publication Data

ISBN: 9781032416311(pbk)
ISBN: 9781003358596 (ebk)
DOI: 10.1201/9781003358596

Typeset in Sabon
by HBK Digital

Table of Contents

List of Tables and Figures

Tables

Figures

Foreword

India as a growing economy has made its significant impact on global affairs in the last few decades. The country has survived many economic, social and political challenges in the past and is determined to become a stronger economy in coming future. The path of this growth is full of challenges of Material research, manufacturing technologies, design and development while ensuring inclusive growth for every citizen of the country. The role of manufacturing in the country's economy and societal development has long been established through their wealth generating capabilities. To enhance and widen our knowledge of materials and to increase innovation and responsiveness to ever-increasing international needs, more in-depth studies of functionally graded materials/ tailor- made materials, recent advancements in manufacturing processes and new design philosophies are needed at present. The objective of this conference is to bring together experts from academic institutions, industries and research organizations and professional engineers for sharing of knowledge, expertise and experience in the emerging trends related to design, advanced materials processing and characterization, advanced manufacturing processes. Since its inception, Yeshwantrao Chavan College of Engineering, Nagpur has been working as an institution committed to contribute in the field of Materials, design, automation and sustainable development through its various academic and non-academic activities. Different departments of YCCE have been conducting various academic conferences, seminars and workshops to discuss on contemporary issues being faced by Design, development and material research sectors and bring together academicians, researchers, technologists, industry experts and policy-makers together on common platforms.

Yeshwantrao Chavan College of Engineering, Nagpur is the pioneer in organizing academic conferences to share and disseminate academic research in the field of engineering & technology with the industry and policy makers since its establishment (1984).The department of mechanical engineering, YCCE is organizing its 1st international conference on recent advances in material, manufacturing, and machine learning processes-2022 (RAMMML-22), April 26–27, 2022 with an objective & scope to deliberate, discuss and document the latest technological and innovative advancements especially in the fields of manufacturing, design and materials engineering.

It was good news to hear from the Conference Organizing Committee that they have received more than 640 papers on different topics related to different subthemes of the conference. These papers cover topics related to manufacturing, design, materials engineering, machine learning, simulation, civil engineering and many more interdisciplinary topics. We believe that this conference will become a common platform to disseminate new researches done by various researchers from different universities/ institutes before industry professionals and policy makers in the government. We hope that these new researches will suggest new directions to innovations in Material research, design, development, manufacturing industries and government policies pertaining to these sectors.

We wish all the best to our conference participants who are the real knowledge champions of their universities/institutes/organizations. We strongly believe that all of us at YCCE will make this conference a good experience for every participant of the conference and this conference will achieve its objectives effectively.

We welcome you all to RAMMML-2022.

Patrons

Chief Patron	Hon'ble Shri. Dattaji Meghe	Chairman, Nagar Yuwak Shikshan Sanstha, Founder Chancellor of Datta Meghe Institute Of Medical sciences.
Chief Patron	Hon'ble Shri. Sagarji Meghe	Secretary, Nagar Yuwak Shikshan Sanstha
Chief Patron	Hon'ble Shri. Sameerji Meghe	Treasurer, Nagar Yuwak Shikshan Sanstha
Chief Patron	Hon'ble Mrs. Vrinda Meghe	Chief Advisor, Nagar Yuwak Shikshan Sanstha
Chief Patron	Hon'ble Dr. Hemant Thakare	COO: Ceinsys Tech. LTD, President, IEI, India
Patron	Dr. U.P. Waghe	Principal, Yeshwantrao Chavan College of Engineering, Nagpur
Patron	Dr. Manali Kshirsagar	Principal, Rajiv Gandhi College of Engineering, Nagpur

Preface

The main aim of the 1st international conference on recent advances in material, manufacturing, and machine learning processes-2022 (RAMMML-22) is to bring together all interested academic researchers, scientists, engineers, and technocrats and provide a platform for continuous improvement of manufacturing, machine learning, design and materials engineering research. RAMMML 2022 received an overwhelming response with more than 640 full paper submissions. After due and careful scrutiny, about 177 of them have been selected for presentation. The papers submitted have been reviewed by experts from renowned institutions, and subsequently, the authors have revised the papers, duly incorporating the suggestions of the reviewers. This has led to significant improvement in the quality of the contributions, Taylor & Francis publications, CRC Press have agreed to publish the selected proceedings of the conference in their book series of Advances in Mechanical Engineering and Interdisciplinary Sciences. This enables fast dissemination of the papers worldwide and increases the scope of visibility for the research contributions of the authors.

This book comprises four parts, viz. Materials, Manufacturing, Machine learning and interdisciplinary sciences. Each part consists of relevant full papers in the form of chapters. The Materials part consists of chapters on research related to Advanced Materials, Ceramics, Shape Memory Alloys and Nano materials, Materials for Aerospace applications, Polymers and Polymer Composites, Glasses and Amorphous Systems, Material characterization and testing, MEMS/NEMS, Bio Materials, Optical/Electronic Materials, Magnetic Materials, 3D Materials, Cryogenic Materials, Materials applications, performance and life cycle etc. The Manufacturing part consists of chapters on Micro/Nano Machining, Metal Forming, Green Manufacturing, Non-Conventional Machining Processes, Additive Manufacturing, Subtractive Manufacturing, Industry 4.0, Sustainable Manufacturing Technologies, Casting Technology, Joining Technology, Plastic processing technology, CAD/CAM/CAE/CIM/HVAC, Product Design and Development, Multi Objective Optimization, Modelling, Analysis and Simulation, Process Monitoring and Control, Vibration Noise Analysis and Control, Thermal Optimization, Energy Analysis etc. The Machine learning part consists of chapters on Machine learning, knowledge discovery, and data mining, Artificial intelligence in biomedical engineering and informatics, Artificial neural networks and algorithms, Knowledge acquisition, representation and reasoning methodologies, Genetic algorithms, Probability-based systems and fuzzy systems, Healthcare process management, Imaging, signal processing and text analysis, Bioinformatics and neurosciences. And the Interdisciplinary part consists of chapters on Condition Monitoring, NDT, Soft Computing, VLSI, Embedded System, Computer Vision, Environment Sustainability, Water Management, Advanced Mechatronics System and Control, Structural and Geo-technical Engineering areas. This book provides a snapshot of the current research in the field of Materials, Manufacturing, Machine learning and interdisciplinary sciences and hence will serve as valuable reference material for the research community.

Details of programme committee

International Advisory Committee

S.No	*Name*	*Details*
1.	Dr. Rajiv Gupta	North Carolina State University, Raleigh, NC, United States
2.	Mr. James Barret	HE Lecturer, Turno College, Plymouth University, England
3.	Abhiram Dapke	Deuce Drone, Albama, USA
4.	Nakul Vadalkar	T&I, Celanese, Germany
5.	Mr. Laxmikant Kolekar	Assistant General Manager-Operations Alam Steel, Al Salmiyah Hawalli, Kuwait
6.	Dr. Naveen Kumar Shrivastava	Birla Institute of Technology and Science, Pilani, Dubai (UAE)
7.	Mr. Aniket Mandlekar	Field Engineer, Intertape polymer group, Canada

National Advisory Committee

S.No	*Name*	*Details*
1.	Dr. Prashant P. Datey	Indian Institute of Technology, Bombay
2.	Dr. Milind Atre	Indian Institute of Technology, Bombay
3.	Dr. Rakesh G. Mote	Indian Institute of Technology, Bombay
4.	Dr. Harekrishna Yadav	Indian Institute of Technology, Indore
5.	Dr. I. A. Palani	Indian Institute of Technology, Indore
6.	Dr. Anand Parey	Indian Institute of Technology, Indore
7.	Dr. Jitendra Sangwai	Indian Institute of Technology, Madras
8.	Dr. Devendra Deshmukh	Indian Institute of Technology, Indore
9.	Dr.Rajesh Ranganathan	Coimbatore Institute of Technology, Coimbatore, Tamilnadu
10.	Dr. Anupam Agnihotri	Director, Jawaharlal Nehru Aluminium Research Development And Design Centre, Bombay
11.	Dr. P. M. Padole	Director, Visvesvaraya National Institute of Technology, Nagpur
12.	Dr. Awanikumar P. Patil	Visvesvaraya National Institute of Technology, Nagpur
13.	Dr. Vilas R. Kalamkar	Visvesvaraya National Institute of Technology, Nagpur
14.	Prof. G. S. Dangayach	Malaviya National Institute of Technology, Jaipur
15.	Dr. Y. M. Puri	Visvesvaraya National Institute of Technology, Nagpur
16.	Dr Rakesh Shrivastava	Former Professor of Mechanical Engineering, Consultant and Trainer, Ex Chairman IEI, Nagpur Chapter
17.	Dr. D. M. Kulkarni	Birla Institute of Technology and Science, Goa
18.	Dr. B. Rajiv	College of Engineering, Pune
19.	Dr. T. N. Desai	Sardar Vallabhbhai National Institute of Technology, Surat
20.	Mr. Saquib Anwar	Manager BEL, Bombay
21.	Dr. Prakash Pantawane	College of Engineering, Pune
22.	Dr. Nalinaksh S. Vyas	Indian Institute of Technology, Kanpur

Conference Chair & Organizing Secretary

S.No	*Commitee*	*Name*	*Details*
1.	Conference Chair	Dr. J.P. Giri	Head of The Department, Department of Mechanical Engineering, Yeshwantrao Chavan College of Engineering, Nagpur
2.	Conference Chair	Dr. R.B. Chadge	Department of Mechanical Engineering, Yeshwantrao Chavan College of Engineering, Nagpur
3.	Organizing Secretary	Dr. S.R. Jachak	Department of Mechanical Engineering, Yeshwantrao Chavan College of Engineering, Nagpur
4.	Organizing Secretary	Dr. S.P. Ambade	Department of Mechanical Engineering, Yeshwantrao Chavan College of Engineering, Nagpur

Contact Persons

Prof. A.P. Edlabadkar

Prof. Neeraj Sunheriya

Organizing Committee Members

Prof. D. I. Sangotra
Prof. N. J. Giradkar
Prof. V. M. Korde
Prof. A. S. Bonde
Dr. S. T. Bagde
Dr. S. S. Chaudhari
Dr. S. V. Prayagi
Dr. A. P. Kedar
Prof. V. G. Thakre
Prof. M. S. Tufail
Prof. P. N. Shende
Prof. A. B. Amale
Prof. G. H. Waghmare
Prof. R. G. Bodkhe
Prof. D. Y. Shahare
Prof. A. P. Edlabadkar
Dr. S. S. Khedkar
Dr. P. D. Kamble
Prof. R. V. Adakane
Prof. D. N. Kashyap
Prof. A. R. Narkhede
Prof. S. P. Kamble
Prof. Y. Y. Nandurkar
Prof. M. M. Dakhore
Prof. P. S. Barve
Prof. N. D. Gedam
Prof. C. A. Mahatme
Prof. P. V. Lande
Prof. G. M. Dhote
Dr. V. R. Khawale
Prof. P. A. Hatwalne
Prof. Dipak M. Hajare
Prof. Praful Shirpurkar
Prof. Ritu Shrivastava
Prof. Sujata A Kimmatkar
Prof. Albela H. Pundkar

1 An investigation on waste plastic materials for hydro carbon fuel production using alternative energy sources

Lavepreet Singh[1,a,], Kaushalendra Dubey[1,b], Shreyansh Gupta[2,c], and Rahul Katiyar[2,d]*

[1]Assistant Professor, Department of Mechanical Engineering, Galgotias, University Greater Noida, Uttar Pradesh, India

[2]Student, Department of Mechanical Engineering, Galgotias University, Greater Noida, Uttar Pradesh, India

Abstract

Plastic is the most often used material in everyday life due to its light weight, durability, and flexibility. There is a deposit of polymer that has incorporated strongly in plastic, causing recycling problems. Plastic degradation has a significant influence on the environment and plants. Plastic pyrolysis is the most effective method for converting wasted solid plastics into hydrocarbon fuels. The materials from spilled plastic and commercial methods for fuel generation were discussed in the present work. The use of alternative energy systems such as solar thermal collectors are able to provide the temperature for solid plastic boiling while reducing the need of high-grade energy in the form of electricity. The solar integrated fuel system helps to reduce hazardous emissions and map the path to a de-carbonized economy.

Keywords: waste plastic, hydrocarbon fuel, plastic pyrolysis, solar thermal collectors, high grade energy.

Introduction

Nowadays plastic is very essential in day-to-day life and its use in the industrial field is steadily increasing. The production and consumption of plastic are increasing at alarming rates as the human population grows, the economy expands, cities continue to grow, and lifestyles change. On the Global basis, plastic production is approximate to be around 300 million tonnes/year, and it is steadily increasing. The vast majority of unused plastic has ended up in landfills; Because of its Plastic waste has caused major environmental concerns due to its vast volumes and disposal issues. In the presence of a catalyst, reprocessed unused plastics are expected to be the most effective method of recovering and using. Significant research have been found by the Ioannis Kalargaris and Guohong Tian.,Shikui Wu, Kaixiong Xu done experiment using from (PEVA)Plastic, where they input as (LDPE) and (EVA) for locating oil during which ensures that the engine runs smoothly Kalargaris et al. (2017). It is observe that trash plastic fuel has properties comparable to fossil fuel like diesel, petrol, natural gas. Farag and Korachy (2017); Wu et al. (2014) experimented with the technique of cracking mixtures of different hydrocarbon feedstock in a steam pyrolysis furnace, the application of trash plastics and petro-oils, is 402°C with a pressure of about 2.0 MPa(max.), and it was discovered that trash plastic and petro-oil cracking has a better efficiency in I.C engines than petroleum Ogunbiyi et al. (2016) (Figure 1.1).

Plastic Waste Is a Serious Issue: Egypt

As per supply chain mapping and evaluation of plastic, "Egypt engenders around twenty million loads of garbage and waste annually, with plastic waste postulated to represent 6 June 1944 out of the entire, distributed over Cairo (60%), Alexandria (16%), the river Delta (19%), and different regions together with Upper Egypt, Suez Canal, and Sinai (5%). Out of the 970 kilotons of plastic waste engendered annually, solely a variety of half-hour is recycled, whereas five-hitter is reused, thirty third is land filled, and thirty second is left to be burned". Overall amplitude of waste of the plastic accounts for 100% of all trash in EGYPT. One-fifth of the total amount of plastic is not mobilized or land filled. This amounts to 1.3 million tonnes in Egypt each year, with Cairo contributing only 0.78 million tonnes. When plastic garbage is burned, it emits hazardous dioxins (very poisonous compounds) that could possibly be ingested by Homo-sapiens, collected in soil and on plants, and collected in surface dihydrogen oxide. Remaining/uncollected

[a]punstu@gmail.com; [b]dubey.kaushalendra@gmail.com; [c]shreyansh_gupta.scmebtech@galgotiasuniversity.edu.in; [d]rahul.katiyar.5264@gmail.com

plastics endanger animals and ocean life Gnansounou and Raman (2019). In the Middle East, there are various effective waste management strategies in use. This investigation focuses on the utilization of plastic trash for generation of fuel. The current approach reduces the amount of solid trash for landfills and the amount of carbonic acid gas emissions caused by plastic garbage in Egypt by roughly 8 May 1945 for the primary year, and by half an hour for the primary 5 years. Furthermore, it may reduce the high demand. Surprisingly, the carbon discharges by this plastic waste based fuel are 93% lower than those generated by fossil fuel like petrol and diesel. The catalyst of the plastic waste is available by using transformation method, some of the catalyst mixture of minerals like mineral, clay, alumina, and silicates in various proportions Syamsiro et al. (2014); Fahim et al. (2021). The three forms of transformation reactions are slow, timesaving and flash shift. The time interval and temperature of the biomass are distinguishing factors. Patni et al. (2013). The primary byproducts of this method are char and gas. Temperature and heating rate, pressure, and duration all influence the proportion of the byproduct. Egypt (2019).

The feasibility of using trash to create a successful fuel processing system by combination of several synthetic polymer matrix composites with various functions from (HDPE), (PET), (PS), and (PP) is calculated in this study. This work successfully disposes of the trash of these undesirable materials which are not easily degradable synthetic polymers and have negative impact on environment & health by converting them to fuel via pyrolysis. This process would be extremely beneficial to the fuel industry because it would make the process more long-lasting. In this context, fuel and bio refinery study will be crucial in future.

Industrial Pyrolysis Process

At first, the plastic trash goes through process like cleaning, drying, and reducing in size; rapid pyrolysis process is available for fuel extraction from unused plastic plastics like HDPE, LDPE, and PET. Food and mouthwash containers include polyethylene terephthalate. Polypropylene can be found in plastic containers like food storage, yogurt, parts of automotive, thermal vests, and diapers disposals etc. Polystyrene can be found in food containers, utensils, and straws etc. To limit the amount of plastic in the reactor, the waste is shredded and crushed into little pieces (1×3 cm^2). To remove any potentially dangerous materials, the plastic parts are washed. The required catalyst is Zeolite Socony Mobil-5 (ZSM-5), a high-silica zeolite that is commonly used in the petroleum industry as a heterogeneous catalyst for hydrocarbon isomerization reactions. The employment of a catalyst is required. However, before breaking, it should be dried in an oven to reduce the moisture content to less than 5.01%, resulting in smaller particle size and a faster reaction rate Juwono et al. (2019). The ratio of plastic to catalyst is 20:2. The inclusion of the catalyst reduces both the process time and the pyrolysis temperatures, resulting in higher conversion rates across a wide range of products. The pyrolysis of low- & high-density polyethylene attempt under thermal pyrolysis environment this process minimizes the process time and improves fuel quality and possibility of wide range of plastic fuel.

Liquid Fuel Synthesis

Before being fed into the pyrolysis reactor, the feedstock is shredded and precisely blended with the catalyst in specific condition which is available in Table 1.1. The reactor is made of stainless steel and has a fixed bed. The heat is supply at rate of 15°C/minutes to 550°C/minutes. During the pyrolysis process, vapour are created first, which are collected in the form of gas in collection chamber, it gets converted into oil after pyrolysis process Khan et al. (2016). Condensation of these compounds yields in the form of fuel,

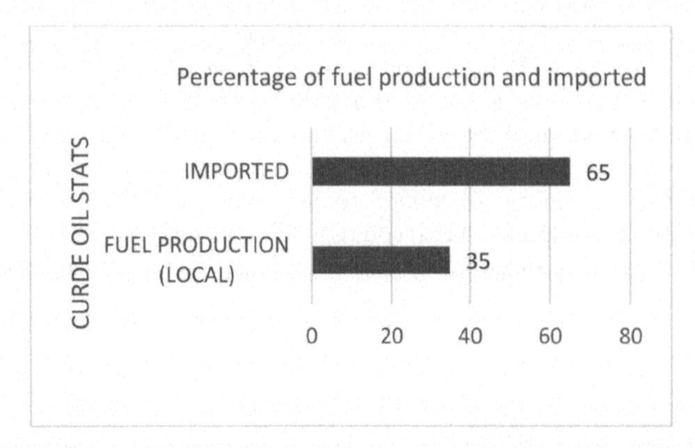

Figure 1.1 Crude oil states (fuel production and import) in Egypt

kerosene and propane gas etc. At refineries, the previously described catalyst is utilized to convert biofuel oil to gasoline and diesel. Ships are supplied with heavy oil. One of the most essential factors in assessing a fuel's efficiency is its heat value. The characteristic difference between diesel, kerosene, bio diesel as shown in Table 1.2. The heat value of the fuel engendered from plastic trash was calculable consistent with the IP 12/58 methodology. Its heat value was 9830 kcal/kg as shown in Figure 1.2. that is proximate to the calorific value of diesel Juwono et al. (2019); Khan et al. (2016); Rosendahl (2017).

Disposing of Plastic Waste Is a Major Issue

Toxic fugitive emissions are eliminated during the polymerization process.

- Burning of plastic garbage in the open is a common occurrence in cities and towns, resulting in toxic emissions such as CO, Cl, HCL, C8H8, C6H6, C4H6, CCL, and C2H4O etc. are polluting the environment.
- Unrecyclable plastic trash, such as metalized pouches and multi-layered sachets, as well as thermoset polymers like SMC/FRP, provides a number of disposal issues.
- Garbage mixed with discarded plastic which clogs recycling and solid waste processing equipment and causes problems in recycling operations.
- According to the centre, over 34 lakh tonnes of plastic garbage were generated in 2019–20, compared to 30.59 lakh tonnes in 2018–19.
- In the last five years, India's plastic trash creation has more than doubled, with an average yearly rise of 21.8%.

Table 1.1 Conditions of catalyst

An Raw material process type	*Raw material quantity(g)*	*Catalyst quantity(g)*	*Raw material ratio*	*Reaction Temp. (C)*	*Heating rate (C/min)*	*Efficiency*
HDPE	5000	500	100%	550°	15	94%
PET	5000	500	100%	550°	15	70%
PS	5000	500	100%	550°	15	80%
Polypropylene	5000	500	100%	550°	15	60%
PP/PET	5000	500	50–50%	550°	15	67%
PS/PP	5000	500	50–50%	550°	15	75%
Mixed	5000	500	25% each	550°	15	85.6–89.5%

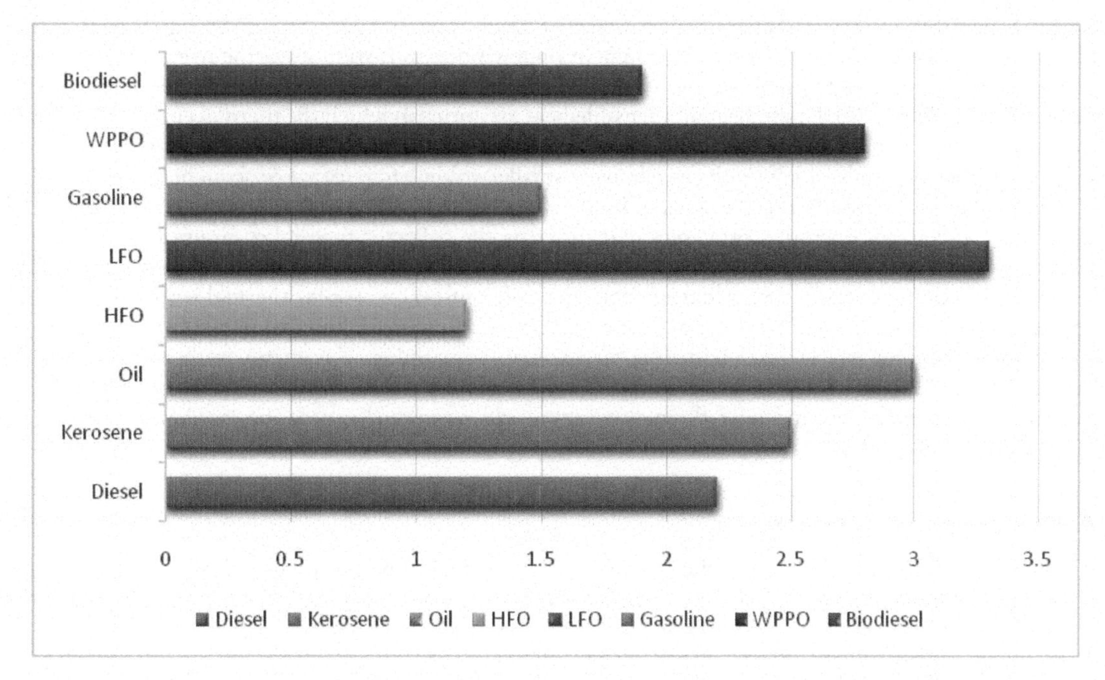

Figure 1.2 Characteristic difference of waste plastic fuel and other fuel Rosendahl (2017)

Demand of Fuel

The current rate of economic expansion is unsustainable unless fossil fuel is conserved. Hahladakis and Velis (2018) reviewed an overview of chemical composition of plastics, its release in environment effect on food chain and environmental impact and its mitigation during use recycling and disposal and found that disposal methods are more preferable by converting into another form of energy.

Technical Detail

In the plastic pyrolysis process the major factor influencing the molecular distribution are Chemical compositions of plastic operating pressure rate of heating temperature of cracking types of catalyst used and type of reactor use in this process. Plastic liquid hydrocarbons at normal temperature and pressure are referred to as liquid fuel. PP, PS, and PE are being prepared as feedstock for the production of liquid hydrocarbons. Plastics are converted to liquid fuel through pyrolysis and condensation of the resultant hydrocarbon. The temperature required for transformation of plastic trash into fuel is 450C°–550C° for liquid fuel production process. The fuel result (a mixture of liquid hydrocarbons) is continuously refined once when the reaction temperature is attained, the plastics trash inside the reactor has tainted adequately to evaporate. A catalyst is used to further crack the evaporated oil; the reactor's hydrocarbons are distilled. Some high-boiling-point hydrocarbons, such as A condenser is a device that condenses kerosene, diesel, and gasoline in water. The liquid collected in the collecting chamber contains, wax, grease, and other impurities; which must be removed using filtration process therefore, the impurities must be removed. In the filtration process, colloidal substances can be removed using filter paper; the filter paper will allow molecules smaller than its pores to pass through, resulting in cleaner fuel. The purified fuel is tested to determine its chemical composition which interprets the quality and properties of the fuel. Roopa and Ravishankar (2016) performed the various experiment on in PP and LLDPE conversion to fuel using clay catalyst with mixing of Bentonite catalyst with polypropylene (PP) and linear low density polyethylene (LLDPE) in a batch reactor in a feed ratio of 2:8 for getting fuel oil on temperature 440–455°C for time three hours & conclude that physical properties using Penske Martin Apparatus of oil samples such as specific gravity, density, fire point and flash pointGhosh and Di Maria (2018) numerous concerns and challenges linked with bio waste management were explored.

Solar Thermal Energy Conversion Systems-Collectors Material, Application and Research Challenges

Energy from the sun is used by the solar thermal system to increase the temperature of the heat carrying fluid which in turn is transferred to another medium of the system or process. Fossil fuels can be replaced by this type of renewable energy solutions and this can prove to be commercially economic alternative for energy generation in presence of competitive market Shahsavari and Akbari (2018); Singh et al. (2021). However, economic feasibility for the development of new projects is still low as the irregular solar energy necessitates an auxiliary thermal storage system which needs abundant investments to be made possible Gautam and Saini (2020). Latent heat and sensible heat are forms in which the thermal energy is stored but the phase change materials (PCM) that use latent heat mechanism, the energy density is greater than that in heat sensible materials which results in relatively smaller volume for storage Schoeneberger et al. (2020).

Solar Thermal Collectors are the devices used for changing solar energy into thermal energy by the application of various operating principles that differ by the collector type taken in use Evangelisti et al. (2019). Availability of space, degree of maturity of the analysed technology, energy requirement and target temperature are the crucial aspects that are considered for installing a solar thermal system in industrial applications Ghazouani et al. (2020); Singh et al. (2021). Here, various low and medium temperature collectors are mentioned which are classified on the basis of use in industry and temperature ranges. Existing method is available which integrated high grade energy source with more than 5 Kw power supply. The present research tittle describe solar thermal energy based waste fuel generation system, so solar energy waste system is available.

A. Low temperature collectors

- Flat Plate Collector (FPC): The thermal energy is intercepted by the plate following which the heat carrying fluid circulates via tubes and reaches the storage system if it is not used directly Hussain et al. (2018).
- Vacuum Tube Collector or evacuated tube collector (ETC)
- Compound Parabolic Collector (CPC)

B. Medium temperature collectors

- Evacuated Flat Plate Collector (EFPC)
- Parabolic Cylinder Collector (PTC)
- Linear Fresnel Collector (LFC)

C. Concentrated collectors

Table 1.2 Simulation of solar collectors with ETC and FPC technologies for industrial purposes and technical, environmental and economic aspects of the application

City/Country located in	*Industries using solar collectors*	*Temperature (°C)*	*Area captured by the collector (m^2)*	*Solar percentage (%)*	*Total fuel saving*	*Reference*
South Africa	Fish Flour Preheating Production	70	384-Flat Plate Collector	81	32061 Litres of fuel oil	Oosthuizen et al. (2020)
Macedonia	Saline Solutions Preparation, Molasses and Ethanol Water Heating Production	95	n/a-Evacuated Tube Collector	n/a	approx. 57 % of fuel oil	Anastasovski (2021)
Reunion Island	Direct Integration of solar collector in the boiler tank for Yogurt-vapour generation	160–170	555-Evacuated Tube Collector	n/a	24% of Fuel	Maillot et al. (2019)
Morocco	Cleaning and drying of fruits, Milk-water heating, cooling and pasteurization	60–90	400-Evacuated Tube Collector	41	77.23 tCO2e/ year	Allouhi et al. (2017)
Ethiopia	Heating and Dying of Clothes-Water	50–90	472-Evacuated Tube Collector	56.3	252.2 tCO2e/ year	Tilahun et al. (2019)

Solar Thermal Technique For Plastic Waste In A Fuel Production Domestic Use

The most known application of solar technology is (PV- photovoltaic) panels, but solar thermal technology is another, often ignored, use of solar energy in low-temperature applications. Solar thermal is most commonly utilized for direct solar water heating and disinfection of water sources stored in dark or translucent containers to be heated in its most basic form. (Solar Cookers International, n.d.), Solar cooking appliances have been used since the late 1700s, because to the rising usage of glass (Solar Cookers International, n.d.). Horace de Saussure, the original conceptualizer came with the most interested design of solar cooker known as 'hot boxes' (solar cookers international, n.d.). Insulated boxes with clear glass coverings and dark metal interiors were used to make these solar-heated containers. Enthused by Saussure Sir John Herschel, In the 1800s, hot boxes were enhanced built one during a South African expedition, but using sand loaded up against the sides of the box. He is reported to have exhibited the capacity to cook an egg in a way similar to today's solar ovens that hold heat and can store heat by relying on the surrounding insulation of bricks and sand. All of these basic solar cookers were able to achieve temperatures over the boiling point, flagging the door for future solar cooker advancements. In poor areas, solar cookers are beneficial, where organizations like CEDESOL in Bolivia and Solar Cookers International are active (Solar Cookers International). To enhance the amount of sunlight that reaches the surface and the temperature of the air within the solar cooker, reflector or focusing lens can be used (such as Fresnel lenses) can be added to the solar cooker, for higher temperature ranges parabolic dish\trough type can be use and for food drying. Low-temperature solar cookers can also be used, water distillation, or purification systems that use direct solar energy without the use of reflection or insulation. Solar water disinfection is as simple as exposing a clear water bottle to direct sunlight for 6 hours, but more effective water treatment requires boiling or distillation, which entails heating water to steam and then re-condensing pollutants that precipitate upon re-condensation. Various tests will be performed to assess the properties of plastic oil, including, viscosity, colour, calorific value, cloud point, flash point, gas chromatography and pour point. Several systems have been developed to transform plastic garbage into gasoline. Using modern plastic pyrolysis technology, the

Table 1.3 Various solar capturing technologies and their costs

Collector type	*Country of origin*	*Cost of collector (USD/m²)*	*Reference*
Compound Parabolic Collector	Italy	131	Bolognese et al. (2020)
	Spain	268–387	Isidoro et al. (2018)
Flat Plate Collector	Zimbabwe	220–347	Hove (2018)
	Mexico	287	Ortega (2018)
	Chile	330–687	Quiñones et al. (2020)
Linear Fresnel Collector	Italy	199	Bolognese et al. (2020)
	Spain	309–506	Isidoro et al. (2018)
Evacuated Tube Collector	Zimbabwe	157–433	Hove (2018)
	Mexico	472	Ortega (2018)
	Chile	460–817	Quiñones et al. (2020)
Parabolic Trough Collector	Chile	379–1263	Quiñones et al. (2020)
	Spain	393–666	Isidoro et al. (2018)
	Italy	262	Singh et al. (2021)
	Mexico	402	Ortega (2018)

problem such as increase landfill size increased agricultural land infertility, issues in domestic waste management operations and processing which caused due to plastic waste can be solved sustainably (Table 1.3).

Parabolic Concentrator

More modern types of solar cookers use curved mirrors to reflect sunlight to a focal point, where a parabolic shape reflects incoming UV radiation to the focus. The most common and mature solar thermal technology is the parabolic trough concentrator, which can also be found in industrial scale solar thermal facilities for high temperature applications. Reflective metal sheets with mirror-like properties are used in parabolic solar thermal collectors to precisely concentrate sunlight to the designed focal points where a receiver unit sits. The entire solar radiation spectrum, including infrared and ultraviolet wavelengths, is present in reflected sunlight. A glass collector tube chosen for its UV penetration properties serves as the focal point of a parabolic solar collector. This collector sleeve houses an absorber tube made of a heat-conductive metal that is dark in color. To reduce heat loss from convective air flow, industrial applications are evacuated between the collector and absorber. To raise temperatures even higher, additional insulation and a heat transfer fluid may be added to the absorber. This fluid is extracted from the collectors and used to generate steam or heat via other conductive processes.

Conclusion

Using modern pyrolysis technology plastics which are non-degradable can be easily converted into high calorific value alternative fuel, which reduces the environment pollution which is produced by burning of plastic, burning of conventional fuel in I.C engines as well as environmental effects such as overheating and green house emission. The pyrolysis process is regarded as an efficient, clean, and highly effective method of dealing with plastic solid waste, as well as a low-cost energy source. Converting plastic waste into fuel in the Indian market would not only solve the country's plastic waste crisis, but would also reduce plastic pollution by avoiding incineration and landfilling, as well as reduce the amount of imported oil barrels. For the burning of the plastic in the conversion chamber the solar energy, the Fresnel lens can be used.

References

Allouhi, A., Agrouaz, Y., Benzakour, M. A., et al. (2017). Design optimization of a multi-temperature solar thermal heating system for an industrial process. Appl. Energy. 206:382–392.

Anastasovski, A. (2021). Improvement of energy efficiency in ethanol production supported with solar thermal energy–a case study. J. Clean. Prod. 278:123476.

Bolognese, M., Viesi, D., Bartali, R. and Crema, L. (2020). Modeling study for low-carbon industrial processes integrating solar thermal technologies. A case study in the Italian Alps: The Felicetti Pasta Factory. Sol. Energy 208:548–558.

Different Plastic Types and How They Are Recycled. 2021. https://www.generalkinematics.com/blog/different-types-plastics-recycled/

Egypt, $ 6.8 Billion 2019 Imported Fuel Bill 2019. https://www.alarabiya.net/aswaq/economy/2019/12/30.

Evangelisti, L., De Lieto Vollaro, R. and Asdrubali, F. (2019). Latest advances on solar thermal collectors: A comprehensive review. Renew. Sustain. Energy Rev. 114:109318.

Fahim, I., Mohsen, O. and ElKayaly, D. (2021). Production of fuel from plastic waste: A feasible business. Polymers 13(6):915.

Farag, M. and Korachy, A. (2017). Plastics Value Chain Mapping and Assessment Technical Report No. 20, USAIDS Strengthening Entrepreneurship and Enterprise Development (SEED). 2017. http://www.seedegypt.org/wp-content/uploads/2019/04/Plastics-Value-Chain-Mapping-and-Assessment.pdf (accessed on 3 February 2021).

Gautam, A. and Saini, R. P. (2020). A review on sensible heat based packed bed solar thermal energy storage system for low temperature applications. Sol. Energy. 207: 937–956.

Ghazouani, M., Bouya, M. and Benaissa, M. (2020). Thermo-economic and exergy analysis and optimization of small PTC collectors for solar heat integration in industrial processes. Renew. Energy 152:984–998.

Ghosh, S. K. and Di Maria, F. (2018). A comparative study of issue, challenges and strategies of bio –waste management in India and Italy. J. Waste Res. Res. 1:8–12.

Gnansounou, E. and Raman, J.K. (2019). Hotspot Environmental Assessment of Biofuels. In Biofuels: Alternative Feedstocks and Conversion Processes for the Production of Liquid and Gaseous Biofuels (pp. 141–162). Academic Press.

Hahladakis, J. N., Velis, C. A. (2018). Review an overview of chemical additives present in plastics: Migration, release, fate and environmental impact during their use, disposal and recycling, J. Hazard. Mater. 344:179–199.

Hove, T. (2018, January). A thermo-economic model for aiding solar collector choice and optimal sizing for a solar water heating system. In Africa-EU Renewable Energy Research and Innovation Symposium (pp. 1–19). Springer, Cham.

Hussain, M. I., Ménézo, C. and Kim, J.-T. (2018). Advances in solar thermal harvesting technology based on surface solar absorption collectors: A review. Sol. Energy Mater. Sol. Cells. 187:123–139.

Isidoro, L.-B., Elena P.-A., Natividad S.-C. and Manuel A.S.-P. (2018). Benefits of medium temperature solar concentration technologies as thermal energy source of industrial processes in Spain. Energies 11(11):2950.

Juwono, H., Nugroho, K. A., Alfian, R., Ni'mah, Y. L. and Sugiarso, D. (2019). New generation biofuel from polypropylene plastic waste with co-reactant waste cooking oil and its characteristic performance. In: Journal of Physics: Conference Series, 1156(1), pp. 012013. IOP Publishing.

Kalargaris, I., Tian, G. and Gu, S. (2017). Experimental evaluation of a diesel engine fuelled by pyrolysis oils produced from low-density polyethylene and ethylene–vinyl acetate plastics. Fuel Process. Technol. 161:125–131.

Khan, M. Z. H., Sultana, M., Al-Mamun, M. R., and Hasan, M. R. (2016). Pyrolytic waste plastic oil and its diesel blend: fuel characterization. J. Environ. Public Health, 2016: 30–36

Maillot, C., Jean, C.-L. and Olivier M. (2019). Modelling and dynamic simulation of solar heat integration into a manufacturing process in reunion island. Procedia Manuf. 35:118–123.

Ogunbiyi, A. W., Openibo, A. O. and Ojowuro, O. M. (2016). Smelting of waste nylon and low-density plastics (ldps) in a fluidized bed system. Procedia Environ. Sci. 35:491–497.

Oosthuizen, D., Goosen, N. J. and Hess, S. (2020). Solar thermal process heat in fishmeal production: Prospects for two South African fishmeal factories. J. Clean. Prod. 253:119818.

Ortega, H. (2018). Energía solar térmica para procesos industriales en México. Estudio base de mercado. CONUEE Y ANES, Ciudad de México.

Patni, N., Shah, P., Agarwal, S., and Singhal, P. (2013). Alternate strategies for conversion of waste plastic to fuels. Int. Sch. Res. Notices. 2013:40–47.

Quiñones, G., Felbol, C., Valenzuela, C., Cardemil, J. M. and Escobar, R. A. (2020). Analyzing the potential for solar thermal energy utilization in the Chilean copper mining industry. Sol. Energy 197:292–310.

Roopa, F. and Ravishankar, R. (2016). Clay catalyst in PP and LLDPE conversion to fuel, Res. J. Chem. Environ. Sci. (RJCES) AELS 4(4S):52–55.

Rosendahl, L., ed. (2017). Direct thermochemical liquefaction for energy applications. United Kingdom: Woodhead Publishing.

Schoeneberger, C. A., McMillan, C. A., Kurup, P., Akar. S., Margolis, R. and Masanet, E. (2020). Solar for industrial process heat: A review of technologies, analysis approaches, and potential applications in the United States. Energy 206:118083.

Shahsavari, A. and Akbari, M. (2018). Potential of solar energy in developing countries for reducing energy-related emissions. Renew. Sustain. Energy Rev. 90:275–291.

Singh, L., Kumar, S., Raj, S. and Badhani P. (2021). Aluminium metal matrix composites: Manufacturing and applications. In IOP Conference Series: Materials Science and Engineering 2021 May 1 (Vol. 1149(1), p. 012025). IOP Publishing.

Singh, L., Kumar, S., Raj, S. and Badhani P. (2021). Development and characterization of aluminium silicon carbide composite materials with improved properties. Mater. Today Proc. 46:6733–6736.

Syamsiro, M., Saptoadi, H., Norsujianto, T., et al. (2014). Fuel oil production from municipal plastic wastes in sequential pyrolysis and catalytic reforming reactors. Energy Procedia 47:180–188.

Tilahun, F. B., Bhandari, R. and Mamo, M. (2019). Design optimization and control approach for a solar-augmented industrial heating. Energy 179:186–198.

Wu, S., Xu, K., Jiang, L. and Wang, L. (2014). The Co-cracking experiment and application route of waste plastics and heavy oil. AASRI Procedia. 7:3–7.

2 Research on the flexible transfer function of DC motor utilizing MATLAB software

Machina Tharuneswar[1,a], *Metta Jagadeesh*[1,b], *Gurmail Singh Malhi*[1,c], *and Md Helal Miah*[2,d,*]

[1]Department of Aerospace Engineering, Chandigarh University, Chandigarh, India

[2]Department of Mechanical Engineering, Chandigarh University, Chandigarh, India

Abstract

MATLAB and SIMULINK provide a wide range of promising tools for various engineering and many other non-engineering fields in enacting and resolving the problem. Some of the far-reaching tools in this software are SYSTEM IDENTIFICATION and LINEARIZATION tools, which proved to help solve many engineering problems. In this paper, it is mentioned regarding the construction of simple real-life physical tools (Paper Armature-Controlled DC Motor). Firstly, the DC motor working principle is discussed with the help of an induced magnetic field to rotate the armature following Fleming's left-hand rule. Then a physical system SIMSCAPE modelling is employed based on essential physical and electrical components already available in the SIMSCAPE library. The electronic components are arranged in series with each other. After the rotational electromechanical converter block, the electrical energy is converted into rotational mechanical energy, and the mechanical blocks perform the remaining task. The system identification method derives a model replicate that is close enough (in this case 84.6% fit) to replace the existing system without knowing the contents. The research has a practical implication to visualise the obtained data and how much it deviated from the required data for the transfer function of DC Motor.

Keywords: Induced magnetic field, linearisation, root mean square error, signal input, step response, transfer function.

Introduction

MATLAB is a high-level programming language that has its software integrated with computational and visualisation, and programming part in a comparatively effortless interface to solve or analyse the various systems according to the defined problem. MATLAB is dedicated to engineering fields and has its application in different purely mathematical problem solving like Fuzzy logic designer, Linear system analysis curve fitting, System identification and Computational Finance, Computational Biology, and Image and Signal processing, to name a few. The above mentioned are in the form of relatively easy applications to access for the problem. In addition to that, it has Simulink and Simscape[tm] as additional applications to solve engineering problems in a very interactive way. Due to this broad spectrum of applications, MATLAB has 4 million users worldwide and counting Chaturvedi (2017).

Mei Li, Chen Lung, and Wenlin Liu (1970) explored a MATLAB system identification Toolbox for the pdf identification and discussed the importance and accuracy of System Identification and the validity of this method on linear and second-order systems De Roeck et al. (2000). This method has also been tested on multi-input-multi-output (MIMO) on all models after the MATLAB update in 2012.M. Ondera and Huba (2006) researched Web-based Tools for Exact Linearization Control Design. It has been stated how one can use linearisation for non-linear control design where most of the computational hardship is put on the computer and reasoning is done by the students to remove the burden and beget this method as comparatively easy to use Ondera and Huba (2006). Tang et al. (2017) examined a speed control of DC motor using MATLAB/SIMULINK and tested the feasibility of the proposal using Arduino and hardware in the loop (HIL) in the pdf 'DC motor speed control based on system identification and PID auto-tuning Tang et al. (2017). Following the previous research work and knowledge, this research illustrates the application of MATLAB/SIMULINK software to identify the most replicable and reliable transfer function for the whole Armature controlled DC motor with fixed voltage parameters.

[a]machinatharuneshwar@gmail.com; [b]Jagadeeshjai334869@gmail.com; [c]gurmailmalhi.me@cumail.in; [d]helal.sau.12030704@gmail.com

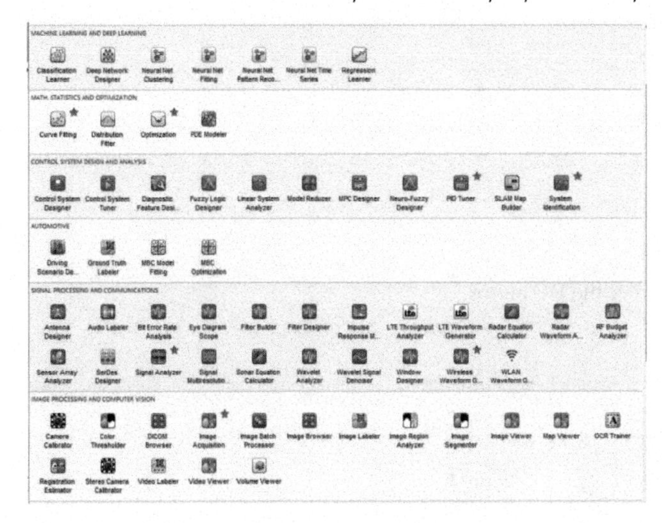

Figure 2.1 Applications of MATLAB software

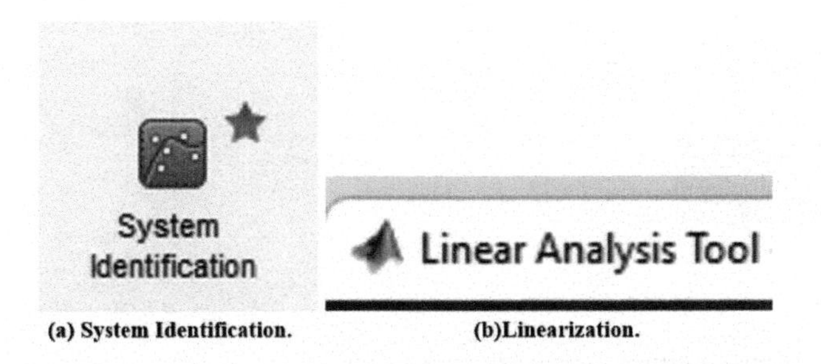

Figure 2.2 System identification and linearisation of MATLAB tools

This paper included the construction of Armature controlled DC motor in SIMULINK and on SIMSCAPE to better understand its working and other properties. After completion of construction of the motor, a signal builder is employed as a voltage regulator to the DC motor input and study its RPM response and initialise SYSTEM IDENTIFICATION by exporting signal builder data and RPM response to workspace and analysing for the transfer function. A similar procedure is followed in LINEARIZATION by specifying linear analysis points on the start and end of the DC motor blocks and solving the transfer function. It subsequently found transfer functions in both ways and used a multiplexer to inspect the signal output variation for the same signal voltage input and determine which process suits best for Armature Controlled DC Motor.

Methodology

A. DC Motor Construction

DC motor works on the principle of converting electrical energy into mechanical energy with the help of an induced magnetic field, which results from current through a conductor coil, which will rotate the armature following Fleming's left-hand rule. The Figure 2.3 represent the diagrammatic representation of the DC motor.

B. Simscape

Modelling a physical system in SIMSCAPE is relatively uncomplicated because all the required essential physical and electrical components are already available in the SIMSCAPE library.

SIMSCAPE handling extends to designing a complete physical system and doing analysis accordingly. The code 'slLibraryBrowser' can be used on the MATLAB command window to access the SIMSCAPE library, or one can type 'simscape' on the command window to access the SIMSCAPE library and to get familiarised with the components and parts that can be accessed for work.

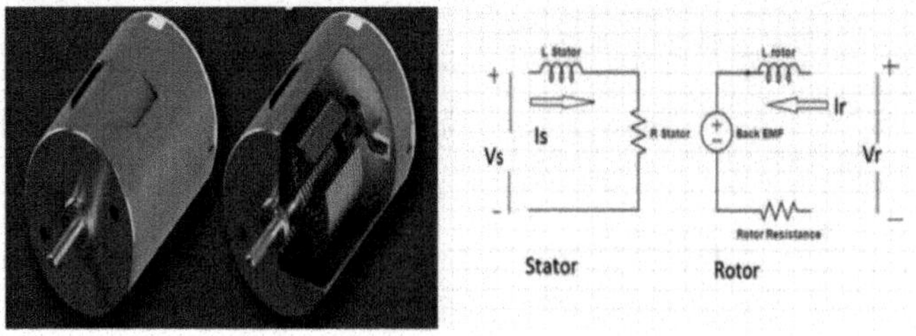

Figure 2.3 The diagrammatic representation of the DC motor

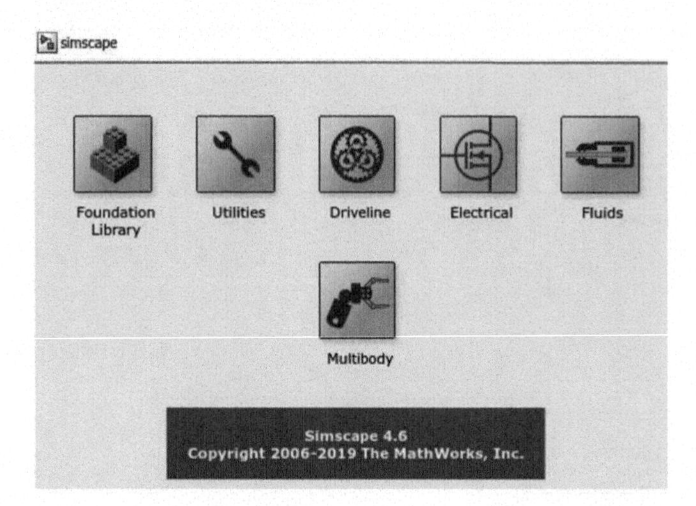

Figure 2.4 Libraries in SIMSCAPE

For the construction of a DC motor, the required components are as follows:

- Resistor.
- Inductor.
- Rotational electromechanical converter block.

The above three components are available in the electrical library of Foundation library in SIMSCAPE. This can also be accessed from the SIMULINK library at the following address 'Simscape/Foundation Library/Electrical/Electrical Elements library. The following mechanical components are required for the construction:

- Rotational damper.
- Inertial block.

The above blocks are available in the Simscape/Foundation Library/Mechanical/Rotational Elements library. The last set of blocks that are required are four connection ports from the 'Simscape/utility library' Watkins et al. (2020). The connection of the blocks can be made by understanding the physical construction of the DC motor itself from the Figures 2.5 and 2.6.

The electrical components in SIMSCAPE are coloured blue and mechanical elements are coloured green. The electrical components have to be arranged in series with each other, as shown in the Figure 2.5.

- Base values of resistance are 1ohm.
- Value of inductance is 0.01H.
- Proportionality constant value of converter is 0.01 rad/s.

After the rotational electromechanical converter block, the electrical energy is converted into rotational mechanical energy, and the mechanical blocks perform the remaining duty. The Block connection of it follows, as shown in the Figure 2.7:

- Damping coefficient is 0.1 V/(rad/s)
- Inertia is 0.01 kg.m^2

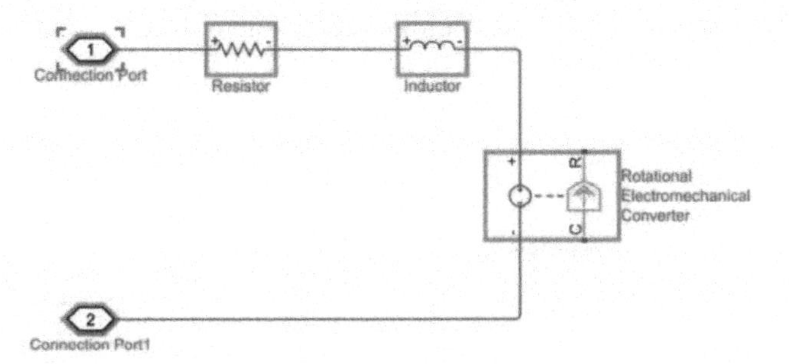

Figure 2.5 The connection of the DC motor blocks

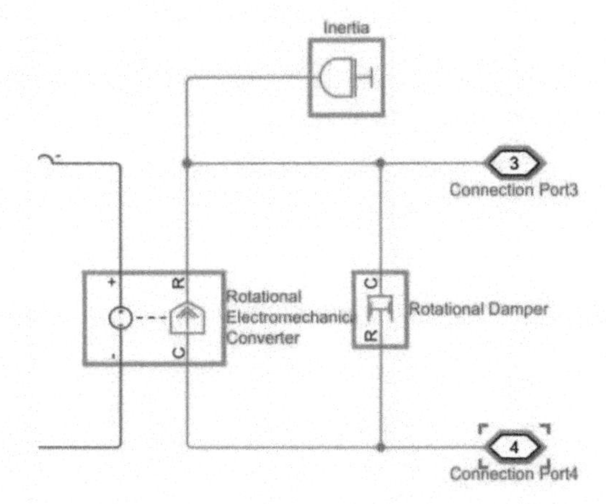

Figure 2.6 The block connection of the motor after the rotational electromechanical converter block

The final arrangement will be as follows:

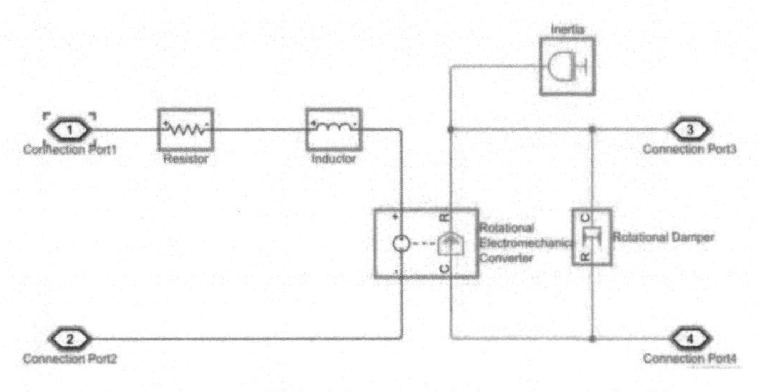

Figure 2.7 Simscape model of dc motor with preassigned values

Simulink

Modelling in SIMULINK is slightly more complex than that of SIMSCAPE. The reason is analogous to SIMSCAPE, and the modelling is done conceptually rather than replicating a physical model. To create a model in Simulink, user need the complete working of the model and all the other physical and surrounding conditions if the system is being influenced. Any conditions that fail to take into account might result in the faulty functioning of the system. For instance, familiarity with its transfer function is necessary to model a DC motor. And to understand its transfer function, the equation behind the working of a DC motor can lend a hand.

A. Deriving functional equations

Newton's laws and Kirchhoff's law can utilise the DC motor system to formulate a set of mathematical equations. Newton's laws for the mechanical (rotational part) and the latter for the electrical part of the system.

After balancing the torques and their consecutive angular acceleration and damping factor accounting friction for the mechanical section, the results are as follows:

$$\frac{d^2\theta}{dt^2} = \frac{1}{j}\left(T - b\frac{d\theta}{dt}\right) \tag{1}$$

Where, J = moment of inertia, T = torque, b = damping coefficient, θ = angular velocity.

After approaching a similar way for the system's electrical components, the resulting equations as follows:

$$\frac{di}{dt} = \frac{1}{L}(V - Ri - e) \tag{2}$$

Where, L = inductance, R = resistance, V = voltage supplied, i = current through conductors, e = backward electromotive force (EMF).

In the conductor coil present in the DC motor, the amount of back EMF (e) generated is proportional to the rotor's angular velocity, which will consequently alter the magnetic flux. Therefore,

$$e = k_b\frac{d\theta}{dt} \tag{3}$$

k_b is proportionality constant.

Substituting equation (6) in equation (4) will result in:

$$\frac{di}{dt} = \frac{1}{L}\left(V - Ri - k_b\frac{d\theta}{dt}\right) \tag{4}$$

Similarly, the torque generated as a result of the induced electromagnetism is proportional to the amount of armature current.

$$T = k_t i \tag{5}$$

Substituting equation (9) in equation (2):

$$\frac{d^2\theta}{dt^2} = \frac{1}{j}\left(k_t i - b\frac{d\theta}{dt}\right) \tag{6}$$

Transfer function application to both equations (4) and (6) will result in ease separation of the input (voltage) and output (angular velocity) with remaining parameters of the system by eliminating the third variable current flowing in the system. The transfer function of a differential function is:

$$\frac{d^n f(t)}{dt^n} = s^n f(s) \tag{7}$$

Using the above reference and applying a transfer function to the above two equations and solving for voltage and angular velocity will give the following result:

$$s \times i(s) = \frac{1}{L}\left[V(s) - R_i(s) - k_b \times \dot{\theta}(s)\right] \tag{8}$$

$$s \times \dot{\theta}(s) = \frac{1}{j}\left[k_t i(s) - b \times \dot{\theta}(s)\right] \tag{9}$$

$$\frac{\dot{\theta}(s)}{v(s)} = \frac{k_t}{\left[(j \times s + b)(L \times s + R)\right] + k_b k_t} \tag{10}$$

$\dot{\theta}$ being angular velocity $\left(\frac{d\theta}{dt}\right)$

B. SIMULINK setup

The final equation (equation 10) has to be transferred into SIMULINK with appropriate mathematical blocks such as 'integration' blocks for integration, 'gain' blocks for performing multiplication and division operations and 'add' blocks for addition and subtraction. The gain and add blocks are available at Simulink/math operations library, and integration blocks are available at Simulink/continuous library. From the above observation, there are two primary equations, which are equations (4) and (6). These equations have a common variable, armature current and made it achievable to eliminate the third variable and equate both equations with getting desired equation for us to understand and analyse the system correctly and efficiently. Before coupling, these two primary equations will become two significant loops in the Simulink. The common variable in both equations or loops (analogously) will be a signal link between the two loops. For this system, the signal flowing is the time differential of current from an in1 block and considering equation (4) for loop one and using gain blocks for inductance and followed by integration block to convert the time differential of the current to time-independent variable and rerouting the current to the input sum block with a negative sign using resistance as a gain block will give the following pattern in Simulink.

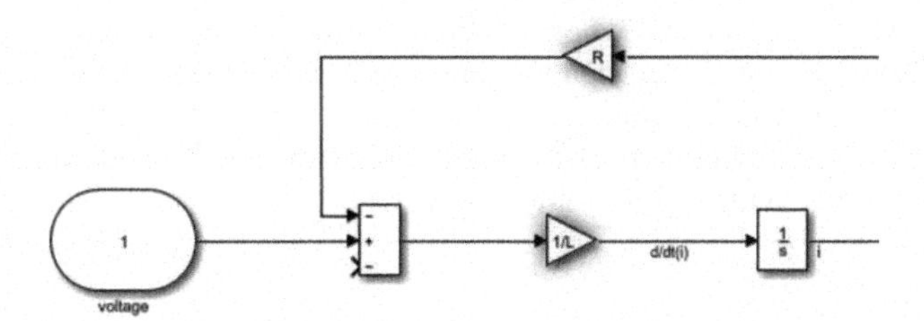

Figure 2.8 Circuit diagram

For the add block, the additional input is for the angular velocity, and the current from the integration block will be connected to the second loop of equation (6) with as in gain block Freidovich and Khalil (2008). In the second loop, the angular acceleration will be integrated similarly to the loop one and then rerouted to the add block with damping co-efficient in the gain block with a negative sign. The result will be as follows:

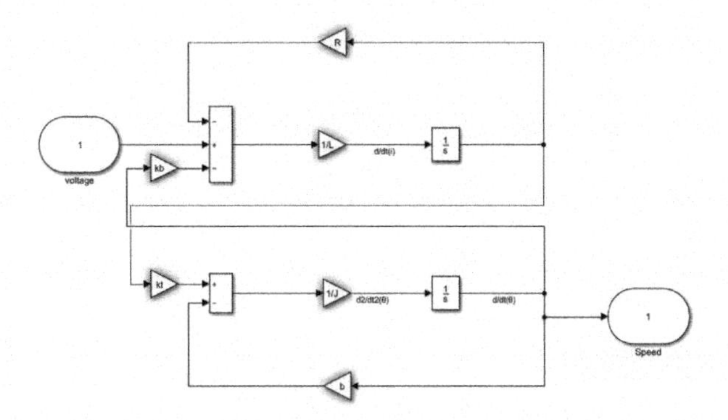

Figure 2.9 The final signal will be converted into angular velocity, this marks the end of primary setup Saghafinia et al. (2013), Weerasooriya and El-Sharkawi (1991)

Assuming reference values for the components of the system be:

- R = 0.5 Ω
- L = 1.5 mH
- K_b = 0.05 Nm/A
- K_t = 0.05 V/rad/s
- b = 0.0001 Nm/rad/s
- J = 0.00025 Nm/rad/s^2.

Assuming reference values for the components of the system be: the red highlight on the constant value mark is that there is no associated value. Assigning the value to the data variables can be done in two ways:

- Directly defining the data variable in the block parameters.
- Exporting the variables to the workspace of the MATLAB.

C. Defining variables directly in Simulink

Double-clicking the blocks that data is assigned will pop a dialogue box called block parameters. In the dialogue box, adding the constant in the place of the variable name is entirely possible.

D. Exporting the variable to MATLAB workspace

Inserting a variable value in the Simulink is relatively easy. Still, when there is a need to change the variables again at some other point of the analysis or if user need to change multiple times, it will be an effort to identify every block and change it individually. In that case, exporting the data variables to the workspace will create a new variable, and assigning/editing the variable will be effortless.

Clicking on the 'create' option on the block parameter dialogue will allow another to 'create new data'. The user can export data variables to the base workspace through this option. After applying the variable created in the base workspace, the variable's name will be visible.

After creating data variables for every constant component of the system, assigning will be as easy as typing the data variable name and its value in the command prompt of MATLAB.

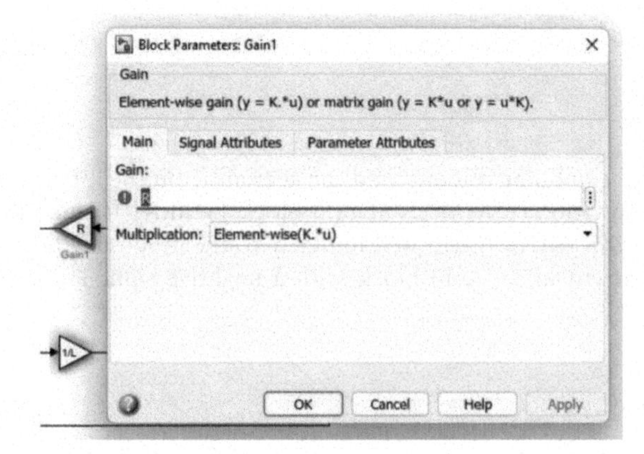

Figure 2.10 Block parameters dialogue box

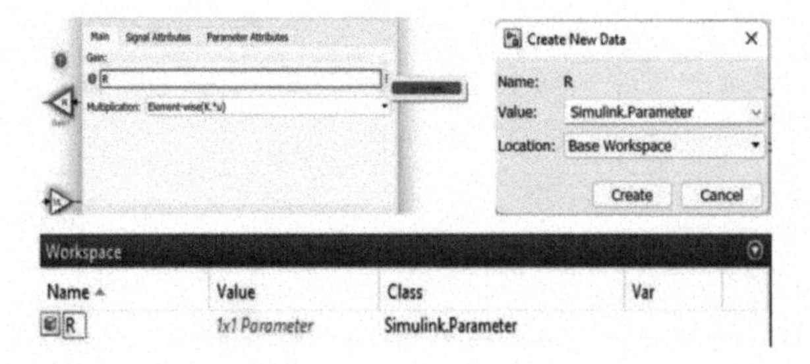

Figure 2.11 Creating data variable in MATLAB base workspace

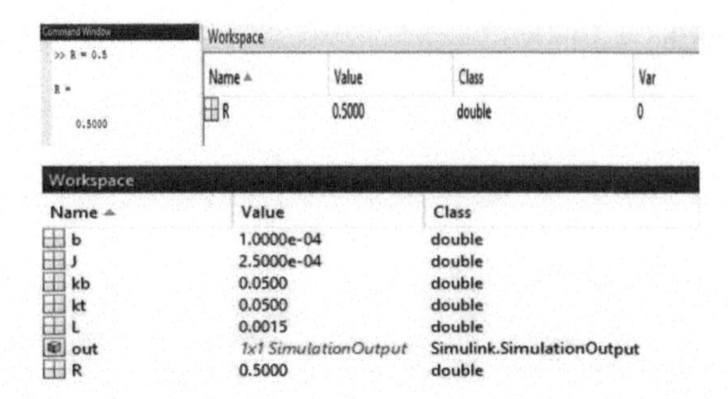

Figure 2.12 The command prompt of the MATLAB

Result and Discussion

The input signal has to be designated with the signal builder (Simulink/source) and to detect the output angular velocity of the system, a scope from Simulink/sink to analyse the system and its working principle. The purpose of the signal builder is to give an input signal to the system, which in this case acts as a supplied input voltage, and the response of the system, which is the angular velocity, is examined through Scope in the form of a plot between time and the response signal.

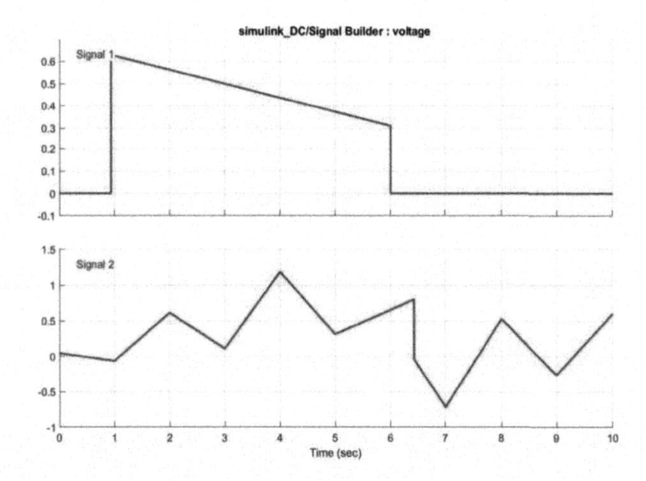

Figure 2.13 Sample signal used for analysis

The above are the signals used as the voltage and are added with Add block's help Miah et al. (2021), Ovando et al. (2007), Patrascoiu (2005), Prasad et al. (2012), Shenoy et al. (2004). A Mux block monitors the input and output signals on the samescope for the scope. There is an inbuilt block 'simout' in SIMULINK to directly export the data to the workspace to monitor the simulation data in the workspace. In the analysis, two simout blocks applied for both voltage and angular velocity data in array form.

Voltage and angular velocity data arrays are exported as 'volt' and 'speed' to the workspace once the simulation is completed.

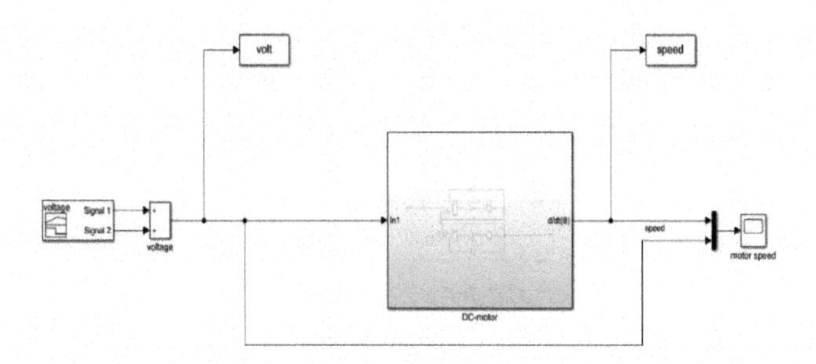

Figure 2.14 Sample circuit design used for analysis

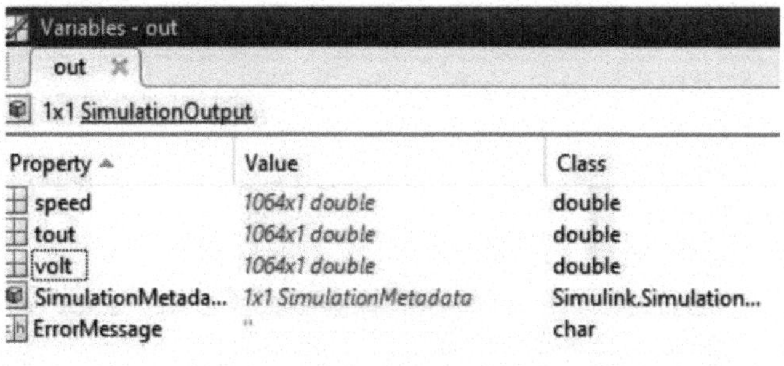

Variables - out

out

1x1 SimulationOutput

Property	Value	Class
speed	*1064x1 double*	double
tout	*1064x1 double*	double
volt	*1064x1 double*	double
SimulationMetada...	*1x1 SimulationMetadata*	Simulink.Simulation...
ErrorMessage	''	char

Figure 2.15 Voltage and angular velocity data arrays

A. System identification

System identification is an inbuilt app in MATLAB used to estimate Mathematical models for Dynamic systems with predefined input and output Ljung and Singh (2012). Then imported time-domain data from the workspace after the simulation and evaluate the system's transfer function for the present system.

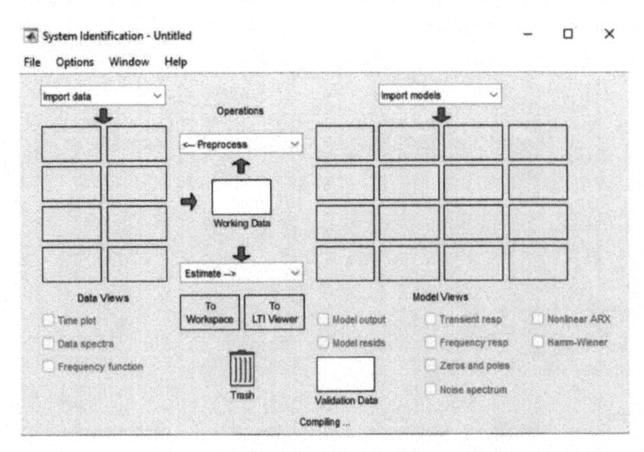

Figure 2.16 Imported time-domain data from the workspace

Estimating the transfer function with the system identification tool starts with launching the tool from the apps and importing the time domain data from the workspace into the System identification tool. After importing both input and output data of the DC Motor, the system can directly estimate the data for the transfer function. This tool gives the user the liberty to choose the holes and zeroes that one need in the system.

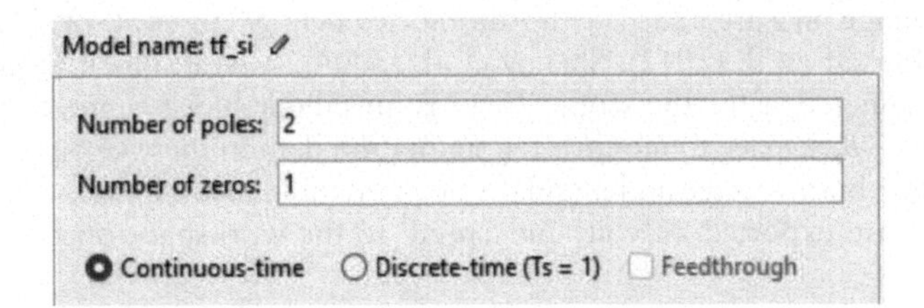

Figure 2.17 The liberty to choose the holes and zeroes

For the system used in this paper, this system identification tool with provided data gave 84.6% match to the data while undergoing 11 iterations. More detailed information is provided in the Table 2.1 below.

Table 2.1 Parameter of the SIMULINK system

Iteration	*Cost ($)*	*Norm of Step*	*First Order Optimality*	*Improvement (%)*		*Bisections*
				Expected	*Achieved*	
0	2993.4	-	1.5e+03	0.0333	-	-
1	408.646	745	2.64e+04	0.0333	86.3	1
2	280.494	1.64e+03	8.97e+04	0.236	31.4	1
3	259.544	4.45e+03	4.49e+04	0.338	7.47	5
4	241.974	1.9e+03	8.41e+03	0.363	6.77	4
5	174.324	391	5.09e+03	0.389	28	3
6	52.7371	41.5	2.15e+03	0.516	69.7	1
7	16.2512	13.1	3.43e+03	1.31	69.2	0
8	16.2113	12.2	1.79e+03	0.0267	0.246	1
9	16.2051	13.3	971	0.0126	0.0382	1
10	16.2009	8.72	696	0.00989	0.0257	2
11	16.2007	10.7	509	0.00782	0.00121	2

The FPE (Final Prediction Error) and MSE (Mean Square Error) obtained during the estimation for the DC Motor system are 16.616 and 16.18, respectively. At the end of the assessment, the transfer function data is transferred into a workspace for further analysis. The final transfer function obtained for the DC Motor data through the system identification tool is given below.

$$tf_{si} = \frac{94.28990s + 0.3634}{s^2 + 4.8555s + 0.008086} \tag{11}$$

This obtained Transfer function through the system Identification tool has to be added to the SIMULINK model linking the same voltage data as the input to monitor its response and calculate its deviation and accuracy according to the original model.

Transfer function can be added directly to the Simulink through the particular block assigned for its sole purpose, which library address is as follows Simulink/continuous/transfer. Scope at the output of the transfer function will give the user a way to visualise the deviation or the accuracy of the same.

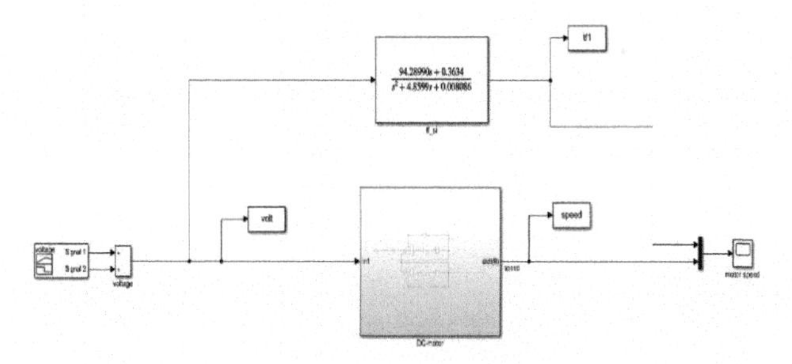

Figure 2.18 Simulink model with additional transfer function from system identification

The replaced transfer function that is obtained through system identification and actual motor speed is as follows.

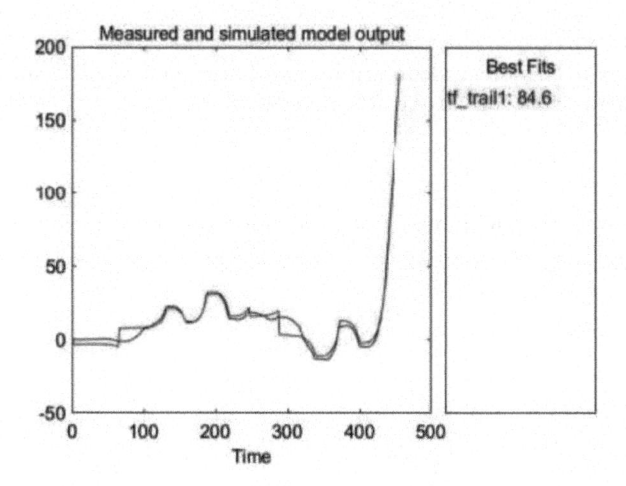

Figure 2.19 Actual speed to tfsi speed plot

B. Linearisation

Linearisation or linearising a system is very handy for solving the linear models for pole placement, loop shaping etc., and linearising tools can be launched from the Simulink by setting linear analysis points for both input and output in proposed DC Motor system Sakunthala et al. (2017).

After initialising the linear analysis points, the analysis tool can be initiated from Simulink/analysis/control design/linear analysis. After launching the analysis for proposed systems for appropriate states will obtain the following linear transfer function.

$$f_{si} = \frac{1.333e05}{s^2 + 333.7s + 6800} \tag{12}$$

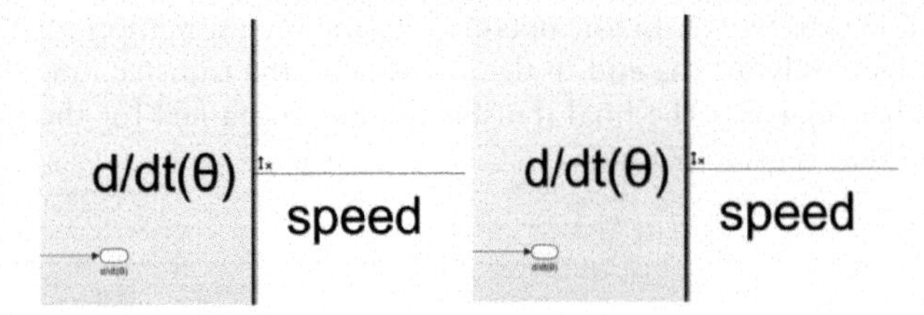

Figure 2.20 The image for the analysis points

The following is the step response for the voltage and speed during linear analysis.

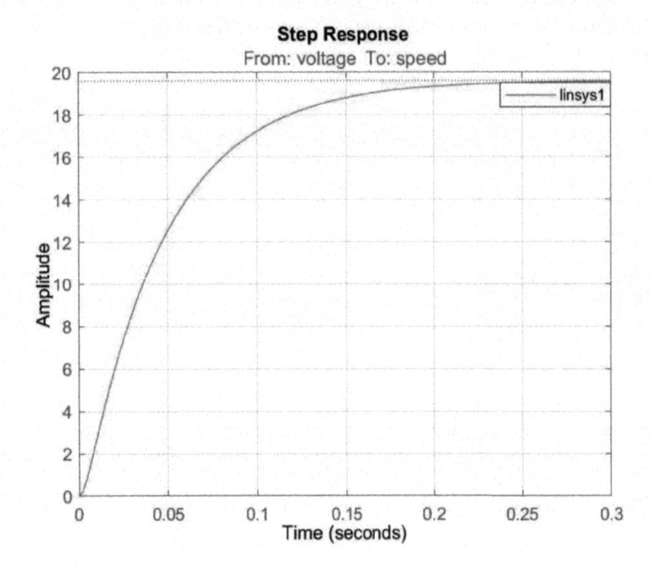

Figure 2.21 Step response

The analysis has both the transfer functions from system identification and linear analysis. The subsequent part analyses both systems to find the best fit for the current DC motor system.

Comparison

The final Simulink set up with all the three systems (DC motor, system identification transfer function, linearisation transfer function) is as follows.

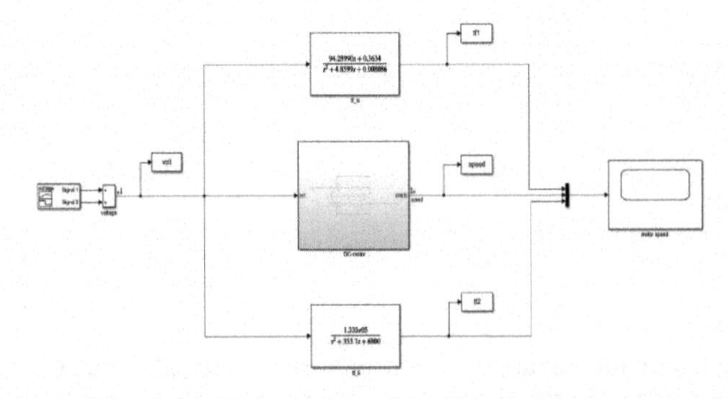

Figure 2.22 Final Simulink setup with data exporting to the workspace

The input voltage data is exported to the workspace in array format under the data variable name 'volt' and DC motor output, system identification output, linearisation outputs are also exported to the workspace for further analysis in the same data format with 'speed', 'tf_1', 'tf_2' respectively. All the data with a timestamp is exported to an excel sheet and uploaded here for further understanding. And comparative data from the scope for all these three subsystems is given below.

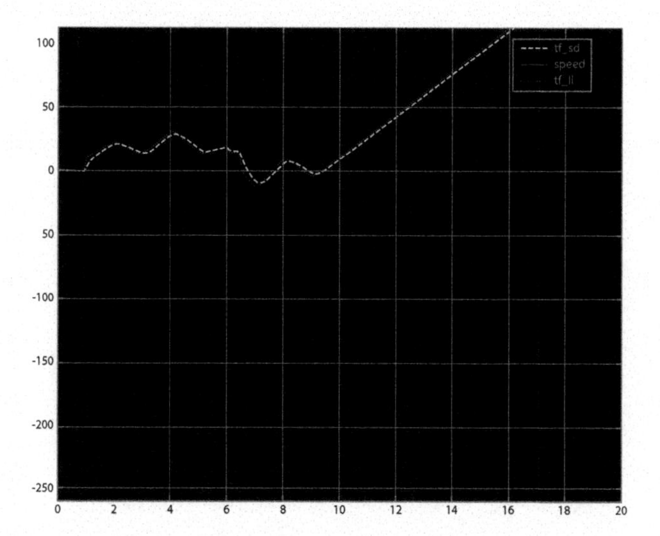

Figure 2.23 The dash-dotted line represents the subsystem plot for the system identification and the dotted line represents the subsystem plot for linerisation

The procedure that is followed to find the best possible DC motor replacement has to be calculated first, and the system with the minor errors is the best match. Root mean square error (RMSE) is employed for this error analysis. The reason for using this method is that squaring the deviation of the observed obtained will preserve both negative and positive errors from subsiding and give inaccurate results. The error calculation is carried out through python after the experimental data is exported into a CSV file and later imported into python. The python script used is linked here if a reference is necessary. After python scripting, the obtained is further imported to MATLAB for detailed plotting and visualising.

D. Data visualisation

The sole aim of this section is to visualise the obtained data and how much it deviated from the required data. During the Python scripting, after getting the required deviation from the data, adding a few lines of script downstream will assist us in obtaining the required data to line plot with error bars for both system identification and linearisation And the data is again imported into MATLAB to plot for visualisation and future usage if applicable.

```
pos_dev1=[]
neg dev1=[]
for i in range(lgt):
  if lis3[i] < 0:
    k=lis3[i]
    neg dev1.append(k)
  else:
    neg dev1.append(0)


  if lis3[i] >= 0:
    k=lis3[i]
    pos_dev1.append(k)
  else:
    k=0
    pos_dev1.append(0)
pos dev2=[]
neg_dev2=[]
  neg2.append(k)
```

```
for i in range(lgt):
  if lis4[i] < 0:
    k=lis4[i]
    neg dev2.append(k)
  else:
    neg dev2.append(0)


  if lis4[i] >= 0:
    k=lis4[i]
    pos_dev2.append(k)
  else:
    k=0
    pos dev2.append(0)
neg1=[]
neg2=[]
for i in range(lgt):
  k=neg_dev1[i]*(-1)
  neg1.append(k)
  k=neg_dev2[i]*(-1)
```

Figure 2.24 Python code for error plot after deviation

The imported data into MATLAB is stored under appropriate data variables for ease of usage, and the following are the error bar plot for both analysed systems Syal et al. (2012).

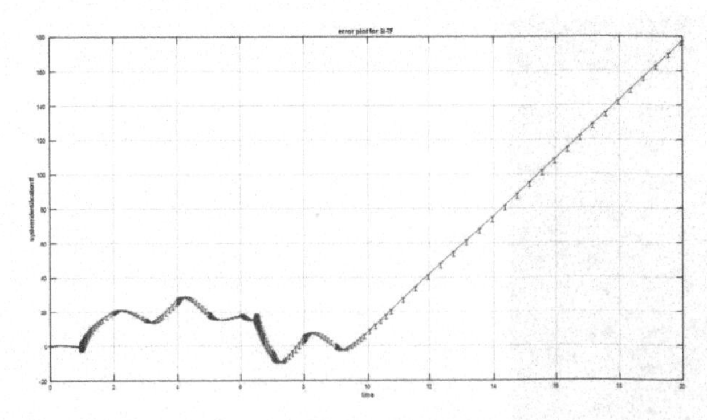

Figure 2.25 Error plot for SI-TF systems

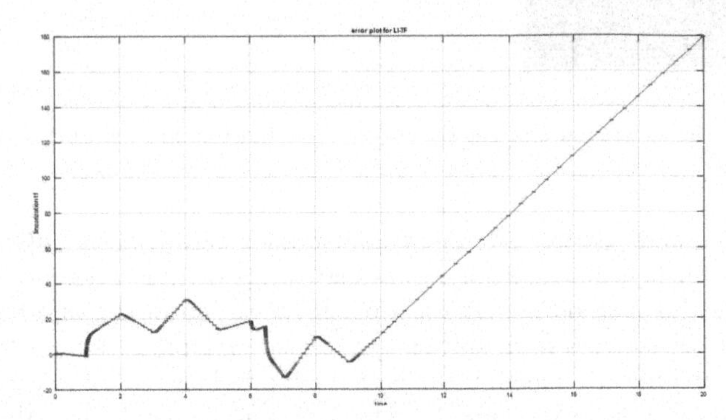

Figure 2.26 Error plot for LI-TF systems

The above plot helps to visualise the amount of deviation observed from actual dc motor data at every point of the plot.

Table 2.2 The MATLAB script for the above plot

% system identification	*% linearisation*
x1=out.tout;	x2=out.tout;
y1=out.tf1;	y2=out.tf2;
pos1=pos_err_si;	pos2=pos_err_li;
neg1=neg_err_si;	neg2=neg_err_li;
errorbar(x1,y1,pos1,neg1)	errorbar(x2,y2,pos2,neg2)
title('error plot for SI-TF');	title('error plot for LI-TF');
xlabel('time');	xlabel('time');
ylabel('system identification tf');	ylabel('linearization tf');

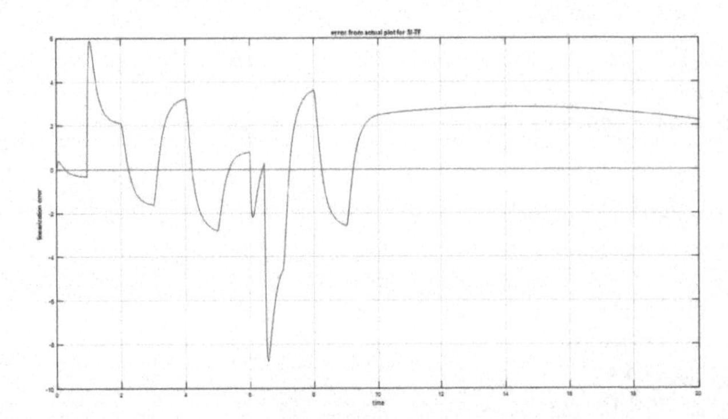

Figure 2.27 Deviation from the actual value plot for SI-TF systems

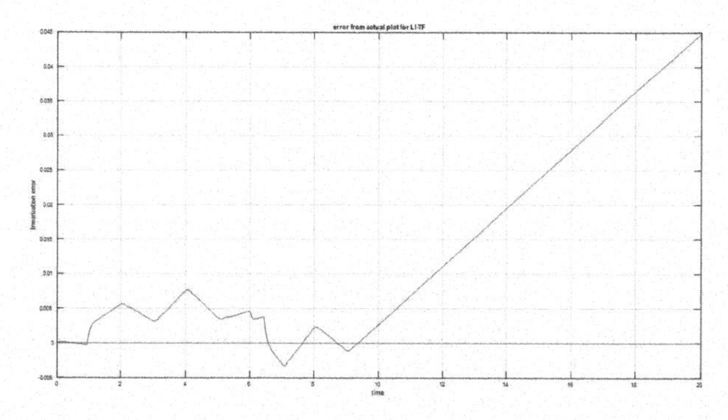

Figure 2.28 Deviation from the actual value plot for LI-TF systems

The above plot shows the difference between both systems' actual and obtained data.

E. RMSE calculation

As mentioned before, the RMSE calculation is computed in Python-3. The empirical formula for calculating this deviation is as follows:

$$RMSE = \sqrt[2]{\frac{\sum_{i=1}^{n}(v_{ai} - v_{oi})^2}{n}} \tag{13}$$

The Python script to employ the above equation is given below:

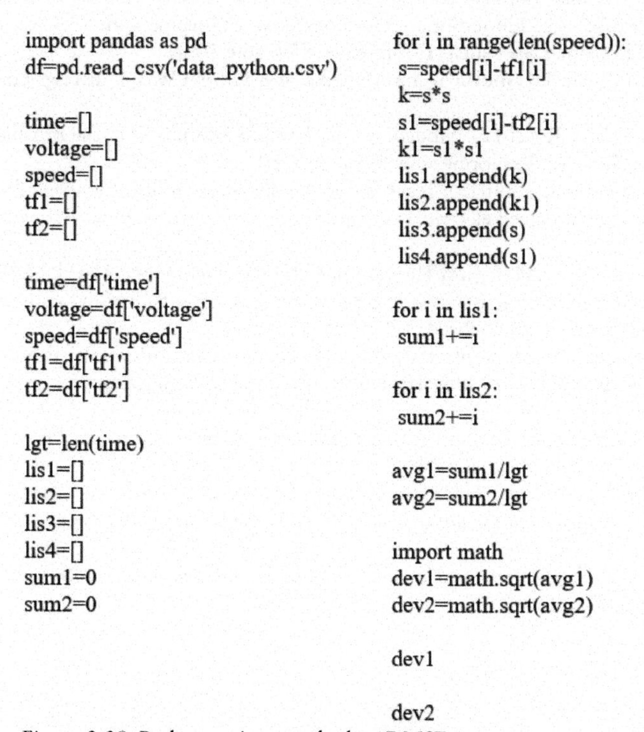

```
import pandas as pd
df=pd.read_csv('data_python.csv')

time=[]
voltage=[]
speed=[]
tf1=[]
tf2=[]

time=df['time']
voltage=df['voltage']
speed=df['speed']
tf1=df['tf1']
tf2=df['tf2']

lgt=len(time)
lis1=[]
lis2=[]
lis3=[]
lis4=[]
sum1=0
sum2=0

for i in range(len(speed)):
 s=speed[i]-tf1[i]
 k=s*s
 s1=speed[i]-tf2[i]
 k1=s1*s1
 lis1.append(k)
 lis2.append(k1)
 lis3.append(s)
 lis4.append(s1)

for i in lis1:
 sum1+=i

for i in lis2:
 sum2+=i

avg1=sum1/lgt
avg2=sum2/lgt

import math
dev1=math.sqrt(avg1)
dev2=math.sqrt(avg2)

dev1

dev2
```

Figure 2.29 Python script to calculate RMSE

Following the above analysis will give the following results:

- RMSE of System Identification Transfer Function is 2.8203543731397973.
- RMSE of lineariston transfer function is 0.007439411386770616.

Conclusion

The system identification method is also referred to as the black box method, for the reason that this process of identification helps derive a model replicate that is close enough (in this case 84.6% fit) to replace the existing system without needing to know the contents of system or the working of it. This procedure purely approximates any system with only input and output data. On the other hand, it linearises the existing systems transfer function to a linear function that is close enough. These transformed linear functions are easy to work within pole placements, loop shaping, etc. It is essential to estimate the system that is replaceable with the original system.

From the Figures 2.25 and 2.26 graphs, it is evident that the error bar magnitude on the system identification transfer function is higher than that of linearisation From Figure 2.15, the maximum and minimum deviation of the observed data for linearisation is comparatively less than that of system identification. The RMSE value for the linearisation is significantly less than System identification. Acknowledging and analysing all the obtained data, the linearised transfer function fits the best compared to system identification for the armature controlled DC motor.

References

Chaturvedi, D. K. (2017). Modeling and simulation of systems using MATLAB® and Simulink®. Boca Raton: CRC Press.

De Roeck, G., Peeters, B. and Ren, W.X. 2000. February. Benchmark study on system identification through ambient vibration measurements. In Proceedings of IMAC-XVIII, the 18th international modal analysis conference, San Antonio, Texas, 1106–1112.

Freidovich, L.B. and Khalil, H.K. (2008). Performance recovery of feedback-linearization-based designs. IEEE Transactions on Automatic Control. 53(10):2324–2334.

Ljung, L. and Singh, R. (2012). Version 8 of the MATLAB system identification toolbox. IFAC Proc. 45(16):1826–1831.

Miah, M. H., Zhang, J., and Chand, D. S. (2021). Knowledge creation and application of optimal tolerance distribution method for aircraft product assembly. Aircr. Eng. Aerosp. Technol. 94(3):431–436. https://doi.org/10.1108/AEAT-07-2021-0193.

Ondera, M. and Huba, M. (2006), June. Web-based tools for exact linearisation control design. In 2006 14th Mediterranean Conference on Control and Automation, 1–6, IEEE.

Ovando, R. I., Aguayo, J., and Cotorogea, M. (2007). June. Emulation of a low power wind turbine with a DC motor in Matlab/Simulink. In 2007 IEEE Power Electronics Specialists Conference, 859–864. IEEE.

Patrascoiu, N. (2005). Modeling and simulation of the DC motor using MatLAB and LabVIEW. Int. J. Eng. Educ. 21(1):49–54.

Prasad, G., Ramya, N. S., Prasad, P. V. N., and Das, G. T. R. (2012). Modelling and Simulation Analysis of the Brushless DC Motor by using MATLAB. Int. J. Innov. Technol. Exploring Eng. 1(5):2120–2025.

Saghafinia, A., Ping, H. W., Uddin, M. N., and Amindoust, A. (2013). Teaching of simulation an adjustable speed drive of induction motor using MATLAB/Simulink in advanced electrical machine laboratory. Procedia Soc. Behav. Sci. 103:912–921.

Shenoy, U. J., Sheshadri, K. G., Parthasarathy, K., Khincha, H. P., and Thukaram, D. (2004). November. MATLAB/PSB based modeling and simulation of 25 kV AC railway traction system-a particular reference to loading and fault conditions. In 2004 IEEE Region 10 Conference TENCON 2004. 100:508511.

Syal, A., Gaurav, K., and Moger, T. (2012). Virtual laboratory platform for enhancing undergraduate level induction motor course using MATLAB/Simulink. In 2012 IEEE International Conference on Engineering Education: Innovative Practices and Future Trends (AICERA), 1–6. IEEE.

Tang, W. J., Liu, Z. T., and Wang, Q. (2017). DC motor speed control based on system identification and PID auto tuning. In 36th Chinese Control Conference (CCC), 64206423. IEEE.

Watkins, C. B., Varghese, J., Knight, M., Petteys, B., and Ross, J. (2020). October. System architecture modeling for electronic systems using mathworks system composer and simulink. In AIAA/IEEE 39th Digital Avionics Systems Conference (DASC), 1–10. IEEE.

Weerasooriya, S. and El-Sharkawi, M. A. (1991). Identification and control of a dc motor using back-propagation neural networks. IEEE Trans. Ene. Conver. 6(4):663–669.

3 Solution of fractional differential equation and its applications in fluid mechanics

Eman Ali Ahmed Ziada[a]

Nile Higher Institute for Engineering & Technology, Mansoura, Egypt

Abstract

This paper aims to solve differential equations with arbitrary orders by using Adomian decomposition method (ADM), and concentrate on giving its applications in fluid mechanics. These applications are Bagley-Torvik equation and Basset problem and discuss applications in other fields such as relaxation-oscillation equation and fractional telegraph equation. The existence and uniqueness of the solution will discuss and the convergence of the series solution will prove.

Keywords: Adomian decomposition method; bagley-torvik equation; basset problem; differential equations of fractional orders; relaxation-oscillation equation; telegraph equation.

Introduction

Fractional Differential Equations (FDEs) have many applications in engineering and mechanics; including fluid flow (Khan et al., 2016; Atangana and Alabaraoye, 2013; Hammad and De la Sen, 2021), fractals theory (Rida and Arafa, (2011; Daraghmeh et al., 2020; Miller and Ross, 1993), viscoelasticity (Podlubny, 1999; Kilbas et al., 2006; El-Salam and El-Sayed, 2007), potential theory (El-Sayed and El-Salam, 2008; Evans and Raslan, 2005; Zwillinger, 1997), chemistry and biology (Mensour and Longtin, 1998; Hefferan and Corless, 2005; El-Sayed et al., 2004; El-Mesiry et al., 2005; Ahmed et al., 2007; El-Sayed, 1993). We use ADM which discussed before in the references (Adomian, 1995; Adomian, 1983; Adomian, 1986; Adomian, 1989; Abbaoui and Cherruault, 1994; Cherruault et al., 1995; Shawaghfeh, 2002; El-kalla, 2008; Momani and Al-Khaled, 2005; Jafari and Daftardar-Gejji, 2006); to solve these types of equations and its application in fluid Mechanics such as Bagley-Torvik equation and Basset problem and other applications such as Relaxation-oscillation equation and fractional telegraph equation. ADM has many advantages such as it used to solve different types of deterministic and stochastic equations if they are linear or nonlinear and it gives an analytic solution for all these types of equations without linearization or discretization. The solution algorithm and the convergence of the ADM series solution will discuss.

Solution Algorithm

Consider the following FDE,

$$ {}_0\mathcal{D}_t^{\sigma_n} y(t) + g(t)y(t) = x(t), \tag{1}$$

With initial conditions

$$ {}_0\mathcal{D}_t^{\sigma_n} y(0) = c_j, \quad j = 1, \ \ldots, \ n. \tag{2}$$

Where,

$$ {}_0\mathcal{D}_t^{\sigma_n} \equiv {}_0\mathcal{D}_t^{\alpha_n} \, {}_0D_t^{\alpha_{n-1}} \cdots {}_0D_t^{\alpha_1},$$

$$ {}_0\mathcal{D}_t^{\sigma_{n-1}} \equiv {}_0D_t^{\alpha_{n-1}} \, {}_0D_t^{\alpha_{n-2}} \cdots {}_0D_t^{\alpha_1},$$

$$ \sigma_n = \sum_{k=1}^{n} \alpha_k, \ 0 \le \alpha_k \le 1, \ k = 1, \ 2, \ \ldots, \ n.$$

The fractional derivative is of Riemann-Liouville sense.

Now we will apply subsequently fractional integration of order $\alpha_n, \alpha_{n-1}, \ldots, \alpha_1$, this reduces the problem (1)–(2) to the following fractional integral equation (FIE):

$$ y(t) = \sum_{j=1}^{n} \frac{c_j}{\Gamma(\sigma_j)} t^{\sigma_j - 1} + \frac{1}{\Gamma(\sigma_n)} \int_0^t (t-\tau)^{\sigma_n - 1} x(\tau) d\tau - \frac{1}{\Gamma(\sigma_n)} \int_0^t g(\tau)(t-\tau)^{\sigma_n - 1} y(\tau) d\tau \tag{3}$$

Where $x(t)$ is bounded $\forall t \in J = [0,T]$, $T \in R^+$, $|g(\tau)| \le M \ \forall \ 0 \le \tau \le t \le T$, M is a finite constant.

[a]eng_emanziada@yahoo.com

Let $y(t)=\sum_{n=0}^{\infty} y_n(t)$ in (3) and applying ADM, we get the following relations,

$$y_0(t)=\sum_{j=1}^{n}\frac{c_j}{\Gamma(\sigma_j)}t^{\sigma_j-1}+\frac{1}{\Gamma(\sigma_n)}\int_0^t (t-\tau)^{\sigma_n-1}x(\tau)d\tau, \tag{4}$$

$$y_i(t)=-\frac{1}{\Gamma(\sigma_n)}\int_0^t g(\tau)(t-\tau)^{\sigma_n-1}y_{i-1}d\tau,\ \ i\ge 1. \tag{5}$$

Finally, the solution is,

$$y(t)=\sum_{i=0}^{\infty} y_i(t) \tag{6}$$

Existence and Uniqueness

Theorem 1:

If $0<\alpha<1$ *where* $\alpha=\frac{MT^{\sigma_n}}{\Gamma(\sigma_n+1)}$, *then the series (6) is the solution of problem (1)–(2) and it is unique.*

Proof *For existence,*

$$\begin{aligned}
y(t)&=\sum_{i=0}^{\infty} y_i(t)\\
&=y_0(t)+\sum_{i=1}^{\infty} y_i(t)\\
&=y_0(t)-\frac{1}{\Gamma(\sigma_n)}\sum_{i=1}^{\infty}\int_0^t g(\tau)(t-\tau)^{\sigma_n-1}y_{i-1}d\tau\\
&=y_0(t)-\frac{1}{\Gamma(\sigma_n)}\int_0^t g(\tau)(t-\tau)^{\sigma_n-1}\sum_{i=1}^{\infty} y_{i-1}d\tau\\
&=y_0(t)-\frac{1}{\Gamma(\sigma_n)}\int_0^t g(\tau)(t-\tau)^{\sigma_n-1}\sum_{i=0}^{\infty} y_i d\tau\\
&=\sum_{j=1}^{n}\frac{c_j}{\Gamma(\sigma_j)}t^{\sigma_j-1}+\frac{1}{\Gamma(\sigma_n)}\int_0^t (t-\tau)^{\sigma_n-1}x(\tau)d\tau-\frac{1}{\Gamma(\sigma_n)}\int_0^t g(\tau)(t-\tau)^{\sigma_n-1}y(\tau)d\tau
\end{aligned}$$

Then the series solution (6) satisfy equation (3) which is the reduced FIE to the problem (1)–(2).

Uniqueness of the solution:
Let y and z are two different solutions to the problem (1)–(2) so

$$\begin{aligned}
|y-z|&=\left|\frac{1}{\Gamma(\sigma_n)}\int_0^t g(\tau)(t-\tau)^{\sigma_n-1}[y-z]d\tau\right|\\
&\le\frac{1}{\Gamma(\sigma_n)}\int_0^t (t-\tau)^{\sigma_n-1}|g(\tau)||y-z|d\tau\\
&\le\frac{M}{\Gamma(\sigma_n)}|y-z|\int_0^t (t-\tau)^{\sigma_n-1}d\tau\\
&\le\frac{MT^{\sigma_n}}{\Gamma(\sigma_n+1)}|y-z|
\end{aligned}$$

Let $\frac{MT^{\sigma_n}}{\Gamma(\sigma_n+1)} = \alpha$ where, $0 < \alpha < 1$ then,

$$|y-z| \le \alpha|y-z|$$
$$(1-\alpha)|y-z| \le 0$$

However, $(1-\alpha)\,|y-z| \ge 0$ and if $(1-\alpha) \ne 0$ then, $|y-z| = 0$ this give $y = z$ and hence the proof is complete.

Convergence

Theorem 2:
The solution (6) of problem (1)-(2) using ADM is converges if $|y_1| < \infty$ *and* $0 < \alpha < 1$ *where* $\alpha = \frac{MT^{\sigma_n}}{\Gamma(\sigma_n+1)}$.

Proof:
Define the Banach space $(C[J],\|\cdot\|)$, is the space of all continuous functions on J taking the norm $\|f(t)\| = \max_{t\in J}|f(t)|$. Define the sequence $\{S_n\}$ as $S_n = \sum_{i=0}^{n} y_i(t)$ the sequence of partial sums from the solution $\sum_{i=0}^{\infty} y_i(t)$.

Taking S_n and S_m as two arbitrary partial sums with $n \ge m$. We are going to show that $\{S_n\}$ is a Cauchy sequence in this Banach space.

$$\begin{aligned}
\|S_n - S_m\| &= \max_{t\in J}|S_n - S_m| = \max_{t\in J}\left|\sum_{i=m+1}^{n} y_i(t)\right| \\
&= \max_{t\in J}\left|\sum_{i=m+1}^{n} -\frac{1}{\Gamma(\sigma_n)}\int_0^t g(\tau)(t-\tau)^{\sigma_n-1} y_{i-1}\,d\tau\right| \\
&= \max_{t\in J}\left|\frac{1}{\Gamma(\sigma_n)}\int_0^t g(\tau)(t-\tau)^{\sigma_n-1}\sum_{i=m}^{n-1} y_i\,d\tau\right| \\
&= \max_{t\in J}\left|\frac{1}{\Gamma(\sigma_n)}\int_0^t g(\tau)(t-\tau)^{\sigma_n-1}\left[S_{n-1} - S_{m-1}\right]d\tau\right| \\
&\le \frac{1}{\Gamma(\sigma_n)}\max_{t\in J}\int_0^t (t-\tau)^{\sigma_n-1}|g(\tau)||S_{n-1} - S_{m-1}|\,d\tau \\
&\le \frac{M}{\Gamma(\sigma_n)}\max_{t\in J}|S_{n-1} - S_{m-1}|\int_0^t (t-\tau)^{\sigma_n-1}\,d\tau \\
&\le \frac{MT^{\sigma_n}}{\Gamma(\sigma_n+1)}\|S_{n-1} - S_{m-1}\| \\
&\le \alpha\|S_{n-1} - S_{m-1}\|
\end{aligned}$$

Taking $n = m + 1$ then,

$$\|S_{m+1} - S_m\| \le \alpha\|S_m - S_{m-1}\| \le \alpha^2\|S_{m-1} - S_{m-2}\| \le \cdots \le \alpha^m\|S_1 - S_0\|$$

From the triangle inequality we have,

$$\begin{aligned}
\|S_n - S_m\| &\le \|S_{m+1} - S_m\| + \|S_{m+2} - S_{m+1}\| + \cdots + \|S_n - S_{n-1}\| \\
&\le \left[\alpha^m + \alpha^{m+1} + \cdots + \alpha^{n-1}\right]\|S_1 - S_0\| \\
&\le \alpha^m\left[1 + \alpha + \cdots + \alpha^{n-m-1}\right]\|S_1 - S_0\| \\
&\le \alpha^m\left[\frac{1-\alpha^{n-m}}{1-\alpha}\right]\|y_1(t)\|
\end{aligned}$$

If $0 < \alpha < 1$ and $n \ge m$ then, $(1-\alpha^{n-m}) \le 1$ so we get,

$$\|S_n - S_m\| \le \frac{\alpha^m}{1-\alpha}\|y_1(t)\| \le \frac{\alpha^m}{1-\alpha}\max_{t\in J}|y_1(t)|$$

Nevertheless, $|y_1(t)| \leq \infty$ and as $m \to \infty$ then, $\|S_n - S_m\| \to 0$ and hence, $\{S_n\}$ is a Cauchy sequence in this Banach space then the series solution (6) is converge.

Estimating Error Analysis

Theorem 3:
The maximum absolute error of the series solution (6) to the problem (1)-(2) is estimating as

$$\max_{t\in J}\left|y(t)-\sum_{i=0}^{m} y_i(t)\right| \leq \frac{\alpha^m}{1-\alpha}\max_{t\in J}|y_1(t)|$$

Proof
Using Theorem 2 we get

$$\|S_n - S_m\| \leq \frac{\alpha^m}{1-\alpha}\max_{t\in J}|y_1(t)|$$

Moreover, $S_n = \sum_{i=0}^{n} y_i(t)$ as $n \to \infty$ then, $S_n \to y(t)$ so,

$$\|y(t) - S_m\| \leq \frac{\alpha^m}{1-\alpha}\max_{t\in J}|y_1(t)|$$

Therefore, the maximum absolute error in the interval J is,

$$\max_{t\in J}\left|y(t)-\sum_{i=0}^{m} y_i(t)\right| \leq \frac{\alpha^m}{1-\alpha}\max_{t\in J}|y_1(t)|$$

Applications

Application (1): Relaxation-Oscillation equation

Consider the following linear FDE,

$$\begin{aligned} &{}_0D_t^{\alpha} y(t) + Ay(t) = f(t),\ t>0, \\ &\quad y^{(k)}(0) = 0, \qquad (k = 0,1,\ldots,n-1), \end{aligned} \tag{7}$$

Where $n - 1 < \alpha \leq n$. For $0 < \alpha \leq 2$, it called the relaxation-oscillation equation. In the references (Podlubny, 1999; Diethelem et al., 2005), numerical methods were using to solve this problem, while in Daftarder-Gejji and Jafari (2005) it is solved by ADM by transform it to a system.

Applying the integrating operator ${}_0D_t^{-\alpha}$ on the both sides of the equation (7) and taking $A = 1$ and $f(t) = H(t)$. Let the integrating operator ${}_0D_t^{-\alpha} \equiv J^{\alpha}$ for simplicity. Then we get,

$$\begin{aligned} J^{\alpha}\left({}_0D_t^{\alpha} y(t)\right) &= J^{\alpha}\left[H(t)\right] - J^{\alpha} y(t), \\ y(t) &= J^{\alpha}\left[H(t)\right] - J^{\alpha} y(t), \end{aligned} \tag{8}$$

Applying ADM to the equation (8), we have,

$$\begin{aligned} &y_0(t) = J^{\alpha}\left[H(t)\right], \\ &y_n(t) = -J^{\alpha} y_{n-1}(t), \quad n \geq 1. \end{aligned} \tag{9}$$

The computation for results at different values of α ($1 \leq \alpha \leq 2$), are shown in Figures 3.1.a–3.1.f (at $n = 50$). We see from these figures that the solution is a damped oscillator whenever $1 < \alpha < 2$, and it will be perfect oscillator if $\alpha = 2$. The results are in good agreement with the analytical solution; obtained from f Green's function solution of two-term fractional differential equation which is

$$y(t) = \int_0^t G_2(t-\tau) f(\tau)\, d\tau, \tag{10}$$

where

$$G_2(t) = t^{\alpha-1}E_{\alpha,\alpha}\left(-t^{\alpha}\right),\ f(t) = H(t).$$

Therefore,

$$y(t) = t^{\alpha}E_{\alpha,\alpha+1}\left(-t^{\alpha}\right) \tag{11}$$

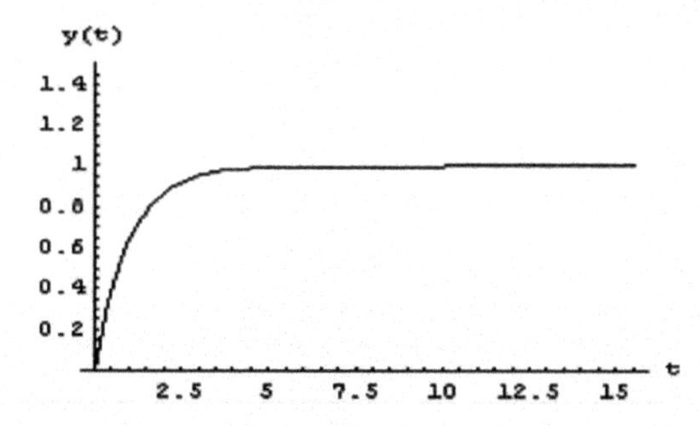

Figure 3.1.a ADM Sol. [$\alpha = 1$]

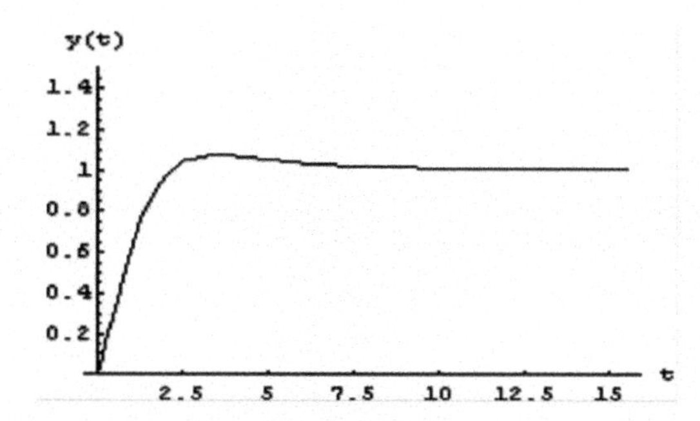

Figure 3.1.b ADM Sol. [$\alpha = 1.2$]

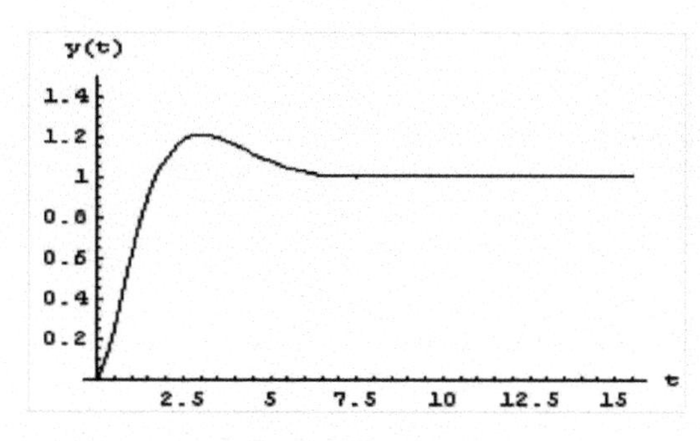

Figure 3.1.c ADM Sol. [$\alpha = 1.4$]

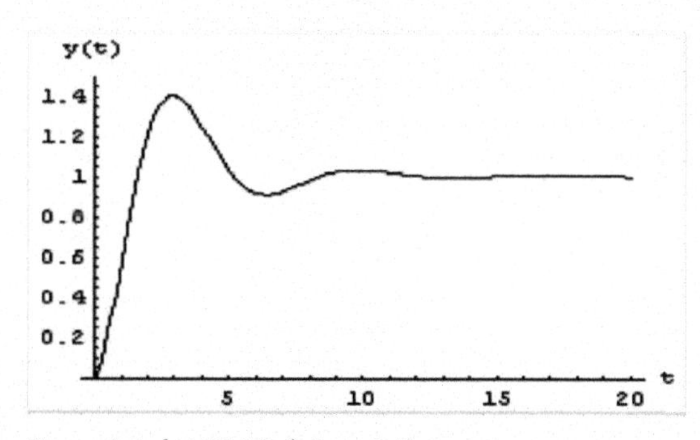

Figure 3.1.d ADM Sol. [$\alpha = 1.6$]

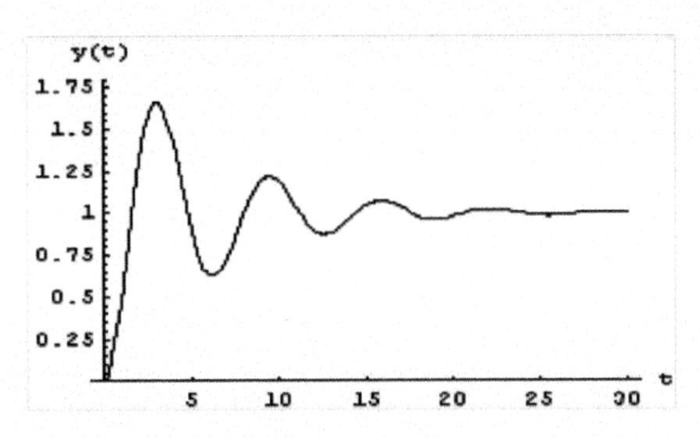

Figure 3.1.e ADM Sol. [$\alpha = 1.8$]

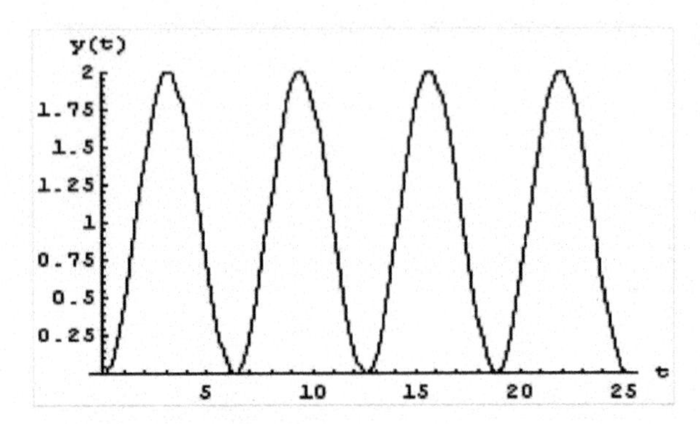

Figure 3.1.f ADM Sol. [$\alpha = 2$]

Let us take the special case when $\alpha = 3/2$, equation (9) will be,

$$y_0(t) = J^{3/2}\left[H(t)\right], \quad y_n(t) = -J^{3/2}y_{n-1}(t), \; n \geq 1. \tag{12}$$

Using the relations (12), the first four-terms of the series solution given by,

$$y(t) = \frac{t^{3/2}}{\Gamma(5/2)} - \frac{t^3}{\Gamma(4)} + \frac{t^{9/2}}{\Gamma(11/2)} - \frac{t^6}{\Gamma(7)} + \cdots. \tag{13}$$

Moreover, the Green's function solution is,

$$y(t) = t^{3/2}E_{3/2,5/2}\left(-t^{3/2}\right) \tag{14}$$

A comparison between ADM and Green's solutions given in Figure 3.1.g (at $n = 50$).

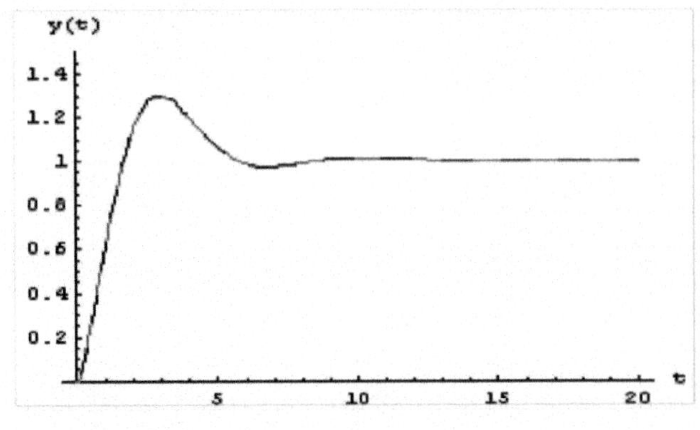

Figure 3.1.g ADM and Green's Sol. [$\alpha = 3/2$]

Application (2): Bagley-Torvik Equation

Consider the equation of motion of the fluid Podlubny (1999),

$$\rho\frac{\partial \upsilon}{\partial t}=\mu\frac{\partial^2 \upsilon}{\partial z^2}, \qquad (15)$$

$$(0<t<\infty,\ -\infty<z<0)$$

Where ρ is fluid density, μ is viscosity and $\upsilon(t, z)$ is transverse velocity a function of a time t and the distance z from fluid plate contact boundary.

This equation is an important application of motion of the fluid. Analytical and numerical solution of this problem given by Podlubny (1999). In (Diethelem et al. ????; Diethelm and Ford, 2002), Diethelm and Ford give a numerical solution of this problem and after that, Trinks and Ruge Trinks and Ruge (2002) discussed the origination of Bagley-Torvik equation and they give a numerical scheme for the solution of this problem. It was solved as a system using ADM in (Momani and Al-Khaled, 2005; Diethelm and Ford, 2002) by Varsha and Shaher.

Now, we will solve it by using the classical

The equivalent ordinary differential equation-to-equation (15) are given in Saha Ray and Bera (2004) as follows,

$$\frac{d^2x(t)}{dt^2}+\frac{B}{A}\frac{d^{3/2}x(t)}{dt^{3/2}}+\frac{C}{A}x(t)=\frac{f(t)}{A}. \qquad (16)$$

Applying ADM to equation (16), we have

$$x(t)=\frac{1}{A}L^{-1}f(t)-\frac{B}{A}L^{-1}\left(D_t^{3/2}\left(\sum_{n=0}^{\infty}x_n(t)\right)\right)-L^{-1}\left(\left(\sum_{n=0}^{\infty}\frac{C}{A}x_n(t)\right)\right), \qquad (17)$$

Where $L\equiv\frac{d^2}{dt^2}$ and we get,

$$x(t)=\frac{1}{A}\frac{d^{-2}}{dt^{-2}}f(t)-\frac{B}{A}\frac{d^{-1/2}}{dt^{-1/2}}\left(\sum_{n=0}^{\infty}x_n(t)\right)-\frac{C}{A}\frac{d^{-2}}{dt^{-2}}\left(\sum_{n=0}^{\infty}x_n(t)\right), \qquad (18)$$

Moreover, from that we get,

$$x_0(t)=\frac{1}{A}\frac{d^{-2}}{dt^{-2}}f(t), \qquad (19)$$

$$x_n(t)=-\frac{B}{A}\frac{d^{-\frac{1}{2}}x_{n-1}(t)}{dt^{-\frac{1}{2}}}-\frac{C}{A}\frac{d^{-2}x_{n-1}(t)}{dt^{-2}},\ n\geq 1. \qquad (20)$$

Let

$$f(t)=\begin{cases}8 & 0\leq t\leq 1,\\ 0 & t>1.\end{cases}$$

Which can written in terms of unit-step function as follows $f(t) = 8(u(t) - u(t - 1))$.

Figure 3.2.a shows ADM Solution which in good agreement with Green's solution given in Podlubny (1999). Taking ($A = 1$, $B = 0.5$, $C = 0.5$ *and* $h = 0.02$) *as taken in* Momani and Al-Khaled (2005). We note that the obtained graph in this case almost coincide with that obtained in Momani and Al-Khaled (2005) by Podlubny by using the numerical methods, this shown in Figure 3.2.b.

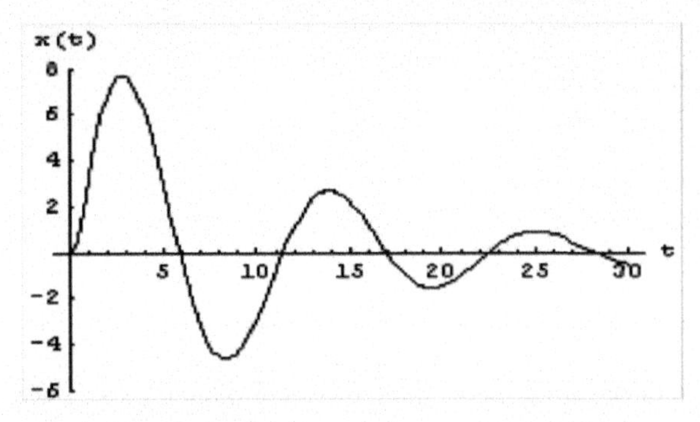

Figure 3.2.a ADM Sol

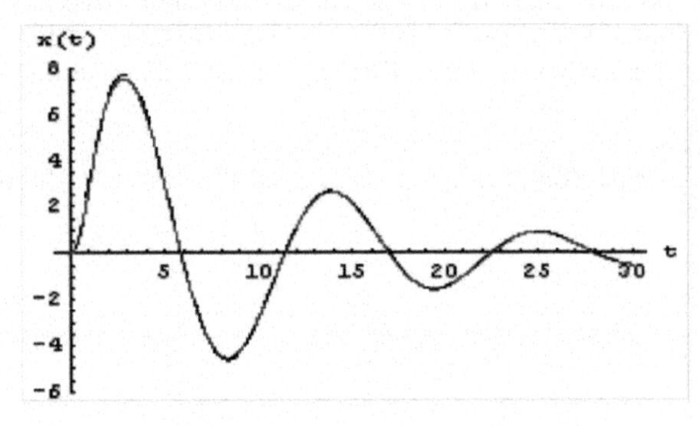

Figure 3.2.b ADM and numerical Sol.

Application (3): Fractional Telegraph Equation

The fractional telegraph equation,

$$(D^{\gamma})^2 u(t) + 2aD^{\gamma}u(t) + bu(t) = 0,$$

$$u(0) = f_1, \quad D^{\gamma}u(0) = f_2, \tag{21}$$

Discussed by Cascaval et al. (2002), the solution given using Laplace transform method. This solution was

$$u(t) = Q_1 E_{\gamma,1}\left(\left(-a+\sqrt{a^2-b}\right)t^{\gamma}\right) + Q_2 E_{\gamma,1}\left(\left(-a-\sqrt{a^2-b}\right)t^{\gamma}\right), \tag{22}$$

Where,

$$Q_1 = \frac{\left(\sqrt{a^2-b}+a\right)f_1 + f_2}{2\sqrt{a^2-b}},$$

$$Q_2 = \frac{\left(\sqrt{a^2-b}-a\right)f_1 - f_2}{2\sqrt{a^2-b}},$$

Now, we will solve it by using ADM, taking
($a = 1, b = 1/2, f_1 = 1, f_2 = 1$), then the equation (21) will be

$$D^{\frac{1}{2}}\left(D^{\frac{1}{2}}u(t)\right) + 2D^{\frac{1}{2}}u(t) + \frac{1}{2}u(t) = 0, \tag{23}$$

$$u(0) = 1, \quad D^{1/2}u(0) = 1.$$

Moreover, its solution by using the Laplace transform will be

$$u(t) = \frac{\left(1+\sqrt{1/2}\right)+1}{2\sqrt{1/2}} E_{1/2,1}\left(\mu_+ t^{1/2}\right) + \frac{\left(-1+\sqrt{1/2}\right)-1}{2\sqrt{1/2}} E_{1/2,1}\left(\mu_- t^{1/2}\right), \tag{24}$$

Where $\mu_{\pm} = -a \pm \sqrt{a^2-b}$.

We begin the solution by operating with $J^{1/2}$ on equation (23),

$$J^{1/2}\left[D^{1/2}\left(D^{1/2}u(t)\right)\right]=-2J^{1/2}\left[D^{1/2}u(t)\right]-\frac{1}{2}J^{1/2}\left[u(t)\right] \tag{25}$$

Using the following formula

$$ {}_{a}^{C}D_{t}^{-\alpha}\left({}_{a}^{C}D_{t}^{\alpha}f(t)\right)=f(t)-\sum_{k=0}^{m-1}f^{(k)}\left(0^{+}\right)\frac{t^{k}}{k!}, \quad (m-1<\alpha\leq m). \tag{26}$$

Equation (25) will be:

$$D^{1/2}u(t)-D^{1/2}u(0)=-2\left[u(t)-u(0)\right]-\frac{1}{2}J^{1/2}\left[u(t)\right], \tag{27}$$

$$D^{1/2}u(t)=3-2u(t)-\frac{1}{2}J^{1/2}\left[u(t)\right], \tag{28}$$

Operating with $J^{1/2}$ again on equation (27),

$$J^{1/2}\left[D^{1/2}u(t)\right]=J^{1/2}[3]-2J^{1/2}\left[u(t)\right]-\frac{1}{2}J^{1/2}\left[J^{1/2}\left[u(t)\right]\right], \tag{29}$$

Using the formula (26), equation (28) will be,

$$u(t)-u(0)=J^{1/2}[3]-2J^{1/2}\left[u(t)\right]-\frac{1}{2}J^{1}\left[u(t)\right], \tag{30}$$

Then,

$$u(t)=1+J^{1/2}[3]-2J^{1/2}\left[u(t)\right]-\frac{1}{2}J^{1}\left[u(t)\right]. \tag{31}$$

Using *ADM to equation (31), we get*

$$u_{0}(t)=1+J^{1/2}[3], \tag{32}$$

$$u_{n}(t)=-2J^{1/2}\left[u_{n-1}(t)\right]-\frac{1}{2}J^{1}\left[u_{n-1}(t)\right], \qquad n\geq 1. \tag{33}$$

From the relations (32) and (33), the series solution is

$$u(t)=1+\frac{6\sqrt{t}}{\sqrt{\pi}}-\frac{\sqrt{t}\left(8+13\sqrt{\pi}\sqrt{t}+4t\right)}{2\sqrt{\pi}}+4t+\frac{56t^{\frac{3}{2}}}{3\sqrt{\pi}}+\frac{25t^{2}}{8}+\frac{2t^{\frac{5}{2}}}{5\sqrt{\pi}}$$
$$-\frac{1}{1680\sqrt{\pi}}\left(t^{\frac{3}{2}}\left(35\sqrt{\pi}\sqrt{t}\left(720+37t\right)+32\left(560+546t+3t^{2}\right)\right)\right)+\cdots$$

A comparison between ADM and Laplace solutions given in Figures 3.3.a–3.3.d. We see from these figures that as we increase the number of terms, the solution will be more accurate. The two curves are identical when in the given interval.

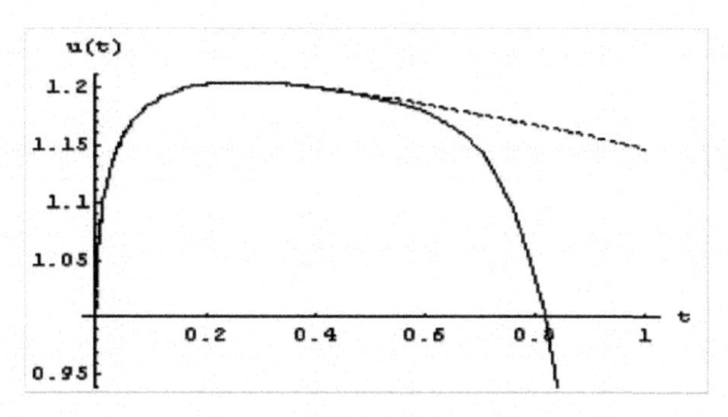

Figure 3.3.a ADM and Laplace Sol. [$n = 20$]

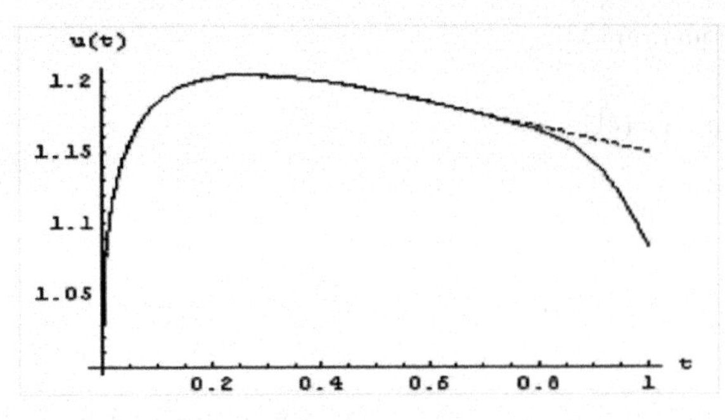

Figure 3.3.b ADM and Laplace Sol. [n = 25]

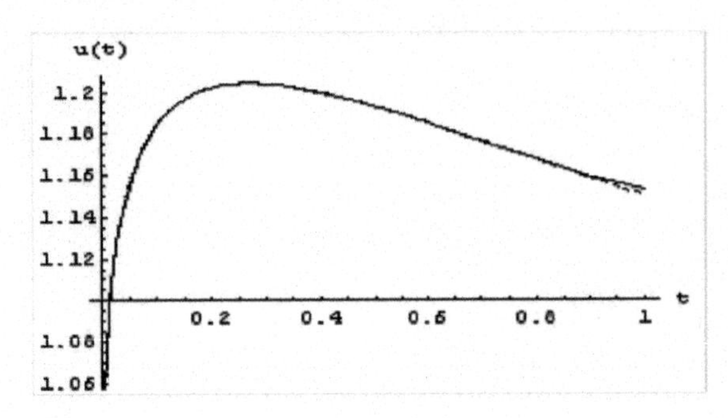

Figure 3.3.c ADM and Laplace Sol. [n = 30]

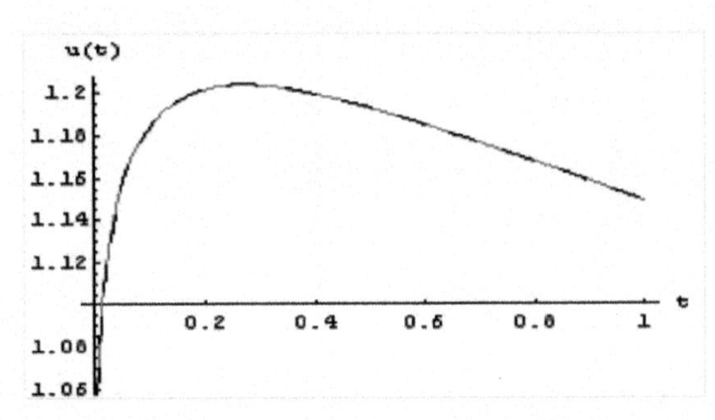

Figure 3.3.d ADM and Laplace Sol. [n = 35]

Application (4): Basset problem

Consider the following linear FDE,

$$Du(t)+2aD^{\frac{1}{2}}u(t)+bu(t)=q(t),\ u(0)=f_1, \tag{34}$$

a is a positive constant. If b = 1, then the linear FDE (34) called *Basset problem, it is a classical problem in fluid dynamics dealing with unsteady motion of a particle accelerating in a viscous fluid under gravity action* Mainardi (1997).

Solution of this problem using Laplace transform method given in Podlubny (1994), when $q(t) = 0$ is,

$$Du(t)+2aD^{\frac{1}{2}}u(t)+bu(t)=q(t),\ u(0)=f_1, \tag{35}$$

Where,

$$C_1=(1+\mu_+)/(\mu_+-\mu_-),\quad C_2=-(1+\mu_-)/(\mu_+-\mu_-),$$
$$\Psi_+=E_{1/2,1}\left(\mu_+\sqrt{t}\right),\quad \Psi_-=E_{1/2,1}\left(\mu_-\sqrt{t}\right),$$
$$\mu_\pm=-a\pm\sqrt{a^2-b}.$$

Now, we will solve it by using ADM when $f_1 = 1$, the equation (34) will be,

$$Du(t) + 2aD^{\frac{1}{2}}u(t) + bu(t) = 0, \quad u(0) = 1, \tag{36}$$

Applying the operator J^1 to equation (36),

$$J^1[Du(t)] = -2aJ^1[D^{1/2}u(t)] - bJ^1[u(t)], \tag{37}$$

By using the formula (26), equation (37) will be

$$u(t) = 1 + \frac{2t^{1/2}f_1}{\Gamma(3/2)} - 2aJ^{1/2}[u(t)] - bJ^1[u(t)], \tag{38}$$

Applying ADM to the problem (38), we get

$$u_0(t) = 1 + \frac{2t^{1/2}f_1}{\Gamma(3/2)}, \tag{39}$$

$$u_n(t) = -2aJ^{\frac{1}{2}}[u_{n-1}(t)] - bJ^1[u_{n-1}(t)], \; n \geq 1. \tag{40}$$

Taking ($a = 1$, $b = 1/2$, $f_1 = 1$) then from the relations (39)-(40) the series solution will be:

$$u(t) = 1 + \frac{4\sqrt{t}}{\sqrt{\pi}} - \sqrt{t}\left(27\sqrt{\pi}\sqrt{t} + 8(3+t)\right)/6\sqrt{\pi} + \frac{1}{120\sqrt{\pi}}\left(t\left(32\sqrt{t}(50+t)\right) + 15\sqrt{\pi}(32+17t)\right) + \ldots \tag{41}$$

A comparison between ADM and Laplace solutions given in Figures 3.4.a–3.4.d.

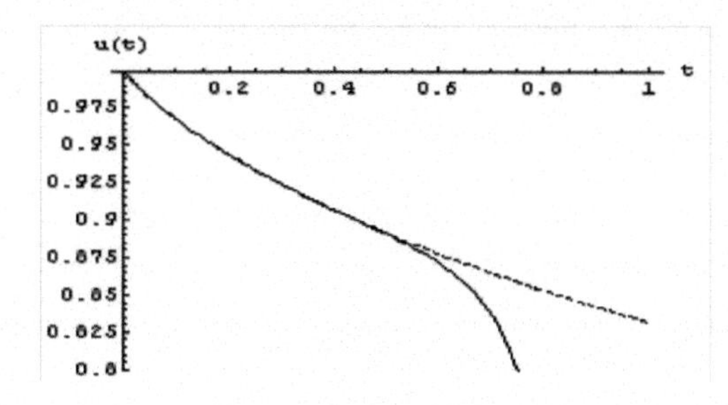

Figure 3.4.a ADM and Laplace Sol. [$n = 20$]

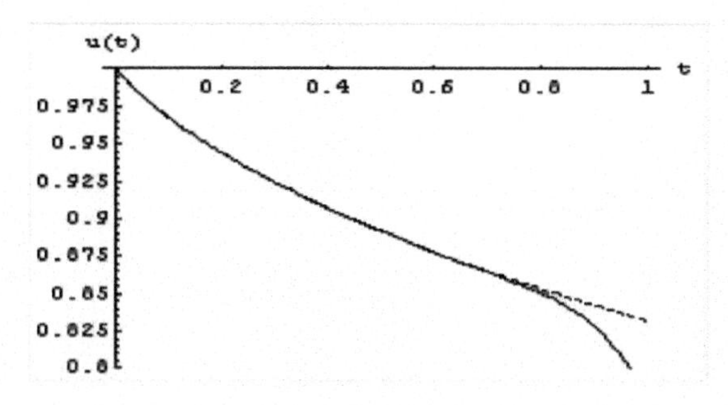

Figure 3.4.b ADM and Laplace Sol. [$n = 25$]

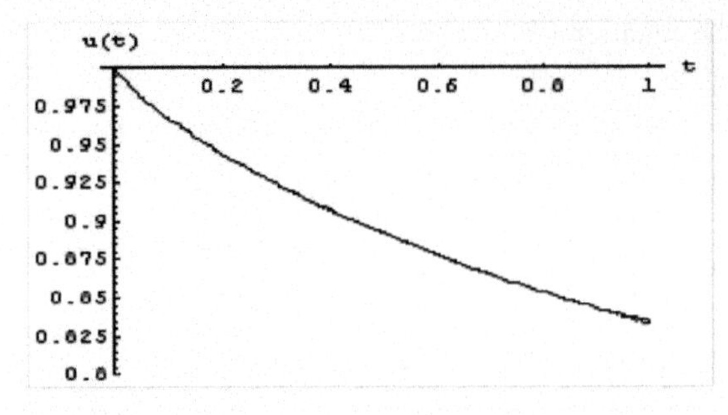

Figure 3.4.c ADM and Laplace Sol. [n = 30]

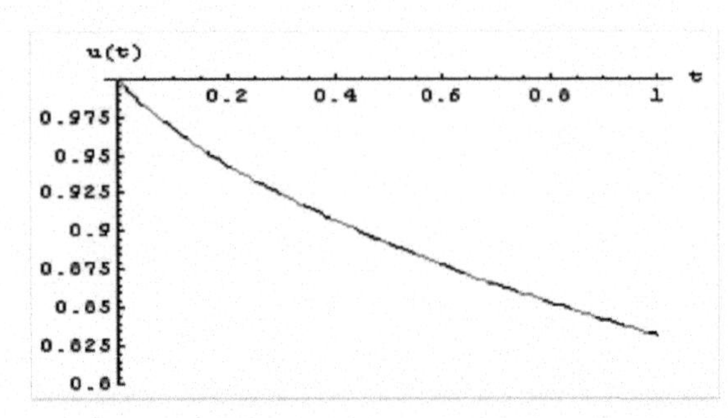

Figure 3.4.d ADM and Laplace Sol. [n = 35]

Now, let turn again to the *Basset problem* (when $a = 1, f_1 = 0, q(t) = 1$)
Its solution by using Laplace transform in Carpinteri and Mainardi (1997) is

$$u(t) = 1 - \left((1-2t)E_{1/2}\left(-\sqrt{t}\right) + 2\sqrt{t/\pi}\right). \tag{42}$$

Moreover, the solution by using ADM will be

$$\begin{aligned} u_0 &= t, \\ u_n &= -2J^{1/2}\left[u_{n-1}(t)\right] - J^1\left[u_{n-1}(t)\right],\ n \geq 1. \end{aligned} \tag{43}$$

From the relation (43), the series solution will be

$$u(t) = t - \frac{8t^{\frac{3}{2}}}{3\sqrt{\pi}} - \frac{t^2}{2} + \frac{1}{30}t^2\left(60 + \frac{64\sqrt{t}}{\sqrt{\pi}} + 5t\right) + \ldots$$

Figure 3.4.e shows the comparison between ADM and Laplace solutions.

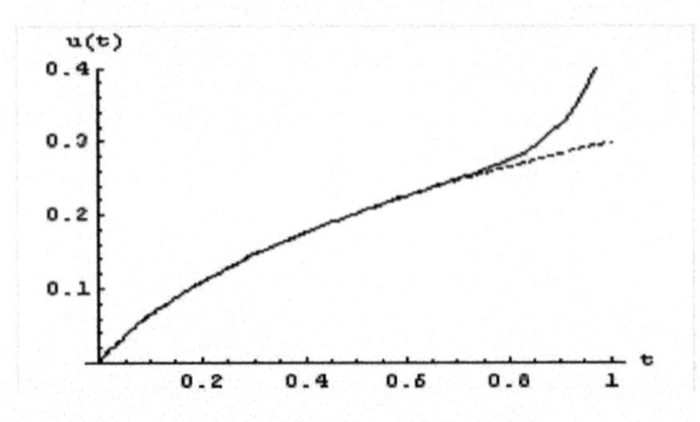

Figure 3.4.e ADM and Laplace Sol. [n = 20]

Conclusions

In this paper, an interesting method (ADM) used to solve four important applications to fractional differential equations used in Fluid Mechanics. These applications are Bagley-Torvik Equation, relaxation-oscillation equation, Basset problem and fractional telegraph equation. This method gives analytical solution and when we comparing ADM solution with Numerical, Laplace and Green's solution methods, we see that the solutions are very enclosed to each other (See Figures 3.1, 3.2, 3.3, 3.4).

References

Abbaoui, K. and Cherruault, Y. (1994). Convergence of Adomian's method applied to differential equations. Comput. Math. Appl. 28:103–109.

Adomian, G. 1983. Stochastic system. Cambridge: Academic Press.

Adomian, G. 1986. Nonlinear stochastic operator equations. New York: Academic Press.

Adomian, G. 1989. Nonlinear stochastic systems: Theory and applications to physics. Dordrecht: Kluwer Academic Publishers.

Adomian, G. (1995). Solving frontier problems of physics: The decomposition method. Dordrecht: Kluwer Academic Publishers.

Ahmed, E., El-Sayed, A. M. A. and El-Saka, H. A. A. (2007). Equilibrium points, stability and numerical solutions of fractional-order predator-prey and rabies models. J. Math. Anal. Appl. 325(1):542–553.

Atangana, A. and Alabaraoye, E. (2013). Solving a system of fractional partial differential equations arising in the model of HIV infection of CD4+ cells and attractor one-dimensional Keller-Segel equations. Adv. Differ. Equ. 94:1–14.

Carpinteri, A. and Mainardi, F. (1997). Fractals and fractional calculus in continuum mechanics. 223–276. Wien and New York: Springer Verlag.

Cascaval, R. C., Eckstein, E. C., Frota, C. L., and Goldstein, J. A. (2002). Goldstein, fractional telegraph equations. J. Math. Anal. Appl. 267:145–159.

Cherruault, Y., Adomian, G., Abbaoui, K. and Rach, R. (1995). Further remarks on convergence of decomposition method. Int J. Bio-Med. Comput. 38:89–93.

Daftarder-Gejji, V. and Jafari, H. (2005). Adomian decomposition method: A tool for solving a system of fractional differential equations. J. Math. Anal. Appl. 301:508–518.

Daraghmeh, A., Qatanani, N., and Saadeh, A. (2020). Numerical solution of fractional differen-tial equations. Applied Math. 11:1100–1115. doi: 10.4236/am.2020.1111074.

Diethelem, K., Ford, N. J., Freed, A. D., and Luchko, Y. (2005). Algorithms for the fractional calculus: A selection of numerical methods. Compt. Methods Appl. Mech. Engrg. 194:743–773.

Diethelm, K. and Ford, N. J. (2002). Numerical solution of the Bagley-Torvik equation. BIT 42:490–507.

El-kalla, I. L. (2008). Convergence of the Adomian method applied to a class of nonlinear integral equations. Appl. Math. Lett. 21:372–376.

El-Mesiry, E. M., El-Sayed, A. M. A. and El-Saka, H. A. A. (2005). Numerical methods for multi-term fractional (arbitrary) orders differential equations. Appl. Math. and Comput. 160(3):683–699.

El-Salam, S. A. A. and El-Sayed, A. M. A. (2007). On the stability of some fractional-order non-autonomous systems. Electron. J. Qual. Theory Differ. Equ. 6:1–14.

El-Sayed, A. M. A. (1993). Linear differential equations of fractional orders. J. Appl. Math. Comput. 55(1):1–12.

El-Sayed, A. M. A. and El-Salam, S. A. A. (2008). On the stability of a fractional-order differential equation with nonlocal initial condition. Electron. J. Qual. Theory Differ. Equ. 29:1–8.

El-Sayed, A. M. A., El-Mesiry, E. M. and El-Saka, H. A. A. (2004). Numerical solution for multi-term fractional (arbitrary) orders differential equations. Comput. Appl. Math. 23(1):33–54.

Evans, D. J. and Raslan, K. R. (2005). The adomian decomposition method for solving delay differential equation. Int. J. Comput. Math. 82:49–54.

Hammad, H.A. and De la Sen, M. (2021). Tripled fixed point techniques for solving system of tripled-fractional differential equations. AIMS Math. 6(3):2330–2343.

Hefferan, J. M. and Corless, R. M. (2006). Solving some delay differential equations with computer algebra. Mathematical Scientist. 31(1):21–34.

Jafari, H. and Daftardar-Gejji, V. (2006). Solving a system of nonlinear fractional differential equations using Adomian decomposition method. J. Appl. Math. Comput. 196(2):644–651.

Khan, N. A., Razzaq, O. A., Ara, A. and Riaz, F. (2016). Numerical Simulation - From Brain Imaging to Turbulent Flows, IntechOpen, London. doi:10.5772/61500.

Kilbas, A. A., Srivastava, H. M. and Trujillo, J. J. (2006). Theory and applications of fractional differential equations. New York: Elsevier.

Mainardi, F. (1997). Fractional calculus: Some basic problems in continuum and statistical mechanics. In Fractals & fractional calculus in continuum mechanics, ed. A. Carpinteri, and F. Mainardi, 291–348. New York: Springer.

Mensour, B. and Longtin, A. (1998). Chaos control in multistable delay-differential equations and their singular limit maps. Phys. Rev. E. 58:410–422.

Miller, K. S. and Ross, B. (1993). An introduction to the fractional calculus and fractional differential equations. New York: Wiley-Interscience.

Momani, S. and Al-Khaled, K. (2005). Numerical solutions for systems of fractional differential equations by the decomposition method. J. Appl. Math. Comput. 162:1351–1365.

Podlubny, I. (1994) The Laplace Transform Method for Linear Differential Equations of the Fractional Order, UEF-02-94, Inst. Exp. Phys, Slovak Acad. Sci., Kosice, pp. 1-32.

Podlubny, I. (1999). Fractional differential equations. New York: Academic Press.

Rida, S. Z. and Arafa, A. A. M. (2011). New method for solving linear fractional differential equations. Int. J. Differ. Equ. 2011:1–8. doi:10.1155/2011/814132.

Saha Ray, S. and Bera, R. K. (2004). Analytical solution of the Bagley Torvik equation by Adomian decomposition method. J. Appl. Math. Comput. 121:331–342.

Shawaghfeh, N. T. (2002). Analytical approximate solution for nonlinear fractional differential equations. J. Appl. Math. Comput. 131:517–529.

Trinks, C. and Ruge, P. (2002). Treatment of dynamic systems with fractional derivatives without evaluating memory-integrals. Comput. Mech. 29(6):471–476.

Zwillinger, D. (1997). Handbook of differential equations. New York: Academic Press.

4 Social network analysis for predicting students' dropouts and performance using e-learning

Samuel-Soma M. Ajibade[1,a], *Oluwadare Joshua Oyebode*[2,b], *Sushovan Chaudhury*[3,c], *Odafe Martin Egere*[4,d], *Gloria Ekene Amadi*[5,e], *and Geovanny Genaro Reivan Ortiz*[6,f]

[1]Department of Computer Engineering, Istanbul Ticaret Universitesi, Istanbul, Turkey

[2]Department of Civil and Environmental Engineering, Afe Babalola University, Ado Ekiti Ekiti State, Nigeria

[3]Department of CSE, University of Engineering and Management, India. ORCID ID: 0000-0002-9336-2654

[4]Department of Marketing, Innovation, Strategy and Operations School of Business, University of Leicester, England, UK

[5]Institute of Climate Change Studies, Energy and Environment, University of Nigeria, Nsukka, Nigeria

[6]Laboratory of Basic Psychology, Behavioral Analysis and Programmatic Development, Catholic University of Cuenca, Ecuador

Abstract

Social network analysis has surfaced as a result of the growing popularity of social networking sites such as Facebook, Instagram, Snapchat and Twitter. Finding the underlying relationships that exist amongst individuals and the causes for their emergence, as well as appraising the sorts of interactions that occur, are examples of tasks that have become increasingly essential in business. Nevertheless, it is not the only situation where social network analysis software could be beneficial. We show that social network analysis may be a useful asset within the educational setting for resolving tough difficulties like establishing a learners' level of cohesion, forum participation, and identifying the most influential students in this study. Furthermore, we show how effective monitoring of social behavioural data, coupled by the use of students' activities, contributes in the construction of further exact outcome and dropout forecasts. Our conclusions are based on a three-year evaluation of an e-learning course delivered at the Federal University of Technology Akure (FUTA).

Keywords: Dropout, e-Learning social network analysis, student performance.

Introduction

In recent years, social network analysis (SNA), which is made up of creating patterns that gives room for the identification of basic interactions that occur amongst users of various, has had a significant impact. 'The presence of social networking sites like, Facebook, Snapchat, Instagram or Twitter, has rekindled interest in this area, there offering techniques used for the advancement of market research based on user activity within those services' Adetola, et al. (2020). The advancement of social media networking platforms has provided unique channels and new chances in order for participation and combination to increase and also provide a good opportunity for changing the way by which people learn Adetola, et al. (2020). 'SNA techniques, on the other hand, do not just focus on social networks, but also on other fields such as marketing (customer and supplier networks) or public safety' (Rehman and Asghar, 2020). 'Education is one of the fields in which they are used' (Desai, et al., 2020). 'SNA provides for the retrieval of numerous metrics from online course student activity, like the learners' cohesion level, their degree of engagement in discussions, or even the recognition of most influential ones and this type of analysis may be useful for teachers in understanding the behaviours of the students and, as a result, achieving better results' (Romero and Ventura, 2010). SNA can also be used to generate new data as attributes, which is then mined using data mining approaches to uncover the patterns of student behaviour. There is a well- specified area in the educational domain known as educational data mining. One of the critical topics explored in this domain is the building of effective outcomes and dropouts predictions to help teachers avoid learners from failing their courses. As a result, classification techniques based on prediction models are commonly used to reveal student behaviour, like the time spent on completing assigned task or activities in discussions forums,

[a]asamuel@ticaret.edu.ng; [b]oyebodedare@yahoo.com; [c]sushovan.chaudhury@gmail.com; [d]ome2@leicester.ac.uk; [e]talk2gloriaekene@gmail.com; [f]greivano@ucacue.edu.ec

resulting in a success, a failure, or a dropout. "Student's academic performance is a significant factor in any educational institution, for that reason it could ensure strategic programme be planned in continuing enlightening or guiding the students for a better performance that may leads them to a better future' (Chweya et al., 2020). SNA introduces a new suitable platform for predictions that has the potential to enhance the efficiency of such models. In this work, which examines an e-learning course given at FUTA, our proposed contribution is identify behavioural patterns and create models that may accurately predict student success and dropouts using SNA. The following is how the paper is structured. In section 2, the literature review in the use of SNA in the education sector is presented while in section 3, the course that is being examined is discussed and also the datasets are presented. In section 4, the experimental result is discussed and lastly section 5 presents the conclusion of this study.

Related Works

SNA is the systematic analysis of social interactions of linked actors. In case of network theory, 'SNA depicts the actors and interactions as a graph or network, with every node representing an individual actor within the network, such as with a person or an organization, and each link representing social interaction among two of those actors, such as friendship. Despite the fact that social networks have been studied for a long time' Scott (2000), 'the recent rise of social network sites such as Facebook, Instagram, Snapchat, and Twitter have resulted in the extraordinary popularity that this field of study now enjoys' (Feng, et al., 2021). 'Social interaction refers to students' ability to perceive themselves as a community supporting positive interdependence. Such an interaction in the learning process occurs when students achieve cooperative tasks and share the sources' (Ajibade et al., 2018). To measure the significance of nodes that exist within a social network, a number of centrality algorithms are being developed. Centrality investigations seek to address the query, 'What nodes inside a social network are most essential?' While there are numerous ways to define relevance, notable nodes are firmly connected to other nodes. 'People with a large number of social network connections are generally thought to be more influential than those with fewer contacts and the degree of a node is the sum of links connected to it regardless of the path of the links, is by far the most basic centrality metric' (Zhang et al., 2020). 'Taking the direction of the links into account, a node possesses indegree and outdegree, which seem to be the amount of inbound and outbound links that are attached to it, however there are some more complicated centrality metrics, such as with a node's betweenness and that is equivalent to the amount of shortest route from across all nodes to all else which pass through this kind of node, along with authorities and hubs' (Sreehari et al., 2019). 'A node is just an authoritative figure only if inbounds links relate it all to nodes with a great volume of outbound links, whilst a node is just a hub if its outgoing links hook up it to nodes with a small number of inbound links' (Freeman, 1977). Centrality measures are actual valuable tool for determining how significant a node is in a network at the level of the node. A few of them, such as betweenness, cannot be quickly estimated due to their nature, so network-level metrics, that can be measured quite effortlessly as well as provide useful information by recognizing the network as whole and, is used to supplement the preceding centrality measures. Network density is one of the network-level performance factors which relates the amount of links that are contained in the network to the maximum permissible number of links. Another interesting network-level statistic is network diameter, which is given as the highest number of network nodes which can only be navigated so as to move from one node to the next. In the field of education, SNA has some applications. Bruun and Evans (2018) utilised quiet a number of numerous regression analysis of the Bonacich centrality in order to analyse the variables which impacts engagement in school communities, such as the age of students or sex of students. Kim (2018) Crespo and Antunes (2015) suggested an approach used to measure each student's global contribution in a teamwork using PageRank algorithm adaptations. 'SNA also provide useful information for educational data mining tasks like performance prediction or dropouts' (Bayer et al., 2012). For example Bayer et al. (2012) predicted dropouts using the following centrality measures: 'degree, indegree, outdegree and betweenness'. It was concluded that such metrics enhance the precision of classification models when compared to using only demographic and academic attributes, such as the age of students', students' gender, or number of completed semesters. "There has been swift information and vast improvement in the technological infrastructural advancement in recent times, hence the use of new media means like social network has become a progressive significant affair, and these have had a good impact on the present-day teaching and learning' (Ajibade et al., 2017).

Research Methodology

The case study makes use of data from an online course called 'Introduction to web design and technologies' which is conducted on Moodle. This course was taught at FUTA for three academic sessions in a row (2016/2017, 2017/2018, and 2018/2019). Web pages, animations, video tutorials, Flash animations were used to create it. The students were asked to carry out eight tasks/assignments as well as a basic final test. Specifically, the forum which was utilized in this research work to create social interactions that exist between students and teachers primarily used by students to make inquiries about how the course was organized, the contents of the course and the set timelines for the course by the teacher to answer every of the students' hesitations and make necessary notices to the students. The average number of students that signed up for this course was 120, and more than half of the students finished the course while the remaining students dropped out. The students' have various backgrounds ranging from engineering to statistics to marketing to economics to computer science to environmental sciences. Three different datasets were created using the Moodle student data and social behaviour from ORA social network analyses. The main activities qualities of students are: (i) duration of class time, (ii) estimated number of sessions done, (iii) average amount spent and total duration of sessions implemented each week, and (iv) estimated number of read and published messages on the discussion board and through email. I degree, (ii) in-degree, (iii) out-degree, (v) betweenness, (vi) hub, and (vii) top3 (viii) percent in top3, which means that the volume of top3 for a given node is set to true. Each of the three datasets contains 245 instances, one per student, and the only difference becoming the values of the class feature. The performance of each student is critical. If a learner decided to drop out, it means he or she failed the course. Regardless of whether he or she dropped out of the course and mixed, if he or she succeeded, failed, or dropped out. This helps to fulfil our main objective of identifying behavioural patterns and create models that may accurately predict student success and dropouts using SNA.

Results

Figure 4.2 illustrates the interactions of network interactions that occur in-between the students and teachers in the 2017/2018 session in FUTA where the course 'Introduction to web design and technologies.'

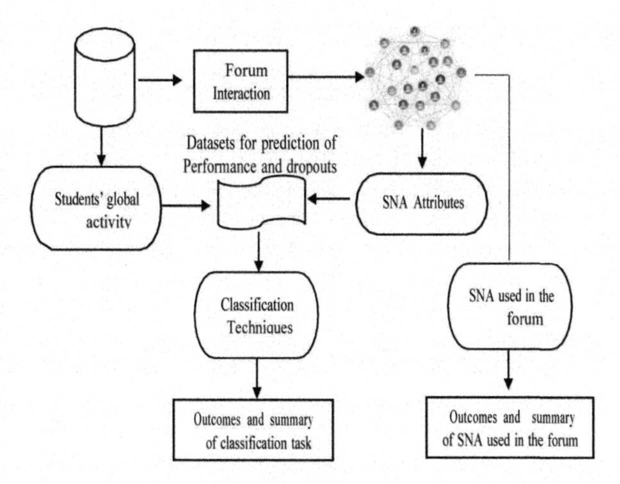

Figure 4.1 The research processes

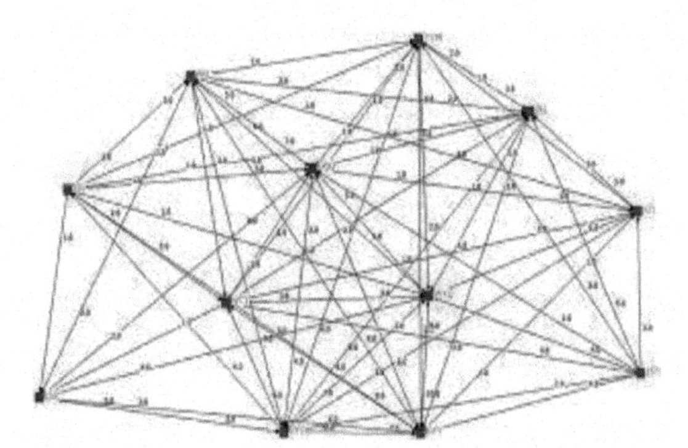

Figure 4.2 Interaction network between the students and teacher of the course "Introduction to web design and technologies"

As shown, the node with the most links corresponds to the major teacher. This means that in this course, the majority of connections in the discussion board happen between both the students and the teachers, while interactions between students are less common. As the teacher pointed out, the discussion group is primarily used in two ways: (i) the learners make inquiries about the course's content, which the teacher should answer. As well as (ii) the teacher post significant notices. This type of interaction is better represented graphically by SNA, which makes it simpler for the teacher to give explanation for them, for example, if the discussion board was used for a specific activity, this could be useful in ensuring its success.

The node centrality attributes that is displayed in Table 4.1 help to clarify the preceding findings. On the one hand, the teacher (node 1) has the most degrees and outdegree attributes. Furthermore, there is a significant outdegree variant that is between the instructor and the 2nd ranked participants. This is also true for the betweenness and hub centrality measures. As a result, it can be deduced that the teacher is the user that responded to many messages that were sent by students in the discussion board. The greatest indegree and authority values, on the other side, correlate to nodes 2, 3, and 6. These learners are the ones who uses the discussion forum and have posted more messages than others. In essence, the learners received the highest marks in the course. As a result of this assessment, it can be ascertained that students who participated frequently in the discussion board are much more likely to receive excellent grade, a fact which will be explored by utilising data mining methodologies. The teacher can gain a better understanding of the students' behaviour and promote participation more in the class. The other two academic sessions, 2016/2017 and 2018/2019, yielded similar results.

First, an investigation was carried out in which SNA attributes were more significant for classifying the three datasets described above. This began by using Weka's attribute selection algorithm, CfsSubSetEval, that chooses the best features for all the classifiers. As an outcome, top3 is by far the most important trait, implying that no other feature is considered more important than top3. Using association rules, the same was confirmed.

A part of the guidelines revealed that if top3 is true, then the students complete the course with a 70% confidence level using the Apriori algorithm and the mixed data as input (discretised with PKI Discretize). On the one hand, it was revealed that the most important feature for this concrete classifier would not be top three, but rather the betweenness and authority centrality measures, utilising Classifier SubSetEval as the attribute selection technique and J48 as the base classifier. The degree, authority, and hub centrality measurements were the most important variables when using naive Bayes as the base classifier. As a result, it was concluded that different features are the most vital for the prediction task for various models. Furthermore, it was investigated on if SNA metrics could be used for performance prediction and dropouts. Six classifiers were built for this task using the three datasets that were described in the previous section, but using social data, not activity data. Bayesian networks, J48, random forests and naive Bayes are the classification techniques selected for this task. The Weka implementations of these algorithms were used, and the main parameters were used. Table 4.2 displays the average accuracy of each algorithm when using ten-fold cross validation.

In Table 4.2, the accuracy for predicting dropouts is greater than 70% in all cases, attaining more than 75% when using Bayesian networks. As regards to performance, all of the algorithms outperformed each other by more than 70%. From the result of this analysis, it can be asserted that the attributes of SNA are able to accurately predict student performance and dropouts. Finally, the accuracy of the classifiers was compared when used with the mixed dataset as an input to determine how the SNA features contribute to better prediction. First, classifiers were built solely based on activity data, and afterwards classifiers were built based on the entire data set, — in other words, social and activity data. Similarly, we investigated on whether to use discretized SNA attributes would improve accuracy. Weka's PKI Discretize algorithm was used to perform the discretisation. The accuracy attained for each case is shown in Table 4.3.

The findings demonstrate that when SNA attributes were used, all the four classifiers improved their accuracy with Bayesian Networks obtaining a substantial improvement of 2.09%. However, when

Table 4.1 Ranking of three nodes for different centrality measures

	Top 1 Node ID	*Value*	*Top 2 Node ID*	*Value*	*Top 3 Node ID*	*Value*
Degree	1	157	3	35	7	32
Indegree	3	30	6	31	7	12
Outdegree	1	142	4	5	23	5
Betweenness	1	402	19	143	12	132
Authority	3	0.87	5	0.66	5	0.32
Hub	1	1.39	21	0.05	3	0.02

Table 4.2 Acuracy percent in performance and dropout of students

Classification Techniques	*Performance*	*Dropout*
Bayesian Networks	74.10	75.30
J48	72.10	72.90
Random Forest	73.35	73.50
Naıve Bayes	70.90	71.00

Table 4.3 Percentage of accuracy in mixed data

Classification techniques	*Activity attributes*	*Activity and SNA attributes*	*Activity and SNA Discretized Attributes*
Bayesian Networks	84.20	86.29	89.31
J48	79.25	80.91	82.22
Random Forest	81.33	82.90	84.66
Näive Bayes	77.68	78.01	78.90

discretized SNA attributes were used, the improvement was 5.11% higher. The accuracy of all other algorithms increased significantly. In summary, we can deduce that the attributes of SNA is able to improve the performance of the students and the dropout prediction.

Conclusion

In this article, SNA techniques were used to analyse interactions that exist between students in a course that was taught at the FUTA over three academic sessions. In addition, data mining techniques was to predict the performance of students and dropouts. In the education sector, SNA is often a strong framework for analysing the social behaviour of students' and their relationships with their teachers. Discussion forums are powerful tools that can be used for this type of analysis. Furthermore, it was discovered that SNA can help the teachers to understand better the way students make use of the discussion forum, noting that they frequently use this tool to make inquiries about the course's contents or its organisation, but it is rarely used to answer questions posted by other students, which are answered by the instructor. In fact, we demonstrated that the students that were able to post most questions and received the most responses are about the same students who received higher grades at the completion of he course. This fact is more accurately reflected by data mining analysis. Based on the classification of the performance of students and their dropouts, it can be concluded that the attributes of SNA are quite beneficial for both goals, as the models of classification of various types of classifiers obtained accuracy values of more than 70% with them. We hope to broaden the scope of our research work in the future. To begin, we want to evaluate the efficacy of our method by utilizing additional e-learning courses with varying characteristics. Despite the fact that the number of instances in educational data mining datasets is typically limited, we intend to explore on courses that have a larger number of students. Subsequently, we want to evaluate some certain SNA metrics and algorithms, such as PageRank, and also data mining techniques other than classification, such as association.

References

Adetola, O., Shamsuddin, S. M., Chweya, R., and Ajibade, S. M. (2020). Social communication of students on social media network platform: A statistical analysis. J. Sci. Eng. Technol. Manag. 2(2): 1–9.

Ajibade, S. S. M., Ahmad, N. B. and Shamsuddin, S. M. (2018). A Study of Online and Face to Face Tutors and Learners. Practices in Collaborative Blended Learning. UTM Computing Proceedings, Innovations in Computing Technology and Applications. Vol.3, page 1–5

Ajibade, S. S. M., Shamsuddin, S. M., and Bahiah, N. B. (2017). Analysis of social network collaborative learning on knowledge construction and social interaction of students. In ASIA International Multidisciplinary Conference (AIMC-2017). Malaysia.

Bayer, J., Bydzovská, H., Géryk, J., Obšıvac, T., and Popelınský, L. (2012). Predicting drop- out from social behaviour of students. In Proceedings of the 5th International Conference on Educational Data Mining. (pp. 103–109).

Bruun, J. and Evans, R. (2018). Network analysis as a research methodology in science education Research. Pedagogika. 68(2):1–17.

Chweya, R., Shamsuddin, S. M., Ajibade, S. M. M. and Moveh, S. (2020). A literature review of student performance prediction in e- learning environment. J. Sci. Eng. Techno. Manag. 1(1):22–36.
Crespo, P. T., & Antunes, C. (2015). Predicting teamwork results from social network analysis. Expert Systems, 32(2), 312–325.
Desai, U., Ramasamy, V. and Kiper, J. D. (2020). A study on student performance evaluation using discussion board networks. In Proceedings of the 51st ACM Technical Symposium on Computer Science Education (pp. 500-506).
Feng, P., Lu, X., Gong, Z., Li, B. and Sun, D. (2021). Social network analysis model for research on organizational structure of the pyramid scheme communication network, MethodsX, Vol 8, 2021, 101259, ISSN 2215-0161, https://doi.org/10.1016/j.mex.2021.101259
Freeman, L. (1977). A set of measures of centrality based on betweenness. Sociometry. 40(1):35–41.
Kim, D. (2018). A Study on the influence of Korean middle school students' relationship through science class applying STAD cooperative learning. J. Techno. Sci. Edu. 8(4):291–309.
Liu, Z., Kang, L., Domanska, M., Liu, S., Sun, J., & Fang, C. (2018, March). Social Network Characteristics of Learners in a Course Forum and Their Relationship to Learning Outcomes. In CSEDU (1) (pp. 15–21).
Rehman, S. U. and Asghar, S. (2020). Online social network trend discovery using frequent subgraph mining. Soc. Netw. Anal. Min. 10(1):1–13.
Romero, C. and Ventura, S. (2010). Educational data mining: a review of the state of the art. IEEE Trans. Syst. Man Cybern. Part C (Applications and Reviews), 40(6):601–618.
Scott, J. (2000). Social network analysis: A handbook. (2nd ed. p. 6). London: Sage.
Sreehari, R., Pillai, R. R. and Indulekha, T. S. (2019). Circuit detection in web and social network graphs. In 2019 2nd International Conference on Intelligent Computing, Instrumentation and Control Technologies (ICICICT). vol. 1, pp. 323-1326). IEEE..
Tabakhi, S. and Moradi, P. (2015). Relevance–redundancy feature selection based on ant colony optimization. Pattern Recognition, 48(9):2798–2811.
Zhang, A. J., Matous, P. and Tan, D. K. (2020). Forget opinion leaders: the role of social network brokers in the adoption of innovative farming practices in North-western Cambodia. Int. J. Agri. Sus. 18(4):266–284.

5 Optimal scheduling methodology for machines, tool transporter and tools in a multi-machine flexible manufacturing system without tool delay using flower pollination algorithm

Sivarami Reddy Narapureddy[1,a], Padma Lalitha Mareddy[2,b], Chandra Reddy Poli[3,c], and Sunil Prayagi[4,d]

[1]Mechanical Engineering Department, Annamacharya Institute of Technology and Sciences, Rajampet, India

[2]Electrical and Electronics Engineering Department, Annamacharya Institute of Technology and Sciences, Rajampet, India

[3]Department of Mathematics, Annamacharya Institute of Technology and Sciences, Rajampet, India

[4]Mechanical Engineering Department, Yeshwantrao Chavan College of Engineering, Nagpur, India

Abstract

The aim of this work is to schedule machines, tool transporter (TT), and tools concurrently in a multi-machine Flexible Manufacturing System (FMS) with the fewest number of copies of each tool variety achievable without tool delay, while also taking tool transfer times into account for minimization of the makespan (MKSN). The tools are kept in a central tool magazine (CTM) which is shared by numerous machines and serves them. The objective is to allocate job-operations (jb-ons) to machines, determine the fewest number of copies of every tool type and to allocate copies and corresponding trips of TT to job-operations for makespan minimization. This work proposes a formulation of nonlinear mixed- integer programming (MIP) to model and flower pollination algorithm (FPA) is employed to work out this problem. The results show that employing an additional copy each for few tool types and a copy each for remaining tool types causes no tool delay as well as reduction in makespan and FPA outperforms Jaya algorithm.

Keywords: Flower pollination algorithm, FMS, makespan, no tool delay, optimisation techniques, scheduling of machines, tool transporter and tools.

Introduction

Manufacturing firms must deal with increasing product complexity, shorter product life cycles, new technologies, global competitive concerns, and a rapidly changing environment. Flexible Manufacturing System (FMS) is primarily established to handle the manufacturing competition. FMS is properly connected manufacturing system that includes multifunctional computer numerical controlled (CNC) machines (MCs) coupled with automated material handling system (AMHS). FMS strives for manufacturing flexibility without compromising product quality. To achieve its flexibility the FMS depends on the flexibility of AMHS, control software, and CNC MCs. Workflow patterns, size, and manufacturing type have all been used to group FMSs into different types. From a planning and control perspective, there are four kinds of FMS: single flexible MC, flexible manufacturing cell, multi-cell FMS, and multi-machine FMS (MMFMS) (Saravanan and Noorul Haq, 2008). Existing FMS installations have already established benefits such as cost savings, increased utilization, lower work-in-process, and so on. SCHG tasks improve resource utilization by lowering the makespan (MKSN). Solving scheduling (SCHG) problems optimally or near optimally is one technique to attain high productivity in FMS.

Tool loading becomes a complex component in the SCHG issues since the tool copies' number is restricted and may be fewer than the MCs' number owing to cost limitations. Job and tool SCHG is a critical issue for industrial systems. Ineffective job SCHG and tool loading planning can result in capital-intensive machinery's underutilization and a high amount of idle time (Shirazi and Frizelle, 2001). Therefore, effective job and tool SCHG allow a production system to maximize MCs usage while reducing idle time. There have been several studies on the MCs and tools concurrent SCHG.

[a]siva.narapureddy@gmail.com; [b]padmalalitha.mareddy@gmail.com; [c]chandramsc01@gmail.com; [d]sunil_prayagi@yahoo.com

Aldrin Raj et al. (2014) employed different algorithms and rules for addressing MCs and tools concurrent SCHG in an FMS with 4 MCs and a CTM for minimisation of MKSN. Özpeynirci (2015) proposed the time-indexed mathematical approach for joint tool and MC SCHG in an FMS for MKSN minimization. Costa et al. (2016) devised the hybrid GA which uses the local search improvement technique to tackle 'p' components SCHG on 'q' MCs considering tool alter activity prompted due to wear of the tool. Sivarami Reddy et al. (2016), Sivarami Reddy et al. (2017) solved the combined tools and MCs SCHG to provide the best sequences in MMFMS that minimise the MKSN. Using the simulated annealing approach, Baykasoglu and Ozsoydan (2017) tackled the indexing of automatic tool changer and tool altering problems in tandem to minimise the non-machining periods in the automatic machining centres. Paiva and Carvalho (2017) solved the problem of job ordering and tool changing by determining the job sequence and tool loading sequence to minimise tool changes. For addressing sequencing problems, an approach that makes use of heuristic, local search methods and graph representation, is applied. Such strategies are used with traditional tooling techniques to solve the issue in local search method with an algorithmic fashion. Gökgür et al. (2018) used models of constraint programming for parallel MCs scenarios to tackle tool SCHG and allocation issues with predetermined tool quantities in the system owing to budgetary constraints to minimise the MKSN. *In the references discussed above, job and tool shift times among MCs are not considered, and operations are only scheduled on specific MCs.*

Sivarami Reddy et al. (2018a) and Sivarami Reddy et al. (2018b) solved the MC and tool combined SCHG problems with one copy of each tool kind considering tool transfer times for minimum MKSN (SMTTATWACT) and made known that tool transfer times had a major influence on the MKSN. In the above references *Job transport times are not considered and operations are scheduled only on specific MCs.*

Few studies are conducted on AGVs, tools, and MCs parallel SCHG with a replica of each tool kind (SMAT). SMAT for minimization of MKSN where operations are planned on specific MCs was addressed by Sivarami Reddy et al. (2019), Sivarami Reddy et al. (2021a), Sivarami Reddy et al. (2021b), and Sivarami Reddy et al. (2021c).

When many jb-ons require the same tool simultaneously, the tool will be assigned to only one operation, causing other operations to wait for it, in SMTTATWACT. The MSN rises as a result of this. Therefore it is important to address machines, TT and tools simultaneous scheduling with the lowest number of copies of each tool kind which are much less than the number of machines without tool delay (SMTTATWLNTC) by sharing the tools among machines to minimise MKSN in an FMS. As it causes no tool delay which in turn reduces MKSN with few additional copies for few tool types. The novelty of the research is addressing the SMTTATWLNTC for minimization of MKSN first time, presenting nonlinear MIP model, FPA is employed to minimise MKSN, and demonstrates its difference from other studies.

The SCHG problem of job shop, and problem of TT SCHG is similar to the lift up and delivery problem, are NP-hard issues. Combing MC SCHG and TT SCHG in an integral way makes the problem a double interlinked NP-hard problem, and moreover, finding out the fewest possible quantity of copies of each tool kind without tool delay and assigning them to jb-ons makes the SCHG problem more multifaceted and also it is anticipated that MC, TT, and tool usage would improve.

Xin-She Yang (2012) developed the FPA. It is based on blooming plants' pollination process, which aims to generate the best population in terms of quantity and quality in order to continue to exist. It's been used to solve engineering optimization issues (Sivarami Reddy et al., 2018b; Ersin and Ali Payidar, 2018; Sivarami Reddy et al., 2021a) with great success. The FPA has benefits over most other meta-heuristic algorithms, including the fact that it only requires two algorithm parameters and is simple to implement.

Problem Formulation

FMS kinds and operations require different setups. The majority of research focuses on specified production systems since having a standard configuration is not practical. The following sections provide details on the system, hypotheses, criterion for purpose, and the issue addressed in this study.

A. FMS environment

To reduce tool inventories, some production systems use common tool storage with one CTM serving numerous MCs. During the machining of the component, the needed tool is shared between MCs or shifted from CTM. The CTM reduces the tool copies in number necessary in the system, lowering tooling costs. TT will switch the tools between the MCs. Automated guided vehicles (AGVs) will perform the job switching between the MCs. The FMS in this study is considered to be made up of 4 CNC MCs, each with its own tool varieties and ATC, as well as a TT and two indistinguishable AGVs. FMS delivers components for manufacture from loading and unloading stations (LUS), and completed components are deposited in LUS and transferred to the ultimate storage facility. Automatic storage and retrieval system (AS/RS) is offered to replenish the work-in-progress. Figure 5.1 depicts its setup.

B. Assumptions

Assumptions mentioned below are made for the problem being studied

- All jobs, TT, and MCs are initially accessible for usage.
- At any particular time, a tool or MC can only do one operation.
- The pre-emption of an jb-on on a MC is prohibited.
- Prior to SCHG any jb-on, the requisite MCs and tools are identified.
- Each job has its own operations set, each of which has processing time.
- The setup time is also factored into the processing time.
- The tools are initially placed in a CTM.
- The time it takes for parts to travel between MCs isn't considered.
- When the jobs arrive, the tools will have enough service life to process the operations assigned to them.
- A flow route layout for the TT is specified, travel times on each path segment are known and it moves along the shortest predetermined paths.
- There is just a TT that transfers tools in the system, and tool swap period across MCs are taken into account.
- TT starts from the CTM initially, returns to the CTM after all their assignments have been done and holds a single unit at a time.

C. Constraints

Process planning information is supplied in terms of precedence restrictions for each task in order to determine the operations' order for optimizing the tolerance stack-up.

D. Problem Definition

Consider a job set 'J' with 'j' job kinds (J_1, J_2, J_3..., J_j) job set's total operations (1,2,3.......N) 'k' tool kinds (t_1,t_2, t_3.......t_k) with little copies of every variety and 'm' MCs (M_1, M_2, M_3..., M_m) in an FMS. The jb-ons are processed in a preset order that is known ahead of time. The jb-ons are processed in a preset order that is known ahead of time. A MC can only carry out one jb-on at a time. A jb-on cannot commence before its preceding operation is accomplished. The arrangement of jb-ons on a MC governs the MC's setup needs. The concurrent SCHG problem is defined as finding the fewest number of copies of each tool variety, allocating copies of tools to jb-ons and sequencing them on MCs without tool delay, every job's commencement and concluding times on each MC and tool copy, and associated flight operations of TT with the empty and loaded flight times of TT for minimization of MKSN in an MMFMS. Because the problem is described clearly, section 4 provides an unambiguous mathematical form.

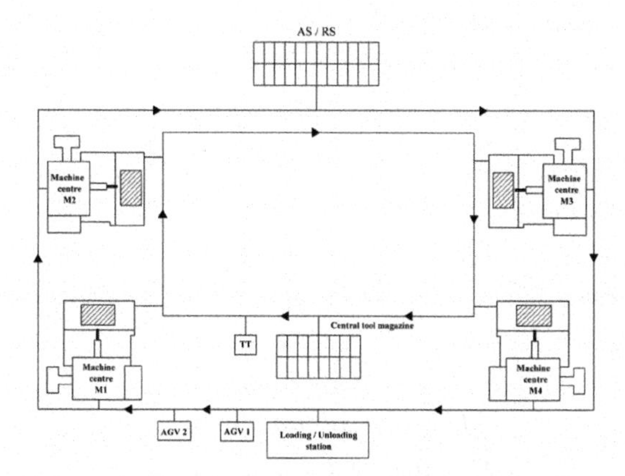

Figure 5.1 FMS environment

Model Formulation

A nonlinear MIP model is offered to visibly illustrate the critical parameters and their impact on the SCHG problem of FMS.
Notations
Subscripts
j job index
i, h jb-ons indices
k tool index
Sets and parameters

J	job set
n_j:	job j's jb-ons
N:	$\sum_{j\in J} n_j$, job set's jb-ons
I:	{1, 2, 3, ----- N}, jb-ons indices set.
I_j:	$\{J_{j+1}, J_{j+2}, \text{------} J_{j+n_j}\}$, the indics connected to job 'j' in 'I', where 'J_j' is jb-ons planned preceding job 'j' and J_1=0.
IS_i:	$I - \{h; h \geq i, i, h \in I_j\}$ set of jb-on indices belong to same job, sans jb-on 'i' , jb-ons succeeding to jb-on 'i'.
IP_h:	$I - \{i; i \leq h, i, h \in I_j\}$ set of jb-on indices belong to same job, sans jb-on 'h' and jb-ons to preceding to 'h'.
ptg_i:	jb- on 'i' processing time.
ctn_i:	jb-on 'i' finish time.
TL:	tool types' set to process the jb-ons
TLCopy:	set of each tool kind copies.
R_k:	set of indices in I coupled with tool kind k in TL, $\forall k \in TL$.
RT_{kj}:	$I_j \cap R_k$ jb-ons index set in I for job j and tool k $\forall k \in TL, \forall j \in J$
b_{kci} :	ready times' set for copies of tool kind 'k' at a MC for jb-on 'i', which is same as ready times at other MCs or CTM plus transport time from those MCs or CTM to this MC.
$TLRM_i$:	$\min(b_{kci})\ \forall k \in TL, \forall c \in TLCopy, \forall i \in R_k$
ct_{ick}:	operation i's finish time with copy c of tool kind k , $\forall i \in I, \forall c \in TLCopy$, and $\forall k \in R_k$
u:	1st jb-on that employs tool k, $u \in R_k, \forall k \in TL$
v:	i's prior jb-on i, $v \in R_k, \forall k \in TL$
L:	TTs present in FMS
a_j:	job j ready time
TM_i:	MC's ready time for jb-on i
ttl_i:	TT loaded flight 'i' tour time together with load and unload times.
ttd_{hi}:	TT empty flight 'i' tour time, commencing at a MC performing jb-on `h` and ending at the MC performing jb-on `i` with the tool that is demanded .
$CTTL_i$:	TT loaded trip 'i' finish time
Q_i:	$\max(ct_{i-1}, TM_i), \forall i \in I$

Decision variables $q_{rs} = \begin{cases} 1 & \text{if } ctn < ctn_s, \text{where } r \text{ and } s \text{ are jb-ons of different jobs} \\ 0 & \text{otherwise} \end{cases}$

$$d_{hi} = \begin{cases} 1 & \text{if the TT is assigned for empty trip between trip 'h' and trip 'i' where the demanded tool is available.} \\ 0 & \text{otherwise} \end{cases} \qquad d_{oi} = \begin{cases} 1 & \text{if the TT starts from CTM to make trip 'i' as its } 1^{st} \text{ assignment} \\ 0 & \text{otherwise} \end{cases}$$

$$d_{ho} = \begin{cases} 1 & \text{if TT returns to the CTM after finishing trip ‘h’ as its final assignment} \\ 0 & \text{otherwise} \end{cases} \qquad ttw = \begin{cases} tttd_{o,u} & \text{if the TT commences from CTM to make trip ‘i’ as its } 1^{st} \text{ accomplishment, otherwise} \\ 0 & \end{cases}$$

A. Mathematical Model

The tool and MC indices are not included in the formulation since every job routing is accessible; nevertheless, the MC index and tool index for every operation index in I are given. There is a one-to-one connection between the jb-ons and the loaded flight. TT loaded flight for operation i is connected for each operation i. The origin of the TT loaded flight i is either a CTM or a MC processing the operation with the tool necessary for operation i and the target is a MC to which operation i is allotted. For operations that are part to the same tool copy and job, the tool must adhere to operations precedence requirements. The objective function for minimisation of MKSN is

$Z = \min(\max(ct_i)), \ \forall i \in I$ subject to	
$Z \geq ct_{N_j+n_j} \quad \forall j \in J$	(1)
$ctn_i - ctn_{i-1} \geq ptg_i + ttl_i,$ $\forall i-1, \ i \in I_j, \ \forall j \in J$	(2a)
$ctn_{N_j+1} \geq ptg_{N_j+1} + ttl_{N_j+1}, \quad \forall j \in J$	(2b)
$ctn_{ick} \neq ctn_{hck}, i \neq h, \ \forall i, \ h \in R_k,$ $\forall c \in TLCopy, \ and \ \forall k \in TL$	(2c)
$\left.\begin{array}{l}(1 + H\ tagd_{rs})\ ctn_r \geq ctn_s + ptg_r - H\, q_{rs} \\ (1 + H\ tagd_{rs})\ ctn_s \geq ctn_r + ptg_s - H(1 - q_{rs})\end{array}\right\}$ $\forall r \in I_j, and \ \forall s \in I_l \ where\ j, \ l \in J, j < l$	(3)
$TM_i < TLRM_i, \ i \in I$ i` is the MC's first scheduled jb-on	(4)
$Q_i - TLRM_i \geq 0 \ \ \forall i \in I$	(5)
$d_{oi} + \sum_h d_{hi} = 1 h \neq i, \quad h, i \in R_k, \ \forall k \in TL$	(6)
$d_{ho} + \sum_i d_{hi} = 1 \ h \neq i, \ h, i \in R_k, \forall k \in TL$	(7)
$\sum_{i \in R_k} d_{oi} \leq L \qquad \forall k \in TL$	(8)
$\sum_{i \in R_k} d_{oi} - \sum_{h \in R_k} d_{ho} = 0, \ \forall k \in TL$	(9)
$CTTL_i - ttl_i \geq ct_v, \ \ \forall i, v \in R_k,$ $u \neq i, \ \ \forall k \in TL$	(10)
$CTTL_i - ttl_i \geq d_{oi}\, ttw + \sum_{h \in R_k, h \neq i} d_{hi} \left(CTTL_h + ttd_{h,v}\right)$ $\forall i \in R_k, \forall k \in TL, \ \ if\, i, \ h \in RT_{kj} \ \ then\ i<h$	(11)
$CTTL_u - ttl_u \geq \sum_{h,u \in R_k, h \neq u} d_{h,u}(CTTL_h + ttd_{ho}), \ \ \forall k \in TL$	(12)

$\max(TM_i, CTTL_i) \le ctn_i - ptg_i, \forall i \in I$	(13)
$ctn_i \ge 0, \ \forall i \in I$ $d_{hi} = 0,1 \quad h \ne i \quad i, h \in R_k, \forall i \in I$	

The first constraint is that MKSN must be more than or equal to the time taken to complete the last operation of all jobs. Operations precedence requirements are given in the constraint 2a. The restriction 2b refers to the time it takes for the jobs' first operations to complete. The constraint 2c applies to operations that are part of the same tool copy.

In the third constraint, H is a big +ve integer, ensuring that no 2 jb-ons owed to the same MC may be completed at the same time. If the 'r' and 's' operations are part of different jobs that require the same MC, then by definition $tagd_{rs}$ is zero. The fourth constraint is that the MC's ready time must be smaller than the $TLRM_i$ because the tool must be transferred from CTM if operation 'i' is the MC's first planned operation or another MC. The fifth restriction is that no tool delay will occur during operation 'i'.

The sixth and seventh constraints indicate that the tools are once placed and unloaded for each operation. The eighth restriction ensures that each TT only enters the system once. The 9th constraint keeps the total number of TTs consistent in the FMS.

The 10th constraint indicates that the TT loaded flight 'i' can commence only when the prior operation 'v' is finished. The 11th limitation specifies that the TT loaded flight 'i' can only commence after the TT empty flight has been finished, and it applies whether it is the TT's first loaded flight or TT loaded flight for operations other than the tools' 1st operation. The 12th restriction stipulates that the TT loaded flight 'u', i.e. the tool's initial operation, can only commence after the TT's empty flight has been completed. The starting timings of the TT loaded trips are tied to the 10th, 11th, and 12th limitations. They agree that the TT loaded flight i cannot begin earlier to the maximum of the empty flight to the preceding operation and the preceding operation's conclusion time. The 13th constraint mentions that the operation 'i' cannot commence earlier to the maximum of the MC's ready time and the TT loaded trip.

However, because of its magnitude and nonlinearity, this formulation is intractable, hence a meta-heuristic technique called FPA is utilised to find near-optimal or ideal solutions.

Because the MKSN must be minimised, a method for calculating the MKSN and determining the fewest number of tool copies for a known schedule must be devised. Figure 5.2 depicts a flow chart for such a calculation.

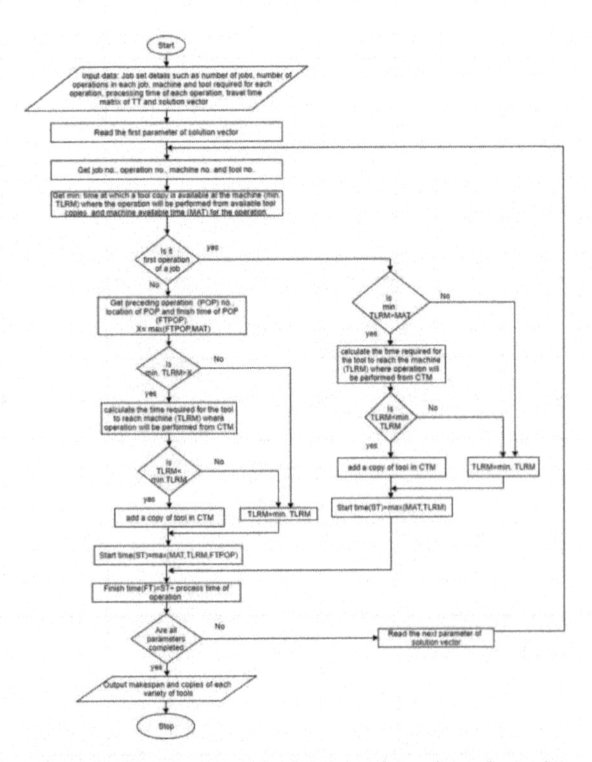

Figure 5.2 For a given schedule, this flow chart shows how to calculate the MSN and the fewest number of tool copies

Input data

Tools will be transferred from one MC to another MC by a TT. The flow path of TT will be determined by FMS configurations that differ topologically from one another. For the simulation, four distinct layout configurations are explored, as described by Bilge and Ulusoy (1995). The 1st 10 job sets utilised by Aldrin Raj et al. (2014) were used for this problem. These issues were created for a variety of t/p ratios (travel times to processing times). Four different layouts (LAOT1, LAOT2, LAOT3, and LAOT4) with three scenarios were utilised to estimate the MSN with increasing processing times. Initial processing times (OLPT), twice the OLPT, and three times the OLPT were employed in cases 1, 2, and 3. For LAOT 1, LAOT 2, and LAOT 3, Case 1 and Case 2 are considered, while for LAOT 4 all cases are included. TT travel times remain same.

The data given below is provided as an input.

- Table 5.1 show the TT travel time matrix, which includes tool load and unload times for various layouts.
- Job set details such as number of jobs, each job's operations, highest operations of a job.
- Every jo-on necessitates the use of a MC (MC matrix),
- Time it takes for each jb to be processed on the MC (process time matrix), and
- Every jb-on demands a tool (Tool matrix)

FPA and ITS Implementation

Xin-She Yang (2012) created the FPA, which is a population-based optimisation approach that simulates floral pollination. Pollination is a plant reproduction physiological phenomenon that makes use of pollinators like birds, bats, bees, and other animals for transferring pollen from one plant to another. Pollination is divided into two types. Abiotic pollination, often known as self-pollination, occurs when pollen carried by wind and does not involve pollinators. Another form is biotic pollination, often known as cross pollination, where pollen is carried from one plant to another. Cross and biotic- pollination transpire over vast distances, and are accomplished by insects such as bees, bats and birds that move large distances. In their activity, bees and birds normally follow the Levy flight. Their movements as a result of this phenomenon should be thought of as discrete jumps that follow the Levy distribution. Insect pollinators, notably honeybees, have also observed the flowers' constancy. Such pollinators tend to prefer certain flower species over others, hence increasing the reproduction of the same species.

The following principles idealise the aforesaid characteristics of the pollination cycle, floral constancy, and pollinator behaviour.

Table 5.1 Travel duration matrix of TT

LAOT 1						*LAOT 2*					
From	*To*					*From*	*To*				
	CTM	*MC1*	*MC2*	*MC3*	*MC4*		*CTM*	*MC1*	*MC2*	*MC3*	*MC4*
CTM	0	04	06	07	08	CTM	0	03	04	06	04
MC1	08	0	04	06	07	MC1	04	0	01	03	01
MC2	07	04	0	04	06	MC2	06	08	0	01	03
MC3	06	06	04	0	04	MC3	04	07	08	0	01
MC4	04	07	06	04	0	MC4	03	06	07	08	0

LAOT 3						*LAOT 4*					
From	*To*					*From*	*To*				
	CTM	*MC1*	*MC2*	*MC3*	*MC4*		*CTM*	*MC1*	*MC2*	*MC3*	*MC4*
CTM	0	01	03	07	08	CTM	0	03	06	07	10
MC1	08	0	01	06	07	MC1	13	0	03	04	07
MC2	07	08	0	04	06	MC2	14	10	0	06	04
MC3	03	04	06	0	01	MC3	08	06	04	0	04
MC4	01	03	04	08	0	MC4	10	10	08	04	0

- *Rule 1:* Cross pollination, also known as biotic pollination, is global pollination in which pollinators follow the Levy distribution, which may be imitated via global search
- *Rule 2:* Self-pollination, also known as abiotic pollination, is a type of local pollination that may be performed by mimicking local search.
- *Rule3:* These pollinators tend to choose to visit specific flower species while avoiding others, resulting in increased reproduction of the same blooms. Constancy in flowers is akin to a chance at reproduction.
- Rule 4: Because pollination might be global or local, a switch probability p is provided to toggle among the two pollination types. The switch probability controls the interplay of global and local pollination.

To develop the updating formulae, the aforementioned steps must be translated into appropriate updating equations.

The following is a mathematical expression of Rules 1 and 3:

$$y_i^{t+1} = y_i^t + L\left(g^* - y_i^t\right) \tag{14}$$

where y_i^t is the solution y_i or a pollen i at iteration t, L is the Lévy fligts-dependent step size and is computed from the Lévy distribution given in '(15)' and g* is the best solution.

$$L \sim \frac{\lambda \Gamma(\lambda) \sin\left(\frac{\pi\lambda}{2}\right)}{\pi} \frac{1}{s^{1+\lambda}} (s \gg s_o > 0) \tag{15}$$

Here $\Gamma(\lambda)$ is gamma function, and this distribution is appropriate for big steps s>0

Rule 2 and Rule 3 are mathematically represented as given below:

$$y_i^{t+1} = y_i^t + \epsilon(y_j^t - y_k^t) \tag{16}$$

y_j^t, y_k^i are pollens from different blooms of the same plant's species. It's a local random walk if pollens are from same population and ϵ is selected from a [0,1] uniform distribution.

FPA is applied for minimisation of MSN. Pollen is a a schedule viable i.e. a solution vector. The vector's parameters match the job set's operations. Every parameter in the solution vector must reflect the jb-on, as well as the tool and MC that are utilised for the jb-on; hence, jb-on, MC, and tool coding is used. The 1st item in the parameter gives the job number, and the sequence in which it appears in the vector specifies the jb-on number. The 2nd item of the parameter represents the MC allocated to carry out the jb-on from the MC matrix, and the 3rd item indicates the tool assigned from the tool matrix to conduct the operation. This coding is significant for checking prior relationships in a vector of job operations.

A. Random Solution Generator (RSG)

The RSG is designed to present initial population. This generator generates a solution vector by adding parameters one by one. An operation has to be eligible for parameter assignment. When all of operation's predecessors have been assigned, it is assumed to be qualified. Qualifying operations are gathered together in a set to be scheduled next. The first operation of each job creates this set in the beginning. Each iteration, a random operation from the set is picked and placed next to the parameter in the solution. Then, for the operation, the MC and tool are selected, and are assigned to parameter to complete the solution. The set is maintained up to date, and the process will keep on if the solution vector is not yet completed.

B. Limit function

It's utilised to guarantee that the new solution's created operations comply with the precedence restrictions requirement. If the requirement of precedence are not followed, the limit function will fix the new solution vector's operations to comply with the precedence requirement restrictions.

The flow chart in Figure 5.3 depicts the FPA application procedures for the problem. In the suggested process, the starting population is produced arbitrarily using RSG. The data discussed under input data section. The coding is done in MATLAB and provides a plan for jb-ons, as well as the assignment of MCs and tools to the related jb-ons, in order to achieve the lowest possible MSN.

Every generation, all population candidates are chosen for substitution. As a result, NP competitions are carried out to choose the members for next generation.

Results and Discussions

The proposed technique has been tested on populations ranging 2 to 12 times the job set operations, and it has been discovered that when the size of population is 10 times the operations of job set, better results are seen. Various combinations of Le'vy flight's step size and switch probability (p) were used; however a combination of 1.5 and 0.8 produced satisfactory results. For most situations, good solutions are obtained between 200 and 250 generations, hence 250 generations is used as the ending threshold. For the MSN minimisation, the MATLAB code for SMTTATWLNTC is run 20 times on each job set stated in section 2.4.3 for a variety of cases and layouts. The MSN best value from 20 runs for different job sets, layout 2 and case 1 with standard deviation (SDV) and mean, the MSN of SMTTATWACT achieved by FPA as recorded in Sivarami Reddy et al. (2018b) achieved by FPA and % decrease in MSN of SMTTATWLNTC over SMTTATWACT, best MSN for SMTTATWLNTC obtained by Jaya algorithm recorded in Venkata Rao (2016) and the lowest possible number of copies for no tool delay for SMATLNTC obtained by two algorithms are given in Table 5.2. The results are similar for other layouts and cases.

A. Gantt chart

The Gantt chart depicts the viability of the best solution for the minimal MSN obtained by FPA for case 1 LAOT 3 of job set 5. The optimal solution is arranged as Table 5.3.

212A 333A 114A 532A 121A 231A 442A 423B 511B 344A 312A 223B 144A

The Gantt chart for the aforementioned solution vector is depicted in Figure 5.4. A jb-on is described by a five-character word. For illustration, in jb-on 5132A, the 1st character '4' specifies the job number, the 2nd character '1' shows the jb-on, the 3rd character '3' denotes the requisite MC, the 4th and 5th characters '2A' denotes tool copy i. e tool type 2 and copy 'A' allocated to the jb-on. The jb-ons allocated to each tool copy and MC are shown on a Gantt chart, along with their start and end periods. TT's empty flights, loaded flights and waiting time are shown in the Gantt chart. M3, M4, M1, and M2 imply MCs, T3A, T4A, T1A, T2A, and T1B denote tool copies in Figure 5.4.
TT loaded trips for operation xxxxx relate to LTT xxxxx.

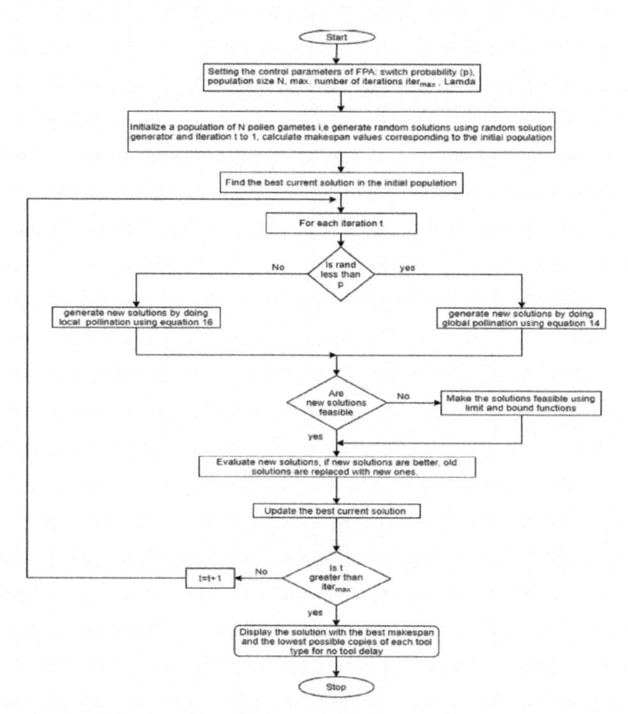

Figure 5.3 Flowchart depicting the FPA application stages

TT empty trips for operation xxxxx relate to ETT xxxxx.
TT waiting time to grab the tool from the MC for operation xxxxx is designated by WPU xxxxx.

Table 5.2 Best MSN of SMTTATWLNTC with mean and SDV, lowest copies for each type of tool obtained by FPA, MSN and lowest copies for each type of tool obtained by Jaya for different job sets, layout 2 and case1

Job set number	*LAOT 2, case1*			*LAOT2, case 1*		*LAOT2, case 1*				*LAOT2, case 1*				
	Best MSN obtained by FPA	*mean*	*SDV*	*MSN of SMTTAT WACT*	*% reduction*	*Lowest copies for each type of tool for minimum MSN*				*Best MSN obtained by Jaya*	*Lowest copies for each type of tool for minimum MSN*			
						T1	*T2*	*T3*	*T4*		*T1*	*T2*	*T3*	*T4*
1	87	87.45	0.5104	95	8.42	3	2	2	1	88	2	1	1	2
2	92	93.6	1.1877	100	8.00	2	2	2	2	97	2	2	2	1
3	91	93.75	1.3717	102	10.78	2	2	3	2	95	2	2	2	2
4	85	85.35	0.7452	98	13.27	2	2	2	1	88	2	3	2	1
5	74	74.45	0.5104	77	3.90	2	1	2	1	75	1	2	2	1
6	99	99	0.0000	99	0.00	1	1	1	1	99	1	1	1	1
7	89	91.25	2.1975	92	3.26	1	1	2	2	94	1	2	2	2
8	135	135.8	1.1517	147	8.16	1	2	2	2	137	2	2	3	2
9	104	104.65	0.8127	129	19.38	1	3	1	3	108	2	3	1	3
10	141	141.8	1.7889	158	10.76	2	3	3	2	147	1	2	3	2

Table 5.3 Optimal solution for job set 5, LAOT 3 of case1

Jb-on	2-1	3-1	1-1	5-1	1-2	2-2	4-1	4-2	5-2	3-2	3-3	2-3	1-3
MC number	1	3	1	3	2	3	4	2	1	4	1	2	4
Tool copy	2A	3A	4A	2A	1A	1A	2A	3B	1B	4A	2A	3B	4A

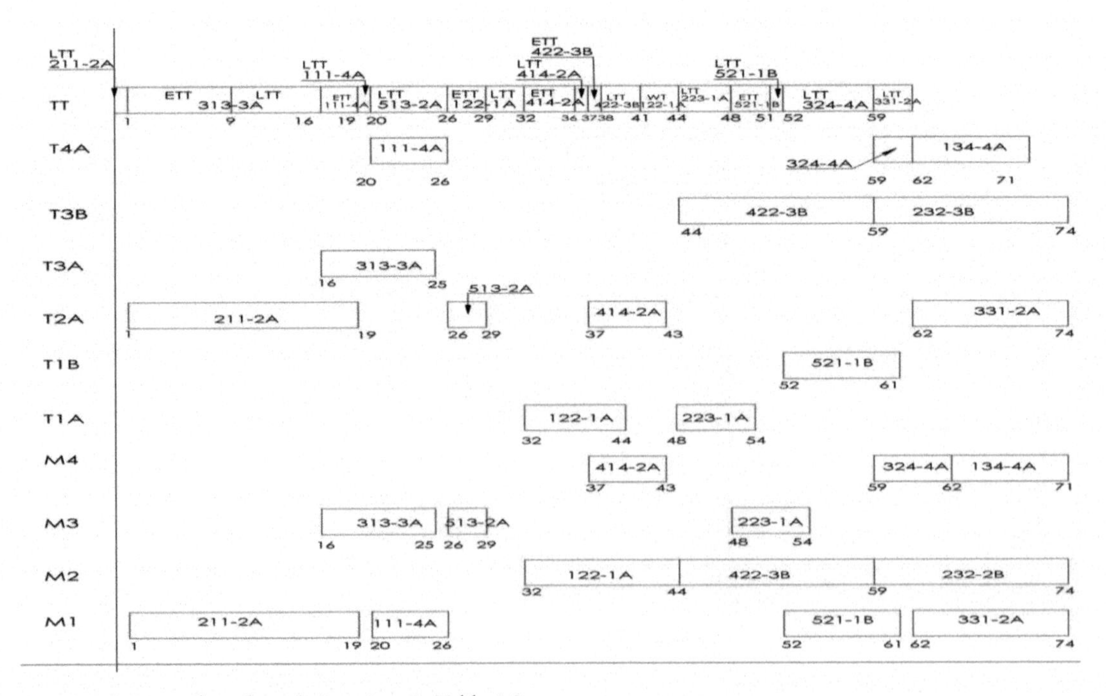

Figure 5.4 Gantt chart for solution given in Table 5.3

TT empty trips for operation xxxxx relate to ETT xxxxx.
TT waiting time to grab the tool from the MC for operation xxxxx is designated by WPU xxxxx.

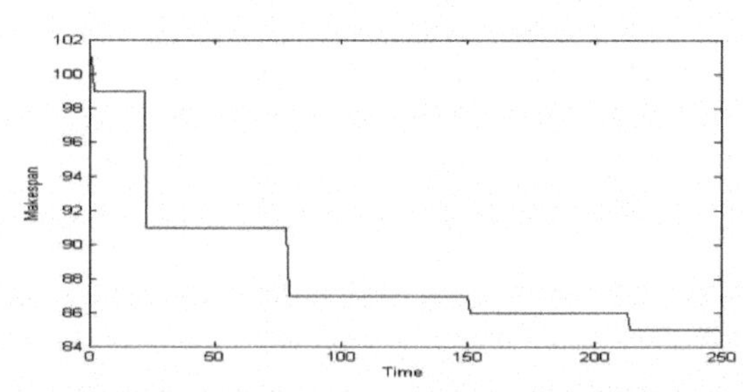

Figure 5.5 Convergence characteristics of the FPA for job set 4, LAOT 2 of case1

B. Convergence characterstics

The FPA convergnce characteristics for case 1 LAOT 2 job set 4 are shown in Figure 5.5. At 214 iterations, the optimum value is85, with a duration per iteration of 0.48722 seconds.

Conclusions

This work offers a nonlinear MIP model for MCs, TT and tools simultaneous SCHG in MMFMS to minimise the MSN with the fewest number of copies of each tool kind for no tool-delay taking into account the tool switch durations between MCs in a MMFMS. This SCHG problem involves evaluating the fewest number of copies of every tool variety without tool delay, allocating suitable tool copy for each jb-on, sequencing and synchronisation of those jb-ons and related trip operations of the TT together with the empty and loaded flights times of the TT for the minimum MKSN. For a given schedule, an algorithm for computing MKSN and the fewest number of copies of every kind of tools without tool delay is devised, and Figure 5.2 depicts its flow chart. The recommended algorithm is tested on job sets mentioned in section III.B. Table 5.2 shows that the FPA is resilient and capable of finding a large number of optimal solutions to the problems, FPA surpasses the Jaya algorithm. It is observed that impact of SMAATWLNTC on reduction in the MSN over SMTTATWACT is significant. In future work with MCs, TT and tool SCHG, subsystems like robots and AS/RS may be integrated.

References

Aldrin Raj, J., Ravindran, D., Saravanan, M. and Prabaharan, T. (2014). Simultaneous SCHG of machines and tools in multimachine flexible manufacturing system using artificial immune system algorithm. Int. J. Comput. Integr. Manuf. 27(5):401–414.

Baykasoglu, A. and Ozsoydan, F.B. (2017). Minimizing tool switching and indexing times with tool duplications in automatic machines. Int. J. Adv. Manuf. Technol. 89:1775–1789.

Bilge, Ü. and Ulusoy, G. (1995). A time window approach to simultaneous scheduling of machines and material handling system in FMS. Oper. Res. 43:1058–1070. https://doi.org/10.1287/opre.43.6.1058

Costa, A., Cappadonna, F. A., Fichera, S. (2016). Minimizing the total completion time on a parallel machine system with tool changes. Comput. Ind. Eng. 91: 290–301.

Ersin, K. and Ali Payıdar, A. (2018). Flower pollination algorithm approach for the transportation energy demand estimation in Turkey: model development and application. Energy Sources, Part B: Econ. Plan. Policy 13(11–12): 429–447.

Gökgür, B., Hnich, B. and Özpeynirci, S. (2018). Parallel machine scheduling with tool loading: A constraint programming approach. Int. J. Prod. Res. 56(16), 5541–5547. https://doi.org/10.1080/00207543.2017.1421781.

Özpeynirci, S. (2015). A heuristic approach based on time-indexed modelling for SCHG and tool loading in flexibl manufacturing systems. Int. J. Adv. Manuf. Technol. 77:1269–1274. https://doi.org/10.1007/s00170-014-6564-2.

Paiva, G. S. and Carvalho, M. M. A. (2017). Improved heuristic algorithms for the job sequencing and tool switching problem. Comput. Oper. Res. 88:208–219.

Saravanan, M. and Noorul Haq, A. (2008). Evaluation of scatter-search approach for scheduling optimization of flexible manufacturing systems. Int. J. Adv. Manuf. Technol. 38:978–986.

Shirazi, R. and Frizelle, G. D. M. (2001). Minimizing the number of tool switches on a flexible machine: An empirical study. Int. J. Prod. Res. 39(15):3547–3560.

Sivarami Reddy, N., Ramamurthy, D. V., Rao, K. P. and Lalitha, M. P. (2016). A Novel metaheuristic method for simultaneous SCHG of machines and tools in multi machine FMS. IET Digital Library International Conference on Recent Trends in Engineering, Science & Technology (pp. 1–6). ISBN-978-1-78561-785-0. **DOI:** 10.1049/cp.2016.1489.

Sivarami Reddy, N., Ramamurthy, D. V., Rao, K. P. and Lalitha, M.P. (2017). Simultaneous scheduling of machines and tools in multi machine fms using crow search algorithm. Int. J. Eng. Sci. Technol. 9(09S): 66–73.

Sivarami Reddy, N., Ramamurthy, D. V., Rao, K. P. (2018a). Simultaneous scheduling of machines and tools considering tool transfer times in multi-machine FMS using CSA. International conference on intelligent computing and applications. Adv. Intell. Syst. Comput. 632:421–432.

Sivarami Reddy, N., Ramamurthy, D. V. and Rao, K. P. (2018b). Simultaneous scheduling of jobs, machines and tools considering tool transfer times in multi-machine FMS using new nature-inspired algorithms. Int. J. Intell. Syst. Technol. Appl. 17(1/2):70–88.

Sivarami Reddy, N., Ramamurthy, D. V., Rao, K. P. and Lalitha, M. P. (2019). Integrated scheduling of machines, AGVs and tools in multi-machine FMS using crow search algorithm. Int. J. Comput. Integr. Manuf. 32(11), 1117–1133. https://doi.org/10.1080/0951192X.2019.1686171.

Sivarami Reddy, N., Ramamurthy, D. V., Lalitha, M. P. and Rao, K. P. (2021a). Minimizing the total completion time on a multi-machine FMS using flower pollination algorithm. Soft Comput. 26,1437–1458. https://doi.org/10.1007/s00500-021-06411-y.

Sivarami Reddy, N., Ramamurthy, D. V., Rao, K. P. and Lalitha, M. P. (2021b). Practical simultaneous scheduling of machines, AGVs, tool transporter and tools in a multi machine FMS using symbiotic organisms search algorithm. Int. J. Comput. Integr. Manuf. 34(2), 153–174. DOI: 10.1080/0951192X.2020.1858503.

Sivarami Reddy, N., Ramamurthy, D. V., Lalitha, M.P. and Rao, K. P. (2021c). Integrated simultaneous scheduling of machines, automated guided vehicles and tools in multi machine flexible manufacturing system using symbiotic organisms search algorithm. J. Ind. Prod. Eng. https://doi.org/10.1080/21681015.2021.1991014.

Venkata Rao, R. (2016). Jaya: A simple and new optimization algorithm for solving constrained and unconstrained optimization problems. Int. J. Ind. Eng. Comput. 7: 19–34.

Yang, X. S. (2012). Flower pollination algorithm for global optimization, in unconventional computation and natural computation. Lect. Notes Comput. Sci. 7445:240–249. https://doi.org/10.1007/978-3-642-32894-7_27

6 Simultaneous scheduling of machines and tools without tool delay using crow search algorithm

Padma Lalitha Mareddy[1,a], *Sunil Prayagi*[2,b], *Sivarami Reddy Narapureddy*[3,c], *Hemantha Kumar, A.*[3,d], *and Lakshmi Narsimhamu K.*[4,e]

[1]Electrical and Electronics Engineering Department, Annamacharya Institute of Technology and Sciences, Rajampeta, India

[2]Mechanical Engineering Department, Yeshwantrao Chavan College of Engineering, Nagpur, India

[3]Mechanical Engineering Department, Annamacharya Institute of Technology and Sciences, Rajampeta, India

[4]Mechanical Engineering Department, Sreevidyanikethan Engineering College, Tirupati, India

Abstract

The aim of this work is to schedule machines and tools concurrently in a multi-machine flexible manufacturing system (MMFMS) with the fewest copies of every tool variety to avoid tool delay, without taking tool shift times into account for minimization of the makespan (MKSN). The tools are initially kept in the central tool magazine (CTM) which is shared by numerous machines and serves them. The objective is to allocate job-operations (jb-ons) to machines, determine the fewest number of copies of every tool type and to allocate copies to jb-ons for MKSN minimization. This work proposes a formulation of nonlinear mixed- integer programming to model and crow search algorithm (CSA) is employed to work out this problem. The results show that employing an additional copy each for few tool types and a copy each for other remaining tool varieties causes no tool delay as well as reduction in makespan and CSA outperforms Jaya algorithm.

Keywords: Crow search algorithm, FMS, makespan, no tool delay, optimization techniques, scheduling of machines and tools.

Introduction

Manufacturing firms must deal with increasing product complexity, shorter product life cycles, new technologies, global competitive concerns, and a rapidly changing environment. Flexible manufacturing system (FMS) is primarily established to handle the manufacturing competition. FMS is properly connected manufacturing system that includes multifunctional computer numerical controlled (CNC) machine (MC) tools coupled with automated material handling system (AMHS). FMS strives for manufacturing flexibility without compromising product quality. To achieve its flexibility the FMS depends on the flexibility of AMHS, control software, and CNC MCs. Workflow patterns, size, and manufacturing type have all been used to group FMSs into different types. From a planning and control perspective, there are four kinds of FMS: single flexible MC, flexible manufacturing cell, multi-cell FMS, and MMFMS (Saravanan and Noorul Haq, 2008). Existing FMS installations have already established benefits such as cost savings, increased utilization, lower work-in-process, and so on. Scheduling (SCHG) tasks improve resource utilization by lowering the MKSN. Solving SCHG problems optimally or near optimally is one technique to attain high productivity in FMS.

Literature Review

Tool loading becomes a complex component in the SCHG issues since the tool copies' number is limited and may be fewer than the MCs' number owing to cost limitations. Job and tool SCHG is a critical issue for industrial systems Ineffective job SCHG and tool loading planning can lead to underutilisation of capital-intensive MCs and a significant amount of inoperative time (Shirazi and Frizelle, 2001). Therefore, effective job and tool SCHG allow a production system to maximize MC usage while reducing idle time. There have been several studies on the MCs and tools concurrent SCHG.

[a]padmalalitha.mareddy@gmail.com; [b]sunil_prayagi@yahoo.com; [c]siva.narapureddy@gmail.com; [d]ahkaits@gmail.com; [e]klsimha@gmail.com

Aldrin Raj et al. (2014) employed different algorithms and rules for addressing MCs and tools concurrent SCHG (SMTWACT) for minimization of MKSN with a replica of every tool kind in an FMS with 4 MCs and a CTM. Özpeynirci (2015) proposed the time-indexed mathematical approach for joint tool and MC SCHG in FMS for MKSN minimisaton. Costa, et al. (2016) devised the hybrid GA which uses the local search improvement technique to tackle 'p' components SCHG on 'q' MCs considering tool alter activity prompted due to wear of the tool. Sivarami Reddy et al. (2016), Sivarami Reddy et al. (2017a), and. Sivarami Reddy et al. (2017b) solved the SMTWACT to provide the best sequences in an MMFMS that minimize the MKSN. Using the simulated annealing approach, Baykasoglu and Ozsoydan (2017) tackled the indexing of automatic tool changer and tool altering problems in tandem to minimize the non-machining periods in the automatic machining centres. Paiva and Carvalho (2017) solved the problem of job ordering and tool changing by determining the job sequence and tool loading sequence to minimize tool changes. For addressing sequencing problems, an approach that makes use of heuristic, local search methods and graph representation, is applied. Such strategies are used with traditional tooling techniques to solve the issue in local search method with an algorithmic fashion. Gökgür et al. (2018) used models of constraint programming for parallel MC scenarios to tackle tool SCHG and allocation issues with predetermined tool quantities in the system owing to budgetary constraints to minimize the MKSN. *In the references discussed above, job and tool shift times among MCs are not considered, and operations are only scheduled on specific MCs.*

Sivarami Reddy et al. (2018a) and Sivarami Reddy et al. (2018b) dealt simultaneous SCHG of MCs and tools without taking into consideration job shift times and taking into account tool shift times between machins with minimm MKSN as an objective and demonstrated *that tool shift times have important effect on MKSN.*

Few studies are conducted on AGVs, tools, and MCs parallel SCHG with a replica of every tool kind (SMAT). SMAT for minimization of MKSN where operations are planned on specific MCs was addressed by Sivarami Reddy et al. (2019), Sivarami Reddy et al. (2021a), and Sivarami Reddy et al. (2021b), and Sivarami Reddy et al. (2021c).

Sivarami Reddy et al. (2021d) and Sivarami Reddy et al. (2022) addressed concurrent SCHG of MCs and tools with a replica of each tool kind and alternate MCs for minimisation of MKSN.

When many jb-ons require the same tool simultaneously, the tool will be assigned to one operation only, causing remaining operations to wait for it, in SMTWACT. The MKSN rises as a result of this. Therefore it is imperative to address concurrent SCHG of MCs, and tools with the fewest of copies of each tool kind which are much less than the number of MCs to avoid tool delay (SMATWLNTC) by sharing the tools amongst MCs to minimize MKSN in an FMS. Because there is no tool delay, MKSN is reduced by a few extra copies for a few tool kinds. The novelty of the study is addressing the SMATWLNTC for minimisation of MKSN first time, presenting nonlinear MIP model, CSA is employed to minimize MKSN, and explains its difference from other studies.

The SCHG problem of job shop is NP-hard issue. Moreover, finding out the fewest possible quantity of copies of each tool kind without tool delay and assigning them to jb-ons makes the SCHG problem more multifaceted and also it is anticipated that MC, and tool usage would improve.

CSA is a met heuristic developed by Askarzadeh (2016). It tries to imitate the cognitive behaviour of crows so as to solve optimization issues. It's been used to solve engineering optimization challenges with great success (Sivarami Reddy et al., 2017a; Upadhyay and Chhabra, 2019; Anter et al., 2019; Sivarami Reddy et al., 2018c; Sivarami Reddy et al., 2019). The CSA has a number of benefits over most other meta-heuristic algorithms, including the fact that it only requires two algorithm parameters to operate, simple, and easy to implement.

Problem Formulation

FMS kinds and operations require different setups. The majority of research focuses on specified production systems since having a standard configuration is not practical. The following sections provide details on the system, hypotheses, criterion for purpose, and the issue addressed in this study.

A. FMS Environment

Figure 6.1 depicts FMS setup. To reduce tool inventories, some production systems use common tool storage with one CTM serving numerous MCs. While machining the component, the needed tool is shared between MCs or shifted from CTM. The CTM reduces the tool copies in number necessary in the system, lowering tooling costs. TT will switch the tools between the MCs. AGVs will perform the job switching between the MCs. The FMS in this study is considered to be made up of 4 CNC MCs,

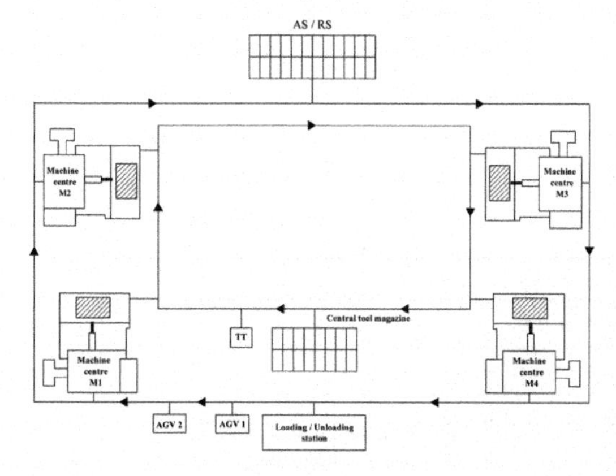

Figure 6.1 FMS environment

each with its own tool varieties and ATC, as well as a TT and two indistinguishable AGVs. FMS delivers components for manufacture from loading and unloading stations (LUS), and completed components are deposited in LUS and transferred to the ultimate storage facility. AS/RS is offered to replenish the work-in-progress.

B. Assumptions

Assumptions mentioned below are made for the problem being considered

- All jobs, and MCs are initially accessible for usage.
- At any particular time, a tool or MC can only do one operation (opn).
- The pre-emption of jb-on on a MC is prohibited.
- Prior to SCHG any jb-on, the requisite MCs and tools are identified.
- Each job has its own operations set, each of which has processing time.
- The setup time is also factored into the processing time.
- The tools are initially placed in a CTM.
- The time it takes for parts to travel between MCs isn't considered.
- When the jobs arrive, the tools will have enough service life to do the operations assigned to them.
- There is just a TT that transfers tools in the system, and tool swap period across MCs are not taken into account

C. Constraints

Process planning information is supplied in terms of precedence restrictions for each task in order to determine the operations' order for optimizing the tolerance stack-up.

D. Problem

Consider a job st J with j job kinds (J_1, J_2, J_3..., J_j) job set's total operations (1,2,3.......N) k tool kinds (t_1,t_2, t_3.......t_k) with few replicas of each variety and m MCs (M_1, M_2, M_3..., M_m) in an FMS. The jb-ons are processed in a preset order that is known ahead of time. The jb-ons are processed in a preset order that is known ahead of time. A MC can only carry out one jb-on at a time. A jb-on cannot commence before its preceding operation is accomplished. The arrangement of jb-ons on a MC governs the MC's setup needs. The concurrent SCHG problem is defined as finding the fewest copies of each tool variety, allocating copies of tools to jb-ons and sequencing them on MCs without tool delay, every job's commencement and concluding tims on each MC and tool copy for minimization of MKSN in an MMFMS. Because the problem is described clearly, section IV provides an unambiguous mathematical form.

Model Formulation

A nonlinear MIP model is offered to visibly illustrate the critical parameters and their impact on the SCHG problem of FMS.

Notations
Subscripts
j job index
i, h jb-ons indices
k tool index

Sets and parameters

J:	job set on hand for processing
n_j:	opns in job j
N:	$\sum_{j\in J} n_j$, job set J total opns.
I:	{1, 2, ------- N}, opns' index set.
I_j:	$\{J_{j+1}, J_{j+2}, \text{------} J_{j+n_j}\}$, the set of indices in I associated with job j, where J_j is jobs' opns planned before job j and J_1=0.
IF_i :	$I - \{h; h \geq i, i, h \in I_j\}$ opns' index set without opn i and same job's following opns to opn i.
IP_h:	$I - \{i; i \leq h, i, h \in I_j\}$ opns' index set without opn h and same job's preceding opns to opn h.
TL:	tool types' set to carry out the jobs' opns.
$TLCopy$::	set of replicas of every tool kind.
R_k:	indices set in I linked with tool type k in TL, $k \in TL$.
RT_{kj}:	$I_j \cap R_k$ the index set of opns in I common for tool k and job j $\forall k \in TL, \forall j \in J$
pt_i:	opn i processing time.
ct_i:	opn i completon time.
b_{kci}:	ready times set for copies of tool type k for opn i, $\forall c \in TLCopy, \forall i \in I$.
S_i:	$\min(b_{kci})\ \forall k \in TL, \forall c \in TLCopy \text{ and } \forall i \in R_k$.
a_j:	job j ready time
TM_i:	opn i's MC ready time
X_i:	$\max(ct_{i-1}, TM_i)$ for $i - 1, i \in I_j, \forall j \in J$.

Decision variables

$$q_{rs} = \begin{cases} 1 & \begin{matrix} \text{if } c_r < c_s, \text{where } r \text{ and } s \\ \text{are operations of different jobs} \end{matrix} \\ 0 & \text{otherwise} \end{cases}$$

A. Mathematical Model

The tool and MC indices are not included in the formulation since every job routing is accessible; nevertheless, the MC index and tool index for every opn are given. For opns that are part to same tool copy and job, the tool must adhere to opns precedence requirements The MKSN minimization objective function is

$Z = \min(\max(ct_i))\ \text{for all } i,\ i \in I$ Subject to the following constraints (cntnt)	
$Z \geq ct_{N_j+n_j} \quad \forall j \in J$ The 1st cntnt ensures that MKSN is higher than or equal to the finish time of last opn of all the jobs.	(1)

$ct_i - ct_{i-1} \geq pt_i,\ \forall i-1, i \in I_j,\ \forall j \in J$ The cntnt set 2a is the opns precedence cntnts.	(2a)
$ct_{N_j+1} \geq pt_{N_j+1},\quad \forall j \in J$ The content set 2b is the cntnt for the finish time of first opns of jobs.	(2b)
$ct_{ick} \neq ct_{hck},\ i \neq h,\ \forall i,h \in R_k, \forall c \in TLCopy,\ and\ \forall k \in TL$ The cntnt set 2c is the cntnt for the opns that belong to the same tool copy.	(2c)
$\left.\begin{array}{l}(1 + H\ tagd_{rs})ct_r \geq ct_s + pt_r - H\,q_{rs} \\ (1 + H\ tagd_{rs})ct_s \geq ct_r + pt_s - H(1 - q_{rs})\end{array}\right\}$ $\forall r \in I_j,\ and\ \forall s \in I_l\ where\ j,\ l \in J,\ j < l$ In the 3rd cntnt, H is a large +ve integer, ensuring that no two opns allotted to the same MC can be processed at the same time.	(3)
$X_i - S_i \geq 0 \quad \forall i \in I$ The 4th cntnt guarantees that there will not be tool delay for opn i.	(4)
$ct_i \geq 0,\ \forall i \in I$ $q_{rs} = 0,1 \quad \forall r \in I_j, \forall s \in I_l\ where\ j,l \in J, j < l$	

However, because of its magnitude and nonlinearity, this formulation is intractable; hence a meta-heuristic technique called CSA is utilized to find near-optiml or ideal solutions.

Because the MKSN must be minimised, a method for calculating the MKSN and determining the fewest number of tool copies for a known schedule must be devised. Figure 6.2 depicts a flow chart for such a calculation.

B. Input data

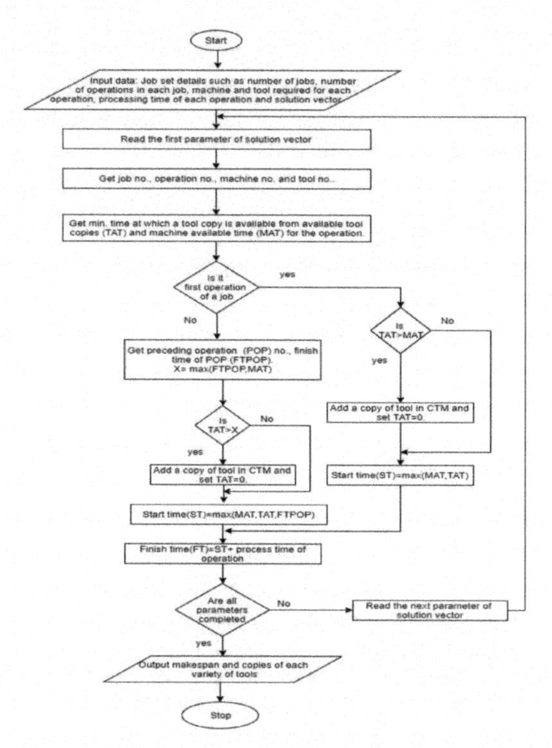

Figure 6.2 For a given schedule, this flow chart shows how to calculate the MKSN and the fewest tool copies to avoid tool delay

Because each job set differs in terms of the amount of jobs, total opns, and MCs, For this problem, Aldrin Raj et al. (2014)'s job sets are used. Each job set has 5 to 10 jobs, and each job's entity contains information about the MC, the jb-on's processing time on those MCs, and the tool required for the jb-on.

The following data is presented as an input.

- Job set details such as total jobs, every job's opns, highest opns of a job.
- Every jo-on necessitates the use of a MC (MC matrix),
- Time it takes for each jb to be processed on the MC (process time matrix), and
- Every jb-on demands a tool (Tool matrix)

CSA and ITS Implementation

Askarzadeh (2016) devised CSA relying on the intelligence of crows. Crows cache their extra food in the environment's hiding spots and retrieve it when it's needed. The flock, crows, and places are referred to as population, searches, and position, respectively, in the CSA. In the search space, each location may be a viable option. The objective (fitness) function is food quality, and the best quality food's location is the problem's global solution. Crows are ravenous because they hunt each other for superior food sources. This implies that each flock member follows the others in order to locate their food effectively. To answer the engineering optimization challenges mentioned below, CSA aims to model this clever behaviour of crows.

Crow i's location at iteraton t is defined as a vector for a flock of N crows, $y^{i,iter} = [y_1^{i,t}, \ldots.. y_d^{i,t}]$ for i = 1, 2, ..., N and t = 1, 2, ..., t_{max}, where d is the problem dimension (decision variables) and the maximum iteratons is denoted by t_{max}. At iteration t, the place of the concealing food that crow i has remembered up to this point is supplied by $m^{i,t}$, so that $m^{i,t} = [m_1^{i,t}, \ldots\ldots. m_d^{i,t}]$. The following approach is used to modify the location of the crow i.

Two situations may arise if crow i attempts to go after crow j to determine the location of its hidden food source $m^{j,t}$.

State 1: Crow j is unaware that crow i is behind it. As a consequence, crow i will come close to crow j's hiding spot, and i's location will be modified as follows: $y^{i,t+1} = r_j \times fl^{i,t} \times (m^{j,t} - y^{i,t}) + y^{i,t}$
where r_j is a uniformly distributed random value betwen 0 and 1, and fl is the crow's flight length (fl) at iteration t.

State 2: Crow j is aware that crow i is following it; as a consequence, crow j will trick crow i by not going to the hiding area, but rather to any random location.

At each CSA' iteration, each crow selects one of the flock crows at random and moves to its hiding site (the best solution found by that crow). This implies that the best locations achieved so far are directly utilized to identify better places in each iteraton of CSA. Crow j's awareness probability (AP) tracked at iteration t can be used to indicate both situations. In CSA, the parameter of awareness probability is primarily responsible for intensification and diversification (AP). The parameter AP allows for a better balance between diversity and intensification. Small AP values indicate local search conduction, while greater AP values indicate global but random search conduction. Finally, both states can be stated in the following way.

$$y^{i,t+1} = \begin{cases} r_j \times fl^{i,t} \times \left(m^{j,t} - y^{i,t}\right) + y^{i,t} \\ r_j \geq AP^{i,t} \quad otherwise \\ a\ random\ position \end{cases} \tag{5}$$

At iteration t, crow j's awareness probability is denoted by $AP^{j,t}$.

A. Random Solution Generator (RSG)

The RSG is designed to present initial population. This generator generates a solution vector by adding parameters one by one. An opn has to be eligible for parameter assignment. When all of opn's predecessors have been assigned, it is assumed to be qualified. Qualifying opns are grouped into a set that will be scheduled next.The first opn of each job creates this set in the beginning. Each iteration, a random opn from the set is picked and placed next to the parameter in the solution. Then, for the opn, the MC and tool are selected, and are allotted to parameter to complete the solution. The set is maintained up to date, and the process will keep on if the vector is not yet completed.

B. Limit function

It's utilised to guarantee that the new solution's created opns comply with the requirement of precedence restrictions. If the requirement of precedence is not followed, the limit function will fix the fresh solution vector's opns to comply with the precedence requirement restrictions.

In the suggested process, the starting population is produced arbitrarily using RSG. The data discussed in section IV.B is provided as an input. The coding is done in MATLAB and provides a plan for jb-ons, as well as the assignment of MCs and tool copies to the related jb-ons, in order to achieve the lowest possible MKSN.

In every generation, all population candidates are chosen for substitution. As a result, NP competitions are carried out to choose the members for next generation.

Results and Discussions

The suggested technique has been tested on populations ranging 2 to 12 times the job set opns, and it has been discovered that when the size of population is 10 times the job set's opns, better results are seen. Various combinations of fl and AP were used; however a combination of 2 and 0.3 produced satisfactory results. For most situations, good solutions are achieved between 8 and 30 generations, hence 30 generations is used as the ending threshold. For the MKSN minimization, the MATLAB code for SMATWLNTC is run 20 times on every job set stated in section IV. B. The best MKSN value from 20 runs for different job sets, with standard deviation (SDV) and mean of SMATWLNTC, the best MKSN of SMATWACT achieved by CSA as recorded in Sivarami Reddy et al. (2018b), and % reduction in MKSN of SMATWLNTC over SMATWACT are specified in Table 6.1. From Table 6.1, it is known the non-zero SDV ranges in the band [1.6330, 3.6530] and one intriguing finding is that the SDV values are quite small in comparison to the size of the mean values. Furthermore, for 19 of the 22 job sets, the SDV is zero. When looking at the 20 runs' final solutions of these problems, one can see that there are several solutions with the identical MKSN value. It implies that there are several optima solutions, but that the suggested CSA can locate them. The % reduction in MKSN of SMATWLNTC over SMATWACT varies from 00.00 to 34.15.

A. Gantt chart

The Gant chart depicts the viability of the best solution for the minimal MKSN obtained by CSA for job set 1. This optimal solution is presented below.

3 3 1A 5 3 2A 1 1 3A 2 1 2B 4 4 3B 1 2 4A 4 2 4A 2 3 3A 3 4 4B 3 1 2A 2 2 1A 1 4 1B 5 1 1B

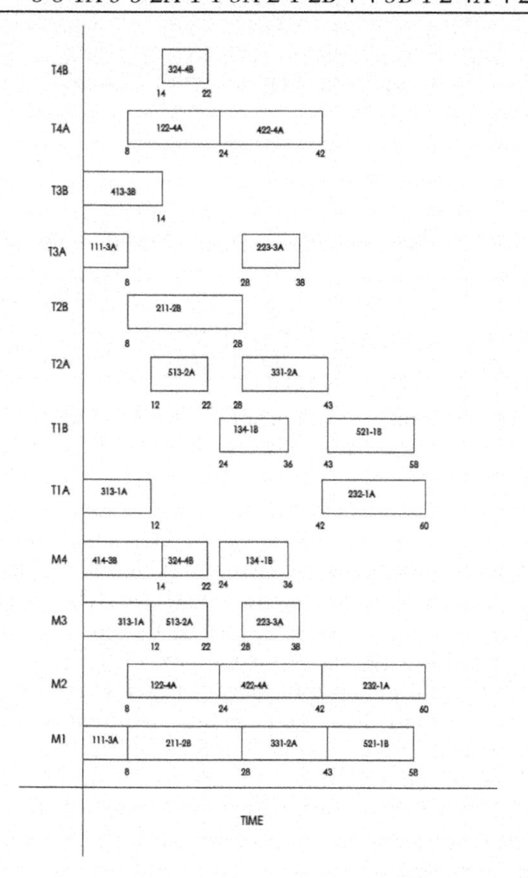

Figure 6.3 Gantt chart for solution given in Table 6.3

Table 6.1 Best MKSN of SMATWLNTC with mean and SDV, best MKSN of SMATWACT acquired by CSA, and % reduction in MKSN of earlier over later

Job set number	*Best MKSN of SMATWLNTC*	*mean*	*SDV*	*Best MKSN of SMATWLNTC*	*Best MKSN of SMATWACT*	*% reduction*
1	60	60	0.0000	60	69	13.04
2	70	70	0.0000	70	80	12.50
3	70	70	0.0000	70	80	12.50
4	54	54	0.0000	54	61	11.48
5	48	48	0.0000	48	48	0.00
6	88	88	0.0000	88	88	0.00
7	66	66	0.0000	66	70	5.71
8	131	131	0.0000	131	131	0.00
9	81	81	0.0000	81	113	28.32
10	112	114	1.6330	112	136	17.65
11	75	75	0.0000	75	93	19.35
12	65	65	0.0000	65	65	0.00
13	92	92	0.0000	92	113	18.58
14	58	58	0.0000	58	70	17.14
15	76	76	0.0000	76	100	24.00
16	71	71	0.0000	71	75	5.33
17	54	54	0.0000	54	61	11.48
18	64	64	0.0000	64	64	0.00
19	70	70	0.0000	70	89	21.35
20	73	73	0.0000	73	92	20.65
21	214	218.3	3.6530	214	325	34.15
22	264	268	3.5901	264	400	34.00

The above solution is arranged as Table 6.3 in the table form. The Gant chart for the aforementioned solution vector is depicted in Figure 6.3. A jb-on is described by a five-character word.

For illustration, in jb-on 5132A, the 1st character '4' specifies the job number, the 2nd character '1' shows the jb-on, the 3rd character '3' denotes the requisite MC, the 4th and 5th characters '2A' denotes tool copy i.e tool type 2 and copy 'A' allocated to the jb- on. The jb-ons allotted to every tool copy and MC are shown on a Gantt chart, along with their start and end periods. M3, M4, M1, and M2 imply MCs, T3A, T4A, T1A, T2A, and T1B denote tool copies in Figure 6.3.

B. Convergence Characterstics

The CSA convergnce characteristics for job set 4 are shown in Figure 6.4. At 8 iteration, the optimum value is 54, with a duration per iteration of 1.414793 seconds.

The best MKSN for SMATWLNTC and the fewest possible copies for no tool delay for SMATWLNTC produced by Jaya algorithm and Teachng-learning-basd optimisation: A novel approach for optimisation problems of constrained mechanical design (TLBO) as reported in Venkata Rao (2016) and Venkata Rao et al. (2011) respectively along with results obtained by CSA are given in Table 6.2.

From Table 6.2, it is clear evident that CSA outperforms the other two algorithms for the job sets which have more opns.

Conclusions

This work offers a nonlinear MIP model for MCs, and tools simultaneous SCHG in MMFMS to minimize the MKSN with the fewest copies of every tool kind for to avoid tool-delay in a MMFMS. This SCHG problem involves evaluating the fewest replicas of each tool variety to avoid tool delay, allocating suitable tool copy for each jb-on, sequencing and synchronization of those jb-ons for the minimum MKSN.

For a given schedule, an algorithm for computing MKSN and the fewest replicas of each kind of tools to avoid tool delay is devised, and Figure 6.2 depicts its flow chart. The recommended algorithm is tested on job sets mentioned in section IV.B. Table 6.1 shows that the CSA is resilient and capable of finding a large number of optimal solutions with identical MKSNs to the problems. It is observed that impact of SMATWLNTC on reduction in the MKSN over SMATWACT is significant which varies from 00.00% to 34.15%. Table 6.2 shows that CSA outprforms the other two algorithms. In future work with MCs and tools SCHG, subsystems like robots and AS/RS may be integrated.

Table 6.2 Best MKSN of SMATWLNTC with fewest copies for each type of tool produced by CSA, Jaya algorithm, and TLBO for various job sets

Job set number	*MKSN by CSA*	*Copies for each variety of tools*								*MKSN by Jaya algorithm*	*Copies for each variety of tools*								*MKSN by TLBO*	*Copies for each variety of tools*							
		T1	*T2*	*T3*	*T4*	*T5*	*T6*	*T7*	*T8*		*T1*	*T2*	*T3*	*T4*	*T5*	*T6*	*T7*	*T8*		*T1*	*T2*	*T3*	*T4*	*T5*	*T6*	*T7*	*T8*
1	60	2	2	2	2	-	-	-	-	60	3	2	2	2	-	-	-	-	60	2	2	2	2	-	-	-	-
2	70	2	2	2	1	-	-	-	-	70	2	2	1	2	-	-	-	-	70	3	3	2	2	-	-	-	-
3	70	2	2	3	2	-	-	-	-	70	2	2	3	2	-	-	-	-	70	2	2	3	2	-	-	-	-
4	54	2	2	2	2	-	-	-	-	54	2	2	2	2	-	-	-	-	54	2	2	2	1	-	-	-	-
5	48	1	1	1	1	-	-	-	-	48	2	3	1	2	-	-	-	-	48	2	3	1	2	-	-	-	-
6	88	1	1	1	1	-	-	-	-	88	1	1	1	1	-	-	-	-	88	1	1	1	1	-	-	-	-
7	66	2	2	2	1	-	-	-	-	66	2	2	3	1	-	-	-	-	66	2	2	3	1	-	-	-	-
8	131	1	1	1	1	-	-	-	-	131	2	2	3	3	-	-	-	-	131	2	2	2	1	-	-	-	-
9	81	2	2	1	3	-	-	-	-	81	2	3	1	3	-	-	-	-	81	2	2	1	3	-	-	-	-
10	112	2	3	3	3	-	-	-	-	114	2	3	3	3	-	-	-	-	114	1	3	4	3	-	-	-	-
11	75	1	2	2	3	-	-	-	-	75	1	2	1	2	-	-	-	-	75	1	2	1	2	-	-	-	-
12	65	1	1	1	1	-	-	-	-	65	2	3	3	2	-	-	-	-	65	2	2	2	2	-	-	-	-
13	92	2	2	3	3	-	-	-	-	92	2	2	3	3	-	-	-	-	92	2	2	3	3	-	-	-	-
14	58	2	3	2	4	-	-	-	-	58	2	3	2	3	-	-	-	-	58	2	3	3	4	-	-	-	-
15	76	2	2	4	2	-	-	-	-	76	1	3	4	2	-	-	-	-	76	1	2	4	2	-	-	-	-
16	71	2	2	3	3	-	-	-	-	71	2	2	2	3	-	-	-	-	71	2	2	2	3	-	-	-	-
17	54	2	2	3	2	-	-	-	-	54	2	2	3	2	-	-	-	-	54	2	2	3	2	-	-	-	-
18	64	1	1	1	1	-	-	-	-	64	2	2	2	1	-	-	-	-	64	2	2	2	1	-	-	-	-
19	70	1	2	4	3	-	-	-	-	70	1	2	4	3	-	-	-	-	70	1	2	4	3	-	-	-	-
20	73	3	3	4	1	-	-	-	-	73	2	4	3	1	-	-	-	-	73	2	4	3	1	-	-	-	-
21	214	4	3	3	4	3	2			219	4	3	4	4	4	3			223	3	3	3	4	3	3		
22	264	3	2	6	4	4	5	4	4	272	4	3	5	3	4	3	5	4	274	4	4	5	4	3	3	4	5

Table 6.3 Job set 1 optimal solution vector

Jb-on	*3-1*	*5-1*	*1-1*	*2-1*	*4-1*	*1-2*	*4-2*	*2-2*	*3-2*	*3-3*	*2-3*	*1-3*	*5-2*
MC number	*3*	*3*	*1*	*1*	*4*	*2*	*2*	*3*	*4*	*1*	*2*	*4*	*1*
Tool copy	*1A*	*2A*	*3A*	*2B*	*3B*	*4A*	*4A*	*3A*	*4B*	*2A*	*1A*	*1B*	*1B*

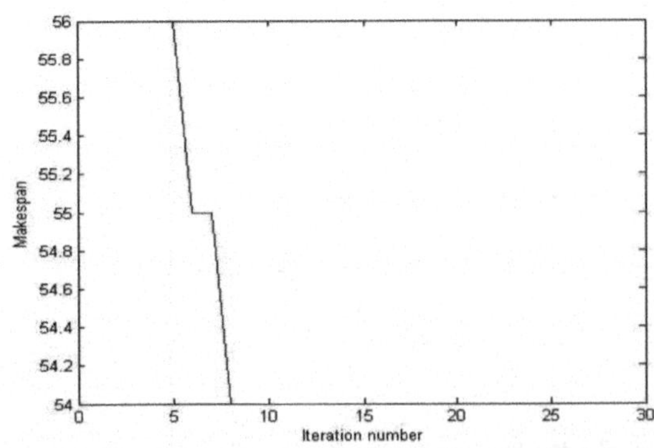

Figure 6.4 Gantt chart for solution given in Table 6.3

References

Aldrin Raj, J., Ravindran, D., Saravanan, M. and Prabaharan, T. (2014). Simultaneous SCHG of machines and tools in multimachine flexible manufacturing system using artificial immune system algorithm. Int. J. Comput. Integr. Manuf. 27(5):401–414. https://doi.org/10.1080/0951192X.2013.834461.

Anter, A. M., Hassenian, A.E., Oliva, D. (2019). An improved fast fuzzy c-means using crow search optimization algorithm for crop identification in agricultural. Expert Syst. Appl. 118:340–354. doi: https://doi.org/10.1016/j.eswa.2018.10.009.

Askarzadeh, A. (2016). A novel metaheuristic method for solving constrained engineering optimization problems.. Comput. Struct. 169:1–12. https://doi.org/10.1016/j.compstruc.2016.03.001.

Baykasoglu, A. and Ozsoydan, F. B. (2017). Minimizing tool switching and indexing times with tool duplications in automatic machines. International J. Adv. Manuf. Technol. 89:1775–1789. DOI 10.1007/s00170-016-9194-z .

Costa, A., Cappadonna, F. A., and Fichera, S. (2016). Minimizing the total completion time on a parallel machine system with tool changes. Comput. Ind. Eng. 91:290–301. https://doi.org/10.1016/j.cie.2015.11.015.

Gökgür, B., Hnich, B. and Özpeynirci, S. (2018). Parallel machine scheduling with tool loading: A constraintprogramming approach. Int. J. Prod. Res. 56(16), 5541–5557. https://doi.org/10.1080/00207543.2017.1421781

Özpeynirci, S. (2015). A heuristic approach based on time-indexed modelling for SCHG and tool loading in flexibl manufacturing systems. Int. J. Adv. Manuf. Technol. 77:1269–1274. https://doi.org/10.1007/s00170-014-6564-2.

Paiva, G. S., and Carvalho, M. A. M. (2017). Improved heuristic algorithms for the job sequencing and tool switching problem. Comput. Oper. Res. 88:208–219. doi: 10.1016/j.cor.2017.07.013.

Saravanan, M., and Noorul Haq, A. (2008). Evaluation of scatter-search approach for scheduling optimization of flexible manufacturing systems. Int. J. Adv. Manuf. Technol. 38:978–986. https://doi.org/10.1007/s00170-007-1134-5

Shirazi, R. and Frizelle, G. D. M. (2001). Minimizing the number of tool switches on a flexible machine: An empirical study. Int. J. Prod. Res. 39(15):3547–3560, DOI: 10.1080/00207540110060888.

Sivarami Reddy, N., Ramamurthy, D. V., Rao, K. P., and Lalitha, M. P. (2016). A Novel metaheuristic method for simultaneous SCHG of machines and tools in multi machine FMS. *IET* Digital Library International Conference on Recent Trends in Engineering, Science & Technology (pp. 1–6). ISBN-978-1-78561-785-0 **DOI:** 10.1049/cp.2016.1489

Sivarami Reddy, N., Ramamurthy, D. V., Rao, K. P. and Lalitha, M. P. (2017a). Simultaneous scheduling of machines and tools in multi machine FMS using crow search algorithm. Int. J. Eng. Sci. Technol. 9(09S):66–73.

Sivarami Reddy, N., Ramamurthy, D. V. and Rao, K. P. (2017b). Simultaneous scheduling of machines and tools to minimize MS in multi machine FMS using new nature inspired algorithms. Manuf. Technol. Today 16(3):19–27.

Sivarami Reddy, N., Ramamurthy, D. V. and Rao, K. P. (2018a). Simultaneous scheduling of machines and tools considering tool transfer times in multimachine FMS using CSA. International Conference on Intelligent Computing and Applications. Adv. Intell. Syst. Comput. 632:421–432.

Sivarami Reddy. N., Ramamurthy, D. V. and Rao, K. P. (2018b). Simultaneous scheduling of jobs, machines and tools considering tool transfer times in multi-machine FMS using new nature-inspired algorithms. Int. J. Intell. Syst. Technol. Appl. 17(1/2):70–88.

Sivarami Reddy, N., Ramamurthy, D.V., and Rao, K.P. (2018c). Simultaneous scheduling of machines and AGVs using crow search algorithm. Manuf. Technol. Today 17(09):12–22. ISSN:0972-7396.

Sivarami Reddy, N., Ramamurthy, D. V., Rao, K. P. and Lalitha, M. P. (2019). Integrated scheduling of machines, AGVs and tools in multi-machine FMS using crow search algorithm. Int. J. Comput. Integr. Manuf. 32(11), 1117–1133. DOI:10.1080/0951192X.2019.1686171, https://doi.org/10.1080/0951192X.2019.1686171

Sivarami Reddy, N., Ramamurthy, D. V., Lalitha, M. P. and Rao, K. P. (2021a). Minimizing the total completion time on a multi-machine FMS using flower pollination algorithm. Soft Comput. 26, 1437–1458. https://doi.org/10.1007/s00500-021-06411-y

Sivarami Reddy N., Ramamurthy, D. V., Rao, K. P. and Lalitha, M. P. (2021b). Practical simultaneous scheduling of machines, AGVs, tool transporter and tools in a multi machine FMS using symbiotic organisms search algorithm. Int. J. Comput. Integr. Manuf. 34(2),153–174. DOI: 10.1080/0951192X.2020.1858503

Sivarami Reddy, N., Ramamurthy, D. V., Lalitha, M. P. and Rao, K. P. (2021c). Integrated simultaneous scheduling of machines, automated guided vehicles and tools in multi machine flexible manufacturing system using symbiotic organisms search algorithm. J. Ind. Prod. Eng. 39(4),317–339. https://doi.org/10.1080/21681015.2021.1991014

Sivarami Reddy, N., Lalitha, M. P., Pandey, S. P. and Venkatesh, G. S. (2021d). Simultaneous scheduling of machines and tools in a multi machine FMS with alternative routing using symbiotic organisms search algorithm. J. Eng. Res. DOI : 10.36909/jer.10653.

Sivarami Reddy, N., Lalitha, M. P., Ramamurthy, D. V. and Rao, K. P. (2022). Simultaneous scheduling of machines and tools in a multi machine FMS with alternate machines using crow search algorithm. J. Adv. Manuf. Syst. https://doi.org/10.1142/S0219686722500305.

Upadhyay, P. and Chhabra, J. K. (2019). Kapur's entropy based optimal multilevel image segmentation using Crow Search Algorithm. Appl. Soft Comput. J. 97: 105522, https://doi.org/10.1016/j.asoc.2019.105522.

Venkata Rao, R. (2016). Jaya: A simple and new optimization algorithm for solving constrained and unconstrained optimization problems. Int. J. Ind. eEng. Comput. 7:19–34.

Venkata Rao, R., Savsani, V. J. and Vakharia, D. P. (2011). Teaching–learning-based optimization: A novel method for constrained mechanical design optimization problems. Comput.-Aided Des. 43(3):303–315.

7 New Tool (DGSK-XGB) for forecasting multi types of gas (T2E3A) based on intelligent analytics

Hadeer Majed, Samaher Al-Janabi[a]*, and Saif Mahmood*

Department of Computer Science, Faculty of Science for Women (SCIW), University of Babylon, Babylon, Iraq

Abstract

Natural gas is one of the alternative energy sources for oil. It is a highly efficient, low-cost, low emission fuel. Natural gas is an essential energy source for the chemical industry and is a critical component of the world's energy supply. It is also considered one of the cleanest, safest, and most beneficial sources of energy available. Therefore, this paper attempts to build a natural gas forecast optimisation model as well as find out what type of gas is connected to the networks using intelligent data analysis. In this study; We will build the HPM-STG system, which consists of integrating two technologies together: XGBoost prediction technology and GSK optimisation technology. This study will solve some problems, including the problem of data coming from natural gas that is collected manually, so the error rate will be because the prediction principle must be correct data, so we will solve it by collecting data through sensors, and the second problem that we will solve is the algorithm problem. Although the XGBoost algorithm is one of the best prediction algorithms, it faces many problems, so its core will be replaced by one of the optimization algorithms, and these two technologies together will give more accurate results.

Keywords: Acetaldehyde, acetone and toluene, ammonia, DGSK-XGB, E2T3A, ethanol, ethylene, GSK, Xgboost.

Introduction

Intelligent data analysis is one of the most important areas in real-world applications as well as computer science. We learn the artificial intelligence-based tools for finding information patterns by intelligently analysing data to provide various techniques of exhibiting throughout the discovery or recovery pattern planning. The outcomes of the data evaluation and processing can be used in the application. As a result, a specific real-world problem must be addressed, realistic data must be assessed, and the best logic approach must be selected. To create a model that can assess data once it has been discovered. The reason for constructing rules, troubleshooting optimization, resulting in data, forecasting results, or providing a concise and relevant summary is the objective of the analysis. To succeed in today's demanding technological environment, gas and oil firms must build a diverse range of hybrid capabilities that allow production processes to interact with information technology. As a result of these requirements, IOFs emerge as a commercial capacity extension, with the leadership controlling the entire value chain rather than just the equipment. Large production environments with a wide assortment of assets open up new possibilities (Li et al., 2021). It improves company operations, such as real-time visibility and process coordination, to bring assets closer to their optimal operating position.

The main problem of this paper is natural gas is the most important source in Iraq's economy and play a prominent role in controlling the country's development in various directions. Therefore, the question of forecasting its production rate for subsequent years is a very important point for drawing plans for a country according to rules and values that are closer to reality. Therefore, this paper attempts to build an optimisation model to predict the gas associated with those networks using Intelligent data analysis. In this study, we will construct a five-step system: (a) Collect data from the Natural Gas network through IOT Platform in real-time, (b) Pre-processing that data based on split it into different intervals, and determining the main limitation and rules on it. (c) build predictive model called (HPM-STG). This predictor is based on Extreme Gradient Boosting (XGboost) typically using Decision Trees (DTs) to produce the predictor. However, it will replace the DTs with a Gaining-Sharing Knowledge-based Algorithm (GSK) since it has the ability to provide more accurate and optimal outcomes than DTs alone. (d) Finally, the HPM-STG outcomes would be assessed using five confusion matrix measures known as "AC, TP, P, F-measure, and Fb". In addition, Cross-Validation will use to validate the accuracy of HPM-STG.

[a]samaher@itnet.uobabylon.edu.iq, [0000-0003-2811-1493]

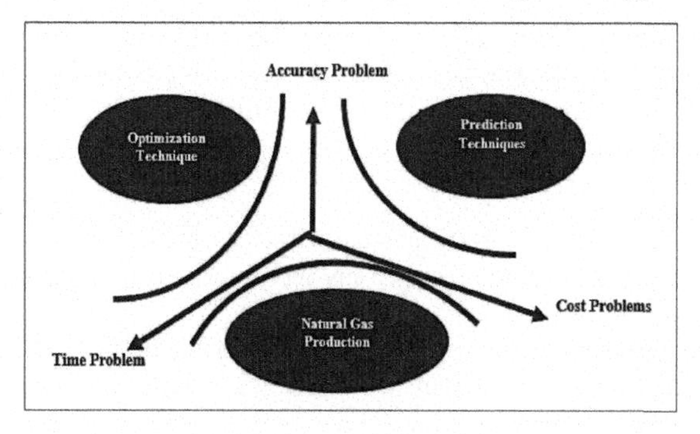

Figure 7.1 Relationships among the main three challenges

Related Work

The issue of perdition the types of Natural gas is one of the key issues related directly to people's lives and the continued of a healthful environment. Since the topic of this research is to find a recent predictive way to deal with types of data that is sensitive and performs within the range of data series, in this part of the thesis, we will try to review the works of past researchers in the same area of our issue and comparing works with seven basis points. (Li et al. 2021) VMD-RSBL prediction technology will be used in conjunction with variable mode decomposition (prediction based on RSBL, SBL with delays that are random and random samples). It would be good to watch how VMD-RSBL corresponds to forecasting workloads in future time series, such as forecasting exchange rate, forecasting loads, and also forecasting wind speed. Simultaneously, we will evaluate the utility of the proposed approach for multivariate price forecasting and give policy recommendations based on the outcomes of the study. Predicting crude oil prices based on raw pricing data is a significant and time-consuming task in our sector. The disparity is due to the fact that we forecast using different methods.

(Wang et al., 2021) By merging machine learning and numerical tank, a program is used to build and analyse data-driven forecasting methods such as GPR, CNN, and SVM models. The GPR and tank models will come after that. The simulation was ran using both the evolutionary method and the standard optimization approach. On the basis of optimization, a data-driven model was constructed A procedure that is both quick and accurate. A replacement for the algorithm of aided numerical simulation optimisation, Our approach is similar in that we applied an Optimization Algorithm, The new production of shale gas machine-based forecasting model learning differs from our business.

(Gupta and Nigam 2020) The use of an artificial neural network to anticipate crude oil prices is a fresh and creative method (ANN). The key benefit of ANN's technique is that it continuously reflects the dynamic pattern of crude oil pricing that has been included during discovery. The optimal delay and delay effect number that governs crude oil prices, Our objective is equivalent to accurate prediction until there is a large and quick change in the real data, at which point it becomes impossible to successfully anticipate the new price. In contrast to our findings, the suggested model successfully accounts for these inclinations.

(Gonzalez et al., 2022) The purpose of the study is to determine if oil and gas production pollutes the environment by examining the impact on primary oil and gas production (the number of drilling sites) and production activities (total volume of oil and gas). This flexibility to geographical, meteorological, environmental, and temporal aspects, as well as continual changes in wind direction as an external source of variation.

(Su et al., 2021) DNN studies the influence of defect size on pipeline failure pressure, suggesting that the deep learning model outperforms empirical equations in terms of prediction accuracy. Simultaneously, a multi-layer ANN deep learning model outperforms a FEM simulation by at least two orders of magnitude. Our method is similar to deep learning, but we employ the FEM model in a different way.

Hybrid Prediction Model for Natural Gas (HPM-STG)

This paper presents the main stages of building the new predictor and shows the specific details for each stage. The HPM-STG is divided into five phases. The first collects data from the natural gas network in real-time utilizing devices linked to the network represented by the Internet of Things. The second stage, pre-processing is the initial phase in the text mining process and plays an essential role in text mining techniques and applications. And that the pre-processing of the data (which was collected through devices, sensors, and tools) is based on dividing it into different periods and conducting some algorithms on it to filter it from impurities, and to determine the main restrictions and rules on it. We are going to build a predictive model called (HPM-STG). Figure 7.2 shows the main stages of HPM-STG and algorithm (1) shows the Building Prediction model.

We can summarise the main steps of this study as follows:

- Collecting data from the natural gas network through the IoT platform using devices, tools, and sensors.
- Through the pre-processing step, combine the datasets. and the basis of my work is the use of forecasting in data mining
- Create a new Hybrid predictor (HPM-STG) by combining the benefits of GSK and XGBoost
- Several metrics will be used to evaluate the prediction results as they are (accuracy, accuracy, reconnection, f-measurement, Fb).

Algorithm#1: Hybrid Prediction Model for Six Types of Gas (HPM-STG)

Input: Stream of real-time data capture from 16 sensors, each sensor, each give 8 features; the total number of features 128 collect from 16 sensors

Output: Predict the six types of Gas (Ethanol, Ethylene, Ammonia, Acetaldehyde, Acetone and Toluene)

// Pre-Processing Stage

1:	***For*** each row in gas dataset
2:	***For*** each column in gas dataset
3:	**Call** Check Missing Values
4:	**Call** Correlation
5:	***End for***
6:	***End for***

// Build DXGBoost-GSK Predictor

7:	***For*** i in range (1: total number of samples in Gas dataset)
8:	Split dataset according to 5- Cross-Validation into Training and Testing dataset
9:	***End for***
10:	**For** each Training part not used
11:	**Call** DXGBoost-GSK //predictive the types of Gas
12:	***End for***
13:	***For*** each Testing part not used
14:	Test stopping condition
15:	***IF*** max error generation < Emax
16:	Go to step 21
17:	***Else***
18:	GO to step 9
19:	***End IF***
20:	End for

// Evaluation stage

21:	***Call Evaluation***
	End HPM-STG

A. The HPM-STG Stages

The initial stage of developing an efficient prediction model in this part is dataset preparation, which comprises Drop Missing Value, Remove Duplication interval, and Five cross Validation. Using the GSK algorithm to forecast six different types of gases. The prediction model (HPM-STG) provides us with accurate findings, and the knowledge sharing acquisition procedure is employed to solve optimization issues in a continuous space. The final stage is to analyse the results using a variety of metrics.

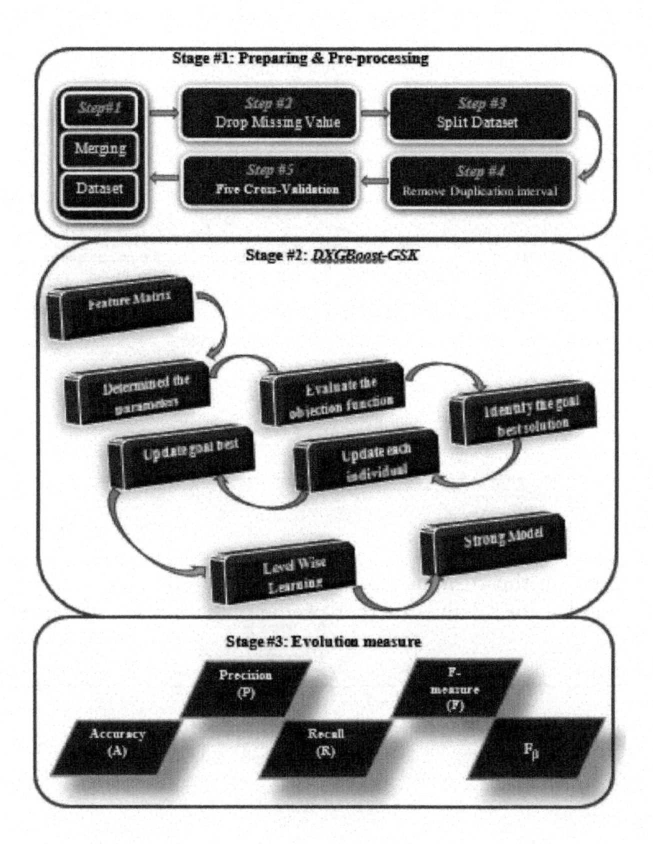

Figure 7.2 Block diagram of prediction model

1) Data Pre-Process Stage:

The information was gathered over a period of several months as explained in algorithm 2.

- The data sets are merging and being stored into a single file.
- Drop Missing values. We utilise the ones in the database and delete the missing values to produce a proper forecast since error values would emerge if it is predicted using default values.
- Finally, for each column in the dataset, apply the correlation.
- Algorithm outlines the stage's major steps

Algorithm#2: Pre-processing		
Input:	A stream of real-time data collected from several sensors	
Output:	The gas optimization approach is applied	
// Checking Missing Value		
1:	**For** each r_i in the dataset	*// i=1… n, n Maximum number of Row*
2:	**For** each c_j in the dataset	*// j=1… m , m Maximum number of Column*
3:	**IF** j=null Then	*//Check missing value*
4:	Delete W[i,j]	

5:	**Else**
6:	V[i,j]=W[i,j]
7:	**End If**
8:	**End For**
9:	**End for**
// Compute Correlation	
10:	For each r_i in dataset
11:	For each c_j in dataset
12:	Compute Pearson Correlation // $C_{r_i, c_j} = \frac{\Sigma(r_i - \bar{r})(c_j - \bar{c})}{\sqrt{\Sigma(r_i - \bar{r})^2 \Sigma(c_j - \bar{c})^2}}$
13:	**End For**
14:	**End For**
15: End Pre-processing	

2) Building HPM-STG predictor:

The goal of developing a prediction model based on the combination of two technologies is to identify gases and then determine the kind of gases found. More details of that predictor with their parameters explain in algorithm 3.

Algorithm#3: DXGBoost-GSK		
Input:	Preprocessed Dataset	
Output:	Predictive Types of Gas	
Initialize Parameter:	**N,K_p, K_r K and p**	**//N=number of individuals in population, kf=junior phase kr=Knowledge ration, and p=Senior Phase, k=find iterations**
// Compute fitness function based on Ackley		
1:	Set Main Parameters: a=20; b=0.2; c=2π	***//i=number of rows***
2:	***For*** **each row in gas dataset**	***//j=d=Total number of features***
3:	***For*** each column in gas dataset	
4:	$\text{Fitness}[i,j] = -a \times \left(-b\sqrt{\frac{1}{d}\sum_{l=1}^{d} v[i,j]^2}\right) - \exp\left(\frac{1}{d}\sum_{l=1}^{d}\cos(c \times v[i,j])\right) + a + \exp(1)$	
5:	***End For***	
6:	***End For***	
7:	Sort Population based on fitness function	
8:	G=0	

// Calculation of the Gained and shared Dimension Phases

9: ***For*** G=1 to Gen^{max}

10: ***For*** each row in gas dataset

11: ***For*** each column in gas dataset

12: $D_{juniorphase} = problemsize \times \left(1 - \frac{G}{Gen}\right)^k$ ***//Junior Phase Equation***

13: $D_{seniorphase} = problemsize - D_{juniorphase}$ ***//Senior Phase Equation***

14: ***IF*** $F\left(v_{i,j}^{new}\right) \leq F\left(v_{i,j}^{old}\right)$ ***//Every vector is updated***

15: $v_{i,j}^{old} = v_{i,j}^{new}$

16: $F\left(v_{i,j}^{old}\right) = F\left(v_{i,j}^{new}\right)$

17: ***End IF***

18: ***IF*** $F\left(v_{i,j}^{new}\right) \leq\leq F\left(v_{best}^{G}\right)$ ***//global best is updated***

19: $x_{best}^{G} = v_{i,j}^{new}$

20: $F\left(x_{best}^{G}\right) = F\left(v_{i,j}^{new}\right)$

21: ***End IF***

22: ***End For***

23: ***End For***

24: ***For*** best individual in population $\left(F\left(x_{best}^{G}\right)\right)$

25: $g_{best}(x_i) = \text{compute the derivative of question}\left(F\left(x_{best}^{G}\right)\right)$

26: $h_{best}(x_i) = \text{compute the second derivative of question}\left(F\left(x_{best}^{G}\right)\right)$

27: Prediction types of Gas:

28:

$$Fit\ types\ of\ Gas = \arg_{\partial \in 0} \sum_{i=1}^{N} \frac{1}{2} h_{best}(x_i)\left[-\frac{g_{best}(x_i)}{h_{best}(x_i)} - \theta(x_i)\right]^2 \cdot {}^{\circ}F_M = \partial\phi_{best}(x)$$

29: ***Return Type of Gas***

30: ***End For***

31: ***End For***

*32: **End DGSK-XGB***

3) Evaluation measures:

The evaluation assesses how successfully program activities meet expected objectives and how much variation in outcomes may be attributable to the program. M&E is crucial because it enables program implementers to make objectively based decisions about program operations and service delivery (Table 7.1).

Table 7.1 Measures of confusion matric

	Prediction value	
Actual value	a (True Positive)	b (False Negative)
	c (False Positive)	d (True Negative)

- Accuracy
  The accuracy of a classification technique or model is defined as the percentage of correct predictions. It is the proportion of "true" observations to all other observations.

$$AC = \frac{a+d}{a+b+c+d} \tag{1}$$

- Precision (P)
  It is the proportion of real positive outcomes to the total number of positive predictions made by the classifier.

$$P = \frac{a}{a+c} \tag{2}$$

- Recall (R)
  Recall relates to how many true positives are correctly anticipated; it is the ratio of positives to the total number of positive class components.

$$TP = \frac{a}{a+b} \tag{3}$$

- F-Measurement (F)
  This measure is based on both measures: precision and recall. That measure compute as eq(4)

$$F = \frac{2 \times Precision \times Recall}{Precision + Recall} \tag{4}$$

- FB
  is the ratio of beta-factor multiplied by Precision and Recall divided by beta-squared multiplied by Precision plus Recall.

$$F_{\beta} = \frac{(1+\beta^2) \times (Precision \times Recall)}{\beta^2 \times Precision + Recall} \tag{5}$$

Implementation and Result HPM-STG

Dataset contains data from 16 chemical sensors that were used in drift correction simulations in a discriminating test comprising 6 gases of varying concentrations. The data was collected at the gas distribution platform facility during a certain period of time (36 months). The measurement system platform allows for the versatility of obtaining desired concentrations of chemicals of interest with high accuracy and reproducibility, reducing common errors caused by human intervention and allowing the measurement system platform to focus solely on chemical sensors for truly meaningful compensation. The resulting data set comprises measurements of six distinct pure gaseous compounds: ammonia, acetaldehyde, acetone, ethylene, ethanol, and toluene.

The results for each step of the HPM-STG are shown in this section. All outcomes will also be justified. Table 7.2 shows the record number, the sample number, the properties number, and the target, as all records are equal to 6 numbers except for the third, fourth, and fifth records consisting of only 5 and missing the number which is 6.

A. Collection of dataset

At this step, the data set is utilised to put the suggested model to the test*.

Table 7.2 The features of samples

Number of records	*Number of samples*	*Number of features*	*# Gas*
1	445	128	6
2	1244	128	6
3	1586	112	5
4	161	112	5
5	197	112	5
6	2300	128	6
7	3613	128	6
8	294	128	6
9	470	128	6
10	3600	128	6

B. Pre-processing

The outcomes of each step of the HPM-STG are shown in this section. Furthermore, all results must be supported by evidence. This section presents the main steps to preprocessing the dataset.

- ***Step #1: Merging***
  In order to build a predictor of high accuracy with minimal computational complications, we will combine the following data from 36 different months together and deal with them as a single block to build the forecaster (Figure 7.3).

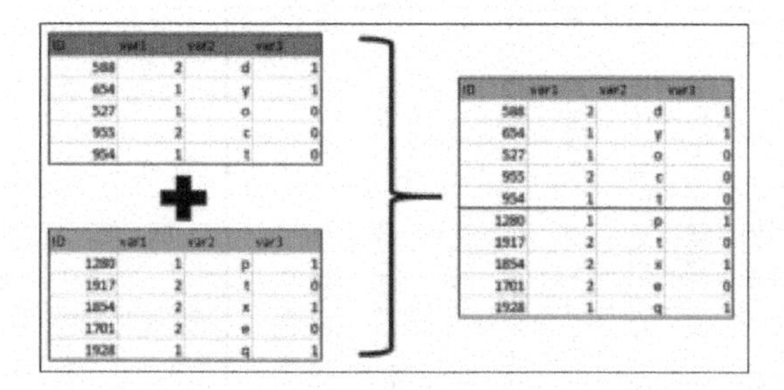

Figure 7.3 The way of merging datasets

After merging all the 10 data together, the record number as a whole will be equal to 13910.

- ***Step #2 Check missing value***
  We check the data if it contains missing values or not. All data has been checked after merging and no missing value appears, since we don't have a problem then there is no need to address it.
- ***Step #3: Correlation***
  The correlation relationship in the database to know the relationship that ties each value with its results, where the darker is closer to the outcomes and the lighter is after, and the checking process is easier for us.

In this part, we show Table 7.3 of the relationship of each sensor with the target and the pictures that illustrate this, in addition to the tables that show each gas with the number of its repetitions.

- The sensors more affect to determine the first gas are (F13, F14) in the first order and in the second-order (F01, F03, F11) while the not important sensors are (F02, F05, F07, F08, F12, F17, F18) therefore to reduce the computation can be neglected.
- The sensors more affect to determine the second gas (F23, F33) in the first order and in the second-order are (F24 , F25 , F34, F35) while the not important sensors are (F22, F28, F32, F38) therefore to reduce the computation can be neglected.
- The sensors more affect to determine the third gas (F43 , F53) in the first order and in the second-order are (F44 , F51) while all other senses are not important therefore to reduce the computation can be neglected.
- The sensors more affect to determine the fourth gas (F63, F73) in the first order and in the second-order are all other

Table 7.3 The first sensor correlation with the target

	F01	*F02*	*F03*	*F04*	*F05*	*F06*	*F07*	*F08*	*F11*	*F12*	*F13*	*F14*	*F15*	*F16*	*F17*	*F18*
F01	1	0.169931	0.98272	0.941909	0.586937	-0.96451	-0.7876	-0.20933	0.87905	0.414247	0.863494	0.84823	0.751252	-0.82618	-0.63253	-0.10768
F02	0.169931	1	0.186405	0.176823	0.075992	-0.17717	-0.12863	0.014044	0.255107	0.815315	0.283978	0.278076	0.208185	-0.25593	-0.19809	-0.01354
F03	0.98272	0.186405	1	0.978705	0.636119	-0.97081	-0.81448	-0.2389	0.84328	0.441371	0.856729	0.854715	0.75637	-0.80844	-0.63048	-0.12525
F04	0.941909	0.176823	0.978705	1	0.741052	-0.94842	-0.83372	-0.30342	0.797083	0.425014	0.819756	0.834129	0.74282	-0.77785	-0.63177	-0.18352
F05	0.586937	0.075992	0.636119	0.741052	1	-0.58034	-0.52862	-0.24899	0.457099	0.188711	0.470362	0.482534	0.511706	-0.44695	-0.37487	-0.16637
F06	-0.96451	-0.17717	-0.97081	-0.94842	-0.58034	1	0.886767	0.342334	-0.84008	-0.43981	-0.84401	-0.84612	-0.72081	0.845855	0.708914	0.233659
F07	-0.7876	-0.12863	-0.81448	-0.83372	-0.52862	0.886767	1	0.678346	-0.65174	-0.3345	-0.6684	-0.7079	-0.63002	0.72306	0.790577	0.570898
F08	-0.20933	0.014044	-0.2389	-0.30342	-0.24899	0.342334	0.678346	1	-0.10021	0.009184	-0.12221	-0.20951	-0.29157	0.24493	0.564364	0.94018
F11	0.87905	0.255107	0.84328	0.797083	0.457099	-0.84008	-0.65174	-0.10021	1	0.567534	0.983361	0.958425	0.843955	-0.96602	-0.7484	-0.12794
F12	0.414247	0.815315	0.441371	0.425014	0.188711	-0.43981	-0.3345	0.009184	0.567534	1	0.618731	0.61855	0.451911	-0.57518	-0.44759	-0.0319
F13	0.863494	0.283978	0.856729	0.819756	0.470362	-0.84401	-0.6684	-0.12221	0.983361	0.618731	1	0.988515	0.867594	-0.96742	-0.76362	-0.14973
F14	0.84823	0.278076	0.854715	0.834129	0.482534	-0.84612	-0.7079	-0.20951	0.958425	0.61855	0.988515	1	0.88824	-0.95996	-0.79348	-0.23194
F15	0.751252	0.208185	0.75637	0.74282	0.511706	-0.72081	-0.63002	-0.29157	0.843955	0.451911	0.867594	0.88824	1	-0.8534	-0.73981	-0.3427
F16	-0.82618	-0.25593	-0.80844	-0.77785	-0.44695	0.845855	0.72306	0.24493	-0.96602	-0.57518	-0.96742	-0.95996	-0.8534	1	0.859647	0.294666
F17	-0.63253	-0.19809	-0.63048	-0.63177	-0.37487	0.708914	0.790577	0.564364	-0.7484	-0.44759	-0.76362	-0.79348	-0.73981	0.859647	1	0.650944
F18	-0.10768	-0.01354	-0.12525	-0.18352	-0.16637	0.233659	0.570898	0.94018	-0.12794	-0.0319	-0.14973	-0.23194	-0.3427	0.294666	0.650944	1
Target	1	1	1	1	1	1	1	1	1	1	1	1	1	1	1	1

- The sensors more affect to determine the fifth gas are (F81, F83, F84, F85, F91, F93, F94, F95) in the first order and in the second-order (F86, F87, F88, F96, F97, F98) while the not important sensors are (F82, F92)) therefore to reduce the computation can be neglected.
- The sensors more affect to determine the six gas (FA1, FA2, FA3, FA4, FA5, FB1, FB2, FB3, FB4, FB5) in the first order while all other senses are not important therefore to reduce the computation can be neglected.

In general, we determined the sensor that have 0.80 or more than as correlated with target as important features.

C. Implementation and Result of DGSK-XGB Stage

Choose the settings that are acceptable for the learning algorithm under consideration. One of the greatest obstacles in science is that XGBoost takes too long to implement and provide results, thus this part discusses how DGSK-XGB handles this problem and overcomes this challenge.

In other words, determining the weights and model number (M) are critical elements that primarily influence DGSK –XGB performance. Table 7.4 displays the primary characteristics of DGSK –XGB.

1) Results of GSK:

This approach, which is based on the aggregation principle, is used to address optimization issues. Following the application of the algorithm, six groups were displayed, each displaying a different gas (Table 7.5).

2) Result XGBoost:

The categorization is determined using this algorithm. After determining the assembly using the previous technique, we will determine its classification in order to determine the classification of each gas (Table 7.6).

D. Evaluation measure

The extent to which changes in outcomes may be ascribed to the program is measured by evaluating how successfully the program activities meet expected objectives. M&E is crucial because it enables program implementers to make objectively based decisions about program operations and service delivery. When

Table 7.4 Table shows the parameters used in DGSK-XGB

Parameter	*Value*
Max depth	6
Learning rate	0.3
n_estimators	100
colsample_bytree	1
Subsample	1
population size	13911
knowledge factor	0.5
knowledge ratio	0.9
Knowledge rate	10
P	0.1
A	20
B	0.2
C	2π

Table 7.5 Results of GSK optimization algorithm represent the number of points in each group

Number of group	*Number of points related to that group*
Group #1	2565
Group #2	2926
Group #3	1641
Group #4	1936
Group #5	3009
Group #6	1833

Table 7.6 Result XGBoost

Target	Average	Initial Residuals	New predictions	New Residuals
1	0.184386457	0.815613543	0.265947811	0.734052189
2	0.210337143	1.789662857	0.389303429	1.610696571
3	0.117964201	2.882035799	0.406167781	2.593832219
4	0.139170441	3.860829559	0.525253397	3.474746603
5	0.216303645	4.783696355	0.69467328	4.30532672
6	0.131766228	5.868233772	0.718589605	5.281410395

Table 7.7 Evaluation measure of confusion matrix-based on HPM-STG

Target	Accuracy	Precision	Recall	F-measure	$F\beta$
1	0.625548127	0.405871511	0.26594781	0.321338302	0.353567872
2	0.59662713	0.489399772	0.19465171	0.27852437	0.287894611
3	0.554416341	0.436073165	0.13538926	0.20662644	0.186880451
4	0.536731124	0.430561448	0.13131335	0.201249339	0.179999917
5	0.543280334	0.491538614	0.13893466	0.216636459	0.206070719
6	0.119764934	1	0.11976493	0.21391085	0.359294803

we create a predictive model, we must assess it using the following metrics: Precision, Accuracy, Recall, F-Measurement, Fb (Table 7.7).

Conclusions

In this study; We will construct a five-step system:(a) Collect data from multi senseor of Gas network through IOT Platform in real-time, (b) Pre-processing that data based on split it into different intervals. (c) build predictive model called (HPM-STG). This predictor is based on Extreme Gradient Boosting (XGboost) typically using Decision Trees (DTs) to produce the predictor. However, it will replace DTs with a Gaining-Sharing Knowledge-based Algorithm (GSK) since it has the ability to provide more accurate and optimal outcomes than DTs alone. (d) Finally, the HPM-STG outcomes would be assessed using five confusion matrix metrics referred to as "AC, TP, P, F-measure, and Fb".

- ***How Gaining-Sharing Knowledge-based (GSK) can be useful in building a new predictor called (HPM-STG)?***
  It had a positive effect because it worked to determine the number of points belonging to each group based on an effective activation function that included both Junior and Senior, and one of its benefits was that it cut the execution time of XGboost, (which had the advantage of requiring many parameters to be specified, such as depth Tree, root selection, and be of great complexity).
- ***How can build an optimal prediction model by replacing the kernel of XGboost with Gaining-Sharing Knowledge-based (GSK).?***
  The basis of XGboost is DT, which has several problems as explained in point one. As a result, in this study, one of the options (HPM-STG) was employed, with GSK as the core at XGboost rather than DT, to minimise time complexity and enhance accuracy while increasing the number of calculations.
- ***Is the assessment measure utilised sufficient to evaluate the results of the suggested predictor?***
  Yes, the confession matrix has five different measures to compute the repost of new prediction XGboost and those measures are sufficient to determine the degree confidence of the predictor.
- ***What are the advantages of developing a predictor using a combination of GSK and XGboost?***
  Through the development of a new predictor known as HPM-STG, which combines GSK and XGboost, where GSK was used to discover the best group of each Gas and the quantity of points associated with it. while XGboost predicts the kind of gas.

References

Gonzalez, D. J., Francis, C. K., Shaw, G. M., Cullen, M. R., Baiocchi, M. and Burke, M. (2022). Upstream oil and gas production and ambient air pollution in California. Sci. Total Environ. 806:150298. https://doi.org/10.1016/j.scitotenv.2021.150298.

Gupta, N. and Nigam, S. (2020). Crude oil price prediction using artificial neural network. Procedia Comput. Sci. 170:642–647. https://doi.org/10.1016/j.procs.2020.03.136.

Li, T., Qian, Z., Deng, W., Zhang, D., Lu, H. and Wang, S. (2021). Forecasting crude oil prices based on variational mode decomposition and random sparse Bayesian learning. Appl. Soft Comput. 113:108032. ISSN 1568-4946, https://doi.org/10.1016/j.asoc.2021.108032.

Su, Y., Li, J., Yu, B., Zhao, Y. and Yao, J. (2021). Fast and accurate prediction of failure pressure of oil and gas defective pipelines using the deep learning model. Reliab. Eng. Syst. Saf. 216:108016. https://doi.org/10.1016/j.ress.2021.108016.

8 Experimental investigation on the frictional losses of a twin cylinder CI engine with pertinent utilization of polymethacrylate blend with conventional lubricants

Rupesh, P. L.[1,a,*], *Raja K.*[1,b], *S. V. S. Subhash*[1,c], *Shiksha Tiwari*[2,d], *K. Rajesh, Babu*[2,e], *and K. Pranay, Chowdary*[1,f]

[1]Department of Mechanical Engineering, Veltech Rangarajan Dr Sagunthala R&D Institute of Science and Technology, Chennai, India

[2]Department of Automobile Engineering, Veltech Rangarajan Dr Sagunthala R&D Institute of Science and Technology, Chennai, India

Abstract

In the developing fields of automobile, a vital role has been played by Internal Combustion Engines. The performance of Internal combustion engines depends on the combustion process that occur in the engine's core. The heat energy delivered in combustion should be converted into practical work in considerable amounts to achieve better engine efficiency. Some of the losses that occur in the engine reduce the conversion percentage of heat into work. The frictional loss contributes a key role in reduction of IC engine's efficiency. As the IC engine consists of several moving components, the friction generated against the total power delivered is assumed to be high. A proper lubricating system with a suitable lubricant and desired properties reduces friction and increases engine performance. In this work, the experimental evaluation on a four-stroke CI engine with a twin-cylinder was carried out using commercial lubricants SAE 40 & SAE 60 and the blend of those conventional lubricants along with viscosity index improver PMA. The current work concentrates on experimental analysis of desired functional parameters such as friction mean adequate pressure, frictional torque, and friction power of the experimentation engine using proposed lubricants. The results of experimentation focus on a selection of suitable lubricants with low frictional mean adequate pressure for efficient engine working. The outcome of the research work indicates that with the addition of PMA with SAE 60, the frictional power loss was slackened from 0.38 kW to 0.21 kW which proves that the addition of PMA is found to be a promising solution to overcome frictional losses.

Keywords: Friction, load, pressure, temperature, viscosity, viscosity index.

Introduction

The design of an automobile engine depends upon the engineer who designs and develops it, by considering some performance variables such as power, exhaust gases emitted, consumption of fuel, the capacity of engine cooling, etc., and some parameters are to be taken care of on reduction of costs and high engine's performance and reliability. The proposed design models of the engine can be evaluated and authenticated on the basis of measurement of flow of air, fuel and the extraction of the combustion process. The quality of an IC engine depends on variation in losses of friction (Duarte et al., 2014; Hernández-Comas et al., 2021; Ma et al., 1997; Senatore et al., 2021; Turnbull et al. 2017). The frictional power delivered in an IC engine arises from the divergence of brake power and indicated power. Due to the wear and tear of the moving parts and components available in the IC engine, a finite quantity of friction heat has been generated, impacting the necessity of cooling the engine. The exhaust gas will carry out most of the heat generated due to friction with an efficient and effective cooling system. The performance evaluation of the engine consists of the investigation of a few parameters such as speed, consumption of fuel and air, power output, etc. (Akl et al., 2021; Derbiszewski et al., 2021; Holmberg et al., 2012; Holmberg et al., 2014; Kamel et al., 2021; Senatore et al., 2021). The three nomenclatures of engine, such as brake power, brake specific fuel consumption, and frictional power, had various significance. If there is any gradual increase of friction power, then the brake power will decrease along with the increase in specific brake fuel consumption (Coy, 1998; Kovach et al., 1982; Priest and Taylor, 2000; Rylski and Siczek, 2020; Sander et al., 2016; Wong

[a]rupeshkumar221@gmail.com; [b]rajamech24@gmail.com; [c]samisubhash@gmail.com; [d]shikshatiwari604@gmail.com; [e]rb843533@gmail.com; [f]Kspc2012@gmail.com

and Tung, 2016). One of the economic aspects of any engine depends on the output delivered by an engine solely dependent on friction power. The sum of work against rubbing friction; accessory and pumping is considered to be total friction work of engine. The pumping loss occurs in gas exchange systems. Rubbing friction occurs in Crankshaft containing Main and front bearings, rear bearing oil sealants. The accessory loss occurs in supplementary components of engine. The deviation of brake power from indicated power output of engine is called as friction power. The performance and efficiency will be determined based upon the friction power loss factors. The lower friction determines the availability of much brake power, and consequently, there observes low consumption of fuel (Akl et al.. 2021; Baker et al., 2017; Knauder et al., 2020; Kovach et al., 1982; Morris et al., 2017; Sander et al., 2016; Wong and Tung, 2016). The elevation in consumption of fuel is due to surge in speed and loss due to friction also surges with engine speed. For design of engine to be effective, the friction power evaluation will be prime consideration. Friction power is used in the evaluation of indicated power and the mechanical efficiency of the engine. It can be determined using (a) Morse test; (b) Willan's line method; (c) Motoring test (Abril et al., 2021; Holmberg and Erdemir, 2017; Holmberg et al., 2014; Pardo García et al., 2021; Priest and Taylor, 2000; Richardson, 2000; Rylski and Siczek, 2020; Valencia Ochoa et al., 2020).

The wear and tear along the sliding parts are the root cause of friction which a proper lubrication system can reduce. The fluid used for lubrication purposes should have low viscosity and high viscosity index and inhibit corrosion (Akl et al., 2021; Derbiszewski et al., 2021; Holmberg and Erdemir, 2017; Holmberg et al., 2012; Holmberg et al., 2014). Some synthetic lubricants are available in which additives can be blended to achieve superior properties (Derbiszewski et al., 2021; Holmberg and Erdemir, 2017; Holmberg et al., 2012; Holmberg et al., 2014; Valencia Ochoa et al., 2020). One of the main properties of lubricant that should be enhanced is viscosity index which the addition of viscosity index improvers can achieve.

A replica of piston ring was designed by Ming-Tang Ma et al. (1997) and a Reynolds equation in two dimensional was incorporated through finite difference to estimate the tribological accomplishment for the piston rings. Veronica Senatore et al. (2021) related tribological etiquette of lubricant greases blend with novel CNT in polyalphaolefin oil and the evaluation also done with addition and elimination of MoS_2. The experimental results reveal that grease blend with MoS_2 and CNT offers a reduced friction coefficient. Akl et al. (2021) described that power loss of engine at a declined rate is assumed to be a important parameter in achieving high engine efficiency. It has been studied those nano lubricants are the most consumable liquid as inherent engine oil for the past few years. Rylski and Siczek (2020) resembles that cooling of oil is a essential work in diesel engines to achieve a greater performance. This study aims to enumerate the values of the flow resistance coefficient, and this flow model describes to study the flow resistance coefficient concerning the change in liquid temperature. Carlos Pardo-García et al. (2021) expelled that usage of Internal combustion engines leads to several serious problems because of compression rings. A framework is suggested and designed by the author to evaluate diesel engines' characteristics on a dynamic, geometric, and operational basis. The assessment of the author reveals that losses that occurred in the course of compression and combustion are assumed to be maximum on the engine's crankcase.

It was observed from the research carried out in the past decade that assessment of frictional loss and its associated parameters are too essential in order to achieve maximum mechanical efficiency. The evaluation of diesel engines' geometric and operational parameters is too dependent on these frictional power characteristics. The importance behind the effective usage of lubricants in reduction of friction losses of the engine was also indicated by several researchers. The appropriate selection of lubricants in reducing the frictional losses is based on the characteristics of lubricants. The present research focuses on all of the above inferences taken from the literature. The current study focuses on the experimental investigation of parameters associated with frictional losses. It was also noted from the literature that usage of nano particles proves to be a costlier effect in the reduction of frictional loss. Therefore, in the present study viscosity index improvers are blended with the conventional lubricants to enhance the viscosity of lubricants. Viscosity is considered to be one of the most influencing parameters for the enhanced performance of lubricant. A twin cylinder diesel engine has been taken for the assessment and reduction of frictional losses with the usage of SAE 40 & SAE 60 blended with PMA in the current work. The current research aims that the experimental investigation of friction mean adequate pressure, frictional power and frictional torque with the usage of conventional lubricants and the blend of PMA in those lubricants. Along with this, the influence of lubricant's viscosity on the frictional parameters is also studied in the present investigation.

Experimentation

A. Experimental Test Rig

The experimentation was accomplished in a Compression Ignition engine with Diesel as fuel along with twin-cylinder setup. The present setup uses a semi-pressure lubrication system through which the lubricant is supplied to the parts in motion. 75% of the lubricant flows in space in the middle of cylinder liner and piston as rubbing friction loss is more compared to other losses. The engine used for experimentation is depicted as an image in below Figure 8.1.

The experimentation was carried out at different loads from 20 to 80 N along with variation in speed from 1000 to 1600 rpm, and the parameters such as fuel consumption, torque were noted as shown in Figure 8.2. The experimentation follows the morse test to determine the friction power of a twin-cylinder engine. The pressure at the suction of air was noted using a manometer. The lubricant used in the present

Figure 8.1 Experimental test rig-4 stroke CI engine

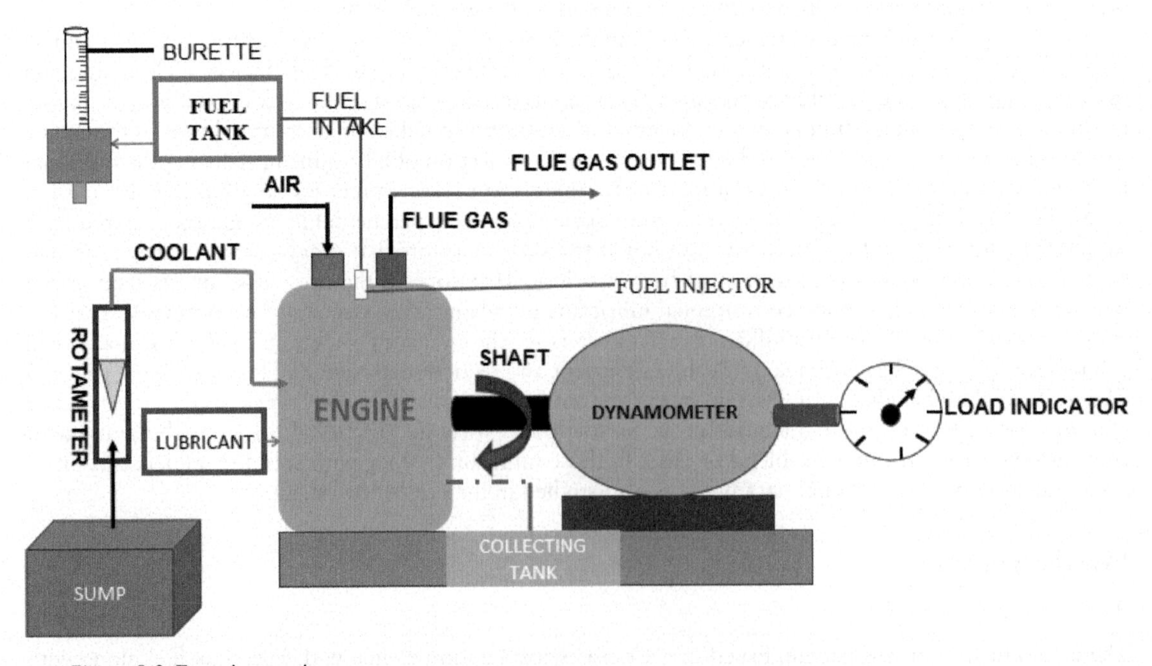

Figure 8.2 Experimentation

experimentation is SAE 40 & SAE 60, along with the blends of these oil with polymethyl acrlyte (PMA). The experimentation was done by changing the lubricants in the lubrication tank periodically after each evaluation was done for higher speed and higher load conditions.

B. Proposed Lubricant Properties

The properties of the proposed lubricants were determined using the redwood viscometer and open cup apparatus. The lubricant's viscosity and density were measured using a redwood viscometer. The open cup apparatus is used to determine flash and fire point of proposed lubricants. Table 8.1 lists the above-mentioned properties of lubricants after evaluation using respective instruments.

Friction power evaluation

The friction power (FP) was evaluated based on the Indicated (IP) and Brake power (BP) evaluated experimentally from the data obtained during the morse test. The deviation of BP from IP leads to evaluation of FP. FP was evaluated at different loads and speeds, as mentioned before. The parameters such as frictional mean adequate pressure and frictional torque are calculated based on the input values of experimentation using the equations from (1) to (6) given below.

$$i_{\text{mep}} = i_{\text{mep gross}} + i_{\text{mep net}} \quad (1)$$

$$T_b = W \times r \quad (2)$$

$$b_{\text{mep}} = \frac{T_b}{V_s} \quad (3)$$

$$f_{\text{mep}} = i_{\text{mep}} - b_{\text{mep}} \quad (4)$$

$$FP = \frac{2\pi N T_f}{60} \quad (5)$$

$$T_f = \frac{60 \times FP}{2\pi N} \quad (6)$$

The above parameters are evaluated through experimentation individually for the lubricants proposed in the current work at variable load and variable speed conditions. The experimentation results used to select suitable lubricant with the reduction in friction mean adequate pressure.

Results & Discussion

The experimental results derived from the above formulae were depicted as plots in the below figures to expose the variation between the friction power delivered for each lubricant proposed in the current study. The variation was also done concerning load and speed to manifest the influence of these parameters on the friction torque and mean effective pressure.

Figure 8.3 depicts the plot of variation between Dynamic viscosity and temperature. The graph shows that the viscosity of lubricants decreases with an increase in temperature. Out of 4 lubricants used in the present study, SAE 60 blend with PMA shows an acceptable variation. There observed a low range of

Table 8.1 Proposed lubricants-characteristics

Proposed Lubricants	*Properties*				
	Kinematic Viscosity (m^2/s)	*Dynamic Viscosity ($N\ s/m^2$)*	*Density (kg/m^3)*	*Flash point (°C)*	*Fire point (°C)*
SAE 40	16.3	0.430	872	240	255
SAE 60	26.1	0.13552	844.2	268	288
SAE 40+PMA	21.9	0.015	900	264	275
SAE 60+PMA	27.9	0.0069	830	275	289

dynamic viscosity for SAE 60 with PMA. This may be due to the presence of PMA as it has high viscosity index.

Figure 8.4 depicts the plot of variation between kinematic viscosity and Temperature. The graph shows that the viscosity of lubricants decreases with an increase in temperature, and it was noticed that SAE 60 blend with PMA shows the low range of kinematic viscosity as PMA's viscosity index is high.

The deviation of friction power generated in the engine with the engine's speed is depicted in Figure 8.5. The plot has observed that the friction power reaches a maximum value at the speed of 1300 rpm for all lubricants at full load conditions.

The variation in the value of friction power ranges from 200 W to 300 W, with the speed range between 1000 and 1600 rpm. The addition of PMA leads to a decrease in the viscosity of SAE 40 & SAE 60, which permits a massive flow of lubricant towards the rubbing friction work. Among blend of PMA with SAE 40 & SAE 60, SAE 60 leads to the low friction power of 200 W at high speed and complete load condition.

Figure 8.6 indicates the response of friction power generated in the engine with the percentage of brake load. The experimentation was carried out at circumstances of half load and full load at speed of 1600 rpm (maintained same). The observation of plot reveals that friction power reaches a maximum value of 180W

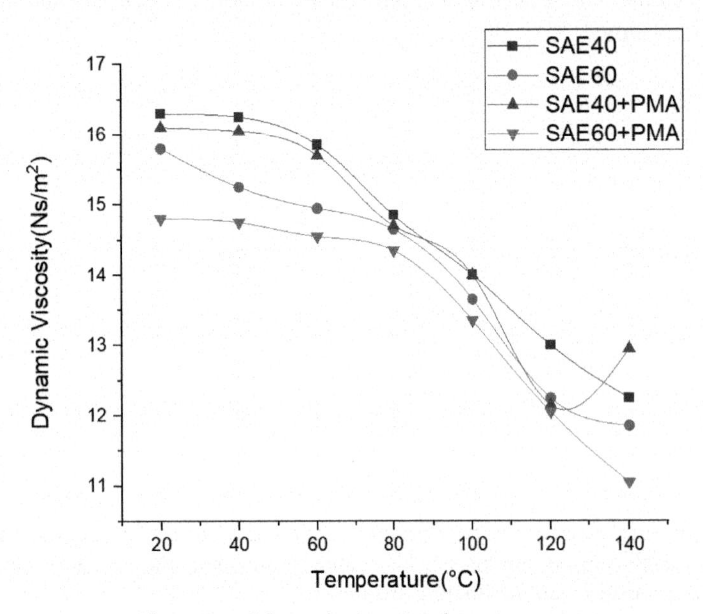

Figure 8.3 Alteration of dynamic viscosity along temperature

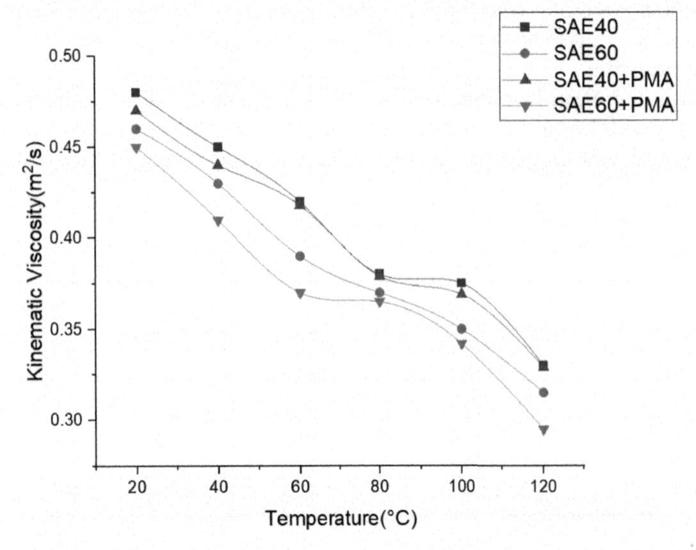

Figure 8.4 Variation of kinematic viscosity with temperature

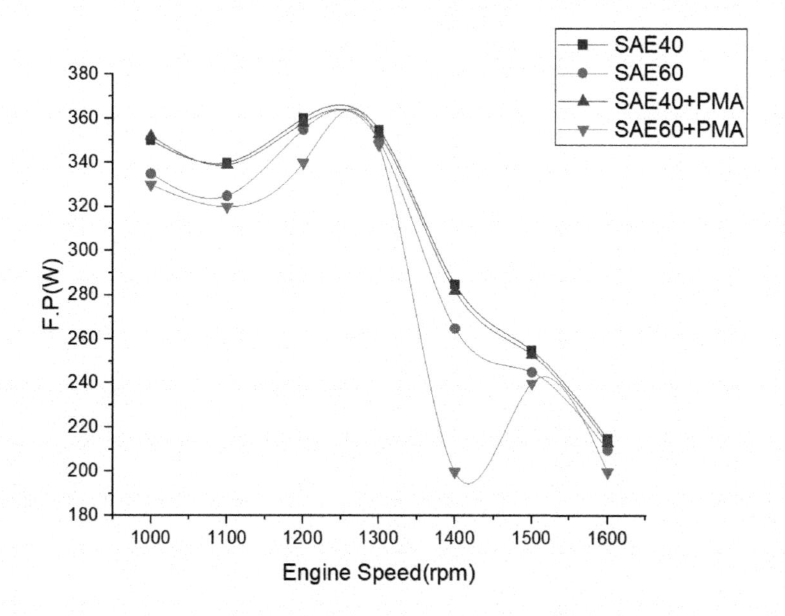

Figure 8.5 Friction power vs speed

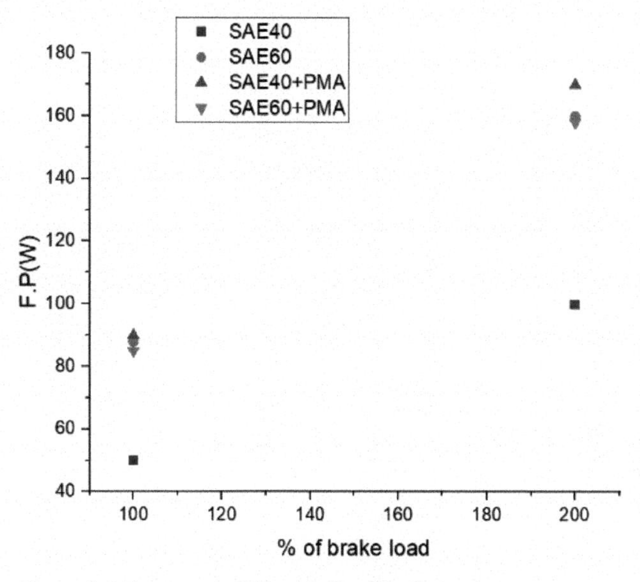

Figure 8.6 Response of FP with % of brake load

at complete load condition for all lubricants. The addition of PMA in SAE 60 leads to low friction power of 160 W at full load conditions. It signifies that the blend of PMA will reduce the heat generated due to friction, thereby reducing the losses.

The f_{mep} is evaluated to study the response of f_{mep} with load and speed as depicted in Figures 8.7 and 8.8, respectively. It was noticed from the plot that f_{mep} declines when load elevates and escalates with elevation in speed. The brake means effective pressure increases as the load increases, and f_{mep} decreases. With the increase in engine speed, FP decreases and observed enhancement in f_{mep}. In both plots, SAE 60 + PMA blend leads to low friction mean effective pressure of 0.6 bar at high load and high-speed conditions.

The torque delivered due to the frictional force reduces along elevation in engine speed, which is plotted as in Figure 8.9. As speed and torque are inversely proportional, there is a slackening of torque with speed enhance. The flash and fire point of the SAE 60 + PMA blend is too high (obtained experimentally), due to which the resistance to the heat produced due to friction becomes high for the blend. As the frictional power loss reduces for SAE60 + PMA, the frictional torque decreases at high engine speed and full load conditions.

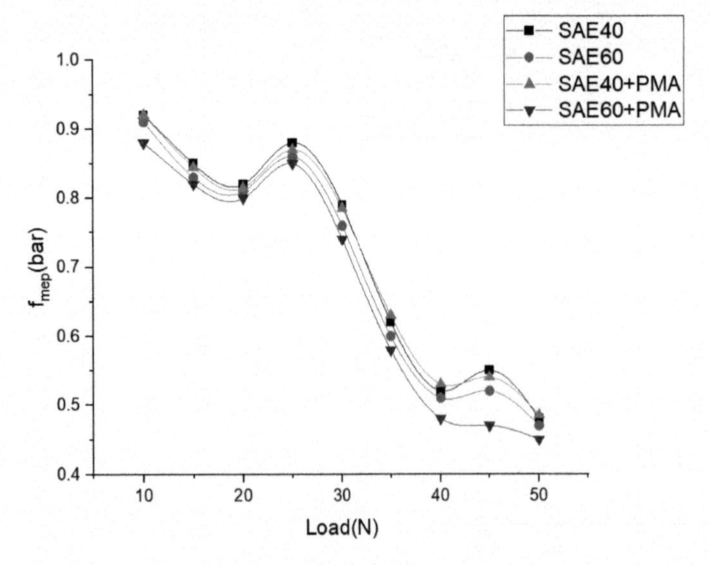

Figure 8.7 Alteration of f_{mep} along load

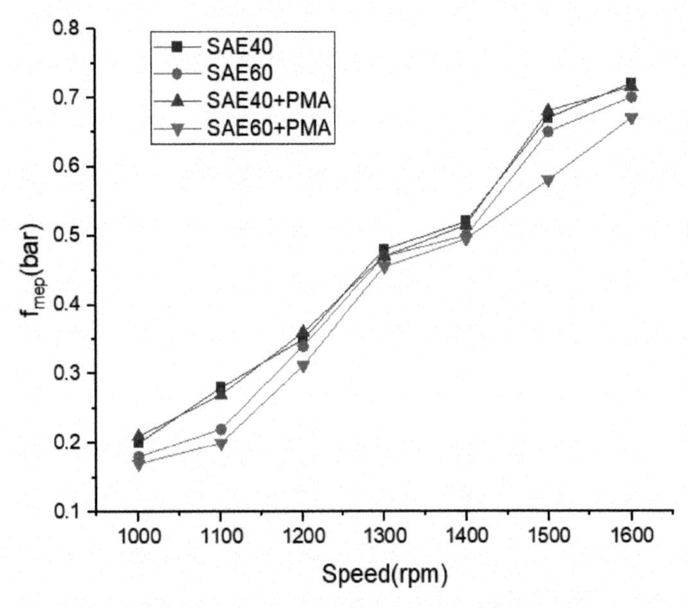

Figure 8.8 Change in f_{mep} with speed

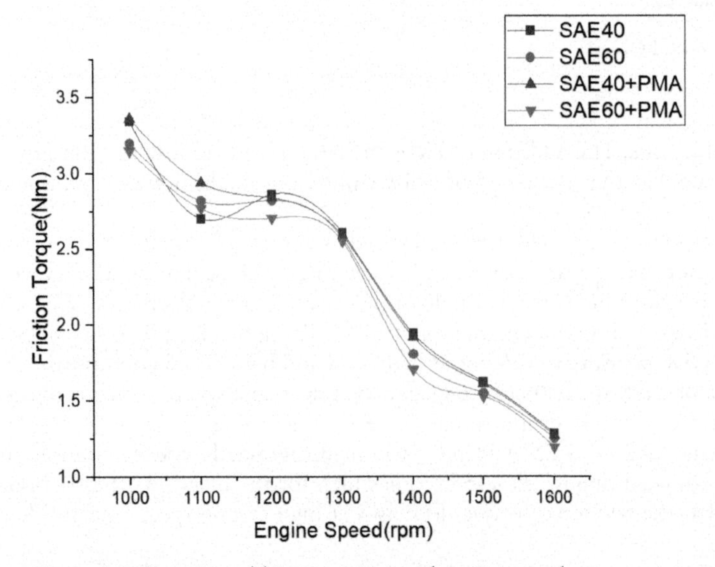

Figure 8.9 Variation of friction torque with engine speed

Conclusion

The experimental evaluation on a four-stroke CI engine with a twin-cylinder was carried out using commercial lubricants SAE 40 & SAE 60 and the blend of those conventional lubricants along with viscosity index improver PMA. The results of the present study deal with the experimental analysis of desired performance parameters such as friction mean effective pressure, frictional power and torque for CI engine using proposed lubricants. The experimentation results also focused on selecting suitable lubricant with low frictional mean effective pressure for efficient engine working. The outcome of research work indicates that the addition of PMA in SAE 60 leads to frictional power (FP) of 0.21 kW along with f_{mep} of 0.65 bar and T_f of 1.23 Nm at a high-speed of 1600 rpm and high load of 50N. The experimentation reveals that the addition of PMA in a high-performance lubricating oil such as SAE 60 increases the engine performance by slackening frictional torque and frictional mean effective pressure.

Acknowledgment

This work supported by the Research & Development Centre of Vel Tech Rangarajan Dr Sagunthala R & D Institute of Science and Technology, Avadi, Chennai, India.

References

Abril, S. O., García, C. P. and León, J. P. (2021). Numerical and experimental analysis of the potential fuel savings and reduction in CO emissions by implementing cylinder bore coating materials applied to diesel engines. Lubricants. 9(2):19.

Akl, S., Elsoudy, S., Abdel-Rehim, A. A. Salem, S., and Ellis, M. (2021). Recent advances in preparation and testing methods of engine-based nanolubricants: a state-of-the-art review. Lubricants. 9(9):85.

Akl, S., Elsoudy, S., Abdel-Rehim, A. A., Salem, S., and Ellis, M. (2021). Recent advances in preparation and testing methods of engine-based nanolubricants: a state-of-the-art review. Lubricants. 9(9):85.

Baker, C., Theodossiades, S., Rahmani, R., Rahnejat, H., and Fitzsimons, B. (2017). On the transient three-dimensional tribodynamics of internal combustion engine top compression ring. J. Eng. Gas Turbines and Power 139(6):1–39.

Coy, R. C. (1998). Practical applications of lubrication models in engines. Tribol. Int. 31(10):563–571.

Derbiszewski, B., Wozniak, M., Grala, L., Waleciak, M., Hryshchuk, M., Siczek, K., Obraniak, A., and Kubiak, P. (2021). A Study on the flow resistance of fluids flowing in the engine oil-cooler chosen. Lubricants 9(8):75.

Duarte, J., Amador, G., Garcia, J., Fontalvo, A., Padilla, R. V., Sanjuan, M., and Quiroga, A. G. (2014). Auto-ignition control in turbocharged internal combustion engines operating with gaseous fuels. Energy. 71:137–147.

Hernández-Comas, B., Maestre-Cambronel, D., Pardo-García, C., Fonseca-Vigoya, M. D. S., and Pabón-León, J. (2021). Influence of compression rings on the dynamic characteristics and sealing capacity of the combustion chamber in diesel engines. Lubricants. 9(3):25.

Holmberg, K. and Erdemir, A. (2017). Influence of tribology on global energy consumption, costs and emissions. Friction. 5:263–284.

Holmberg, K., Andersson, P., and Erdemir, A. (2012). Global energy consumption due to friction in passenger cars. Tribol. Int. 47:221–234.

Holmberg, K., Andersson, P., Nylund, N. O., Mäkelä, K., and Erdemir, A. (2014). Global energy consumption due to friction in trucks and buses. Tribol. Int. 78:94–114.

Kamel, B. M., Tirth, V., Algahtani, A., Shiba, M. S., Mobasher, A., Hashish, H. A,. and Dabees, S. (2021). Optimization of the rheological properties and tribological performance of SAE 5w-30 base oil with added MWCNTs. Lubricants. 9(9):94.

Knauder, C., Allmaier, H., Sander, D. E., and Sams, T. (2020). Investigations of the friction losses of different engine concepts: Part 3: Friction reduction potentials and risk assessment at the sub-assembly level. Lubricants 8(4):39.

Kovach, J. T., Tsakiris, E. A. and Wong, L. T. (1982). Engine friction reduction for improved fuel economy. SAE Technical Paper 820085. https://doi.org/10.4271/820085.

Ma, M. T., Sherrington, I., Smith, E. H., and Grice, N. (1997). Development of a detailed model for piston-ring lubrication in IC engines with circular and non-circular cylinder bores. Tribol. Int. 30(11):779–788.

Morris, N., Mohammadpour, M., Rahmani, R. and Rahnejat, H. (2017). Optimisation of the piston compression ring for improved energy efficiency of high-performance race engines. Proc. Inst. Mech. Eng. Part D: J. Automob. Eng. 231(13):1806–1817.

Pardo García, C., Rojas, J. P., and Orjuela Abril, S. (2021). A numerical model for the analysis of the bearings of a diesel engine subjected to conditions of wear and misalignment. Lubricants. 9(4):42.

Priest, M. and Taylor, C. M. (2000). Automobile engine tribology—approaching the surface. Wear. 241(2):193–203.

Richardson, D. E. (2000). Review of power cylinder friction for diesel engines. J. Eng. Gas Turbines Power. 122(4):506–519.

Rylski, A. and Siczek, K. (2020). The effect of addition of nanoparticles, especially ZrO2-based, on tribological behavior of lubricants. Lubricants 8(3):23.

Sander, D. E., Allmaier, H., Priebsch, H. H., Witt, M., and Skiadas, A. (2016). Simulation of journal bearing friction in severe mixed lubrication–Validation and effect of surface smoothing due to running-in. Tribol. Int. 96:173–183.

Senatore, A., Hong, H., D'Urso, V. and Younes, H. (2021). Tribological behavior of novel CNTs-based lubricant grease in steady-state and fretting sliding conditions. Lubricants. 9(11):107.

Turnbull, R., Mohammadpour, M., Rahmani, R., Rahnejat, H., and Offner, G. (2017). Coupled elastodynamics of piston compression ring subject to sweep excitation. Proc. Inst. Mech. Eng. Part K: J. Multi-Body Dyn. 231(3):469–479.

Valencia Ochoa, G., Cárdenas Gutierrez, J., and Duarte Forero, J. (2020). Exergy, economic, and life-cycle assessment of ORC system for waste heat recovery in a natural gas internal combustion engine. Resources. 9(1):2.

Wong, V. W. and Tung, S. C. (2016). Overview of automotive engine friction and reduction trends–Effects of surface, material, and lubricant-additive technologies. Friction 4(1):1–28.

9 Intermittent renewable energy sources for green and sustainable environment – a study

Pratheesh Kumar S.[a,*], *Dinesh R.*[b], *Balaganesh S.*[c], *and Vijayakumar V.*[d]
[1]Department of Production Engineering, PSG College of Technology, Coimbatore, Tamil Nadu, India

Abstract

Variable renewable energy sources, also known as intermittent renewable energy resources, are renewable energy sources that are difficult to dispatch due to their fluctuating nature. This study primarily focused on intermittent renewable energy sources and its power storage mechanisms during unfavorable climatic conditions. This also gives an overview on the need for intermittent renewable energy sources for given and sustainable environment. It ensures stable power supply and fuel diversification, which improves energy security and reduces the risk of fuel spills. Intermittent renewable resources also contribute to the conservation of the country's natural resources. This resource's most efficient application is turning it into chemical energy that can be stored in batteries. Industry technology must continue to evolve, building on previous triumphs, in order to improve dependability, boost capacity factors, and lower production costs. Renewable energy is significant because it has the ability to provide a constant supply of electricity without depleting natural resources. While lowering the need for imported fuels, there is also a lesser danger of environmental issues such as oil spills and minor emissions issues. Renewable energy could cover the electricity needs for years to come with reliable supplies and fuel variety. Diversification helps to lessen reliance on imported fuels. It generates a lot of economic development and opportunity by increasing the long-term viability of these resources, therefore reducing their volatility.

Keywords: Green environment, intermittent renewable energy, solar and wind energy, sustainable environment, technology trends.

Introduction

Wind and solar energy are two examples of renewable energy sources whose output is intermittent or variable by nature. Numerous storage methods for renewable energy sources are examined in this article. It is possible to diversify fuel supply and reduce reliance on other oil-producing countries, such as Iran, the United Arab Emirates, and others, by storing fossil fuel reserves over an extended period of time. When the weather is inclement, it is possible to utilise a range of energy sources. The reliance on fossil fuels such as coal and crude oil may be reduced by utilizing intermittent renewable energy sources, thereby lowering pollution from their combustion and promoting a more environmentally friendly and long-term environmental scenario. Themes discussed are potential future improvements to intermittent renewable energy sources, as well as the opportunities and constraints connected with their proper use. Solar energy is a fweorm of energy that is generated directly from the sun. Humans benefit from this because it is active and archived for future use. As the sun's light reaches longer distances, solar energy is becoming an increasingly common source of electricity.

Sun cells (photovoltaic cells) are solar energy conversion devices that directly convert solar energy to electricity (photovoltaic cells). When light contacts the junction of a metal and a semiconductor (such as silicon), or the junction of two distinct semiconductors, a very small electric potential can be generated. In isolation, solar cells typically produce less than two watts of electricity. Solar electric plants and large residential arrays can generate hundreds or even thousands of kilowatts of energy by bringing together a large number of individual cells, such as solar-panel arrays. Solar photovoltaic systems can be combined to create substantial amounts of electricity Wolak (2021). A second possibility is to construct smaller solar photovoltaic systems for use in micro grids or for individual consumption. Increased access to electricity in developing countries using solar photovoltaics and micro grids is a feasible option. When solar panels

[a]spratheeshkumarth@gmail.com; [b]dineshrjrj@gmail.com; [c]balaganesh160404@gmail.com; [d]vijayyajiv379@gmail.com

are placed, the orientation of the panels has a considerable effect on their performance. Calculate the solar panel's tilt angle (the angle formed by the horizontal ground and the solar module) by using location's latitude as a reference point and the solar panel's length Mamun et al. (2021). Batteries are essential for off-grid solar systems as a backup source of electricity. Due to the connection of this battery to a solar inverter, it can be charged by solar panels or the electrical grid, depending on the situation. A cable is required to connect both the positive and negative terminals of the battery to the inverter. To connect the inverter to the grid, it only needs to be connected to the main power switchboard and granted authorization to draw electricity from the grid. Additionally, the output connector is attached to the circuit board that is responsible for supplying the residence with electrical power. An excellent illustration of this is depicted in Figure 9.1 Rajendran et al. (2016) which shows the vast majority of solar energy storage batteries now in use. Lead acid batteries are the most often utilised type of battery in photovoltaic systems and are also the least expensive type of solar storage battery to purchase upfront when compared to other types of batteries. While lead acid batteries have a lower energy density, a higher efficiency, and a higher maintenance demand than other battery types, they have a longer life and cost less. These deep cycle batteries are frequently the cheapest deep cycle battery technology in terms of cost per amp-hour and cost each kWh cycle, as well as the cheapest deep cycle battery technology in terms of overall cost.

Due to the fact that lead acid batteries are the most often used type of battery in the great majority of rechargeable battery applications, they have the unique advantage of a well-established and developed technical basis Bashir et al. (2017).

Wind energy is the process of collecting wind energy and converting it to mechanical energy that is then utilised to spin electric generators, also known as wind energy collection and transformation. Wind energy, which has a lower environmental impact than fossil fuel combustion, is gaining popularity as a clean, renewable energy source. Wind turbines are equipment that harness the kinetic energy of the wind and convert it to electrical energy that may be used for a variety of purposes. When the wind is harnessed, wind turbines may generate power. Wind-powered turbines utilise the force of the wind to propel propeller-like blades around a rotor, which drives a generator, which generates electricity.

The blade length of a wind turbine is the most critical factor in determining the quantity of electricity it can generate (El-Ahmar et al., 2017; Kumar, 2018). Today, horizontal-axis wind turbines (HAWTs) generate the maximum share of wind energy. HAWTs are three-bladed wind turbines that are located upwind of the tower and generate the vast majority of the energy. Horizontal-axis turbines have blades resembling those on aero plane propellers and are frequently supplied with three blades, as indicated in Figure 9.2(a) El-Ahmar et al. (2017). Electricity is generated using horizontal-axis turbines. In contrast to Vertical-axis wind turbines (VAWTs), which are affixed to a tower and must be oriented in the direction of the wind, these turbines are free-standing. Smaller turbines are led by a simple wind vane, whereas larger turbines are guided by a wind sensor and a tilt mechanism, among other things. As illustrated in Figure 9.2(b) El-Ahmar et al. (2017), vertical-axis turbines create electricity using blades positioned at the top and bottom of a vertical rotor, respectively. Vertical-axis wind turbines (VAWTs) are wind turbines whose main rotor shaft is oriented vertically in proportion to the direction of the wind Castellani et al. (2019). As a result of this design, the turbine does not need to be oriented straight at the wind to function, which is especially advantageous in places with a wide

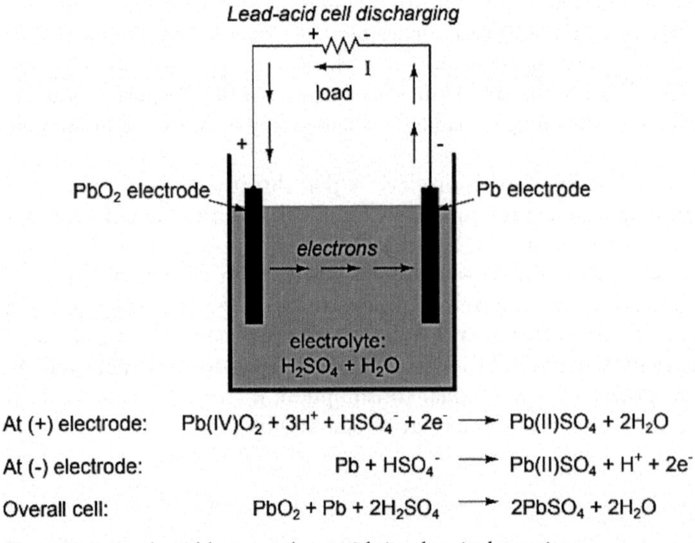

Figure 9.1 Lead-acid battery along with its chemical reaction

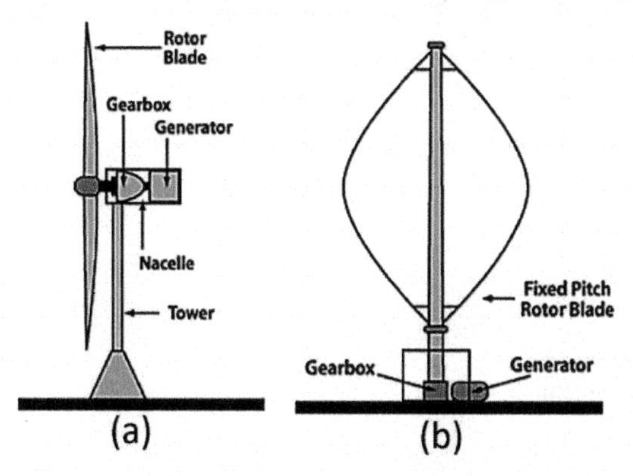

Figure 9.2 (a) Horizontal axis, (b) Vertical axis

variety of wind directions. Additionally, when a turbine is integrated into a structure, it is more advantageous because the turbine is fundamentally less steerable when integrated. Additionally, by directly connecting the rotor assembly to the ground-based gearbox, the generator and gearbox can be positioned near the ground, making maintenance easier. At the moment, just a few vertical-axis wind turbines are operational worldwide. Electricity generation from fossil fuel power plants is the leading source of carbon dioxide emissions. Additional greenhouse gas emissions, such as carbon dioxide, are major contributors to climate change and global warming. Natural gas, coal, and other fossil fuels have a substantially higher carbon footprint than alternative energy sources. Switching to renewable energy sources for power production will benefit the environment by delaying and reversing climate change. Droughts, floods, and storms induced by global warming will be reduced if the consequences of climate change are slowed and finally reversed. Increasing the number of utility-scale energy systems produced can boost economic growth and create jobs in the installation and manufacturing industries. Therefore, study in this field is essential for the implementation of these energy sources that benefit the society and the present living environment. Adequate studies on energy sources facilities for intermittent sources of energy such as solar and wind energy is not primarily discusses in research domain. This study aims to address the various energy storage system that could be utilised for the storage of intermittent energy from wind and solar sources.

Energy Storage and its Applications

Energy technologies such as solar photovoltaics, wind power, or hydropower are unable to store and deliver electricity to the system; therefore, batteries must be utilised to store and supply electricity to the grid Barton and Infield (2004). It is a relatively young technical field that has grown rapidly in response to growing global demand for more environmentally friendly energy sources, which has fueled the development of large-scale renewable energy storage systems.

A. Energy storage in photovoltaic systems

Chemical energy can be transformed to electrical energy via modern technology such as fuel cells and batteries. On the other hand, when recharged, rechargeable batteries convert electrical energy to chemical energy, which can subsequently be utilised to power new devices. As illustrated in Figure 9.3 (Singh et al., 2014) it is possible to capture a greater proportion of the energy generated by the photovoltaic system by using batteries rather than solar panels. When the solar panels' output is insufficient to meet the grid's demands, batteries can be employed to supplement the panels' output. Numerous variables, like a drop in electricity output caused by harsh weather, an increase in energy consumption above usual, or other anomalies with the photovoltaic power gathering equipment, could be at the root of this issue (Malvoni et al., 2020; Shiva Kumar and Sudhakar, 2015). As discussed previously, batteries assist in establishing the direct current (DC) operating voltages of auxiliary components when utilised in conjunction with a solar-powered system. However, they are prohibitively expensive, and each system must be constructed to a sufficiently high standard to justify the increased expenses. Due to the additional equipment necessary for photovoltaic systems with battery backup, such as inverters, batteries, and charge controllers, photovoltaic systems with battery backup are significantly more expensive to construct than conventional solar systems. To design and install these systems properly, a load assessment is required, as is the usage of specialised subpanel wiring that is only possible with these systems. The average daily electrical demand and the amount of days that the battery can store are also factors that affect the battery's capacity Hlal et al. (2019).

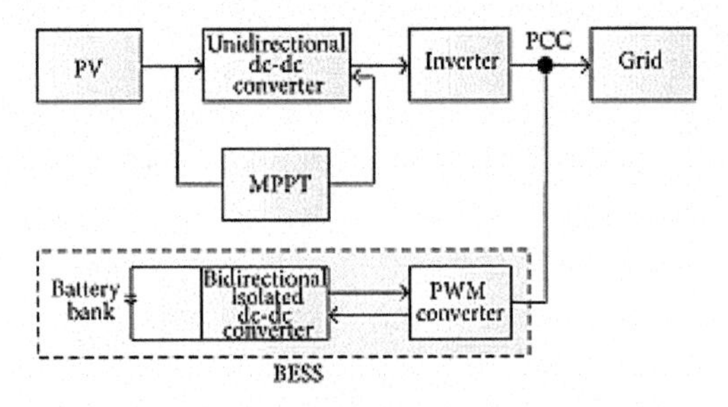

Figure 9.3 Schematic diagram of energy storage in PV system

B. Energy Conversion in Wind Turbine Systems

Batteries are commonly used to store energy in solar energy applications, and they can also be used to store wind energy when it is available. Lead acid batteries' high electrical output, trickle charging capability, and high charging efficiency make them an excellent choice for a wide variety of applications. Wind turbines can push a significant amount of compressed air into above-or below-ground tanks or tunnels, which can then be used to generate electricity. If compressed air is required, it can be expanded directly into a compressed air motor. Apart from that, hydrogen fuel cells have a high capacity for energy storage in their batteries. A hydrogen generator uses wind turbine energy to electrolyse water, which is then stored and converted back to electricity when electricity is required via a fuel cell power system (Alves, 2008). The energy generated by wind turbines is used to electrolyse water in a hydrogen generator. Despite the fact that pumped storage is typically associated with hydroelectric generating systems, it has not yet been used to generate wind energy. Theoretically, it is conceivable to use surplus energy to pump water up to an elevated reservoir and then use that energy to power a water turbine when needed (Sun et al., 2015).

C. Battery Storage System

Wind and solar energy can be utilised to charge a battery storage system. A battery storage system is a form of energy storage system. While algorithms coordinate energy production, intelligent battery software determines whether energy should be stored for reserve purposes or released to the grid, and computerised control systems determine whether energy should be stored or released to the grid (Hesse et al., 2017; Nadeem et al., 2019). Additionally, clever battery software chooses when and how much energy to keep for reserve purposes vs when and how much energy to release to the grid. Compressed air energy storage devices are available in the following configurations: The energy required to compress and store air in these machines, which are frequently housed in large chambers, is greater than the energy required to operate the device. The compressed air is released and directed via an air turbine, which produces energy as a by-product of the compressed air discharge. To transport concrete blocks up a building, a mechanical gravity energy storing device is necessary. Because energy must be generated when it is required, the concrete blocks are lowered, allowing gravity to work its magic and generate power. Chemical energy is generated in flow batteries, which are essentially rechargeable fuel cells, when two chemical components dissolve in liquids contained within the device and separated by a reactive membrane (Arabkoohsar, 2021; Chalamala et al., 2014). Flow batteries are a rechargeable type of fuel cell.

D. Energy storage in distributed systems

Uninterruptible Power Supplies (UPS) are a form of energy storage device that can be employed in the event of a power outage. UPSs are power sources that automatically switch to battery backup in the case of a power breakdown. A UPS is similar to a power strip in that it can connect multiple devices simultaneously and often has built-in surge protection and noise filtering. When a voltage drop occurs, the UPS detects it and promptly changes to battery backup mode, averting damage. Following that, UPS-connected components will continue to work for a predefined amount of time until normal power is restored or the system can be appropriately shut down, depending on the configuration. A rectifier converts alternating current (AC), which alternates in direction on a periodic basis, to direct current (DC), which flows in a single direction Zahedi (2002). The inverter is in charge of performing the inverse of the preceding function. Uninterruptible power supply systems are another critical application of Energy Storage System (ESS) in renewable microgrids, and specifically in renewable microgrids. For example, cloud cover and shade, dusk, and a lack of wind can disturb the production of power from solar panels and wind turbines, resulting in decreased output. On the other hand, backup generators can take up to ten minutes to reach full power after being switched off. It is likely that the microgrid will encounter some type of outage during this time period. If the microgrid is equipped with this technology, in the event of a power outage, the load can be provided by the ESS Anaba and Olubusoye (2021).

Overview on Intermittent Renewable Energy Sources

A. Advancements in solar energy

i. Passivated emitter rear cell (PERC):

Since its inception in the 1980s as a thin-film technology, PERC has evolved into an add-on technology aimed at increasing the efficiency of first-generation solar cells. Two layers of crystalline silicon are stacked one on top of the other on the rear of the cell using this production procedure. They assist electron transport across the cell's membrane by interacting with it. Additionally, they serve to reflect light that has gone through the cell for the first time back onto the surface, where it can be transformed into usable electrical energy for the cell.

ii. Heterojunction technology (HJT):

In the 1980s, the Japanese company Sanyo (later Panasonic) invented heterojunction technology, which enabled photovoltaic (PV) cells to boost their efficiency and power output (HJT). Numerous solar businesses began conducting HJT research when the technology's patent expired in 2010. The objective was to increase the efficiency of their photovoltaic solar panels.

iii. Building-integrated photovoltaics (BIPV):

The most prevalent type of building-integrated photovoltaic (BIPV) system is solar shingles, which became commercially available in 2005. Solar cells are installed on the roof in place of asphalt shingles to help in saving money on energy bills. Each of them is capable of producing 60 watts of power. Certain varieties of shingles are capable of producing up to 100 watts of energy from a single shingle.

iv: Multi-junction solar cells:

Multiple-junction solar cells, sometimes referred to as stacked cells, have a better efficiency than single-junction solar cells, which accounts for their higher cost. Approximately 45% efficiency levels have been consistently attained in laboratory settings up to this point. They are currently being employed commercially in the field of space technology, where they appear to be a possible component of the clean energy revolution. On the other hand, despite their rising popularity in the office, multi-functional devices remain prohibitively expensive and unavailable for domestic usage. It is based on the fact that various semiconductive materials absorb solar energy at different wavelengths, resulting in a multi-junction photovoltaic cell with a higher efficiency Ehrler et al. (2020). Simply stacking two, three, or more components does not suffice to create a cohesive thing. This is often not the case, as their structural similarities are too great. To yet this year, researchers have developed two independent ways for making stacked cells function properly. To begin, they may be able to create tunnel connections between the layers, allowing electrons to travel across. Alternatively, they may combine a variety of semiconductive materials into a single device via chemical bonding. Energy flows through the chemical bonds, which cause the materials to become electrically connected as a result of the chemical bonds produced between their constituent elements.

v. Concentration photovoltaic cell (CPV):

Concentrators like as lenses and mirrors are used to focus and direct light to a small photovoltaic cell, which is made possible by the employment of optical collectors such as lenses and mirrors. Typically, solar cells designed for space or military use and built to survive extreme temperatures are employed in this application. At the moment, multi-junction solar cells are being employed to maximise solar energy efficiency (about 30% in the field, but up to 43% in the laboratory). To maximise direct irradiance while minimizing thermal damage to the materials, a double-axis solar tracking device with active cooling must be used. To function effectively, the light rays entering the lens must always be perpendicular to the lens.

vi. Perovskite solar cells:

In chemistry, they are referred to as perovskites. This is an exceedingly broad term that refers to a large number of chemical compounds that have structural features with a naturally occurring substance called calcium titanate. They are inexpensive and easy to create, which makes them desirable. Numerous them are outfitted with exceptional photovoltaic capacity. In 2012, researchers created the world's first thin-film perovskite with a 10% efficiency, the first of its kind. Due to significant technical advancements, the figure has been increased to 25% when tested in a laboratory setting. Tandem cells, which utilise 'perovskite on silicon' technology, have shattered efficiency records at a rate of 29.1%, with no indication of slowing Kothandaraman et al. (2020). It took 300 hours in total to restore the prototype's efficiency to its prior level. Even in isolation, this longevity is remarkable given that existing perovskites are not at all heat stable.

B. Advancements in wind energy

i. Taller towers:

Wind turbines and their supporting towers are growing in size and height as they are constructed throughout the world. GE Renewable Energy's offshore wind turbine will generate 12 MW, 13 MW, or 14 MW, depending on the model. A prototype is now being built in the Netherlands' port of Rotterdam. It will be unveiled in 2020. When compared to the largest wind turbine currently in service, it produces approximately one-third more electricity.

ii. Increased Energy Production

It is feasible to boost the system's output capacity by using larger blade diameters and taller turbines. To put this in context, a typical wind turbine a decade ago produced 1.5 megawatts of energy. Their capacity has increased throughout the years, with General Electric's Halide X being the largest. On the other hand, several are significantly smaller in stature. According to a recent report from the Department of Energy's Wind Exchange platform, which gives access to scientific and wind energy data, the average nameplate capacity of newly constructed land-based wind turbines in the United States was 2.55 MW in 2019.

iii. Onsite Assembly

Despite the fact that the larger blades are spinning high in the air, those monstrous blades are causing gridlock on the ground below. Due to their size and design, they require the use of specialised trucking routes and authorization applications, as well as the use of the appropriate vehicles, equipment, and specialists. Even if they are travelling through highways and interstates at all hours of the day or night, these enormous pieces of equipment can generate traffic congestion and bottlenecks. The blades are constructed from segmented components that can be combined to form a single unit to facilitate transportation. Despite its prototype status, the commercial world has showed interest in this notion. A pair of blades, each around 70 to 80 meters in length, would be transported in two sections for on-site assembly.

iv. Onsite Construction:

Wind turbine towers in the United States are typically constructed of tubular steel and supported by concrete foundations. On the other hand, several manufacturers are experimenting with the construction of fully concrete towers in order to increase the longevity of their structures. The tower's construction will necessitate the manufacture of concrete pieces on-site. At the moment, due to transportation limits, the base's breadth cannot exceed 4.5 meters. As a result, the turbine's height is limited.

v. Turbine Recycling Growing:

Wind turbine blades are expected to generate 43.4 million metric tons of rubbish by 2050, according to a 2017 study undertaken by academics at the University of Cambridge Institute for Manufacturing Liu and Barlow (2017). Taking the predicted numbers into account, more wind turbine manufacturers are studying ways to recycle their equipment once it reaches the end of its useful life.

C. Challenges of Intermittent Renewable Energy Sources

The fact that the power system is built around massive, regulated electric generators makes deploying variable energy sources problematic. Due to the grid's limited storage capacity, it is critical to maintain a continuous balance between supply and demand in order to avoid a blackout or other cascading disaster (Hirsch et al., 2018; Kwasinski et al., 2012). Additionally, intermittent renewable energy sources complicate matters by deviating from normal methods for planning the day-to-day operation of the electric grid. They create varying amounts of energy throughout the day, necessitating adjustments to the grid operator's day-ahead, hour-ahead, and real-time operations. The grid operator's day-ahead plan must contain generators capable of rapidly adjusting their output in order to compensate for variations in solar energy generation during daylight hours. Additionally, power plants that operate 24 hours a day may be forced to shut down during the day to allow solar energy to be used in place of fossil-fuel electricity. Additionally, in addition to the natural oscillations induced by sunrise and sunset, solar panel output can fluctuate dramatically depending on the amount of cloud cover present. Due to the unpredictability of clouds, estimating the amount of additional electric generation necessary over the next hour of the day is more challenging. As a result, calculating precisely what each generator's output should be during the load-following phase of the system becomes more complex. Wind and solar energy production variations affect the second-to-second balance of total electric supply and demand, as well as the hourly load following the grid planning phase.

D. Potentials of intermittent renewable energy sources

Wind and solar energy have shown tremendous potential in terms of reducing carbon dioxide emissions linked with the electric power business. However, the extensive use of renewable energy sources has been questioned on numerous occasions in the past due to its intermittent nature. The existing grid integration ratings has been examined and the massive quantity of data collected on the behaviour and reliability of intermittent renewable energy sources using this paradigm. Recent simulation results indicate that the modelling methodologies need to characterise the system are strongly dependent on the energy penetration of intermittent technologies. As a result, the modelling approaches demonstrate markedly different behaviour in regimes with low and high energy penetration, respectively. On the basis of both penetration regimes, a paradigm is proposed for incorporating grid integration findings into evaluations of decarbonisation plans (Al-Shetwi et al., 2020; Hart et al., 2012).

The well-being of the people living in a society is greatly improved because of the reduction in emission of greenhouse gases. It's usage increases the energy diversity and minimises the reliance on fossil fuels which enhances the economical development of the society. It helps prevent the depletion of natural available resources that are utilised for energy. By utilizing the intermittent energy sources, the effect of adverse change in climate prevalent in the environment can be minimised or even completely discarded thus preserving the environment. Intermittent renewable energy sources (IRES) are renewable energy sources that cannot be dispatched because of their fluctuating nature, such as wind and solar power. Solar and wind energy can be used by the utilization of specific energy storage devices such as photo voltaic (PV) cells, wind turbines and so on. The utilization of these sources of energy enables us to diversify the energy consumption for the human needs. This in turn reduces the pressure on the utilization of non- renewable sources.

Conclusion

When used to address some of the world's most serious problems, such as global warming, air and water pollution, soil erosion, and soil erosion, carbon-neutral or near-carbon-neutral energy sources can contribute to a more environmentally conscious society, as well as a more environmentally conscious society. Solar, wind, and hydroelectric power are all examples of renewable and clean energy sources to consider. Renewable energy sources, particularly in developing nations, are becoming increasingly important in the development of technical infrastructure around the world, particularly in developing countries. Most developing countries are concentrating their efforts on this issue in order to encourage economic growth while simultaneously protecting the environment and safeguarding human health.

References

Al-Shetwi, A. Q., Hannan, M. A., Jern, K. P., Mansur, M., and Mahlia, T. M. I. (2020). Grid-connected renewable energy sources: Review of the recent integration requirements and control methods. J. Clean. 253:119831.

Alves, M. (2008). Hydrogen energy: Terceira island demonstration facility. Chem. Ind. Chem. Eng. 14:77–95.

Anaba, S. A. and Olubusoye, O. E. (2021). Electricity generation from renewable resources Encyclopedia of the UN Sustainable Development Goals. Springer Nature. 338–350.

Arabkoohsar, A. (2021). Classification of energy storage systems. Mech. Energy Storage Technol. 1–12.

Barton, J. P. and Infield, D. G. (2004). Energy storage and its use with intermittent renewable energy. IEEE Trans. Energy Convers. 19:441–448.

Bashir, N., Sardar, H. S., Nasir, M., Hassan, N. U., and Khan, H. A. (2017). Lifetime maximization of lead-acid batteries in small scale UPS and distributed generation systems. In IEEE Manchester Powertech, Manchester, United Kingdom.

Castellani, F., Astolfi, D., Peppoloni, M., Natili, F., Buttà, D., and Hirschl, A. (2019). Experimental vibration analysis of a small scale vertical wind energy system for residential use. Machines. 7:35.

Chalamala, B. R., Soundappan, T., Fisher, G. R., Anstey, M. R., Viswanathan, V. V., and Perry, M. L. (2014). Redox flow batteries: An engineering perspective. Proc. IEEE Inst. Electr. Electron. Eng. 102:976–999.

Ehrler, B., Alarcón-Lladó, E., Tabernig, S. W., Veeken, T., Garnett, E. C., and Polman, A. (2020). Photovoltaics reaching for the shockley–queisser limit. ACS Energy Lett. 5:3029–3033.
El-Ahmar, M. H., El-Sayed, A.-H. M., and Hemeida, A. M. (2017). Evaluation of factors affecting wind turbine output power. Nineteenth International Middle East Power Systems Conference (MEPCON). IEEE. 1471–1476.
Hart, E. K., Stoutenburg, E. D., and Jacobson, M. Z. (2012). The potential of intermittent renewables to meet electric power demand: Current methods and emerging analytical techniques. IEEE Inst. Electr. Electron. 100:322–334.
Hesse, H., Schimpe, M., Kucevic, D., and Jossen, A. (2017). Lithium-ion battery storage for the grid—A review of stationary battery storage system design tailored for applications in modern power grids. Energies 10:2107.
Hirsch, A., Parag, Y., and Guerrero, J. (2018). Microgrids. A review of technologies, key drivers, and outstanding issues. Renew. Sustain. Energy. 90:402–411.
Hlal, M. I., Ramachandaramurthy, V. K., Sarhan, A. Pouryekta, A., and Subramaniam, U. (2019). Optimum battery depth of discharge for off-grid solar PV/battery system. Energy Storage. 26:100999.
Kothandaraman, R. K., Jiang, Y., Feurer, T., Tiwari, A. N., and Fu, F. (2020). Near infrared transparent perovskite solar cells and perovskitebased tandem photovoltaics. Small Methods. 4:2000395.
Kumar, A. (2018). College Motihari, M. S., India, M. Z. U. K. University of Management & Technology, Sialkot Campus, Pakistan, and Gyancity Research Lab, Motihari, India. Wind Energy 2:29–37.
Kumar, B. S. and Sudhakar, K. (2015). Performance evaluation of 10 MW grid connected solar photovoltaic power plant in India. Energy Rep. 1:184–192.
Kwasinski, A., Krishnamurthy, V., Song, J., and Sharma, R. (2012). Availability evaluation of micro-grids for resistant power supply during natural disasters. IEEE Trans. Smart Grid. 3:2007–2018.
Liu, P. and Barlow, C. Y. (2017). Wind turbine blade waste in 2050. Waste Manag. 62:229–240.
Malvoni, M., Kumar, N. M., Chopra, S. S., and Hatziargyriou, N. (2020). Performance and degradation assessment of large-scale grid-connected solar photovoltaic power plant in tropical semi-arid environment of India. Sol. Energy 203:101–113.
Mamun, M. A. A., Islam, M. M., Hasanuzzaman, M., and Selvaraj, J. (2021). Effect of tilt angle on the performance and electrical parameters of a PV module. Comparative indoor and outdoor experimental investigation. Energy Built Environ. 3(3):278290.
Nadeem, F. Hussain, S. M. S., Tiwari, P. K., Goswami, A. K., and Ustun. T. S. (2019). Comparative review of energy storage systems, their roles, and impacts on future power systems. IEEE Access. 7:4555–4585.
Rajendran, S., Rathish, R. J., Prabha, S. S., and Anandan, A. (2016). Green electrochemistry a versatile tool in green synthesis: An overview. Port. Electrochim. Acta. 34:321–342.
Singh, R., Taghizadeh, S., Tan, N. M. L., and Pasupuleti, J. (2014). Battery energy storage system for PV output power leveling. Adv. Power Electron. 2014:1–11.
Sun, H. Luo, X., and Wang, J. (2015). Feasibility study of a hybrid wind turbine system – Integration with compressed air energy storage. Appl. Energy. 137:617–628.
Wolak, F. (2021). Long-term resource adequacy in wholesale electricity markets with significant intermittent renewables. Cambridge: National Bureau of Economic Research.
Zahedi, A. (2002). Energy, people, environment. Development of an integrated renewable energy and energy storage system, an uninterruptible power supply for people and for better environment. In IEEE International Conference on Systems Man and Cybernatics, San Antonio.

10 Evolution, revolution and future of plastics for an eco-friendly environment

Pratheesh Kumar S.[a,*], *Dinesh R.*[b], *Balaganesh S.*[c], *and Vijayakumar V.*[d]
Department of Production Engineering, PSG College of Technology, Coimbatore, Tamil Nadu, India

Abstract

Plastic is a material that contains both a required element and an organic compound in sufficient amounts. Plastics have become an integral part of our everyday lives as their use has increased. It has grown to be one of the most commonly used materials on the planet since its inception. Without a question, plastics have exceeded and supplanted a wide variety of conventional materials, becoming ubiquitous in contemporary culture. The objective of this study is to gain knowledge about the evolution, revolution, and future of plastics. We now understand how plastics can reach previously unfathomable levels of degradation. The purpose of this study is to look at how plastic has evolved over time. To understand the market impact of plastics, it is necessary to evaluate their properties in a variety of industries. Property information is used to complete the analysis. Plastics can be expected to contribute to a wide variety of societal and technological improvements in the future. As a result of this inquiry, plastic has been maintained in the environment. Despite their shortcomings, plastics are an integral part of our future. By examining the current state of the market for plastics, one can forecast the future. Plastics' long-term viability has begun to dwindle as environmental concerns have grown. This study could be used in a variety of ways to aid in the forecasting of the plastics industry's future. This page also helps to explain the evolution, application, and features of plastics.

Keywords: Ecosystem, environment, plastics, polymers, technology.

Introduction

Plastics assist us in reconciling modern expectations in today's environment. Cast, pressedor extruded plastic materials include films, fibres, plates, tubes, bottles, and boxes are used in commercial product market. Cellulose, coal, gas, salt, and crude oil are all used to make plastics. Plastics are a man-made or semi-man-made substance that is used in a variety of applications. Everywhere there is plastic. Plastics simplify, secure, and improve the quality of our lives. Plastic is used in almost every part of humans life. Toys, displays, information technology tools, and medical equipment are all constructed of plastic. Humans have long strived to create materials that outshine natural ones. Natural substances such as shellac and chewing gum were used to create plastics. The next stage was to modify natural substances such as rubber, nitrocellulose, collagen, and galalite. Modern polymers were developed more than a century ago. Alexander Parkes developed Parkesine in 1855. Celluloid is the modern term for this material. PVC was originally synthesised between 1838 and 1872, Jia et al. (2017). Bakelite was invented in 1907 by Leo Baekeland. The following listrepresents plastic's evolution.

- 1862 Birth of plastic
- 1862 Cellulose (John Hyatt)
- 1880 Celluloid (X-ray/motion picture films)
- 1907 Bakelite (Leo Baekland)
- 1914 PVC (Eugen Baumann and Waldo Semon)
- 1921 Injection moulding (John Wesley Hyatt)
- 1922 Super polymers

[a]spratheeshkumarth@gmail.com; [b]dineshrjrj@gmail.com; [c]balaganesh160404@gmail.com; [d]vijayyajiv379@gmail.com

- 1933 Polythene (Hans von Pechmann)
- 1935 Aircraft Cockpit
- 1938 Epoxy resins (Paul Schlack)
- 1935 Plastic in healthcare industry
- 1940 Plastic in war
- 1946 Plastic in banking & finance
- 1950 Plastic in clothing industry
- 1956 Plastic in automobiles
- 1960 Plastic in furniture
- 1970 Plastic in IT &Tele-communications
- 1980 Plastic in Nanotechnology
- 1990 Bio plastics
- 2000 Single use plastics
- 2021 Recycling of plastics

In 1869, in response to a $10,000 offer from a New York corporation, John Wesley Hyatt produced the first synthetic polymer. Hyatt combined cellulose and camphor to create a material that could be moulded to resemble tortoiseshell, horn, linen, and ivory, this acted as a game changer. Finally, human production was liberated from natural limits. Metal, stone, bone, tusk, and horn are all considered rare. Now we have the ability to manufacture new materials. Human beings and the environment both benefited. It was hailed as the saviour of elephants and tortoises. Plastics can help save the natural environment from human exploitation. Natural resource scarcity forced humanity to develop new materials, which alleviated social and economic restraints. As affluence increased, celluloid became more affordable. The revolution in plastics begun.

Leo Hendrik Baekeland, a Belgian immigrant, invented the first synthetic polymer in 1907. Wire insulation is necessary in electric motors and generators. He discovered a sticky mass comprised of wood flour, asbestos, or slate dust that could be used to create sturdy and fire-resistant 'composite' materials. Pressure tanks were employed to eliminate bubbles and achieve a smooth, homogeneous result during the synthesis of the novel substance. It was mostly employed in electrical and mechanical components until the 1920s, when it became widespread in commerce and jewellery. Bakelite, a thermosetting synthetic, was another early pioneer.

Polymers are chains of big molecules that are covalently linked. The number of molecular mass units varies between hundreds and millions (as opposed to the tens of atomic mass units commonly found in other chemical compounds). Molecules' size, state, and structure all contribute to the polymers' unique features.

Bakelite was created in 1907 by Leo Baekeland, which is a synthetic equivalent to shellac, a natural insulator. Bakelite was a tough, heat-resistant material that could be mass-produced mechanically. Bakelite, with its "thousand applications," could be formed into nearly anything. As a result of Hyatt and Baekeland's success, additional polymers joined celluloid and Bakelite. In contrast to Hyatt and Baekeland, contemporary research is focused on developing novel polymers with application potential.

Thermoplastics include polyethylene and polystyrene. A heated foamed polystyrene cup, for example, can be formed. Molecules of thermoplastic polymer pass through one another. Small, branched, or linear molecules are easily separated and therefore highly mobile. Plants are abundant in thermoplastics. Polyethylene (LDPE and HDPE), polypropylene (PP), polyvinyl chloride (PVC), and polystyrene (PS) are the most commonly used. Polymers for utility wire and low-load applications. Thermoplastic polymers provide support for textiles. Molding is possible using thermoplastic materials (through heat). Due to polymer chain breaking, thermally deteriorated thermoplastic polymers should not be recycled.

However, there is no recycling. Thermosets during the early stages of thermosetting resin manufacturing, an insoluble network is formed. It is heated to the point where a single big molecule forms. Cross-linking occurs when epoxy polymers are heated to form a golf club laminate. Indeed, re-heating the material may degrade it. Thermosetting compounds become harder when exposed to heat. One type is thermosetting polymers. It is either a liquid or a pliable solid. Heat cures/hardens polymers, resulting in the formation of covalent bonds. Apart from external sources, heat can be generated chemically. Accelerating the cure rate using a catalyst or a hardener. Injection moulding, extrusion moulding, compression moulding, and spin casting are all methods for shaping heat-set polymers.

Starch, cellulose, oils, and lipids are used to make bioplastics. Others are partially biobased, with the carbon coming from renewable sources. Additionally, the proportion of biobased components and the conditions under which they degrade may vary depending on the chemical makeup, crystallinity, and surrounding environment Muniyasamy and John (2017). Recently, biodegradable bioplastics have gained appeal. Market share for these innovative polymers will be determined by the public's commitment to environmental protection, notably waste reduction and fossil fuel conservation. Biobased and biodegradable polymers must overcome higher prices and limited production capacity in order to obtain a major market share.

Role of Plastics in Daily Life

Plastics are extremely adaptable materials with a wide range of consumer and industrial applications. Plastics are flexible and lightweight. Several polymers can be manufactured to conduct electricity in addition to providing thermal and electrical insulation. They are resilient and resistant to a variety of substances that degrade other materials in harsh settings. Certain varieties are translucent for optical purposes. They can be formed into complex shapes, allowing for the incorporation of a wide variety of materials and functionalities. By adding reinforcement fillers, colours, foaming agents, flame retardants, and plasticisers to plastics, certain physical qualities can be achieved. Theoretically, plastics can have any combination of qualities.

There are two forms of plastic processing. Thermoplastic materials may be repeatedly softened and moulded using heat and pressure. Thermoset materials undergo a chemical process that is irreversible. Fabrics, steering wheels, bumpers, light lenses, fuel and water tanks, furnishings and fittings, and foams and insulation are all manufactured using PVC resins. Dishes, glasses, and jugs Plastics processing is a high-stress job. Resin is softened by heat and pressure. Resin is used to sculpt the completed product. Compression and injection moulding are the most frequently used methods. The influence of plastics on daily life is seen in Figure 10.1, Szlachetka et al. (2021).

A. Plasticulture Applications

India saw its first Green Revolution in the 1970s. India achieved agricultural self-sufficiency through increasing the use of fertilisers, insecticides, and pesticides. India today sustains approximately 16% of the world's population Lakshmana (2013). 58.2% of the population is employed in agriculture. While agriculture contributes approximately 17% of India's GDP, it utilises nearly all available water. Plasticulation has the potential to usher in a "Second Green Revolution" in Indian agriculture. Numerous crop enhancement strategies are included in Table 10.1, Markarian (2005).

PE is the most often used polymer in agricultural and related applications (LLDPE, LDPE, HDPE). It improves yield and quality while cutting costs. Table 10.2 summarises water savings, water efficiency, and

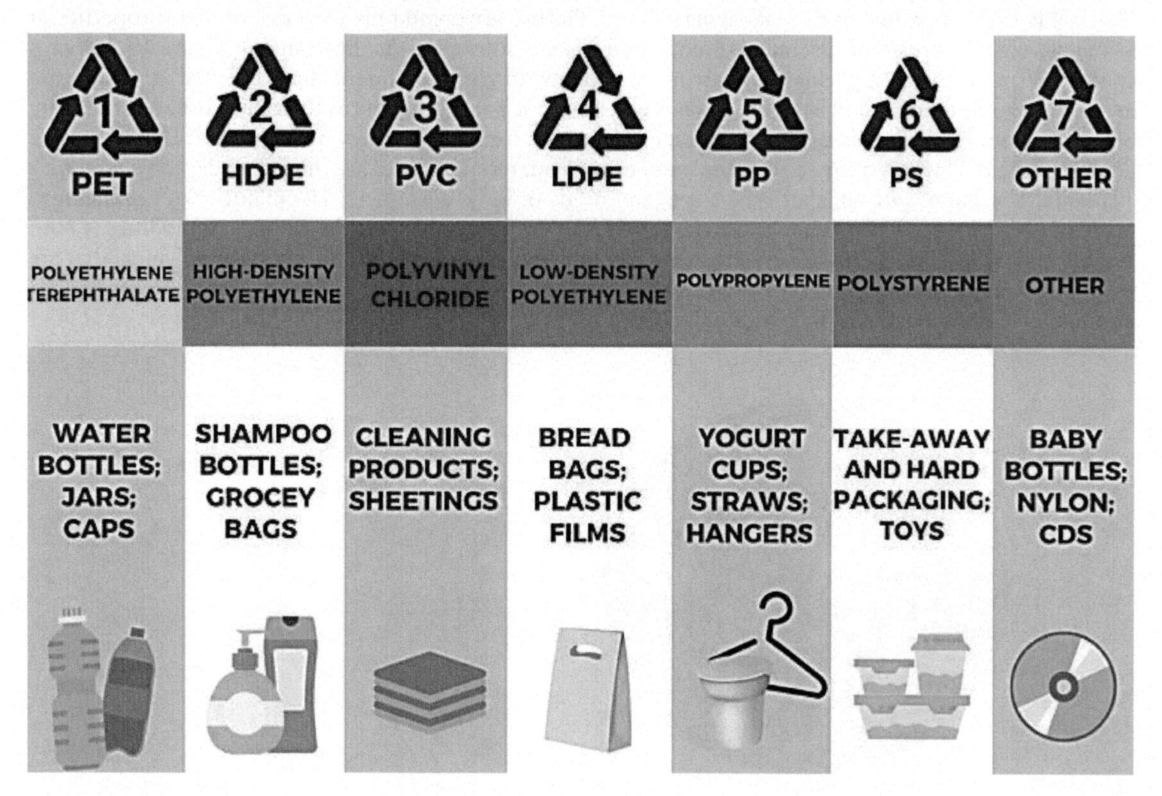

Figure 10.1 Plastics in our daily life

Table 10.1 Applications of plastics in plasticulture

Applications	Description
Linings for ponds and reservoirs	Plastic film is used to line canals, ponds, and reservoirs to keep out seepage. A lack of water for drinking and farming will not be a result of this.
Mulching with plastic	A variety of materials, such as plastic film, straw, dry leaves, hay, and stones, are used to cover the soil around a plant as part of the mulching process. A protective covering that keeps moisture in the soil and keeps it from escaping into the atmosphere
Plastics in green house construction	Glass or plastic film is used to protect a framed construction. It serves as a radiation filter that allows plants to be grown in a controlled environment only in certain wavelengths of light.
Tunnel made of plastic	It enhances the plant's ability to take in carbon dioxide, resulting in a higher yield as a result.

Table 10.2 Potential benefits of plasticulture

Plasticulture applications	Savings in water usage (%)	Efficiency in water usage (%)	Efficiency in fertiliser usage (%)
Sprinkler irrigation	30–50	35–60	30
Farm, pond lined with plastic film	100	40–60	Not available
Drip irrigation	40–70	30–70	30
Shade nets	30–40	30–50	Not available
Green house	60–85	20–25	36
Tunnel	40–50	20–30	Not available
Plastic mulching	40–60	15–20	26

fertiliser efficiency. Each application can save between 30% and 100% of the water consumed. It is highly advantageous to have agricultural ponds with zero-water-loss plastic film walls. Additionally, it may save money.

B. Food processing

Plastisol is frequently used in the packaging of food. Plastics are commonly used due of their properties. It is a lightweight, corrosion-resistant, and moisture-resistant substance. As illustrated in Figure 10.2 (Evans et al., 2020) new Indian guidelines that are more in line with global standards have resulted in an increase in packaging requirements. Plastics are a common choice for packaging due to their low cost and simplicity of sterilisation. Plastics make excellent packaging materials for a wide variety of businesses. They are either flexible or rigid. Plastic is a low-cost, useful, and durable material. Plastics are the most adaptable materials available. Their aim is to safeguard the assets and needs of large customers. The majority of packaging is made of plastics since it consumes low energy and easily transportable. Packaging for commodities is both flexible and rigid. Plastic films are used to create flexible packaging. PVC and PE are both polymers that are used in multilayered laminated sheets. Polyethylene or polypropylene accounts for around 62% of flexible packaging (Tajeddin and Arabkhedri, 2020).

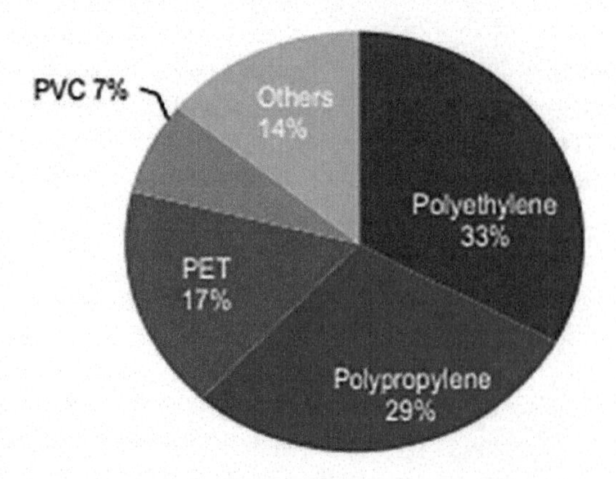

Figure 10.2 Plastics in flexible packaging

C. Automobile industries

It is used in instrument panels, wire wrapping, pipelines, and doors of automobiles. Acronyms ABS stands for Acetate Butadiene. ABS is made by combining styrene, acrylonitrile, and polybutadiene (acrylonitrile-butadiene-styrene). The presence of styrene renders the plastic impenetrable. Superior mechanical, chemical, and/or thermal qualities are provided by high performance polymers. Automobiles are increasingly made of high-performance plastics. These comprise more than 70% of all high performance polymers used in automobiles. Plastic sheet dividers are durable and lightweight, most often used in colours. Transparent, coloured, scratch-resistant, and combustible polymers are all possible with these polymers. Commercial and municipal bus companies are erecting sturdy plastic barrier screens to prevent the spread of COVID-19. Awe-inspiring design, available in virtually any colour Durability, It readily forms. Resistant to water and corrosion. acoustic insulating material is a more fuel-efficient design. It is available in lubrication and cleaner resistance varieties. Additionally, there are flammability classes for vehicles.

D. Electrical and electronic appliances

From simple cables to smartphones, modern electronics rely on a new class of polymers. Plastics are used by innovators due to their versatility. Plastics are used by designers on electrical and technological devices. To ensure the safety of electrical devices, we incorporate insulating materials into them. Thermoset polymers are used to manufacture switches, lamps, and electrical equipment handles. To prevent leaking and conduction, plastic is used to cover the majority of hair dryers, electric coffee makers, juicers, and razors (Muthamilselvan and Mondal, 2021). Plastics are poor conductors of heat. Polymers can be used to reduce the temperature in ovens, kettles, and frying equipment and to protect consumers from burns. Numerous electrical equipment incorporate plastics to lower their weight, allowing users to carry them more easily and decreasing power consumption. In comparison to metal, plastic is impervious to water, hygienic, and simple to clean and maintain. Plastique is an anti-grease material. Plastics are more affordable to manufacture than other materials. After use, the plastic components of a product can be recycled.

Medical industries

In medical applications, plastic fibres and resins such as PVC, PP, PE, PS and nylon are the most frequently used materials. To restore damaged arteries, a flexible plastic prosthesis might be employed. Hearing loss patients now have the option of receiving plastic implants in their ears to recover their hearing. Plastic medical equipment has made a significant contribution to the medical field. This equipment, on the other hand, is entirely composed of single-use plastics. General medical supplies include catheters, syringes, IV kits, and tubing. MRIs, ECG monitors, and surgical instruments are all examples of medical equipment. Orthopedics: Prostheses for the shoulder, elbow, hip, and knee are common. This category includes implants that deliver drugs to a specific location. Recent years have seen a rise in interest and demand for regulated medication administration into the body. Bacteria may wreak havoc on the teeth and gums. Dental implants may be beneficial in certain circumstances. Heart uses: Fat deposits and other factors contribute to the occlusion of valves and arteries. Replacement or unblocking of devices may resolve the issues.

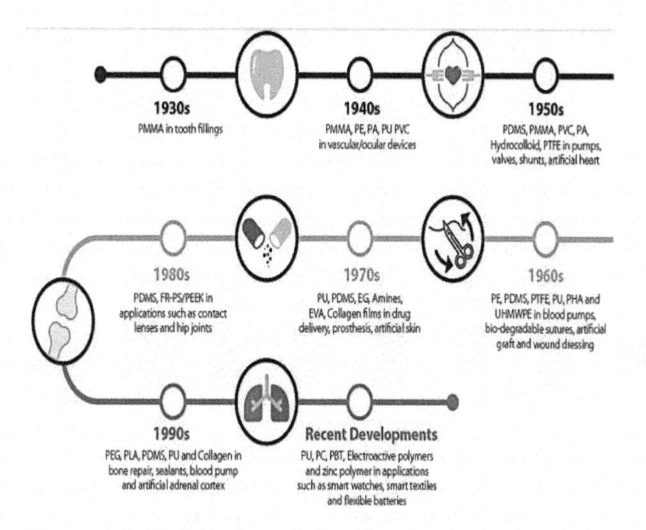

Figure 10.3 High-level trend of polymers in medical applications

Gloves, sterile syringes, sticky bandage strips, IV bags, IV tubing, and heart valves are all made of plastic. Plastic is an excellent material for medical packaging. It protects against pollution due to its exceptional barrier properties. The advancements in plastics enable unique procedures. A patient's requirements may be met by 3-D printing an artificial plastic heart or bacteria-resistant polymers. The progress of plastics in medicine is depicted in Figure 10.3 (Basmage and Hashmi, 2020).

Plastics Market Share

The global plastics industry is expected to grow at a compound annual rate of 3.4% from 2021 to 2028, reaching USD 579.7 billion in 2020, Szabo (2001). Demand is expected to be driven by the construction, automotive, and electronics industries. Automotive manufacturers have started using plastic components instead of metal ones like aluminium and steel in order to meet stricter fuel efficiency and carbon emission regulations. Plastic consumption has increased significantly in emerging nations such as Brazil and China. Laxer FDI regulations and increasing demand for public and industrial infrastructure all contribute to the market's expansion. Federal governments have increased building investment to meet growing infrastructure demands. Infrastructure and building spending will increase plastic consumption, particularly in Asia. Metallic materials have a greater specific gravity. They can save up to 80% of weight and 30% to 50% on individual components in the automotive and construction industries. Figure 10.4 (Czerniawski, 2007; Geyer et al., 2017) shows the plastics industry segmented by product.

A. Application insights

Polyethylene accounts for approximately 25% of all sales in 2020. It is a substance that is frequently found in containers, bottles, bags, films, and geomembranes. PE polymers are categorised according to their molecular weight as HDPE, LDPE, or LLDPE. Lubricants, waxes, and plastics all contain PE polymers. Consumption of bottled food and other liquid food items is predicted to increase as a result of the restraint effort to tackle COVID-19. ABS is an extremely lucrative market sector. ABS is gaining popularity in consumer goods and electronics due of its high stiffness, strength, and dimensional stability. Acidic meals and extreme heat have no effect on it. ABS is easily injection moulded and recycled due to its liquefaction. ABS's low melting point precludes its use in high-temperature situations. ABS is a plastic material that is used in LEGO kits and computer keyboards. ABS is a thermoplastic material that is used to make electrical enclosures, automobile trim components, and safety helmets.

By 2020, injection moulding will account for approximately 43% of total revenues. It is used to create customised plastic components. The procedure is paused while the plastic components are extracted from the moulds. We employ a plastic material and an injection moulding equipment. Inject and cool chilled plastic into the cavity of a mould. It is used in automobile components, containers, and medical devices. This is an application for a market. It is used to fabricate thermoplastic polymer films. It is used in the

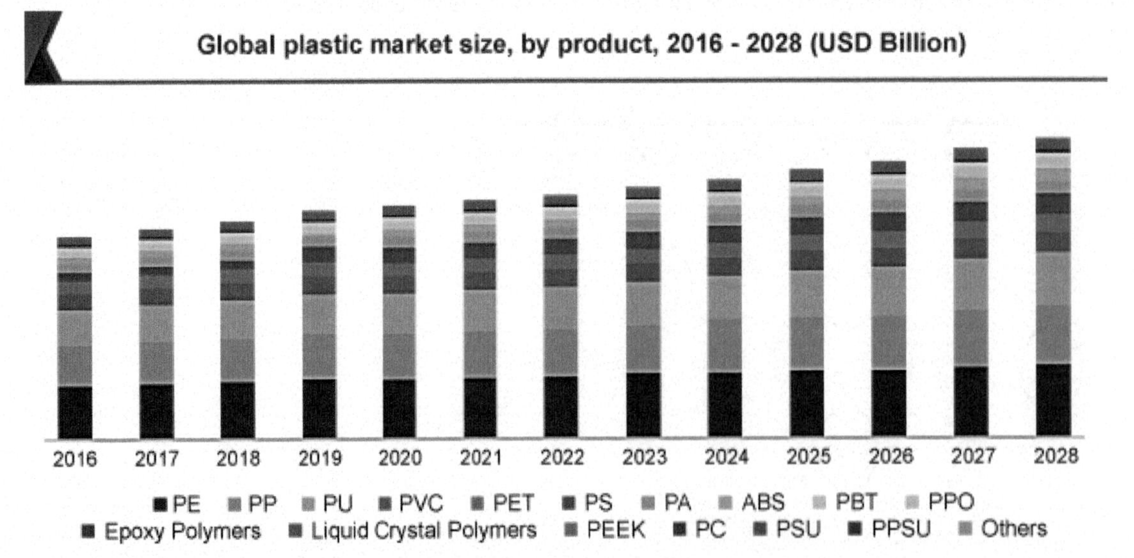

Figure 10.4 Plastic market in terms of products

manufacture of PVC and other thermoplastics. As a result, you can enhance or emboss the surface characteristics, as well as laminate in-line. As a result, the packaging industry continues to flourish.

By 2020, packaging will generate 36% of total revenue with a high-growth in end-use industrial products. Plastics have historically been used in packaging. Bioplastics have transformed the packaging of food, pharmaceuticals, and beverages. PET and PC are used in consumer, appliance, toy, and garment packaging. The appliance packaging market will experience significant growth. The COVID-19 pandemic is unlikely to have a significant effect on packaging. As industrial output has fallen, so has demand for luxury goods and some B2B transit packaging.

B. Regional insights

In 2020, Asia Pacific (including China) will account for more than 44% of global sales. Plastic consumption is expected to increase as the vehicle, building, packaging, and electrical and electronic sectors rise. Western technology transfer has boosted auto manufacture in India and China in recent years. The industry is likely to profit from Taiwan, China, and South Korea. The chemical industry in India is a significant manufacturer of plastic. Rapid urbanisation facilitates the expansion of infrastructure. The country's plastic demand is fuelled by the growing automotive and electronics industries, as well as a desire for lightweight components to increase vehicle efficiency and reduce electronic component weight.

C. Demand overview

Plastic raw materials are produced in a wide range of forms to meet the needs of a wide range of industries. The most often used polymers are commodity, engineering, and specialty plastics. Commodity plastics dominate the plastics business and, by extension, the plastics industry itself in terms of petrochemicals. Most polymers fall into one of these four categories: Polyethylene, Polypropylene, Polyvinyl Chloride, and Polystyrene. In terms of mechanical and thermal properties, engineering and specialty plastics are unique. Plastics like polyoxymethylene (POM), polymethyl methacrylate (PMMA), and polycarbonate (PC) are just a few examples. Polyethylene (PE) comes in three main forms: LDPE, HDPE, and LLDPE, all of which are currently available on the market. There are many basic chemicals utilised in the manufacturing of PVC, PS, ABS, and PC, such as bensene. One of the fastest developing industries in India is plastics. Figure 10.5 shows a rise from FY08's 6 MnTPA (million tonnes per year) to FY13's 8.5 MnTPA (million tonnes per year).

As a raw material, polyethylene (PE) is widely used in Indian industry. In FY13, there were 3.6 million TPA of demand, an annual growth rate of 8%. Between 2008 and 2013, Polypropylene (PP) showed a 2% annual rise in consumption. Between FY13 and FY14, annual PVC use increased by 10%, from 1.2 million tonnes to 2.1 million tonnes. Polystyrene (PS) rose from 94,000 TPA to 141,000 TPA to 250,000 TPA in FY13 at a rate of 3% per year. PE and PP are the two most commonly used polymers on Earth, accounting for 43% and 24% of total plastic use, respectively. Figure 10.5 shows that HDPE now makes up 20% of all PE consumption for the first time. LLDPE production is predicted to rise as a result of the material's increased usage.

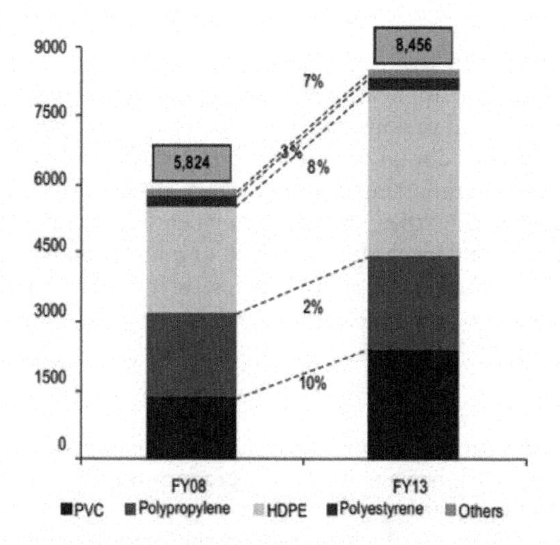

Figure 10.5 Demand growth of plastics

Polystyrene/ethylene propylene styrene is a synthetic polymer used in technical applications. Plastics are unrivalled when it comes to design flexibility. Plastics are extremely resource-efficient because of their inertness, corrosion resistance, bioinertness, high thermal/electrical insulation, and nontoxicity. "Compounds comprising carbon or hydrogen," according to PlasticsEurope's definition (2008). Renewable resources like sugar and corn can also be used in place of traditional fossil fuels. In terms of oil and gas use, plastics account for around 4% of global output, and their energy consumption is about equal to that amount (Russo et al., 2021). As a carbon sink, plastics can be reused and recycled multiple times.

Future of Plastics

A. PLA (Poly-lactic Acid) in Medical Applications

In medicine, PLA is the most often utilised biopolymer due to its biocompatibility and biodissolvability via ester backbone hydrolysis. A hydrolytic degradation of pharmaceuticals is the most typical approach. Zhang and colleagues synthesisednanodiamond-enhanced TECs (2011). To make the composites, PLA was dissolved in chloroform and ND-ODA was added. Both chloroform-dissolved PLA solution and ND-ODA dispersion solution had to be evaporated in order to accomplish this. Furthermore, it is safe to utilise on osteoblasts from mice. PLA-PEG and PLA-p-dioxanone-PEG are the two BMP carriers (BMPs). Components of biologics that promote bone growth. Bone graft replacements and biomaterials are utilised in conjunction with them to aid in bone repair. PLA-derived bone, on the other hand, was extremely sparse. Copolymers of low molecular weight were created. Recombinant BMP2 was delivered via a PLA scaffold (rhBMP2).

Polylactic acid (PLA) and its copolymers are used in surgeries. Li et al. investigated the treatment of ureteral damage with PLA ureteral stents (2011). They're an excellent choice because they're both biodegradable and reversible. As a result, PLA stents appear to be a treatment option for ureteral injury that needs to be explored further. A study by Qin et al. found that PLA has better mechanical qualities than ABS for decreasing postoperative adhesions, including higher tensile strength and Young's modulus. Using the PLA surgical bandage may help reduce wound failure after third molar extraction, according to Brekke.

In orthopaedics, biodegradable polymers are utilised to minimise the need for a second surgical treatment to remove unnecessary hardware.. PLA polymers are necessary for biodegradable suture anchors, screws, and pins. Absorbable screws and pins with high abrasion resistance are often used in therapeutic applications. When using PLA for bone fixation, it may be necessary to increase its impact tensile strength and fracture modulus. PLA copolymers have been touted for their biocompatibility. Pla/PGA copolymer pins cause aseptic cavities in roughly six out of every 120 patients.

B. PLGA (Poly lactic-co-glycolic acid) used in bone tissue engineering

The GF/PL scaffold was used by Kim et al. (2006) to build a polymeric/nano-HA composite. In GF/PL scaffolds, cell proliferation, alkaline phosphatase activity, and in vitro mineralisation are all increased. For his 2011 bionic scaffold, Ebrahimian-Hosseinabazi used TIPS and a nano-biphasic component (nBCP), both stiffened with powdered HA. nBCP composite scaffolds have a yield strength and Young's modulus of between 20 and 30% of their weight.

C. PLGA in dentistry

PLGA materials have proven their worth in a variety of dental applications. To treat periodontal disease, to make buccal mucosa or indirect pulp-capping treatments, and to construct bone fixation screws, these are employed (Paknejad et al., 2017; Pilipchuk et al., 2015). Periodontal disease can be effectively treated with PLGA implants, discs, and dental films, which have been demonstrated to increase antibiotic localisation while minimising systemic adverse effects. Bone regeneration textiles, such as highly degradable PLGA and SiO2 - CaO gel nonwoven fabrics, developed an apatite crystal layer after one week of interaction with simulated bodily fluid. When growth factors are included into the PLGA microparticles, direct pulp capping of mechanically exposed teeth utilising PLGA composites is also possible.

D. PCL Poly (ε-caprolactone) in drug-delivery systems

PCL is a promising choice for controlled drug administration due to its high drug permeability, biocompatibility, and ability to totally excrete after biosorption. PCL has a slower biodegradation rate than other polymers, making it appropriate for long-term drug delivery systems. The breakdown kinetics of PCL can also be altered by combining it with various polymers. Along with PCL content, the percentage of drug in microcapsules and the microcapsules' size determine the release rate of PCL drug. In order to improve stress resistance, crack resistance, and the pace of medicine delivery, PCL has been combined with other polymers. Peptide and protein delivery systems that can be controlled by PCL have been the focus of recent study (Jain et al., 2013; Shantha Kumar et al., 2006). Poly caprolactone–poly L-lactic acid (PCL–PLLA), poly caprolactone–poly L-lactic acid (PCL–DLLA), and poly caprolactone–TMC rods were examined in vitro and in vivo by Lemmouchi and colleagues for the release of isometamidium chloride and ethidium bromide.

E. PCL applied in tissue engineering

When it comes to tissue replacement, preservation, or improvement, tissue engineering is a collaborative endeavour between biologists and engineers (including bone, cartilage, and blood vessels). Tissue regeneration necessitates certain structural and mechanical properties. Tissue engineering is also a term that refers to the use of artificially generated cells in a support system to perform biological activities (including an artificial liver, or pancreas). Since its inception, tissue engineering has evolved into its own set of application tactics and tissue replacement methods. To make tissues in the lab using a combination of extracellular matrixes, physiologically active chemicals, and cells (commonly referred to as 'scaffolds'), scientists can now combine these elements. Because of its low melting point and exceptional rheological and mechanical qualities, PCL is a popular biomaterial in cardiovascular and bone tissue engineering. Scaffolds made of PCL are extremely strong and lightweight. It is possible to fabricate a wide range of scaffolds from PCL, a bioresorbable polymer that may be treated in virtually any way. As far as implantable polymeric scaffolds go, the most coveted attribute is the capacity to replace deteriorated polymer with genuine tissue created by cells (Abruzzo et al., 2014; Shoichet, 2010).

F. Petrochemical industries

Chemicals play a significant role in the economic well-being of all nations. Thirteen percent of India's industrial value added is derived from chemicals. Due to Asia's growing contribution to the chemical industry, India is increasingly becoming an important commercial location worldwide. All industries depend on chemicals, hence they are essential to a country's economic growth. During the next five years, it is predicted that the chemical industry in India will increase at an annual pace of 8%. Polymers, synthetic fibres, and surfactants are all manufactured in India, which accounts for 20% of the global market Banerjee et al. (2014). Everything from textiles to construction materials to furniture to autos to packaging to medical devices, electronics, and electrical goods all rely on petroleum-based products. As a result, we use more petroleum. India's long-term potential is demonstrated by the country's low per capita plastic item usage when compared to other affluent countries.

India's packaging industry has seen a dramatic increase in plastic usage with a per capita plastic products consumption of 9.7 kg/person, shown in Figure 10.6. However, agriculture has not fully realised the promise of plastics. Plastics are utilised in agriculture at an 8% global rate, but at a 2% rate in India (Jalil et al., 2013). India's target of boosting manufacturing's share of GDP from 16% to 25% by 2022 bodes well for the expansion of the petrochemical industry. Urbanisation and rising income levels all contribute to increased India's petrochemical supply and demand. Plastics have a significant role in India's petrochemical industry. According to futurist Hammond (2007), the rate of technological progress is accelerating, to the point where the first three decades of this century will appear to be a century by 2030. It is being utilised to boost the insulation of homes, as well as reusable electronic graphic medium for books and periodicals, as well as packaging that constantly monitors for spoiling.

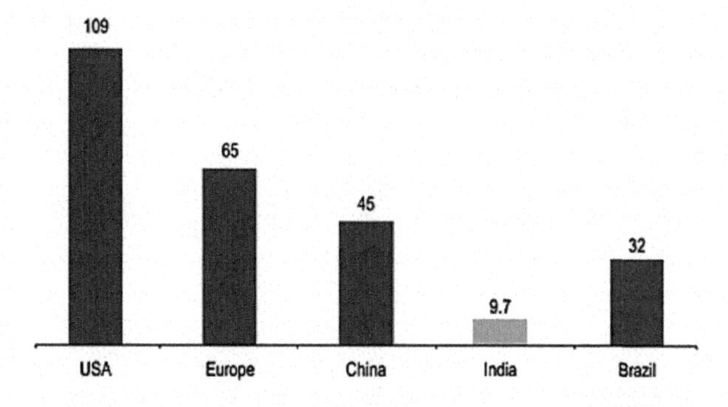

Figure 10.6 Per capita plastic products consumption(kg/person)

Conclusion

Research on plastics' characteristics and functions is conducted. Due to their greater biocompatibility and bioresorbability biodegradable polymers such as PCL, PLA, and PLGA can be used in medical applications such as tissue engineering, medicine delivery, and biomedical devices. Despite their diverse biological applications, biopolymers require minor chemical and physical changes to improve mechanical properties and allow for complete implant site absorption. It is less expensive to develop biocompatible, less crystalline modified, and mixed biomaterials. It will require time and study to create a method for generating biocompatible bioplastics that can be used for next-generation implantation in a variety of biomedical sectors. The end result is an incomprehensible plastic. Due to environmental concerns, plastic usage has grown perplexing. It will be upgraded when technology advances more rapidly. Without a doubt, plastics will evolve in all possible ways in future.

References

Abruzzo, A., Fiorica, C., Palumbo, V. D., Altomare, R., Damiano, G., Gioviale, M. C., ... , and Lo Monte, A. I. (2014). Using polymeric scaffolds for vascular tissue engineering. Int. J. Polym Sci. 1:1–9.

Banerjee, T., Srivastava, R. K., and Hung, Y. T. (2014). Chapter 17: Plastics waste management in India: an integrated solid waste management approach. In Yung-Tse Hung, Lawrence K Wang and Nazih K Shammas, eds. Handbook of environment and waste management: Land and groundwater pollution control, World Scientific. 1029–1060.

Basmage, O. M. and Hashmi, M. S. J. (2020). Plastic products in hospitals and healthcare systems. Encyclopedia of renewable and sustainable materials. 1:648–657.

Czerniawski, B. (2007). Analysis of plastics packaging domestic market. Polimery 52(11/12):811–819.

Evans, D. M., Parsons, R., Jackson, P., Greenwood, S., and Ryan, A. (2020). Understanding plastic packaging: The co-evolution of materials and society. Glob. Environ. Change 65:102166.

Geyer, R., Jambeck, J. R. and Law, K. L. (2017). Production, use, and fate of all plastics ever made. Sci. Adv. 3(7):e1700782.

Jain, A., Jain, A., Gulbake, A., Shilpi, S., Hurkat, P. and Jain, S. K. (2013). Peptide and protein delivery using new drug delivery systems. Crit. Rev.™ Ther Drug Carrier Syst. 30(4):293–329.

Jalil, M. A., Mian, M. N. and Rahman, M. K. (2013). Using plastic bags and its damaging impact on environment and agriculture: An alternative proposal. Int. J. Learn. Dev. 3(4):1–14.

Jia, P., Zhang, M., Hu, L., Wang, R., Sun, C. and Zhou, Y. (2017). Cardanol groups grafted on poly(vinyl chloride)-synthesis, performance and plasticization mechanism. Polymers (Basel) 9(11):621.

Lakshmana, C. M. (2013). Population, development, and environment in India. Chinese J. Popul. Resour. Environ. 11(4):367–374.

Markarian, J. (2005). Plasticulture comes of age. Plast. Addit. Compd. 7(1):16–19.

Muniyasamy, S. and John, M. J. (2017). Biodegradability of biobased polymeric materials in natural environments. In Handbook of composites from renewable materials. eds. V. K. Thakur, M. K. Thakur, M. R. Kessler, United States: Scrivener Publishing LLC. 625–654.

Muthamilselvan, T. and Mondal, T. (2021). Thermally conductive plastics for electronic applications. In Sereni, J. G. R., ed. Reference Module in Materials Science and Materials Engineering. Netherland: Elsevier.

Paknejad, Z., Jafari, M., Nazeman, P., Rad, M. R., and Khojasteh, A. (2017). Periodontal and peri-implant hard tissue regeneration. In Biomaterials for oral and dental tissue engineering, 405–428. Woodhead Publishing.

Pilipchuk, S. P., Plonka, A. B., Monje, A., Taut, A. D., Lanis, A., Kang, B., and Giannobile, W. V. (2015). Tissue engineering for bone regeneration and osseointegration in the oral cavity. Dent. Mater. 31(4):317–338.

Russo, S., Valero, A., Valero, A., and Iglesias-Émbil, M. (2021). Exergy-based assessment of polymers production and recycling: An application to the automotive sector. Energies 14(2):363.

Shantha Kumar, T. R., Soppimath, K., and Nachaegari, S. K. (2006). Novel delivery technologies for protein and peptide therapeutics. Curr. Pharm. Biotechnol. 7(4):261–276.

Shoichet, M. S. (2010). Polymer scaffolds for biomaterials applications. Macromolecules 43(2):581–591.

Szabo, F. (2001). The world's plastic industry. Int. Polym. Sci. Technol. 28(11):1–9.

Szlachetka, O., Witkowska-Dobrev, J., Baryła, A. and Dohojda, M. (2021). Low-density polyethylene (LDPE) building films–Tensile properties and surface morphology. J. Build. Eng. 44:103386.

Tajeddin, B. and Arabkhedri, M. (2020). Polymers and food packaging. In Tajeddin, B. and Arabkhedri, M, eds. Polymer science and innovative applications, 525–543. London: Elsevier.

11 Effect of fine recycled aggregates on strength and durability properties of concrete – A review

Mohd Imran[a,*], *Mandeep Kaur*[b], *and Humaib Nasir*[c]

School of civil engineering, (Research Scholar & Assistant Professor), Lovely Professional University, Jalandhar, India

Abstract

This research discusses the current literature on the fine aggregates derived from building and demolition waste as a partial or total alternative for fine natural aggregates in the manufacture of concrete. The analysis involves an overview of current regulations as well as preliminary research on the subject. As we are taking the steps into the future, we are wearing out the natural resources that will once day perish completely if we do not take some necessary actions. This lack of resources is only because of the construction industry, and it is also responsible for the environmental imbalance. To address this issue and to overcome this concern, we will be using the construction wastes and turning the wastes into Fine recycled Aggregate manufactured from old concrete. Fines are a natural byproduct of the milling process. and many other small elements which are too small for identification. Taking the strength parameter for the Fine recycled concrete aggregate from the brick aggregate concrete, it was found to be more significant that stone aggregate concrete. One of the Interesting observations that was highlighted is that the rehydrated strength was more if Portland cement and fly ash was mixed to already preheated Fine recycled aggregate concrete. So, the proper treatment of the construction wastes (i.e., Fines recycled concrete aggregate) can result into many uses which will cover the environmental concerns and construction industry will also be benefitted since many of the good results in compression, tension and torsion were obtained and are very good for practical uses.

Keywords: Compressive strength, durability, fine recycled aggregates, recycling, sustainability.

Introduction

Fine aggregates are basically the sand particles when sieved through the 3/8-inch sieve passes completely. These are generally the crushed stones or natural sand. And the fine recycled concrete are the small particles or crushed stones which are not produced naturally but they are produced through the multiple crushing by jaw crushers and many other machines and instruments and finally we get the desired product as Fine recycled aggregates. As solid innovation develops and climate disintegrates, it is presently certain that the limitless utilisation of development materials, with beginning expense being the pervasive determination model is a relic of the old times.

Ecological concerns indicated by new legislation and business patterns have driven the solid company to minimise its natural impact, primarily by reducing CO2 discharges and common asset consumption. Hu et al. (2013). There are numerous sources of recycled fine aggregates, including construction and demolition waste, as well as concrete waste from natural disasters, concrete waste from burning of buildings and so many others as well Kirthika et al. (2020). As humanity is taking its steps into the future by making many beautiful structures and buildings but at the same time it is alarming to know that how the nature is being affected by all the construction that is going around. By all the construction going around, the need of raw materials as aggregates (course or fine aggregates) is also increasing at a very high pace. That's why the depletion of these resources has taken it pace also at an alarming rate because it's a natural resource which is very crucial for the construction purpose and many of the properties of concrete are dependent on it. So, in order to overcome this depletion, the earth of its resources completely, Researchers came up with an idea of making the fine recycled aggregates out of concrete wastes from numerous sources.

[a]emibhatt@gmail.com; [b]Mandeep.kaur@lpu.co.in; [c]Humaib.18648@lpu.co.in

Various examinations on the assessment of reused total properties have been done from alternate points of view, identified with the quality and the wellspring of the waste cement. Nonetheless, the vast majority of them assess the coarse portion, and the outcomes acquired regard to the qualities and properties of these totals are fundamentally the same as one another. By doing so, the Earth is also recovering. But now it was still a mystery for researchers that if the idea will work the way they have thought it would. So, much of the tests and analysis on these recycled fine aggregates were still be done Velay-Lizancos et al. (2018), Kou and Poon (2009a). Upon putting the recycled fine aggregate through a series of tests and analysis it was found out that the concrete mix prepared from recycled fine aggregates is behaving almost the same ways as the conventional concrete does. All the tests and analysis are necessary because many of the countries are still unknown to the idea and some countries even have completely prohibited the use of fine recycled aggregates.

For the sustainable development of our future, it is very important that we take care that the earth remains green forever Velay-Lizancos et al. (2018). And the arising problem like the earth is lacking its resources and its environment is changing should be solved and overcome any obstacle coming in the way of doing that. By taking the recycled fine aggregates in our construction works many of the properties of Recycled aggregate need to be analysed and tested in order to actually be sure that it is really beneficial to use the recycled fine aggregates into the construction works. Many of the properties that definitely changes are its Physical properties, Fresh Properties and Hardened Properties Vinay Kumar et al. (2018).

The strength as alone will be tested many times and with different procedures and methods so that we will finally be sure that the recycled fine aggregates that we are using in the mix will live to our expectations. Same goes for the case of durability as well. Generally, strength is tested by taking one as the reference mix and other mixes will be having some percentage of recycled fine aggregate in them. If the water cement ratio is also managed in the same way and strength on different days such as 7 days 14 days and 28 days can be tested out Lin et al. (2004). The density and water absorption alone has a huge impact on the concrete mix. It can be said that this impact is around 70–80% in the volume terms. The studies that are done earlier has revealed that the decrease in the strength is in the range 1530% for recycled concrete aggregate if the replacement is done with natural aggregates. And recently one of the findings were saying that the compressive strength was reduced by about 10% as compared to that of a conventional concrete. And for the compressive strength of about 55 MPa Wang et al. (2019b), it is mandatory that the 100% replacement of recycled aggregate with natural aggregates should be done for the required compressive strength.

But however, apart from the studies made on Recycled concrete aggregate, the studies made on the fine recycled concrete for the strength of the mix, it was maximum reduction of mechanical strength was about 5% for the replacement of 30% of natural aggregate with recycled fine aggregate and about 10% of reduction in compressive strength as compared to that of conventional concrete for the replacement of 100% with natural recycled fine aggregate Wang et al. (2019a). Unlike many characteristics, Carbonation and chloride penetration plays an important role in the long run of durability. There are other characteristics also which contribute to the durability such as water absorption and capillarity and others Basheer et al. (2001). In most of the cases it is seen that water. A chemical-based phenomenon in which the alkalinity of the concrete is reduced very drastically when the carbon dioxide sinks into the concrete. By doing so, carbon dioxide reacts with calcium hydroxide and the reduction happens Pan et al. (2017). The composition of the cement type aggregate and the porosity of cement are the factors on which carbonation is dependent. Durability is reduced because of these two factors which are Carbonation and the chloride penetration. By performing various tests and analysis for the durability property. It can be said that in most of the cases an increase in the water absorption is observed. This increase is observed because of the porous nature of the fine recycled aggregates. As compared to the conventional concrete only a little increase is observed so it does not bother much to construction. Since very little change are observed in strength and durability of the mixes prepared from the fine recycled aggregates. Simultaneously, development industry continually seeing worth added source for reused totals, including utilisation of RCA as immediate substitution to essential totals got from characteristic assets, where, in fact and monetarily conceivable. It is a high time to change the tradition of construction which are not sustainable. If we take up the use of fine recycled aggregates, it will automatically heal Earth and the atmosphere of the earth. And the sustainable construction will be adopted Evangelista and De Brito (2007a).

Methodology and Extraction of Fine Recycled Aggregate

The methodology followed in this review study is that already published literature about recycled concrete aggregate and about recycled fine aggregates is being studied individually and extensively. So that each aspect of mechanical and durability properties of concrete made with Recycled concrete aggregates is evaluated properly and promising results and conclusion can be made about the use of fine recycled concrete aggregates.

Normally, FRA is an undesired consequence of CDW crushing, but with high pollutant content determined that FRA might be used provided that these pollutants are screened throughout the CDW production process. It is also thought that the CDW's origins and interactions with the FRA arrangement have a substantial impact on the FRA arrangement. The expectation that the comminution cycle of FRA will be comparable to that of crushed FNA is a fundamental condition for its advancement (Baldusco et al., 2019); Rodrigues et al., 2013); Evangelista and De Brito, 2014). A novel approach at the FRCA used a standard jaw-smasher with those given utilizing a model Smart Crusher and acquired molecule size conveyances with the last technique substantially the same as those of FNA., showing that the creators figured out how to isolate the FNA from the concrete glue, particularly for the totals above 0.5 mm (Florea et al., 2012). The model smasher produced several times as many fines' totals under 0.5 mm, the majority of which consisted of concrete adhesive, which the developers eventually used to substitute concrete. The demonstration that the FRA derived from the model was free of concrete glue and was conducted efficiently using differential filtering calorimetry, which demonstrated that many -SiO2 precious stones turned to -SiO2, in comparison to quartz found in common sand. A survey of the handling flowcharts of the various Portuguese reuse plants was done, and it was discovered that in the majority of them the RA created are exposed distinctly to essential squashing, with not many utilising optional pulverising.

FRA improved their features by using mineral comminution and partition methods, bringing them closer to those of the FNA. A three-stage crushing method was employed for this, beginning with crushing at the reuse plant, then pounding with a jaw-smasher, and finally pounding with a vertical pivot impactor. According to the data, this method eliminates mortar from parts larger than 0.15 mm and is more productive at greater rates. The FRA was subjected to partition efficiency analyses that took into account both thickness and attractiveness. The findings of the compound investigations, scanning electron microscopy (SEM), XRD, and FTIR demonstrated that the FRA produced was rounder and had less following mortar content.

A. Research objectives

With the trend of making concrete mixes with recycled concrete aggregates, this review is based on:

- Examine the published literature about the use of fine recycled aggregates in reinforced applications, as well as the maximum optimum replacement of fine recycled aggregate (FRA) with fine natural aggregate (FNA).
- To investigate the effect of replacing FNA with FRA on the mechanical and durability qualities of concrete.

Strength Parameters of Fine Recycled Concrete Aggregate

A. Compressive strength

Kou and Poon (2009a) prepared 13 mixes with two of them being as reference mixes and other 11 mixes with different replacement ratios and water to binder ratios and these 13 mixes were under three series namely series 1, series 2 and series 3. In the mixes of replacement ratio of 25% of series 1, when w/b ratio of 0.53 was introduced, a decrease in the compressive strength was observed.

Khatib (2005) prepared 9 mixes with fine recycled aggregates ratio ranging from 0 to 100%. The fine recycled aggregates were obtained from crushed concrete (CC) and crushed bricks (CB) and in both the RFA form CC and CB compressive strength was observed to decrease after 28 days. Behera et el. (2019) in order to study the strength with age development, the compressive strength of SCC mixes was evaluated at various ages. Comparing the strength of all the mixes, the compressive strength increased by age at a higher rate than at an early stage. A mixture showed a slight reduction in strength of up to 8% by 50% RRA and ternary FA and SF combinations. 50% RFA substitution can be used to substitute NFA for SCC production, with low compressive strength compensation. At 100% replacement for RFA, even in latter ages, the Compressive strength fell significantly regardless of binding blends. Carro-López et al. (2015) investigated the effect of time on the rheology of auto compacting concrete of incorporation of fine recycled aggregates (at 15, 45 and 90 min). At 0 per cent, 20 per cent, 50% and 100 per cent of natural aggregates were replaced by recycled sand. The 20% suitable alternative mix maintained the required performance and filling capacity. This trend was induced by an increase in yield stress and an increase in plastic viscosity. The compressive strength of the 50 and 100% replacement mixes dropped significantly, while the compressive strength of the 20% replacement mix dropped below 10%.

Evangelista and de Brito (2007b) studied the use of recycled fine concrete aggregates in the production of structural concrete to replace, natural fine aggregates, either partially or completely (sand). The findings revealed that using fine recycled concrete aggregates has no effect on the mechanical qualities of concrete and can increase replacement ratios by up to 30%. The compressive strength was assessed in two stages and in the first stage, the compressive strength increases but there is a loss in the second stage and the loss of strength was justified by the higher percentages of FRA used in the mixes. Fan et al. (2016) analysed two types of samples where production of fine recycled aggregates using crushed concrete waste has been subject to two methods: first, the production of coarse and fine aggregate (R1), and secondly, the production of only fine aggregates (R2).

Furthermore, the compressive strength demonstrated that, under the identical conditions of replacement ratio, water cement ratio, and curing period, the compressive strength of the concrete specimen including R2 is greater than the compressive strength of the concrete specimen containing R1. Shui et al. (2008) intended to use FRCA as the primary component of the construction mortar and used thermal treatment Techniques for the dehydrated and processed phases of pre-heated ferric acid (FRCA) were used in thermo-gravimetrical differential scanning calorimetry) and X-ray diffracting. Results showed the rehydration reactivity of the preheated FRCA and fly ash and Portland cement can be added to FRCA pre-heated to improve its rehydration strength significantly.

Kou and Poon (2009b) Comparison with river sand, crushed fine petroleum (CFS) and fine recycled aggregate (FFRA) of the properties of concretes prepared with use of fine aggregates. Results showed that the compressive strength and the drying shrinkage decreased at fixed water–cement ratios as the furnace bottom ash (FBA) content increased. The compression strength was lowered by the FRA and the drying of the concrete was increased. Results also showed that the fine aggregates for concrete production are suitable for both FBA and FRA in Figure 11.1. Zega and Di Maio (2011) presented the durable behavior of different RFA (0%, 20% and 30%) structural concrete is investigated. The compressive strength of recycled fine aggregate at 20 and 30% concretes was like that of concrete made of 100% natural fine aggregate. RC concrete has a lower effective water/cement ratio than CC concrete.

Pereira et al. (2012a) analysed the two types of ultrasonic superplasticisers of latest generation were used. They differed in terms of water reduction capacity and robustness, and they examined each composition's workability, density, and compressive strength. And the results revealed that there were compressive force gains in the FRC, and the more water there was, the less super-plasticiser there was, and these strength improvements are inextricably linked to a decrease in the water/cement ratio. Gonçalves et al. (2020) produced two types of commercially available reactive MgO_s (10%, 15 and 20% by weight) and finely recycled concrete were used as a partial cement substitute for silicious sand specimens (50 and 100%, by volume). The results indicated an overall decline in performance related to mechanical using fine recycled concrete aggregates, but increased retraction behavior in all MgO containing specimens was observed. Fonseca et al. (2011) experimented to assess the mechanical performance of concrete made of gross recycling aggregates from crushed concrete by differing curing conglomerates. The main result can be stated that regarding mechanical performance is that recycled aggregate concrete (RAC) is affected roughly in the same way as conventional concrete by curing conditions. Koulouris et al. (2004) produced concrete was made from a natural gravel mixture and a recycled aggregate content of up to 100%. Fresh property results showed no effect due to the coarse RCA inclusion with ±20 mm resistant tolerance results. The engineering properties showed similar performances between two types of concrete (strength development, drying shrinkage, static elasticity modulus, flexural strength and swelling). The compressive strength had no effect in the two series of mixes that he prepared, and the average value was about 36 MPa and shown in Figure 11.2.

Anastasiou et al. (2014) experimented and implemented a program. To study the feasibility of making concrete including a significant volume of industrial by-products and secondary components, the strength development method of construction and demolition waste (CDW) aggregates appears to have dropped considerably, but in comparison to CDW limestone combinations, the CDW in combination with steel slag appears to be superior. Katz (2003) tested concrete with a compressive strength of 28 MPa over 28 days was crushed for new concretes at 1, 3 and 28 days and simulated the precast concrete. And resulted in recycled OPC cement made from aggregates crushed at age 1 days a slightly higher compressive strength was observed and there was no significant difference between the ages of crushed 3 and 28 days. Khoshkhenari et al. (2014) studied the influence of fine 0–2 mm aggregate in the compressive and partitioning tensile strengths of recycled concrete aggregate at normal and high strengths (RCA). The same gradation of RCAs was used to replace two regular and high strength concrete compositions. The compressive and dividing tensile strengths were measured after 3, 7, and 28 days. The results of the tests suggest that ground and fine RCAs produced by a parent concrete with a compression strength of 30 MPa have an absorption propensity that is approximately 11.5 and 3.5 times more than regular ground and fine additives. Table 11.1 compiles all of the compressive strength data, which is displayed in Figure 11.1.

B. Split-Tensile Strength

SC Kou and Poon (2009a) prepared 13 mixes with two of them being as reference mixes and other 11 mixes with different replacement ratios and water to binder ratios and these 13 mixes were under three series namely series 1, series 2 and series 3. And an increase in the mixes of series 2 and series 3 mixes was observed after 28 days.

Behera et al. (2019) also measured tensile strength at 28 days and 56 days and the modulus of rupture was measured at age 28 days and was seen that no effect on tensile strength property was investigated at 50% replacement of RFA. Pereira et al. (2012b) analysed the effect of superplasticisers on the mechanical properties, there was a decrease in the relative performance of concrete made of recycled aggregates.

The identical concrete with admixtures outperformed reference mixes without or with a less active superplasticiser in terms of mechanical performance. Former concrete is treated with superplasticisers to lower the water-cement ratio.

The strength gain in the splitting tensile strength was observed by the researcher and the reason was incorporation of Super-plasticisers. Evangelista and de Brito (2007b) investigated the use of recycled fine concrete aggregates in the construction of structural concrete to partially or completely replace natural fine aggregates (sand). The findings revealed that using fine recycled concrete aggregates has no effect on the mechanical qualities of concrete and can increase replacement ratios by up to 30%. The replacement ratio increases as the recycled aggregates have a more pore structure and therefore the splitting tensile strength decreases. Split-tensile strength values are given in Table 11.2 and illustrated in Figure 11.2.

C. Flexural strength

Evangelista and de Brito (2017) tested flexural behavior of the reinforced concrete beams made by replacing natural by recycled fine-aggregates and the results obtained indicated that the use of FRA in structural parts did not significantly affect their flexural performance even if their bearing capacity is somewhat less than conventional concretes. Koulouris et al. (2004) produced concrete was made from a natural gravel mixture and a recycled aggregate content of up to 100%. Fresh property results showed no effect due to

Table 11.1 Compressive strength values

Reference	*Compressive strength (MPa)*			
	7 days	*28 days*	*56 days*	*90 days*
Kou and Poon (2009a)	34	44.5	-	56.5
Khatib (2005)	30.4	39.2	-	50.9
Behera et al. (2019)	-	38.4	-	60.72
		44.85		70.82
Evangelista and de Brito (2007b)	-	61.3	-	-
Fan et al. (2016)	47.53	53.27	-	-
Shui et al. (2008)	8.2	8.3	-	-
Kou and Poon (2009b)	40	53	60	
Zega and Di Maio (2011)	-	42.7	-	-
Pereira et al. (2012a)	54.5	65.4	67.2	-
Carro-López et al. (2015)	-	54	-	-
Fonseca et al. (2011)	42.8	49.8	61.5	-
Koulouris et al. (2004)	-	37	-	-
Anastasiou et al. (2014)	-	48.2	-	-
Katz (2003)	23.4	30.5	-	38.7
Khoshkenari et al. (2014)	63.9	81.3	-	-

Table 11.2 Split tensile strength values

Reference	*Split-tensile strength (MPa) at 28 Days*
Kou and Poon (2009a)	3.5
Behera et al. (2019)	4.99
Pereira et al. (2012b)	4.5
Evangelista and de Brito (2007b)	3.65

Table 11.3 Flexural strength values

Reference	*Flexural Strength (KN) at 28 Days*
Evangelista and de Brito (2017)	73.6
Koulouris et al. (2004)	57

the coarse RCA inclusion with ±20 mm resistant tolerance results. The engineering properties showed similar performances between two types of concrete (strength development, drying shrinkage, static elasticity modulus, flexural strength and swelling). The flexural strength almost had no effect in the two series of mixes that were prepared and the average value 5.41 MPa. Flexural strength is tabulated in Table 11.3 and illustrated in Figure 11.3.

Durability Parameters of Fine Recycled Concrete Aggregates

Carriço et al. (2020) analysed the fresh and hardened properties of concrete through thermogravimetry and found that there was a weak exothermic peak associated to a C2S allotropic transformation and the mechanical strength was low to a 20% thermo activated recycled cement (TRC) substitution and only decreased by 12%. Velay-Lizancos et al. (2018) studied concretes with different replacement ratios of natural aggregates by the recycled aggregates which were 8 per cent, 20 per cent and 31 per cent and found that negative effect on the development of the E-modulus of the first 12 hours on the recycled aggregate compared to the reference mixture.

Kalinowska-Wichrowska et al. (2020) made the research on the basis of two variables that determine the recycling process. The test results showed that the correct thermal treatment of concrete scrap allows a fine fraction of a high quality. This fine fraction can be used in mortars and concrete as a partial substitute for cement.

Güneyisi et al. (2016) investigated the rheological and fresh properties of self-compacting concrete (SCC) made with recycled concrete aggregate (RCA) as coarse and fine aggregates. Self-compatibility characteristics of the concretes are remarkably improved by the replacement levels of CRCA and FRCA. The results show that Herschel-Buckley and modified Bingham models provide well defined rheological representations in Figure 11.3 for SCC with RCA. Tiwari et al. (2016) reviewed the feasibility of a wide variety of industrial by-products such as bottom ash, waste foundry sand, copper slag, plastic waste, recycled rubber waste, crushed glass aggregate. Šefflová and Pavlů (2017) analysed the durability and FRA came from crushed structures of old concrete. A total of four concrete mixes were prepared. The first blend was a natural sand reference. Natural sand was replaced with FRA in different substitution rates in other concrete mixtures, specifically 10%, 20% and 30%. Results stated that FRA concrete can be used in the same applications as conventional concrete, depending on the durability.

A. Water Absorption of FRA

Carro-López et al. (2015) investigated the effect of fine recycled aggregates on the rheology of auto compacting concrete over time (at 15, 45 and 90 min). At 0%, 20%, 50% and 100% of natural aggregates were replaced by recycled sand. The 20% replacement mix maintained the appropriate performance and filling capacity and an increase in the water absorption was seen due to the suction capillarity of the pores. Pan et al. (2017) attempted to improve the carbonation effectiveness by raising the amount of carbonate compounds in the RFA demolition using Calcium Hydroxide (CH). Before carbonation, the RFAs were pre-soaked with CH, and their properties were superior to those of carbonation alone, including a lower crush value, water absorption, and powder content. The results of the experiments revealed that by treating RFAs under appropriate conditions, the crush value, water absorption, and powder content were greatly reduced. Fan et al. (2016) also analysed water absorption in the two samples first; coarse and fine aggregate (R1), and second, only fine aggregates (R2). R1 has a higher water absorption rate than R2, implying that R1 should be used in concrete, in the case of R2 both specimens have the same substitute ratio, results in greater water absorption than the one obtained. Zega and Di Maio (2011) presented the durable behavior of different RFA (0%, 20% and 30%) structural concrete. And the results of recycled concrete's durable behaviour were as good as conventional and a very little increase in water absorption was seen in his study. Gonçalves et al. (2020) produced two types of commercially available reactive MgO_s (10%, 15 and 20% by weight) and finely recycled concrete were used as a partial cement substitute for silicious sand specimens (50 and 100%, by volume). The results indicated an overall decline in performance related to both

Table 11.4 Water absorption at 24 hours

Reference	*Water Absorption in 24 h (%)*
Carro-López et al. (2015)	9.3
Pan et al. (2017)	4.35
Fan et al. (2016)	8.9
Zega and Di Maio (2011)	8.5
Gonçalves et al. (2020)	7.2
Anastasiou et al. (2014)	8
Evangelista and de Brito (2010)	13.1

durability using fine recycled concrete aggregates, but increased retraction behaviour in all MgO containing specimens was observed.

Anastasiou et al. (2014) experimented and implemented a program. The use of fine building and demolition waste aggregates increases the pore content of concrete and reduces strength and durability, while its combination with steel slag partially recovers resistance and the loss of durability, according to the results of an investigation into the possibility of producing concrete incorporating a large volume of industrial by-products and secondary materials. This poor performance could be attributed to an increase in water absorption as a result of the increased amount of FRA employed in the mixes.

Evangelista and de Brito (2010) also analysed the water absorption by two methods (by Immersion and through capillarity) and a significant rise in the water absorption was seen in the both samples of FRA and with 30% replacement of FRA. Water absorption values are tabulated in Table 11.4 and illustrated in Figure 11.4.

B. Density of FRA

Khatib (2005) prepared nine mixes with fine recycled aggregates ratio ranging from 0 to 100%. The fine recycled aggregates were obtained from crushed concrete (CC) and crushed bricks (CB) and average value of density for the mixes at 28 days is 2362.33 kg/m^3. Carro-López et al. (2015) investigated the effect of fine recycled aggregates on the rheology of auto compacting concrete over time (at 15, 45 and 90 min). At 0 per cent, 20 per cent, 50% and 100 per cent of natural aggregates were replaced by recycled sand. The 20% replacement mix maintained the appropriate performance and filling capacity.

After 28 days both fresh density and hardened density decreased. Zhao et al. (2013) discovered a linear association between the mean size of four granular classes and the concentration of hardened cement paste. Special methods for the measurement of the cement paste content have been developed based on salicyl acid dissolution. Results demonstrate a strong correlation between bound water and density of FRCA and the cement paste content. Density Values are tabulated in Table 11.5 and illustrated in Figure 11.5.

C. Carbonation

Pan et al. (2017) attempted to improve the carbonation effectiveness by raising the amount of carbonate compounds in the RFA demolition using Calcium Hydroxide (CH). Before carbonation, the RFAs were pre-soaked with CH, and their properties were superior to those of carbonation alone, including a lower crush value, water absorption, and powder content. The carbonation settings had a good influence on CO2 curing, and the optimal state was found.

Zega and Di Maio (2011) presented the durable behavior of different RFA (0%, 20% and 30%) structural concrete. Carbonation evaluation on prismatic specimens subjected to a natural urban-industrial setting (75 x 75 x 300 mm). The site is located on the outskirts of the Argentina village of La Plata, with a 1000 mm/year precipitation regime, 78% average relative humidity, and minimum and highest average

Table 11.5 Density values of FRA

Reference	*Density (kg/m^3)*
Khatib (2005)	2340
Carro-López et al. (2015)	2100
Zhao et al. (2013)	2068

Table 11.6 Carbonation depths values

Reference	*Carbonation depth (mm)*
Pan et al. (2017)	-
Zega and Di Maio (2011)	2
Koulouris et al. (2004)	23

temperatures of 5°C and 30°C, respectively, and with time almost no change in carbonation was seen after 620 days.

Koulouris et al. (2004) produced concrete was made from a natural gravel mixture and a recycled aggregate content of up to 100%. RCA concrete's durability performance was evaluated, with results for RCA hardness equivalent concretes being very similar to NA concrete (near surface absorption, penetration of carbonation (CO2), freese thaw scaling, chloride input resistance and abrasion). Carbonation depth values are given in Table 11.6.

D. Resistance to Chloride ion Penetration

Kou and Poon (2009a) prepared 13 mixes with two of them being as reference mixes and other 11 mixes with different replacement ratios and water to binder ratios and these 13 mixes were under three series namely series 1, series 2 and series 3. Resistance to chloride ion penetration was observed to increase due to the fact incorporation of the fine recycled aggregates of size less than 0.30 mm. Anastasiou et al. (2014) experimented and implemented a program. To examine the feasibility of building concrete with a high volume of industrial byproducts and secondary materials, as well as the strength development of construction and demolition debris (CDW) aggregates appears to have dropped considerably, but in comparison to CDW limestone combinations, the CDW-steel slag combination appears to be superior and the concentration of chloride measured for the various concrete mixtures at 40 – 50 mm depth. The content of chloride for all mixtures is acceptable. Chloride ion penetration is apparently increased slightly by HCFA and CDW, while chloride resistance decreases further in combination with HCFA and CDW. Chloride intake resistance showed no change in the substitution of limestone with EAF slag aggregates.

E. Ultra-Sonic Pulse Velocity

Khatib (2005) prepared 9 mixes with fine recycled aggregates ratio ranging from 0 to 100%. The fine recycled aggregates were obtained from crushed concrete (CC) and crushed bricks (CB) and average value of ultra-sonic pulse velocity at days was found to be 4.854 km/s.

Results and Disscussions

The mechanical properties like Compressive strength, flexural strength, and split tensile strength appear to have declined as fine recycled materials were replaced with fine natural aggregates. Kou and Poon (2009a) discovered a decrease in compressive and split-tensile strength without the incorporation of fly ash, and found that the most promising values of compressive strength and split-tensile strength were obtained at the replacement of 25 to 50% of fine recycled aggregates with fine natural aggregates. Khatib (2005) also observed the similar pattern for long term strength. When the replacement of class M sand was done with class CC sand the reduction in long term strength could reach up to 30% and only 15% loss in long term strength was observed if the replacement was only 25%.

Similarly, compressive strength, split tensile strength, and flexural strength findings were obtained and presented in Figures 11.1–11.3, by other researchers, if percentage of FRA keeps increasing, then strength will keep on decreasing and the appreciable values of mechanical properties are obtained with the percentage replacement of fine recycled aggregates in the range of 25% to 55%.

And, the durability properties of the concrete made with fine recycled aggregates has also been affected as the mechanical properties have. This is because of the increase in the water absorption values and decrease in the density of the concrete made with fine recycled aggregates. 100% replacement of fine recycled aggregates have serious effects on the durability of the concrete made with FRA, however if the replacement if about 30%, the durability features have no negative consequences and can be used for structural concrete Evangelista and de Brito (2010).

When FRA is used in higher percentages, the durability properties such as water absorption, density, carbonation depths, resistance to chloride ion, and ultrasonic pulse velocity suffer.

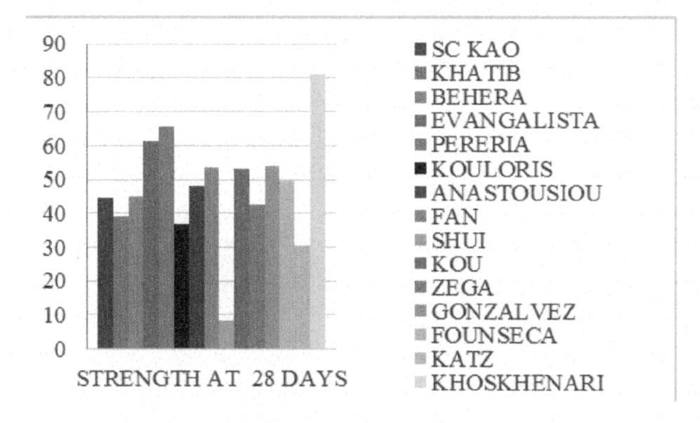

Figure 11.1 Compressive strength values as per different researches

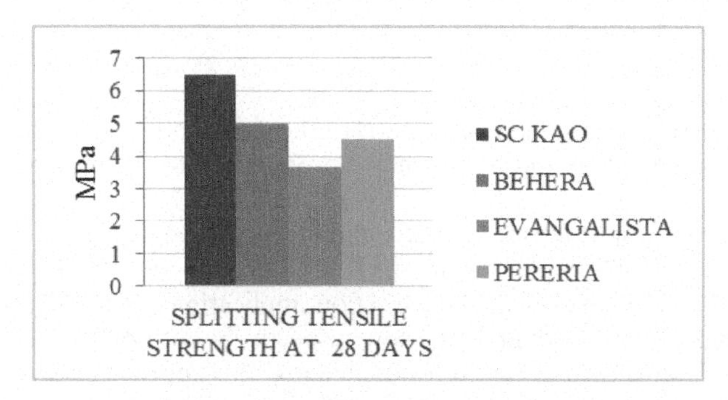

Figure 11.2 Split-tensile strength values as per different researches

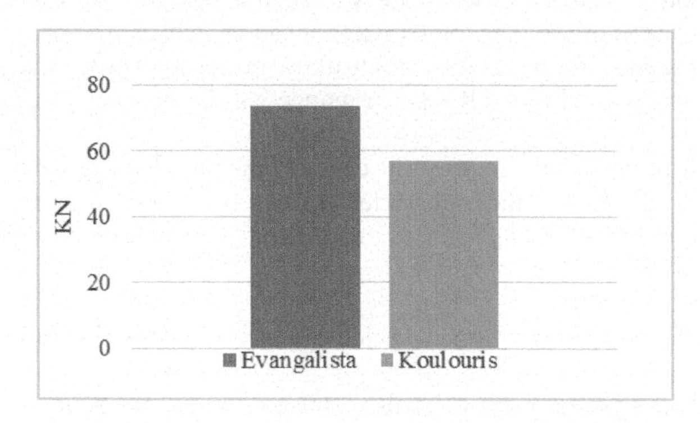

Figure 11.3 Flexural strength at 28 days for Evangelista and de Brito (2017) and Koulouris et al. (2004)

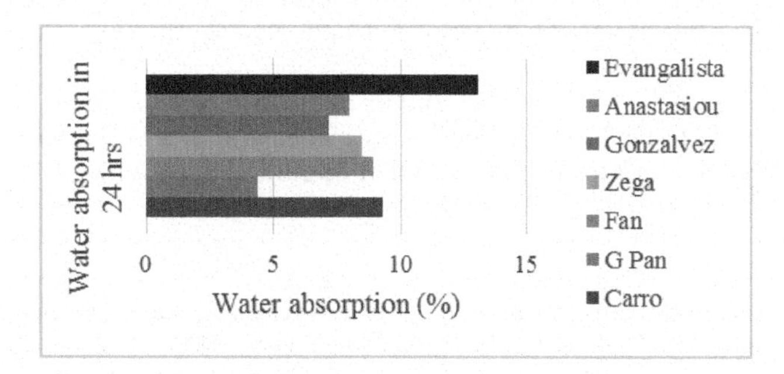

Figure 11.4 Water absorption values comparison

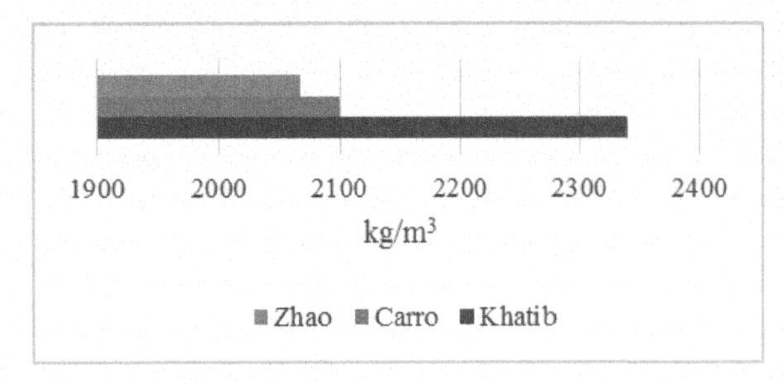

Figure 11.5 Density of FRA

However, appreciable results are obtained with a replacement percentage of 30% FRA with FNA and can be used in structural concrete. Water absorption is one of the durability qualities and density of concrete made with fine recycled aggregate are depicted in the Figures 11.4 and 11.5.

Conclusions

Although, it is true that the use of FRA is completely restricted in many parts of the globe and in some countries the use of FRA is allowed to be used to some extent and by the major proportions of the researcher community.

Mechanical parameters like compressive strength, split-tensile strength, and flexural strength were lower than in typical concrete, but it was due to the 100% replacement of FRA in the concrete mixes and the obtained values were safe to be applied to some parts of the construction applications. Similarly, the durability properties like water absorption and density were high and low respectively but not out of the standards ranges given by different literature of various countries. It is possible that this approach is overly conservative in terms of the usage of FRA in construction. If approached, all of the results observed and cited in this review work would be obtained, taking into consideration all the needs and obstructions to the current lack of resources to the construction purposes to use the FRA in such a way that maximum performance of the concrete made from FRA even exhibiting at some of stages can be achieved. Also, the results so far have shown that it can be said to use the FRA mixes into practical use rather than only talking about this in the theory. From all the results it is proved that FRA performance will be beneficial in the construction works.

There are obstacles and difficulties in every situation, same goes for the case of Fine recycled aggregates. But that doesn't mean they cannot be overcome. Therefore, these obstacles must be overcome so that the use of FRA is recommended and can be used on the larger scale. Some of the research gaps that needs to be identified and further to be studied are as follows:

- First, there should be a basic standard for the assessment of fine recycled aggregates' water absorption (FRA).
- There are many properties that are studied, and proper research is also done on some of the properties but, there are still some properties left out that needs further studying and further research and investigation especially about the density and durability of the concrete made from fine recycled aggregates.
- Since there are many methods and techniques that are adopted by many researchers for FRA to be used in the concrete mixtures. There should be some replacements ratios to be adopted for the experiments and what type of recycled stuff is in the FRA actually.
- Although there are many studies and research done on the FRA and its use in different situations and in different places many more things are still to be discovered like many mechanical properties, Stress-strain curves of FRA concrete, and many relationships regarding to that of durability parameters.

References

Anastasiou, E., Georgiadis Filikas, K., and Stefanidou, M. (2014). Utilization of fine recycled aggregates in concrete with fly ash and steel slag. Constr. Build. Mater. 50:154–161. doi: 10.1016/j.conbuildmat.2013.09.037.

Baldusco, R., Nobre, T. R. S., Angulo, S. C., Quarcioni, V. A., and Cincotto, M. A. (2019). Dehydration and rehydration of blast furnace slag cement. J. Mater. Civ. Eng. 31(8):04019132. doi: 10.1061/(asce)mt.1943-5533.0002725.

Basheer, L., Kropp, J., and Cleland, D. J. (2001). Assessment of the durability of concrete from its permeation properties: A review. Constr. Build. Mater. 15(2–3):93–103. doi: 10.1016/S0950-0618(00)00058-1.

Behera, M., Minocha, A. K., and Bhattacharyya, S. K. (2019). Flow behavior, microstructure, strength and shrinkage properties of self-compacting concrete incorporating recycled fine aggregate. Constr. Build. Mater. 228:116819. doi: 10.1016/j.conbuildmat.2019.116819.

Carriço, A., Real, S., Bogas, J. A., and Costa Pereira, M. F. (2020). Mortars with thermo activated recycled cement: Fresh and mechanical characterisation. Constr. Build. Mater. 256:119502–119502. doi: 10.1016/j.conbuildmat.2020.119502.

Carro-López, D., González-Fonteboa, B., De Brito, J., Martínez-Abella, F., González-Taboada, I., and Silva, P. (2015). Study of the rheology of self-compacting concrete with fine recycled concrete aggregates. Constr. Build. Mater. 96:491–501. doi: 10.1016/j.conbuildmat.2015.08.091.

Evangelista, L. and De Brito, J. (2007a). Environmental life cycle assessment of concrete made with fine recycled concrete aggregates. In Port. SB 2007 - Sustainable Construction, Materials and Practices: Challenge of Industry for the New Millenium, 789–794.

Evangelista, L. and de Brito, J. (2007b). Mechanical behaviour of concrete made with fine recycled concrete aggregates. Cem. Concr. Compos. 29(5):397–401. doi: 10.1016/j.cemconcomp.2006.12.004.

Evangelista, L. and de Brito, J. (2010). Durability performance of concrete made with fine recycled concrete aggregates. Cem. Concr. Compos. 32(1):9–14. doi: 10.1016/j.cemconcomp.2009.09.005.

Evangelista, L. and De Brito, J. (2014). Concrete with fine recycled aggregates: A review. Eur. J. Environ. Civ. Eng. 18(2):129–172. doi: 10.1080/19648189.2013.851038.

Evangelista, L. and de Brito, J. (2017). Flexural behaviour of reinforced concrete beams made with fine recycled concrete aggregates. KSCE J. Civ. Eng. 21(1):353–363. doi: 10.1007/s12205-016-0653-8.

Fan, C. C., Huang, R., Hwang, H., and Chao, S. J. (2016). Properties of concrete incorporating fine recycled aggregates from crushed concrete wastes. Constr. Build. Mater. 112:708–715. doi: 10.1016/j.conbuildmat.2016.02.154.

Florea, M. V. A. and Brouwers, H. J. H. (2012). Recycled concrete fines and aggregates: The composition of various size fractions related to crushing history. Proc. Int. Conf. Build. Mater. 18:1034–1041.

Fonseca, N., De Brito, J., and Evangelista, L. (2011). The influence of curing conditions on the mechanical performance of concrete made with recycled concrete waste. Cem. Concr. Compos. 33(6):637–643. doi: 10.1016/j.cemconcomp.2011.04.002.

Gonçalves, T., Silva, R. V., de Brito, J., Fernández, J. M., and Esquinas, A. R. (2020). Mechanical and durability performance of mortars with fine recycled concrete aggregates and reactive magnesium oxide as partial cement replacement. Cem. Concr. Compos. 105:1–10. doi: 10.1016/j.cemconcomp.2019.103420.

Güneyisi, E., Gesoglu, M., Algin, Z., and Yazici, H. (2016). Rheological and fresh properties of self-compacting concretes containing coarse and fine recycled concrete aggregates. Constr. Build. Mater. 113:622–630. doi: 10.1016/j.conbuildmat.2016.03.073.

Hu, J. Wang, Z., and Kim, Y. (2013). Feasibility study of using fine recycled concrete aggregate in producing self-consolidation concrete. J. Sustain. Cem. Mater. 2(1):20–34. doi: 10.1080/21650373.2012.757832.

Kalinowska-Wichrowska, K., Kosior-Kazberuk, M., and Pawluczuk, E. (2020). The properties of composites with recycled cement mortar used as a supplementary cementitious material. Materials (Basel), 13(1):64. doi: 10.3390/ma13010064.

Katz, A. (2003). Properties of concrete made with recycled aggregate from partially hydrated old concrete. Cem. Concr. Res. 33(5):703–711. doi: 10.1016/S0008-8846(02)01033-5.

Khatib, J. M. (2005). Properties of concrete incorporating fine recycled aggregate. Cem. Concr. Res. 35(4):763–769. doi: 10.1016/j.cemconres.2004.06.017.

Khoshkenari, A. G., Shafigh, P., Moghimi, M., and Bin Mahmud, H. (2014). The role of 0-2mm fine recycled concrete aggregate on the compressive and splitting tensile strengths of recycled concrete aggregate concrete. Mater. Des. 64:345–354. doi: 10.1016/j.matdes.2014.07.048.

Kirthika, S. K., Singh, S. K. and Chourasia, A. (2020). Performance of recycled fine-aggregate concrete using novel mix-proportioning method. J. Mater. Civ. Eng. 32(8):04020216. doi: 10.1061/(asce)mt.1943-5533.0003289.

Kou, S. C. and Poon, C. S. (2009a). Properties of concrete prepared with crushed fine stone, furnace bottom ash and fine recycled aggregate as fine aggregates. Constr. Build. Mater. 23(8):2877–2886. doi: 10.1016/j.conbuildmat.2009.02.009.

Kou, S. C. and Poon, C. S. (2009b). Properties of self-compacting concrete prepared with coarse and fine recycled concrete aggregates. Cem. Concr. Compos. 31(9):622–627. doi: 10.1016/j.cemconcomp.2009.06.005.

Koulouris, A., Limbachiya, M. C., Fried, A. N. and Roberts, J. J. (2004). Use of recycled aggregate in concrete application: Case studies. In Proceedings of the international conference on sustainable waste management and recyling construction and demolition waste, (pp. 245–257).

Lin, Y. H., Tyan, Y. Y., Chang, T. P. and Chang, C. Y. (2004). An assessment of optimal mixture for concrete made with recycled concrete aggregates. Cem. Concr. Res. 34(8):1373–1380. doi: 10.1016/j.cemconres.2003.12.032.

Pan, G., Zhan, M., Fu, M., Wang, Y. and Lu, X. (2017). Effect of CO2 curing on demolition recycled fine aggregates enhanced by calcium hydroxide pre-soaking. Constr. Build. Mater. 154:810–818. doi: 10.1016/j.conbuildmat.2017.07.079.

Pereira, P., Evangelista, L. and De Brito, J. (2012a) The effect of superplasticisers on the workability and compressive strength of concrete made with fine recycled concrete aggregates. Constr. Build. Mater. 28(1):722–729. doi: 10.1016/j.conbuildmat.2011.10.050.

Pereira, P., Evangelista, L. and De Brito, J. (2012b). The effect of superplasticisers on the mechanical performance of concrete made with fine recycled concrete aggregates. Cem. Concr. Compos. 34(9):1044–1052. doi: 10.1016/j.cemconcomp.2012.06.009.

Rodrigues, F., Carvalho, M. T., Evangelista, L., and De Brito, J. (2013). Physical-chemical and mineralogical characterisation of fine aggregates from construction and demolition waste recycling plants. J. Clean. Prod. 52:438–445. doi: 10.1016/j.jclepro.2013.02.023.

Šefflová, M. and Pavlů, T. (2017). The durability of fine recycled aggregate concrete. Adv. Mater. Res. 1144:59–64. doi: 10.4028/www.scientific.net/amr.1144.59.

Shui, Z., Xuan, D., Wan, H., and Cao, B. (2008). Rehydration reactivity of recycled mortar from concrete waste experienced to thermal treatment. Constr. Build. Mater. 22(8):1723–1729. doi: 10.1016/j.conbuildmat.2007.05.012.

Tiwari, A., Singh, S., and Nagar, R. (2016). Feasibility assessment for partial replacement of fine aggregate to attain cleaner production perspective in concrete: A review. J. Clean. Prod. 135:490–507. doi: 10.1016/j.jclepro.2016.06.130.

Velay-Lizancos, M., Martinez-Lage, I., Azenha, M., Granja, J., and Vazquez-Burgo, P. (2018). Concrete with fine and coarse recycled aggregates: E-modulus evolution, compressive strength and non-destructive testing at early ages. Constr. Build. Mater. 193:323–331. doi: 10.1016/j.conbuildmat.2018.10.209.

Velay-Lizancos, M., Vazquez-Burgo, P., Restrepo, D., and Martinez-Lage, I. (2018). Effect of fine and coarse recycled concrete aggregate on the mechanical behavior of precast reinforced beams: Comparison of FE simulations, theoretical, and experimental results on real scale beams. Constr. Build. Mater. 191:1109–1119. doi: 10.1016/j.conbuildmat.2018.10.075.

Vinay Kumar, B. M., Ananthan, H., and Balaji, K. V. A. (2018). Experimental studies on utilization of recycled coarse and fine aggregates in high performance concrete mixes. Alexandria Eng. J. 57(3):1749–1759. doi: 10.1016/j.aej.2017.05.003.

Wang, Y., Liu, F., Xu, L., and Zhao, H. (2019a). Effect of elevated temperatures and cooling methods on strength of concrete made with coarse and fine recycled concrete aggregates. Constr. Build. Mater. 210:540–547. doi: 10.1016/j.conbuildmat.2019.03.215.

Wang, Y., Zhang, H., Geng, Y., Wang, Q., and Zhang, S. (2019b). Prediction of the elastic modulus and the splitting tensile strength of concrete incorporating both fine and coarse recycled aggregate. Constr. Build. Mater. 215:332–346. doi: 10.1016/j.conbuildmat.2019.04.212.

Zega, C. J. and Di Maio, Á. A. (2011). Use of recycled fine aggregate in concretes with durable requirements. Waste Manag. 31(11):2336–2340. doi: 10.1016/j.wasman.2011.06.011.

Zhao, Z., Remond, S., Damidot, D., and Xu, W. (2013). Influence of hardened cement paste content on the water absorption of fine recycled concrete aggregates. J. Sustain. Cem. Mater. 2(3–4):186–203. doi: 10.1080/21650373.2013.812942.

12 Analysis of a rocket nozzle using method of characteristics

Vinayak H. Khatawate[a,*], *Aniruddha Mallick*[b], *Dhaval Bhanushali*[c], *and Greeshma Gala*[d]

Department of Mechanical Engineering (Associate Professor) Dwarkadas J. Sanghvi College of Engineering, Mumbai, India

Abstract

A nozzle is an integral part of a rocket that is responsible for increasing the thrust that a rocket can gain from a unit weight of propellant burned. In this paper, a minimum length nozzle is designed by using the Method of Characteristics (MOC) approach. For performing calculations, compressible isentropic flow of the gases is considered. A thermo-structural analysis is conducted so that the design of the nozzle can be validated for its safety. Three materials for the nozzle are considered and compared based on the stresses and deformations obtained. The analysis is performed using Finite Element Analysis. Two cooling methods, regenerative and film cooling are generally used for rocket cooling. A CFD analysis of film cooling is performed to verify the reduction in temperature.

Keywords: Film cooling, method of characteristics, minimum length nozzle, nozzle, regenerative cooling.

Introduction

Propulsion is the act of changing the motion of a body with respect to an inertial reference frame (Sutton and Biblarz, 2017). A rocket engine is a propulsion system which provides the force to a rocket to move it from rest and accelerate it enough to overcome retarding forces and eventually leave the atmosphere. Rocket propulsion works on the principle of Newton's third law.

A rocket engine nozzle is a converging-diverging type of nozzle which expands and accelerates the combustion gases produced by the burning of the propellants. The exhaust gases exit the nozzle at supersonic velocities.

Materials used for a rocket nozzle vary according to different requirements. They generally range from graphite, tungsten, aluminium alloys and many other combinations of materials which can withstand such high temperatures.

A rocket generally runs with 3500 K combustion temperature. Therefore, the structure of the rocket depends upon its cooling parameters too. The primary objective of cooling in rocket is to prevent the chamber and the nozzle walls from becoming too hot, so that they will be able to withstand the imposed stresses. Majorly two types of cooling are used for rocket nozzles, regenerative and film cooling.

Methodology

Design of Nozzle using method of characteristics

Supersonic nozzles are of 2 types: gradual-expansion and minimum-length nozzles. The former type is commonly used in supersonic wind tunnels where the flow needs to be stabilised before the test is performed. Rocket nozzles, on the other hand, have to be smaller in size due to length and weight constraints. Hence minimum-length type of nozzles are employed.

A minimum length supersonic nozzle is designed by using the method of characteristics (MOC) (Khan et al., 2013). At the throat, the flow is sonic due to which the subsonic portion i.e., the converging section can be studied independently. The main difficulty is experienced in the diverging section of the nozzle where supersonic flow exists.

[a]vinayak.khatawate@djsce.ac.in; [b]animallick236@gmail.com; [c]1802dhavaldb@gmail.com; [d]greeshma.gala@gmail.com

For uniform flow at the exit, the number of Mach wave reflections and intersections needs to be minimised (known as the non-simple region). These Mach waves are known as characteristics. These lines in supersonic flow are oriented in definite directions of propagations of pressure waves.

Nozzle length is kept as short as possible such that rapid expansion takes place. The expansion section fans out from an abrupt corner (at the throat). This phenomenon is typically modelled as a continuous series of expansion waves, each turning the airflow in small amount along with the contour of the channel wall. Prandtl-Meyer function governs these expansion waves.

When the exit pressure is equal to the outside pressure, optimal expansion takes place. In this type of expansion, MOC is used to design a supersonic nozzle which is shock free. The contour is obtained in such a way that shockwaves are eliminated as no pressure waves will be interfering with one another.

Only two systems of waves are encountered by the fluid element in a minimum length nozzle, right-running and left-running.

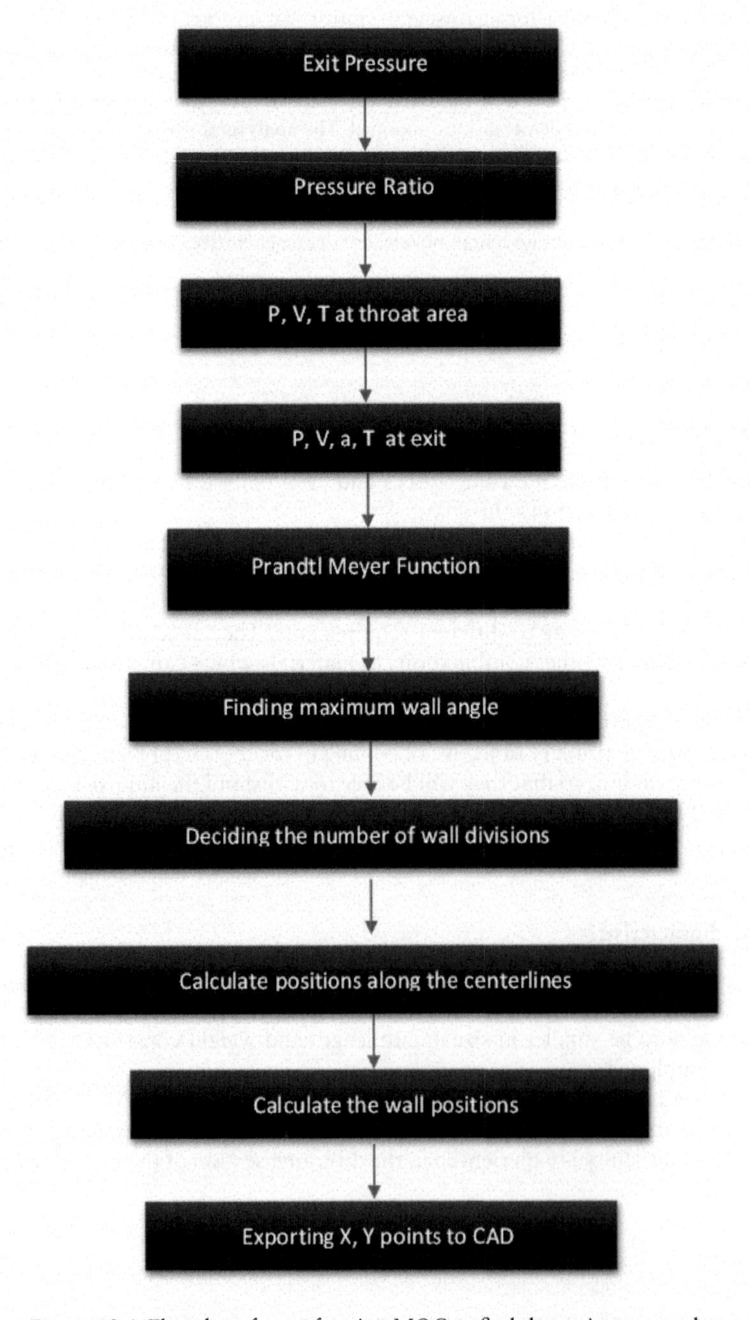

Figure 12.1 Flowchart for performing MOC to find the optimum nozzle contour

The angle through which sonic flow (M = 1) expands to obtain a given Mach number is called Prandtl-Meyer angle. It is the output of the Prandtl-Meyer function. The methodology to be followed for obtaining the nozzle contour is illustrated in the form of a flowchart in Figure 12.1.

Using MATLAB to generate 3-D points

As an example, the following input parameters are taken as mentioned in Table 12.1. A conventional method to obtain the desired Nozzle CAD model is to make use of the Rocket Propulsion Analysis software (RPA) (Baxi 2021). For the purpose of this paper, a MATLAB code is executed to use the input parameters and give an output of the 3D points.

The exit pressure is calculated from empirical relations which are derived as a function of altitude. At an altitude of 7500 m, the exit pressure is 39.356 kPa.

$$\text{Pressure ratio} = \frac{\text{Outside Pressure}(P_2)}{P_1} = 0.0174$$

$$\text{Temperature ratio} = (\text{Pressure ratio})^{\frac{\gamma-1}{\gamma}} = 0.314$$

$$\text{Critical}(\text{throat})\text{temperature} = \frac{2.T_1}{\gamma+1} = 1000\,K$$

$$\text{Critical (throat) pressure} = \left(\left(\frac{2}{\gamma+1}\right)^{\frac{\gamma}{\gamma-1}}\right) \times 2.068(\text{for air}) = 1.0925 \text{ kPa}$$

$$\text{Critical (throat) velocity} = \sqrt{\frac{2.\gamma.R.T_1}{\gamma+1}} = 704.98\,m/s$$

$$\text{Exit velocity} = \sqrt{\frac{2\gamma RT_1}{\gamma-1}.(1-\text{Temp ratio})} = 1430.2\,m/s$$

$$\text{Exit temperature} = T_1.(\text{Pressure ratio})^{\frac{\gamma-1}{\gamma}} = 376.829\,K$$

$$\text{Exit sound speed} = \sqrt{\gamma.R.Exit\,temp} = 432.76\,m/s$$

$$\text{Exit Mach} = \frac{\text{Exit velocity}}{\text{Exit sound speed}} = 3.305$$

Table 12.1 Input parameters

Parameters	*Value*
Chamber pressure (Pa)	2.27E+06
Chamber temperature (K)	1200
Thrust (N)	4000
Altitude (m)	7500
Coefficient of heats	1.4
R (J/kg-K)	355
Throat radius (mm)	35

Thus, the exit Mach number is found from the input parameters. In a minimum length nozzle, the maximum wall angle downstream of the throat is equal to one-half the Prandtl-Meyer angle, obtained for the design exit Mach number. A 1-dimensional mesh is obtained in the form of wall divisions.

As the mesh is refined, the result of nozzle height and length becomes steadier (Ali Md. et al. 2012). Angle increment is calculated as a function of the maximum wall angle and the number of wall divisions is taken as twice the maximum wall angle.

The Prandtl-Meyer function is:

$$PM = \sqrt{\frac{\gamma+1}{\gamma-1}}.atan\sqrt{\frac{\gamma-1}{\gamma+1}.(M^2-1)} - atan\sqrt{M^2-1}$$

PM (Exit Mach) = 55.307°

$$\text{Maximum wall angle} = \frac{1}{2}.\{PM(Exit\ Mach)\} = 27.65°$$

Angle increment = (90°-Max angle)-floor(90°-Max angle)=0.35°

Number of wall divisions = 2*maximum wall angle≈56

The minimum length is achieved with the mesh generated by 56 characteristic lines. The reflection section is shown in Figure 12.2 while the expansion section can be seen in Figure 12.3, which ultimately gives the positions of points along the wall. The reflection section begins from the length of 0.823 mm and stretched to the length of 48.79 mm. The expansion waves are completely cancelled at the length of 137.042 mm and height of 63.304 mm.

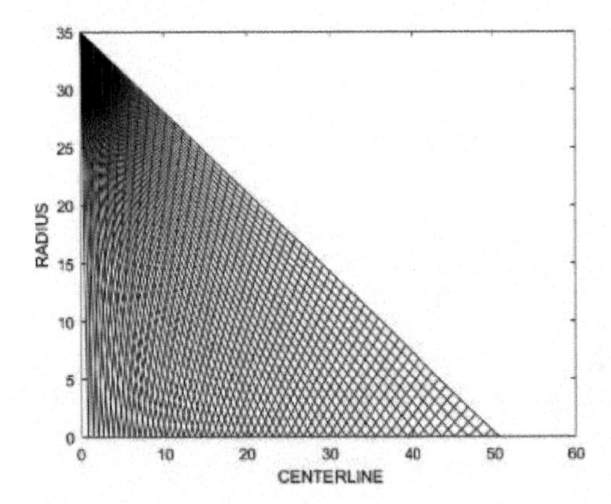

Figure 12.2 Finding the positions along the centerlines

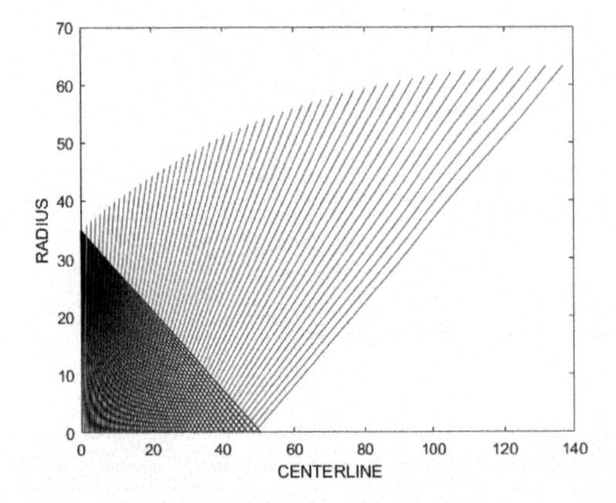

Figure 12.3 Finding the wall positions

Figure 12.4 Nozzle CAD

3D Modelling using Solidworks

On generating the points for the nozzle curve, the CAD model was created by importing the excel file in Solidworks. The final CAD is obtained as seen in Figure 12.4.

Selection of Material

Considering the temperature and stresses generated, three materials are taken into consideration: Silicon Carbide, Inconel Alloy 625 and Aluminium alloy. The properties of the three materials are listed in Table 12.2. Ceramics like Silicon Carbide (SiC) can withstand extremely high temperatures as compared to Aluminium. On the other hand, Aluminium alloys and Inconel Alloy have found widespread application in the aerospace industry due to their reduced weight and relatively low manufacturing costs. The chemical composition of Inconel Alloy 625 gives it excellent corrosion resistance. With the advent of Additive manufacturing (AM), ceramics can be used for improving the thermal performance of nozzles.

Finite Element Analysis

A. Thermal-Structural Analysis

This is an example of a one-way FSI (Fluid Structure Interaction) study. It involves carrying out CFD analysis on the nozzle geometry so that the temperatures acting on the body is obtained. Figure 12.5 shows the geometry used for this analysis.

For the CFD analysis, inviscid flow is selected so that the effect of shear stresses along the wall can be ignored. Air is set as an ideal gas and a density-based solver is used. Figure 12.6 shows the temperature contour generated. The maximum and minimum temperatures obtained are 922.04 °C and 4.0528 °C.

It is then imported into ANSYS Mechanical so that structural analysis can be carried out. The Mechanical nodes get mapped to the CFD domain and the remaining nodes are set to an average temperature of 491.18^{0}C. Planes are created at particular distances from the throat section so that the converging and diverging sections can be sliced into multiple bodies. The area ratio is found at each section and the pressure ratio is found using isentropic flow calculator (Angle).

The pressure load values are applied to the sliced parts as a boundary condition (surface effect). The mesh size is reduced to 2 mm to refine the mesh. Fixed support is added to the edge of converging side, as seen in the setup shown in Figure 12.7, and then by solving, the result of thermo- structural analysis is obtained.

Table 2 Material properties

Properties	*Materials*		
	Silicon carbide	*Inconel Alloy 625*	*Aluminium alloy*
Mass Density (kg/m^3)	3100	8440	2574
Thermal Conductivity (W/mK)	120	24	253
Specific Heat (J/KgK)	750	632.5	982
Compressive Strength (MPa)	3900	927.3	280
Yield Strength (MPa)	550	424.35	276
Poisson's Ratio (μ)	0.14	0.334	0.33
Modulus of Elasticity (GPa)	410	140.95	68.3

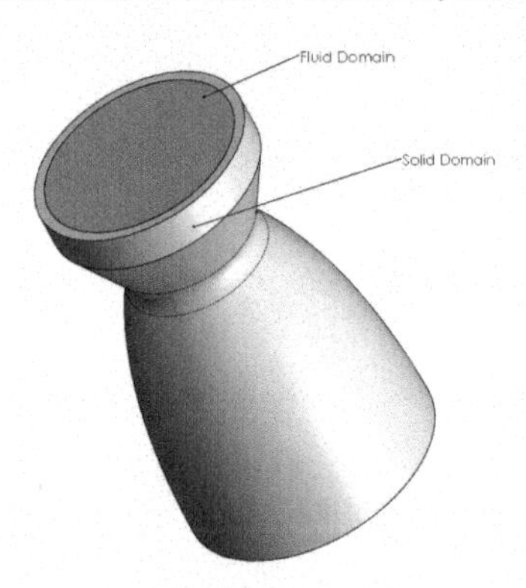

Figure 12.5 Defining the domain

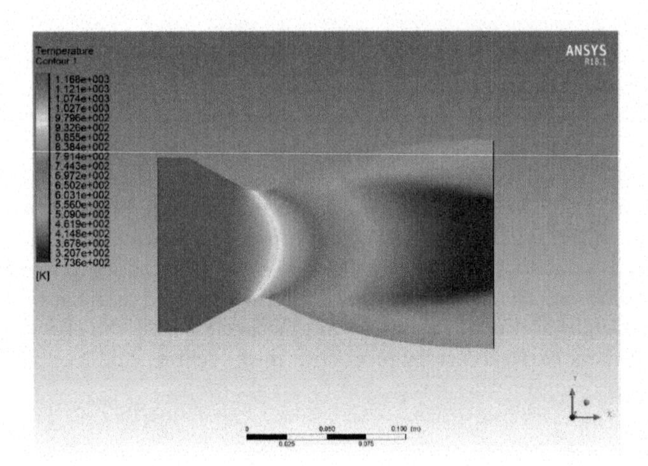

Figure 12.6 Temperature contour obtained by inviscid flow

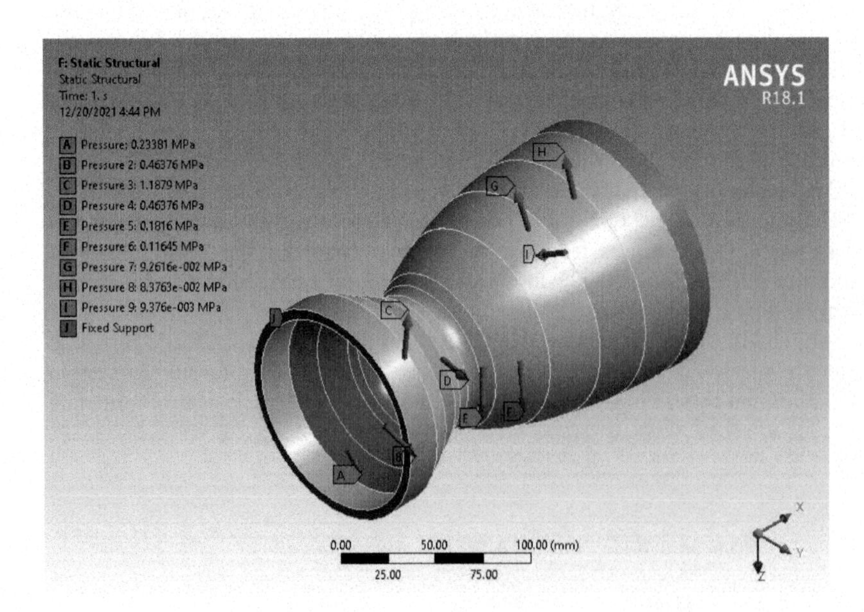

Figure 12.7 Setup of pressure loads and fixed support for static-structural analysis

B. Modal and Harmonic Analysis

After completing the thermo-structural analysis, the solution is used to perform modal analysis (in this case, pre-stressed modal analysis). The output of this analysis are the natural frequencies and their corresponding mode shapes. This analysis can be performed in conjunction with Harmonic analysis, where the peak responses of stress, directional displacement and strain can be found, near the natural frequencies. This is known as mode-superposition method. In harmonic analysis it is assumed that all input loads are sinusoidal and they have the same frequency. Transient effects are neglected along with any start-up vibrations.

C. Impact of cooling

At the throat, the highest temperatures are observed. High thermal stresses are generated which if it exceeds the yield strength, can cause catastrophic failure. Cooling mechanisms are needed to limit the temperatures obtained and this increases the lifetime of nozzle component. Two major forms of cooling used are regenerative cooling and film cooling.

Regenerative cooling is the standard cooling system for almost all modern main stage booster and upper state engines. Gaseous film cooling is used to find out the reduction in temperature in the nozzle. Hydrogen gas is used as the coolant, supplied at a temperature of -30C and an assumed velocity of 40 m/s.

Using CFD analysis, two geometric models are compared, one where film cooling is incorporated and another where no cooling is present. The area average of static temperature is compared from both cases to find the effectiveness of cooling in the design.

As seen in Figure 12.9, face meshing has been carried out by dividing the surface body into multiple portions, followed by edge sizing by giving an input of 100 number of divisions to obtain a fine mesh. An edge sizing with a bias factor of 4 is set to the throat section to increase the accuracy of results.

For this problem, the SST solver was used to model turbulence such that the effect of viscosity is considered. Sutherland's law of viscosity is applied to model the viscosity of air with the wall. Also, species transport is considered to model the combustion process using a reduced H_2-O_2 kinetic model. A slip wall condition is applied in the film cooling geometry from where the coolant is injected(Amato, 2016).

Results and discussion

A. Thermal-structural analysis

For structural and thermal analysis, the stresses and temperature points are taken into consideration. Figures 12.10–12.15 show the result of thermo-structural analysis carried out by applying the properties of three materials respectively. In solution information, equivalent stress (Von-Mises stress) and deformation are obtained and mentioned in Table 12.3. From the three materials, Silicon carbide deforms the least under the same conditions.

According to this analysis, Inconel Alloy 625 performs similar to Aluminium Alloy yet it is a superior material as it exhibits higher strength at cryogenic temperatures. This analysis verifies the thermal performance of ceramics. The physical properties of the SiC-nozzle depends on the AM method used.

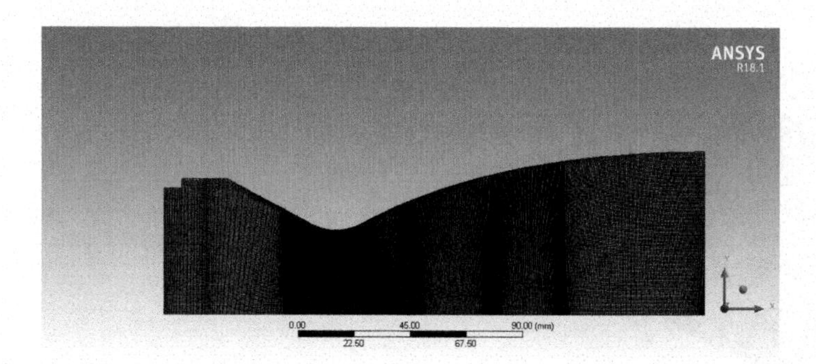

Figure 12.8 Meshing for the nozzle geometry with coolant-injector for film cooling

Table 12.3 Results of thermal-structural analysis

Materials	*Solution information*	
	Max. equivalent stress (GPa)	*Max. total deformation (mm)*
Silicon Carbide	6.818	1.2508
Inconel Alloy 625	2.671	1.2683
Aluminium Alloy	2.568	2.4302

B. Silicon carbide

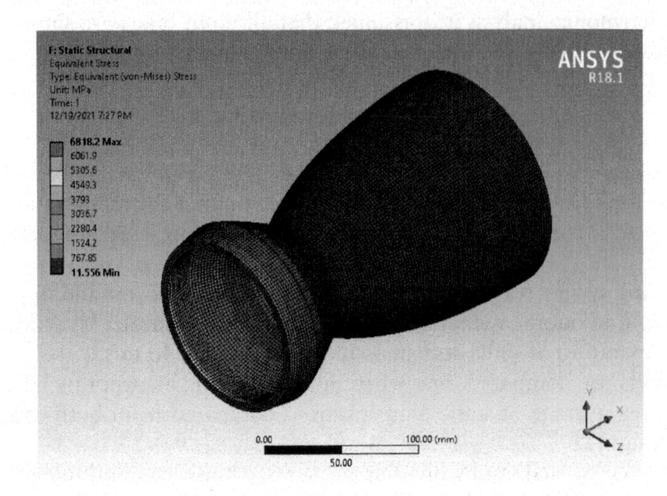

Figure 12.9 Equivalent stress

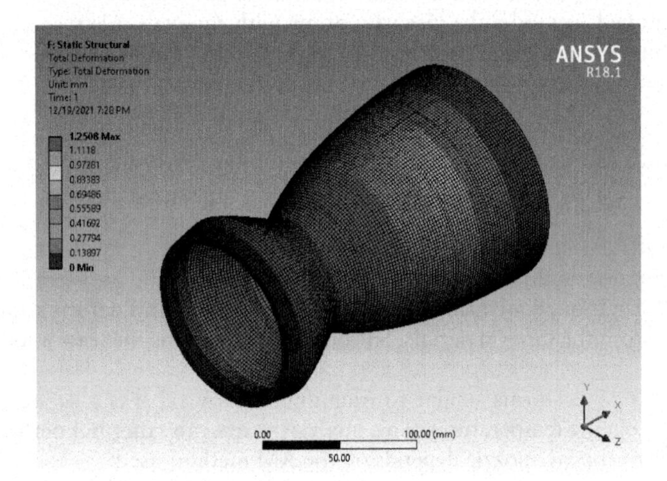

Figure 12.10 Total deformation

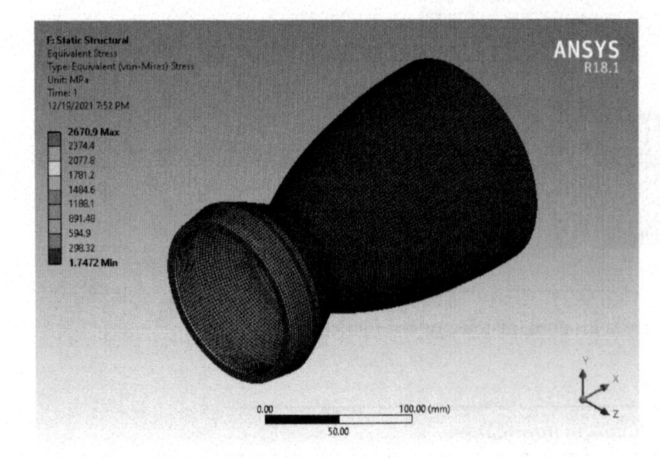

Figure 12.11 Equivalent stress

C. Inconel Alloy 625:

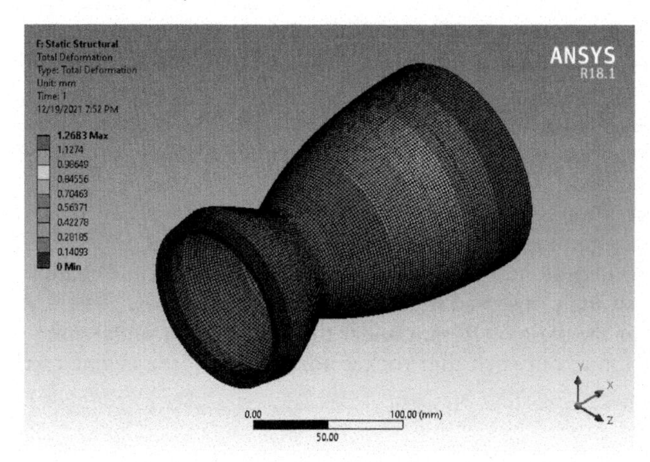

Figure 12.12 Total deformation

D. Aluminium alloy:

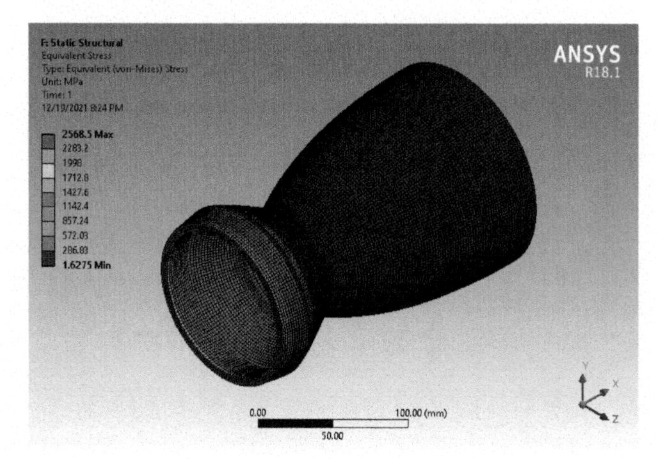

Figure 12.13 Equivalent stress

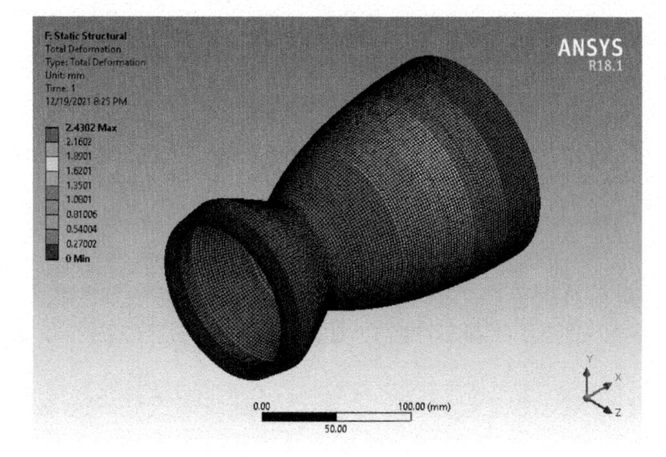

Figure 12.14 Total deformation

E. Modal and Harmonic Analysis

Six modes are found using the pre-stressed Modal analysis, as shown in Table 12.4.

To define the sinusoidal vibration environment, an example is taken from(Falcon User's Guide 2020). In the axial direction, acceleration loading is applied, as seen in Figure 12.15. The frequency range is entered in the analysis setting, in this case it is taken from 800 to 2000 Hz. The frequency range of the acceleration loading and analysis setting is selected so that the effect of the first four modes can be captured in the harmonic analysis (using the mode-superposition method).

The frequency response for directional deformation in the x-direction is plotted in Figure 12.16. The analysis occurs over 50 solution intervals. The magnitude plot shows that the axial deformation reaches its first peak value at 880 Hz, followed by a second peak at 940 Hz, after which the deformation steadily increases. There is no appreciable change in phase angle observed over this range of frequencies.

This analysis reveals that the nozzle deforms at the scale of 10 mm under the effect of sinusoidal vibrations. Appropriate damping coefficients can be found out such that rocket nozzle as well the engine can sustain vibrations.

Table 12.4 Results of modal analysis

Mode	*Frequency (Hz)*
1	900.24
2	900.44
3	1010.5
4	1010.8
5	2192.8
6	2315.2

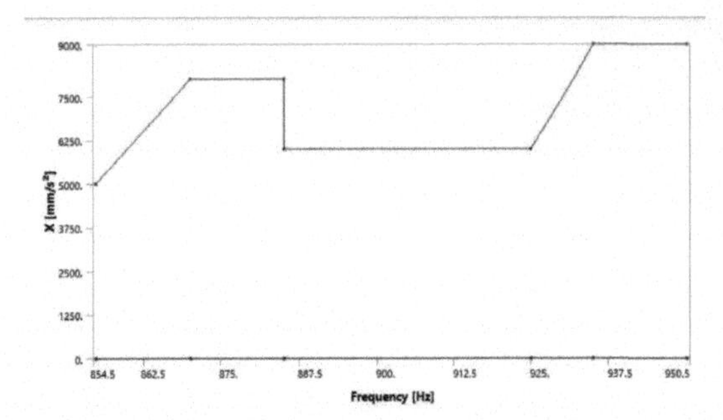

Figure 12.15 Graph of acceleration loading in the axial direction

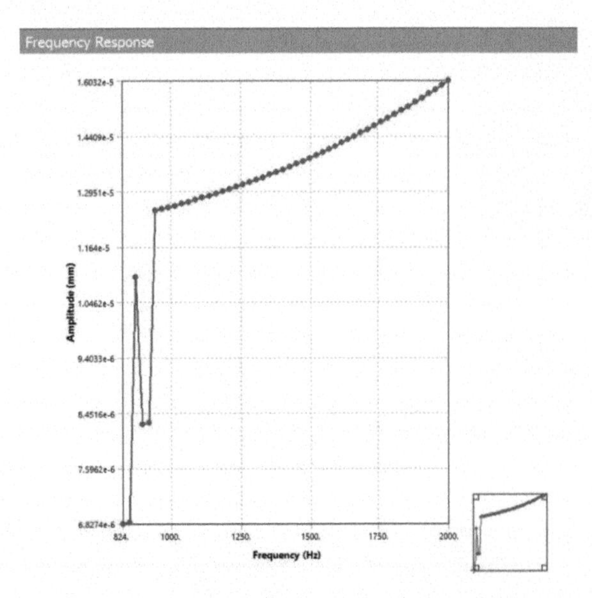

Figure 12.16 Frequency response of directional deformation (x-direction)

F. Film cooling

The static temperature contours with and without film cooling are shown in Figures 12.17 and 12.18.

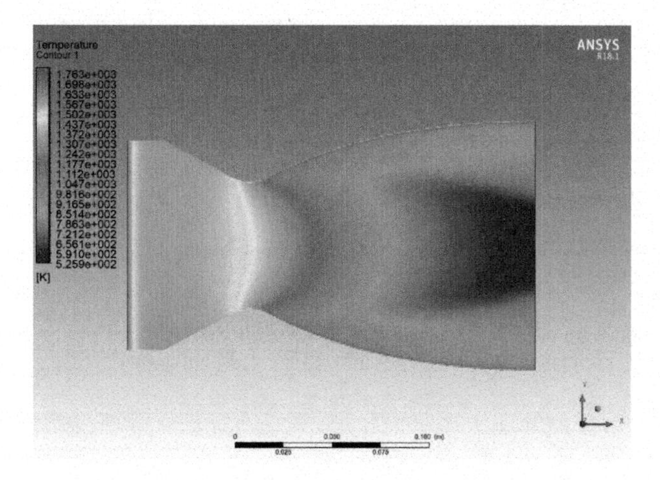

Figure 12.17 Static temperature contour without film cooling

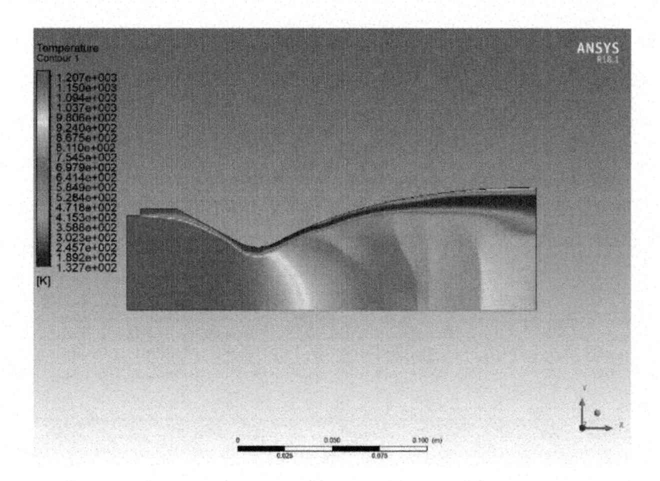

Figure 12.18 Static temperature contour with film cooling

Using area average of temperature as a deciding factor, without film cooling the value obtained is 820.88 K and with film cooling, the value obtained is 673.61 K. This shows that film cooling is somewhat effective in reducing the temperature of the nozzle.

Conclusion

A minimum length nozzle was designed using MATLAB and SOLIDWORKS and its analysis is successfully carried out using FEA. Thermo-structural analysis of three different materials is carried out to determine which material gives the best results. Harmonic analysis gives information about the frequency response such that the location where maximum deformation and stress occurs can be found, using the contour plot option.

Using ANSYS Fluent, it is seen how film cooling is partially helpful in reducing the wall temperature (as is evident upon comparing the contours of static temperature) and reducing the overall temperature of the nozzle (using area average).

In practice, regenerative cooling is coupled with film cooling and several other cooling methods such as ablative cooling to reduce the induced thermal stresses along the nozzle wall (Shine and Nidhi, 2018).

References

Ali, M. H., Mashud M., Al Bari A., and Islam M. U. (2012). Numerical Solution for the Design of Minimum Length Supersonic Nozzle. ARPN. J. Eng. Appl. Sci. 7(5):605–612.

Amato N., Leylegian J., and Naraghi M. 2016. CFD Analysis of Film Cooling and Heat Transfer in a Bipropellant Rocket Nozzle, Incorporating Chemically Reacting Flow. Propulsion and Energy Forum: 52nd AIAA/SAE/ASEE Joint Propulsion Conference, Salt Lake City, UT, USA, July 25–27, 2016.

Baxi, P., Jain, R., Dhadke, R., Chhabra, Y., and Khatawate, V. H. 2021. Design and Analysis of Bell-Parabolic De Laval Rocket Exhaust Nozzle. 4th Biennial International Conference on Nascent Technologies in Engineering (ICNTE): 1–6.

Falcon User's Guide, SpaceX, 2020. https://www.grc.nasa.gov/www/k-12/airplane/pranmyer.html, https://www.spacex.com/media/falcon_users_guide_042020.pdf

Khan M., Sardiwal S., Sharath M., and Chowdary D. (2013). Design of a Supersonic Nozzle using Method of Characteristics. Int. J. Eng. Res. Technol. 2(11):19–24.

Prandtl-Meyer Angle, Glenn Research Center, NASA.

Shine S. and Nidhi S. 2018. Review on film cooling of liquid rocket engines. Propuls. Power Res. 7(1):1–18.

Sutton, G. and O. Biblarz. 2017. Rocket propulsion elements. Hoboken, New Jersey: Wiley & Sons, Inc.

13 Building integrated system to generation DC-power based on renewable energy

Ihab Al-Janabi[1], *Samaher Al-Janabi*[2,*], *Monaria Hoshmand*[1], *and Saeed Khosroabadi*[1]

[1]Faculty of Electrical Engineering, Imam Reza International University, Mashhad, Iran

[2]Department of Computer Science, Faculty of Science for Women (SCIW), University of Babylon, Babylon, Iraq

Abstract

This paper builds an integrated system based on different types of hardware (i.e., CPU, GPU, and FPGA) to increase the production of electrical power (DC-power) in less time based on solar plants; That system collects data in real-time and pre-processing it. Then create two embedded intelligence models (i.e., linear and nonlinear); after that implementation, both based on CPU, GPU, and FPGA). The main point of that paper compares the performance of two types of smart micro-grid systems the first is based on linear embedded intelligence (LEI) while the other is based on nonlinear embedded intelligence (NEI) to determine which one is more efficient in a generation the max DC-power in less time. As a result, in both models, the FPGA gives a time of implementation less than CPU and GPU. Also; there is a high correlation between (DC power and AC power), (DC power and irradiation), (DC power, AC power, module temperature, and ambient temperature). NEI model requires preparing multi parameters but their results are best than LEI. Finally, the features more effective in generating max DC-power are AC-power, irradiance, and temperature.

Keywords: FPGA, linear embedded intelligence model, max Dc power, non-linear embedded intelligence model, renewable energy.

Introduction

Renewable energy is a scientific term given to energy that is generated without polluting the environment. It is a type of inexhaustible energy, and it is called renewable energy because it comes from natural sources (wind, sun, water), its advantages renewable energy is available in most countries of the world. Do not pollute the environment, Maintains the general health of living organisms. Economical in many uses, ensuring continued availability and existence. Uncomplicated techniques are used. And it is constantly renewed. In addition to generating electrical energy from nature or sources that are not harmful to the environment, in some countries, such as Iraq the amount of energy they produce is insufficient, which makes them need to import energy from other countries.

On the other hand, some countries produce energy in a greater quantity than they need at certain times and do not use all the produced energy, but at other times energy is not produced with a need for it so in this paper, we will store the excess energy from some times and use it when needed there are several types of renewable energy: solar energy, bioenergy, wind energy hydropower, sustainable biofuel energy geothermal or geothermal energy and tidal energy. Wind energy is generated from wind turbines that are in a large open area. Wind energy is renewable, local energy that does not cause global warming and does not produce polluting gases such as carbon dioxide, nitric oxide or methane, so its harmful impact on the environment is minimal. the cost of wind is relatively high and requires many standards and lands to be used as wind fields that can be used for other purposes such as agriculture, grazing. Hydropower is generated from waterfalls, water rapids, and dams. It is also having limited power and it's relatively expensive. As for solar energy, it is one of the most suitable types of renewable energy in the Arab world due to the emergence of the sun for long periods and at a limited cost. Where solar cells are used, which are photovoltaic converters direct sunlight into electricity, which are semiconducting and light-sensitive devices surrounded by a front and back conductor envelope for electricity, types of the solar cell, crystalline silicon cells.

Proposed system

The proposed system includes three stages; first collection and pre-processing related to collecting the values from the sensors in real-time for both datasets (i.e., solar plant and weather), dropping any record that

asamaher@itnet.uobabylon.edu.iq

has missing values then merging both that dataset based on (primary key); the second stage focuses on split the stream of the dataset into multi-interval and remove the duplication. The final stage focuses on building the model from linear and nonlinear and compares both for implementation it with three techniques CPU, GPU, and FPGA. Figure 13.2 shows a block diagram of the proposed system.

A Wireless Sensor Network (WSN) is a network in which nodes are distributed randomly or symmetrically, depending on the requirements, to detect data and perform various actions. WSN is one of the world's largest and most widely utilized networks. Solar energy harvesting is highly common in WSN and is also utilized for energy gathering. In these WSN applications, the wireless sensor node's life cycle and performance, as well as communication channels, are critical. A sensor node is made up of four major components: a sensing unit, a transceiver unit, a processing unit, and a power unit, as well as supplementary application-specific components including a mobilizer, location detecting system, and power generator.

Sensors are one of the IoTs support technology which contains various kinds of device sensing and actuators, sensors help systems to operate in a real-time environment. Sensors are used to collect a huge amount of data which is used as an input to large data applications, therefore sensors technology is considered as an essential tool for computing and analysing data, that functions related to IoT. The analysis of huge data resulting from sensing units is performed by using Data Mining and Machine Learning for building models, specifying patterns, relation creation, and deploying the outcomes for support and make-decision.

The external effects detection on electronic sensors is divided into active and passive sensing. The active sensing will alter in reaction to external effects and generate a change in current or voltage when an excitation signal is used. In passive sensing, there is no response to external effects. Passive temperature sensing is commonly utilized such as thermocouples. Another type of sensor, which is based on electronic circuit systems, is suitable, inexpensive, and simple but it lacks when used to monitor hostile conditions such as oxidizing fluids, high temperatures, or high pressure, and they need a large amount of space.

TA40 and TA60 Sensor

The TA40 and TA60 Series air temperature sensors are effective and out-quality electronic sensors, building to determine ambient air temperature. The enhanced form using great confidence elements produces more active and precise temperature measures, the standard ranges temperature measure is (-30–50°C),. The main features of this sensor

- Microprocessor accuracy.
- Strong signal no wost for 250 m distance.
- Exchanging battery beside no need software modification
- Waterproof, strong installation, semi-conductor component

Pyranometer

A pyranometer is a sensor that converts the global solar radiation it receives into an electrical signal that can be measured. The measurement of the sun's radiation on the earth is referred to as global solar radiation. It refers to direct and diffuse radiation, the direct radiation is the amount of solar energy that reaches to surface directly from the sun above the pyranometer sensor while diffuse radiation is the reach of solar energy horizontally, that sun's light is diffused by the atmosphere, also measured by pyranometer sensor. The main features of this sensor

- Measuring both direct and diffuse radiation.
- Global radiation range (0–1400 W/m^2).

AcuDC240

AcuDC240 is a measurement of DC power, which is used for measuring controlling, getting, and saving data measured in real-time. Used for measuring significant parameters in an effective way such as current, power, energy, and voltage. The main features of this sensor

- Display data in real-time.
- Measures with high accuracy.
- The input range for direct voltage measurement is 0-1000V and for current is 0-±10A.

RCB56A1

RCB56A1 set is for the accuracy measurement: DC, AC, pulse, and irregular wave current in real-time. The main features of this sensor

- Great accuracy, regular linearity
- Wide bandwidth
- Quick reply
- overload ability

Implementation

The integration system contains multi-stages, *first stage* captures data in real-time from multi-sensors. Then merging between solar plant dataset and weather datasets. After that checking the missing values by dropping any record have missing value. *The second stage* called pre-processing includes adding some features

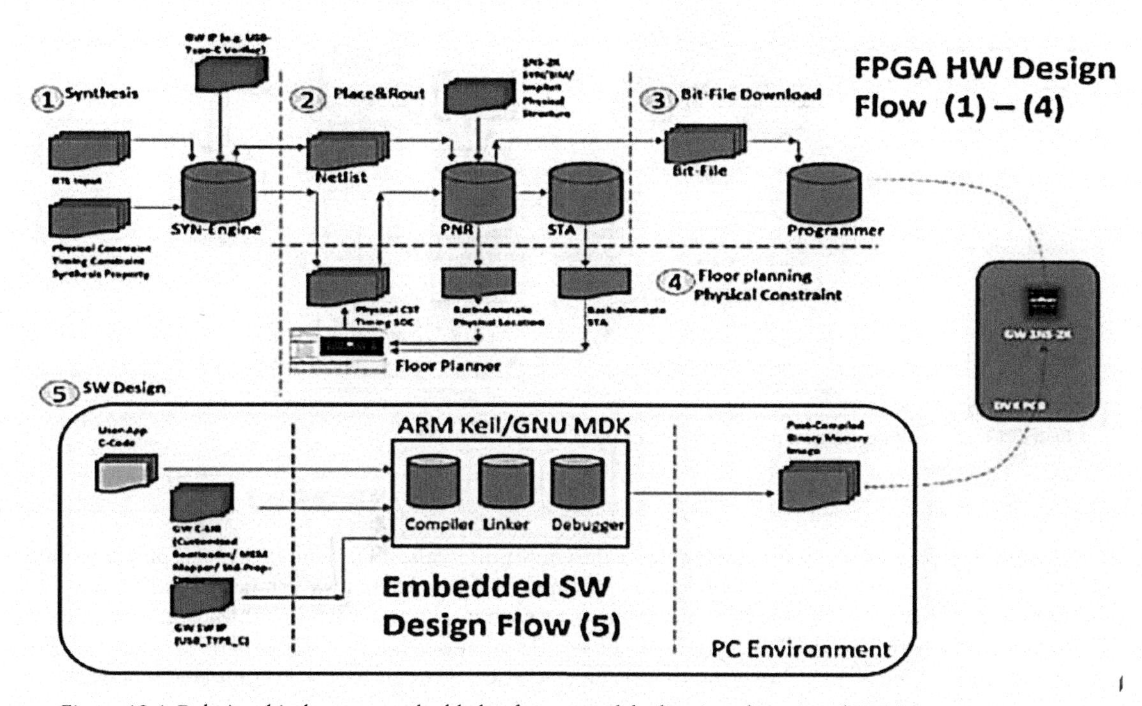

Figure 13.1 Relationship between embedded software models (linea\nonlinear) and FPGA

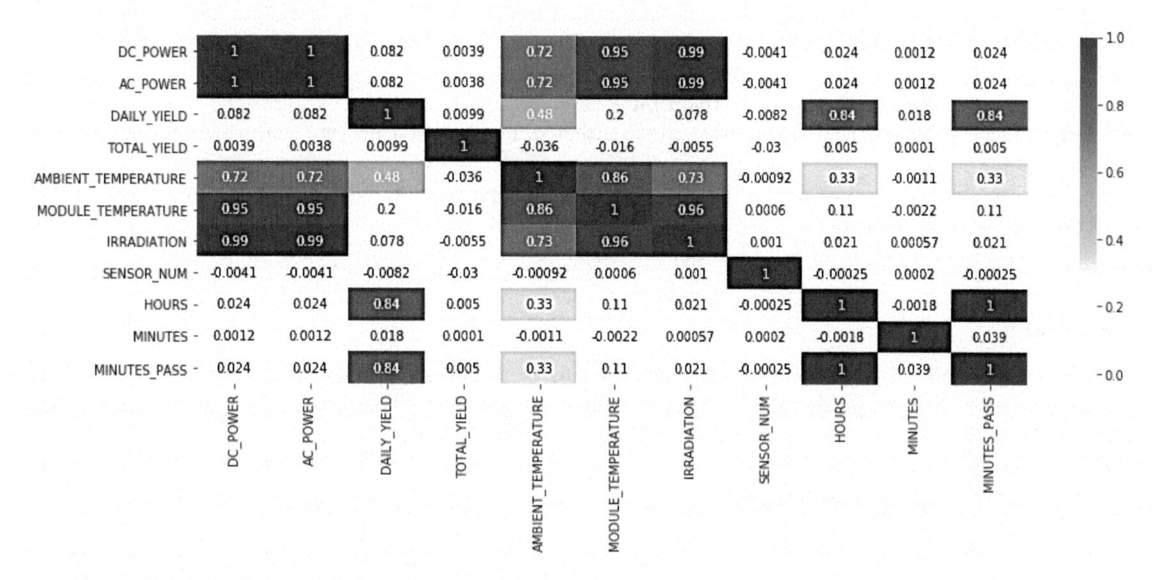

	DC_POWER	AC_POWER	DAILY_YIELD	TOTAL_YIELD	AMBIENT_TEMPERATURE	MODULE_TEMPERATURE	IRRADIATION	SENSOR_NUM	HOURS	MINUTES	MINUTES_PASS
DC_POWER	1	1	0.082	0.0039	0.72	0.95	0.99	-0.0041	0.024	0.0012	0.024
AC_POWER	1	1	0.082	0.0038	0.72	0.95	0.99	-0.0041	0.024	0.0012	0.024
DAILY_YIELD	0.082	0.082	1	0.0099	0.48	0.2	0.078	-0.0082	0.84	0.018	0.84
TOTAL_YIELD	0.0039	0.0038	0.0099	1	-0.036	-0.016	-0.0055	-0.03	0.005	0.0001	0.005
AMBIENT_TEMPERATURE	0.72	0.72	0.48	-0.036	1	0.86	0.73	-0.00092	0.33	-0.0011	0.33
MODULE_TEMPERATURE	0.95	0.95	0.2	-0.016	0.86	1	0.96	0.0006	0.11	-0.0022	0.11
IRRADIATION	0.99	0.99	0.078	-0.0055	0.73	0.96	1	0.001	0.021	0.00057	0.021
SENSOR_NUM	-0.0041	-0.0041	-0.0082	-0.03	-0.00092	0.0006	0.001	1	-0.00025	0.0002	-0.00025
HOURS	0.024	0.024	0.84	0.005	0.33	0.11	0.021	-0.00025	1	-0.0018	1
MINUTES	0.0012	0.0012	0.018	0.0001	-0.0011	-0.0022	0.00057	0.0002	-0.0018	1	0.039
MINUTES_PASS	0.024	0.024	0.84	0.005	0.33	0.11	0.021	-0.00025	1	0.039	1

Figure 13.2 Correlation of dataset after merging datasets

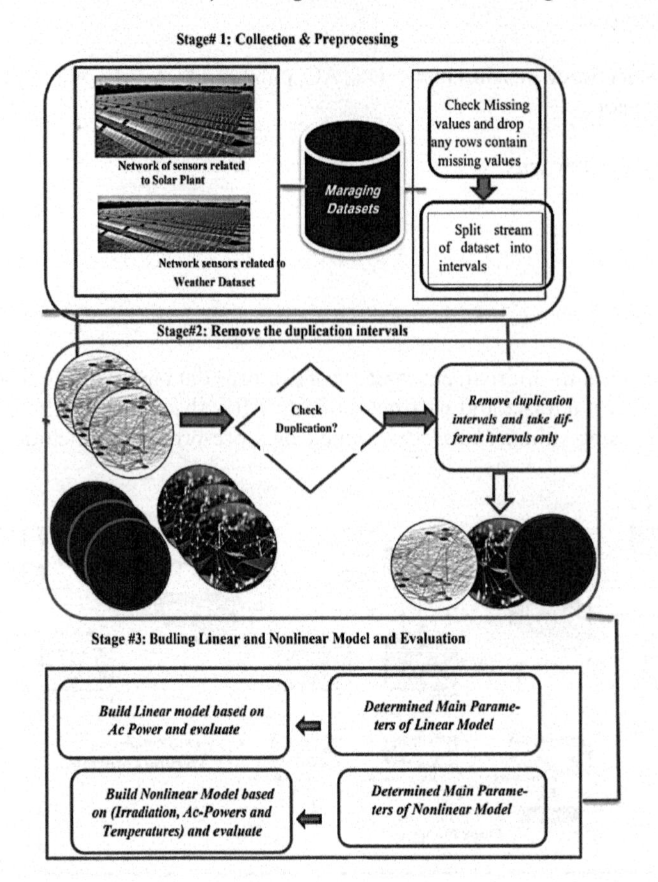

Figure 13.3 Block diagram of proposed system

that are useful in prediction, splitting the dataset into multi-intervals, then removing the duplication interval. Find the correlation among the features as explained in Fig. 3. In the third stage create LEI, and NEI models before that determining the main parameters for each model as shown in Table 13.2. Based on LEI The Relationship between DC-Power and (AC-Power\ Irradiations \Dc-Power) explain in Figures 13.5–13.7 sequentially. While; Fig. 8 shows the Dc-power generation through NEI. The comparison between the two models LEI and NEI is present in Fig. 9.

The final stage includes the implementation of the design models on three different hardwares and compares the execution time for each one as explained in Table 13.3. As a result; we can summarisation the point achieves in that paper as follow:

- Collecting the data from sensors in real-time
- Merging between solar -plant dataset and weather datasets based on two features (i.e. plant-id and date-time).
- Check missing values and drop them if found Delete duplicate intervals, and add some features that are important in prediction.
- building LEI and NEI models and evaluating them based on two points performance and time.

Conclusion

This paper presents an integrated system based on Different types of hardware to evaluation including CPU, GPU, and FPGA to predict the DC-power from the solar plant. The data contains two sets of information about two solar power plants that were acquired over 34 days:

- Power generation data include seven features (i.e., Source-key, Plant-ID, DC-power, AC-power, Yield delay, Total Yielded, Date-Time)

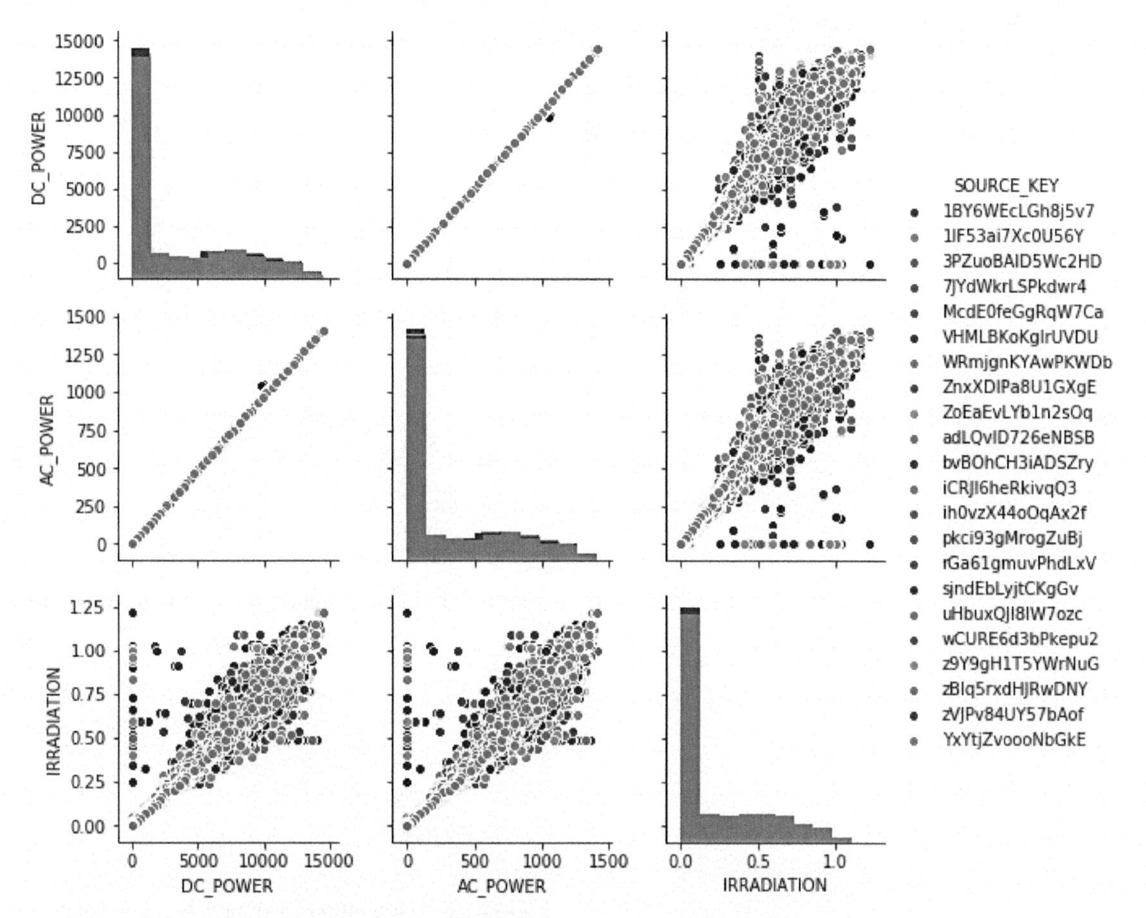

Figure 13.4 relationship among DC-power and (AC-power\irradiations\DC-power)
'popt=', array ([1.11067030e+01, -8.36585233e-10, -2.88700293e+08, -6.56012551e-02]))
'pcov=', array ([[2.95118747e-01, -3.37619553e-09, 1.17373200e+09, 6.74016006e-03], [-3.37619550e-09, 7.98285837e-14, -2.73849543e+04, -4.92535159e-10], [1.17373199e+09, -2.73849543e+04, 9.39432660e+21, 1.69319923e+08], [6.74016006e-03, -4.92535159e-10, 1.69319923e+08, 1.56495347e-04]]))

- Weather data contain six features (i.e., Source-key, Plant-ID, Irradiation, Date-Time, Temperatures, Mobility Temperatures)

After collecting both datasets in real-time maraging in the signal dataset based on both features (Date-Time and Plant-Id) to become that dataset contains nine features rather than 13. After that checking, if any record of that dataset has missing values drop it. In the second stage split the dataset into multi-intervals the time of each interval take 15 minutes; in this stage focus on removing the duplication interval and take only the different intervals to build the model. In general, the total number of intervals is 51 while after remove duplication remained only 22 intervals. Finally building two models (linear and nonlinear) and implementation based on three types of technique (CPU, GPU, and FPGA). As a result, in both models, the FPGA gives the time more reduce compare with CPU and GPU. We can summarisation the main advantages of Models as follow:

- The high correlation between DC power and AC Power
- The high correlation between power and irradiation
- Correlation between DC power, AC power, and module temperature and Ambient Temperature
- Correlation between daily yield and ambient temperature
- A nonlinear model require multi parameters must preparing but their results are best than a linear model
- Multi features affect in generated max DC-power include (AC-power, irradiance and temperature)

Table 13.2 Parameters effect in both linear and nonlinear model

Model	*Equations*	*Parameters*	*Values of P*	*Results*
Linear Model	$P(t)=a+b\cdot E(t)$	$P(t)$:DC power $E(t)$: irradiance a,b : random coefficients	a=rnd() b=rnd()	*Figure 13.5* Generation DC-power based on AC-power *Figure 13.6* Generation DC-power based on irradiations *Figure 13.7* Generation DC-power based on temperatures
Nonlinear Model	$P(t)=aE(t)$ $(1-b(T(t)+E(t)$ $800(c-20)-25)-$ $d\ln(E(t)))$	irradiance $E(t)$, Temperature $T(t)$ coefficients a,b,c,d popt, pcov	'popt 'pcov'	*Figure 13.8* Generation DC-power based on NLM

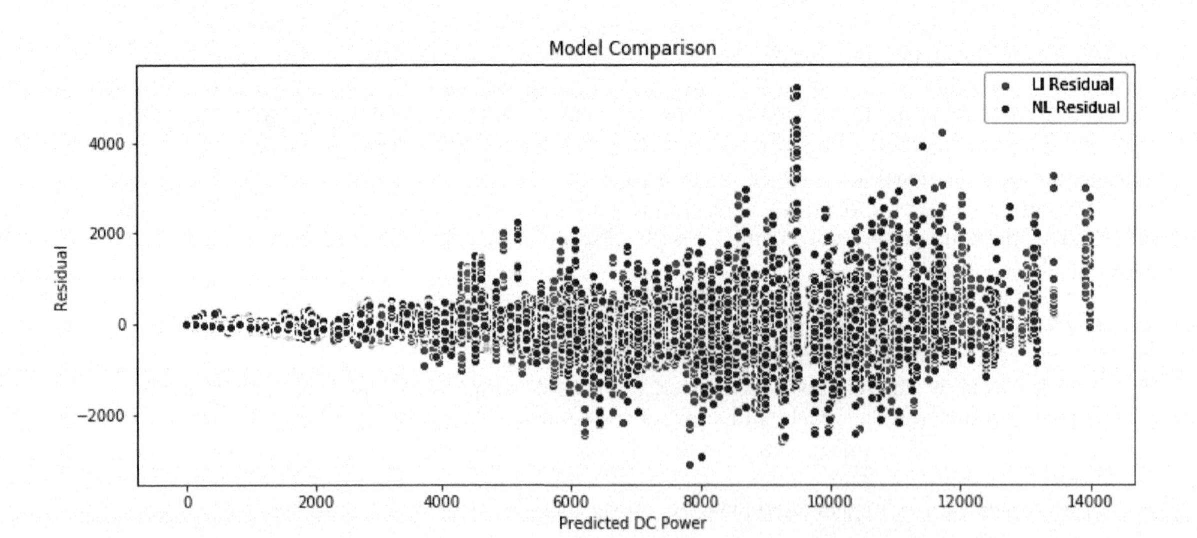

Figure 13.9 Compare between LM and NLM in generation DC-power

Table 13.3 Compare between LEI and NEI Model from time computation

Name of model	*CPU*	*GPU*	*FPGA*
LEI MODEL	12.597	5.992	1.004
NEI MODEL	23.994	17.023	1.962

References

Al-Janabi, S. (2015). A Novel Agent-DKGBM Predictor for Business Intelligence and Analytics toward Enterprise Data Discovery. JUBPAS. 23(2):482507.

Al-Janabi, S., Alkaim, A. F, Al-Janabi, E., and Aljeboree, A. (2021). Intelligent forecaster of concentrations (PM2.5, PM10, NO2, CO, O3, SO2) caused air pollution (IFCsAP). Neural Comput. Appl. 33:14199–14229. https://doi.org/10.1007/s00521-021-06067-7

Al-Janabi, S., Alkaim, A.F., and Adel, Z. (2020). An Innovative synthesis of deep learning techniques (DCapsNet & DCOM) for generation electrical renewable energy from wind energy. Soft Comput. 24:10943–10962. https://doi.org/10.1007/s00500-020-04905-9

Ang, K. L.-M. and Seng, J. K. P. (2021). Embedded Intelligence: Platform Technologies, Device Analytics, and Smart City Applications. IEEE Internet Things J. 8(17):13165–13182. doi:10.1109/jiot.2021.3088217

Basaran, K., Özçift, A., and Kılınç, D. (2019). A New Approach for Prediction of Solar Radiation with Using Ensemble Learning Algorithm. Arab. J. Sci. Eng. 44:7159–7171. doi:10.1007/s13369-019-03841-7

Bibri, S. E. (2018). The IoT for smart sustainable cities of the future: An analytical framework for sensor-based big data applications for environmental sustainability. Sustain. Cities Soc. 38:230–253. doi:10.1016/j.scs.2017.12.034

Cotfas, L. A., Delcea, C., Roxin, I., Ioanăş, C., Gherai, D. S., and Tajariol, F. (2021). The Longest Month: Analyzing COVID-19 Vaccination Opinions Dynamics from Tweets in the Month following the First Vaccine Announcement. IEEE Access. 9:3320333223. doi: 10.1109/ACCESS.2021.3059821.

Dhanraj, Joshuva, A., Ali, Mostafaeipour., Karthikeyan, Velmurugan., Kuaanan, Techato., Prem K. Chaurasiya, P. K., Jenoris, M., Solomon, Anitha Gopalan, A., and Khamphe, Phoungthong. (2021). An Effective Evaluation on Fault Detection in Solar Panels. Energies. 14:22:7770. https://doi.org/10.3390/en14227770

Dogan, A. and Birant, D. (2020). Machine Learning and Data Mining in Manufacturing. Exp. Syst. Appl. 114060. doi:10.1016/j.eswa.2020.114060

Gao, T. and Lu, W. (2020). Machine learning toward advanced energy storage devices and systems. iScience. 101936. https://doi.org/10.1016/j.isci.2020.101936

Hao, J. (2020). Deep Reinforcement Learning for the Optimization of Building Energy Control and Management. PhD Thesis, University of Denver.

Hochreiter, S. and Schmidhuber, J. (1997). Long Short-Term Memory. Neural Comput. 9(8):1735–1780. doi:10.1162/neco.1997.9.8.1735

Hossny, K., Magdi, S., Soliman, A. Y., and Hossny, A. H. (2020). Detecting explosives by PGNAA using KNN Regressors and decision tree classifier: A proof of concept. Prog. Nuc. Energy. 124:103332. doi:10.1016/j.pnucene.2020.103332

Kabir, E., Kumar, P., Kumar, S., Adelodun, A. A., and Kim, K. H. (2018). Solar energy: Potential and future prospects. Renew. Sustain. Energy Rev. 82:894900. https://doi.org/10.1016/j.rser.2017.09.094

Kaur, H. and Buttar, A. S. (2019). A review on solar energy harvesting wireless sensor network. Int. J. Comput. Sci. Eng. 7(2):398404. doi: https://doi.org/10.26438/ijcse/v7i2.398404

Khan, A., Sohail, A., Zahoora, U., and Qureshi, A. S. (2020). A survey of the recent architectures of deep convolutional neural networks. Artif Intell. Rev. 53:5455–5516. https://doi.org/10.1007/s10462-020-09825-6

Lamb, J. J. and Pollet, B. G. (2020). Micro-Optics and Energy. doi:10.1007/978-3-030-43676-6

Mathew A., Amudha P., and Sivakumari S. (2021). Deep Learning Techniques: An Overview. Advanced Machine Learning Technologies and Applications, AMLTA 2020. Advances in Intelligent Systems and Computing. In: ed. A., Hassanien, R., Bhatnagar , and A. Darwish, 1141. Singapore: Springer. https://doi.org/10.1007/978-981-15-3383-9_54

Megahed, T. F., Abdelkader, S. M., and Zakaria, A. (2019). Energy Management in Zero-Energy Building Using Neural Network Predictive Control. IEEE Internet Things J. 6(3):53365344. doi: 10.1109/JIOT.2019.2900558.

Miljkovic, D. (2017). Brief review of self-organizing maps. 40th International Convention on Information and Communication Technology, Electronics and Microelectronics (MIPRO). doi:10.23919/mipro.2017.7973581

Rustam, F., Khalid, M., Aslam, W., Rupapara, V., Mehmood, A., and Choi, G. S. (2021). A performance comparison of supervised machine learning models for Covid-19 tweets sentiment analysis. Plos one. 16(2):e0245909. https://doi.org/ 10.1371/journal.pone.0245909.

Sah, D. K., and Amgoth, T. (2020). Renewable Energy Harvesting Schemes in Wireless Sensor Networks: A Survey. Information Fusion. 63:223247. doi:10.1016/j.inffus.2020.07.005

Samaher Al-Janabi, S. and Ayad Alkaim, A. 2022. A novel optimization algorithm (Lion-AYAD) to find optimal DNA protein synthesis. Egyp. Info. J. 23(2): 271290. https://doi.org/10.1016/j.eij.2022.01.004.

Schmidhuber, J. (2015). Deep learning in neural networks: An overview. Neural Networks. 61:85117. https://doi.org/10.1016/j.neunet.2014.09.003

Sodhro, A. H., Pirbhulal, S., Luo, Z., and De Albuquerque, V. H. C. (2019). Towards an optimal resource management for IoT based Green and sustainable smart cities. Journal of Cleaner Production, 220, 1167-1179. https://doi.org/10.1016/j.jclepro.2019.01.188.

Tan, P. N., Steinbach, M., and Kumar, V. (2016). Introduction to data mining. Pearson Education India: University of Minnesota.

Wang, J., Chen, Y., Hao, S., Peng, X., and Hu, L. (2019). Deep learning for sensor-based activity recognition: A survey. Pattern Recognition Letters, 119, 3-11. https://doi.org/10.1016/j.patrec.2018.02.010

Xiangdong Zhang, X., Gunasekaran Manogaran, G., and BalaAnand Muthu, B. A. (2021). IoT enabled integrated system for green energy into smart cities. Sustain. Energy Technol. Assess. 46:101208. https://doi.org/10.1016/j.seta.2021.101208.

Zhu, J., Wang, H., Zhang, Z., Ren, Z., Shi, Q., Liu, W., and Lee, C. (2020). Continuous direct current by charge transportation for next-generation IoT and real-time virtual reality applications. Nano Energy. 73:104760. https://doi.org/10.1016/j.nanoen.2020.104760

14 An innovative predictor (ZME-DEI) for generation electrical renewable energy from solar energy

Zainab K. Al-Janabi and Samaher Al-Janabi[a,*]

Department of Computer Science, Faculty of Science for Women University of Babylon, Babylon, Iraq

Abstract

Due to the technological development that occurred in different areas of life, the world and the environment became more vulnerable to different types of pollution, such as the increase in the emission of carbon dioxide and gases that contain different toxicity rates, especially in industrial areas and dense residential areas. Therefore, renewable energy (RE) is the best solution to this problem. In general; RE is infinite and natural energy which can be called also environmentally friendly, clean energy, or green energy that meets a clean, pollution-free climate, and supplies people with a novel cost-efficient kind of energy. There are multi types of renewable energy resources (RER), biomass energy (bioenergy) it is the transformation of biomass into the interesting format, like heat, electricity, and liquid fuels. Geothermal energy is an efficient extraction of the earth renewable energy. In hydropower, falling water transformed into energy by utilization turbines. Marine energy (ocean) extract from six roots: waves, tidal range, tidal currents, ocean thermal energy conversion, and salinity gradients, each has the various root and needs various technics for transformation. Solar energy extraction requires, utilisation of sun's energy to fit hot water by solar thermal systems or electricity by solar photovoltaic (PV) and concentrating solar power (CSP) system. This paper designs an integration model called "Sero to Max Energy Predictor Model Based on Deep Embedded Intelligence Techniques (ZME_ DEI)" based on the concept of deep embedded intelligent techniques. To determine the main rules for each unit that are effective in generating the maximum electrical energy based on the nature of each dataset. ZME-DEI model contains multi-stages, first stage includes many steps; first step captures data in real-time from multi-sensors. The second step makes merging between solar plant and weather datasets. Then the third step checks missing values. While; second stage considered of pre-processing contains four steps such as deleting duplicate, adding some features that are useful in prediction, split capture datasets after merging into multi-intervals each interval containing the reading through 15 minutes. In the third stage, the ZME- DEI model creates concerning knowledge constraints and adopts gradient boosting techniques through replace the kernel of GBM represent by DT by new kernel based on multi-objective functions. The dataset divides into two subsets using ten cross-validation methods, training used to construct the ZME-DEI model, and a testing set for evaluating it. The final stage includes evaluation results based on three measures such as coefficient of determination (R2), root mean square error (RMSE) and mean error (ME), furthermore max energy generation (MEG).

Keywords: Deep embedded intelligent, gradian boosting machine, multi objective optimization, optimisation, renewable energy.

Introduction

The natural source of energy can cover 50% of the total need the world of energy if it used in the prefect way. The natural source can split into solar, wind, water, act. Machel mentions about the use of resource energy could reduce premature mortality rate, waste job days, and minimise the total costs for healthcare (Vinoth et al., 2020). Thus, the changing to RE, economic, reduce air pollution with harmful gases in addition to that could assist to employ a large number of workers.

The energy that is out from RER is changeable (non-dispersible, interrupted, and unreliable) (Omar et al., 2014) also the lack of sufficient spaces and extensive experience in this field, in addition to the lack of an extensive feasibility study for the implementation of this technology for electric power generation in Iraq, are the obstacles facing this technology. We focus in this work on the two renewable energy resources, (1) solar, (2) wind energy, the reason for using (1) is unlimited, free, and do not cause air or water pollution the reason of (2) is potentially cheap construction and also do not cause air or water pollution (Omar et al., 2014). This study offers, RE filed, and the characteristics of the significant resources, design an optimiser based on embedded intelligence system EI-FPGA integrated circuit, to maximise the generation of electric power, and the work applies on smart microgrid model.

Embedded intelligent system (EI) is a nascent project domain, combining machine learning algorithms such as ("machine learning and neural networks, deep learning, expert systems, fuzzy intelligence, swarm

[a]samaher@itnet.uobabylon.edu.iq

intelligence, self-organizing map and extreme learning") and smart resolution-produce abilities into movable and implanted devices or systems (Seng et al., 2021). IoT is the contraction of appliances, software, sensors, operators, and physical objects are embedded in the WSN, vehicles, house devices, and other outputs that aid these objects to transmissions and data sharing. The IoT is rapidly growing with the recent evolutions in wireless technology and embedded devices, with lower energy Microcontrollers that have been evaluated that are typical for IoT's remotely located in separated areas to bind and operate for the large period that need not repairing (Yasin et al., 2021). Many implementations that could take advantage of EI are in independent systems and edge computing. The challenges of edge computing for independent driving systems to have the ability to carry out great computational power to satisfy the energy-efficient field that is responsible for ensuring the safeness systems on independent cars. The edge computing systems for independent driving want to own the ability to handle vast data from multi-sensing and safeness systems in existent time. We mention various challenges that should handle to satisfy efficient EI application in hardware like, the requirements for high Computational processing (it is significant since algorithms for EI become so complicated; cost efficiency(to satisfying the need to cost-effective embedded devices against its low batteries power); and scalability to adjust various nets, dimensions and topologies(to solutions detection at EI appliances that make the algorithms and techniques able to be achieved inside architectures whose malleable and able to be scale to conform the computational demands and hardware resources whose ready for applicable) (Seng et al., 2021).

'Many countries in the world suffer from a clear lack of electricity production, which has led most countries to turn towards producing energy from natural sources or environmentally friendly sources that do not cause the emission of carbon dioxide gas while not causing pollution to the environment. The problem of producing electrical energy from environmentally friendly sources with high efficiency and low cost is one of the most important challenges in this field' (Al-Janabi et al., 2020). Therefore; this paper will design an integration model based on the concept of deep embedded intelligent techniques. To determine the main rules (constraints) for each unit that are effective in generating the maximum electrical energy based on the nature of each dataset. Added to that, *Zero to Max Energy Predictor Model Based on Deep Embedded Intelligence Techniques (ZME- DEI).*

The reminded of this paper is organised as follow: section number one shows the introduction, the second section presents the related work, the third section explains the main tools, the fourth section explain the main stages related to ZME-DEI, fifth section implementation of (ZME-DEI), section sixth show how ZME-DEI suitable for the increase in the production of electrical power compared to other comparable techniques? The final section presents the main conclusion of that work.

Related Works

Many works attempt to handle the same problem present in section number on and the authors of those works used different techniques as follow:

Soydan (2020) present method to find the perfect location for installing the solar energy plants (SEP), based on analysis eleven layers include ("sunshine duration, solar radiation, slope, aspect, road, water sources, residential areas, earthquake fault line, mine areas, power line, and transformers") through geographic information system (GIS) and analytical hierarchy process (AHP). The result finds the best location for 80% of SEP and fails to find the best location of 20% from SEP. These results evaluated based on consistency ratio (CR) come from divided consistency indicator (CI) on random indicator (RI). This work differs from the evaluation measures (EM) and technologies of our work while using the same type of renewable energy resource (RER) (i.e., solar plant).

Guozhou et al., (2021) presented a method to build a dynamic system for generating renewable energy (RE) in real-time with considering the uncertain state, for this hybrid-wind-solar model (HM) was built based on optimal control and multi targets, based on analysing historical data (HD) of daily wind and solar energy through deep reinforcement algorithm (DNN). The model could be used to reduce costs. This result is evaluated by control policy and cost (C) This work is different from our work by using uncertainty and measures while similar by using solar and wind plants and dynamic systems.

Ahmed et al., (2020) present method to build math model for maximise supply to satisfy demand, based on optimiser for deterministic and stochastic methods to simulate micro-grid models (MG). The result is evaluated by the minimum cost energy and total cost when the PV solar used, This work different from our work in using DSM and technologies used, while similar in using VP solar plant and predicate economic feasibility(EF).

Razmjoo et al., (2021) presented model of hybrid based on gathering data from the meteorological organisation (MO) and handled using HOMER software. The result is evaluated by pointers like policies and investments that indicate minimum emissions, maximum energy and good investment. This work is different by using hybrid model and evaluation measures while similar by using the same plant such as PV solar and wind.

Bahareh et al., (2021) presented method to find the best ordering of three RER to be the first one to take on consider and with little constraints, based on massive analysis of five groups ("1. Economic and financial, 2. Social, cultural, and behavioral, 3. Political & regulatory, 4. Technical, and 5. Institutional") using AHP method, The result is solar PV that has first priority with little constrains and then, wind and biomass. This work similar with our work by using the same plants accept one, and different in the measures and techniques.

Kaabeche and Bakelli, (2019) presented method to find the best performance of four algorithms like("ant lion optimiser algorithm (ALO), grey wolf optimiser algorithm (GWO), krill herd algorithm (KH) and JAYA algorithm') according to less unit electricity cost (UEc) corresponding to simple less power supply probability (LPSp),that applied to solve optimisation problems, based on analysing data set for daily climate, and building hybrid solar/wind system that, and using three battery technologies, for forecasting the good performance many and massive analysis have been used considering the effect of battery life, depth of discharging(DOD) and approximate battery cost on UEc. The result show that JAYA algorithm is the best. This work different from our work by using different alg. and measure while similar in using the solar PV and wind resource.

Shin'ya et al., (2021) proposed model to minimise cost generation by arrange the RER around places connecting and exchanging power through WAN to handle current changing, analysing 7 area included 42 city based on GA algorithm, use heat storage tank (HST) as an energy storage instead of battery that is expensive, the result is minimum Cg and maximum system efficiency evaluated through capacity and economic efficiency. This work is different from our work in evaluation measures and techniques while similar in using the solar and wind plant.

Aziah et al., (2020) presented method to find the best place to build the appropriate type of RER for these places based on analysing the Sarawak map by using clustering and selection and then image segmentation included (color thresholding, circular though transform, and K-means) and regional techniques. The result determine nine from 420 place that is suitable for building hybrid energy (solar PV and hydropower) these result evaluated by HOMER program and the total cost. This work different from our work in the technique used and measures while similar in using the solar PV plant.

Utkucan (2020) suggested a method for predicting total renewable and hydro energy installed capacity (TR-HEIC) and electricity generation (EG) in Turkey from 2019 to 2030, based on a model for analysis data set from 2009 to 2018 through the ("fractional nonlinear grey Bernoulli") model (FANGBM). The result predicts that the TR-HEIC and EG will be minimised from 2019 to 2030 this is evaluated by the mean absolute percentage error (MAPE) and the accuracy (A). This work is similar to our work by using the accuracy measure and different in using hydropower and hydro energy and other measure and techniques.

Ning et al. (2021) presented method to minimise the TC for setting MG included S&D technique and the energy wasted, a model based on analysis unrealistic dataset through hybrid gravitational search and pattern search (GSA-PS) algorithm. The result finds that using such technologies is high performance than others and the production cost of (GSA-PS) algorithm greatly less. This work is different from our work in using HRE, unrealistic data and different techniques while similar in using micro-grid.

Main Tools

A. Internet of Things (IoT)

In SC, IoTs provided excellent facilities for linking different smart appliances with the internet furthermore, IoTs provided facilities in many disciplines such as academia, industry, healthcare, and business (Xiangdong et al., 2021). In an intelligent city climate, a smart grid is attached to the IoT's platform which stand-alone mode and pervasive everywhere, that assist to specify the delivery and distribution of energy (Megahed et al., 2019).

The development of SC in actual-time is very complex since, the small-scale, energy consumption, and ability of appliance's sensors. Furthermore, the hardness of controlling and detecting battery lifespan of various sensors-base appliances. Therefore, appears of needing to ease the SC program that is achieved by IoT (Sodhro et al., 2019).

IoT is a network of interconnected objects or appliances that can interact via the Internet to gather, transport, and share data. While the IoT creates a lot of data and requires a lot of capacity to distribute it for various applications like SC, cloud computing services aid by storing and centralisation processing data (Ang and Seng, 2021).

Suffering from the slow process of gathering and sending data for processing in a centralised location, before that the need for handle and analysis raw data to obtain pure data which useful for a specific implementation, those all are challenges need to solve. But in decentralised processing and enables IoT technologies and mobile computing, edge computing is required for processing data close to IoT appliances or objects sources rather than sending data to cloud server for analysis (performed in real-time), so this will reduce latency and bandwidth usage. Another benefit is that the energy needs for sending and receiving are also reduced between appliances and servers (Ang and Seng, 2021).

B. Solar plants

Solar energy is the most extensively utilised environmental energy source for its ready availability and steady energy scavenging capabilities throughout the day, but it has the disadvantage of being inaccessible at night or in adverse weather conditions. If energy is retrieved properly, harvested energy can even persist through the night. Solar energy is among the most well-established, well-known, and advantageous renewable energy sources (Kaur and Buttar, 2019).

Recently the dependence on clean energy generation from solar and wind becoming the main goal to satisfy, because of the obvious shortage in Fossil Fuel Resources (FFR) and environmental degradation (Megahed et al., 2019).

Fast alteration in the photovoltaic industry, technology, and institutional teams over the last decennium has changed PV's economical feasibility and significant transformation of the energy industrial competitive. Now, solar energy and photovoltaics become universal, many billion-dollar industries giving minimum power cost to millions of individuals across the world in the markets which wide and rising (Vinoth et al., 2020).

The collection and usage of light and/or heat energy created by the Sun also uses passive and active methods required to maximise the gain, which is considered the entirety of the solar energy harvesting concept. Mostly, active technology is divided into two classes photovoltaic and thermal technology.

Recently, photovoltaic active technology which uses semiconductor materials to convert photons into electrons is a popular choice. Solar thermal technology converts solar energy into thermal energy for home and/or commercial uses as drying, heating, pooling, cooking, and so on. On a larger scale, Concentrated Solar Thermal (CST) and Concentrated Solar power meet heating needs, while CSP technologies are used to generate energy. CSP uses big mirrors before converting it to heat energy (Kabir et al., 2018).

C. Sensor

A wireless sensor network (WSN) is a network in which nodes are distributed randomly or symmetrically, depending on the requirements, to detect data and perform various actions. WSN is one of the world's largest and most widely utilised networks. Solar energy harvesting is highly common in WSN and is also utilised for energy gathering (Gao and Lu, 2020). In these WSN applications, the wireless sensor node's life cycle and performance, as well as communication channels, are critical. A sensor node is made up of four major components: a sensing unit, a transceiver unit, a processing unit, and a power unit, as well as supplementary application-specific components including a mobiliser, location detecting system, and power generator (Kaur et al., 2019).

Sensors are one of the IoTs support technology which contains various kinds of device sensing and actuators, sensors help systems to operate in a real-time environment. Sensors are used to collect a huge amount of data which is used as an input to large data applications, therefore sensors technology is considered as an essential tool for computing and analysing data, that functions related to IoT. The analysis of huge data resulting from sensing units is performed by using Data Mining and Machine Learning for building models, specifying patterns, relation creation, and deploying the outcomes for support and make-decision (Bibri, 2018).

The external effects detection on electronic sensors is divided into active and passive sensing. The active sensing will alter in reaction to external effects and generate a change in current or voltage when an excitation signal is used. In passive sensing, there is no response to external effects. Passive temperature sensing is commonly utilised such as thermocouples. Another type of sensor, which is based on electronic circuit systems, is suitable, inexpensive, and simple but it lacks when used to monitor hostile conditions such as oxidizing fluids, high temperatures, or high pressure, and they need a large amount of space (Lamb et al., 2020).

D. Embedded Intelligent (EI)

The EI is a nascent project domain, combining machine learning algorithms and smart resolution-produce abilities into movable and implanted devices or systems (Seng et al., 2021). There exist many challenges to do to achieve efficient EI realisation in a device such as a requirement for high computational processing, effective cost, and scalability to accommodate different networks sizes and topologies. The computations challenge is distinguished as the complexity of Embedded Intelligent algorithms and techniques. For example, there is a need for high computations in Deep Neural Networks (DNNs) that use multilayer networks to extract high-level features. The effective cost is required for satisfying the request of cost-effective embedded devices with a limited energy supply. The scalability challenges permit the algorithms and techniques to be implemented within architectures that are adaptable and scalable to fit the target computational conditions and hardware (Seng et al., 2021).

E. Prediction Techniques of Deep Embedded Intelligent from side Data Mining

Prediction is the primary method for energy harvesting systems. Since renewable energy is uncontrollable, so prediction estimation gives approximate efficient outcomes. Solar energy harvesting is highly common in WSN and is also utilised for energy gathering (Gao and Lu, 2020).

As a result of the existing and need for analysing huge data to extract the interest information that is used in various applications, DM has the main role in this domain. DM has more attention in knowledge extraction from the huge amount of data (Tan et al., 2016).

1) Decision tree (DT)

The DT is a series of logical and mathematical processes based on probabilistic in determining the class label of test records. DT principal working as looking on and comparing the consistency of specific attributes and the threshold node. And it makes classification by assigning the class that occurs more frequently to the test records (Hossny et al., 2020). DT collects data in the form of a tree and maybe reformed as a collection of discrete rules. The building block of decision trees includes a root node and a lot of internal branching nodes. The class shown in a leaf node corresponds to an example, whereas features and branching reflect the consistency of features that leads to the classification being performed in internal nodes. The good way of building DT from the training set reflects the effectiveness of DT (Rustam et al., 2021).

2) Random forest (RF)

The RF is used for classification and regression problems. RF is an ensemble model that utilises bagging techniques which creates many trees and makes voting. Creating a large number of trees will fit prediction accuracy. RF can handle over-fitting. RF is described as the type of the various prediction trees (Rustam et al., 2021). RF is more robust in the election of the training set as compared to the decision tree classifier, RF is hard to interpret, however, its hyper-parameters can be turned with simple (Cotfas et al., 2021).

3) XGBoost (eXtreme Gradient Boosting)

The XGBoost is an advanced version of gradient boosting tree growth to concentrate on computational speed and model efficiency. To fit training data XGBoost assume an initial value always 0.5. It builds its regression tree using the individual tree, that fits residuals. A leaf node is used to begin the tree, all residuals are put on a leaf then computing similarity score, and then compute the gain to specify how to split. Based on these scores and the gain, the XGBoost chooses the largest gain assumed threshold for pruning (Hao, J., 2020). The XGBoost has much popularity as considered a tree-based model. The XGBoost provides a speed boost since trains the number of poor students (DT) parallelly, different from the gradient that does this sequentially. The XGBoost can handle over-fittings, which is unfeasible with Gradient Boosting also Adaboost classifiers. The XGBoost can be implemented on a distributed system also can process larger datasets, as a result, it has scalability features. It helps to reduce the loss and enhance accuracy by utilizing a Log Loss function (Rustam et al., 2021).

4) Extra Tree Classifier(ETC)

The ETC is an ensemble learning model same as RF. Increases prediction accuracy by using the meta-estimator, which trains on different samples of the dataset a large number of weak classifiers (i.e. DT). The way of building a tree is different between ETC and RF. ETC creates DT using the original training sample, whereas RF uses samples from the entire dataset. At each iteration, the tree is given a random sample of attributes from the dataset on every test node. Based upon Gini Index criteria the optimal feature must be determined by DT to partition data. Several de-correlated DTs are created because of provided random sample (Rustam et al., 2021).

5) Tree Net

Tree Net is a specific processing loop of classification and regression tree (CART), including forwarding strategy for providing a series of binary trees. Do not use randomness in selecting variables and bootstrap sampling for transformation to the optimal accuracy. The number of iterations is reduced, to control over-fitting. At each iteration, the current tree is used to fit a residual of the previous one. Tree Net performance is influenced by the restricted wrong size of a tree changes since it does not handle the relationship between variables. Furthermore, with a small number of samples, a weak prediction can occur. Additionally, no need for data conversion or avoidance of outliers and interaction effects among predictors automatically managed (Al-Janabi, S., 2015).

F. Prediction Techniques of Deep Embedded Intelligent from side Neural Network

Deep neural networks (DNN), are the fundamental building blocks of deep learning. Those techniques have allowed significant growth in many fields like sound and image processing, as well as facial recognition, speech recognition, automated language processing, computer vision, text classification (Schmidhuber, J., 2015).

Main prediction techniques from side Neural network and it categorised as deep embedded intelligent prediction techniques.

1) Convolutional Neural Networks (CNN)

The CNN uses multi-stage to extract many features that could automatically recognise the representation from data. Show high ability in machine vision techniques and image processing. Can exploit the local or time correlation between data. Consists of a set of convolution layers, non-linear processing units, and subsampling layers. The CNN is a feed-forward algorithm, has a hierarchical learning model, multiple tasks, and sharing weight. It's lacking interpretation and explanation. Not dealing with noise, and may lead to misclassification. Needing a huge of training data to learn. Selection of the hyper-parameter and the little change in its values can affect CNN performance. There are various CNN architectures like LeNet, AlexNet, VGGNet, GoogleNet, ResNet, ZFNet. (Khan et al., 2020).

2) Long Short Term Memory (LSTM)

The LSTM is a spatial kind of Recurrent Neural Network (RNN), results from back-propagation of errors that is infinite inside the cell the ability of LSTM to bridge the long temporal interval. It can deal with noise and distributed forms and continuous values. Can be segeneralised well. The time complexity for LSTM is O(1) for each weight and step, so it is considered local in both time and place. Suffer from problems that same to that face the feed-forward. Do not rely on random weight estimation but use small weight initialisation (Hochreiter and Schmidhuber, 1997; Al-Janabi et al., 2021).

3) Multinomial Naive Bayes (MNB)

The NBC is a probability classification based on the strong assumption classify data that attributes are autonomous from each other. Although its simplicity, these Bayes theorem is quick, accurate, and reliable in a variety of natural language processing (NLP) classification jobs (Cotfas et al., 2021). Handle noise point. Can process missing values during model construction. Robust to unrelated features. Correlated features can damage the performance (Tan, 2016; Ardianto et al., 2020).

3) Support Vector classifier (SVC)

The SVC is a linear supervised learning algorithms model, which gives data points to each object within n-dimension, the variable n denotes the attribute's number. SVC finds the best hyper-plane that separates between the points, so performs a binary classification that suffers few from over-fitting. Besides, it can perform multiple classifications by combining multi-binary classification functions. Furthermore, it can perform other tasks like regression and outlier detection (Rustam et al., 2021; Cotfas et al., 2021).

4) Self-organising map (SOM)

The SOM employs unsupervised, competitive learning to output low dimensional, discretised descriptions about given high dimensional data, and maintains similarity relations among the given data items at the same time. That low-dimensional representation is known as a feature map. SOM is an individual-layer NN including units set together with an n-dimensional grid. Hexagonal grids, some three or higher dimensional spaces are used from Multiple applications. SOMs provide low-dimension projection images to distribution high-dimensional data, under preserving similar relations between data items, two-dimensional and rectangular grids, are the most applications used (Miljkovic, 2017; Mahdi and Al-Janabi, 2019).

Build Max Energy Predictor Model Based on Deep Embedded Intelligence Techniques (ZME-DEI)

This section presents a predictor named zero To Max Energy Predictor Model Based on Deep Embedded Intelligence Techniques (ZME_ DEI) to predict DC-POWER generation.

The ZME-DEI model contains multi-stages, stage number one captures data in real-time from multi-sensors. The second step makes merging between solar plant and weather datasets. Then the third step checks missing values.

In stage two pre-processing contains four steps such as deleting duplicate features, adding some features that are useful in prediction, split capture datasets after merging into multi-intervals each interval containing the reading through 15 minutes.

In the third stage, the ZME- DEI model creates concerning knowledge constraints and adopts gradient boosting techniques through adding multi-objective functions to develop it. The dataset divide into two subsets using ten cross-validation methods, training used to construct the ZME-DEI model, and a testing set for evaluating it.

Table 14.1 Deep embedded intelligent prediction techniques from side data mining

Techniques	*Advantage*	*Disadvantage*
DT [Rustam et al., 2021]	• Simple and easy to build and interpret. • Multi-stage decision-making. • Its performance depends on the way of building on the training set.	• Only one feature is examined at each node. • Suitable only for a limited number of features.
RF [Cotfas et al., 2021] [Rustam et al., 2021]	• Used to solve both classification and regression problems. • Reduces the over-fitting problems. • Robust with selecting training samples. • Provide more accurate results as compared to another classifier.	• Hard to interpret, however, its hyper-parameter is an easy turn.
Xgboost [Rustam et al., 2021]	• Used for classification and regression. • Popular and fits the number of DT parallelly, unlike GBC which does this sequentially. • Control over-fitting • Process large datasets. • Scalability feature. • Use log loss function to enhance accuracy	• Steps backward require to fix over-fitting. • The computational time required for large datasets.
ETC [Rustam et al., 2021].	• Higher prediction accuracy by implementing meta-estimator. • Generate DT from the original training sample. • Same to RF classifier in which both ensembles learning model. • Different from RF in the way of building trees. • Based on math Gini index criteria, it selects the best feature to split the data.	• Several de-correlated DTs are created because of provided random sample.
Tree Net [Al-Janabi, S.,2015].	• Using forwarding strategies to create binary trees. • Do not use randomness. • Control the growing tree. • Avoid over-fitting. • No need for transformation or outlier prevention. • Dealing with the interaction between predictors automatically.	• Weak prediction can occur if the number of samples is small. • Does not consider the relationship between variables.
GBM [Al-Janabi, S., 2015].	• One of the more powerful ML algorithms for prediction. • Uses an additive or average model for boosting the error (loss function). • Uses many weak learner models for each step, the new model tries to minimize the error of the previous model. • The accuracy is the best because its uses a learning rate value (< 1) for reducing the contribution of each BRT.	• Work in a sequential style therefore it is slow in processing and analyzing data. • High computation and time complexity.

Table 14.2 Deep embedded intelligent prediction techniques from side neural network

Prediction t.	*Advantage*	*Disadvantage*
CNN [Khan et al., 2020]	Using multi-stage extraction features. Exploit the local or time correlation between data. Hierarchical learning model. Multiple tasks, and sharing weights.	Hard to interpret and explain. Unhanding noise, and may lead to misclassification. Needing a huge of training data to learn. Hyper-parameter choosing is more sensitive. Ineffective when used to estimate the object's location, orientation, and pose.
LSTM [Al-Janabi et al., 2021]	The ability to bridge the long temporal interval. Can deal with noise, distributed forms, and continuous values. Can be generalised. The time complexity is O(1) for each step. Local in both position and time and space.	Suffer from problems that same to that face the feedforward. do not rely on random weight estimation. using small weight initialisation.
MBC [Cotfas et al., 2021]	Simple, quick, accurate in NLP classification tasks. Handle noise. Robust for unrelated features. Process missing value.	Correlated features can damage its performance.
SVM [Rustam et al., 2021]	binary classification and multi-classification. finding fittest hyper-plane. regression and outlier detection task.	suffer few from over-fitting.
SOM [Miljkovic D., 2017]	low dimensional, discretised descriptions about given high dimensional data, and maintains similarity relations among the given data items at the same time.	Weak prediction can occur if the number of samples is small. Does not consider the relationship between variables.

The final stage includes evaluation results based on three measures such as coefficient of determination (R2), Root Mean Square Error (RMSE), and Mean Error (ME), furthermore Max Energy Generation (MEG), and Accuracy(A). In Figure 14.2 there is illustrated the sequential stages of the ZME-DEI block diagram, and in (1) the algorithm for building it. The summarisation of this study can observe below:

- Collecting the sensor's reading from solar and weather plants.
- Merging between datasets.
- Check missing values that affect prediction
- Delete duplicate features, and adding some features that are important in prediction, and split readings into intervals, all this presents in pre-processing stage.
- ZME-DEI new predictor building that related to gradient boosting techniques with multi-objective function to obtain good accuracy results.
- Evaluation results of ZME-DEI based on five measures.

Solar power generation forecasting has recently been developed using a variety of intelligent-based forecasting methodologies. However, several parameters like humidity, ambient temperature, Global Horizontal Irradiance (GHI kW/m^2), Direct Normal Irradiation (DNI kW/m^2), cloud variation, seasonal change, and so on impact the accuracy of PV output forecasts. To test the accuracy of the Renewable Energy Forecast (REF) function (F_{t+1}) in predicting renewable energy at time t. The traditional model comprises independent factors for renewable energy forecasting ahead of time (t + 1), whereas the suggested cascaded auxiliary model identifies the model with observation variables and renewable energy forecasting ahead of time (t + 1) (Amir and Khan, 2021).

$$\mathrm{h}(\mathrm{argmin}_h L(y_{t+1}), \mathrm{argmin}_g L(\mathrm{REF}_{t+1}) \quad (1)$$

$$h(\mathrm{argmin}_h L(y_{t+1}), R^*(\mathrm{F}_{t+1}) \quad (2)$$

$$h(F_{t+1}), \mathrm{R}^*(\mathrm{F}_{t+1}) \quad (3)$$

As a result, a comparison is made between the two models to anticipate overall power generation Y_{t+1}. The observable value confidence intervals and the predicted renewable energy prediction. The optimum function f for estimating renewable energy forecasts (REF) is:

$$f(F_{t+1}) = f^* = \mathrm{argmin}_f L(y_{t+1}) \quad (4)$$

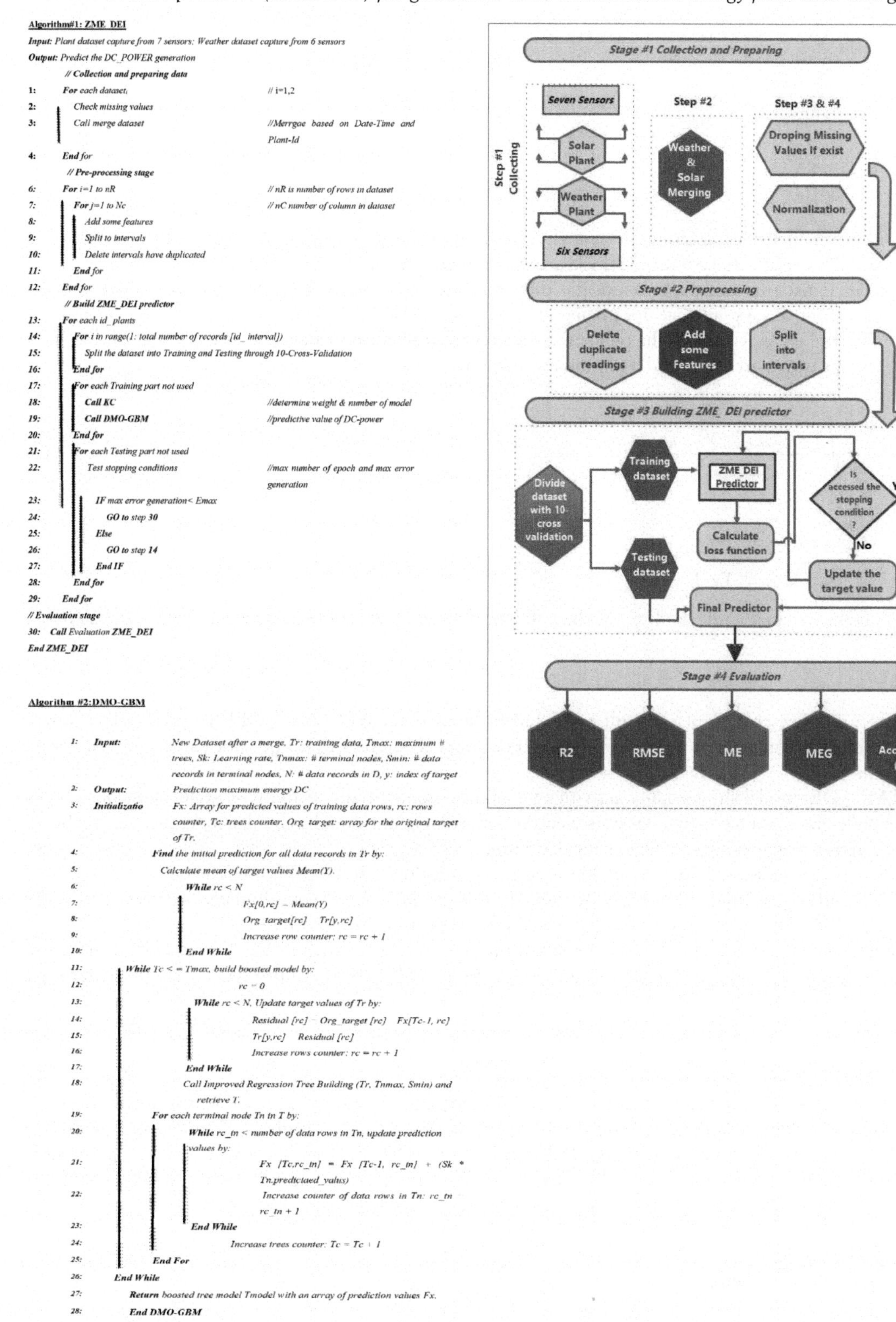

Algorithm#1: ZME DEI
Input: Plant dataset capture from 7 sensors; Weather dataset capture from 6 sensors
Output: Predict the DC_POWER generation
// Collection and preparing data
1: For each dataset // i=1,2
2: Check missing values
3: Call merge dataset //Merrgae based on Date-Time and Plant-Id
4: End for
// Pre-processing stage
6: For i=1 to nR // nR is number of rows in dataset
7: For j=1 to Nc // nC number of column in dataset
8: Add some features
9: Split to intervals
10: Delete intervals have duplicated
11: End for
12: End for
// Build ZME_DEI predictor
13: For each id_plants
14: For i in range(1: total number of records [id_ interval])
15: Split the dataset into Training and Testing through 10-Cross-Validation
16: End for
17: For each Training part not used
18: Call KC //determine weight & number of model
19: Call DMO-GBM //predictive value of DC-power
20: End for
21: For each Testing part not used
22: Test stopping conditions //max number of epoch and max error generation
23: IF max error generation< Emax
24: GO to step 30
25: Else
26: GO to step 14
27: End IF
28: End for
29: End for
// Evaluation stage
30: Call Evaluation ZME_DEI
End ZME_DEI
Algorithm #2:DMO-GBM
1: Input: New Dataset after a merge, Tr: training data, Tmax: maximum # trees, Sk: Learning rate, Tnmax: # terminal nodes, Smin: # data records in terminal nodes, N: # data records in D, y: index of target
2: Output: Prediction maximum energy DC
3: Initializatio Fx: Array for predicted values of training data rows, rc: rows counter, Tc: trees counter. Org target: array for the original target of Tr.
4: Find the initial prediction for all data records in Tr by:
5: Calculate mean of target values Mean(Y).
6: While rc < N
7: Fx[0,rc] – Mean(Y)
8: Org target[rc] Tr[y,rc]
9: Increase row counter: rc = rc + 1
10: End While
11: While Tc < = Tmax, build boosted model by:
12: rc = 0
13: While rc < N, Update target values of Tr by:
14: Residual [rc] – Org_target [rc] Fx[Tc-1, rc]
15: Tr[y,rc] Residual [rc]
16: Increase rows counter: rc = rc + 1
17: End While
18: Call Improved Regression Tree Building (Tr, Tnmax, Smin) and retrieve T.
19: For each terminal node Tn in T by:
20: While rc_tn < number of data rows in Tn, update prediction values by:
21: Fx [Tc,rc_tn] = Fx [Tc-1, rc_tn] + (Sk * Tn.predictaed_valus)
22: Increase counter of data rows in Tn: rc_tn rc_tn + 1
23: End While
24: Increase trees counter: Tc = Tc + 1
25: End For
26: End While
27: Return boosted tree model Tmodel with an array of prediction values Fx.
28: End DMO-GBM
Stage #1 Collection and Preparing
Step #1 Collecting
Seven Sensors
Solar Plant
Weather Plant
Six Sensors
Step #2
Weather & Solar Merging
Step #3 & #4
Droping Missing Values If exist
Normalization
Stage #2 Preprocessing
Delete duplicate readings
Add some Features
Split into intervals
Stage #3 Building ZME_DEI predictor
Divide dataset with 10-cross validation
Training dataset
Testing dataset
ZME_DEI Predictor
Calculate loss function
Is accessed the stopping condition ?
Yes
No
Update the target value
Final Predictor
Stage #4 Evaluation
R2
RMSE
ME
MEG
Accuracy (A)

Figure 14.1 Block diagram of ZME DEI construction

$$R^2_{adj} = 1 - \left(\frac{\sum_{i=1}^{N} (y_i - \hat{y}_i)^2}{\sum_{i=1}^{N} (y_i - \bar{y})^2} \right) X N - 1/N - P - 1 \quad (5)$$

$$RMSE = \frac{1}{N} \sum_{i=1}^{N} (y - y)^2 \quad (6)$$

The R2 value (coefficient of determination is directly reliant on the variance of the dependent variable) and the adjusted of its value are two forms of error measurement (Amir and Khan, 2021).

The RMSE is usually used to measure the difference between predicted values of the model and the actual values from the system that does being created. The RMSE is determined as the square root about the mean squared error. R2 shows the percentage variety of prediction values. The rate of the R2 is between 0 and 1. The measures are described as following formulas:

$$RMSE = \sqrt{\frac{\sum_{i=1}^{N} (y_{mod,i} - y_{obs,i})^2}{N}} \quad (7)$$

$$R^2 = 1 - \frac{\sum_{i=1}^{N} \left(y_{obs,i} - y_{mod,i}\right)^2}{\sum_{i=1}^{N} \left(y_{obs,i} - \overline{y_{obs}}\right)^2} \quad (8)$$

where y_{obs} is the original value, and y_{mod} is the value of the model at a time/place i (Basaran et al., 2019).

Results

This section will present the main result extracted from each stage of (ZME_DEI) to predict DC-power which is the maximum energy generation from Renewable Resources that does not cause environmental degradation.

The ZME_ DEI model shows many activities which flow sequentially in stepwise style; stage one shows real-time collecting data from two datasets weather and solar plant each having basic features then checking if the dataset contains missing values for dropping.

Stage two pre-processing contains: (a) Merging between two datasets. (b) Using correlation to the final dataset. (c) Splitting readings into intervals every fifteen minutes. (e) Delete duplication intervals. In the third stage, ZME_ DEI model is constructed based on gradient boosting techniques.

The final stage includes evaluation results based on three measures such as coefficient of determination (R2), Root Mean Square Error (RMSE), and Mean Error (ME), furthermore Max Energy Generation (MEG), and Accuracy (A).

A. Collecting datasets

This step shows reading datasets weather and solar plant each having important features, the information of weather plant containing ('Date_Time, Plant_Id, Source_Key, Ambient_Temperature, Module_ Temperature, and Irradiation') which has 3182 entries, 0 to 3181. And seven sensors for solar plant such as ("Date_Time, Plant_Id, Source_Key, Dc_Power, Ac_Power, Daily_Yield, and Total_Yield") which has 71358 entries, 0 to 71357, then apply of describing function to each one and the result is:

a. Merging two datasets
b. Compute the correlation

To find which parameters in the dataset affect much on the target DC_power the coefficient of correlation method is used to find out the relationship between those parameters. The result shows that the parameters (Ac_Power, Module Temperature, and Irradiation) have high effects on our target.

B. Splitting dataset into intervals and deleting duplications

According to the name of sensors (Source_Key), the data rows are split into intervals every fifteen minutes, then applying compre among intervals with deleting of duplicates, and keeping the difference only.

C. Apply the DMO-GBM

Before push DMO-GBM to predict the Dc-Power need to determined the parmaters such as Goal and features have important affect to take desions that extraction from the privous stages; as explain in the following steps and Table III.

Step1: Create a Base Model/Average Model.

$$\hat{Y} = \frac{1}{n}\sum_{i=1}^{n} y(i)$$

- Set 1.40 as a predicted value for all records.
- Compute the Residuals (Loss function) values

$$\text{Residuals} = y(i) - \hat{Y}$$

Step2:
Create a second decision tree model RM1 to fit on Residuals.
Compute the mean for each node
Mean of Residuals (Node1) = 2.26

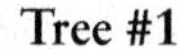
Tree #1

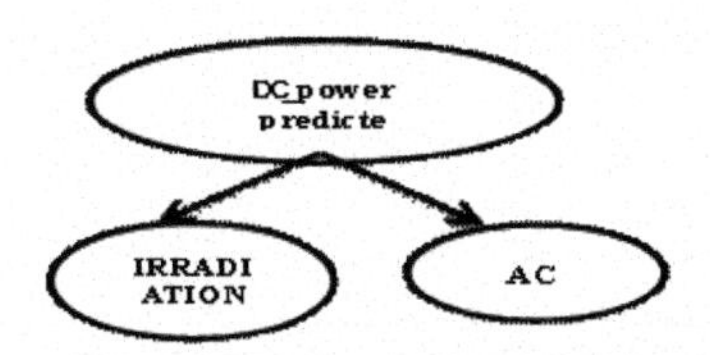

Mean = 1.566 Mean = 3.3

- RM1 for record split by Node 2 = $\hat{Y} + \alpha \times$ mean(Node 2) = 1.9 + (0.1 × 1.566) = 2.056
- RM1 for record split by Node 2 = $\hat{Y} + \alpha \times$ mean(Node 2) = 1.9 + (0.1 × 3.3) = 2.23

Step3:
- Create a third decision tree model RM2.
- Residuals in Tree2= y(i) – RM1

Tree#2

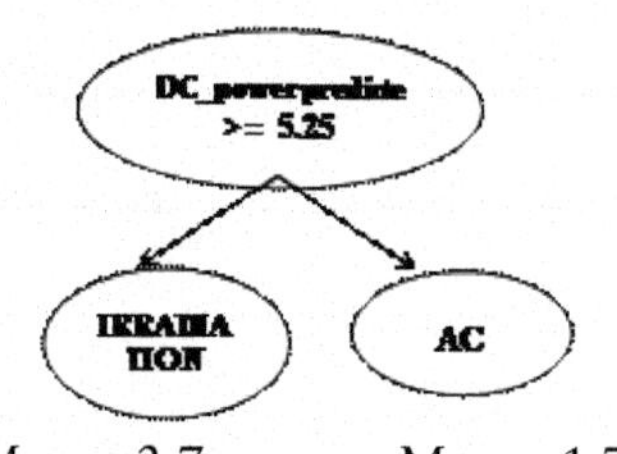

Mean= 2.7 Mean= 1.77

Mean of Residuals (Node1) = 2.33
RM2 for record split by Node 2 =RM1+ $\alpha \times$ mean (Node 3)
RM2 for record split by Node 3 =RM1+ $\alpha \times$ mean (Node 3)

D. Evaluation

This stage is used for evaluation results based on three grid measures such as coefficient of determination (R2), Root Mean Square Error (RMSE), and Mean Error (ME), furthermore Max Energy Generation (MEG) and Accuracy(A). Results of all masseurs present in table 9.

Conclusion

This paper produces a predictor named Sero to Max Energy Predictor Model based on Deep Embedded Intelligence Techniques (ZME- DEI) to predict DC-power which is the maximum energy generation from

Table 14.3 Sample of dataset based on the features most important

Ac	*MODULE_TEMPERATURE*	*IRRADIATION*	*DC-Power*
3.585714286	52.12607433	0.848850921	37.14286
5.1625	53.64847207	0.900069661	53.5
5.585714286	54.3455892	0.93859401	58
5.628571429	56.65431327	0.912929421	58.42857
5.25	56.80053313	0.966185457	54.375
0	58.69564164	0.975161304	0
4.228571429	59.36714047	0.967890089	43.85714
5.671428571	58.6883928	0.966922209	58.57143
2.366666667	56.83325393	0.89780282	24.5
5.425	56.27651418	0.902461681	56.125

Table 14.4 Results of DMO-GBM

Ac	*MODULE_TEMPERATURE*	*IRRADIATION*	*DC_power predicter*	*Target (residuals)*
3.585714286	52.12607433	0.848850921	1.40	4.1
5.1625	53.64847207	0.900069661	1.40	1.9
5.585714286	54.3455892	0.93859401	1.40	1.9
5.628571429	56.65431327	0.912929421	1.40	1.9
5.25	56.80053313	0.966185457	1.40	0.9
0	58.69564164	0.975161304	1.40	1.9
4.228571429	59.36714047	0.967890089	1.40	0.1
5.671428571	58.6883928	0.966922209	1.40	1.9
2.366666667	56.83325393	0.89780282	1.40	0.9
5.425	56.27651418	0.902461681	1.40	7.1

Table 14.5 Results after applying RM2

Ac	*MODULE_ TEMPERATURE*	*IRRADIATION*	*Target Residuals*	*RM2*
3.585714286	52.12607433	0.848850921	3.77	2.4
5.1625	53.64847207	0.900069661	2.056	2.32
5.585714286	54.3455892	0.93859401	2.056	2.32
5.628571429	56.65431327	0.912929421	2.056	2.23
5.25	56.80053313	0.966185457	1.056	2.23
0	58.69564164	0.975161304	2.23	2.5
4.228571429	59.36714047	0.967890089	0.23	2.4
5.671428571	58.6883928	0.966922209	2.056	2.32
2.366666667	56.83325393	0.89780282	1.056	2.32
5.425	56.27651418	0.902461681	6.77	2.5

Table 14.6 Results of evaluation measures for training and testing datasets

Rate of Training Dataset	*Rate of Testing Dataset*	*Results of the Training Dataset*			*Results of the Testing Dataset*		
		R2	*RMSE*	*Accuracy*	*R2*	*RMSE*	*Accuracy*
80% =19 intervals	20% =5 intervals	0.423	0.497	0.766	0.167	0.328	0. 620
60% =14 intervals	40% =10 intervals	0.562	0.526	0.853	0.433	0.334	0.741
50% =12 intervals	50% =12 intervals	0.301	0.792	0.870	0.010	0.421	0.800
40% =5 intervals	60% =19 intervals	0.296	0.787	0.891	0.020	0.203	0.825
20% =10 intervals	80% =14 intervals	0.097	0.0167	0.945	0.136	0.190	0.883

Renewable Resources that does not cause environmental degradation. ZME- DEI model shows many stages; can summarisation the main benefit of this research as follow; today the friendly environment energy such as solar plant energy become the main source of energy and very important in multi countries for the following reasons: Economic and political independence: Renewable energy is good for maintaining the local economy, as reliance on imported fossil fuels leads the country to be subject to the economic and political goals of the supplying country. As for renewable energy represented in wind, sun, water, and organic materials, it exists all over the world. On the other hand, renewable energy needs more labour compared to other energy sources that rely mostly on technology, where there will be workers to install solar panels, and technicians to maintain Wind farms, and other jobs that increase employment. Low prices: Renewable energy is witnessing a continuous decrease in costs, despite the progress made in its development, as the equipment used in it has become more efficient, and technology and engineering work has also become more advanced in this field, unlike gas, fossil fuels, and other energy sources that despite its advantages are that prices fluctuate periodically. Improving public health: Coal and natural gas plants lead to air and water pollution, which leads to many health problems; Such as breathing disorders, nervous problems, heart attacks, cancer, premature death, and other serious problems, and it is noteworthy that the majority of these negative health effects, resulting from water and air pollution, are not caused by the use of renewable energy technologies, as wind, solar, and energy systems hydroelectricity all generate electricity without any emissions causing air pollution, although some types of renewable energy can cause pollution; Such as geothermal energy systems and biomass, but the total polluting emissions in them are generally much less than the total emissions from power plants that use coal and natural gas. Inexhaustibility: Renewable energy is inexhaustible, compared to other energy sources; Such as coal, gas, and oil, and this means that they are always available, such as: the sun, which produces energy, and falls within the natural cycles, and this makes renewable energy an essential element in a sustainable energy system that is capable of development and development without risking, or harming future generations.

References

Ahmed, M. A., Haidar, A. F., Andreas, H.)2020). Sustainable energy planning for cost minimization of autonomous hybrid microgrid using combined multi-objective optimization algorithm. Sustain. Cities Soc. . 62:102391. https://doi.org/10.1016/j.scs.2020.102391.

Al_Janabi, S. (2015). A Novel Agent-DKGBM Predictor for Business Intelligence and Analytics toward Enterprise Data Discovery. JUBPAS. 23(2):482–507.

Al-Janabi, S. and Mahdi, M. A. (2019). Evaluation prediction techniques to achievement an optimal biomedical analysis. Int. J. Grid Util. Comput. 10(5):512–527.

Al-Janabi, S. and Alkaim, A. F. (2020). A nifty collaborative analysis to predicting a novel tool (DRFLLS) for missing values estimation. Soft Comput. 24(1):555–569.

Al-Janabi, S., Alkaim, A. F., and Adel, Z. (2020). An Innovative synthesis of deep learning techniques (DCapsNet & DCOM) for generation electrical renewable energy from wind energy. Soft Comput. 24:10943–10962. https://doi.org/10.1007/s00500-020-04905-9

Al-Janabi, S., Alkaim, A., Al-Janabi, E., Aljeboree, A., and Mustafa, M. (2021). Intelligent forecaster of concentrations (PM2. 5, PM10, NO2, CO, O3, SO2) caused air pollution (IFCsAP). Neural. Comput. Appl. 131.

Amir, M., and Khan, S. Z. (2021). Assessment of renewable energy: status, challenges, COVID-19 impacts, opportunities, and sustainable energy solutions in Africa. Energy Built Environ. 3(3):348–362. https://doi.org/10.1016/j.enbenv.2021.03.002

Ang, K. L.-M., and Seng, J. K. P. (2021). Embedded Intelligence: Platform Technologies, Device Analytics, and Smart City Applications. IEEE Internet Things J. 8(17):13165–13182. doi:10.1109/jiot.2021.3088217

Ardianto, R., Rivanie, T., Alkhalifi, Y., Nugraha, F. S., and Gata, W. (2020). Sentiment Analysis on E-Sports For Education Curriculum Using Naive Bayes and Support Vector Machine. J. Ilmu komput. Inf. 13(2):109–122.

Aziah Khamis, A.,Tamer Khatib, T., Nur Amira, N., Haziqah Mohd Yosliza, H., Aimie Nazmin Azmi, A. N.,)2020(, Optimal selection of renewable energy installation site in remote areas using segmentation and regional technique: A case study of Sarawak, Malaysia. Sustain. Ener. Technol. Assess. 42:100858. https://doi.org/10.1016/j.seta.2020.100858.

Bibri, S. E. (2018). The IoT for smart sustainable cities of the future: An analytical framework for sensor-based big data applications for environmental sustainability. Sustain. Cities Soc.38:230–253. doi:10.1016/j.scs.2017.12.034

Cotfas, L. A., Delcea, C., Roxin, I., Ioanăş, C., Gherai, D. S., and Tajariol, F. (2021). The Longest Month: Analyzing COVID-19 Vaccination Opinions Dynamics from Tweets in the Month following the First Vaccine Announcement. IEEE Access. 9:3320333223. doi: 10.1109/ACCESS.2021.3059821.

Das, H. S. and Roy, P. (2019). A deep dive into deep learning techniques for solving spoken language identification problems. In Intelligent Speech Signal Processing (pp. 81–100). Cambridge:Academic Press.

Dhanraj, J. A., AMostafaeipour, A., Velmurugan, K., Techato, K., Chaurasiya, P. K., Solomon, J. M., Gopalan, A., and Phoungthong, K. (2021). An Effective Evaluation on Fault Detection in Solar Panels. Energies. 14(22):7770. https://doi.org/10.3390/en14227770

Dogan, A., and Birant, D. (2020). Machine Learning and Data Mining in Manufacturing. Expert Systems with Applications, 114060. doi:10.1016/j.eswa.2020.114060

Gao, T. and Lu, W. (2020). Machine learning toward advanced energy storage devices and systems. iScience. 13, 24(1):101936. doi.org/10.1016/j.isci.2020.101936

Hao, J. (2020). Deep Reinforcement Learning for the Optimization of Building Energy Control and Management. Phd Thesis. University of Denver.

Hossny, K., Magdi, S., Soliman, A. Y., and Hossny, A. H. (2020). Detecting explosives by PGNAA using KNN Regressors and decision tree classifier: A proof of concept. Prog. Nuc. Energy. 124:103332. doi:10.1016/j.pnucene.2020.103332

Kabir, E., Kumar, P., Kumar, S., Adelodun, A. A., and Kim, K. H. (2018). Solar energy: Potential and future prospects. Renewable and Sustainable Energy Rev. 82:894900. https://doi.org/10.1016/j.rser.2017.09.094

Kaur, H. and Buttar, A. S. (2019). A review on solar energy harvesting wireless sensor network. Int. J. Comput. Sci. Eng. 7(2):398–404. doi: https://doi.org/10.26438/ijcse/v7i2.398404

Khan, A., Sohail, A., Zahoora, U., and Qureshi, A. S. (2020). A survey of the recent architectures of deep convolutional neural networks. Artif Intell Rev. 53:5455–5516. https://doi.org/10.1007/s10462-020-09825-6

Lamb, J. J. and Pollet, B. G. ((2020). Micro-Optics and Energy. doi:10.1007/978-3-030-43676-6

Liu, C., Wang, C., and Luo, J. (2020).Large-Scale Deep Learning Framework on FPGA for Fingerprint-Based Indoor Localization. IEEE Access. 8:6560965617. doi: 10.1109/ACCESS.2020.2985162.

Mahdi, M. A. and Al-Janabi, S. (2019). A novel software to improve healthcare base on predictive analytics and mobile services for cloud data centers. In International conference on big data and networks technologies. (pp. 320–339). Springer.

Oryani, B., Yoonmo K., Y., Shahabaldin, R. and Afsaneh, S.)2021). Barriers to renewable energy technologies penetration: Perspective in Iran. Renew. Ener.. 174:971–983. https://doi.org/10.1016/j.renene.2021.04.052.

Razmjoo, A., Gakenia Kaigutha, L., Vaziri Rad, M. A., Marzband, M., Davarpanah, A., and M. Denai, M. (2021). A Technical analysis investigating energy sustainability utilizing reliable renewable energy sources to reduce CO_2 emissions in a high potential area. Renew. Energy. 164:46–57. https://doi.org/10.1016/j.renene.2020.09.042.

Samaher Al-Janabi, S., and Ayad Alkaim, A. 2022 A novel optimization algorithm (Lion-AYAD) to find optimal DNA protein synthesis, Egyptian Informatics Journal, 2022, https://doi.org/10.1016/j.eij.2022.01.004.

Yin, L. and Zhang, Y. (2020) ,Village precision poverty alleviation and smart agriculture based on FPGA and machine learning, Microprocessors and Microsystems. 103469.

Zhang, G., Hu, W., Cao, D., Liu, W., Huang, R., Huang, Q., Chen, Z., and Blaabjerg, F.)2021(. Data-driven optimal energy management for a wind-solar-diesel-battery-reverse osmosis hybrid energy system using a deep reinforcement learning approach. Energy Conve. Manag. 227:113608. https://doi.org/10.1016/j.enconman.2020.113608.

15 Application of additive manufacturing in orbital-bone fracture reconstruction surgery

Deepak Shirpure[1,a,*], *Prajwal Gedam*[2,b], *Jayant Giri*[2,c], *Aniket Mandlekar*[1,d], *Atharva Wankhede*[2,e], *Atharva Chaudhari*[2,f], *Mohanish Khotele*[2,g], *and Rajkumar Chadge*[2,h]

[1]Concordia University, Gina Cody school of Engineering, Canada

[2]Yeshwantrao Chavan College of Engineering, Nagpur, India

Abstract

Restoration of orbital cavity generated due to injuries causes a deep blowout fracture which is challenging for surgeons to operate due to its complex anatomy. This research study focuses on the application of rapid prototyping for the repair of the orbital fracture using the intraoperative navigation system. The study was conducted on five patients who had undergone fracture in the orbital wall cavity. Using the computer tomography (CT) scan of the patient's head and capturing images as DICOM (Digital imaging and communication in medicine) which is reconstructed using the 3-D printing technique. Ophthalmic examination is carried out on surgical results for assessment. The application of 3-D printing for the diagnosis of orbital wall fracture proves to be a beneficial tool for the assessment of a patient's anatomical condition to improve clinical outcomes.

Keywords: Additive manufacturing, DICOM, orbital cavity, 3-D printing.

Introduction

Orbital fracture contributes 4–16% of the facial medical traumas which is caused by buckling or hydraulic mechanism due to vehicle accidents or injuries Darwich et al. (2020). The blunt impact blow-out the orbital wall bone which sometimes causes hernia and displacement of the eyeball in the maxillary sinus cavity, which leads to diplopia (double vision) or an enophthalmos (posterior displacement of the eye). Orbital inferior wall located at infraorbital groove and canal is majorly contributed towards orbital cavity fractures. Computer tomography is the preferred image capturing modality for bones and offers quicker scans (Osti et al., 2019). The CT scans were saved as (.DCM) files. We had used InVesalius open-source DICOM to 3-D volume generation software for this study. The software offers a user-friendly interface with two types of segmentation features viz thresholding segmentation and watershed segmentation. InVesalius has inbuilt export to standard tessellation language (STL) feature which is uncommon in comparative freeware software (Matsiushevich et al., 2018).

The application of additive manufacturing specifically 3-dimensional printing (3-DP) technology in orbital fracture repair surgeries will bring up substantial efficiency. 2-D computer tomography (CT) images do not provide a clear picture of the bones which were affected by the injury. 3-D printed model of the patient-specific injury gives a clear vision of the length and angle of the fracture which removes the ambiguity among surgeons (Dubois et al., 2016). The 3-D printed model is also used to determine the surgical tool's usage location. It also makes the patient aware of the operative procedure to be undergone on them. Hence 3-D printing of the concerned region via DICOM files provides excellent tangible information and helps plan the procedure of operation. Rapid prototyping (RP) or 3-D printing is extensively applied in a variety of medical fields especially in orthopedics, craniofacial, plastic, and reconstructive surgeries. Rapid prototyping is applicable for the generation of the implant which will substitute for the original orbital broken bone. Publications focusing on 3-D printing applications have increased over the recent decade since the availability of fast operating, easily accessible and cost-effective printers. The availability

[a]shirpuredeepak07@outlook.com; [b]prajwalpg55@gmail.com; [c]jayantpgiri@gmail.com; [d]asmandlekar@gmail.com; [e]atharvawankhede1515@gmail.com; [f]rbchadge@rediffmail.com

of cost-effective and robust materials has paved the way in the medical domain. DICOM 3-D printing is a research hotspot nowadays. Complex anatomical part makes it difficult to operate hence the involvement of 3-D printing will provide efficient results.

Methodology

Our present research study explores the possibilities and implementations of the 3-D printed model of the medial orbital fracture for clinical and surgical practice purposes or in preoperative planning. The methodology proposed with the use of our method is summarised in the given flowchart in Figure 15.1.

A) Patients and methods

Orbital fracture contributes 40% of maxillofacial injuries (Cha1 et al., 2016). A prospective study is conducted, this research focuses on the study of the patients who had taken a blow and developed a fracture in the craniofacial region and reconstruction of the same using rapid prototyping technique. The concern regarding the fracture can be stated by indications like restriction in the extraocular muscles, an excessive fracture which is 2 mm wide or enophthalmos (fracture of more than 2 mm), and increment of more than 5% in the orbital cavity (Cha1 et al., 2016). A total of five patients have undergone blowout fracture reconstruction with the application of RP technology and the intraoperative navigation system. Each patient took a CT scan of the head section. An ophthalmologist made preoperative evaluation for diplopia, enophthalmos, oculomotor movement dysfunction, and other vision-related issues. Enophthalmos is the difference of 2 mm or greater difference on the Hertel Exophthalmometer (Cha1 et al., 2016). Primary diplopia is severe while other diplopia was stated as mild. Each patient, post the operation received a CT scan. The injured orbit is compared with the contralateral. After the one-week ophthalmologic evaluation was conducted for the complications like diplopia, postoperative enophthalmos, and oculomotor dysfunction.

This study was conducted on the sample data of five patients which is available on the TCIA portal. The metadata of patients like Injury sustained by them accompanied with retrobulbar haemorrhage or not and other fracture involvement is summarised in Table 15.1.

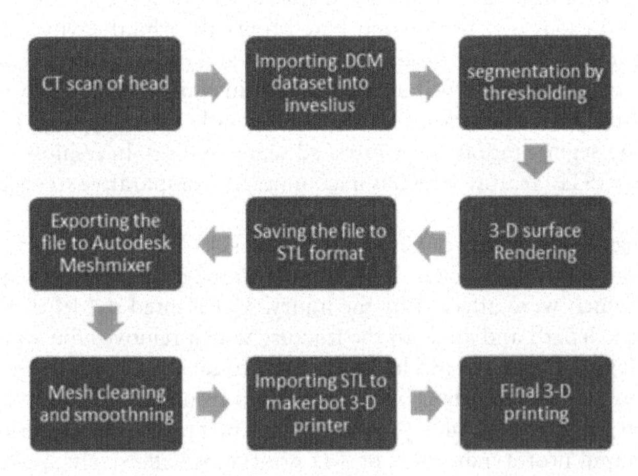

Figure 15.1 Workflow of the complete 3-D printing procedure

Table 15.1 Summary of the patients

Patient	*Age*	*Gender*	*Injury (mechanism)*	*Time to repair(day)*	*Orbits involved*	*Retrobulbar Haemorrhage*	*Other fracture*
1	64	M	Fall	8	Inferior	No	None
2	55	F	Work Injury	7	Medial	No	None
3	58	M	Accident	9	Inferomedial	Yes	Nasal bone
4	71	M	Slip down	6	Inferomedial	No	Nasal bone
5	42	F	Accident	8	Medial	Yes	None

B) Segmentation and mesh reconstruction

This method had performed on the data of five patients sample taken from the The Cancer Imaging Achieve (TCIA) portal. The five patients had undergone CT scans of the skull for the confirmation of the orbital fracture of the bone. The scans taken were exported as a DICOM file. We have adopted a semiautomatic segmentation process in this study. Dataset of CT scan comprised of 47 85 image slices. DICOM images of size 512*512 were imported into the InVesalius slicing software. InVesalius is open-source PAC software that offers thresholding-based segmentation in which the minimum and maximum thresholding values were set as 210 and 1537 HU for the extraction of bones out of the complete head skull structure. The thresholding outline was checked for coverage in axial, coronal, and sagittal views before surface rendering the model. The generated model was checked for complete surface rendering before exporting it to the STL file.

The generated STL file is needed reconstruction of triangular facets for the smooth surface and to rectify mesh queries like degenerated facets or excessive parts extraction. This process is to be carried out with the guidance of a medical expert by a technical operator. STL file is imported to Autodesk Meshmixer for the mesh reconstruction and to optimise the voids created during 3-D surface rendering in the software. For the orbital region of interest selection, the whole model of the upper skull and bottom area from the nasal bone is cropped in Autodesk Meshmixer for convenience in 3-D printing. The cropped model needs further pre-processing before exporting it to a 3-D printer. The open voids created due to cropping the model were filled and with the inspector tool in Meshmixer, the mesh query issues were rectified.

C) 3D printing technique

CT scan images obtained from the imaging modality are converted into a solid model with the help of PACS software. In this study, we have used InVesalius open-source software. DICOM files of 0.5 mm in thickness were imported into InVesalius and a 3-D virtual model is created. As the CT scan is taken of the patient's head, the patient's anatomical condition is assessed and searched for the correct orientation for cropping thus the required orbital region of interest is extracted and utilised for printing. The virtual model is prepared by the information provided in terms of Hounsfield unit intensities and hence we had provided the intensities of 210 HU to segment the mandible structure. The location and size of the defect are examined through preoperative knowledge of the condition and to do advanced planning (Figures 15.2–15.4 and Table 15.2).

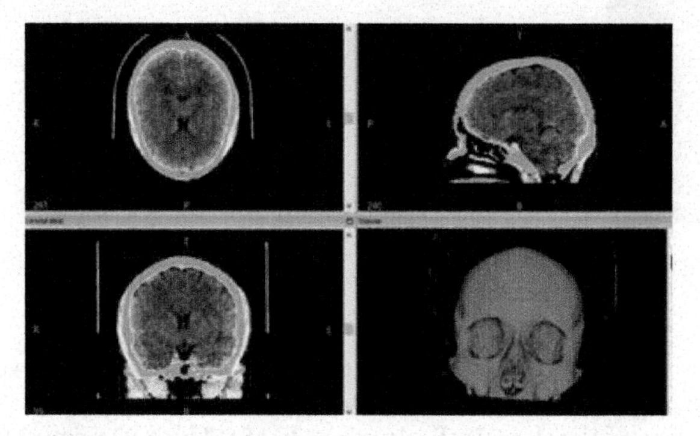

Figure 15.2 Thresholding segmentation of skull in InVesalius

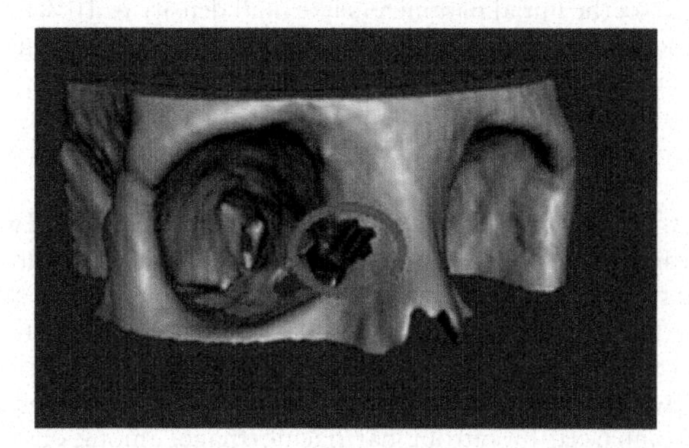

Figure 15.3 Mesh reconstruction of the cropped model in Meshmixer

Table 15.2 Parameters used for 3-D printing

Sr. No	*Parameters used*	*values*
1	Filament material	PLA
2	Filament diameter	1.75 mm
3	Layer thickness	0.3 mm
4	Infill density	10%
5	Nozzle diameter	0.4mm
6	Material required	50.4g
7	Dimensions	127*43.5*85.8 (mm)
8	Printing time	240 minutes

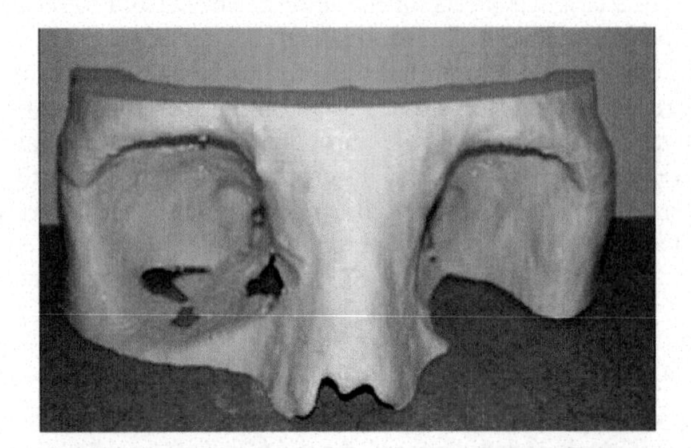

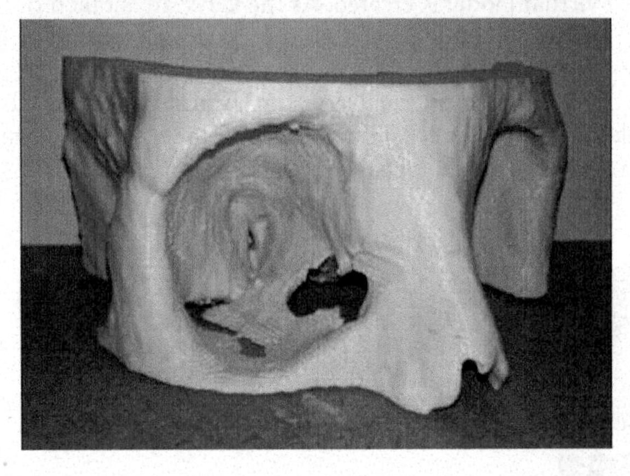

Figure 15.4 3-D printed orbital wall fracture by FDM

To search for the defect in the orbital cavity, a virtual model of injured orbital cavity bone was overlapped with the image of the uninjured cavity through mirroring this resulted in the visibility of the 3-D mark of an interface that was not overlapped because of the defects (Park et al., 2015). The 3-D virtual file is exported into a STL file. We have used fused deposition modelling and a Makerbot 3-D printer. STL is imported into the Makerbot integrated software and set the initial parameters like infill density to 10% and layer thickness as 0. 3 mm, with the scale of 1:1 and of dimensions 127 mm in length, 43.5 mm in height, and 85.8 mm in width. Taking these specifications, the printing took 240 minutes.

Results

The patients (three males and two females) from the patient's group were aged between 42 and 71. Two patients suffered from inferomedial blowout fracture. Three patients suffered from orbital wall fractures. The reason for the fracture was a motor accident in two cases and a fall in one case, a work injury, and a slip down in the one-one case. The operations were carried out at an average of 6.75 ± 1.58 days after meeting with an injury. The mean time of operation was 118 ± 29.2 minutes, and the duration of follow-up lasted 530 weeks. Three patients had developed diplopia two patients had enophthalmos. Four patients developed extraocular dysfunctional movement two had accompanied by retrobulbar haemorrhage. among one of the two patients with retrobulbar haemorrhage is accompanied by traumatic optic neuropathy. The patient with retrobulbar haemorrhage had no entrapment of muscle in the radiological examination carried out earlier, and hence operation was carried out post observing process of the observation period. Most of the patients (4 out of 5) not experienced complications like diplopia or extraocular movement dysfunction.

Patient no. 3 (Table 15.1) had a fractured bone of about 25 mm wide we have 3-D printed the model for the analysis of the depth and curvature in the anatomy. The dimensional data is further used in the generation of the implant which will fill the cavity generated due to the accident.

Discussion

3-D printing comes under the umbrella of Additive manufacturing, which uses CAD (computer-aided design) and CAM (computer-aided manufacturing) to deposit the material layer by layer to manufacture a solid part. By this approach, a complex part can be created with ease with a variety of materials ranging from plastics to ceramics and metals. 3-D printing is a hotspot in the medical field with its wide usage in the structurally complex craniofacial field. Rapid prototyping is applicable for medical implants and reconstruction of the defective regions because of its capability of creating complex contours parts with ease which eventually reduces surgery and anaesthesia duration (Pang et al., 2018). The dimensional accuracy is very high which makes it promising in the medical field. The availability of a wide range of robust materials like acrylonitrile butadiene styrene (ABS), polylactic acid (PLA), and nickel-titanium alloys enable to select the and imply the right material for implant or replication of the part (Shahrubudina et al., 2019). 3-D printing creates complex body parts flexibly unlike subtractive manufacturing.

The RP technique is showing promising results in the craniofacial field, which is geometrically complex (Mankovich et al). In 1994 first used the rapid manufacturing approach to print skull models to find the donor of calvarial bones and they succeeded in finding one with a similar anatomical model as that of the patient. RP model for orthognathic surgery is used for the design of osteotomy and printed a pre-bent plate of titanium by using the mandible of the model to improve accuracy in the surgical process. As per the recent studies, the titanium-based material implant can be used for human body implants for the reconstruction of maxillary and calvarial defects with minimal effects.

The orbital cavity has complex anatomical structure owing to this fact RP technology in orbital cavity reconstruction surgery is advantageous. The RP of the orbital volume involves 1) CT scan 2) pre-processing of CT data and virtual model generation 3) Final 3-D printing this process took approximately 240 minutes. With the help of a virtual 3-D model, one can locate and determine the size of the orbital cavity and anatomy of the patient, but the 3-D printed model provides six degrees of freedom for the deep analysis by the doctors. The RP technique in medicine needs expertise in medical software we have used open-source segmentation software for the reduction in overall RP cost. FDM printing with PLA material provides effective and affordable RP of the model. The complete model required approximately 50 grams of material costing less than $ 5 USD. The continuous development of cost-effective and easy-to-operate 3-D printers increases the potential of the major application of this technology in medicine.

Our present study has some shortcomings, it was practiced on only five patients' data also present method has a few drawbacks (1) the segmentation process must be carried out in the presence of an anatomical expert in that field. (2) The time required for building the model makes it infeasible for trauma cases where surgery is needed to be done within an hour. (3) The close cooperation requirement among several people sum up the additional cost of the surgery.

Conclusion

The complex shape of orbital fracture is in the blind zone hence it makes it difficult for the surgeons to operate without prior planning. The application of rapid prototyping not only makes surgeries less prone to errors by providing tangible information. The only limitation of this study is the long printing time requirement which is not favourable for trauma cases by considering their complexities. The segmentation of ROI plays prime importance and requires close monitoring of technical people hence more focus is required here to minimise the maximum errors. In conclusion, the inclusion of the RP technique in medical orbital surgical procedures will provide a good understanding of anatomical structure, helps in reducing operating time. The RP technique is having the potential in restoring orbital cavity volume.

References

Cha, J. H., Lee, Y. H., Ruy, W. C., Roe, Y., Moon, M. H., and Jung, S. G. (2016). Application of rapid prototyping technique and intraoperative navigation system for the repair and reconstruction of orbital wall fractures. Arch. Craniofac. Surg. 17(3):146–153.

Darwich, A., Attieh, A., Khalil, A., Sza´vai, S., and Nazha, H. (2021). Biomechanical assessment of orbital fractures using patient-specific models and clinical matching. Oral Maxillofac. Surg. 122(4):e51–e57.

Dubois, L., Jansen, J., Schreurs, R., Habets PEM, Reinartz SM, Gooris PJJ, and Becking AG. (2016). How reliable is the visual appraisal of a surgeon for diagnosing orbital fractures? J. Craniomaxillofac. Surg. 44(8):1015–1024

Matsiushevich, K., Belvedere, C., Leardini, A., and Durante, S. (2018). Quantitative comparison of freeware software for bone mesh from DICOM fifiles. J. Biomech. 84:247–251.

Osti, F., Santi, G. M., Neri, M., Liverani, A., Frizziero, L., and Stilli, S. (2019). CT conversion workflow for intraoperative usage of bony models: From DICOM data to 3D printed models. Appl. Sci. 9:708. https://doi.org/10.3390/app9040708

Pang, S. S. Y., Fang, C., and Chan, J. Y. W. (2018). Application of three-dimensional printing technology in orbital floor fracture reconstruction. Trauma Case Rep. 17:23–28.

Park, S. W., Choi, J. W., Koh, K. S., and Oh, T. S. (2015). Mirror-imaged rapid prototype skull model and pre-molded synthetic scaffold to achieve optimal orbital cavity reconstruction. J. Oral. Maxillofac. Surg. 73(8):1540–53.

Shahrubudina, N., Leea, T. C. and Ramlana, R. (2019). An overview on 3D printing technology: Technological, materials, and applications. Procedia Manufacturing: n. pag., DOI:10.1016/J.PROMFG.2019.06.089, Corpus ID: 202096072. 35:1286–1296.

16 CFD analysis of vortex tube using multiple nozzles

Neeraj Sunheriya[1,a,*], *Amey Bhoyar*[2,b], *Jayant Giri*[1,c], *R. B. Chadge*[1,d], *Chetan Mahatme*[1,e], *and Pratik Lande*[1,f]

[1]Department of Mechanical Engineering, YCCE, Nagpur, Inida

[2]Technische Hochschule Deggendorf, Germany

Abstract

A vortex tube is a device that splits the high-pressure airflow at room temperature into two low-pressure streams, one at a higher temperature than the inlet and the other at a lower temperature than the inlet. Even though its simplicity in operation, the physics involved in this temperature separation is extremely complex. Also, there are various parameters such as inlet pressure, number of nozzles, length of the tube, diameter of the tube, and hot end and cold end mass fractions that affect the performance of the tube. The main aim of this work is to study the energy separation phenomenon in the tube, find the effect of the number of nozzles on the performance, the effect of change in effective tube length, effect of inlet pressure. For this purpose, instead of widely used vortex tubes in which we use cone-shaped obstruction for separation of hot and cold streams, we have used a valve that closes vertically to obstruct the flow for experimental verification. The model of this is designed on the computer using ANSYS WORKBENCH, meshed with the help of software ICEM, and then analysed with the help of solver FLUENT. The results are compared with the data obtained by performing experiments on the actual model. Then both the results are compared by plotting them on the graph. After verification, intense computational fluid dynamics analysis is carried out to find the optimum parameters.

Keywords: Computational fluid dynamics, nozzle, numerical analysis, Vortex tube.

Introduction

A vortex tube is a simple device consisting of one or more inlet nozzles, vortex chamber, hot-end tube, cold-end tube, and hot-end control valve. In a vortex tube, a pressurised gas (generally air) enters the vortex chamber tangentially through one or more inlet nozzles. The tangential entry of the pressurized fluid into the vortex chamber gives rise to a vortex or swirling motion inside the tube. The swirling flow is along the circumference of the tube (also called a swirl chamber). This flow further goes on to strike the valve. The central portion is at lower pressure due to tangential injection allowing the reverse flow of the incoming gas. As its kinetic energy is less, which contributes to the total temperature, it is cold as compared to the circumferential flow. For a given inlet pressure and temperature, the temperatures of hot and cold streams can be varied by varying the fraction of hot and cold streams with the help of a valve. In general, the vortex tube is designed to obtain either (i) the maximum temperature difference or (ii) the maximum efficiency.

The input nozzle draws compressed air (Figure 16.1). Figure 16.2 represents the side view of experimental setup. Figure 16.3 represents the front view of experimental setup. The vortex motion inside the tube is generated by swirl generators at the intake plane. A temperature separation forms as the vortex goes along the tube. The tube's exterior is filled with hot air, while the inner core is filled with cold air. The heated air is subsequently permitted to depart the tube by the cone valve at the far end, while the cold air outlet is located adjacent to the inlet plane. The Ranque-Hilsch effect refers to the consequent radial temperature separation inside the vortex tube.

A. Novelty of the Work

The vortex tube is currently used in many industrial applications such as cleaning of gas, separation of gases, cooling of electronics components, cooling of electronic cabinets, cooling food, dehumidification of gas, chilling environment chamber, cooling of fireman suits, liquefaction of natural gas, due to its features of simple in construction with non-moving parts, compactness, light-weight, robustness, reliability, low

[a]neeraj.sunheriya@gmail.com; [b]ameybhoyar97@gmail.com; [c]Jayantpgiri@gmail.com; [d]rbchadge@rediffmail.com; [e]chetanmahatme@gmail.com; [f]pratiklande@gmail.com

Figure 16.1 Schematic diagram of vortex tube (De Vera, 2010)

cost, low maintenance required, durability, safety, instantaneous cooling or heating, ventilation in addition to cooling/heating, ease in adjusting output temperature and environment friendly. But, its low thermal efficiency is a main limiting factor for its application. Also, the noise and availability of compressed gas may limit its application.

B. Objectives of the work

The primary objectives of this research can be summarised below.

- To develop a Computational fluid dynamics (CFD) simulation model for a Vortex tube. To use CFD tool for obtaining the Maximum temperature difference between the hot and cold ends of the vortex tube.
- Study the effect of variation in various physical parameters and thermodynamic parameters, mass flow rates which affect the performance of Vortex tube.
- Suggest the best set of parameters for enhanced performance.
- Experimentally validate the results.

Literature Survey

Vortex tube has been an area of interest for many because of its usefulness and also because the perfect physics behind its operation is not completely known yet. Some papers have been published that state their results based on experimental studies, while some authors have worked on understanding the physics of vortex tubes with the help of CFD as a tool. Out of these publications, the following papers proved much useful. De Vera (2010) gave a brief introduction to the basic theory and concept of vortex tubes with its principal of working. It introduces numerical methods i.e. mathematical representation of flow in vortex tube with some empirical relations. It also explains temperature separation in the tube.

Wua et al. (2007) dealt with an experimental study related to a modification in various parameters that affect the performance of vortex tubes. The paper gives an idea about how various parameters like inlet pressure, valve opening, temperature, and the number of nozzles affect the performance of the tube. Eiamsa-ard and Promvonge (2007) numerically simulated the thermal separation in a Ranque–Hilsch vortex tube. The modelling of turbulence for compressible, swirling flows used in the simulation is discussed. The work has been carried out to provide an understanding of the physical behaviour of the flow, pressure, and temperature in a vortex tube. Farouk et al. (2009) simulated the CFD model of vortex tube in 2-D and studied the effect of large eddies on the cold and hot end temperatures. Simulations were conducted for different cold mass fractions by changing the hot end pressure. The effect of the cold mass fraction on the temperature separation was studied.

Xue et al. (2009) presented a critical review of current explanations of the working concept of a vortex tube. Hypotheses of pressure, viscosity, turbulence, temperature, secondary circulation, and acoustic streaming were discussed in the work. Zin et al. (2010) used CFD as a tool to study mainly the effect of inlet pressure on the performance of vortex tubes. It also gives a slight outline regarding the generation of mesh on the vortex tube model. Authors have also shown the results that are plotted in graphical form for better representation. Pouraria and Zangooe (2011) carried out a numerical investigation to study the effect of using a divergent vortex tube which gives better results than the conventional tube and to find the optimum angle of divergence. The existence of heat and work transfer inside the tube was investigated. Numerical results indicated that an increase in divergent tube angle increases by cooling performance of the vortex tube. Pourmahmoud et al. (2012) studied the effect of helical nozzles on both energy separation and refrigeration phenomena in the Ranquee-Hilsch Vortex Tube (RHVT) by CFD techniques. It was observed that the radial gap of the inlet of the helical nozzle from the vortex chamber stands as a significant designing parameter. Shamsoddini and Khorasani (2012) used CFD simulations to provide an interpolation table and defined two correlations for the velocity magnitude to the normal velocity with a simple thermodynamic approach. The results of the proposed algorithm were verified with the experimental and numerical studies. This approach was found to be a proper method to optimise the cooling performance of a vortex tube at different operating conditions. Reasonable results were obtained by applying appropriate constraints.

However many researchers have done various studies shown above, not check the performance of vortex tubes with different numbers of nozzles. In the proposed work CFD analysis of vortex tube with a different number of nozzles is done and the behaviour of the different flow parameters is checked numerically.

Experimental Setup

Tubes with lengths of 300 mm, 100 mm, and 70 mm are studied in this project. For each length mentioned above one nozzle, two nozzles, and three nozzles are used to study their effects on the performance of the Vortex Tube. The basic dimensions of the vortex tube e. g. diameter, the thickness of the tube, and dimensions of the nozzle are kept constant.

The tube with a total length of 385 mm is used in the actual setup having an internal diameter of 16.7 mm for the flow of air. The material of the tube is stainless steel. The inlet nozzle has an outside diameter of 22 mm whereas the internal diameter is 18 mm as shown. It is attached to the tube tangentially so that when air enters the tube it produces a swirl motion. An obstruction is provided with the tube, which is at a 35 mm distance from the cold end and the inlet is 5.5 mm from this obstruction. Obstruction is provided so that air is forced to flow towards the hot end and the air returning from the valve after striking it is centrally separated from the main flow. Obstruction also has a central hole of 5 mm so that the central cold flow is taken to the cold end. The valve on which the swirling airstrikes are provided is 300 mm from the

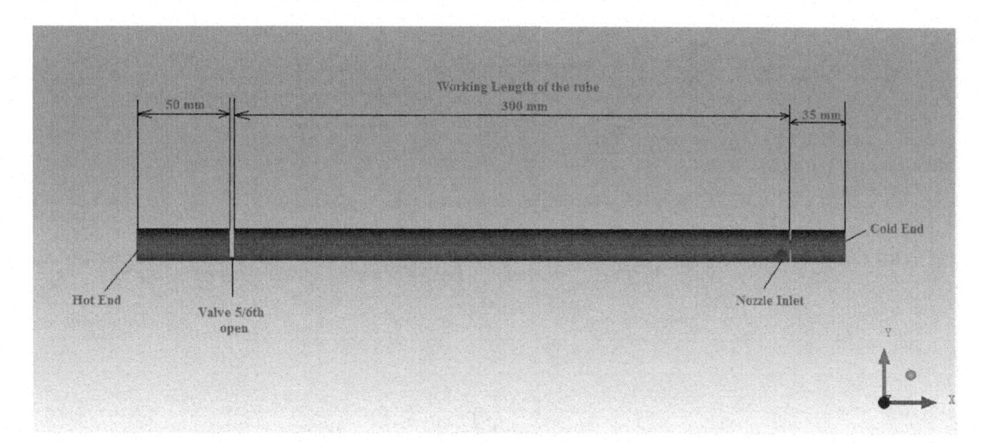

Figure 16.2 Side view of experimental setup

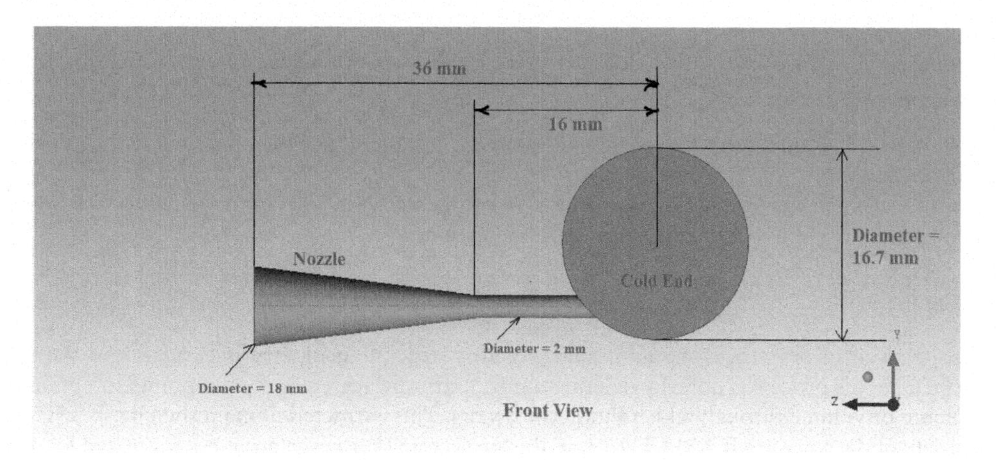

Figure 16.3 Front view of experimental setup

obstruction. Thus the total working length of the tube becomes 300 mm. Further, there is a hot end where we collect hot air on the outer periphery of the tube at a 50 mm distance from the valve. The valve has a diameter same as that of the tube and is kept 5/6 open as shown, which experimentally gave optimum results in terms of temperature difference. This arrangement is for the sake of validation with the actual result.

Besides this, geometries with tube lengths of 100 mm and 70 mm working lengths are also created. In addition to this, each geometry is studied for one nozzle, two nozzles, and three nozzles cases. Hence total of 9 different geometries is prepared with the help of ANSYS WORKBENCH to study the effect of change in length and number of nozzles on the performance of the vortex tube.

Numerical Analysis

The numerical simulation of the vortex tube has been created by using the FLUENTTM software package. The flow is assumed as a 3-D, steady-state and employs the standard k-ε turbulence model. This is because the simulations that were done earlier suggested that the results obtained with the help of this turbulence model are quite close to the actual results. The compressible turbulence flows in the vortex tube are governed by the conservation of mass, conservation of momentum, and energy equations, which are given by

$$\frac{\partial}{\partial xj}(\rho_{uj}) = 0:$$

Continuity equation:

$$\frac{\delta u}{\delta x} + \frac{\delta v}{\delta y} = 0$$

Momentum equation in X-direction:

$$\frac{\delta}{\delta t}(\rho u) + \frac{\delta}{\delta x}(\rho uu) + \frac{\delta}{\delta y}(\rho vu) = -\frac{\delta p}{\delta x} + \frac{\delta}{\delta x}\left\{(2\rho(v + v_t))\frac{\delta u}{\delta x}\right\} + \frac{\delta}{\delta y}\left\{\rho(v + v_t)\left(\frac{\delta u}{\delta y} + \frac{\delta v}{\delta x}\right)\right\}$$

Momentum Equation in Y-direction:

$$\frac{\delta}{\delta t}(\rho v) + \frac{\delta}{\delta x}(\rho uv) + \frac{\delta}{\delta x}(\rho vv) = \frac{\delta p}{\delta y} + \frac{\delta}{\delta x}\left\{\rho(v + v_t)\left(\frac{\delta u}{\delta x} + \frac{\delta v}{\delta y}\right)\right\} + \frac{\delta}{\delta y}\left\{2\rho(v + v_t)\frac{\delta v}{\delta y}\right\}$$

Energy equation:

$$\frac{\delta}{\delta t}(\rho c_p T) + \frac{\delta}{\delta x}(\rho c_p uT) + \frac{\delta}{\delta y}(\rho c_p vT) = \frac{\delta}{\delta x}\left\{\rho c_p(a + a_t)\frac{\delta T}{\delta x}\right\} + \frac{\delta}{\delta y}\left\{\rho c_p(a + a_t)\frac{\delta T}{\delta y}\right\}$$

A. Pre Processing

Geometry creation

This process consists of designing the desired geometry on the computer and then meshing it with the help of meshing software.

The Vortex tube geometry is first drawn with the help of the software ANSYS WORKBENCH (release 13). These geometries are shown in the figures under this section.

Meshing

The next step involved in pre-processing is meshing. In this step, the geometry is cleaned and the fluid domain is extracted. This process is done by deleting unnecessary surfaces, curves, and points to obtain the exact fluid domain or volume through which fluid flow occurs. This extracted domain then meshes. This is done with the help of the software ICEM CFD 13. In this project work, the surface mesh is created by using all triangular patch independent types. Volume meshing is tetrahedral type. This mesh is often of very low

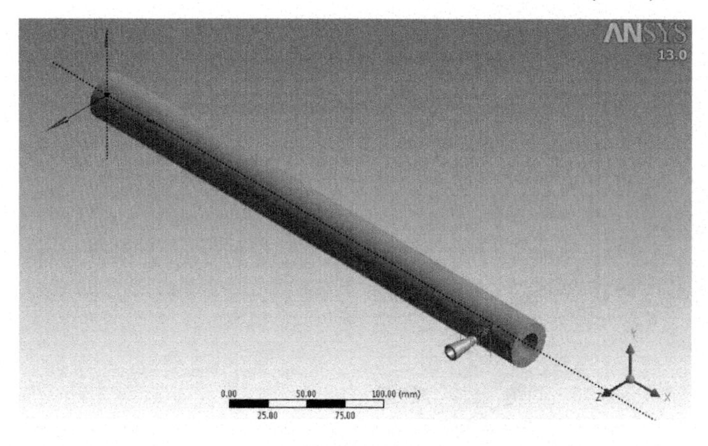

Figure 16.4 Basic geometry of available vortex tube setup

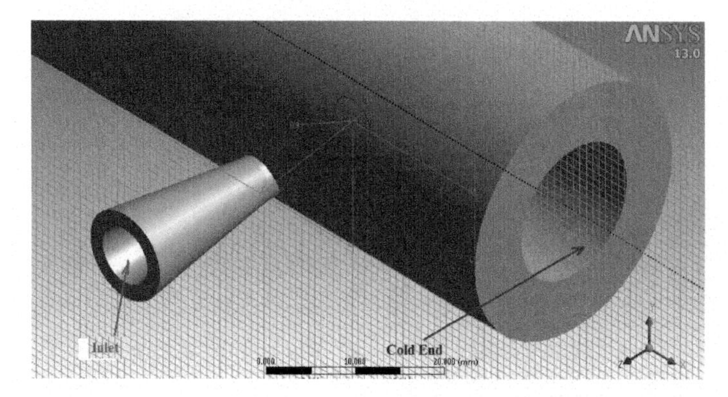

Figure 16.5 Inlet nozzle and cold end

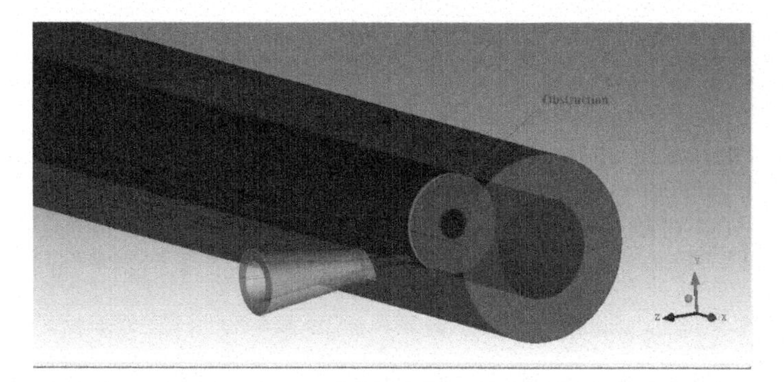

Figure 16.6 Transparent view of cold end showing obstruction

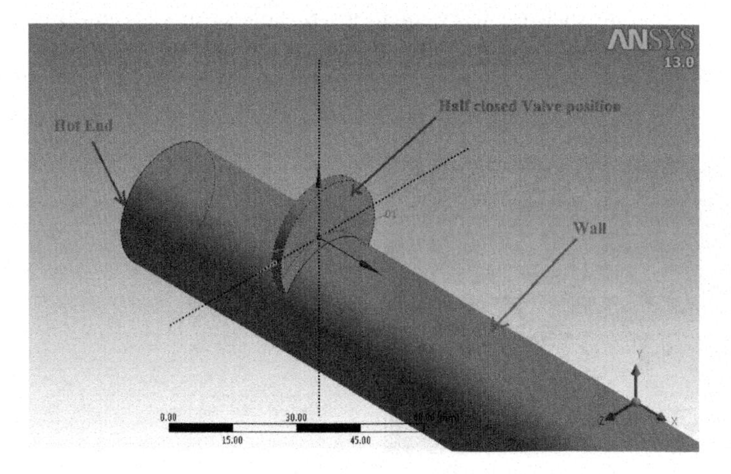

Figure 16.7 Showing hot end, half closed valve position, wall

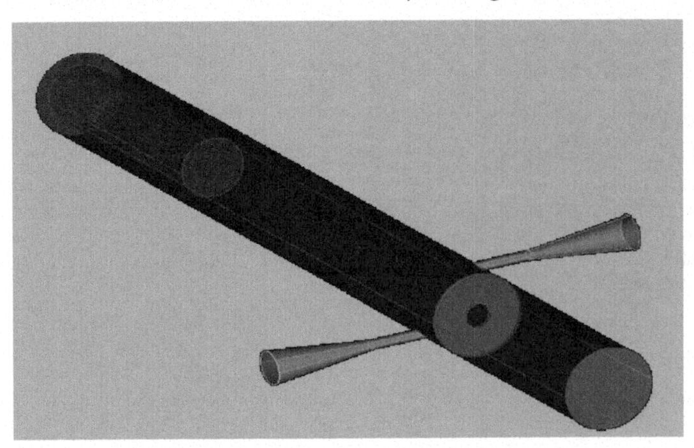

Figure 16.8 Geometry of two nozzles 100 mm vortex tube

Figure 16.9 Geometry of three nozzles 70 mm vortex tube

quality. To obtain better results during the actual solution through solver, a good quality mesh is required. Thus this raw mesh is smoothened. The boundaries are tagged or named e. g. inlet1, nozzle1, wall, hot end, cold end, valve, etc. This mesh is then exported to a suitable solver for further processing. The meshed fluid domains are shown in the figures that follow.

B. Solver

The Mesh that is created in the pre-processing step is then fed to a suitable solver for the actual solution of the problem. ANSYS FLUENT 13 software package was used to obtain the solution in this project work. The mesh is read in this solver. Various equations such as energy equations, turbulence models, multi-phase, and species that are involved in flow are defined according to the physics of the problem. In this project work, a density-based solver was used because of the highly compressed flow involved. Material air is

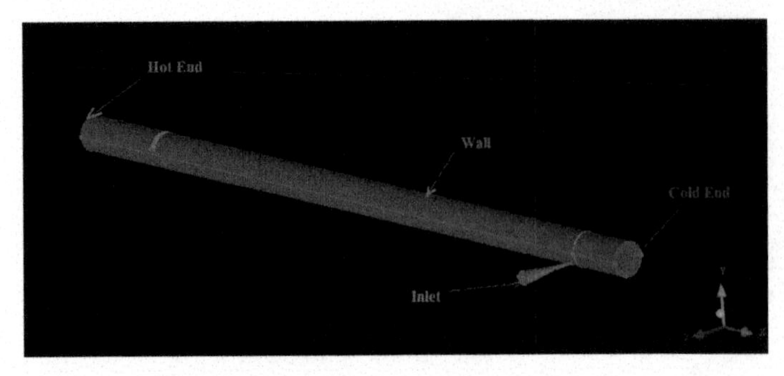

Figure 16.10 Extracted fluid domain in ICEM

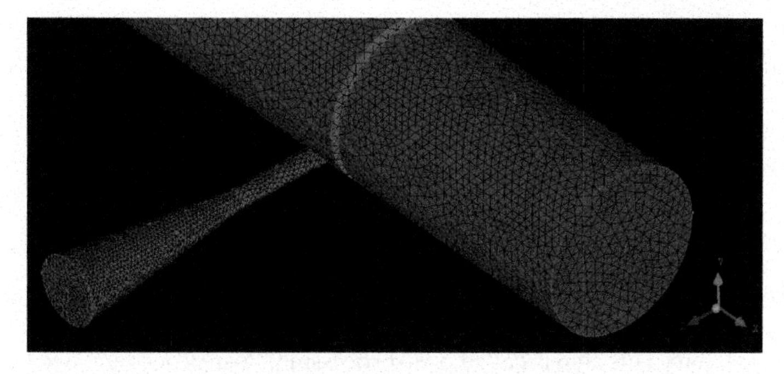

Figure 16.11 Surface meshing of all tri patch independent type

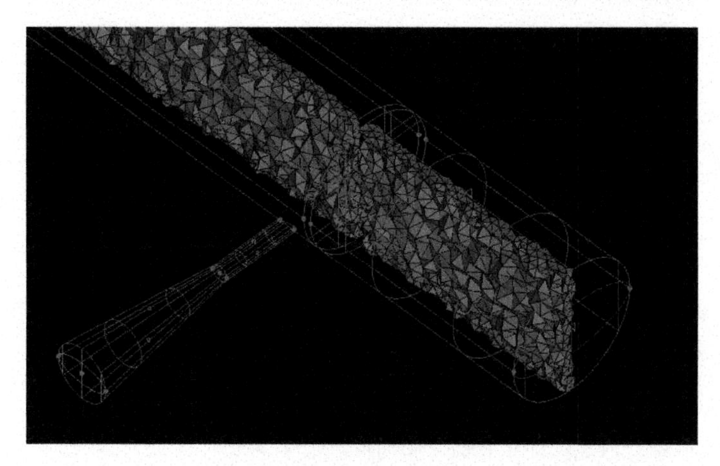

Figure 16.12 Volume meshing (Tetra type) at a cut plane

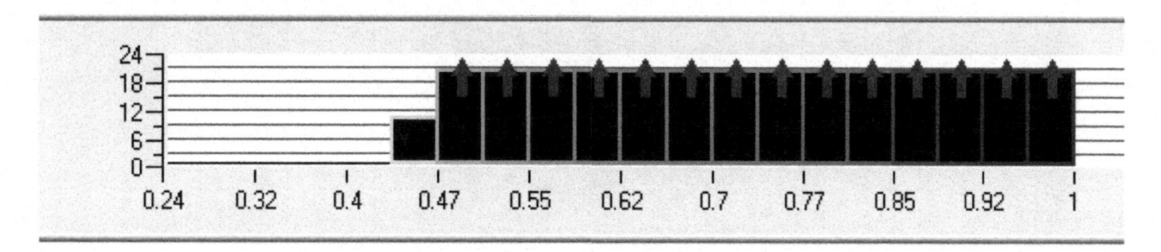

Figure 16.13 Mesh quality after mesh smoothening (Min 0.45 was reported)

necessarily an ideal gas. Energy Equation kept on because of the energy transfers and temperature difference within the fluid domain. The standard k-ε model is used for Turbulence as it has shown agreement with the experimental studies in the studies done by Abdol Reza Bramo (Pourmahmoud et al., 2012).

Default values in Standard values were used. Values were

- Cmu = 0.09
- C1-Epsilon =1.44
- C2-Epsilon = 1.92
- TKE Prandtls No. = 1

The boundary conditions are defined e.g. Pressure inlet, Pressure outlet, wall, etc. Suitable input parameters such as pressure, velocity, and temperature at particular surfaces according to their type are defined in this step.

- In this work, the Boundary conditions used are,
- Inlet of nozzle = Pressure Inlet at atmospheric temperature (295 K)
- Hot end, cold end = Pressure Outlets at atmospheric pressure and temperature (295 K)
- All other surfaces were taken as walls at 295 K with no heat transfer.

The method of solution which is also known as the scheme is defined as a first-order upwind or second-order upwind scheme. For this project, the second-order upwind scheme was used. The solution is then initialised, and the monitors are set to set the convergence criteria. The equations as mentioned in the above paragraph are then solved for the desired number of iterations till a converged solution is obtained. The solver solves the equation at the centroid of each cell and stores it in form of numerical data.

C. Post-processing

The solver step generates a large amount of numerical data but it is of no use as it is not understandable to a human. Thus for easy understanding, this data is then represented in the form of vectors, contours, graphs, path lines of the particles, surface integrals, etc. This step is called post-processing. Post-processing need not be done in separate software as ANSYS FLUENT itself provides the tools related to post-processing. Thus the data is represented in suitable form as per required.

Validation Of CFD Model

Verifying numerical predictions with experimental data is an important aspect of any modelling study. The overall objective is to demonstrate the accuracy of the CFD model so that they may be used with confidence for simulation and that the results be considered for decision-making in design.

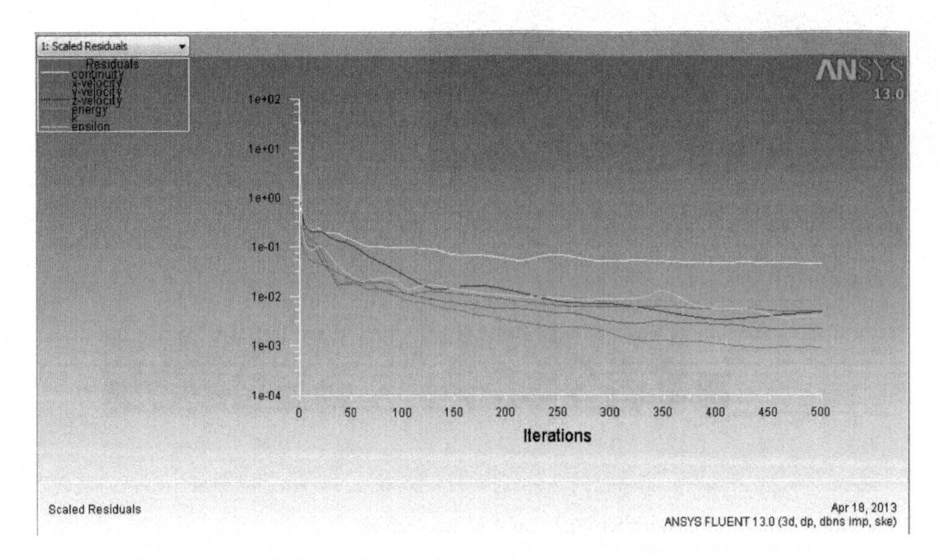

Figure 16.14 Iterative solution of the problem with convergence

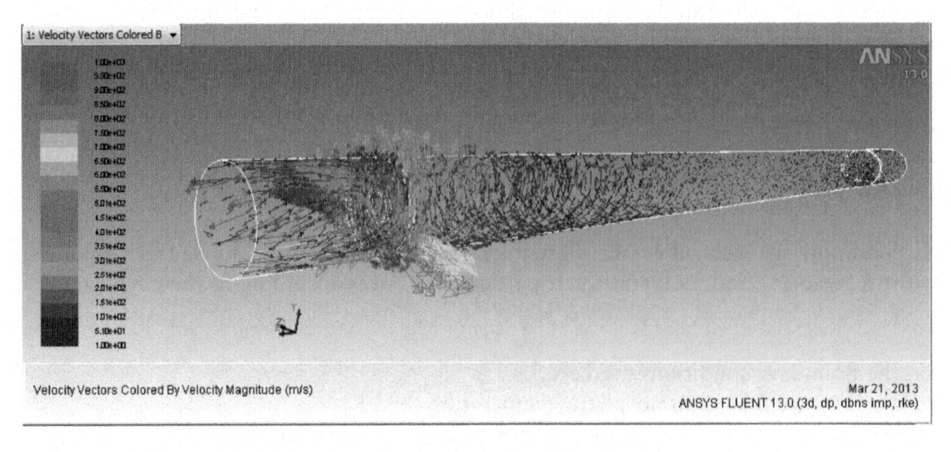

Figure 16.15 Example of post processing (vectors of velocity in m/s)

A. Validation of smooth tube

For validation modelling of the tube is done using the ANSYS WORKBENCH software for geometry and then meshing it in ICEM. Tetrahedral meshing is used for meshing the volume. Specification of tube:-

- Working length – 300 mm
- External diameter – 31.25 mm
- Internal diameter – 16.7 mm

The atmospheric temperature was taken as 305 K as in the actual experimental case. Table 16.1 represents the Validation of results for 300 mm tube length with 5/6th opening of the valve at 305 K temperature.

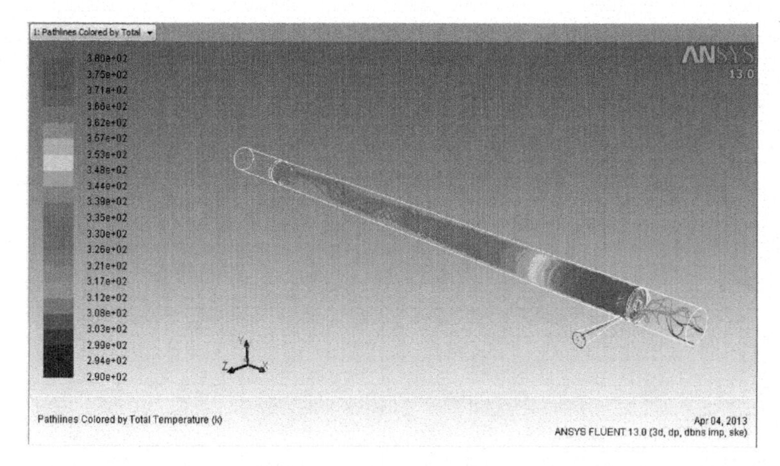

Figure 16.16 Path-lines showing swirl motion of air along the tube

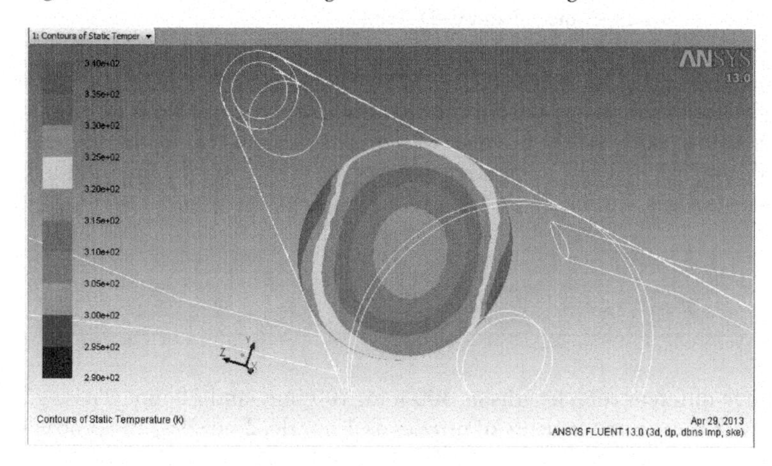

Figure 16.17 Temperature contour (in K) at 10 mm from inlet

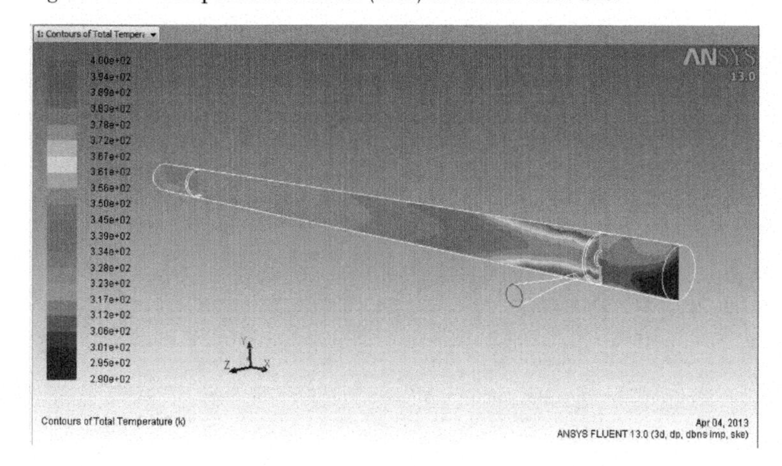

Figure 16.18 Temperature contour (in K) at 10 mm from inlet

Table 16.1 Validation for 300 mm tube, 5/6th opening of the valve, 305 K

Pressure (bar)	*Exp. Values ΔTExp (K)*	*CFD Values*			*% error*
		Tc (K)	*Th (K)*	*ΔT (K)*	
1	2	304.137	306.0103	1.8733	-6.763465542
2	4	303.437	308.1975	4.7605	15.97521269
3	7	302.3240	309.871	7.54696	7.247421478
4	13	297.0621	313.7149	16.6528	21.93504996
5	17	297.3240	320.8684	23.5444	27.79599395

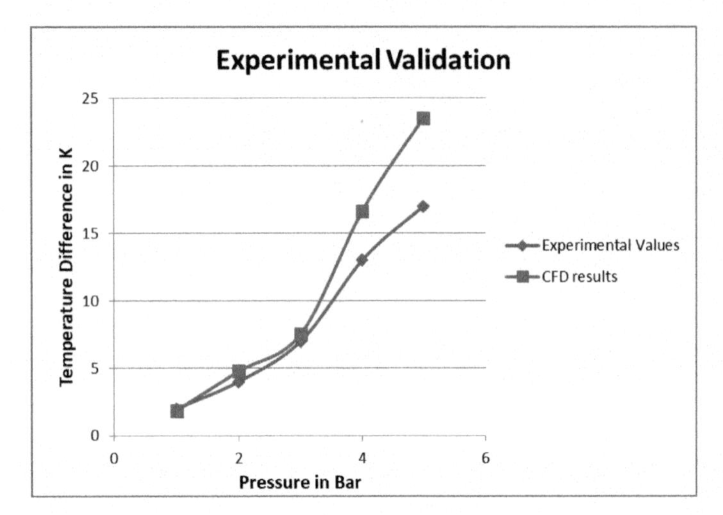

Figure 16.19 Graph of ΔT vs P to compare experimental and CFD results

As seen in the above table, the error goes on increasing, but it is within an acceptable range. Error is because, in CFD analysis, the tube is assumed to be perfectly adiabatic whereas there is a certain amount of heat loss in the actual case. It is also observed that the presence of moisture in air decreases the temperature difference in the actual case while in CFD analysis; we assume that dry air is entering the tube. Air in CFD analysis is also assumed to be an ideal gas. Thus CFD results were found close to actual results and thus validated.

CFD Results and Analysis

Analysis

The analysis is carried out on three different tube lengths as 300 mm, 100 mm, and 70 mm. These geometries are then further provided with three arrangements of nozzles as 1 nozzle, 2 nozzles, and 3 nozzles.

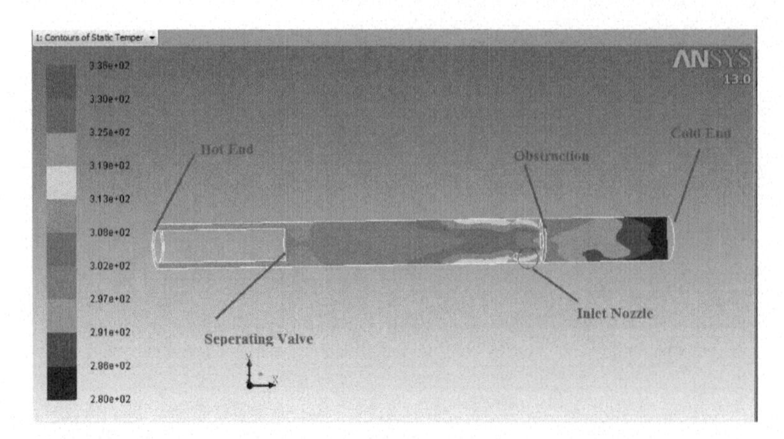

Figure 16.20 Temperature contours side view single nozzle 100 mm tube

Thus a total of nine different geometries are formed. Each of these geometries was solved for different values of inlet pressure ranging from 1 bar (Gauge) up to the gauge pressure which does not show significant variation from the previous input pressure in terms of cold end temperature and thus in temperature difference between the hot and cold end. The temperature contours at the central plane of some of these cases are plotted as shown in this section.

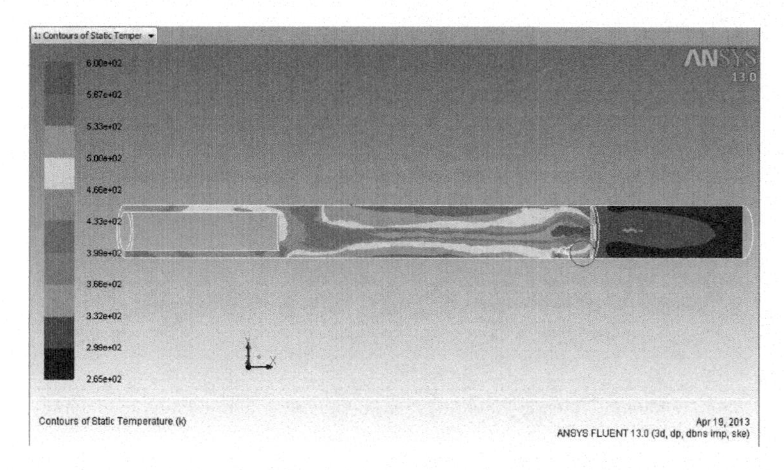

Figure 16.21 Temperature contours (in K) side view with two nozzles in 100 mm tube

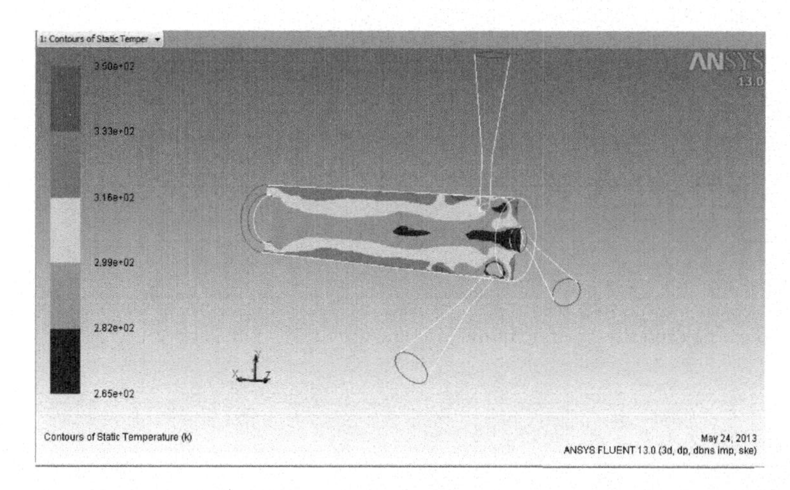

Figure 16.22 Temperature contours (in K) side view with 3 nozzles in 70 mm tube

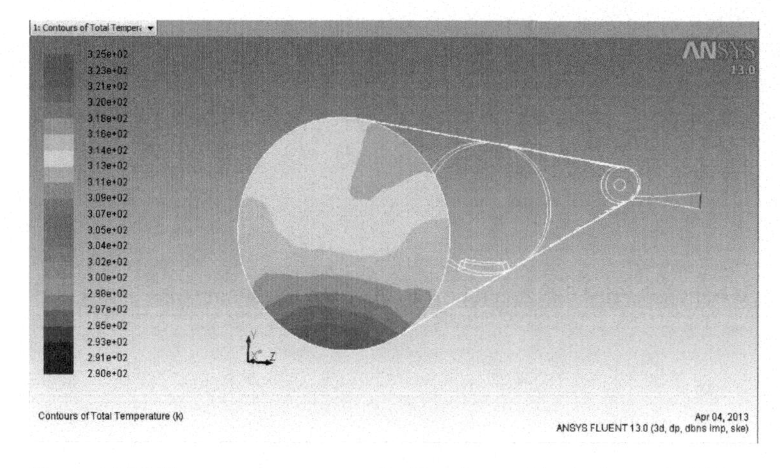

Figure 16.23 Temperature contours (in K) at hot end in single nozzle 300 mm tube

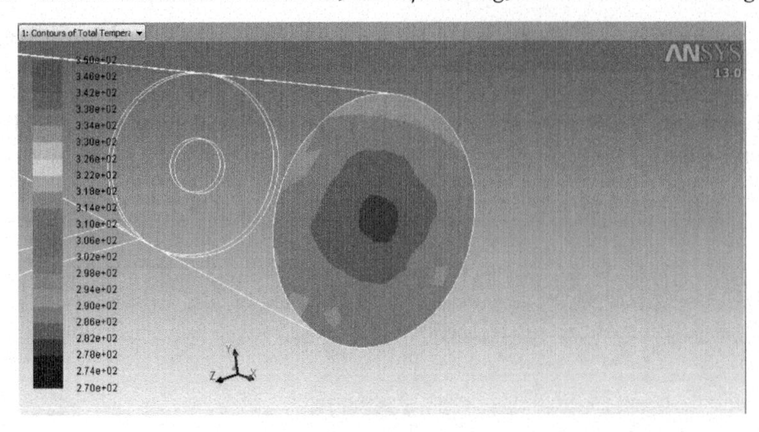

Figure 16.24 Temperature contours (in K) at cold end in single nozzle 100 mm tube

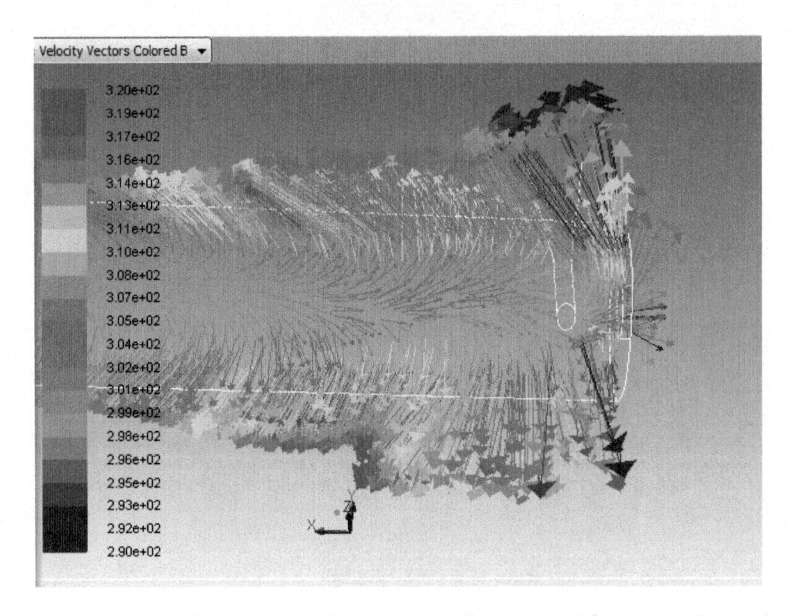

Figure 16.25 Velocity vectors showing centrally separated flow in single nozzle 100 mm tube

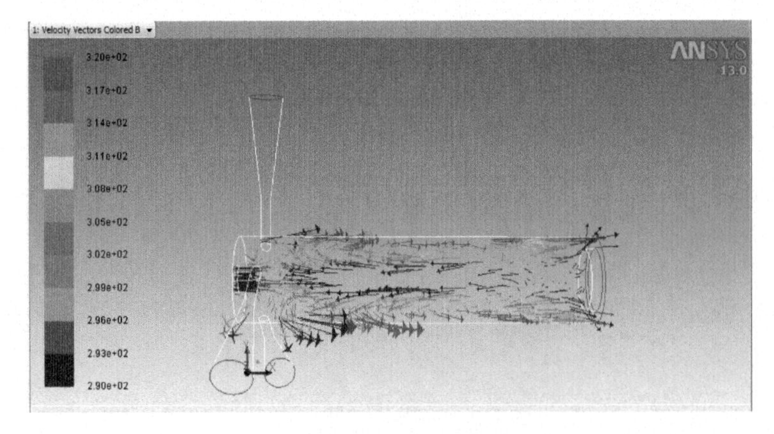

Figure 16.26 Vectors showing centrally separated flow in a tube with three nozzles 70 mm tube

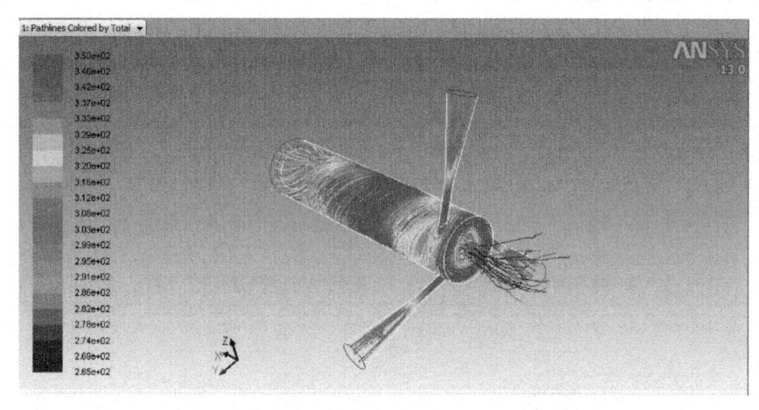

Figure 16.27 Path lines showing swirling flow in 3 nozzles 70 mm tube

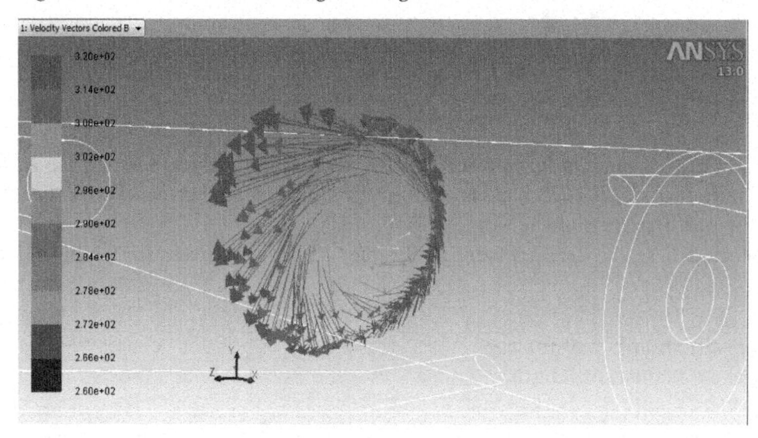

Figure 16.28 Swirl motion at a section in two nozzles 70 mm tube

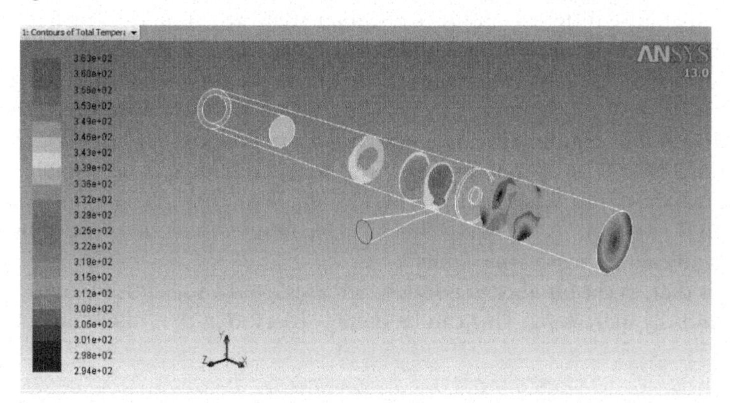

Figure 16.29 Total temperature contours at different sections in one nozzle 100 mm tube

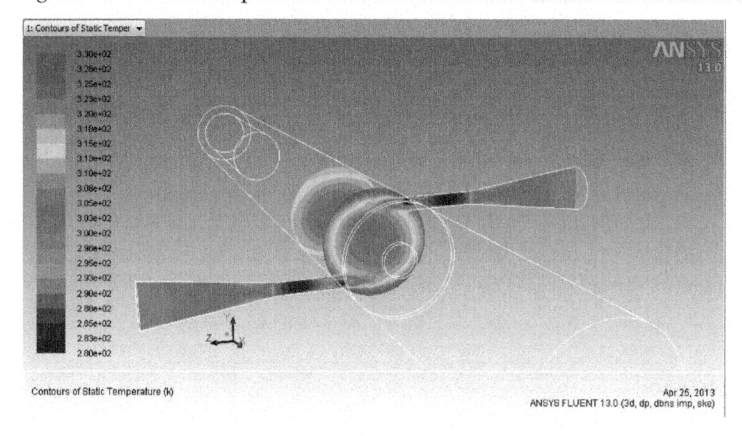

Figure 16.30 Temperature contours at the inlet of 2 nozzles 100 mm tube

The CFD temperature contours give a rough idea about the distribution of temperature on the surface. Contours are diagrammatic representations and thus are known as qualitative results. But to obtain an average temperature on the surface, these contours are not helpful. This is calculated with the help of the surface integral tool available in the reports option in the post-processor. This tool integrates the required parameter, temperature in this case, over the selected surface, to give one specific number which represents its average temperature on that surface. This is also called a quantitative result. This is shown below.

```
>
> Reading "H:\Ansys models\Short tube\New mesh\Short Tube coarse mesh_008_data_SOU_10.dat"...
Done.

              Area-Weighted Average
                Static Temperature                     (k)
-------------------------------- --------------------
                         cold_end            270.06668
                          hot_end            313.64481
                ---------------- --------------------
                              Net            284.27444
```

Figure 16.31 Results obtained by integrating temperature over the hot and cold end

Thus, with this tool, we calculate the average value of temperature over cold and hot ends. These values are then saved in tabular form. These results are then used to plot graphs that give a better idea of the variation of various parameters on the performance of the Vortex Tube.

The various graphs that are plotted here are Temperature difference ΔT in K vs Pressure in the bar for the following cases:

1. Variable tube lengths for the same number of nozzles.
2. A different number of nozzles at a constant length.

Variable tube lengths for the same number of nozzles

As seen in the above graphs, the temperature difference follows a similar trend when plotted against inlet pressure which is the gauge pressure in the bar, for all tube lengths and irrespective of the number of nozzles. It is observed that at low pressures, like 2 to 3 bar, the temperature difference between the hot and cold end is not significant. But after 3 bars it shows a significant increase in the temperature difference. This is visible as the slope of the curve is high in this region. This rise continues up to about 6 bar after which the temperature difference does not change significantly and whatever rise that is observed is very small. This might be due to the choking phenomenon of the nozzle, i.e. no rise in mass flow even after the increase in inlet pressure. Thus after about 6 to 7 bar, there is no significant increase in inlet mass flow and thus the temperature difference shows no significant increase after 7 bar.

One more point to be noted here is that, as the number of nozzles increases from 1 to 3, it is observed that the temperature difference is continuously increasing. This can be better observed in the following graphs.

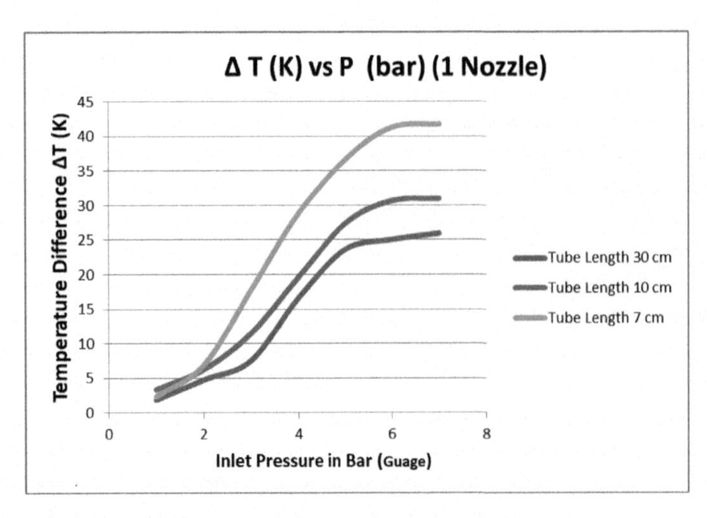

Figure 16.32 ΔT vs P for different tube lengths with 1 nozzle

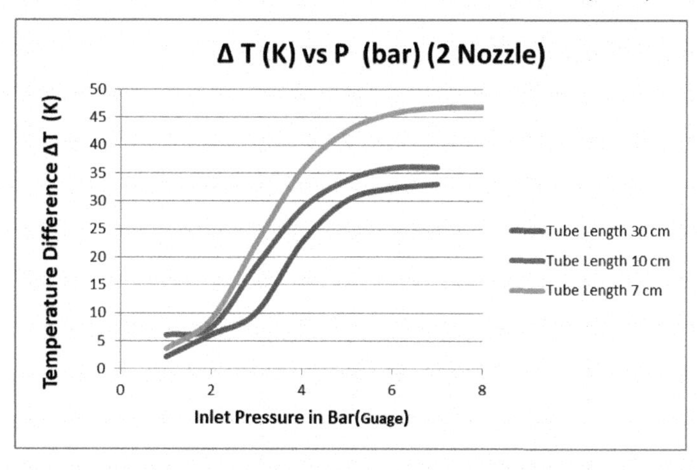

Figure 16.33 ΔT vs P for different tube lengths with 2 nozzles

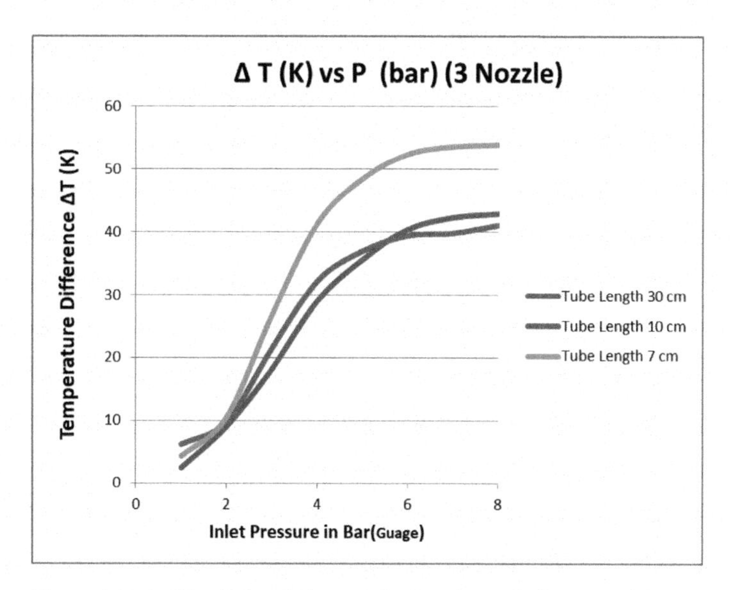

Figure 16.34 ΔT vs P for different tube lengths with three nozzles

Different number of nozzles at a constant length

Three graphs are plotted for temperature difference vs inlet pressure for constant working tube lengths with a different number of nozzles. The trend followed by the temperature difference is the same as that in the previous case. But as it can be seen from the graphs, the more the number of nozzles, the more the temperature difference. This is because as the number of nozzles increases, more mass of air is injected into the tube which gives rise to an intense vortex. This vortex is sustained up to a larger distance from the inlet. This results in more space for inner cold flow to expand within the core and thus decreases its pressure and thus temperature. While on the circumference of the tube, an intense vortex results in highly pressed air along the wall causing high pressure and thus high temperature. As a result, the temperature difference between the hot and cold end increases.

Actual hot and cold end temperature vs inlet pressure graph

The above graphs are drawn for ΔT vs P. But to get a better idea of the exact values of temperature that are obtained at hot and cold ends under various conditions, the following graphs are plotted.

The graphs are self-explanatory as it can easily be observed that, in all cases, cold end temperature rapidly decreases between 2 to 6 bar and then becomes almost steady. On the other hand, the hot end temperature rapidly increases between 2 to 6 bar and it also does not show significant variation. Sub-zero temperatures were reported at the cold end of the 70 mm tube at higher pressures for 2 and 3 nozzles, at an inlet temperature of 295 K.

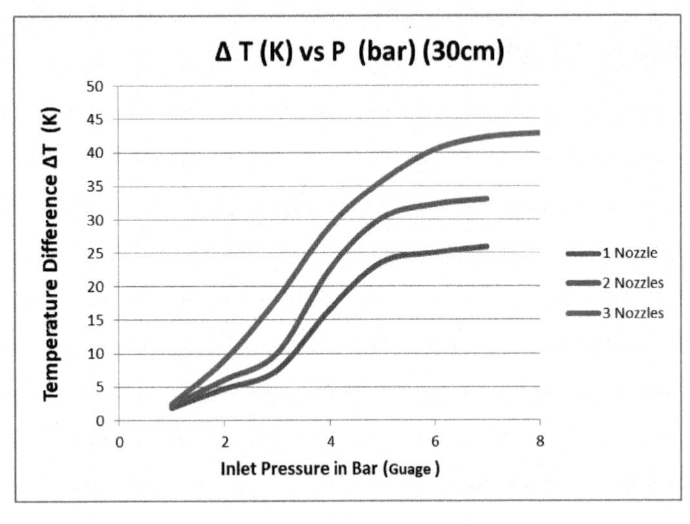

Figure 16.35 ΔT vs P for 30 cm tube with different number of nozzles

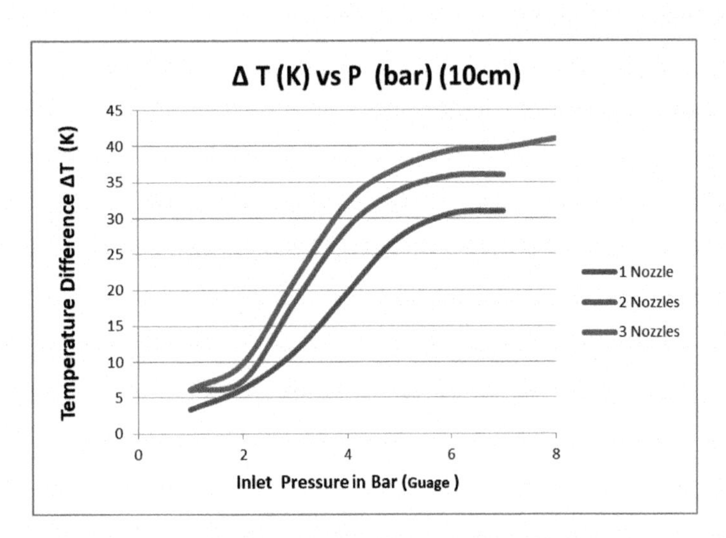

Figure 16.36 ΔT vs P for 10 cm tube with different number of nozzles

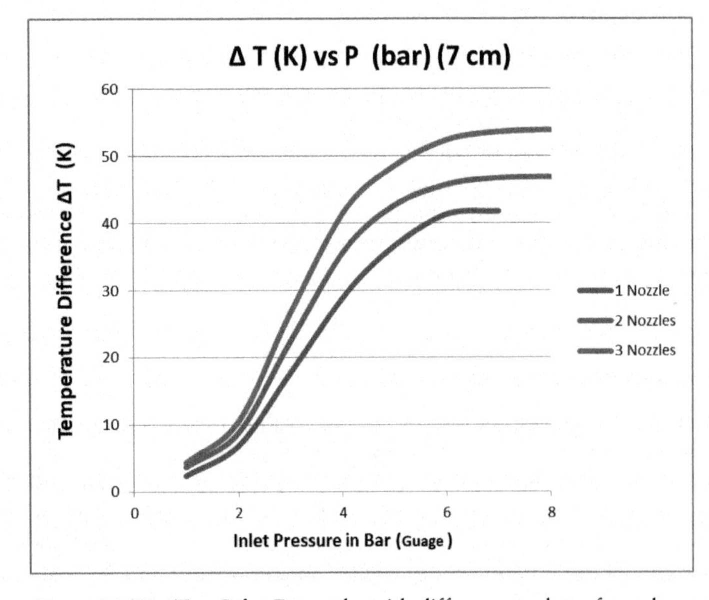

Figure 16.37 ΔT vs P for 7 cm tube with different number of nozzles

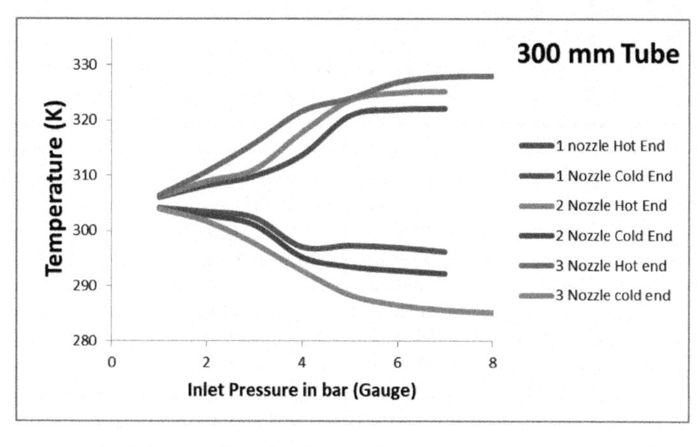

Figure 16.38 Hot and cold end temperature variation for 30 cm tube

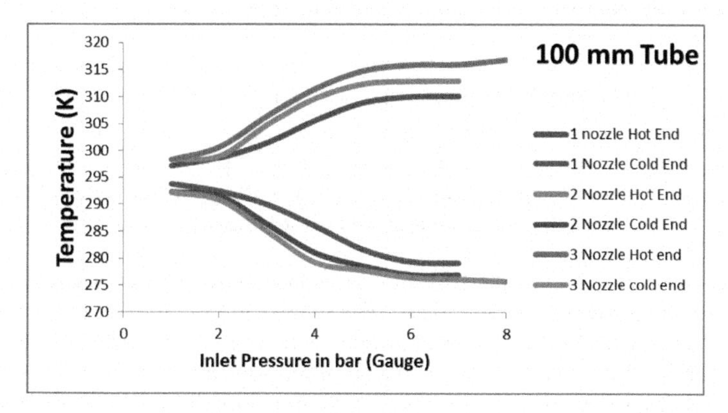

Figure 16.39 Hot and cold end temperature variation for 10 cm tube

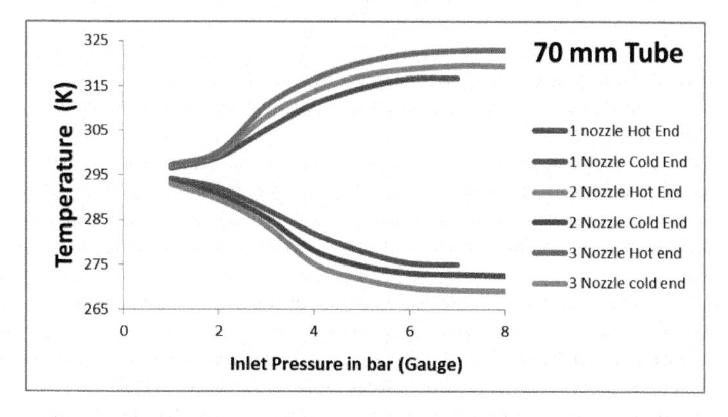

Figure 16.40 Hot and cold end temperature variation for 7 cm tube

Conclusions

The objective of the study was to develop a CFD simulation model for a Vortex tube and to analyse the performance by varying different parameters in a tube. The purpose of the work was to optimise the performance, using a CFD calculation which is an easier, accurate, cheaper, and faster method to analyse as compared to the experimental due to practical constraints.

In this study, the parameters affecting the performance of the vortex tube such as inlet pressure, number of nozzles, and tube length are studied. For this purpose, geometries with tube lengths 300 mm, 100 mm, and 70 mm were prepared on ANSYS WORKBENCH. These geometries were further provided with 1, 2,

Table 16.2 Maximum and minimum temperatures at different conditions

Tube length (mm)	*Number of nozzles*	*Pressure (bar)*	*Hot end temp.* T_h *(0C)*	*Cold end temp.* T_c *(0C)*
300	1	7	49.1676	23.2318
	2	7	52.266	19.2118
	3	8	55.071	12.184
100	1	7	37.185	6.2003
	2	7	40.001	3.9597
	3	8	43.912	2.8358
70	1	7	43.75	2.0113
	2	8	46.416	-0.449
	3	8	49.981	-3.9

and 3 nozzles to create nine different geometries. These were then studied for different inlet pressures. Table 16.2 represents the Maximum and minimum temperature values at different conditions.

It was observed that

- As the pressure increases, the temperature difference between the hot and cold ends of the tube continuously increases till an optimum value after which the temperature difference remains unchanged.
- The intensity of the impact of the swirl flow on the valve has a greater role to play in the temperature difference between the two ends. As this impact is more severe in the tube with a shorter length, better results are obtained for shorter tubes. But it is also mentionable that after some optimum tube length, it may not affect the temperature difference.
- The number of nozzles attached to the tube is directly affecting the inlet mass flow rate. Also due to more number of tangential nozzles, intense swirl motion is developed in the tube and thus its impact on the valve at the other end is severe. Thus with more number of nozzles, a greater temperature difference is obtained between hot and cold ends. However, there are practical limitations on many nozzles. Also after some optimum number of nozzles, the temperature difference obtained might not be sufficiently large to explain the investment for more number of nozzles.

References

De Vera, G. (2010). The Ranque-Hilsch vortex tube. Int. J. Refrig. Air Conditioning. 36:629–956.

Eiamsa-ard, S. and Promvonge, P. (2007). Numerical investigation of the thermal separation in a Ranque–Hilsch vortex tube. Int. J. Heat Mass Transfer. 37:156–162.

Farouk, T., Farouk, B., and Gutsol, A. (2009). Simulation of gas species and temperature separation in the counter-flow Ranque–Hilsch vortex tube using the large eddy simulation technique. Int. J. Heat and Mass Transfer. 52:3320–3333.

Pouraria, H. and Zangooe, M. R. (2011). Numerical investigation of vortex tube refrigerator with a divergent hot tube. Energy Procedia. 14:1554–1559.

Pourmahmoud, N., Hassanzadeh, A., and Moutaby, O. (2012). Numerical analysis of the effect of helical nozzles gap on the cooling capacity of Ranquee-Hilsch vortex tube. Int. J. Refrig. 35:1473–1483.

Pourmahmoud, N., Hassan Zadeh, A., Moutaby, O., and Bramo, A. (2012). Computational Fluid Dynamics Analysis of Helical Nozzles effects on the energy separation in a vortex tube. Thermal Science. 16(1):151–166.

Shamsoddini, R. and Khorasani, A. F. (2012). A new approach to study and optimize cooling performance of a Ranquee-Hilsch vortex tube. Int. J. Refrig. 35:2339–2348.

Wua, Y. T., Dinga, Y., Jia, Y. B., Maa, C. F., and Geb, M. C. (2007). Modification and experimental research on vortex tube. Int. J. Refrig. 30:1042–1049. (April edition).

Xue, Y., Arjomandi, M., and Kelso, R. (2009). A critical review of temperature separation in a vortex tube. Exp. Therm. Fluid Sci. 34:1367–1374.

Zin, K. K., Hansske, A., and Ziegler, F. (2010). Modeling and optimization of the vortex tube with computational fluid dynamic analysis. Energy Res. J. 1(2):193–196.

17 Data pre-processing using descriptive statistics for mathematical modeling

Pallavi Jayant Giri[1,a,*] *and Prashant Patil*[2,b]

[1]Assistant Professor, Laxminarayan Institute of Technology, Nagpur, Maharashtra, India

[2]Technische Universität Ilmenau , Germany

Abstract

This paper presents a thorough insight into the role of descriptive statistics for linear and nonlinear mathematical models for hydrological applications. The Pench reservoir is used for the case and after sample size calculation; readings are recorded from the actual site. Overall data is recorded for 19 independent and six dependent variables for 15 years. Before modelling descriptive statistics like standard deviation, variance, skewness, and kurtosis is calculated for all dependent and independent variables. Results reflect that data is normally distributed and fit for mathematical modelling. Further reliability analysis using chronbatch alpha is checked and the coefficient is more than 0.5 which is desirable for data validity. Hotelling's t square test and significance level hint toward the sustainability of the research hypothesis and lead to rejecting the null hypothesis. A definite underlying relationship between independent and dependent variables is then presented based on the goodness of fit of multiple regression and artificial neural network models. Good coefficient of correlation, coefficient of determination, and statistically significant models hint toward the effectiveness of descriptive statistics for modelling hydrological application. The methodology may be further useful for different hydrological applications.

Keywords: Descriptive statistics, hotelling t–square, kurtosis, reliability, skewness.

Introduction

Quantitative summarisation of different features of data essentially requires descriptive statistics Illustrious representation for inferential statistics is possible using descriptive statistics. Descriptive statistics is required to draw conclusion based on the set of data and it is not based on probability theory only. Simple summery of samples and observation could be drawn based on descriptive statistics, it does includes with initial description of data which may be a part of extensive statistical analysis.

Most of the modelling approaches are dependent on precise data collection, and proper pre-processing of data to handle specific uncertainties and noise in the data. Elaborated on the basic statistical inference for physical models and phenomena. Mapping a physical system is always a complex scenario. Uncertainty and errors are usually induced in the system of measurement and data collection process. Classic statistical inference is an essential part of the modelling process, failing to do that may lead to wrong interpretations and conclusions. Singh (2018), hydrological models are complex, sometime interdependency of variables that may have an impact on the overall hydrological model needs clear introspection. Historical perceptive for efficient modelling of the hydrological phenomenon is also important, which reflects data trends and their behaviour over the period. Model construction, calibration, and data processing are extremely important to draw suitable fruitful conclusions for hydrological applications.

Garcia MH (ed) (2008), measurement and modelling of sedimentation-related problems of hydrological applications are data-driven and it is important to pre-processed data before modelling it. Garcia MH (ed) (2008), explored the details of modelling and error measurement for accurate prediction for hydrological applications. National Institute of Hydrology Roorkee (2017), provided current and future trends for modelling hydrological systems. Hydrological models are built on dominant hydrological processes and depend on various factors. The different modelling and optimisation approaches depend on the precision of data collection and its corresponding pre-processing. The complex hydrological models like rainfall-runoff, watershed modelling, flood management system and water resource management depend on precise and noise-free data, corresponding data trends, and its behaviour with other independent and dependent variables. Dozier et al. (2017), focuses on and discussed the importance of data pre-processing for the optimization of water resources applications. Dwivedi et al. (2017), signify and revealed the details about

[a]pallavijgiri@gmail.com; [b]prashantpatil18011996@gmail.com

data-driven spatial models for geospatial application, geospatial is an integral part of hydrological applications and depends on huge datasets. Kurt K. Benke and Robinson (2017) pierson criteria of statistical correlation was used for robust regression analysis based on its ability to handle outliers from the data. Hossein Tabari (2019) elaborated the need of innovative method of statistical and stochastic modelling to handle the hydrological extremes.

In the present research, study data is collected extensively from the actual dam operation site. Recording of the logbook maintains at the dam site and other records maintained at the public works department is taken into consideration for the consolidation of data.

Study Area

A. Pench Dam Details

The name of the dam in the National Register of Dams is Kamthikhairy Dam and, the Official Designation of Kamthikhairy (Pench) Dam Irrigation Project is "Kamthikhairy (Pench) Dam, D - 01100" and in India-WRIS (Water Resources Information System) is JI00281. Pench Irrigation Project is constructed as part of an irrigation project by the Government of Maharashtra. It was completed in 1976. The dam is constructed for irrigation and supplies water to two districts of Maharashtra, Nagpur, and Bhandara. Pench irrigation project is a multipurpose project with the objective of domestic, industrial, and irrigation water supply. The catchment area of Pench Hydro Electric Project at Totladoh is 4273 sq. km. and of Pench Dam at Navegaon Khairy is 388 sq. km. The gross storage capacities are 1241 MCM and 230.032 MCM respectively. The dam is an Earth fill dam. The length of the dam is 2880 m (9448.82 ft) having the main dam, overflow section, and one sub bund. One Ogee-type gated spillway is provided with 16 gates of size 12 x 8 m.

B. Sample Size Calculation for Mathematical Modeling

The sample size is important for any empirical study where the target is to have inferences about a population from a sample. Statistical validity of data is an extremely important terminology while getting conclusive output from the study .Initially, the appropriate sample size is calculated using the following standard relationship to collect data for the present investigation.

$$N= \sigma 2 * p*(1 - p) / E^2$$

Where N is the sample size, σ = 1.96 which is the standard value from the Z table related standard deviation corresponding to 95% confidence interval, p = percentage picking a choice, expressed as a decimal, E = confidence interval, expressed as a decimal. By considering p=0.5, the minimum number of data points which is required to collect is 384.

Data is collected based on the minimum sample size required for modelling, a total of 2295 readings are taken into consideration for modelling of reservoir operation. Four-time periods have been considered for the collection of data namely Time periods T, T-5, T-10, and T-15. Figure 17.1 shows the Study area of Pench Dam.

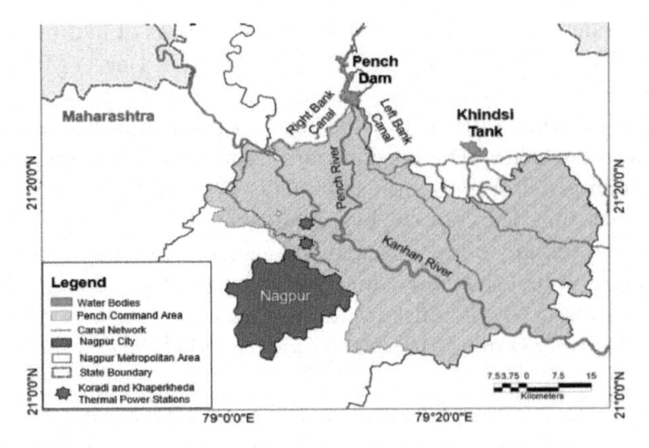

Figure 17.1 Study area Pench Dam (Source Maharashtra state irrigation department)

C. List of independent and dependent variables are shown in Tables 17.1 and 17.2 respectively.

Table 17.1 List of the independent variable

Code	*Description*	*Code*	*Description*
x_1	Temperature	x_{11}	Total Content
x_2	RL(Opening cond^n)	X_{12}	Drainage Losses
x_3	Capacity (Gross storage)	X_{13}	Evaporation Depth
x_4	Area (Opening cond^n)	X_{14}	Evaporation Losses Capacity
x_5	Live Storage	X_{15}	Release RBC
x_6	% Live Storage	X_{16}	Release LBC
x_7	Rainfall	X_{17}	Day's Outflow
x_8	Total Rainfall	X_{18}	Total Day's Outflow (Cumul.)
x_9	Day's Inflow	X_{19}	Live Capacity(Before Spill)
x_{10}	Total Day's Inflow (Cumulative)		

Table 17.2 List of the dependent variable

Code	*Description*	*Code*	*Description*
Y1	Live capacity	Y4	Area
Y2	Reservoir level	Y5	Deficit capacity/spill
Y3	Gross capacity	Y6	Discharge through all gates

Result and Discussion

A. Descriptive Statistics for Time Period T

The descriptive table exhibits the sample size, mean, standard deviation, and standard error of skewness and kurtosis for the data set of time period T. The sample means disbanding around the 639.60 standards by what appears to be a small amount of variation. The skewness and kurtosis values reported in the statistics table are all quite within a specified range, showing that the distributions of these variables are definitely towards normality. From Table 17.3–17.6, it is observed that there is small deviation in standard error for skewness and kurtosis which confine the sample size with 2295,2129,1980,1830 samples fit for the mathematical modelling .Data set is statistically valid to established the relationship between 19 input variables and 5 output variables.

Table 17.3 Descriptive statistics of all variables for time period T

	N Statistic	*Minimum Statistic*	*Maximum Statistic*	*Mean Statistic*	*Std. Deviation Statistic*	*Variance Statistic*	*Skewness*		*Kurtosis*	
							Statistic	*Std. Error*	*Statistic*	*Std. Error*
x1	2295	26.00	42.50	29.9672	3.26794	10.679	2.365	.051	5.264	.102
x2	2295	314.35	325.10	323.6408	1.79010		3.204	.051	4.833	.102
							−2.110			
x3	2295	61.83	232.55	199.4084	36.16034	1307.571	−1.611	.051	2.306	.102
x4	2295	8.60	25.34	23.1176	2.86245	8.194	−2.189	.051	5.181	.102
x5	2295	11.83	182.55	149.4084	36.16035	1307.571	−1.611	.051	2.306	.102
x6	2295	6.57	101.41	82.9953	20.08689	403.483	−1.611	.051	2.306	.102
x7	2295	0.00	143.50	7.2571	15.33046	235.023	3.734	.051	17.562	.102
x8	2295	0.00	1560.00	639.6023	428.70063	183784	.138	.051	−1.051	.102
x9	2295	0.00	326.25	6.8811	15.75168	248.115	10.806	.051	171.327	.102
x10	2295	0.00	3038.65	437.2944	626.48430	392482	2.726	.051	7.185	.102
x11	2295	12.73	500.37	156.2957	41.11805	1690.694	−.290	.051	6.152	.102
x12	2295	.00	.08	.0072	.00164	.000	38.138	.051	1699.879	.102
x13	2295	.70	12.00	3.3376	2.32850	5.422	2.376	.051	5.296	.102
x14	2295	.01	3.50	.0770	.09581	.009	21.344	.051	718.403	.102
x15	2295	0.00	4.69	.8836	.55645	.310	1.334	.051	4.405	.102
x16	2295	0.00	9.25	2.6301	2.73280	7.468	.381	.051	−1.562	.102
x17	2295	.03	11.11	3.5978	2.93117	8.592	.417	.051	−1.413	.102
x18	2295	.78	3021.70	439.8880	616.92723	380599	2.641	.051	6.840	.102
x19	2295	11.83	499.74	152.6979	41.60332	1730.837	−.284	.051	6.062	.102

Table 17.4 Descriptive statistics of all variables for time period T-5

	N	*Minimum*	*Maximum*	*Mean*	*Std.*	*Variance*	*Skewness*		*Kurtosis*	
	Statistic	*Statistic*	*Statistic*	*Statistic*	*Deviation Statistic*	*Statistic*	*Statistic*	*Std. Error*	*Statistic*	*Std. Error*
x1	2129	26.00	42.40	29.6293	2.65666	7.058	2.568	.053	7.585	.106
x2	2129	315.37	325.10	323.7018	1.74098	3.031	-2.099	.053	4.687	.106
x3	2129	61.83	232.55	200.7731	35.51566	1261.362	-1.643	.053	2.374	.106
x4	2129	9.93	25.34	23.2131	2.78385	7.750	-2.188	.053	5.110	.106
x5	2129	11.83	182.55	150.7731	35.51566	1261.362	-1.643	.053	2.374	.106
x6	2129	6.57	101.41	83.7534	19.72877	389.224	-1.643	.053	2.374	.106
x7	2129	0.00	143.50	7.8110	15.78136	249.051	3.592	.053	16.227	.106
x8	2129	0.00	1555.00	642.7862	415.19720	172388.714	.161	.053	-1.013	.106
x9	2129	0.00	326.25	7.0930	16.31222	266.089	10.453	.053	159.883	.106
x10	2129	.76	3021.52	426.8852	611.28435	373668.559	2.792	.053	7.515	.106
x11	2129	12.73	500.37	157.8728	40.79437	1664.181	-.217	.053	6.558	.106
x12	2129	.00	.08	.0072	.00169	.000	37.617	.053	1626.886	.106
x13	2129	.70	12.00	3.0957	1.89448	3.589	2.594	.053	7.710	.106
x14	2129	.01	3.50	.0721	.09362	.009	24.577	.053	854.790	.106
x15	2129	0.00	4.69	.8552	.53814	.290	1.245	.053	3.856	.106
x16	2129	0.00	7.68	2.5872	2.70224	7.302	.406	.053	-1.534	.106
x17	2129	.03	9.43	3.5217	2.88709	8.335	.431	.053	-1.400	.106
x18	2129	5.37	2978.62	428.3388	601.04713	361257.652	2.712	.053	7.189	.106
x19	2129	11.83	499.74	154.3511	41.27878	1703.937	-.213	.053	6.467	.106

Table 17.5 Descriptive statistics of all variables for time period T–10

	N	*Minimum*	*Maximum*	*Mean*	*Std.*	*Variance*	*Skewness*		*Kurtosis*	
	Statistic	*Statistic*	*Statistic*	*Statistic*	*Deviation Statistic*	*Statistic*	*Statistic*	*Std. Error*	*Statistic*	*Std. Error*
1980	26.00	42.30	29.3815	2.20290	4.853	2.652	.055	9.816	.110	
x2	1980	316.42	325.10	323.7575	1.69361	2.868	−2.077	.055	4.464	.110
x3	1980	72.78	232.55	201.9589	34.85133	1214.615	−1.665	.055	2.388	.110
x4	1980	11.38	25.34	23.2977	2.70515	7.318	−2.174	.055	4.953	.110
x5	1980	22.78	182.55	151.9589	34.85133	1214.615	−1.665	.055	2.388	.110
x6	1980	12.66	101.41	84.4121	19.35973	374.799	−1.665	.055	2.388	.110
x7	1980	0.00	143.50	8.1806	16.18233	261.868	3.508	.055	15.353	.110
x8	1980	0.00	1552.00	650.1757	401.01311	160811.518	.184	.055	−.978	.110
x9	1980	0.00	326.25	7.2973	16.87774	284.858	10.113	.055	149.366	.110
x10	1980	8.85	2999.46	422.3888	601.33033	361598.165	2.821	.055	7.599	.110
x11	1980	24.45	500.37	159.2553	40.50843	1640.933	−.133	.055	6.965	.110
x12	1980	.00	.08	.0072	.00174	.000	37.156	.055	1561.339	.110
x13	1980	.70	12.00	2.9172	1.56816	2.459	2.688	.055	10.017	.110
x14	1979	.01	3.50	.0672	.08803	.008	30.761	.055	1174.962	.110
x15	1980	0.00	4.69	.8381	.54401	.296	1.333	.055	4.112	.110
x16	1980	0.00	7.36	2.5439	2.65972	7.074	.428	.055	−1.491	.110
x17	1980	.03	9.43	3.4564	2.85522	8.152	.455	.055	−1.346	.110
x18	1980	9.35	2941.02	422.9702	590.67989	348902.738	2.749	.055	7.313	.110
x19	1980	22.78	499.74	155.7989	40.99962	1680.969	−.132	.055	6.867	.110

B. Descriptive statistics for time period T

Reliability analysis consent to study the properties of measurement scales and the items that compile the scales. The Reliability Analysis procedure calculates a number of commonly used measures of scale reliability and also endows with information about the relationships amid individual items on the scale. Intraclass correlation coefficients can be used to compute inter-rater reliability estimates. Eligibility of dataset for mathematical modelling is confirmed by chronbatch alpha values of 0.776 and 0.886 for standardise items. Internal consistency of data set for time period T is also justified through Cronbach alpha for time period T. Tables 17.7 and 17.8 shows the Reliability statistics and Hotelling's t-squared test data respectively.

Table 17.6 Descriptive statistics of all variables for time period T–15

	N	*Minimum*	*Maximum*	*Mean*	*Std.*	*Variance*	*Skewness*		*Kurtosis*	
	Statistic	*Statistic*	*Statistic*	*Statistic*	*deviation Statistic*	*Statistic*	*Statistic*	*Std. Error*	*Statistic*	*Std. Error*
1830	26.00	40.20	29.1878	1.84769	3.414	2.469	.057	10.091	.114	
x2	1830	316.42	325.10	323.7970	1.65123	2.727	–2.058	.057	4.366	.114
x3	1830	72.78	232.55	202.8938	34.19886	1169.562	–1.678	.057	2.391	.114
x4	1830	11.38	25.34	23.3591	2.63297	6.933	–2.157	.057	4.873	.114
x5	1830	22.78	182.55	152.8938	34.19886	1169.562	–1.678	.057	2.391	.114
x6	1830	12.66	101.41	84.9315	18.99729	360.897	–1.678	.057	2.391	.114
x7	1830	0.00	121.00	8.5012	16.18239	261.870	3.321	.057	13.476	.114
x8	1830	19.00	1549.00	657.0928	385.66533	148737.743	.210	.057	–.944	.114
x9	1830	0.00	326.25	7.4934	17.52051	306.968	9.750	.057	138.564	.114
x10	1830	15.20	2962.51	416.8597	589.27257	347242.160	2.852	.057	7.708	.114
x11	1830	26.10	500.37	160.3877	40.30247	1624.289	–.026	.057	7.395	.114
x12	1830	.00	.08	.0072	.00179	.000	36.605	.057	1490.129	.114
x13	1830	.70	10.40	2.7789	1.31154	1.720	2.514	.057	10.389	.114
x14	1829	.01	3.50	.0647	.08907	.008	32.127	.057	1217.087	.114
x15	1830	0.00	4.69	.8172	.54091	.293	1.399	.057	4.535	.114
x16	1830	0.00	7.36	2.4899	2.60581	6.790	.454	.057	–1.436	.114
x17	1830	.03	9.43	3.3790	2.79761	7.827	.477	.057	–1.286	.114
x18	1830	13.31	2903.82	416.9148	578.41089	334559.156	2.790	.057	7.475	.114
x19	1830	22.78	499.74	157.0087	40.79812	1664.487	–.029	.057	7.287	.114

Table 17.7 Reliability statistics

Time period	*Cronbach's Alpha*	*Cronbach's Alpha Based on Standardised Items*	*N of Items*
T	0.776	0.886	24
T-5	0.677	0.877	24
T-10	0.677	0.888	24
T-15	0.677	0.889	24

Table 17.8 Hotelling's t-squared test

Time period	*Hotelling's T-squared*	*F*	*df1*	*df2*	*Sig*
T	17418252983114704.000	750049347250677.800	23	2271	.000
T-5	18298567962235628.00	787364827731044.200	23	2106	.000
T-10	14201965891989854.000	610608987662816.100	23	1956	.000
T-15	13821823168067518.000	593716407609407.800	23	1806	.000

C. Hotelling's t-ssquare test for all time period

The results of the test for time period T are represented in the table for five output variables. Here we get a test statistic of 593716407609407.800 with 1806 degrees of freedom, 1806 coming from the 19 independent variables. The p-value for the test is less than 0.0001 indicating that we can reject the null hypothesis and ultimately indicate that the data set is ready for mathematical modelling to establish the research hypothesis. In another way there is a definite possibility exists of a relationship between 19 independent variables and five dependent variables for the T-15 data set. Similarly for all other time periods also data set is eligible for modelling.

After detailed descriptive statistics, it is evident that the data is valid and suitable for the modelling process. The details about the independent variables and dependent variables hint toward modelling using the linear method and nonlinear method. Following are the results of the modeling.

D. Summary of mathematical models

After detail statistical analysis ,based on the validity check and eligibility of the linear and nonlinear mathematical models for all periods based on the 19 independent variables and five dependent variables are formulated and depicted in the tables.

Table 17.9 Summary of regression analysis models

Modeling technique	*Model Name (based on dependent variables)*	*Coefficient of correlation(R)*	R^2	*Standard Error of estimate*
Regression analysis of basic variables (Time period T)	Live Capacity	0.956	0.913	2.9375
	Reservoir Level	0.967	0.953	0.2878
	Gross Capacity	0.957	0.915	2.9375
	Area	0.968	0.937	0.4490
	Deficit Capacity/Spill	0.946	0.8949	1.9634
Regression analysis of basic variables (Time period T-5)	Live Capacity	0.959	0.916	2.9924
	Reservoir Level	0.945	0.893	0.2962
	Gross Capacity	0.956	0.913	2.9924
	Area	0.946	0.894	0.4625
	Deficit Capacity/Spill	0.959	0.919	2.0350
Regression analysis of basic variables (Time period T-10)	Live Capacity	0.945	0.893	3.0476
	Reservoir Level	0.984	0.968	0.3053
	Gross Capacity	0.925	0.855	3.0476
	Area	0.934	0.872	0.4760
	Deficit Capacity/Spill	0.945	0.893	2.1052
Regression analysis of basic variables (Time period T-15)	Live Capacity	0.952	0.906	3.1132
	Reservoir Level	0.941	0.8841	0.3157
	Gross Capacity	0.956	0.913	3.1132
	Area	0.946	0.895	0.4919
	Deficit Capacity/Spill	0.958	0.917	2.1797

Table 17.10 Summary of regression analysis model for total discharge from all gates

Modelling technique	*Model name (based on dependent variables)*	*Coefficient of correlation(R)*	R^2	*Standard error of estimate*
Regression analysis of basic variables (Time period T)	Total Discharge from all the gates	0.939	0.881	355.8587

Table 17.11 Summary of ANN models for basic variables and all dependent variables

Modelling technique	*Model*	*Coefficient of correlation (R)*	R^2	*SSE*		*RMSE*	*Relative error*	
				Training	*Prediction*		*Training*	*Prediction*
ANN Models for basic variables (Time period T)	Live Capacity	0.998	0.996	0.257	0.124	1.493	0.007	0.009
	Reservoir Level	0.994	0.988	0.851	0.251	0.221	0.029	0.026
	Gross Capacity	0.999	0.998	1.648	0.342	1.3351	0.012	0.005
	Area	0.992	0.984	0.654	0.223	0.3578	0.024	0.018
	Deficit Capacity/Spill	0.99931	0.9986	0.164	0.094	2.09	0.028	0.036
ANN Models for basic variables (Time period T-5)	Live Capacity	0.99918	0.9983	0.158	0.145	1.899	0.005	0.011
	Reservoir Level	0.99243	0.9849	0.659	0.237	0.2406	0.026	0.029
	Gross Capacity	0.99907	0.9981	0.171	0.073	2.2395	0.005	0.005
	Area	0.99107	0.9822	0.777	0.327	0.36503	0.033	0.029
	Deficit Capacity/Spill	0.99936	0.9987	0.222	0.101	2.2475	0.041	0.046
ANN Models for basic variables (Time period T-10)	Live Capacity	0.99901	0.9980	0.545	0.288	1.8019	0.017	0.018
	Reservoir Level	0.99021	0.9805	0.780	0.437	0.2579	0.031	0.037
	Gross Capacity	0.99915	0.9983	0.321	0.124	1.3622	0.010	0.008
	Area	0.9906	0.9812	0.926	0.317	0.33327	0.035	0.030
	Deficit Capacity/Spill	0.99939	0.9987	0.054	0.048	1.8649	0.011	0.020
ANN Models for basic variables (Time period T-15)	Live Capacity	0.998	0.996	0.446	0.100	1.9926	0.015	0.008
	Reservoir Level	0.98898	0.978	1.062	0.302	0.3070	0.047	0.030
	Gross Capacity	0.99868	0.9973	0.304	0.101	1.9280	0.010	0.008
	Area	0.9868	0.9737	0.690	0.345	0.5242	0.029	0.041
	Deficit Capacity/Spill	0.99942	0.9988	0.690	0.061	1.2397	0.029	0.041

Table 17.12 Summary of ANN model for total discharge from all gates

Modelling technique	*Model*	*Coefficient of Correlation (R)*	R^2	*SSE*		*RMSE*	*Relative Error*	
				Training	*Prediction*		*Training*	*Prediction*
ANN models for basic variables	Total discharge from all the gates	0.999	0.998	0.041	0.116	18.244	0.008	0.043

Tables 17.9–17.12 depict assorted models with the name of the modelling technique, the dependent variables along with the coefficient of correlation and RMS error for all the models as well as the coefficient of determination. Pre-processing of data and detailed descriptive statistics which leads to handling the variables for mathematical models have reaped good results.

The correlation coefficient obtained for various models using different modelling techniques signifies that ANN models with basic variables are superior to Linear Regression modelling. Even the Regression model tapped the specific benefit of predictive modelling for the present investigation to predict dependent variables abidingly well. But the function-fitting machine learning process of ANN outperformed Regression Analysis. ANN models have been exhaustively checked with all possible combinations of transfer functions of input and output layers. Further, the overall effect of various optimisation algorithms is also tested with specific zeal. ANN modelling tested for the present investigation is a comprehensive effort to test the underlying relationship of complex hydrological operations.

The following figures depict the effectiveness of function fitting and machine learning of Regression with the actual values of Dependent variables. A closed association of curves shows that the coefficient of correlation for all models is very good. Comparison of predicted values of Live capacity, RL, gross capacity, area, deficit capacity and dischrage (Regression, ANN) with actual values are shown in Figures 17.2, 17.3, 17.4, 17.5, 17.6 and 17.7 respectively.

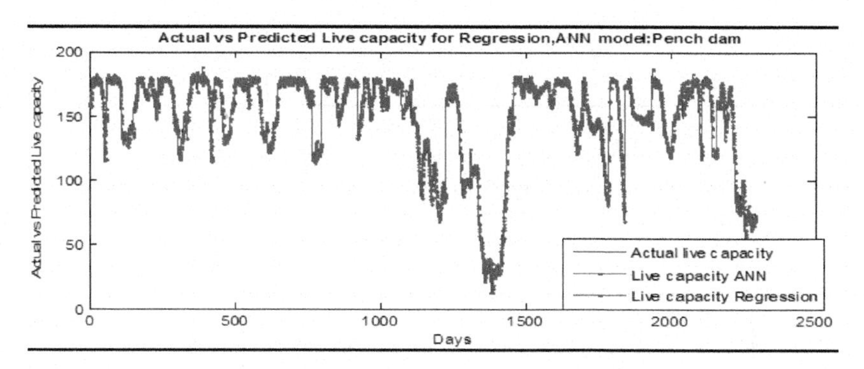

Figure 17.2 Comparison of predicted values of live capacity (regression, ANN) with actual

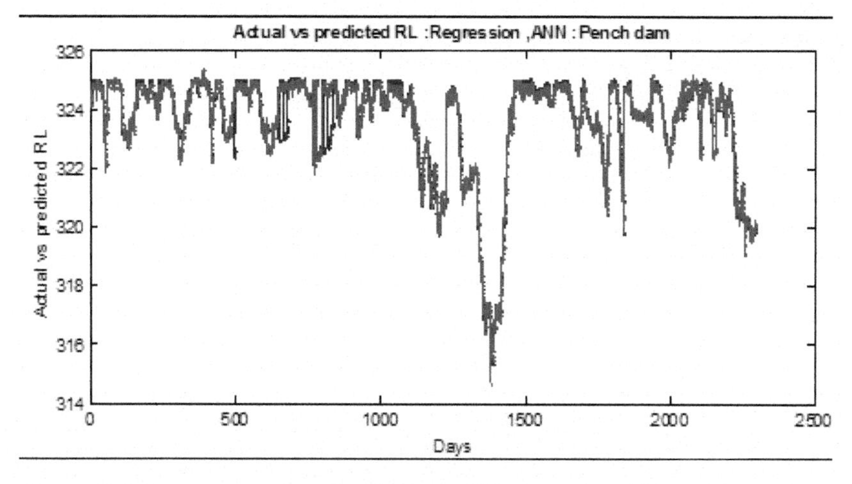

Figure 17.3 Comparison of predicted values of RL (regression, ANN) with actual

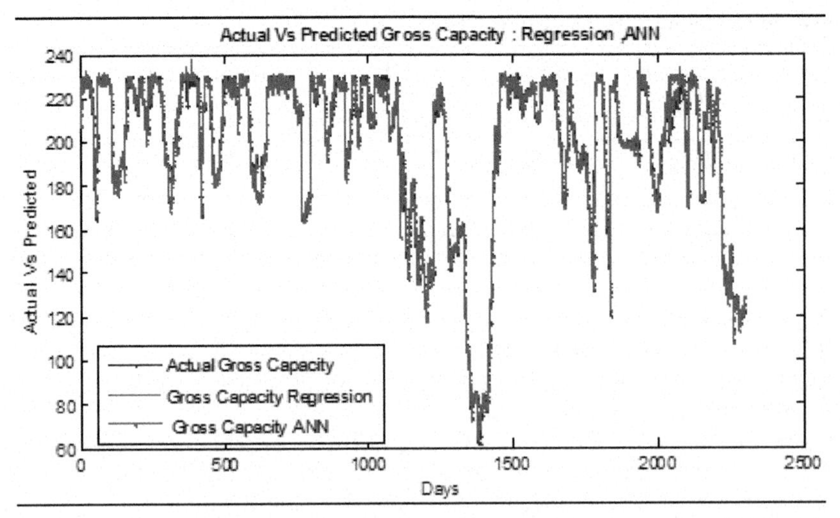

Figure 17.4 Comparison of predicted values of gross capacity (regression, ANN) with actual

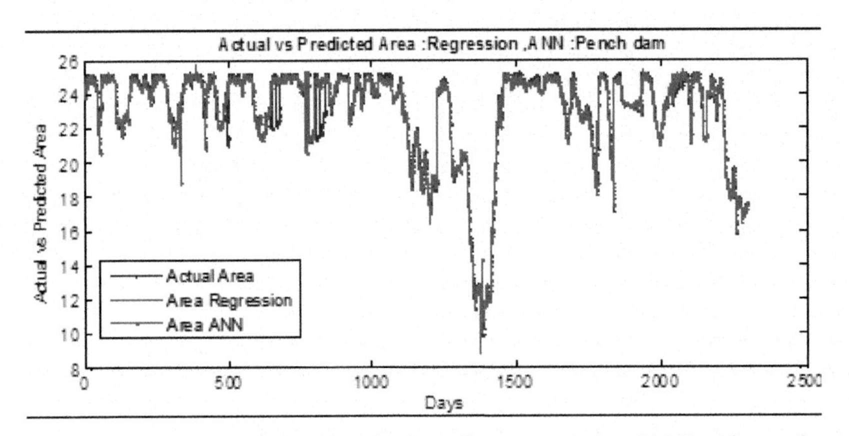

Figure 17.5 Comparison of predicted values of area (regression, ANN) with actual

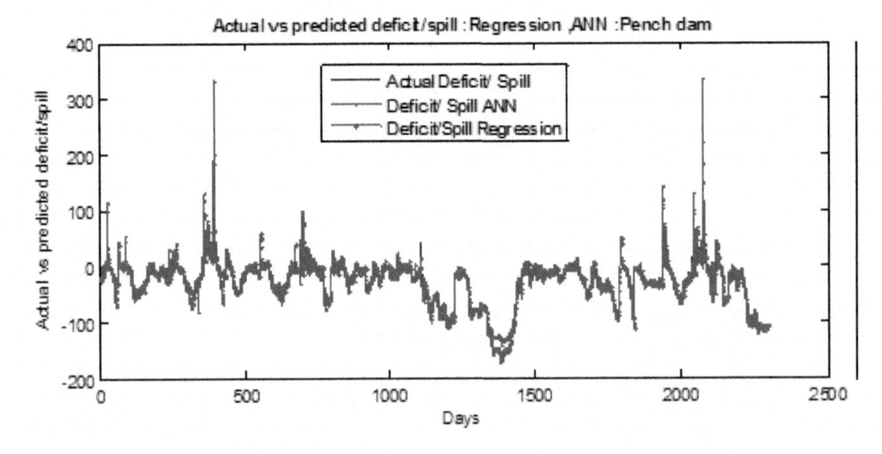

Figure 17.6 Comparison of predicted values of deficit capacity/spill (regression, ANN) with actual

Closed associativity of predicted values of regression models and ANN models and the overall goodness of fit of all models hints towards proper pre-processing of data. The methodology adopted to record the variables and sample size calculation to ensure the statistical validity of the model is well validated. However the same process may be replicable for a similar type of hydrological model.

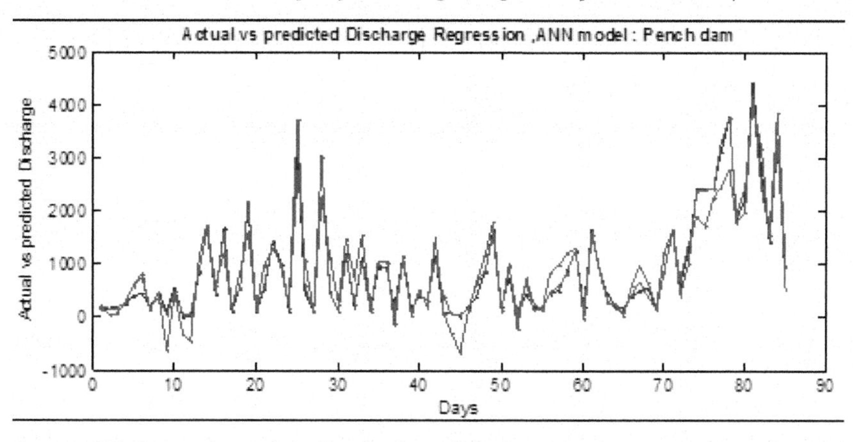

Figure 17.7 Comparison of predicted values of discharge (regression, ANN) with actual

Conclusion

An attempt of revealing the effect of proper and systematic descriptive statistics and data validity before linear and nonlinear mathematical modelling is presented in this paper. Overall it is reflecting conclusive output for regression and ANN models. The goodness of fit for both the modelling techniques signifies that the pre-processing of data before actual mathematical modelling has a productive impact on the outputs of models. Predictability and underlying relationship between independent variables and dependent variables related to reservoir operation are thus established properly. This study will be replicable to a similar type of hydrological model.

References

Benke, K. K. and Robinson, N. J. (2017). Quantification of uncertainty in mathematical models: The statistical relationship between field and laboratory pH measurements. Appl. Environ. Soil Sci. 2017:12. Article ID 5857139. https://doi.org/10.1155/2017/5857139.

Dozier, A., Arabai, M., Labadi, J. and Fontane, D. (2017). Optimization approaches for integrated water resources management. Chapter 24. In Handbook of applied hydrology, ed. V. P. Singh, 24-1–24-7. New York, N Y: McGraw-Hill Education.

Dwivedi, D., Dafflon, B., Arora, B., Wainwright, H. M. and Finsterle, S. (2017). Spatial analysis and geostatistical methods. Chapter 20. In Handbook of applied hydrology, ed. V. P. Singh, 20-1–20-9. New York: McGraw-Hill Education.

Garcia, M. H. (ed.) (2008). Sedimentation engineering: Processes, measurements, modeling, and nature. ASCE Manuals and Reports on Engineering Practice no. 110. (pp. 1132). Virginia:ASCE, Reston

National Institute of Hydrology Roorkee (2017). Hydrological Modeling – Current Status and Future Directions. Center for Excellence in Hydrological Modeling at NIH National Hydrology Project National Institute of Hydrology Roorkee.

Singh, V. P. (2018). Hydrologic modelling: progress and future directions. Geosci. Lett. 5:15. https://doi.org/10.1186/s40562-018-0113-z.

Tabari, H. (2019). Statistical analysis and stochastic modelling of hydrological extremes. Water 11:1861. https://doi.org/10.3390/w11091861.

18 Prediction of response parameter for turning AISI 4340 steel using hybrid taguchi-ANN method (HTAM)

Prashant D. Kamble[1,a,*], *Atul C. Waghmare*[2,b], *Ramesh D. Askhedkar*[3,c], *and Shilpa B. Sahare*[1,d]

[1]Department of Mechanical Engineering, Yeshwantrao Chavan College of Engineering, Nagpur, India

[2]Department of Mechanical Engineering, KDK College of Engineering, Nagpur, India

[3]Department of Mechanical Engineering, Ex-Professor, VNIT, Nagpur, India

Abstract

In this study, the investigation for turning AISI 4340 steel is experimentally done. The main purpose of this work is to minimise the surface roughness by using Hybrid Taguchi Artificial Neural Network Method(HTAM). L_{27} orthogonal array is adopted by the Taguchi technique. CNC lathe machine is used for experimentation. The readings are recorded for each level of noise factor (spindle vibration). S/N ratio and main effect plot are used to find the optimal setting. An artificial neural network (ANN) is further implemented to analyse and predict the output parameter. The response is predicted by adopting the additive model and it comes to 0.947 um. The result revealed that the most significant parameter is feed rate and least significant parameter is Tool type. The optimal setting obtained is A_3 B_3 C_3 D_2 and E_1. A comparison of experimental and predicted values shows there is a minimum variation. The conformity test is also implemented to find the range of acceptable experimental values.

Keywords: ANN, surface roughness, taguchi method, turning process.

Introduction

In any metal removing process especially in turning, quality of the product (surface finish) and productivity (Material Removal rate) are the most important terms. But these are contradictory in nature. It becomes important to find the breakeven point where better quality and productivity are gained.

Turning process is the most extensively used metal removal process. There is always a chance of improvement in such processes to make it better and better. Advancement in technology can be used to get the precious value of responses. Artificial neural network is the advanced tool with help of which prediction of output parameters is done in advanced level. The main goal of this study is to forecast the surface roughness by ANN while processing of AISI4340 steel and to find the optimal setting to get best surface finish.

Many work is done on optimization of machining process. Few literature reviews is mentioned here. As per Arjun Joshy et al. (2019), the ANN can be used for predicting the output parameters like surface roughness. In this work, back propagation neural network (BPNN) approach was made in MATLAB software. According to Karkalos and Markopoulos (2017), Tool life is the crucial factor in turning processes. In this Study, ANN is applied for prediction of tool life. Further it is mentioned that the application of ANN method along with Taguchi method can be used. Sreenivasulu, Reddy (2013), in this research work, Ra in CNC face milling is predicted by neural network modelling. The feed forward artificial neural networks is used for this purpose. As per Sanjeev Kumar et al. (2014), a hybrid Taguchi artificial neural network (ANN) can be practiced to obtain the best setting of control parameters. Surface integrity is tested by SEM. Asiltürk and Çunkaş (2011), in this work, the response is investigated by conventional factorial design. For prediction, ANN is implemented. Yalcin et al. (2013), this work proposed the investigation on effect of cutting parameters on Cutting Force, Surface Roughness and Temperatures by using ANN. As per Davim et al. (2008), the ANN can be used for the prediction of surface roughness. Back-propagation training algorithm is used for this purpose. L_{27} orthogonal array is implemented for experimentation. Result revealed that cutting speed is the most significant factor. Mia and Dhar (2016), stated that ANN is the suitable method for the prediction of average surface roughness for turning. The regression coefficient was more than 0.997 which is the good one. Kishore et al. (2018) used ANN model to predict the response parameter. It is found that the predicted valued by ANN is closely nearer to experimental values. Analysis of variance

[a]drpdkamble@gmail.com; [b]dracwaghmare@rediffmail.com; [c]r.askhedkar@rediffmail.com; [d]mrsspkamble@gmail.com

is implemented for percentage contribution of input parameters. Kamble et al. (2021), this paper Surface Roughness and Tool Wear are optimized by hybrid Taguchi-Grey relational analysis (HTGRA) method. Surface Roughness and Tool Wear are optimized to 0.958 um and 0.0401 mm.

After studying above literatures, the research gap is found as follows:-

- Very few researches used the noise factor for optimization.
- AISI 4340 steel is not used any researchers as a workpeice material although it is having the huge application in industries especially in aerospace and hydraulic industries.

Experimentataion

CNC Lathe Machine (Spinner 15) is used for performance. The CNC machine and experimental set up is shown in Figures 18.1 and 18.2.

The Table 18.1. Shows the input factors and their different values.

L_{27} OA (Orthogonal Array) is adopted by Taguchi Approach. Table 18.2 shows the design matrix.

A. Measurement of Output Parameters

The surface roughness (SR) is measured by MITECH MDT310 Portable Surface Roughness Tester. Individual values of roughness are recorded at 1.7 m/s^2, 4.3 m/s^2 and 6.9 m/s^2 of spindle vibration. Figure 18.3 show the recording of roughness. Table 18.3 shows the recorded output at each level of noise factor and its average value.

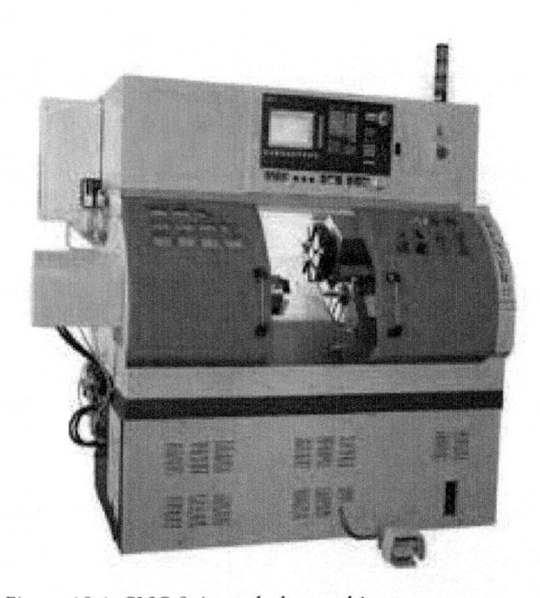

Figure 18.1 CNC Spinner lathe machine

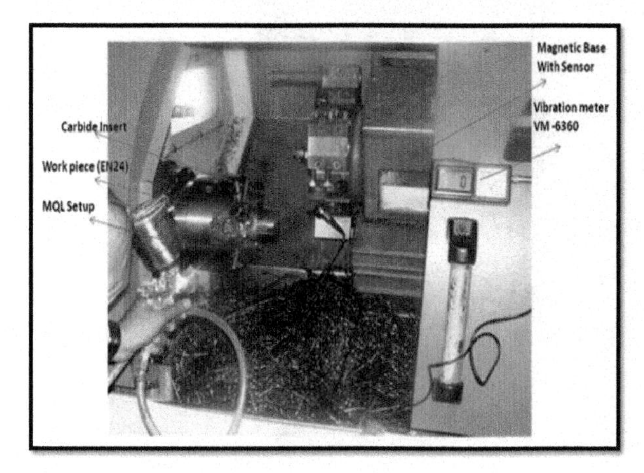

Figure 18.2 Experimental setup

Table 18.1 Control parameters and their number of levels

Process Parameter	*Level*		
	Low	*Mid*	*High*
CE (Cutting Environment)	1	2	3
NR (Nose radius)(mm)	1	2	3
FR (Feed rate) (mm/rev)	1	2	3
DOC (Depth of cut) (mm)	1	2	3
TT (Tool Type)	1	2	3

Table 18.2 L27 orthogonal array design matrix

RUN	*CE*	*NR*	*FR*	*DOC*	*TOOL TYPE*
1	DRY	0.4	0.15	0.5	UNCOATED
2	DRY	0.4	0.25	1	PVD
3	DRY	0.4	0.35	1.5	CVD
4	DRY	0.8	0.15	1	CVD
5	DRY	0.8	0.25	1.5	UNCOATED
6	DRY	0.8	0.35	0.5	PVD
7	DRY	1.2	0.15	1.5	PVD
8	DRY	1.2	0.25	0.5	CVD
9	DRY	1.2	0.35	1	UNCOATED
10	WET	0.4	0.15	0.5	UNCOATED
11	WET	0.4	0.25	1	PVD
12	WET	0.4	0.35	1.5	CVD
13	WET	0.8	0.15	1	CVD
14	WET	0.8	0.25	1.5	UNCOATED
15	WET	0.8	0.35	0.5	PVD
16	WET	1.2	0.15	1.5	PVD
17	WET	1.2	0.25	0.5	CVD
18	WET	1.2	0.35	1	UNCOATED
19	MQL	0.4	0.15	0.5	UNCOATED
20	MQL	0.4	0.25	1	PVD
21	MQL	0.4	0.35	1.5	CVD
22	MQL	0.8	0.15	1	CVD
23	MQL	0.8	0.25	1.5	UNCOATED
24	MQL	0.8	0.35	0.5	PVD
25	MQL	1.2	0.15	1.5	PVD
26	MQL	1.2	0.25	0.5	CVD
27	MQL	1.2	0.35	1	UNCOATED

Figure 18.3 Surface roughness measurement

Table 18.3 Average SR and its signal/noise ratio

RUN	*SR_NF1*	*SR_NF2*	*SR_NF3*	*Avg*	*S/N*
1	7.40	7.44	10.07	8.30	–18.3834
2	5.79	5.81	7.04	6.21	–15.8636
3	4.51	4.53	5.49	4.84	–13.7044
4	6.27	6.30	8.53	7.03	–16.9452
5	4.83	4.85	5.87	5.18	–14.2887
6	2.93	2.94	3.56	3.14	–9.9410
7	4.96	4.99	5.43	5.13	–14.1991
8	3.52	3.54	4.47	3.84	–11.6875
9	2.07	2.08	2.52	2.23	–6.9513
10	6.67	6.70	8.11	7.16	–17.0969
11	5.23	5.25	6.35	5.61	–14.9780
12	3.78	3.80	4.60	4.06	–12.1633
13	5.50	5.52	5.43	5.48	–14.7814
14	4.21	4.23	4.47	4.30	–12.6739
15	2.31	2.32	2.81	2.48	–7.8905
16	4.35	4.37	5.29	4.67	–13.3826
17	2.90	2.92	3.53	3.12	–9.8784
18	1.46	1.47	1.77	1.57	–3.8945
19	6.20	6.23	5.43	5.96	–15.4989
20	4.61	4.63	4.47	4.57	–13.1956
21	3.16	3.18	3.85	3.40	–10.6198
22	5.04	5.07	6.13	5.41	–14.6686
23	3.60	3.61	4.37	3.86	–11.7326
24	1.70	1.70	2.06	1.82	–5.2012
25	3.73	3.75	5.43	4.31	–12.6813
26	2.13	2.14	4.47	2.91	–9.2782
27	0.96	0.97	1.16	1.03	–0.2554

Data Analysis

Following steps can be sued for data analysis.

a) S/N ratio formula Selection.
Nature of response decides the equation. The low SR is desirable, therefore equation of Lower-the-Better S/N ratio is adopted.

$$S/N = -10\log\left(\left(\frac{1}{n}\right)\sum y^2\right) \tag{1}$$

b) S/N ratio calculation. Table 18.3 shows SR and its S/N ratio.
c) Creation of main effect. The average S/N is shown in Table 18.4.
d) Forecasting of optimum level of input parameters.

A. Average Signal to Noise

The average of S/N ratio is computed for each level of all the parameters and are shown in the response Table 18.4. Figure 18.4 shows the residual plot for Signal to noise ratio.

B. Forecasting of optimum level of input parameters

The mean S/N for SR is observed in Figure 18.5. From fig, it is realised that the maximum mean S/N ratio of CE is MQL. This is at 3rd level. Therefore, this is the optimal level of CE. In the same manner, the maximum

Table 18.4 Average S/N ratios

Level	*CE*	*NR*	*FR*	*DOC*	*TT*
1	–13.552	–14.612	–15.293	–11.651	–11.197
2	–11.860	–12.014	–12.620	–11.282	–11.926
3	–10.348	–9.134	–7.847	–12.827	–12.636
Delta	3.204	5.477	7.446	1.546	1.439
Rank	*3*	*2*	*1*	*4*	*5*

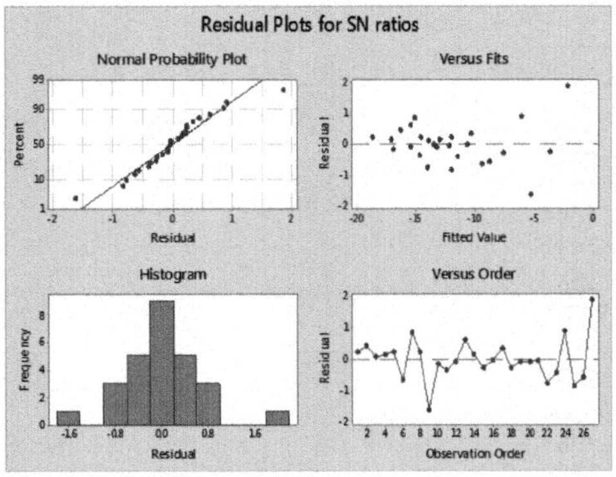

Figure 18.4 Residual plot

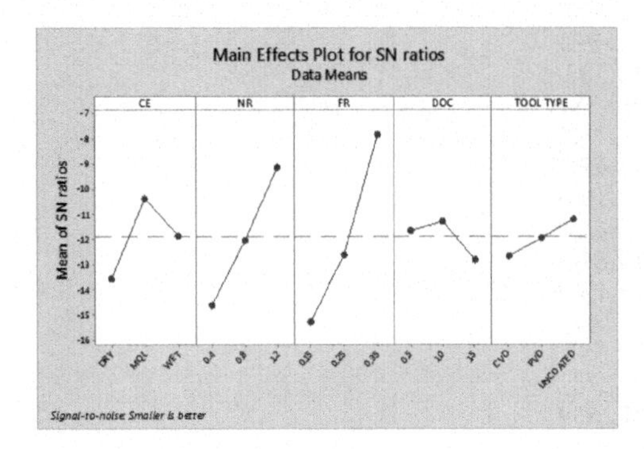

Figure 18.5 Main effect plot

mean S/N ratio remaining input parameters are identified. Hence 'A_3 B_3 C_3 D_2 & E.' is the optimal setting to get the best SR.

C. Projected surface roughness

Additive model helps to project response at optimal level

$$\mu_{pred} = \bar{Y} + \sum(\bar{Y}_i - \bar{Y}) \tag{2}$$

$$\bar{Y}_{surface\ roughness} = Y_{surface\ roughness} + \bar{Y}(A_3 - Y_{surface\ roughness}) + (B_3 - Y_{surface\ roughness}) + \bar{Y}(C_3 - Y_{surface\ roughness}) + \bar{Y}(D_2 - Y_{surface\ roughness}) + (E_1 - Y_{surface\ roughness})$$

$$\bar{Y}_{surface\ roughness} = 4.36 + (3.70 - 4.36) + (3.20 - 4.36) + (2.73 - 4.36) + (4.35 - 4.36) + (4.40 - 4.36) = 0.947\text{um}$$

Conformity Test

Confidence interval finds the range of acceptance. This test is practiced to identify the experimental output parameter at optimal setting is within the range given by confidence interval or not.

A. CI (Confidence Interval)

Mathematically CI is calculated by equation 3, if confidence level= 95 %

$$CI = \sqrt{\frac{F_a(1, f_e)V_e}{n_{eff}}} \tag{3}$$

The range obtained by confidence interval is as follows

$0.256 < 0.321 < 0.386$

Table 18.5 shows the confirmatory experiments for SR (um).

Prediction Using Artificial Neural Network (ANN)

Steps:-

1) Data Input:-
   a) *Normalisation of input data*:- Before data input, the normalisation of data is done by following equation

$$X_i(k) = \frac{\max Y_i(k) - Y_i(k)}{\max Y_i(k) - \min Y_i(k)} \quad (4)$$

Where, $Y_i(k)$ = actual value, $\min Y_i(k)$ = minimum value, $\min Y_i(k)$ = maximum value

2) Training ANN
   Figure 18.6 shows the Architecture of ANN. It also shows the input layer, hidden layer and out put layer.

In the Figure 18.9 the dotted line reflects smooth fit. The line in colour blue shows a direct fit. Training model of ANN is embodied by Projected value = 0.94 * trial value + 0.019. This displays best relation in experimental and projected values.

Figure 18.10 shows dotted line and red line. The equation obtained is:-

Projected value = 0.88 * trial value + 0.076.

Validation of ANN model

In Figure 18.11, the correspondence of trial and projected values of output during confirmation. The equation obtained is.

Projected value = 0.93 * trial value + 0.027

Result and Discussion

The developed ANN is trained for the dependent and independent factors. The standard for ending the training hinge on the no. of epochs. 1000 epochs are implemented here. The network simulation is

Table 18.5 Confirmatory experiment for SR (um)

Surface Roughness (um)			
Sample	*NF1 1.7 m/s²*	*NF2 4.3 m/s²*	*NF3 6.9 m/s²*
1	0.906	0.919	1.094
2	0.908	0.913	1.149
3	0.906	0.916	1.070
Average	0.907	0.916	1.104
Total Average	0.98		

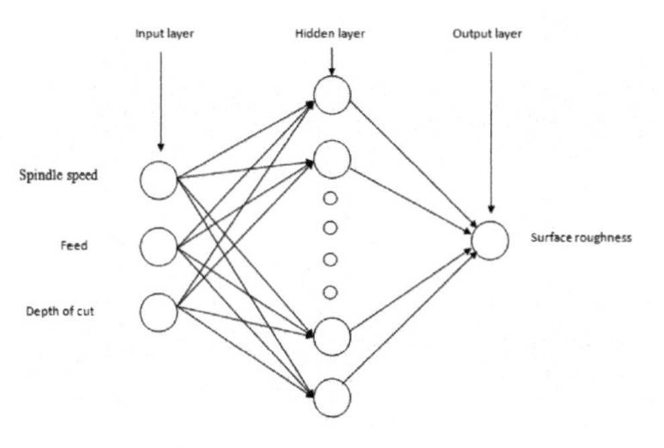

Figure 18.6 Architecture of ANN

Table 18.6 Normalisation of input data

RUN	*SR*	*Nomalised Value*
1	8.30	0.00
2	6.21	0.29
3	4.84	0.48
4	7.03	0.17
5	5.18	0.43
6	3.14	0.71
7	5.13	0.44
8	3.84	0.61
9	2.23	0.84
10	7.16	0.16
11	5.61	0.37
12	4.06	0.58
13	5.48	0.39
14	4.30	0.55
15	2.48	0.80
16	4.67	0.50
17	3.12	0.71
18	1.57	0.93
19	5.96	0.32
20	4.57	0.51
21	3.40	0.67
22	5.41	0.40
23	3.86	0.61
24	1.82	0.89
25	4.31	0.55
26	2.91	0.74
27	1.03	1.00

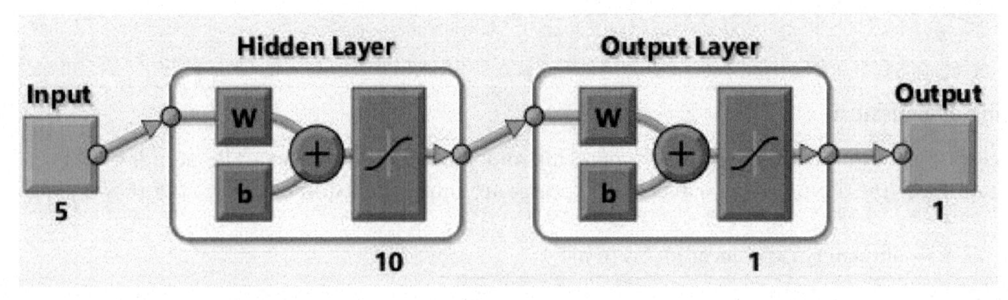

Figure 18.7 ANN network

accompanied for nondependent the aim values. The training of network is done for input while test readings. Then comparison of input and the target with real output is made. Finally the comparison is done for predicted (Ra) experimental (Ra). A minimal variations is obtained after comparison. The Table 18.7 and Figure 18.12 shows the experimental vs predicted surface roughness.

Conclusion

The Taguchi method is effectively implemented along with the ANN method for turning process.
Following are the conclusions:

1) The surface roughness is minimised to 0.947 um.
   The optimal setting obtained is
   A_3 B_3 C_3 D_2 and E_1.
2) Feed Rate is the most and Tool type is the least important factors.
3) The conformity test showed that the experimental and predicted values are within the range of confidence interval.
4) Very less variation in the trial values and projected values obtained by ANN is shown.

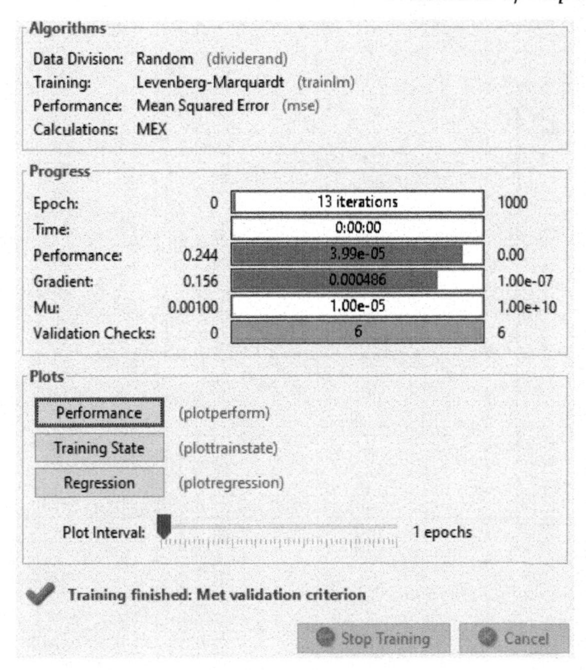

Figure 18.8 Neural network training (nntraintool)

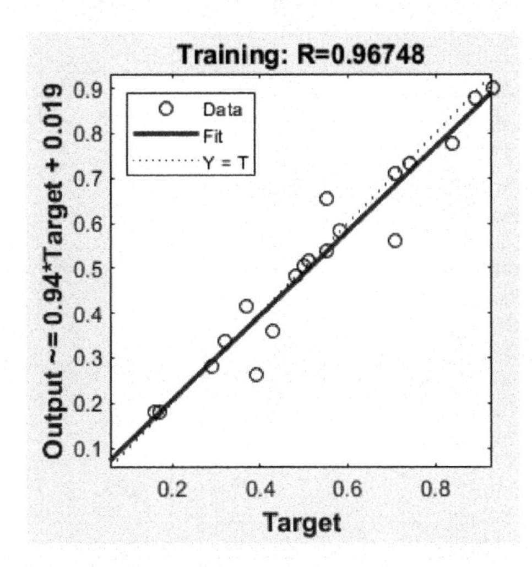

Figure 18.9 Training

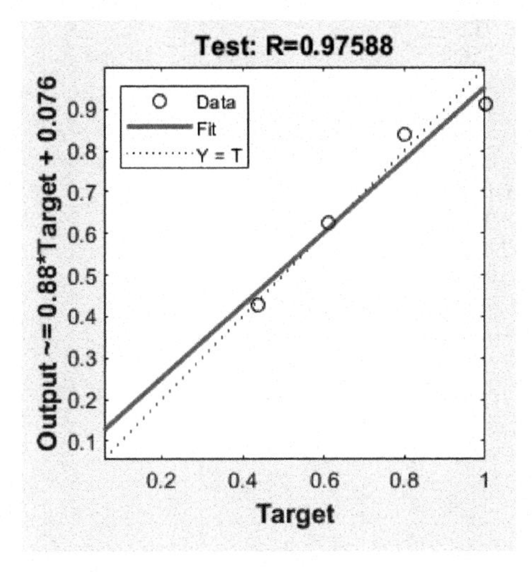

Figure 18.10 Testing

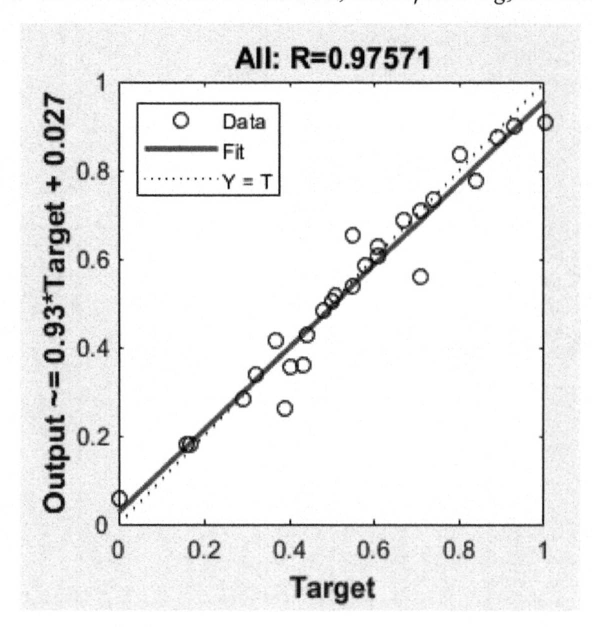

Figure 18.11 Validation

Table 18.7 Experimental vs predicted surface roughness

RUN	*SR*	*Nomalised value*	*Predicted value by ANN*
1	8.30	0.00	0
2	6.21	0.29	0.280569
3	4.84	0.48	0.46439
4	7.03	0.17	0.164472
5	5.18	0.43	0.416016
6	3.14	0.71	0.686911
7	5.13	0.44	0.425691
8	3.84	0.61	0.590163
9	2.23	0.84	0.812683
10	7.16	0.16	0.154797
11	5.61	0.37	0.357968
12	4.06	0.58	0.561138
13	5.48	0.39	0.377317
14	4.30	0.55	0.532114
15	2.48	0.80	0.773984
16	4.67	0.50	0.48374
17	3.12	0.71	0.686911
18	1.57	0.93	0.899756
19	5.96	0.32	0.309594
20	4.57	0.51	0.493415
21	3.40	0.67	0.648212
22	5.41	0.40	0.386992
23	3.86	0.61	0.590163
24	1.82	0.89	0.861057
25	4.31	0.55	0.532114
26	2.91	0.74	0.715935
27	1.03	1.00	0.96748

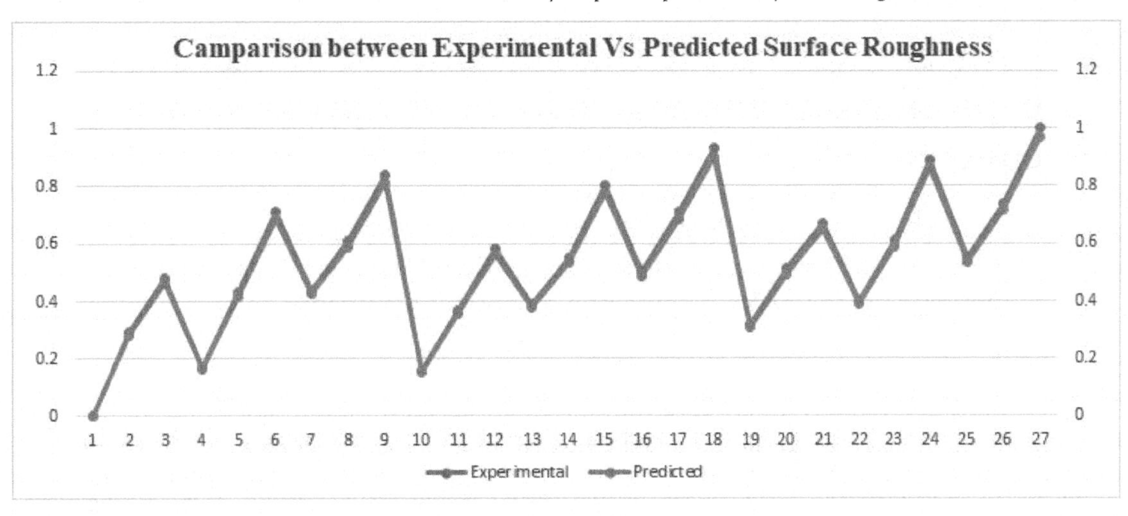

Figure 18.12 Experimental vs predicted surface roughness

Table 18.8 The optimal setting

Process parameter	*Level*		
	Low	*Mid*	*High*
CE	1	2	3
NR	1	2	3
FR	1	2	3
DOC	1	2	3
TT	1	2	3

References

Asiltürk, I. and Çunkaş, M. (2011). Modeling and prediction of surface roughness in turning operations using artificial neural network and multiple regression method. Expert Syst. Appl. 38(5):5826–5832.

Davim, J. P., Gaitonde, V. N., and Karnik, S. R. (2008). Investigations into the effect of cutting conditions on surface roughness in turning of free machining steel by ANN models. J. Mater. Proc. Technol. 205(1-3):16–23.

Joshy, A., Dsouza, R., Muthirulan, V., and Sachidananda, K. H. (2019). Experimental analysis on the turning of aluminum alloy 7075 based on taguchi method and artificial neural network. J. Eur. Syst. Autom. 429–437.

Kamble, P. D., Waghmare, A. C., Askhedkar, R. D., and Sahare, S.B. (2021). Application of hybrid Taguchi-Grey relational analysis (HTGRA) multi-optimization technique to minimise surface roughness and tool wear in turning AISI4340 steel. J. Phys. 1–7.

Karkalos, E. and Markopoulos, A. P. (2017). Applicability of ANN models and Taguchi method for the determination of tool life in turning. 21st Innovative Manufacturing Engineering & Energy International Conference (pp. 1–6).

Kishore, D. S. C. et al. (2018). Investigation of surface roughness in turning of in-situ Al6061-TiC metal matrix composite by Taguchi and prediction of response by ANN. Mater. Today: Proc. 5(9):18070–18079.

Kumar, S. et al. (2014). A hybrid Taguchi-artificial neural network approach to predict surface roughness during electric discharge machining of titanium alloys. J. Mech. Sci. Technol. 28(7):2831–2844.

Mia, M. and Dhar, N. R. (2016). Prediction of surface roughness in hard turning under high pressure coolant using artificial neural network. Measurement 92:464–474.

Reddy, S. (2013). Optimization of surface roughness and delamination damage of GFRP composite material in end milling using Taguchi design method and artificial neural network. Procedia Eng. 64:785–794.

Sahare, S. B., Untawale, S. P., Chaudhari, S. S. Shrivastava, R. L., and Kamble, P. D. (2018). Optimization of end milling process for Al2024-T4 aluminum by combined Taguchi and artificial neural network process. In Soft Computing: Theories and Applications, 525–535. Singapore: Springer.

Yalcin, U., Karaoglan, A. D., and Korkut, I. (2013). Optimization of cutting parameters in face milling with neural networks and Taguchi based on cutting force, surface roughness and temperatures.. Int. J. Prod. Res. 51(11):3404–3414.

19 Critical study on the characterstics of industrial byproducts as fine aggregate on the performance of concrete

B. V. Bahoria[1,a], *R. S. Berad*[1,b], *and S. A. Pande*[2,c]

[1]Department of CivilEngineering, Yeshwantrao Chavan College of Engineering, Nagpur, India

[2]Department of Physics, Laxminarayan Institute of Technology, Nagpur, India

Abstract

The quality of materials, their performance, and their safety are necessary for making sure they meet specifications, perform flawlessly, and are not unsafe. In the past, it was believed that fine aggregate smaller than 75 μm (No. 200 sieve) would affect the strength of the concrete. However, recent testing showed that fine aggregate can add strength and durability to concrete. This isn't always necessarily the case with manufactured best aggregates. At the same time in the direction of changing well known specifications to permit a higher percent of micro fine aggregates, there is necessity to find a way of examining whether these micro fines will have detrimental effects. In this article technique such as X-ray tomography, X-ray diffraction, DTA and SEM-EDS has been discussed for studying the micro fines in order to find the effects on concrete. Easy tests for micro fines that can be utilised as a criterion for their incorporation or prohibition were reviewed in addition to thoroughly defining the aggregates utilising improved methodologies. For any such test to be significant there ought to be a robust interrelationship between its effects and urban overall performance. The test results of X-ray diffraction, SEM EDS for natural fine aggregate, quarry prey dirt and waste low density polyethylene (LDPE) have been discussed. This study thoroughly describes micro fines and assesses simple tests for forecasting total concrete performance.

Keywords: Characterisation techniques, fine aggregates, micro fines.

Introduction

Incorporating micro fines in construction fascinates as the huge quantity in which it is generated and need to be avoided while gradation of aggregate. The number is foreseen to escalate else keeping in view the prospective plans for overall development of the country. The Pulverisation unit in our country is evaluated to generate revenue of Rs. 5000 crore per year equal to about $1 billion) and is thus a financially viable sector. Recently, massive endeavour are carried out to investigate the use of modern trash-debris in the construction of cement. Prey dirt is granulating waste that has been crushed during the parent stone's destructive procedure. It has recently got a lot of attention for being used as a strong filler material rather than fine total. In addition to increasing the characteristics of cement in both new and solidified forms, effective utilisation of these ingredients will reduce environmental load, waste board costs, and substantial creation costs. The solidity of cement is commonly tested in less-than-ideal conditions without focusing. The strength of cement is likely the most important quality since it is critical that a structure is able to stand out the state it has been designed for its entire designed life (Bahoria et al., 2013a, 2013b). The lack of sturdiness might may be the reason for external natural sources or internal causes inside the substance itself (Bahoria et al., 2014; Vijayalakshmi et al., 2013). Various types of side-effects, such as reused considerable total, prey dirt, fine fly debris, and residues, few types of fabricated totals, has been considered by several scientists when it comes to optional materials employed as total fillers in concrete. Several test evaluations demonstrated that fractional supplanting of fine total with squashed stone residue at various rates resulted in the development of the further developed. It has a tendency to be visible that the impregnation of cement multiplied altogether whilst the tension surpasses as much as 40% of a definitive energy of the significant examples (Sobhani et al., 2012). Tests due to useful solidifying substances in concrete display a multiplied competition of cement; in every other evaluate it turned into determined that the lessening with inside the coefficient of porousness occurred with time, that's relied upon due to steady hydration in the instance simply as dormant pore-obstructing (Thomas and Partheeban, 2010). The deficiency of the significant sturdiness is probably made through the seriousness of the weather which its miles exposed or through internal modifications with inside the evolved significant itself (Song et al., 2010). Stability evaluation of significant concerning fuel line penetrability take a look at may be a becoming method for estimating the pervasion

[a]boskey.bahoria@gmail.com; [b]rupaliberad@gmail.com; [c]sap7001@gmail.com

homes of cement (Shi et al., 2009). The enlargement of miniature fillers display similarly evolved pore filling affects at decrease water-to-cowl percentage and moreover display higher execution in fuel line porousness. It thoroughly can be visible that the porousness of cement diminishes on addition of silicon dioxide as much as 8%, contributes the rheological of cement denser (Binici et al., 2008). Likewise, the beyond exploratory evaluate connotes that the solidness of cement is based upon the drawn out effects of significant supportability closer to adverse ecological situations simply as for the duration of purposeful burdens whilst the development is located getting used. It has a tendency to be summed up from the beyond investigations that the significant creation placed getting used is uncovered to starting stacking due to useless weight which in turns produces breaking. This calls for checking out porousness of cement in targeted on situation to evaluate the vital exhibition of cement below broke situation. The lineation of the micro fines can also additionally have an impact at the converting water call for of the concrete mix. Naturally occurring sand generally have non-angular to round debris, even synthetic fine aggregate possess the physical properties of having lineation. Flakiness and elongation features. Distinct variation in combinations may affect the affinity to water in concrete (Hudson, 1997). Particles having excessive tapering features such as too angular, elongation considerably have a significant effect on concrete handling property as compared to ill graded coarse aggregates. While using uneven formed synthetic likely aggregates, it has been well proven that by use of well graded aggregates the affinity for water demand can be controlled. The extra abnormal the micro fine debris is formed, the extra packing development micro fines provide (Kronlof, 1994).

Some wonderful consequences of inclusive of best fillers in combinations are: smaller water requirement because of stepped forward particle packing; elevated energy because of smaller water requirement and stepped forward interplay among the ingredients; reduced impregnation; and improved resilience.

It is now no longer the best stuff managing the quandary supplied with the aid of using dirt of fracture. Many international locations along with India all have multiplied the permissible restriction of micro fines to 20% of its incorporation in designing concrete. However, so as for the USA to boom the permissible confines on micro fines, it's far essential to apprehend the results that micro fines have on concrete, which calls for each feature of the micro fine be recognised. Properties which probably affect concrete comprises of the physical features and its mineralogical composition an intense study about micro fine combination strategies and its findings for their incorporation in mix design had been taken into account.

Tests for Microstructure Properties of Fines

The characterisation of micro fine geology can be performed in different ways. In order to know the compounds & deposits diffraction can be effectively used in the form of powdered samples such as fine dust. The existence or absence of clay can be found out by it. On the other hand, for high resolution recognition of elements and the compounds SEM along with EDS can be used in 2-D cross-sections of samples.

X-ray diffraction rays in the material under consideration scattered by atoms in a manner that indicates spacing of the lattice elements as the X-rays are in the state, it will reflect positive obstruction giving a wavelength peak as in X-ray diffraction sample. The spacing's between the upper and lower planes of the material can be found by taking measurement of distance between corresponding points of two consecutive waves for large range of angles. The unknown substances can be identified by the powder diffraction pattern which is accounted by a diffractometer along with the spacing between the upper and the lower planes of relative the severity of the diffraction lines. The data recorded is then contrasted with the typical pattern lines existing for available blends (PDF) database". The database can be used for the quality comparison for the identification of the components in the specimen to be analysed.

SEM in combination with EDS serves to be an impressive technique for vivid recognition of a particle that cannot be seen even under an optical microscope. The surface of the specimen is focused by an electron beam in SEM. As the beam is exposed to the surface of the specimen the negatron are dispersed or absorbed; these responses are taken into account which gives SEM images. By this principle the good conductive object can be imperceptibly tested. By X-rays released by the sample EDS explores the elements present in a given sample. Due to characteristic energy position possessed by the element it has a specific characteristic emission from the electron. The X-ray photons exposed by the specimen are collected by EDS and at each emission voltage it is converted to a number of "counts". "Total number of counts and the amount of element present directly proportional". To identify the possible reasons responsible for the performance of the fracture dust in concrete maximised depiction taken with SEM can be utilised. For better correlation of the factors responsible and the performance data available there are some improved computerised image analysis techniques that can be exploited (Masad et al., 2001).

Portrayal representation can be considered as an effective measure for physical characteristics such as shape as the output are likely to be not affected by the positioning of the particle which may happen while preparing the specimen. The size distribution of the particles by portrayal representation can be considered as inaccurate due to the possible positioning of the particle. Similar to the micro fines, the void ratio of the material is influenced by shape and lineation of particles, as is the case with the fracture dust when it is cohesion less (Sukumaran and Ashmawy, 2001) for quantifying the image parameters there are several techniques.

To elaborate the surface structure of the particle there are some basic terminologies often used. The particle shape or the form in which it exist that is the first-level surface structure property, is termed as the total form of a particle, and is not independent on evenness and surface texture. Angular shape or evenness, that forms the second level property, depicts quantity and harshness of the edges of the particle surface. The outer feel or the irregularity forms the third level property that depicts the quantity, dimension and severity of the sharpness along the edges of the particles and its surface.

The complete set of particle with all the three level of properties in the same plane forms a good correlation with compacted void contents of the fine aggregate. The particle shape, evenness, irregularity these three indications can be expressed as aspect ratio, lineation, and Coarseness (Kuo, 2002). The following equations 1–3 represent the calculation of these surface indications. The permissible longitudinal and lateral, $Perimeter_{convex}$ which is circumference of the polygon and $Perimeter_{ellipse}$ is the circumference of ellipse under consideration.

$$Aspect\ ratio = Length/Width \tag{1}$$

$$Lineation = \left\{ \frac{Perimeter_{convex}}{Perimeter_{ellipse}} \right\} \tag{2}$$

$$Coarseness = \left\{ \frac{Perimeter}{Perimeter_{convex}} \right\} \tag{3}$$

'The ratio of longitudinal to lateral dimension will be near to unity for similar dimensional or rounded aggregates, and will have larger values for elongated and flat particles. The lineation for either a circle or an ellipse will be unity. For angular particles, lineation will be larger than 1. Therefore, larger values of lineation indicate a higher degree of lineation Coarseness will increase when the surface texture of the aggregate increases (and thus its perimeter) while having the same bounding polygon (and thus it's Perimeter convex)' (Kuo, 2002).

By using the conventional black and white images was used to identify the particle shape characteristics as described by radius method which was used to find the lineation index. The radius method for calculating particle angularity measures the difference between the particle radius, Rθ, at each angle θ and the radius of an equivalent ellipse, REEθ, in accordance with Equation 4.

$$Lineation\ index = \sum_{\theta=0}^{355} \frac{[R\theta - REE\theta]}{REE\theta} \tag{4}$$

In the similar manner shape index was calculated from a planer particle. The particle radius, R_θ, and an incremental angle change, Δθ, of 4°. Shape index is as given in equation 5.

$$Shape\ Index = \sum_{\theta=0}^{\theta=360-\Delta\theta} \frac{[R\theta + R\Delta\theta]}{R\theta} \tag{5}$$

Fourier series is applied to analyse shape, lineation, and coarseness of the aggregate using one general by merely altering the variables the parameters can be found out.

In short, the aggregate surface characteristics can be portrayed by the general Fourier equation but which differ by frequency magnitude Masad et al. (2001) as shown in Figure 19.1. The three particles in the figure resembles the same shape signature number due to similar form. Particle B has a much higher lineation as compared to A or C, and particle C has a higher texture designation reflecting the small changes in the surface that occur at much smaller intervals than those that occur for the lineation and form designation.

X-Ray tomography can also be employed for studying three-dimensional analysis of aggregates which may be applicable for classifying micro fines provided there exists a correlation. For computational modelling of the concrete, X-ray tomography can be used instead of individual variable.

The variation in the material properties due to temperature changes can be studied by thermal analysis. In this method the temperature of the concerned specimen and a dormant recommended material such as αAl_2O_3 is compared in DTA, the thermal variation between the samples is noted as they are both heated. During an endothermic reaction, the temperature of the sample will straggle as compared to the temperature of the reference material. In DTA representation temperature is plotted on the x-axis whereas the differential temperature on the y-axis. The peaks' pointing downward depicts an endothermic reaction whereas that showing up depicts an exothermic reaction. The data from the analysis can be used to The DTA data can be recommended predominantly for the determination of particular information as the kind of reaction whether endothermic or exothermic has taken place as shown in Figure 19.2, Ramachandran (2001) which elaborates the typical peaks of known impurities found to be hazardous to concrete which may lead to durability issues.

Research Analsis

In order to check the feasibility of the micro fine there are simple tests to evaluate their performance when mixed in concrete. According to the different test methodology the tests were completed on the substantial in accordance to the standards. The nominal and high grade of concrete as M20, M30, and M40 were tested according to standard codes. The prey dirt and waste LDPE was utilised as a substitution of normal sand in shifting rates as 0,25,50,75,100 overall alongside 2,4,6,8% of waste plastic step by step have been utilised to get ready changed cement. To concentrate on the mineralogy, portrayals of materials (normal sand, quarry dust, squander LDPE) were tried for XRD, SEM-EDS, Functionality testing have been done on normal concrete and altered cement concrete.

X-ray crystallography

The investigation of the mineralogical properties of micro fines should be possible in numerous ways. Dispersion of X-beam can distinguish mixtures and metals in prey dirt tests alongside micro fine materials. The indistinguishable can be utilised to recognise the existence or nonappearance. SEM - EDS can be a

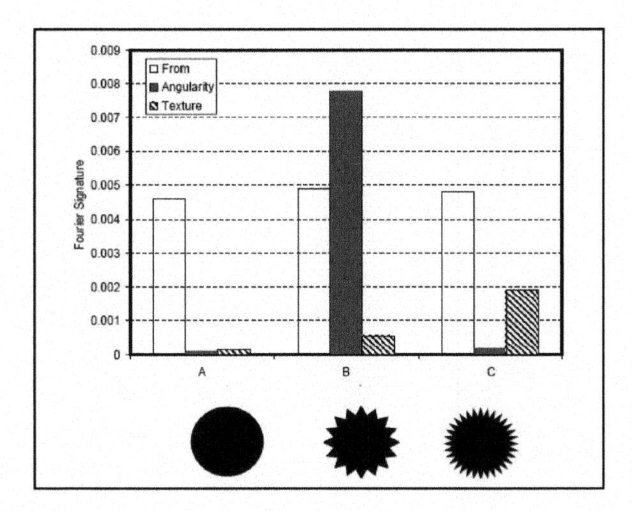

Figure 19.1 Image analysis by fourier series (Masad et al., 2005)

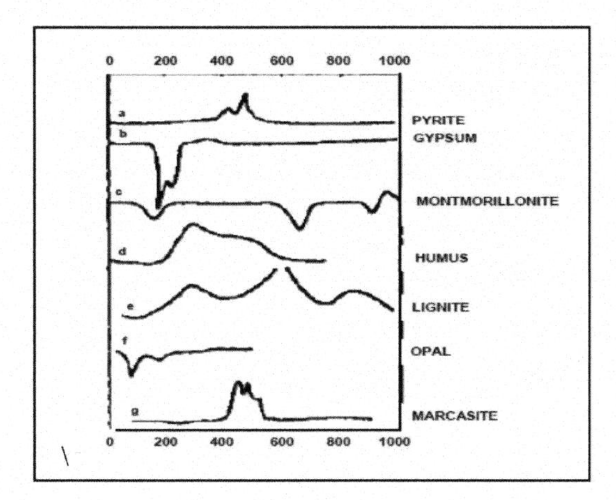

Figure 19.2 Thermal analysis of aggregates (Ramachandran, 2001)

hopeful gadget for outwardly analysing a molecule that is too little to even consider being observed in light magnifying instrument. Checking electron magnifying instruments (SEM) utilise a light emission to photo tests with definitiveness down to the nanoscale. Negatrons are transmitted from a fibre and collimated straight forwardly into a pillar inside the electron source. Electrically powered particle can be analysed minutely along these lines (Sarkar et al., 2001). EDS distinguishes the variables found in an example under the reason of recognizing the X-beams dispersed. Because of the presentation capacity of each detail, it has a particular electron discharge. The example emanates X-beam antiproton particles, collected by EDS and changed over into a total series of 'tally' for every discharge potential drop. The overall assortment of records that have a chosen detail is promptly corresponding to how much that gift given inside the article. Expanded photos utilizing SEM can be tried to find various components that can likewise be connected with the general exhibition of micro fine constituents' present concrete. Database Chatterjee (2001) this method can make a subjective assurance of the elements and associations inside the examined matter. The results from X-ray diffraction show that a number of minerals are present in the micro fines. Each micro fine was analysed to match the peak intensities found for the individual angles to those for known minerals in a database. A graph of typical analysis is shown in Figures 19.3, 19.4, 19.5 and 19.6 where the microphone peak intensity data is shown on the top graph and the known data is shown in Table 19.1 which lists the minerals that matched X-ray intensities for each of the micro fines. As can be seen, similar materials are signified by similar composition. Figure 19.3 indicates that the natural sand was primarily composed of quartz along with the presence of calcite. It could be concluded that practically identical substances are described through a similar piece, sand and (NS01) comprised specifically of geode in all with the existence of chalk as apparent from Figure 19.4

From Figures 19.5 and 19.6 the prey soil and dolerite or hold onto rock (TR02) carry graphite close by calcium oxides. The littlest garbage is routinely seen in which dirt and diverse hazardous product are

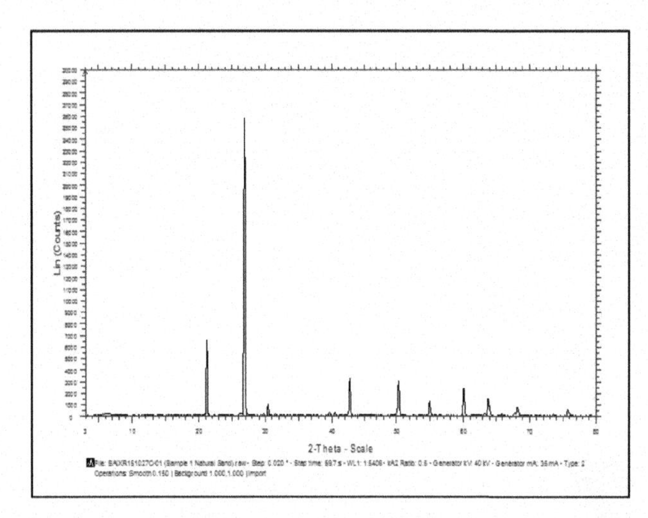

Figure 19.3 Natural fine aggregate sample

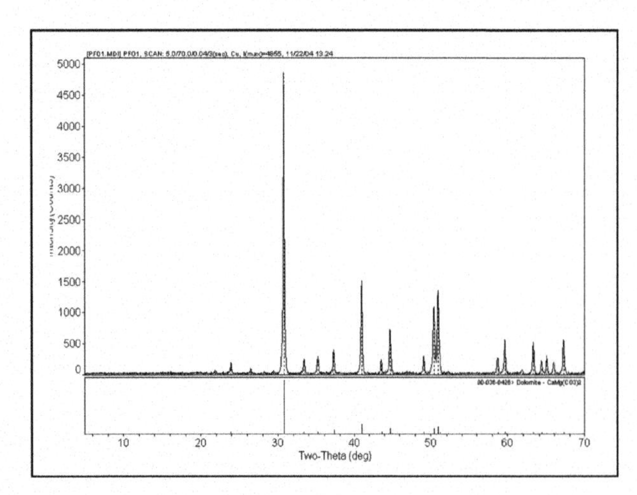

Figure 19.4 Classic graph of microfine (ICAR-107)

noticed. To choose particular miniature fines contained dirt/diverse hazardous matter, tests containing useless or discarded objects of substantially smaller microns have been extricated the utilisation of a deposition chamber as utilised inside side the hydrometer test. In the wake of agreeing to 6 hours, tests have been wiped out from the zenith of the settling chamber. The one with micro fine textures, then, at that point, 2 μm was situated on a laminated plate and permitted to wither. These examples have been then revealed to the X-beam diffusion. The outcomes acquired are portrayed in Table 19.1.

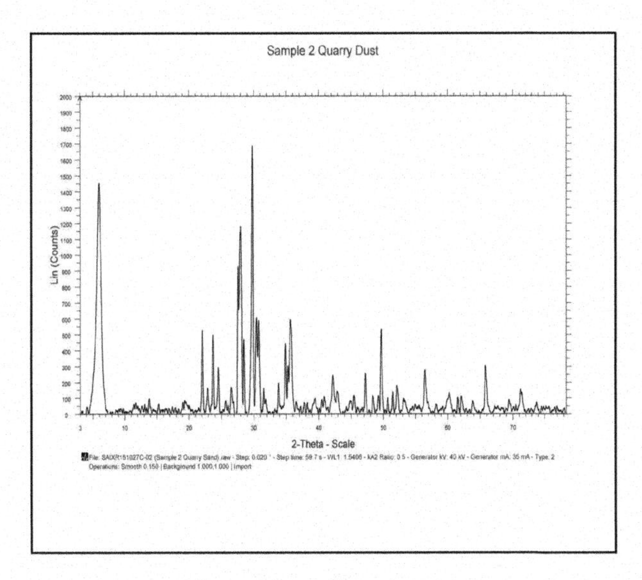

Figure 19.5 Dust fracture sample

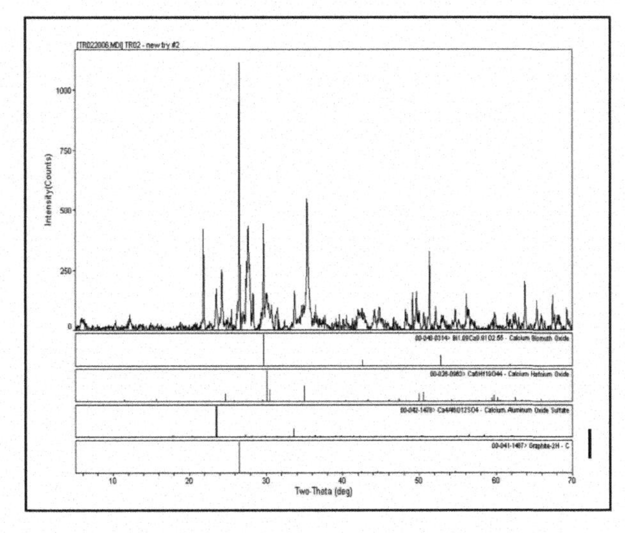

Figure 19.6 Classic graph of TR02 (ICAR-107)

Table 19.1 Results of less than two micrometer matter

Aggregate	*Deposits*
NS-01	Chalk
PF-01	Geode
TR-01	Chalk
TR-02	Clonochlore Nimite

The electron of TR02 incorporates each of the deposits inside the chlorite group which are very much like biotite. According to muds, biotite is one of the phyllosilicate of deposits. Unwanted results on the overall in general presentation of cement were resolved on account of the existence of biotite. In any case, the results are generally relying upon the type of biotite and presently at this point not on the quantity of biotite Muller (1971).

Figure 19.7 shows the dispersion of waste LDPE sample whereas Figure 19.8 recommends the Classic dispersion for LDPE. It very well might be noticeable that LDPE is to a limited extent glasslike and partially undefined shape as a result of the ways of life of sharp slim diffraction pinnacles and huge top. The d-dividing at $2\theta = 20°$ is 5.379 Å.

Figure 19.8 shows that X-ray dispersion for LDPE 2.49 phr and five phr composite recommends a translucent example shape with the d-separating charge at $\theta = 22.86°$ and $2\theta = 22.32°$ are 3.053 Å and 5.163 Å individually. The top movements to a superior viewpoint as contrasted and the LDPE, which compare to the hole among interlayer, diminishes. The diminishing d-dispersing expense is at filler stacking 2.49 phr. It is obvious that the filler stacking will build the compound arise as extra translucent in view of the predominant of sharp slim diffraction tops.

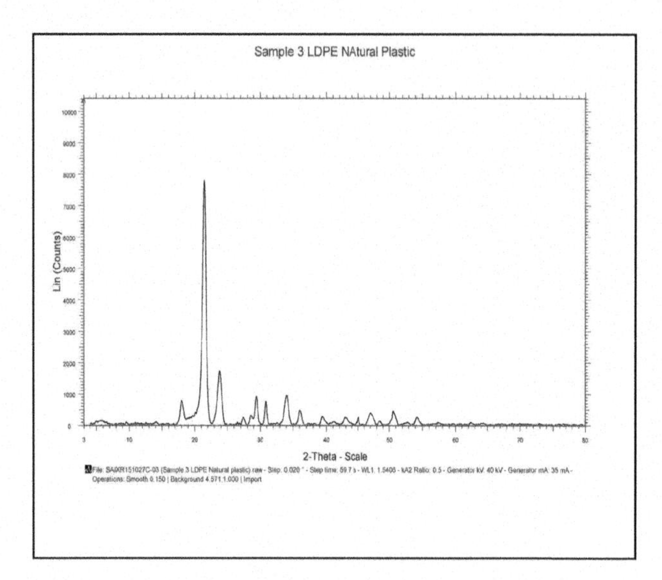

Figure 19.7 LDPE (waste plastic) sample

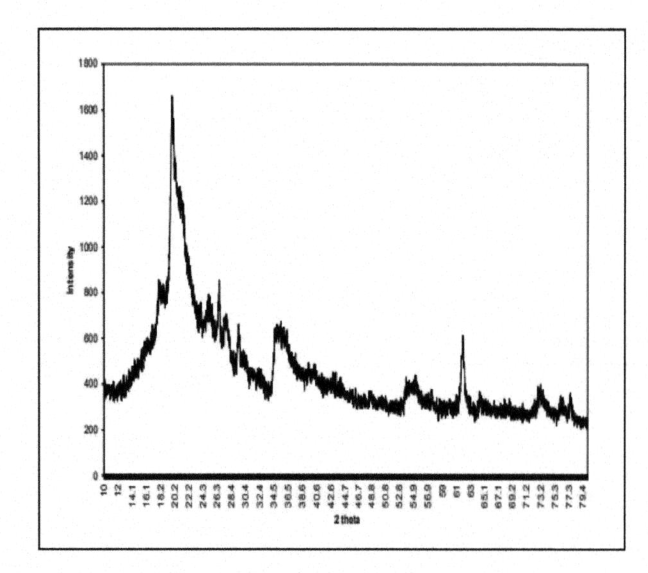

Figure 19.8 Classic dispersion of LDPE (Supri, 2008)

SEM–EDS

EDS recognises elements present in a specimen by the dispersion of rays scattered by the specimen. The specimen disperse antiproton rays, accumulated through EDS and converted into a series of 'tally' at each emission voltage as shown in Figures 19.9, 19.10 and 19.11 and data as depicted in Table 19.2. The tally range is directly proportional to the amount of that matter given away within the item. High resolution depiction by SEM possibly be used to find number of metals to correlate with the execution of specific micro fines.

Refromed concrete

The major axis measures about 1.56 μm and minor axis about 1.45 μm as in Figure 19.12. The major axis length and minor axis length is found to be approximately equal, and the elongation parameter which is the ratio of length of axes was about 0.99. From this it was concluded that the particles are spherical. The surface area of the particle was found to be 1.87 square microns and the circumferential parameter counted about 4.77 μ. The 0.98 was the roundness; it showed that the particles are of spherical nature. This

Table 19.2 Metals through in EDS analysis

Metal	*Weight- percentage*	*Atomic percentage*
C K	3.019	5.099
O K	62.01	72.16
Al K	0.30	0.23
Si K	32.12	21.29
Ca K	0.64	0.22
	Prey Dirt	
Metal	*Weight- percentage*	*Atomic percentage*
C K	1.579	4.98
O K	39.04	60.02
Na K	1.64	1.433
Mg K	1.099	1.98
Al K	4.24	4.55
Si K	16.64	13.09
Ca K	5.89	2.89
Ti K	2.48	1.12
Fe K	19.18	6.67

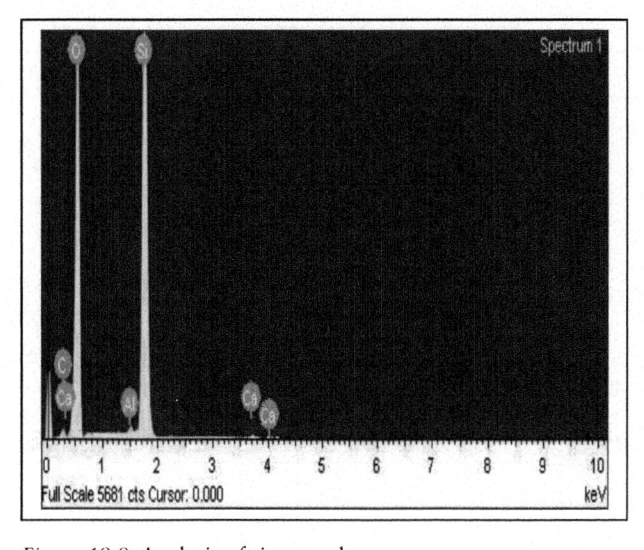

Figure 19.9 Analysis of river sand

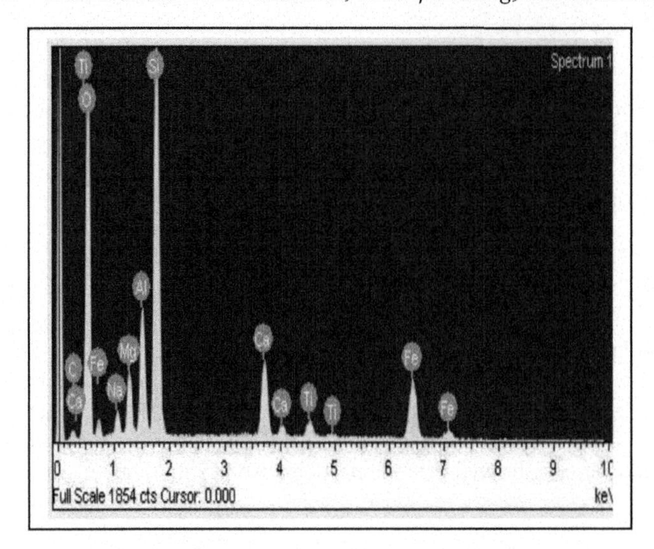

Figure 19.10 Analysis of prey dirt

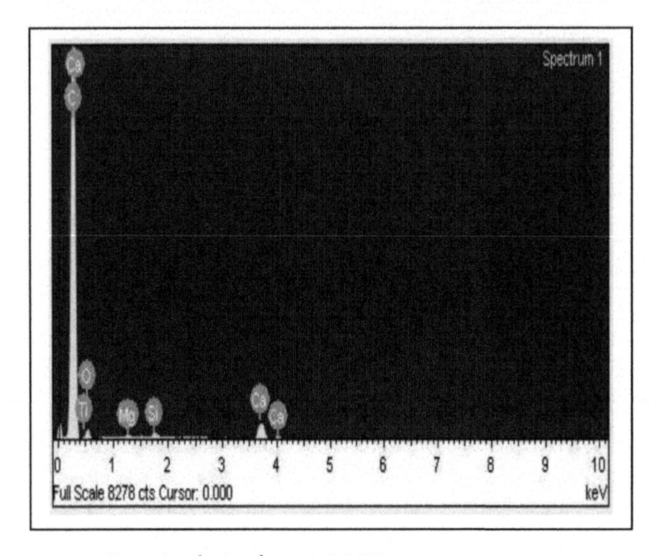

Figure 19.11 Analysis of waste LDPE

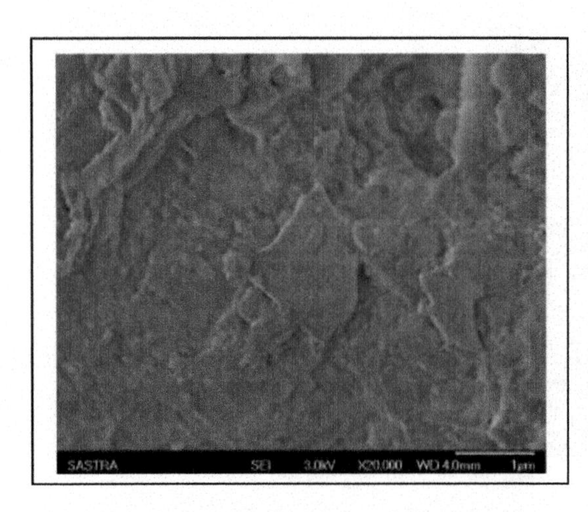

(a) Conventional Concrete

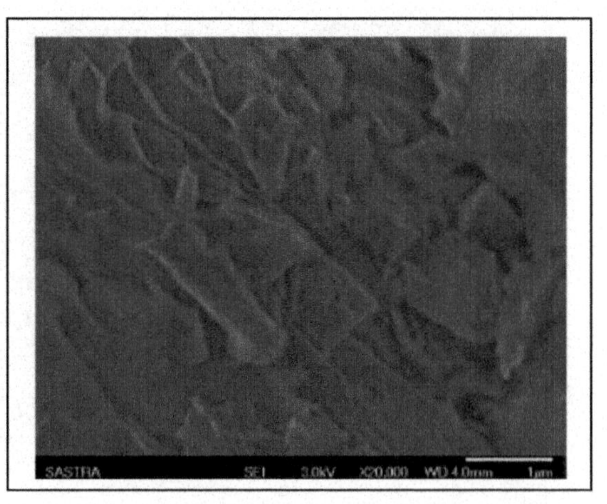

(b) Refromed concrete

Figure 19.12 Images of concrete samples with river sand, prey dirt, LDPE by SEM

particularly aided in enhancing workability parameter. The modified and conventional concrete represents SEM depiction of the specimen with noble matter at 20000x range. The length of major axis was 1.88 μm and that of the minor axis was 0.56 μm. The prolongation measured about 3.4 which show that particles are widened. The surface area counted about 1.02 square microns and the circumference measured about 4.85 microns. The particle sphericity measured about 0.56, this revealed about the lineation of the particles. The measured characteristics enabled in forming improved dense structure which in turn enhanced the strength and longevity properties. The presence of silica, calcium and oxides were found and the reaction of calcium with silica and oxides resulted in hydrated form of calcium silicates. This hydrated calcium silicates were actually responsible for attaining strength at premature stage.

Conclusion

1. From the critical review of the tests carried out to study the microstructure properties of finer materials, it was concluded that the particle size is not the essential parameter that can decide whether the particle is harmful but it depends on the mineralogical properties of the particles. The type of clay and its quantity is a key factor. The reason of increased strength was due to the presence of compounds of calcium & silica which reduced the demand of water to admixture and thereby drying depreciation. It can be safely concluded that the particle size is not the primary factor affecting the performance of concrete as the non-clay particle substitution were underlying the clay deposits.
2. Further in addition to this by using a suitable admixture in concrete the demand of admixture due to the effect of clays can be minimised without compromising the performance of concrete.
3. The modified and conventional concrete depicts SEM images of the specimen with noble matter at 20000x resolution. The length of major axis was 1.88 μm and that of the minor axis was 0.56 μm. The prolongation measured about 3.4 which show that particles are widened. The surface area counted about 1.02 square microns and the circumference measured about 4.85 microns. The particle sphericity measured about 0.56, this revealed about the lineation of the particles. The measured characteristics enabled in forming improved dense structure which in turn enhanced the strength and longevity properties.
4. EDS study of LDPE revealed that there is no such delirious element affecting concrete and can be utilised effectively to enhance the strength and durability parameters.
5. The SEM images depict that LDPE with membrane, fibre like pattern which means it is not having porous arrangement. Owing to fibre like arrangement, it contributed in enhancing strength.

Acknowledgement

I, would like to extend my gratitude to Dr.D.K.Parbat, Dr. P.B. Nagarnaik Dr. A.M. Pande, for their valuable and timely guidance for the successful completion of the research work. Yeshwantrao Chavan College of Engineering, (An Autonomous Institute) Wanadongri Nagpur for the provision of necessary research facility. Siddheswar's crushing plant & Chemech Pvt.Ltd, MIDC Nagpur for the supply of study material.

References

Bahoria, B. V. Parbat, D. K., Nagarnaik, P. B. and Waghe, U. P. (2013a). Experimental study of effect of replacement of natural sand by PREY DIRT by PREY DIRT and waste plastic on compressive strength of M20 concrete. In 4th Nirma university international conference on engineering, Ahemdabad, Gujrat, India.

Bahoria, B. V., Parbat, D. K., Nagarnaik, P. B. and Waghe, U. P. (2013b). Comprehensive literature review on use of waste product in concrete. Int. J. Appl. Innov. Eng. Manag. 2(4):387–394.

Bahoria, B. V., Parbat, D. K., Nagarnaik, P. B. and Waghe, U. P. (2014). Sustainable utilisation of quarry dust and waste plastic fibers as a sand replacement in conventional concrete. In ICSCI 2014 ASCE India section (Oct 17–18, 2014), Hitex, Hyderabad, Telangana, India.

Binici, H., Shah, T., Aksogan, O. and Kaplan, H. (2008). Durability of concrete made with granite and marble as recycle aggregates. J. Mater. Proc. Technol. 208(1–3):299–308.

Chatterjee, A. K. (2001). Petrographic and technological methods for evaluation of concrete aggregates. In Analytical techniques in concrete science and technology, eds. V. E. Ramachandran, and J. J. Beaudoin, (Chapter 8), New York: William Andrew Publishing/Noyes Publications.

Hudson, B. (1997). Manufactured sand – destroying some myths. Quarry, 57–62.

Kronlof, A. (1994). Effect of very fine aggregate on concrete strength. Mater. Struct. 27:15–25.

Kuo, C.-Y. (2002). Correlating permanent deformation characteristics of hot mix asphalt with aggregate geometric irregularities. J. Test. Eval. JTEVA 30(2):136–144.

Masad, E., Olcott, D., White, T. and Tashman, L. (2001). Correlation of fine aggregate imaging shape indices with asphalt mixture performance. Transp. Res. Rec. 1757:148–156. Paper No. 01-2132.

Muller, O. H. (1971). Some aspects of the effect of biotiteceous sand on concrete. Civ. Eng. S. Afr. 13(9):313–315.

Ramachandran, V. S. (2001). Thermal Analysis. In Handbook of analytical techniques in concrete science and technology, eds. V. E. Ramachandran, and J. J. Beaudoin. New York (N Y): William Andrew Publishing/Noyes Publications.

Sarkar, S. L., Aimin, X. and Jana, D. (2001). Scanning electron microscopy, X-Ray microanalysis of concretes. In Handbook of analytical techniques in concrete science and technology, eds. V. E. Ramachandran and J. J. Beaudoin, (Chapter 7). New York: William Andrew Publishing/Noyes Publications.

Shi, H.-S., Xu, B.-W. and Zhou, X.-C. (2009). Influence of mineral admixtures on compressive strength, gas permeability and carbonation of high performance concrete. Constr. Build. Mater. 23(5):1980–1985.

Sobhani, J., Najimi, M. and Pourkhorshidi, A. R. (2012). Effects of retempering methods on the compressive strength and water permeability of concrete. Sci. Iran. 19(2):211–217.

Song, H.-W., Pack, S.-W., Nam, S.-H., Jang, J.-C. and Saraswathy, V. (2010). Estimation of the permeability of silica fumecement concrete. Constr. Build. Mater. 24(3):315–321.

Sukumaran, B. and Ashmawy, A. K. (2001). Quantitative characterisation of the geometry of discrete particles. Geotechnique 51(7):619–627.

Supri, A. G., Salmah, H., and Hazwan, K. (2008). Low Density Polyethylene-Nanoclay Composites: The Effect of Poly(acrylic acid) on Mechanical Properties, XRD, Morphology Properties and Water Absorption, Malaysian Polymer Journal (MPJ), 3(2):39–53.

Thomas, F. K. and Partheeban, P. (2010). Study on the effect of granite powder on concrete properties. Proc. Inst. Civ. Eng. 163(2):63–70.

Vijayalakshmi, M., Sekar, A. S. S. and Prabhu, G. G. (2013). Strength and durability properties of concrete made with granite industry waste. Constr. Build. Mater. 46:1–7.

20 Cupronickel composites: An overview of recent progress and applications

Ajay D. Pingale[1,a], *Ayush Owhal*[1,b], *Anil S. Katarkar*[2,c], *Sachin U. Belgamwar*[1,d], *and Jitendra S. Rathore*[1,e]

[1]Department of Mechanical Engineering, Birla Institute of Technology and Science, Pilani, Rajasthan, India

[2]Department of Mechanical Engineering, National Institute of Technology Agartala, Tripura, India

Abstract

This study presents an overview of the various types of composites and their fabrication methods. The study then presents a summary discussion on the fabrication of various cupronickel composites using electrodeposition method and powder metallurgy method. Cupronickel composites are fabricated by adding reinforcing material into the cupronickel alloy matrix, which has led to a significant interest due to their superior hardness, wear resistance, tensile strength, macrofouling resistance, corrosion resistance anti-bacterial properties. Due to these properties, cupronickel composites are employed in various engineering applications such as condenser tubes, cooling circuits, heat exchangers, microelectronics, and condensers. Based on the critical assessment of the literature, existing challenges with possible solutions and future research opportunities were discussed.

Keywords: Electrodeposition, hardness, powder metallurgy, reinforcement, sintering temperature, wear resistance.

Introduction

The need for superior mechanical, tribological, electrical, optical and corrosion properties has resulted in extensive research work in the development of metal matrix composites (MMCs). The MMCs have been extensively used in several industries such as aerospace, chemical, automobile and marine due to their enhanced mechanical and physical properties. The MMCs are composites embedding reinforcement into a metal matrix to obtain desirable functional properties which are not offered by conventional unreinforced monolithic metal or metal alloy counterparts. In far back as 1200 B.C, Egyptians and Hebrews have fabricated synthetic composite by the addition of straw as reinforcement in bricks to enhance their mechanical properties. Wattle and daub is 6000 years old synthetic composite material and has been used in many historic buildings. Fiberglass is the first modern composite material and is extensively used in car bodies, building panels, sports equipment and boat hulls. Drivers for improved composite materials are weight reduction, cost reduction as well as to improve performance by enhancing resistance to fatigue, corrosion, and mechanical damage. To meet the requirement of a particular application, the composite material can be fabricated by selecting an appropriate reinforcing element and matrix material. Composite materials include reinforced concrete, reinforced plastics, ceramic matrix composite, metal matrix composites, plywood, etc. Composite materials have been commonly used for bridges, buildings and structural material for storage tanks and bathtubs. Also, composite materials have been used for industrial applications due to their high strength, less expensive, or lightweight than traditional materials. Composite materials are widely employed in advanced engineering applications such as sensing, computation, communication, and actuation.

A. Matrix

In composite material, the matrix serves different functions such as transfer load between the reinforcement, binds the reinforcement, provides the composite component its net shape and protects the reinforcement from mechanical and environmental damage. A composite matrix may be ceramic, polymer, metal, or

[a]ajay9028@gmail.com; [b]ayushowhal@gmail.com; [c]anil.katarkar@gmail.com; [d]sachinbelgamwar@pilani.bits-pilani.ac.in; [e]jitendrarathore@pilani.bits-pilani.ac.in

carbon. The polymer matrix is widely used in aerospace applications due to its good mechanical, electrical, and chemical properties. Ceramic and metal matrices are widely used in automobile and aerospace applications due to their high mechanical strength. The carbon matrix is commonly used in high-temperature applications due to its superior thermal resistance.

B. Reinforcement

The reinforcement element is generally used to enhance the desired properties of the matrix material. Reinforcements are categorised into different groups: whiskers, short fibres, continuous fibres, platelets, and particles. The enhancement in the properties of the composites depends upon the type of reinforcement, the orientation of reinforcement and the geometry of reinforcement. The roles of reinforcement are to carry the load from the matrix and strengthen the composite by improving its overall properties. Reinforcements are also used for the special purpose of resistance to corrosion, heat resistance, heat conduction, improve the strength and provide rigidity. As per the functional requirements, a specific reinforcement could be selected for the fabrication of composites.

Classification of Composites

A composite material consists of two basic parts: Matrix and reinforcement. Composite materials can be classified by their matrix type as shown in Figure 20.1. A composite matrix may be ceramic, polymer, metal, or carbon. Therefore, a composite material could be categorized as ceramic matrix composites (CMCs), polymer matrix composites (PMCs), carbon-carbon composites (CCCs), or metal matrix composites (MMCs).

A. Ceramic Matrix Composites

The CMCs are composite materials and key materials for advanced energy systems. They generally consist of ceramic fibres or whiskers reinforced in a ceramic matrix, developing a ceramic fibre-reinforced material. CMCs have been developed to overcome the brittleness problem of unreinforced ceramic materials. SiC/SiC, C/C, Al_2O_3/Al_2O_3 and C/SiC are the most commonly used CMCs in several industrial applications. Applications for CMCs are being considered for the recirculating fan, Gas-fired radiant, burner tubes, canned motor, filtration, and heat exchanger (Krenkel, 2008).

B. Polymer Matrix Composites

The PMCs are composite materials comprised of a specific type of fibre, bound together by the polymer matrix to accomplish desired properties. The fibre of PMCs consists of aramid, fiberglass, and graphite.

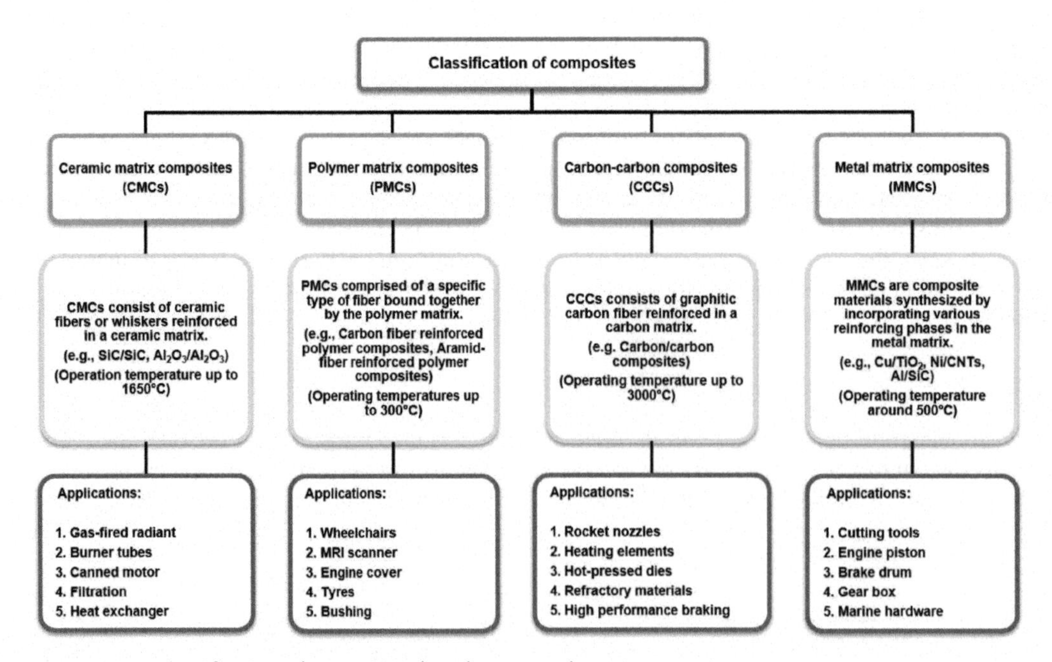

Figure 20.1 Classification of composites based on type of matrix

The PMCs are easy to fabricate compared to metal-matrix, ceramic-matrix, and carbon-matrix. The PMCs have many advantages such as good abrasion and corrosion resistance, lightweight, high strength, and high stiffness along the direction of reinforcement. Nowadays, PMCs are widely used in automobiles, aircraft, marine structures, and other moving structures.

C. Carbon-carbon Composites

The CCCs consist of graphitic carbon fibre reinforced carbon matrix. CCCs are commonly used for high strength and modulus of rigidity. Also, these composites are lightweight and can withstand up to 3000°C. The CCCs are widely used in aircraft, rocket nozzles, space shuttle nose tip, F1-racing cars and train brakes due to their remarkable properties such as high thermal and abrasion resistance, high electrical conductivity, low density and high strength.

D. Metal Matrix Composites

The MMCs are composite materials synthesised by incorporating various reinforcing phases in the metal matrix. Numerous matrices and reinforcing materials used for the fabrication of MMCs are represented in Figure 20.2. The MMCs are the potential contestants for operation in complex service conditions such as marine, nuclear power plants, automobile, chemical and infrastructure. In MMCs, the main matrix materials may be Ni, Cu, Al, Mg, and Ti. The main reinforcements used are alumina, carbide, and silicon. MMCs with lightweight and high strength have been developed for satellites, aircraft, missiles, jet engines, and high-speed machinery. Presently, MMC is used in diesel engine piston developed by Toyota, which shows high wear resistance and high-temperature strength (Office of Technology Assesment, 1988). The MMCs are most commonly used in several engineering applications and are the focus of this thesis. Presently, particulate reinforced MMCs have attracted considerable attention from researchers worldwide due to their low cost, ease of synthesis, and near-isometric enhancement in the overall properties (Katarkar et al., 2021; Pingale et al., 2020c; 2021b). Also, the incorporation of nanoparticles in the metal matrix has shown significant enhancement in the resulting composite's mechanical, tribological, electrical, and optical and corrosion properties. In the past few decades, various methods have been employed to fabricate MMCs. The classification of MMCs fabrication method is represented in Figure 20.3. Among these methods, powder metallurgy and electrochemical deposition have been extensively used to prepare various composites due to some good features such as simplicity in operation, low cost and scalability (Katarkar et al., 2021; Owhal et al., 2021).

Need for cupronickel composites

A wide range of objects be they components, tools, sub-assemblies, machines, or entire plants are made from different types of materials. In most cases, the lifetime of these materials is strongly affected by external factors and the operational environment. The surface of the components is usually damaged due to the

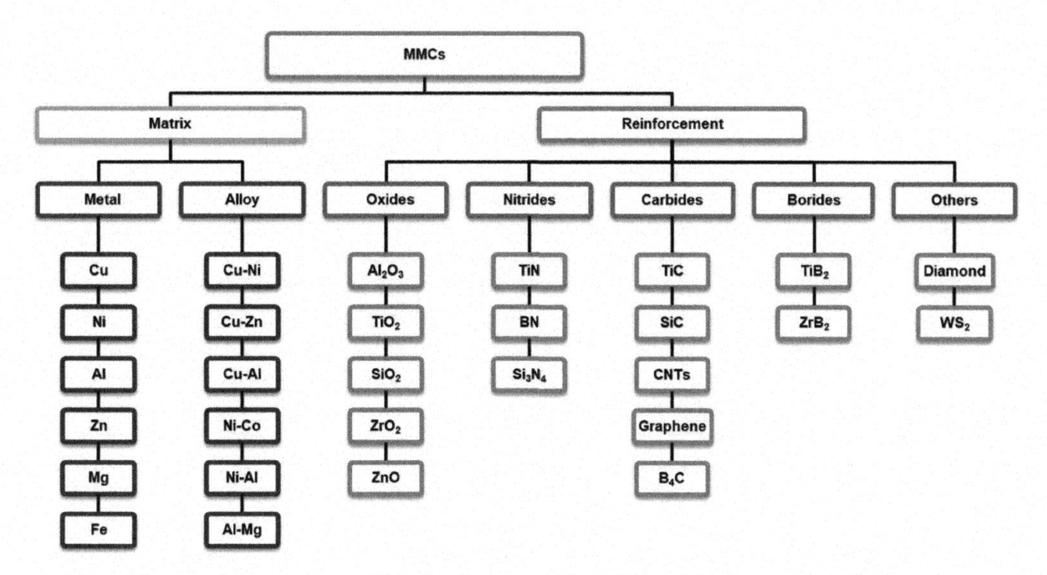

Figure 20.2 Numerous matrices and reinforcing materials used for the fabrication of MMCs

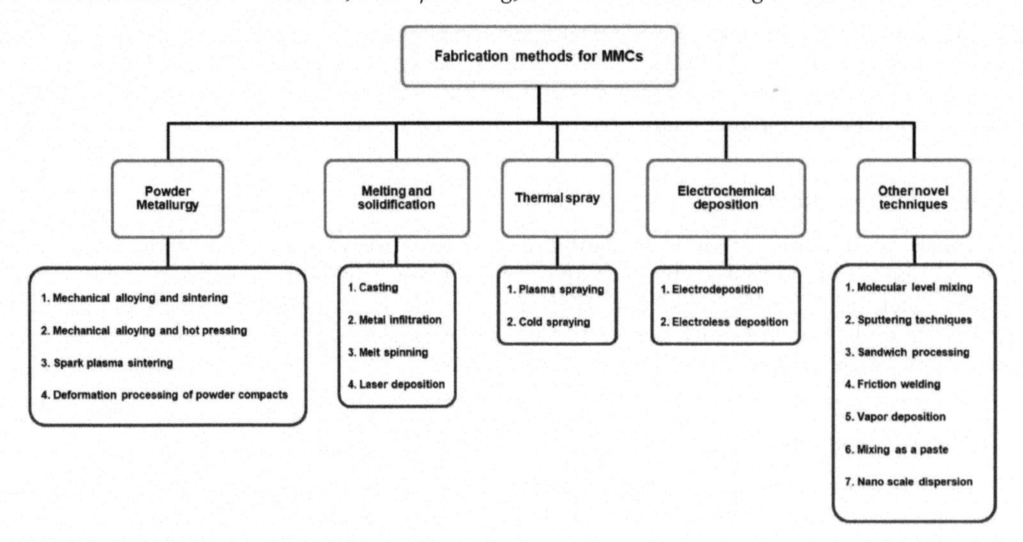

Figure 20.3 Fabrication methods for MMCs

Figure 20.4 Some applications of cupronickel alloy

mechanical interaction between the surfaces of the components in contact with each other as well as electrochemical reactions with the environment (Kanani, 2004). The damaging changes at the surface arise due to oxidation, erosion, electrochemical corrosion, scaling, cavitation, weathering, microbiological damage and wear. Alloy deposition is an old technique with the same scientific principles as individual metals electrodeposition (Schlesinger and Paunovic, 2011). The interest in the utilization of alloy coating is increased due to the wide range of possible alloy combinations and the related possible applications. Alloy coatings have superior properties in certain composition ranges than those of individuals metal coatings. They can be harder, better resistance against corrosion, stronger and tougher, more wear resistance and superior in magnetic properties. Over the past few years, cupronickel alloy coatings have attracted considerable scientific and industrial interest (Metikoš-Huković et al., 2011). Due to this, cupronickel alloy coatings are widely used to reduce corrosion and wear rate. Hence, it finds large number of applications in piping, condensers and heat exchangers in seawater systems, desalination plants, marine hardware, boat hulls, oil rigs and platforms, seawater intake screens and fish farming cages (Ngamlerdpokin and Tantavichet, 2014; Goranova et al., 2014; Calleja et al., 2012; Ahmed et al., 2008; Thurber et al., 2016; Dai et al., 2016). Some applications of cupronickel alloys are shown in Figure 20.4. The recognition to increase the lifetime of engineering components that are exposed to wear and the corrosive environment is yet unresolved. Hence, in order to achieve further enhancement of mechanical, wear and anti-corrosion properties, there is a need to fabricate cupronickel composites.

Electrodeposition of cupronickel composites

Several reinforcing elements such as graphene, carbon nanotubes, Al2O3, TiO2, SiC and Y2O3 have been added into a cupronickel alloy matrix to enhance mechanical, tribological and anti-corrosion properties of resulting composite coatings (Pingale et al., 2020b; 2020d; 2021a). Bath composition and operating conditions employed for preparing Cu-Ni alloy composite coatings is given in Table 20.1. According to obtained results, the microhardness and wear resistance of Ni-Cu/Al_2O_3 composite coating are increased about 2.4 and 3.75 times, respectively, compared to pure Ni-Cu alloy coating. The incorporation of Al_2O_3 nanoparticles increased the microhardness of the coating from 331 HV to 570 HV. The wear rate for Ni-Cu/Al_2O_3 composite coatings was decreased with an increase in Al_2O_3 nanoparticles bath (Alizadeh and Safaei, 2018). the corrosion resistance and hardness of Ni-Cu/ZrO_2 composite coatings have been enhanced compared to pure Ni and Ni-Cu alloy coatings. The presence of ZrO_2 in the Ni-Cu alloy matrix results in an increase in microhardness about 100 HV compared to pure Ni-Cu alloy coating. The hardness of Ni-Cu/ZrO_2 composite coatings is increased with a decrease in the duty cycle (Li et al., 2019). The duty cycle and current density significantly affect the corrosion resistance of the Ni-Cu/TiN-ZrO_2 nanocomposite coating compared to frequency (Li et al., 2019). Ni-Cu/Y_2O_3 nanocomposite coatings are prepared using the electrodeposition method to enhance the corrosion resistance performance in the marine environment (Safavi et al. 2021). The obtained results show that Ni-Cu-Y_2O_3 (4 g/L) composite coating shows higher anti-corrosion preperformance mainly due to the higher content of Y_2O_3 nanoparticles and the least microstructural defects such as pores, voids and micro-cracks. SiC has been widely used in photo-induced wettable composite coatings due to a highly effective photocatalyst under UV or visible light irradiation. Ni-Cu-SiC composite coatings have been synthesized on the Mg-Li alloy using the electrodeposition method and stearic acid modification to enhance corrosion resistance and hydrophobicity (Ji et al. 2018). Wettability transition test and corrosion test were carried out to analyse the coating samples. Surface wettability tests revealed that the contact angle for Ni-Cu-SiC composite coating was increased with an increase in SiC concentration in the electrolyte bath up to 20 g/L and decreased with further increase in SiC concentration in the electrolyte bath. Corrosion study showed that the high hydrophobic state of coatings results in enhancement in the corrosion resistance of coatings.

Table 20.1 Bath composition and operating conditions used for preparing cupronickel composite coatings

Coating	*Bath composition*	*pH*	*Temp. (°C)*	*Goal*	*Ref.*
Cu-Ni/Al_2O_3	105 g/L $NiSO_4.7H_2O$, 25 g/L $CuSO_4.5H_2O$, 59 g/L $Na_3C_6H_5O_7$, Al_2O_3 (0-30 g/L)	4	35	Enhancing hardness, tribological and anti-corrosion performance	(Alizadeh and Safaei, 2018)
Ni-Cu/TiN	160 g/L $NiSO_4.7H_2O$, 15 g/L $CuSO_4.5H_2O$, 20 g/L H_3BO_3, 8 g/L TiN, 1 g/L $C_7H_5NO_3S$, 0.05 g/L $NaC_{12}H_{25}SO_4$	4.3	50	Enhancing corrosion resistance by ultrasonic agitation	(Li et al., 2019)
Ni-Cu-ZrO_2	70.96 g/L $NiSO_4.7H_2O$, 15 g/L $CuSO_4$, 42.6 g/L Na_2SO_4, 63 g/L Lactic acid, (0-16 g/L) ZrO_2	8-9	30	Increasing microhardness and corrosion resistance properties	(Li et al., 2019)
Ni-Cu/ TiN-ZrO_2	160 g/L $NiSO_4.7H_2O$, 15 g/L $CuSO_4.5H_2O$, 20 g/L H_3BO_3, 6 g/L ZrO_2, 4 g/L TiN, 1 g/L $C_7H_5NO_3S$, 0.05 g/L $NaC_{12}H_{25}SO_4$	4.2	50 ± 1	Investigating effect of pulse parameters on microstructure, morphology and anti-corrosion performance	(Li et al., 2019)
Ni-Cu-Y_2O_3	150 g/L $NiSO_4.7H_2O$, 30 g/L $CuSO_4.5H_2O$, 50 g/L $Na_3C_6H_5O_7$, 30 g/L H_3BO_3, 0.4 g/L $NaC_{12}H_{25}SO_4$, 1 g/L $C_7H_5NO_3S$, (0-5 g/L) Y_2O_3	4-5	60	Exploring the effect of Y_2O_3 on microstructure, morphology and anti-corrosion behaviour	(Safavi et al., 2021)
Ni-Cu-SiC	300 g/L $NiSO_4.7H_2O$, 1.3 g/L $CuSO_4.5H_2O$, 19.4 g/L NH_4HF_2, 35 g/L H_3BO_3, 0.4 g/L $NaC_{12}H_{25}SO_4$, (0, 10, 20 and 30 g/L) SiC	4.4-4.8	50	Improving hydrophobicity and corrosion resistance	(Tian et al., 2016)

Table 20.2 Fabrication and process parameters employed for preparation of cupronickel composites by powder metallurgy method

Composite	*Processing route*	*Compression pressure (MPa)*	*Sintering temp. (°C)*	*Goal*	*Ref.*
$Cu\text{-}Ni/Al_2O_3$	In-situ chemical reaction + compaction + sintering	700	900	Improving microhardness by increasing interfacial bonding between Cu-Ni and Al_2O_3 and	(Ali et al., 2022)
$TiB_2\text{-}B_4C/Cu\text{-}Ni$	Ball milling + spark plasma sintering	30	1850	Enhancing flexural strength and fracture toughness	(Wu et al., 2017)
Cu-Ni-graphite	Powder mixture + compaction + sintering	600	800	Improving hardness and wear resistance	(Wang et al., 2017)
$Cu_{0.7}Ni_{0.3}(Al_2O_3)$	Ball milling + compaction + microwave sintering	350	300	Enhancing tribological and anti-corrosion properties	(Baghani et al., 2017)
$Cu\text{-}Ni\text{-}Al_2O_3$	Chemical reaction + compaction + sintering	15	1000	Increasing homogeneous distribution Al_2O_3 into the Cu-Ni alloy matrix	(Ramos et al., 2017)
Cu-Ni-graphite	Ball milling + compaction + sintering	600	900	Enhancing flexural strength and microhardness	(Wang et al., 2017)

Fabrication of cupronickel composites using powder metallurgy method

Recently, several cupronickel composites were fabricated using powder metallurgy method to meet the requirement of advanced engineering applications. Fabrication and process parameters employed for the preparation of cupronickel composites by powder metallurgy method are listed in Table 20.2. Cu-Ni-graphite composites have been fabricated using powder metallurgy method (Powder mixture + compaction + sintering) to enhance the tribological performance of Cu-Ni alloy (Wang et al. 2017). It is reported that the addition of graphite into the Cu-Ni alloy matrix significantly improved hardness and reduced the friction coefficient of resulting composites. Also, $Cu\text{-}Ni/Al_2O_3$ composites were prepared using in situ chemical reaction technique followed by powder metallurgy method (Ali et al., 2022). However, the addition of ceramic particles into the cupronickel alloy matrix will cause excessive brittle phases forming, which is undesirable and reduces the strength of the resulting composites (Wu et al., 2017).

Summary

Several reinforcing elements such as Al_2O_3, SiC, TiO_2, Y_2O_3, ZrO_2, TiN, MMT and Cr have been used to synthesize cupronickel composite coatings. However, the incorporation of ceramics can lead to problems with brittleness and delamination, resulting in short service life. Therefore, there is a need to explore a promising reinforcing element for fabricating cupronickel composite coatings. Several attempts have been made towards the improvement in the uniform dispersion of carbon nanomaterials without damaging their structure. Still, the uniform dispersion of carbon nanomaterials in the metal matrix without damaging its structure is a major challenge in the field of composite synthesis. Therefore, a modified method to overcome the above challenges needs to be investigated. Study on computational techniques in cupronickel composites is not yet reported. Computational techniques will help in predicting the properties of the composites without even fabrication.

References

Ahmed, J., Ramanujachary, K. V., Lofland, S. E., Furiato, A., Gupta, G., Shivaprasad, S. M., and Ganguli, A. K. (2008). Bimetallic Cu-Ni Nanoparticles of Varying Composition (CuNi3, CuNi, Cu3Ni). Colloids Surf. A: Physicochem. Eng. Asp. 331(3):206–12. https://doi.org/10.1016/j.colsurfa.2008.08.007.

Ali, M., Sadoun, A. M., Abouelmagd, G., Mazen, A. A., and Elmahdy, M. (2022). Development and Performance Analysis of Novel in Situ Cu–Ni/Al2O3 Nanocomposites. Ceramics Int. January. https://doi.org/10.1016/j.ceramint.2022.01.287.

Alizadeh, M. and Safaei, H. (2018). Characterization of Ni-Cu Matrix, Al2O3 Reinforced Nano-Composite Coatings Prepared by Electrodeposition. Appl. Surface Sci. 456:195–203. https://doi.org/10.1016/j.apsusc.2018.06.095.

Calleja, P., Esteve, J., Cojocaru, P., Magagnin, L., Vallés, E., and Gómez, E. (2012). Developing Plating Baths for the Production of Reflective Ni–Cu Films. Electrochimica Acta. 62:381–89. https://doi.org/10.1016/j.electacta.2011.12.049.

Dai, P. Q., C. Zhang, J. C., Wen, H. Rao, C., and Wang, Q. T. (2016). Tensile Properties of Electrodeposited Nanocrystalline Ni-Cu Alloys.J. Mater. Eng. Perform. 25(2):594–600. https://doi.org/10.1007/s11665-016-1881-2.

Goranova, D., Georgi, A., and Rashko, R.. (2014). Electrodeposition and Characterization of Ni-Cu Alloys. Surf. Coat. Technol. 240:204–210. https://doi.org/10.1016/j.surfcoat.2013.12.014.

Ji, P., Long, R., Hou, L., Wu, R., Zhang, j., and Zhang, M. (2018). Study on Hydrophobicity and Wettability Transition of Ni-Cu-SiC Coating on Mg-Li Alloy. Surf. Technol. 350: 428–35. https://doi.org/10.1016/j.surfcoat.2018.07.038.

Kanani, N. 2004. *Electroplating:* Basic Principles, Processes and Practice. Unitesd Sates: Elsevier Science. https://books.google.co.in/books?id=Ii1e-pp1pq0C.

Katarkar, A. S., Majumder, B., Pingale, A. D., Belgamwar, S. U., and Bhaumik, S. (2021). A Review on the Effects of Porous Coating Surfaces on Boiling Heat Transfer. Mater. Today: Proc. 44:362–67. https://doi.org/10.1016/j.matpr.2020.09.744.

Katarkar, A. S, Pingale, A. D., Belgamwar, S. U., and Bhaumik, S. (2021). Experimental Study of Pool Boiling Enhancement Using a Two-Step Electrodeposited Cu–GNPs Nanocomposite Porous Surface With R-134a. J. Heat Transfer. 143(12):121601. https://doi.org/10.1115/1.4052116.

Krenkel, W. 2008. Ceramic Matrix Composites: Fiber Reinforced Ceramics and Their Applications. Online Library: Wiley & Sons. https://books.google.co.in/books?id=j81hCENfG60C.

Li, B., Du, S., and Mei, T. (2019). Pulse Electrodepsoited Ni-Cu/TiN-ZrO 2 Nanocomposite Coating: Microstructural and Electrochemical Properties. Mater. Res. Expr. 6(9):096433. https://doi.org/10.1088/2053-1591/ab31e9.

Li, B., Mei, T. Li, D., and Du, S. (2019).Ultrasonic-Assisted Electrodeposition of Ni-Cu/TiN Composite Coating from Sulphate-Citrate Bath: Structural and Electrochemical Properties. Ultrason. Sonoch. 58:104680. https://doi.org/10.1016/j.ultsonch.2019.104680.

Li, B., Mei, T., Li, D., Du, S., and Zhang, W. (2019). Structural and Corrosion Behavior of Ni-Cu and Ni-Cu/ZrO2 Composite Coating Electrodeposited from Sulphate-Citrate Bath at Low Cu Concentration with Additives. J. Alloys Compd. 804:192–201. https://doi.org/10.1016/j.jallcom.2019.06.381.

Metikoš-Huković, M., Babić, R., Škugor Rončević, I., and Grubač, Z. (2011). Corrosion Resistance of Copper-Nickel Alloy under Fluid Jet Impingement. Desalination. 276(1–3):228–32. https://doi.org/10.1016/j.desal.2011.03.056.

Mohammad, B., Aliofkhazraei, M., and Poursalehi, R. (2017). Low Temperature Microwave Sintering of Cu 0.7 Ni 0.3 (Al 2 O 3) Nanocomposite. Powder Metallur. 60(1):73–83. https://doi.org/10.1080/00325899.2016.1275097.

Ngamlerdpokin, K. and Tantavichet,. N. (2014). Electrodeposition of Nickel-Copper Alloys to Use as a Cathode for Hydrogen Evolution in an Alkaline Media. Int. J. Hydro. Energy. 39(6):2505–15. https://doi.org/10.1016/j.ijhydene.2013.12.013.

Office of Technology Assesment, U. S. Congress. 1988. Advanced Materials by Design. OTA-E-351. U.S. Government Printing Office.

Owhal, A., Pingale, A. D., Khan, S., Belgamwar, S. U., Jha, P. N., and Rathore, J. S. (2021). Electro-Codeposited γ-Zn-Ni/Gr Composite Coatings: Effect of Graphene Concentrations in the Electrolyte Bath on Tribo-Mechanical, Anti-Corrosion and Anti-Bacterial Properties. Transac. IMF. 99(6):324–31. https://doi.org/10.1080/00202967.2021.1979815.

Pingale, A. D., Belgamwar, S. U., and Rathore, J. S. (2020a). Effect of Graphene Nanoplatelets Addition on the Mechanical, Tribological and Corrosion Properties of Cu–Ni/Gr Nanocomposite Coatings by Electro-Co-Deposition Method. Trans. Indian Inst. Met. 73(1):99–107. https://doi.org/10.1007/s12666-019-01807-9.

Pingale, A. D., Owhal, A., Belgamwar, S. U., and Rathore, J. S. (2021a). Electro-Codeposition and Properties of Cu–Ni-MWCNTs Composite Coatings. Trans. IMF. 99(3):126–32. https://doi.org/10.1080/00202967.2021.1861848.

Pingale, A. D., Owhal, A., Katarkar, A. S., Belgamwar, S. U., and Rathore, J. S. (2021a). Recent Researches on Cu-Ni Alloy Matrix Composites through Electrodeposition and Powder Metallurgy Methods: A Review. Mater. Today: Proc. 47:3301–3308. https://doi.org/10.1016/j.matpr.2021.07.145.

Pingale, A. D., Belgamwar, S. U., and Rathore, J, S. (2020c). The Influence of Graphene Nanoplatelets (GNPs) Addition on the Microstructure and Mechanical Properties of Cu-GNPs Composites Fabricated by Electro-Co-Deposition and Powder Metallurgy. Mater. Today: Proc. 28:2062–2067. https://doi.org/10.1016/j.matpr.2020.02.728.

Ramos, M. I., Suguihiro, N. M., Brocchi, E. A., Navarro, R., and Solorzano, I. G. (2017). Microstructure Investigation of Cu-Ni Base Al2O3 Nanocomposites: From Nanoparticles Synthesis to Consolidation. Metall. Mater. Trans. A. 48(5):2643–2653. https://doi.org/10.1007/s11661-017-4000-6.

Safavi, M. S., Fathi, M., Mirzazadeh, S., Ansarian, A., and Ahadzadeh, I. (2021). 'Perspectives in Corrosion-Performance of Ni–Cu Coatings by Adding Y 2 O 3 Nanoparticles. Surf. Eng. 37(2):226–235. https://doi.org/10.1080/02670844.2020.1715543.

Savage, G. and G M Savage, G. M. (1993). Carbon-Carbon Composites. Londan: Chapman & Hall. https://books.google.co.in/books?id=_qF2BD9sI-wC.

Schlesinger, M. and Paunovic, M. 2011. Modern Electroplating. The ECS Series of Texts and Monographs. Wiley. https://books.google.co.in/books?id=j3OSKTCuO00C Tian, W. -M., Li, S. M., Wang, B., Chen, X., Liu, J. -H., and Yu, M. (2016). Graphene-Reinforced Aluminum Matrix Composites Prepared by Spark Plasma Sintering. Int. J. Miner. Metall. 23(6):723–729. https://doi.org/10.1007/s12613-016-1286-0.

Thurber, C. R., Ahmad, Y. H., Sanders, S. F., Al-Shenawa, A., D'Souza, N., Mohamed, A. M. A., and Golden, T. D. (2016). Electrodeposition of 70-30 Cu–Ni Nanocomposite Coatings for Enhanced Mechanical and Corrosion Properties. Curr Appl Phys. 16(3):387–96. https://doi.org/10.1016/j.cap.2015.12.022.

Wang, Y., Gao, Y., Li, Y., Zhang, C., Huang, X., and Zhai, W. (2017). Effect of Milling Time on Microstructure and Mechanical Properties of Cu–Ni–Graphite Composites. Mater. Res. Exp. 4(9):096506. https://doi.org/10.1088/2053-1591/aa84a7.

Wang, Y., Gao, Y., Sun, L., Li, Y., Zheng, B., and Zhai, W. (2017). Effect of Physical Properties of Cu-Ni-Graphite Composites on Tribological Characteristics by Grey Correlation Analysis. Results Phys. 7:263–71. https://doi.org/10.1016/j.rinp.2016.12.041.

Wu, Z., Zhang, J., Shi, T., Zhang, F., Lei, L., Xiao, H., and Fu., Z. (2017). Fabrication of Laminated TiB 2 -B 4 C/Cu-Ni Composites by Electroplating and Spark Plasma Sintering. J. Mater. Sci. Technol. 33(10):1172–76. https://doi.org/10.1016/j.jmst.2017.05.012.

———. 2020b. "Synthesis and Characterization of Cu–Ni/Gr Nanocomposite Coatings by Electro-Co-Deposition Method: Effect of Current Density." *Bulletin of Materials Science* 43 (1): 66. https://doi.org/10.1007/s12034-019-2031-x.

———. 2021b. "Effect of Current on the Characteristics of CuNi-G Nanocomposite Coatings Developed by DC, PC and PRC Electrodeposition." *JOM* 73 (12): 4299–4308. https://doi.org/10.1007/s11837-021-04815-7.

———. 2020d. "A Novel Approach for Facile Synthesis of Cu-Ni/GNPs Composites with Excellent Mechanical and Tribological Properties." *Materials Science and Engineering: B* 260 (July): 114643. https://doi.org/10.1016/j.mseb.2020.114643.

———. 2021b. "Facile Synthesis of Graphene by Ultrasonic-Assisted Electrochemical Exfoliation of Graphite." *Materials Today: Proceedings* 44 (November): 467–72. https://doi.org/10.1016/j.matpr.2020.10.045.

21 The significance of fine aggregates on strength and shrinkage properties of mortar (1:4)

Aaquib. R. Ansari[a] *and Anmol W. Dongre*[b]

Department of Civil Engineering, G. H. Raisoni Institute of Engineering & Technology, Nagpur, India

Abstract

The experimental research is done to understand the effect of sand particle size on strength and shrinkage properties of mortar. In order to examine the same, an exploratory study is carried out. Cement Sand mortar (1:4) was made by keeping the water to cement ratio 0.5 for all combinations mixes. 06 mortar mixes were produced and moulded in the form of 70.6 cubic meter cubes in accordance with IS-10086-1982. Water absorption was evaluated following 30 and 90 days. For calculating drying shrinkage, mortar bar of size 25 cm × 2.5 cm × 2.5 cm were prepared as per ASTM. Strength against the compression was calculated after curing mortar samples for 7, 28 and 90 days. Mortar Shrinkage was measured after three, seven, and thirty-five days of humidity curing. Compressive strength for mortar mix M44 was highest among the others. Similarly other parameters like water consumption and shrinkage were within the permissible limits for the same mix.

Keywords: Cement mortar, compressive strength, drying shrinkage, moisture absorption, sand gradation.

Introduction

In this research different mortar mixes were prepared for the different proportion of the sand. The sand was classified according to the particle diameter. Sand passing 0.236 cm standard sieves and sustained on 0.6 mm standard sieve is coarse sand and portion passed through 0.6 mm is classified as fine sand. Total six mortar mixes were prepared by taking different percentages of fine and coarse sand including the conventional mortar. Water cement ratio for all type of mixes were kept constant as 0.5. Compressive strength, water absorption and drying shrinkage for all the samples were observed and compared carefully. For calculating drying shrinkage, mortar bar of size 25 cm × 2.5 cm × 2.5 cm were prepared as per American standard for testing of materials (ASTM).

A. Literature Review

Five alternative mix compositions were investigated, with cement sand ratio (1:3) and lime concentrations ranging from 10 to 75%. Mechanical strength, stiffness as well as Ultrasonic pulse velocity (UPV) have been assessed related to lime content. Results indicate decrease in mechanical strength and stiffness by 14 and 12% respectively for every 10% increase in lime (Ramesh et al., 2019). The approach of multivariate adaptive regression splines is being used to determine the shrinkage characteristics of concrete. The same includes some of the major factors such as humidity, water-to-cement ratio and CS at about one month. (**Kaveh** et al. 2018). Electron microscope used to test the crack formation of restricted contraction has been used to assess the width of crack of dried contraction. This material investigates the degree of drying shrinkage and fracture width (Yatagan 2015). Purpose of this research is to discover and investigate the possibilities of re-cycling a substantial impact of plastic bag trash. It will also investigate the mechanical characteristics and resilience of mortars including plastic bag wastes. As a fine aggregate, it is mixed with a varying amount of sand (10, 20, 30 and 40%. Research analysis indicate that the inclusion of the same is allowed for a reduction of 18–23% in the mechanical properties of mortars containing ten and 20% waste, respectively, while being close to the traditional one (Ghernouti and Rabehi, 2012). In this study waste ceramic is used as binder and sand. One conventional mortar mix using cement, sand while other one was made by replacing cement with ceramic particle by 40% and sand is replaced completely by ceramic particle, their findings suggests that combination of ceramic powder and fine particles of ceramic lowered the flow ability of mortar. It was also observed that strength (compressive and tensile) increases with increase in waste ceramic. Drying shrinkage reduces for ceramic mortar as compared to conventional one

---

[a]aaquibansari12@gmail.com; [b]anmolwdongre@gmail.com

(Hosseini et al., 2019). Combination of 108 samples of 2.5 × 2.5 × 27.5 cm prisms and ninety eight samples of 7.5 cm × 15.0 cm cylinders of masonry mortar were casted and tested from active building projects in Riyadh. For the identical curing circumstances and with the same mixture proportions, white sand creates mortar with a greater elastic modulus with lesser shrinkage (Schutter and Pope, 2004).

B. Gap Identification

After reading the literature related to the properties of mortar using different techniques. I have identified that the effect of the different percentages of sand according to the coarse and fine fraction classification has not been studied. Using the fine and coarse sand one can analyse the variation in strength and drying shrinkage properties of mortar. Durability of mortar also can be studied by using different salt solution and also different diluted acids like HCL, H_2SO_4.

C. Novelty of work

Strength and shrinkage properties of conventional and modern mortars with and without pozzolanic materials such as fly ash and rice husk ash have been studied worldwide by using river sand without taking any specific percentage of sand particle size. In this research particle size of natural sand has been identified using two standard size sieves. Standard sieve size 2.36 mm and 600 μ is being used to define the coarse and fine fraction of sand to be used in mortars

Experimental Procedure

A. Ingredients and Related Characteristics

OPC: The cement containing zero pozzolanic material is tested as per standards in India, verifying IS-8112 (2013) and outcome of the test are tabulated below.

Table 21.1 OPC's chemical structure (43 grade)

	Values Obtain	*Standard Criteria*
Silica soluble in %	21.5	---
Alumina proportion %	5.0	---
ferrous oxide	2.9	---
% Cao	62	
Magnesium percentage	3.2	Limited to 6%
Insoluble residue	3.1	Limited to 4%
% Sulphor trioxide	3.0	Limited to the value 3.5%
% loss of ignition	2.2	
Lime saturation factor	1.057	Limited to 5%
Proportion of alumina to iron oxide	0.72	In between 0.66 to 1.02
Tricalcium aluminates	8.5	More than 0.66 % Limited to
chloride percentage	0.5	the value 0.1 %

Table 21.2 OPC (43 grade) physical characteristics

Test conducted	*Values Obtain*	*Standard criteria*
m^2/Kg (Fineness)	260	More than 200
Compressive strength (Days)		
03	35.5	More than 23
07	42	More than 33
28	54	The value below Min 43/Max 58
Setting time (minutes)		
Initial	180	Not less than30
Final	300	Not more than 600
Soundness		
Le-chatelier expn. (mm)	0.00	Not more than 10 mm
Autoclave expn. (%)	0.033	Not more than 0.8%
a) % Standard Consistency	26.8	
b) Test temperature (°C)	287+/-2	
c) Fly ash content (%)	2.9	Limited to 5%

Fine aggregate (river sand):- Fine Fraction is selected depending on particle diameter. Fine sand is river sand that is been furnace dried and it has gone via both 0.236 cm and 0.6 mm standard sieve. The next kind is gritty sand (coarse), which is natural sand and passes through a 2.36 mm IS sieve and remains on a 0.6 mm Standard sieve in an oven-dry condition.

Portable water: The portable water was needed for both the mortar's mixing and curing.

B. Various mortar mixtures for (1:4)

Table 21.3 shows the different mortar mixes with sand NOTE: M41, M42, M43, M44, M45, M46 stands for mixes for (1:4).

C. Sample Preparation Process

Six different (1:4) mixtures were cast. For all six cement mortars, the w/c ratio was kept consistent at 0.5. To guarantee mix homogeneity, the sand cement was completely mixed. Natural oil or light cup greasing was applied lightly to the inside sides of the sample mould. Excess liquid or lubricant was cleaned from inside faces as well as the top and bottom surfaces of every mould, after they were joined. Similarly drying shrinkage bar were prepared using standard mould size of 25 cm × 2.5 cm × 2.5 cm and they are kept in the moisture chamber as shown in Figure 21.3.

Table 21.3 Different mortar mixes (1:4)

Mix type	*Percent cement*	*Ratio W/C*	*Sand (%)*	
			Coarse	*Fine*
M41	100	1/2	100	0
M42	100	1/2	All passing standard sieve of 2.36 mm	
M43	100	1/2	70	30
M44	100	1/2	60	40
M45	100	1/2	50	50
M46	100	1/2	0	100

Figure 21.1 Hand mixing of mortar mixes

Figure 21.2 Mortar cubes for M43

Figure 21.3 Drying shrinkage bar

Testing Methodology

A. Compression Test

The compression test was done according to IS: 2250-1981 for 7D, 28D and 90D of curing. Due care is to be taken that the cube shall be centrally aligned with the centre of plates bearing the testing machine. Only auxiliary steel plates must be used between the faces of the specimen and steel plates of the Compression machine. Average of three cube samples was taken as Compressive strength.

$$\text{Compressive strength} = \frac{\text{Failure Load}}{\text{Area}}\left(\frac{\text{N}}{\text{sq.mm}}\right)$$

B. Drying Shrinkage test

The drying shrinkage test was conducted as per IS 4031 (part-10) and also the help of ASTM C157-C 157-M 08 which is normal test method for finding change in length of set cement mortar and concrete. Mould size 25 cm × 2.5 cm × 2.5 cm was selected as per requirement and the test values are measured for 7, 35, and 90 days in terms of length change of specimens in %.

$$\Delta D_x = \frac{\text{Fianl CDR} - \text{Initila CDR}}{\text{G}} \times 100$$

ΔDx = Change in length of specimen at any age in %. CDR = Variance between the specimen's comparator value and the reference bar at any period L_g = Length of gauge (0.25 m)

Figure 21.4 Compression test

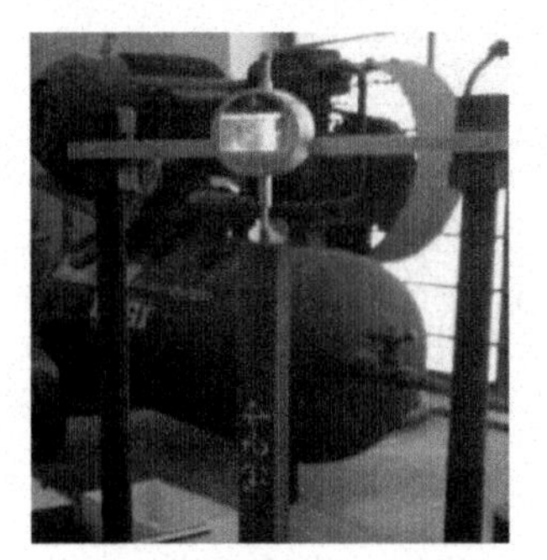

Figure 21.5 Drying shrinkage test

C. Moisture Uptake Evaluation

Following moulding, the cubes were completely submerged for 28 and 90 days to cure then are weighed, and the resulting weight was recorded as the cube's wet weight (WW). All samples then were oven dried at 110°c until the bulk became constant before being examined again. This weight was observed as the cube's dry weight (DW). Now, these dry cubes were immersed in water for a day to cure, and the wet weight was taken as the wet weight.

Weight was taken as the wet weight.

$$\%WA = \frac{WW - WD}{WD} \times 100$$

Where, WW and WD are wet weight and dry weight of cube respectively.

Results and Discussion

This portion of research discloses the brief observation on the values obtained from the test results are discussed to understand the variations in the outcome

A. Compressive Strength

Compressive strength for 7, 28 and 90 days are shown below.

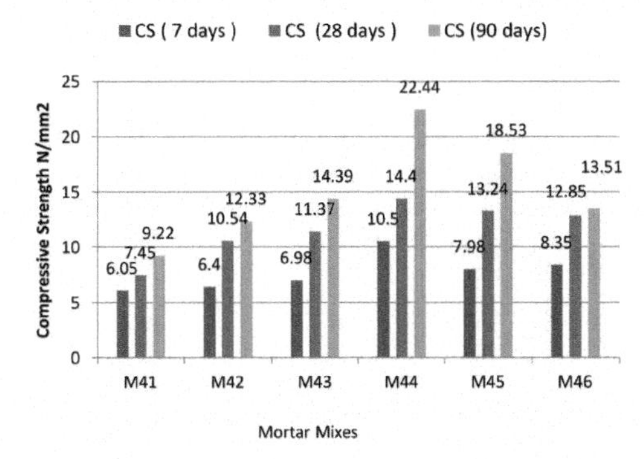

Figure 21.6 Strength results for 07, 28 and 90 days

Strength results for all mortar types after certain duration of water curing is shown in "Figure 21.6". The compressive strengths of M41 are least for 7, 28 and 90 days. Similarly maximum compressive strengths are obtained for the M44. The compressive strength of M41 for 28 days is minimum 7.45 N/mm^2 which is less than 7.5 N/mm^2 as per IS-2250. Hence the value is not acceptable.

B. Drying Shrinkage test

Drying shrinkage of the samples M41, M42 and M43 are shown in the figure.

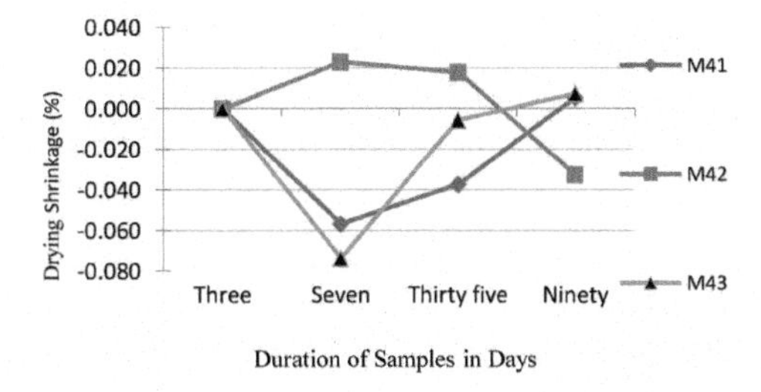

Figure 21.7 Shrinkage variation of M41, M42 and M43

Drying shrinkage values are the characteristic values of mortar which shows the shrinkage and expansion properties of mortar shown in above graph in Figure 21.7.

Mortar mix M41 shows the shrinkage after 7 and 35 days of moisture curing. On the other side M42 shows expansion at same ages.

C. Water Absorption

After period of 30 days curing M46 shows the maximum water absorption 7.94% and M41 have the minimum value 4.77%. M44 shows maximum value of 6.87% after 90 days. As the duration of curing increases the absorption starts decreasing.

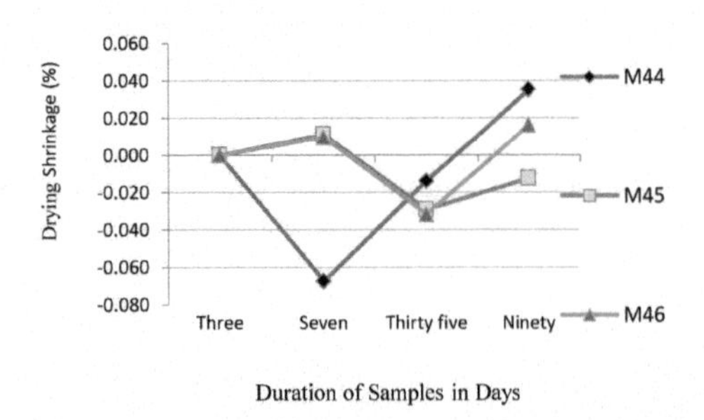

Figure 21.8 Shrinkage variation of M44, M45 and M46

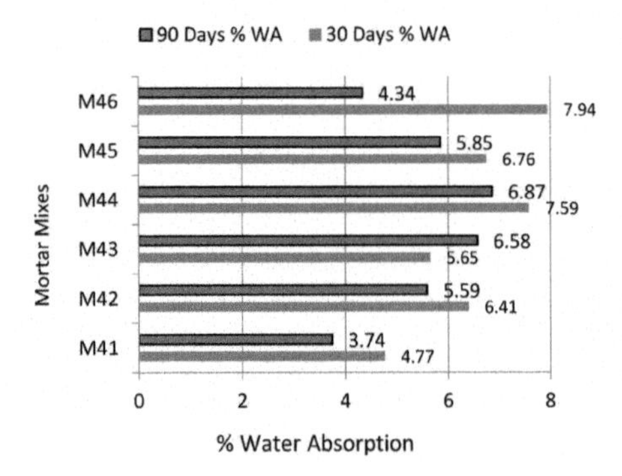

Figure 21.9 Percentage moisture absorbed 30 and 90 days

Conclusions

- Strength of Compression for cement and sand mortar varied with sand particle size in mortar mixtures.
- M41 (100% CS + 0% FS) demonstrated the lowest compressive strength values after seven, Twenty eight and ninety days of curing.
- The CS of M41 for 28 days is minimum 7.45 N/mm^2 which is less than 7.5 N/mm^2 as per IS-2250. Hence the value is not acceptable.
- Maximum compressive strengths are obtained for the M44 for all ages.
- Water absorption values are much higher for 30 day curing for all six mortar mixes.
- M45 showed the expansion initially after 7 days and then showed continuous shrinkage after 35 and 90 days.
- M41 (1:4) showing the minimum water absorption value at 30 and 90 days i.e. 4.77% and 3.74% respectively and M46 has maximum absorption level at 30 days as 7.94%.

References

Aaquib, A., Nandurkar, B., Bhagat, R., Raut, J., Ganvir, V., Agrawal, V., Kedar, A., and Sahare, P. Feb 2018 Influence of Sand Particles on Strength and durability of Mortar (1:3). Int. J. Res. Appl. Sci. Eng. Technol. *6(2).*

Amjad, M. A. (1999). Elasticity and Shrinkage of Cement: Sand Mortar Produced in Riyadh. *JKAU: Eng. Sci.* 11(2):91105.

ASTM C157/C 157M-08, Standard test method for length change of hardened Hydraulic cements Mortar and concrete.

Canova1, J. A., NetoG, G. de. A., and Bergamasco, R. (2009). Mortar with unserviceable tire residues. J. Urban Environ. Eng. 3(2):63–72. ISSN 1982-3932.
Choi, S. J., Lee, S. S., and Monteiro, P. J. M. Effect of Fly Ash Fineness on Temperature Rise, Setting, and Strength Development of Mortar.. Journal of ASCE.
Ghernouti, Y. and Rabehi, B. (2012). Strength and Durability of Mortar Made with Plastics Bag Waste (MPBW). Int. J. Conc. Struct. Mater. 6(3):145–153.
Hosseini, H. M., Lim, N. H. A. S., Tahir, M. M., Alyousef, R., Samadi, M. (2019). Performance Evaluation of green mortar comprising of ceramic waste. SN Appl. Sci. 1:557. https://doi.org/10.1007/s42452-019-0566-5
Kaveh, S. M., Hamze-Ziabari1 L., and Bakhshpoori, T. (2018). Estimating drying shrinkage of concrete using multivariate adaptive regression splines approach. Int. J. Optim. Civil Eng. 8(2):181194
Malathy, R. and Subramanian, K. (2007). Drying shrinkage of cementitious composites with mineral admixtures. Indian J. Eng. Sci. 14:146150.
Potty, N. S., Vallyutham, K. (2014). Properties of Rice Husk Ash (RHA and MIRAH) mortars. Res. J. Appl. Sci. Eng. Technol. 7(18):38723882.
Ramesh, M., Azenha, M., and Lourenço, P. B. (2019). Mechanical properties of lime–cement masonry mortars in their early ages. Material. Structure, Construction. 52:13.
Schutter, G. De. and Pope, A. M.(2004). Quantification of the water demand of sand in mortar. Construct. Build. Mater. 18(7):517521.
Yatagan, M. S. (2015). The Investigation of the Relationship between Drying and Restrained Shrinkage In View of the Development of Micro Cracks. GSTF J. Eng. Technol. 3(3).
Zhang, W., Zakaria, M., and Hama, Y. (20122013). Influence of Aggregate Materials Characteristics on the Drying Shrinkage Properties of Mortar and Concrete. J. Construct. Build. Mater. 49:500510.

22 Optimisation of sewing workstation parameters to reduce low back compressive load for sewing operator

Somdatta Tondre[1,a], *Tushar Deshmukh*[2,b], *and Rajesh Pokale*[1,c]

[1]Department of Mechanical Engineering, P.R. Pote College of Engineering and Management, Amravati, India

[2]Department of Mechanical Engineering, Prof. Ram Meghe Institute of Technology and Research, Badnera, Amravati, M.S., India

Abstract

Sewing machine operators experiences work related musculoskeletal disorders induced because of constrained working postures. The aim of present study was to optimize workstation parameters (sewing needle distance, sewing table top inclination and sewing table top height) to reduce low back compressive force/load for sewing operator. Twenty female sewing machine operators carry out sewing task under nine set of experimental arrangement by varying workstation design parameters. Trunk forward flexion was measured and the low back compressive force was estimated under different experimental trials. Trunk posture was strongly related to low back compressive force. The experimental results shows, sewing needle distance and sewing table top inclination parameters had significant effect on low back compressive force ($p < 0.008$, $p < 0.019$ resp.) of sewing operator. The load on musculoskeletal system due to forward flexed trunk posture could be reduced by improving the design parameters of sewing workstation.

Keywords: Low back compressive force, sewing machine operator, sewing workstation, trunk posture.

Introduction

A. Prevalence of musculoskeletal discomfort among seated operators

It is worldwide accepted and known that, constrained awkward postures and repetitive nature of work results musculoskeletal disorders in seated workers (Li, et al., 1995; Ariens, et al., 2000; Delleman and Dul 2002). Musculoskeletal discomfort as consequence of poor constrained work postures associated with reduction in efficiency and performance (Corlett, 1981; Bhatnager et al., 1981). Some studies demonstrated association between prevalence of low back pain and history of sedentary work type (Vihma et al., 1982). Sewing operators exposed to high risk of musculoskeletal disorders has been recorded in various studies (Vihma et al., 1982; Keyserling et al., 1982; Punnett, et al., 1985; Westgaard and Jansen 1992). Sewing machine operators suffers from musculoskeletal discomfort on different body regions due to awkward constrained work postures (Vihma et al., 1982; Wick and Drury, 1986; Blader, et. al., 1991). According to Herbert, et.al. (2001), garment workers have reported high complaints of upper body work-related musculoskeletal disorders compared to other industrial workers. Afonso et al. (2014) reported the prevalence of work related musculoskeletal disorders among sewing machine operators in Portugal to be 76%. Whereas Akodu et al. (2013) concluded that the prevalence of work related musculoskeletal disorders in Lagos, Nigeria was 92.0% and the lower back pain was reported in 78% of the cases. Similarly, higher prevalence of musculoskeletal discomfort in sewing machine operators than other workers have been reported (Punnett, et al., 1985; Westgaard and Jansen, 1992). The sewing work demands long hours of sitting with forward inclined posture and repetitive work. The sewing work requires a coordination of body parts (feet, hands, and eye), which involves static and rhythmic muscular activity. The stressful constrained awkward postures adopted for work treated as health risk (Brider, 2003). Prolonged seated work owes to musculoskeletal discomfort, which results absenteeism and medical burdon (Duquette et al., 1992).

Working posture is measure of musculoskeletal stresses induced due to workstation and is very useful for evaluation of workstation (Beatriz et.al., 1997). Rapid Upper Limb Assessment (RULA) and Rapid Entire Body Assessment (REBA) are the validated tools used to analyse working posture for ergonomic risk assessment at workplaces. RULA was developed for the evaluation of postures adopted and muscle actions of operators whose repetitive tasks are associated with upper limb disorders whereas REBA was developed for evaluation of postures adopted and muscle actions of operator whose repetitive tasks are associated with the entire body disorder. Many studies reveals need for improvement of working posture, studies shows stressful postures adopted at work has relationship with muscular pain in various body

[a]tondresomdatta@gmail.com; [b]tushar.d69@gmail.com; [c]rajeshpokale79@gmail.com

parts (Aaras and Stranden, 1988). Suitable neutral working posture always reduces musculoskeletal load which leads to improve work performance. The precautionary steps are very important to improve constrained awkward working posture and to promote occupational health (Mattila and Vilkki, 1999). The well designed workstation not only improves working postures but also reduces body stresses (Kroemer 1988; 1997). In general a well-designed workstation assures health and well-being with improvement in functional efficiency, whereas a poorly designed workstation contributes risk of health complaints or occupational injuries, which ultimately results poor task performance (Clark, 1996; Ayoub, 1973; Kadefors, 1998).

B. Biomechanics of spinal loading

Earlier researchers documented, less muscular tension may develop on the spine in erect posture compared to forward flexed inclined posture (Colombini et al., 1986). The back extensor muscles responsible for maintaining postural stability of human body. The extensor muscles resist the forces due to forward flexion of trunk like Guy ropes on one side of mast (Brider, 2003).

C. Estimation of spinal compression

The seated sewing operator represents simplistic biomechanical cantilever model (Figure 22.1(a)). The lumbar spine is like a fulcrum for upper human body. The backside erector spinal muscles of lumbar spine having lever arm of 5 cm. The total spinal compressive load is sum compressive load developed by upper body weight and the compressive load created by back extensor muscles for postural stability. The load moment developed in forward bending posture is the product of load due to upper body weight and moment arm due to upper body weight, whereas in erect posture, the weight body above spine imposes only load. The counter moment developed by back extensor muscles is due to back muscle force and the back muscle lever arm (Brider, 2003).

In forward flexed posture, the movement of centre of gravity of upper body in forward direction of lumbar spine increases load moment, whereas the back extensor muscles resists postural and task load moments. In forward-flexed posture, the upper body weight produces two components of force i.e. compressive force and shear force shown in Figure 22.1(b) (Hamill et al., 2015).

A simple biomechanical model was be used to calculate compressive load of sewing operator.

Force is expressed as product of mass and acceleration:

$$f = m \times a \tag{1}$$

whereas,
m = Mass of the body (kilograms),
a = Acceleration(meter per second)
f = Force (Newton's)

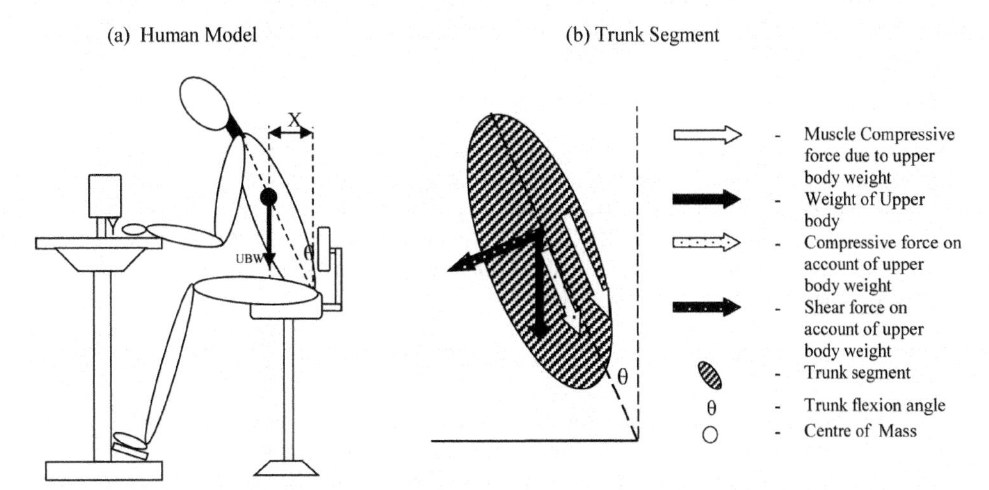

Figure 22.1 Biomechanical model of sewing operator (a) Human model (b) Trunk segment. (UBW - Upper body weight passing through centre of mass of upper body, θ = Trunk flexion angle, X = Perpendicular distance L5/S1 to vertical line passing through centre of mass)

Acceleration due to gravity on earth is 9.81 m/s^2, hence mass 1 kg results force of 9.81 N on earth surface. The total compressive force developed at spine of seated operator without carrying load in hands;

C_t =(Compressive force on account of upper body weight) + (Compressive force on account of back muscle contraction needed to maintain posture) (2)
(Brider, 2003).

Earlier researchers reported need to redesign of sewing workstations for improving working posture of sewing operators. Even after implementing certain modifications in sewing machine workstation, it is observed that sewing operators still adopt forward bending posture of neck and trunk, which makes the back support less effective. This study aimed to modify conventional sewing machine workstation so as to improve trunk posture effectively and to investigate the effect of workstation parameters on low back compressive force between the spinal discs L5/S1 of sewing operator. It is seen from the previous researchers' work, a gap was noticed on the correlation between workstation parameters and low back compressive force between the spinal discs L5/S1 of sewing operator. A prototype was developed by modifying a traditional lock stitch sewing machine workstation with incorporating the workstation design principles. The experimental work performed on modified workstation to study the effect of workstation parameters on low back compressive load.

The features of workstation design parameters as follows.

(1) Provision of sliding arrangement was made for sewing desk to move in fore/aft direction.
(2) Provision of slope arrangement was made for sewing table top to make adjustable between 0 and 10^0.
(3) Sliding arrangement was provided to change height of sewing table top.

Materials and Methods

A. Subjects

Twenty female sewing machine operators (average age 30 year with range 22–40 year; average height 1600 mm with range 1550–1650 mm, average weight 48 kg with range 5 kg) of Amravati region working in garment industry participated in the experimental trials. Subjects under study were well experienced in field sewing work (average work experience 5 year, range 2.5–10). The majority of Indian household women prefer sewing profession as a source of livelihood. Selected subjects for study were experienced with normal health condition (Tondre and Deshmukh, 2019).

B. Design of Experimentation

Genichi Taguchi proposed quality engineering method to improve quality characteristic in most economic way. Taguchi's methodology provides best result with comparatively few numbers of experimental trials (Roy 2001). In this study Taguchi method was used for experimentation purpose to study effect of workstation parameters on low back compressive load of sewing operator.

C. Selection of control factors/parameters and its levels

The sewing workstation design parameters identified after critically analysing working posture of sewing operator and brain storming session. The sewing needle distance, sewing table top inclination and sewing table top height were workstation parameter with three levels for the experimental work. The level for sewing needle distance was 0 mm, 70 mm and 140 mm respectively. Whereas 0 degree, 5 degree and 10 degree and 762 mm, 787 mm and 813 mm was the level for sewing table top inclination and sewing table top height respectively (Tondre and Deshmukh, 2019).

D. Selection of orthogonal array

Orthogonal array is specially constructed tables for experimentation purpose in quality engineering methodology proposed by Genichi Taguchi. The orthogonal array table's makes the design of experiments very simple and efficient and the desired results can be obtained within less number of experiments compared with other experimentation techniques. Table 22.1 shows L-9 orthogonal array used for this experimental work. Total nine experimental trials were performed by referring L-9 orthogonal array (Tondre and Deshmukh, 2019).

Table 22.1 Experimental design (orthogonal array L-9)

Experiment No	*Control Factors (level)*		
	Sewing needle distance (mm)	*Sewing table top inclination (degree)*	*Sewing table top height (mm)*
1	0	0	762
2	0	5	787
3	0	10	813
4	70	0	787
5	70	5	813
6	70	10	762
7	140	0	813
8	140	5	762
9	140	10	787

E. Signal to Noise Ratio (S/N)

The S/N ratio measures sensitivity of the quality characteristic within controlled environment. Signal means the desirable mean result for dependent variables, whereas noise means undesirable/disturbance in getting desirable result, the noise affects the output due to external independent variables (Roy 2001).

The S/N ratio defined as:

S/N ratio (Smaller the better),

$$\eta = -10\ \mathrm{Log}_{10} \frac{1}{n} \sum_{i=1}^{n} \mathrm{yi}^2 \quad (3)$$

S/N ratio (Larger the better),

$$\eta = -10\ Log_{10} \frac{1}{n} \sum_{i=1}^{n} 1/yi^2 \quad (4)$$

Where, n = no. of experiments, and yi = Dependent variable.

In this study low back compressive force between the spinal discs L5/S1 of sewing machine operators is output quality characteristics.

F. Experimental Procedure

Taguchi approach with standardized form of design of experiment (L-9 orthogonal array) was used for experimentation. The sewing operators performed sewing task on a nine set of experiments on modified adjustable sewing workstation. Each experimental trial was of 45 minutes with breaks of 15 minutes after experimental trial. Time of 45 minute found to be sufficient for the sewing operator to adjust themselves to work with sewing workstation. Every experiment represent one of nine experimental trials, there were maximum three to four experimental trials per day. Prior to the experimental trial the each operator select pedal and chair position according to convenience and the experiments were performed as per experimental design table. Total 180 experiments were conducted for twenty subjects (Tondre and Deshmukh, 2019).

G. Trunk Flexion Measurement Method

Working posture of sewing operator at each experimental trial was recorded through video system. The working posture of sewing operator during the experimentation was largely considered to be in two dimensional sagittal plane. Reflective markers were sticked at the ear (tragus), the neck (C7), the hip (greater trochanter), the shoulder (greater tubercle) and at the elbow (lateral humeral epicondyle) (Figure 22.3). The left hand plays dominant role in sewing operation hence reflective stickers were fixed on the left portion of the body (Delleman and Dul, 2002). The marking positions were determined with reference to the neutral posture, which is mentioned in the study performed by Delleman and Dul (2002). The screen shots of the working postures from video based system were taken on the basis of maximum duration spend of adopted posture while sewing operation. On the basis of the marking points on the image (Figure 22.2), trunk flexion defined as the angle made by the line passing through point on neck (M2) and point on hip i.e. greater trochanter (M3) with the vertical line. A positive angle of flexion indicates inclination of trunk flexed forward. To measure trunk flexion angle, vertical axis was taken as reference i.e. zero degree. Positive angle of flexion indicates forward bending and negative angle of flexion indicate backward bending. In this

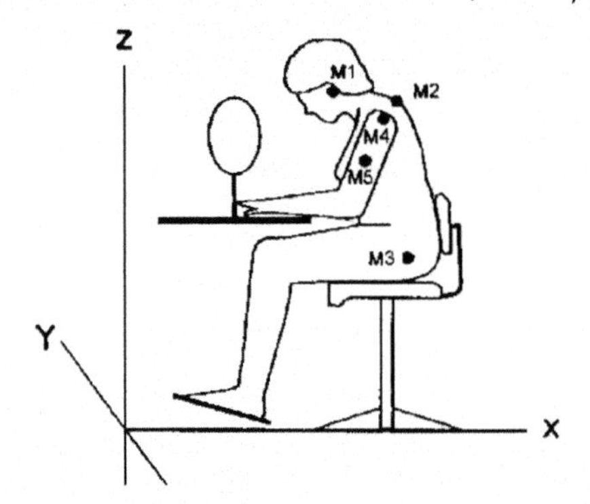

Figure 22.2 Marking points for working posture measurement (Delleman and Dul, 2002)

study negative value of trunk flexion is taken as zero as the back support was provided in the seating chair (Tondre and Deshmukh, 2019).

H. Procedure for estimation of total compressive force

In the present study, force due to body load was taken as upper body weight (UBW) of the sewing operator seating on a chair with back support. Earlier researcher investigated that, weight proportion of upper body to total weight of human body varies between 5565%. Hence for clarity of presentation the upper body weight was assumed to be 60% of total body weight of human being (Ray et. al., 1981). Similarly, Reference (Chaffin et al., 1999). investigated that the upper body centre of mass distance from L5/S1 found to be 23% of the total height of human body.

The total compressive force was calculated stepwise as given above

1) Measure angle of trunk flexion under nine set of experimental condition.
2) Calculate Compressive force on account of upper body weight

   Compressive force = (Force because of upper body weight) × COS (angle of flexion)

   Compressive force = UBW. × 9.81) × COS(θ) (5)

3) Calculate turning moment due to upper body weight.

   Turning Moment (N-m) = UBW × X (6)

   X = Perpendicular distance between line of action of UBW and L5/S1.

   X = Distance of Centre of mass to L5/S1 × SIN (θ)

   Distance of Centre of Mass to L5 to S1 = 0.23 × Total height of Subject

4) Calculate muscle force to counterbalance turning moment due to upper body weight.

   Muscle Force (N) = Turning Moment / 0.05 (7)

   Muscle leverage = 0.05 m

5) Calculate total compressive force on account of upper body weight.

   Total Compressive Force = (Compressive force + Muscle force) (8)

I. Data analysis

The experimental results obtained converted to S/N ratio to measure quality characteristics. Higher S/N ratio relates to better quality characteristics. Statistical analysis of variance (ANOVA) was performed to study the significance of each workstation parameters on low back compressive force. The optimal combination of sewing workstation design parameters were predicted by analyzing S/N ratio and ANOVA results. Selected level of significance was 0.05 (Tondre and Deshmukh, 2019).

Results

A. Total Compressive Force

The Experimental Results of S/N ratio and mean value of Total compressive force presented in Table 22.2. The range analysis was done to investigate influence of every workstation design parameter to minimize total compressive force. Table 22.3 indicates the results of range analysis. The sewing needle distance and sewing table top inclination shows significant effect on trunk flexion angle ($p < 0.008$, $p < 0.019$ resp.), whereas height of sewing table top shows less significant effect (Figure 22.3). From ANOVA table, the contribution sewing needle distance shows 65.84%, sewing table top inclination shows 28.84% and for sewing table top height was 4.74% (Table 22.4).

The optimum level of parameters was examined in terms of their contributions to the low back compressive force of sewing operators. The minimum value of compressive force was 300.48 Newton found in experiment nine as shown in Table 22.2. The proposed optimum workstation parameters with minimum compressive force on L5/S1 spinal segment was shown in Table 22.5.

Table 22.2 S/N ratio for total compressive force (smaller the better)

Expt. No.	*Independent variables*			*Mean trunk flexion (degree)*	*Mean total compressive force in newton*	*Standard deviation (Newton)*	*S/N ratio*
	Sewing needle distance (mm)	*Sewing table top inclination (degree)*	*Sewing table top height (mm)*				
1	0	0	762	17.55	927.83	101.52	-59.40
2	0	5	787	13.35	782.49	90.52	-57.92
3	0	10	813	11.85	732.08	96.21	-57.36
4	70	0	787	10.55	686.45	92.18	-56.81
5	70	5	813	10.35	677.74	83.34	-56.68
6	70	10	762	5.3	497.12	81.23	-54.04
7	140	0	813	9.2	639.79	109.57	-56.24
8	140	5	762	2.8	407.60	99.92	-52.45
9	140	10	787	0	300.48	19.12	-49.57

Table 22.3 Rangeanalysis for total compressive force in response table for means (smaller the better)

Level	*Sewing needle distance*	*Sewing table top inclination*	*Sewing table top height*
1	814.1	751.4	610.9
2	620.4	622.6	589.8
3	449.3	509.9	683.2
Delta	364.8	241.5	93.4
Rank	1	2	3

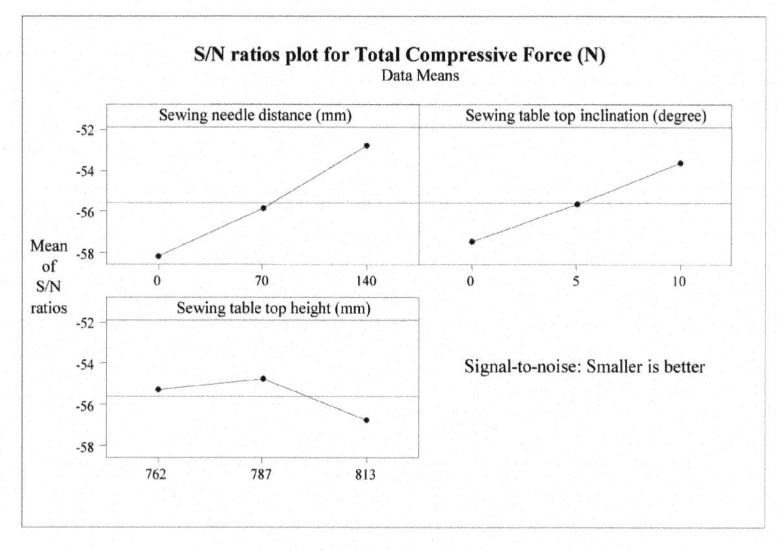

Figure 22.3 S/N ratios plot for total low back compressive force

Table 22.4 Anova table for total compressive force

Control factor/parameter	*Degree of freedom*	*Seq sum square*	*Adj sum square*	*Adj mean square*	*F-value*	*P-value*
Sewing needle distance	2	199919	199919	99959	117.88	0.008
Sewing table top inclination	2	87582	87582	43791	51.64	0.019
Sewing table top height	2	14401	14401	7200	8.49	0.105
Error	2	1696	1696	848		
Total	8	303598				

S = 29.1201 R-Sq = 99.44% R-Sq(adj) = 97.77%

Table 22.5 Proposed optimum sewwing workstation parameters and their level

Sr. No.	*Factor*	*Level*	*Value*
1	Sewing needle distance	3	140
2	Sewing table top inclination	3	10
3	Sewing table top height	2	787

Discussion

The study focused into the ways in which sewing workstation parameters influences sewing machine operator's posture. As per biomechanical principle, musculoskeletal loads may be calculated on the basis of moment of force due to postural angle; hence working postures become useful measure of musculoskeletal load and is helpful for evaluating the workstation design (Mastalerzi and Palczewska, 2010). The type and magnitude of stresses depend upon work environment. The sewing operator has to perform sewing task with constrained awkward sitting posture for long duration due to which from postural stresses induced. The reason of high postural stress is improper workstation design i.e. neglecting ergonomic principles of workstation design. The study demonstrated that sewing workstation design parameters significantly affect postural variables. As the sewing table top moves in fore/aft direction i.e. from 0 to 140 mm towards the sewing operator, the vision of the sewing operator to perform sewing task became clearer without need to bend the trunk segment in forward direction. Similarly sewing table top inclination found to be highly significant to sewing operator's trunk posture. The better view of sewing needle/task results improvement in trunk posture. In this study low back compressive force was estimated to quantify the load on L5/S1 segment of trunk. During forward leaning trunk posture, the stress induced was much more compared to natural upright trunk posture. The results obtained were in agreement with the findings of earlier researchers (Schuldt et.al., 1986; Maria et.al., 1997; Colombini, et. al., 1986). The relationship between the trunk flexion and estimated low back compressive force seen to be straightforward. The low back compressive force developed at L5/S1 body segment of the sewing operators could be minimised by adjusting the fore/aft sewing needle distance from 0 to140 mm towards the sewing operator and by inclining sewing table top from 0 to 10^0 towards the sewing operator. This alteration will be helpful to maintain natural upright trunk posture so that, posture adopted will minimise the postural stress at trunk segments.

Conclusion

Experimental result shows that, in sewing operation the determinant factor for trunk posture is the sewing needle distance and sewing table top inclination. On the basis of the results, guidelines have been developed to improve trunk posture of sewing operator: (a) the fore/aft sewing needle distance should be adjusted between 0 to 140 mm towards sewing operator; (b) a sewing table top should be inclined from 0 to 10^0 towards the sewing operator. The new workstation design incorporated in the sewing machine workstation would be valuable for garment industries and for self employed sewing operators. The main conclusion can be drawn from this study is the alterations in the sewing workstation design parameters could be helpful for reducing trunk flexion and consequently low back postural stress of sewing machine operators. Health related issues of seated workers at a work such as prevalence of musculoskeletal disorders and burden of settlement of medical claims against musculoskeletal disorders may be solved by implementing the principles of ergonomics for designing and modifying workstation. This inference may possibly useful for other sedentary workplaces also.

References

Aaras, A., Stranden, E. (1988). Measurement of postural angles during work. Ergonomics. 31(6):935944.

Afonso, L., Pinho, M. E., and Arezes, P. 2014. Prevalence of WMSD in the sewing sector of two companies of the footwear industry. In Occupational Safety and Hygiene II, I ed. 610614. London: Taylor and Francis.

Akodu, A., Ba, T. and Adebisi, O. A. (2013). Prevalence, pattern and impact of musculoskeletal disorders among sewing machine operators in Surulere local government area of Lagos state, Nigeria. Indian J. Physi. Occu. Ther. 8(2):1520.

Ariens, G. A., Van Mechelen, W., Bongers. P. M., Bouter, L. M., and Vander Wal, G. 2000. Physical risk factors for neck pain. Scand J Work Environ Health. 26(1):719.

Ayoub, M. M. 1973. Workplace design and posture. Human Factors 15(3):265268.

Bhatnager, V., Drury, C. G., and Schiro, S. G. (1981). Posture, postural discomfort, and performance. Human Factors. 27(2):189199.

Blader, S., Barck-Hoist, U., Danielsson, S., Ferhm, E., Kalpamaa, M., Leijon, M., Lindh, M., and Markhede, G. (1991). Neck and shoulder complaints among sewing-machine operators. Appl. Ergonomics. 22(4):26.

Brider, R. S. 2003. Introduction to Ergonomics. Taylor and Francis publication 2nd Edition.

Chaffin, D. B., Andersson, G. B., and Martin, B. J. (1999). Occupation Biomechanics, New York (N Y): John Wiley & Sons, Inc.

Clark, D. R. 1996. Workstation evaluation and design. In Occupational ergonomics: theory and practice. ed. A. Bhattacharya, J. D. McGlothlin, 226. New York, (N Y): CRC Press.

Colombini, D., Occhipinti, E., Frigo, C., A. Pedotti, A. Grieco. (1986). Biomechanical, electromyo-graphical and radiological study of seated postures. In The ergonomics of working postures: models, methods and cases. ed. N. Corlett, J. Wilson, and I. Manenica, 26. London: Taylor & Francis.

Corlett, E. N. 1981. Pain, posture and performance. In Stress, Work Design and Productivity, ed. E. N., Corlett, and Richardson, 2742. London: John Wiley & Sons

Delleman, N. J. and Dul, J. (2002). Sewing machine operation: workstation adjustment, working posture, and workers perceptions. Int. J. Indus. Ergonomics. 30:341353.

Duquette, J., Lortie, M., and Rossignol, M. 1992. Perception of troublesome factors for the back associated with the workplace: a study of aircraft assemblers. In Advances in Industrial Ergonomics and Safety IV, ed. S., Kumar. London: Taylor & Francis.

Hamill, J., Kathleen, M. K., and Timothy, R. 2015. Biomechanical Basis of Human Movement, Wolters Kluwer, 4 th Edition, Walnut Street, Philadelphia, 2015. http://www.sfu.ca/~leyland/Kin201%20Files/Spinal%20 Biomechanics. pdf

Herbert, R., Dropkin, J., Warren, N., D Sivin, D., Doucette, J., Kellogg, L., Bardin, J., Kass, D., and Zoloth, S. (2001). Impact of a joint labour management ergonomics program on upper extremity musculoskeletal symptoms among garment workers. Appl. Ergonomics. 32:453460.

Kadefors, R. 1998. An integrated approach in the design of workstations In Encyclopaedia of occupational health and safety, ed. J. M., Stellman. 5660. Geneva, Switzerland: International Labour Office.

Keyserling,W. M., Donoghue, J. L., Punnet, L., and Miller, A. B. 1982. Repetitive trauma disorders in the garment industry Report. Boston: Department of Environmental Health Sciences, Harvard School of Public Health.

Kroemer, K. H. E. 1988. VDT workstation design. In Handbook of human-computer interaction, ed. Helander, 521539. North-Holland, Amsterdam: Elsevier Science.

Kroemer, K. H. E. 1997. Design of the computer workstation. In Handbook of human-computer interaction. ed. M., Helander, T. K., Landauer, P., Prabhu P. 1395–1414. The Netherlands: Elsevier Science, North-Holland, Amsterdam.

Li, G., Haslegrave, C. M., and Corlett, N. (1995). Factors affecting posture for machine sewing tasks. Appl. Ergonomics. 26:3546.

Villanueva, M. B., Jonai, H., Sotoyama, M., Hisanaga, N., Takeuchi, Y., and Saito, S. (1997). Sitting Posture and Neck and Shoulder Muscle Activities at Different Screen Heights Settings of the Visual Display Terminal. Industrial Health 35(3):330336.

Mastalerzi, A. and Palczewska, I. (2010). The influence of trunk inclination on muscle activity during sitting on forward inclined seats. Acta Bioeng. Biomech. 12(4):1924.

Mattila, M., Vilkki, M. 1999. OWAS Methods. In *The occupational ergonomics handbook*, ed. W. Karwowski, W. S. Marras , 447–59. Boca Raton: CRC Press LLC.

Punnett, L., Robins, J. M., Wegman, D. H., and Keyserling, W. M. 1985. Soft tissue disorders in the upper limbs of female garment workers. Scand. J. Work Environ. Health. 11:417425.

Ray, C. G., Sen, R. N., Nag, P. K. (1981). Relationship Between Segmental and Whole Body Weights and Volumes of Indians. J. Human. Ergol. 10:3548.

Roy, R. K. 2001. Design of Experiments Using the Taguchi Approach: 16 Steps to Product and Process Improvement. John Wiley & Sons, Inc.

Schuldt, K., Ekholm, J., Harms-Ringdahl, K. (1986). Effects of changes in sitting work posture on static neck and shoulder muscle activity. Ergonomics. 29(12):15251537.

Tondre, S. and Deshmukh, T. (2019). Guidelines to Sewing Machine Workstation Design for Improving Working Posture of Sewing Operator. Int. J. Indus. Ergonomics. 71:37–46.

Vihma, T., Nurminen, M., and Mutanen, P. 1982. Sewing-machine operators' work and musculoskeletal complaints. Ergonomics. 25(4):29529.

Westgaard, R. H. and Jansen, T. (1992). Individual and work related factors associated with symptoms of musculoskeletal complaints, II Different risk factors among sewing machine operators. Br. J. Ind. Med. 49:154162.

Wick, J. and Drury, C. G. 1986. Postural change due to adaptations of a sewing workstation. In The Ergonomics of Working Postures, ed. E. N., Corlett, J. R. Wilson, and Manenica, I. 375379. London: Taylor & Francis.

23 Monitoring and correction of power factor for residential loads

Y. Y. Dhage[a], *R. R.Tambekar*[b], *R. B. Lasurkar*[c], *L. Dhole*[d], *P. J. Kathote*[e], *and Atul Lilhare*[f]

Department of Electrical Engineering, YCCE, Nagpur, Nagpur, India

Abstract

This paper is deal with the concept of power factor improvement technique specially for resisdential load. Fundamentals of power system is used in this paper to continuously monitor the power factor of residential load. For monitoring purpose Ardunino Uno (microcontroller board) is used because of its easy to use and lot of facility available for programmer. For sensing purpose, Voltage and current sensor are used instead of of CT and PT. This sensors are cheper and accurate in measurement. The said system is tested with MATLAB software and also implements with the help of embedded system. The said system is able to monitor automatically the variation in power factor (0 to 1) and calculate the parameters accordingly to improve it. It results are tested and found to be satisfactory (near to unity) as per standard defined for power factor improvements.

Keywords: Embedded system, MATLAB, Power Factor.

Introduction

Today, we are observing an ever-increasing demand for electrical energy if we can contribute a smaller device to improve power factor and reduce the loss in any sector then it can give a little boost to the idea of saving and using the energy that we have more effectively and efficiently. Generally, the use of inductive load in the system increases the reactive load in the system. So that current lag behind the voltage, it occurs lagging power factor due to this the efficiency of the system get reduces and electricity bill gets increase Jarad et al. (2018). Also, due to significant phase difference between voltage and current at the load terminal, system draw a power at low power factor, it means it also have a distorted waveform.

So many methods are developed now a days in order to monitor and control the power factor of resisdenial load. In Kukde and Lilhare (2017), one of the most important converter is used for water pumping application which is further utilised for the development power factor correction in water pumping application of resisdential load as mentioned by Lilhare et al. (2021).

A well system is developed to achieve high power factor in single phase system by operating in continuous conduction mode which work on the principal of avaregae current mode control. In Patil et al. (2019), a microcontroller device is used which ensure automatic correction of the power factor without operator. The triac based power factor corrector for singlr phase domestic load is proposed by Ali et al. (2018). Other approach by Rakiul Islam et al. (2016) is to use a PFC device for each load means in multiple way in order to acieve a desired power qualityin the area of residential load. Instaed of above a concept of filter and capacitor compensator (MPF/C) Green Plug is used for effective energy conservation, for two phase, three wire household load (Sharaf and Chhetri, 2006). An open source energy monitoring library were implemented in order to monitor the energy consumption of a system and automatically improve its power factor (Kabir et al., 2017).

With above discussion , it observed that there is still some scope to work in the area of residential load for power factor improvement. There is need of development of cheper and effective power factor correction device which will be able to maintain unity power factor at residential load. This paper deals with automatic monitoring and improvement of power factor with the help of embedded system. It simple consist of Arduino board , capacitor bank and relays. Instead of CT and PT , voltage and current sensors are used. This sensors are cheper and have better accuracy as compared to CT and PT. This helps to reduced the cost of device and lighter in weight. A programmed is developed for switch in and switch out of capacitor in

[a]yogeshdhage726@gmail.com; [b]rutujat2001@gmail.com; [c]rohitlasurkar@gmail.com; [d]dholeleena00@gmail.com; [e]pranaykathote1@gmail.com; [f]0000-0001-5350-7532

the system so that the power factor should be mainteain as per definition of electricity board. Care is taken in regards of lagging and leading term. The said work is divided in five section and explained in following manner.

Topology of Proposed System

Figure 23.1 consists of inductive load, caoacitive bank, current sensor,voltage sensor, grid supply and PFC unit for monitoring and maintain desire power factor of resisdential load. Here inductive load is considered as it is the main component of resisdential load. This load draw reactive power from the electric supply which leads to poor power factor. In order to supply reactive power , a capacitive bank is used so that it can fulfill the requirement of reactive power of load and maintained power factor of the system. For monitoring the power factor of the system, there in need of current and voltage sensing and therefore current and voltage sensor are used in the system. For deciding the function of capacitor switch in and out there is need of manupulationof voltage and current so that desired power factor can be achieved. So for this work PFC is used, which is defined as power factor correction.

Capacitor Selection

The system which is shown in Figure 23.1, supply voltage is considered as 220V rms and for inductive load , the value of resistance is considred as R=10Ω and L=0.31831H.The system frequency is 50Hz.

A. The Reactive Power Rating of the Inductive Load

In given system, first we need to determine the actual reactive power for load. Once reactive power is know then operating power factor of the load can be determined.

For given inductive load, the impedance can be calculated as

$$Z = R + jX_L \tag{1}$$

$$Z = 10 + j(2\pi \times 50 \times 0.31831) = 100.49\angle 84.28°\Omega$$

The current drawn by the load is

$$I = \frac{V}{Z} \tag{2}$$

$$I = \frac{220\angle 0°}{100.49\angle 84.28°\Omega} = 2.19\angle -84.28° A$$

Nature of the current is lagging as the phase angle is negative.Similerly, the complex power of the load is calculated as

$$S = VI^* \tag{3}$$

$$S = 220\angle 0° \times 2.19\angle -84.28° = 481.8\angle -84.28° VA$$

$$S = 48.02 - j479.401 VA$$

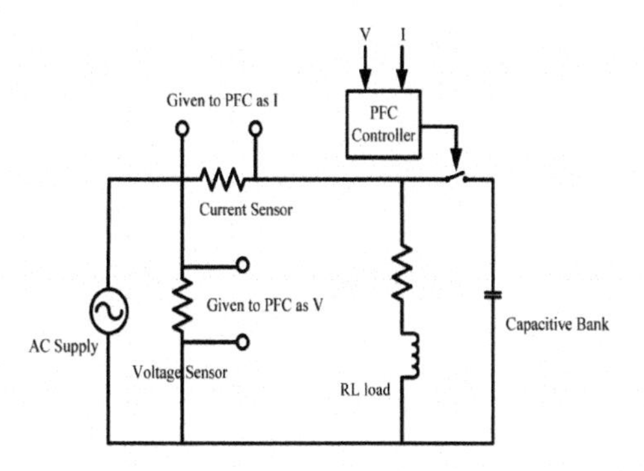

Figure 23.1 Power factor monitoring and correction unit for residential load

Once complex power is evaluated, power factor can be calculated as shown in equation in(4).

$$p.f = \frac{P}{S} = \frac{48.02}{481.8} = 0.0996 \tag{4}$$

B. Capacitive Bank for Unity Power Factor Operation

For unity power factor operation, the complex power should be equal to active power which means that total reactive power drawn by the system is zero. By submitting the imaginary part of load current is zero, the power factor of the system can be improved. So reactive power required to fed by capacitor is

$$Q_C = 479.401\,Var$$

With the help of equation (5), capacitive reactance can be calculated, once the value of reactance known, required value of capacitor is determined from equation (6).

$$Q_C = \frac{V^2}{X_c} \tag{5}$$

$$X_C = \frac{V^2}{Q_C} = \frac{220^2}{479.401} = 100.96\Omega$$

$$X_C = \frac{1}{2\pi fC} \tag{6}$$

$$C = \frac{1}{2\pi fX_C} = \frac{1}{2\pi \times 50 \times 100.96} = 31.52\mu F$$

Hardware Blockdiagram and Control Strategy

Figure 23.2 shows a schematic for impletation of power factor monitoring and correction unit which simply consist of d.c.regulated power supply , Arduino Uno board, current sensing, voltage sensing, relay driver circuit, inductive load , capacitive bank and LCD for display. Regulated power supply is required for Arduino Uno board as well as for other control unit. Mathematical expression are evaluated with the help controller knows as Arduino Uno. Controller ensure the proper switching of capacitor in and out so that the desire power factor is to be maintained. The value of voltage and current is sense by the circuit which is describe as follows.

A. Voltage Sensing Circuit

With the help of step down transformer, step down voltage is given to zero crossing detector (i.e. LM339). LM339 is comparator specially designed for low level sensing and memory application. Its features are high gain and wide bandwidth characteristics. It will detect a sign wave zero crossing from positive half

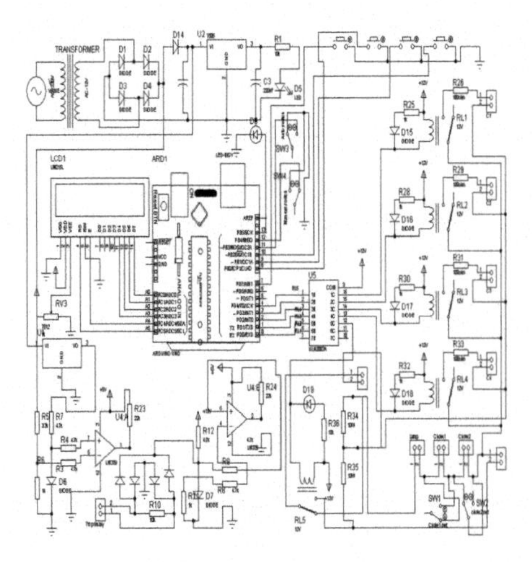

Figure 23.2 Schematic of power factor monitoring and correction unit for residential load

cycle to negative half cycle. This generated signal is then fed to controller as sense voltage signal as shown in Figure 23.3.

B. Current Sensing Circuit

Similerly, bias resistor sensed the current and convert it into voltage as controller only work on voltage signal. This generated voltage is passed through zero crossing detector so that zero crossing from positive half cycle to negative half cycle can be detected. This generated signal is then fed to controller as sense current signal as shown in Figure 23.4.

C. Controller Circuit

A microcontroller is the heart of the system. It will sense the phasor term of voltage and current from sensing unit. After receiving information it will manipulate phase angle and phase difference between this two waves.This information information display on LCD screen. A microncontroller detect the angle between current and voltage signal and Once angle information is confirmed , a value of power factor is obtained. Depend upon the power factor value , controller will take a decision of switch in or out of capacitor in such way that the system power factor will remain at unity value.

D. Relay Driver Circuit

Relay driver as shown in Figure 23.5, ensure connection or disconnection of capacitor in the circuit. Relay work is totally depend on the signal send by the controller. It means relay on its own can not controlled the switching of capacitor in the system.

Simulation Result of the Proposed System

The said system is tested with the help MATLAB software. By considering the all parameter as mentioned in above parameter design , the results are obtained.

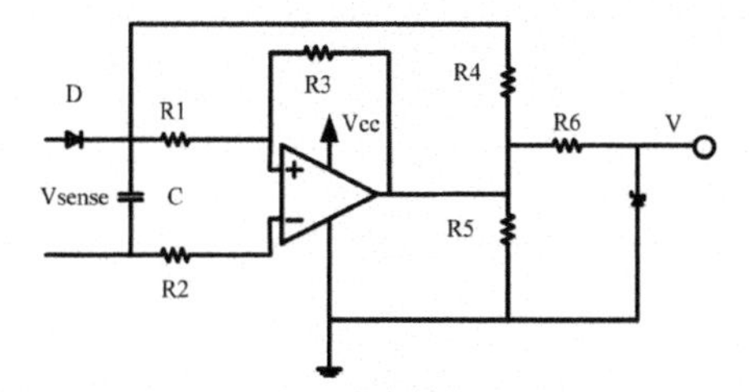

Figure 23.3 Voltage sensing circuit for PFC unit for residential load

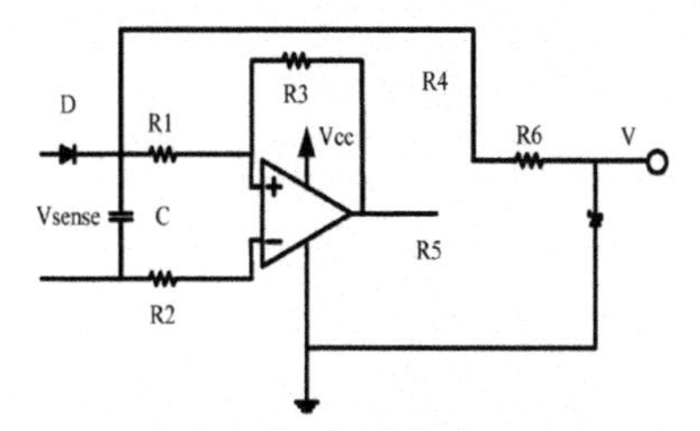

Figure 23.4 Current sensing circuit for PFC unit for residential load

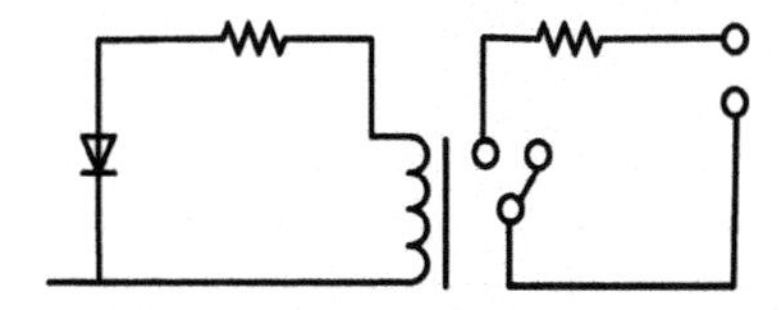

Figure 23.5 Relay driver circuit for PFC unit for residential load

Figure 23.6 shows a result for supply voltage and current. The supply voltage of 220V is applied to the load and measured value is found to be accurate as shown in Figure 23.6. At 0.5 sec, there is change in magnitude of load current due to capacitor switch in the system. At initial stage, due to inductive load, system draw a current which is lag to the supply voltage as shown in Figure 23.7. This phase angle is sensed by the PFC unit , so it switch on the capacitor in circuit at 0.5 sec in order to achieve the unity power factor as shown in Figure 23.6. It can be observe that , magnitude of current is decresed and its in phase with the supply voltage as shown in Figure 23.8.

In Figure 23.9, the active and reactive power are shown. It can be seen that, before switching capacitor in the circuit, the system draws a huge reactive power from the supply.But after switching the capacitor in the circuit, reactive power drawn by the system is zero. Due that system shows a improvement in power factor as shown in Figure 23.10.

In Figures 23.11 and 23.12, hardware circuit and its operation shown through LCD display. It can be observe that, the said system able to maintained unity power factor on source side.

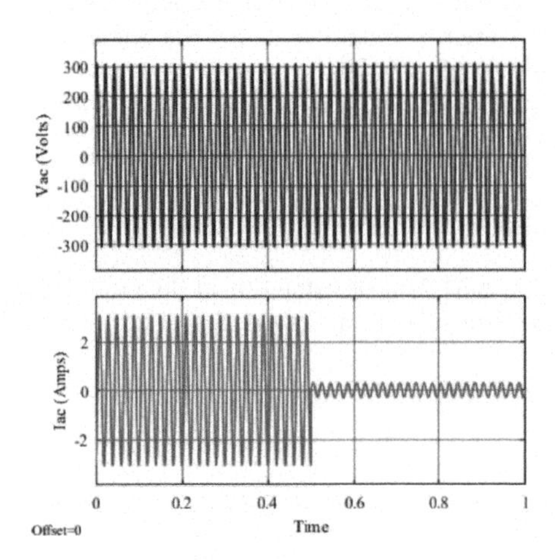

Figure 23.6 Supply voltage and current of PFC unit for residential load

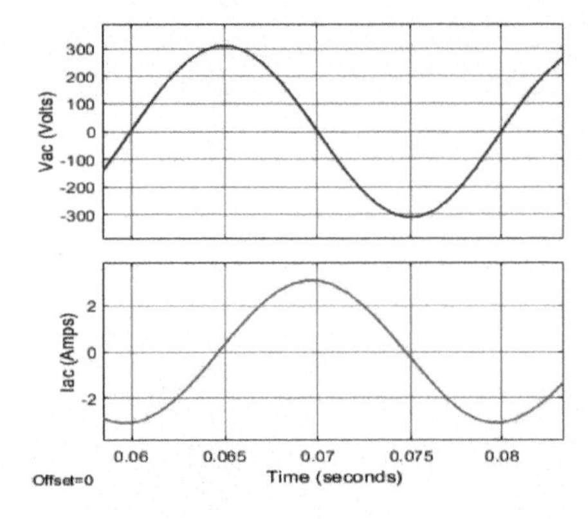

Figure 23.7 Supply voltage and lagging current of PFC unit for residential load

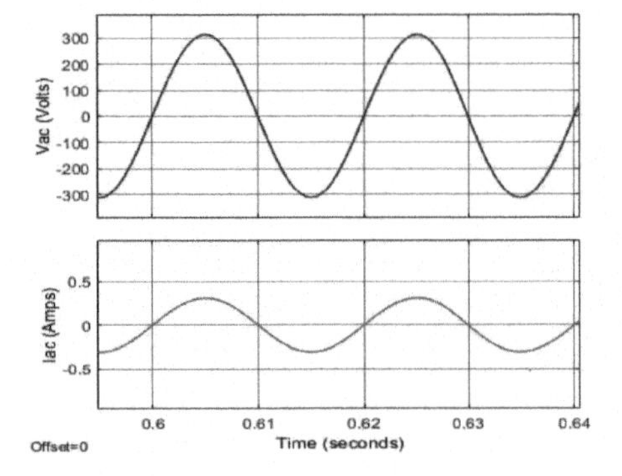

Figure 23.8 Supply voltage and in phase current of PFC unit for residential load

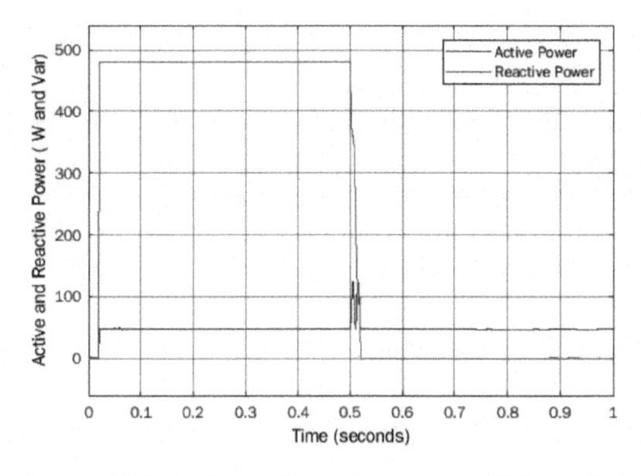

Figure 23.9 Active and reactive power of PFC unit for residential load

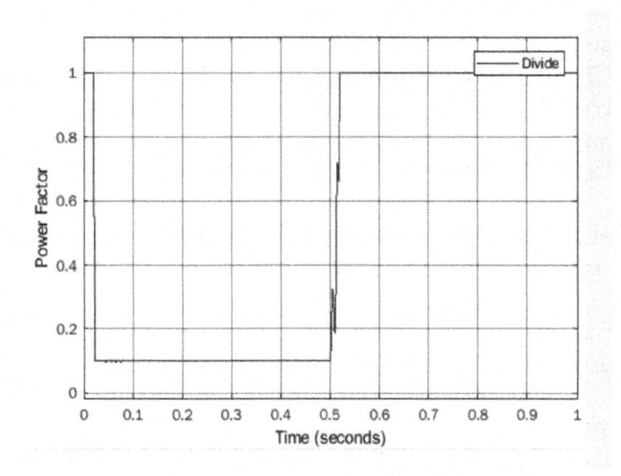

Figure 23.10 Power factor of PFC unit for residential load

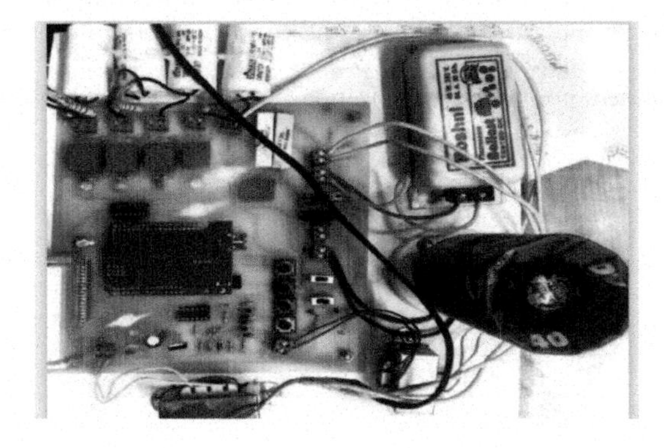

Figure 23.11 Hardware set up of PFC unit for residential load

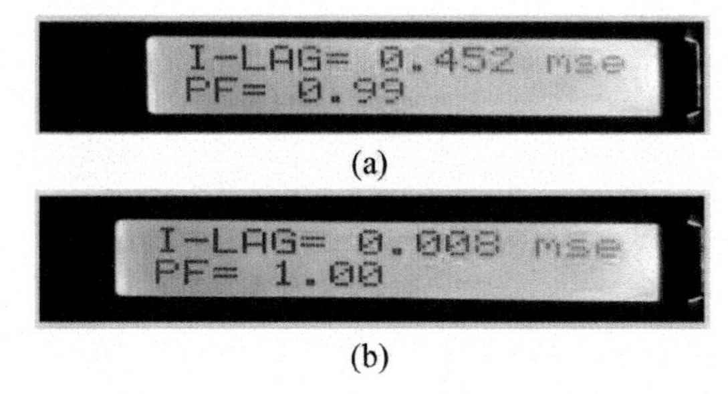

Figure 23.12 Power factor on LCD display of PFC unit for residential load

Conclusion

The said system is able to correct the unity power factor on residential side. The components used for desiging power factor correction device are cheper as compared to traditional devices. It also more sensitive to change in load so that it will auto correct the power factor. Power factor correction unit will be beneficial for the residential sector. It will ensure the continuity of the power supply at unity power factor. Automation mechanism is quite simple and easy to install. This unit is cost effective and also require less space for installation.

Acknowledgment

Authors thanks for the support of funding by Electrical Engineering Department, YCCE, Nagpur in their project work.

References

Ali, W., Farooq, H., Jamil, M., Rehman, A. U., Taimoor, R., and Ahmad, M. (2018). Automatic power factor correction for single phase domestic loads by means of arduino based triac control of capacitor banks. In 2018 2nd international conference on energy conservation and efficiency (ICECE), (pp. 72–76). doi: 10.1109/ECE.2018.8554986.

Jarad, S. B., Lohar, V. D., Choukate, S. P., and Mangate, S. D. (2018). Automatic optimization and control of power factor, reactive power and reduction of THD for linear and nonlinear load by using arduino uno. In 2018 Second international conference on inventive communication and computational technologies (ICICCT), (pp. 1128–1132). doi: 10.1109/ICICCT.2018.8473191.

Kabir, Y., Mohsin, Y. M., and Khan, M. M. (2017). Automated power factor correction and energy monitoring system. In 2017 second international conference on electrical, computer and communication technologies (ICECCT), (pp. 1–5). doi: 10.1109/ICECCT.2017.81179693.

Kukde, H. and Lilhare, A. S. (2017). Solar powered brushless DC motor drive for water pumping system. In 2017 international conference on power and embedded drive control (ICPEDC) (pp. 405-409). doi: 10.1109/ICPEDC.2017.8081123.

Lilhare, A., Kadwane, S. G., and Fulzele, P. (2021). Grid supported solar water pump system. In 2021 innovations in power and advanced computing technologies (i-PACT), (pp. 1–5). doi: 10.1109/i-PACT52855.2021.9696889.

Patil, V., Dhuri, P. B., Kushe, P. P., Mondkar, P. A., and Acharekar, S. (2019). Minimizing penalty in industrial power consumption by using apfc unit: A review. In 2019 IEEE international conference on system, computation, automation and networking (ICSCAN), (pp. 1–5). doi: 10.1109/ICSCAN.2019.8878863.

Rakiul Islam, S. M., Maxwell, S., Hossain, M. K., Park, S.-Y., and Park, S. (2016). Reactive power distribution strategy using power factor correction converters for smart home application. In 2016 IEEE energy conversion congress and exposition (ECCE), (pp. 1–6). doi: 10.1109/ECCE.2016.7855377.

Sharaf, A. M. and Chhetri, R. (2006). A novel household efficient power/energy utilization and power quality compensator devices. 13–27. 10.1109/LESCPE.2006.280353.

24 Mathematical Modelling different parameters with friction stir processing for future applications

Bazani Shaik[1,a,*], *P. Siddik Ali*[2], *and M. Muralidhara Rao*[1]

[1]Professor, Ramachandra College of Engineering, Eluru, Andhra Pradesh, India

[2]Assistant Professor, Ramachandra College of Engineering, Vatluru, Eluru, Andhra Pradesh, India

Abstract

Friction stir processing is a very promising method widely joining varieties of metals in other relatively marine, ship-building, automotive industries, aeronautical and heavy machinery industries due to the following advantages i.e. low porosity, fewer tendencies to cracking and fewer defects. Research investigates the parameters on aluminium alloys by using friction stir processing on based with to improve strength and quality with the help of mathematical modelling maximum tensile strength of 167 MPa measured on the basis of ASTM-E8 on specimens and analysis for carrying and used design of experiments and mathematical modelling, the relations with empirical process useful for the development for automated design and used future growth of industrial applications needed.

Keywords: Aluminum alloys, automated design, dissimilar welding, machinery industries, mathematical modeling.

Introduction

Friction stir process is currently very useful for ship manufacturing and industry-oriented aircraft and automotive for butt, lap with spot-on (Patel et al., 2018) dissimilar joining of applicability Al-alloys and other materials of Mg- alloys, the production of mass of light transportation systems and fuel consumption has significantly reduced (Dorbane et al., 2016). Studied resistance of ironing with process aluminum alloys are increased to improve the silicon oxide nano particles for the limit of iron (El-Batahgy et al., 2016). Studied mechanical properties and microstructural evaluation of AZ31B of sheets has 3 mm thickness welded of optimum conditions. The material of work pieces for joining are used friction stir processing with tool shown in Figure 24.1 (Goebel et al., 2016; Ni et al., 2015). Studied of tempered steel with quench property is feasible of tensile strength 1635 Mpa and research focus of different types of high carbon steels and medium are accepted successfully of friction stir welds Das and Toppo (2018). joining of Al6061 or NiTip composite with the distribution of homogeneous particle without product interface reaction is prepared successfully by friction stir. Dalwadi et al. (2018) processing took place combination of good damping with thermal physical properties on the treatment of heat process in the composite (Dragatogiannis et al., 2015; Ma et al., 2017). Shaik et al. (2019a) AL-Li 2099T86 of stress corrosion cracking applications and developments of new alloy on aircraft industries are identified. Shaik et al. (2019c) aluminium- lithium alloys with the substitute of high strength aluminum alloys on spacecraft manufacturing and launchers. The properties of strength, toughness, stiffness adopted with aluminium alloys. Shaik et al. (2018) The aluminium-lithium alloys advanced taken place with stress corrosion cracking on structural space applications. The parameters are used for welding has cohesive band and circular shape and path studied of tool intention Shaik et al. (2019b,c,d).

Materials and methods

Friction stir process mainly involves the basic need with materials and methods influences with welding of dissimilar AA7075T651 and AA6082T651 with having thickness of 6 mm and by using advanced numerically controlled stir process are carried out experiments on the basis of lot of literature survey and trail error methods on input parameters varying with proportionate condition done at Annamalai university. Chemical compositions with base material shown Table 24.1. The specimens of the plate taken dimensions on the basis of gap is 100 mm × 50 mm × 6 mm. the dimensions cut by the edges with smooth areas to do easily joining process of butt welding for the two dissimilar aluminium alloys are placed advancing side

[a]dr.bazani.sk@gmail.com

and retreating side are shown Figure 24.2 for the fixed clamps will be adjusted for specimens. The designed tool with advanced condition material taken as M2-Grade SHSS tool diameter of shoulder is 18 mm and length of probe is 6 mm. After the friction stir processing the weld zone appears perfectly, for the testing of the welding specimens are taken as standards of ASTM-E8 and tensile test specimens Figure 24.3 and specimens after testing Figure 24.4. The combination and particular diameter of standards specimens are taken for the impact strength shown in Figure 24.5. The AA7075T651 advancing side and AA6082T651 in retreating side to have the proper joining of materials and for the improvement of mechanical properties.

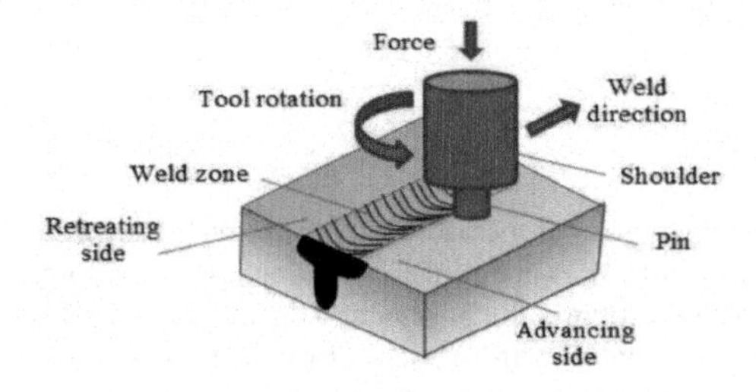

Figure 24.1 Friction stir welding process

Table 24.1 Chemical compositions AA7075T651 and AA6082T651

Elements	*Si*	*Fe*	*Cu*	*Mn*	*Mg*	*Cr*	*Ni*	*Zn*	*Ti*	*Al*
AA7075-T651	0.12	0.2	1.4	0.63	2.53	0.2	0.004	5.62	0.03	89.26
AA6082-T651	1.05	0.26	0.04	0.68	0.8	0.1	0.005	0.02	0.01	97.03

Figure 24.2 Weld position of dissimilar aluminium alloys of friction stir welding

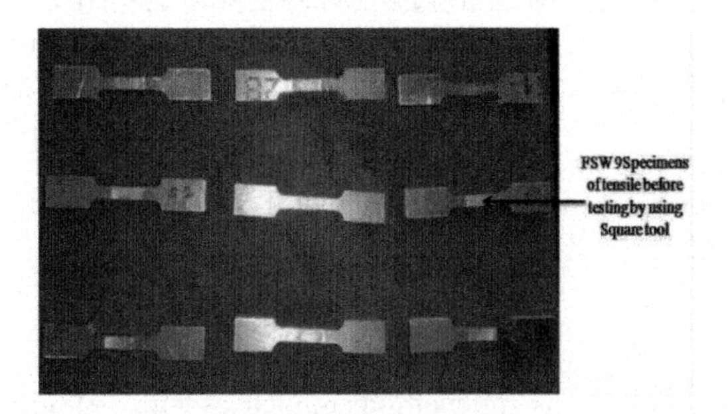

Figure 24.3 Specimens of tensile test before testing with ASTM E8

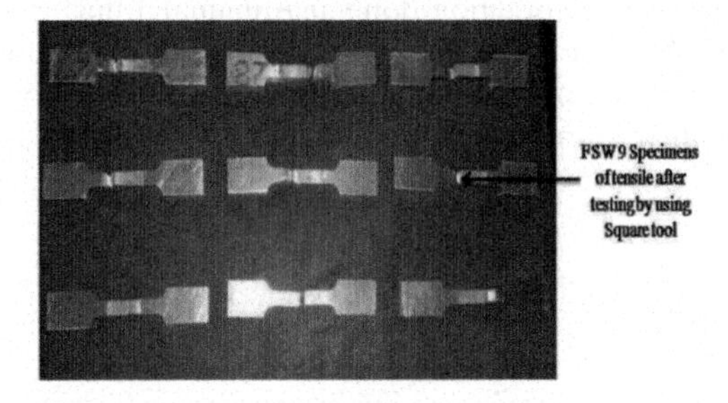

Figure 24.4 Specimens of tensile test after testing

The advanced methodology applied for different parameters to obtained easy way of influencing the properties of mechanical by using dissimilar welding of notations and units are described Table 24.2 and Table 24.3 experimental design of Taguchi model input parameters and output parameters.

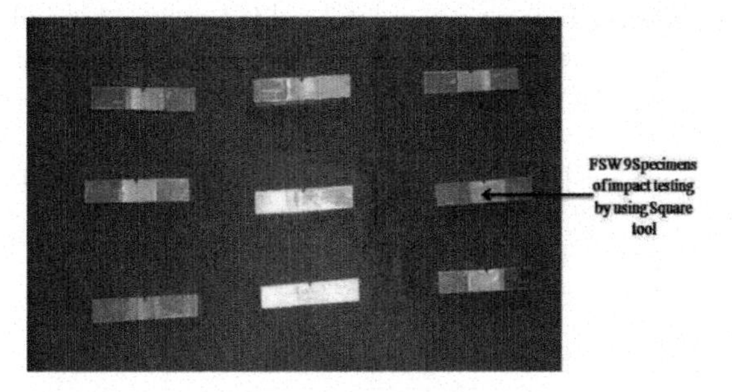

Figure 24.5 Specimens of impact test

Table 24.2 Input variables for actual and coded

Sno	*Parameters*	*Notation*	*Unit*	*Levels*		
				1	*2*	*3*
1	Welding speed	WS	mm/min	40	50	60
2	Rotational speed	RS	rpm	1150	1250	1350
3	Axial Force	AF	KN	9	10.5	11

Table 24.3 Experimental design of taguchi model

Input process parameters					*Output responses*		
Exp No	*Rotational speed (rpm)*	*Welding speed (mm/min)*	*Tilt angle (degree)*	*Axial force (KN)*	*Tensile strength (MPa)*	*Impact strength (J)*	*Elongation (%)*
1	1150	40	1	10	162.00	10.55	9.60
2	1150	50	2	11	158.99	10.31	9.41
3	1150	60	3	12	155.00	9.00	8.50
4	1250	40	2	12	171.00	12.20	10.80
5	1250	50	3	10	164.99	11.16	10.05
6	1250	60	1	11	158.00	9.30	6.78
7	1350	40	3	11	174.99	13.10	12.15
8	1350	50	1	12	173.00	13.03	11.25
9	1350	60	2	10	167.00	11.30	10.10

Design of Expert

The design of experts in series with the test for the researcher useful for changes in input variables on a processor system is shown in Figure 24.6 due to the effect of variables of responses measured. The applicability of computer simulation models and physical on the factorial designs took place sensitively for the estimation of the combination of effect for two or more factors.

The design of experiments and methods of the traditional difference taken place approach in a better way of values on variables of parallel and it does not cover main effects on the variables on the different interactions and the possibility of approach for identifying optimal values on the variables of combination with experimental runs. The design of experiments is carried in four phases are screening, panning, optimisation, and verification.

Increased based on the tool welding speed varies the strength with respect to the elongation has improved the maximum extent depends up on the rotational speed. The Figure 24.7 shows the increases of rotational speed depends up on the heat increases at the welding zone area. The friction coefficient decreases with the melting condition. The friction stir process region intricate the fine particles will be distributed in the uniform portion. The effect of tool stirred the position on the flow of metal optimum depends up on the increase of tensile strength.

Percentage elongation along transverse direction obtained from the tensile test plotted against the welding speed. The plates Figure 24.8 shows welded with the rotational speed is 1250 rpm and weld speed 40 mm/minutes. while the plates welded 1150 rpm and 60 mm/minutes. The influence shows the properties of higher heat input on the basis influenced elongation.

Effect Figure 24.9 shows the manner of the position at the bottom area of the welded part and it will be increased the position of tool speed with respect to the material and designed shoulder based. The region of the position will be making difference between the tool changes the yield strength to improve the microstructure with ductility.

Influence with axial force on tensile strength Figure 24.10 shows the significance of friction stir processing at the joining area. The joint taken place the position of rotational speed is 1250 rpm and tensile strength 164.99 MPa and the welding speed takes the major role due to increasing of force is 12 KN has the strength will be superior at the position of part counter.

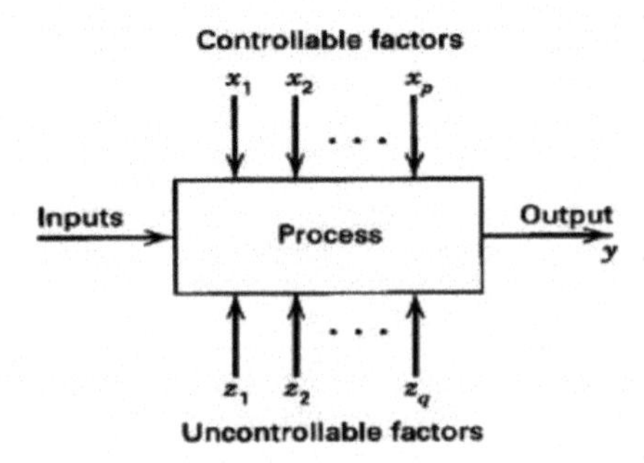

Figure 24.6 Process model of the design of expert

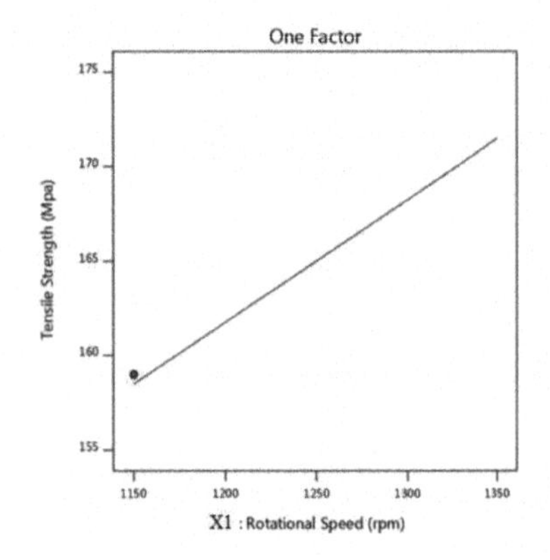

Figure 24.7 Effect of rotating speed on tensile strength

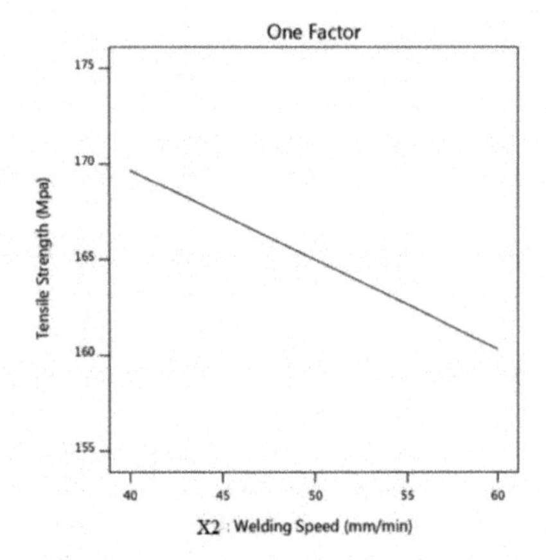

Figure 24.8 Effect of welding speed on tensile strength

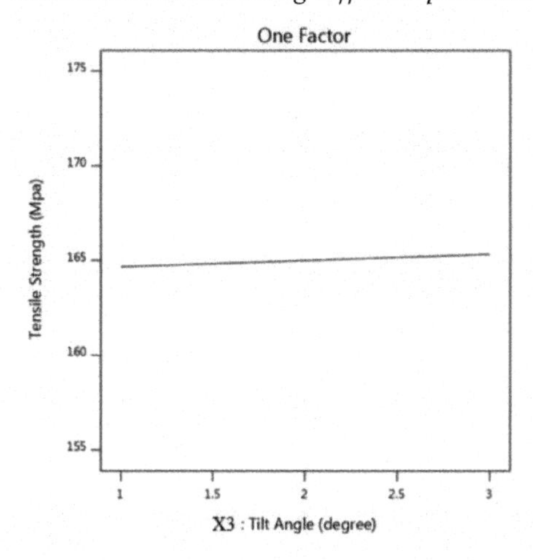

Figure 24.9 The effect of tilt angle on tensile strength

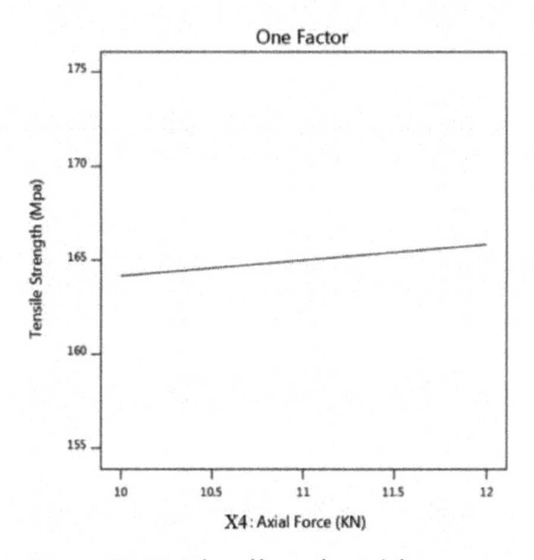

Figure 24.10 The effect of axial force on tensile strength

Conclusions

The present investigation shows the aluminium alloys with application of Taguchi design of experiments helped us in conducting the experiments in an effective manner without losing accuracy. Two-dimensional plotted parameter using Design-Expert software. The tensile strength is increasing values and tensile strength is decreasing with the increase in the weld speeds. The Impact strength increases, when there is an increase in the values, whereas the impact strength tends to decrease with the increase in the weld speeds. The elongation also results presented in the work are analysed on the basis of analysis process conducted with microstructures with different zones on thermo mechanical treatment zone has higher plasticity due to eutectic constituents Cu-Al precipitation on rolled condition and parent metal has rolled temper condition. Research Development done on the different materials that which help for the future applications to apply mathematical modelling concepts with quality for future applications.

References

Dalwadi, C. G., Patel, A. R., Kapopara, J. M., Kotadiya, D. J., Patel, N. D., and Rena, H. G. (2018). Examination of mechanical properties for dissimilar friction stir welded joint of Al alloy (AA-6061) to PMMA (Acrylic). Mater. Today: Proc. 5:4761–4765.

Das, U. and Toppo, V. (2018). Effect of tool rotational speed on temperature and impact strength of friction stir welded joint of two dissimilar aluminum alloys. Mater. Today: Proc. 5:6170–6175.

Dorbane, A., Ayoub, G., Mansoor, B., Hamade, R. F., Kridli, G., Shabadi, R,. and Imad, A. (2016). Microstructural observations and tensile fracture behavior of FSW twin-roll cast AZ31 Mg sheets. Mater. Sci. Eng. A 649:190–200.

Dragatogiannis, D. A., Koumoulos, E. P., Kartsonakis, I., Pantelis, D. I., Karakizis, P. N., and Charitidis, C. A. (2015). Dissimilar friction stir welding between 5083 and 6082 Al alloys reinforced with tic nanoparticles. Mater. Manuf. Processes. 31(16). doi:10.1080/10426914.2015.1103856.

El-Batahgy, A.-M., Miura, T., Ueji, R., and Fujii, H. (2016). Investigation into the feasibility of FSW process for welding 1600 MPa quenched and tempered steel. Mater. Sci. Eng. A 651:904–913.

Goebel, J., Ghidini, T., and Graham, A. J. (2016). Stress-corrosion cracking characterisation of the advanced aerospace Al-Li 2099-T86 alloy. Mater. Sci. Eng. A. 673:16–23. DOI: http://dx.doi.org/10.1016/j.msea.2016.07.013.

Ma, Z. Y., Feng, A. H., Chen, D. L., and Shen, J. (2017). Recent advances in friction stir welding/processing of aluminum alloys: Microstructural evolution and mechanical properties. Crit. Rev. Solid State Mater. Sci. 43(4). DOI:10.1080/10408436.2017.1358145.

Ni, D. R., Wang, J. J., and Ma, Z. Y. (2016). Shape memory effect, thermal expansion and damping property of friction stir processed NiTip/Al composite. J. Mater. Sci. Technol. 32(2):162–166. http://dx.doi.org/doi: 10.1016/j.jmst.2015.12.013-October 2015.

Patel, A. R., Drupal, J. et al. (2018). Investigation of mechanical properties for hybrid joint of aluminium to polymer using friction stir welding. Mater. Today: Proce. 5(2):4242–4249

Shaik, B., Gowd, G. H., and Prasad, B. D. (????). Parametric investigations on friction stir welding of aluminium alloys. Emerging Trends in Mechanical Engineering, Lecture Notes in Mechanical Engineering, ISSN 2195-4364, ISBN 978-981-32-9931-3, 333–345. https://doi.org/10.1007/978-981-32-9931-3_33.

Shaik, B., Gowd, G. H., and Prasad, B. D. (2018). Experimental investigations on friction stir welding process to join aluminum alloys. Int. J. Appl. Eng. Res. 13(15):12331–12339. ISSN 0973–4562.

Shaik, B., Gowd, G. H., and Prasad, B. D. (2019a). An optimization and investigations of mechanical properties and microstructures on friction stir welding of aluminium alloys. Int. J. Mech. Prod. Eng. Res. Dev. 9(1):227–240. ISSN(P): 2249-6890; ISSN(E): 2249–8001.

Shaik, B., Gowd, G. H., and Prasad, B. D. (2019b). Experimental and parametric studies with friction stir welding on aluminium alloys. Mater. Today: Proc. 19:372–379. https://doi.org/10.1016/j.matpr.2019.07.615.

Shaik, B., Gowd, G. H. and Prasad, B. D. (2019c). Investigations and optimization of friction stir welding process to improve microstructures of aluminium alloys Taylor & Francis, Online, May, 2019. https://doi.org/10.1080/23311916.2019.1616373.

Shaik, B., Gowd, G. H., and Prasad, B. D. (2019d). Investigations on friction stir welding process to optimise the multi responses using GRA method. Int. J. Mech. Eng. Technol. (IJMET), 10(03):341–352. ISSN Print: 0976-6340 and ISSN Online: 0976-6359.

25 Assessment of dimensional overcut induced during EDM of Die Steel D3

Ajaj Khan[1,a], Abdul Jabbar Ansari[1,b], and Syed Asghar Husain Rizvi[2,c]

[1]Mechanical Engineering Department, Bansal Institute of Engineering and Technology, Lucknow, India

[2]Mechanical Engineering Department, Khwaja Moinuddin Chishti Language University, Lucknow, India

Abstract

Dimensional overcut is basically experienced while performing precise machining on EDM as a result of extra sparks striking the wall of the workpiece. EDM being a machine tool for complex cutting uses complex shaped tool. While using a cylindrical tool, a radial defect is noted which result in excess in dimension as compared to the size of the tool. This excess of dimension is assessed using Scanning Electron Microscopy. Response Surface approach was employed for designing the trials. ANOVA revealed that peak current (p value = 0.0009) crucially dominates the radial overcut while pulse on time have least influence. Validation run was performed to assess the validity of the model and it was found that the residual error was 1.80%. Thus, the model developed is accurate to predict the value of radial overcut.

Keywords: ANOVA, EDM, SEM, radial overcut.

Introduction

EDM is a process of non-conventional machining of metals utilising Spark erosion for removing the material from the workpiece. Die Steel D3 has various industrial applications and thus widely used at commercial level. During spark erosion, there are various defects that are observed at microscopic level. These defects general include oversizing, deposition of recast layer, microcracks, subsurface damage, burnouts, etc. By using proper set of EDM parameters, these defects can be reduced and better machining characteristics can be achieved.

Rashid et al. (2019) investigation to assess the effect of conductive coating over a EDMed Aluminium nitride ceramic found that lower level of discharge energy was suitable for machining micro holes in ceramics. Though it takes more machine time, the surface produced is rough.

Pradhan (2018) in the research to optimise material removal rate, rate of tool wear and the over cut produced during EDM revealed that peak current was the major influencing factor with highest contribution towards the performance measures followed by pulse duration, duty factor and voltage.

In another investigation carried by Bhaumik and Maity (2019) to assess the influence of electrode material during EDM of Titanium alloy found that the copper electrode was suitable to produce better surface finish and minimum overcut. It was also concluded that thin and uniformly distributed white layer was obtained when Titanium is machine and using copper electrode. Further, they recommended copper electrode over other chosen electrodes in order to achieve better precision and higher level of finish of the machine surface.

Kubade and Jadhav (2012) during their investigation on AISI D3 to assess the rate of electrode wear, degree of material removal and over cut found that peak current was the crucial parameter for all the performance measures. They suggested that a lower extent of EDM variables must be used for achieving superior machining characteristics.

Moreover, Muthukumar et al. (2014) performed mathematical modelling while undergoing electro discharge machining on Incoloy 800 for the overcoat produced. Their study revealed that peak current and voltage possess high significance over the radial overcut. The developed mathematical model was found to have good agreement with the predictions.

In another investigation carried by Singh et al. (2013) during machining of aluminium alloy using rotating electrode established that current and of duration crucially influence radial overcut. Radial overcut is found to have an increasing trend with both current and duration. With off duration, the radial overcut tends to decrease.

[a]er.azazkhan@gmail.com; [b]techno.abdul@gmail.com; [c]sahr.me@gmail.com

Rizvi et al. (2018) during their investigation for analysing tool wear and recast layer develop the mathematical model found that the pulse current was major influencing factor for tool wear and recast layer thickness. They also concluded that duty factor is insignificant for the performance measures. Their developed model shows that AISI 4340 can be used commercially for industrial application.

Modi and Agarwal (2019) during their research with powder mixed EDM for analysing surface roughness and rate of material removal found that powder mixed dielectric fluid enhances the rate of material removal and minimises the surface the roughness of machined surface. They concluded that by adding powder of chromium, higher rate of metal removal is obtained and rougher surfaces is obtained. When aluminium powder is added, an inverse effect is observed.

In another research carried by Rizvi and Agarwal (2016) to assess rate of metal removal, surface roughness, developed residual stresses and cracks during electro discharge machining of AISI 4340 found that current and on duration are the crucial variables for all performance measures. They also concluded that the surface crack density tends to increase at lower level of current. A higher on duration widens up the rupture. Further they concluded that the internal stress developed fluctuates when the induration is varied.

Alhodaib et al. (2021) carried research on Nimonic 90 with powder mixed EDM and found that powder in dielectric reduces the roughness of the machined surface as well as the recast deposited post machining. On duration was crucial parameter for both the responses.

Further, Muthuramalingam et al. (2019) carried out research for developing a model to predict white layer thickness. Their developed model was having an accuracy of 97%. Kumar et al. (2021) conducted research on Inconel 825 to assess crack density and recast layer. They found that the crack density was majorly dominated by current and on duration while the formation of recast layer depends upon onlu on duration.

Moreover, Balamurugan and Sivasubramanian (2020) conducted research to find suitable electrode for machining and found that copper with brass coating produces higher rate if material removal. They also concluded that copper with brass coating leads to lower rate of tool wear.

Based on above survey, it was discovered that there were no studies conducted on dimensional overcut of Die Steel D3. The majorly selected EDM parameters are peak current, pulse on time and voltage and same will be selected for the present research.

Material and Methods

The material chosen for the present research is die Steel D3 Which is real hardened alloy Steel possessing high chromium and carbon. It finds its industrial application as dies punches tools. The specimens were turned through lathe into cylindrical shape 16 mm and thickness 10mm. The tool material used is copper having a diameter of 5 mm and length of 10 mm. The following Table 25.1 shows the chemical composition of the work material.

The trails were designed based on response surface approach using twenty set of experiments. Three factors viz. peak current, pulse on duration and voltage and their three levels were selected to assess their influence on radial overcut. The following Table 25.2 shows the EDM parameters and their level selected for the present modelling.

Table 25.1 Chemical composition of work material

Material	*% Composition*
Iron	86.5
Nickle	0.0689
Manganese	0.269
Chromium	11.05
Carbon	2.07
Silicon	0.191
Copper	0.00367
Vanadium	0.0218
Molybdenum	<0.002

Table 25.2 EDM parameters and their levels

Parameter/level	*Level 1*	*Level 2*	*Level 3*
Peak Current (A)	1	5	9
Pulse on Time (μsec)	10	15	20
Voltage (V)	90	120	150

The experiments with selected EDM parameters were performed on Electronica ZNC EDM. After machining the work on EDM, focused electron beam microscopy i.e., SEM was executed for evaluating the diameter of machined hole. The machined hole diameter was compared with diameter of tool. Radial overcut was calculated by the formula shown in Equation1 below.

$$\text{Radial Overcut} = \text{Diameter of Machined Hole} - \text{Diameter of the Tool} \quad (1)$$

The following Figure 25.1 on the next page shows the EDM machining setup with Die Steel D3 specimen and Copper tool.

Figure 25.1 Machining of die steel D3 specimen using copper tool on EDM

Results and Discussions

The dimensional overcut caused due to excess sparks experienced by the wall of machined hole was assessed through scanning electron microscopy and value of each radial overcut at different set of EDM parameters was tabulated as shown in following Table 25.3.

Table 25.3 Values of radial overcut corresponsing to different set of EDM parameters

Exp. No.	*Peak Current (Ip) (in Ampere)*	*Pulse on Time (Ton)(in μsec)*	*Voltage (V)*	*Radial Overcut (mm)*
1	1	10	90	0.101
2	9	10	90	0.211
3	1	20	90	0.097
4	9	20	90	0.171
5	1	10	150	0.075
6	9	10	150	0.189
7	1	20	150	0.075
8	9	20	150	0.118
9	1	15	120	0.065
10	9	15	120	0.095
11	5	10	120	0.108
12	5	20	120	0.165
13	5	15	90	0.124
14	5	15	150	0.162
15	5	15	120	0.112
16	5	15	120	0.109
17	5	15	120	0.107
18	5	15	120	0.113
19	5	15	120	0.107
20	5	15	120	0.108

Table 25.4 Lack of fit test for radial overcut

Source	*Sequential p value*	*Lack of Fit p value*	*Adjusted R^2*	*Predicted R^2*	
Linear	0.0079	<0.0001	0.4217	0.2101	**Suggested**
2FI	0.65	<0.0001	0.3698	0.4149	
Quadratic	0.0677	<0.0001	0.5861	0.7594	**Suggested**
Cubic	<0.0001	0.0351	0.9903	1.3326	

Table 25.5 Sequential model sum of square (SMSS) test for radial overcut

Source	*Sum of Square*	*DOF*	*Mean Square*	*F Value*	*P value*	
Mean vs Total	0.2909	1	0.2909			
Linear vs Mean	0.0148	3	0.0049	5.62	0.0079	Suggested
2FI vs Linear	0.0016	3	0.0005	0.5612	0.65	
Quadratic vs 2FI	0.0062	3	0.0021	3.26	0.0677	Suggested
Cubic vs Quadratic	0.0062	4	0.0016	105.6	<0.0001	
Residual	0.0001	6	0			
Total	0.3198	20	0.016			

The radial oversized hole obtained was measured using electron microscopy. To determine the appropriate polynomial equation for radial overcut, lack of fit and SMSS tests were performed as shown in the Tables 25.4 and 25.5 below. Both suggested quadratic equation for the present model.

Further, ANOVA was performed for the present quadratic model and is depicted in Table 25.6. From the ANOVA, the model F-value of 3.99 shows the significance of the model. It is known that a p-value less than 0.05 points out the significant model terms. In this case current was perceived as crucial term for radial overcut having a *p*-value of 0.0009. The ANOVA also depicts that the other electrical parameters have less significant influence over radial overcut. The lack of fit *F*-value of 187.82 its crucialness. Only 0.01% possibility exist for an estimate this high can arise because of noise.

Figure 25.2 shows the response surface plots for Radial Overcut with different EDM parameters. Figure 25.2 (i) considers current and on duration and reports that major variation of overcut is observed with current and almost minimum influence is depicted with on duration. A similar curve was obtained with peak current and voltage as parameters in Figure 25.2 (ii). Peak current again was elucidated as crucial parameter for radial overcut while voltage has least influence. The Figure 25.2 (iii) considers on duration and voltage as varying parameters and the radial overcut initially decreases and then increases with these parameters. A higher level of peak current and lower level on duration and voltage must be selected so as to achieve minimum radial overcut. The curved plot represents the presence of quadratic term in the model.

The following equation Equation 2 is in actual factor terms that is used for predicting responses at selected level of parameters.

$$\begin{aligned} Radial\,Overcut(mm) = {} & 0.637528 + 0.043562 \times Ip - 0.022825 \times Ton - 0.007696 \\ & \times V - 0.000669 \times Ip \times Ton - 0.000028 \times Ip \times V - 0.000022 \\ & \times Ton \times V - 0.002088 \times Ip^2 + 0.000924 \times Ton^2 + 0.000033 \times V^2 \end{aligned} \tag{2}$$

Table 25.6 Analysis of variance for radial overcut

Source	*Sum of Square*	*DOF*	*Mean Square*	*F Value*	*P value*	
Model	0.0226	9	0.0025	3.99	0.0209	Significant
A-Peak Current	0.0138	1	0.0138	21.87	0.0009	
B-Pulse on Time	0.0003	1	0.0003	0.5345	0.4815	
C-Voltage	0.0007	1	0.0007	1.15	0.3092	
AB	0.0014	1	0.0014	2.27	0.1625	
AC	0.0001	1	0.0001	0.1448	0.7115	
BC	0.0001	1	0.0001	0.1448	0.7115	
A^2	0.0031	1	0.0031	4.88	0.0517	
B^2	0.0015	1	0.0015	2.33	0.1579	
C^2	0.0024	1	0.0024	3.83	0.079	
Residual	0.0063	10	0.0006			
Lack of fit	0.0063	5	0.0013	187.82	<0.0001	Significant
Pure Error	0	5	6.67E-06			
Cor Total	0.0289	19				

The Figure 25.3 on the following page depicts SEM figures of machine Sample 1 in Figure 25.3 (a) (peak current = 1A; voltage = 90V; pulse on time = 10 µsec) and Sample 6 in Figure 25.3 (b) (peak current = 9A; voltage = 150V; pulse on time = 10 µsec).

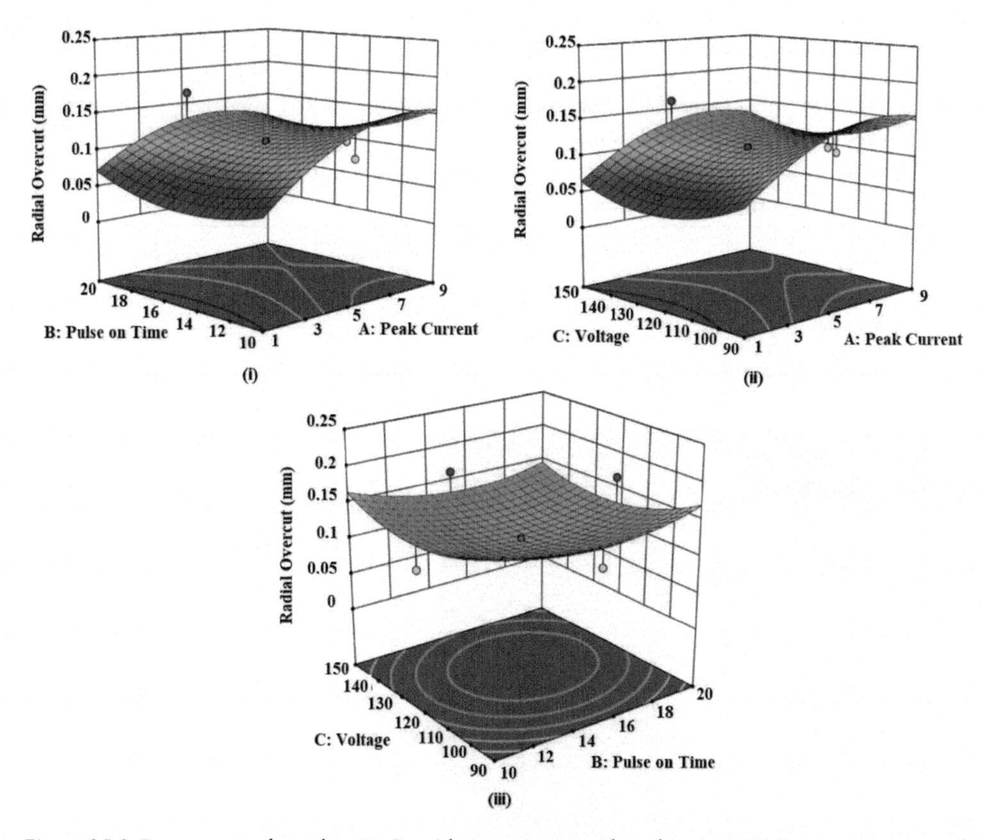

Figure 25.2 Response surface plots (i) Considering current and on duration; (ii) Between current and voltage; (iii) Between on duration and voltage

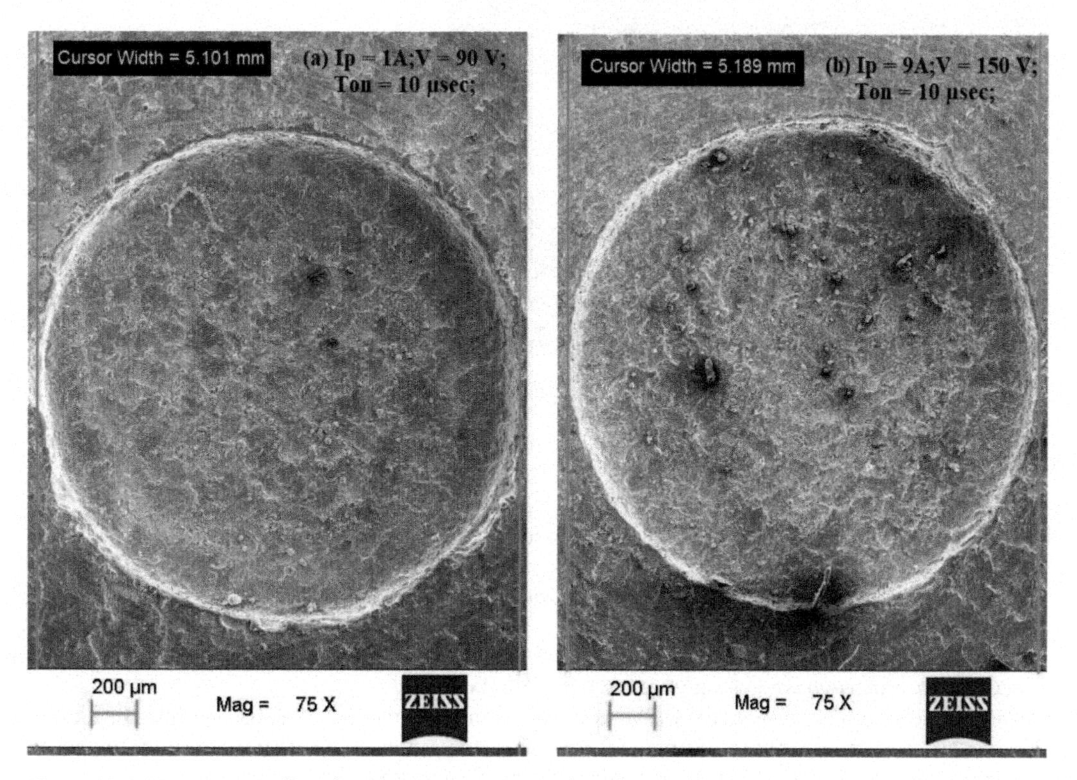

Figure 25.3 SEM images of machined specimen (a) Sample 1 (b) Sample 6

Validation of Developed Model

The developed model was validated in order to find that the present model is suitable in forecasting the radial overcut or not. It compares the validated radial overcut with the predicted radial overcut. The model is considered to be accurate when the residual error falls under 5%. Figure 25.4 shows the SEM image of machined specimen for validation run.

The predicted value of radial overcut was 0.110964 mm. the validated value for the set of EDM parameters was estimated as 0.109 mm. Thus, the residual error was found as 1.80% hence indicating the model as valid. Table 25.7 depicts the model validation.

Table 25.7 Model validation run for radial overcut

S.No.	*Input Parameters*	*Predicted Radial Overcut (mm)*	*Validated Radial Overcut (mm)*	*% Residual Error*
1	Ip = 5A; Ton = 15μsec; Voltage = 120V	0.110964	0.109	1.80%

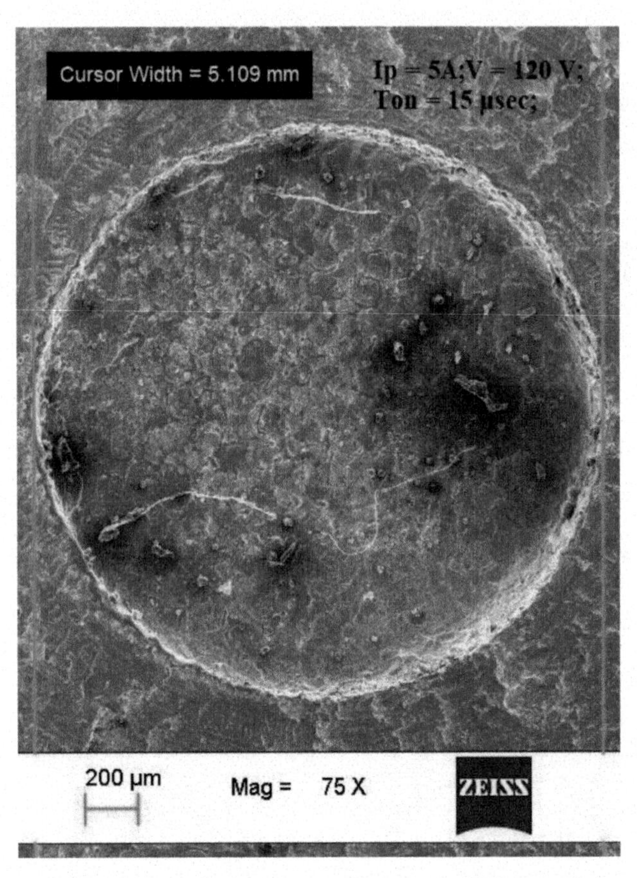

Figure 25.4 SEM images of machined specimen for validation run

Conclusion

Based on the above research performed on EDM of Die Steel D3 using Copper tool using response surface methodology, the following conclusions were drawn for the developed radial overcut:

1. The Lack of fit and SMSS test suggested Quadratic equation for the present developed model of Radial Overcut.
2. Current was perceived as crucial term for radial overcut having a p-value of 0.0009 while the other electrical parameters have less significance over radial overcut.
3. To achieve lower radial overcut, the EDM parameters must be set to Ip = 1A, Ton = 15μsec and V = 120V.
4. The curved response surface plot obtained represents the presence of quadratic term in the model.
5. The residual error found as 1.80% indicates that the developed model as valid.

References

Alhodaib, A., Shandilya, P., Rouniyar, A. K., and Bisaria, H. (2021). Experimental investigation on silicon powder mixed-EDM of nimonic-90 superalloy. Metals 11:1–17.

Balamurugan, G. and Sivasubramanian, R. (2020). Prediction and analysis of electric discharge machining (EDM) die sinking machining of ph 15-5 stainless steel by using taguchi approach. METABK 59:67–70.

Bhaumik, M. and Maity, K. (2019). Effect of electrode materials on different EDM aspects of titanium alloy. Silicon 11:187–196.

Kubade, P. R. and Jadhav, V. S. (2012). An experimental investigation of electrode wear rate (EWR), material removal rate (MRR) and radial overcut (ROC) in EDM of high carbon-high chromium steel (AISI D3). Int. J. Eng. Adv. Technol. 1:135–140.

Kumar, P., Gupta, M., and Kumar, V. (2021). Experimental investigation of surface crack density and recast layer thickness of WEDMed inconel 825. J. Comput. Appl. Res. Mech. Eng. 11:205–216.

Modi, M. and Agarwal, G. (2019). Effect of aluminium and chromium powder mixed dielectric fluid on electrical discharge machining effectiveness. Adv. Prod. Eng. Manag. 14:323–332.

Muthukumar, V., Rajesh, N., Venkatasamy, R., Sureshbabu, A., and Senthilkumar, N. (2014). Mathematical modeling for radial overcut on electrical discharge machining of incoloy 800 by response surface methodology. Procedia Mater. Sci. 6:1674–1682.

Muthuramalingam, T., Saravanakumar, D., Babu, L. G., Phan, N. H., and Pi, V. N. (2019). Experimental investigation of white layer thickness on EDM processed silicon steel using ANFIS approach. Silicon 11:1–7.

Pradhan, M. K. (2018). Optimisation of EDM process for MRR, TWR and radial overcut of D2 steel: A hybrid RSM-GRA and entropy weight based TOPSIS approach. Int. J. Ind. Syst. Eng. 29:273–302.

Rashid, A., Bilal, A., Liu, C., Jahan, M. P., Talamona, D., and Perveen, A. (2019). Effect of conductive coatings on micro-electro-discharge machinability of aluminum nitride ceramic using on-machine-fabricated microelectrodes. Materials 12:1–19.

Rizvi, S. A. H. and Agarwal, S. (2016). An investigation on surface integrity in EDM process with a copper tungsten electrode. Procedia CIRP 42:612–617.

Rizvi, S. A. H., Agarwal, S., and Bharti, P. K. (2018). Modelling of tool wear and recast layer thickness in die sinking EDM process. In Euspen's 18th international conference & exhibition, (pp. 1–2).

Singh, M. P., Kalra, C. S., and Singh, S. (2013). An experimental investigation of radial overcut during machining of Al/Al2O3 MMC by Rotary EDM. Int. J. Emerging Technol. Adv. Eng. 4:617–619.

26 Databases for machine learning: A journey from SQL to NOSQL

Dipali Meher[1,a], *Baljeet Kaur*[2,b], *Alaknanda N. Pawar*[3,c], *and Sheetal Parekh*[4,d]

[1]Department PES Modern College of Arts, Science and Commerce, Ganeshkhind, Pune, India

[2]Symbiosis Institute of Computer Studies and Research, Symbiosis International (Deemed University), Pune, India

[3]Department, Bhosala Military College, Rambhoomi Nashik, India

[4]Associate Architect, EXFO India Pvt. Ltd., Magarpatta, Pune, India

Abstract

Before 1990 relational databases were dominant and before that data was stored in punch cards. It is impossible to imagine any software without databases. After 1990, all organisations either small or large, started using internet and over the last 10 years, cloud came into picture. Now data grows into petabytes and terabytes i.e., big data. It was difficult to query faster on these big databases. Relational databases were not capable of handling this burden. Today, more than 80% of the data is unstructured data. Relational databases use structured data. Getting business insights from unstructured data was a big challenge for database administrators and researchers. Relational databases fail to provide clustering concept used in management of unstructured data. Though they strongly provide concurrency and data recovery techniques, but they were unable to handle customised clustering techniques for firing and execution of queries in efficient manner. At this point, evolution of NOSQL i.e. Not Only Structured Query Language concept made it possible to query more and more data with better efficiency than structured data. Carlo Strozzi has documented the term NOSQL in 1998 for handling of clustering concept. The problem of 2k came into picture in the year of 1999. So many developers worry about economic future of internet. In the year 2000, web properties were increased in scale. In 2006 Google has published his research paper on BigTable distributed database structure and soon in 2007 Amazon has published his research paper about Dynamo database handling technique. Soon then user of various NOSQL databases or technologies handling for unstructured data has been exploded and became the backbone of Machine Learning. This paper aims to give pitfalls of relational databases with their advantages and disadvantages to describe the databases journey from SQL to NOSQL databases as they provide handling of customised clustered data over 21st century web estate and almost all of them are used in Data Science for Machine Learning concept.

Keywords: Clustering in NOSQL, databases for machine learning, impedance mismatch, JSON, NOSQL, RDBMS.

Introduction

In the era of mainframes relational databases played active role before internet, cloud and big data, Capacity of relational databases was improved by updating servers, processors, storages so that these databases can scale vertically (Sadalage and Fowlerm, 2012). NOSQL databases emerged from exponential growth of WWW and various online applications over internet (Karwan and Shakir, 2019). Google had out Bigtable research paper in 2006 Chang et al. (2006) and Dynamo research paper by Amazon in 2007 DeCandia et al. (2007) which is starting era of NOSQL databases. From this point onwards, NOSQL databases are developed to meet requirements of new generation of enterprise database requirements i.e., big data. Relational databases were the only choice in front of project development before 1990s (Till and Riede, 2002). In 1990s Object oriented databases emerged which changed the world of database and project development. In 2007, NOSQL databases evolved in surprise way (Chen and Lee, 2019). NOSQL databases provide agility and, they can operate at any scale. They are non-relational databases having huge capacity of storage, retrieval and recovery of data. Real-time web applications and big data uses NOSQL databases. They are called Not Only SQL means they sometimes supports SQL. Examples of different NOSQL databases includes OrientDB MarkLogic, Aerospike, FairCom, c-treeACE, Google Spanner, etc (Sethi et al., 2014).

[a]mailtomeher@gmail.com; [b]baljeet.rayat@gmail.com; [c]alaknanda.pawar@bmc.bhonsala.in; [d]sheetalbhalgat@gmail.com

The Value of Relational Databases

Relational databases worked as backbone for any software development process. When data storage and retrieval concept is there relational databases are always there. MySQL, PlPgSQL, PostgresSQL, (Vershinin and Mustafina, 2021) are some of the examples of relational databases. Core benefits of relational databases can be listed below.

A. Use of Joins

Relational databases provide meaningful information by joining tables. Joining allows user to understand how tables in a database relate to each other. For joins all relational databases are normalised as in any form such 1NF till BCNF (Boyce Codd Normal Form). Normalisation is used to avoid database duplication, and cardinalities can be used to go for normalisation process.

B. Persistent Data Provision

Relational databases are keeping data in persistent mode. As computer architectures provide two types of memory main memory(volatile) and secondary memory(backup). Retrieval data from databases is also at very fast speed. By using any kind of memory structure and provision of various recovery mechanisms such as deferred update and immediate update user can get consistent database at any given point of time.

C. Concurrency Control Mechanism

The ability of the database system to allow multiple users to affect and run multiple transactions in an interleaved manner with the concept of serialisability. These multiple users are working on different parts of database. When many users are working on same databases their operations should be interleaved in such a way that to maintain the serialisability. There should be coordination among these interleaved fashions of execution transactions (Lam et al., 2002). Example. Two users at different locations try to book same hotel room. These room booking operations are to be executed in interleaved manner such that if a room is booked by one user another user can see it is booked. It is difficult to manage concurrency control mechanisms by expert programmers. Many enterprise applications have lots of users working on same database concurrently so there will be nig change of errors occurred a run time situation. Due to transaction concept in relational databases all concurrency control mechanisms are handled very carefully and systematic manner in real time databases, but these mechanisms are complex.

D. Integration

Enterprise applications is a combination of many multiple applications written by different terms in collaboration. These teams may have some organisation limitations. All these small multiple applications must use same data and made update to this database with visibility of all updates to all users (Ziegler and Dittrich, 2007). To achieve this concept shared database integration techniques is used. In this technique,-data of multiple applications will be shared in single database. A centrally available single database allows all applications in one enterprise to communicate with each others and see each others database in easiest way. Concurrency control mechanism provided by relational database allows to handle this in similar way as it handales several users in a sole application.

E. Standard Model of RDBMS

Relational databases were succeeded because by learning simple or basic model any one can start for applying it for project development. There are many differences between relational models, but basic idea remains same. Different vendors have different products and languages like PlPgSQL, MySQL, ORACLE but the transaction concept remains same in all languages.

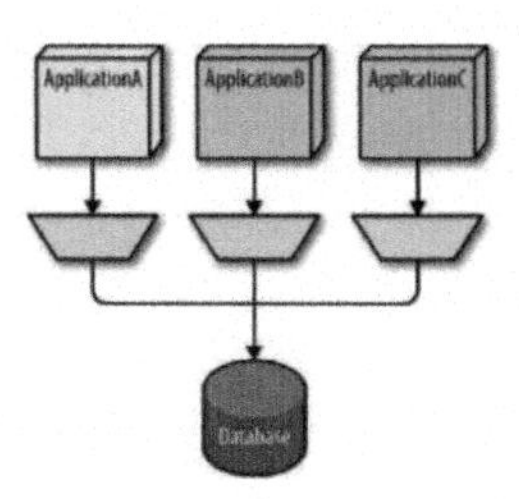

Figure 26.1 Database integration

Limitations of RDBMS

As there are limitations of RDBMS so there is emergence of NOSQL. In case of RBMS, DBA must define database structure and schema (ER diagram) to start with any model or design process. RDBMS provides ACID properties (atomicity, consistency, isolation, durability) which are useful for transaction management and recovery of databases from failure. But these properties give performance overhead and more downtime as shown in Figure 26.2.

NOSQL contains semi structured data, less functionality than compared to RDBMS but gives high performance. In current ers of programming languages , most of them are store data in JSON format which is not supported by RDBMS (as queries in RDBMS like create , insert, update and delete does not support JSON format).

Impedance Mistmatch

In early days relatinal databases provide many advantages but while using them they also provide many frustations to its users. Impedance mismatch is the main frustation provided by relational databases to database administrators and application developers Burger and Riley (1974).

In relational databases data stored in tables and tuples which ocntians single record for that relation. Tuple can be considered as name value pair. Relational databases use SQL query language and output of query is subset of relations. Various relational algebra notations are is used to take instances from relations such as project, selcet. When relational databases are developed developers have not throught of their complexcity which will be faced in future. Consider a simple example of bill where order is placed which contains data about payment details made. In this example data from different tables like orer, customer, item card will be taken which replicate the data in bill table. If data which is taken from different tables will be stored on different places then bringing that data to a single place becomes tedious task. Object oriented databases solve such segregated data issue, which will be the leading for software development in coming centuries. This database is used by object oriented programming languages, but they sometimes create anomalies. Relational databases have integrated themselves to so that these were used for programmers professionally. Impedance mismatch is nothing but keeping the required data for joins on same place where the joins were made. Relational databases failed to support this feature. Hybernate and iBATIS are object relational mapping frameworks. They will be implemented with well-known mapping patterns but still mapping (i.e. bringing tables required for join on single place) issue is not solved. These frameworks try to solve the impedance mismatch issues but mapping patterns in them are hard to use and they were suffered by issues of query performance. There is need for such databases which will solve impedance mismatch issue and thought process for NoSQL data structures were started. Following Figure 26.3 explains this concept.

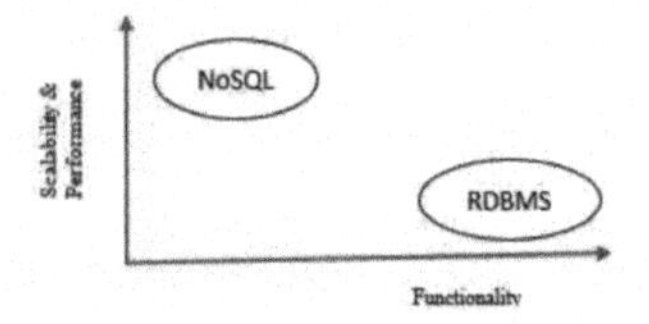

Figure 26.2 RDBMS verses NOSQL

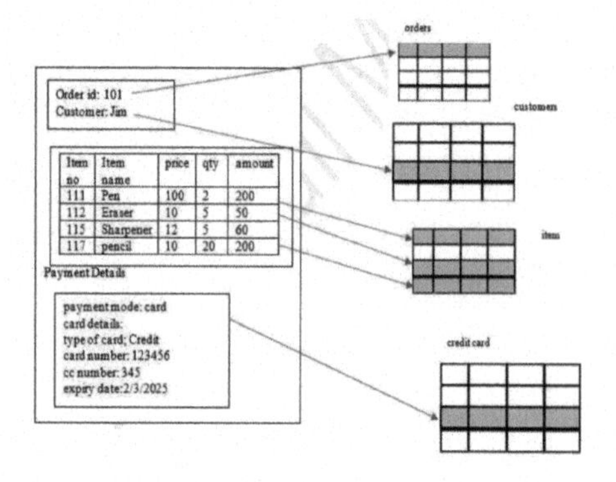

Figure 26.3 Impedance mismatch example

Application, Integration and Combination of Databases

SQL databases always provides integration mechanism between different applications under same roof. So, SQL databases are called as an integration database. They integrate and assimilate with numerous applications developed by dissimilar teams and store their data in shared database. To maintain consistency of database communication between teams has to be increased Karcher et al. (2018). This is also known as shared database instance. One big drawback of this shared database instance is its complex structure as many applications are integrated together to work on shared instance. If any application wants to do changes to its data storage mechanism, then it needs to connect with applications using that shared instance. Also, applications working on shared instance have different structural(physical and logical) and performance needs so indexing technique used also becomes problematic. Each application has its own development team which implies that shared database cannot trust on these applications in data updating to preserve integrity concept so database itself must take this responsibility on its own shoulder. Another approach than shared instance is treating database as an application database. In this approach, database is directly accessed by single application code developed by single team. The team which accesses database needs to know the structure and database schema. In this situation both the database and application code can be controlled by single team and an integrity can be done with application code.

In the year 2000 interoperability concept for web applications was run with HTTP and web services uses SQL with shared techniques. (They are known as service-oriented architectures) (Lewis and Smith, 2008). Web services give more flexibility to database structure which is being exchanged between different web applications. The web services mostly use XML or JSON structure (Bourhis et al., 2020). In case of remote communication, number of round trips required for data exchange will turn to single request response. Once a decision that application database is used, developer will get freedom to choose database. There is always a decoupling between database used by application and services used by that application to communicate with outside world (outside databases/applications). The outside world (application or database) does not care about how the internal database is stored. Security feature of relational databases is less useful in application databases due to encapsulation feature. Relational databases become dominance to application structure also.

Attack of Clusters and Changed Way

The problem of 2k came into picture in year 1999. So many developers worry about economic future of internet. In the year 2000, web properties were increased in scale (Kaur, 2021). There were many dimensions to this scale as noted below:

A. Tracking

All web sites started detailed user tracking activity to which user visits and what user searched for and what website user is investing for.

B. Increase in Links(hyperlinks), Social networks, logs, mapping data of web pages

The amount of Web page structure is increased by giving more hyperlinks, increase in social media websites, increase in web logs and mapping of web logs to trace the user's activity.

C. Increase in Visitors to Website

With the above two scaling the amount of web data has started increasing with number of users and visitors using internet (and hence traffic to internet is increased).

To manage the amount of scaling dimensions in data and traffic purpose, more computing resources are required. There were two choices to cope with this situation. Either to make changes that is up or to don't make anything let it go on i.e., out. Up means use large storages, hardware resources, bigger machine in terms of storage, processors, and disks. But these bigger machines are expensive as expenses are increased because of size of data will also increase in future. One alternative to this is smaller machines with low storage capacity, processors can be used on clusters. Cluster provides commodity usage of hardware and finally cheaper for scaling. These clusters are resilient (strong) as if any individual machine fails in that cluster other can be built up to overcome failures by providing high reliability. So, using a clustering provides many advantages like reliability, scalability, less expensive, resilient but when large databases are shifted to clusters, a new problem raises. Relational databases are not intended to run on clusters. Relational databases like Oracle RAC or Microsoft SQL Server work on shared disk concept. These databases have a cluster aware file system that writes to a highly available disk subsystem. This means that cluster has a disk subsystem which may undergo failure runtime. Relational databases may run at separate servers with different sets of data known as sharding of database. This sharding concept separates the load but applications which uses this concept have preserve path and DBA has to track which database server dialog for which bit of data. Sharding loses the concept of querying, applying transactions, referential integrity, and consistency constraints (i.e., ACID properties). ACID consists of atomicity, consistency isolation and durability. When clustering is introduced, licensing of relational databases must be done to overcome technical problems. Commercial databases are running on single server, so by clustering pricing is increased which reflects negotiations overhead on purchase department. Clustering and relational databases have changed the database storage technique of commercial databases. Google and Amazon are the two best companies dealing with large size of clustering. These companies are capturing large amount of data. These companies are rich with strong technical components required for clustering and were large growing companies in the world. Due to this, these companies are motivated to bring such a database which is unstructured and must be used in clusters.

In the year 2006, BigTable Distributed data Structure concept is published by Google which is its BigTable distributed structure database. 'BigTable is a distributed storage system for managing structured data that is designed to scale to a very large size: petabytes of data across thousands of commodity servers.' BigTable stores rows by single key and stores data in rows connected by column families. So related data is accessed using IDs rather than using joins in RDBMS. Successively in 2007, Amazon had released a paper describing Dynamo data storage application. The Dynamo database is structured as follows:

"Dynamo is used to manage the state of services that has very high reliability requirements and need tight control over the tradeoffs between availability, consistency, cost – effectiveness and performance." This paper state that primary key is used to store Amazon data and consistent use of hashing was done to divide and distribute data. Also, way of using object versioning to maintain consistency across data centres. This paper was the first paper to label use of key value structure globally at Amazon.

The above two papers have changed the way of thinking how huge data is handled using a simple clustering technique. Both Amazon and Google operate on scales. But such technique may not be useful to average organisations. The organisations want to capture and process data but with RDBMS it was not possible. So, Google and Amazon have leaked more information about the technique, process used for data storage. Based on the techniques used by both, other organisations also started designing the database in the same way in the world of clusters. The threat of RDBMS is phantom problem but now the threat forms cluster become a serious gives emergence to NoSQL databases.

The Emergence of Nosql

Carlo Strozzi in 1998was first documented the term NOSQL (Bach and Werner, 2014). He has visited to San Francisco to meet some people to talk together around the trivial(lightweight) relational database created by him. In his database, ASCII files are used to store database, tabs are used to separate each tuple which is represented by a line. At that time, relational databases were dominant. He has used the term NoSQL because he used shell scripts to access database created by him rather than SQL to access RDBMS. The original meaning of NoSQL was instead of using SQL use the query mechanism from developers end (in Carlo it was UNIX open-source environment). So, the use of No SQL terms had put a strong frustration on relational database developers. Developers found that using NoSQL is better than firing complex queries on relational database using joins. Such meeting in San Francisco come and went and developers continued to develop better query environment for complex queries involving joins. One example is using Hibernate library in java in which developers don't have to worry so much about how the underlying database is structured — developers just call functions on objects.

In this situation cost stocks i.e., debugging such complex queries was very hard as it requires cost for development, administration, and testing. One abstraction like library concept is used but relational databases don't work well with this library concept also. Exiting relational databases were not working well with explosion and growth of www and internet. By that time in 1991, other things were happening like in 1991 a public web page was created. It was just before 7 years of that Carlos San Francisco meet. In the year 1994 Yahoo and Amazon were founded. Google was there before but was not founded till. Remember that before Google AltaVista (by and shut down by Yahoo) and Ask Jeeves were served the purpose what Google has thought for. Then in 1997, XML was introduced for system- to-system communication. In 1999, XSLT specification was released. In the year 1997, web is newborn baby and people are thinking for how to make use of web for money generation process. In 2009 many NoSQL databases have emerged like Riak, MongoDB, HBase, Accumulo, Hypertable, Redis, Cassandra. In 2007 to 2009, Neo4j was emerged.

Then two software developers, Eric Evans from Rackspace, and Johan Oskarsson from company Last.fm have organised first modern NoSQL meet up (Vyawahare, 2018). They have created #NoSQL tag to distribute title of NoSQL among social media. With the statement 'This meetup is about 'open source, distributed, non-relational database' they have invited people from all over world. The meeting was worth with included speakers from social networks like LinkedIn, Facebook, Powerset, Stumbleupon, ZVents, and couch.io who were discussed Voldemort, Cassandra, Dynamite, HBase, Hypertable, and CouchDB, respectively. This meeting first time come up to a fruitful discussion about different approaches to non-relational databases and to brand them as NoSQL(either NoSQL or NOSQL). The clear thing is that NoSQL databases don't use SQL always but sometimes. Some of the databases may have query language to look similar with SQL. Example Cassandra's CQL. In NoSQL then term No stands for Not Only. All these NOSQL databases are generally open-source projects. This is distinguishing NoSQL databases from other databases. NoSQL databases run on clusters which effect on data modelling and approach to handling consistency (Diogo et al., 2019). As in RDBMS due to ACID properties consistency is surely handled. In NoSQL databases due to clustering handling consistency become a tedious and complex job. However not all databases of NoSQL run on clusters. Graph databases are also type of NoSQL databases that have same properties as like relational databases and provides different data model to handle data with complex relationships. Graph databases are famous for the quote "odd man out fish in NOSQL pond". NoSQL databases founded for the needs of the early 21st century web estates. NoSQL databases are schemaless (without schema) means allowing user to freely add fields to database records without defining structure of database first (Nayak et al., 2013). This is useful when user wants to deal with nonuniform data. Polyglot persistence is the terms used by NoSQL for storing data. It is to use many data storage technologies, selected based on the method data is being used by individual applications or components of a single application. Different kinds of data give best results for different purposes. Polyglot persistence means picking the right database technology or programming language for the right use case. Applications should be written in a combination of languages to take advantage of the fact that language supports. Dissimilar languages are suitable for tackling dissimilar problems. To use this concept in work, organisations need to change from integration databases to application databases. Following Table 26.1 shows a road map for Emergence of NOSQL (Gessert et al., 2017).

However, JSON format is having many features over XML.JSON format is light weight. JSON format is language independent. JSON format is easy to read and write. JSON is text based human readable data exchange format. JSON objects have standard structure that make easy for developers to read and write code. In case of working with AJAX, quick access time for loading data and asynchronous way is required, due to light weight feature of JSON all these tasks can be done with negligible downtime (Sahatqija et al., 2018). JSON format is language independent, meaning that it works well with all modern programming languages. Suppose in case developer want to change server-side programming language then developer can do it as JSON structure is same for all languages (Karande, 2018). Following Table 26.2 shows the difference between RDBMS and NOSQL (Győrödi et al., 2015).

Table 26.1 Emergence of NoSQL: A road map

Year	*Roadmap*
Before 1991	Oracle RAC or Microsoft SQL Alta Vista, Ask Jeeves
1991	Creation of public webpage
1994	Foundation of Yahoo, Amazon
1997	XML
1998	Google officially launched by Larry Page and Sergey Brin Carlo Strozzi at Sanfrascio used term NoSQL
1999	XSLT, 2k problem
2006	Google Research Paper
2002	First Neo4j Version
2007	Amazon Research Paper Amazon DynomoDB, Neo4j open-source graph database
2008	Cassandra wide column store by Apache software Foundation
2009	Frst modern NoSQL meet by Eric Evans from Rackspace and Johan Oskarsson from Last.fm MongoDB by MongoDB Document store Riak, MongoDB, HBase, Hypertable, Redis
2011	Couchbase document stare by Couchbase

Table 26.2 RDBMS verses NOSQL

RDBMS	*NODSQL*
Rigid	Flexible
Looks at Parts	Looks at Whole
Relational	Object-Oriented
Mature	Emerging
Stable	Scalable
Consistent	Eventually Consistent
ACID properties	Cap Theorem
Uses structured Query Language	Uses No declarative query language
Predefined Schema	No Predefined Schema
Data and its relationships are stored in separate tables	Key-value pair storage column storage, document storage, graph databases
Structured data	Unstructured data
Tight Consistency	Prioritised high performance and high availability and scalability
Vertically scalable	Horizontally scalable
Applications: enterprise data marts	Example: Facebook Gaming
Example: PLPgSQL, MySQL, ORACLE	Example, MongoDB, Riak, Cassandra, Neo4j

JSON Format

All NOSQL databases uses this format. It stands for Java Script Object Notation. AS like XML to transfer data between client and server and vice versa this format is used. Example Var Dipali = {"firstname": "Dipali", "lastname":" Meher", "Age":39}

Reasons for Using NOSQL

NOSQL databases are built for 21st century web estates as web contains unstructured data. They are schema less databases as no schemas were designed before its use. Size of data: When side of data is big (generally in pera bytes). Relationships in the data is not important. Unstructured data changes in and ended time.

Constraints and joins support are not required at database level. Data growth rate is continuous and high, and user need to scale the regular handling of database. In relational databases table is created, schema is defined, and different datatypes of fields are set before inserting data into database. In NoSQL all database creation, insertion and updating can be done on fly. NoSQL databases runs at very fast speed. NoSQL databases cluster design is preferred data which is access together is stored together. NoSQL databases gives better control over availability of data in database. Data structures used on NoSQL languages are flexible. The most important rise in NoSQL database id Polyglot persistence and Polyglot programming.

There are various types of NOSQL databases fall into four categories as Key-value databases(e.g. Redis, Memcached), Document data bases(e.g. MongoDB), Column Family Databases(Amazon Simple DB, Cassandra) and Graph Databases(Neo4j, OrientDB, Flock DB).

Barriers to NOSQL

NoSQL databases doesn't provide support for low level query languages. They do not have standardised interfaces for accessing, updating and maintain databases. Every database has its own query language. Vast earlier investments in existing relational databases will become barrier to NoSQL as shifting from SQL to NoSQL requires changing in Database to JSON structure. NoSQL databases lacks for ACID properties of transactions. NoSQL databases faces stale read problem: As NoSQL databases offers an eventual consistency in which database changes are spread (propagated) to all nodes so queries might not result in returning updated data. Queries may return data the is not accurate according to given time. This refers to stale read problem. NoSQL databases may exhibit lost write form of data loss. Data consistency is bigger challenge to NoSQL databases. NoSQL databases does not support joins. NoSQL databases don't have constraints like RDBMS while crating databases hence cannot go for support of transactional concepts. Following table shows types of NOSQL databases.

Advantages and Disadvantages of NOSQL Databases

A. Advantages of NOSQL Databses

High Scalability (horizontal scaling): NoSQL databases provides sharding and replications concepts. By which data can be partitioned and placed on multiple machines for high availability of database in case of any node failure. Vertical scaling can also be done in NoSQL databases by adding more resources to multiple machines (Karande, 2018). Due to scalability concepts NoSQL databases handle huge amount of data in well-organised manner. Replication feature of NoSQL databases makes it highly available because in case of failure of any node data become available as number of replicas of databases leads to previous consistent state of database.

B. Disadvantages of NOSQL Databases

There are many disadvantages of NoSQL databases too. NoSQL databases provides narrow focus as they are designed for storage considerations. These databases provide very little functionality. RDBMS is better choice for Transaction Management databases then NoSQL. NoSQL databases are open-source databases. Yet there is no reliable standard for NoSQL databases. Aim of Big data management tools is to manage the large amount data in simple way. It seems to be not that much easy. Data management in NOSQL is complex as compared to RBDMS. It might be hectic to manage NoSQL databases in daily transactions. GUI tools to access the NOSQL databases is not flexible and clearly available in the market. Regular backup of NoSQL database is a weak point as example in MongoDB has no approach to take backup of data in consistent manner. Document databases contains database documents in large size requiring high network bandwidth, speed to manage them.

Use of NOSQL Databases for Machine Learning

As various relational databases for example MYSQL, PostgreSQL are used in execution and development of machine learning algorithms.

In this big data era, to store unstructured data various NOSQL databases were used. Machine learning algorithm works on test and train databases. Bigdata itself contains data in petabytes and due to CRUD (Create, Update, Read, Delete) operations available for NOSQL databases they are relatively better than relational databases. Indexing structure of relational databases gives pain to DBA whereas using large structure and optimisation capability in NOSQL indexing will be handled in efficient manner. All NOSQL databases itself supports the concept of bigdata. The MongoDB database with IaaS (Infrastructure as a service) like AWS (Amazon Web Service) is used in every organisation for e- business. In now a day we have seen Data Science concept has camping in picture with its huge advantage to any industry as well as in research, NoSQL data stores can bec ome substitute to relational databases that focus on minimizing latency while restricting the types of operations that can be performed.

Data validation for machine learning is imperative. It is not optional. Data validation consists of testing accuracy and quality of source data to remove anomalies. Correct data should be served to training and testing process of ML. Most of the time it is seen that there are errors in training dataset while building model but at the time of real use of model there should not be any error. Data errors should be detected and removed at initial stage to avoid overfitting errors. As there are various challenges for data validation and researcher must think and use most of its time in this process while building machine learning model. There are various approaches available for it. For example, unit test approach by Amazon, data Scheme approach by Google. According to stack overflow system report generated in 2019 NOSQL databases will be the most used databases for machine learning. Again, from NOSQL databases Redis is the most loved database and MongoDB is the Most wanted database. UCI, GitHub, Netflix, Instagram are the main important repositories from where database can be downloaded for ML.

The topmost databases will be used in machine learning will be as follows:

Apache Cassandra: it is open source and highly scalable NOSQL DB used by GitHub, Netflix, Instagram, Reddit etc. Hadoop integration part in Cassandra allows Map reduce concept execution properly. Fault tolerance and elastic scalability are the best features provided by this database.

Couchbase: open-source, distributed, NoSQL document-oriented DB. This database allows data operations per milliseconds and has powerful query engine for SQL query execution. This DB provides powerful APIs for multiple programming languages hence supports for polyglot programming and allows to used cloud technology.

1) DynamoDB: This DB is provided by Amazon having built in security features. Due to its high availability and durability and high performance and scaling it is used by every single user.
2) Elasticsearch: This open-source DB provides Apache Lucene indexing techniques and works for all types of data as text, numeric, geospatial, structured, and unstructured type. Other than speed, scalability, and resiliency it provides features like data rollup, index management, infrastructure monitoring etc.
3) MLDB: This is open-source DB widely used for ML and comprehensively used for SQL as well as RDBMS.

4) MongoDB: This is document database management system used JSON like structure and allows data flexibility and distributed data management.
5) Redis: This is open source in memory data structure and allows various data structures and provides advantage like recovery from automatic failover.

All above databases will be used in Machine Learning as well as Data Science for knowledge generation and interpretation.

Conclusion

NOSQL databases have changed the way of thinking data in structured way. Unstructured data can be managed using different NOSQL technologies. Almost all information that is generated on web is unstructured and handling these databases with NoSQL has enabled Machine learning algorithms to give better insights to fuel business. But still there are some situations where NOSQL database cannot be used as when relationships in data are not important. The customisation storage of data can be helpful to retrieve, manage storage data from the way of firing queries. This paper has highlighted the pitfalls of relational databases and advantages of NOSQL databases for 21st century web estate databases. Relational Databases were born before internet, cloud, big data. They are required to scale up vertically to meet up the requirement of big data. NOSQL databases evolved as an output of internet and web applications. NOSQL databases like MONGODB, DynamoDB, MLDB can expand horizontally to meet the competitive requirement of today's business problems. Distributed NOSQL databases are agile and match up the requirement of Machine Learning Process.

References

Bach, M. and Werner, A. (2014). Standardization of NoSQL database languages. In: Kozielski, S., Mrozek, D., Kasprowski, P., Małysiak-Mrozek, B., Kostrzewa, D., eds. Beyond Databases, Architectures, and Structures. BDAS 2014. Communications in Computer and Information Science, 424. Springer, Cham. https://doi.org/10.1007/978-3-319-06932-6_6

Bourhis, P., Reutter, J. L., and Vrgoč. D. (2020). JSON: Data model and query languages. Inf. Syst. 89:101478. ISSN 0306-4379, https://doi.org/10.1016/j.is.2019.101478.

Burger, C. P. and Riley, W. F. (1974). Effects of impedance mismatch on the strength of waves in layered solids. Exp. Mech. 14:129–137. https://doi.org/10.1007/BF02322835.

Chang, F., Dean, J., Ghemawat, S., Hsieh, W. C., Deborah, A. (2006). BigTable: A distributed storage system for structured data. In OSDI 2006 - 7th USENIX symposium on operating systems design and implementation, (pp. 205–218).

Chen, J.-K. and Lee, W.-Z. (2019). An introduction of NoSQL databases based on their categories and application industries. Algorithms 12:106. 10.3390/a12050106.

DeCandia, G., Hastorun, D., Jampani, M., and Kakulapati, G. (2007). Dynamo: Amazon's highly available key-value store. Ope. Syst. Rev. - SIGOPS. 41:205–220. 10.1145/1294261.1294281.

Diogo, M., Cabral, B., and Bernardino, J. (2019). Consistency models of NoSQL databases. Future Internet. 11:43. 10.3390/fi11020043.

Gessert, F., Wingerath, W., Friedrich, S., Norbert, R. (2017). NoSQL database systems: A survey and decision guidance. Comput. Sci. Res. Dev. 32:353–365. https://doi.org/10.1007/s00450-016-0334-3.

Győrödi, C., Gyorodi, R., and Sotoc, R. (2015). A comparative study of relational and non- relational database models in a web- based application. Int. J. Adv. Comput. Sci. Appl. 6:254. 10.14569/IJACSA.2015.061111.

Karande. N. (2018). A survey paper on NoSQL databases: Key-value data stores and document stores. Int. J. Res. Adv. Technol. 6(2):153–157.

Karcher, S., Willighagen, E. L., Rumble, J. and Ehrhart, F. et al. (2018). Integration among databases and data sets to support productive nanotechnology: Challenges and recommendations. NanoImpact, 9:85–101. ISSN 2452-0748, https://doi.org/10.1016/j.impact.2017.11.002.

Karwan, J., and Shakir, A. (2019). development history of the world wide web. Int. J. Sci. Technol. Res. 8:75–79.

Kaur, H. (2021). Analysis of nosql database state-of-the-art techniques and their security issues. Turkish J. Comput. Math. Edu. (TURCOMAT) 12(2):467–71. https://doi.org/10.17762/turcomat.v12i2.852.

Lam, K-.Y., Kuo, T.-W., Kao, B., Lee, T., and Cheng, R. (2002). Evaluation of concurrency control strategies for mixed soft real-time database systems. Inf. Syst. 27(2):123–149. ISSN 0306-4379. https://doi.org/10.1016/S0306- 4379(01)00045-X.

Lewis, G. A. and Smith, D. B. (2008). Service-oriented architecture and its implications for software maintenance and evolution. Front. Software Maintenance, 1:1–10. doi: 10.1109/FOSM.2008.4659243.

Nayak, A., Poriya, A., and Poojary, D. (2013). Article: Type of nosql databases and its comparison with relational databases. Int. J. App. Inf. Syst. 5:16–19.

Sadalage, P. J. and Fowlerm M. (2012). NoSQL, distilled: A brief guide to the emerging world of polyglot persistence. Vasa., Pearson Education.

Sahatqija, K., Ajdari, J., Zenuni, X., Raufi, B. and Ismaili, F. (2018). Comparison between relational and NOSQL databases. 0216-0221. 10.23919/MIPRO.2018.8400041.

Sethi, B., Mishra, S., and Patnaik, P. K. (2014). A study of NoSQL database. Int. J. Eng. Res. Technol. 3(4):14–21.

Till, A. and Riede, K. (2002). RDBMS -an introduction to relational database management systems. Advanced User Guide. 10.13140/RG.2.1.1510.3520.

Vershinin, I. S. and Mustafina, A. R. (2021). Performance analysis of postgre SQL, MySQL, Microsoft SQL server systems based on TPC-H Tests. In 2021. International Russian automation conference (RusAutoCon), (pp. 683–687). doi: 10.1109/RusAutoCon52004.2021.9537400.

Vyawahare, H. (2018). Brief review on SQL and NoSQL. Int. J. Trend Specific Res. Dev. ISSN 2456–6470. 2:968–970..

Ziegler, P. and Dittrich, K. R. (2007). Data integration — problems, approaches, and perspectives. In Conceptual modelling in information systems engineering, J. Krogstie, A. L. Opdahl, S. Brinkkemper, eds. Berlin, Heidelberg: Springer. https://doi.org/10.1007/978-3-540-72677-7_3.

27 Optimum solution of goal programming problem by modified Charnes penalty method

Monali G. Dhote[1,a,*] *and Girish M. Dhote*[2,b]

[1]Department of Applied Mathematics and Humanities Yeshwantrao Chavan College of Engineering, Nagpur, India

[2]Department of Mechanical Engineering Yeshwantrao Chavan College of Engineering, Nagpur, India

Abstract

Present article shows an attempt which deals to explain goal programming by using Modified Charnes Penalty Method. Many times, this method involves less or at the most an equal number of iteration because this method helps to skip additional calculations of net evaluation, so one can save important time and hence improved solution has been achieved. Therefore, the mentioned procedure is going to be a new approach which provides outcomes instantly. It is better understood by resolving a cyclic problem.

Keywords: Goal programming problem, less iteration, linear programming, optimal solution, proposed algorithm, save time.

Introduction

It is impossible many times to fulfil the definite précised goals in given constraints for several queries in any organisation. Then these queries convert as one amongst maximising degree of attainment of those goals. Goal programming (GP) has an awareness which illuminate these queries of satisfying (probably differing) goals moreover seeing that feasible once a number of them have a top priority as compare to others. Fundamentally if there is only single goal then linear Programming technique is applicable, like maximising the profit or minimising the cost. And wherever the system might have more than single (probably differing) goals. For example, there might have a collection of goals in an Industry, such as stability of employment, excessive quality of your goods, maximisation of profit, minimising overtime or price, and so on. Thus, in these circumstances, different technology is desired which appears for a negotiated solution that carried on the relatively equal importance of every objective. So, one can conclude that GP is the renowned technique which helps to minimise the deviations from the goal assigned by the management. To find out an accurate goal for every objective, the process is like to first formulate an objective function for each objective, afterward as per the basic approach of GP, find solution that minimises the summation of deviations from its corresponding goals.

Significantly in GP the interest has been improved in the latest past, as has its concrete execution. In 1955, Charnes, Cooper along with Ferguson was first utilised this concept of GP, though Charnes and Cooper were the first who materialised the real name of GP in 1961. They found that GP is an extension of Linear Programming which originated and arrived in the same way. Afterward this attitude was expanded and amplified by their student along with other investigators, particularly Ijiri (1965); Jaaskelanen (1969); Lee and Clayton (1970); Lee (1972); Ignizio (1976), who pointed out that actual real-world problems invariably involve non-deterministic system for which a variety of conflicting inconsistent objectives exist.

Dauer and Krueger (1977) have expanded an Iterative Approach for GP. Kornbluth and Steuer (1981) examined GP through linear fractional criteria. Romero (1991) pursued the seminal works. Scniederjans (1995) has mentioned the GP methodology and applications. A discussion about Fuzzy GP approach for finding the results of Bi-Level programming problems has been made by Moitra and Pal (2002). Pramanik and Kumar (2006) applied Fuzzy GP tool to multi-level programming problems. In 2009, to work out on decentralised bi-level multi-objective programming problems, Baky derived Fuzzy GP algorithm. A latest-book by Jones and Tamiz (2010) givesa ample overview GP. In 2014, an Approximation algorithm has been expanded by Khobragade et al. (2014) which is designed for best possible solution to the LPP. Again in 2017 Alternative Approach has been designed by Birla et al. (2017) to support of answering Bi-Level Programming Problems. Putta and Khobragade (2019) extracted most favourable solution of Goal and Fractional Programming Problem. New Approximation Algorithm has been suggested by Dhote and Ban (2021), to find optimum solution of GP problem.

---

[a]thakaremonali@gmail.com; [b]girishdhote@gmail.com

In today's dynamic business setting, most of the time organisations have multiple conflicting objectives to realise. Not solely do corporations explore for profit and revenue maximisation or price diminution however produce other non-profit goals to cater to love social responsibilities, publicity, industrial and worker relations, etc. underneath such things, GP assumes utmost importance and is a strong quantitative technique capable of handling multiple call criteria. So, in this article a new algorithms and methodology is included and calculated to find a fundamental or prime solution for GP problem depended on maximising the profit also minimising the cost. This proposed method which is very easy to handle and simple to understand, also decreases quantity of iterations, save precious time and obtained optimum solutions. By this modified technique one can find the optimum solution with less or even equal numbers of iterations than established method and gives improved outcomes as compared to existing methods which was suggested previously. There are many inventions in LP, but GP is still somewhere needs more attention, so it is important to focus on some new techniques while studying GP. In this article small attempt has been perform to find improved optimum solution for GPP.

Artificial Variable

To discover the initial basic feasible solution, first place the given GPP towards its ideal format and afterwards insert a positive variable to the left side of all equations that requires most essential sorting basic variables. And the newly added positive variable is called artificial variables.

Charnes Penalty Method

A Charnes suggested that a very high penalty can be paid for introducing the artificial variables in the constraints of a stated example, by giving an extremely large negative charge (penalty) to the artificial variables of the problem. So the objective function can be represented like

$$Z = C_{ori}X_{ori} + 0.x_{sla} + 0.x_{sur} - Mx_{art}$$

Where (−M) is the huge cost, unspecified but sufficiently large, assign to each artificial variable.

The artificial variable has been purposely introduced to get an initial basic feasible solution and so one would like to get rid of these variables once the purpose has been achieved. Thus, the objective of huge penalty cost is to ensure that all artificial variables will be determined to zero as Z is optimised by using Simplex method. The method of solving a LPP by maximising Z where high penalty cost has been assigned to the artificial variables is known as the method of penalty or the big M method.

In this paper, we suggest modified Charnes Penalty Method and use it to solve GPP.

Proposed Modified Charnes Penalty Method for GPP

Charnes Penalty technique adds following stages to work out on GP problem.

Stage (1). Prefer min $\sum x_{ij}$, $x_{ij} \geq 0$ to enter vector.

Stage (2). Insert an artificial variable $A_1 \geq 0$ to the left side of all equation having no observable initial basic variables.

Stage (3). Prefer highest coefficient of decision variables.

i). For only one highest coefficient, the element containing to this row and column transfer in pivotal.

ii). And for more than one highest coefficient, tie breaking method is applicable.

Stage (4). As a minimum one artificial variable should present in the basis with zero value, then the existing optimum solution is deteriorated.

Stage (5). After skipping accompanying row and column, go on stage 3 for remaining elements moreover to get optimal solution or to get the indication for unbounded solution, replicate the same procedure.

Stage (6). After neglecting all rows and columns, optimal solution exists.

Solved Problem

Problem-1

TATA Motors and tyres produce vehicle parts. There are two Units for production. Unit 1 has the rate of production as 50 parts per hr. whereas it is 60 parts per hr. for unit 2. TATA Motors has made an agreement to deliver 1200 parts daily to Ambika Motors. Presently, the usual operation phase for all units is 8 hours. The fabrication in charge of the TATA Motors is aimed to find out the best every day functional hours for the both units in order to achieve the following goals:

i: Fabricate and deliver 1200 parts daily.

ii: Daily overtime operation hours should be limited to 2 or 3 hours.

iii: Reduce underutilisation of the usual daily operation hours of both units. Allot different loads on given productivity rate.

iv: Reduce the possible operation hours of both units for every day. Allot different loads of given rate of overtime. The expenditure of operation has supposed to be matching for both production units.

Do the formulation as GP model and solve.

Solution:

Assume that x_1 *as well as* x_2 represents production rate of unit 1 along with 2 correspondingly. Following equations shows, the constraints and goals of the problem:

$$Maximize\ \ Z = 50x_1 + 60x_2 \tag{1}$$

$$x_2 \leq 11, x_1 \leq 8$$

Formulation of equation (1) as GP model is:

$$Minimize\ \ Z = p_1 d_1^- + p_2 d_2^+ + p_3(5d_1^- + 6d_2^-) + p_4(6d_1^+ + 5d_2^+) \tag{2}$$

Subject to the constraints:

$$50x_1 + 60x_2 = 1200 \tag{3}$$

$$x_1 + d_1^- - d_1^+ = 8$$

$$x_2 + d_2^- - d_2^+ = 11$$

$$x_1, x_2, d_1^+, d_1^-, d_2^-, d_2^+ \geq 0$$

d_1^-, d_2^- are quantities that underachieve required objective.

d_1^+, d_2^+ are quantities that overachieve required target.

Since there does not exist columns of an identity matrix which can be assumed for the opening basis matrix.

So, we include requisite identity column to complete the identity column [1 0 0] as the new column.

Clearly, this amount to adding an artificial variable $A_1 \geq 0$ to first constraints.

So, we get initial basic feasible solution as

$$A_1 = 1200, d_1^- = 8, d_2^- = 11.$$

Hence, reformulation of given example as GP model is:

$$Minimize\ \ Z = p_1(-A_1) + p_2 d_2^+ + p_3(5d_1^- + 6d_2^-) + p_4(6d_1^+ + 5d_2^+) \tag{4}$$

Subject to the constraints:

$$50x_1 + 60x_2 + A_1 = 1200 \tag{5}$$

$$x_1 + d_1^- - d_1^+ = 8$$

$$x_2 + d_2^- - d_2^+ = 11$$

$$x_1, x_2, d_1^+, d_1^-, d_2^-, d_2^+, A_1 \geq 0$$

Table 27.1 Initial table

			0	0	$5p_3$	$6p_4$	$6p_3$	$(p_2 + 5p_4)$	$-p_1$
C_B	y_B	x_B	x_1	x_2	d_1^-	d_1^+	d_2^-	d_2^+	A_1
$-p_1$	A_1	1200	50	60	0	0	0	0	1
$5p_3$	d_1^-	8	1	0	1	−1	0	0	0
$6p_3$	d_2^-	11	0	1	0	0	1	−1	0
			↑						↓

As min $\sum x_{ij} = 51$

Therefore, the column vector x_1 jump to the next step as well as the column vector A_1 departs.

Table 27.2 Introdece x_1 and drop A_1

			0	0	$5p_3$	$6p_4$	$6p_3$	(p_2+5p_4)	$-p_1$
C_B	y_B	x_B	x_1	x_2	d_1^-	d_1^+	d_2^-	d_2^+	A_1
0	x_1	24	1	6/5	0	0	0	0	1/50
$5p_3$	d_1^-	−16	0	−6/5	1	−1	0	0	−1/50
$6p_3$	d_2^-	11	0	**1**	0	0	1	−1	0
				↑			↓		

As min $\sum x_{ij} = 1$
Therefore, the column vector x_2 jump to the next step as well as the column vector d_2^- departs.

Table 27.3 Introdece x_2 and drop d_2^-

			0	0	$5p_3$	$6p_4$	$6p_3$	(p_2+5p_4)	$-p_1$
C_B	y_B	x_B	x_1	x_2	d_1^-	d_1^+	d_2^-	d_2^+	A_1
0	x_1	54/5	1	0	0	0	−6/5	6/5	1/50
$5p_3$	d_1^-	−14/5	0	0	1	−1	6/5	−6/5	−1/50
0	x_2	11	0	1	0	0	1	−1	0

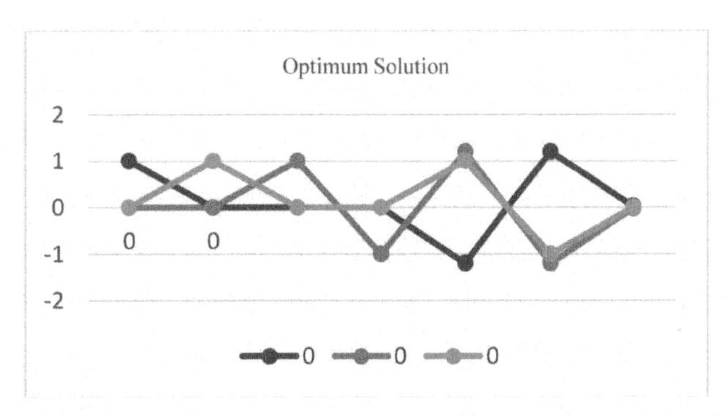

Figure 27.1 Reduction of operation hours for every day

Optimum solution is

$$x_1 = 54/5,\ x_2 = 11,\ d_2^- = d_3^- = d_4^- = 0 \tag{6}$$

Problem-2
Ali Chicken shop is occupied in making Butter Chicken and Chili Chicken, which gives a unit profit of Rs. 400 and Rs. 300, respectively. Shop needs uncooked material which has restricted delivery of 300 kg per month. Both Chicken requires uncooked material one kg per unit. Further, it is known that to prepare Butter Chicken shop require 2 hours while Chili Chicken needs one hour. The shop contains a usual preparation ability of 400 hours per month. The advertising department has assured that according to the existing conditions, the maximum number of Butter Chicken and Chili Chicken which can be sold each month is 150 kg and 350 kg, respectively. Following goals have been placed by the supervisor of the shop as per their significance priorities:

i: Ignore under consumption of common productive capacity.
ii: Retail the maximum number of Butter Chicken and Chili Chicken. But, since both chickens give up profit in the ratio 400:300, the supervisor decided to give the importance to the sale of the chicken in the same ratio.
iii: Minimise overtime of preparation capacity.

Do the Formulation to achieve required goal furthermore solve.

Solution:-

Assume x_1, x_2 represents profit of Butter Chicken as well as Chili Chicken respectively.

$$Minimize \quad Z = P_2 d_1^- + P_2(4d_2^- + 3d_3^-) + P_3 d_1^+ \quad (1)$$

Subject to the constraints:

$$x_1 + 3x_2 \le 300 \quad (2)$$

$$2x_1 + x_2 = 400$$

$$x_1 \le 150$$

$$x_2 \le 350$$

$$x_1, x_2 \ge 0$$

Here an artificial variable $A_1 \ge 0$ is introduced in second constraints.
And initial basic feasible solution is $A_1 = 400, d_1^- = 300, d_2^- = 150, d_3^- = 350$.
The above can be written as

$$Minimize \quad Z = P_2 d_1^- + P_2(4d_2^- + 3d_3^-) + P_3 d_1^+ - A_1 \quad (3)$$

Subject to the constraints:

$$x_1 + x_2 + d_1^- - d_1^+ = 300 \quad (4)$$

$$2x_1 + x_2 + A_1 = 400$$

$$x_1 + d_2^- = 150$$

$$x_2 + d_3^- = 350$$

$$x_1, x_2, d_1^+, d_1^-, d_2^-, d_3^-, A_1 \ge 0$$

Table 27.4 Initial table

			0	0	p_1	p_3	$4p_2$	$3p_2$	−1
C_B	y_B	x_B	x_1	x_2	d_1^-	d_1^+	d_2^-	d_3^-	A_1
p_1	d_1^-	300	1	**1**	1	-1	0	0	0
−1	A_1	400	2	1	0	0	0	0	1
$4p_2$	d_2^-	150	1	0	0	0	1	0	0
$3p_2$	d_3^-	350	0	1	0	0	0	1	0
				↑	↓				

As min $\sum x_{ij} = 3$
Therefore, the column vector x_2 jump to the next step as well as the column vector d_1^- departs.

Table 27.5 Introdece x_2 and drop d_1^-

			0	0	p_1	p_3	$4p_2$	$3p_2$	−1
C_B	y_B	x_B	x_1	x_2	d_1^-	d_1^+	d_2^-	d_3^-	A_1
0	x_2	300	1	1	1	−1	0	0	0
−1	A_1	100	**1**	0	−1	1	0	0	1
$4p_2$	d_2^-	150	1	0	0	0	1	0	0
$3p_2$	d_3^-	250	−1	0	−1	1	0	1	0
			↑						↓

As min $\sum x_{ij} = 2$
Therefore, the column vector x_1 jump to the next step as well as the column vector A_1 departs.

Table 27.6 Introdece x_1 and drop A_1

			0	0	p_1	p_3	$4p_2$	$3p_2$	−1
C_B	y_B	x_B	x_1	x_2	d_1^-	d_1^+	d_2^-	d_3^-	A_1
0	x_2	300	0	1	2	−2	0	0	−1
0	x_2	100	1	0	−1	1	0	0	1
$4p_2$	d_2^-	50	0	0	1	−1	1	0	−1
$3p_2$	d_3^-	350	0	0	−2	2	0	1	350

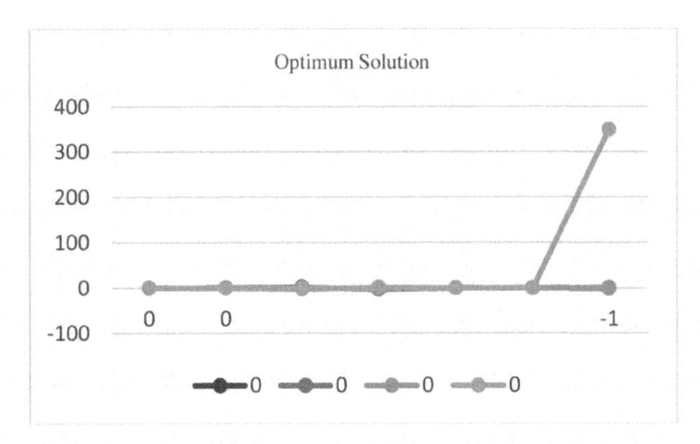

Figure 27.2 Reduction of over time

Optimum solution is

$$x_1 = 100,\ x_2 = 200,\ d_2^- = 50,\ d_3^- = 350 \tag{5}$$

Problem-3

A Cotton textile industry produces two types of fabrics, Denim and French terry. Denim gives actual profit of Rs. 80 per unit, along with that of French terry gives actual profit of Rs. 40 per unit. Industry has taken a decision to produce Rs. 900 in upcoming week. In addition, the administration wishes toward achieve sales amount for two fabrics correspondingly close to 17 and 15. Do the Formulation to achieve the goal and solve.

Solution:-

Assume that x_1 *and* x_2 are two types of fabrics, Denim and French terry correspondingly.
Then, the given problem is expressed as

$$Maximize\ Z = 80x_1 + 40x_2 \tag{1}$$

$$x_1 \le 17, x_2 \le 15 \text{ and } x_1 \ge 0, x_2 \ge 0$$

Formulation of (1) as GP model is:

$$Minimize\ Z = d_1^- + d_2^- - A_1 \tag{2}$$

Subject to the constraints:

$$80x_1 + 40x_2 + A_1 = 900 \tag{3}$$

$$x_1 + d_1^- = 17$$

$$x_2 + d_2^- = 15$$

$$x_1, x_2, d_1^-, d_2^-, A_1 \ge 0$$

Here we introduced an artificial variable $A_1 \ge 0$ in the first constraint.
So, initial basic feasible solution is $A_1 = 900, d_1^- = 17, d_2^- = 15$.

Table 27.7 Initial table

			0	0	1	1	1	−1
C_B	y_B	x_B	x_1	x_2	d_1^-	d_1^+	d_2^-	A_1
−1	A_1	900	80	40	0	0	0	1
1	d_1^-	17	1	0	1	−1	0	0
1	d_2^-	15	0	1	0	0	1	0
				↑				↓

As min $\sum x_{ij} = 40$

Therefore, the column vector x_2 jump in to the next step as well as the column vector A_1 departs.

Table 27.8 Introdece x_2 and drop A_1

			0	0	1	1	1	−1
C_B	y_B	x_B	x_1	x_2	d_1^-	d_1^+	d_2^-	A_1
0	x_2	45/2	2	1	0	0	0	1/40
1	d_1^-	17	1	0	1	−1	0	0
1	d_2^-	−15/2	−2	0	−1	1	1	−1/40
		↑			↓			

As min $\sum x_{ij} = 1$

Therefore, the column vector x_1 jump to the next step as well as the column vector d_1^- departs.

Table 27.9 Introdece x_1 and drop d_1^-

			0	0	1	1	1	−1
C_B	y_B	x_B	x_1	x_2	d_1^-	d_1^+	d_2^-	A_1
0	x_2	23/2	0	1	−2	2	0	1/40
0	x_1	17	1	0	1	−1	0	0
1	d_2^-	53/2	0	0	2	2	1	−1/40

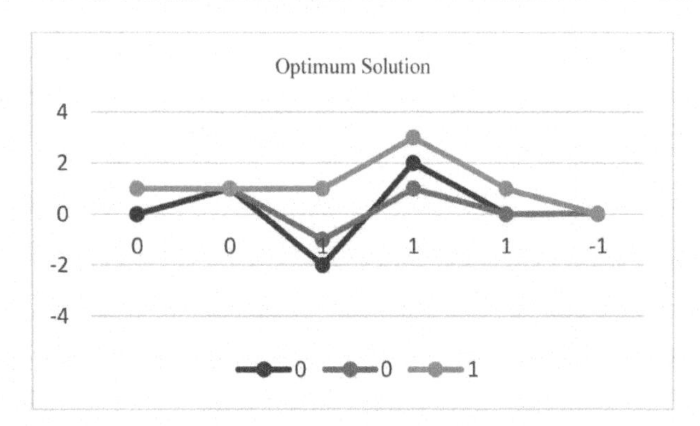

Figure 27.3 Increment of sale amount

Optimum solution

$$x_1 = 17,\ x_2 = 23/2,\ d_2^- = 53/2 \tag{4}$$

Problem-4

The following garments are available in two shops. The table present the production periods in minutes per unit for the two shops.

Garments	*Shop 1*	*Shop 2*
Sari	7	5
Shirts	8	4

The everyday production slices for both garments are 70 and 50 units, correspondingly. Each shop works 8 hours daily. Eventually, though not pleasing, it is compulsory to both shops to acquire the production quota. Do the Formulation to achieve required goal and solve.

Solution:-
Assume that x_1 *and* x_2 are clothes of shop 1 as well as 2 correspondingly.
Then, the problem is expressed as:

$$Minimize\ Z = \{ d_1^+, d_2^+, d_2^- \} \tag{1}$$

$$x_1 \le 70, x_2 \le 50 \text{ and } x_1 \ge 0, x_2 \ge 0$$

$$7x_1 + 5x_2 = 480$$

$$8x_1 + 4x_2 = 480$$

The formulation of the problem is:

$$Minimize\ Z = d_1^+ + d_2^+ + d_2^- - A_1 - A_2 \tag{2}$$

Subject to the constraints:

$$x_1 + d_1^- - d_1^+ = 70 \tag{3}$$

$$x_2 + d_2^- - d_2^+ = 50$$

$$7x_1 + 5x_2 + A_1 = 480$$

$$8x_1 + 4x_2 + A_2 = 480$$

$$x_1, x_2, d_1^+, d_1^-, d_2^+, d_2^-, A_1, A_2 \ge 0$$

Here we introduced an artificial variable $A_1 \ge 0$ in the second and third constraints.
And initial basic feasible solution is $d_1^- = 70, d_2^- = 50, A_1 = 480, A_2 = 480$

Table 27.10 Initial table

			0	0	0	0	1	1	−1	−1
C_B	y_B	x_B	x_1	x_2	d_1^-	d_1^+	d_2^-	d_2^+	A_1	A_2
0	d_1^-	70	1	0	1	−1	0	0	0	0
1	d_2^-	50	0	1	0	0	1	−1	0	0
−1	A_1	480	7	5	0	0	0	0	1	0
−1	A_2	480	8	4	0	0	0	0	0	1
				↑					↓	

As min $\sum x_{ij} = 10$
Therefore, the column vector x_2 jump to the next step as well as the column vector A_1 departs.

Table 27.11 Introdece x_2 and drop A_1

			0	0	0	0	1	1	−1	−1
C_B	y_B	x_B	x_1	x_2	d_1^-	d_1^+	d_2^-	d_2^+	A_1	A_2
0	d_1^-	70	1	0	1	−1	0	0	0	0
1	d_2^-	−46	−7/5	0	0	0	1	−1	−1	0
0	x_2	96	7/5	1	0	0	0	0	1/5	0
−1	A_2	96	**12/5**	0	0	0	0	0	−4/5	1
			↑							↓

As min $\sum x_{ij} = 17/5$
Therefore, the column vector x_1 jump in to the next step as well as the column vector A_2 departs.

Table 27.12 Introdece x_1 and drop A_1

			0	0	0	0	1	1	−1	−1
C_B	y_B	x_B	x_1	x_2	d_1^-	d_1^+	d_2^-	d_2^+	A_1	A_2
0	d_1^-	30	0	0	1	−1	0	0	4/5	−1
1	d_2^-	10	0	0	0	0	1	−1	−22/15	7/12
0	x_2	40	0	1	0	0	0	0	2/3	−7/12
0	x_1	40	1	0	0	0	0	0	−1/3	5/12

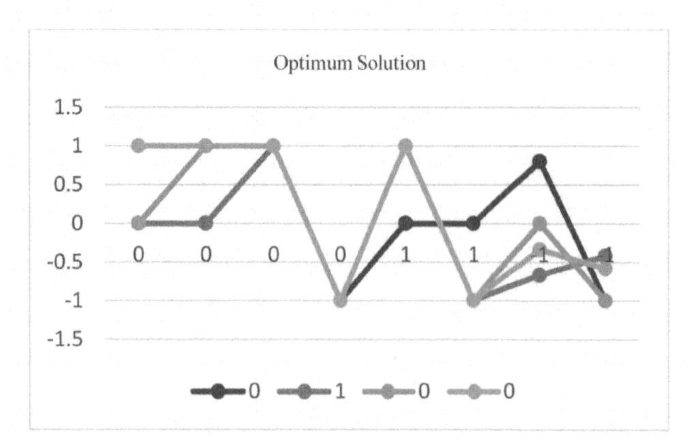

Figure 27.4 Increment of production quota

Optimum solution is

$$x_1 = 40,\ x_2 = 40,\ \ d_1^- = 30,\ d_2^- = 10 \tag{4}$$

Problem-5

Indian Life Insurance Company has two marketing officers Mr. Patil and Mr. Bhole, assigned for marketing of insurance policy. Mr. Patil can visit places at the rate of 40 per hour, with an exactness of 97%. Mr. Bhole visits daily 30 places per hrs with an exactness of 95%. Salary of Mr. Patil is Rs.5 per hrs, and that of Mr. Bhole is Rs. 4 per hrs. A fault done by any officer expenses Rs. 3 to the company.There are simply nine officers with Mr. Patil and eleven officers with Mr. Bhole presented in the company. The company desires to allot work to the available marketing officers so as to reduce the entire cost of the marketing. Do the formulation as a GP model and solve.

Solution:-

Assume that x_1 *and* x_2 denotes Mr. Patil and Mr. Bhole respectively
Then, the above problem is expressed as:

$$Minimize\ Z = 0x_1 + 0x_2 \tag{1}$$

$$4x_1 + 3x_2 \geq 25,\ x_1 \leq 9 \text{ and } x_2 \leq 11$$

The formulation of the problem is:

$$Minimize\ Z = P_1 d_1^- + P_2(d_2^- + d_3^-) + P_3 d_2^+ + P_4 d_3^+ \tag{2}$$

Subject to the constraints:

$$4x_1 + 3x_2 + d_1^- - d_1^+ + A_1 = 25 \tag{3}$$

$$x_1 + d_2^- - d_2^+ = 9$$

$$x_2 + d_3^- - d_3^+ = 11$$

$$x_1, x_2, d_1^-, d_2^-, d_3^-, d_1^+, d_2^+, d_3^+, A_1 \geq 0$$

Here we introduced artificial variable $A_1 \geq 0$ in the 1st constraint.
And initial basic feasible solution is $A_1 = 25,\ d_2^- = 9,\ d_3^- = 11$

Table 27.13 Initial table

			0	0	P_1	0	P_2	P_3	P_2	P_4	−1
C_B	y_B	x_B	x_1	x_2	d_1^-	d_1^+	d_2^-	d_2^+	d_3^-	d_3^+	A_1
−1	A_1	25	4	3	1	−1	0	0	0	0	1
P_2	d_2^-	9	1	0	0	0	1	−1	0	0	0
P_2	d_3^-	11	0	1	0	0	0	0	1	−1	0
				↑							↓

As min $\sum x_{ij} = 4$
Therefore, the column vector x_2 jump to the next step as well as the column vector A_1 departs.

Table 27.14 Introdece x_2 and drop A_1

			0	0	P_1	0	P_2	P_3	P_2	P_4	−1
C_B	y_B	x_B	x_1	x_2	d_1^-	d_1^+	d_2^-	d_2^+	d_3^-	d_3^+	A_1
0	x_2	25/3	4/3	1	1/3	−1/3	0	0	0	0	1/3
P_2	d_2^-	9	1	0	0	0	1	−1	0	0	0
P_2	d_3^-	8/3	−4/3	0	−1/3	1/3	0	0	1	−1	−1/3
			↑				↓				

As min $\sum x_{ij} = 1$
Therefore, the column vector x_1 jump to the next step as well as the column vector d_2^- departs.

Table 27.15 Introdece x_1 and drop d_2^-

			0	0	P_1	0	P_2	P_3	P_2	P_4	−1
C_B	y_B	x_B	x_1	x_2	d_1^-	d_1^+	d_2^-	d_2^+	d_3^-	d_3^+	A_1
0	x_2	−11/3	0	1	1/3	−1/3	−4/3	4/3	0	0	1/3
0	x_1	9	1	0	-	0	1	−1	0	0	0
P_2	d_3^-	44/3	0	0	−1/3	**1/3**	4/3	−4/3	1	−1	−1/3
						↑				↓	

As min $\sum x_{ij} = 0$
Therefore, the column vector d_1^+ jump to the next step as well as the column vector d_3^- departs.

Table 27.16 Introdece d_1^+ and drop d_3^-

			0	0	P_1	0	P_2	P_3	P_2	P_4	−1
C_B	y_B	x_B	x_1	x_2	d_1^-	d_1^+	d_2^-	d_2^+	d_3^-	d_3^+	A_1
0	x_2	11	0	1	0	0	0	0	1/3	−1/3	2/3
0	x_1	9	1	0	0	0	1	−1	0	0	0
0	d_1^+	44	0	0	−1	1	4	−4	3	−3	1

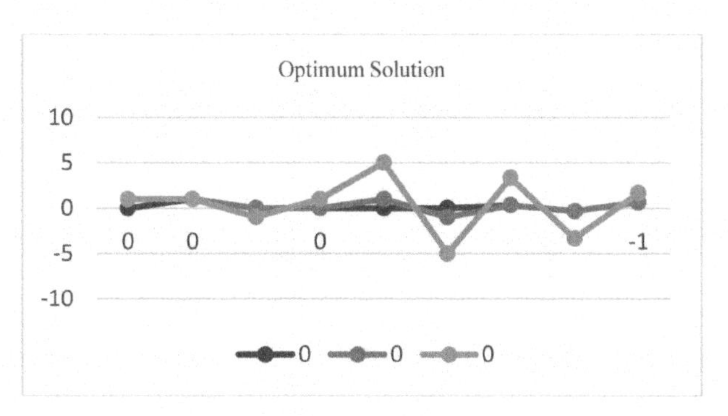

Figure 27.5 Reduction of cost of marketing

Optimum solution is

$$x_1 = 9, x_2 = 11, d_1^- = 44 \tag{4}$$

Conclusion

Modified Charnes Penalty Method has been advised in present article in order to explain GPP. As per the general observation, this method offered reduction of quantity of iterations, done calculations efficiently furthermore given optimum solutions. Therefore, the mentioned procedure is strongest mode which provides outcomes instantly.

References

Baburao, P. and Khobragade, N. W. (2019). Optimum solution of Goal and Fractional Programming Problem. Int. J. Manag. Technol. Eng. 9:142–152.

Birla, R., Agarwal, V. K., Khan, I. A., and Mishra, V. N. (2017). An Alternative Approach for Solving Bi-Level Programming Problems. Am. J. Oper. Res. 7:239–247.

Charnes, A., Cooper, W. W., Ferguson, R. (1955). Optimal estimation of executive compensation by linear programming. Manag. Sci. 1:138–151.

Charnes, A. and Cooper, W. 1961. Management Models and Industrial Applications of Linear Programming. New York (N y): John Wiley.

Dauer, J. P. and Krueger, R. J. (1977). An Iterative Approach to Goal Programming.Operat. Res.. Q. 28(3):671–681.

Dhote, M. G. and Ban, C. P. (2021). Optimum Integer Solution of Goal Programming Problem by Modified Gomory. Constraint Technique TURCOMAT. 12:3443–3452.

Ibrahim, A. B. (2009). Fuzzy goal programming algorithm for solving decentralised bi-level, multi- objective programming problems. Fuzzy Sets Syst. 160:2701–27–13.

Ignizio, J. P. 1976. Goal programming and extensions. Lexington, MA: Lexington Books.

Ijiri, Y. 1965. Management Goals and Accounting for Control. Amsterdam: North-Holland Publishing Company.

Jaaskelainen 1969. Linear Programming and Budgeting. New York: Petrocelli-Charter. Accounting and Mathematical Programming. New York (N Y): Helsinki School of Economics.

Jones, D. F. and Tamiz, M. 2010. Practical Goal Programming. New York (N Y): Springer Books.

Khobragade N. W., Vaidya, N. V., and Lamba, N. K. (2014). Approximation algorithm forOptimal solution to the linear programming problem. Int. J. Math. Operat. Res. 6(2):139–154.

Kornbluth, J. S. H. and Steuer, R. E. 1981. Goal programming with linear fractional criteria. Eur. J. Operat. Res. 8:58–65.

Lee, S. M. and Clayton, E. R. 1970. A Goal Programming Model for Academic Planning. Management Sciences American Meeting (11th), Los Angeles, California.

Lee, S. M 1972. Goal programming for decision analysis. Philadelphia: Auerbach Publishers.

Moitra, B. N. and Pal, B. B. 2002. A fuzzy goal programming approach for solving bi-level programming problems. In Lecture Notes in ArtificialIntelligence, ed. N. R. Pal, M. Sugeno, 2275(91–98). Berlin, Heidelberg: Springer.

Pramanik, S. and Kumar, R. T. (2006). Fuzzy goal programming approach to multi-level programming problems. Eur. J. Operat. Res. 176:1151–1166.

Romero, C. 1991. Handbook of critical issues in goal programming. Oxford: Pergamon Press.

Scniederjans, M. J. 1995. Goal programming methodology and applications. Boston: Kluwer publishers.

28 A review on implementation of artificial intelligence techniques with smart grid

Ankur Kumar Gupta and Rishi Kumar Singh*

Department of Electrical Engineering, Maulana Azad National Institute of Technology, Bhopal, M.P., India

Abstract

The existing power grid in today's scenario is facing a huge increase in demand. But it could not affect its performance in a wide scale. This is due to various technologies equipped in the system through various components present in distribution as well as generation side. These technologies include advanced communication system, metering system, data management system, the energy efficient system, smart management system, and cyber security of electrical systems. Artificial intelligence (AI) is used for these systems widely. It is used to handle large amount of data through cloud computing technologies and data base management system to support energy management process of different equipments in distribution. Some of the examples in electricity sector are smart grids (SG), advance load forecasting, analysis of fault, protection of equipment etc. With these new advance systems the operation of grid is enhanced. This paper gives an inclusive review of implementation of Artificial Intelligence technologies to improve performance of the system and support all the associated parameters of power systems.

Keywords: Artificial intelligence, distributed grid Intelligent energy management, market model for power, Smart grid, smart storage system.

Introduction

The increase power demand will exhaust the conventional sources therefore these resources can be replaced by renewable sources of energy. The renewable energy sources are adopted by smart grid in new existing power system. According to ICT techniques, handling the consumer/retailer data may reconstruct the existing grid Johannesen et al. (2019). According to power market scenario the role of Artificial intelligence is to analyse all kind of incentives made by private agencies, monitoring the automation technologies adopted by the companies and monitoring the carbon emissions problem. In current scenario, the energy market sector is becoming more complex, so to simplify the system smart tools are used to manage the whole process more effectively and accurately. Artificial Intelligence uses different machine learning algorithms which after implementation solve the system problems and helps in load forecasting Neves et al. (2018). Earlier due to lack of advancement in technologies many operations were executed in offline mode.

Secondly, the optimization of load using Artificial Neural Network (ANN) leads to reduction in cost. The management of micro grids can be done through genetic algorithms (Ahmad and Chen, 2018). The limitation of traditional grid system leads to their failure. This can be overcome by AI application in grid. Cyber-attacks could also be reduced using this approach (Utkarsh et al., 2016). This paper discusses the importance of AI and its usage technique in smart grids in energy generation and storage sector of power systems.

Integration of Artificial Intelligence

Conventional power systems already use mathematical algorithm approaches and schematics for observation and control for decades, especially in well-structured communication and monitoring techniques. To construct and implement a model additional assumptions and computations are required. This could be done by deep learning (DL) methods. Because DL is possible to extract compact and appropriate data that is accountable for substantial grid operation. The deep reinforcement learning (DRL) improves the decision-making process by utilizing the data gathered rather than merely believing on current technologies.

Load forecasting and fault detection are useful for grid power analysis as it shows the relationships between the inputs and outputs. DL as an algorithmic technique, has achieved remarkable outcomes in the several sectors of electricity grids due to deep neural networks (DNN) exceptional feature of engineering and representation ability. It is a powerful tool that can be used to retrieve nonlinear features from the data. Due to DNNs outstanding value design and presentation capability, DL has achieved remarkable results in several sectors of power grids as an analytical technique. System failure, efficiency, accurate and efficient fault detection assisting in isolation and rapid network recovery can be done with AI. Figure 28.1 represents a block diagram of Artificial Intelligence framework. Figure 28.2 shows the application of AI in various sectors of power systems.

Future Topology

Smart grid is used to overcome the problems in existing electrical grid infrastructure by using modern tools such as intelligent control techniques (ICT) (Ford et al., 2014). It is an existing energy network that can cost-effectively combine all users' actions to ensure a low-loss, economically and supply security (Vaccaro and Zobaa, 2013). A focus on eight areas for smart grid standardisation has been presented by Asare-Bediako et al. (2013). The intelligence creates a grid that monitors and controls devices on the consumer/distribution side and responds to them (Ma et al., 2013). The intelligent a technology in their first layer includes the smart energy meter (SEM), intelligent based inverters, and EV chargers, and also governs the energies in a home automation system. With the support of connected phones or digital switches, the second layer completes community-level goal such as group congestion traffic, exchange of information and grid stability improvement. Grid intelligence developed at the system level observes and controls the devices throughout the distributed side as per its response to the information. Figure 28.3 shows various stages in power grid and topology of AI usage in it.

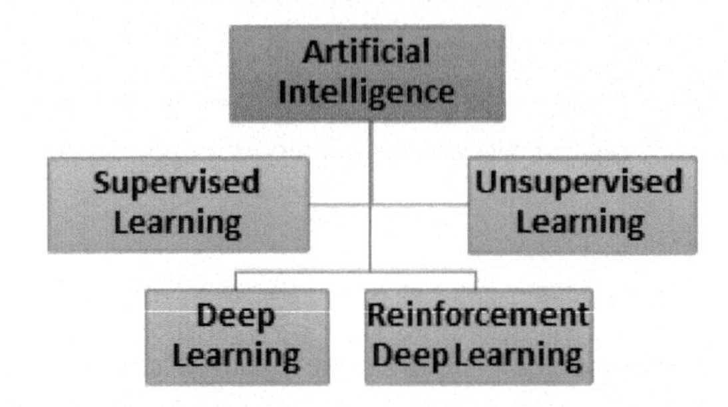

Figure 28.1 Block diagram of AI framework

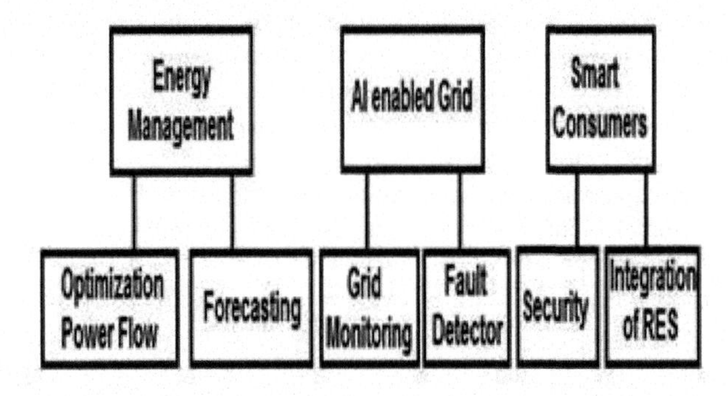

Figure 28.2 Applications of AI techniques used in grid

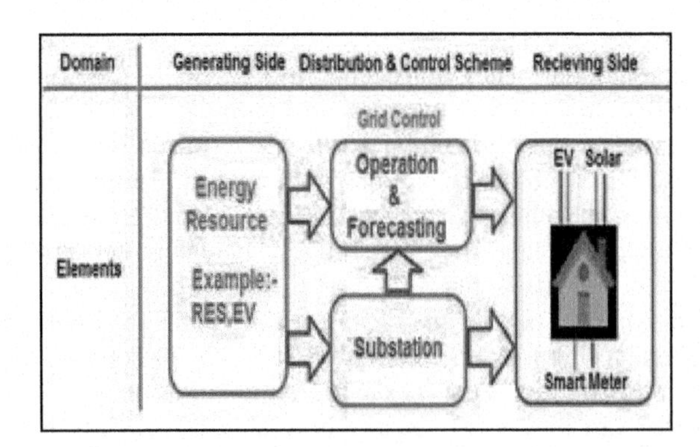

Figure 28.3 Topology of AI in grid

In smart grid, due to bidirectional flow of electricity the consumers or enablers can utilise it or can share it with the local grids through advance metering techniques (Colson and Nehrir, 2011).The local consumers can easily trade with local body and can decide their tariffs without investing in RES units (Darab et al., 2019). With Ai along with algorithms helps in forecasting the demand of consumers with accuracy and detection of fault as shown in Figure 28.2. Energy meter optimisation strategies based on different sorts of load situations are discussed by Blake and Sullivan (2018). Management of distributed grid can help in following fields of systems: energy management sector, monitoring and controlling, and fault detection. Controlling voltage through online mode is one of the important aspect (Chen et al., 2009). Most advanced techniques, decentralised scheme, intelligent, algorithms frameworks have been explained by the researcher.

Information gathering and control techniques were handled by centralised scheme controllers (Al-Alawi et al., 2007), which increase the central controller's workload (communication and calculation), putting the system at danger. Many scholars recommended decentralised control mechanisms to safeguard the system and control activities to collect local information to solve this challenge. Many power sectors previously uses the supervisory control and data acquisition system control energy resources, but these systems were not feasible because of security and slow down the operations (Foruzan et al., 2018). Therefore, all the associated systems became low productive due to involvement of human interfacing operations and hence interconnection of grid becomes more complex and a high precise network is required to collect the processed data.AI and its algorithms based technologies are used to store the data from the security point of view (Duan et al., 2019). This paper explained the process of AI techniques used in distribution grid operators in energy market for trading of energy at local model.

Integration and Implementation of Renewable Energy Sources

Because of global warming scenario, there is a shift towards renewable energy sources. These sources utilised in the power sector use certain technologies for integration with grid. An operator of power systems thus faces the challenge of electrical power quality and reliability. Due to awareness towards AI, other operations adopt base value of automation technique in automation as well as control strategy. Implementation of AI technologies to identify the occurrence of fault is very beneficial. AI technologies have been implemented to all types of renewable energy sources for analysing and distribution in various sectors of power systems. A survey concluded exponential increases in supply and demand management. An efficient supply system requires combination of renewable energy sources (RES) into conventional system. Figure 28.4 shows total generation year wise from 2009–10 to 2021–22. Figure 28.5 shows generation through each source until 30-11-2021. As per the graph, contribution of hydro sector at 1190 GW, solar (green) consumed at 586 GW. A literature survey on implementation of AI techniques in renewable energy sources has been provided in Table 28.1.

Energy market terms has been finalised by all power sector regulation under direction of CERC. Power forecasting can be done by Artificial Intelligence algorithms. Load forecasting can be monitored by neural

Table 28.1 Literature survey

Reference	*Objective*	*Method*	*Limitation*
Johannesen et. al. (2019)	Load forecasting	linear regression algorithms	load demand is not considered
Neves et al. (2018)	Isolated microgrid	GA	PV not considered
Ahmad and Chen (2018)	Load forecasting	linear regression model	long-term forecasting
Utkarsh et al. (2016)	Reduction of Active Power	Genetic Algorithms	issues regarding security
Ford et al. (2014)	Energy fraud detection	ANN	Non-technical losses
Vaccaro and Zobaa (2013)	Voltage regulation	ANN	fast-switching devices
Asare-Bediako et al. (2013)	Load forecasting	ANN	demand side management
Ma et al. (2013)	Voltage profile improvement	GA	load demand is not considered
Colson and Nehrir (2011)	Micro-grid	Multi-agent system	agent algorithm
Darab et al. (2019)	Fault location detection	ANN	reduction reliability
Blake and Sullivan (2018)	Power forecasting	ANN	charging/discharging
Chen et al. (2009).	Energy market	ANN	Issue of operation
Al-Alawi et al. (2007)	Emission of fuel	ANN	DER not mentioned

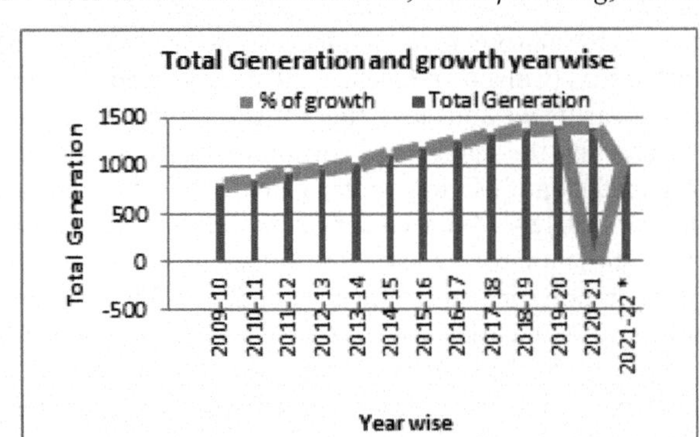

Figure 28.4 Current statics of total generation of country throughout the year since 2009–10 to 2021–22

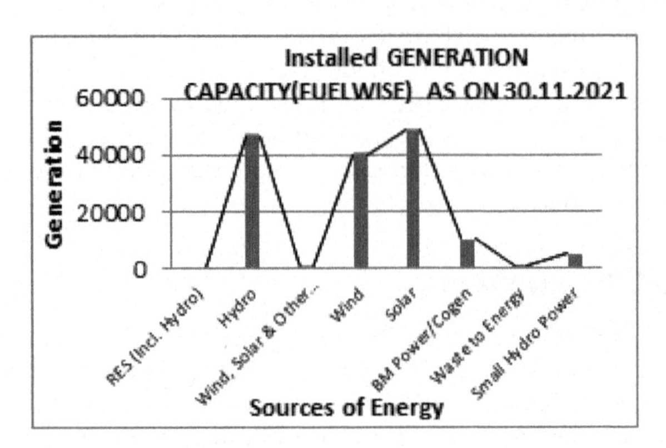

Figure 28.5 Installed capacity fuel wise as 30.11.2021

networks using genetic algorithms. The transformation of conventional power system into micro energy oriented with scheme of centralization can be done through AI. Optimization of power flow in PV system can be done easily. Traditional grid has analysed less data resources as compared to smart grid. The components in SG can do the transformation and operation at the same time while handling the different strategies for operation of distribution grid. The RES supplies to all the components for grid reliability. The existing grid which is a centralised system will provide policies for Smart grid system. SG also provides supports to reduce the burden on the low voltage operated grid using Point of Common Coupling (PCC). SGs along with artificial neural network are used to forecast various parameters. The algorithm used helps in real time distribution in power systems. Energy management scheme (EMS) is totally based on an improved communication technology is aided by distributed grid intelligence. An intelligent, cooperative design can benefit from the advantages from energy resources and services. Optimisation process, the setup of additional resources introduced to the system, and the detection and recovery from irregularities could all be aided by intelligent algorithms. The addition of distributed generation units towards the smart grid architecture provides new aspects to the smart grid architecture while, previously, the grid has acted as a sink for generations in most parts of the globe and has limited ability to accept new resource penetration.

Integration of Energy Storage System

Energy storage system (ESS) is playing a vital role in field of smart grid along with future prospective planning. These ESS act as a backup to usage of renewable sources of energy, assuring that clients have electricity access at all times. They assist with the local administration of a DC micro grid by increasing performance and reducing expenses. It also assists in reducing peak energy demand on system. ESS are executed during off-peak time hours, so that it helps to maintain homeostasis and delivers it to clients at peak times, reducing the pressure on the main power grid and boosting economic rewards. Batteries are the backbone of alternative fuel vehicles and devices, enabling us to improve the environment while remaining connected. The sizable implementation of ESS in the power grid provides power to 60% of the population by 2030, reducing carbon emissions in the electricity sector and including transport about 30%. Traditional systems work only on data and energy storage. The ESS technique is an important factor which can reshape the current grid's structure and operation. It can superior energy to all system components at different levels, guaranteeing stability of the grid. Predicting voltage and frequency is indeed very helpful in the SG concept as it ensures grid reliability. A real-time distribution strategy AI based for addressing energy demands by trying to charge and drain the ESS for user with worldwide Energy storage system is attracting the focus of researchers.

Energy Management System

The principal goal of power system is to ensure a stable and sustainable supply. Managing unpredictable aspects, coordinating numerous resources, and developing advanced algorithms are all required for smooth operation of the system. The aim of Energy Management Scheme is to integrate the numerous distributed energy resources (DER) in order to accomplish various operational goals. A unique characteristic of DER is extreme variability and low predictability which makes this aim difficult to achieve. Furthermore, to fully leverage their benefits, such as smoothing output, volatility, and stability along with improving profitability, the ESS requires real-time forecasting scheduling technique. To begin, create a smart load algorithm that takes temperature, time, pricing, and peak demand into account. Following that, artificial neural networks trained on actual data are being used to approximate the suggested adaptive model approach. To estimate the home actual power, a neural artificial network is implemented into the EMS. In a competitive electrical economy with several consumers and a service provider, different contractors may own distributed energy resources connected to distribution networks, and their goals may vary. Individuals and communities want to attain their own goals, such as higher revenue or assuring electric grid, without concern for others. Concerns for privacy and security make it impossible to access the precise details of different clients, resulting in search of solutions for model-based energy management systems. Taking into account the MG energy market Q-learning algorithms can be used that offer multi-agent-based energy EMS topologies for microgrids (Foruzan et al., 2018). Each stakeholder introduces energy storage and load consumers.

The assessment of the electrical sector with several actors lends itself well to multi-agent. When many microgrids are linked to a distribution network a Distributed generation technique based on artificial intelligence was proposed (Duan et al., 2019). The RL technique was used to optimise selling value on an ongoing bidding basis in order to make the most money from supplying electricity while lowering the distribution network operator's demand-side intensity ratio (DSO). The DNN approach would be used to address the link of hourly retail prices and global distribution costs between DSOs and interconnecting mini-grids. As per current scenario, the function of agent is to first produce the price relative data and then proceed to DNN that feeds reverse to particular agents. The transferred power is used to compute the instant reward for current actions.

Figure 28.6 represents an energy demand forecasting technique that uses DSO model. Every micro grid administrator strives to reduce its operational expenses under provisional pricing, resulting in the micro grid economic load dispatch (ED) model. Power dissipation in distribution networks are produced by voltage magnitude of bus and reactive power source. Volt-VAR optimisation (VVO) aims to improve operating effectiveness of system by regulating bus voltage and lowering power outage. VVO is a MINLP issue having integer variables with non - linear load factors, which is computationally difficult. Despite the fact that relaxation techniques help to simplify the problem, the solution procedure remains inefficient. In Bose (2017) multi-agent Bayesian Neural is used to perform VVO in unbalanced distributed system in order to reduce computing cost and accomplish real-time optimisation. The agents here attempt to optimise the settings of smart stabilisers, voltage regulators, and shunt capacitors in a coordinated manner to manage the voltage levels and eliminated voltage drop at the same time.

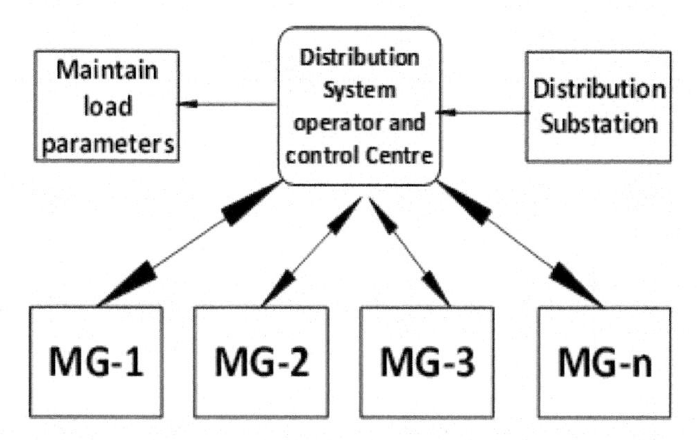

Figure 28.6 Energy demand forecasting through DSO model

Demand response (DR) for EMS is applicable to programs that motivate end-users and decrease energy requirements within immediate term in response to a pricing mechanism on an ad hoc basis energy. The grid power administrator may begin a trade or a catalyst. Due to economic programming, consumers can buy fixed or time-varying cash rewards and incentives for reducing their energy use at rush hours and eventualities. In the smart grid, a scheme for incentive-based demand systems and practices, can enable the load management operator to generate specific consumption encroachments where customer's privacy is protected when load management settlements are made. The approach safeguards private information by segregating real identification and fine-grained billing data, i.e., the DR must focus on learning one of its two and cannot conflate the two together. From a statistical strategy based on linear equations to a computer method capabilities forecasting model, there are many techniques for forecasting electricity price. End-users change their electricity consumption patterns above their trends in conjunction with energy prices over the time span specified by the utility, or even in order to get monetary support to risk voltage stability associated to high demand. An hour-ahead DR technique implemented to combine learning technique as well as neural network can minimise the unpredictability in forecasting electricity tariff with respect to stake holders (Chen et al., 2015). AI approaches may be used to simulate load and demand predictions in the context of subscribers and utility datasets.

Control of Operation of Power Grid

A Microgrid improves electricity dependability, durability, efficiency, and cost-savings by being a controlled entity with inverter-interfaced DERs and a varied customer base. Layered control architecture of the grid is shown in Figure 28.7.

In the secondary control, a microgrid regulator is installed to reduce frequency profile and voltage variations caused by the main control. Throughout there is a central controller located at the point of connection, this control scheme is responsible for monitoring all the microgrids interconnected to the transmission network. At this level, the power management layer is responsible for managing the whole system's economic dispatch. Hierarchical control provides a smoother transition from grid-connected mode to transient stability mode with really no disruptions in load supply and underlies electric capability. Generally, MG control approaches have been model-based, involving a thorough understanding of system interplay. Two problems are mentioned in this case. They are as follows:

i. There is a Loss of Comprehensibility

A DNN scheme is programmed for estimating the value that indicates the power systems dynamics Chen et al. (2015). For example: While training, the agent functions act as a black box with a control scheme. The system will not available throughout this phase, where the intakes, variables, and outcomes are believed to have a relationship.

ii. Exploratory Problem

On the one side, the trained agent must make the final choice given the existing strategy, it is achievable; nonetheless, there really is no guarantee of managerial consistency. The analysis and implementations of DRL are currently supported by machine learning, such as probabilities and statistical concept.

A predefined distribution governs the dataset being used to educate an agent. Such that, if trained agent is used to generate control commands, the unidentified inputs should lie under or be quite near to the same retraining data distribution, due to the varying techniques, otherwise the critical factor would be

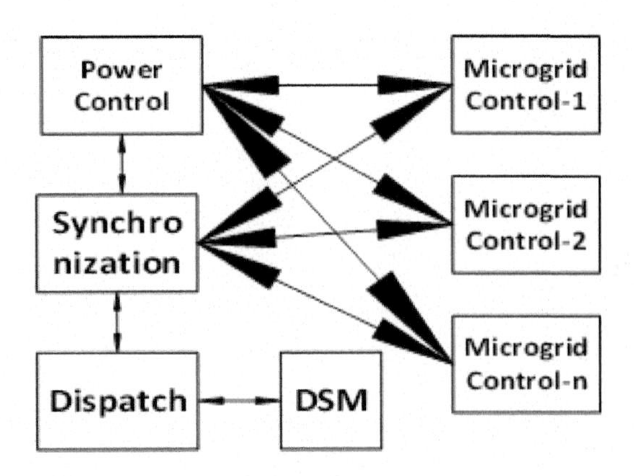

Figure 28.7 Layered control architecture of grid

threatened. This assumptions are not worthy for control problems since big disruptions are unavoidable. In the event of a major breakdown, the agent might issue a regulated order that creates unstable issues. Since the networked micro grids (NMGs) can support one other by transferring energy, they can also give extra dependability and resilience benefits. As a result, the investment in reserves is reduced. Linkage of multiple MGs also allows for involvement in energy and associated service markets. When viewing the NMGs as a whole, the minimum capacity requirements are decreased.

Risks Associated with Networking of Microgrids

There are various risks associated with microgrid networking, as well as several risks induced by microgrid communication networks. All these risks are shown in Figure 28.8.

Within NMGs, one is the collaboration of diverse owners. Individual micro grids inside NMGs may be owned by various parties, such as residential, powers. Various organisations have multiple objectives so they attempt to maximise their profits without looking to others advantages or overall system's performance. NMG efficiency is degraded due to instability problems. A network operating system (NMG) is a technological system with different participants, a competitive market, and insufficient system intelligence gathering. We believe that use of multi-agent DRL and use of DL are potential answers to the problems mentioned. Every Micro Grid can be act as an agent acting capable of communicating with other entities and improving its own intelligence. All techniques involve proper understanding of the process, which is difficult to gain in practical mode however; an NMG is managed by communications network. The communication between stations results in cyber-attack. With above scenario we conclude that NMG is just a combination of cyber cum physical network with the features of multi-agent, market mode as well as inadequate exchange of information. The SG normally falls into a certain distribution under normal circumstances, which may be recorded through the use of DNNs (Omitaomu and Niu, 2021). The gathered genuine data can be utilised to train DNNs with methods. And, if a failure occurs, the trained artificial neural network can identify anomalous operational conditions. The SG normally falls into a certain distribution conditions under normal circumstances, which may be recorded through the use of DNNs. The gathered genuine data can be utilised to train DNNs.

Optimisation of Grid

The best reconfigured network model suitable for different goals as well as for different objectives is to be found out using optimisation techniques.

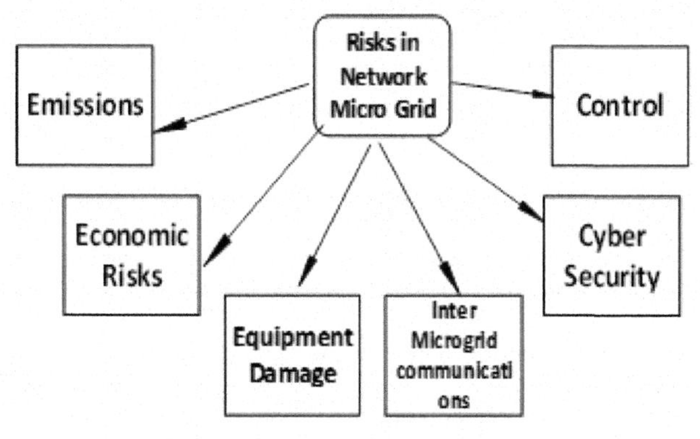

Figure 28.8 Risk associated with grid

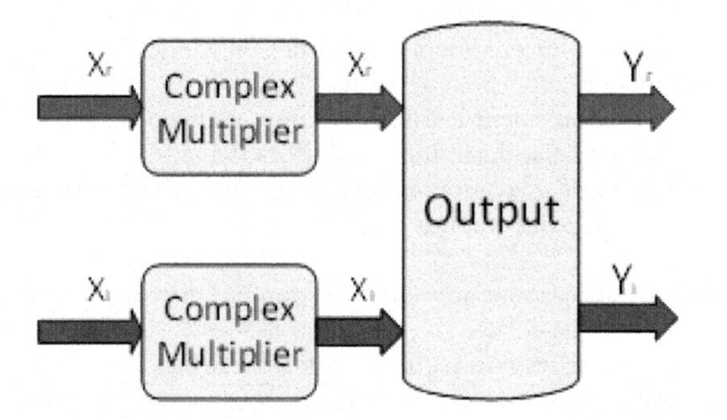

Figure 28.9 Reconfigurable complex twiddle factor multipliers

DRL is well suited for solving the technical and official protocol (TOP), which could be separated into two sub-problems:

- upper-level microgrid
- lower-level microgrid

A unique method for creating moderate FFT twiddle factor multipliers is shown in Figure 28.9. Initially construct cost-effective trigonometric formulations for twiddle factor coefficients. Using the exchange of mathematical formulations and Computer Science and Designs commons sub expressions, we are enabled to generate FFT twiddle factor multipliers with large coefficient vocabulary size at cheaper capital costs than prior approaches published in the literature.The proposed method will expand the range of low-complexity FFT processors' applications in digital signal processing. This FFTs are used to provide the momentum needed to keep the frequency of microgrid stable as ESS are critical island mode for MGs.

Conclusion

The use of Artificial Intelligence in power grids has been seen in this paper. The different topologies of Artificial Intelligence Techniques for RES integration with gridand in energy storage and management have been seen. AI implementations benefit the power grids. AI approach uses suitable algorithm to utilise data. Modern power systems have a high level of uncertainty, a lot of nonlinearity, and a lot of data which can be used for forecasting therefore AI techniques are appropriate for them. This research provides an outline of AI applications in modern systems, such as predicting and fault detection, autonomous power network management, and demand - side management. The difficulties with micro grids are examined, as well as potential AI solutions are suggested. Artificial Intelligence methods can be helpful in minimising power losses which further improves power quality and reliability of the system making the grid smart. Artificial intelligence can help with better and more automated distribution resource management. Cyber security, machine learning, information theory, or intelligence identification techniques are also discussed which is required for security of power grids by cyber-attacks.

References

Ahmad, T. and Chen, H. (2018). Utility companies strategy for short-term energy demand forecasting using machine learning based models. Sustain. Cities Soc. 39:401–417.

Al-Alawi, A., Al-Alawi, S. and Islam, S. (2007). Predictive control of an integrated PV-diesel water and power supply system using an artificial neural network. Renew. Energy 32:1426–1439.

Asare-Bediako, B., Kling, W. L., and Ribeiro, P. F. (2013). Day-ahead residential load forecasting with artificial neural networks using smart meter data. In Proceedings of the IEEE grenoble conference, grenoble, (16–20 June 2013), France.

Blake, S. T. and Sullivan, T. J. D. (2018). Optimization of distributed energy resources in an industrial microgrid. Proc. CIRP. 67:104–109.

Bose, B. K. (2017). Artificial intelligence techniques in smart grid and renewable energy systems—some example applications. Proc. IEEE 105(11):2262–2273. doi: 10.1109/JPROC.2017.2756596.

Chen, C., Duan, S., Cai, T., Liu, B., and Yin, J. (2009). Energy trading model for optimal microgrid scheduling based on genetic algorithm. In Proceedings of the IEEE international power electronics and motion control conference, (17–20 May 2009), Wuhan, China.

Chen, C., Wang, J., Qiu, F., and Zhao, D. (2015). Resilient distribution system by microgrids formation after natural disasters. IEEE Trans. Smart Grid 7(2):958–966.

Colson, C. and Nehrir, M. H. (2011). Algorithms for distributed decision-making for multi-agent microgrid power management. In Proceedings of the IEEE power and energy society general meeting, (24–28 July 2011), Detroit, MI, USA.

Darab, C., Tarnovan, R., Turcu, A., and Martineac, C. (2019). Artificial intelligence techniques for fault location and detection in distributed generation power systems. of the IEEE International Conference on Modern power systems (MPS), (21–23 May 2019), ClujNapoca, Romania.

Duan, J., Yi, Z., Shi, D., Lin, C., Lu, X., and Wang, Z. (2019). Reinforcement-learning-basedoptimal control of hybrid energy storage systems in hybrid AC–DC microgrids. IEEE Trans. Industr. Inform. 15(9):5355–5364.

Ford, V., Siraj, A., and Eberle, W. (2014). smart grid energy fraud detection using artificial neural networks. In Proceedings of the IEEE symposium on computational intelligence applications in smart grid (CIASG), (9–12 December 2014) Orlando, FL, USA.

Foruzan, E., Soh, L. K., and Asgarpoor, S. (2018). Reinforcement learning approach for optimal distributed energy management in a microgrid. IEEE Trans. Power Syst. 33(5):5749–5758.

Johannesen, N. J., Kolhe, M., and Goodwin, M. (2019). Relative evaluation of regression tools for urban area electrical energy demand forecasting. J. Clean. Prod. 218:555–564.

Ma, H., Chan, K. W., and Liu, M. (2013). An intelligent control scheme to support voltage of smart power systems. IEEE Trans. Ind. Inform. 9:1405–1414.

Neves, D., Pina, A., and Silva, C. A. (2018). Comparison of different demand response optimization goals on an isolated microgrid. sustain. Energy Technol. Assess. 30:209–215.

Omitaomu, O. A. and Niu, H. (2021). Artificial intelligence techniques in smart grid: a survey. United States: N. Web. doi:10.3390/smartcities4020029.

Utkarsh, K., Trivedi, A., Srinivasan, D., and Reindl, T. (2016). A consensus based distributed computational intelligence technique for real-time optimal control in smart distribution grids. IEEE Trans. Emerging. Top. Computer. Intell. 1:51–60.

Vaccaro, A. and Zobaa, A. (2013). Voltage regulation in active networks by distributed and cooperative meta-heuristic optimisers. Electr. Power Syst. Res. 99:9–17.

29 Smart micro-grid model to generated renewable energy based on embedded intelligent and FPGA

Ihab Al-Janabi[1] *and Samaher Al-Janabi*[2,a]

[1]Faculty of Electrical Engineering, Imam Reza International University, Mashhad, Iran

[2]Department of Computer Science, Faculty of Science for Women University of Babylon, Babylon, Iraq

Abstract

The problem of producing electrical energy from environmentally friendly sources with high efficiency and low cost is one of the most important challenges in this field. Therefore, to avoid those limitations need efficiency technique to prediction max energy from the solar energy and find useful knowledge help the country to apply or not of this type of project. The important of this paper in real life can summarisation as: economic and political independence: renewable energy is good for maintaining the local economy, as reliance on imported fossil fuels leads the country to be subject to the economic and political goals of the supplying country. As for renewable energy represented in wind, sun, water, and organic materials, it exists all over the world. On the other hand, renewable energy needs more labour compared to other energy sources that rely mostly on technology, where there will be workers to install solar panels, and technicians to maintain Wind farms, and other jobs that increase employment. Low prices: renewable energy is witnessing a continuous decrease in costs, despite the progress made in its development, as the equipment used in it has become more efficient, and technology and engineering work has also become more advanced in this field, unlike gas, fossil fuels, and other energy sources that despite its advantages are that prices fluctuate periodically. Improving public health: coal and natural gas plants lead to air and water pollution, which leads to many health problems. Such as breathing disorders, nervous problems, heart attacks, cancer, premature death, and other serious problems, and it is noteworthy that the majority of these negative health effects, resulting from water and air pollution, are not caused by the use of renewable energy technologies, as wind, solar, and energy systems Hydroelectricity all generate electricity without any emissions causing air pollution, although some types of renewable energy can cause pollution; Such as geothermal energy systems and biomass, but the total polluting emissions in them are generally much less than the total emissions from power plants that use coal and natural gas. Inexhaustibility: renewable energy is inexhaustible, compared to other energy sources, such as coal, gas, and oil, and this means that they are always available, such as: the sun, which produces energy, and falls within the natural cycles, and this makes renewable energy an essential element in a sustainable energy system that is capable of development and development without risking, or harming future generations.

Keywords: Embedded intelligent system, FPGA, LEIM, NEIM, renewable energy, solar plant.

Introduction

The exponential growth of technology in the world in many different fields increases the need for energy. Energy sources are classified into three categories (fossil fuels, nuclear resources, and renewable sources); most energy sources today make the world face a major turning point represented by climate change. Most Energy sources threaten the environment and human health because it produces toxic substances, generate harmful emissions, and requires using of huge quantity of water and some others are limited, their price is variable All of these reasons and many others lead to needing to reduce our consumption of these carbon-intensive fuels and avoid the consequences of environmental hazards by relying on the use of renewable energy sources (low-carbon and environmentally friendly) such as solar energy, wind, water, bioenergy, and geothermal energy Al-Janabi and Alkaim (2022).

The Basic Concepts of This Work

This section presents the main theoretical background related to the problem statement that is displayed in chapter one

A. Renewable Energy Sources (RES)

RES are friendly environment energy (alternative energy) Al-Janabi, et al. (2021) that reduces the harmful impacts on the environment (Hossny, et al. 2020; Kabir, et al. 2018; and Kaur and Buttar 2019). This

[a]samaher@itnet.uobabylon.edu.iq

concept of energy is linked to the energy that is obtained from natural sources that produce enormous amounts of energy and is capable of regenerating naturally; the resources of friendly environment energy are driven by the wind, hydropower, ocean waves, biomass from photosynthesis, and direct solar energy; friendly environment energy has many advantages where it considers non-polluting, sustainable, (one-time installation), economic, ubiquitous, safe and it offers a wide variety of options and also there are drawbacks to some of the sources (Maybe more costly or its effects by some environmental influences); These sources can using the devices such as sensors to facilitate the dealing with it [Al-Janabi et al. 2020;]. Figure 29.1 shown main types of Renewable energy.

B. Internet of Things (IoTs)

IOTs has emerged as enormous technology that finds new fields that have a significant impact on our lives. The main objective of the IoT is to provide an environment that enables the objects (such as sensors) and living beings to coordinate together to collect and share information [Al-Janabi et al. 2021]. IoT as a combination of

- Information Technology.
- Operational Technology.

Where to provide the IoT functionality need to the following elements (identity for each object, sensors or actuators, Communications Devices, Compute Devices, Services IoT) as shown in Figure 29.2.

The main advantages of the IoT in the real world are improved data collection, improved customer interaction, reduced waste, and Technology optimisation. Also, there are disadvantages of using the IoT such as Flexibility, Privacy, and Complexity. IoT uses sensors and actuators to interact and control the environment to achieve the specific purpose; The use of sensors and devices generate a huge amount of data about the specific features in the environment [Mathew, et al. 2021], these data must process and discover a useful pattern from using it to modify the environment, the discovering of data pattern can perform by intelligent data analysis.

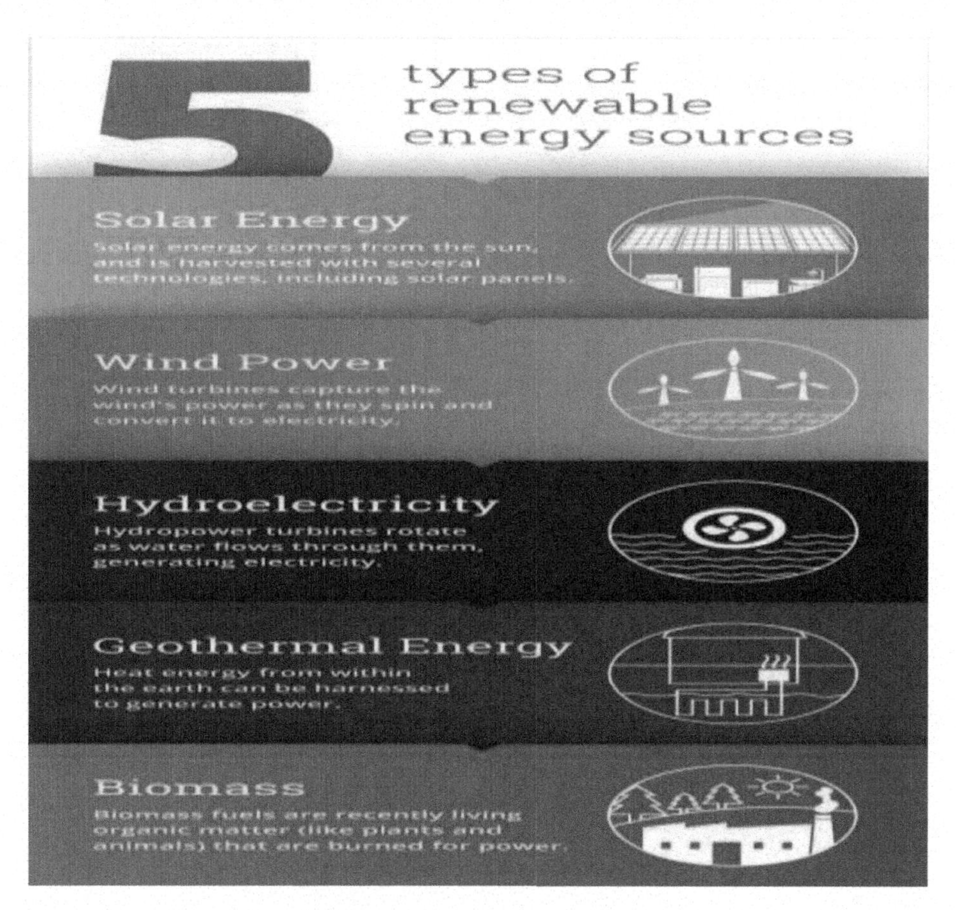

Figure 29.1 Types of renewable energy resources

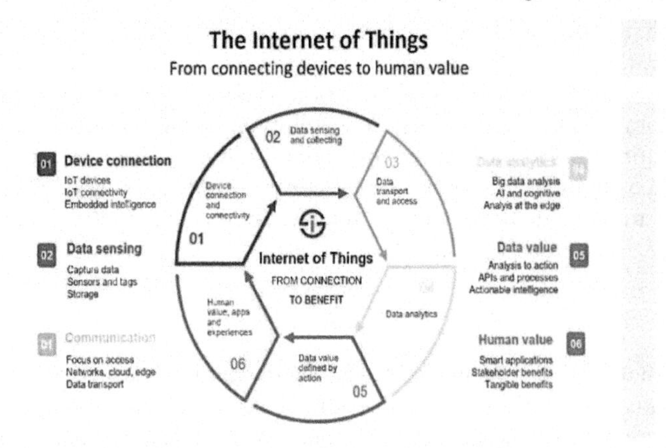

Figure 29.2 Internet of things

C. Intelligent Data Analysis (IDA)

IDA is the process concerned with the analysis of data effectively to discover the meaningful information huge data to using them to make a better decision [Al-Janabi 2021 and Al-Janabi and Mahdi 2019] and reduce the risks. Data analysis consists of the following major steps [Samaher, et al. 2014] as shown in Figure 29.3:

- Data collection: acquiring data, and transforming it into a suitable format for analysis.
- preparation and data exploration: examination specific characteristics
- Data analysis by machine learning algorithms and the algorithms of deep learning.

Intelligent data analysis (IDA) has emerged as the incorporation of numerous fields such as AI, machine learning, statistics, and pattern recognition. It aims to create predictive models based on the analysis of the data [Basaran and Kılınç 2019; Bibri 2018; Cotfas, et al. 2021; Dhanraj, et al. 2021]. Some of these models' predictions are explainable but not very effective and the accuracy is low. The others model their predictions are difficult to explain, complex, and the accuracy is high. Intelligent data analysis models and machine learning models suffer from trade-off the accuracy and explain ability, this gap is reduced with the upgrading of IDA.

D. An Embedded Intelligence System (EIS)

EIS is an electronic system, usually considered as part of the intelligent devices that designed specifically to perform specific functions; EIS aims to combine intelligent decision-making capabilities with machine learning into embedding system [Amir and Khan 2021; Ang and Seng 2021] as shown in Figure 29.4. Based on the hardware the ES can group as

- FPGA (Field-programmable gate arrays). Figure 29.5 explain design of FPGA
- A microcontroller (Arduino is the most well-known microcontroller system)
- The microprocessors.
- System-on-chip devices.

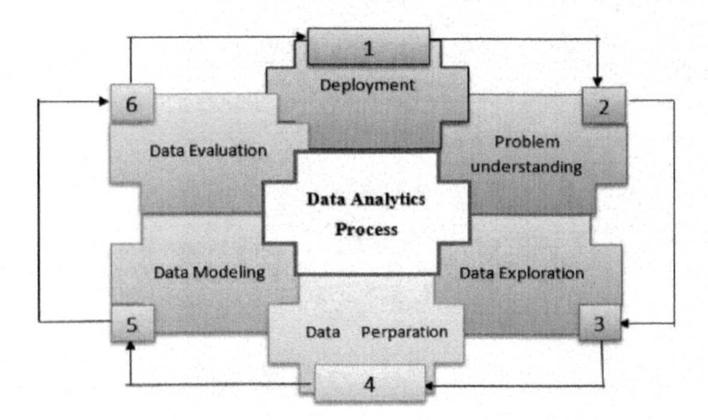

Figure 29.3 Main stages of data analysis (Al-Janabi, et al., 2021; Khan, et al., 2020)

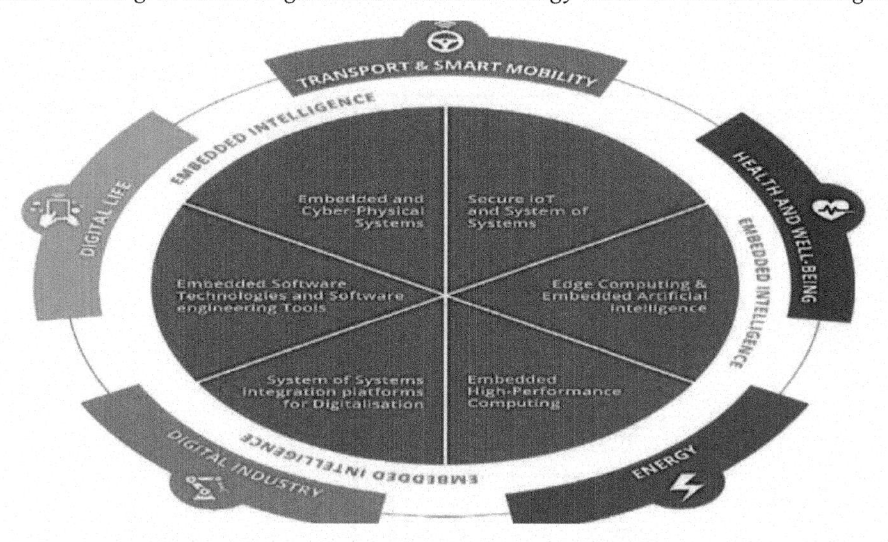

Figure 29.4 Embedded intelligence system

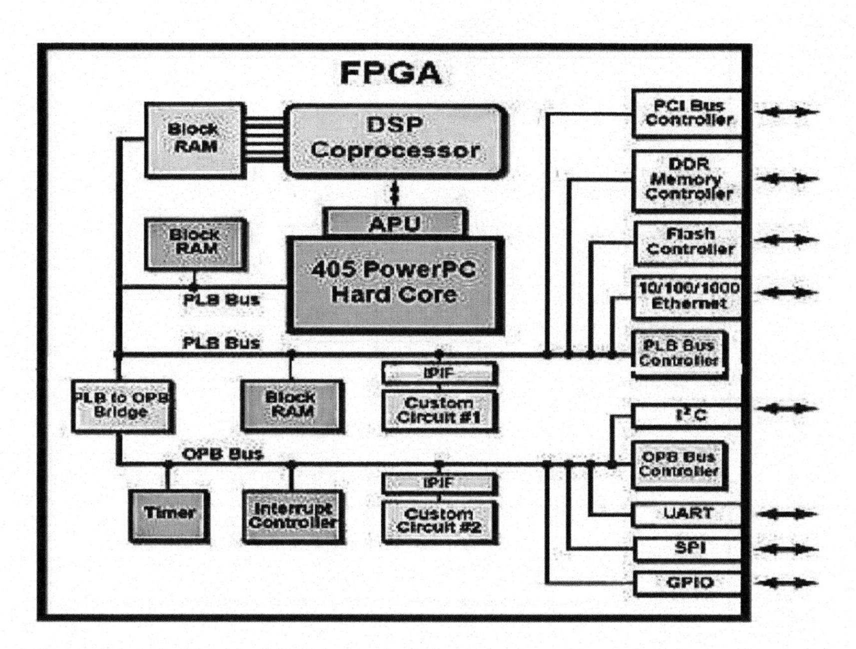

Figure 29.5 Structure of FPGA circuit

The system can control and implement the tasks with some intelligence by exploiting the main characteristic of ES that is based on using the microcontroller sensors, Wi-Fi adapter, GPS, and other devices (Al_Janabi and Hussein, 2020).

The Smart Micro Grid Model (SMGM) Stages

The integration system called Smart Micro-Grid Model (EI-FPGA) contains multi-stages, ***first stage*** captures data in real-time from multi-sensors related to datasets of solar plant and weather. Then merging between solar plant dataset and weather datasets. After that checking the missing values by dropping any record have missing value.

The second stage called pre-processing includes adding some features that are useful in prediction, splitting the dataset into multi-intervals, then removing the duplication interval. Find the correlation among the features.

In the ***third stage*** create linear embedded intelligent (LEI), and non-linear embedded intelligent (NEI) models before that determining the main parameters for each model.

The ***final stage*** includes the implementation of the design models on three different hardware (CPU, GPU, and FPGA) and compares the execution time for each one. As Figures 29.6 and 29.7.

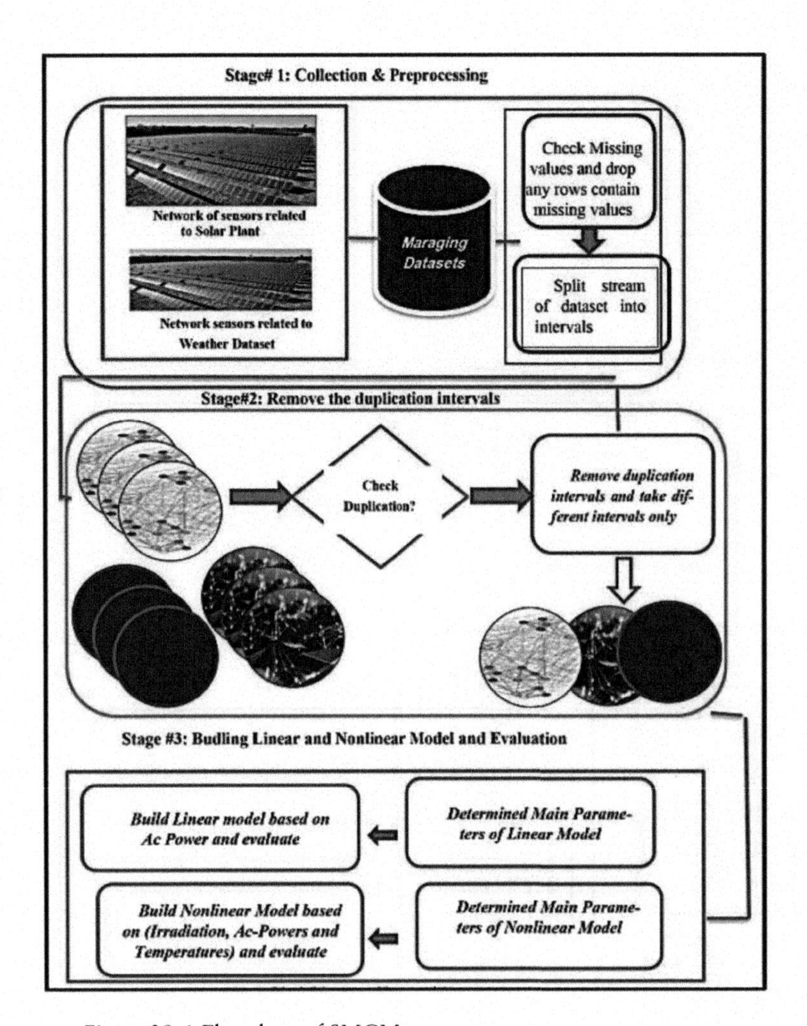

Figure 29.6 Flowchart of SMGM

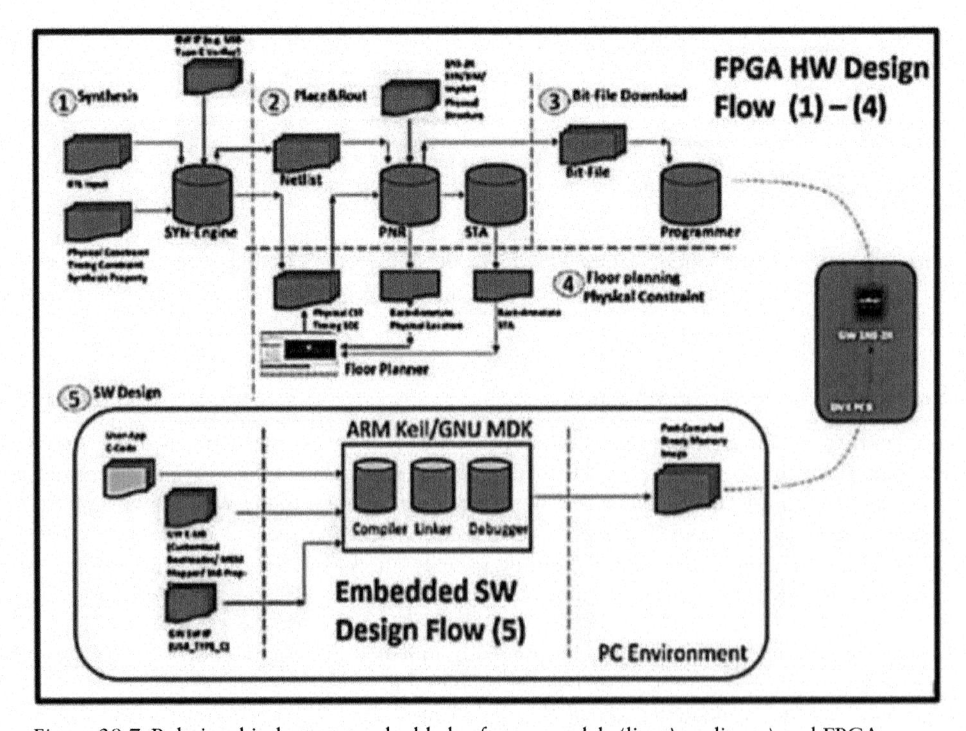

Figure 29.7 Relationship between embedded software models (linea\nonlinear) and FPGA

```
Algorithm  : EIM_FPGA
Input: Plant dataset capture from 7 sensors; Weather dataset capture from 6 sensors
Output: Predict the DC_POWER generation
        // Collection and preparing data
1:    For each datasetK                          // K=1,2
2:      Call Merge datasets                      //Merrgoe based on Date-Time and Plant-Id
3:      Check Missing values
4:    End for
        // Pre-processing stage
5:    For i=1 to nR                              // nR is number of rows in dataset
6:      For j=1 to Nc                            // nC number of column in dataset
7:        Add new features
8:        Split to intervals
9:        Delete duplicated intervals
10:     End for
11:   End for
        // Build EIM_FPGA predictor
12:   For each iterval_plants
13:     For I in range(1: total number of records [Interval_plants])   //Training Dataset
          // Determine main parameters of LEI model
14:       Call Build Linear embeded Intelligent Model       //Generated DC-Power
          // Determine main parameters of NLEI model
15:       Call  Build  NonLinear                            //Generated DC-Power
          embeded Intelligent Model
          // Test stopping conditions
16:       IF max error generation< Emax
17:         GO to step 23
18:       Else
19:         GO to step 12
20:       End IF
21:     End for
22:   End for
// Evaluation stage
23:  Call Evaluation EIM-FPGA, EIM-CPU, and EIM-GPU     // Time and Dc-Power
End EIM_FPGA
```

A. Data Preprocessing Stage

After datasets collection through multi sensor. That dataset needed to handle it before building the predictor as follow.

- Merage the dataset of solar plant with the dataset of weather and save it in separated file hold the name of this Plant-ID.
- After that, checking the missing values through drop each row has one or more missing values.
- Find the correction among the features and the target (i.e., DC-power)

Finally, split the dataset into multi-intervals (i.e., steam of data split based on time each 15 minutes into interval) and remove the duplication from intervals by save only the different intervals (different reading from sensors). Algorithm 2 explains the main steps of that stage.

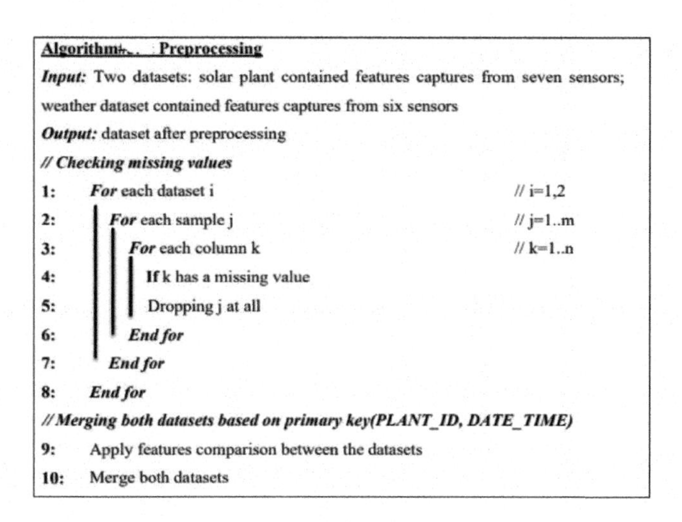

```
Algorithm  : Preprocessing
Input: Two datasets: solar plant contained features captures from seven sensors;
weather dataset contained features captures from six sensors
Output: dataset after preprocessing
// Checking missing values
1:    For each dataset i                               // i=1,2
2:      For each sample j                              // j=1..m
3:        For each column k                            // k=1..n
4:          If k has a missing value
5:          Dropping j at all
6:        End for
7:      End for
8:    End for
//Merging both datasets based on primary key(PLANT_ID, DATE_TIME)
9:    Apply features comparison between the datasets
10:   Merge both datasets
```

```
// Apply correlation
11:  For each sample in the dataset
12:  |  For each column in the dataset
13:  |  |  Compute Correlation      // R_C,S = covr(c,s)/(σ_c σ_s) = Exp(c−c̄)(s−s̄)/(σ_c σ_s)
14:  |  End for
15:  End for
// Split dataset into multi intervals
16:  Build empty set called Q
17:  For i=1 to the total number of intervals   // time of intervals equal 15 minutes
18:  |  W[i]=read samples
19:  End for
20:  For i=1 to the total number of intervals
21:  |  If w[i] = w[i+1]
22:  |  |  Delete w[i+1]                         // delete duplications
23:  |  Else
24:  |  |  Q= w[i]
25:  |  End if
26:  End for
27:  End preprocessing
```

B. Prediction Stage

This stage considers the core of SMGM that include build predicting part of both models linear and non-linear. Algorithm 3 and 4 explain the main steps of that models.

```
Algorithm#__: Linear Embedded Intelligent Model (LEIM)
Input: different intervals results from preprocessing stage with the important features determined through correlation
Output: Predicte the Dc-Power
// Split intervals into training and Testing dataset
1:   For each id_ interval
2:   |   For i in range(1: total number of records [id_ interval])
3:   |   |   Split the dataset into Training and Testing
4:   |   End for
5:   End for
     For i= 1 total number of samples in training dataset
//  |  Build Linear Embeded intelligent model
6:  |  df_train = df_plant1[df_plant1["DATE_STR"].isin(train_dates)]
7:  |  reg.fit(df_train[["IRRADIATION"]], df_train.DC_POWER)
8:  |  prediction = reg.predict(df_plant1[["IRRADIATION"]])
9:  |  df_train["Prediction"]=reg.predict(df_train[["IRRADIATION"]])
10: |  df_train["Residual"]=df_train["Prediction"]-
11: |  df_train["DC_POWER"]
12: |  df_plant1["Prediction"]
13: |  =reg.predict(df_plant1[["IRRADIATION"]])
14: |  df_train["DC_POWER"].plot()
15: |  plt.show()
16: |  print('orginal DC_POWER')
17: |  df_plant1["Prediction"].plot(color='r')
18: |  plt.show()
19: |  print('prediction DC_POWER based on Irr')
20: |  df_plant1["Residual"]=df_plant1["Prediction"]-
21: |  df_plant1["DC_POWER"]
22: |  df_plant1["Residual_abs"] = df_plant1["Residual"].abs()
23: |  df_plant1["Residual_abs"].plot()
24: |  plt.show()
25: |  print('Residual_abs of LEIM  based on Irr')
26:    End For
27:  End LEIM
```

```
Algorithn.    . Nonlinear Embedded Intelligent Model (NEIM)
Input: different intervals results from preprocessing stage with the important
features determined through correlation
Output: Predicte the Dc-Power
// Split intervals into training and Testing dataset
1:  For each id_ interval
2:      For i in range(1: total number of records [id_ interval])
3:          Split the dataset into Training and Testing
4:      End for
5:  End for
    For i= 1 total number of samples in training dataset
//  Build NonLinear Embeded intelligent model
6:  def func(M, a, b, c, d):
7:     x,y = M
8:     x=x*1000
9:     y=y*1000
10:    return (a*x*(1-b*(y+x/800*(c-20)-25)-d*np.log(x+1e-10))
11: p0 = [1.,0.,-1.e4,-1.e-1]
12: popt, pcov = curve_fit(func, (df_train.IRRADIATION,
    df_train.MODULE_TEMPERATURE), df_train.DC_POWER,
13: p0, maxfev=5000)
14: df_train["Prediction_NL"] =func((df_train.IRRADIATION,
    df_train.MODULE_TEMPERATURE), *popt)
15: df_train["Residual_NL"] = df_train["Prediction_NL"] -
    df_train["DC_POWER"]
16: df_plant1["Prediction_NL"] = func((df_plant1.IRRADIATION,
    df_plant1.MODULE_TEMPERATURE), *popt)
17: df_plant1["Residual_NL"] = (df_plant1["Prediction_NL"] -
18: df_plant1["DC_POWER"])
19: df_plant1["Residual_NL"].plot()
20: plt.show()
21: End For
22: End NEIM
```

C. Evluation Stage

An integrated system based on different types of hardware (i.e., CPU, GPU, and FPGA) to increase the production of electrical power (DC-power) in less time based on solar plants.

Implmentation and Results of (EIM-FPGA)

This section focuses on the implementation of the integration system contains multi-stages, ***First stage*** captures data in real-time from multi-sensors. Then merging between solar plant dataset and weather datasets. After that checking the missing values by dropping any record have missing value.

The second stage called pre-processing includes adding some features that are useful in prediction, splitting the dataset into multi-intervals, then removing the duplication interval. Find the correlation among the features as explained in Figure 29.8.

In the third stage create LEI, and NEI models before that determining the main parameters for each model. Based on LEI The relationship between DC-power and (AC-power\ irradiations \DC-power) explain in Figures 29.9, 29.10, and 29.11 sequentially. While; Figurers 29.12 and 29.14 showed the DC-power generation through LEIM-FPGA for training and testing datasets. In addition; Figurers 29.13 and 29.15 showed the DC-power generation through NEIM-FPGA for training and testing datasets. The comparison between the two models LEIM-FPGA and NEIM-FPGA is present in Figure 29.16.

The final stage includes the implementation of the design models on three different hardware and compares the execution time for each one. As a result; we can summarisation the point achieves in that paper as follow:

- Collecting the data from sensors in real-time
- Merging between solar -plant dataset and weather datasets based on two features (i.e., plant-id and date-time).
- Check missing values and drop them if found
- Delete duplicate intervals, and add some features that are important in prediction.
- Building LEI and NEI models and evaluating them based on two points performance and time.

A. Description of Dataset

This step shows reading datasets weather and solar plant each having important features, the information of weather plant containing ("Date_Time, Plant_Id, Source_Key, Ambient_Temperature,

Table 29.1 Merging between the weather and plant datasets

Date_Time	*Plant_Id*	*Source_Key*	*Dc_Power*	*Ac_Power*	*Daily_Yield*	*Total_Yield*	*Date_Time*	*Plant_Id*	*Source_Key*	*Ambient_ Temperature*	*Module_ Temperature*	*Irradiation*
15/05/2020 00:15	4135001	3PZuoBAID5Wc2HD	0	0	0	6987759	15/05/2020 05:45	4135001	HmiyD2TTLFNqkNe	24.28921113	23.09669193	0.000862721
15/05/2020 00:15	4135001	7JYdWkrLSPkdwr4	0	0	0	7602960	15/05/2020 06:00	4135001	HmiyD2TTLFNqkNe	24.08844607	22.2067566	0.005886957
15/05/2020 00:15	4135001	McdE0feGgRqW7Ca	0	0	0	7158964	15/05/2020 06:15	4135001	HmiyD2TTLFNqkNe	24.01163527	22.35345867	0.022281607
15/05/2020 00:15	4135001	VHMLBKoKgIrUVDU	0	0	0	7206408	15/05/2020 06:30	4135001	HmiyD2TTLFNqkNe	23.97673127	22.893282	0.049409724
15/05/2020 00:15	4135001	WRmjgnKYAwPKWDb	0	0	0	7028673	15/05/2020 06:45	4135001	HmiyD2TTLFNqkNe	24.21899	24.44244393	0.095394454
15/05/2020 00:15	4135001	ZnxXZZZa8U1GXgE	0	0	0	6522172	15/05/2020 07:00	4135001	HmiyD2TTLFNqkNe	24.5373984	27.18565287	0.141940443
15/05/2020 00:15	4135001	ZoEaEvLYb1n2sOq	0	0	0	7098099	15/05/2020 07:15	4135001	HmiyD2TTLFNqkNe	24.8159595	28.88847786	0.154712676
15/05/2020 00:15	4135001	adLQvlD726eNBSB	0	0	0	6271355	15/05/2020 07:30	4135001	HmiyD2TTLFNqkNe	24.98878987	29.6056438	0.148799153
15/05/2020 00:15	4135001	3PZuoBAID5Wc2HD	0	0	0	6987759	15/05/2020 05:45	4135001	HmiyD2TTLFNqkNe	24.28921113	23.09669193	0.000862721
15/05/2020 00:15	4135001	7JYdWkrLSPkdwr4	0	0	0	7602960	15/05/2020 06:00	4135001	HmiyD2TTLFNqkNe	24.08844607	22.2067566	0.005886957

Table 29.2 Split dataset into multi intervals and remove the duplication

Intervals	*# Samaples related to interval*	*Average*
Interval #1	3276	0.045909
Interval #2	3276	0.045909
Interval #3	3248	0.045517
Interval #4	3247	0.045503
Interval #5	3246	0.045489
Interval #6	3246	0.045489
Interval #7	3246	0.045489
Interval #8	3246	0.045489
Interval #9	3246	0.045489
Interval #10	3246	0.045489
Interval #11	3246	0.045489
Interval #12	3246	0.045489
Interval #13	3245	0.045475
Interval #14	3242	0.045433
Interval #15	3240	0.045405
Interval #16	3235	0.045335
Interval #17	3233	0.045307
Interval #18	3231	0.045279
Interval #19	3231	0.045279
Interval #20	3231	0.045279
Interval #21	3231	0.045279

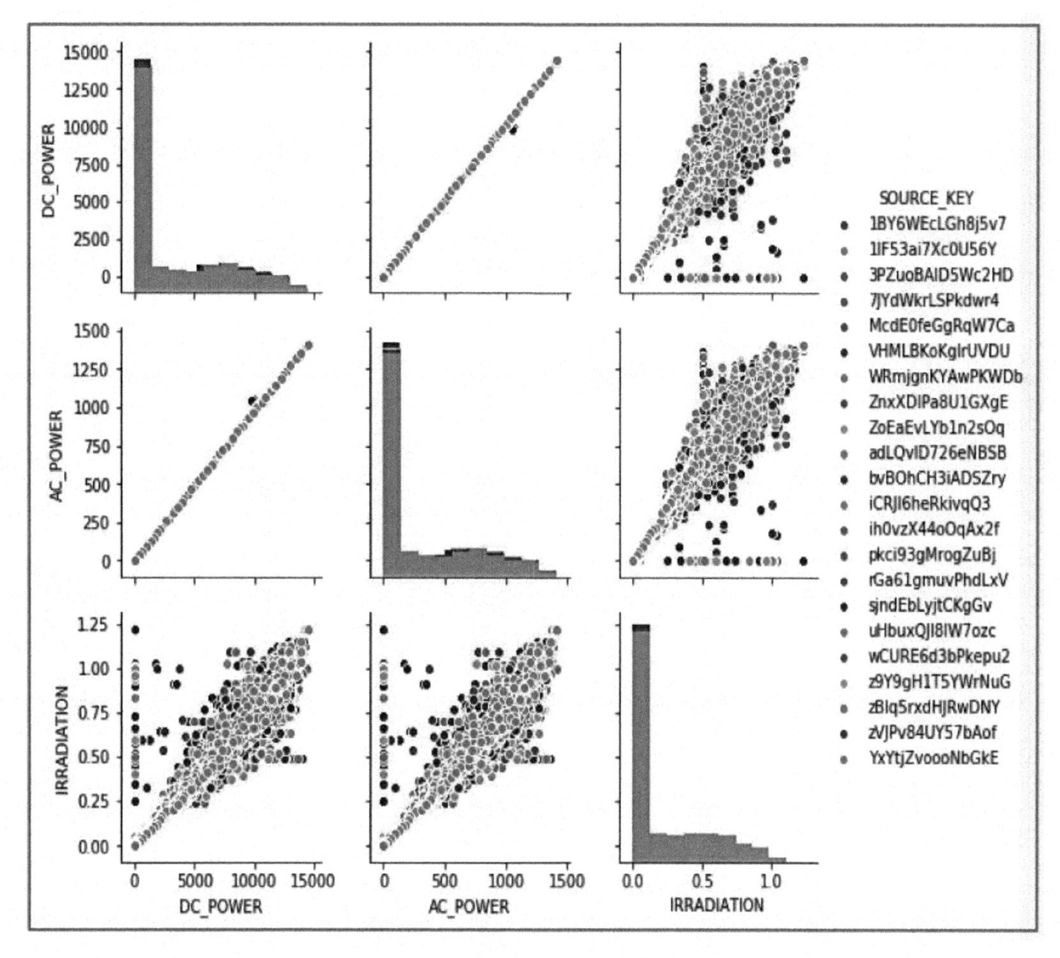

Figure 29.8 Relationship among DC-power and (AC-power\Irradiations\DC-power)

Table 29.3 Parameters effect in both linear and nonlinear model from

Model	*Equations*	*Parameters*
Linear Model	$P(t) = a + b \cdot E(t)$	$P(t)$:DC power $E(t)$: irradiance a, b: random coefficients
Nonlinear Model	$P(t) = aE(t)(1 - b(T(t) + E(t)800(c - 20) - 25) - d\ln(E(t)))$	irradiance $E(t)$, Temperature $T(t)$ coefficients a, b, c, d popt, pcov

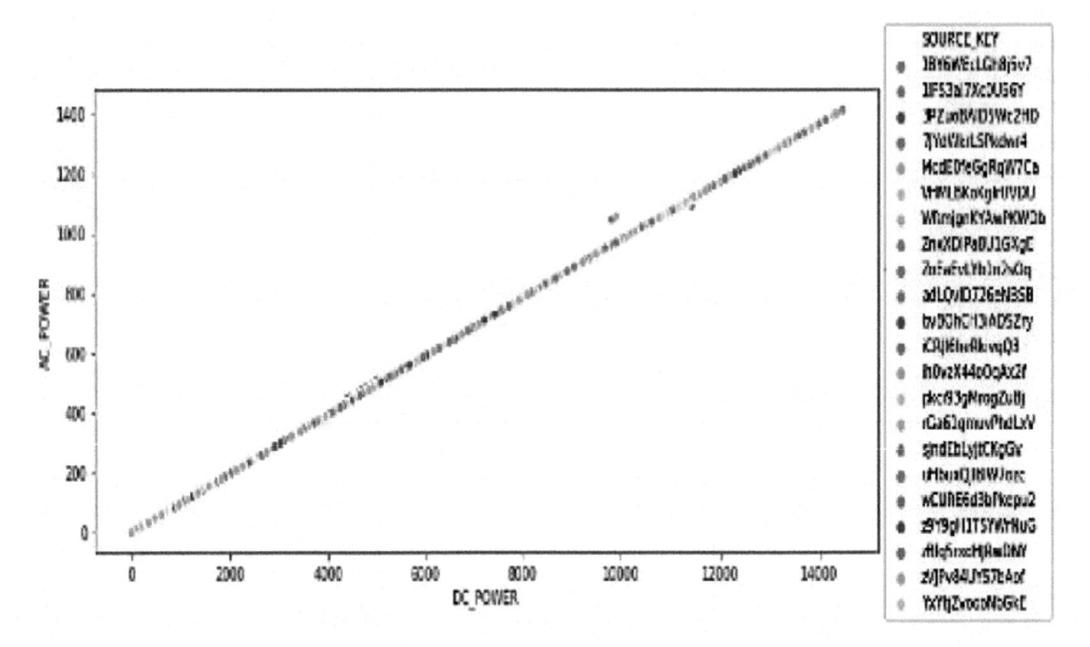

Figure 29.9 Generation DC-power based on AC-power

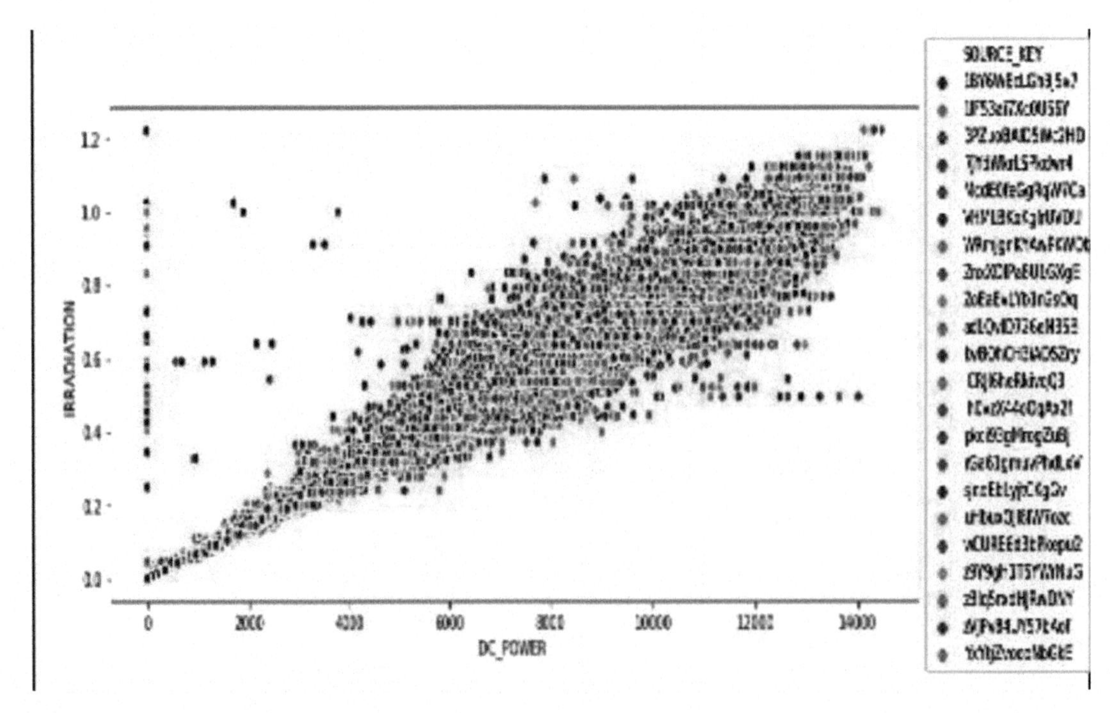

Figure 29.10 Generation DC-power based on irradiations

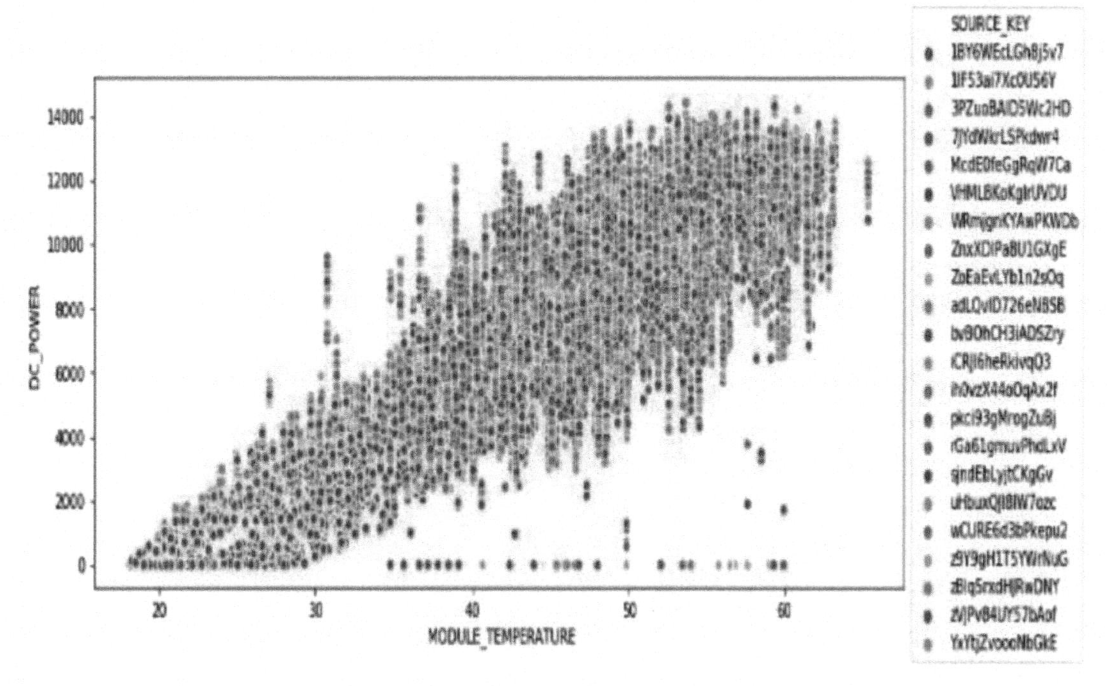

Figure 29.11 Generation DC-power based on temperatures

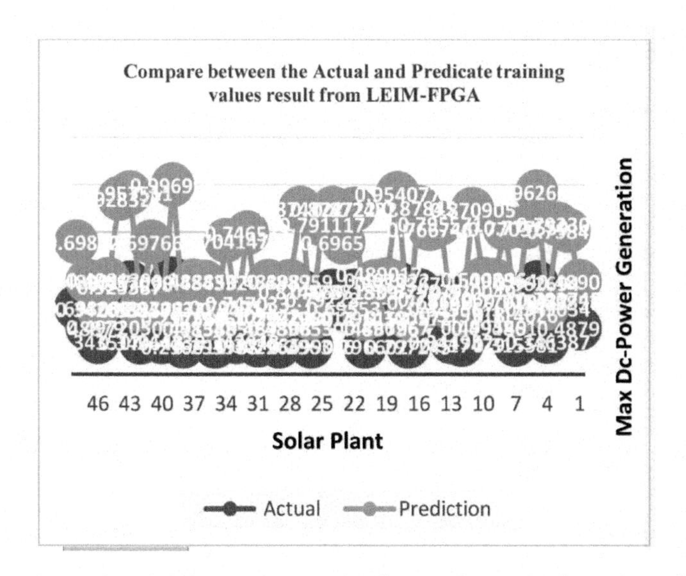

Figure 29.12 DC-power generation from LEIM-FPGA for training dataset

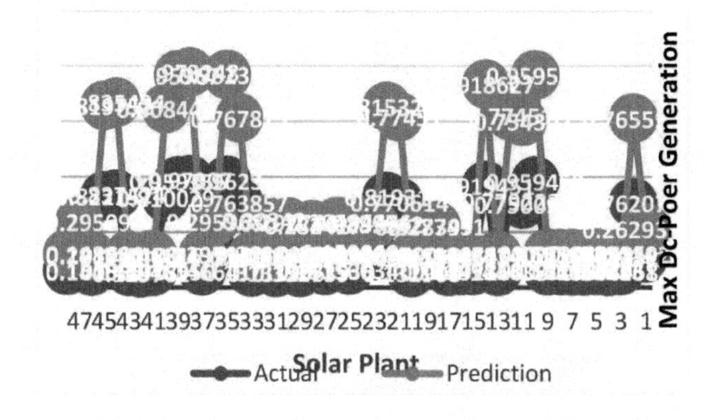

Figure 29.13 DC-power generation from NEIM-FPGA for training dataset

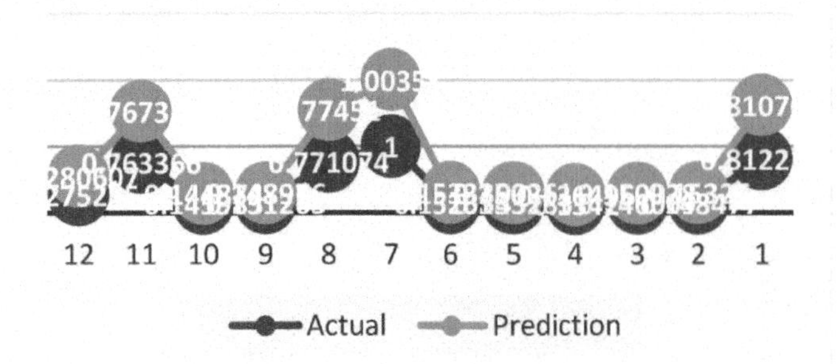

Figure 29.14 DC-power generation from LEIM-FPGA for testing dataset

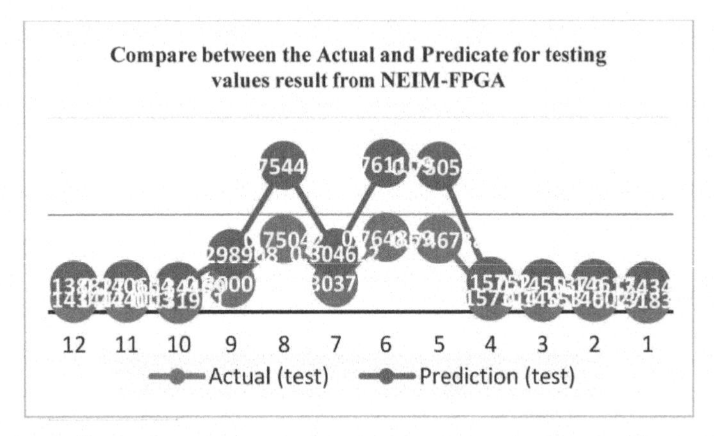

Figure 29.15 DC-power generation from NEIM-FPGA for testing dataset

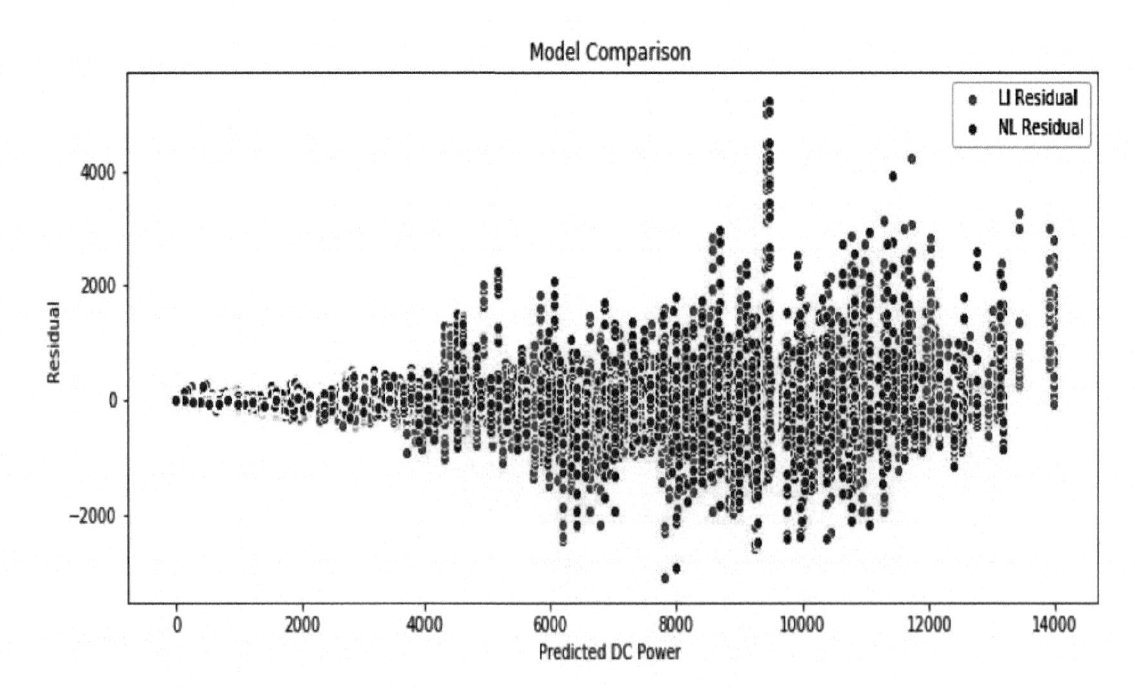

Figure 29.16 Compare between LEIM-FPGA and NEIM-FPGA from their predicted DC-power

Module_Temperature, and Irradiation") which has 3182 entries, 0 to 3181. And seven sensors for solar plant such as ("Date_Time, Plant_Id, Source_Key, Dc_Power, Ac_Power, Daily_Yield, and Total_Yield") which has 71358 entries, 0 to 71357, then apply of describing function to each one and the result is:

B. Results of Preprocessing Stage

In this section, will explain the results of each step of preprocessing stages.

1) Merging Datasets
   To apply the merge, firstly adjust Date_Time format because the solar dataset has '%d-%m-%Y %H:%M' formatting while the weather dataset has '%Y-%m-%d %H:%M:%S' formatting then drop unnecessary columns and merge both data frames along DATE_TIME.
2) *Compute the Correlation among the features* and Target
   To find which parameters in the dataset affect much on the target DC_power the coefficient of correlation method is used to find out the relationship between those parameters. The result shows that the parameters (Ac_Power, Module_Temperature, and Irradiation) have high effects on our target.
3) *Splitting Dataset into multi Intervals and Remove Duplications*
   According to the name of sensors(Source_Key), the data rows are split into intervals every fifteen minutes, then applying compression between intervals deleting duplicates, and keeping the difference only.

C. Build Models

The types of models as equation with main parameters related to that model explain in Table 29.3.
Values of Parameters for each Model:

- a=rnd( )
- b=rnd( )
- **popt=** array ([1.11067030e+01, -8.36585233e-10, -2.88700293e+08, -6.56012551e-02]))
- **pcov=**array([[2.95118747e-01,-3.37619553e-09,1.17373200e+09,6.74016006e-03],[-3.37619550e-09, 7.98285837e-14, -2.73849543e+04, -4.92535159e-10], [1.17373199e+09, -2.73849543e+04, 9.39432660e+21, 1.69319923e+08], [6.74016006e-03, -4.92535159e-10, 1.69319923e+08, 1.56495347e-04]]))

D. Evaluation the Models

The both suggest models test through three types of hardware to compute the time of implementation it theses hardware parts called CPU, GPU and FPGA. In general FPGA approve their ability to implementation the both model in time less than other two other hardware parts. As shown in Table 29.4.

Dicussion

This paper presents an implementation of integrated system based on Different types of hardware to evaluation including CPU, GPU, and FPGA to predict the Dc-power from the solar plant. The data contains two sets of information about two solar power plants, that were acquired over 34 days:

- Power generation data include seven features (i.e., Source-key, Plant-ID, Dc-Power, Ac-Power, Yield delay, Total Yielded, Date-Time)
- Weather data contain six features (i.e., Source-key, Plant-ID, Irradiation, Date-Time, Temperatures, Mobility Temperatures)

After collecting both datasets in real-time maraging in the signal dataset based on both features (Date-Time and Plant-Id) to become that dataset contains nine features rather than 13. After that checking, if any record of that dataset has missing values drop it. In the second stage split the dataset into multi-intervals

Table 29.4 Compare between LEIM and NEIM from time computation

Name of Model	*CPU*	*GPU*	*FPGA*
LEI MODEL	12.597	5.992	1.004
NEI MODEL	23.994	17.023	1.962

the time of each interval take 15 minutes; in this stage focus on removing the duplication interval and take only the different intervals to build the model. in general, the total number of intervals is 51 while after remove duplication remained only 22 intervals. Finally building two models (linear and nonlinear) and implementation based on three types of technique (CPU, GPU, and FPGA). As a result, in both models, the FPGA gives the time more reduce compare with CPU and GPU. We can summarisation the main advantages of Models as follow:

- The high correlation between DC power and AC power
- The high correlation between power and irradiation
- Correlation between DC power, AC power, and module temperature and ambient temperature
- Correlation between Daily Yield and Ambient Temperature
- A nonlinear Model require multi parameters must preparing but their results are best than a linear model
- Multi features affect in generated max DC-power include (AC-Power, irradiance, and temperature)

Conclusions and Futuer Works

The conclusions of this work are summarised as follows:

- It solves one of main problem related to generate the renewable energy through building-integrated software to generate max renewable energy from friendly environment resources
- Using field-programmable gate array (FPGA) with the embedded intelligent system (FPGA-EI) rather than CPU–EI or GPU-EI
- Find the solution of the three challenges represent in the triangular problem that has three different sides though (reduce time of implantation problem this achieve using FPGA, increase the accuracy of results by remove the samples that have missing values and determined the main features impact to the DC-Power by correlation problem, finally, reduce the computation by remove duplication from intervals and work on the different intervals only)
- Design integration system as software to generate max renewable energy from a solar plant with a new part of Hardwar represented by FPGA.
- The pre-processing stage involves three stages; first stage involves managing both dataset, checking of missing value, then split the stream of dataset into multi intervals and remove the duplication this stage enhance the performance of design software and reduce the computation of total system.
- There are many embedded intelligent models to find the best solutions, but the linear and nonlinear model was chosen to work on the best features that have 0.95 correlation with the PC-power this lead to gives high accuracy in the results when the number of feature is few.
- Nonlinear embedded intelligent model (NEIM-FPGA) is consider as pragmatic model in generated the max DC-Power while it require multi parameters must preparing but their results are best than a linear embedded intelligent model (LEIM-FPGA).

While, the following point represent some of good ideas for future work.

- Using other models based on Regression rather than LEIM or NLEIM or sms such as whale optimisation algorithm (WOA) or particle swarm optimisation (PSO) or genetic algorithm (GA) (Kaghed, et al., 2006).
- Investigation impact of the location of solar plant on the amount of max DC-Power generation from this system through add new part to hardware represent by based on global positioning system (GPS)
- Verification from the prediction results based on other evaluation measures such as (Accuracy, Recall, Precision, F, and FB) (Dogan and Birant, 2020; Al-Janabi, et. al., 2021).
- Test the model on the new dataset that contain other concentrations rather than these used in this study.

References

Al-Janabi, S. (2015). A novel agent-DKGBM predictor for business intelligence and analytics toward enterprise data discovery. J. Babylon Univ./Pure Appl. Sci. 23(2):482–507.

Al-Janabi, S., Alkaim, A., Al-Janabi, E. et al. (2021). Intelligent forecaster of concentrations (PM2.5, PM10, NO2, CO, O3, SO2) caused air pollution (IFCsAP). Neural. Comput. Appl. 33:14199–14229. https://doi.org/10.1007/s00521-021-06067-7.

Al-Janabi, S., Alkaim, A. F. and Adel, Z. (2020). An Innovative synthesis of deep learning techniques (DCapsNet & DCOM) for generation electrical renewable energy from wind energy. Soft Comput. 24:10943–10962. https://doi.org/10.1007/s00500-020-04905-9.

Al-Janabi, S. and Alkaim, A. (2022). A novel optimization algorithm (Lion-AYAD) to find optimal DNA protein synthesis, Egyptian Inf. J. https://doi.org/10.1016/j.eij.2022.01.004.

Amir, M. and Khan, S. Z. (2021). Assessment of renewable energy: Status, challenges, COVID-19 impacts, opportunities, and sustainable energy solutions in Africa. Energy Built Environ. https://doi.org/10.1016/j.enbenv.2021.03.002.

Ang, K. L.-M. and Seng, J. K. P. (2021). Embedded intelligence: platform technologies, device analytics, and smart city applications. IEEE Internet of Things J. 8(17):13165–13182. doi:10.1109/jiot.2021.3088217.

Basaran, K., Özçift, A. and Kılınç, D. (2019). A new approach for prediction of solar radiation with using ensemble learning algorithm. Arab. J. Sci. Eng. doi:10.1007/s13369-019-03841-7.

Bibri, S. E. (2018). The IoT for smart sustainable cities of the future: An analytical framework for sensor-based big data applications for environmental sustainability. Sustain. Cities Soc. 38:230–253. doi:10.1016/j.scs.2017.12.034.

Cotfas, L. A., Delcea, C., Roxin, I., Ioanăş, C., Gherai, D. S. and Tajariol, F. (2021). The longest month: analyzing COVID-19 vaccination opinions dynamics from tweets in the month following the first vaccine announcement. IEEE Access 9:33203–33223. DOI: 10.1109/ACCESS.2021.3059821.

Dhanraj, J. A., Mostafaeipour, A., Velmurugan, K., Techato, K., Chaurasiya, P. K., Solomon, J. M., Gopalan, A. and Phoungthong, K. (2021). An effective evaluation on fault detection in solar panels. Energies 14(22):7770. https://doi.org/10.3390/en14227770.

Dogan, A. and Birant, D. (2020). Machine Learning and data mining in manufacturing. Expert Syst. Appl. 114060. doi:10.1016/j.eswa.2020.114060.

Hao, J. (2020). Deep reinforcement learning for the optimization of building energy control and management. (Doctoral dissertation, University of Denver).

Hochreiter, S. and Schmidhuber, J. (1997). Long short-term memory. Neural Comput. 9(8):1735–17 doi:10.1162/neco.1997.9.8.1735.

Hossny, K., Magdi, S., Soliman, A. Y. and Hossny, A. H. (2020). Detecting explosives by PGNAA using KNN Regressors and decision tree classifier: A proof of concept. Prog. Nucl. Energy 124:103332. doi:10.1016/j.pnucene.2020.103332.

Kabir, E., Kumar, P., Kumar, S., Adelodun, A. A. and Kim, K. H. (2018). Solar energy: Potential and future prospects. Renew. Sustain. Energy Rev. 82:894–900. https://doi.org/10.1016/j.rser.2017.09.094.

Kaur, H. and Buttar, A. S. (2019). A review on solar energy harvesting wireless sensor network. Int. J. Comput. Sci. Eng. 7(2):398–404. DOI: https://doi.org/10.26438/ijcse/v7i2.398404.

Khan, A., Sohail, A., Zahoora, U. et al. (2020) A survey of the recent architectures of deep convolutional neural networks. Artif. Intell. Rev. 53:5455–5516. https://doi.org/10.1007/s10462-020-09825-6.

Lamb, J. J. and Pollet, B. G. (Eds.). (2020). Micro-optics and energy. doi:10.1007/978-3-030-43676-6.

Mahdi, M. A. and Al-Janabi S. (2020) A novel software to improve healthcare base on predictive analytics and mobile services for cloud data centers. In Big data and networks technologies. BDNT 2019. Lecture notes in networks and systems, ed. Y. Farhaoui, (vol 81). Springer, Cham, DOI: https://doi.org/10.1007/978-3-030-23672-4_23.

Mathew, A., Amudha, P. and Sivakumari, S. (2021). Deep learning techniques: An overview. In Advanced machine learning technologies and applications. AMLTA 2020. advances in intelligent systems and computing, A. Hassanien, R. Bhatnagar, and A. Darwish, eds. (vol 1141). Singapore: Springer. https://doi.org/10.1007/978-981-15-3383-9_54.

Miljkovic, D. (2017). Brief review of self-organizing maps. In 2017 40th International convention on information and communication technology, electronics and microelectronics (MIPRO). doi:10.23919/mipro.2017.7973581.

30 Fossil energy formation and its significance in earth's energy resources

Pratheesh Kumar S.[1,a], *Rajesh R.*[2,b], *Jayasuriya J.*[3,c], *Lokes Arvind K.*[3,d], *Santhosh B.*[3,e], *and Madhan Kumar S.*[3,f]

[1]Assistant Professor Department of Production Engineering, PSG College of Technology Coimbatore, India

[2]Assistant Professor (Sr.Gr), Department of Production Engineering, PSG College of Technology Coimbatore, India

[3]ME Manufacturing Engineering Department of Production Engineering, PSG College of Technology Coimbatore, India

Abstract

The study of various energy sources, their production and distribution, as well as their appraisal and selection, production and consumption, orderly replacement, and development prospects, is known as energy science. There has been one enormous life explosion and five great extinctions throughout the planet's history. They are concerned with the emergence, evolution, and global spread of several organic-rich layers, all of which are critical for the development, evolution, and distribution of fossil fuels. To get a better grasp of the earth's past and present, as well as human behavior and aptitude, a systematic energy discipline must be established. Sustainable and renewable energy resources of 150 Petawatt hours (PWh) is required as the major investment for transformation from fossil fuels to non renewable resources. The considerable energy research and are particularly interested in the evolutionary connections between energy and the environment. Fossil fuel production and development, as well as the discovery and utilisation of energy in both the Earth's interior and outer space, require a diverse variety of energy scientific fields. The best practices followed in this process is that it has been found that new energy demands the production of energy through green technology that has no effect on environment. Energy science would have a profound effect on the disciplinary system, development, energy transition clarity, carbon-neutral geological exploration, and the creation of a habitable world.

Keywords: Earth habitability, energy science, fossil energy, life explosion, mass extinction.

Introduction

Human civilisations require energy in order to survive and progress. Since the industrial revolution began in the 1760s, fossil fuels have accounted for the majority of human energy use. Wind, solar, water, and nuclear energy generation all increased considerably throughout the twentieth century as a result of technical developments. Natural gas and renewable energy production have increased from 21% in 1965 to 40% in 2019. Electricity is recommended for the terminal's energy use.

According to the International Energy Agency, electricity penetration increased from 15% in 2000 to 20% in 2019 and is forecast to reach 36% by 2040. China's current energy production structure includes only "three small and one large" new energy sources, necessitating a shift away from a "one large and three small" coal-dominated structure toward a "three small and one large" new energy-dominated structure in order to achieve "energy independence" and carbon neutrality through new energy.

Coal, oil, and natural gas, like the three-stage geobiological process of biological primary productivity, sedimentary organic matter, and burial organic matter, are the outcome of species-environment interactions. Over the course of the biosphere's existence, there have been five catastrophic extinctions and one life boom. By shifting away from a "one large and three smaller" coal-dominated structure to a "three small and one larger" new energy-dominated structure, China can achieve its stated goal of "energy independence," as well as its stated goal of "carbon neutrality."

Biomass, sedimentary organic matter, and burial organic matter all result from species-environment interactions, just like the three stages of the geobiological process described above. There have been five mass extinctions and one life boom in the biosphere's existence. Disruptions to the biosphere's delicate balance have occurred despite the Cambrian period's richness of species. When new energy sources are developed, consumption will move from high-carbon sources to low-carbon and finally zero-carbon ones. Despite the

[a]spk.prod@psgtech.ac.in; [b]rrh.prod@psgtech.ac.in; [c]21mp01@psgtech.ac.in; [d]20mp31@psgtech.ac.in; [e]21mp02@psgtech.ac.in; [f]21mp31@psgtech.ac.in

abundance of diversity during the Cambrian period, catastrophic extinctions upset the fragile biological equilibrium of the biosphere. Consumption will shift away from high-carbon sources and toward low-carbon and, eventually, zero-carbon sources when new energy sources enter development.

Lynch, 1999 mentioned that conventional oil and gas reserves will deplete in the second half of this century, but it is not the final end of oil and gas. Bockris, 2007 commented that pessimistic notion that lack of energy would lead to the demise of the high-technology countries is a misleading exaggeration which would never happen at all. Gavin and Andrew, 2010 reported that Worldwide gas fields, oil wells and coal mines supply 21 Gb/year (3.066 GTOE), 29 Gb/year (4.234 GTOE), and 27 Gb/ year (3.942 GTOE), respectively. Fossil fuel reserves are in a fixed quantity subject to decline in coming decades. Nick et al. 2010 suggested that even if the oil is going peak or decline, it would not deplete in the next 50–60 years. Jacobson and Deluchhi, 2011 suggested that global energy demands are met through solar, wind and water watts. Global energy report 2021 indicated that the current total primary energy supply (TPES) of 12 BTOE should increase by 17 BTOE (22.6 TW) in 2035. Patzek and Croft (2010) showed that global coal consumption rate of 103 MTOE/year is higher than global coal reserves increase rate of 19.2 MTOE/year. Thus, coal is also subject to peak, decline and depletion. Rutledge, 2011 presented there has been considerable increase in production of coal from 2400 in 2000 to 3450 MTOE in 2010 and is likely to peak at 3650 MTOE by 2035 and decline to 2700 MTOE in 2050 and 1350 MTOE by 2070. Perez 2009 indicated that World power demand is hardly 17 TW per year, which can be supplied with sun or wind alone. Discovery of fusion and CO_2 based fuel would be the ultimate answer. All fossil fuel reserves, known to humankind, are equal to 20 days' sunshine. Omar et al. (2014) found that renewable energy resources have the virtual potential to supply thousands time more energy than the current global demand of 17.12 TW-year or 150 PWh. Dewangan et al. (2021) developed a simple solar water heater (SSWH) with technical effective parameters and cost-effective solar thermal system. The technology discussed is widely followed in many countries for domestic and industrial purposes.

A. Formation of Organic RichStrata

As seen in Figure 30.1, variations in sea level, sudden climate shifts, oxygen depletion in the atmosphere, hypoxia, and biological extinction all have an impact on the formation and distribution of organic-rich strata across time.

The organic matter content of shale gas has an effect on the amount of gas that can be generated and stored. As a result of understanding how organic matter accumulates in shales, potential places for shale gas exploration and production can be identified. TOC, SiO2, and Ba indicate the formation of organic matter in shales; organic matter preservation (Mo, U, and Th) is related with reducing environments; and detrital dilution markers are also significant (Al2O3, K2O, TiO2, and Zr). Organic-rich strata are formed as a result of oxidation processes, glacial interglacial cycles, and volcanic eruptions.

B. Sources of Organic Carbon Accumulation

The amount of oxygen in the atmosphere has a direct effect on biological evolution. Significant oxidation events, which also impact organic matter accumulation, have an effect on the number of photosynthetic organisms.Oxidation episodes occurred during the Paleoproterozoic and Neoproterozoic periods, coinciding with the transition of prokaryotes into eukaryotes and unicellular animals into multicellular creatures, illustrating the critical role of oxygen in biological evolution Goncalves-Araujo (2016). The oxygen concentration in the atmosphere grew from less than 0.1% to 1–15% during the early Paleoproterozoic ("great oxidation event") (2.35–2.45 Ga) (Decker and van Holde. 2014; Wang et al., 2018). Following that, the oxygen concentration on Earth remained rather constant. Marine life development is aided by a combination of growing terrestrial waste, rapid nutrient circulation, and a return to more normal temperatures

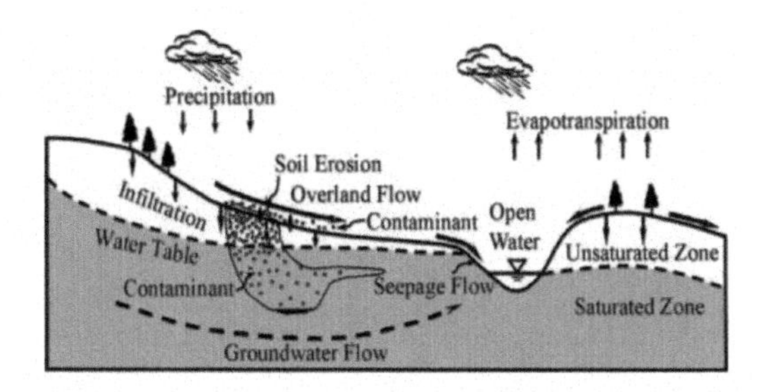

Figure 30.1 Schematic on erosion phenomenon

following the glacial age. When a volcano erupts, nutrient-rich gas is released into the atmosphere. Volcanic activity has a range of beneficial effects on productivity, including:

1. Iron, phosphorus, nitrogen, silicon, and manganese are all found in volcanic ash, and they can stimulate algae and other organisms, resulting in an increase in primary production.
2. Salt films like sulphide and halide form when volcanic ash reacts with gases like SO2, HCl, and HF. These salt films dissolve quickly in water and significantly boost the nutritional value of seawater.
3. At the bottom of the euphotic zone, a layer of volcanic ash prevents sunlight from reaching it, killing life and conserving biological waste.
4. Clay minerals such as montmorillonite and chlorite can be formed by volcanic ash in saltwater. These minerals interact with organic matter to increase enrichment, deposition, and adsorption of organic matter.

Anatahan Volcano outbreak in 2003 enhanced the productivity of the Pacific Ocean's surface, resulting in the growth of algae covering around 4800 km^2. The 2008 outbreak of the Kasatochi Volcano resulted in a phytoplankton coloration of approximately (1.5–2.0) 106 km^2 (Langmann et al., 2010). An increase in oxygen content is required to keep the Cambrian life boom going. The statement of organic matter and the evolution of species are both affected by the oxygenation of the atmosphere.

Organic carbon accumulation, redox state alterations, volcanic activity, and other events all lead to biological species extinction. A water body's ability to produce and store biological matter is influenced by its redox conditions and its productivity at its surface. The Ordovician mass extinction occurred concurrently with a large rise in global sea level and temperature. At the bottom of the sea shelf, sulphide anoxic water formed, giving excellent conditions for the preservation of biological remnants. Elimination of predators, combined with the rapid expansion of algae such as planktons and other species, resulted in a massive increase in the sea surface's primary output. The formation of organic materials contributed in the synthesis of organic-dense black shale (Lee, 1993; Mani et al., 2017; Petsch et al., 2005). Several volcanic eruptions occurred during the mass extinction, sending large amounts of nutrients into the water but also obstructing light and creating hypoxia at the bottom, allowing organic waste to accumulate.

Major Biologicalevents

On Earth, numerous significant biological events have occurred, consisting of a single biological explosion and five distinct major extinctions. These occurrences have a significant impact on how the ecology of the planet evolves. They were able to have a significant impact on organic matter production and preservation by influencing the establishment of organic-rich layers. Researchers determined how organic-rich strata are spread in relation to Earth system evolution after researching the planet's environment.

A. Cambrian LifeExplosion

The creation of a metazoans predominate in the marine ecology is one of the ecological consequences of the Cambrian life explosion. During the first brief geological age (514–541 Ma) of accelerated metazoan expansion, at least twenty metazoans and six biogroups perished.During the Cambrian epoch's increasing primary productivity and material foundation for organic matter enrichment, bottom hydrothermal activity and upwelling ocean currents resulted in the conveyance of necessary trace elements for mass reproduction (Hou et al., 2017). Due to oxygen scarcity during this era, broad and thick organic-rich shale layers replaced the original carbonate rocks, resulting in the Niutitang formation of carbon and

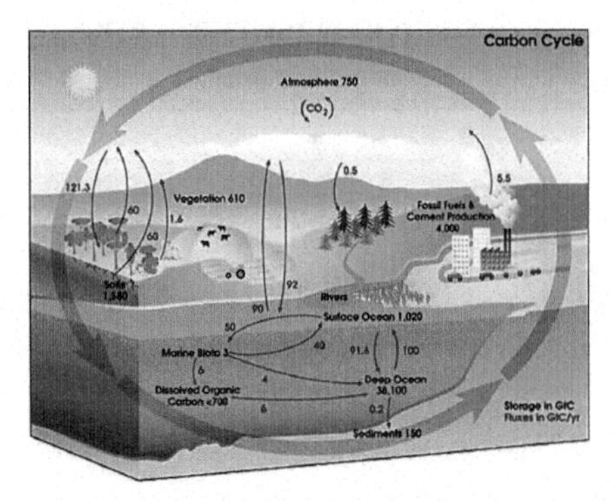

Figure 30.2 Microbial transformations of carbon

silicon-rich black shales. Anoxic ocean bottom water aided organic material retention, resulting in the production of high-concentration carbon-rich shales on the ocean's surface. Across the Yangtze Plateau, the Cambrian and Niutitang formations are creating deep black shale strata with thicknesses ranging from 50–600 metres, a distribution area of 90–104 square kilometres, TOC levels ranging from 0.5–9.0%, and Ro values ranging from 1–5% (mature stage). In these southern China shale layers, shale gas production is a distinct possibility. Organic molecules have been found in the Cambrian shale layers of European alum shale, Australian Arthur shale, and Russian Olenyok oil shale.

B. Ordovician MassExtension

Around 61% of biological genera and 86% of biological species perished during the Ordovician-Silurian mass extinction as a result of climate change, glacier formation, and sea level rise.

During the first stage, the environment became substantially colder and the sea level fell significantly. All of the ocean currents were full with life, and the water itself was overflowing with nutrients. Ice sheets, sea levels and shallow waters became unsuitable for cold-water species as the Earth's climate warmed, culminating in mass exodus. Paleotemperature shifts are the most important factor in this great catastrophe. When graptolites bloom, they deposit organic-rich black shale, which is the most important drilling stratum in southern China (Pan et al., 2021; Wang et al., 2017). The Ordovician-Silurian period produced black shales that can be found all over the planet.

C. Late Devonian MassExtension

As illustrated in Figure 30.3, the Late Devonian transition between the Frasnian and Famennian periods resulted in the loss of at least 75% of species and dramatic changes in community organisation, having a catastrophic impact on life on Earth (Qie et al,. 2019). Seawater stratification and hypoxic water diffusion to shallow water areas led to hypoxia and an increase in buried organic carbon on the continental shelf, as well as an overall drop in atmospheric CO_2. During glacial eras, global temperatures began to drop fast, resulting in large extinctions of tropical species and the destruction of existing coral reefs. As the seas cooled in the mid-latitudes, they began to absorb more CO_2 from the atmosphere while creating less, because cooler water is more soluble in CO_2. As a result of this ideal composition, CO_2 levels were roughly one-third lower throughout the ice ages. Species thrived and died during the F-F era, providing a stable supply of organic materials for deposition. The preservation of sedimentary organic materials was helped by rising sea levels and anoxic settings (Erwin, 1997).

D. Permian MassExtension

Permo-Triassic transition was the time of greatest mass extinction since the Phanerozoic. Due to this calamity, 90% of marine and 70% of terrestrial species are estimated to have died. During the Paleozoic-Mesozoic transition period, massive climatic catastrophes occurred, including supercontinent convergence, supervolcanic eruptions, and sea level changes, as well as ocean hypoxia and temperature oscillations International Energy Agency (2021).

Asteroids, volcanic eruptions, sea level decreases, ocean hypoxia, high heat and acidity, and other factors have all had a role in the extinction of animal and plant species on Earth. Because of massive CO_2 emissions, widespread wildfires, a lack of ocean circulation, increased acidity and hypoxia, and rapid temperature rises, the mass extinction is inexorably linked to the creation of Siberian igneous rocks. The average temperature of the Earth increased during the Carboniferous, Permian, and Triassic epochs, as did the extinction and replacement of marine and land plants. Throughout the Carboniferous–Permian period, the world's first significant coal accumulation phase, a swamp forest dominated by lycopods served as the principal environment for coal formation, as the Permian is one of the most organically rich strata on the globe DiMichele et al. (2001).

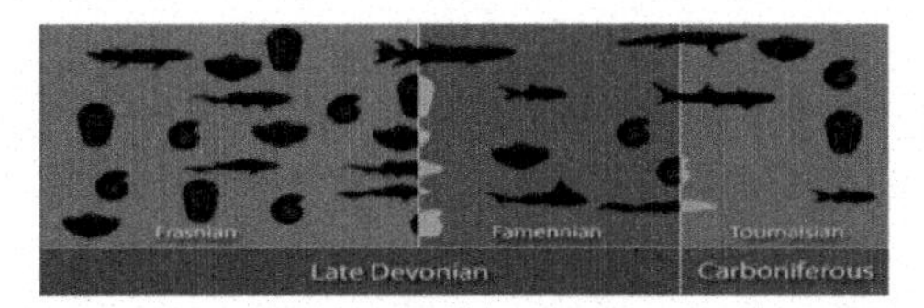

Figure 30.3 Late devonian mass extinction

E. Triassic MassExtension

Around 76% of all marine and terrestrial species, as well as around 20% of all taxonomic groups, perished in the Triassic Period's global extinction catastrophe (252 million to 201 million years ago). The end-Triassic extinction was caused by large-scale and widespread volcanic eruptions. More than 76% of Earth's marine and terrestrial species perished as a result of ocean acidification and global warming that began around 200 million years ago (Hoegh-Guldberg and Bruno, 2010).

Ecosystems can adapt to climate change, as evidenced by the enormous extinction of marine habitats during the Triassic-Jurassic period, which wiped out 52% of genera and 76% of species. Oxygen and CO_2 levels were at their lowest during the Mesozoic period. Because of the greenhouse effect, which has been exacerbated by increasing CO_2 and CH_4 levels, extreme weather events have become more frequent. This has led to an increase in terrestrial input and the eventual collapse of the marine ecosystem, while simultaneously enhancing the water cycle and surface weathering. The broad deposition of organic-rich black shale was promoted by anoxia, which happened regularly prior to and during the extinction event. Shales containing organic matter, such as Montney shale in Canada, Emm shales in European, and Kockatea shales Australia, date from the late Triassic and were found in abundance on the Palaeo-Tethy continental shelf, which is thought to have been influenced by a warm, moist climate called a greenhouse.

F. Cretaceous MassExtension

Sixteen percent of marine creature families and 47% of marine organism genera were destroyed by an asteroid impact and volcanic eruption in the late Cretaceous age. Increasing CO_2 levels in the atmosphere are a result of large-scale volcanic eruptions and collisions with celestial bodies, which have resulted in acid rain and the degradation of the ecological environment. Because of the lack of biological photosynthesis, marine invertebrates and vertebrates have experienced a significant decline in diversity.

During the Cretaceous–Paleogene period, the climate and climatic circumstances, as well as the velocity of species turnover, significantly changed. When organic-rich black shales were discovered, an entire species went extinct. Anoxic conditions during the Middle Cretaceous epoch are the principal source of hydrocarbons on Earth. Many coal deposits were made in the Jurassic and Cretaceous periods in the Xinjiang region, Northern Shaanxi province and Inner Mongolia region of China and Northeast China (Thomas, 2020).

Major Biologicalevents

A range of events, including biological, environmental, and geological processes, as a result, organic-rich strata are formed, as shown in Figure 30.4 (Grotzinger and Jordan, 2014). As a result, it's the result of several complex systems interacting through time, including those in our biosphere and solar system. Advances in Earth system research have made it possible to study the formation of organic-rich strata and the interplay between human activities and energy development.

A. Sources of Fossil Energy

Nonrenewable fossil fuels such as oil, coal, and natural gas were developed as a result of prehistoric plant and animal decomposition followed by burial beneath successive layers of rock. The temperature and pressure conditions generated when organic matter is buried for a lengthy period of time result in the

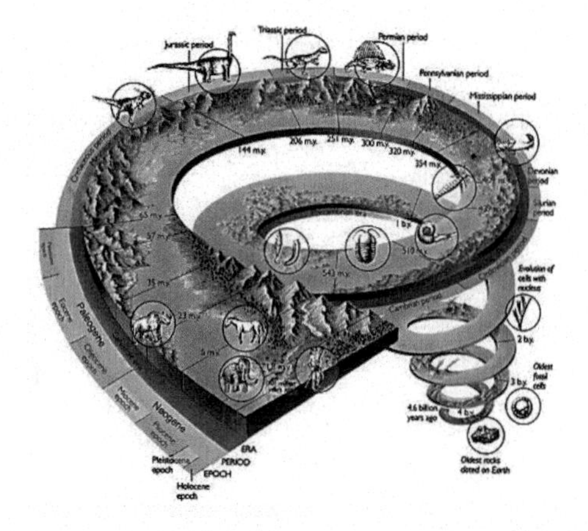

Figure 30.4 Evolution diagram of earth and energy

formation of a variety of fossil fuels. These fossil fuels are the result of a three-phase geobiological process that began with biological primary productivity and ended with burial organic matter, in which organisms and their habitats coevolved. Fossil fuels provide 82% of global energy (Abas et al., 2015; Gürsan and de Gooyert, 2021). They are the primary source of energy. The majority of confirmed fossil fuel reserves are held by non-OECD countries. Because they take billions of years to form, fossil fuels aren't something that can be replenished. Once these resources are spent, they cannot be refilled. A major contributor to climate change, carbon dioxide is produced mostly by the burning of fossil fuels, which has a substantial impact on the environment and human health. People all over the world are looking for environmentally acceptable, renewable energy sources as a result of these concerns. Furthermore, when traditional fossil fuel reserves have run out, firms have been compelled to seek out more difficult deposits. You might want to explore elsewhere if you're seeking for a more environmentally friendly option.

B. Birth and Development of Earth System

Since the Earth's creation 4.6 billion years ago, the Sun has been the primary force behind its evolution (Shikazono, 2012). The paleo-ocean was generated by the old atmosphere's CH_4, ammonia, CO_2, water vapor, and other elements at 4.4 Ga, creating the ideal conditions for the origin of life. Biodiversity and fossil fuel production can be traced back to the appearance of prokaryotes at 3.5–3.8 Ga. Human evolution entered a new phase with the appearance of Homo sapiens some six million years ago. A wide variety of fossil fuels became increasingly important as human civilisation progressed. As we move away from fossil fuels and toward renewable forms of energy, this is a tipping point.

C. World's Dependence on Fossil Energy

Humans are changing the Earth's biosphere at a breakneck pace, and their impact is bigger than ever. The Anthropocene emphasises the impact of human activities on climatic and environmental change Zhang et al. (2015). The fourth industrial revolution, as well as the sixth scientific and technological revolution, have both begun. We have increased our ability to manage nature and utilise its resources as a result of scientific and technical developments. Our energy sources have progressed from highly polluting to low-polluting to carbon-free over time. The use of resources to extract energy is being phased out in favour of technology-driven 'energy generation.' The eventual transition to artificial energy is seen as a metaphor for artificially caused nuclear fusion. Energy research in the Anthropocene epoch must consider the Earth's past and present, energy origins and evolution, as well as human behaviour and capacities. This is especially important given recent climate change tendencies.

Major Biologicalevents

Research and development in the field of fossil energy benefits the general public, as shown in Figure 30.5 (Zou et al., 2016). By undertaking research and management, removing market barriers to ecologically responsible fossil fuel usage, and cooperating with industry to commercialise fossil energy technology, we contribute to the development of information and policy choices that benefit the public.

A. Earth Science Development

Discipline splits and fusions have resulted in Earth system science. In the realm of energy research, there is no exception. The majority of today's energy studies start with resource use and then move on to more

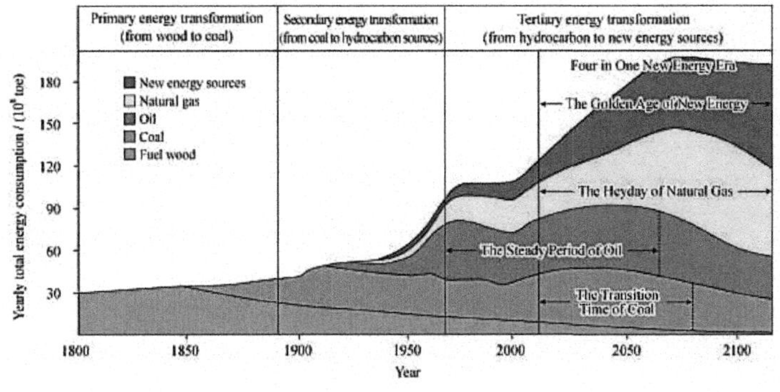

Figure 30.5 Energy revolution: Fossil energy era to a new energy era

in-depth topics like coalfield geophysics, petroleum geosciences, unconventional petroleum geology, and alternative energy sources. These ideas can be used to explain nature, evolution, and energy growth principles. These fields tend to focus on a single energy source, which means they haven't addressed how Earth, energy, and humans interact and evolve together at many spatial and temporal scales during Earth evolution. The birth of humanity ushered forth a new era of energy consumption.

The generation and use of energy from natural resources is required for human activities. Fossil fuels have been a key source of energy and a driving force for human progress since industrialisation. Because of resource depletion, pollution, and global warming, as well as energy inequity, humanity and the Earth's ecology are unable to coexist peacefully.

B. Hydrogen Fuel

To achieve zero-emission mode of transportation, hydrogen will be necessary. This might turn our fossil-fuel economy into one based on hydrogen. As shown in Figure 30.6, there are various methods for producing hydrogen, but only a few are environmentally friendly Lynch (1999). Water electrolysis can be used to generate energy that is both ecologically beneficial and long-lasting. There are both advantages and disadvantages to using hydrogen fuel cell or IC engine as the primary fuel.

Like gasoline engines, hydrogen internal combustion engines employ oxygen in the air as a catalyst to produce expanding hot gases that drive the engine's physical components directly. Water vapour and a trace amount of nitrous oxides are the only leftovers. The efficiency rate is now hovering around 20%, which is far from ideal. An electrochemical process is used to generate power in a hydrogen fuel cell, which requires hydrogen and oxygen. Because of their water-rich output, fuel cells are projected to be widely deployed in the near future. Fuel cell technology will contribute in the fight against global warming by reducing greenhouse gas emissions as it develops. The scarcity of hydrogen is one of the most significant barriers to broad use of hydrogen fuel cells. Despite the abundance of hydrogen on the earth, the technologies for creating, storing, and transporting hydrogen are prohibitively expensive, and they can release up to 10% hydrogen gas while in use. The current method of hydrogen production is reliant on natural gas, which releases CO_2. As a result, greenhouse gases continue to be released into the environment even when renewable energy is generated. It's also important to remember that fuel cell technology is still in its early stages.

C. Hydrogen Fuel

As raw materials, natural resources can be employed in chemical or industrial manufacturing to produce material products such as metals and plastics, or they can be converted into energy-carrying molecules. Thus, when it comes to acquiring energy, the term 'energy' refers to the total amount of substances that have evolved or arrived from space that may be digested and used to generate energy by humans.

Scientific investigation of how the Earth's natural resources are linked to human civilisation is one focus of energy study. The two pillars of energy research are: (1) energy production inside the Earth system; (2) energy use as feedback to the Earth's climate. If we think about how Earth and humanity are intertwined, this is an appropriate representation of Earth-human interplay. As long as we have technology, humans and energy are inextricably linked. In order to replace and transform carbon and hydrogen energy and to construct a green and harmonious development law between the energy system and the habitable Earth, it is the purpose of energy research to create a symbiotic distribution link between fossil and non-fossil new

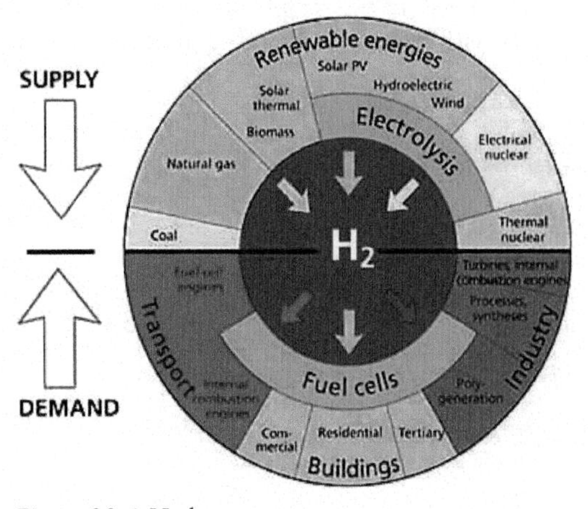

Figure 30.6 Hydrogen economy

energy. In this study, the production and distribution of fossil energy are evaluated in relation to Earth's evolution and energy formation, as well as human development, rather than just looking at fossil energy generation and dispersion over time.

D. EnergyResearch Framework

A wide range of activities related to energy research are included, including the discovery and utilisation of deep-earth and deep-space resources as well as energy strategy and planning. These findings will help us better understand how organisms respond to changes in the Earth's environment, as well as how changes in the behaviour of living things are affected by these changes in the environment.

Renewable energy sources that originate from natural or man-made processes are studied as a way to connect the dots between natural resource transformation technologies and the energy-intensive activities of modern society. Aside from a search for alien energy sources, human efforts have been made to investigate the Earth's mantle and core and its interior and outer space. Planned development of a new energy system aims to create a 'green home and a habitable Earth' while also considering global warming, carbon emissions reductions, and carbon neutrality. Energy planning, development architecture, and structural optimisation in the Earth system are all examined in this study.

E. Significance of EnergyResearch

Certainly, energy science has a major role to play in expanding the disciplinary system, accelerating energy development, defining the energy transition path, and creating a habitable Earth. The energy development law is defined by systematic evolution, setting the foundation for other energy-related disciplines based on Earth system evolution. In particular, renewable and low-carbon energy sources such as wind and solar power are being promoted as a main strategy. There must be a greater emphasis on bringing together the development of fossil and new forms of energy, as this is crucial to accelerating the transition from high carbon to low carbon and carbon-free.

All aspects of the long-term and current carbon cycles are examined in depth using sedimentology and geochemical techniques. As a result of the research, there will be significant changes in the field of carbon neutral geology.

Conclusion

Changes in the Earth's environment caused by important events in the planet's history are intimately linked to an organism's existence or extinction. The emergence, growth, and extinction of organisms are examples of biological occurrences. Volcanic eruptions and other natural disasters provide the raw materials for fossil energy synthesis, therefore biological activities are crucial to its development and spread. Another benefit of a healthy ecosystem is the availability of medicinal plants. Ecosystem deterioration has a negative impact on human health. Efforts to protect endangered animals promote human well-being. Fossil fuel generation and geological processes are incomprehensible without a thorough comprehension of biological activities. The chemical industry, which will continue to play a vital role in human progress for decades to come, relies heavily on raw materials derived from fossil fuels.

The development of the Earth arrangement serves as a lens through which to view the field of energy studies as a whole, with particular attention paid to the interactions between energy and the planet, the environment, and humanity. There is a wide range of energy sources that are studied in the field of energy science. These sources include coal, oil, natural gas, and nuclear power. New energy sources, new fossil fuels discoveries, exploration in Earth's depths for fossil fuels, as well as planning for future energy needs all fall under energy research. Energy research is influenced by environmental changes as well as scientific and technological breakthroughs. Changes in the primary energy source will force us to reconsider our priorities in the areas of the environment, energy, and people.

References

Abas, N., Kalair, A. and Khan, N. (2015). Review of fossil fuels and future energy technologies. Futures 69:31–49.

Bockris, J. O. M. (2007). Will lack of energy lead to the demise of high-technology countries in this century. Int. J. Hydrog. Energy 32:153–158.

Decker, H. and van Holde, K. E. (2014). Oxygen and the evolution of life, 2011th ed. Berlin, Germany: Springer.

Dewangan, S. K., Sunheriya, N., Ravikiran, T. and Ravikiran, C. A. (2021), Review of component wise performance enhancement techniques for simple solar water heater. DOI:10.1080/15567036.2021.1954728.

DiMichele, W. A., Pfefferkorn, H. W. and Gastaldo, R. A. (2001). response of late carboniferous and early permian plantcommunities to climate change. Annu. Rev. Earth Planet. Sci. 29:461–487.

Erwin, H. (1997). The late Devonian mass extinction: The frasnian/famennian crisis. George R. mcghee, Jr. Q. Rev. Biol. 72:192–192.

Gavin, B. and Andrew, W. (2010). Less is more: Spectres of scarcity and the politics of resource access in the upstream oil sector. Geoforum, 41:565–576.

Goncalves-Araujo, R. (2016). Tracing environmental variability in the changing Arctic Ocean with optical measurements of dissolved organic matter. Thesis PhD, University Bremen.

Grotzinger, J. and Jordan, T. H. (2014). Understanding earth, 1st ed., New York, NY: Bedford/Saint Martin's.

Gürsan, C. and de Gooyert, V. (2021). The systemic impact of a transitionfuel: Does natural gas help or hinder the energy transition. Renew. Sustain. Energy Rev. 138.

Harper, T. and Servais, T. (2013). Early palaeozoic biogeography and palaeogeography. Geol. Soc. Lond. Mem. 38.

Hoegh-Guldberg, O. and Bruno, J. F. (2010). The impact of climate change on the world's marine ecosystems. Science 328:1523–1528.

Hou, X.-G. et al. (2017). The cambrian fossils of chengjiang. China. Chichester, UK: John Wiley & Sons, Ltd.

International Energy Agency (2021). Introduction. In World Energy Outlook. OECD.

Jacobson, M. Z. and Deluchhi, M. A. (2011). Providing all global energy with wind, water and solar power. Part 1: Technologies, energy resources, quantities and areas of infrastructure and materials. Energy Policy 39:1154–1169.

Langmann, B., Zakšek, K. and Hort, M. (2010). Atmospheric distribution and removal of volcanic ash after the eruption of Kasatochi volcano: A regional model study, J. Geophys. Res. 115.

Lee, C. (1993). Early organic evolution: Iplications for mineral and energy resources. Geochim. Cosmochim. Acta 57:2665–2666.

Lynch, M. C. (1999). Oil scarcity, oil crises, and alternative energies — Don't be fooled again. Appl. Energy 64:31–53.

Mani, D., Kalpana, M. S., Patil, D. J. and Dayal, A. M. (2017). Organic matter in gas shales, in shale gas. Elsevier.

Nick, A. O., Oliver, R. I. and David, A. K. (2010). The status of conventional world oil reserves—Hype or cause for concern? Energy Policy 38:4743–4749.

Omar, E., Haitam, A. R. and Frede, B. (2014). Renewable energy sources: Current status, future prospects and their enabling technology. Renew. Sust. Energy Rev. 30:748–764.

Pan, S. et al. (2021). Major biological events and fossil energy formation: On the development of energy science under the earth system framework. Pet. Explor. Dev. 48:581–594.

Petsch, S. T., Edwards, K. J. and Eglinton, T. I. (2005). Microbial transformations of organic matter in black shales and implications for global biogeochemical cycle. Palaeogeogr. Palaeoclimatol. Palaeoecol. 219:157–170.

Qie, W. MA, X., XU, H., QIAO, L., LIANG, K., GUO, W., SONG, J., CHEN, B. and LU, J. (2019). Devonian integrative stratigraphy and timescale off China. Sci. China Earth Sci. 62:112–114.

Rutledge, D. (2011). Estimating long-term world coal production with logit and probit transform. Int. J. Coal Geol. 85:23–33.

Shikazono, N. (2012). Evolution of the earth system in introduction to earth and planetary system science. Tokyo, Japan: Springer.

Thomas, L. (2020). Coal geology, 3rd ed., Hoboken, NJ: WileyBlackwell.

Wang, X., Zhao, W., Zhang, S., Wang, H., Su, J., Canfield, D. E., and Hammarlund, E. U. (2018). The aerobic diagenesis of Mesoproterozoic organic matter. Sci. Rep. 8:13324.

Wang, Y., Li, X., Dong, D., Zhang, C. and Wang, S. (2017). Main factors controlling the sedimentation of high-quality shale in the Wufeng–Longmaxi Fm, Upper Yangtze region. Nat. Gas Ind. B 4:327–339.

Zhang, S., Wang, X., Hammarlund, E. U., Wang, H., Costa, M. M., Bjerrum, C. J., Connelly, J. N., Zhang, B., Bian, L., and Canfieldb, D. E. (2015). Orbital forcing of climate 1.4 billion years ago. Proc. Natl. Acad. Sci. U.S.A. 112:1406–1413.

Zou, C., Zhao, Q., Zhang, G. and Xiong, B. (2016). Energy revolution: From a fossil energy era to a new energy era. Nat. Gas Ind. B 3:1–11.

31 Advances in reclaimed asphalt pavement-geopolymer as subbase or base layer or aggregate in pavements – a review

Tehmeena Bashir[1,a], *Humaib Nasir*[2,b], *and Mandeep Kaur*[2,c]

[1]Research Scholar, School of Civil Engineering, Lovely Professional University, Punjab, India

[2]Assistant Professor, School of Civil Engineering, Lovely Professional University, Punjab, India

Abstract

India is a developing country and has the world's second-largest street network. With a sudden increment of urbanization, the interest in the improvement of the streets is likewise expanding. The Reclaimed Asphalt Pavements(RAP) is the awesome financially savvy (decrease in expense is around 25 to 30% by reusing the aggregates created at the same site) option in contrast to normal assets we could have for the development of streets to address the issues, diminish the negative ecological impact and utilization of common aggregates. As immense no. of development, fix and upkeep of streets are going on; there is an adequate measure of RAP delivered. This paper audits the advancement in an examination on RAP material utilised as a substitute of fine aggregates and coarse aggregates blended in with geopolymer. The water assimilation of RAP is diminished because of the covering of the black-top on the aggregates. Studies have uncovered that the presentation of asphalt by utilizing around 30% RAP material is like that of asphalt built with common aggregates without RAP materials. Objective: To explore the progressing examination of utilizing RAP materials with and without fortification, as an option for characteristic assets as base, sub-base, or fine and coarse aggregates for adaptable asphalt development. To assess the diverse physical and mechanical properties of the same, for example, functionality, compressive strength, and flexural strength. So forth Tests: UCS, XRD, SEM, Gradation, specific gravity, LA abrasion, etc. are done to check the physical, mechanical and mineralogical properties of the RAP and RAP-FA geopolymer.

Keywords: Physical and mechanical properties, RAP coarse aggregate, RAP fine aggregate, RAP-FA geopolymer, reclaimed asphalt pavement (RAP).

Introduction

Being a developing country, India has the world's second-largest street network. It has about 5 million km of the long road network as of April 2019 which includes 2.7% of national Highways, state highways 3.14%, district roads 10.03% and rural roads 70.23%. Most of these roads are flexible pavements that require regular maintenance. India being a developing country is having many road development projects in progress. The other reason for the requirement of the development of roads is the rapid urbanization that is taking place in India being the populous country. The development of the roads needs a huge amount of natural resources. Due to the huge demand for resources, natural resources are exhausting rapidly. As the demand cannot be fulfilled by the available amount of resources in the coming years, there is the requirement of alternative materials. There are many industries that produce huge amounts of waste that can be reused in many ways. A huge no. of people are already researching the industrial wastes (rice husk, slag, fly ash, etc.) to be used in the construction of different structures. Other than industrial wastes, construction and demolition wastes are getting huge attention. Reclaimed asphalt pavement (RAP) and recycled aggregate concrete (RAC) are the most often used construction and demolition wastes for the building or restoration of structures. The removed and demolished concrete waste from the structures, when used as coarse aggregates in the construction, is known as recycled aggregate concrete. The removed bituminous surface is crushed and reused in the new pavements is called reclaimed Asphalt Pavement (RAP). The physical and mechanical features of recovered asphalt pavement when blended with various natural aggregates and geopolymers are reviewed in this work. In the road construction, maintenance, and repair process, a large amount of RAP will be produced with a certain amount of other waste materials such as fine or coarse aggregates or dust, causing a certain percentage of environmental degradation and

[a]Tammytais77@gmail.com; [b]humaib.18648@lpu.co.in; [c]mandeep.kaur@lpu.co.in

waste production. Because of the scarcity of landfill space, it creates a waste management issue. Instead of dumping RAP on the open ground which increases the pollution and ecological imbalance, the material should be reused to minimise the negative environmental impact of the waste and degradation of the natural resources. During the refurbishment, rehabilitation, or reconstruction operations, the large amount of existing bituminous pavements are generally crushed/removed or they get removed on their own, the reason being their durability. The removed surface is first separated according to its quality and then used accordingly. The high-quality RAP is mostly utilised in manufacturing of hot mix asphalt (HMA), while the remaining quality is used in the building of road bases and sub-bases for flexible pavements, as well as recycled coarse aggregate in rigid pavements. The RAP cannot be used for construction by replacing natural aggregate fully but it can be used after mixing with the natural aggregates in different quantities such that it may be used in the construction of several layers of roadways. RAP can be reinforced or stabilised by adding geopolymers or natural aggregates before being utilised as a sub-base, base, or coarse aggregate in different types of pavements. Geopolymers to be used can be polymer geogrid, bamboo geogrid, polymer geocells, bamboo geocells, fly ash, or slag. Recycling of Asphalt pavements is used in many developed countries successfully. Whereas in India RAP hasn't been given much importance in the 20th century due to many reasons such as unavailability of the updated technology and types of machinery or high technology instruments and the poor economic conditions. Because of these reasons the cost of the processing of the RAP is much higher than the new asphalt mix. However, many people in India started researching the RAP in the late 20th century and early 21st century. In India, RAP is used in two different ways, first is the extraction of bitumen from demolished roads and using same with coarse aggregate in asphalt mix separately and the second is, as aggregate in HMA with natural aggregates. The research is going on for the third method, RAP to be used as aggregate in the rigid pavement. While as in other developed countries such as the USA, the RAP is used in all three methods. It is recorded, in many projects of the National highway development plan in India, 50% of RAP has been used as partial replacement of wet mix macadam and granular sub-base. The recycling of crushed bituminous materials has gained popularity in India in recent times because of numerous successful tests on selected projects.

The Advantages of using RAP -

- Cost-effective
- Reduction in degradation of natural aggregates
- Less impact on the environment
- Less production of waste at construction sites.

Less damage to the roads as no need of transportation of RAP produced at the same site. Due to the minimum need for transportation, energy and fuel are also saved.

The removed bituminous surface from old pavements and then reused in the new pavements is known as reclaimed Asphalt Pavements (RAP). The RAP generated at the site can be used at the same time at the same site or it can be stockpiled and reused accordingly. The stockpiling is done after grading the RAP. The gradation of the RAP is done as different sites and structures require different grades and sizes of the RAP. The demolished bituminous material is pulverised in a crusher to get the required grades then only it can be used. There are different sources of RAP, some of them are mentioned below:

- Generation of wastes from HMA plant
- Milling of Hot Mix Asphalt layer generates RAP.
- Demolition or destruction of the older pavements.
- Repair, rehabilitation of the pavements.
- Destruction due to the disasters like flood, earthquake, tsunami etc
- Deterioration of pavement due to moisture content.

The existing researches on RAP mixtures and RAP used in different new bituminous mixtures evaluated its potential of multiple recycling. The authors determined that various factors and additives, such as size, gradation, bitumen amount, rejuvenator, and so on, impact the mechanical characteristics of various RAP mixtures Antunes et al., 2019. The strength and microstructural development of RAP stabilised with a slag-based geopolymer (RAP) (S) as the pavement base was investigated using various tests such as unconfined compressive strength (UCS), X-ray diffraction (XRD) and scanning electron microscopy (SEM) to analyse and evaluate the strength of the geopolymer-RAP-S when a liquid alkaline activator (NaOH/Na2SiO3 ratio) was mixed. The results showed that to produce RAP along with 20% S geopolymers as raw material with UCS value of 7 days to meet the least necessary strength specified by Thailand's Ministry of Roads The ratio of NaOH / Na2SiO3 is suitable as 60:40. The increase in Na2SiO3 decreases the rate of geopolymerisation formation which leads to the reduction of UCS Hoy et al. 2018. XRD and SEM analysis show that the development of the strength of the RAP-FA mixture occurs during the geopolymerisation in the mixture, due to reaction between alumina and silica in Fly Ash, and Calcium present in RAP leads to C-S-H gel formation. From TCLP results it was concluded that FA can decrease the leachability in the RAP-FA blend. (Horpibulsuk et al., 2017). To investigate the strength development of samples, XRD and SEM was done to measure the Unconfined Compressive Strength and leaching behaviour of the heavy metal by preparing two different samples–RAP-FA blend and geopolymer. It was concluded that with an increase of Na2SiO3, the geopolymerisation formation rate decreases, which leads to the reduction of UCS and vice versa. The leaching of the heavy metal decreases due to the geopolymer binding. (Hoy et al., 2016b. Some paper worked on the approaches to increase the amount of RAP from allowable10-20% to 40% and above in asphalt mixtures. The problems and distresses in the pavement caused by high RAP concentration in mixes, as well as measures to improve it, were investigated (Zaumanis and Mallick, 2015). To reduce creep deformation, one study advocated using new polymeric alloy (NPA) geocell contained RAP. The creep deformation of geocell was investigated by Tensile Creep Test. Nineteen static plate-loading experiments were used to evaluate the applied vertical stress-displacement results, the connection between creep strain and time, and the parameters affecting creep behaviour of RAP bases. When compared to unreinforced RAP, the initial deformation and creep rates of RAP base were reduced by 18–73% and 6–60%, respectively. With a decrease in vertical stress and a rise in NPA, creep deformation was reduced (Thakur et al., 2013).

Material and Methodology

The data was collected by doing a thorough evaluation of experimental studies published in peer-reviewed journal papers. For this project, only articles published in English were considered. The selected peer-reviewed publications were then read, catalogued, and summarised in the database while ensuring that no data was replicated across numerous articles. The experimental parameters for each study were reviewed in accordance with a data collection procedure. The hardened properties of RAP concrete samples used in the experiments, as well as the mix design parameters and the RAP properties were extracted and recorded in the data.

- A literature assessment was done to assess the present state of practice in the use of RAP as a base and/or sub base material, as well as the consequences of its usage. The review focused on peer-reviewed research and literary sources.
- During the database's creation, keywords relating to the present topic were utilised to find relevant peer-reviewed literature. The keywords such as RAP, geopolymer, Pavements, sub base/base materials, physical properties, mechanical properties, Fly ash geopolymer, etc. were used to get the related papers.
- Before deciding whether or not to put an item in the database, the significance of the keywords was evaluated by a screening procedure. The keywords and articles were compared to themes related to RAP-geopolymer materials and RAP concrete mechanical characteristics.
- The unrelated papers were eliminated and related ones were downloaded and reviewed.

According to most of the research papers reviewed, the samples of RAP and Fly ash geopolymer collected from the mill and lignite based power plant were used. Alkaline solution as activator having composition of sodium hydroxide (NaOH) and sodium silicate (Na_2SiO_3) with a 10M concentration (Horpibulsuk et al., 2017). Different samples of RAP-FA geopolymer is prepared by mixing RAP, FA and alkaline Solution with different ratio of components of L that is NaOH/Na2SiO3 to be 50:50, 60:40, 90:10 and 100:0. The replacement by FA is in ratios of 10%, 20% and 30% by weight of RAP. Samples are soaked for 2 hours in water after curing for 7 and 28 days at room temperature, and then air-dried for 1 hour before the UCS Test (Kallas, 1985).

A. Wetting and Drying (W-D) Test

After 28 days of cure, the samples are immersed in water for 5 hours. Then it was oven-dried for 42 hours at 70°C before being air-dried for 1 hour. This is an example of a w-d cycle. UCS was evaluated at 1 to 20 cycles and compared to samples that did not undergo a w-d cycle (Horpibulsuk et al., 2017).

B. Mineralogical and Microstructural Analyses

The microstructure and mineralogical changes of UCS samples during w-d cycles tests were investigated using XRD and SEM investigations, which were conducted on small pieces extracted from the broken parts of UCS samples to investigate the microstructure and mineralogical changes individually (Horpibulsuk et al., 2017).

C. Toxicity Characteristic Leaching Procedure (TCLP) Test

The United States Environmental Protection Agency (EPA) uses this test (Method 1311) to determine if solid waste is hazardous. For a range of heavy metals, tests were conducted on specimens containing 100% RAP, a blend of RAP-FA, and RAP-FA geopolymer.

Data Collection

A. Physical Properties of RAP

1) *Particle size distribution:* In RAP aggregates the majority of the particles are finer as it goes through different processes like crushing and milling (Arulrajah et al., 2014). Due to the processing of RAP, the gradation of RAP is noted to be denser and finer than NA. In comparison with NA, RAP fine particles were coarser and coarse aggregates were finer. However, the fine particles were well-graded than NA due to the fulfillment of the gradation gap due to the dust particles present in large amounts in RAP (Chyne and Sepuri, 2019; Hossiney et al. ????; Su et al., 2009).
2) *Specific Gravity and Unit Weight:* The number of aggregates used in a particular design mix for concrete pavement may be determined using specific gravity. The specific gravity of RAP is smaller than that of NA (Zaumanis and Mallick, 2015; Thakur et al., 2013). The RAP unit weight is affected by moisture content and type of the aggregates in the pavement. Unit weight range of RAP was observed to be 1940 to 2300 kg/m3 in previous researches (Al-Mufti and Fried, 2018; Chyne and Sepuri, 2019; Franke and Ksaibati, 2015; Hassan et al., 2000; Smith, 1980).
3) *Moisture content:* The moisture content of 5% or more was observed in RAP still in storage due to its exposure to rain Decker et al. (1996). Which can increase to 7–8% at times of heavy rainfall (Thakur et al., 2010). The properties are shown in Table 1, and compared values of Specific gravity and water absorption of coarse NA, fine and coarse RAP are shown in Table 2 (Franke and Ksaibati, 2015).
4) *LA Abrasion:* It was observed that RAP doesn't have uniform hardness as the abrasion resistance value was found to be less than 38% for the coarse RAP (Delwar et al., ????).
5) *California Bearing Ratio:* According to the findings of a literature review of several previous research, the California bearing ratio (CBR) of 100% RAP is unsuitable for use as the base of flexible pavement in India as per IRC standards. When RAP is combined with crushed stone aggregates in various proportions and stabilised with minor quantities of cement, its soaking CBR value improves from 20% to more than 100%, making it ideal for use as a subbase/base of flexible pavement.

B. Mechanical Properties of RAP

1) *Compressive Strength:* Many research have shown that increasing the replacement quantity of RAP reduces the compressive strength of concrete due to the poor link between cement and RAP caused by the asphalt adhering to aggregates (Hossiney et al., ????; Solanki and Dash, 2016; Tia et al., ????). Previous studies have found a 65% and an 80% drop in strength in concrete containing coarse and

Table 31.1 Properties of RAP

S. No.	*Properties*	*Range*
1	Unit weight (Kg/m^3)	19002250
2	Compacted unit weight (Kg/m^3)	15001950
3	Asphalt content	56%
4	California bearing Ratio (CBR) 100% RAP	2025%
5	Asphalt penetration(%) at 25°C	1080

Table 31.2 Properties of NA and coarse RAP and fine RAP (Delwar et al., ????)

S. No.	*Aggregates used*	*Water absorption*	*Specific gravity (g/cc)*	*Apparent specific gravity (g/cc)*
1	Coarse natural aggregate	0.32%	2.68	2.63
2	Fine RAP	1.01%	2.161	2.11
3	Coarse RAP	1.06%	2.42	2.48

fine RAP particles, respectively, when compared to a standard concrete mix after 28 days of curing. (Franke and Ksaibati, 2015; Mahmoud et al., 2013; Mathias et al., 2009; Okafor, ????; Singh et al., 2019; Tia et al., ????). When both coarse and fine RAP are utilised in the concrete, the strength is less than 44Mpa after 28 days of curing (Al-Mufti and Fried, 2018).

a) *Flexural Strength:* The flexural strength was found to behave similarly to the compressive strength (Solanki and Dash, 2016). When compared to normal mix concrete, concrete with coarse RAP and concrete with coarse and fine RAP used combined has a strength loss of up to 35% and more than 45%, respectively. In concrete containing RAP, the elasticity modulus was reduced, but load absorption was improved over conventional concrete (Al-Mufti and Fried, 2018; Franke and Ksaibati, 2015; Fried, 2018; Hossiney et al., ????; Huang et al., 2006; Said et al., 2018; Tia et al., ????).

2) *Split Tensile Strength:* Its behavior was observed to be the same as compressive strength Singh et al. (2019). In many previous types of research, the concrete with coarse RAP performs way better than concrete with fine RAP and concrete with both fine and coarse RAP (Franke and Ksaibati, 2015; Huang et al., 2005; Mahmoud et al., 2013; Mathias et al., 2009; Okafor ????; Said et al., 2018; Singh et al., 2018; Singh et al., 2019; Tia et al. ????).
3) *Permeability:* Because coarse and fine RAP are finer and coarser than NA the concrete becomes permeable when RAP fine is used in large quantity as the voids may be more than the NA concrete. It is proposed that Fly Ash can be used in the mix that can fill the voids to lower the permeability. Another solution can be heating RAP, due to which the Asphalt film on aggregates melts down to fill the voids. Concrete having fine and coarse RAP combined had a higher permeability than concrete containing coarse and fine RAP separately (Al-Mufti and Fried, 2018; Sargious and Mushule, ????; Seferoglu et al., 2018; Wartman et al., 2004).
4) *Wetting-drying cycled strength:* For several cycles w-d(C), the UCS of the geopolymer and mix of RAP + 20% FA in varied proportions increased with increasing C, up to $C = 6$, and then decreased when $C > 6$. Based on the cyclic wd data, it can be stated that the RAP + 20% FA blend provides pretty decent durability. The RAP-FA blend's durability can be increased by using FA geopolymer, especially for samples with a NaOH / Na2SiO3 ratio of 100: 0 (Horpibulsuk et al., 2017).
5) *Mineralogical and microstructural changes:* When using a 50:50 alkali solution of NaOH/Na2SiO3, the RAP + 20% FA geopolymer exhibits worse endurance against w-d cycles than when utilised in a 100:0 ratio (Horpibulsuk et al., 2017).
6) *Toxicity characteristic leaching procedure results:* The material is assigned an unsafe waste under the US EPA if an identified metal is available at quantity >100 times higher than drinking water norms. When settled with 20% FA, the investigation results demonstrate that RAP is precisely and monetarily feasible for usage in essential asphalt applications. Horpibulsuk et al. (2017)

Data Analysis and Result and Discussion

- To analyse the impact of the RAP (coarse or fine) on the water absorption of aggregate, the water absorption as a result of different sized RAP replacement is appeared in Figure 31.1.
- It can be seen that every one of the explored concretes have a comparable graph. The Figure 31.1 shows that the water absorption of the coarse RAP is the highest among all three and Coarse natural aggregate has the lowest among all.
- Figure 31.2 gives further information of specific and apparent specific gravity of coarse RAP, coarse NA and fine RAP. This is a result of the comparable specific and apparent specific gravity variety for every one of the tried concretes with various size of RAP.
- It can be analysed how coarse RAP has the specific gravity close to the natural coarse aggregate. The fine aggregate has the lowest value for both properties among the three.
- RAP has a somewhat lower specific gravity than NA.

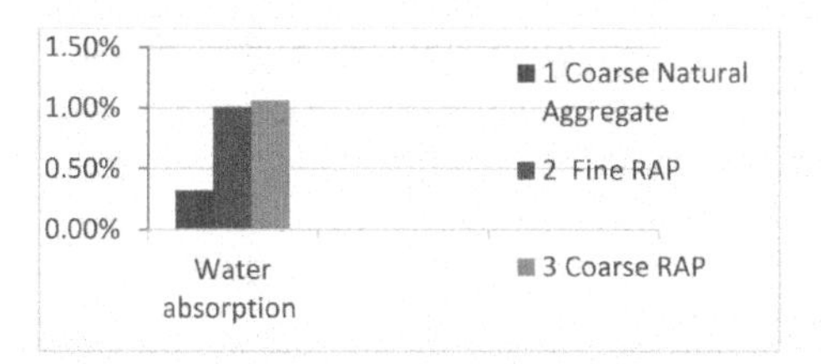

Figure 31.1 Water absorption of coarse RAP, coarse NA and fine RAP

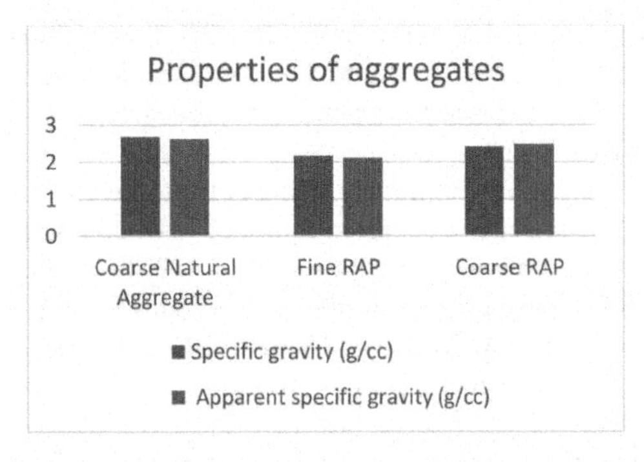

Figure 31.2 Sp. gravity and apparent sp. gravity of coarse RAP, coarse NA and fine RAP

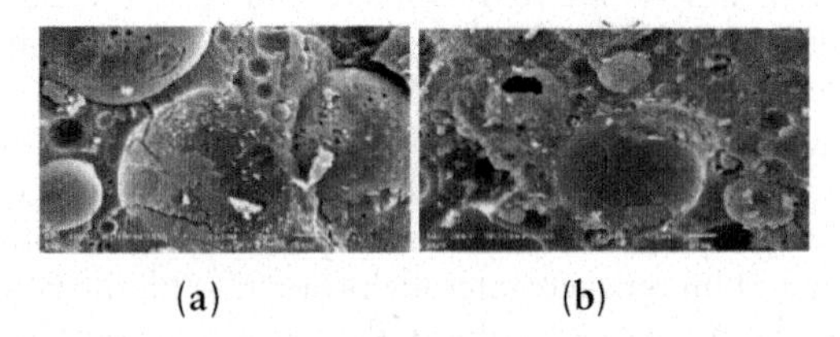

Figure 31.3 SEM images of sample at C=12 (a) with NaOH/Na_2SiO_3 ratio 100:0 (b) NaOH/Na_2SiO_3 ratio 50:50
Source: S Horpibulsuk et al 2017 IOP Conf. Ser.: Mater. Sci. Eng. 273 012005

- The RAP-FA geopolymer has a higher UCS than the RAP-FA mix for alkali solution ratios less than 10:90. When the FA replacement rate surpasses 20%, the UCS value increases just somewhat.
- Within the first w-d cycle, both the RAP+20% FA geopolymer and blend lose a significant amount of weight, which then steadily increases with increase in cycles.
- At a NaOH/Na2SiO3 ratio of 100:0, there are the fewest cracks on the surface of the RAP+20% FA geopolymer, whereas at a ratio of 50:50, there are more cracks.
- The geopolymerisation products in the sample with NaOH/Na2SiO3 = 50:50 are greater than those in the sample with NaOH/Na2SiO3 = 100:0, according to SEM images Figure 3 of RAP+20% FA geopolymer at alkali solution of both ratios. The same was detected in XRD patterns. Micro-fractures are readily evident in RAP+20% FA geopolymer samples as a result of moisture loss, which results in external surface cracks and a loss of strength for C > 6, and especially C = 12.
- From the laboratory investigation, Arshad 2020 found that UCS value increases with increase in Cement dosage, curing time as well as the replacement by fresh granulated material.

Conclusion

- The RAP-FA (compacted) mix's 7-day UCS at Optimum Water Content fulfills the strength requirements for base course specified by open road specialists for both 20% and 30% FA replacement. When the FA replacement extent outperforms 20%, the UCS improves, indicating that this is the appropriate mix. RAP-UCS FA's and strength may be improved by using FA-geopolymer (Horpibulsuk et al., 2017).
- RAP aggregates have somewhat worse mechanical and physical qualities than normal aggregates. The reason for this is that the presence of asphalt particles adhering to the aggregates necessitates the milling and crushing of RAP (Chyne and Sepuri, 2019; Horpibulsuk et al., 2017).
- According to the TCLP findings, RAP-FA mix and geopolymer examine may be safely used as a supportable asphalt base application because these materials provide no significant ecological or filtration risk to the soil and groundwater sources. Furthermore, the FA used can reduce metal fixation leachability from the RAP-FA blend. This examination concludes that these reused materials can be used as they are sustainable and emit very few ozone-depleting greenhouse gases. Horpibulsuk et al. (2017)
- Addition of RAP results in poor workability because of the shape of aggregates which is mostly found to be angular and irregular.
- As the amount of RAP increases, the strength of the samples decreases due to a poor link between the asphalt layer adhering to the RAP aggregate and the cement paste. However, because it is free of excessive dirt and asphalt layer, treated RAP outperforms untreated RAP.
- The SEM and XRD investigation of RAP-FA blends exhibit improvement of C-A-S-H gel, hence its UCS increases after some time. Horpibulsuk et al. (2017)

- The split tensile strength and flexural strength of the samples with RAP shows the same pattern as that of the compressive strength. However, because of the binder adhered to the RAP aggregates, the load absorption of the RAP samples increased as compared to the samples made using NA. The reason being the decrease in elasticity modulus of sample when RAP content is increased.
- The permeability of sample of concrete with RAP is higher, which implies a higher opportunity for corrosion of the reinforced material when utilised. The parameters that impact the corrosion of the sample are the gradation of the RAP aggregates and their fineness modulus. The use of fly ash, which increases the permeability of concrete with RAP aggregates, might be a solution to this problem (Avirneni et al., 2016; Hoy et al., 2016a; Saha and Mandal, 2017).

References

Al-Mufti, R. L. and Fried, A. N. (2018). Non-destructive evaluation of reclaimed asphalt cement concrete. Eur. J. Environ. Civil Eng. 22(6):770–782.

Antunes, V., Freire, A. C., and Neves, J. (2019). A review on the effect of RAP recycling on bituminous mixtures properties and the viability of multi-recycling. Constr. Build. Mater. 211:453–469.

Arshad, M. (2020). Laboratory investigations on the mechanical properties of cement treated RAP-natural aggregate blends used in base/subbase layers of pavements. Constr. Build. Mater. 254:119234. ISSN 0950-0618. https://doi.org/10.1016/j.conbuildmat.2020.119234.

Arulrajah, J., Piratheepan, J., and Disfani, M. M. (2014). Reclaimed asphalt pavement and recycled concrete aggregate blends in pavement subbases: Laboratory and field evaluation. J. Mater. Civil Eng. 58:245–257.

Avirneni, D., Peddinti, P. R. T., and Saride, S. (2016). Durability and long-term performance of geopolymer stabilised reclaimed asphalt pavement base courses. Constr. Build. Mater. 121(15):198–209.

Chyne, J. A. and Sepuri, H. (2019). A Review on Recycled Asphalt Pavement in cement concrete. Int. J. Latest Eng. Res. Appl. 4: 9–18.

Decker, D. S. and Young, T. J. (1996). Handling RAP in an HMA Facility. In Proceedings of the Canadian Technical Asphalt Association, Edmonton, Alberta.

Delwar, M., Fahmy, M., and Taha, R. (1997) Use of reclaimed asphalt pavement as an aggregate in portland cement concrete. Mater. J. 94(3): 251–256.

Franke, R. and Ksaibati, K. (2015). A methodology for cost-benefit analysis of recycled asphalt pavement (RAP) in various highway applications. Int. J. Pavement Eng. 16(7):1–7.

Hassan, K. E., Brooks, J. J., and Erdman, M. (2000). The use of reclaimed asphalt pavement (RAP) aggregates in concrete. Waste Manag. Series 1:121–128.

Horpibulsuk, S., Hoy, M., Witchayaphong, P., Rachan, R. and Arulrajah, A. Recycled asphalt pavement – fly ash geopolymer as a sustainable stabilized pavement material Published by IOP Publishing Ltd. (2017). IOP Conference Series: Materials Science and Engineering, International Conference on Informatics, Technology and Engineering (InCITE 2017), Bali, Indonesia. 273:24–25.

Hossiney, N., Tia, M., and Bergin, M. J. (2020) Concrete containing RAP for use in concrete pavement. Int. J. Pavement Res. Technol. 3(5):1–13.

Hossiney, N., Wang, G., Tia, M., and Bergin, M. (2020). Evaluation of concrete containing recycled asphalt pavement for use in concrete pavement. In Transportation Research Board 87th Annual Meeting Transportation Research Board, Issue 08-2711.

Hoy, M., Arulrajah, A., and Mohajerani, A. (2018). Strength and microstructural study of recycled asphalt pavement: slag geopolymer as a pavement base material. J. Mater. Civil Eng. 30:1–11. 10.1061/(ASCE)MT.1943-5533.0002393.

Hoy, M., Horpibulsuk, S., and Arulrajah, A. (2016a). Strength development of recycled asphalt pavement – fly ash geopolymer as a road construction material. Constr. Build. Mater. 117:209–219.

Hoy, M., Horpibulsuk, S., Rachan, R., Chinkulkijniwat, A., and Arulrajah, A. (2016b). Recycled asphalt pavement – fly ash geopolymers as a sustainable pavement base material: Strength and toxic leaching investigations. Sci. Total Environ. 573:19–26.

Huang, B., Shu, X., and Burdette, E. G. (2006). Mechanical properties of concrete containing recycled asphalt pavements. Mag. Concr. Res. 58(5):313–320.

Huang, B., Shu, X., and Li, G. (2005). Laboratory investigation of portland cement concrete containing recycled asphalt pavements. Cement Concr. Res. 35(10):2008–2013.

Kallas, B. F. and United States. (1985). Federal Highway Administration. Offices of Research, Development, and Technology. & Asphalt Institute. Flexible pavement mixture design using reclaimed asphalt concret. [Washington, D.C.] : Springfield, VA: U.S. Department of Transportation, Federal Highway Administration, Research, Development, and Technology; National Technical Information Service [distributor], https://nla.gov.au/nla.cat-vn3880233.

Mahmoud, E., Ibrahim, A., El-Chabib, H., Chowdary, V., and Patibandla, V. (2013). Self-consolidating concrete incorporating high volume of fly ash, slag, and recycled asphalt pavement. Int. J. Concr. Struct. Mater. 7(2):155–163.

Mathias, V., Sedran, T., and de Larrard, F. (2009). Modelling of mechanical properties of cement concrete incorporating reclaimed asphalt pavement. Road Mater. Pavement Des. 10(1):63–82.

Okafor, F.O. (2010). Performance of recycled asphalt pavement as coarse aggregate in concrete. Leonardo Elec. J. Pract. Technol. 9(17):47– 58.

Saha, D. C. and Mandal, J. N. (2017). Laboratory investigations on reclaimed asphalt pavement (RAP) for using it as Base Course of flexible pavement. Procedia Eng. 189:434–439. ISSN 1877-7058.

Said, S. E. E. B., Khay, S. E. E., and Louliz, A. (2018). Experimental investigation of PCC incorporating RAP. Int. J. Concr. Struct. Mater. 12:8.

Sargious, M. and Mushule, N. (2011). Behaviour of recycled asphalt pavements at low temperatures. Cana. J. Civil Eng. 18:428–435.

Seferoglu, A. G., Seferoglu, M. T., and Akepmar, M. V. (2018). Investigation of the effect of recycled asphalt pavement material on permeability and bearing capacity in the base layer. Adv. Civil Eng. 2018:6, Article ID 2860213.

Singh, S., Ransinchung R. N. G. D., and Kumar, P. (2018). Performance evaluation of RAP concrete in aggressive environment. J Mater. Civil Eng. 30(10).

Singh, S., Ransinchung, R. N. G. D., and Kumar, P. (2019). Feasibility study of RAP aggregates in cement concrete pavements. Road Mater. Pavement Des. 20(1):151–170.

Smith, R. W. (1980). State-of-the-art hot recycling. transportation research board, Record No. 780. In Proceedings of the National Seminar on Asphalt Pavement Recycling, Washington, DC.

Solanki, P. and Dash, B. (2016). Mechanical properties of concrete containing recycled materials. Adv. Concr. Constr. 4(3):207–220.

Su, K., Hachiya, Y., and Maekawa, R. (2009). Study on recycled asphalt concrete for use in a surface course in airport pavement. Resour. Conserv. Recycl. 54(1):37–44.

Thakur, J. K., Han, J., and Parsons, R. L. (2013). Creep behavior of geocell-reinforced recycled asphalt pavement bases. J. Mater. Civil Eng. 25:1533–1542.

Thakur, S. C., Han, J., Chong, W. K., and Parsons, R. L. (2010). Laboratory evaluation of physical and mechanical properties of recycled asphalt pavement. In Geo Shanghai 2010 International Conference Paving Materials and Pavement Analysis.

Tia, M., Hossiney, N., Su, Y. M., Chen, Y., and Do, T. A. (2012). Use of reclaimed asphalt pavement in concrete pavement slabs. Tallahassee 321. https://rosap.ntl.bts.gov/view/dot/25372.

Wartman, J., Grubb, D. G., and Nasim, A. (2004). Select engineering characteristics of crushed glas. J. Mater. Civil Eng. 16:526–539.

Zaumanis, M. and Mallick, R. B. (2015). Review of very high-content reclaimed asphalt use in plant-produced pavements: State of the art. Int. J. Pavement Eng. 16:(1):39–55.

32 Study on fresh properties of self-compacting concrete blended with sugarcane bagasse ash, metakaolin and glass Fibre

Monali Wagh[a] *and U. P. Waghe*[b]

Department of Civil Engineering, Yeshwantrao Chavan College of Engineering, Nagpur, India

Abstract

Fill the spaces, pass through, and resist segregation are all characteristics of self-compacting concrete. In crowded reinforcement, SSC flows under its own weight and spreads evenly. The aggregates used in SCC are tiny. In India, there is a large amount of industrial and agricultural waste, posing disposal issues. Pozzolanic properties are found in several industrial and agricultural wastes. So, the utilisation of these materials as a supplemental cementitious material in SCC is one of the solutions to prevent environmental threats. The purpose of this research is to determine and evaluate the fresh and strength qualities of SCC manufactured using sugarcane bagasse ash (SBA) and metakaolin (MK) as replacements for cement in a 0 to 20% combined incorporation with 0.1% glass fibre. When the amount of SBA and MK grows up to 15% owing to partial cement replacement, the fresh qualities of SCC improve, but this improves with the inclusion of glass fibre.

Keywords: pozzolanic material, metakaolin, self-compacting concrete, workability.

Introduction

SCC stands for self-compacting concrete and is defined by its filling and passing capacity and isolation resistance. SSC flows under its own weight and spread smoothly in congested reinforcement. In SCC small size aggregates are used. When compared to the SCC without supplementary cementitious material, the cracking and spalling process in the SCC with supplemental cementitious material was less severe as compared to the SCC without material. Upto the 40% replacement of cement by SBA and MK, shows the positive result on spalling effect (Larissa et al., 2020). Upto the 15% of MK in SCC gives the positive results on compressive strength (Gill and Siddique, 2017). With 5% of addition of SBA in SCC improves the CS, STS, WAb, UPV (Zareei et al., 2018). Upto 35% of cost reduces with the use of SBA in SCC (Akram et al., 2009). Modulus of rupture and STS enhanced with the addition of glass fibre in SCC (Ahmad et al., 2017).

The purpose of this research is to employ SBA and MK as an ecologically acceptable supplemental cementitious material in the creation of SCC. In the research, the effect of adding glass fibre in prepared SCC is investigated. The rheological and hardened characteristics of the suggested SCC are investigated using an experimental approach. This study would add to a rising of knowledge concerning the utilisation of agricultural wastes in ecologically friendly concrete manufacturing.

Materials Used:

Ordinary Portland Cement: In order to manufacture self-compacting concrete, grade 43 OPC cement was employed. IS 8112-2013 was followed for the cement. Ultra Tech cement was used for all the SCC combinations.

Coarse and Fine Aggregate: The sand and coarse aggregate utilised in the experiments met the requirements of IS 383. The coarse aggregates were sieved with 10mm and 20mm in size, and the fine aggregates was 4.75 mm sieve. Fine aggregate, 10 mm coarse aggregate, and 20mm coarse aggregate have specific gravity of 2.680, 2.860, and, 2.900, respectively.

Bagasse-Ash (SBA): The SBA used in this study came from the sugarcane industry in Tumsar, India's Devhala region. X-ray fluorescence (XRF) was employed to investigate the characterisation of SBA, as shown in Table 32.2. XRF testing was carried out at IBM Nagpur. SBA has a maximum concentration of silica 55% SiO_2, followed by Al_2O_3 37%, TiO_2 2.530%, and Fe_2O_3 1.150 %, according to the results of an

[a]wagh.monali04@gmail.com; [b]udaywaghe@yahoo.com

Table 32.1 Development of self-compacting concrete using supplementary cementitious material

Ref. NO.	*Waste (Ash) material used for production of Self Compacting Concrete and % used (by weight)*	*Various Fresh test conducted*	*Discussion*	*Mechanical Properties Test of SCC*	*Discussion*
Le et al. (2018)	SBA and BFS used as replacement of cement in 10, 20 & 30% individual and collectively	SF, VF, T50SF,LB	replacement of cement by SBA or BFS in mixtures -lesser flowability	density, CS, UPV, SR, WAb and electrical resistivity	CS with 30% SBA and 30% BFS in SCC was equivalent to the control mix. SR was improved by SBA and BFS. At 28 days, SCC samples exhibited a minimal corrosion rate.
Hamza Hasnain et al. (2021)	By weight, sand is replaced with RHA and BA in amounts of 0, 10, 20, and 30%.	T50SF, J-ring, LB and VF	use of SBA & RHA, enhance the viscosity	density, WAb, CS, STS and SR	density decreases and CS increases either SBA & RHA. SR increases due to pozzolanic activity of ashes.
Larissa et al. (2020)	Sugarcane bagasse ash and metakaolin are used to replace 20% of the cement each.	SF, J-ring, LB and VF, visual stability index.	Rheological properties of SCC are within the range of EFNARC.	mass loss, CS, UPV, capillarity absorption, void index ,XRD and immersion,	SCC with 40% SBA and metakaolin as a cement substitute is more resistant to high temperatures, cracks less, and loses less strength.
Karahan et al. (2012)	metakaolin content replaced by 0%, 5%, 10% and 15% by weight of binder and and fly ash by 20%, 40%, and 60%.	SF, T50, VF and LB	The inclusion of metakaolin enhances the density. The fresh characteristics of developed SCC are unaffected.	CS, FS, STS and sorptivity, absorption, bond strengths, porosity, and RCPT	strength properties increase by small amount
Hassan et al. (2012)	Developed SCC containing 0%, 3%, 5%, 8%, 11%, 15%, 20%, and 25% MK by weight of cement	SF, VF, T50SF, LB	MK shows higher passing ability	CS, Drying shrinkage, Freezing and thawing, Salt scaling, RCPT	Strength properties and durability of SCC are enhanced by increasing amount of MK. MK is beneficial to reduce chloride-permeability.
Ofuyatan et al. (2019)	Silica fume (SF) and metakaolin (MK) was used to substitute the Portland cement to 5%, 10%, 15%, 20% and 25%	SF, VF, T50SF, LB	Mix with metakaolin had good workability properties. By increasing metakaolin %, increase workability.	CS, FS, STS	With the accumulation of MK to SCC, the hardened characteristics improve. The maximum CS is achieved when MK is used to replace 15% of cement.
Ozcan and Kaymak (2018)	Metakaolin and Calcite replaced by 10, 15, 20 cement	SF, VF, T50SF, LB	Rheological properties of SCC with MK are within the range of EFNARC.	CS, Abrasion Test	Utilising MK and C together increased CS
Khotbehsara et al. (2017)	Pumice and Metakaolin is replaced by 0 to 15% by weight of cement	SF, visual stability index, T50SF, VF, and LB	With the addition of Pumice and MK, improve SCC's workability.	CS, Water Absorption, Density, Electrical Resistivity	The usage of pumice and MK instead of cement resulted in a higher CS decrease in all mixes' water absorption.
Panda et al. (2017)	Cement replaced by metakaolin by 5-20%	SF, T50SF, J-Ring, LB, VF and U-Box tests	increase in MK content improves the workability of CC and SCC	CS, STS and FS, Carbonation depth, WAb and density	MK replaced 10% of the cement and increased the hardened characteristics.

SF, Slump-Flow; VF, V funnel test; T500SF, T50 slump Flow; LB, L-Box-test; CS, Compressive Strength; STS, Split Tensile Strength; FS, Flexural Strength; WAb, Water absorption; SF, Slump Flow; UPV, Ultrasonic Pulse Velocity; SR, Sulphate Resistance; XRD, X-ray Diffraction; RCPT, Rapid Chloride Permeability

X-ray fluorescence study (Wagh and Waghe, 2022). SBA is a grey colour (Figure 32.1). Bagasse ash from sugarcane has a specific gravity of 1.9. SEM analysis reveals that SBA has an uneven form (Figure 32.2). The result of the EDS analysis for SBA is shown in Figure 32.5. The concentration of silica is higher than the other elements, according to EDS analysis.

Metakaolin (MK): MK is off-white powder (Figure 32.3), procured from Apple Cheme Pvt. Limited, Nagpur. X-ray fluorescence (XRF) was performed on MK at IBM, Nagpur and the results are shown in Table 32.2. MK has higher content of SiO_2 i.e. 74.41%. The specific gravity of MK was found to be 2.6. SEM testing revealed that MK has an even form as shown in Figure 32.4. MK is in form of powder and colour was off White. 50% particles are below the 1.72 μ. The result of the EDS analysis for MK is shown in Figure 32.6. The concentration of silica and alumina are higher than the other elements, according to EDS analysis.

Table 32.2 X-ray fluorescence analysis of SBA and MK composition of chemical

Materials	*Chemical composition (%)*								
	Na_2O	MgO	Al_2O_3	SiO_2	SO_3	K_2O	P_2O_5	SrO	BaO
SBA (Wagh and Waghe, 2022)	0.18	0.19	37	55	0.15	0.15	0.11	0.02	0.02
Metakaolin	0.14	2.05	1.11	74.41	1.64	4.66	2.5	0.01	0.01
CeO_2	TiO_2	Cr_2O_3	Fe_2O_3	MnO_2	CaO	Cl	ZrO_2	Bi_2O_3	CuO
0.050	2.530	0.030	1.150	0.020	1.30	0.020	0.040	0.030	
	0.12	0.01	1.15	0.11	2.75	0.43			0.01

Figure 32.1 Photographs of the SBA sample used in this study

Figure 32.2 SEM image for SBA

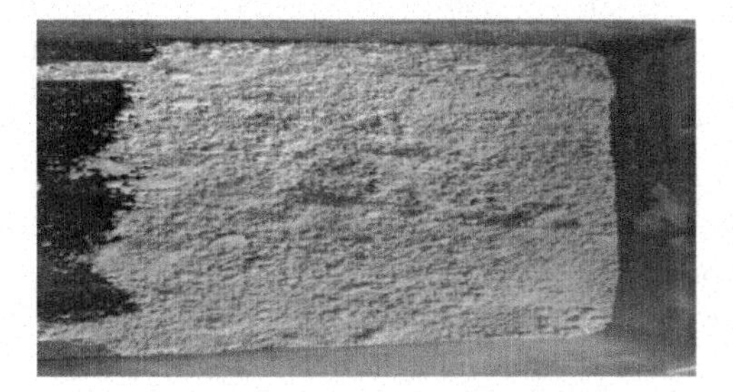

Figure 32.3 Photographs of the MK sample

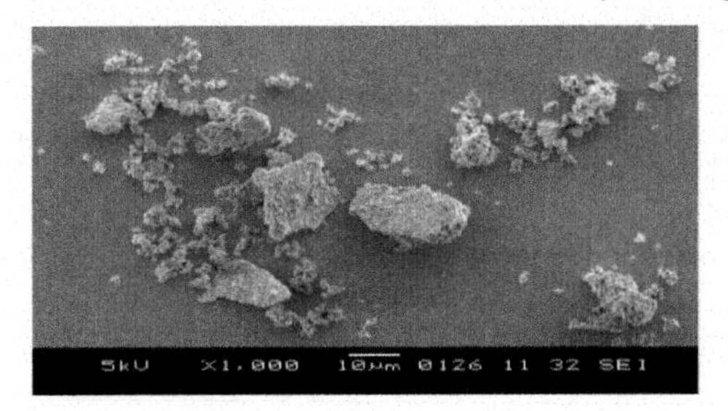

Figure 32.4 SEM image for MK

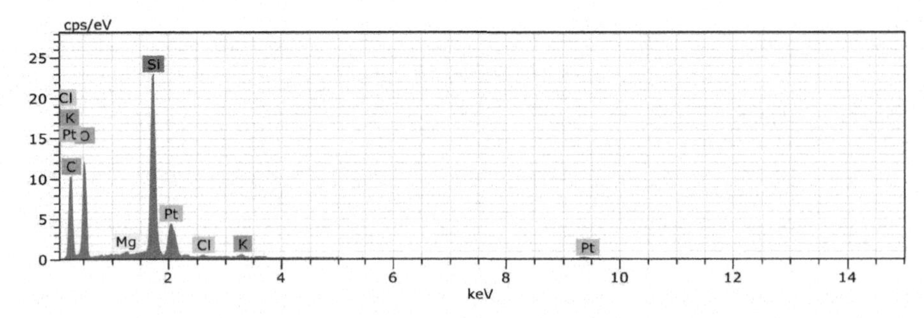

Figure 32.5 Energy dispersive spectroscopy (EDS) report for SBA

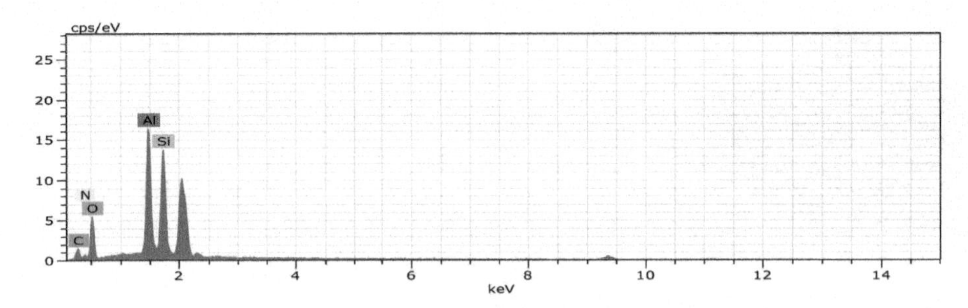

Figure 32.6 Energy dispersive spectroscopy (EDS) report for MK

Table 32.3 Proportion of developed SCC in the mix

Sr. No.	*Mix Type*	*Mix-ID*	*% SBA*	*% MK*	*Aggregates (kg/m³)*		*Cement (kg/m³)*	*Superplasticizer(%)*	*Glass Fibre %*
					Coarse	*Fine*			
1	Single	SCCB0M0	0	0	796.6	958.60	530	1.2	0
2	Tertiary	SCCB10M5	10	5	796.7	915.97	450.5	1.5	0
3	Tertiary	SCCB10M10	10	10	796.63	911.19	424	1.5	0
4	Tertiary	SCC1B0M0	0	0	796.6	927.07	530	1.3	0.1
5	Tertiary	SCC1B10M5	10	5	796.7	915.97	450.5	1.5	0.1
6	Tertiary	SCC1B10M10	10	10	796.63	911.9	424	1.5	0.1

Superplasticizer: SCC employed Viscoflux 5507 as a superplasticizer. At 27°C and a pH of more than 6, the relative density is 1.11. ASTMC494 Type A and F, as well as IS 9103--1999, were used to certify this product (Wagh and Waghe, 2022).

Viscosity Modifying Agent (VA): In SCC, AC-Gel-Build was utilised as the VA. The density ratio is 1.0., and the pH ranges from 5 to 8. IS 9103-1999 was used to validate this product (Wagh and Waghe, 2022).

Experimental Programme

Table 3 shows the mix proportioning data for self-compacting concretes including bagasse ash, metakaolin, and glass fibre, as well as their designations. The images showing the fresh characteristics of SCC are shown in Figures 32.7, 32.8, and 32.9. The cube, beam, and cylinder are displayed in Figure 32.10. The SCC mixes contains 0% and 10% of SBA incorporation with 0, 5 and 10% of MK as a supplementary cementitious material designated as SCCB0M0, SCCB10M5, SCCB10M10 respectively. 0.1% of glass fibre is added and it designated as SCC1B0M0, SCC1B10M5, SCC1B10M10. In this study, an over-all of 6 trial mixes were created by combining various percentages of SBA, MK, and glass fibre. For each mix proportion, 9 cubes of 150 mm by 150 mm by 150 mm and 9 beams of 100 mm by 100 mm by 500 mm are constructed. To achieve exceptional workability and flowability, Viscoflux, concrete additive, was used. The quantity of superplasticizer in the binder component was kept constant at 1.2 to 1.5% by weight of binding material. A VMA dosage of 0.3 to 0.35% by entire weight of the binder components is used in this project. In the fresh test, the slump-flow, L box, V funnel, and T50 slump flow tests were all evaluated. In the hardened condition, compressive strength (CS) and flexural strength (FS) were measured at 7, 14 and 28 days.

Figure 32.7 Slump flow test

Figure 32.8 L-Box test

Figure 32.9 V-Funnel test

Figure 32.10 Prepared cube, beam and cylinder

Result and Discussion

Slump Flow Test: The slump flow diameter for all the SCC mixes was within the range of EFNARC of SCC, however the slump flow for developed SCC mixes blended with SBA, MK and glass fibre ranged between 650 mm and 750 mm as shown in Figures 32.11, 32.12 and 32.13. With the use of SBA and MK collectively as a supplementary-cementitious material, increases the flowability upto 20% of replacement. For the new mix SCCB10M5, the slump values increase from 693 mm to 712mm as compare to control mix. For the new mix SCC B10M10, the value reduces to 702 mm as compare to SCCB10M5. With the use of 0.1% of glass fibre, the flowability reduces but within the range of EFNARC as shown in Figure 32.13. With the accumulation of 0.1% of glass fibre, the slump flow reduces from 693 to 675 for control mix, 712 to 700mm for SCCB10M5 and 702 to 691mm for SCCB10M10.

T-50 Slump Flow Test: The newly created SCC mixes had a slump flow diameter of 670 to 720mm, showing good flowability. T-500 slump flow testing is the simplest method for assessing SCC flowability. The time required for the SCC mixes to blow-out 500 mm, according to EFNARC guidelines (EFNARC, 2002), differed from 2 to 5 seconds. The T-500 slump flow time decreases from 3.6 to 3.5 for the mix type SCCB10M5 which shows the greater flowability. For the mix SCCB10M10, the time increases to 4 but within the range of EFNARC. With the addition of 0.1% of glass fibre in SCCB0M0, the slump flow time increases by 35.83%. 14.28% increment is recorded in the sump flow time for the SCCB10M5 after accumulation of 0.1% of glass fibre. Supreme time of slump flow is recorded for SCC1B0M0. Addition of SBA and MK collectively in SCC is beneficial as shown in Figures 32.12 and 32.13.

V-Funnel Test: The V funnel test timing was observed in the series of 7 to 8 seconds without glass fibre, as illustrated in Figure 32.16. The V funnel timing was observed in the series of 10 to 12 seconds with the

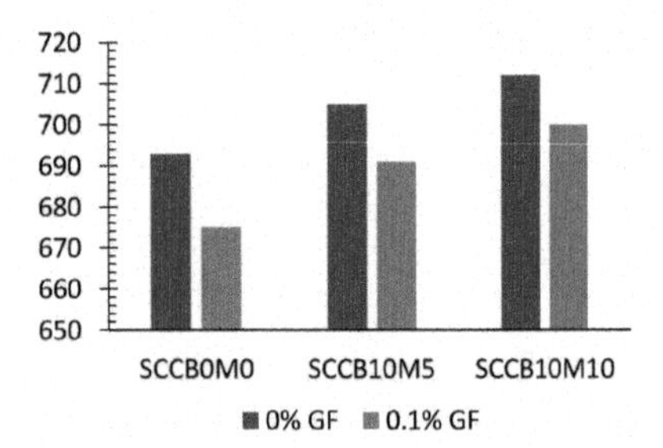

Figure 32.11 Slump flow chart of prepared SCC with and without glass fibre

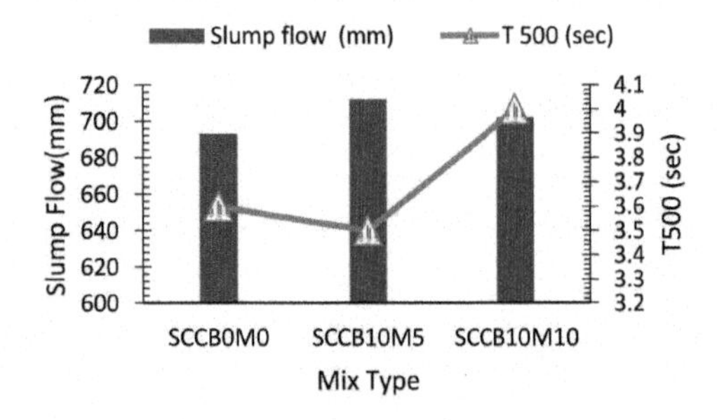

Figure 32.12 T-50 slump flow and slump flow of developed SCC without glass fibre

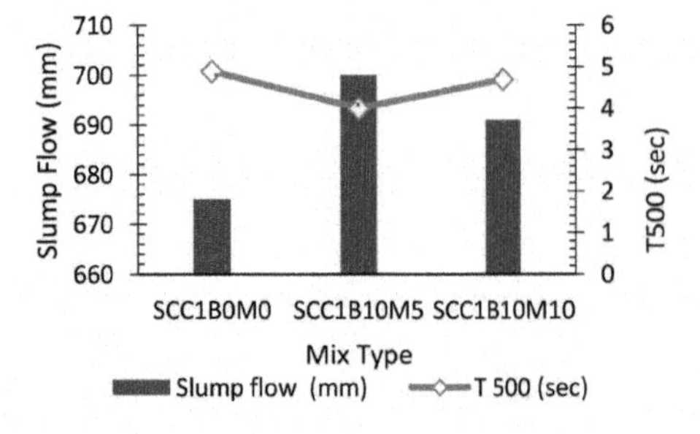

Figure 32.13 T-50 slump flow and slump flow of developed SCC with glass fibre

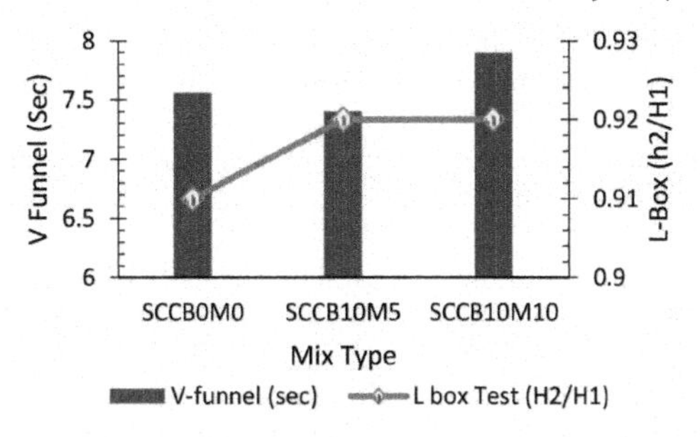

Figure 32.14 V-Funnel and L-Box of prepared SCC without glass fibre

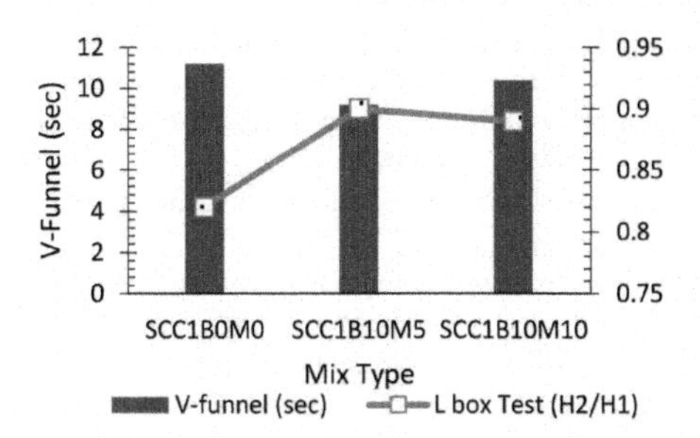

Figure 32.15 L-Box and V-Funnel of prepared SCC with glass fibre

0.1% of glass fibre, as illustrated in Figure 15. Mixes made with the SBA, MK and 0.1% of glass fibre, all had V-funnel durations that were within the permitted range by increasing little amount of superplasticizer. With the addition of glass fibre in various blended mixes, the V-funnel time increases, it means reduces the flowability of SCC. Best suitable combination for V-funnel timing is SCCB10M5.

L Box Test: The L Box test examines whether or not, a product can pass through tightly packed reinforcement on its own weight. The H_2 by H_1 ratio were in the series of 0.8 to 0.92 for SCC blended with SBA, MK and 0.1% of glass fibre as given in Figures 32.14 and 32.15.

Conclusion

1. The usage of SBA, MK as a cement alternative in the creation of SCC might be beneficial.
2. As glass fibre is added to the SCC mix, the fresh qualities of SCC are significantly reduced, but they stay within the EFNARC range.
3. Filling and passing ability of SCC blended with bagasse ash and Metakaolin improved by increasing the slight dose of water reducing admixture.
4. For SCCB10M5, the slump flow increases from 693 to 712mm, T-5oo and V-funnel sec reduces from 3.6 to 3.5 and 7.56 to 7.4sec respectively.
5. With the addition of glass fiber, the slump flow for SCCB0M0, SCCB10M5 and SCCB10M10 reduces from 693 to 675mm, 712 to 700mm and 702 to 691 mm respectively.
6. Most suitable combination for SCC in terms of workability is SCCB10M5 which gives the maximum workability by increasing the slight dose of water reducing admixture.

References

Ahmad, S., Umar, A., and Masood, A. (2017). Properties of normal concrete, self-compacting concrete and glass fibre-reinforced self-compacting concrete: an experimental study. Procedia Eng. 173:807–813. doi:10.1016/j.proeng.2016.12.106.

Akram, T., Memon, S. A., and Obaid, H. (2009). Production of low cost self compacting concrete using bagasse ash. Constr. Build. Mater. 23(2):703–712. doi: 10.1016/j.conbuildmat.2008.02.012.

Gill, A. S. and Siddique, R. (2017). Strength and micro-structural properties of self-compacting concrete containing metakaolin and rice husk ash. Constr. Build. Mater. 157:51–64. doi: 10.1016/j.conbuildmat.2017.09.088.

Hamza Hasnain, M., Javed, U., Ali, A., and Saeed Zafar, M. (2021). Eco-friendly utilization of rice husk ash and bagasse ash blend as partial sand replacement in self-compacting concrete. Constr. Build. Mater. 273: 121753. doi:10.1016/j.conbuildmat.2020.121753.

Hassan, A. A. A., Lachemi, M., and Hossain, K. M. A. (2012). Cement & concrete composites effect of metakaolin and silica fume on the durability of self-consolidating concrete. Cem. Concr. Compos. 34(6):801–807. doi: 10.1016/j.cemconcomp.2012.02.013.

Karahan, O., Hossain, K. M. A., Ozbay, E., Lachemi, M., and Sancak, E. (2012). Effect of metakaolin content on the properties self-consolidating lightweight concrete. Constr. Build. Mater. 31:320–325. doi: 10.1016/j.conbuildmat.2011.12.112.

Khotbehsara, M. M., Mohseni, E., and Ozbakkaloglu, T. (2017). Durability characteristics of self-compacting concrete incorporating pumice and metakaolin. J. Mater. Civ. Eng. 29(11):1–9. doi: 10.1061/(ASCE)MT.1943-5533.0002068.

Larissa, L. C., Marcos, M. A., Maria, M. V., de Souza, N. S. L., and E. C. de Farias. (2020). Effect of high temperatures on self-compacting concrete with high levels of sugarcane bagasse ash and metakaolin. Constr. Build. Mater. 248:118715. doi: 10.1016/j.conbuildmat.2020.118715.

Le, D. H., Sheen, Y. N., and Lam, M. N. T. (2018). Fresh and hardened properties of self-compacting concrete with sugarcane bagasse ash–slag blended cement. Constr. Build. Mater. 185:138–147. doi: 10.1016/j.conbuildmat.2018.07.029.

Ofuyatan et al. (2019) Incorporation of silica fume and metakaolin on self compacting concrete incorporation of silica fume and metakaolin on self compacting concrete. International Conference on Engineering for Sustainable World, Journal of Physics: Conference Series, Conf. Ser. 1378 042089 doi: 10.1088/1742-6596/1378/4/042089.

Ozcan, F. and Kaymak, H. (2018). Utilization of metakaolin and calcite : working reversely in workability aspect — as mineral admixture in self-compacting concrete. 2018.

Panda, K. C. (2017). Effect of metakaolin on the properties of conventional and self compacting concrete. Adv. Concr. Constr. 5(1):31–48. doi:10.12989/acc.2017.5.1.31.

Wagh, M. and Waghe, U. P. (2022). Development of self-compacting concrete blended with sugarcane bagasse ash. Mater. Today Proc. 60(3):1787–1792. https://doi.org/10.1016/j.matpr.2021.12.459. doi:10.1016/j.matpr.2021.12.459.

Zareei, S. A., Ameri, F., and Bahrami, N. (2018). Microstructure, strength, and durability of eco-friendly concretes containing sugarcane bagasse ash. Constr. Build. Mater. 184:258–268. doi: 10.1016/j.conbuildmat.2018.06.153.

33 Transfer Learning for Glaucoma Detection

Shreshtha Gole[a], *Neeraj Rangwani*[b], *Prachi Yeskar*[c], *Aishwarya Vyas*[d], *Parag Jibhakate*[e], *and Kanchan Dhote*[f]

Department of Electronics Engineering, Shri Ramdeobaba College of Engineering and Management Nagpur, India

Abstract

The early detection of glaucoma is discussed in this paper. A comparison of two different transfer learning algorithms is carried out. The morphological markers of glaucoma, which predict the start of abnormalities, were quantified using machine learning algorithms. Glaucoma screening is an expensive, time-consuming, and human-error-prone technique. In developing and poor areas, there are fewer eye experts. This project will save time, money, and resources by making glaucoma screening more accessible to the general public. Because glaucoma is the leading cause of blindness in the United States, it is critical to diagnose it early. With this initiative, we hope to make a difference in society. Both the algorithms used in this methodology are the advanced implementation of Transfer Learning and Deep Learning. With the help of this a comparison between shallow and deep models are done to find the efficiency of both the algorithms on same dataset.

Keywords: Blindness, glaucoma, machine learning, transfer learning.

Introduction

This research analyses the two most effective algorithm strategies for detecting glaucoma in Machine Learning. Glaucoma is one of the most serious undiscovered disorders on the planet. It might cause irreparable harm to the eyes if not detected early. Let us attempt to comprehend what glaucoma entails. Glaucoma is a neurodegenerative eye disease that causes optic neuropathy and visual abnormalities due to optic disc cupping and nerve fibre destruction. It is a major cause of blindness all around the world. For open-angle glaucoma, lowering intraocular pressure (IOP) is a realistic, evidence-based treatment (OAG). When glaucoma symptoms appear and the disease progresses, a correct diagnosis of glaucoma can be difficult at first. Routine glaucoma testing is therefore necessary and recommended. A study of colour fundus pictures, which may show glaucomatous optic neuropathy, including rim shrinkage and compression, reduction, cutting, high cup-to-disc rate, disc bleeding, and a layer of the retinal nerve fibre, is the most basic diagnostic technique for identifying glaucoma (Huang et al., 2017).

Glaucoma is marked by the dysfunction and loss of retinal ganglion cells, which results in a loss of visual space and structural alterations in the optic nerve's head. When persons with the disease are not diagnosed early enough, they lose a lot of function. As a result, early discovery of glaucoma allows for early treatment, which can help to prevent visual loss. Due to those eyes, the disc shape characteristic, and the visual field feature, diagnosing myopic glaucoma in individuals with brain disorders such as brain tumours is difficult. Medical practitioners could benefit from a learning model for a more effective glaucoma detector (Fu et al., 2018). Glaucoma should be diagnosed through a thorough study of machine learning. There are two methods for detection. Using transfer learning with deep learning is one of the methods.

Literature Review

We used several feature extraction and dimensionality deduction investigations based on the architecture to identify the critical components and isolate them for future investigation. We came to the conclusion that an automated system of algorithms may be used to detect glaucoma (Quigley and Broman, 2006).

Glaucoma detection was achieved by combining convolutional and recurrent neural networks. By extracting specific temporal features from the accessible retinal movies and images, a pre-trained CNN model was created. To attain the requisite accuracy, the network was trained using these two methods. The best accuracy was achieved by combining VGG-16 with LSTM (Krizhevsky et al., 2012).

[a]goless@rknec.edu; [b]rangwanink@rknec.edu; [c]yeskarps@rknec.edu; [d]vyasav@rknec.edu; [e]jibhakateps@rknec.edu; [f]dhotek@rknec.edu

The categorisation of glaucoma was done utilising three transfer learning approaches using feature descriptors and vector machines as support, as well as ensemble methods in the techniques indicated. In the strategies utilised for the provided data set, the ensemble method fared the best. Deep learning algorithms were utilised by the authors in (Szegedy et al., 2015) to detect clinical glaucoma. They retrieved annotated images and created a set of heterogeneous 3D optical coherence tomography scans for the early identification of glaucoma using that architecture (Szegedy et al., 2016).

Image processing, deep learning, and machine learning techniques were used to create the application. The input image is validated using the Le-Net architecture, and the Region of Interest (ROI) is detected using the brightest spot algorithm. Furthermore, U-Net architecture is used to segment the optic disc and optic cup, and SVM, Neural Network, and Ad boost classifiers are used to classify the data (Shinde, 2021).

The paper compares and contrasts various state-of-the-art deep learning approaches, including Xception, Inception, DenseNet, ResNet, and VGG. Furthermore, criteria like as precision, recall, and accuracy are used to compare techniques. The findings of this study could be used to develop handheld glaucoma diagnostic tools that medical practitioners and researchers can use to analyse retinal pictures and predict glaucoma. As a result, diagnosis using computer-aided diagnosis (CAD) systems with imaging modalities would function better in the presence of lighting disturbances, as well as minimise diagnostic time and cost when compared to traditional instruments such as tonometers and pachymeters for retinal examination (Thakur and Juneja, 2021).

A comprehensive study of early glaucoma analysis is conducted using various techniques such as Machine Learning, Convolutional Neural Networks, Transfer Learning, and Deep Learning (Janani and Rajamohana, 2021).

The approach is tested using two fundus image datasets for the same people, which were collected using a smartphone and retinography, respectively. Using both databases, we attain 100% accuracy, demonstrating the resilience of our strategy. Furthermore, when the Samsung-M51 and Samsung-A70 smartphone devices are used, the detection takes 0.027 and 0.029 seconds, respectively. In remote clinics or places with limited access to fundus cameras and ophthalmologists, our suggested smartphone software provides a cost-effective and widely accessible mobile platform for early glaucoma screening (Mrad et al., 2021).

Classification of Glaucoma

Predicated on how intraocular pressure increases glaucoma, they are divided into two types:

A. Open Angle Glaucoma

This type of glaucoma is also generally called chronic glaucoma. It is one of the most common types of glaucoma. The liquid is gone long ago or when the drainage system is in the eye closed over time. As a result, the liquid/ fluid cannot escape the eye. This results in an increase in intraocular pressure. The progression is slow, and cases are unfit to prevent vision loss until the complaint has increased dramatically. A lot more than 80 glaucoma cases suffer from open-angle glaucoma. This kind of glaucoma generally responds well to medicines that can be fluently detected and treated early.

B. Close Angle Glaucoma

This is also called acute small-angle glaucoma. It is rare and severe. About 10 glaucoma cases are affected by unrestricted-angle glaucoma. Closed-angle of glaucoma occurs in people with veritably little space between the cornea and iris. As the eye grows, the pupil grows, leading to this small space's restriction. Thus, water inflow is restricted. This leads to a rapid-fire increase in intraocular pressure. Because of the unforeseen increase in pressure within the eye, glaucoma is more severe. It is treated surgically, creating a pathway fluid inflow between the cornea and iris.

The early detection of glaucoma could be done in different ways. Some of them are:-

1. **Examination of increased intraocular pressure:** In most cases, glaucoma exists without an increase in IOP; in that case, IOP measurement methods fail to detect the disease.
2. **Unusual viewing environment:** Visual space testing requires special equipment that is usually only available in tertiary hospitals if they have a fundus camera and OCT (Optical Coherence Tomography). Therefore, this method is not suitable for glaucoma tests.
3. **Examination of changes in retinal structure:** Internal surface of the eye, the retina contains layers of light-sensitive neurons. Eye diseases such as diabetic retinopathy, macular degeneration, and glaucoma

affect the neuron layer. Therefore, images of the eye retina fundus are analysed to discover glaucoma. There are two main problems in glaucoma-recognition using images of the fundus (Li et al., 2018):

a. ***Extrusion of texture element in retina images:*** use of image elements (pixel intensity, texture, spectral features, histogram model parameters) in binary categories between glaucomatous and healthy subjects. These features are commonly computerised at the image level. In these ways, selecting features and stages of the strategy is complex and challenging.
b. ***Exposure of structural element in retina images:*** This strategy is based on the clinical indication of vertical cup size to disk, disk width, and peripapillary atrophy.

Implementation

There are several clinical ways to diagnose glaucoma. Still, population growth and rising obesity rates mean that the number of eye doctors needed for a direct diagnosis is a limiting factor. Therefore, the system automatically detects the features of pathological cases, which can be of great benefit (Simonyan and Zisserman, 2014; Targ et al., 2016). There are many natural ways to diagnose glaucoma. One of the main symptoms of glaucoma is the location of the cup. The cup expands to take up more disk space. Many methods have been developed to determine optical cup enlargement. A large percentage indicates an outbreak of the disease. However, the data showed that physical detection and prognosis for glaucoma are tedious and subtle, and it depends entirely on the experience of the expert. In the last few decades, significant effort has been made to find and predict cases of glaucoma using many machine learning methods. Some sensory network methods are decision trees based on the ID3 algorithm, vector support machine, naive Bayes classifier for nearby neighbours, and Canny edge detector. Figure 33.1 represents the flow of work in the implementation part of the work.

1. VGG-16

The architecture of the VGG-16 algorithm is depicted in Figure 33.2. Convolution, max pooling, ReLU, and SoftMax are all levels in this stack. Each layer denotes a different level of significance for the image in question. It's the CNN structure that's been used (Li et al., 2018).The original pixel values based on three colour channels R, G, and B [224 × 224 × 3] will be stored in the input image of width 224 height 224 for the glaucoma detection system.

A. Convolution Layer

The convolution level is the main component of CNN and controls the object problem. The critical layer uses a set of source input images and uses a set of filtered filters that can ultimately create 2D animation cards. CNN allows the system to capture the image function due to weight distribution to lower the

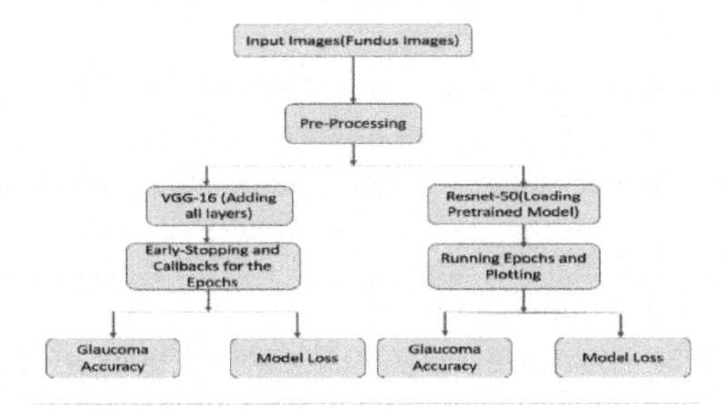

Figure 33.1 Flowchart for VGG-16 and Resnet-50 implementation

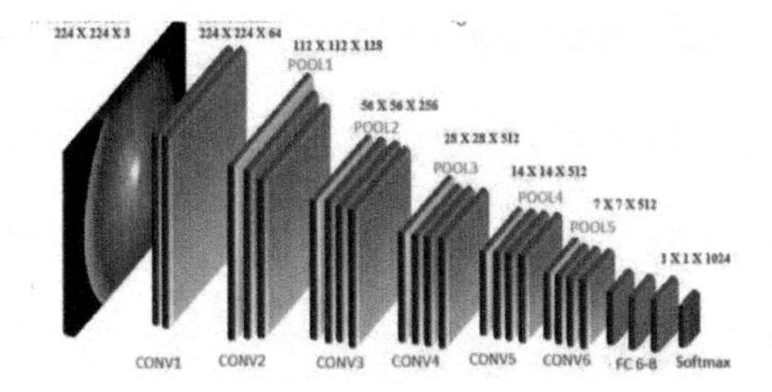

Figure 33.2 VGG-16 architecture

integral cost. In a glaucoma detection system, there are 13 convolutional layers, and a convolutional filter is 3 × 3 layer and after the data size is [7 × 7 × 512]. The image is transferred to the first stack of the first layer of the smallest layer of size 224 × 3, and the activation of the ReLU is continued. Each layer has 64 filters. The conversion step is paused with one pixel and ends with another. This parameter stores position adjustment and the output card is the same as the image input size. The activation card is then sent to the upper surface area of the two pixels in the 2 × 2-pixel window. This allows the activation size to be reduced twice. So, at the end of the first stack, the activation size is 112 × 112 × 64.

B. Max Pooling Layer

The pooling layer in CNN architecture is used to abstract image features. The most common form of a pooling layer uses a 2 × 2 filter in two steps to select 25% of the activations of the activation map in the convolutional layer. Each max-pooling operation picks the most significant number in the 2 × 2 area. So, the data size will be [16 × 16 × 12]. Applying a pooling layer reduces the size of objects and compute parameters in the network, controls overfitting, and improves overall network performance. When merging functions, the maximum for that group is still 4, so we get the same result in the integrated version. This process gives convolutional neural networks the possibility of being "spatially distributed." Because max-pooling lowers the resolution of a given output of the convolutional layer, the web will see a large area of the image at a time, reducing the number of parameters in the network and reducing the computational load. Max-pooling also helps minimise overfitting. The intuition for max-pooling works is that the network will look for a specific function for a particular image.

C. Rectified Linear Unit

The Rectified Linear Unit (ReLU) is a type of activation function that computes the weighted sum of input values and determines whether or not to keep it. In comparison to other activation functions, ReLU takes fewer computer resources. The ReLU function is easy to use and does not require any complicated arithmetic or calculations. Therefore, it may take less time to train or run the model. Another vital property that is considered an advantage is its sparsity. The ReLU gives a zero output for every negative input; it is very likely that a given block will never be executed, resulting in a sparse network. Neural networks are trained using gradient descent. Gradient descent consists of backpropagation steps and is a chain rule that changes weights to reduce losses after each epoch. It is important to note that derivatives play an essential role in updating weights. As the number of layers increases, the slope continues to decrease. This will lower the gradient values of the initial layers, and these layers will not be able to train correctly. In other words, the gradient tends to be zero because the depth and activation of the network shift the value towards zero. This is called the vanishing gradient problem. ReLU, on the other hand, does not suffer from this problem because the gradient does not stagnate or 'saturate' as the input grows. For this reason, models using the ReLU activation function converge faster.

D. SoftMax

A SoftMax layer is an activation function that converts a fully linked layer's output into probabilities that add up to one. It returns a vector containing the possible output classes and their associated probability. The SoftMax function's main purpose is to convert a fully-connected layer's (unnormalised) output of K units (which is represented as a vector of K elements) to a probability distribution (which is commonly expressed as a vector of K elements, each of which is between 0 and 1). (a probability). All of these factors add up to one (a probability distribution). It's mostly used to fit zero and one into the output of neural networks. It represents the 'probability' of certainty. Figure 33.3 represents the architecture of the ResNet-50 algorithm.

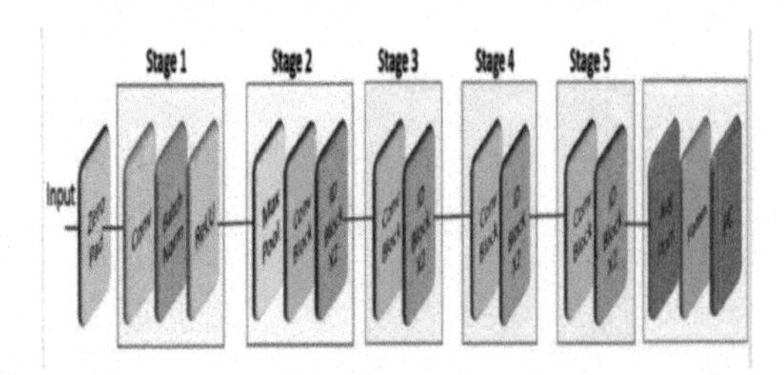

Figure 33.3 ResNet-50 architecture

2. ResNet-50

a. Identity Block

Figure 33.4 shows the identity block, which is the first block in the ResNet-50 architecture, followed by the skip connection. The identity block is the standard block used in ResNet architecture and has similarities to the case where the input activation has the same dimension as the output activation.

b. Convolutional Block

Figure 33.5 represents the convolutional block, which consists of layers similar to VGG-16, with a normalisation layer to normalise each image. In ResNet-50 architecture, a "shortcut" or a "skip connection" allows the gradient to be directly backpropagated to earlier layers. It is divided into five stages, and each stage has different blocks or combinations of blocks.

c. Residual Block

The ResNet is much deeper than VGG-16, and the model size is less than the cost of reducing the size. Face recognition, based on the deep learning architecture of the neural network, can cast the height width (32, 3) to the channel width (32, 3). The 2D component convolution has 64 filter shapes 77 and uses 22 steps. For the ResNet-50 model, a 1x1 wrapper is used to reduce the channel depth, then rebuilds the 10-layer bottleneck block to reduce the computational load, simply replacing each 2-layer residual block. ResNet-50 is a deep learning neural network residual model with 50 layers (Al-Bander et al., 2017; Fu et al., 2018). This architecture introduces the concept of a "residual network." This network uses skip connection technology. The benefit of adding this type of skip connection is that it skips to normalisation if the whole tier achieved architecture performance. This allows the training of intense neural networks without the problem of vanishing gradients. The convolutional and Identity residual blocks are present as sub- residual blocks in the architecture (Ahn et al., 2018).

d. Dataset

Table 33.1 displays the information about the dataset which was used for this project. Figure 33.6 shows sample images from the dataset used for algorithms implementation. This Dataset is taken from open source available on the internet. More images are also in consideration for better accuracy in further iterations.

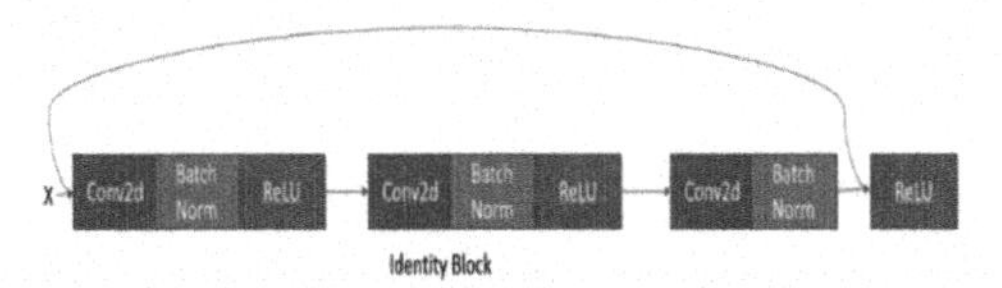

Figure 33.4 Identity block in ResNet-50

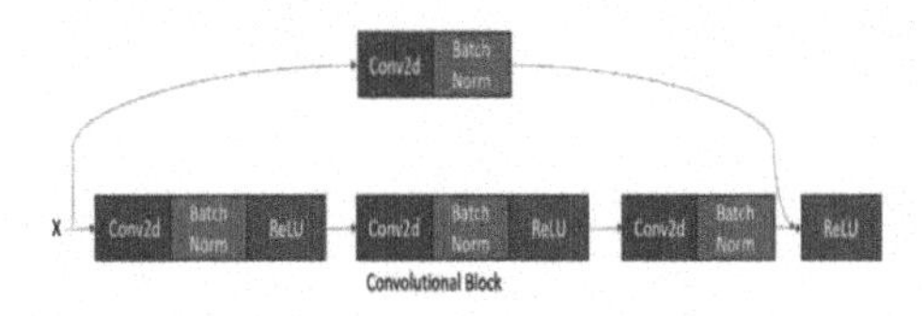

Figure 33.5 Convolutional block in ResNet-50

Table 33.1 Dataset used for implementation

Type of Images	*Glaucoma*	*Nonglaucoma*
Training Images	233	486
Validation Images	73	136

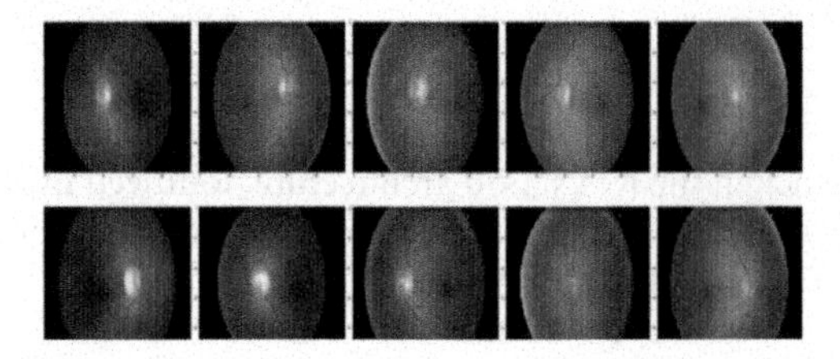

Figure 33.6 Sample of fundus images dataset used in implementation

Results and Conclusion

Various transfer learning algorithms were applied to a dataset of images for detecting the presence of glaucoma. Algorithms like VGG-16 and ResNet-50 were trained on the training data. An accuracy of 94.64% was obtained for ResNet-50 on training data and 90.44% on validation data. The least accuracy of 82.77% was obtained for the VGG-16 on training, data making it the least favourable for detection of glaucoma. Table 33.2 is a representation of the accuracies of the algorithms in percentage.

Figure 33.7 displays the accuracies and loss for the VGG-16 algorithm. Figure 33.8 displays the accuracies and losses for the ResNet-50 algorithm in a visual form. Since the dataset is varying, there seems to be a variating graph for both VGG-16 and ResNet -50. The training accuracy in both cases is more or equal to the validation accuracy since the dataset is more for training. The Losses are in the declining trend with the number of epochs which shows the consistency and closeness to an ideal situation. Here the testing loss is more than training for the same reason stated above.

Table 33.2 Representation of accuracies of algorithms

Comparative results of algorithms	*Training accuracy*	*Validation Accuracy*	*Loss*
VGG-16	82.77%	82.77%	0.39
ResNet-50	94.64%	90.44%	0.24

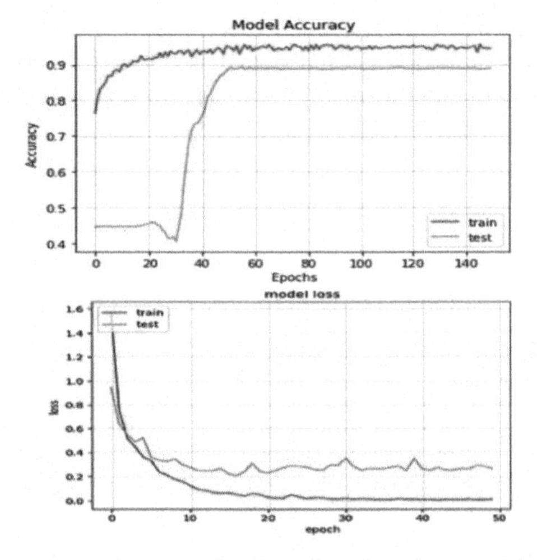

Figure 33.7 Visualised results of accuracy and loss for the VGG-16 algorithm

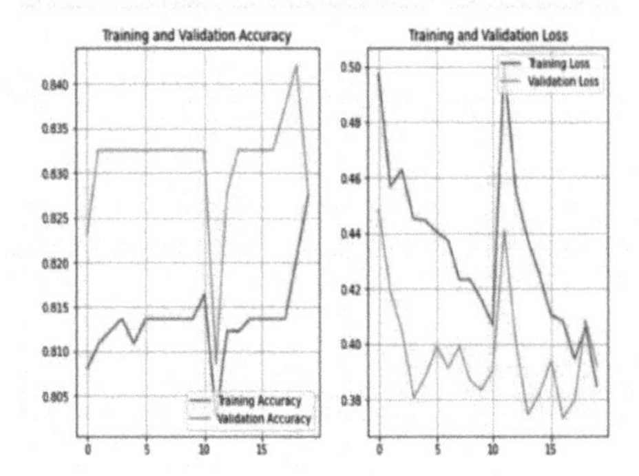

Figure 33.8 Visualised results of accuracy and loss for the Resnet-50

Discussion

Various techniques are used for glaucoma detection, but the concept of gradient comes into play while evaluating various algorithms for the model. Those problems were dealt with in several ways and allowed tens of networks to coalesce. The Training data was vast compared to validation data, so the precision is more. With the help of the ResNet-50 algorithm, the problem of the gradient is resolved with improved accuracy. The residual network is better for the Glaucomic fundus image than the VGG-16 algorithm. In both the algorithms, if we use transfer learning, we get better accuracy than building the model and then training the dataset. With each iteration or epoch, the accuracy varies, which shows the variation in the dataset, resulting in close to the real-time fundus images present. Sometimes overfitting issues arise in these algorithms because of a shortage of dataset or more layers, especially in higher layers algorithms like Resnet-50 and onwards. In VGG-16, that issue does not arise because of its compact architecture. Also, an early stopping feature is available in that algorithm to avoid redundancy in the runtime for epochs.

References

Ahn, J. M., Kim, S., Ahn, K.-S., Cho, S.-H., Lee, K. B., and Kim, U. S. (2018). A deep learning model for the detection of both advanced and early glaucoma using fundus photography. PloS One. 13(11):e0207982.

Al-Bander, B., Al-Nuaimy, W., Al-Taee, M. A., and Zheng, Y. (2017) Automated glaucoma diagnosis using deep learning approach. In 2017 14th International Multi- Conference on Systems, Signals & Devices (SSD), pp. 207–210, IEEE.

Chen, X., Xu, Y., Wong, D. W. K., Wong, T. Y., and Liu, J. (2015). Glaucoma detection based on deep convolutional neural network," in 2015 37th annual international conference of the IEEE engineering in medicine and biology society (EMBC), pp. 715–718, IEEE.

Christopher, M., Belghith, A., Bowd, C., Proudfoot, et al. (2018). Performance of deep learning architectures and transfer learning for detecting glaucomatous optic neuropathy in fundus photographs. Scientific Rep. 8(1):16685.

Fu, H., Cheng, J., Xu, Y., et al. (2018). Disc-aware ensemble network for glaucoma screening from fundus image," IEEE transactions on medical imaging. 37(11):2493–2501.

Huang, G., Liu, Z., Van Der Maaten, L., and Weinberger, K. Q. (2017). Densely connected convolutional networks. In Proceedings of the IEEE conference on computer vision and pattern recognition, pp. 4700–4708.

Janani, R., Rajamohana, S. P. (2021). Early detection of glaucoma using optic disc and optic cup segmentation: A survey. Mater. Today Proc. 45(2):2763–2769.

Krizhevsky, A., Sutskever, I., and Hinton, G. E. (2012). Imagenet classification with deep convolutional neural networks. Adv. Neural Inform. Process. Syst. 1:1097–1105.

Li, Z., He, Y., Keel, S., Meng, W., Chang, R. T., and He, M. (2018). Efficacy of a deep learning system for detecting glaucomatous optic neuropathy based on color fundus photographs. Ophthalmology. 125(8):1199–1206.

Mrad, Y., Elloumi, Y., Akil, M., and Bedoui, M. H. (2021). A fast and accurate method for glaucoma screening from smartphone-captured fundus images. IRBM, 137–144.

Quigley, H. A. and Broman, A. T. (2006). The number of people with glaucoma worldwide in 2010 and 2020. Br. J. Ophthalmol. 90(3):262–267.

Sharanya S. (2019). Glaucoma detection using machine learning. Int. J. Sci. Res. 7:583.

Shinde, R. (2021). Glaucoma detection in retinal fundus images using U-net and supervised machine learning algorithms. 5:100038.

Simonyan, K and Andrew. (2014). Very deep convolutional networks for large-scale image recognition. CoRR abs/1409.1556: n. pag.

Szegedy, C., . Liu, W., Jia, Y., et al. (2015). Going deeper with convolutions. In Proceedings of the IEEE conference on computer vision and pattern recognition, pp. 1–9.

Szegedy, C., Vanhoucke, V., Ioffe, S., Shlens, J., and Wojna, Z. (2016). Rethinking the inception architecture for computer vision. In Proceedings of the IEEE conference on computer vision and pattern recognition, pp. 2818–2826.

Targ, S., Almeida, D., and Lyman, K. (2016). Resnet in resnet: Generalizing residual architectures. CoRR abs/1603.08029 doi: https://doi.org/10.48550/arXiv.1603.08029:1–4.

Thakur, N. and Juneja, M. (2021). Early stage prediction of glaucoma disease to reduce surgical requirement using deep-learning. Mater. Today. Proc. 45(6): 5660–5664.

34 Experimental evaluation and correlation of plasma transferred arc welding parameters with hardfacing defects

Sachin Kakade[1,a], *Ajaykumar Thakur*[1,b], *Sanjay Patil*[2,c], *and Dhiraj Deshmukh*[3,d]

[1]Department of Mechanical Engineering, Sanjivani College of Engineering, Kopargaon, India

[2]Department of Quality Control, KOSO India Pvt. Limited, Nashik, India

[3]Department of Mechanical Engineering, MET's, Institute of Engineering, Nashik, India

Abstract

Component's which are used in severe working conditions must be given the utmost attention. The surface treatments have been used in the oil, gas, and petrochemical sectors to apply different materials and metals on component. The welding technique may be used to apply a layer of corrosion and wear-resistant material on top of a substrate material. The poor welding procedures or weld flaws including porosity, blowholes and undercuts were responsible for component failure. In the case of an overlay process, weldment properties such as penetration, bead shape, reinforcing, penetration depth, and HAZ play a critical role in structural integrity and joint homogeneity. The substrate material's basic composition, bonding, and surface qualities all have a significant impact in extending the life of the components. The hard surface failure may be avoided if process capabilities are properly controlled. To retain desired mechanical and tribological characteristics during hardfacing, required the proper control over input process parameters. Considering all of these aspects, this article takes an experimental method to explaining the correlation of process parameters with surface defects in during plasma transferred arc welding techniques (PTAW). From this investigation, it is found that the working range of process parameters to deposit Colmonoy-4 on SS grade 410 could be a transferred arc current between 100 A and 180 A, powder feed rate between 5 g/min and 13 g/min, welding speed between 70 mm/min and 190 mm/min, oscillation speed 450 mm/min to 650 mm/min and plasma gas flow rate between 1.5 L/min and 3.5 L/min

Keywords: Cracks, hardfacing, PTAW, surface defects.

Introduction

Abrasion, wear, and corrosion degrade most components used in power, nuclear, and marine applications. In these sectors, abrasion, wear, and corrosion that cause material deterioration via hazardous chemical reactions (Balasubramanian et al., 2009; Deng et al., 2010; Deshmukh and Kalyankar, 2018, 2021; Flores et al., 2009a; Kakade and Thakur, 2021). This not only degrades the appearance but also creates hazardous conditions, raising the cost of repair and reconstruction (Flores et al., 2009a,; 2009b; Singh et al., 2015). The coating application technique is depending on the material's characteristics and behaviour in severe environments. The most common and practical method of protection of materials in extreme conditions is to cover them with a highly resistant metallic coating (Bharath et al., 2008). Choosing the right procedures is crucial for long-term coating qualities. A surface layer is also applied to the components utilising overlay methods. Overlay is the process of welding corrosion and wear resistant materials onto a substrate. Higher energy supplied during the deposition resulted fine microstructures and improved mechanical, corrosion, and tribological properties because of slowdown in heating and quenching from the melt (Bourithis et al., 2002). Welding creates a strong metallurgical bond between the coating and the base components. As a consequence, overlay techniques are often utilised to prolong component life by enhancing the substrate surface's corrosion and wear resistance. Around the globe, researchers are studying coating characteristics and material behaviour to determine their applicability and maintainability for varied material compositions and combinations. Porosity, blowholes, undercuts, and irregular heat-affected zones are some of the problems caused by faulty welding procedures. The failure of the part was caused by the incorrect welding properties, including penetration, bead shape, reinforcement, and depth of penetration, and HAZ (Sada et al., 2020). Literature shows the application of plasma transferred arc welding (PTAW) technique. Tembhurkar et al. (2021) examined the influence of fillers and no fillers on the welding of austenitic

[a]sachinkakade2107@gmail.com; [b]ajay_raja34@yahoo.com; [c]sanjay.patil@koso.co.in; [d]dhirgajanan@gmail.com

stainless steels 316L and 430 ferritic stainless steels. Madadi et al. (2012) and Deng et al. (2010) commented that, a better material coating is needed for exposed components in the petrochemical, marine, oil, nuclear, aerospace, and power industries to prevent component failure. Ambade et al. (2021) studied the effect of welding passes over microstructure, mechanical and corrosion properties of ferritic stainless steel 409M. Mandal et al. (2020) have reported weld bead characteristics and dilution can be affected by process conditions. The process factors may be controlled to produce surface layers of the desired thickness with free of fractures, and the desired mechanical, wear, and tribo-corrosion characteristics. It is critical to understand how specific factors affect low distortion, low porosity, and crack free surfaces with optimal hardness, penetration, and reinforcement for a given material composition. The objective of this paper is to understand the PTAW process, main influencing process parameters and identify their extreme working ranges by experimental method. This will help researchers to correlate the working ranges of PTAW process parameters for various materials. Hence, in this context, the correlation of PTAW welding hardfacing process parameters on surface defects are explored.

Material and Method

The substrate material was a rectangular plate of stainless-steel (SS) grade 410 with dimensions of 130 mm × 100 mm × 30 mm, and the hardfacing (coating) material was powder of Colmonoy-4 (nickel-based alloy) used for the experimentation. This nickel-based alloy has good wear resistance, metallurgical and physical qualities, and can sustain hardness up to 600°C. It is also resistant to oxidation. Tables 34.1 and 34.2 show the chemical compositions of SS grade 410 and Colmonoy-4, respectively.

A. PTAW technique

The plasma transferred arc welding (PTAW) technique rely on non-consumable tungsten electrodes, which may be found within the torch. The PTAW uses powder as a filler material and argon as an inert gas for transport to the arc area. The PTAW process has shown to be an effective alternative to current thermal spraying techniques for surface modification or treatment (Bharath et al., 2008; Deng et al., 2010). At high temperatures, the PTAW overlay has excellent quality, competitive wear, and corrosion resistance, as well as better microstructure and hardness stability (Deng et al., 2010). Thermal sprayed coatings have a greater manufacturing cost and poorer productivity than PTAW overlays. A metallurgical contact exists between the substrate and the coating in PTAW overlays, which makes them more impact resistant than laser-induced overlays (Flores et al., 2009a, b). overlay procedures varied greatly in welding efficiency and weld plate dilution rates. Hardfacing's composition and qualities are greatly influenced by the amount of dilution it receives. Controlling dilution or penetration is critical during the overlay process when minimal dilution is desired (Balasubramanian et. al, 2009). In PTAW coatings, the dilution is greatly reduced. High temperature stability, superior quality, and competitive wear resistance are all features of PTAW (Deng et al., 2010). Protecting against extreme service life conditions including heat, abrasion, corrosion, erosion, adhesion and abrasive wear are some of the most probable uses for PTAW. It may also be used to rebuild damaged or worn-out pieces. Quality welds are dependent on a number of factors including shielding gas flow rate, the distance from the nozzle to the work piece and welding current, and weld speed. Deposition of better alloys on substrate materials in order to improve corrosion and wear resistance has lately become more common using the PTAW technique. Alloys based on Co and Ni are widely used in a variety of industries, including chemical and fertiliser facilities, nuclear and steam power plants, and pressure vessels, since they are easy to work with and effective. Furthermore, the overlay process is examined in relation to its ability to provide a robust and lasting covering that is devoid of faults. We want to find out what causes the surface faults that appear during processing, and we also want to find a solution to the problems associated with the PTAW overlay. It is utilised to identify the cause of surface defects by the use of a process parameter variation.

Table 34.1 Chemical composition materials

Elements (wt. %)	C	Cr	Fe	Ni	Si	P	Mn	S	B
SS grade 410 (Substrate)	0.15	13.5	Bal	0.75	1	0.04	1	0.03	--
Colmonoy-4 (hardfacing)	0.30	7.5	2.5	84.5	3.5	--	--	--	1.7

B. Experimentations

A PTAW machine developed by Primo automation Ltd., for KOSO India Pvt. Limited Ambad, Nashik, provides the equipment for the investigations as shown in Figure 34.1. Parts such as valve seats, collars and sleeves have been coated using this process. It has also been used to surface valves and components. Multitrack layers of deposition are created on a 30 mm thick plate by altering different process parameters. With the groove 1G position of the electrode negative (DCEN) in accordance with ASME21 welding process standard (WPS) We use a shielding flow rate of 15 litres per minute, a powder feeding flow rate of 3 litres per minute, and an industrially pure argon (99.99%) flow rate to power our 4-mm diameter tungsten electrodes a 25-mm diameter torch orifice, and 99.99% pure argon. Welding speed (mm/min), filler material flow rate, i. e. powder feed rate (g/min), the oscillation speed of the torch (mm/min), and plasma gas were selected for the studies. In order to analyse and examine the impact of process parameters on defects, visual inspections and dye penetration tests are performed.

Results and Discussion

Hardfacing surface characteristics must be understood and correlated in order to produce the desired surface characteristics. This may be achieved by studying and comparing the impact of various process variables on the weld bead shape and surface properties. But the most influential process parameters must be identified and the operating range of every input process parameter constrained. In this investigation, various experiments were carried out to identify the most important processing parameters and their operating ranges. Process parameters are varied between the given ranges in order to prepare the substrate material for deposition. The geometry, look, and imperfections of the weld bead's surface are examined using a dye penetration test. The following results were obtained and discussed for individual parameters about hardfacing operation and surface defects.

A. Effect of transferred arc current

Substrate penetration and partial powder melting can be detected if transferred arc current is less than 100 Amp. There is an undercut and spatter on the weld bead surface when the transferred arc current is larger than 180 Amp. In order to produce an arc, constrictor nozzles need heat energy that rises with the amount of current being transmitted. In the studies, it is revealed that heat has a significant impact on the melting of substrate materials. The melting of the substrate material rises as the current increases, resulting in the full fusing of the powder delivered at that time with the substrate surface, resulting in increased deposition and undercuts as shown in Figure 34.2. A decrease in current results in a drop in temperature and hence a reduction in bonding between the overlying materials.

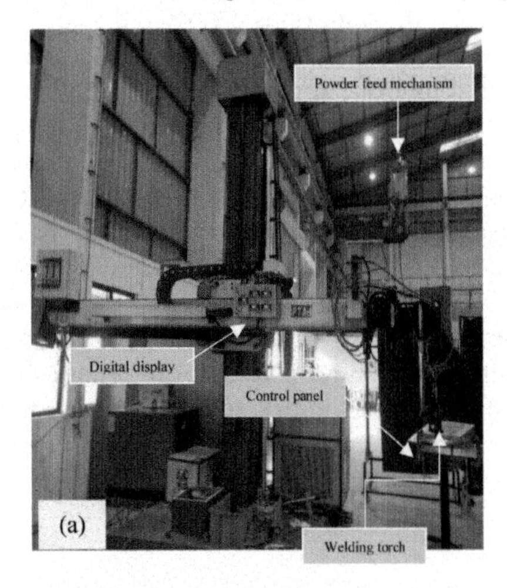

Figure 34.1 Actual experimental setup

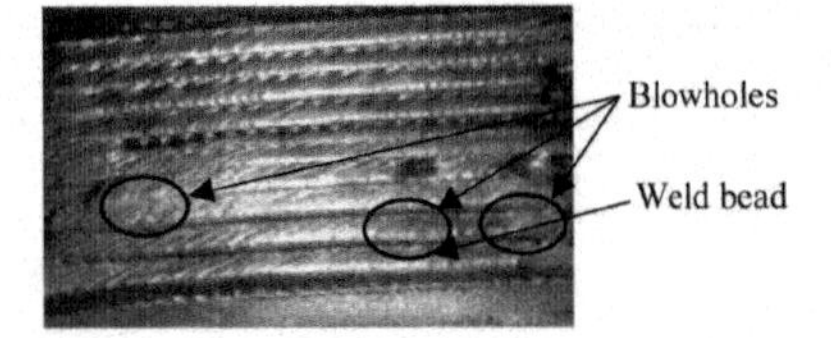

Figure 34.2 Overheating of substrate material showing blowholes on hardfaced surface

B. Effect of powder feed rate

The melting of base material and tungsten electrode overheating occurred when the powder feed rate was less than 05 g/min. The breadth and strength of deposition increase when the powder feed rate approaches 13 g/min. Incomplete melting of powders resulted in a non-smooth weld bead production as shown in Figure 34.3(b). It's possible that the substrate material will be partially melted because of the large quantity of heat energy being used at the time. When the powder supply is adequate at given current, this results in less substrate material melting. The melting of substrate material reduces as the powder feed rate increases, it is found that the cross-sectional area of deposition increases. Spatter and undercutting on the surface, as well as an uneven layer of deposition, are the results of a lower powder feed rate and a larger current, as seen in Figure 34.3(b). On the other hand, higher powder feed rates, result in less bonding in the deposited layer since they lower the substrate material's melting temperature as shown in Figure 34.3(a).

C. Effect of welding speed

The welding speed is the speed at which the welding torch move on the work piece. Under 70 mm/min welding speed results in excess deposition of weld metal and increased reinforcing height. Welding speed greater than 190 mm/min minimise the time for deposition, resulting in an extremely thin layer of material that has an irregular pattern. In addition, the covered surface shows reduced bonding and partial penetration at greater transit speeds. Deposition was broken short by splatter as shown in Figure 34.4 (a, b). Thus, the proper welding speed is required to ensure that the substrate material and the overlying layer are bonded together properly.

D. Effect of oscillation speed

The weld bead was not as smooth when the torch oscillation speed was less than 450 mm/min. At 650 mm/min oscillation speed resulted in broader weld bead and lower bead diameter. A increase in dilution with increasing torch oscillation was noticed during experiments. Hence a reduction in reinforcement with a greater torch oscillation. The mean width of the torch's oscillations grows as the oscillation speed increases. Using the same oscillation speed and powder feed rate, a thin deposit layer with spatter and blowholes may be visible. The low oscillation speed is reflected in the uncertain reinforcement deposition and thin, uneven weld beads. The risk of blowholes in the coating increases as the oscillation speed increases, as seen in Figure 34.5.

Figure 34.3 (a) Over deposition of powder, (b) Weld bead not smooth due to lower powder

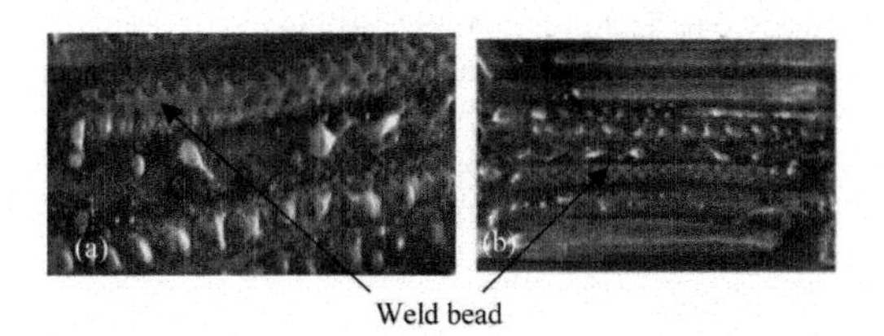

Figure 34.4 (a) Thin an interrupted pattern of weld bead, (b) Irregular pattern of weld bead

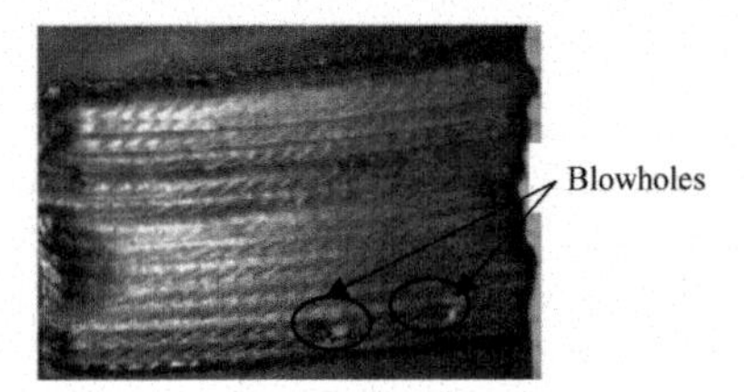

Figure 34.5 Blowholes on the hardfaced surface

E. Effect of plasma gas

Lack of plasma gas flow shortens the operating life of the constrictor nozzle, resulting in premature failure of nozzle. Hardfacing material penetrates more deeply into the substrate when the plasma gas flow rate is over 3.5 L/min. Consequently, the hardfaced surface qualities deteriorate. The bonding between the substrate and the hardfaced material is also affected by a reduced plasma gas flow rate of less than 1.5 L/min.

Conclusions

Surface degradation failure is a major concern for a variety of industrial equipment, including those in the power, marine, oil, and chemical industries. In present work, the significant process parameters and their working ranges for SS grade 410 deposited by Ni based Colmonoy-4 material are investigated. Observations made during and after the hardfacing process were also taken into consideration. This study examined the relationship between plasma transferred arc welding's most critical process parameters and surface flaws and it has contributed following important points:

1. Process factors have a significant impact on the plasma transmitted arc welding overlay process, according to the results of the current research.
2. Weld speed, powder feed, oscillation speed, transferred arc current, plasma gas and other process parameters, as well as post- and pre-heat treatments, must all be carefully monitored and controlled.
3. The input process parameters are properly controlled and systematic studies are performed in an attempt to analyses and explore the sources of surface defects in this investigation.
4. In this investigation, the working ranges of PTAW process parameters are drawn as, i) transferred arc current between 100 A to 180 A, ii) powder feed rate between 5 g/min to 13 g/min, iii) welding speed between 70 mm/min to 190 mm/min, iv) oscillation speed 450 mm/min to 650 mm/min and v) plasma gas flow rate between 1.5 L/min to 3.5 L/min. As described in section III, using extreme levels of process parameters results in surface flaws.
5. In welding procedures, the qualities of welded connections are typically more important than those of the main materials used in the process. As a result, in order to achieve better weld quality, it is crucial to use correct welding process parameters.

References

Ambade, S., Tembhurkar, C., Patil, A., and Meshram, D. B. (2021). Effect of number of welding passes on the microstructure, mechanical and intergranular corrosion properties of 409M ferritic stainless steel. World J. Eng. 19:368–374.

Balasubramanian, V., Lakshminarayanan, A. K., Varahamoorthy, R., and Babu, S. (2009). Application of response surface methodolody to prediction of dilution in plasma transferred arc hardfacing of stainless steel on carbon steel. J. Iron Steel Res. Int. 16:44–53.

Bourithis, E., Tazedakis, A., and Papadimitriou, G. (2002). A study on the surface treatment of "Calmax" tool steel by a plasma transferred arc (PTA) process. J. Mater. Process. Technol. 128:169–177.

Chetan, T., Kataria, R., Ambade, S., Verma, J., Sharma, A., and Sarkar, S. (2021). Effect of fillers and autogenous welding on dissimilar welded 316L austenitic and 430 ferritic stainless steels. J. Mater. Eng. Perform. 30:1444–1453.

Deng, H., Shi, H., and Tsuruoka, S. (2010). Influence of coating thickness and temperature on mechanical properties of steel deposited with Co-based alloy hardfacing coating. Surf. Coat. Technol. 204:3927–3934.

Deshmukh, D. D., and Kalyankar, V. D. (2018). Recent status of overlay by plasma transferred arc welding technique. Int. J. Mater. Prod. Technol. 56:23–83.

Deshmukh, D. D., and Kalyankar, V. D. (2021). Analysis of deposition efficiency and distortion during multitrack overlay by plasma transferred arc welding of Co–Cr alloy on 316L stainless steel. J. Adv. Manuf. Syst. 20:705–728.

Flores, J. F., Neville, A., Kapur, N., and Gnanavelu, A. (2009a). Erosion–corrosion degradation mechanisms of Fe–Cr–C and WC–Fe–Cr–C PTA overlays in concentrated slurries. Wear 267:1811–1820.

Flores, J. F., Neville, A., Kapur, N., and Gnanavelu, A. 2009b. An experimental study of the erosion–corrosion behavior of plasma transferred arc MMCs. Wear 267:213–222.

Kakade, S., and Thakur, A. (2021). comparative analysis and investigations of welding processes applied for hardfacing using AHP. Int. J. Modern Manuf. Technol. XIII: 53–63.

Madadi, F., Ashrafizadeh, F., and Shamanian, M. (2012). Optimization of pulsed TIG cladding process of stellite alloy on carbon steel using RSM. J. Alloys Comp. 510:71–77.

Mandal, S., Kumar, S. and Oraon, M. (2020). Process Parameter Effects over Bead Properties during Material Deposition of PTAW Process. Mater. Sci. Forum 978:55–63.

Ravi Bharath, R., Ramanathan, R., Sundararajan, B., and Bala Srinivasan, P. (2008). Optimization of process parameters for deposition of Satellite on X45CrSi93 steel by plasma transferred arc technique. Mater. Design 29:1725–1731.

Sada Oro-Oghene, S., and Achebo, J. (2020). Optimisation and prediction of the weld bead geometry of a mild steel metal inert gas weld. Adv. Mater. Process. Technol.: 1–10.

Singh, K. A., Madhusudhan Reddy, G., and Srinivas Rao, K. (2015). Pitting corrosion resistance and bond strength of stainless-steel overlay by friction surfacing on high strength low alloy steel. Defence Technol. 11:299–307.

35 Lead generation strategy for an organisation through salesforce experience cloud: a case study for YCCE

Ganesh Yenurkar[1,a], *Ganesh Khekare*[2,b], *Nupur Bagul*[3,c], *Saylee Prakashe*[3,d], *Punam Bandhate*[1,e], *and Shreya Pillai*[1,f]

[1]Assistant Professor, Department of Computer Technology, Yeshwantrao Chavan College of Engineering, Nagpur, India

[2]Assistant Professor, Department of Computer Science & Engineering, Parul University, Vadodara, Gujarat, India

[3]Student, Department of Computer Technology, Yeshwantrao Chavan College of Engineering, Nagpur, India

Abstract

The main purpose of this paper is to create a Social Media Marketing Plan for YCCE College based on data collected from seniors who have been accepted to YCCE College and are meeting with an academic advisor to register for freshmen level YCCE college curriculum. People from many walks of life use social media in a number of ways on a daily basis. Facebook, Instagram, YouTube, and Twitter, among other social media sites, have grown to the point that they can no longer be ignored. Customers in higher education are demanding more attention and prompt service, thus proactive colleges are relying on technology – customer relationship management systems – to meet this need. Social media has been included to the new social CRM component, which focuses on consumer interaction. Customer engagement is the centre of the new social CRM dimension, and social media technologies have transformed how businesses and customers interact. The customer journey through digital advertising leads to the discovery of new prospects or the engagement of inactive users. It activates CRM data for audience targeting and assists with user integration. CRM can now contact the right people, target prospects with personalized content, and invite interested students to relevant events, among other things. The CRM keeps track of every encounter with the 'customer' in one spot. CRM will assist us in achieving our email marketing objectives through advertising. It will coordinate all promotions with our efforts in social, mobile, sales, and customer service. It will use Google Ads, Social Studios, Instagram and YouTube to engage customers. The CRM system will assist the institution in focusing on the appropriate recruitment initiatives and organizing communications. Students might benefit from this during the admissions process. This initiative will send students and parents email alerts about current and forthcoming events to their registered email addresses.

Keywords: Customer email alerts, salesforce, social media.

Introduction

To enhance enrolment at Yeshwantrao Chavan College of Engineering (YCCE) by utilising social media and other technology. YCCE College will employ social media particularly for its recruitment efforts. Facebook, Instagram, YouTube, and Twitter, for example, have all developed to the point where they can no longer be ignored. People of various ages and demographics use social media on a daily basis. As a result, it makes sense to employ social media as a marketing tool. From the standpoint of both students and parents, these social media channels appear to be of little or no use while looking for college information. With the aid of our project, we can solve this problem.

Clients in higher education are demanding greater attention and prompt service; therefore, forward-thinking colleges are turning to technology – customer relationship management systems – to meet this need. Customer data from numerous channels is combined into a single database, allowing for simpler information access, message tailoring and customisation, and timely distribution to students. This information contains the lead's/name, customer's gender, educational background, phone, email, marketing materials, social media, and any other pertinent information. Every interaction with the "customer" is recorded in the CRM.

Related Study

Because of the rapid speed of digital disruption, SMEs have innovated and grew by strategically merging social media with customer interaction initiatives. Based on the limitations and gaps discovered, this study

[a]ganeshyenurkar@gmail.com; [b]khekare.123@gmail.com; [c]nupurbagul09@gmail.com; [d]sayleeprakashe@gmail.com; [e]punambandhate99@gmail.com; [f]shreyapillai2000@gmail.com

had developed a social CRM research integration model that clearly depicted the research state of socialised CRM. It predicted future research directions for socialized CRM principles, empirical studies, and localisation studies, as well as innovative and constructive ideas. The study's findings have reflected the most commonly utilised subcategory for each categorisation, as well as trends and tendencies. Social CRM is a novel and effective technique to deal with a disruptive shift. Social CRM (Social Customer Relationship Management) is a novel concept that mixes traditional customer relationship management with social media. The major purpose of this research was to conduct a review of scholarly studies on the advantages of social CRM for firms and the factors that have affected adoption. This study presented a new model for social CRM that incorporates a new construct of customer engagement efforts as well as adjustments to other constructs to account for the effect of social media technologies on CRM. The new dimension of social CRM has centred on customer engagement, and social media technologies have dramatically changed how organisations and customers interact.

Alt, R. and Reinhold, O. (2012, 2016) were provided a two way communication channel for customers with new business trends by using various social media. To enhance the customers experience through social media collaboration and dialogues were added by Baird and Parasnis (2011). The characteristics, functionalities, and challenges of the internet of things could be used to add the novel advantage into the CRM system was provided by Khekare et al. (2021). The customer centric automated CRM system and its management discovered by Nyoman et al. (2014) and Trainor et al. (2014), which would able to search the connection between customer and their profiles. The real world challenges of social CRM and its overview for identification, construction, and estimation suggested by Pinheiro et al. (2017) and Social Media Regional Overview 2016. The business strategy for social CRM added the knowledge of customers' behaviour, their attitude and with their mood by Woodcock et al. (2011). Topic identification in twitter dataset was provided by Yenurkar et al. (2021). In future perspective, the CRM may also be used for searching of books in library as RFID was used in existing systems added the advantage by Yenurkar et al. (2017). The performance of big data machine learning and data mining algorithms through quality of service parameters was studied by Yenurkar and Mal (2021).

Methodology

Table 35.1 depicts the social media connect strategy, which would provide the objectives and outcomes in existing research. The flow chart for the project is given Figure 35.1, and it begins with a social media advertisement and then redirects to our community website by clicking on the link provided in the online advertisement. After logging in, all visitors to the website can self-register. The students then have the option of registering or not. When a student registers, he or she becomes a subscriber and will get detailed email alerts with all important information about the admissions process and upcoming college activities. If a student visits the page but does not register, we treat them as a contact and send them college marketing through email. Figure a depicts the project's flow chart, which begins with a social media advertisement and subsequently redirects to our community website by clicking on the link supplied in the online ad. All visitors to the website can self-register after logging in. The pupils can then chose whether or not to register. If a student registers, he or she becomes a subscriber and will get comprehensive email notifications with all relevant information regarding the admissions process and college events. If a student does not register but only sees the page, we consider them a contact and send them college marketing through email.

The fundamental activities that are conducted in four phases to persuade readers by comparing with other research studies and giving viable answers or beneficial ideas are shown below in Figure 35.2:

A) Data fetching
B) Data Manipulating
C) Data filtering
D) Data analysis

A. Data Fetching

We fetch the data of all visited students along with the all registered students with their personal information and only email id will be fetched of those students who have not registered.

B. Data filtering

When working with data it has to be filtered to get the registered students data from all the visited students' data.

Table 35.1 Social media connect strategy

Ref. No.	*Objective*	*Advantage*	*Outcome*	*Highlight*
Malthouse et l. (2013)	Managing Customer Relationships in the Social Media Era: Introducing the Social CRM House	they are already familiar with from planning more traditional marketing activities Company manages customer connections in order to optimise lifetime value.	Technologies have also enabled consumers to filter out advertising and CRM messages, compare prices with competitors from anywhere with mobile devices, and distribute positive or negative brand messages to a global audience Consumers block advertising and CRM communications using mobile devices.	The premise of CRM is that the firm could, and should, manage relationships with its customers to maximize lifetime value, an objective that benefits only the firm. Social media and other new technologies have empowered consumers. 3 key factors: • Empowering culture • Relevant skillsets • Operational excellence.
Sashi (2012)	Customer engagement, buyer-seller relationships, and social media	Social media to aid possible combinations with other digital and conventional non-digital media has been planned.	Parts of social media such as YouTube, Twitter, and Facebook were integrated with banner advertising, TV advertising, and personal selling.	Customer provide superior value relative to competitors by • Generating • Disseminating
Kietzmann et al. (2011)	Social Media? Get Serious! Understanding the Functional Building Blocks of Social Media	Firms monitor social media activity which differs in function and impact, and develops a consistent social media strategy.	Organisations, communities, and people all experience significant and pervasive changes as a result of social media.	7 building blocks: • Identity • Conversations • Sharing • Presence • Relationships • Reputation • Groups
Gallaugher and Ransbotham (2010)	Social Media and customer dialog Management at Starbucks	• Understand the support, planning, coordination, and execution of social media activities.	The investigation includes monitoring all of Starbucks' public-facing social media outlets, including MyStarbucks Idea.	Interactions as a framework with 3 components: • Megaphone • Magnet • Monitor
Kupper et al. (2014)	Performance Measures for Social CRM: A Literature Review	Social CRM into 4 parts: • Infrastructure • Process • Customer • Organisational performance.	The search phrases are not all-encompassing and possibly miss assemblies.	2 stage multi-method approach: • Systematic and rigorous literature review • Sorting procedure

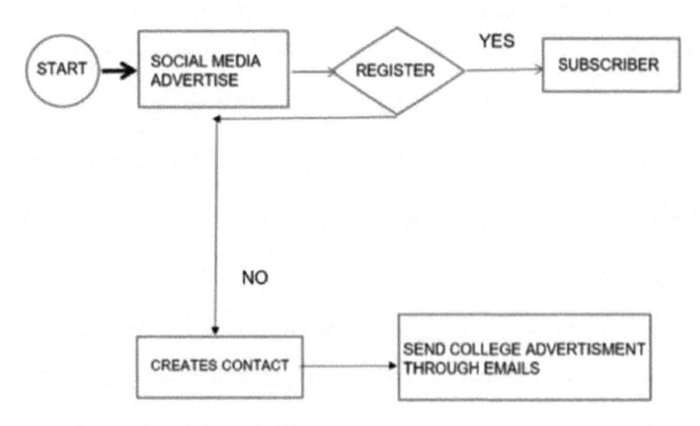

Figure 35.1 Workflow model

C. Send email notification

Now those students who have registered will get the detailed email about the college admission process and college events (basically updates about college related activities).

D. Analysis of data

When working with data it has to be filtered in various parts. Visited students are divided into two parts i.e., registered students and non-registered students and accordingly emails will be sent.

Community Cloud has been renamed Experience Cloud since it can be used to create portals, support centres, forums, websites, mobile apps, and manage content in addition to communities.

Experience Cloud is a Salesforce-integrated platform that allows us to develop online communities for customer, partner, and employee communication. This platform enables us to create a branded environment with several customisation options. The Experience cloud setup decribes how to enable experience cloud in Salesforce go to setup -> Digital Experience -> Settings

->Setup->Digital Experience->All Sites->New -> Template The template which was chosen is Customer Service Portal Template.

The technique for making a **New Contact** is that, first, go to the Contacts app and tap the Add New Contact icon. Select an account to which the contact should be linked. Fill in the information for the contact.

To finish the modifications and create the new contact, tap the Done or Save button.

Now go to the website created (Practice) and select administration.

Administration->Members

Select Customer Community Plus Login User and Add it in the Selected Profiles.

Figure 35.3 explains how to save the settings and then again switch to Salesforce Classic.

Select Manage External User > Enable User. Administration->Members

Select Customer Community Plus Login User and Add it in the Selected Profiles.

Thus users are created.

MODULES

Modules of the project are:-

A. *Social Media Integration for Post posting.*
B. *Survey Module for students and parent using community portal.*
C. *Email and SMS marketing for leads.*
D. *Analysis modules for leads.*

Module A: Social media integration for post posting.

For authentication and authorisation, Google APIs employ the OAuth 2.0 protocol. Google supports OAuth 2.0 in a number of circumstances, including web server, client-side, installation, and applications with restricted input.

The authorisation process starts with the application sending a web service request for an authorisation number to a Google URL. The programme displays the response to the user, which contains parameters such as a URL and a code.

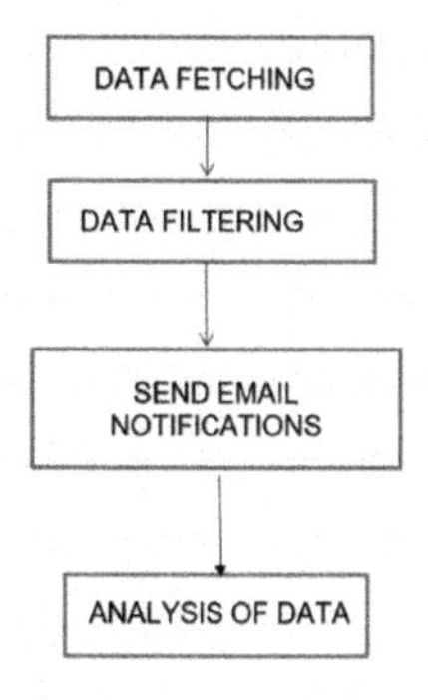

Figure 35.2 Architecture

Figure 35.3 Enabled user

The user gets the URL and code from the device and then transfers it to a device or computer with more input options. The user opens a browser, goes to the URL supplied, logs in, and enters the code. Meanwhile, the application checks a Google URL at regular intervals. When the user grants access, the Google server responds with an access token and a refresh token. The access token should be used to access a Google API, while the refresh token should be retained for future use. When the access token expires, the software uses the refresh token to request a new one.

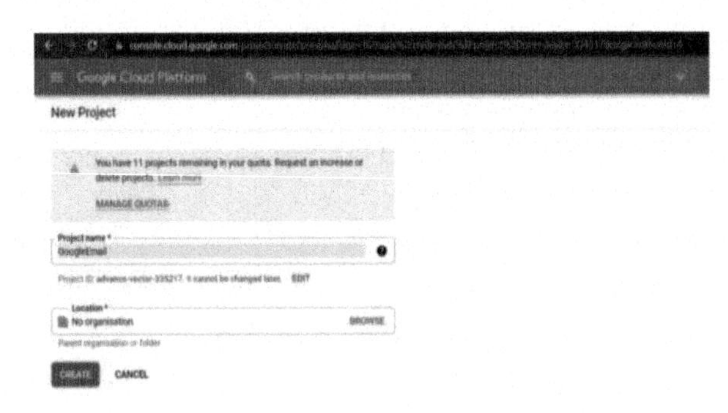

Figure 35.A1 New project

The Above given Figure 35.A1 explains how to create a new project. Also project name and location is given to specify. OAuth Consent Screen shows the Edit App Registration where we can create an app and Select OAuth 2.0 client id.

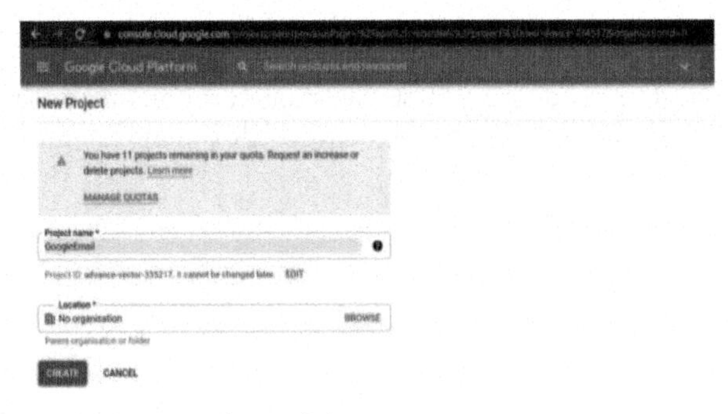

Figure 35.A2 Authorized domains

Figure 35.A2 explains about the Authorized Domains where we can add domains and later need to add the information of the developer i.e. email address.

Non-Sensitive scopes helps us to add the scopes needed for the project. Also the user-facing description is given after the addition of the scopes. Credentials list the API keys, OAuth 2.0 Client IDs and Service Accounts.

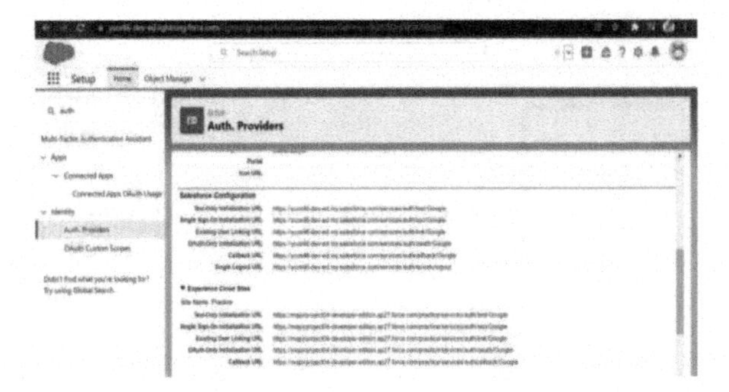

Figure 35.A3 Auth providers

The figure shows how to create a new credential in Google developer console as “LeadCollectionAuth”.

The above Figure 35.A3 shows the links provided under Auth Providers section. Next, we need to add the call back URL of Salesforce configuration and experience cloud sites in Redirect Uri.

Client Id and server describes how to copy the client id and the client server to the Salesforce Auth provider.

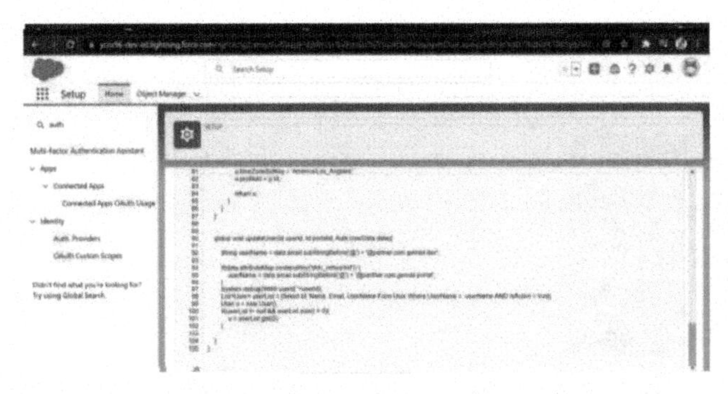

Figure 35.A4 Code (i)

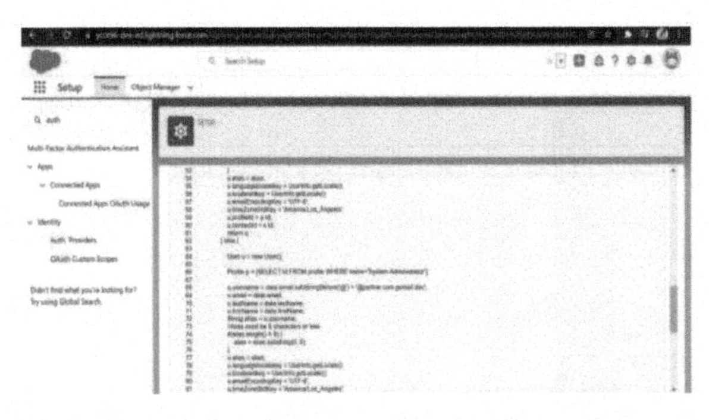

Figure 35.A4 Code (ii)

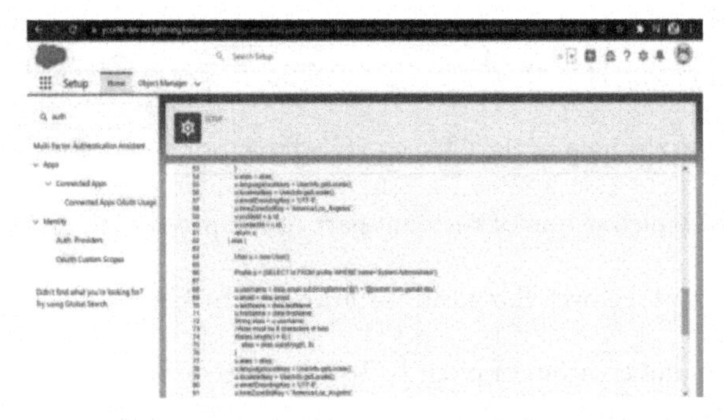

Figure 35.A4 Code (iii)

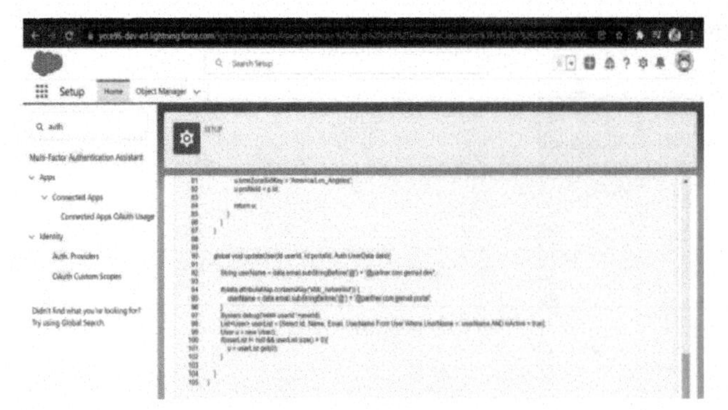

Figure 35.A4 Code (iv)

The above Figures 35.A4 (i) (ii) (iii) (iv) illustrates the whole code which has been written for the making of login page. For this, from Registration Handle select GoogleEmailHandler and add the code.

Add Domain helps us to know how to add domains.. Now from quick find type "My Domain" under company settings and go to Authentication configuration and add Google.

Module B: survey module for students and parent using community portal

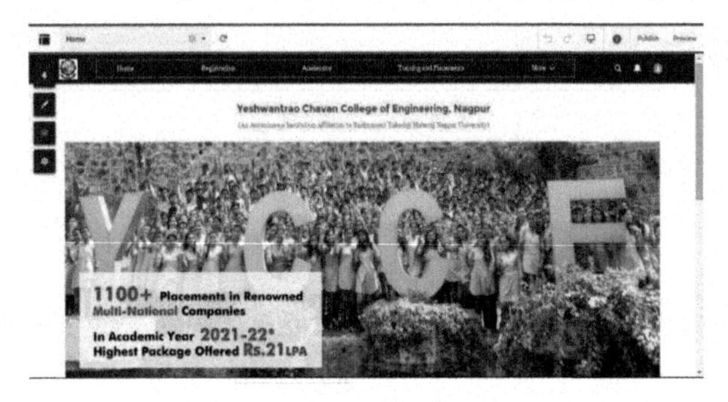

Figure 35.B1 Home page

The above Figure 35.B1 illustrates the Home Page of our website.

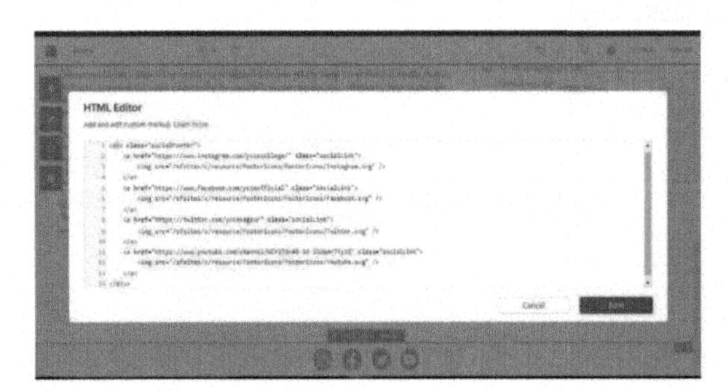

Figure 35.B2 HTML code

Figure 35.B2 demonstrates the HTML Code. So using HTML we have created the footer for our website using HTML Editor Component.

Figure 35.B3 demonstrates the CSS Code. So with the help of the CSS, we styled the footer of our website using CSS Editor.

Figure 35.B4 describes the creation of the flow which is one of the vital part of the project. To create a flow:-

Setup->Flows->New Flow->Screen Flow->Next->Freeform For Lead Capture Screen Flow we need to add a Screen to Collect User Input.

Create Record of the Object depicts the editing and creation of records. To create records, add a Create Records Element.

Figure 35.B5: We have created a 'Lead Capture Screen Flow' using flow builder to capture lead data.

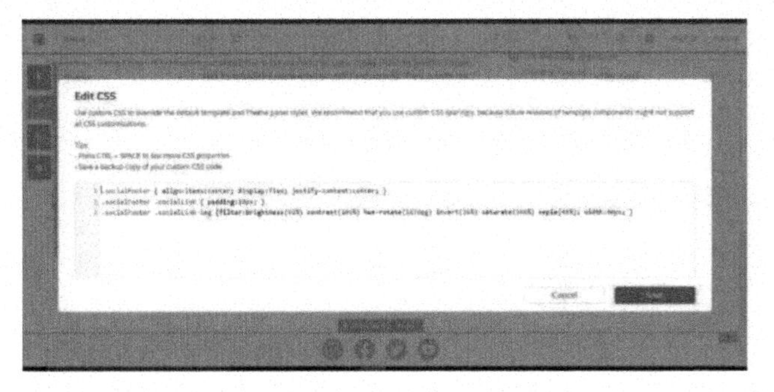

Figure 35.B3 CSS editor

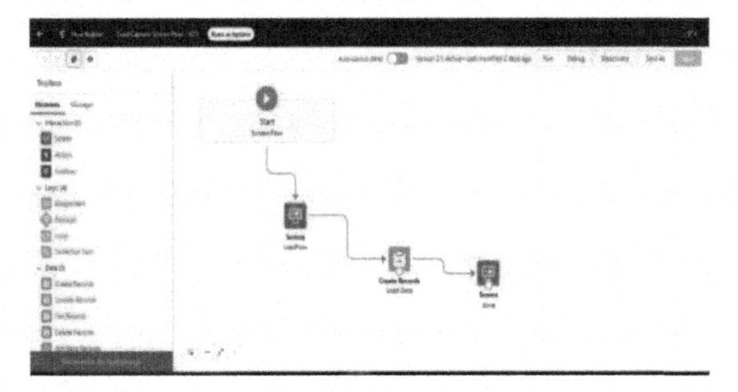

Figure 35.B4 Flow

Figure 35.B5 Lead capture screen flow

All of the records have been saved in a lead record. After filling out the form produced by the flow builder, the data will be sent into the lead object.

Module C: Email and SMS marketing for leads

Figure 35.C1 shows that from the lead object page we can directly send the mail to the leads (students, subscribers, etc.)

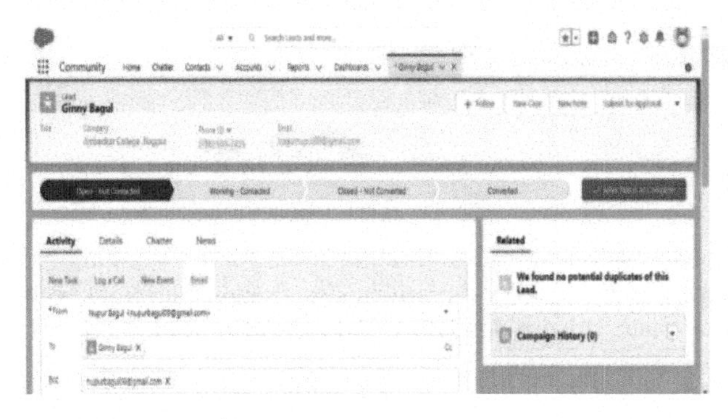

Figure 35.C1 Leads

Figure 35.C2 illustrates the whole content present in the mail. Thus, it shows how successfully email is sent to the user.

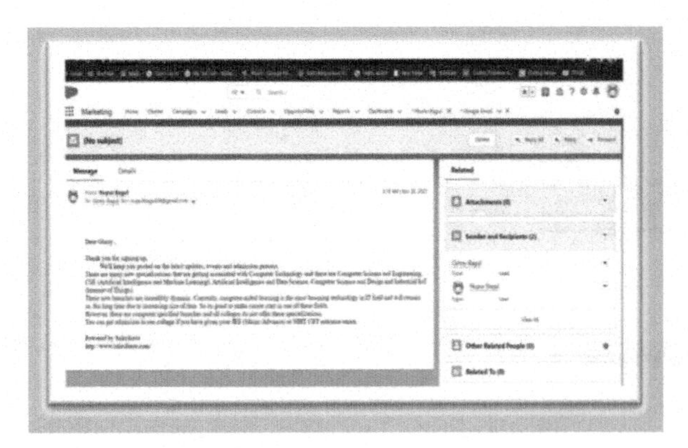

Figure 35.C2 Email

Module D: Analysis modules for leads

To begin collecting basic data from a website, follow these steps:

1) Login up for an Analytics account or sign in if you already have one: Go to google.com/analytics to learn more:
   Choose one of the following options:
   a) Click Start for free to establish an account.
   b) Click Sign in to Analytics to sign in to your account.
2) A property represents a website or app and is the location where data from the site or app is collected in Analytics.
3) In the property, create a reporting view. Views allow us to filter the website so that we may collect data in our Analytics property.

The Above Figure 35.D1 illustrates the graph which shows the input data about the users, sessions, bounce rate and session duration.

Property settings describes the property settings. In this property name, default URL, view and industry category is shown.

The community which we have created has been successfully linked with the Google Analytics.

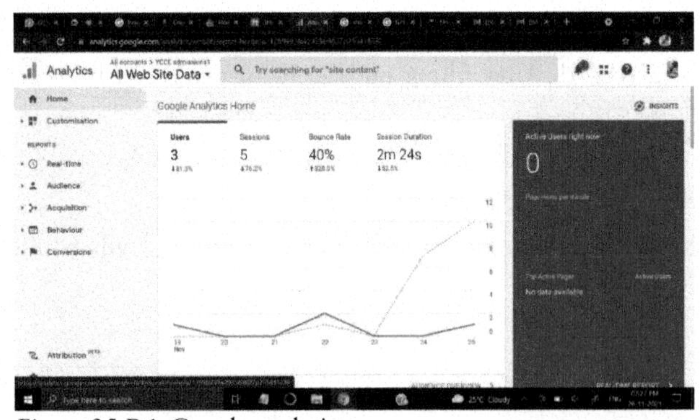

Figure 35.D1 Google analytics

Result

Social media is one of the most often utilized advertising and marketing tools for establishing a loyal following for a product. Companies and goods are promoted via platforms such as Instagram, Facebook, LinkedIn, and others. Getting the most out of social media might help you make more money. Because social media is generally free, you can establish a vast portfolio with little money. This study provided an in-depth look into Salesforce and social media marketing. It also included a step-by-step tutorial for integrating social media with Salesforce.

Figure 35.V.1 assesses the graph depicting the number of users, new users, sessions, number of sessions per user, page views, pages / sessions, average session duration, bounce rate, language, city, location, and device type.

Figure 35.V.2 shows a table with numerical figures for users, new users, bounce rate, pages/sessions, and average time on page. Goal conversion rate, goal completions, and goal value are all factors to consider.

The Above Figure 35.V.3 demonstrates the number of users, number of sessions, percent of bounce rate and the session duration.

Figure 35.V.4 (ii) illustrates the table and the pie chart. This figure describes the table showing the users logged in from various cities and pie chart showing new visitors and returning visitors.

Graph details in numbers displays the total number of users, new users, and sessions, as well as the number of sessions per user, page views, pages / sessions, average session duration, bounce rate, language, city, and device type.

Figure 35.V.5 demonstrates the graph showing the comparison that how many users have logged in, on which date. The figure also shows the number of active users.

Discussion

The project has been created with the help of Salesforce by creating an experience cloud so the problem definition is "Lead generation strategy for an organisation through Salesforce experience cloud: A case study for YCCE". The project is especially for our college purpose.

Why Salesforce has been used?? There are many reasons behind that.

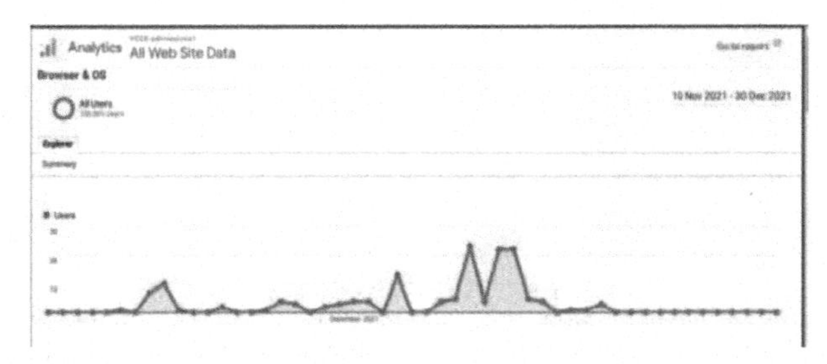

Figure 35.V.1 Graph 1

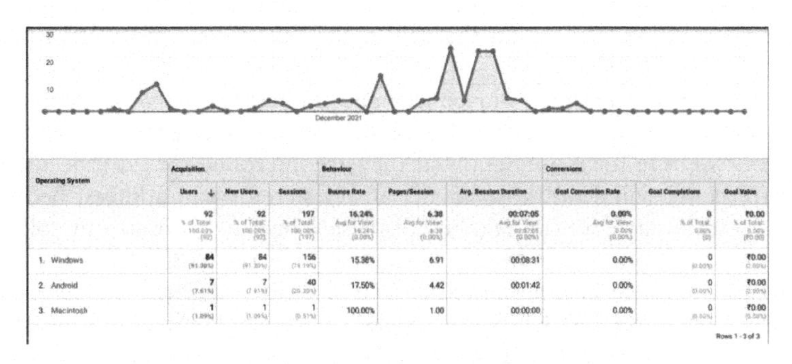

Operating System	Acquisition			Behaviour			Conversions		
	Users	New Users	Sessions	Bounce Rate	Pages/Session	Avg. Session Duration	Goal Conversion Rate	Goal Completions	Goal Value
	92	92	197	16.24%	6.38	00:07:05	0.00%	0	₹0.00
1. Windows	84	84	156	15.38%	6.91	00:08:31	0.00%	0	₹0.00
2. Android	7	7	40	17.50%	4.42	00:01:42	0.00%	0	₹0.00
3. Macintosh	1	1	1	100.00%	1.00	00:00:00	0.00%	0	₹0.00

Figure 35.V.2 Table

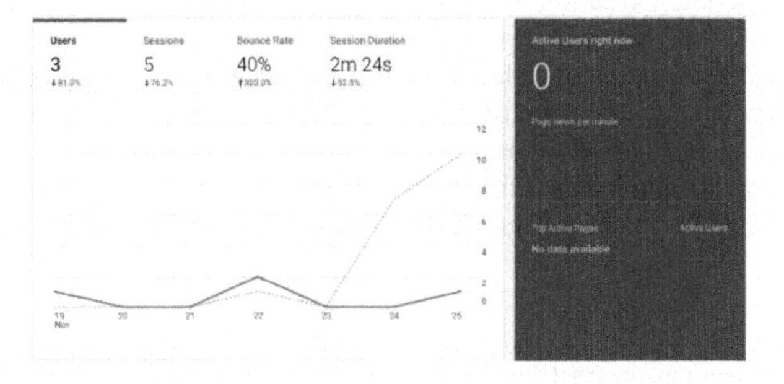

Figure 35.V.3 Graph 2

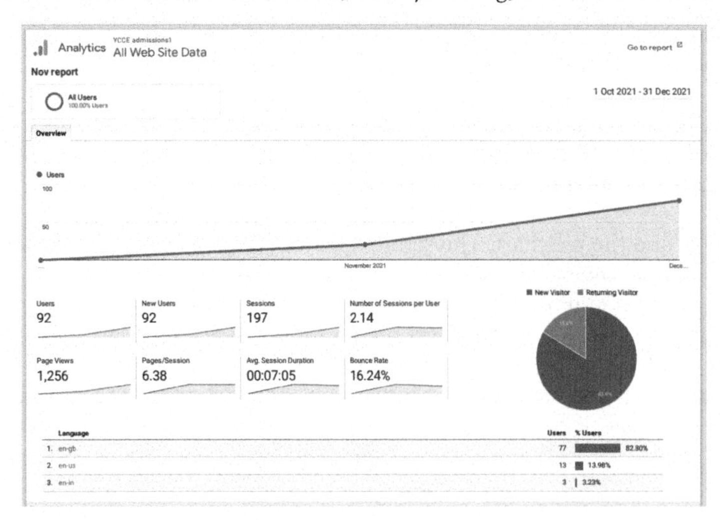

Figure 35.V.4(ii) Table and pie chart

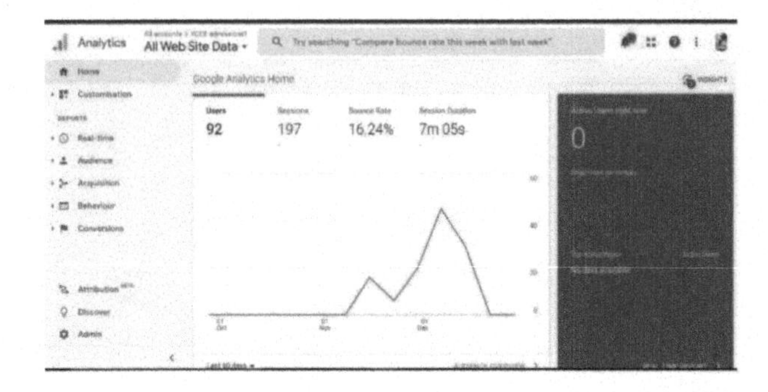

Figure 35.V.5 Graph of logged in users and date

Salesforce's main purpose is to handle customer data. Salesforce offers a user-friendly interface for accessing customer details contained in objects (tables). So this will make this project more efficient in many ways.

The project helps in establishing relationship between our college and interested eligible students and their parents. With the help of the project, college website is linked to different social media platforms so the interested students can visit our website for more details. If they are interested they can become a subscriber and then through the project we will send those emails on registered email address. Because practically everyone is on social media these days, we can take use of these venues to promote our college.

The major goal of this project is to develop a Social Media Marketing Plan for YCCE College based on information acquired from seniors who have been admitted to YCCE College and are meeting with an Academic Advisor to register for freshman level YCCE college coursework. People of various ages and demographics use social media on a daily basis. Facebook, Instagram, YouTube, and Twitter have all developed to the point that they can no longer be ignored.

CRM can now contact the right people, target prospects with personalized content, and invite interested students to relevant events, among other things. The CRM keeps track of every encounter with the "customer" in one spot. CRM will assist us in achieving our email marketing objectives through advertising. It will coordinate all promotions with our efforts in social, mobile, sales, and customer service. It will use Google Ads, Social Studios, Instagram, and YouTube to engage customers. The CRM system will assist the institution in focusing on the appropriate recruitment initiatives and organizing communications. Students might benefit from this during the admissions process. This initiative will send students and parents email alerts about current and forthcoming events to their registered email addresses.

Conclusion

As a result, our website serves as a useful communication tool for our colleges and institutions. Every lead receives one-on- one contact that helps them progress along their individual path.

Our college's CRM system will assist us in focusing on the correct recruitment initiatives and organizing our interactions. Students might benefit from this during the admissions process. This initiative will send students and parents email alerts about current and forthcoming events to their registered email addresses. As a result, our website serves as a useful communication tool for our colleges and institutions. Every lead receives one-on-one contact that helps them progress along their individual path.

References

Alt, R. and Reinhold, O. (2012). Social customer relationship management (Social CRM): Application and technology. Bus. Inf. Syst. Eng. 4:287–291. https://doi.org/10.1007/s12599-012-0225-5.

Alt, R. and Reinhold, O. (2016). Social customer relationship management: Grundlagen anwendungen und technologien, Springer gabler. ISBN 978-3-662-52789-4 ISBN 978-3-662-52790-0 (eBook), DOI: 10.1007/978-3-662-52790-0.

Baird, C. H. and Parasnis, G. (2011). From social media to social customer relationship management. Strategy Leadersh. 39(5):30–37.

Gallaugher, J. and Ransbotham, S. (2010). Social media and customer dialog management at starbucks. MIS Q. Exec. 9(4):197–212.

Khekare, G., Verma, P., Dhanre, U., Raut, S. and Yenurkar, G. (2021). Analysis of internet of things based on characteristics, functionalities, and challenges. International J. Hyperconnectivity and the Internet of Things (IJHIoT), 5(1):44–62, IGI Global 2021/1/1.

Kietzmann, J. H., Hermkens, K., McCarthy, I. P., and Silvestre, B. S. (2011). Social media? Get serious! Understanding the functional building blocks of social media. Bus. Horiz. 54(3):241–251.

Kupper, T., Jung, R., Lehmkuhl, T., Walther, S. and Wieneke, A. (2014). Performance Measures for Social CRM: A Literature Review. 27th Bled conference.

Malthouse, E. C., Haenlein M., Skiera, B., Wege, E. and Zhang, M. (2013). Managing customer relationships in the social media Era: Introducing the social CRM house. J. Interact. Mark. 27(4):270–280.

Nyoman, K., Iping, S. and Nur, M. (2014). Social CRM using web mining. International Conference on Information Technology Systems and Innovation (ICITSI). pp. 264–268.

Pinheiro, M., Jacob, A., Reinhold, O. Santana, A. and Lobato, F. (2017). Social CRM: Biggest challenges to make it work in the real world. In Business Information Systems Workshops Springer International Publishing.

Sashi, C. (2012). Customer engagement buyer-seller relationships and social media. Manag. Decis. 50(2):253–272.

Social Media Regional Overview. WeAreSocialSG January 2016. https://wearesocial.com/sg/.

Trainor, K. J., Andzulis J. M., Rapp, A. and Agnihotri, R. (2014). Social media technology usage and customer relationship performance: A capabilities- based examination of social CRM. J. Bus. Res. 67(6):1201–1208.

Woodcock, N., Green, A. and Starkey, M. (2011). Social CRM as a business strategy. J. Database Mark. Cust. Strategy Manag. 18(1):50–64.

Yenurkar, G. K., Nasare, R. K. and Chavhan, S. S. (2017). RFID based transaction and searching of library books. In 2017 IEEE International Conference on Power, Control, Signals and Instrumentation Engineering (ICPCSI), pp. 1870–1874, doi: 10.1109/ICPCSI.2017.8392040.

Yenurkar, G. and Mal, S. (2021). Performance analysis of big data based mining and machine learning algorithms: A review. Turk. Online J. Qual. Inq. 12(7):9437–9752, Science Research Society.

Yenurkar, G., Nagapure, D., Marathe, K., Vaidya, V. and Chaudhary, G. (2021). Topic Detection in Twitter Dataset Using Python. International Conference on Advanced Computing and Communication Technology Tirunelveli, India, 06th & 07th, May 2021.

36 Innovative model development for enhancing the serviceability of flexible pavement using fea

Meghana Dongre[1,a], *Yogesh Kherde*[2,b], *Devendra Y. Shahare*[2,c], *Aniket Pathade*[3], *and Snehal Banarse*[4]

[1]Civil Engineering Yeshwantrao Chavan College of Engineering Nagpur, Maharashtra, India

[2]Yeshwantrao Chavan College of Engineering Nagpur, Maharashtra, India

[3]Jawaharlal Nehru Medical College, Datta Meghe Institute of Medical Sciences, Sawangi(M), Wardha, Maharashtra, India

[4]Spruha Engineering Services, Amravati, Maharashtra, India

Abstract

In most urban cities, rigid pavement is getting applicable, but it has certain disadvantages like the use of more concrete creates a bad impact on the environment and human health. As when cement is manufacture that calcining process the major source of emission of SOx, CO2, NOx and particulate matter. Those gases and particulate matter emission from cement manufacturing plant are degrading air quality and thus it produces air pollution. Aim of project is to find out cost effective structural solution for serviceability improvement of flexible pavement. To estimate design features requirements of precast panel i.e., (length, width, height and shape) as per IRC specifications. To model large precast element by means of a relevant finite element analysis. To analyse and design of a homogenous open ended rectangular element subjected to a combination of compressive and lateral live loads. To carry out comparative analysis of flexible pavement with and without using precast panel.

Keywords: FEA, Flexible pavement, Precast Panel, Cost-effective.

Introduction

Highway pavement consists of rigid and flexible pavement. The major structural element for rigid pavement is Portland cement concrete (PCC). Large use of cement concrete is directly proportional to increase in production of cement, which ultimately create adverse effect on environment and human health. In contrast to that, flexible pavement consists of several layers composed of natural materials, mainly soil and gravel. Subgrade, sub-base, base, and bituminous layers make up the flexible pavement structure of highways. The load act on bituminous layer is transfer to base then base to sub base then sub base to subgrade. A permanent purpose of a subgrade is to spread the vehicle weight delivered to the earth. The ultimate goal is to ensure that wheel load stress is transmitted to the bottom layer by grain to grain transmission in the granular structure. As considering urban road condition the width of carriageway is 3.5 m for single lane road as per IRC:86 (2018). Drainage hole is provided at 1 m to reduce poor water pressure.

Several variables contribute to the development of stresses. The most typical failures of flexible pavements are fatigue cracking, rutting, and heat cracking. Sub-grade rutting can occur as a result of inadequate compaction during construction, resulting in long-term secondary settling. Horizontal tensile strain at the bottom of asphaltic concrete causes fatigue cracking in flexible pavement. So, the durability and sustainability of flexible pavement is depending on various factors like thickness of layer, material properties and environmental condition (Abed and Al-Azzawi, 2012). There is a pressing need to expand our road network to all terrains, and because flexible pavement is the preferred pavement type, there is an immediate need to switch at high-preference and new and alternative materials or technologies that can be applied to flexible pavement and provide a long-term solution. So providing precast concrete panel to flexible pavement is the sustainable solution (Melaku and Qui, 2015).

In flexible pavement, precast concrete panels are utilised for two main purposes: foundation reinforcement and stability. Restricted depth precast panel is placed at the bottom of sub base layer or unbounded layer of flexible pavement system to improve load bearing capacity of flexible pavement under repetitive vehicle load. Precast panel is used in subgrade stabilisation to provide a construction platform over a weak subgrade to hold equipment and ease pavement building without excessive subgrade deformation. In between the base layer and the sub base layer of flexible pavement, a precast concrete panel is used to disperse shear stresses caused by automobile loads (A-Alzaawi, 2012). The purpose of a project at the

[a]meghnadongare1996@gmail.com; [b]ypkherde@ycce.edu; [c]devshahare0501gmail.com

University of Minnesota is to establish an analytical approach for measuring load transfer in concrete using a non-destructive testing approach for pavement joints. The fundamental analysis entailed dynamically loading the joints and doing a frequency response analysis various joint conditions for load transfer, ranging from total to partial load transmission, were analysed using a three-dimensional finite element approach (Koubaa and Krauthammer, 1990). From the research published by Bala and Kennedy (1986), to determine overlay required and pavement performance, observed a three-dimensional finite element analysis was used to incorporate pavement deflections. The surface deflections observed by a deflector graph were used to calibrate the finite element analysis response predictions. A parametric research was used to establish changes for the calibration procedure Dynamic test findings for in situ samples acquired in the lab were compared to the qualities expected for the materials in the research. Research publish by Saad et al. (2006) gives the benefits of providing geosynthetic in three location in flexible pavement at base-asphalt, in base-sub base and in sub base layer. Result gives large reduction in fatigue strain and highest decrease in rutting strain carried out in finite element program ADINA using 3-D modelling (Saad et al., 2006).

Aim and objective

The research gap of paper is finding a structural solution of serviceability in flexible pavement. Under vehicle's moving load flexible pavement get fails. It occur fatigue crack, rutting and thermal cracking. The aims and objective of paper is,

- To find out cost effective structural solution for serviceability improvement of flexible pavement.
- To estimate design features requirements of precast panel i.e., (length, width, height and shape) as per IRC specifications.
- To model large precast element by means of a relevant Finite Element computer program (ANSYS APDL).
- To analyse and design of a homogenous open ended rectangular element subjected to a combination of compressive and lateral live loads.
- To carry out comparative analysis of flexible pavement with and without using Precast panel.

Methodology

Flexible pavement section consists of sub-base layer, base layer and bituminous layer. The most frequent flexible pavement failures are fatigue cracking, rutting, and heat cracking. Fatigue cracking in flexible pavement is caused by horizontal tensile strain at the bottom of the asphaltic concrete. Pavement degradation is the process through which the pavement develops distress (defects) as a result of the combined impacts of traffic and environmental factors. So, precast pavement technology is a novel and unique construction process that may be utilised to address the demand for structural reinforcement. Strengthening and provide durability to the flexible pavement. When a precast concrete panel is installed, it separates subgrade soil into aggregate base or sub-base and restricts soil particle migration from the subgrade. Precast panel Confines base course for increasing mean stress and also increase in shear strength. After providing precast concrete panel at below the sub-base layer Precast panels are used to provide a platform over a weak subgrade to hold the pavement and simplify pavement building without causing excessive subgrade deformation. In between the base layer and the sub base layer of flexible pavement, a precast concrete panel is used to disperse shear stresses caused by automobile load.

The thickness of the flexible pavement model is chosen as per codal provision IRC:37 (2018). A flexible pavement section is modelled and analysed as a multi-layered structure is subjected to static class AA loading conditions as per IRC:6 (2014). The flexible pavement with or without precast panel are modelling and analysis is done with application software of FEA i.e. ANSYS APDL, considering two separate cases, one considering pavement layers with Precast panel and the second considering without Precast panel presented in Figure 36.1.

A. Precast Concrete Panel

Precast concrete panel technology recently gained lots of attraction from the construction industry. Also it is more suitable for repair work of pavement and strengthening of flexible pavement. In the precast system, the panels are constructed remotely on the casting area. Because these panels are made in a controlled atmosphere, a better grade of concrete may be used in construction.

The optimum panel size is taken as 4 m to achieve a perfect equilibrium between temperature stresses and easy erection of panel in flexible pavement. The thickness of panel is taken as 100 mm after 100% stabilisation of CBR value and 45% improvement of CBR cement stabilisation by IRC:50 (1973). As single lane width for urban road pavement is 3.5 m as per IRC:86 (2018), so the size of panel is 3.5 m × 4 m × 0.1 m presented in Figure 36.2. The main aim of the project is to reduce cement consumption because cement is harmful to environment and human health. The panel is designed as U shape open ended rectangular panel. U shape panel is used to strengthening the flexible pavement. This U-shaped concrete panel will assist in the sub-grade strengthening by providing a platform to accept weight transmitted from the flexible pavement structure's uppermost layer. This panel will act as a barrier between the undisturbed soil below and the pavement construction above.

B. Finite Element Model Parametric Study

The finite element tool is a popular structural analysis tool that is utilised all over the world. But when analysing on finite element tool several factors are considered in mind. Firstly, the method is approximate analysis method whose accuracy is depends on degree of mesh discretisation. Secondly accuracy in result is depending on influence of problem behaviour included in analytical idealisation (Melaku and Qui, 2015).

In finite element analysis, determined the stress, strain and deformation with effect of load which is acting on flexible pavement with and without precast panel. For analysis, the elements used for discretisation are SOLID95, It is a higher-order variant of SOLID45, a 3-D 8-node solid element. It can handle irregular forms with less loss of precision. Curved borders are well-suited to SOLID95 components. There are 20 nodes that make up the element. In analysis material properties of precast panel (density23kN/m^3, E27.5Mpa, ν0.205), sub-base course (density23.5kN/m^3, E110Mpa, ν0.4), base course (density23.5kN/m^3, E1650Mpa, ν0.35), and binder course (density22.8kN/m^3, E2200Mpa, ν0.35) has been taken in account. For analysis, hexagonal and quadric meshing is done with mesh size 200 and 100 respective for the different layers of the flexible pavement. As a considering boundary conditions, all sides of the flexible pavement are treated as fixed when analysing flexible pavement. And considering loading condition, the pressure is delivered at the top of the pavement surface, which has a contact area of 0.85 × 3.6 m and a pressure intensity of 35 tonnes.

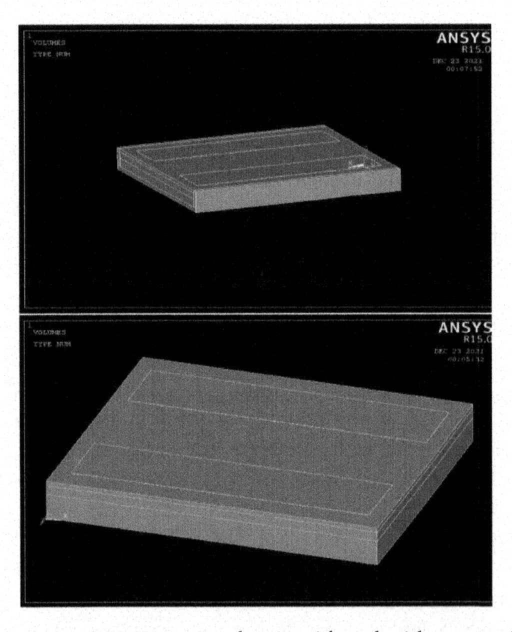

Figure 36.1 Pavement layers with and without panel

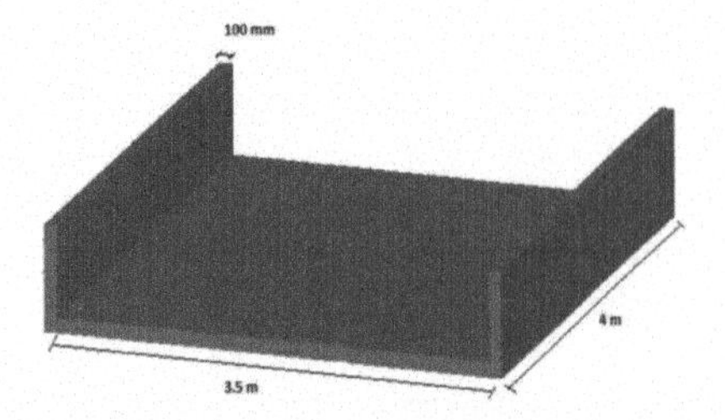

Figure 36.2 Precast concrete panel

Material Model

A. Calculation of Layers Thickness Reduction by CBR Method

For designing flexible pavement two methods are used one is IRC method and second is CBR method. The CBR technique is repeated with different values of soil CBR, taking the original CBR of sub-grade as 3%, to determine the decrease in layer thicknesses by enhancing the CBR value of sub-grade. Using the same procedure as before, layer thicknesses are calculated for this increased CBR of 6%. In accordance with IRC 50, the soil stabilisation approach may enhance the CBR of sub-grade by up to 45%, hence for a 45% improved CBR, the CBR technique is used to calculate layer thicknesses, and the results are summarised as present in table no 36.1.

The thickness of flexible pavement is calculated on the basis CBR value of soil. For sub-grade soil CBR value is taken as 3%. After 100% improvement of original subgrade CBR value is taken as 6% so the thicknesses of layers are reduced. Furthermore, material attributes are employed in the ANSYS study of flexible pavement and the layer model is derived from literature as shown in Table 36.2.

Loading System

As per IRC:6 (2014), class AA loading is adopted up to certain municipal limit, certain existing industrial are a, along with specified highway and bridges heavy loads. In IRC:6 (2014), clause no 204.1, the maximum wheel load is taken as 20 tons for single axle wheel load or 40 tons for bogie of two axes wheel load with spacing not more than 1.22 m centre to centre. As per IRC clause no 204.3, the load is check under

Table 36.1 Comparison of layer thicknesses with different CBR values

Sr. No.	*Description*	*Thickness of sub-base layer*	*Thickness of base layer*	*Thickness of bituminous layer*
1.	For original CBR value of sub-grade i.e. 3%	400 mm	150 mm	70 mm
2.	After 100 % improvement in original CBR i.e. 6%	200 mm	150 mm	70 mm
3.	After improvement of CBR by cement stabilization i.e. 45 % as per IRC-50	Total thickness of pavement = 100 mm		

Table 36.2 Properties of material

Material	*Origional depth (mm)*	*Improved depth (mm)*	*Density (kn/m3)*	*Poisons ration (v)*	*Elastic modulus (mpa)*
Sub-base course	400	200	23.5	0.40	110
Base course	150	150	23.5	0.35	1650
Binder course	70	70	22.8	0.35	2200

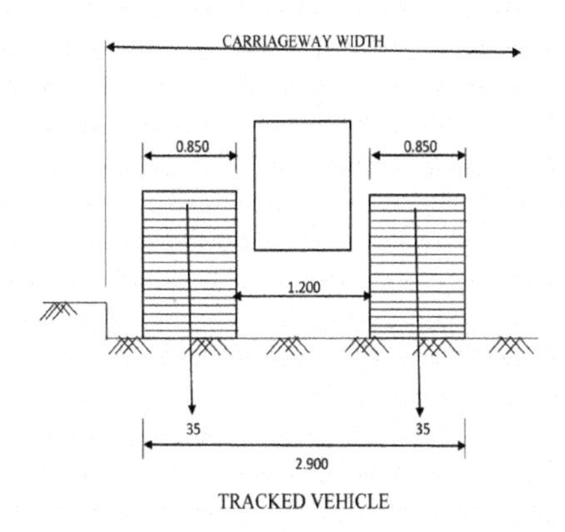

Figure 36.3 Class AA tracked and wheeled vehicles

class 70R loading from which we get the contact area of wheel is 0.85 mm × 3.6 m and the wheel load of 35 tonns on each wheel shown in Figure 36.3 (IRC:6, 2014).

Result and Discussion of Finite Element Analysis

A. Deformation of flexible pavement

Total deformation of flexible pavement with and without precast panel is calculated on application software of FEA i.e. ANSYS APDL. The total deformation of flexile pavement with panel is 0.319771 mm and total deformation of flexible pavement without panel is 0.572988 mm as shown in Figure 36.3. The comparison of flexible pavement with and without panel is shown in Figure 36.4. As in Figure 36.4 show a red colour portion showing the extreme stress act on which portion flexible pavement will be fails. When adding a panel to the flexible pavement is most distortion is reduced when the precast panel is put at the bottom of the sub base foundation layer. Such reduction of deformation is up to 47% which indicates improved serviceability presented in Figure 36.5. Providing precast concrete panel to flexible pavement is reduce the von-misses' stresses of pavement up to 31% indicates the improved strength and durability of panel as shown in Figure 36.5.

B. Flexible Pavement Stress-Strain State

Flexible pavement stress strain analysis is performing on finite element analysis tool. This result indicates impact of vehicle load on road pavement. Elemental stress occur with panel is 2.85641 MPa and without panel is 4.18868 MPa presented in Figure When analysing elemental strain, the strain value with panel is 0.001811 and stain value without panel is 0.01904 is shown in Figure 36.7 and comparison is shown in Figure 36.8. The stress analysis of pavements with and without precast panels indicates that the largest stress occurs at the vehicle's tyre contact region, while the least stress occurs at the panel's centre. Placing precast concrete panel below sub base layer leads to decrease von misses elastic strain. The reduction of resulting strain is up to 5% indicates the improved serviceability.

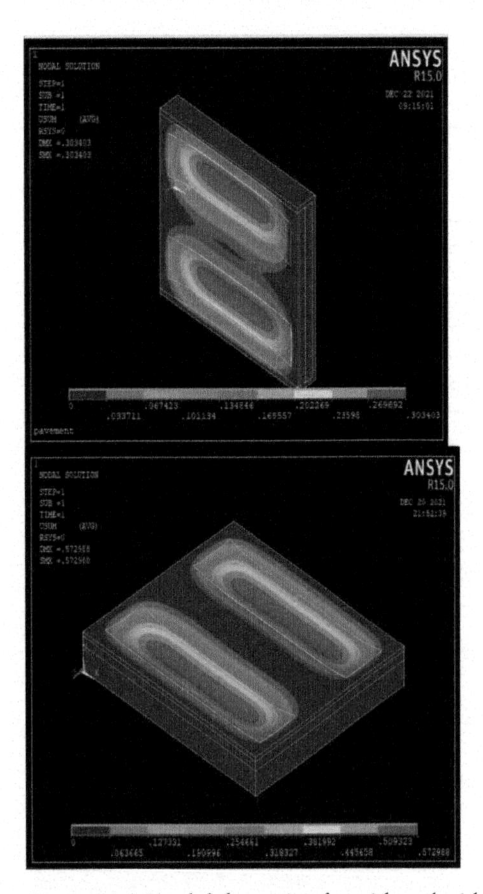

Figure 36.4 Total deformation for with and without panel

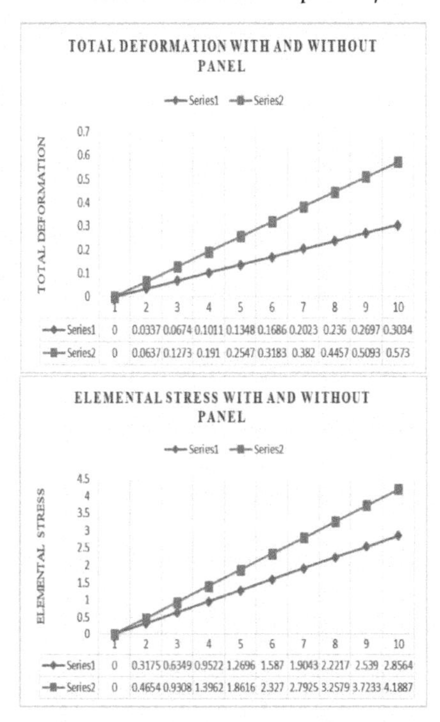

Figure 36.5 Comparison of total deformation and elemental stress

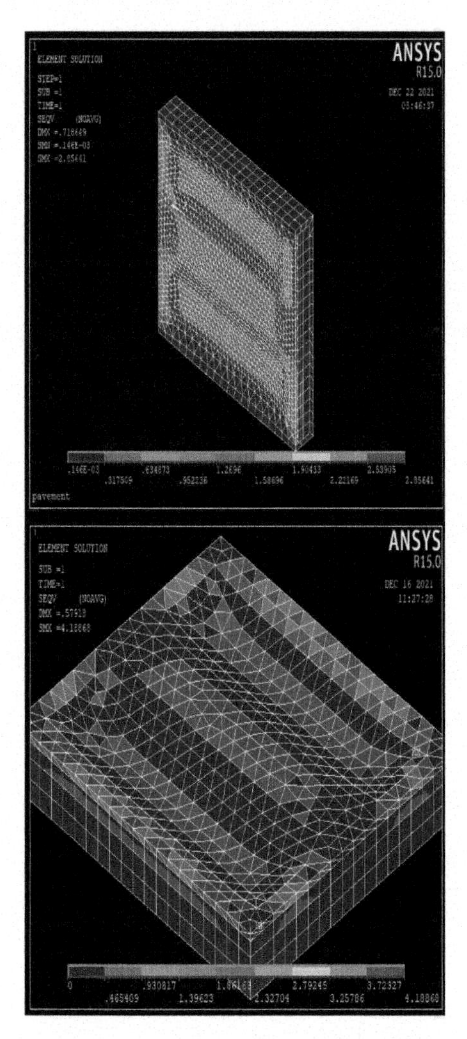

Figure 36.6 Elemental stress for with and without panel

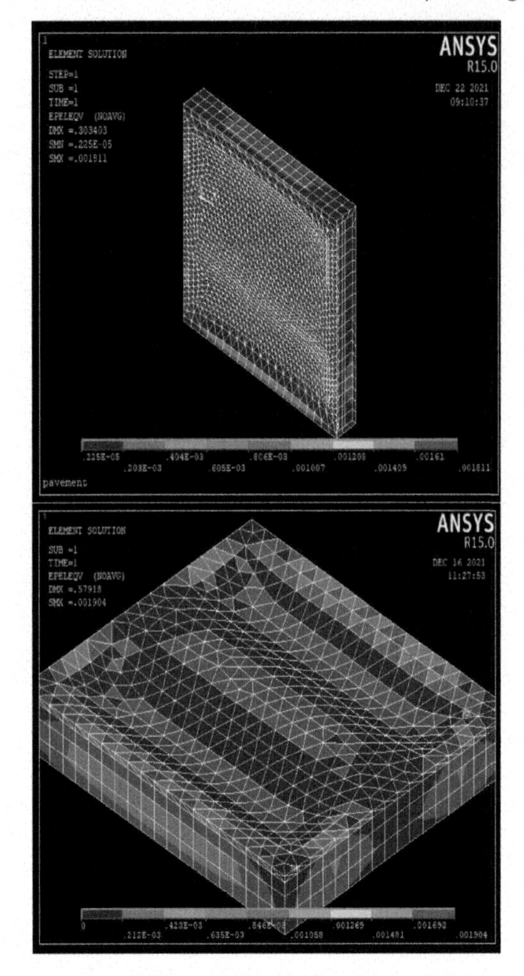

Figure 36.7 Elemental strain for with and without panel

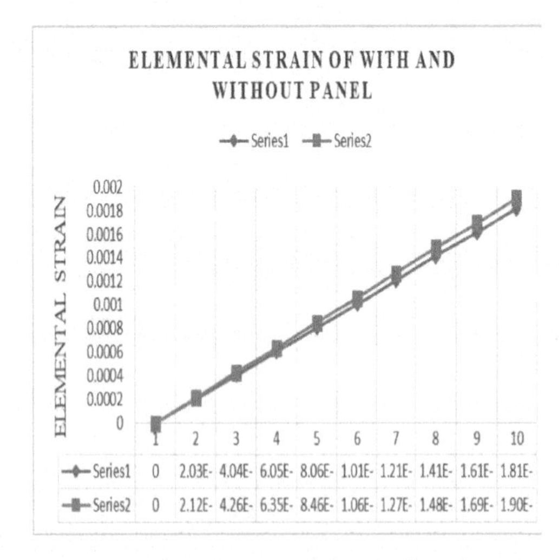

Figure 36.8 Comparison of elemental strain

Table 36.3 Results of flexible pavement with and without panel

SR NO	Result	With panel	Without panel	% Errors
1	Deformation (mm)	0.303403	0.572988	47%
2	Elemental stress (MPa)	2.85641	4.18868	31%
3	Elemental strain	0.001811	0.001904	5%

Conclusion

From the finite element analysis, precast concrete panel is used for improvement of flexible pavement. The result shows that deformation, stress and strain of flexible pavement is reduced by providing precast panel in the flexible pavement. When a precast panel is put at the above the sub grade layer and below the base layer, it reduces vertical deflection, improving the overall performance of the flexible pavement. Following conclusion are drawn from model parametric study,

- CBR technique is used in this project for various sub-grade CBR values. The entire thickness of the pavement decreases as the CBR of the sub-grade improves, according to the results of this approach. As a result of the foregoing findings, it can be determined that increasing the CBR of the sub-grade to 45% reduces the overall thickness by 17%.
- The biggest decrease in deformation occurs when precast panels are positioned at the bottom of the sub foundation layer. Such reduction of deformation reaches to 47% which indicates improved serviceability.
- Providing precast concrete panel to flexible pavement is reduce the von misses stresses of pavement up to 31% indicate improved strength and durability. The stress analysis of pavements with and without precast panels indicates that the largest stress occurs at the vehicle's tyre contact region, while the least stress occurs at the panel's centre.
- Placing precast concrete panel below sub base layer leads to decrease von misses elastic strain. The reduction of resulting strain is up to 5% indicate improved serviceability of flexible pavement.

References

A-Alzaawi, A. A. (2012). Finite element analysis of flexible pavements. strengthen with geogrid. ARPN. J. Eng. Appl. Sci. 7(10).

AASHTO, 1993. AASHTO Guide for Design of Pavement Structures. (pp.: 640). Washington: AASHTO.

Abed, A. H. and Al-Azzawi, A. A. (2012). Evaluation of rutting depth in flexible pavements by using finite element analysis and local empirical model. Am. J. Eng. Appl. Sci. 5(2):163–169.

Abu-Farsakh, M. Y., Gu, J., Voyiadjisb, G. Z., and Chena, Q. (2014). Mechanistic–empirical analysis of the results of finite element analysthe is on the flexible pavement with geogrid base reinforcement. Int. J. Pavement Eng. 15(9):786–798.

Bala, K. V. and Kennedy, C. K. 1986. The Structural Evaluation of Flexible Highway Pavements Using the Deflectograph. Proceedings, 2nd International Conference on the Bearing Capacity of Roads and Airfields, Plymouth, England. 1618.

Koubaa, A. and Krauthammer, T. 1990. Numerical Assessment of Three-Dimensional Rigid Pavement Joints Under Impact Loads. Minneapolis: University of Minnesota.

IRC:37 (2018). Guidelines for the design of flexible pavements (fourth revision).

IRC:6 (2014). Standard specifications and code of practice for road bridges.section 2 loadstresstrese.

IRC:50 (1973). Recommended design criteria for the use of cement-modified soil in road construction.

IRC 86: (2018). Geometric design standards for urban roads in plains.

Li, S. and Hu, C. (2018). Finite element analysis of GFRP reinforced concrete pavement under static load. IOP Conf. Series: Earth Environ. Sci. 113:012188.

Li1a, S. and Hu1b, C. 2018. Finite element analysis of GFRP reinforced concrete pavement under static load. IOP Conf. Series: Earth and Environm. Sci. 113:012188.

Melaku, S. and Qui, H. (2015). Finite Element Analysis of Pavement Design Using ANSYS Finite Element Code. The Second International Conference on Civil Engineering, Energy and Environment.

Saad, B., Mitri, H., and Poorooshasb, H. 2006. 3D FE Analysis of Flexible Pavement with Geosynthetic Reinforcement. J. Transp. Eng. 132(5).

37 Research on key charrecteristics of aircraft product assembly process to improve assembly accuracy

Md Helal Miah[1,a], *Dharmahinder Singh Chand*[2,b], *and Gurmail Singh Malhi*[2,c]

[1]Department of Mechanical Engineering, Chandigarh University, Chandigarh, India

[2]Department of Aerospace Engineering, Chandigarh University, Chandigarh, India

Abstract

The aircraft assembly process mainly refers to the drilling and riveting/joint process of the aircraft product/components. It's necessary to assemble the aircraft with high accuracy to ensure the safety of the aerospace vehicle. In this research, factor affecting aircraft assembly accuracy is discussed. And the efficient process is demonstrated to improve the aircraft component's assembly accuracy based on experts' knowledge information. At first, the assembly accuracy requirements of different aircraft components assembly are provided. According to the required accuracy of aircraft assembly, finishing the drilling and riveting/joint operation is not easy at all. Because various factors affect the aircraft components assembly and manufacture processes, such as the transmission dimension and geometry errors. The accumulation of transmission errors will cause multiple inconsistencies. These inconsistencies are ultimately manifested as component shapes inconsistencies and component joint inconsistencies. This article carries out the precaution, awareness, and effective process of the aircraft component adjustment in the jig to avoid surface failure due to excessive clamping force, drilling, and riveting process. Then summarizes the aircraft components adjustment, clamping, drilling, and riveting processing methods during aircraft components assembly and manufacturing. Finally, these methods have been applied in aircraft parts assembly, which represents an efficient assembly finishing of the aircraft components.

Keywords: Adjustment of aircraft components, aircraft assembly process, drilling and joint, interchange coordination, surface failure.

Introduction

The quality of aircraft assembly depends on good component assembly finishing and surface finishing Najmon et al. (2019) and MIAH et al. (2022). The key process of assembling aircraft fuselage panels, fuselage frames, bulkhead, primary control system, secondary control system, and engine nacelle is drilling and riveting Durham (2014). Component assembly finishing refers to after the riveting process of the aircraft components assembly is completed and the final product structure is completely closed Mei et al. (2019). The aircraft component joint holes and surface finishing are processed based on the component shape in the drilling and riveting process. Also, the joint holes have to meet the requirements of the drawings or will meet the needs of the joint specifications NanYan et al. (2018). Regarding aircraft assembly, the drawing requirement focuses on the riveting errors, improving the manufacturing accuracy of components, the coordination accuracy between product parts, and realizing the processing method of component interchange. It's a difficult task to perform the components assembly finishing for some aircraft products. Therefore, it's a challenging task for the modern aircraft industry to improve the aircraft's assembly accuracy and components coordination accuracy Zhu et al. (2019).

Significant of the Components Assembly Finishing

A. Assembly error analysis

According to the modern aircraft manufacturing technology level, the intelligent manufacturing method is used generally Zhang et al. (2019). The coordination system is divided into absolute coordination system and relative coordination system. The workpiece line intersecting point, intersecting plane point, and datum hole determine the absolute coordinates system. The absolute coordinates system, coordination route, workpiece line intersecting point, intersecting plane point, and datum hole determine the relative coordinates system Zhu and Gao (2009). In terms of coordinates system, the transmission of dimension

[a]helal.sau.12030704@gmail.com; [b]sdsinghchand25@gmail.com; [c]gurmailmalhi.me@cumail.in

and geometry may have certain transmission errors Velex et al. (2016). The accumulation of transmission errors will cause various inconsistencies. These inconsistencies are ultimately manifested as component shapes inconsistencies and component joint inconsistencies. This research paper illustrates the coincidence of the hole function coordinates method as an example.

The general coordination route between the component shape and the assembly tooling design is shown in Figure 37.1.

$$\sum \Delta\varepsilon = \Delta\varepsilon_1 + \Delta\varepsilon_2 + \Delta\varepsilon_3 + \Delta\varepsilon_4 - (\Delta\varepsilon_5 + \Delta\varepsilon_6) \quad (1)$$

In the formula:

$\Delta\varepsilon_1$, $\Delta\varepsilon_2$, $\Delta\varepsilon_5$, $\Delta\varepsilon_6$ are the transmission errors of each link.

$\Delta\varepsilon_3$ is the part manufacturing error.

$\Delta\varepsilon_4$ is the riveting assembly error.

$\sum\Delta\varepsilon$ is the coordination error between the shape of the component and tooling design. The allowable range of these errors is shown in Table 37.1.

The general coordination route of the butt joints on the two parts A and B that are connected to each other. In the Table 37.2. $\Delta\varepsilon_{10} \sim \Delta\varepsilon_{16}$ is the transmission error of each link $\Delta\varepsilon_A$ and $\Delta\varepsilon_B$ are the accumulated errors of the parts A and B joints, $\Delta\varepsilon_{AB}$ is the coordination error of the parts A and B joints, A and B are the $\Delta'\varepsilon_A$ and $\Delta'\varepsilon_B$ accumulated error after finishing machining of part A and B joint, $\Delta'\varepsilon_{AB}$ is the coordination error after finishing machining of part A and B. Taking the coordinated route of the JL-8 fuselage and wing as an example, the allowable $\Delta\varepsilon_{10} \sim \Delta\varepsilon_{16}$ range is shown in table 37.2.

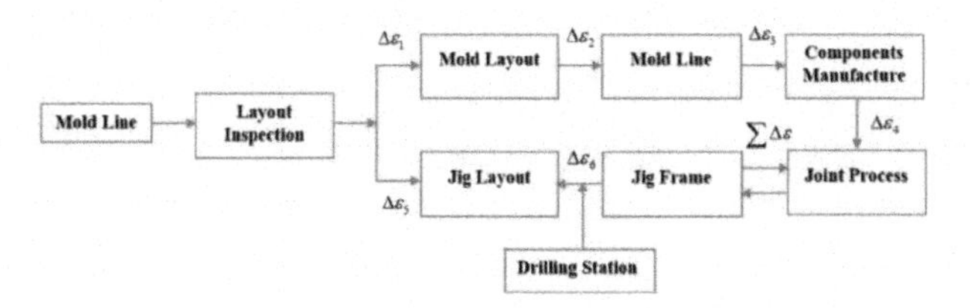

Figure 37.1 General coordination route of components assembly process

Table 37.1 $\Delta\varepsilon_1 \sim \Delta\varepsilon_6$ transmission error range

Error	$\Delta\varepsilon_1$	$\Delta\varepsilon_2$	$\Delta\varepsilon_3$	$\Delta\varepsilon_4$	$\Delta\varepsilon_5$	$\Delta\varepsilon_6$
Upper Limit	0	0	+0.3	+0.5	0	+0.1
Lower Limit	−0.15	−0.1	0	−0. 8	−0.1	0

Table 37.2 $\Delta\varepsilon_{10} \sim \Delta\varepsilon_{16}$ transmission error range

Error	$\Delta\varepsilon_{10}$	$\Delta\varepsilon_{11}$	$\Delta'\varepsilon_{11}$	$\Delta\varepsilon_{12}$	$\Delta\varepsilon_{13}$	$\Delta\varepsilon_{14}$	$\Delta\varepsilon_{15}$	$\Delta'\varepsilon_{15}$	$\Delta\varepsilon_{16}$
Value	±.033	±.5	±.075	±.033	±.033	±.033	±. 5	±.075	±.033

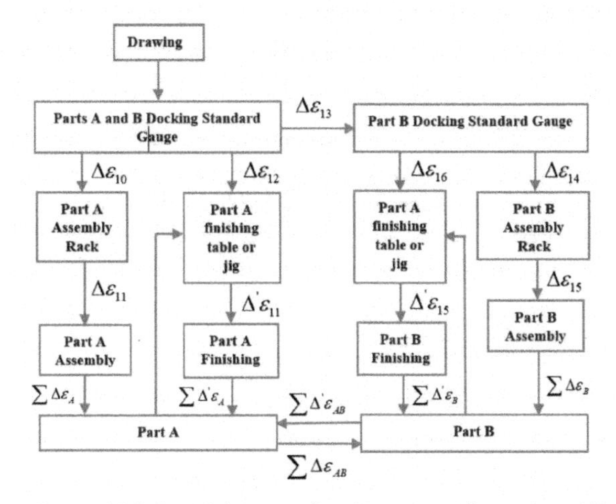

Figure 37.2 Butt-joint coordination route of part A and B

$$\sum \Delta\varepsilon_A = \Delta\varepsilon_{11} + \Delta\varepsilon_{12} \quad (2)$$

$$\sum \Delta\varepsilon_B = \Delta\varepsilon_{13} + \Delta\varepsilon_{14} + \Delta\varepsilon_{15} \quad (3)$$

$$\sum \Delta\varepsilon_{AB} = \sum \Delta\varepsilon_A - \sum \Delta\varepsilon_B \quad (4)$$

$$\sum \Delta'\varepsilon_A = \Delta'\varepsilon_{11} + \Delta\varepsilon_{12} \quad (5)$$

$$\sum \Delta'\varepsilon_B = \Delta\varepsilon_{13} + \Delta\varepsilon_{16} + \Delta'\varepsilon_{15} \quad (6)$$

$$\sum \Delta'\varepsilon_{AB} = \sum \Delta'\varepsilon_A - \sum \Delta'\varepsilon_B \quad (7)$$

According to the probability method (take M = 0.5), the coordinated error value of the centre distance between the main joint holes of the JL-8 fuselage and wing is calculated as shown in Table 37.3.

The above-mentioned $\Sigma\Delta\varepsilon$, $\Sigma\Delta\varepsilon_A$, $\Sigma\Delta\varepsilon_B$ and $\Sigma\Delta\varepsilon_{AB}$ finally appear as:

- The shape of the component part is larger than that of the card board, which is compulsive when the cardboard is closed, the partial shape of the component is smaller than that of the card board, and there is a gap between the shape and the card board.
- There is a certain position deviation between the parts A and B joints and their respective assembly frame joint positioning parts.
- The parts A and B cannot be connected smoothly, and the parts are not interchangeable.
- The position of the component joint and the component shape is not coordinated with each other.

$\Delta\varepsilon_4$, $\Delta\varepsilon_{11}$ and $\Delta\varepsilon_{15}$ are the effects the main factor of the value $\Sigma\Delta\varepsilon$, $\Sigma\Delta\varepsilon_A$, $\Sigma\Delta\varepsilon_B$ and $\Sigma\Delta\varepsilon_{AB}$. For the riveting assembly of any part, these errors are inevitable, so the parts must be finished.

B. The role of assembly finishing

Part finishing has the following functions:

- When the parts are adjusted on a dedicated finishing table or an assembly frame with inspection card, the parts are in a free state, and the shape is not forced, and it objectively reflects the state of the parts, and the overall shape can be adjusted to the best form relative to the theoretical position.
- Make the height of the joint position in line with the joint positioning parts on the finishing table or assembly frame, significantly reducing the accumulated error $\Sigma\Delta\varepsilon_A$ and $\Sigma\Delta\varepsilon_B$. As shown in Table 37.3, $\Sigma\Delta'\varepsilon_A << \Sigma\Delta\varepsilon_A$ and $\Sigma\Delta'\varepsilon_B << \Sigma\Delta\varepsilon_B$.
- Make the relative position of the component joint and the shape highly coordinated.
- Make the parts interchangeable. After the parts are processed by the same method on the same finishing equipment, the shape and size of the parts are consistent.
- Favourable conditions are created for the smooth docking of components A and B. As shown in Table 37.3, $\Sigma\Delta'\varepsilon_{AB} << \Sigma\Delta\varepsilon_{AB}$.

The finishing of the components eliminates the riveting assembly errors caused by the deformation of the riveting assembly and the influence of human operation $\Delta\varepsilon_{11}$ and $\Delta\varepsilon_{15}$. The docking coordination error of components A and B only depends on the coordination route itself, which greatly improves the coordination accuracy between components.

Component finishing can only adjust the shape error caused by the component assembly link and eliminate the joint hole position error caused by the component assembly link, but cannot eliminate the transmission error of each coordination link. To ensure the smooth page 3 of 5 11/15/2021 docking between components, it is necessary to design a reasonable process coordination route and carry out a reasonable process tolerance distribution.

Table 37.3 $\Delta\varepsilon_{10} \sim \Delta\varepsilon_{16}$ accumulated and coordinated error calculation values

Error	$\Sigma\Delta\varepsilon_A$	$\Sigma\Delta\varepsilon_B$	$\Sigma\Delta\varepsilon_{AB}$	$\Sigma\Delta'\varepsilon_A$	$\Sigma\Delta'\varepsilon_B$	$\Sigma\Delta'\varepsilon_{AB}$
Calculated	±0.655	±0. 678	±1.138	±0.098	±0.115	±0.170

Components Finishing Method

The finishing method of the parts can be divided into two types: finishing the joint hole and the final installation of the joint.

A. Finish the joint hole

This is the most widely used type of assembly finishing method. The main point is that when the part is adjusted on a special assembly finishing table or an assembly frame equipped with an inspection card, the part is adjusted according to the part joint positioning piece. The joint hole is processed to the final size to achieve the purpose of assembly finishing parts, so the joint hole and hole end face should be left with a machining allowance. This method is suitable for the assembly finishing of parts such as joint holes that are convenient to leave a machining allowance and have processing structural passages Liu (2012).

B. Final installation of the connector

Some component joints are inconvenient to make up during the component assembly stage, such as the head withbearing, pulley plate, slide groove, etc. Nekoz et al. (2018). The final installation of the joint is an effective way to solve the assembly finishing of such parts. As shown in Table 37.2, $\Delta\varepsilon_{11}$ and $\Delta\varepsilon_{15}$ are caused by factors such as assembly deformation, poor structural manufacturability, and low level of operation during the assembly process. After the structure is closed, no further errors will be caused. When connecting the joint with the component structure, it can also play a role in the assembly finishing of the component. The specific process is as follows: in the riveting assembly process, use process rivets and process bolts with a diameter smaller than that of the connecting parts shown in the Figure 37.2. to temporarily fix the part joints and the structure. When the structure is assembled and the part structure is completely closed, the process rivets or process bolts can be removed. Open all the cards in the assembly frame, use the inspection card to adjust the shape of the parts to the best state; install the joint on the structure according to the joint positioning piece on the assembly frame, and ensure the position error of the joint relative to the joint positioning piece zero, to achieve the purpose of assembly finishing parts.

The characteristic of this kind of method is that the assembly finishing process is carried out on the assembly frame, no special finishing table is needed, and special tools and powerheads are often not needed.

Structural Requirements for Assembly Finishing

The prerequisite for the parts to be finished is that the structure of the parts must be finished. To finish machining the joint hole, its structure should meet the following conditions:

- There is an assembly finishing allowance for the joint hole and the end face of the joint hole, and its value should be greater than the sum of the accumulated error of the joint and the joint displacement caused by the adjustment of the shape.
- Allow the machining tool to enter the structure to be finished.

To adopt the assembly finishing method of the final installation of the joint, its structure should meet the following conditions:

- There are design compensations such as overlap, gap, or padding between the joint and the component structure so that the relative position between them is adjustable.
- When the structure has access, the connection between the joint and the structure can be ordinary bolts, rivets, etc. When the structure has no access, the connection should be a single-sided connection, such as pull nails, high shear rivets, etc.
- • To use inspection card as shown in Figure 37.3 to measure the actual position of curvature or synthetic component of aircraft in the fixture.

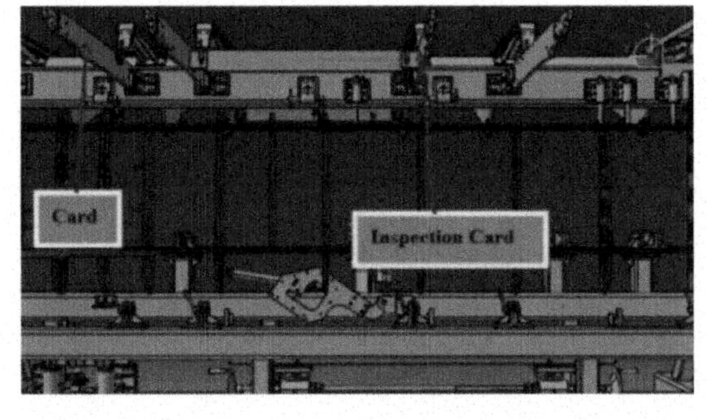

Figure 37.3 Inspection card to measure the component's position

Adjustments During Assembly Finishing

A. Shape adjustment

The shape can be used as a reference or the joint can be used as a reference to adjust the shape to ensure the coordination accuracy of the relative position of the joint and the shape. Under normal circumstances, it is very difficult to adjust the shape after the structure is closed. Even if it can be done, it is easy to damage the surface quality. The number of rivets will loosen, and the finishing of the joint is much simpler and easier than correcting the shape. Therefore, when parts are finished, the shape should be used as the adjustment standard to finish the joint. Occasionally, when the part structure does not have the joint finishing conditions and is a thin-walled structure, the shape is corrected based on the joint Mei and Maropoulos (2014).

B. Adjustment method

When large parts such as fuselage and wings are mass-produced, they are usually finished on a dedicated assembly finishing table. When adjusting the shape of the part, the level measuring point indicator on the assembly finishing table can be used to measure. So that the level measuring point meets the requirements of the liquid measurement chart and the process tolerance distribution. For the parts that are finished on the assembly frame, two or three sets of working cards can be replaced with inspection cards. Also, the shape of the parts can be adjusted to make the gap between the two shape faces and the inspection card as even as possible. Even if the shape of the part is compared with the inspection card, the shape error is in the smallest state. For parts that require only a one-sided appearance, such as guard plates and speed reducers. the parts can be adjusted to make the appearance as close as possible to the outermost two working pallets in the assembly frame, abandoning the remaining working pallets, and performing assembly finishing after clamping Yuan et al. (2017).

Clamping During Assembly Finishing

It should be noted that during the assembly finishing and clamping of parts, as shown in Table 37.2, $\Delta'\varepsilon_{11}$ and $\Delta'\varepsilon_{15}$ are calculated values of hole-shaft fit under normal conditions, and do not include errors caused by clamping deformation and other factors. In fact, these errors are even larger than $\Delta'\varepsilon_{11}$ and $\Delta'\varepsilon_{15}$, so the parts are required to be in a free state during assembly finishing. However, in order to ensure the stability of the parts relative to the position of the assembly finishing table or assembly frame, the parts must be clamped. For aircraft components, especially small aircraft components, the structural rigidity is low, which will cause elastic deformation of the components during clamping. In the clamping state, the clamping is loosened after processing, and the part rebounds, causing the part joint to shift relative to the position of the positioning part. Therefore, it is very important to control and detect the elastic deformation of the part caused by the clamping, such as a small aircraft When the wing is finished, the position of the wing joint caused by the clamping deformation is 0.2 ~ 1 mm Zhang et al. (2021), which makes the docking of the wing and fuselage difficult.

A. Control of clamping deformation

The control of clamping deformation is to make the clamping deformation as small as possible or even basically eliminate it Mannan and Sollie (1997). The following clamping methods can be used to control the clamping deformation:

- The joints, beams, rigid ribs, and rigid frames of the clamping components are strongly rigid parts. Also, the joints, beams, rigid ribs, and rigid frames need to hold against six degrees of freedom with less deformation during numerical machining operation.
- The clamping part is as close as possible to the assembly finishing part.
- As far as possible, design a clamp that clamps up and down or left and right at the same time.

B. Detection of clamping deformation

In order to effectively control the clamping deformation of the components, it is necessary to measure the clamping deformation accurately and effectively Walsh et al. (2020). The specific method is independent of the parts and clamping device. Fix the dial indicator (as shown in Figure 37.4) for testing, touch the probe of the dial indicator to the clamping position and the clamping shape. Finally, adjust the clamping device so that it can be appropriately clamped. For components, the deformation value indicated by the dial indicator

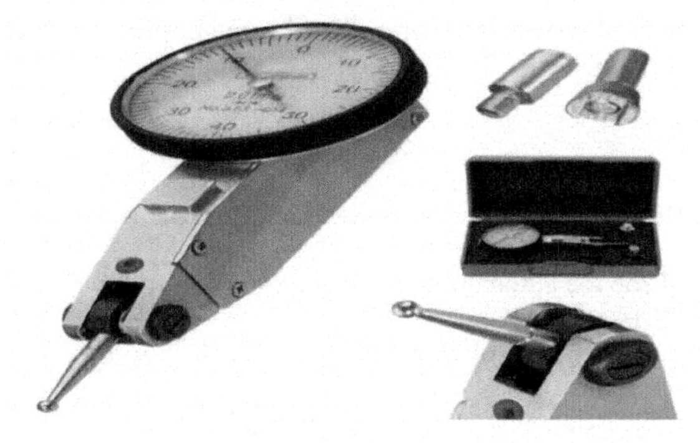

Figure 37.4 Dial indicator for clamping deformation measurement

should be as zero as possible to improve the conformity of the joint and the joint positioning part after the complete aircraft components assembly.

Component Joints Methods

The components joint methods have two goals. One is to process the additional joints to the size required by the final drawing or to install the final size of the joint on the structure. The second is to control the factors that affect the quality of $\Delta'\varepsilon_{11}$ and $\Delta'\varepsilon_{15}$ finishing, and as much as possible to reduce the finishing error and as shown in Table 37.2, so that the finished joint fits the finishing joint positioning requirements. Pay attention to the following points during processing:

- Choose the appropriate cutting tool, cutting tolerance, rotational speed, and feed rate. In general, the feed rate should not be too fast because the rigidity of the aircraft parts is low. Therefore, it needs to select appropriate cutting tool rotational speed and feed rate ratio for efficient cutting operation.
- The joint positioning tooling or guide should be as thick as possible, and the cantilever length should be as short as possible to increase the aircraft component's rigidity.
- Try to avoid the horizontal state processing mode of the cutting tool, and the vertical state processing mode should be adopted.
- If the joint is a body-pressure bushing structure, the body should be processed for assembly finishing, but the bushing should not be processed. The processing machine body can use the eccentric bushing to compensate for the hole position error, and there is a remedy when the hole diameter is out of tolerance.
- For the assembly process, the final process is joint. Generally, this process is performed after the drilling process of the connection hole diameter and is more straightforward than the connection hole calibration process. It is essential to make two positioning holes smaller than the actual connection hole diameter and ensure higher accuracy. Use temporary fastener to joint aircraft components through positioning hole, and check the joint position to ensure the other positioning deviation are correct. Then complete the joint process of aircraft assembly. Finally, remove the temporary fastener and install the two connectors with positioning holes made first.

Conclusion

This paper mainly illustrated the factor affecting the assembly process, especially aircraft component positioning and clamping during clamping riveting/joining. The following item needs to be optimised during the drilling and riveting process for efficient assembly finishing:

- Reduce the error caused by the frame and skin due to the tight-fitting.
- Minimise the shape error of the frame parts manufacturing.
- Minimise the assembly error of the frame.
- Minimise the deformation produced by excessive clamping force.
- Select the proper ratio of feed rate and rotational speed during machining.
- Try to employ the vertical state processing mode of the cutting tool.

Acknowledgement

Prof. Dharmahinder Singh Chand instructed and conducted research. Then Mr. Md Helal Miah did the experiment, analysed the result, designed the model, and wrote the paper. Prof. Gurmail Singh Malhi from

Chandigarh University, India helped to improve the final research paper and add some important points for this research. All authors had approved the final version.

References

Alexandr, N., Oleksandr, V., and Alexandr, B. (2018). Durability of cutter assemblies and its causative factors. *Foods Raw. Mater.* 6(2):358–369.

Durham, B. J. (2014). Determining appropriate levels of robotic automation in commercial aircraft nacelle assembly. PhD diss., Massachusetts Institute of Technology, 2014.

Liu, C. H. (2012). Safety improving technology for finish machining of wing-fuselage joints. In *Advanced Materials Research*, (vol. 566, pp. 263–266). Trans Tech Publications Ltd.

Mannan, M. A. and Sollie, J. P. (1997). A force-controlled clamping element for intelligent fixturing. *CIRP Ann.* 46(1):265–268.

Mei, B., Zhu, W., Zheng, P. and Ke, Y. (2019). Variation modeling and analysis with interval approach for the assembly of compliant aeronautical structures. *Proc. Inst. Mech. Eng. Part B: J. Eng. Manuf.* 233(3):948–959.

Mei, Z. and Maropoulos, P. G. (2014). Review of the application of flexible, measurement-assisted assembly technology in aircraft manufacturing. *Proc. Inst. Mech. Eng., Part B: J. Eng. Manuf.* 228(10):1185–1197.

Miah, M. H., Zhang, J. and Chand, D. (2022). Knowledge creation and application of optimal tolerance distribution method for aircraft product assembly. *Aircr. Eng. Aerosp. Technol.* 94(3):431–436. https://doi.org/10.1108/AEAT-07-2021-0193.

Najmon, J. C., Raeisi, S. and Tovar, A. (2019). Review of additive manufacturing technologies and applications in the aerospace industry. *Addit. Manuf. Aerosp. Ind.* 1:7–31.

NanYan, S., ZiMeng, G., Li, J., Liang, T. and Zhu, K. (2018). A practical method of improving hole position accuracy in the robotic drilling process. *Int. J. Adv. Manuf. Technol.* 96(5-8):2973–2987.

Parietti, F. and Asada, H. H. (2014). Supernumerary robotic limbs for aircraft fuselage assembly: Body stabilization and guidance by bracing. In *2014 ,IEEE International Conference on Robotics and Automation (ICRA)*, (pp. 1176–1183). IEEE.

Velex, P., Chapron, M., Fakhfakh, H., Bruyère, J. and Becquerelle, S. (2016). On transmission errors and profile modifications minimising dynamic tooth loads in multi-mesh gears. *J. Sound Vib.* 379:28–52.

Walsh, W., Abotula, S. and Konda, B. (2020). Ring expansion testing innovations: Hydraulic clamping and strain measurement methods. In *International Pipeline Conference*, 84461:V003T05A009. American Society of Mechanical Engineers.

Yuan, L. I., Zhang, L. and Yanzhong, W. A. N. G. (2017). An optimal method of posture adjustment in aircraft fuselage joining assembly with engineering constraints. *Chinese J. Aeronaut.* 30(6):2016–2023.

Zhang, L., Zhou, L., Ren, L. and Laili, Y. (2019). Modeling and simulation in intelligent manufacturing. *Comput. Ind.* 112:103–123.

Zhang, W., An, L., Chen, Y., Xiong, Y. and Liao, Y. (2021). Optimisation for clamping force of aircraft composite structure assembly considering form defects and part deformations. *Adv. Mech. Eng.* 13(4): doi:1687814021995703.

Zhu, S. and Gao, Y. (2009). Noncontact 3-D coordinate measurement of cross-cutting feature points on the surface of a large-scale workpiece based on the machine vision method. *IEEE Trans. Instrum. Meas.* 59(7):1874–1887.

Zhu, W., Li, G., Dong, H. and Ke, Y. (2019). Positioning error compensation on two-dimensional manifold for robotic machining. *Robot. Comput. Integr. Manuf.* 59:394–405.

38 Performance analysis of conventional inverters driven SVM-PMSM drive

Rakesh G. Shriwastava[1,a], *Nilesh C. Ghuge*[2], *D. D. Palande*[2], *and Ashwini Tidke*[1]

[1]Electrical Engineering Department, MCOERC, Nashik (M.S.), India

[2]Mechanical Engineering Department, MCOERC, Nashik (M.S.), India

Abstract

This paper focus on performance analysis of conventional inverters driven space vector modulation-permanent magnet synchronous motors (SVM-PMSM) drive using microcontroller in automotive application. The purpose of this paper is to decrease the ripples in torque of Micro controllers based SVM-PMSM drive. The innovative method consists of conventional inverters, SMPS, PMSM motor and microcontroller. It is used to maximise fundamental component of torque also. The both the results of the suggested Micro controller are compared on the basis of torque and speed and it proves that reduces the torque ripple and improve the dynamic response of the system in comparative analysis. This paper organised introduction, mathematical model of the PMSM, proposed microcontroller technique, hardware and simulation results and conclusion in different sections.

Keywords: Clock generator, conventional inverters, microcontroller, permanent magnet synchronous motors, switched-mode power supply, voltage source inverter.

Introduction

Electrical machines are the heart of industrial application. The generators convert energy transformation to charge the batteries for motor operation. Required torque developed by motors for drive the wheels. The permanent magnet, switched reluctance, and induction motors are used in industrial application (Bose, 2002; Leonhard, 1990; Ehsani et al., 1997; Jung et al. (2001). This paper explains simulation and hardware implementation of voltage source inverter (VSI) fed permanent magnet synchronous motors (PMSM) drives for automotive application. For the speed and torque control, VSI fed PMSM drive is one of the method used. The VSI fed PMSM drive system is designed, simulated and implemented using Microcontroller. A hardware setup is designed and implemented based on a Four-pole 0.25 HP PMSM. The control hardware consists of, Main power circuit, Control circuit, Isolator and driver circuit and PMSM (Zhong et al., 1997; Zhong et al., 1999; Rahman et al., 1998; Buja et al., 2004; Casadei et al., 2002; Rahman et al., 2004; Li, 2011). The fixed dc voltage is obtained from three phase rectifier circuit and used shunt capacitor for filter purpose. The output of three phase rectifier circuit is given to MOSFET bridge inverters in power circuit (Li et al., 2007). The function of control circuit is to turn the MOSFET by using Gating pulses Isolator and driver circuit is used for proper isolation by using opt isolator (Shriwastava et al., 2020; Shriwastava et al., 2019). The steps involved in any drive system are design, building its hardware and its testing. Once the system is design, depending on designed values of various components, the system hardware is developed. Once the assembly of hardware is completed then it is tested by carrying out various experiments on the developed system to test whether the system is working satisfactory or not. If it is not working properly then the error must be removed and the system must be updated and modified. The drive developed here is operating variable frequency mode. The theme of the paper is simulation and hardware analysis of torque minimisation of PMSM drive for automotive application.

Mathematical Model

A. PMSM modelling in rotor reference frame

The induced voltage in the direct-axis winding:

$$u_d = R_d i_d + \frac{d\lambda_d}{dt} - \omega_r \lambda_q \qquad (1)$$

Where, '*id*' and '*Rd*' are called direct-axis stator current and resistance respectively

---

[a]rakesh_shriwastava@rediffmail.com

The induced voltage in the quadrature-axis winding:

$$u_q = R_q i_q + \frac{d\lambda_q}{dt} - \omega_r \lambda_d \tag{2}$$

Where, '*Rq*' and '*iq*' are called the quadrature -axis resistance and current of stator.

$$\lambda_d = L_d i_d + \lambda_m \tag{3}$$

λd = flux-linkage in the direct-axis of stator in webers, λ_m is the PM rotor flux

$$\lambda_q = L_q i_q \tag{4}$$

λq = flux-linkage in the quadrature-axis stator (Wb)

In this case of quadrature-axis, there are no magnets so λm is absent.

Considering round rotor PMSM, we have

$$L_d = L_q \tag{5}$$

The PMSM torque equation is given by:

$$T_e = \frac{3}{2}\frac{p}{2}\left(\lambda_d i_q - \lambda_q i_d\right) \tag{6}$$

Substituting for '*λd*' and '*λq*' in the torque equation of PMSM,

$$T_e = \frac{3}{2}\frac{p}{2}\left[\left(\lambda_d i_d + \lambda_m\right) i_q - L_q i_q i_d\right] \tag{7}$$

$$T_e = \frac{3}{2}\frac{p}{2}\left[\left(L_d - L_q\right) i_d i_q + \lambda_m i_q\right] \tag{8}$$

The two components of torque developed are:

$$Reluctance\,torque = \frac{3}{2}\frac{p}{2}\left(L_d - L_q\right) i_d i_q \tag{9}$$

$$field\,torque = \frac{3}{2}\frac{p}{2}\lambda_m i_q \tag{10}$$

$$T_e = \frac{3}{2}\frac{p}{2}\lambda_m i_q \tag{11}$$

In a round rotor PMSM the electromagnetic torque present is the field torque present due to the PM flux linkage, λm. For a chosen PMSM, the PM rotor flux-linkage *(λm)* and the number of poles (*p*) is constant. Hence, for the round-rotor PMSM, the electromagnetic torque equation is

$$T_e = Kti_q \tag{12}$$

Where, Kt = Torque constant

$$K_t = \frac{3}{2}\frac{p}{2}\lambda_m \tag{13}$$

Therefore, electro-magnetic torque is given by

$$T_e = T_l + B\omega_m + J\frac{d\omega_m}{dt} \tag{14}$$

ωm = rotor's mechanical speed, *ωr* = rotor's electrical speed

Simulation Model and Results

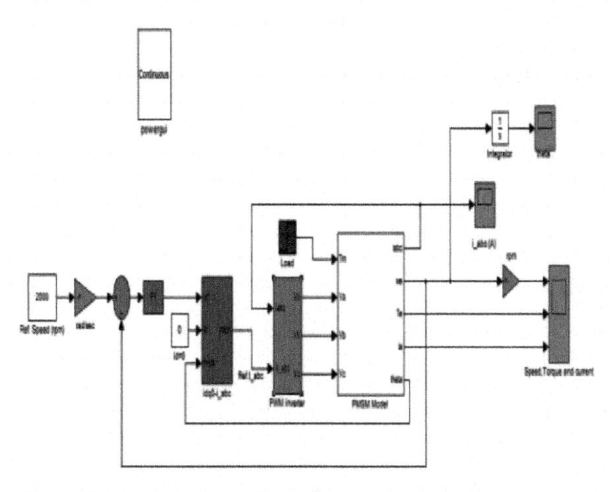

Figure 38.1 Simulation diagram

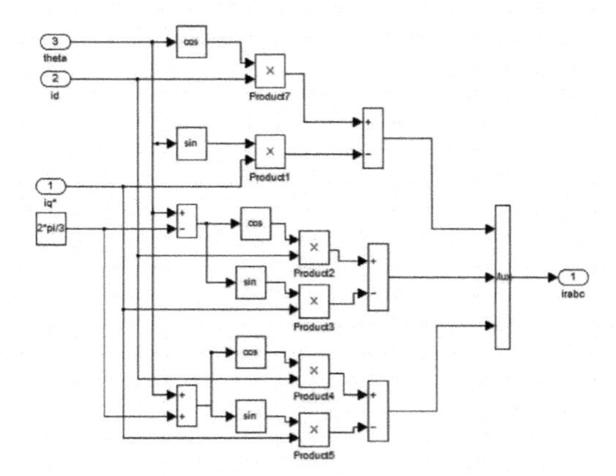

Figure 38.2 D-axis and Q-axis currents model

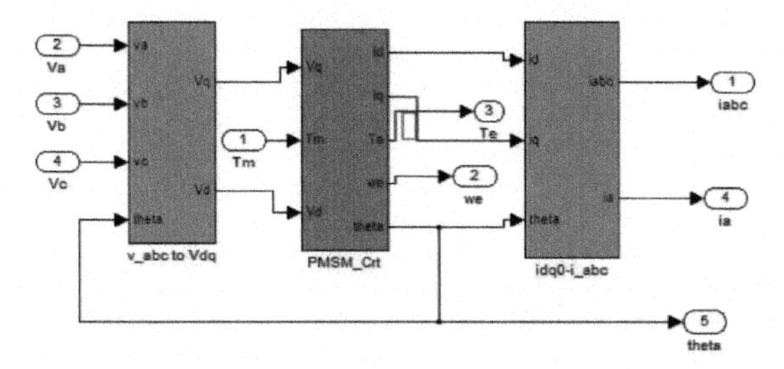

Figure 38.3 PM synchronous motor model

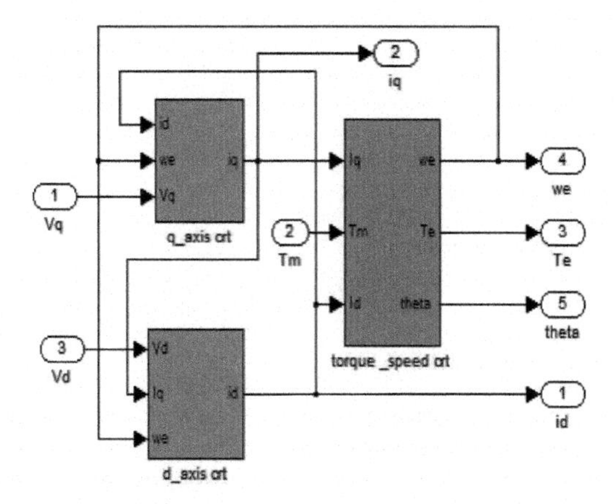

Figure 38.4 PMSM characteristics diagram

The simulation response of Voltage source inverter-fed PMSM drives at different speed range is shown in Figure 38.5–38.7.

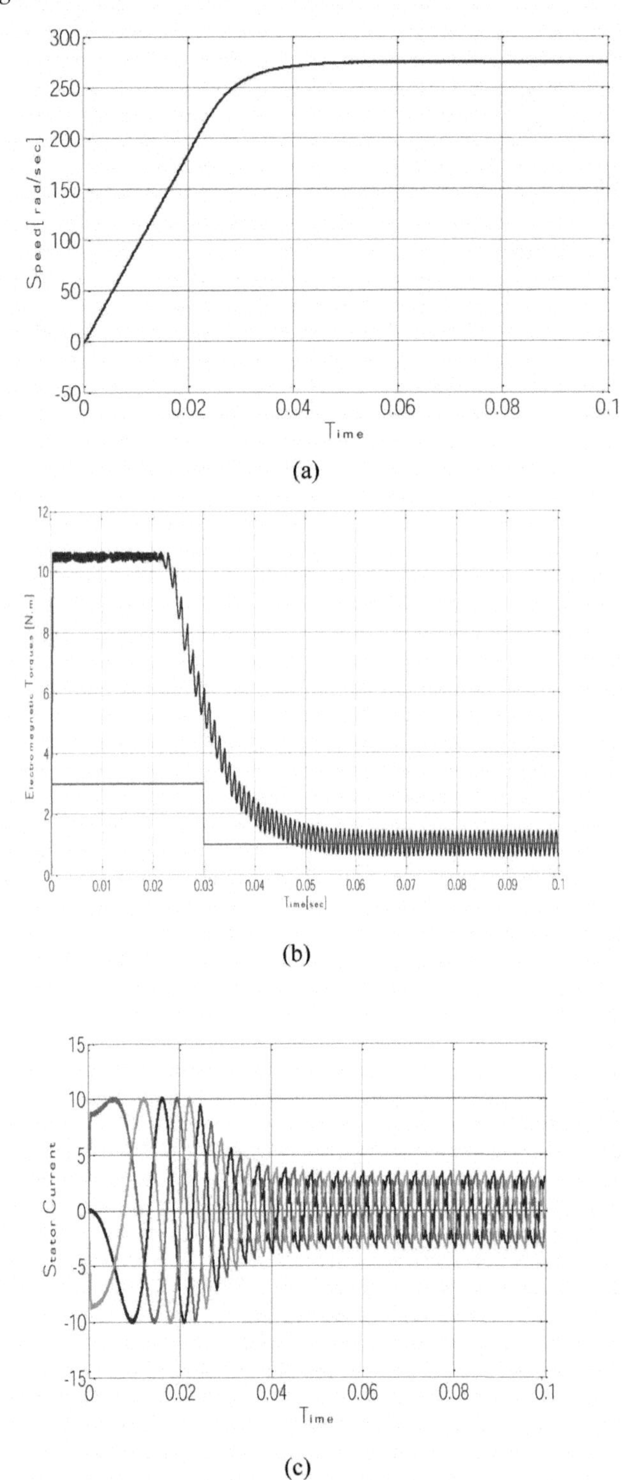

Figure 38.5 (a) Speed response (b) Torque response (c) Current response

Hardware Design

Initially the drive operation was worked on dummy load (three phase star connected lamps) by connecting the motor to the drive. Various outputs are taken out from the control card on the drive. Firstly the frequency knob on the drive was kept at its minimum position. The converter card was then switched on. The power supply status was indicated by all the LEDs inside the unit. By using CRO probes, the waveforms

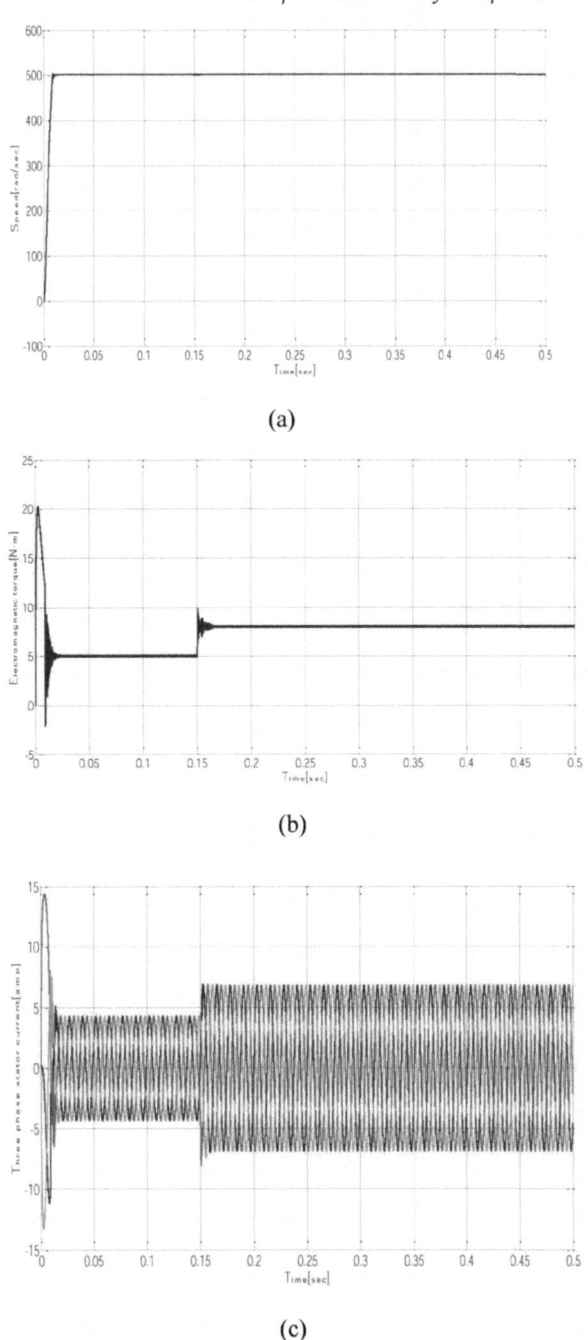

Figure 38.6 (a) Speed response (b) Torque response (c) Current response

at various test points were observed. While observing the waveforms on CRO, care was taken to use the unearthed CRO only. The basic timer IC 555 output frequency was observed at the test point. This frequency can be changed by changing the position of frequency knob. The pulses for various gate drives were observed at the test points. The output frequency was observed at the test point. The minimum and maximum frequency was noted and tabulated in the result table. The frequency knob position was slowly changed along with the set speed pot. Slowly the lamps were turned on as the soft start in the result table. The output line and phase voltage waveforms for the three phases were observed on the CRO at the corresponding test points. After observing the satisfactory performance of the control card then the motor was connected to the drive. The dummy lamp load was now removed and the motor was connected in the output connector. Thus the motor runs at variable speed depending on the frequency. After testing the various test points the dummy star connected load is removed and across the output of the bridge inverter PMSM is connected. Then gradually increasing the load on the shaft of the motor for different values of frequency speed of the motor is measured.

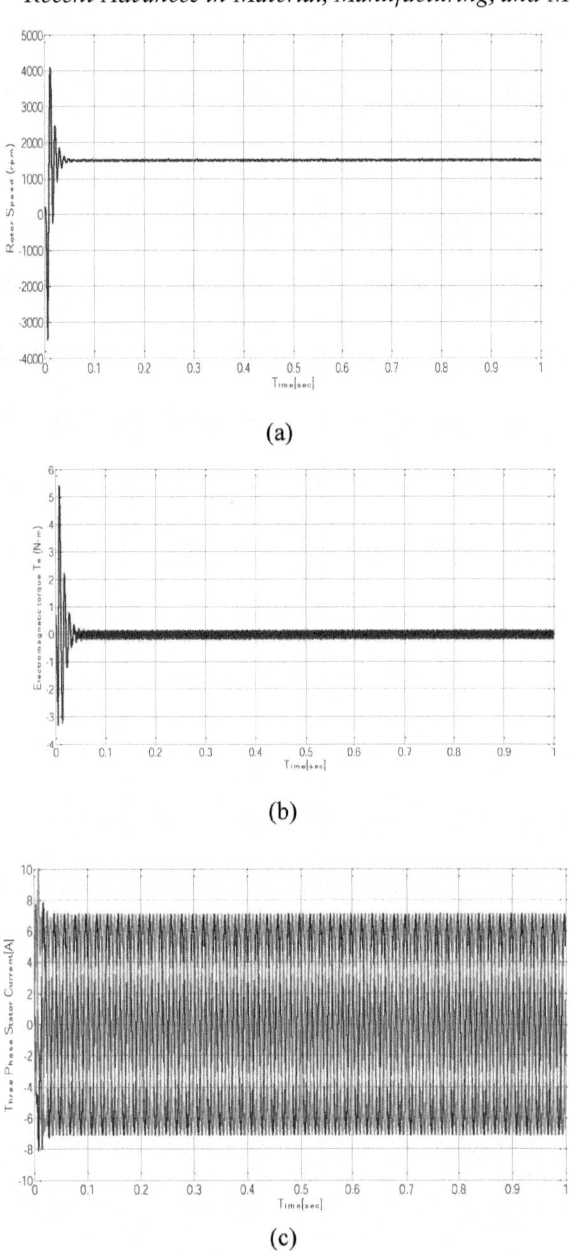

Figure 38.7 (a) Speed response (b) Torque response (c) Current response

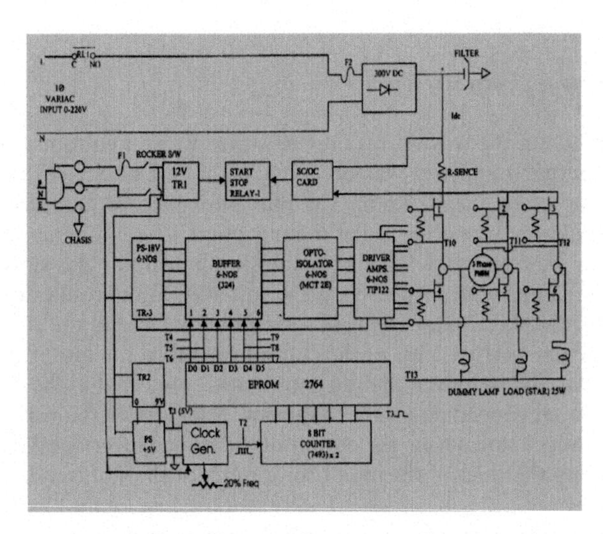

Figure 38.8 Block diagram of conventional inverters driven PMSM drive

Table 38.1 Motor speed frequency variation

Sr.No.	*Time (m.s.)*	*Frequency (Hz)*	*Actual Speed (rpm)*	*Measured Speed (rpm)*	*Voltage (Volts)*
1	30	33.4	995	1005	260
2	25	40.1	1250	1219	265
3	22	45.5	1462	1370	265
4	20	50.2	1530	1518	265
5	18	55.6	1765	1672	265
6	17	60	1785	1795	265

Table 38.2 Motor speed variation by load at 33.4 Hz

Sr. No.	*Weight (gm)*	*Frequency (Hz)*	*Actual Speed (rpm)*	*Expected Speed (rpm)*	*Torque (N-m)*	*Output Power (W)*
1	500	33.4	1110	995	0.15715	16.59
2	1000	33.4	1110	995	0.2743	35.28
3	1500	33.4	1110	995	0.4814	48.99
4	2000	33.4	1110	995	0.6186	65.45
5	2500	33.4	1110	995	0.8357	79.98
6	3000	33.4	1110	995	0.9829	99.49

Table 38.3 Motor speed variation by load at 50 Hz

Sr. No.	*Weight (gm)*	*Frequency (Hz)*	*Actual Speed (rpm)*	*Expected Speed (rpm)*	*Torque (N-m)*	*Output Power (W)*
1	500	50.2	1500	1540	0.16715	25.89
2	1000	50.2	1500	1540	0.3043	49.96
3	1500	50.2	1500	1540	0.4814	76.45
4	2000	50.2	1500	1540	0.6886	99.79
5	2500	50.2	1500	1540	0.7857	125.51
6	3000	50.2	1500	1540	0.9229	148.78

Table 38.4 Motor speed variation by load at 60 Hz

Sr. No.	*Weight (gm)*	*Frequency (Hz)*	*Actual Speed (rpm)*	*Expected Speed (rpm)*	*Torque (N-m)*	*Output power (W)*
1	500	60	1700	1770	0.157	29.64
2	1000	60	1700	1770	0.324	58.16
3	1500	60	1700	1770	0.491	89.73
4	2000	60	1700	1770	0.658	115.33
5	2500	60	1700	1770	0.795	142.90
6	3000	60	1700	1770	0.922	172.49

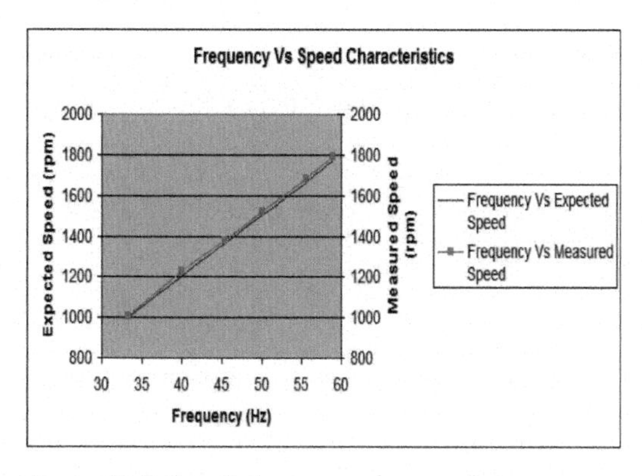

Figure 38.9 Speed- frequency characteristics

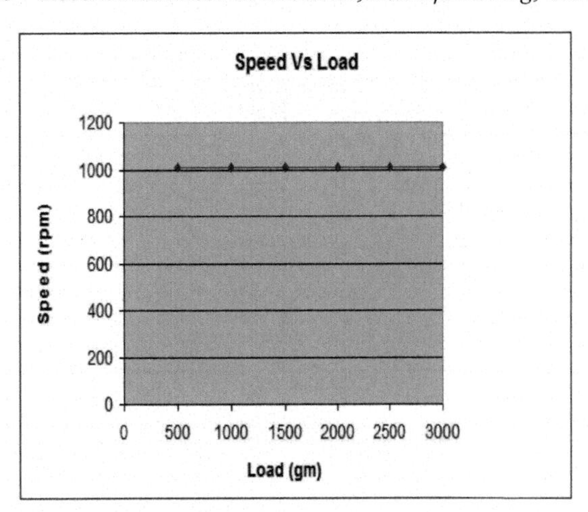

Figure 38.10 Load-speed characteristics at 33.3 Hz

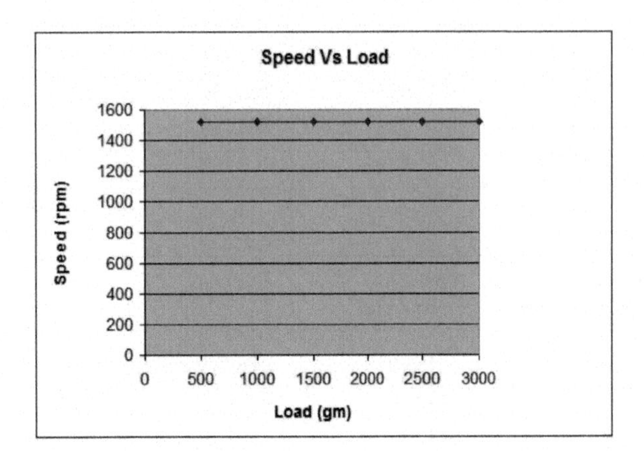

Figure 38.11 Load-speed characteristics at 50 Hz

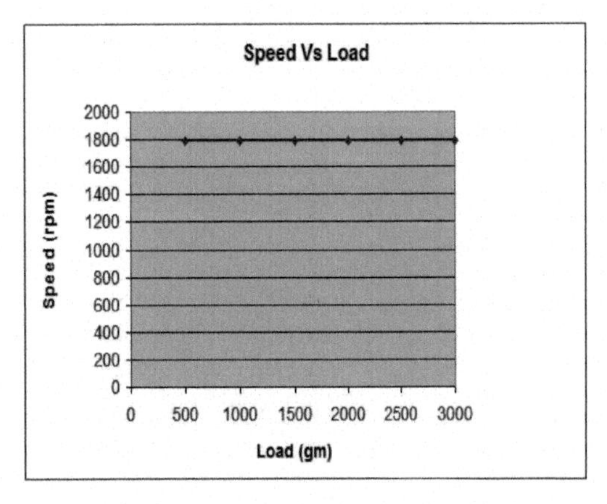

Figure 38.12 Load-speed characteristics at 59 Hz

Torque ripples analysis of VSI-fed PMSM drives at different speed range calculated by formula
Torque ripple (%) = $(Tmax - Tmin)/Tavg \times 100$

Table 38.5 Torque ripples analysis at different speeds

Sr.No	*Controller speed*	*(%)Torque ripple*
1.	1000 rpm	28.38%
2.	1500 rpm	22.41%
3.	1800 rpm	18.49%

Conclusion

This paper focus on simulation and hardware analysis of conventional inverters driven PMSM drive using microcontroller. A detailed Simulink model of torque minimisation of PMSM drive has being developed. It is observed that, at constant frequency, speed remains constant irrespective of load. The motor runs at synchronous speed. Speed also gets changed accordingly the inverter frequency. The overall motor performance can be very well judged from the performance characteristic shown. Table 38.5 shows that torque ripples analysis at different speed. Finally we conclude that simulation results and hardware implementation shall give better performance analysis. Hence it is used in automotive application.

References

Borse, P. S. (2020). Analytical evaluation of FOC and DTC induction motor drives in three levels and five levels diode clamped inverter. In 2020 International Conference on Power, Energy, Control and Transmission Systems (ICPECTS). Chennai, India, 2020, (pp. 1–6). doi: 10.1109/ICPECTS49113.2020.9337015.

Bose B. K. (2002). Modern power electronics an AC drives. Prentice-Hall, Upper Saddle River.

Buja, G., Kazmierkowski, M. P. (2004). Direct torque control of PWM inverter-fed AC motors-a survey. IEEE Trans. Ind. Electron. 51:744–757.

Casadei, D. (2002). FOC and DTC: Two viable schemes for induction motors torque control. IEEE Trans. Power Electron. 17:779–787.

Ehsani, M., Rahman, K., and Toliyat, H. (1997). Propulsion system design of electric and hybrid vehicles. IEEE Trans. Ind. Electron. 33:19–27.

Gaidhani, T. (2020). HVDC fault current reduction through MMC and DCCB coordination. In 2020 International Conference on Power, Energy, Control and Transmission Systems (ICPECTS), (pp. 1–6), doi: 10.1109/ICPECTS49113.2020.9337020.

Heydari, F., Sheikholeslami, A., Firouzjah K. G,. and Lesan, S. (2010). Predictive field-oriented control of PMSM with space vector modulation technique Front. Electr. Electron. Eng. China. 5(1):91–99.

Kazmierkowski, M. P. and Malesani, L. (1998). Current control techniques for three-phase voltage-source PWM converters: A survey. IEEE Trans. Ind. Electron. 45(5):691–703.

Leonhard, W. (1990). Control of electrical drives. Verlag: Springler, Berlin, Heidelberg GmbH.

Li, Y. H. et al. (2007). Simulation study on the effect of voltage Vector on torque in direct torque control system of permanent magnet synchronous motor. In Proc. 2007 IEEE Industrial Electronics and applications Conf. (pp. 15211525.

Li, Y. H. et al. (2008). A novel switching table to suppress unreasonable torque ripple for the PMSM DTC drives. In Proc. 2008 IEEE Electrical Machines and Systems Conf., (pp. 972–977).

Li, Y. H. et al. (2009). The control of stator Flux and torque in the surface permanent magnet synchronous motor direct torque control system. In Proc. 2009 IEEE Industrial Electronics and applications Conf., (pp. 2004–2009).

Li, Y. H. et al. (2010a). A simplified voltage vector selection strategy for the permanent magnet synchronous motor direct torque control drive with low torque ripple and fixed switching frequency. In Proc.2010 IEEE Electrical Machines and Systems Conf., (pp. 674–679).

Li, Y. H. et al. (2010b). The comparison of control strategies for the interior PMSM drive used in the electric vehicle. In Proc. the 25th World Electric Vehicle Symposium and Exposition. paper 4C-7.

Matale, N. P. (2020). Alleviation of Voltage Sag-Swell by DVR Based on SVPWM Technique. In 2020 International Conference on Power, Energy, Control and Transmission Systems (ICPECTS), Chennai, India, 2020, pp. 1-6, doi: 10.1109/ICPECTS49113.2020.9336972.

Merzoug, M. S. and Naceri, F. (2008). Comparison of field-oriented control and direct torque control for permanent magnet synchronous motor (PMSM) world academy of science. Eng. Technol. 45:1797–1802

Park, J. W, Koo, D. H, Kim, J. M., and Kim, H. G. (2001). Improvement of control characteristics of interior permanent-magnet synchronous motor for electrical vehicle. IEEE Trans. Ind. Appl. 37:1754–1760.

Paturca, S. V. (2006). Direct torque control of permanent magnet synchronous motor (PMSM) - an approach by using Space Vector Modulation (SVM). Proc. of the 6th WSEAS/IASME In Conf. on Electric Power Systems, High Voltages, Electric Machines, (pp. 16–18).

Rahman, M. F., Zhong, L., and Lim, K.(1998). A direct torque-controlled interior permanent magnet synchronous motor drives incorporating field weakening. IEEE Trans. Ind. Appl. 34:1246–1253.

Rahman, M. F., Haque, Md. E., Tang, L., Zhong, L., and Zhong, L. (2004). Problems associated with the direct torque control of an interior permanent-magnet synchronous motor drive and their remedies. IEEE Trans. Ind. Electron. 51:799–809.

Shriwastava, R. G., Bhise, D. R., and Nagrale, P. (2019). Comparative analysis of FOC based three level DCMLI driven PMSM Drive. Published. In IEEE Explore ICITAET 2019 Conference, record no ISBN-978-1-7281-1901-4 (pp. 26–31)

Shriwastava, R. G., Daigavane, M. B., and Wagh, N. B. (2018a). Comparative analysis of three phase 2-Level VSI with 3-Level and 5-Level DCMLI Using CB-SVM. Published in IEEE International Conference on Electrical, Electronics, Computers, Communication, Mechanical and Computing (EECCMC-2018), at Priyadarshini Engineering College, Chettiyappanur, Vaniyambadi - 635751, Vellore District, Tamil Nadu, India. Conference records no 978-1-5386-4304-4/18/$31.00 ©2018 IEEE.

Shriwastava, R. G., Daigavane, M. B., and Wagh, N. B. (2018b). Performance investigations of PMSM drive using various modulation techniques for three level NPC Inverter. Published in 5th International Conference on Computing for Sustainable Global Development (INDIACOM-2018), IEEE Conference at Bharati Vidyapeeth's Institute of Computer Applications and Management (BVICAM), New Delhi with Conference record no INDIACom-2018; ISSN 0973-7529; ISBN 978-93-80544-28-1 Page no-1038-1042

Shriwastava, R. G., Wagh, N., and Shinde, S. K. (2020). Implementation of DTC-controlled PMSM driven by a matrix converter. Int. J. Recent Technol. Eng. (IJRTE), 8(5):4420–4424. doi:10.35940/ijrte.E6641.018520,.

Shriwastava, R. G., Thakre, M. P., Patil, M. D., and Bhise, D. R. (2021). Performance analysis of CB-SVM three-phase five-level DCMLI fed PMSM drive. published in Turkish J. Comput. Math. Edu. 12(12):887–889.

Sun, D., Weizhong, F., and Yikang, H. (2001). Study on the direct torque control of permanent magnet synchronous motor drives. IEEE/ ICEMS. 1:571–574.

Zambada, J. (2007). Microchip corporation, sensorless field oriented control of PMSM motors. Microchip Technology Inc, www.microchip.com.

Zhong, L., Rahman, M. F., Hu, W. Y., and Lim, K. W. (1997). Analysis of direct torque control in permanent magnet synchronous motor drives. IEEE Trans. Power Electron. 12(3):528–536.

Zhong, L., Rahman, M. F., Hu, W. Y., Lim, K. W., andRehman, M. A.. (1999). A direct torque controller for permanent magnet synchronous motor drives. IEEE Trans. Energy Convers. 14(3):637–642.

39 Enviro-economic study of sustainable multi-crystalline silicon based solar photovoltaic powered irrigation system for Indian rural areas

Kaushalendra Kumar Dubey[1,a], *Rohit Tripathi*[2,b], *Kapil Rajput*[1,c], *Lavepreet Singh*[1,d], *RS Mishra*[3,e], *and Gunjan Aggarwal*[4,f]

[1]Department of Mechanical Engineering, Galgotias University, Greater Noida, Uttar Pradesh, India

[2]Department of Electronics Engineering, J. C. Bose University of Science and Technology, YMCA Faridabad, Haryana, India

[3]Department of Mechanical Engineering, Delhi Technological University, Delhi, India

[4]Department of Physics, Sharda University, Greater Noida,Uttar Pradesh, India

Abstract

Irrigation is a significant cultural practice and labour exhaustive task of agriculture sector. Today irrigation entirely reliant on electric and diesel pump based in India. Both energy generation techniques consume fossil fuel for power production and responsible for severe impact on environment through carbon emission with other toxic discharge. The enviro-economic analysis is important for its economic evaluation, climatic impact, and assessment of recovery of capital and energy investment. The present work deals the energy consumption, carbon emission, pump performance with available solar energy and system payback from traditional fuel-based irrigation system. The employment of 05 KW of photovoltaic technology-based DC solar powered irrigation systems (SPIS) is consider for 01 hectare of farming land irrigation for rice and wheat crop production. SPIS able to save 23 unit of electricity daily wise and reduces 04 and 13 ton of CO_2 from grid power and diesel power-based pumping set respectively in 08 months of seasonal irrigation cycle. This study estimates the proposed SPIS generate the extensive economy in terms of carbon credits and valuable reduction in harmful carbon emission. The SPIS payback is counted 6.5 years where as energy payback is evaluated with respect of grid connected pumping system is 0.9 years. The sale of farming surplus energy is newly added approach for other farmers is also incorporated. Results of this work helps in the implementation of solar photovoltaic based all range of pumping system for urban to rural areas irrigation needs

Keywords: Carbon emission; conventional power generation; solar photovoltaic energy; stand-alone solar irrigation; system payback time.

Introduction

The agriculture in India is important and primary sector of rural livelihood areas. India has approximately 179.7 million hectares agriculture land area which is seventh largest in the world (https://data.worldbank.org/indicator). India is the world's top producer of pulses, jute, millet and among the top three in the production of wheat, rice, potato, tomato, onion. The net irrigated area in India is about 6.8 million hectares (www.financialexpress.com; Siebert et al., 2010). The various sources of irrigation in India are canals, tube wells, other wells, tanks and other sources. Tube wells and canals are accounting for about 45% and 24% of total irrigation respectively. Due to massive involvement of tube wells in irrigation, about 30 million irrigation pump sets are in use (www.fao.org/nr/water/aquastat/irrigationmap/ind; www.ceicdata.com/en/india/electricity). The current power consumption by the agriculture sector in India according to CEI is 207791 GWh which is about 4.12% more than year of 2018 power consumption (mospi.nic.in/sites/default/files/Statistical_year_book_india; and mospi.nic.in/statistical-year-book-india). Around 70% of these pumps are based on electricity and remaining 30% are based on diesel. Agriculture sector consumed 7123 thousand tons of high-speed diesel (HSD), light diesel oil (LDO) and furnace oil for plantation in was 44 000 and 243 000 tonnes respectively (https://indiaenergyportal.org; Agarwal and Jain, 2015). Solar energy is one of the most vital renewable sources of energy and can be attractive option for irrigation. The Ministry of new renewable energy (MNRE) reported that government of India (GOI) has been installed 181,000 solar water pumps across the country by October 2019. And also launched KUSUM (Kisan Urja

[a]dubey.kaushalendra@gmail.com; [b]rohittripathi30.iitd@gmail.com; [c]kapilrajput.iet@gmail.com; [d]punstu@gmail.com; [e]rsmishra@dtu.ac.in; [f]gunjanaggarwal188@gmail.com

Suraksha Evam Utthaan Mahabhiyaan) plan for providing 7.5 HP of solar pumping for rural areas farmers (MNRE (2017a); mercomindia.com; Ministry of New & Renewable Energy, 2017). The heat loss factor, PV production factor, degradation loss factor, and other important quality parameters were studied in a recent research on photovoltaic systems. All of these variables are affected by the PV system's array. Increase in the internal temperature of the solar cell as a result of prolonged exposure to high ambient temperatures in the summer. The impact of packing factors on solar collectors is also addressed as an important element. For 1 m2 of aperture area, the optimal packing factor for collector is determined to be 0.5. When the packing factor is reduced, the thermal and total energy will rise (Tripathi et al., 2020; Tiwari et al., 2020).The proposed research work deals the energy and carbon emission estimation of 5 HP of irrigation pumping system for one hectare of agriculture land. The sizing of SPIS is depend on the flow rate through the pump, types of crop production and its irrigation cycle. The wheat and rice crop farming have been considered for this analysis, because both crop takes five month for complete irrigation, and estimation of carbon emission from five months will give remarkable results. Wheat and rice crop production is common in entire Indian farming practices. The effect of cooling case on different parameters of irrigation pumps like flow rate, pump efficiency, solar panel heat transfer rate, etc. have been discussed. The solar irradiation also essential factor for pumping head delivery (Tiwari et al., 2020; 2021; Yadav et al., 2020).

A. Solar photovoltaic irrigation pump set and its balance of system (BOS)

The various range of irrigation pump are using for different flow rate and head discharge level. Generally 01 HP to 05 HP of submersible pump set is applicable for irrigation of domestic kitchen garden to farming land. Table 39.1 is present the communally used pump set and its specifications.

The high grade energy based pumping set is now occupied by the solar powered pump set due to clean energy based operation and zero toxic emission. The Figure1shows the layout of SPIS with its all components. The specification of 05HP capacity of SPIS is available in Table 39.2.

Table 39.1 Size of irrigation pump set (www.shasyadhara.com)

S.No	*Size of pump set*	*Agriculture application*
1	1 HP	Applicable for drip irrigation, irrigation for small nurseries, kitchen garden, green houses and low discharge head water lifting capacity
2	2 HP	Applicable for drip irrigation, irrigation for small nurseries, Kitchen garden, green houses and low discharge head water lifting capacity
3	3 HP	Applicable for drip irrigation system. Sprinkler irrigation system is suitable for much range of (medium and big) nurseries, vegetable and fruit farm and medium head water discharge.
4	5HP	Applicable for drip irrigation system and sprinkler irrigation, The motor pump is suitable for ground water level high. It can irrigate a farming land area with high head discharge.

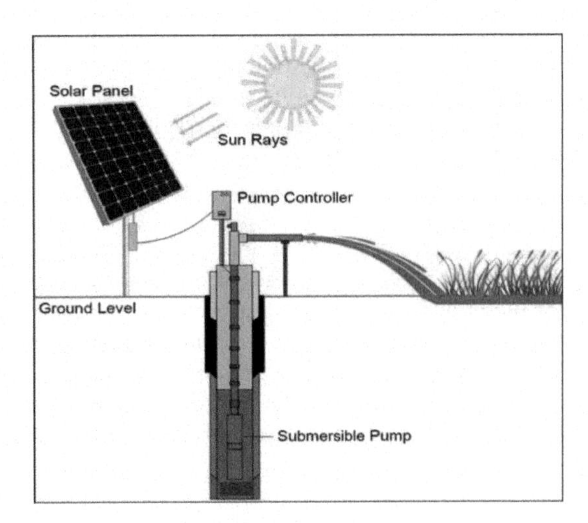

Figure 39.1 Solar powered irrigation pump set (www.electronicsforu.com)

Table 39.2 Specification of 05 HP DC SPIS (www.indiagosolar.in)

S.No	Components	Specification and unit
1	Water pump type	Submersible DC powered
2	Capacity of water pump	5 HP
3	Solar panel wattage	5 KWp
4	Balance of system of solar powered DC pumping set	Solar module or panel, DC pump set, charge controller, solar inverter and structure.
5	Discharge (Lt/day)	42,000 – 105,000 (depends on head)
6	Peak voltage or V_max	110 V DC
7	Voltage_open circuit or VOC	90140 V DC
8	Current_max input	8.2 Amps
9	Voltage_output	3085 V
10	Power_input	900 W DC

B. Sizing of solar photovoltaic module for pumping system

The size of SPV module for irrigation pump set MNRE-GOI recommends 900 Wp of SPV for 1 HP pump set (MNRE (2017a).

The SPV power estimation as follow-

The 1 HP is equivalent to 750 Watt of power Considering losses -5%-wiring loss,5%-thermal loss.5%-radiation loss and 10% of energy used for 1st start of pump motor (as per MNRE estimation)

So total SPV power required = 1HP+25% OF 1HP

1 HP pump set is required=950 Wp of SPV or 1KWp of SPV

For 5 HP pumping set SPV size should be 5 KW

C. Sizing of pump set

The pump size is depending on the flow rate of water for proper irrigation of farming land. The flow rate of pump is estimated by the maximum water flow rate as per crop development requirement and irrigation needs. The proper growth of crop needs water requirement as per the season within the fix days of irrigation cycle and hours of pumping operation daily basis (Kumar, 2010). The pump flow rate is estimates as: (Pump flow rate

$$\frac{\text{water requirement in season} \times \text{agriculture land area}}{\text{No of cycle of irrigation per season} \times \text{No of days in irrigation cycle} \times \text{Total hours of irrigation in 01 day}}$$

For wheat-pump flow rate is 30.6 m3/hour
For rice-pump flow rate is 23.15 m3/hour
(The daily 6 hours of time taken for irrigation)
Overall pump efficiency (η_spvpump)

$$= \frac{Pump\,electricity\,supply\left(\frac{KWh}{day}\right)}{Average\,solar\,radition\,on\,panel\,DNI\left(\frac{KWh}{day}\right)}$$

D. Enviro-Economic Study of 05HP SPIS system

The energy expenses during the product development, resource exhaustion, its climatic impact, health-safety of society and waste management, all environmental issues are prominently discussed. The enviro-economic

Table 39.3 Crop water requirement (Kumar, 2010)

Crop	*Water requirement in one season (mm)*	*Irrigation cycles per season*	*No of days in irrigation cycle*
Rice	1250	30	3
Wheat	550	6	5

analysis is covers all factors of economic evaluation and helps to develop life cycle assessment model of system. The remarkable factors have been considered for the enviro-economic study of SPIS.

1. Components performance in terms of SPV efficiency, DC pump efficiency, diesel setup efficiency, etc.
2. Carbon emission and energy loss due to grid and diesel based pumping system.
3. SPIS payback and its energy payback estimation.
4. Energy distribution for irrigation within cycle.
5. Annual energy consumption and saving evaluation.

Energy generation from 01 KWp of SPVas follows-

Assumptions: Lifetime-20 years
Degradation-0.7%per year
Installation- In India (So assume 300 sunshine days)
Peak sunshine hours- 5.5 hours/day
So KWH generated per day would be-1000W * 5.5 = 5.5 KWh
KWH Generated per year would be- 5.5*300 = 1650 KWh

(Average1300-1500KWh of energy is generated yearly by 01 KWh of SPV module as per DNI available in India)
With degradation Average Annual Generation would be- 1500* (1-.007) = 1490 KWh
So energy generated in 20 years would be- 1490*20 = 298 MWh or 300MWh

For 05 KW of SPV energy produced in year = 8167KWh

E. Solar pumping performance and system payback

The available solar incident energy is main source of SPIS. The intensity of solar radiation enhances the solar power as input for pump operation. Table 4 gives the performance of irrigation pump as per available solar energy in terms of KWh/m^2 per day at rural area of Sidhi disctrict, MP state of India. The DNI refers from http://www.synergyenviron.com/tools/solarirradiance/india/madhya-pradesh/sidhi.

Carbon economy- (CDM model)

The CDM model analysis for carbon emission estimation and carbon economy from 1 hectare farming land is based on the type of crop production, its season-wise irrigation time. The two main crop production has considered rice and wheat. Rice takes 30 cycles in season for irrigation and three days consumes one cycle, whereas Wheat takes six cycles in season for irrigation and five days consumes one cycle with assumption of six hours daily pump work. The CDM analysis deals five months of irrigation for combined rice and wheat.

Carbon emission and its ton of CO_2 (TCO_2) cost accounts significant contribution for system employability and its economic aspect. As per EPA, the CO_2 emission from coal fired thermal power plant for one

Table 39.4 Available solar power and required pump power for irrigation of 01 hectare land

Crop production	*Month*	*DNI (KWh/m^2) per day*	*Solar SPV power (KWh)*	*Farmer irrigation energy (pump power)(KWh)*	*Farmer monthly surplus energy (KWh) (Solar SPV power-farmer irrigation energy)*
Wheat	Jan	5.29	132.25	49.26	82.98
	Feb	6.47	161.75	60.25	101.49
	Mar	6.65	166.25	61.92	104.32
	Apr	6.02	150.5	56.06	94.43
	May	6.5	162.5	60.53	101.96
	Jun	4.45	111.25	41.44	69.80
	Jul	2.58	64.5	24.02	40.47
	Aug	2.84	71	26.44	44.55
	Sep	4.66	116.5	43.39	73.10
Rice	Oct	6.07	151.75	56.52	95.22
	Nov	5.87	146.75	54.66	92.08
	Dec	4.94	123.5	46.00	77.49

unit (1 KWh) of power is about 707 gram of CO_2. [7.07×10^{-4}metric tons CO_2/unit (www.epa.gov/energy; Dubey et al., 2021). Table 5 estimate the carbon credit from SPIS and emission through grid and diesel based pumping set.

(Only CO_2 emission is considered, no other any greenhouse gas emissions)

Carbon emission by appliances = 707 gm of CO_2/unit × Energy consumption by appliances in unit (KWh)

The possible cost of TCO_2 globally is about 20 USD/TCO_2 as per World Bank data

The comprehensive evaluation of carbon emission and carbon-credit

Table 39.5 Carbon credit estimation

Carbon credit by SPV plant against grid power	*Carbon credit by SPV plant against diesel set*
Total energy consumed by 5 HP pump = 3750 W = 3.75 KW For six hours daily work = 22.50 KW Unit consumption by the pump = 22.5 KWH Carbon emission through the pump daily-(750 gram CO_2 emission per unit) = 17 Kg/day For eight month of seasonal irrigation (240 days of irrigation) = 4080 Kg = 4 ton of CO_2) The SPV plant can save 4 ton of CO_2 emission and able to gain carbon economy as = 04 × 20 USD = 80 USD = 6000 INR in 08 month from 01 hectare farming land.	Ten litres of diesel consumes for one hectare of area irrigation for three hours (for six hours daily diesel pump set running, 1820 litres of diesel consumed) and one litre of diesel emits 2.64 kg of CO_2 20 litres of diesel produce-53 kg of CO_2 daily, For eight months of seasonal irrigation The carbon emission by the diesel pump set = 13 ton of CO_2 emission in eight months of irrigation. SPV plant can save 13 TCO_2 in eight months of seasonal irrigation and able to earn carbon economy as =13 × 20 USD = 260 USD = 19500 INR
Carbon credit earned 6000 INR by 1 hectare farming land in five months of seasonal irrigation against grid power based irrigation system.	Carbon credit earned 19500 INR by 1 hectare farming land in five months of seasonal irrigation against diesel power based irrigation system.

SPIS Payback and energy payback-

The recovery of initial cost of system installation is important for replacement of existing and high grade energy consumption system. The proposed 5 HP of SPIS is operating for 1 hectare land of wheat and rice farming. Eight months of irrigation is considering as per irrigation cycle of crop production. Taking the all factors of payback time and determine by following arithmetic steps.

$$Payback\ period = \frac{Cost\_installation}{Saving\_yearly}$$

$$Cost\ installation - Cost\_SPIS - Cost\_financial\ incentives$$

$$\text{Cost average electricity} = \frac{\text{Cost\_annual electricity}}{Annual\ electricity\ used}$$

$$\text{Yearly savings} = \text{Cost average electricity} \times \text{Energy yearly produced by SPV}$$

$$\text{Energy payback} = \frac{Energy\ consumed\ by\ exixting\ system}{Energy\ produced\ by\ solar\ based\ system\ in\ year\left(\frac{KWh}{year}\right)}$$

Table 39.6 Payback time of 5 HP-SPIS

System	*Installation cost*	*Total electricity used*	*Annual cost of used electricity (energy tariff @ 6.5 INR per KWh)*	*Energy produced by 05KW of SPV of SPIS (5KW*6 hours of sunshine daily*degradation factor of 7% in annually* 240 days of irrigation)*	*Yearly energy saving due to SPIS*	*Payback time*	*Energy payback time (Energy_ consumption by grid connected pump/Energy produced by SPIS)*
5 HP-SPIS	300000 INR	5520 KWh	35889 INR	7150 KWh	46473 INR	6.5 Years	0.9 year

Result and Discussion

The employment of 5 HP capacity based SPIS for irrigation is able to replace grid connected and diesel power based pumping set and SPIS specially applicable for un-electrified village, remote areas. Table 7 concludes the all possible operating condition and resultant parameters of SPIS.

The irrigation for wheat crop production is recommended in January to April month whereas Rice crop production irrigation routine from month of October to December yearly. Figure 39.2 explains the available average solar energy in terms of DNI of farming location for all months as input source and required pump energy as output. The surplus energy of farmer can be utilised for energy sale to others for irrigation. The solar energy and pump power is minimum during the monsoon months (July-September) of Indian climate.

Table 39.7 Operating and resultant parameters with cost

Parameters	*Values with unit*	*Parameters*	*Values with unit*	*Parameters*	*Values with unit*	*Parameters*	*Values with unit*
Farming land area	1 hectare	SPV lifetime	20 years	Diesel pump set mechanical efficiency	40%	Yearly energy saving by SPIS	46473 INR
Crop Production	Wheat and Rice	SPV capacity	5 KW, Poly crystalline module.	Diesel pump electric motor efficiency	35%	Yearly save grid energy by SPIS	5.5MWh
DC type pump capacity	5HP	SPV plant	Standalone pumping system (without grid connected)	Diesel pump-set lifetime	7 years	Yearly save diesel by SPIS	4800 Lt
Expected life of pump	10 years	SPV module efficiency	15%	Density of diesel	835 kg/m3	Carbon emission through diesel pump set in 08 month of irrigation	13 TCO_2
Mechanical Efficiency of pump	45%	Framework of SPV module installation	10 module framework	Heating value of diesel	46 MJ/kg	Water flow rate as per irrigation cycle	Wheat@30.6 m3/hour and rice @23.15 m3/hour
Electric motor efficiency	80%	Payback of SPIS and Energy payback	6.5 years and 0.9 year	Carbon emission through grid connected pump in 08 month of irrigation	4 TCO_2	Average surplus energy for sale to other farmer	81.29 KWh

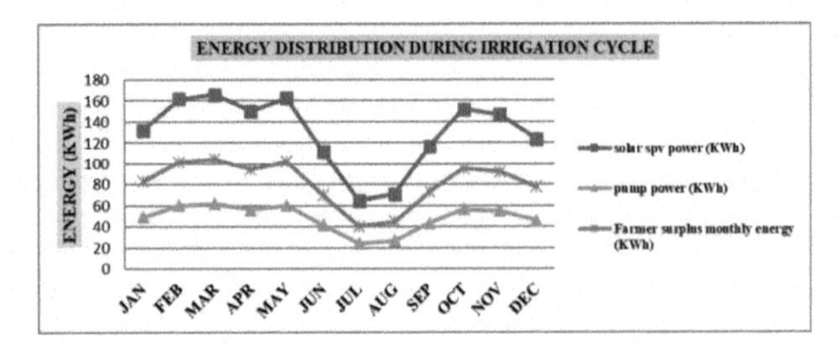

Figure 39.2 Energy distribution during eight months of irrigation cycle

Conclusion

The irrigation is essential activity of agriculture sector. The fossil fuel power operated irrigation systems are responsible for the toxic emissions and also create severe impact on environment. The solar power based irrigation system can provide reliable, sustainable and cost-effective solution for farmers and built the clean and green agriculture infrastructure. The present enviro-economic study of 05 HP DC solar photovoltaic powered pumping system is considered for wheat and rice farming in 01 hectare land. The wheat crop require 30.6 m3/hour of water flow rate in 05 days of 06 irrigation cycle in 06 hours of daily pump set working whereas rice crop needs 23.15 m3/hour of flow rate in three days of 30 irrigation cycle with same working hours of pump operation. The major highlights of this analysis is concluded below-

1) The high-grade energy based irrigation pumping system is consumes 23 units of electricity and 20 litres of high density diesel when operate with electricity or diesel engine respectively. Both power generation system are responsible for accountable carbon and another toxic discharge in to the environment.
2) During the 08 month of complete irrigation for both crop emits 4 tons and 13 tons of CO_2 from grid and diesel powered pumping set respectively. This considerable emission from only 1 hectare of agriculture land with single unit of 5 HP DC pump set.
3) The solar powered irrigation system able to save severe carbon emission and also earn significant economy in terms of carbon economy as 60 USD and 160 USD in 5 month of irrigation cycle, which is indicates the building of special economic infrastructure for agriculture globally.
4) India has provided 1.81 lacs of solar pumping set to farmers in past year. This volume has tremendous potential to generate green economy in terms of earning of carbon-credits in global market. (One 05 HP pump set consumes 6000KWh yearly in India, and discharge 4.5 TCO_2 yearly).
5) The extensive review of Indian irrigation and solar energy potential estimated that solar powered water pumping sets can save at least 250 billion litres of diesel combustion per year.
6) The investment recovery in terms of SPIS payback and energy payback are 6.5 years and 0.9 year, which is remarkable recovery of solar powered pumping system.
7) The energy distribution during the irrigation cycle helps to farmers for utilization of surplus energy. The available SPV is more than 60% of pumping operation power and this difference is valuable energy gain for other farmers, who have not solar powered irrigation pumping set. It can be develop irrigation energy sale concept by the farmers and for the farmers.

References

Agrawal, S. and Jain, A. 2015. Solar Pumps for Sustainable Irrigation: A Budget Neutral Opportunity'. New Delhi: Council on Energy, Evironment and Water.

Dubey K. K., Nandan, S., and Agarwal, G, and Trivedi, R. 2021. Energy-Economic Study of Smart Lightning Infrastructure for Low Carbon Economy. Advances in Industrial and Production Engineering. Lecture Notes in Mechanical Engineering. (pp. 219226). Singapore: Springer. https://www.springer.com/gp/book/9789813343191, Print ISBN: 978-981-334-319.

https://www.financialexpress.com/budget/india-economic-survey-2018-for-farmers-agriculture-gdp-msp/1034266/ (Accessed 15 April 2020)

http://www.fao.org/nr/water/aquastat/irrigationmap/ind/IND-gmia.pdf'(Accesse 18 April 2020)

https://www.ceicdata.com/en/india/electricity-consumption-utilities/electricity-consumption-utilities-agriculture (Accessed 26 May 2020)

http://www.indiaenergyportal.org/subthemes_link.php?text=agriculture&themeid=15 (Accessed 10 May 2020)

https://www.shasyadhara.com/best-motor-pump-for-crop-irrigation-submersible-pumps-for-2-acre-3-acre-4-acre-and-5-acre-agriculture-land-price-of-1-hp-2-hp-3-hp-and-5-hp-motor-pump/(Accessed 26 May 2020)

https://www.indiagosolar.in/shop/solar-water-pump/5hp-solar-water-pumpset/ (Accessed 3 June 2020)

https://www.electronicsforu.com/technology-trends/tech-focus/solar-powered-irrigation-systems (Accessed 3 June 2020)

http://www.ceew.in/sites/default/files/CEEW-Solar-for-Irrigation-Deployment-Report-17Jan18_0.pdf (Accessed 5 June 2020)

https://www.solarexpertsindia.com/solar-water-pump-1hp-2hp-3hp-5hp-7hp-10hp-price/(Accessed 7 June 2020).

https://data.worldbank.org/indicator/AG.LND.AGRI.K2 (Accessed 1 June 2020)

http://mospi.nic.in/sites/default/files/Statistical_year_book_india_chapters/Ch_12_SYB2017.pdf (Accessed 1 June 2020)

http://mospi.nic.in/statistical-year-book-india/2017/181(Accessed 1 June 2020)

https://www.epa.gov/energy. (Accessed 6 June 2020)

http://www.synergyenviron.com/tools/solar-irradiance/india/madhya-pradesh/sidhi (Accessed 4 June 2020)

https://archive.indiaspend.com/cover-story/how-agriculture-consumes-23-of-indias-electricity-picks-7-of-tab-96206. (Accessed 7 June2020)

Kumar, R. 2010. Irrigation in Wheat Crop, Agropedia. Available at: http://agropedia.iitk.ac.in/content/irrigation-wheat-crop.

MNRE (2017a) Annual Report 2016-2017 Ch. 4: National Solar Mission. New Delhi. (Accessed 20 May 2020)

https://mercomindia.com/over-181000-solar-water-pumps-installed (Accessed 20 May 2020)

Ministry of New & Renewable Energy (2017. Benchmark Cost for 'Off-grid and Decentralised Solar PV Applications Programme' for the year 2017-2018. New Delhi: Government of India.

Siebert, S., Burke, J., Faures, J. M., Frenken, K., Hoogeveen, J., Döll, P., and Portmann, F. T. (2010). Groundwater use for irrigation – a global inventory. Hydrol. Earth Syst. Sci. 14:1863–1880.

Tiwari, A. K., Sontake, V. C., and Kalamkar, V. R. (2020). Enhancing the performance of solar photovoltaic water pumping system by water cooling over and below the photovoltaic array. J. Solar Energy Eng. 142:2.

Tiwari A., Tripathi R., Tiwari G. N. 2020. Improved Analytical Model for Electrical Efficiency of Semitransparent Photovoltaic (PV) Module. In: Advances in Energy Research. ed. S., Singh, V., Ramadesigan. 1. Singapore: Springer. https://doi.org/10.1007/978-981-15-2666-4_10.2020.

Tiwari, A. K., Kalamkar, V. R., Pande, R. R., Sharma, S. K., Sontake, V. C., and Jha, A. (2021). Effect of head and PV array configurations on solar water pumping system. Mater. Today: Proceed. 46:54755481.

Tripathi, R., Bhatti, T. S., and Tiwari, G. N. 2020. Effect of packing factor on electrical and overall energy generation through low concentrated photovoltaic thermal collector in composite climate condition. Mater. Today: Proceed. 31(2):449453.

Tripathi R., Sharma R., and Tiwari G. N. 2020. (Experimental Study on PV Degradation Loss Assessment of Stand-Alone Photovoltaic (SAPV) Array in Field: A New Simplified Comparative Analytical Approach. In: Advances in Greener Energy Technologies. Green Energy and Technology. Ed. A. Bhoi, K., Sherpa, A., Kalam, G. S., Chae. Singapore: Springer. https://doi.org/10.1007/978-981-15-4246-6_40.24.2020

Yadav, D. H., Tiwari, A. K., and Kalamkar, V. R. (2020). Social and economic impact assessment of solar water pumping system on farmers in Nagpur District of Maharashtra State of India. In Advances in Applied Mechanical Engineering. (pp. 1926). Singapore: Springer.

40 Optimising feature selection technique through machine learning for intrusion detection system

Pandurang V. Chate[a], *Sunil B. Mane*[b], *and Rohini Y. Sarode*[c]
Department of Computer Engineering and IT, College of Engineering Pune, Pune, India

Abstract

The development in Computer Science and Telecommunication Engineering has given tremendous opportunities to humanity to be prosperous at all levels in their life. However, as per the rule of a coin, great inventions come with great hurdles, and hence today's network communications, traffic brings a number of vulnerabilities, attack possibilities to the resources and services. One can easily understand that most of the attacks are being spread through computer networks mainly. Considering this factor into account, this research work tries to integrate the machine learning domain with network security and aims to increase the accuracy of prediction along with reducing the computational time (designing a solution of high accuracy with minimum computation) by learning and testing the packet instances of different network attacks which were captured in CIC-IDS2017 dataset, generated by The Canadian Institute for cyber security in 2017. The goal of this study is to create a machine learning model that can predict network intrusions with high accuracy, as well as to propose a feature selection technique that removes noise from the dataset and selects the most relevant input features for the machine learning model to work efficiently.

This research proposes another optimal feature selection strategy for reducing the independent feature set while maintaining or improving the machine learning model's prediction accuracy. As an outcome, a 99.97% accuracy score was acquired, with the number of input characteristics reduced from 78 to 26.

Keywords: Feature selection, gain ratio, intrusion detection system, machine learning, mutual information; network security, Pearson's correlation.

Introduction

Intrusion detection is an important part of the network defence process since it alerts security administrators to malicious behaviours including intrusions, assaults, and viruses. IDS is a must-have line of defence for vital network protection against these ever-increasing invasive activities Pachghare et al. (2012). As a result, IDS research has grown throughout the years to offer improved IDS solutions. However, most of the research performed in the past contained old datasets which can give minimum results when deployed in real-time situations due to an inadequate number of features or properties of the network attacks. Hence this project aims to use the latest, possible, open-sourced dataset named as CIC-IDS2017 dataset (Sharafaldin et al., 2018) referred from the University of New Brunswick (UNB) platform.

Detecting an intrusion is a mechanism using which monitoring and analysis of network traffic take place. While monitoring network traffic, if there is any malicious activity observed then the alert message or response will be triggered to the concerned entity (Abbas et al., 2021). Based on the monitoring technique (Pachghare, 2019), intrusion detection systems are divided into two categories. 1) Network intrusion detection system (NIDS), which analyses all the network incoming-outgoing traffic and detects any malicious packet present based on the already defined criteria. 2) host intrusion detection system (HIDS), which works off the host, monitors the system events, and audits the event logs. It then compares the logs with the safe snapshot of the system and helps in detecting intrusion. The results of the proposed model can help in a network intrusion detection system as the dataset contains all the possible properties to capture any malicious network traffic.

In the proposed model, we have analysed different supervised machine learning algorithms with the chosen dataset and strived to get optimum results in the aspect of a maximum number of correct predictions while feeding machine learning (ML) model with as minimum as possible input features. As a result, we have achieved 99.97% accuracy with the top 26 input features through Random Forest (RF) classifier ML model optimised from 78 input features based on the arithmetic manipulations of mutual information (MI) and gain ratio (GR) values of each input feature. The steps taken to accomplish the task of intrusion detection includes,

- Data gathering
- Data cleaning (Pre-processing)

---

[a]chatepv20.comp@coep.ac.in; [b]sunilbmane.comp@coep.ac.in; [c]rys.comp@coep.ac.in

- Feature engineering
- Hypothesis setting (ML models)
- Hypothesis testing (Evaluation)

The remaining sections of this paper describe the following highlights. Section 2 highlights past research work on the same grounds. Section 36 respectively describe an overview of the dataset used, proposed methodology, system implementation, and result analysis and system comparison, and validation. In section 7, the conclusion has been depicted.

Literature Review

Intrusion detection systems have been already proposed by the great minds from researcher society in the past timeline. However, most of the research performed had used and referred quite old and outdated datasets such as KDD'99 dataset created by DARPA in 1999 year (Tahe et al., 2019; Meyer and Labit, 2020). Hence proposed model aimed to use the latest available dataset with latest attributes of the normal and malicious network traffic to cope up recent intrusion attacks.

Past research achieved (Meyer and Labit, 2020) has proposed to detect DOS, Probe, U2R, R2L attacks of intrusion. However, the proposed model work aimed to detect the frequent, common network security attacks such as DOS, DDOS, Brute force, SQL Injection, Cross-Site Scripting (XSS), Botnet attacks which are frequently occurring in the real world as the computing and Web technology improved in recent times.

The author in Abbas et al. (2021) has discussed different feature selection techniques such as Anova, chi-square from general feature selection methods and applied through SVM and RF. The author has obtained 99.67% accuracy after analysing different supervised models on the CICIDS-2017 dataset. Every ML model should give optimum results based on the solution need and to achieve the same the number of input features selected should be minimum. Numerous researchers have tried different approaches for optimising feature selection methods. Adhao and Pachghare (2020) have used Principal Component System and Genetic Algorithm (PCS&GA) on the CICIDS-2017 dataset and had achieved 99.53% accuracy through Decision Tree Classifier ML model. In the proposed system by Vijayanand and Devaraj (2020) had discussed the use of the Whale Optimization Algorithm (WOA) and genetic algorithm (GA) methods to select input features. The author was able to obtain 95.91% accuracy through the Support Vector Machine (SVM) ML model.

Nebrase Elmrabit et al. (2020) have used different machine learning algorithms along with a Deep Learning algorithm (DNN) to detect anomalies inside the network through the CICIDS2017 dataset. The authors were able to achieve a 99.9% accuracy score. However, the authors had to use all 78 input features without a proper feature selection technique which could have achieved the same accuracy score but with minimum computation. Similarly, Kayvan Atefi1 et al. (2019) have specifically used KNN and DNN machine learning models with their respective accuracy scores of 88.2% and 92.9% but without feature filtering technique which could have acquired the best relevant input features to avoid computational hurdle.

Adhao and Pachghare (2019) have proposed one of the finest feature selection techniques in which a feature can be categorised into an important feature, secondary feature, and unimportant feature based on the behaviour of the ML model by removing that selected feature against the accuracy score. The authors had achieved a 96.1 % accuracy score with the best 15 relevant features through features selected by the above method.

The proposed model has considered the need for improvement in selecting only the best input features to train the ML model and hence implemented the ML model with the best relevant input feature set to achieve optimal performance. Different authors (Adhao and Pachghare, 2020; Siddiqi and Pak, 2020; Vijayanand and Devaraj, 2020; Prasad et al., 2020; Sharafaldin et al., 2018) have achieved different accuracy scores as per implemented feature selection technique. Compared to them, this research work has experimented with different arithmetic operations on Mutual Information and Gain Ratio Values of input features with Pearson's Correlation factor to optimise feature selection criteria which ultimately improves the computational cost while training any ML model. As a result, these experiments have given the best feature selection technique that has reduced the input feature set count from 78 to 26 with increasing the prediction accuracy to 99.97% which is considerably better (Adhao and Pachghare, 2020; Siddiqi and Pak, 2020; Vijayanand and Devaraj, 2020; Prasad et al., 2020; Sharafaldin et al., 2018) as shown in Table 40.4.

Dataset Used

In this project, we have attempted to use the CIC-IDS2017 dataset (Sharafaldin et al., 2018) which was generated in 2017 year with an intention to let researchers study on the network the security attacks with the help of the maximum number of available properties to tell the difference between what is normal and what is malicious network packets. The abstractive view of dataset shows that, data collector has performed experiments in the well-equipped lab environment and has captured daily network packets from Monday till Sunday for a week. In the acquired instances, researcher have collected data for Benign, SFTP, SSH, BForce, DoS, Heartbleed Attacks, Slowris, Slowhttptest, Hulk, Golden Eye, Web Infiltration attacks, Web BForce, XSS, SQL Injection, DDOS, Botnet, Portscans network packets and hence this has been considered for intrusion detection dataset due to its nature on realistic ground.

In this project work, all the frequent occurring attacks have been considered and categorised into DOS, DDOS, WEB attack, BOTNET, BruteForce (SFTP, SSH) attacks as shown in Figure 40.1. 70% of the entire dataset has been considered for training the ML models and remaining 30% is for testing the model.

Proposed Methodology

There have been several methodologies proposed by the research community with their possible best efforts and as a part of the same community, every researcher tries and enhances already developed method, and hence this project has aim to achieve the same. This study provides a technique for obtaining a minimum set of input features by combining and analysing the information gain, mutual information, and gain ratio values of each input feature using Pearson's correlation factor. We have used different ML models and different data analysis techniques to validate the best working ML model. The following hypothesis have been observed one after another and finally best has been selected based on the high correct predictions and minimum number of independent features.

- Logistic Regression ML model (LR).
- Decision Tree ML model (DT).
- RF-ML model
- Naïve Bayes (NB) ML model
- Support Vector Machine ML model (SVM).

The Figure 40.2 represents the diagrammatic process of the designed system model. This project has used different feature engineering techniques to increase the intrusion prediction rate.

As shown in the system model working diagram (Figure 40.2), the feature selection technique has helped in increasing the prediction rate and reducing a significant number of input features which will ultimately help in improving execution time while the data learning phase.

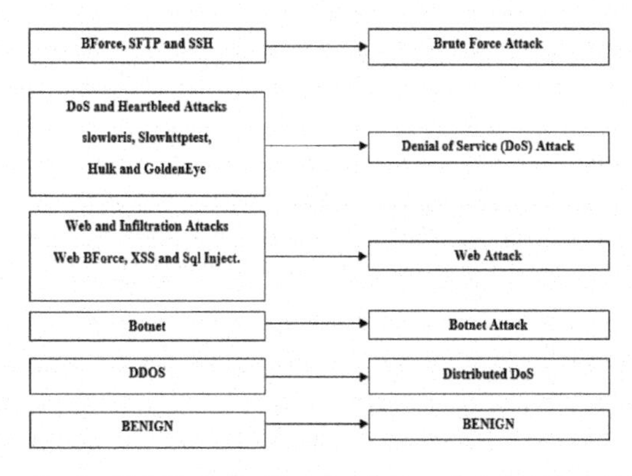

Figure 40.1 Attacks mapping from sub types into main

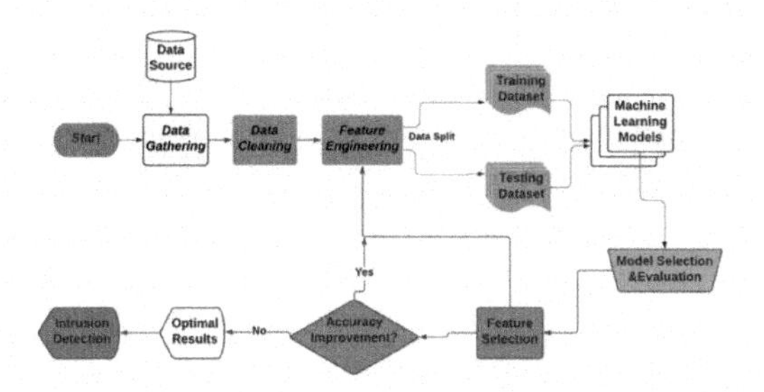

Figure 40.2 System model of proposed methodology

It can be noted that the model training phase continues until the model evaluation gives overall accuracy improvement and it is achieved through two feature selection techniques which are as follows.

- Mutual information (MI) and information gain (IG).
- Gain ratio (GR).

The detailed working principle of above information theory techniques has been explained under section 5 of this paper.

System Implementation and Result Analysis

System implementation has been done through following steps.

1) DATA GATHERING: This project has referred CIC-IDS2017 dataset Sharafaldin et al. (2018) which was generated in 2017 year with the intention to let researcher study network the security attacks with the help of maximum number of available properties about network packets to distinguish between normal and abnormal network packets. The dataset used contains random instances of network traffic packets.
2) DATA CLEANING (PREPROCESSING): Data pre-processing is the second step in the system implementation in which raw data gathered from the sources has been cleaned, impurities, null values, blank values, infinite values have been arranged in such a way that machine learning model could learn about the pattern inside the data. Since the number of null values were very less in quantity, we have neglected their presence through removal technique. After handling all such impurities in the dataset, this project has boiled at a point to use 69 total features in which 68 are independent features and 1 is dependent one whereas in the raw dataset, the count of all features is 79.
3) FEATURE ENGINEERING: The output of data pre-processing step will be used as input to the Feature Engineering step in which data scaling (normalisation) can be done and feature selection can also be done so that unnecessary features from independent feature set can be ignored. The need for data normalisation arises in the LR and SVM machine learning models as their working principle demands equal dominance of the values present in the dataset. This basically helps in preventing bias while learning data patterns. In this project the normalisation technique used is Min-Max scaling. The mathematical model for this normalisation technique can be shown below.
MIN-MAX SCALING: Min-max normalisation (also known as feature scaling) is a linear adjustment that is applied to the original data. This method obtains all scaled data in the [0,1] range. The following is the formula for doing this:

$$X_{scaled} = \frac{X - X_{\min}}{X_{\max} - X_{\min}} \quad (1)$$

Where X is every single instance from each feature from the independent feature set, X_{min} and X_{max} are the minimum and maximum values, respectively from each feature from the independent feature set.

The relationships between the original input values are preserved while using min-max normalisation. This method suppresses the effect of outliers by creating bounded range inside the dataset and smaller deviations of feature values.
4) FEATURE SELECTION: The feature selection technique helps in reducing the number of input features by removing the noise from the dataset which will be used to train the ML model. The result of this solution improves the execution of the time model training phase and hence improves overall efficacy. The working principle of this technique describes that the chosen number of input features (reduced) will give the same or high predication rate as of selecting all the input features at the time of training phase. There are numerous feature selection techniques available, but the proposed model focuses on below two methods with a combination of Pearson's Correlation technique. These two techniques work on the teaching of Shannon's Entropy principle.

To understand above techniques, we must understand Shannon's Entropy principle. Equations (2–5) represent formulae for entropy, mutual information, information gain and gain ratio, respectively.

SHANNON'S ENTROPY:

$$H(X) = \sum_{i=1}^{n} Pi \times \log\left[\frac{1}{Pi}\right] \quad (2)$$

Where X = features, P = Probability of each feature

The basic working principle of entropy is a measurement of impurity, randomness, or more precise measurement of information in the set of data on which entropy is applied. Hence entropy helps in gaining the information from features and ultimately in the selection of important features from the independent feature set.

a) Mutual information (MI) and information gain (IG): Halimaa and Sundarakantham (2019). The mutual information of two variables that are random in nature is a measurement of the mutual reliance between them in information theory and probability theory. It estimates the 'amount of knowledge' gleaned from seeing one random variable while monitoring another random variable. Mutual information and entropy of any random variable, a fundamental concept in information theory that measures the predicted 'amount of information' carried in a random variable, have a strong relationship. Variable is nothing but a feature in our case. The mutual Information between two random variables X and Y is as follows,

$$I(X;Y) = H(X) - H(X|Y) = H(Y) - H(Y|X) \tag{3}$$

Where H(X) and H(Y), respectively, are entropies, while H(X|Y) and H(Y|X), respectively, are conditional entropies of features X and Y. Similarly, the information gain measures the entropy reduction based on the given value of a feature (Brownlee, 2019). Value-wise, both MI and IG give same output if applied on the same data (Brownlee, 2019).

$$IG(X,a) = H(X) - H(X \mid a) \tag{4}$$

Where IG(X, a) = information for the dataset X for variable a for a random variable, H(X) = Entropy for the dataset before any split and H(X|a) = conditional entropy of dataset given variable a.

b) Gain ratio (GR): Gain ratio (Priyadarsini et al., 2011) helps in reducing the bias that occurred in information gain information selection criteria as sometimes information gain chooses a feature with the high number of values over the high number of distinct values which reduces the chances of actual information gain of the input feature in association with the target feature.

$$GR = \frac{IG(X,a)}{H(X)} \tag{5}$$

c) Pearson's correlation: It basically checks the relation between dependent and independent features of the dataset. It helps finding the strong or weak association giving values in range −1 to +1. Value close to 1 denotes that the two features have high correlation among themselves.

5) HYPOTHESIS SETTING (ML MODELS): The output of the Feature Engineering step containing the ready-to-model data has been used as a training dataset as input in the current step. Considering the result of a set of experiments with different supervised ML models while consideration of high true correct predictions (high correct accuracy) and the minimum set of independent features, the RF ML models have been considered as the best prediction ML model.

The splitting criteria, number of decision trees in Random Forest, depth of decision tree have been decided through experiments performed on the chosen dataset. The experiments have been performed on a system with Windows 10 OS, RAM 4GB, Intel(R) Core (TM) i5-5200U CPU @ 2.20GHz, Jupyter tool configurations.

6) HYPOTHESIS TESTING (EVALUATION): The optimised model (RF) has been evaluated through below three techniques.

i) Accuracy score: In which the ML model has been tested on the dataset which is not used for the training purpose. The mathematical expression used for accuracy score is,

$$\text{Accuracy Chkirbene et al., 2020} = \frac{TP + TN}{TP + TN + FP + FN} \tag{6}$$

Where TN denotes true negative, TP denotes true positive, FN denotes false negative, and FP denotes false positive predictions.

ii) Confusion matrix (Chkirbene et al., 2020)

iii) K – Fold cross validation

7) RESULT ANALYSIS: Table 40.1 describes the results of chosen five ML classifiers in terms of overall accuracy for 10 folds of dataset considering all 68 input features. SVM gives 99.97% accuracy for all 68 independent features but fails to perform well with a reduced set of independent features. However,

the RF model increases accuracy score with the reduced feature set. Hence RF is considered as the best model and used for further input feature optimisation experiments.

Figure 40.3 shows graphical representation of Table 40.1 content.

Table 40.2 shows the results of applied feature selection techniques on the best ML model is as follows.

As shown in Table 40.2, we have calculated MI and GR values for each independent feature. Then arithmetic operations were performed on them giving a different set of values. After analysation, it became clear that GR-MI values with the use of Pearson's Correlation factor of value 0.91 have reduced the count of 68 independent features to 26 independent features with an accuracy of 99.97% through RF-ML model. It clearly indicates that the removal of noise from the dataset helps in improving accuracy as well as minimising the number of independent features.

Table 40.3 shows the top 26 features which are obtained using different feature selection experiments on the CICIDS-2017 dataset.

Table 40.1 The performance examination results

Supervised ML classification model	*10-fold cross validation accuracy score (in %)*
NB	52.38
LR	67.56
DT	99.89
RF	99.91
SVM	99.97

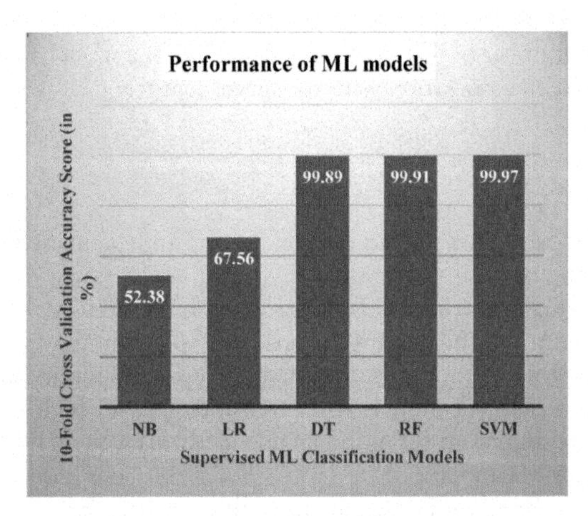

Figure 40.3 ML model performance using 10-fold cross validation

Table 40.2 Feature selection results

Experimentation	*Reduced features*	*Accuracy in %*
Without use of GR and MI values	68	99.91
GR values	51	99.90
MI values	46	99.93
GR+MI values	46	99.93
MI-GR values	43	99.93
*GR*MI values*	54	99.93
MI/GR values	53	99.93
GR/MI values	48	99.95
GR-MI values	39	99.96
GR-MI values and Pearson's correlation	26	99.97

Table 40.3 Reduced set of independent features

Sr. No.	*Feature*	*Sr. No.*	*Feature*
1	FIN_Flag_Count	14	Active_Max
2	URG_Flag_Count	15	Active_Mean
3	RST_Flag_Count	16	Flow_IAT_Min
4	SYN_Flag_Count	17	Bwd_IAT_Min
5	ACK_Flag_Count	18	Idle_Mean
6	PSH_Flag_Count	19	Subflow_Fwd_Packets
7	Active_Std	20	Fwd_IAT_Min
8	Down_Up_Ratio	21	Destination_Port
9	Bwd_Packet_Length_Min	22	Bwd_Packet_Length_Std
10	Idle_Std	23	Init_Win_bytes_forward
11	Min_Packet_Length	24	Bwd_Header_Length
12	Fwd_Packet_Length_Min	25	Bwd_IAT_Std
13	min_seg_size_forward	26	Init_Win_bytes_backward

System Comparison and Validation

1) SYSTEM COMPARISON: Table 40.4 and Figure 40.4 shows the comparison between proposed model and past work.
2) SYSTEM VALIDATION: The results of the proposed model have been validated through K-fold cross-validation in which the ML model will be evaluated or executed on different folds of a provided dataset instead of on the only single set of instances. The result of validation will be mean of different tests performed using the K-fold cross-validation process. As described in section 5, the results are considerably good in terms of overall positive and negative predictions.

Table 40.4 Comparison between proposed model and different research results

Reference	*Number of features (Out of 78)*	*Classifier*	*Accuracy in %*
Iman Sharafaldin et al. (2018)	78	RF	98
Adhao and Pachghare (2020)	40	DT	99.53
Vijayanand and Devaraj (2020)	35	SVM	95.91
Siddiqi and Pak (2020)	35	Deep Neural Network	99.73
Prasad et al. (2020)	40	Bayesian RS	98.08
Proposed Model	26	RF	99.97

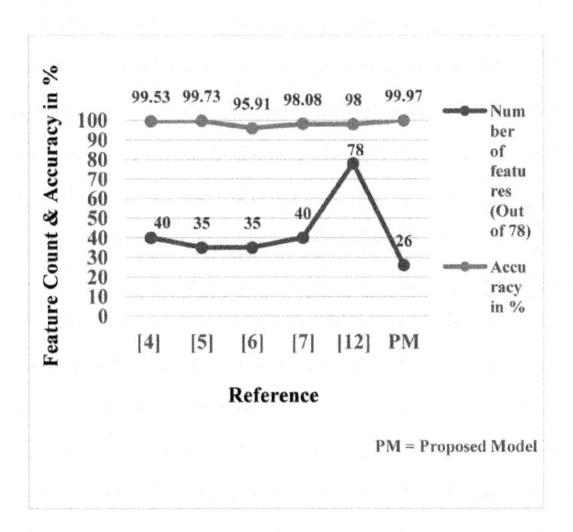

Figure 40.4 Graphical representation of Table 40.4

Conclusion

Detection of intrusion in the network to achieve optimum network security can and has been achieved with the integration of the latest computational technologies, ML algorithms, AI, etc. great implementations. Hence considering this fact, proposed model has used one of the frequently used techniques called ML to enhance network security by increasing the detection rate of malicious network packets while focusing on reduction in the number of independent features. From the technical perspective, the proposed model has experimented with the set of values of mutual information, information gain, gain ratio, and Pearson's Correlation factor and found out that if a selection of the best subset of input features is necessary, then arithmetic operations on their values can give reduced input feature set while improving intrusion prediction accuracy score which can create impact in the research community while searching a method to reduce computational load while training and deploying ML models. In the future, authors would like to apply the proposed approach with different datasets to ensure proposed feature selection technique helps in reducing execution load of underlying computational unit of computer system.

References

Abbas, A., Khan, M. A., Latif, S., Ajaz, M., Shah, A. A. and Ahmad, J. (2021). A new ensemble-based intrusion detection system for internet of things. Arab. J. Sci. Eng. 47:1–15.

Adhao, R. B. and Pachghare, V. K. (2019). Performance-based feature selection using decision tree. In 2019 international conference on innovative trends and advances in engineering and technology (ICITAET), (pp. 135–138). IEEE.

Adhao, R. and Pachghare, V. (2020). Feature selection using principal component analysis and genetic algorithm. J. Discret. Math. Sci. Cryptogr. 23(2):595–602.

Atefi, K., Hashim, H. and Kassim, M. (2019). Anomaly analysis for the classification purpose of intrusion detection system with K-nearest neighbors and deep neural network. In 2019 IEEE 7th conference on systems, process and control (ICSPC), (pp. 269–274). IEEE.

Brownlee, J. (2019). e-blog created on October 16, 2019 in Probability. Information gain and mutual information for machine learning. https://machinelearningmastery.com/information-gain-and-mutual-information/.

Chkirbene, Z., Eltanbouly, S., Bashendy, M., AlNaimi, N. and Erbad, A. (2020). Hybrid machine learning for network anomaly intrusion detection. In 2020 IEEE international conference on informatics, IoT, and enabling technologies (ICIoT), (pp. 163–170). IEEE.

Elmrabit, N., Zhou, F., Li, F. and Zhou, H. (2020). Evaluation of machine learning algorithms for anomaly detection. In 2020 international conference on cyber security and protection of digital services (Cyber Security), (pp. 1–8). IEEE.

Halimaa, A. and Sundarakantham, K. (2019). Machine learning based intrusion detection system. In 2019 3rd International conference on trends in electronics and informatics (ICOEI), (pp. 916–920). IEEE.

Meyer, M. L. B. and Labit, Y. (2020). Combining machine learning and behavior analysis techniques for network security. In 2020 international conference on information networking (ICOIN), (pp. 580–583). IEEE.

Pachghare, V. K. (2019). Cryptography and information security. PHI Learning Pvt. Ltd.

Pachghare, V. K., Khatavkar, V. K. and Kulkarni, P. A. (2012). Pattern based network security using semi-supervised learning. Int. J. Inf. Network Security 1(3):228.

Prasad, M., Tripathi, S. and Dahal, K. (2020). An efficient feature selection based Bayesian and Rough set approach for intrusion detection. Appl. Soft Comput. 87:105980.

Priyadarsini R. P., Valarmathi, M. L. and Sivakumari, S. (2011). Gain ratio based feature selection method for privacy preservation. ICTACT J. Soft Comput. 1(4):201–205.

Sharafaldin, I., Lashkari, A. H. and Ghorbani, A. A. (2018). Toward generating a new intrusion detection dataset and intrusion traffic characterization. ICISSp, 1:108–116.

Siddiqi, M. A. and Pak, W. (2020). Optimising filter-based feature selection method flow for intrusion detection system. Electronics 9(12):2114.

Tahe, K. A., Jisan, B. M. Y. and Rahman, M. (2019). Network intrusion detection using supervised machine learning technique with feature selection. In 2019 International conference on robotics, electrical and signal processing techniques (ICREST), (pp. 643–646). IEEE.

Vijayanand, R. and Devaraj, D. (2020). A novel feature selection method using whale optimization algorithm and genetic operators for intrusion detection system in wireless mesh network. IEEE Access 8:56847–56854.

41 Evaluation of prediction performance for K-Nearest neighbour, random forest algorithm and extreme gradient boosting algorithms in the forecasting of fault diagnosis for roller bearings

Ayyappa, T.[1,a], *I.V. Manoj*[2,b], *Shridhar Kurse*[3,c], *Jayant Kumar*[4,d], *and Tilak Bhattacharyya*[5,e]

[1]Senior AI & ML Engineer, TPRI Technologies Pvt Ltd, Bangalore, India

[2]Department of Mechanical Engineering, National Institute of Technology Karnataka, Surathkal, Karnataka, India

[3]Department of Mechanical Engineering, New Horizon College of Engineering, Bengaluru, India

[4]CEO&MD, TPRI Technologies Pvt Ltd, Bangalore, India

[5]Delivery Manager, TPRI Technologies Pvt Ltd, Bangalore, India

Abstract

Condition monitoring for ball bearings has become a necessity in identifying the early faults thereby necessary precautions can be implemented before the complete failure of the machine. In this paper, three different robust classification algorithms are applied in the fault prediction of rolling bearings for which the data is taken from a bearing datacentre of Case Western Reserve University (CWRU). These include the K-Nearest Neighbour algorithm (KNN), Random Forest algorithm (RF) and Extreme Gradient Boosting algorithm (XGBoost). All of the algorithm performances were decent but the XGBoost is comparatively more efficient as it has better performance measures like Accuracy, Precision, Recall and F1 scores. All the three dataset from different domains like time, wavelet energy and wavelet entropy were used for training, testing and prediction by these algorithms. In the comparison of the performance of each algorithm, it was seen that XGBoost showed the best performance with 99.56% for the prediction of faults in all the domains.

Keywords: Accuracy, K-NN, Precision, Prediction, Recall, RF, XGBoost.

Introduction

Since the bearing is an important machine elements, monitoring these elements is necessary for the detection of the faults. The failure of such elements consequences to devastating damages on the overall system (Manoj et al. 2016; Kankar et al., 2011). This not only applies to the simplest of roller bearing but also to the most complex magnetic bearings. Condition monitoring can be carried out using vibration analysis, temperature analysis or acoustic analysis. These diagnoses from the conditional monitoring indicate occurs of faults in the bearing due to certain vibrational characteristics exhibited change in their reference levels (Zhao and Yan, 2013; Cheng et al., 2021). The datasets obtained from conditional monitoring can be used as a powerful pattern recognition tool.

Many artificial intelligence techniques like convolutional neural network, K-Nearest Neighbour algorithm, artificial neural network, Random Forest (RF) algorithm, AdaBoost, Xgboost etc. have been used for different predictions (Manoj and Narendranath, 2021a; Manoj and Narendranath, 2021b; Nguyen et al., 2019; Manoj et al., 2022). Application of AI models has been of keen interest to researchers for fault detection. The deep learning models like different neural networks perform efficiently for imbalanced fault diagnosis of rotating machinery Liu et al. (2018). Ayas and Ayas (2022) formulated a novel deep residual learning with batch normalization (BN) which provided a higher accuracy in fault finding during condition motoring of bearing, than other models. The average classification accuracy of the model was 99.98%. Karabacak and Özmen (2022) utilised k-NN, servo vector mechanism (SVM) and artificial neural network (ANN) on compressed signal processing (CSP) for conditional monitoring. The findings indicated that ANN with CSP gave exceptional results during worm gear condition monitoring for variable operating conditions. It was also observed that CSP features perform well in time and frequency domain features.

[a]thalawarayyappa@gmail.com; [b]vishalmanojvs@gmail.com; [c]skurse@gmail.com; [d]jayant.kumar@technopro.com; [e]Tilak.Bhattacharyya@technoproindia.com

Sarothi et al. (2022) have explored eleven well-established ML models for forecasting the strength of bearing here the RF model had a prediction that was nearing experimental results. Vives (2022) have used KNN and SVM methodology in fault diagnosis for bearings used for installation of high power wind turbines. It was concluded that by using Ml models there were reducing costs and time, making them accurate and reliable. Buchaiah and Shakya (2022) demonstrated the superiority of SVM is shown by comparing it with the KNN and the Naive Bayes method in bearing fault diagnosis. Brito et al. (2022) have used unsupervised models for fault detection in rotating machinery where the performance parameter F1 was derived to have the maximum value of 96.72% for Ensemble, k-NN and Cluster-based Local Outlier Factor (CBLOF) models in mechanical fault detection. Rathore and Harsha (2022) implemented k-NN model with online degradation state classification technique. They concluded that it was found to be a promising model as the area under the curve used to estimate the remaining useful life of bearing was 0.94. Kumar and Hati (2022) proposed a method employing conventional models and convolutional neural network (CNN) for the prediction of bearing faults and broken rotor bar detection. The python software with the Keras and TensorFlow packages were used for the models. It was seen that the dilated CNN gave better accuracy in prediction than the other fault detection models. Zhong and Ban (2022) employed two ensemble learning models such as bagging and Adaboost strategies for crack detection in rotatory elements for nuclear applications. The RF and Adaboost tree were used as it gave promising prediction results with small data. It was also seen that these models mitigate the negative effects of noise during machine learning process. Zhu et al. (2022) found that RF was more accurate compared to conventional machine learning methods, such as Backpropagation Neural Network (BPNN) and SVM. The RF model was stronger robustness as it requires lesser training data and the parameter tuning faster. Doua and Zhoua (2016) employed Probabilistic Neural Network (PNN), K-NN algorithm, Particle Swarm Optimization optimised Support Vector Machine (PSO-SVM) and a Rule-Based Method (RBM) algorithm and a new Rule Reasoning Mechanism (RRM) for fault detection of rotating equipment. It was found that PSO-SVM ranked first which was followed by RRM although RRM was most appropriate for real-time applications. Yu et al. (2022) proposed an error check model that was established on the RF which gave the efficiency of 94.8% outperforming linear discriminant analysis, decision trees, supporting vector machine and Nave Bayes during the prediction of defect identification of rotating bars. Nishat Toma and Kim (2020) have utilised the discrete wavelet transform (DWT) for feature extraction in fault diagnosis. The authors have predicted using ensemble machine learning techniques like RF and extreme gradient boosting (XGBoost). It was observed that authors proposed a method which gave an accuracy slightly greater than 99%. Minhas and Singh (2021) have used improved multiscale permutation entropy method and XGBoost was employed. It was observed that the proposed method exceeds accuracy of 3% to 18%. Trizoglou et al. (2021) have made use of XGBoost and LSTM in effective monitoring strategies. It was concluded that XGBoost was efficient than LSTM in predictive accuracy as it needs smaller training times and a smaller sensitivity to noise.

From the above literature, it can be seen that deep learning algorithms like different neural networks was more efficient in the imbalanced dataset. In this paper, the comparison of three algorithms KNN, RF and XGBoost was performed for fault detection. The dataset was borrowed from Case Western Reserve University (CWRU) where there were some of the faults introduced and this was used for the fault prediction. Different evaluation has been performed using the algorithm with the help of performance measurements such as accuracy, precision, recall and f-1 score. These algorithms are used for time domain, wavelet energy domain and wavelet entropy domain data. Extreme Gradient Boosting algorithm (XGboosting) was found to be more efficient in the prediction, implementation of the standardised dataset in all three domains.

Methods

A. K-Nearest Neighbour Algorithm (K-NN)

K-NN is a type of representative based learning, the function of this algorithm is to approximate locally and then the total computation is carried out by delaying the function evaluation. The training samples are created in P- dimensional space with P features from the sample and also the testing samples are performed in the same manner K nearest neighbour is to be chosen accordingly and then by computing the Euclidean distance between the training and testing samples. The qth training sample and testing sample are defined as an Equation 1 Doua and Zhoua (2016)

$$\mathrm{ED_q} = \sqrt{\sum_{p=1}^{P}\left(TE_p - TR_{qp}\right)^2} \tag{1}$$

B. Random Forest (RF)

The RF can be understood as a combination of a series of predictor trees where each tree depends on an independent sampled arbitrary vector with this concept gets applied to all forest trees. As the number of decision trees increases in the random forest, errors of the forest approaches a limit value. An error in the forest depends on the ability of the specific trees in the forest and their correlation. Utilizing a random set of features which would be split to generate error rates and also they are stable about noise. The increase in the splitting of the features was carried out with the internal calculations of intensity, correlation, and track error. To gauge variable importance, internal forecasts are also used (Yu et al., 2022; Nishat Toma and Kim, 2020).

C. Extreme Gradient Boosting Algorithm (XGBoost)

The XGBoost algorithm is based on the decision-tree ensemble machine learning that utilises a framework of gradient boosting. The decision tree-based algorithms are examined as the best-in-class when we consider the data with structured whereas the artificial neural network outperforms when it comes to unstructured data. XGBoost is a scalable and accurate implementation of gradient boosting machines. This algorithm was main formulated for the increasing the model efficiency and speed of computing. It was the best algorithm among other boosted trees algorithms that demonstrated to push the limits of computing power (Minhas and Singh, 2021; Trizoglou et al., 2021).

Experimentation and Data Analysis

The Case Western Reserve University (CWRU) bearing data centre have collected information through experimentation which was used for prediction. A range of 0.021, 0.007, 0.014 and 0.028 faults were manually applied using electro-discharge machining to the bearing. These faults were introduced in form of indentations on the ball, outer race and inner race (in inches) respectively. Figure 41.1 shows the test rig that was used to generate data. With the aid of the accelerometer, the data was gathered from which the dataset was bored for our analysis. Identify and classify the faults in the bearing according to the plot obtained. From the test rig which has the capacity of 2hp load and with nine classes of faults in the ball, inner race and outer race with three each different faults were recorded namely ball defect (0.007), ball defect (0.014), ball defect (0.021), inner race defect (0.007), inner race defect (0.014), inner race defect (0.021), normal, outer race defect (0.007), outer race defect (0.014), outer race defect (0.021) Ayas and Ayas (2022). All defects are in inches.

Measurement of Performance Characteristics

To evaluate the performance of the model we have used Accuracy, Precision, Recall and F-1 score metrics. The models were built using python programming. The different performance index were used to rate the models. These performances were measured using a confusion matrix derived from the models. The confusion matrix gives the performance of the model during training and testing of data Different characteristics were calculated based on four parameters false positive (γ), true positive (α), true negative (β), and false-negative (φ) which are shown in Table 41.1.

Figure 41.1 Test-rig consisting Zhu et al. (2014)

Table 41.1 Representation of general confusion matrix

		Predicted Class	
Actual Class		*Positive*	*Negative*
	Positive	α	φ
	Negative	γ	β

Accuracy (A): Accuracy is the ratio between correctly expected observation and overall observation.

$$A = \frac{\alpha + \beta}{\alpha + \beta + \gamma + \varphi} \tag{2}$$

Precision (P): Precision is the right positive observation expected to the total positive predictive observation.

$$P = \frac{\alpha}{\alpha + \gamma} \tag{3}$$

Recall (R): Recall is the ratio of correctly expected positive observations to all real-yes-class observations.

$$R = \frac{\alpha}{\alpha + \varphi} \tag{4}$$

F-1 score (f): The F-1 score is a performance characteristic derived from weighted average the Precision and Recall. Both false positives and false negatives are considered so that all the false cases are evaluated. So f1 is generally more useful than accuracy and precision, especially if you have an uneven spreading of classes.

$$f = \frac{2 \times (R \times P)}{(R + P)} \tag{5}$$

Results and Discussion

The data was borrowed by the diagnosis of rolling bearings from Case Western Reserve University (CWRU) Bearing Datacenter. We have used Python for coding with the help of dictionaries such as NumPy, Pandas, Seaborn, Matplotlib and SKlearn for training and testing the data which is imported. The KNN, RF and XGBoost algorithms were coded with the help of python dictionaries for all three domains. When we apply the KNN, RF and XGBoost algorithm to the dataset with 2300 features that need to be split into samples for training and testing. In this paper, we have taken 80 per cent of the data to be training samples which were 1840 features and the remaining 20 per cent for testing data that comes to be 640 features. Figure 41.2(a) shows the time domain that possesses a data set of 2300 rows and 10 columns. Whereas Figure 41.2 (b) and (c) shows the wavelet energy and wavelet entropy domains have 2300 rows and 9 columns respectively. The wavelet energy and wavelet entropy domain data had 9-time domain characteristics i.e. maximum (V1), minimum (V2), mean value (V3), standard deviation (V4), root mean square value (RMS) (V5), skewness (V6), kurtosis (V7), crest factor (V8), and shape factor (V9). Thus, we get the 2300 × 9 feature matrix size. In the case of time-domain data, the feature matrix adds an extra column to the default shape so that the final function size matrix comes to be 2300 × 10 Zhu et al. (2014).

A. Comparisons of K-NN, RF and XGBoost Algorithm with and without Standardization

From the above dataset models using k-NN, RF and XGBoost Algorithm were built and the overall accuracy was calculated for both with and without standardised data set. Figure 41.3 shows the accuracy percentage for all the datasets using all the three models in all the three domains. Observing the train and test efficiency we can conclude that the dataset has to be standardised to get more efficiency from the models in all three domains. So the dataset was standardised and used for T1 and T2 of models. It was seen that the RF model and XGBoost model gives 100% training efficiency in all the domains but the test

```
In [6]: data.head(15)
Out[6]:
```

	max	min	mean	sd	rms	skewness	kurtosis	crest	form	fault
0	0.35986	-0.41890	0.017840	0.122746	0.124006	-0.118571	-0.042219	2.901946	6.950855	Ball_007_1
1	0.46772	-0.38111	0.022255	0.132488	0.134312	0.174699	-0.081548	3.482334	6.035202	Ball_007_1
2	0.46855	-0.43809	0.020470	0.149651	0.151006	0.040339	-0.274069	3.102819	7.376926	Ball_007_1
3	0.58475	-0.54303	0.020960	0.157067	0.158422	-0.023266	0.134692	3.691097	7.558387	Ball_007_1
4	0.44685	-0.57891	0.022187	0.138189	0.139922	-0.081534	0.402783	3.193561	6.312085	Ball_007_1
5	0.43726	-0.44435	0.021119	0.138763	0.140328	-0.131329	-0.168557	3.115990	6.644538	Ball_007_1
6	0.45353	-0.49129	0.021464	0.138461	0.140082	-0.114175	0.308107	3.237609	6.526352	Ball_007_1
7	0.43955	-0.45228	0.020880	0.150120	0.151526	-0.021955	-0.272298	2.900820	7.263886	Ball_007_1
8	0.49233	-0.37217	0.020244	0.145361	0.146729	0.074174	-0.421814	3.355377	7.248013	Ball_007_1
9	0.37154	-0.49087	0.018105	0.136393	0.137556	-0.136237	-0.097889	2.701005	7.597902	Ball_007_1
10	0.38761	-0.34776	0.017720	0.126351	0.127557	-0.054869	-0.350092	3.038718	7.198563	Ball_007_1
11	0.47606	-0.40763	0.017677	0.136432	0.137540	0.038106	0.170824	3.461258	7.780711	Ball_007_1
12	0.48232	-0.45706	0.017111	0.141645	0.142839	0.057154	-0.027017	3.376676	8.347949	Ball_007_1
13	0.40346	-0.39366	0.018397	0.127684	0.128971	-0.010814	0.186028	3.128291	7.010409	Ball_007_1
14	0.47689	-0.38344	0.017997	0.136440	0.137589	-0.004357	-0.099890	3.466056	7.645231	Ball_007_1

```
data.head(15)
```

	V1	V2	V3	V4	V5	V6	V7	V8	fault
0	16.415	14.855	0.18270	0.029149	0.003602	0.005503	0.001981	0.000701	Ball_007_1
1	21.237	15.495	0.17681	0.026414	0.002804	0.005116	0.001718	0.000421	Ball_007_1
2	26.286	20.195	0.18248	0.029666	0.002424	0.005042	0.000993	0.000080	Ball_007_1
3	30.537	20.655	0.16648	0.029571	0.003195	0.005692	0.001971	0.000722	Ball_007_1
4	21.279	18.536	0.20850	0.041363	0.010714	0.010341	0.005792	0.004666	Ball_007_1
5	22.614	17.390	0.24450	0.041405	0.012424	0.012567	0.007475	0.006094	Ball_007_1
6	20.931	18.950	0.23606	0.039485	0.009946	0.010888	0.005576	0.004547	Ball_007_1
7	27.318	19.491	0.17240	0.029425	0.003658	0.005503	0.001775	0.000851	Ball_007_1
8	25.914	17.962	0.17703	0.027165	0.002926	0.005512	0.001902	0.000643	Ball_007_1
9	22.329	16.161	0.20133	0.036026	0.007951	0.009089	0.004150	0.002968	Ball_007_1
10	18.100	14.993	0.18382	0.030779	0.004342	0.006478	0.002328	0.001019	Ball_007_1
11	22.550	15.996	0.16354	0.024904	0.001883	0.004652	0.001239	0.000166	Ball_007_1
12	24.291	17.270	0.18565	0.028921	0.002436	0.005625	0.001514	0.000318	Ball_007_1
13	18.874	14.971	0.18416	0.028087	0.002341	0.005243	0.001095	0.000158	Ball_007_1
14	21.904	16.652	0.17892	0.027085	0.002226	0.004919	0.000998	0.000091	Ball_007_1

(a)

```
In [4]: data.head(15)
Out[4]:
```

	V1	V2	V3	V4	V5	V6	V7	V8	fault
0	62.583	59.330	0.26518	1.6009	0.001467	0.018364	0.032765	0.060336	Ball_007_1
1	85.084	53.239	0.28289	1.6765	0.001716	0.017286	0.029688	0.062880	Ball_007_1
2	80.503	84.168	0.30216	1.6733	0.001349	0.013060	0.028566	0.059382	Ball_007_1
3	101.400	71.313	0.26996	1.3808	0.001791	0.015103	0.027356	0.058245	Ball_007_1
4	75.326	71.221	0.28575	1.6650	0.001243	0.014989	0.025672	0.060098	Ball_007_1
5	73.079	75.345	0.29058	1.6312	0.001123	0.015766	0.025455	0.060939	Ball_007_1
6	78.557	65.909	0.31294	1.6543	0.001350	0.013835	0.033252	0.064605	Ball_007_1
7	98.706	67.136	0.29753	1.4711	0.001077	0.013382	0.026059	0.054039	Ball_007_1
8	86.290	75.069	0.26507	1.3614	0.001172	0.016527	0.025535	0.055497	Ball_007_1
9	92.747	53.161	0.25046	1.5497	0.001207	0.014631	0.028356	0.061824	Ball_007_1
10	80.551	49.547	0.27721	1.6239	0.000972	0.017911	0.027389	0.064057	Ball_007_1
11	74.988	69.488	0.24063	1.4948	0.001383	0.013861	0.024958	0.057927	Ball_007_1
12	93.597	56.078	0.29305	1.7098	0.001413	0.016097	0.028236	0.064438	Ball_007_1
13	72.209	56.315	0.29700	1.6386	0.001157	0.013853	0.028008	0.061578	Ball_007_1
14	83.883	61.529	0.26991	1.4515	0.001299	0.013431	0.027479	0.061899	Ball_007_1

Figure 41.2 Dataset used for (a) Time-domain (b) Wavelet energy domain and (c) Wavelet entropy domain

efficiency decreases for the test dataset. Comparing all the three-domain, the maximum training accuracy was 98.15%, 100% and 100% for the k-NN, RF and XGboost models respectively. The maximum accuracy in testing was 97.17%, 99.34% and 99.56% for the k-NN, RF and XGBoost model for time, wavelet energy and wavelet entropy domains.

B. Confusion and Performance Matrix for the Standardised data

After the training (T1) and testing (T2) of standardised data, the confusion matrix was generated. Figures 41.4, 41.5 and 41.6 shows the confusion matrix for different algorithms in all three domains respectively. From the confusion matrix, all the other performance parameters like accuracy, recall and f1 score were calculated. Table 41.2 shows different performance parameters that were tabulated to analyse the efficiency of the models. From the performance tables, Table 41.3 was concluded where it was seen that the k-NN model needs more intervention maintenance and checking whereas RF and XGBoost are more friendly models that can be easily implemented. The evaluation matrix like Mean absolute error (MAE), Root mean squared error (RMSE) and R squared values for all the models were calculate in Figure 41.7. It was seen that XGBoost wave entropy has least MAE and RMSE values whereas the R2 values are higher.

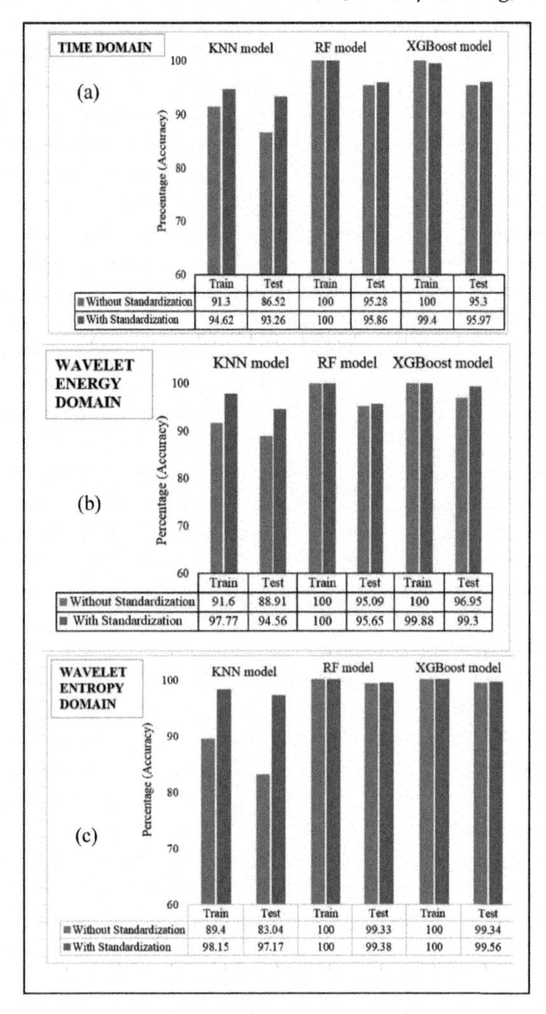

Figure 41.3 Dataset used for (a) Time domain (b) Wavelet energy domain and (c) Wavelet entropy domain

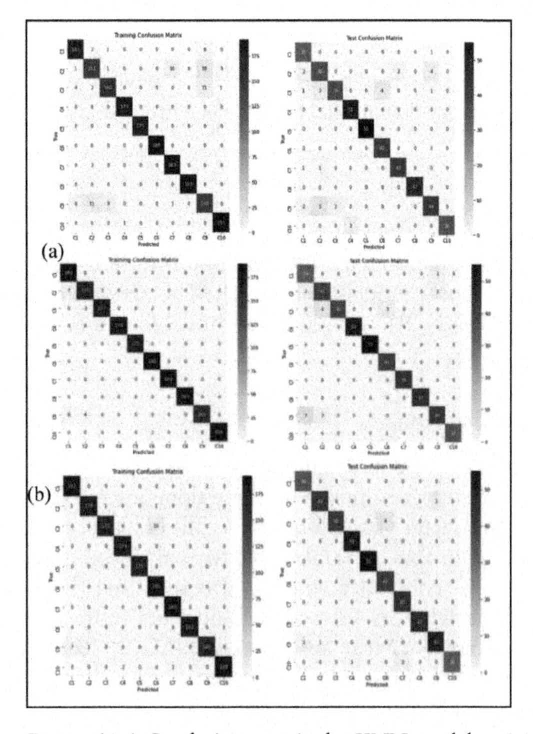

Figure 41.4 Confusion matrix for KNN model at (a) Time domain (b) Wavelet energy domain and (c) Wavelet entropy domain for standardised data set

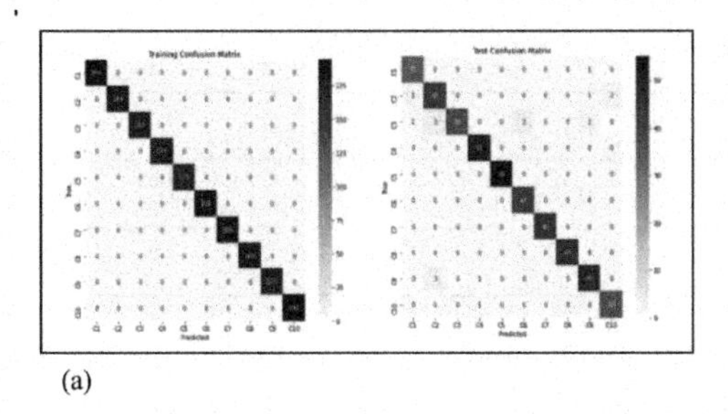

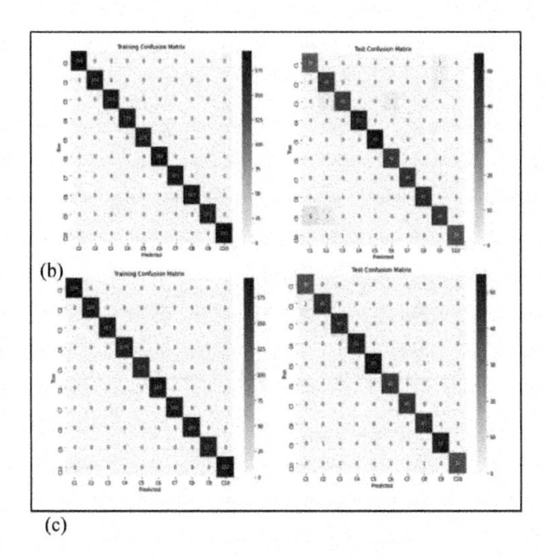

Figure 41.5 Confusion matrix for RF model at (a) Time domain (b) Wavelet energy domain and (c) Wavelet entropy domain for standardised data set

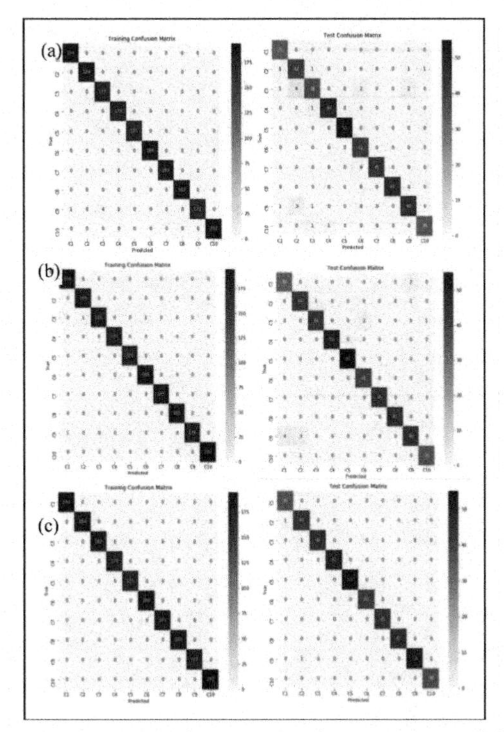

Figure 41.6 Confusion matrix for XGBoost model at (a) Time domain (b) Wavelet energy domain and (c) Wavelet entropy domain for standardised data set

Table 41.2 Performance of the algorithms for different models in all the three domains using standardised data

Method	*Precision*		*Recall*		*f1-score*	
	T1	*T2*	*T1*	*T2*	*T1*	*T2*
			Time domain			
KNN	0.95	0.93	0.95	0.93	0.95	0.93
RF	1.00	0.96	1.00	0.96	1.00	0.96
XGBoost	0.99	0.95	0.99	0.95	0.99	0.95
			Wavelet energy domain			
KNN	0.98	0.95	0.98	0.95	0.98	0.95
RF	1.00	0.96	1.00	0.96	1.00	0.96
XGBoost	1.00	0.96	1.00	0.96	1.00	0.96
			Wavelet entropy domain			
KNN	0.98	0.97	0.98	0.97	0.98	0.97
RF	1.00	0.99	1.00	0.99	1.00	0.99
XGBoost	1.00	0.99	1.00	0.99	1.00	0.99

Table 41.3 Decision on monitoring for different models all three domains with standardised data based on performance

Method	*Intervention*	*Maintenance*	*friendly*
	Time domain		
KNN	Needed	Medium	No
RF	None	Medium	Yes
XGBoost	None	Medium	Yes
	Wavelet energy domain		
KNN	None	Medium	No
RF	None	Medium	Yes
XGBoost	None	Medium	Yes
	Wavelet entropy domain		
KNN	Needed	Medium	No
RF	None	Medium	Yes
XGBoost	None	Medium	Yes

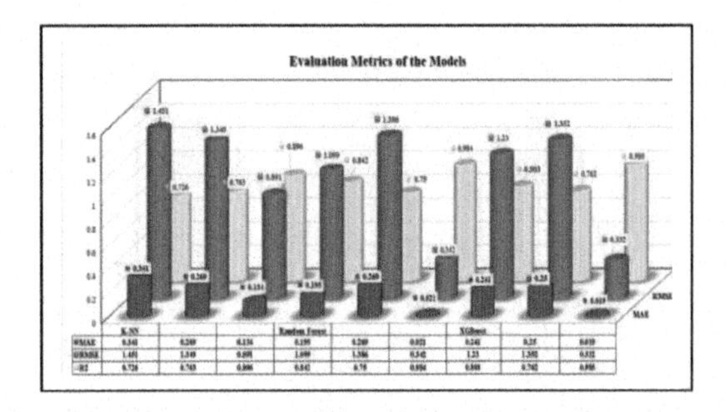

Figure 41.7 Evaluation metrics for prediction models

Conclusions

The data set was used with and without standardization and all the predictive models were used for fault prediction. All the performance parameters were calculated and conclusions were drawn as

1. The results of the wavelet entropy of the XGBoost algorithm were found to be 99.56% for testing samples which is better than the time domain and wavelet energy domain with standardisation of the data.
2. Without standardisation of dataset, XGBoost give higher accuracy of 95.30% in the time domain, 96.95% in wave energy and 99.34% in wave entropy domains in testing samples of dataset.
3. Although during tuning of hyper-parameters, the XGBoost algorithm needs more attention in programming than the KNN and RF. XGBoost models have proved to be accurate which was observed in evaluation metrics.
4. It can be concluded that the XGboost and RF models are friendlier than k-NN models as the maintenance is low and the intervention required is none for fault detection in the mechanical systems.

References

Ayas, S. and Ayas, M. S. (2022). A novel bearing fault diagnosis method using deep residual learning network. Multimed. Tools Appl. 81:22407–22423. https://doi.org/10.1007/s11042-021-11617-1.

Brito, L. C., Susto, G. A., Brito, J. N., and Duarte, M. A. V. (2022). An explainable artificial intelligence approach for unsupervised fault detection and diagnosis in rotating machinery. Mech. Syst. Signal Process. 163:108105. ISSN 0888-3270. https://doi.org/10.1016/j.ymssp.2021.108105.

Buchaiah, S. and Shakya, P. (2022). Bearing fault diagnosis and prognosis using data fusion based feature extraction and feature selection. Measurement 188:110506. https://doi.org/10.1016/j.measurement.2021.110506.

Cheng, Y., Lin, M., Wu, J., Zhu, H., and Shao, X. (2021). Intelligent fault diagnosis of rotating machinery based on continuous wavelet transform-local binary convolutional neural network. Knowl. Based Syst. 216:106796. https://doi.org/10.1016/j.knosys.2021.106796.

Doua, D. and Zhoua, S. (2016). Comparison of four direct classification methods for intelligent faultdiagnosis of rotating machinery. Appl. Soft Comput. 46:459–468.

Kankar, W. P. K., Sharma, S. C., and Harsha, S. P. (2011). Fault diagnosis of ball bearings using machine learning methods. Expert Syst. Appl. 38:1876–1886. doi: 10.1016/j.eswa.2010.07.119.

Karabacak, Y. E. and Özmen, N. G. (2022). Common spatial pattern-based feature extraction and worm gear fault detection through vibration and acoustic measurements. Measurement 187:110366. ISSN 0263-2241, https://doi.org/10.1016/j.measurement.2021.110366.

Kumar, P. and Hati, A. S. (2022). Dilated convolutional neural network based model for bearing faults and broken rotor bar detection in squirrel cage induction motors. Expert Syst. Appl. 191:116290. ISSN 0957-4174. https://doi.org/10.1016/j.eswa.2021.116290.

Liu, R., Yang, B., Zio, E., and Chen, X. (2018). Artificial intelligence for fault diagnosis of rotating machinery: A review. Mech. Syst. Signal Process. 108:33–47. https://doi.org/10.1016/j.ymssp.2018.02.016.

Manoj, I. V. and Narendranath, S. (2021a). Evaluation of WEDM performance characteristics and prediction of machining speed during taper square profiling on Hastelloy-X. Aust. J. Mech. Eng. DOI: 10.1080/14484846.2021.1960670.

Manoj, I. V. and Narendranath, S. (2021b). Slant type taper profiling and prediction of profiling speed for a circular profile during in wire electric discharge machining using Hastelloy-X. Proc. Inst. Mech. Eng. Part C: J. Mech. Eng. Sci. 235:5511–5524.

Manoj, I. V., Soni, H., Narendranath, S., Mashinini, P. M., and Kara, F. (2022). Examination of machining parameters and prediction of cutting velocity and surface roughness using RSM and ANN using WEDM of Altemp HX. Adv. Mater. Sci. Eng. 2022:1–9. https://doi.org/10.1155/2022/5192981.

Manoj, I. V., Srihari, P. V., Kulkarni, S. S., Kumar, K. S., and Bharatish, A. (2016). Assessment of thermal effects on the levitation speed of bump foil bearings made of low cost spring steel. Measurement 92:453–463.

Minhas, A. S. and Singh, S. (2021). A new bearing fault diagnosis approach combining sensitive statistical features with improved multiscale permutation entropy method. Knowl. Based Syst. 218:106883. ISSN 0950-7051. https://doi.org/10.1016/j.knosys.2021.106883.

Nguyen, H., Bui, X., Bui, H., and Cuong, D. (2019). Developing an XGBoost model to predict blast-induced peak particle velocity in an open-pit mine: a case study. Acta Geophys. 67:477–490. https://doi.org/10.1007/s11600-019-00268-4.

Nishat Toma, R. and Kim, J. M. (2020). Bearing fault classification of induction motors using discrete wavelet transform and ensemble machine learning algorithms. Appl. Sci. 10(15):5251. https://doi.org/10.3390/app10155251.

Rathore, M. S. and Harsha, S. P. (2022). Prognostic analysis of high-speed cylindrical roller bearing using weibull distribution and k-nearest neighbor. ASME J. Nondestruct. Eval. Diagn. Progn. Eng. Syst. 5(1):011005. https://doi.org/10.1115/1.4051314.

Sarothi, S. Z., Ahmed, K. S., Khan, N. I., Ahmed, A., and Nehdi, M. L. (2022). Predicting bearing capacity of double shear bolted connections using machine learning. Eng. Struct. 251:113497. ISSN 0141-0296. https://doi.org/10.1016/j.engstruct.2021.113497.

Trizoglou, P., Liu, X., and Lin, Z. (2021). Fault detection by an ensemble framework of extreme gradient boosting (XGBoost) in the operation of offshore wind turbines, Renew. Energy 179:945–962. ISSN 0960-1481. https://doi.org/10.1016/j.renene.2021.07.085.

Vives, J. (2022). Vibration analysis for fault detection in wind turbines using machine learning techniques. Adv. Intell. Syst. Comput. 2:15. https://doi.org/10.1007/s43674-021-00029-1.

Yu, L., Shen, H., Zhen, H., Gu, D., Tong, T., Shen, P., and Chen, H. (2022). Design of error check model for electric energy metering device based on random forest. Energy Syst. https://doi.org/10.1007/s12667-021-00498-w.

Zhao, D. and Yan, J. (2013). Ant colony clustering analysis based intelligent fault diagnosis method and its application to rotating machinery. Pattern Analy. Appl. 16:19–29. doi: 10.1007/s10044-012-0289-3.

Zhong, X. and Ban, H. (2022). Crack fault diagnosis of rotating machine in nuclear power plant based on ensemble learning. Ann. Nucl. Energy 168: ISSN 0306-4549, https://doi.org/10.1016/j.anucene.2021.108909.

Zhu, K., Song, X., and Xue, D. (2014). A roller bearing fault diagnosis method based on hierarchical entropy and support vector machine with particle swarm optimization algorithm. Measurements 47:669–675. doi: 10.1016/j.measurement.2013.09.019.

Zhu, M., Yang, Y., Feng, X., Du, Z., and Yang, J. (2022). Robust modeling method for thermal error of CNC machine tools based on random forest algorithm. Intell. Manuf. https://doi.org/10.1007/s10845-021-01894-w.

42 Fabrication and characterization of nanoencapsulated phyto-compounds

Prasad Sherekar[1,a], *Roshni Rathod*[2,b], *Harsha Pardeshi*[2,c], *Sanvidhan G. Suke*[2,d], *and Archana Dhok*[3,e]

[1]Department of Biotechnology, Priyadarshini Institute of Engineering and Technology, Nagpur, India

[2]Department of Biotechnology, Priyadarshini College of Engineering, Hingna Road, Nagpur, India

[3]Jawaharlal Nehru Medical College, Datta Meghe Institute of Medical Sciences (Deemed to be Uniersity), Wardha, India

Abstract

Obstruct pharmacokinetics and low bioavailability of active phytoconstituents; diosgenin and emodin are major limitations for their therapeutic success in several inflammatory diseases. Nanoencapsulation of both drugs will promisingly overcome these limitations. Herein, biodegradable/biocompatible polymer; poly(lactic-co-glycolic acid) (PLGA) has been used to prepare nano encapsulated diosgenin (DG)n and emodin (ED)n via a modified solvent-emulsion-diffusion-evaporation method. Functional stability of prepared nano-drugs and in vitro physiological characterization including mean particle sizes distribution, polydispersity index, surface zeta potential, and morphological examinations were performed. Both nano-drug formulations demonstrated the presence of functional drug within nanoparticles, 50 to 150 nm sizes with homogeneous particle distribution, negative surface zeta potential stability below -25 mV, and uniform spherical morphology. Obtained resulting characteristics reveal the efficacy of (DG)n and (ED)n to be a suitable nano-drug delivery modality with improved pharmacological strength.

Keywords: Diosgenin, drug delivery, emodin, nanoencapsulation, poly(lactic-co-glycolic acid).

Introduction

Nanoparticular drug delivery is known to be an approach, technology, formulation, and widely studied system to transport a broad range of synthetic drugs, bio-molecules, and natural compounds into the body. This drug delivery system has a long frame of unique properties and potential that can be used to enhance the therapeutic effects of drugs through improving concentration and absorption of the drug at pathological sites (Samanta et al., 2016; Silva et al., 2017; Mahmood et al., 2017). The purpose of nanoparticles utilization is lessening toxicity and reducing unfavourable side effects of drugs by focusing to definite site of action, through altered pharmacokinetics, lowering dosing frequencies by controlling drug release, and enhancing shelf life by improving stability (Raval et al., 2019). However, the characterization of pharmaceutical nanoparticles formulations, especially the nanostructure examination during formulation, their stability evaluation, and biological transportation through the body, are important characteristics for the success of this drug delivery approach (Mahmood et al., 2017).

In this study, particulate drug carrier, poly(lactic-co-glycolic acid) (PLGA) copolymer was used for nanoencapsulation of herbal steroidal sapogenin and anthraquinone such as diosgenin and emodin respectively. Diosgenin (DG) extensively occurred in *Dioscoreaceae* and *Agavaceae* family is the sugar-free (aglycone) hydrolysis product of dioscin which has anti-oxidative, anti-inflammatory, and other pharmacological activities in the attenuation and cure of variety of cancers, cardiovascular diseases, atherosclerosis, and neurological diseases. Despite the potential pharmacological activities, the pharmacokinetic properties and clinical applications of DG have been severely hampered due to its hydrophobic nature, insolubility in water, poor bioavailability, and fast biotransformation in physiological conditions (Cai et al., 2020; Xu et al., 2012). Several nanotechnology-based formulations including DG nanocrystals, eight-arm-PEG-DG conjugate nanoparticles demonstrated the improved bioavailability, pharmacokinetic profile, and enhanced solubility of DG which further significantly enhanced therapeutic efficiency to prevent and treatment of arterial and venous thrombus (Cai et al., 2020).

Another active phytoconstituent, emodin (ED) (1, 3, 8- trihydroxy-6-methyl anthraquinone) from *Rheum officinale* Baill and *Aloe barbadensis* Makino, showed strong antifibrotic (Yang et al., 2016; Di et al., 2015)

Senior Research Fellowship Project funded by Indian Council of Medical Research (ICMR), Government of India.

[a]sherekar.vprasad@gmail.com; [b]rathodroshni992@gmail.com; [c]harshapardeshi2@gmail.com; [d]sgsuke@hotmail.com; [e]drarchanadhok@gmail.com

anti-inflammatory, anti-oxidant, anti-cancerous, anti-tuberculosis (Shia et al., 2010; Liu et al., 2012) activities in different *in vivo* animal models. Despite its favorable pharmacological activities, ED has demonstrated exceptionally low, (<1%) and (<3%) oral bioavailability in rats and rabbits respectively since its wide glucuronidation (Di et al., 2015). Fast and extensive metabolism of ED in the small intestine leading to the production of emodin glucuronides after oral administration. Whereas, ω-OHE and ω-OHE sulfates/glucuronides production following intravenous administration (Shia et al., 2010; Liu et al., 2012) which are obstructing its medical applications. However, emodin glucuronidation was inhibited with the use of piperine which partially improved *in vivo* bioavailability and pharmacokinetics of emodin (Di et al., 2015).

The pharmaceutical nanocarriers have a great benefit against present approaches to avoid the blood-brain barrier and GI tract, to advance the bioavailability of drugs (Samanta et al., 2016). Several biodegradable natural and synthetic polymers such as collagen, fibrin, chitosan, dextran, polylactic acid (PLA), PLGA, polyglycolic acid (PGA), polyesters, etc., have been used in drug delivery (Jeon and Park, 2009). Among these polymers, FDA approved PLA and PLGA co-polymer have been successfully developed for clinical applications because their physical properties allow them to make micro and nano-drug particles and can be easily controlled to obtained desirable pharmacokinetic and biodegradable properties (Jeon and Park, 2009; Lee et al., 2016). This colloidal nanoparticle system can deliver both hydrophilic and hydrophobic drugs which demonstrate improved pharmacokinetic properties such as late or controlled absorption of the drugs, limited bio-distribution, and an amplified retention time at the site of action, specificity to the affected tissues, and minimise side effects and toxicity of drug (Garcia, 2014).

Hence, the present study aimed was to synthesise biodegradable PLGA nanoparticles, loaded with diosgenin and emodin to improve their bioavailability for potent therapeutic delivery. Also, the characterization of prepared nano-drug particles was done to validate their physiological constancy.

Materials and Methods

A. Nanoencapsulation of DG and ED

Pure herbal compounds, Diosgenin (DG) (HiMedia, India) and Emodin (ED) (SRL, India) were separately encapsulated in PLGA co-polymer (85:15) L:G ratio (Sigma-Aldrich) using modified single solvent-emulsion-diffusion-evaporation method (Samanta et al., 2016; Rafiei and Haddadi, 2017). Briefly, 100 mg of PLGA dissolved in 5 mL of ethyl acetate (w/v) and 5 mg of DG and ED were also dissolved separately in 1 mL of ethyl acetate (w/v). A 1 mL of 2.2% (w/v) didodecyl dimethyl ammonium bromide (DMAB) (Sigma-Aldrich) stabilizing solution was prepared in distilled water. Non-ionic detergent (Tween 20) 2 to 3 drops, PLGA, and DMAB solutions was mixed and placed on a magnetic stirrer. The drug solution was then drop-wise added over the mixture with constant stirring speed. The whole emulsified suspension was then shaken vigorously followed by homogenization (REMI, RO-127A) for 5 min at 8000 rpm. The homogenised suspension was left on a magnetic stirrer for 2 hours for complete removal of organic solvent. After consecutive centrifugation (REMI, C-24BL) at 17,000 x g and washing steps with phosphate-buffered saline (PBS), the pellet of resulting nano-drug formulations (DG)n and (ED)n were obtained and resuspended in 2 ml of (0.025M) PBS followed by lyophilization (MAC Lyophilizer, MSW-137) and stored at -20°C temperature for future use.

B. Nano-Drugs Characterization

Fourier transformed-infrared spectroscopy (FT-IR):

The encapsulation chemistry of prepared nanoparticles (DG)n and (ED)n were analysed through FT-IR spectroscopy (IRAFFINITY-IS) and comparative expression of both nanoparticles with pure Diosgenin and Emodin were obtained (Ghosh et al., 2015; Hou et al., 2017; Suke et al., 2018).

Field emission scanning electron microscopy (FE-SEM):

Lyophilised nanoparticles were coated with a thin layer of platinum, and subjected to high-resolution electron image scanning across the surface of nanoparticles. The size, topography, and morphological evaluation of empty PLGA nanoparticles and (DG)n or (ED)n using FE-SEM (JSM-7610F) were confirmed (Samanta et al., 2016; Hussein et al., 2013; Mahmood et al., 2017).

Particle size, polydispersity index, and stability

Nanoparticles were suspended in water (25°C) and were subjected to mean diameter/size distribution profile, polydispersity index (PDI), and zeta potential (ZP) analysis using Zetasizer (Malvern Nano-ZS 90) (Samanta et al., 2016; Rafiei and Haddadi, 2017).

Result and Discussion

Diosgenin and emodin both are active phytoconstituents having antioxidant, anti-inflammatory, and several pharmacological properties against a wide range of diseases (Cai et al., 2020; Di et al., 2015; Shia et al., 2010). Moreover, the poor bioavailability and solubility of these drugs might be improved with their nanoencapsulation subject to appropriate characterization (Mahmood et al., 2017). Functional stability, physiological properties, size distribution, and morphology are the major characteristics of nano-particles that relate to the success percent of nanoparticulate drug delivery.

A. FT-IR Analysis

The FT-IR spectrum for pure diosgenin and (DG)n are depicted in Figure 42.1A. The FTIR of DG showed peaks at 3664 (O-H stretching), 2358 (O=C=O stretching), 1764 (C=O stretching), 1548 (N-O stretching), 675 (aromatic C=C bending), and 514 (C-Cl stretching). The similar peak which appears both in DG and (DG)n were 2358, 1764, and 1548 cm^{-1} confirming the existence of functional diosgenin in (DG)n nanoparticles (Ghosh et al., 2015). Similarly, the FT-IR data (Figure 42.1B) showing O=C=O stretching, C=O stretching and aromatic C-H bending at 2358, 1764, 1479 cm^{-1} along with 453 cm^{-1} presence in pure emodin and were also in (ED)n nanoparticles (Hou et al., 2017) From FTIR results, functional groups and chemical stability of both pure drugs with their nano-formulations remains functional and unchanged.

B. Surface Morphology Using FE-SEM Examinations

The FE-SEM images (Figure 42.2) demonstrated the sizes of the prepared nano-formulations were ranges between 50 to 150 nm for (DG)n and 50 to 100 nm for (ED)n which are appropriate and facilitating drug delivery in target organs like lungs and liver with different roots of administration (Samanta et al., 2016; Mahmood et al., 2017). Moreover, both nano-drug formulations showed uniform spherical morphology with more or less uniform particle size distribution.

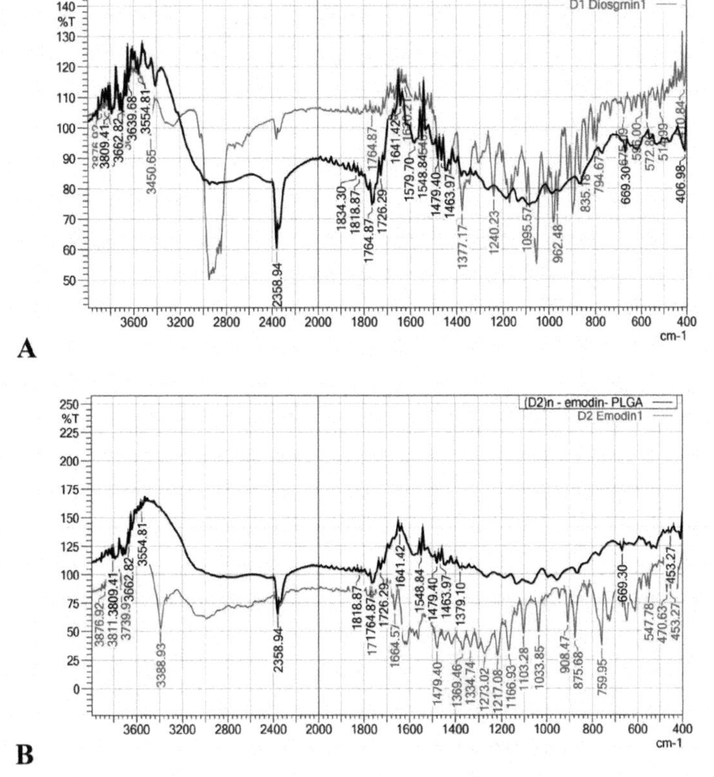

Figure 42.1 FT-IR spectrum of A: Diosgenin & (DG)n and B: emodin & (ED)n

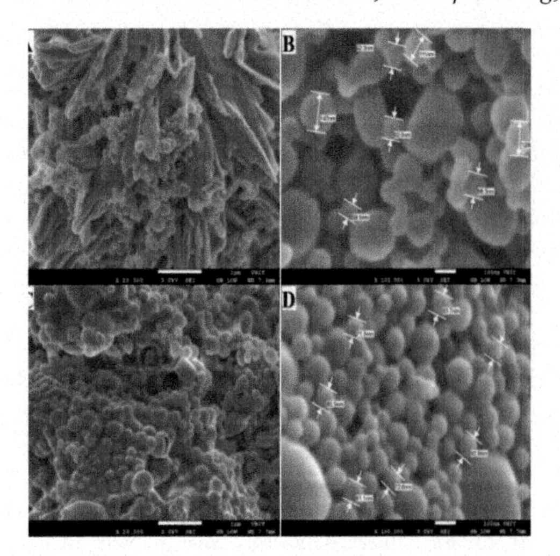

Figure 42.2 FE-SEM images showing nanoencapsulation of diosgenin (A & B) and emodin (C & D) in PLGA

C. Physiological Characteristics and Zeta Stability Evaluations

The average particle size (Z_average in nm) distributions and polydispersity index (PDI) for nano-formulations were depicted in Figure 42.3 and Table 42.1. DG-loaded PLGA nanoparticles had a mean diameter of 267.66±17.78 nm and PDI of 0.62±0.1 showing a slightly polydistributed nanoparticle population. Whereas, 200.5±13.14 nm average size and 0.54±0.05 of PDI were recorded for ED-PLGA nanoparticles exhibited a uniform particle distribution. Negative zeta potential of −14.46±3.15 mV and −15.43±2.75 mV observed for (DG)n and (ED)n respectively proves its stability below −25 mV (Table 42.1) (Samanta et al., 2016; Rafiei and Haddadi, 2017).

The physical characters including shape, size/diameter distribution, PDI, and zeta stability play an important role in determining biotransformation and therapeutic potential of nano-drug upon administration to the body Rafiei and Haddadi, 2017; Mahmood et al., 2017). The resulting nanoencapsulation method and characteristics observations for both fabricated nano-formulations are in desirable ranges which are suitable for drug delivery through any roots of administration given herbal nano-drug treatment for various diseases.

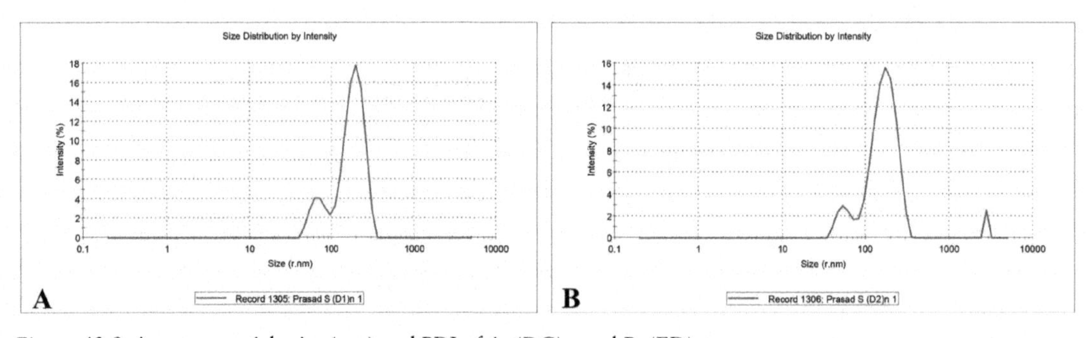

Figure 42.3 Average particle size (nm) and PDI of A: (DG)n and B: (ED)n

Table 42.1 Nanoparticle composition, size, PDI and stability

Nano-drug formul-ations	*Nano-drug characteristics*		
	Mean particle size (nm)	*Polydispersity index (PDI)*	*Zeta potential (mV)*
(DG)n	267.66±17.78	0.62±0.1	−14.46±3.15
(ED)n	200.5±13.14	0.54±0.05	−15.43±2.75

(Results were expressed as mean ± SD of 3 replicates of each formulation)

Conclusion

Prepared PLGA nanoparticles were used to encapsulate bioactive steroidal molecule diosgenin and phyto-anthraquinone emodin for a sustained-release drug delivery system. In this study, the functional stability, mean particle diameter size, PDI, and negative surface zeta potential value which are the important characteristics for the success of nano-drugs delivery are found in a desirable range, which further will be in favour of a long-circulating period in the blood and controlled absorption at target tissues for encapsulated active phytoconstituents. However, more studies are essential including *in vivo* pharmacokinetic, biocompatibility, and bioavailability evaluation for underlying its therapeutic potential.

Acknowledgment

Prasad Sherekar thanks Indian Council of Medical Research for SRF (No. 3/1/2(4)/COPD/2020-NCD-II).

References

Cai, B., Zhang, Y., Wang, Z., Liu, G., Chun, C. J., and Li, J. (2020). Therapeutic potential of diosgenin and its major derivatives against neurological diseases: Recent advances. Oxid. Med. Cell. Longev. 16. Article ID 3153082.

Di, X., Wang, X., Di, X., and Liu, Y. (2015). Effect of piperine on the bioavailability and pharmacokinetics of emodin in rats. J. Pharm. Biomed. Anal. 115:144–149.

Garcia, F. M. (2014). Nanomedicine and therapy of lung diseases. Einstein 12(4):531–533.

Ghosh, S., More, P., Derle, A., Kiture, R, Kale, T., Kundu, G. C., Chopade, B. A. and Kale, S (2015). Diosgenin functionalised iron oxide nanoparticles as novel nanomaterial against breast cancer. J. Nanosci. Nanotechnol. 15:9464–9472.

Hou, X., Wei, W., Fan, Y., Zhang, J., Zhu, N., Hong, H., and Wang, C. (2017). Study on synthesis and bioactivity of biotinylated emodin. Appl. Microbiol. Biotechnol. 101(13):5259–5266.

Hussein, A. S., Abdullah, N., and Ahmadun, F. R. (2013). In vitro degradation of poly (D, L-lactide-co-glycolide) nanoparticles loaded with linamarin. IET Nanobiotechnol. 7(2):33–41.

Jeon, O. and Park, K. (2009). Biodegradable polymers for drug delivery systems. Encyclopedia of Surface and Colloid Science (2nd edn), (pp. 1–15). Taylor and Francis, USA.

Lee, B. K., Tun, Y., and Park, K. (2016). PLA micro- and nano-particles. Adv. Drug Deliv. Rev. 107:176–191.

Liu, W., Feng, Q., Li, Y., Ye, L., Hu, M., and Liu, Z. (2012). Coupling of UDP-glucuronosyltransferases and multidrug resistance-associated proteins is responsible for the intestinal disposition and poor bioavailability of emodin. Toxicol. Appl. Pharmacol. 265:316–324.

Mahmood, S., Mandal, U. K., Chatterjee, B., and Taher, M. (2017). Advanced characterizations of nanoparticles for drug delivery: Investigating their properties through the techniques used in their evaluations. Nanotechnol. Rev. 6(4):355–372.

Rafiei, P. and Haddadi, A. (2017). Docetaxel-loaded PLGA and PLGA-PEG nanoparticles for intravenous application:- pharmacokinetics and biodistribution profile. Int. J. Nanomedicine 12:935–947.

Raval, N., Maheshwari, R., Kalyane, D., Youngren-Ortiz, S. R., Chougule, M. B., and Tekade, R. K. (2019). Importance of physicochemical characterization of nanoparticles in pharmaceutical product development. In Basic fundamentals of drug delivery (pp. 369–400). Elsevier.

Samanta, A., Bandyopadhyay, B., and Das, N. (2016). Formulation of catechin hydrate nanocapsule and study of its bioavailability. Med. Chem. (Los Angeles) 6:399–404.

Shia, C. S., Hou, Y. C., Tsai, S. Y., Huieh, P. H., Leu, Y. L., and Chao, P. D. L. (2010). differences in pharmacokinetics and ex vivo antioxidant activity following intravenous and oral administrations of emodin to rats. J. Pharm. Sci. 99:2185–2195.

Silva, A. L., Cruz, F. F., Rocco, P. R. M., and Morales, M. M. (2017). New perspectives in nanotherapeutics for chronic respiratory diseases. Biophys. Rev. 9:793–803.

Suke, S. G., Sherekar, P., Kahale, V., Patil, S., Mundhada, D., and Nanoti, V. M. (2018). Ameliorative effect of nanoencapsulated flavonoid against chlorpyrifos-induced hepatic oxidative damage and immunotoxicity in Wistar rats. J. Biochem. Mol. Toxicol. 32(5):e22050.

Xu, T., Zhang, S., Zheng, L., Yin, L., Xu, L., and Peng, J. (2012). A 90-day subchronic toxicological assessment of dioscin, a natural steroid saponin, in Sprague–Dawley rats. Food Chem. Toxicol. 50:1279–1287.

Yang, T., Wang, J., Pang, Y., Dang, X., Ren, H., Liu, Y., Chen, M. and Shang, D. (2016). Emodin suppresses silica-induced lung fibrosis by promoting Sirt1 signaling via direct contact. Mol. Med. Rep. 14:4643–4649.

43 Influence of process parameters on electric discharge machining of DIN 1.2714 steel

Harvinder Singh[1,a], *Santosh Kumar*[2,b], *and Swarn Singh*[2,c]

[1]PhD Reseasch Sscholar Chandigarh University, Graruan Mohali Punjab India, Department of Mechanical Engineering, Chandigarh Group of Colleges Mohali, India

[2]Department of Mechanical Engineering, Chandigarh Group of Colleges Mohali, India

Abstract

Electrical discharge machining (EDM) is a widespread process which works very effectively in the machining of difficult-to-cut materials and alloys in the die and aerospace industries with high dimensional accuracies. DIN 1.2714 steel is widely used in forging tools, dies of all sorts, hot forging, pressing tools, moulds, bushings and piercers etc. Thus, there is great need on carrying research on this material so that EDM parameters are to be optimised to get higher MRR. Hence, the aim of the current investigation is to analyse the influence of distinct parameters (current (I), Voltage (V), Pulse-on-time (T_{on}), Pulse-off-time (T_{off}) and flushing pressure (F_p)) on metal removal rate (MRR), of DIN 1.2714 steel. Further, to conduct the experiment, L_{27} orthogonal array (OA) was chosen for the five machining parameters at 3 levels each. The experiments were executed in a random order with two successive trials. Further, the analysis of variance (ANOVA) was executed and the optimal levels for maximising the responses were recognised. The percentage contributions of I, V, T_{on}, T_{off} and F_p on MRR are 82.28%, 4.01%, 13.21%, 0.015% and 0.070% respectively. In addition, it was found that at optimum parameters settings MRR improved by 1.15 times.

Keywords: Electrical discharge machining, material removal rate, surface roughness.

Introduction

In the manufacturing business, machining is the most widely used metal shaping procedure (Pandey and Shah, ????). Years after years, global investment in metal-machining processes continues to rise. Machining is more expensive than other production techniques like as casting, shaping and moulding, but it is frequently justified due to the need for precision. Another reason for its appeal is its adaptability; by adjusting the standard electrode, sophisticated and free form structures with numerous characteristics across a wide size range may be created more inexpensively, rapidly, and easily. There has been a rise in the availability and usage of difficult-to-machine materials, due to the need of high strength materials for diverse applications. For machining, non-traditional machining procedures are required. EDM is utilised for machining electrically conductive materials (Chigal et al., 2013). EDM is a thermo-electric method in which we can remove material by using a controlled spark generating process. It is widely utilised non-traditional machining methods in the industry today. EDM is widely utilised in the mould and die business, as well as in the production of automotive, aerospace, and surgical components. Fragile and thin parts can be machined without damage because there is no mechanical contact b/w the w/p and tool.

The mechanisms behind the development of MRR are very dynamic, complex and process dependent. Several factors influence the MRR obtained in an EDM operation. These can be categorised as controllable factors (Current, Voltage, Spark-on, Spark-off, Duty cycle, Arc gap and Flushing pressure) etc and uncontrollable factors (geometry and material properties of both tool and work-piece) (Rohith et al., 2019; Singh et al., 2021).

There are so many non-conventional machining methods utilised for the machining of hard material or alloys viz; Ultrasonic machining (USM), Water Jet machining (WJM), Abrasive Jet machining (AJM), Chemical machining (CHM), Electro-chemical machining (ECM), Electro chemical grinding (ECM), Ion beam machining, (IBM)., Plasma arc machining (PAM)., Laser beam machining (LBM), and Electrical discharge machining (EDM), etc. (Lee and Li, 2001; Kumar et al., 2020; Singh et al., 2020; Ramasawmy and Blunt, 2002; Kumar et al., 2018; Mahajan et al., 2021; Leao and Pashby, 2004; Kumar et al., 2021). Various authors (Liao et al., 2004; Kansal et al., 2005; Keskin et al., 2006; Kurita and Hattori, 2006; Abbas et al., 2007; Kumar et al., 2007; Zarepour et al., 2007; Kiyak and Cakir, 2007; Kanagarajan et al., 2008; Chiang, 2008; Hascalyk and Caydas, 2007; Pradhan et al., 2009; Kumar et al., 2009; Tzeng, 2008; Kumar et al., 2021; Pradhan and Biswas, 2009; Habib, 2009; Wang and Lin, 2009) have machined distinct

[a]harvinder.coeme@cgc.edu.in; [b]santosh.3267@cgc.edu.in; [c]swarn.coeme@cgc.edu.in

material by using distinct EDM parameters. Their results revealed that by optimizing the parameters of EDM, MRR and other parameters can be improved. Hence, in the present work EDM process has been taken up for the increase of MRR of **DIN 1.2714 steel.** As the EDM process is most productive process, the study is expected to be quite beneficial. Hence EDM ZNC (Z-axis Numerically Controlled) Model S50; make: Sparkonix has been adopted for the current study to analyse the influence of process parameters on the MRR of **DIN 1.2714 steel.**

Experimental Details

The experiments have been performed on Electric Discharge as shown in Figure 43.1.

A. Work-Piece Material

DIN 1.2714 steel has been selected for the study work with copper as an electrode material. Actual image of the work-piece (after machining) is represented in Figure 43.2.

The hardness of DIN 1.2714 steel was 388-390 HBW, while the chemical composition of DIN 1.2714 steel as per Spark Emission Spectrometer is given below in the Table 43.1.

B. Tool Material

The tool material chosen for this study is pure red copper having cylindrical in shape with 12 mm diameter and 30mm length as shown in Figure 43.3.

Figure 43.1 Sparkonix S50 ZNC EDM

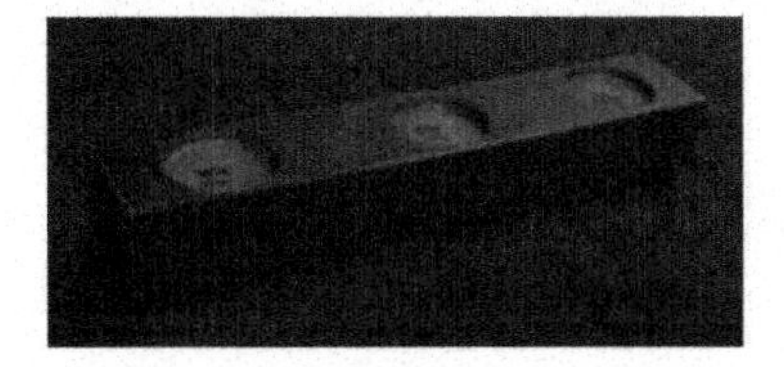

Figure 43.2 Work-piece after EDM process

Table 43.1 Chemical configuration of the DIN 1.2714 steel

Element Name	*Concentration in %*
Nickel (Ni)	1.58
Silicon (Si)	0.28
Carbon (C)	0.59
Phosphorus (P)	0.013
Manganese (Mn)	0.83
Chromium (Cr)	1.13
Molybdenum (Mo)	0.46
Vanadium (V)	0.07

Figure 43.3 Image of the tool (electrode) utilised for machining

Design of Experiments

A series of experiments have been conducted on ZNC S50 EDM machine and the kerosene oil is utilised as a dielectric fluid to investigate the influence of distinct machining parameters of EDM process on MRR. The test was conducted on DIN 1.2714 steel work-piece material with commercially available copper tool and kerosene oil was utilised as a dielectric medium. The machining time for each experiments was varies from 2 to 5 minute.

Selection of Basic Levels

The distinct applied variables and their range are depicted in the Table 43.2.

A. Constant Parameters

The constant parameters during machining operation were:

Tool material = Pure Red Copper
Depth of cut = 2.0 mm
Work-piece material = DIN 1.2714 steel

B. Performance Measure

In the present study the MMR was selected as response parameters.

The MRR (*gms/min*) is determined by the following method:

$$MRR = \frac{W_1 - W_2}{t} \tag{1}$$

Where,

W_1: is the initial weight of w/p (*gram*);
W_2: is the final weight of w/p (*grams);*
t: is the entire time for machining (*mins)*.

Result and Discussion

In this work, the outcomes of the experiments are studied using the S/N ratio and ANOVA. Based on the results of the S/N ratio and ANOVA, optimal settings of the machining parameters for MRR is obtained and verified. The machine was set for a cut of 2mm each during the experiment.

Table 43.2 Process variables and levels

S. No.	*Parameters (Units)*	*Description*	*Levels*		
			1	*2*	*3*
1	Gap Voltage (V)	A	50	60	70
2	Pulse Current (A)	B	9	12	15
3	Pulse Off Time (μs)	C	15	45	90
4	Pulse On Time (μs)	D	90	120	150
5	Flushing Pressure (kg/cm^2)	E	0.25	0.50	0.75

Material Removal Rate (MRR)

The experiments have been conducted according to the Orthogonal Array (OA) selected for the investigation i.e. L27. Each trial is repeated twice for better results. The results of the MRR are shown in the Table 43.3. Where trail number gives the experiment number and R1, R2 gives the repetitions of the same trail.

Here, R1 and R2 means repetitions of the same trail

The objective function in this work is maximisation of MRR, the ratio of signal-to-noise (S/N ratio) defined according to the Taguchi method, "higher is better" (Equation 2) is used.

$$S/N_{HB} = -10\log\left(\frac{1}{r}\sum_{i=1}^{r}\frac{1}{y_i^2}\right) \tag{2}$$

Where,

y = observation response data

r = total number of observation

y_i = i[th] response

However, the average value and main effect of S/N ratio for each level of the parameters are summarised and called the average and main effects of S/N data table for MRR is listed in Table 43.4.

A. *Influence of distinct parameters on MRR:*

"The greater the level of S/N in the Taguchi technique, the better the whole performance, hence the factor levels with the maximum S/N ratio value should always be chosen. As a result, to construct the data, the

Table 43.3 Experimental results with S/N ratio (MRR)

Trial No.	*Material removal rate (gms/min)*		*Average* $(R_1+R_2)/2$	*S/N Ratio (db)*
	R_1	R_2		
1	0.096332	0.089573	0.092953	-20.6522
2	0.093401	0.098151	0.095776	-20.3830
3	0.104381	0.094862	0.099622	-20.0628
4	0.136642	0.139545	0.138094	-17.1979
5	0.14282	0.139362	0.141091	-17.0120
6	0.138358	0.137851	0.138105	-17.1959
7	0.16504	0.154107	0.159574	-15.9560
8	0.143423	0.148116	0.14577	-16.7300
9	0.16143	0.1588	0.160115	-15.9122
10	0.067985	0.065232	0.066609	-23.5348
11	0.067371	0.063649	0.06551	-23.6844
12	0.067738	0.06508	0.066409	-23.5606
13	0.144766	0.143475	0.144121	-16.8255
14	0.154746	0.147008	0.150877	-16.4359
15	0.159634	0.148591	0.154113	-16.2601
16	0.182435	0.168617	0.175526	-15.1332
17	0.171526	0.181828	0.176677	-15.0673
18	0.190116	0.158541	0.174329	-15.2796
19	0.072205	0.076722	0.074464	-22.5729
20	0.066727	0.072134	0.069431	-23.1888
21	0.074778	0.069073	0.071926	-22.8829
22	0.096694	0.092736	0.094715	-20.4773
23	0.094017	0.098605	0.096311	-20.3335
24	0.10353	0.092645	0.098088	-20.2076
25	0.189565	0.170918	0.180242	-14.9176
26	0.17185	0.173827	0.172839	-15.2475
27	0.186588	0.173423	0.180006	-14.91170

Table 43.4 Average value and main effects (S/N data: MRR)

Process Parameters Designation	*Process Parameter*	*Average Values of Material Removal Rate*		
		Level 1	*Level 2*	*Level 3*
A	Voltage	–17.900	–18.420	–19.415
B	Current	–22.280	–17.994	–15.462
C	T_{on}	–18.622	–18.584	–18.529
D	T_{off}	–20.044	–18.392	–17.299
E	Flushing	–18.675	–18.575	–18.475

average for each experimental level was determined by utilising the highest S/N ratio at the level for each variable, and S/N ratio graphs were made on this basis.

1) *Influence of voltage*
   Figure 43.4 represents the influence of voltage on MRR

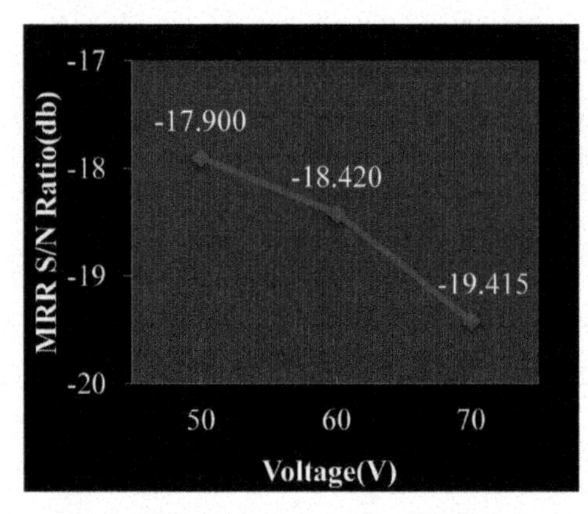

Figure 43.4 Influence of voltage on MRR

This graph reveals that with the enhance in the gap voltage, the MRR is decreasing from level 1 to level 3 rapidly as shown in Figure 4. This might be due to a widening of the space between the workpiece and the tool electrode. By way of this distance enhanced the electrons does not strike the w/p with its complete force.

2) *Influence of current*
   Figure 43.5 indicates the influence of current on MRR

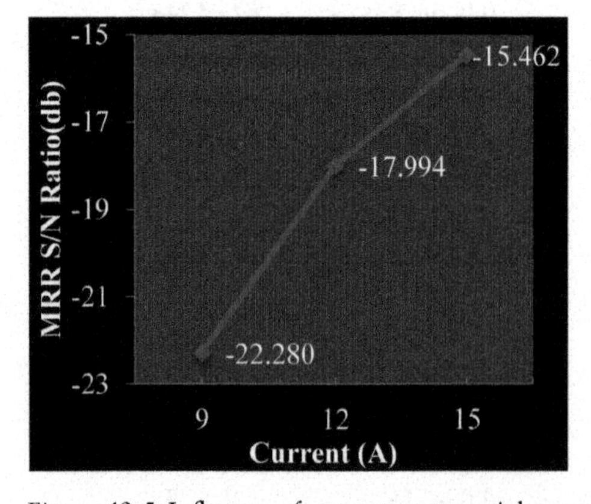

Figure 43.5 Influence of current on material removal rate

The Figure 43.5 clearly displays that increasing the current value raises the MRR. Maximum MRR is obtained at level 3 i.e. at current value 15A. At the higher current value the discharge intensity increases which results in greater MRR. The S/N ratio plot also depicts that the MRR is supplementary for 15A with level 3.

3) *Influence of pulse-off-time*

Figure 43.6(a) and (b) signifies the influence of T_{off} on MRR

The MRR rises with the increase in T_{off} and maximum MRR is obtained at level 3 as presented in Figure 43.6(a). This may be because of enhancing in the non-machining time with the rise in the T_{off}. It represents that more pulse off time is needed for increases MRR because more pulse-off-time improves the debris removal b/w the work-piece and tool for better flow of current and hence improves machining.

Figure 43.6(b) depicts that within the rise in the T_{on} the MRR also enhances and its maximum value takes place at level 3. The MRR is reliant on the energy input, which is directly proportional to the T_{on} *i.e.* longer pulse duration.

4) *Influence of flushing pressure*

Figure 43.7 shows the influence of F_p on Material Removal Rate.

With changes in flushing, the MRR does not alter appreciably. Even though there is a gradual increasing trend in MRR. An optimal flushing is found to be at level 3 of the experiment as shown in Figure 43.7.

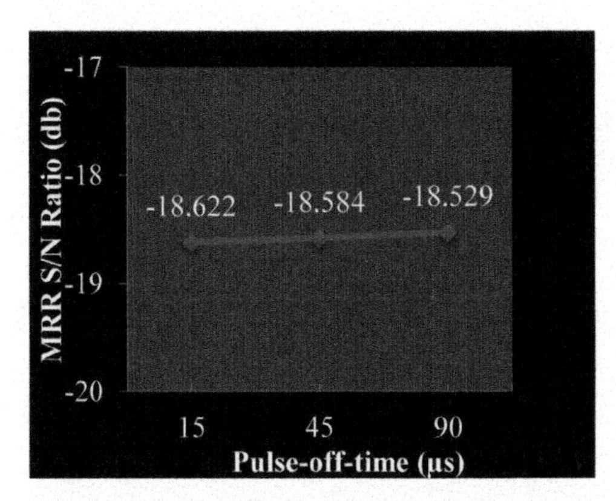

Figure 43.6(a) Influence of T_{off} on MRR

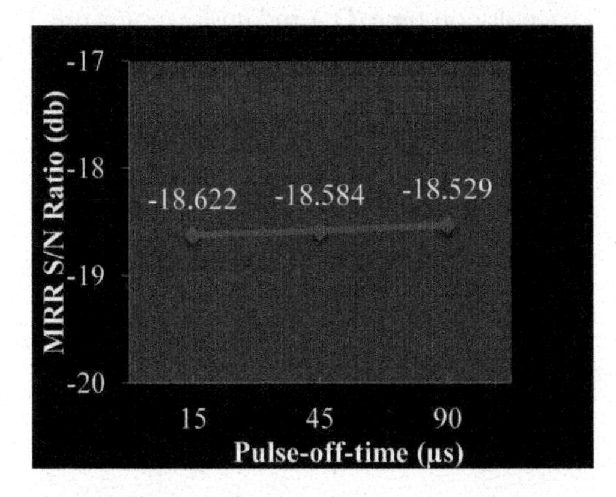

Figure 43.6(b) Influence of T_{on} on MRR

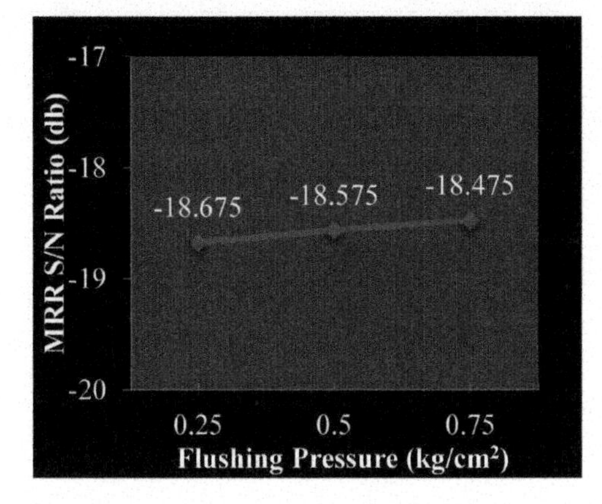

Figure 43.7 Effect of F_p on MRR

5) *Influence of most significant variables on MRR*
The influence of most significant parameters on MRR is represented in Figure 43.8.

B. Computation of Sum of Square:

In ANOVA, overall sum of squares is determined by:

$$SS_T = \sum_{i=1}^{N} (\eta_i - \bar{\eta})^2 \quad (3)$$

Here, N is the no. of orthogonal array trials, which in this case is 27, I is the S/N ratio for the ith trial, and M is the complete mean S/N ratio".

The overall sum of squared deviations (SST) is divided into 2 parts: the sum of square deviation owing to each process variable (SSP) and sum of square error (SSE).

The total of each process parameter's squared variances may be computed as:

$$SS_P = \sum_{j=1}^{t} \frac{(S\eta_j)^2}{t} - \frac{1}{m}\left[\sum_{i=1}^{m} \eta_i\right]^2 \quad (4)$$

Here, p indicates one of the experiment parameters; j: level no. of this variable p, t: the repetition of each level of the parameters p, $S\eta_j$: sum of the S/N ratio containing this variable p at level j.

C. Computation of Degree of Freedom:

The total degree of freedom can be found by:- $D_T = n - 1$, here *n:* the no. of experiments. *i.e.* $D_T = 27 - 1 = 26$

The DOF of the tested variable, $D_p = t - 1$

Where, *t* is the repetition of each level,of the variable *p*.

D_A = Degree of freedom for voltage = Number of repetitions at each level –1, so,

$D_A = 3 - 1 = 2$, similarly for current (D_B), pulse-off-time (D_C), pulse-on-time (D_D) and flushing pressure (D_E).,

$D_B = 3 - 1 = 2, D_C = 3 - 1 = 2, D_D = 3 - 1 = 2, D_E = 3 - 1 = 2$

Degree of error (De) can found by:-

$D_e = D_T - D_A - D_B - D_C - D_D - D_E$

$= 26 - 2 - 2 - 2 - 2 - 2 = 16$

D. Computation of Mean Square or Variance:

The mean square or variance of the parameter tested is $V_p = SS_p/D_p$

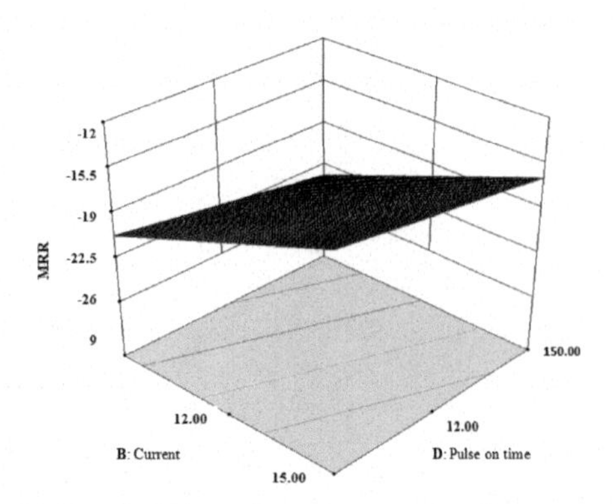

Figure 43.8 Influence of most important variables on MRR

Variance for the parameter voltage (V_A)	$= SS_A/D_A$	$= 10.6718/2$	$= 5.336$
Variance for the parameter current (V_B)	$= SS_B/D_B$	$= 213.9677/2$	$= 106.984$
Variance for the parameter pulse-off-time (V_C)	$= SS_C/D_C$	$= 0.0389/2$	$= 0.0195$
Variance for the parameter pulse-on-time (V_D)	$= SS_D/D_D$	$= 34.3630/2$	$= 17.1815$
Variance for the parameter flushing pressure (V_E)	$= SS_E/D_E$	$= 0.1824/2$	$= 0.0912$
Variance for the error(V_e)	$= SS_e/D_e$	$= 0.8080/2$	$= 0.0505$

E. Computation of F-Ratio:

"The F-value is equal to the mean of squares variation divided by mean of the squared error" i.e. ($F_p = V_p/V_e$).

F-value for the parameter voltage (F_A)	$= V_A/V_e$	$= 5.336/0.0505$	$= 105.663$
F-value for the parameter current (F_B)	$= V_B/V_e$	$= 106.984/0.0505$	$= 2118.49$
F-value for the parameter pulse-off-time (F_C)	$= V_C/V_e$	$= 0.0195/0.0505$	$= 0.3861$
F-value for the parameter pulse-on-time (F_D)	$= V_D/V_e$	$= 17.1815/0.0505$	$= 340.223$
F-value for the parameter flushing pressure (F_E)	$= V_E/V_e$	$= 0.0912/0.0505$	$= 1.8059$

F. Computation of Percentage Contribution:

%age contribution (ρ) of each of the design variable is given by $\rho_p = SS_p/SS_T$.

Percentage contribution of voltage (ρ_A)	$= SS_A/SS_T$	$= 10.6718/260.03$	$= 4.01$
Percentage contribution of current (ρ_B)	$= SS_B/SS_T$	$= 213.968/260.03$	$= 82.28$
Percentage contribution of pulse-off-time (ρ_C)	$= SS_C/SS_T$	$= 0.0389/260.03$	$= 0.015$
% contribution of pulse-on-time. (ρ_D)	$= SS_D/SS_T$	$= 34.3630/260.03$	$= 13.21$
% contribution of flushing pressure (ρ_E)	$= SS_E/SS_T$	$= 0.1824/260.03$	$= 0.070$

Table 43.5 shows the F-test results for these control variables to give an indication of the effect of each of the experimental factors on the MRR. Note that the significance level is 0.05. The result shows that the theoretical boundary value is F 0.05(2, 16) = 3.63.

It can be found that current is the most imperative variable for affecting the MRR along with voltage and pulse-on-time {F 0.05(2, 16) = 3.63} while the other parameters (T_{on}, F_p) are not having much effect on MRR. The T_{on} and F_p in the range given by Table 43.3 has an insignificant influence on MRR. The results of the ANOVA (see Table 43.3) also reveal that discharge current, which reached 82.28%, made the major contribution to overall performance. The % contribution for V, T_{on}, T_{off}, and F_p are lower at 4.01%, 13.21%, 0.015% and 0.070% respectively.

Table 43.5 Result of the ANOVA for MRR

EDM process parameters	*Degree of freedom*	*Sum of Square*	*Mean square or Variance*	*F-ratio*	*Contribution(%)*
Voltage(V)	2	10.6718	5.336	105.6634	4.01*
Current(A)	2	213.9677	106.984	2118.495	82.28*
T_{on}(μs)	2	0.0389	0.0195	0.3861	0.015
T_{off}(μs)	2	34.3630	17.1815	340.2277	13.21*
Flushing (kg/cm^2)	2	0.1824	0.0912	1.8059	0.070
Error	16	0.8080	0.0505		
Total	26	260.03			

*Indicates the significant factor

G. *Confirmation Tests:*

The last step is to forecast and verify the development of the performance characteristic by utilising the optimal level of the process variables once the optimal level of the process variables has been chosen. Using the optimal level of the process variables, the estimated S/N ratio can be calculated as:

$$\hat{\eta} = \eta_m + \sum_{i=1}^{q} (\bar{\eta}_i - \eta_m) \tag{5}$$

Where,

η_m: Entire mean of S/N ratio,

$\bar{\eta}_i$: mean S/N ratio (at the optimal level)

q: no. of the process variables that influence the performance representative.

The expected S/N ratio utilising the optimal machining variables for MRR can be attained and the corresponding MRR can also be considered by employing below equation

$$\hat{\eta} = \eta_m + (\bar{\eta}_1 - \eta_m) + (\bar{\eta}_2 - \eta_m) + (\bar{\eta}_3 - \eta_m) + (\bar{\eta}_4 - \eta_m) + (\bar{\eta}_5 - \eta_m)$$

$$= -18.5786 + (-17.900 + 18.5786) + (-15.462 + 18.5786)$$

$$+ (-18.529 + 18.5786) + (-17.299 + 18.5786) + (-18.475 + 18.5786)$$

$$= -13.3506$$

Put $\eta = -13.3506$ in equation "higher is better" will get the MRR value corresponding to estimated S/N ratio utilising the optimal machining parameters ($\hat{\eta}$):

$$-13.3506 = -10 log(1/y^2)$$

$$1.33506 = log(1/y^2)$$

$$1.33506 = (1/y^2)$$

$$\frac{1}{y^2} = 21.630$$

$$y^2 = 0.0462$$

$$y = 0.2149$$

It will be forecast for MRR value for optimal machining variables, now take the initial cutting parameters as A2B2C2D2E2 *i.e.* V = 60V, I= 12A, T_{OFF} = 45μs, T_{ON} = 120μs and F_P = 0.50kg/cm^2. The optimal cutting variables are A1B3C3D3E3 *i.e.* V = 50V, I = 15A, T_{OFF} = 90μs, T_{ON} = 150 μs and F_P = 0.75kg/cm^2 respectively and conduct two trials as per above values of parameters keeping the previous trial conditions remain same.

Equivalent to initial cutting factors as A2B2C2D2E3 and optimal setting variable as A1B3C3D3E3, measured MRR values (*gms/min*) are 0.151 and 0.192 respectively. The S/N ratio ratios can be measured to corresponding measured MRR values by using Equation (5.1) as follows:

For **A2B2C2D2E2**

S/N ratio = −10 log (1/0.151^2) = −16.420

Similarly for, **A1B3C3D3E3**

S/N ratio = −10 log (1/0.192^2) = −14.334

From, the initial cutting setting to the ideal cutting variables, the S/N ratio is increased. = [−14.334 + 16.420] = 2.036

The findings of the confirmation trial employing MRR's optimal machining settings are given in Table 43.6. There is a high level of agreement between the projected and actual machining performance.

Table 43.6 Result of the confirmation trial

	Initial cutting Parameters	*Optimal cutting variables*	
		Prediction	*Experiment*
Level	A2B2D2	A1B3D3	A1B3D3
MRR (*gms/min*)	0.151	0.2149	0.192
S/N ratio (db)	–16.420	–13.354	–14.334
Enhancement of S/N ratio		2.086	

Table 43.4 shows the growth in S/N ratio as an individual performance feature. The MRR is raised by 1.15 times based upon the confirmation test, and the experimental findings corroborate the prior variable design for optimal machining variables with diverse performance variables in the EDM.

H. Mathematical Model:

The final response equation for MRR is as follows after removing the non-significant terms:
(In coded terms)

$$MRR = -18.58 + 0.68 \times A_1 + 0.16 \times A_2 - 3.70 \times B_1 + 0.58 \times B_2 + 1.28 \times D_1 + 0.19 \times D_2 \quad (6)$$

Where, Parameter A depicts Voltage and subscripts 1, 2 and 3 are levels
Similarly parameters B depicts Current and subscripts 1, 2 and 3 are levels
Parameters D depicts Pulse-on-time and subscripts 1, 2 and 3 are levels
A_1 = 50Volts A_2 = 60Volts, B_1 = 9Amperes, B_2 = 12Amperes, D_1 = 90μs and D_2 = 120μs.

Conclusion

- The experimental results revealed that the discharge current is the main variable along with T_{on} and V among 5 controllable factor (I, V, $T_{on,}$ T_{off} and F_p) that influence the MRR in electric discharge machining of DIN 1.2714 steel.
- The percentage contributions of I, V, $T_{on,}$ T_{off} and F_p are 82.28%, 4.01%, 13.21%, 0.015% and 0.070% respectively.
- The optimal parametric conditions are: I = 15A, V = 50V, T_{on} = 150μs, T_{off} = 90μs and F_p = 0.75 kg/cm^2 to obtain maximum MRR for specific test range.

References

Abbas, N. M., Solomon, D. G., and Fuad Bahari M. (2007). A review on current research trends in electrical discharge machining (EDM). Int. J. Mach. Tools Manuf. 47:1214–1228.

Chiang, K.-T. (2008). Modeling and analysis of the effects of machining parameters on the performance characteristics in the EDM process of Al2O3+TiC mixed ceramic. Int. J. Adv. Manuf. Technol. 37:523–533.

Chigal, G., Saini, G., and Singh, D. (2013). A study on machining of al 6061/Sic (10%) composite by electro chemical discharge machining (ECDM) process. Int. J. Eng. Res. Technol. (IJERT) 2(1):1–12.

Habib, S. S. (2009). Study of the parameters in electrical discharge machining throughresponse surface methodology approach. Appl. Math. Model. 33:4397–4407.

Hascalyk, A. and Caydas, U. (2007). Electrical discharge machining of titanium alloy (Ti–6Al–4V). Appl. Surf. Sci. 253:9007–9016.

Kanagarajan, D., Karthikeyan, R., Palanikumar, K., and Paulo Davim, J. (2008). Optimization of electrical discharge machining characteristics of WC/Co composites using non-dominated sorting genetic algorithm (NSGA-II). Int. J. Adv. Manuf. Technol. 36:1124–1132.

Kansal, H. K., Singh S., and Kumar, P. (2005). Parametric optimization of powder mixed electrical discharge machining by response surface methodology. J. Mater. Proc. Technol. 169:427–436.

Keskin, Y., Halkacy, H. S., and Kizil M. (2006). An experimental study for determination of the effects of machining parameters on surface roughness in electrical discharge machining (EDM). Int. J. Adv. Manuf. Technol. 28:1118–1121.

Kiyak, M. and Cakir, O. (2007). Examination of machining parameters on surface roughness in EDM of tool steel. J. Mater. Process. Technol. 191:141–144.

Kumar, A., Kumar, R., Kumar, S., and Verma, P. (2021). A review on machining performance of AISI 304 steel. Mater. Today Proc. 56:9281–9286. https://doi.org/10.1016/j.matpr.2021.11.003.

Kumar, B. V. M., Ramkumar, J., Basu, B., and Kang, S. (2007). Electro-discharge machining performance of TiCN-based cermets. Int. J. Refract. Met. Hard Mater. 25:293–299.

Kumar, S., Kumar, M. and Handa, A., (2018). Combating hot corrosion of boiler tubes- a study. J. Eng. Fail. Anal. 94:379–395. https://doi.org/10.1016/j.engfailanal.2018.08.004.

Kumar, S., Kumar, M., and Jindal, N. (2020). Overview of cold spray coatings applications and comparisons: A critical review. World J. Eng. 17(1):27–51. DOI: 10.1108/WJE-01-2019-0021.

Kumar, S., Mahajan, A. Singh, H., and Kumar, S. (2021). Friction stir welding: types, merits & demerits, applications, process variables & effect of tool pin profile. Mater. Today Proc. 56:3051–3057 https://doi.org/10.1016/j.matpr.2021.12.097.

Kumar, S., Singh, R., Singh, T. P., and Sethi, B. L. (2009). Surface modification by electrical discharge machining. J. Mater. Process. Technol. 209:3675–3687.

Kurita, T. and Hattori, M. (2006). A study of EDM and ECM/ECM-lapping complex machining technology. Int. J. Mach. Tools Manuf. 46:1804–1810.

Leao, F. N. and Pashby, I. R. (2004). A review on the use of environmentally-friendly dielectric fluids in electrical discharge machining. J. Mater. Proc. Technol. 149:341–346.

Lee, S. H. and Li, X. P. (2001). Study of the effect of machining parameters on the machining characteristics in electrical discharge machine of tungsten carbide. J Mach. Technol. 115:344–358.

Liao, Y. S., Huang, J. T., and Chen, Y. H. (2004). A study to achieve a fine surface finish in Wire-EDM. J. Mater. Proc. Technol. 149:165–171.

Mahajan, A., Singh, H., Kumar, S., and Kumar, S. (2021). Mechanical properties assessment of TIG welded SS 304 joints. Mater. Today Proc. 56:3073–3077, 1–5. https://doi.org/10.1016/j.matpr.2021.12.133.

Pandey, P. C., and Shah, H.S. (1981). Modern machining processes. New Delhi: Tata McGraw Hill.

Pradhan, B. B., Masanta, M., and Sarkar, B. R. (2009). Investigation of electro-discharge micro-machining of titanium super alloy. Int. J. Adv. Manuf. Technol. 41:1094–1106.

Pradhan, M. K. and Biswas, C. K. (2009). Modeling and analysis of process parameters on surface roughness in EDM of AISI D2 tool steel by RSM approach. World Acad. Sci. Eng. Technol. 57:1–10.

Ramasawmy, H. and Blunt, L. (2002). 3D surface characterisation of electropolished EDMed surface and quantitative assessment of process variables using taguchi methodology. Int. J. Mach. Tools Manuf. 42:1129–1133.

Rohith, R., Shreyas, B. K. S., Kartikgeyan, S., Sachin, B. A. and Umesha, K. R. (2019). Selection of non-traditional machining process. Int. J. Eng. Res. Technol. (IJERT) 8(11):1–5.

Singh, S., Kumar, H., Kumar, S., and Chaitanya, S. (2021). Systematic review on recent advancements in abrasive flow machining (AFM). Mater. Today Proc. 56:1–9. https://doi.org/10.1016/j.matpr.2021.12.273.

Singh, S., Kumar, S., Singh, S., Kumar, R., and Sidhu, H. S. (2020). advance technologies in fine finishing and polishing processes: A study. J. Xidian Univ. 14(4):1387–1399. https://doi.org/10.37896/jxu14.4/161.

Tzeng, Y.-F. (2008). Development of a flexible high-speed EDM technology with geometrical transform optimization. J. Mater. Process. Technol. 203:355–364.

Wang, C.-C. and Lin, Y. C. (2009). Feasibility study of electrical discharge machining for W/Cu composite. Int. J. Refract. Met. Hard Mater. 27:872–882.

Zarepour, H., Tehrani, A. F., Karimi, D., and Amini, S. (2007). Statistical analysis on electrode wear in EDM of tool steel DIN 1.2714 used in forging dies. J. Mater. Proc. Technol. 187–188:711–714.

44 Stacked classifier for network intrusion detection system

D. P. Gaikwad[1,a], *D. Y. Dhande*[2,b], *and A. J. Kadam*[1,c]

[1]Department of Computer Engineering, AISSMS College of Engineering, Pune, India

[2]Department of Mechanical Engineering, AISSMS College of Engineering, Pune, India

Abstract

Intrusion Detection System is useful to monitor and analyses computer system in network and produces alert about malicious activity. However, existing intrusion detection systems do not offer acceptable accuracy and offers high false positive rate. To overcome these problems, many researchers have proposed hybrid approach of intrusion detection system. Recently, an ensemble method such as Bagging, Boosting and AdaBoost methods of machine learning is being widely used to reduce false positive rate with high accuracy.

In this paper, a novel stacked classifier has proposed for network intrusion detection system. Appropriate selection of base classifiers and Meta classifier is very important aspect in stacking based ensemble. BayesNet, PART rule learner and J48 Decision tree have used as base classifier. These three different base classifiers have stacked using Logistic regression Meta classifier. Novelty of the proposed intrusion detection system is that new stacking method with appropriate base classifiers have used to implement detection system. Base learners and Meta classifier have trained and tested using NSL-KDD datasets. Experimental results exhibit that the proposed stacked classifier beats its base learners and existing intrusion detection systems on test, training datasets and cross validation. Proposed stacked classifier also offer better false positive, precision and recall values than its base learners and existing intrusion detection systems.

Keywords: Base classifier, ensemble, J48, naïve bayes, PART, stacked classifier.

Introduction

Now-a-days, worldwide usages of computer networks have increased rapidly. Cyber adversaries try to exploit delicate points of network to abolish important information of organisations (DeWeese, 2009; Eom et al., 2012; Vatis, 2001). Therefore, network intrusion detection systems are widely proposed by researchers to spot and identify invaders in computer network. These intrusion detection systems are broadly divided into anomaly and misused based detection techniques (DeWeese, 2009; Eom et al., 2012; Vatis, 2001; Lee et al., 1999; Moustafa and Slay, 2014; Valdes and Anderson, 2019). Signature based detection technique is used to detect known attacks. It offers advanced accuracy and inferior false positive rates than the Anomaly detection technique. Anomaly based detection technique identify unknown attacks (Ghosh et al., 1998). Anomaly detection technique is being widely used to detect and identify new attacks in networks. Many researchers have proposed anomaly based detection technique to develop intrusion detection systems (IDS). Presently, machine learning (ML) and data mining methods become a famous tool for IDS. Deep learning technique of ML also is taking lead in developing IDS. Individual classifier of ML is not capable to offer higher classification accuracy and lower false positives. Therefore, researchers are offering ensemble classifier for IDS. Ensemble is a method of combining individual learners that offers improved classification accuracy with less value of false positive. Homogenous, heterogeneous, lazy and eager classifiers can be used to develop ensemble classifiers. In ensemble method, selection of individual classifiers is very important exercise (Moustafa and Slay, 2016). Specifically, Bagging, AdaBoost and Stacked methods of ensemble are being used for intrusion detection system. They offer very admirable accuracies with very low false positive rates. For getting precise great accuracy and small false positive rate, researcher are involved in researching suitable base learners. Stacking of similar classifiers sometimes will not give good classification due to their similar performances. The over these research gap heterogeneous classifiers have been used in this research work.

In this paper, a new stacked ensemble classifier has suggested for intrusion detection system. Three appropriate heterogeneous base classifiers have chosen for stacked. One decision tree, one rule learner and Bayes Net classifiers have selected for stacked. Logistic regression based Meta Classifier has proposed for stacked base learners. All Base learners and stacked classifier have trained and tested using NSL-KDD dataset. Rest

[a]dp.g@rediffmail.com; [b]dp.g@rediffmail.com; [c]dp.g@rediffmail.com

part of the paper is planned as below. In section 2, some standard existing intrusion detection systems have discussed. Section 3 is dedicated to discuss the proposed methodology for intrusion detection system. In section 4, the proposed system architecture of intrusion detection system has presented. Section 5 discusses experimental results and analysis. Lastly, the paper has concluded in section 6.

Literature Survey

Ensemble is a method of ML in which base classifiers are used to build new classifier to reduce false positives rates. Researchers have used different homogenous and heterogeneous base classifiers to build ensemble classifiers. In this section, some latest research in intrusion detection system have presented in short. Mohamed and Ejbali (2021) have developed intrusion detection system using Reinforcement learning algorithms. AE-deep SARSA algorithm has implemented to propose new intrusion detection system that provides a high prediction performance for U2R and R2L type attacks. This system offers 81.24% classification accuracy. Shone et al. (2018) have proposed a novel feasible intrusion detection technique using deep learning method. Auto encoder has used to learn unsupervised features. Non-symmetric deep auto encoders have stacked to propose a novel model. Model trained using KDDCup-99 and NSLKDD datasets using GPU. They have obtained very promising results over existing proposals. Abbas et al. (2012) has proposed an ensemble-based intrusion detection model. They have combined naïve Bayes, logistic regression and Decision tree heterogeneous classifiers. These base learner s and ensemble classifier have trained and tested using CICIDS-2017 dataset. This model detects all kinds of attacks with major accuracy with small computational power. This system offers maximum 88.96% accuracy. Nagpal et al. (2021) have proposed IDS using optimised Support vector Machine. They have used Big Bang Big Crunch optimisation to optimise parameters of Support vector machine. They have used KDD-99 dataset set to train and test model. Training and testing dataset has pre-processed to reduce dimension by fusing ranked features to decrees false positive rate and training time. Mogg et al. (2021) have proposed Decision Tree based IDS. Genetic Algorithm is utilised to obtain optimal features of attack. Model is trained using NSL-KDD dataset. Teardrop and Nmap attacks are classified as benign. Krishna et al. (2021) has proposed two-stage classification for intrusion detection. This proposal works in two phases. In first phase, CISCIDS-2017 dataset have pre-processed to clean and normalise dataset. In second phase, two-stage classification is used to detect attacks. In first stage, Fast k Nearest Neighbor classifier is used to decide whether request is legitimate or attack. In second stage, multi classification used to decide type of attack. Least Variance Feature Elimination method is used to reduce size of dataset.

Raymond Mogg et al. (2021) have suggested an intrusion detection tool which is based on Decision Tree. They have used NSLKDD dataset for training Decision trees. In this proposal, Genetic algorithm select relevant feature and explore the feasibility of producing interpretable dodging assaults against IDSes. Investigational results presented attacks that like to a given seed attacks are classified as kind attack. Mebawondu et al. (2020) have proposed Artificial Neural Network (ANN) based IDS. Authors have used UNSW-NB15 dataset for training ANN. Continuous attributes have discretised in binary before training ANN. They have used Gain ratio for ranking attributes. The offered model shows positive correlation value 0.57 and gives an accuracy of 76.96%. Jabbar et al. (2021) has suggested an ensemble classifier for IDS using Random Forest and Average One-Dependence Estimator. They have detected that Random forest improved accuracy with less error rate. The offered ensemble classifier offers 90.52 % accuracy with False Positive rate 0.14. Pooja and Shrinivasacharya (2019) has developed an automated technique for network based IDS. They have used UNSW-NB15 complex and KDDCUP-99 datasets for training system. LSTM deep learning model has trained. LSTM offers higher accuracy both datasets. They have suggested Convolution Neural network for intrusion detection in network. Xu et al. (2021) have proposed novel interpretable IDS. In this proposal, Additive Decision Tree is used as classifier that offers better predictive performance and interpretability. Model was trained and tested on UNSW-NB15 dataset. Aleesa et al. (2021) has proposed collaborative intrusion detection system using deep learning using RNN, ANN and DNN. Improved UNSW-NB15 dataset have used for training purpose. The performances of Binary and Multi-class models have measured with UNSW-NB15 dataset. Meftah et al. (2019) proposed two stage network intrusion detection. Random Forest and Recursive Feature Elimination have used to select relevant. They have done binary classification using Support vector machine (SVM), Logistic Regression and Gradient Boost Machine. The output of this first stage classification feed to multinomial classifiers such as C4.5, Naïve Bayes and multinomial SVM. This two-stage classifier offers 86.04% accuracy. Kumar and Raaza (2018) have proposed Linear Regression and Random Forest for intrusion detection system. UNSW-NB15 dataset is utilised for training and testing classifiers. They have used WEKA tool to find relevant features. Beloucha et al. (2018) have evaluated different classifiers for intrusion detection system Apache Spark. SVM, Decision Tree and Naïve Bayes and Random Forest classifiers have trained and evaluated using UNSW-NB15 dataset. Authors have found that Random Forest offer highest 97.49% accuracy and 93.53% sensitivity.

Zhao et al. (2009) have proposed intrusion detection system using fusion theory. Different detection measures have fused to drop false positive and false negative rates. System consists of Data fusion-based and abstracting process for intrusion detection model. Fused method offers higher detection rate. Zareapoor and Seeja (2015) have proposed feature selection methods for classification of emails. Authors have used PCA and LSA for feature extraction for emails. Authors have found that Latent Semantic Analysis is the greatest method which beats other methods. Srikanth and Shashi (2019) have proposed two new metrics for differentiating the profiles in recommender systems. It clusters the shills to distinguish from usual users. It is reliable to separate the violence profiles inserted through any of the attack methods. This system identifies the shilling profiles. Akhter et al. (2019) have proposed ML based method to spot cyber harassment activities in social media. Multinomial Naïve Bayes have used as a classifier to categorise the type of harassment. They have designed fuzzy rule sets to specify the power of dissimilar kinds of bullying. Chahal and Kaur (2016) have planned a hybrid system for network IDS. In first phase, K-mean algorithm is used to cluster the data. Adaptive-SVM is used to categorise packets. NSL-KDD dataset have used to train model. This hybrid model offers 98.47% accuracy on training dataset.

Methodology

In this section, methodology of intrusion detection system have discussed in detail. NSL-KDD dataset have utilised for training and testing individual base learners and the proposed stacked classifier. NSL-KDD dataset is publically available which is widely adopted by researchers for IDS. This dataset has desired samples of the whole KDD dataset. This dataset is developed form of KDDCup99 Dataset in which redundant samples have removed to prevent biased result (Meftah et al., 2019). NSL-KDD dataset includes 42 features with a class label attribute. The available NSL-KDD dataset have pre-processed to refine training dataset and testing datasets. This refined training dataset comprises of 67,343 normal samples and 58,638 anomaly samples. The dimension of training dataset is 125,981 samples.

Combination of strong base learner through Meta learning is called as stacking. In stacking method, each base learner is trained on whole training dataset. The selected Meta classifier or Super Learner is trained on outcomes of all base learners. Outputs of base learners are considered as features of Meta classifier. Stacked is used to improve predication accuracy by reducing generalisation error. In stacked or voting regressor, we can use same or different types of base learners. Meta Classifier is trained to combine predications of its all base learners. Stacked absorbs the best combination of the individual results. Each base learner works independently with different hypothesis. This method of ensemble is used when we do not get proper prediction of base learners with different regression results. The final result of Meta classifier or Super learner is an average of predication results of base learners. Basically, there are two methods of stacked one is averaging and second is weighted averaging of base learner's predications.

Architecture of Proposed Stacked Classifier

In this section, the system architecture of the proposed intrusion detection has presented. The selection of base learners and Meta classifier or Super classifier is important work in stacking method. Three strong heterogeneous base learners have selected based on their advance features and classification accuracies. PART, J48 and BayesNet classifiers have selected as a base learners for training. These base classifiers are also called as 'zero Level' classifiers. Logistic regression classifier have selected as a Meta Classifier or 'One Level' classifier. Algorithm 1 describes steps of stacking of base learners. According to Algorithm 1, initially three base learners are trained using same training NSL_KDD dataset. The predication values are collected to form feature matrix. This feature matrix is combined with original response vector to form 'level one' dataset. This 'one level' dataset is feed to Meta Classifier for training. Based on predications of Base learners, Meta classifier decides final predication values on test dataset. In this method, same training dataset is used to train all base classifiers; C_1, C_2 and C_3. These base classifiers produces predictcations; P_1, P_2 and P_3 respectively. Selection of best classifiers is main contribution of this research work.

Algorithm 1: Stacking of Strong base learner s.
Input: L-Base learners, P- Base learner s Predications.
Step1: Selection of base and Meta Classifier.

- Select base learner s list L = {Bayes Net, J48 and PART}
- Select a Meta learner or Super Learner

Step 2: Train the ensemble

- Train base learners on same training dataset.
- Do k-fold cross validation of each Base learner
- Assemble cross validated predicted values p1, ..., pL
- Assemble N predicted values of L learner and form feature matrix of N × L size
- Create 'Level One' dataset n {[p1] ... [pL][y] // Feature matrix of N × L and along with original response vector y
- Train the Super learner or Meta learner on the level-one dataset.

Step 3: Predict on new data

- Collect estimates from the base learners.
- Feed base predictions into the Meta learner

Output: Prediction of the ensemble classifier.

The architecture of the stacked classifier have depicted in Figure 44.1. BayesNet, PART and J48 are used as base classifiers. These base classifiers have purposefully selected form different classification categority to combine their advantages. The predictions of these base classifiers have fed to Logistic Regression clasifier for stacking base classifiers. Finally, Logistic regression classifier produces final prediction of rest sample.

Experimental Results and Analysis

Initially, base learners are trained and tested. These base learners have stacked using Logistic Regression based Meta Classiifer. The all classifiers are trained and tested on Lenovo laptop with 8 GB Ram, CPU @ 2.50 GHz and coding is done in JAVA language with WEKA tool. Performances of classifiers have evaluated in term of precision, accuracy and recall using followings equations [1-3].

$$Accuracy = \frac{TP_c + TN_c}{TP_c + TN_c + FP_c + FN_c} \tag{1}$$

$$P_c = \frac{TP_c}{TP_c + FP_c} \tag{2}$$

$$R_c = \frac{TP_c}{TP_c + FN_c} \tag{3}$$

Where, TP_c presents True positives) and TN_c is True Negative. FP_c is False positives and FN_c present False negative. The performances of classifiers have measured in terms on accuracy, false positive rate, precision, recall and F-score. The quality of any IDS is evaluated in term of accuracy on test dataset as this accuracy is measured on unknown patterns. The experimental results on test dataset have listed in Table 44.1 and Table 44.2. From Tables 44.1 and 44.2, it can be perceived that overall performance of the proposed stacked classifier is better than its all base classifiers on test dataset. The classification accuracy, recall and

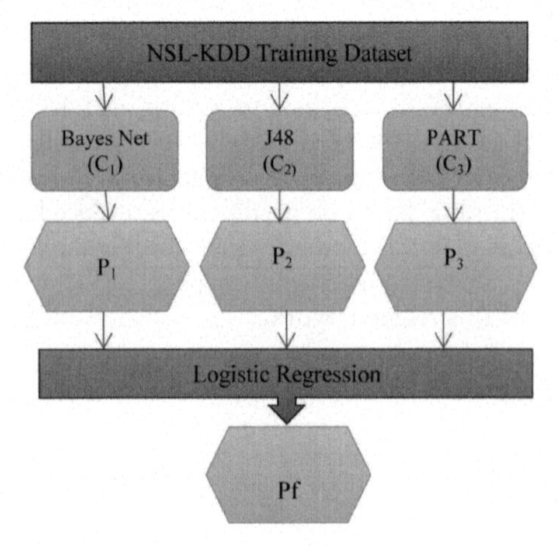

Figure 44.1 System architecture of the proposed stacked classifier

precision values are higher than its all base classifiers. The proposed stacked classier also exhibit better accuracy than existing system proposed in Mebawondu et al. (2020) on test dataset. The proposed stacked ensemble classifier takes more time to train than its base classifiers.

The experimental results on training dataset have listed in Tables 44.3 and 44.4. In following Tables 44.3 and 44.4, the performances of proposed stacked classifier have presented on training dataset. From Tables 44.3, 44.4 and Figure 44.2, it can be determined that the performance of the proposed stacked classifier is better than its two base classifiers in term of accuracy, false positive rate, and recall and precision values. It can be observed that BaysNet offers better accuracy than stacked classifier on training dataset. The proposed stacked ensemble classifier have compared with three existing intrusion detection system proposed by Jabbar et al. (2021)and Beloucha et al. (2018) and Chahal and Kaur (2016) on training dataset. It is can be concluded that proposed stacked ensemble outperforms these classifiers. But, stacked classifier takes more time for training proposed model.

The results of classifiers have presented in Tables 44.5 and 44.6 on 10-fold cross validation. From Table 44.3, it can observe that the output results of the proposed stacked classifier are marginally better than its base classifiers. The classification accuracy, false positive rate, recall and precision values are higher than its base classifiers on 10-fold cross validation.

Table 44.1 Classification accuracy and false positive rate of classifiers on test dataset

Classifier	*Test accuracy*	*False Positives*
Proposed Stacked Ensemble	82.12 %	0.142
Bayes Net	74.43%	0.200
PART	81.26 %	0.149
J48	81.53 %	0.146
Mebawondu et al. (2020)	76.96%.	NA

Table 44.2 Recall, precision and *F*-score values of classifiers on test dataset

Classifier	*Precision*	*Recall*	*F-Score*
Proposed Stacked Ensemble	0.861	0.821	0.821
Bayes Net	0.822	0.744	0.739
PART	0.856	0.813	0.812
J48	0.858	0.815	0.815

Table 44.3 Classification accuracy and false positive rate of classifiers on training dataset

Classifier	*Training Accuracy*	*False Positive*
Proposed Stacked Ensemble	99.93 %	0.001
Bayes Net	97.18 %	0.031
PART	99.93 %	0.001
J48	99.91 %	0.001
Jabbar et al. (2021)	90.52 %	NA
Beloucha et al. (2018)	97.49%	NA
Chahal and Kaur (2016)	98.47%	NA

Table 44.4 Recall, precision and *F*-score values of classifiers on training dataset

Classifier	*Precision*	*Recall*	*F-Score*
Proposed Stacked Ensemble	0.999	0.999	0.999
Bayes Net	0.973	0.972	0.972
PART	0.999	0.999	0.999
J48	0.999	0.999	0.999

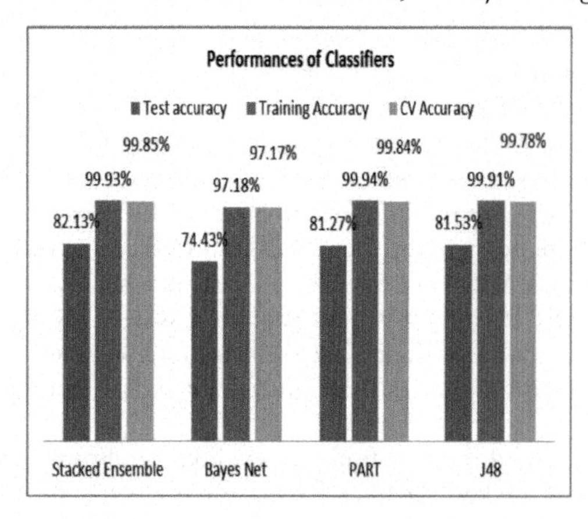

Figure 44.2 Performances of base classifiers and stacked classifier on training dataset

Table 44.5 Accuracy and false positive rate of classifiers on cross validation

Classifier	*CV Accuracy*	*False Positive*
Proposed Stacked Ensemble	99.84%	0.001
Bayes Net	97.17%	0.032
PART	99.83%	0.002
J48	99.78%	0.002

Table 44.6 Precision, recall and *F*-score values of classifiers on cross validation

Classifier	*Precision*	*Recall*	*F-Score*
Proposed Stacked Ensemble	0.999	0.998	0.998
Bayes Net	0.972	0.972	0.972
PART	0.998	0.998	0.998
J48	0.998	0.998	0.998

Conclusions with Future Scope

Network intrusion detection system is a tool that is used to capture and inspect packets in computer network. It is used to detect and identify to decide incoming packet is normal or abnormal. Specifically, this tool is used inspect internal packets. There are many methods to identify packets in network. MLis leading statistical approach that is widely used in intrusion detection system. Many researchers are being used ensemble scheme of MLto implement IDS. Bagging, AdaBoost and Stacked are three main methods of ensemble classification.

In this paper, a stacked method of ensemble has proposed for intrusion detection system. Three suitable heterogeneous base learners have selected for stacked purpose. One decision trees, one rule learner and BayesNet have stacked using Logistic Regression based Meta classifier. These base learners and stacked classifier have trained and tested using NSL-KDD dataset. Experimental results exhibit that proposed stacked classifier beats all its base learners and existing classifiers due to heterogeneous base classifiers. Overall, the proposed stacked classifier outperforms its base learners on test dataset, training dataset and cross validation. Stacked classifier also offer better accuracy, false positive rate than some existing intrusion detection system. Limitation of this work is that training and testing dataset used for building and testing model are old. In future, new primarily and secondary dataset will be used for real time intrusion detection system.

References

Abbas, A., Khan, M. A., Latif, S., Ajaz, M., Shah, A. A., and Ahmad, J. (2012). A new ensemble-based intrusion detection system for internet of things. Arab. J. Sci. Eng. https://doi.org/10.1007/s13369-021-06086-5.

Akhter, A., Acharjee, U. K., and Polash, M. A. (2019). Cyber bullying detection and classification using multinomial naïve bayes and fuzzy logic. Int. J. Math. Sci. Comput. 4:1–12. Published Online November 2019 in MECS (http://www.mecs-press.net), DOI: 10.5815/ijmsc.2019.04.01.

Aleesa, A. M., Yonis, M., Mohmmed, A., and Sahar, N. M. (2021). Deep intrusion detection system with enhanced UNSW-NB15 dataset based on deep learning techniques. J. Eng. Sci. Technol. 16(1):711–727. © School of Engineering, Taylor's University, 2021.

Beloucha, M., El Hadaj, S., and Idhammad, M. (2018). Performance evaluation of intrusion detection based on machine learning using Apache Spark. The first international conference on intelligent computing in data sciences. Procedia Comput. Sci. 127:1–6.

Chahal, J. K., and Kaur, A. (2016). A hybrid approach based on classification and clustering for intrusion detection system. Int. J. Math. Sci. Comput. 4:34–40. Published Online November 2016 in MECS (http://www.mecs-press.net) DOI: 10.5815/ijmsc.2016.04.04.

DeWeese, S. (2009). Capability of the People's Republic of China (PRC) to conduct cyber warfare and computer network exploitation. Darby, PA: DIANE Publishing.

Eom, J.-H., Kim, S.-H., and Chung, T.-M. (2012). Cyber military strategy for cyberspace superiority in cyber warfare. In 2012 International conference on cyber security, cyber warfare and digital forensic (Cyber Sec), IEEE.

Ghosh, A. K., Wanken, J., and Charron, F. (1998). Detecting anomalous and unknown intrusions against programs. In computer security applications conference, 1998. proceedings. 14th Annual. IEEE.

Jabbar, M. A., Aluvalub, R., and Reddy, S. (2021). RFAODE: A novel ensemble intrusion detection system. In Seven international conference on advances in computing & communications, August (2017), Cochin, India.

Krishna, K. V. Swathi, K., Rama Koteswara Rao, P., and Basaveswara Rao, B. (2021). TSC: A two-stage classifier for network intrusion detection system on green cloud. J. Green Eng. (JGE) 11(2):1500–1510..

Kumar, S. P. and Raaza, A. (2018). Study and analysis of intrusion detection system using random forest and linear regression. Period. Eng. Nat. Sci. 6(1):197–200. ISSN 2303-4521.

Lee, W., Stolfo, S. J. and Mok, K. W. (1999). A data mining framework for building intrusion detection models. In proceedings of the 1999 IEEE symposium on security and privacy.

Mebawondu, J. O., Alowolodu, O. D., Mebawondu, J. O., and Adetunmbi, A. O. (2020). Network intrusion detection system using supervised learning paradigm. Sci. Afr. 9:e00497.

Meftah, S., Rachidi, T., and Assem, N. (2019). network based intrusion detection using the UNSW-NB15 Dataset. International Journal of Computing and Digital Systems ISSN (2210-142X). Int. J. Com. Dig. Sys. 8(5):478–487.

Mogg, R., Simon, E., and Kim, D. S. (2021). A framework for generating evasion attacks for machine learning based network intrusion detection systems. TechRxiv, Preprint, https://doi.org/10.36227/techrxiv.15164463.v1.

Mohamed, S. and Ejbali, R. (2021). Adversarial multi-agent reinforcement learning algorithm for intrusion detection system. Int. J. Inf. Technol. Secur. 13(3):87.

Moustafa, N. and Slay, J. (2014). UNSW-NB15 dataset for network intrusion detection systems. 2015 Military Communications and Information Systems Conference (MilCIS), 10–3 12 November 2015, Canberra, ACT, Australia, 1–6. Retrieved from http://www.cybersecurity.unsw.adfa.edu.au/ADFA% 20NB15%20, Datasets, 2014.

Moustafa, N. and Slay, J. (2016). The evaluation of network anomaly detection systems. Statistical analysis of the UNSW-NB15 data set and the comparison with the KDD99 data set. Inf. Secur. J. A Global Perspect. DOI: 10.1080/19393555.2015.1125974.

Nagpal, M., Kaushal, M., and Sharma, A. (2021). A feature reduced intrusion detection system with optimised SVM using big bang big crunch optimization. Wireless Pers. Commun: Int. J. 122:1939–1965.. https://doi.org/10.1007/s11277-021-08975-2, 2012.

Pooja, T. S. and Shrinivasacharya, P. (2019). Evaluating neural networks using Bi-directional LSTM for network IDS in cyber security. Procedia Manuf. 7(3):524–539.

Shone, N., Ngoc, T. N., Phai, V. D., and Shi, Q. (2018). A deep learning approach to network intrusion detection. IEEE Trans. Emerg. Top. Comput. Intell. 2(1):41–50.

Srikanth, T. and Shashi, M. (2019). New metrics for effective detection of shilling attacks in recommender systems. Int. J. Inf. Eng. Electronic Bus. 4:33–42. published Online July 2019 in MECS (http://www.mecs-press.org).

Valdes, A. and Anderson, D. (2019). Statistical methods for computer usage anomaly detection using NIDES (Next-Generation Intrusion Detection Expert System). In Proceedings of the third international workshop on rough sets and soft computing (RSSC94), (pp. 306–311). San Jose, CA: USW, 2019.

Vatis, M. A. (2001). Cyber-attacks during the war on terrorism: A predictive analysis. DTIC Document. Hanover. NH: Institute for Security Technology Studies at Dartmouth College.

Xu, W., Fan, Y. and Li, C. (2021). I2DS: Interpretable intrusion detection system using auto-encoder and additive tree. Hindawi, Secur. Commun. Netw. 2021:9. Article ID 5564354 https://doi.org/10.1155/2021/5564354, 2021.

Zareapoor, M. and Seeja, K. R. (2015). Feature extraction or feature selection for text classification: a case study on phishing email detection. Int. J. Inf. Eng. Electronic Bus. 2:60–65. Published Online March 2015 in MECS (http://www.mecs-press.org/).

Zhao, X., Jiang, H. and Jiao, L.Y. (2009). A data-fusion-based method for intrusion detection system in networks. Int. J. Inf. Eng. Electronic Bus. 1:32–40. published Online October 2009 in MECS (http://www.mecs-press.org/).

45 Paper classification of dental images using teeth instance segmentation for dental treatment

D. P. Gaikwad[a], *V. Joshi*[b], *and B. S. Patil*[c]

Department of Computer Engineering, AISSM College of Engineering, Pune, India

Abstract

In orthodontic treatments, segmenting each tooth and classifying it into different sub classes is very important task for dentist. Many researchers are proposing different algorithms for segmentation of teeth for classification. These segmentation results are then transformed with mathematical models into 3-D meshes that are then used for preparing dental casts. Dentist can use these results for teeth restoration and other dental practices. Localising each tooth and classifying it is most challenging task. Since teeth are irregular and obscure these methods fail to scale these on diverse teeth models. In this paper, a novel technique of instance segmentation of teeth images clicked from different views, successfully localising each tooth and their respective contours have proposed. The proposed method for classification and segmentation of each tooth solves the problems related to misclassification, crowding teeth and broken or missing teeth. This model can be scaled to new datasets and performs explicably well on new data points. Experimental results exhibit that the proposed algorithm outperforms all the start-of-art methods with average precision for segmentation by 88% and classification score by 92%. The model performs efficiently for different geometric deformities and varying degrees of tooth arrangements.

Keywords: Mask, orthodontic, R-CNN, segmentation, teeth segmentation, x-rays.

Introduction

Segmentation of different organs is vital task in medical treatment, as it is the fundamental process of identifying and localising each organ and its boundaries. In this technique we train the model to localise and identify the boundaries of a given object. The output of these models is a set of data points known as 'mask' which gives us the points on the image where that object is located. Thus, we get contours for that object. There has been lot of research in segmentation in medical practice. Ronneberger et al. (2015) have trained a U-NET convolutional network in order to segment transmitted light microscopy images of neuronal structures in electronic microscopic cells. In Sirinukunwattana et al. (2016) authors proposed a new neighboring ensemble predictor (NEP) coupled with convolutional neural network to make detection and classification of nuclei in histological images of routine colon cancer. Ait Skourt et al. (2018) have proposed technique for lung CT image segmentation using the U-net for cancer detection. Structure Correcting Adversarial Network is used for segmentation of lungs and heart boundary (Dai et al., 2017). Instance segmentation is process where we localise as well as classify it by labeling it into classes. Thus, there are mainly two problems we solve through instance segmentation. Localise the object and second is we identify its class. Instance segmentation is very difficult task in medical imaging as feature extraction is very complex, in Figure 1 we can observe that each tooth has its boundaries and classified in four quadrants, where every quadrant has a set of central incisors, lateral incisor, canine the premolars and molars. Thus, through this paper we successfully localise and predict these labels for each tooth with best accuracy. Segmentation of teeth to create 3D anatomy structure of patients with variety of deformities has always been challenging. Several commercials CAD software for orthodontics predicts tooth segmentation on its own. However, they fail as the structure become abstruse and the variation starts increasing. Many different techniques process different types of images for segmentation. Anchor based object detection is used in many ways for segmentation of teeth, but all of them are on teeth X-Rays proposed (Jader et al., 2018). They use Mask R-NNN on dental X-Rays for panoramic, bitewing and periapical X-Rays. But these segmentations are mostly not useful as their teeth contours are concurrent to X-ray images, thus not predicting the exact surface boundaries for 3-D dental frame preparation. Another method Tian et al. (2019) uses orthodontic CAD systems to create 3D tooth segmentation. It uses 3-D CNN's framework to create segmentation of teeth but fails to generalise the model on different attributes of teeth and its complexities. Lee et al. (2021) authors have provided a method for segmentation of images from laser scanner by detecting features on

[a]dp.g@rediffmail.com; [b]vibhav031998@gmail.com; [c]bharatp_04@rediffmail.com

two range images computed from 3-D-image. It uses orthodontic techniques like dental arch detection and two-step curve fitting. Based on survey, this is first project that makes use of MASK R-CNN for creating 2-D mask of such images. More importantly this works on varied types of teeth frameworks not depending in any kind of external modifications. The classification and segmentation scores solve the problems related to misclassification, crowding teeth and broken or missing teeth. This model can be scaled to new datasets and performs explicably well on new data points as the hyper parameters tuned in this model balance the bias-variance tradeoff. Rest part of the paper is prepared as below. Section 2 discusses existing proposals in short. In section 3, the proposed method for teeth segmentation and classification have discussed. Section 4, experimental and analysis have discussed. Finally, in section paper is concluded.

Literature Survey

Kronfeld et al. (2010) have proposed different techniques for segmentation of teeth are proposed on different form of data. The active contour model for representing the image contours with a snake model estimates the teeth mean curvature. The contours obtained are transformed into a snake by minimizing the energy function. Wu et al. (2014) proposal fails to capture all vertices as the transition between teeth and gum is hard to predict. It uses segmentation pipeline on scanned dental meshes with specific designed energy function. To extract boundary it uses mean curvature thresh-holding which allows roughly estimated mess features to be transformed. Zhao et al. (2005) proposal solves the crowding problems between the teeth as the separation technique refines the boundaries using contour by making them smooth a robust algorithm on scanned dental mesh images is presented in this project. The contour lines are sampled and the boundaries are detected from these contour lines in the concave regions. The segmentation is a harmonic field, which is crucial in isolating the tooth from the dental model. This field is able to detect the shape variation around the cutting boundaries. Lin et al. (2016) and Long et al. (2015) proposed Semantic Segmentation is a process where every part of the image will be assigned a label Propose fully connected network for sematic segmentation where conditional random fields (CRFs) with CNN-based pairwise potential functions to label each pixel in image with a class. Luc et al. (2016) propose the generative adversarial networks which transform the z sample from a fixed (e.g. standard Gaussian) distribution pz (z), to approximate the distribution of training samples x using deterministic differentiable deep network g (•). Senthilkumaran (2012) have proposed other methods that work on dental X-Rays images use different techniques like anchor base approaches on X-Rays. The mask region proposal network (RCNN) model trained on 1500 panoramic images gives good accuracy on training, but works only in specific views and predicting contours on X-Rays is not useful. Different techniques of segmentation have evaluated on teeth dataset like cluster-based, threshold-based, region-based, boundary based and watershed-based. It formulates a segmentation method using ROI's region of interest method in (Wang et al., 2016). This paper provides a different approach in dental radiography for 400 cephalometric radiographs. It tries to solve major challenges that are classifying anatomical landmarks and segmenting tooth structures on different radiographs.

Kim et al. (2020) have proposed a method to reconstruct teeth areas using a generative adversarial network (GAN). This method fails to take into the complete model of teeth, as intraoral scan data consists of single unit and cannot be used to create a complete 3-D model (Fan et al., 2015). In this paper, the correlation clustering segmentation technique have proposed which transforms the segmentation problem in terms of graph labeling addressing weights to salient and non-salient edges. Based on the study of segmentation, a programmed segmentation and classification method for teeth models from 2-D images clicked from phone camera have presented. .Major studies fail to provide segmentation approach on simple RGB images. There are lots of complexities for this project such as variations in patient to patient teeth framework, various restorations done on the teeth. Existing methods are not efficient for segmentation and classification of teeth images such as shown in Figures 45.1 and 45.2. To overcome these limitations, a unique architecture for segmentation of RGB images of patients has proposed.

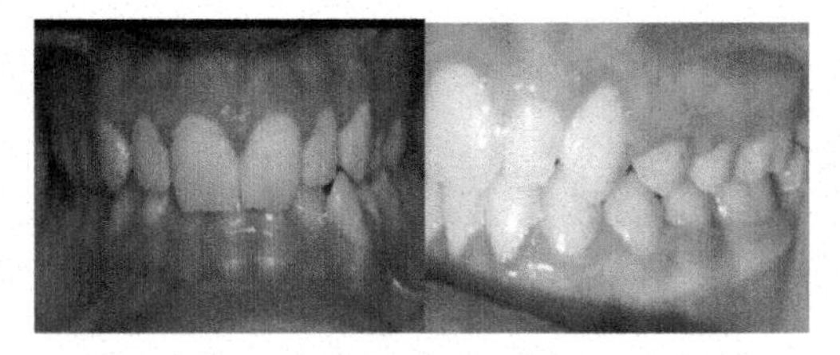

Figure 45.1 Images of teeth from different angles

Proposed Region and Anchor Based Segmentation of Teeth

Usually, in neural network layers incrementally are added to increase accuracy. But the problem of fading gradients started in deeper networks began to converge, and accuracy began to decrease. This was not caused by over fitting but by the layers being stacked on top of each other. To overcome this problem, He et al. (2016), Girshick (2016) have proposed the deep residual learning structure. There was a jump connection or a shortcut linking used to learn uniqueness mapping.

In this paper, ResNet have used as a base architecture of classifier with RPN to produce Regions of Interest. In Figure 45.3, the proposed architecture for teeth segmentation and classification has shown.

The region suggestion network is used in RCNN and Faster RCNN to forecast the probability of existence of the object location. In this work, convolutional neutral network have used as a region suggestion networks to predicts anchors that mark along with the slip of the fold neural networks. The threshold value is defined for scores and Region of Interest of 2000 anchors height for further processing of the mask. These anchors act as a class bounding box repressors. Non-appropriate and appropriate collaborative training is used to train RCNN and RPN. The Region of Interest pooling layer is used for feature mapping from the RPN network. In faster RCNN, a work-study approach of is used to train the Network. In first step, the RPN is formed individually. In second step, separate discovery network is formed which is based on the suggestions made in step 2. The third step only refines the RPN without affecting another network. In the fourth step, Convolutional neural network and RCNN are trained. View anchors in RPN are shaped for keeping objects of different sizes and aspect ratios. Twenty thousand anchors of each image are transmitted for Object bounding box repressors. It is found that there is no exciting suppression with the threshold. The bounding Box with the highest score is retained according to the number of points. Finally, limit box and points for them have obtained. Faster RCNN works great with MS COCO and PASCAL VOC recordings for the object acknowledgement. Using RCNN is much faster and is used for factual purposes.

The new architecture has used in this work for mask prediction with Bbox, additional parameter of Mask and class ranking value. Sharing technique with Convolutional Neural network and suggesting regions is used to reduce a lot of compute time and increase the performance. The feature pyramid network was a further enhancement of the faster RCNN architecture. Instance segmentation is the succeeding goal after semantic segmentation. In instance segmentation, each class example must be notable and a binary mask must be specified for each instance (Lin et al., 2017). This job is difficult because there may be multiple instances of the same class name in near proximity. For preserving space, the evidence about the example and the prediction mask have obtained. The class evaluation and the bounding Box are obtained from mask RCNN.

For object recognition, faster RCNN have used for recognition of object using Region of Interest. In instance segmentation, each mask is linked to the clusters of classes. Each example of each class has a dissimilar class, so the aim was to locate each autonomous object and class associated with it. The RCNN mask is the same as the faster RCNN except that the segmentation mask is enabled with each Region of Interest with a bounding box and class trust value. In Mask RCNN, RoIAlign have chosen instead of RoIPool to restore pixel-to-pixel alliance and preserve spatial positions. The proposed RCNN mask outperforms all other models on COCO dataset trained on GPU with 200 ms. The output of Mask RCNN is a bit

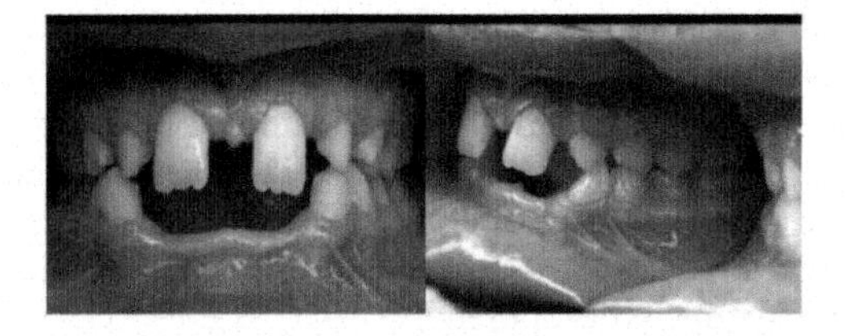

Figure 45.2 Teeth with different deformities

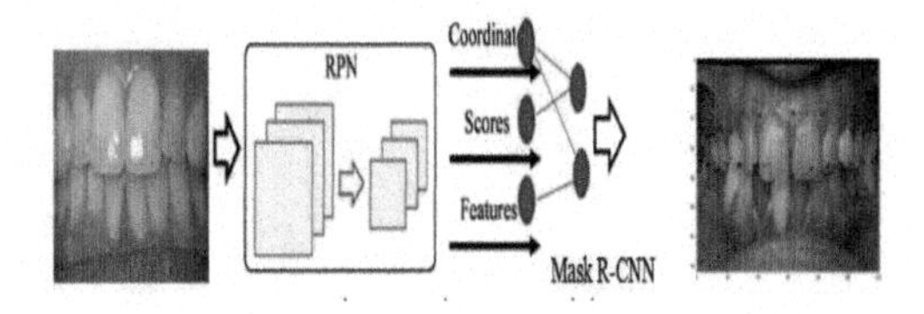

Figure 45.3 Proposed mask RCNN on teeth dataset

mask for each Region of Interest. Pixel by pixel, the mask abstracts the spatial structure. RoIPool is used to abstract functionality map that is quantised and RoI is changed into a discrete granularity of the feature map. In the RCNN mask, RoIAlign have utilised to decrease this quantification and correctly line up feature maps. This can be compared to RoIWrap that performs bilinear resampling. Derivatives of ResNet 50 are used as the support in mask RCNN design. Additional Mask RCNN is the FPN Feature Pyramid Network which uses top to bottom approach with adjacent connections. Therefore, the use of ResNet and FPN provides decent accuracy and fast GPU processing.

Experimentation Results and Analysis

A. Data Pre-Processing

First the teeth images were labeled in LabelBox online software with mask and its classes. Each teeth boundary was drawn and the images were converted to labeled images. First iteration of 1500 RGB images was labeled into following classes.

'premolar', 'molar', 'lateral_incisor', 'central_incisor', 'canine'

Others were labeled only as teeth and its boundary. This dataset was converted to the COCO data set format. The following parameters need to be passed to the model in form of COCO datasets.

- **"bbox": bbox,**
- **"bbox_mode": BoxMode.XYWH_ABS,**
- **"segmentation": segm,**
- **"category_id": classes.index(anno['value']),**
- **"category_id": m[anno['value']],**
- **"iscrowd": 0**

The mask contours from labeled dataset is converted in (x, y) points format which is list of these contour points of these submasks. Dataset Catalog and Metadata Catalog are arranged which holds this COCO data and is delivered to function used by our model. Then scan every pixel and only take the Boolean positive value of mask as it has the mask of that label of class. Afterward receiving these contours, height and width have found which are required in COCO format. The segmentation property of COCO has obtained by converting these contours to the polygon data structure. These polygons are changed to segmentation in which every x and y coordinate of the mask is involved. The bounding box has obtained by parsing these contours, then the bounding box mode is defined which determine the type of format.

B. Training the MASK RCNN-50 FPN Model

The data is trained in different Mask R-CNN back bone architectures like ResNet 101 and ResNet 50 with FPN, DC5 and C4 .Each data pipeline for model consists of 1500 images labeled in different ways for iteration. For iteration, the dataset is trained in new dataset with different labels. The hypermeters for the mask

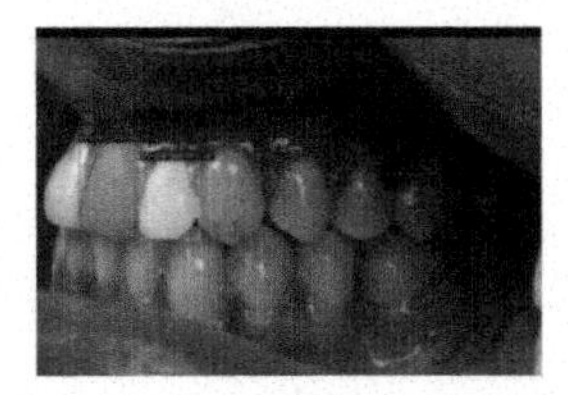

A. The mask for central incisor and lateral incisor

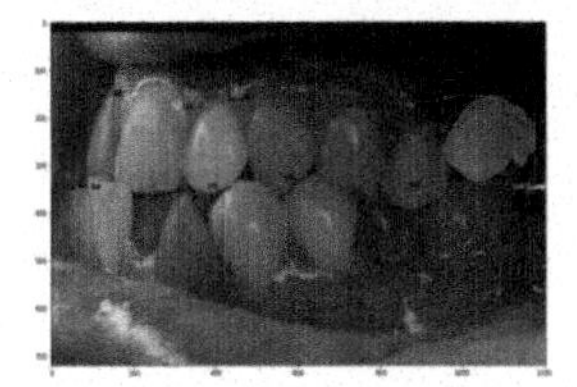

B. Every tooth labeled with mask

Figure 45.4 Shows the mask and classes labeled in LabelBox

RCNN model a tuned accordingly by balancing the variance and bias. As shown in Table 45.1 we have average precision which demonstrates segmentation score and class score which shows the classification score. The best Average Precision is for teeth label as this signifies a generic boundary for each tooth. This shows that the proposed model identifies each contour and successfully localises each tooth. Other labels like Central Incisor, Lateral Incisor, Canine, and Premolar, Molar show exceptional results for both Average Precision and class score. According to Table 45.1, it can be concluded that the proposed model overcomes the challenges for variation in teeth structure and other deformities.

As demonstrated in Figure 45.3 the teeth segmentation gives near perfect boundaries for teeth labels from all angles. All the separation between teeth and gums is clearly distinguished with high-definition contours predicted for each tooth. In Figure 45.4, the mask for central incisor and lateral incisor and every tooth labeled with mask have shown. In Figure 45.5, result of segmentation for model trained on all classes have shown. In Figure 45.8, final prediction on images with deformities have shown. Figure 45.6 demonstrates results for central incisor and lateral incisor. Each sub class of teeth is predicted from every angle the image is taken. The model is trained with a lot of variation and calibrating the respective hyper parameters which gives the highest accuracy. The Figure 45.7 shows the smoothing of accuracy vs. epochs graph for mask RCNN model. The accuracy stables after 400 epochs and remains constant for every range of epochs representing very low over fitting. Thus on any new image taken at any angle the accuracy remains same.

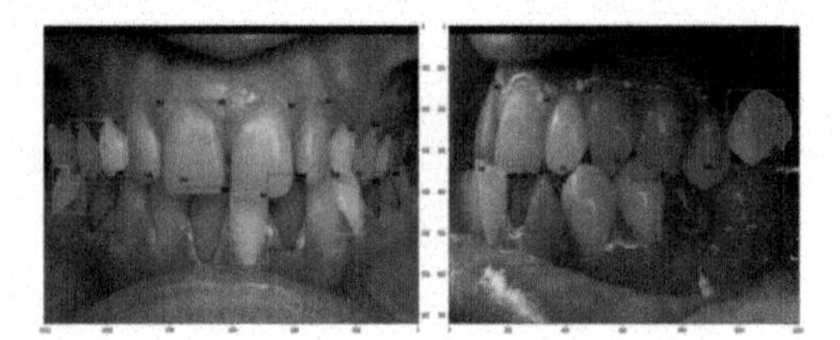

Figure 45.5 Result of segmentation for model trained on all classes

Table 45.1 Average precision for segmentation and class score for classification for respective labels

Label	*Average precision (AP)*	*Class score*
Teeth	95	98
Central incisor	90	92
Lateral incisor	91	92
Canine	95	96
Premolar	89	90
Molar	91	82

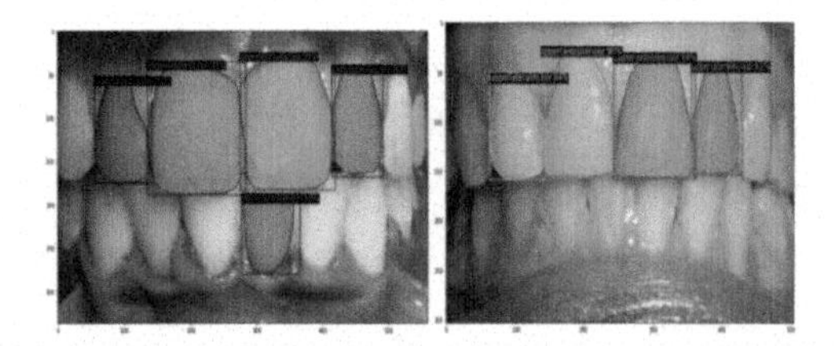

Figure 45.6 Result of segmentation for model trained only on class's upper central incisor and lateral incisor

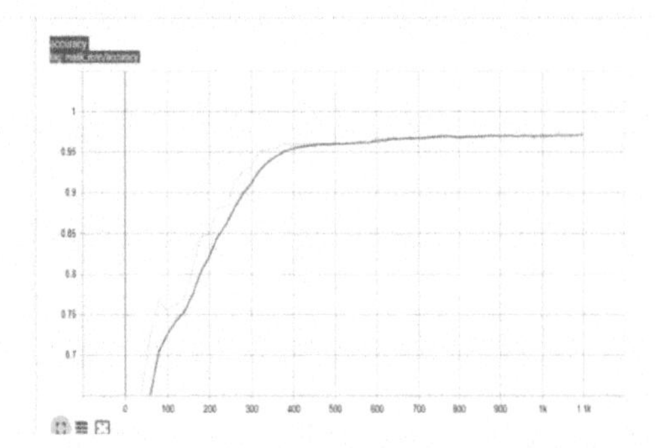

Figure 45.7 Accuracy for Mask R-CNN per epoch

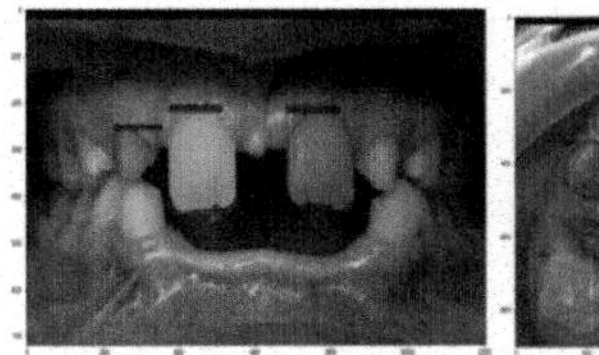
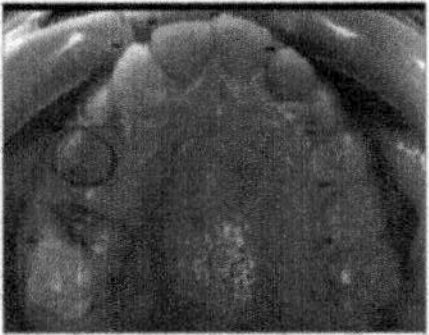

Figure 45.8 Prediction on images with deformities

As Shown in Figure 45.7 the proposed model performs explicably well on all deformities like missing teeth or patient-to-patient variation. Every tooth structure is predicted with good accuracy for all types' teeth. This model is able to handle all distortion for teeth without compromising on accuracy.

Conclusion

Teeth segmentation is very challenging for RGB based dental images, which have been pursuit for many researchers for decades. Many different ways are proposed on different kind of datasets but none of them pass in generalising the model to all test cases. For dental practice the teeth segmentation is very crucial part for many different applications. The proposed model with Mask R-CNN region-based architecture gives exception results on all edge cases like deformities and variations. We are able to transform the labeled data to COCO format which is then fed to the Mask RCNN model. The model acquires almost 90 % accuracy on segmentation with ResNet 50 FPN as best backbone architecture. The output from the model is used to derive the contours for the teeth which then can be used for teeth restoration and for preparations of dental casts. Currently, the model can improve in identifying the quadrant of the teeth. In the future with more images in all angles of moth, we can distinguish every tooth in each quadrant of the teeth. Considering all challenges, our model solves these obscure tasks with highly trained model. This proposed method will accelerate segmentation in orthodontic medical imaging.

References

Ait Skourt, B., El Hassani, A., and Majda A. (2018). Lung CT image segmentation using deep neural networks. Procedia Comput. Sci. 127:109–113.

Dai, W., Doyle, J., Liang, X., Zhang, H., Dong, N., Li, and Xing, E. P. (2017). SCAN: Structure correcting adversarial network for organ segmentation in chest x-rays. ArXiv, Corpus ID: 5767685.

Fan, R., Wang, C. C. L., and Jin, X. (2015). Multiregion segmentation based on compact shape prior. IEEE Trans. Autom. Sci. Eng. 12(3):1047–1058.

Girshick, R. (2016). Fast R-CNN: Fast region-based convolutional networks for object detection. In IEEE international conference on computer vision, (pp. 1440–1448).

He, K., Ren, S., Zhang, S., and Sun, J. (2016). Deep residual learning for image recognition. ArXiv: 1512.03385. Histology Images. IEEE Trans. Med. Imaging 1196–1206.

Jader, G., Fontineli, J., Ruiz, M., and Lima, K. A. B. (2018). Deep instance segmentation of teeth in panoramic x-ray images. In 31st SIBGRAPI Conference on Graphics, Patterns and Images, 10.1109/SIBGRAPI.2018.00058.

Kim, T., Cho, Y., Kim, D., Chang, M., and Kim, Y. J. (2020). Tooth segmentation of 3D scan data using generative adversarial networks. Appl. Sci. 10:490. https://doi.org/10.3390/app10020490.

Kronfeld, T., Brunner, D., and Brunnett, G. (2010). Snake-based segmentation of teeth from virtual dental casts. Comput. Aided Des. Appl. 7(2):221–233.

Lee, J., Chung, M., Lee, M., and Shina, Y.-G. (2021). Tooth instance segmentation from cone-beam CT images through point-based detection and gaussian disentanglement. Republic of Korea: Seoul National University. arXiv: 2102.01315v1.

Lin, G., Shen, C., Van den Hengel, A., and Reid, I. (2016). Efficient piecewise training of deep structured models for semantic segmentation. In Proceedings of the IEEE conference on computer vision and pattern recognition, (pp. 3194–3203).

Lin; T. -Y, Dollár, P., Girshick, R., He, K., Hariharan, B., and Belongie, S. (2017). Feature pyramid networks for object detection. In IEEE conference on computer vision and pattern recognition (CVPR), (pp. 936–944), Honolulu, HI, 2017. doi: 10.1109/CVPR.2017.106.

Long, J., Shelhamer, E., and Darrell, T. (2015). Fully convolutional networks for semantic segmentation. In Proceedings of the IEEE conference on computer vision and pattern recognition, (pp. 3431–3440).

Luc, P., Couprie, C., Chintala, S. and Verbeek, J. (2016). Semantic segmentation using adversarial networks. arXiv preprint arXiv: 1611.08408.

Ronneberger, O., Fischer, P. and Brox, T. (2015). U-Net: Convolutional networks for biomedical image segmentation. In international conference MICCAI 2015: medical image computing and computer-assisted intervention (MICCAI), (pp. 1–8).

Senthilkumaran, N. (2012). Genetic algorithm approach to edge detection for dental x-ray image segmentation. Int. J. Adv. Res. Comput. Sci. Electron. Eng. 1(7):5236–5238.

Sirinukunwattana, K., Raza, S. E. A, Tsang, Y. W., Snead, D. R. J., Cree, I. A., and Rajpoot, N. M (2016). Locality sensitive deep learning for detection and classification of nuclei in routine colon cancer histology images. IEEE Trans. Med. Imaging 35:(5):1196–1206.

Tian, S., Dai, N., Zhang, B., Yuan, F., Yu, Q., and Cheng, X. (2019). Automatic classification and segmentation of teeth on 3D dental model using hierarchical deep learning networks. IEEE Access 7:84817–84828. 10.1109/ACCESS.2019.2924262.

Wang, C. W., Huang, C. T., Lee, J. H., Li, C. H., Chang, S. W., Siao, M. J., Lai, T. M., Ibragimov, B., and Lindner, C. (2016). A benchmark for comparison of dental radiography analysis algorithms. Med. Image Anal. 31:63–76.

Wu, K., Chen, L., Li, J., and Zhou, Y. (2014). Tooth segmentation on dental meshes using morphologic skeleton. Comput. Graph. 38:199–211.

Zhao, M., Ma, L., Tan, W. and Nie, D. (2005). Interactive tooth segmentation of dental models. Engineering in medicine and biology society. In 27th annual international conference of the. IEEE, (pp. 654–657).

46 Recent Practices and Development in the Wire EDM process

Dipak P. Kharat[a], *M. P. Nawathe*[b], *and C. R. Patil*[c]

Research Scholar, Department of Mechanical Engineering, Prof. Ram Meghe Institute of Technology and Research, Badnera, Amravati, India

Abstract

For cutting conductive metals of any hardness or for cutting metals that are difficult or impossible to cut using existing methods, the wire electric discharge machining (WEDM) technique has become increasingly popular in recent years. It is possible to use these machines to cut elaborate designs or delicate geometries that would be impossible to achieve with typical cutting methods. There has been an explosion in the manufacturing capacity of machine tools over the last decade, but data on their use and development are still few. One of the key objectives of this research is to consolidate the work of several scientists in this field over the last several decades. It would be helpful to new researchers in assessing the scope of research and the availability of materials for research based on wire EDM. In search of the ideal operating circumstances, many scholars and engineers have been working for many years now. In order to construct an empirical response model, we used RSM and a central composite rotatable design of second order. WEDM process parameters are expected to evaluate the cutting rate, surface roughness, gap current and dimensional deviation throughout the machining process. Cutting rate, surface roughness, gap current and dimensional change can all be modelled using a response surface methodology. Plans for the rotatable central composite half-fractional second order For example, pulse on/off times, spark gap set voltages, and peak currents are varied to examine how these effects the response.

Keywords: Cutting rate, deviations, optimisation, Response surface technique, WEDM.

Introduction

Wire electric discharge (WEDM) and other non-traditional material removal methods are frequently used to make components with complex forms and profiles (Manivannan and Kumar, 2016). There are too many variables and stochasticity for optimal performance to be achieved. Our goal is to discover the relationship between process metrics and their controllable input parameters, and then to identify which input parameters should be combined best in order to achieve the highest potential process performance (Manivannan and Kumar, 2016; Meena et al., 2017; Chandramouli and Eswaraiah, 2017). A mathematical model based on the response surface modelling (RSM) technique is used to correlate various WEDM parameters and performance measures (Marichamy et al., 2017).

WEDM is an innovative modification on traditional EDM technique in which an electrode is employed to trigger the sparking process immediately after start-up (Khullar et al., 2017). A continuously travelling wire electrode composed of thin copper, brass, or tungsten wire with a diameter of 0.05-0.30 mm is used in WEDM, whereas a thin copper or brass wire electrode is used in WEDM. WEDM can produce extremely small corner radii when used in conjunction with thin copper, brass, or tungsten wire electrode. Because the wire is held in tension by way of a mechanical tensioning device, there is a lower chance of incorrect parts being manufactured (Kumar and Mondal, 2021). In the course of the WEDM process, the material erodes ahead of the wire and there is no direct contact between the work piece and the wire, resulting in the elimination of mechanical stresses that would otherwise develop throughout the machining procedure.

Revision of the Literature

Throughout this article, the most recent studies and developments in the WEDM process will be reviewed in depth. A study of the literature has been carried out in order to address the approaches and research in the field of environmental decision making (WEDM). The research includes a thorough examination of relevant literature, which is made available in its entirety in the report. It also discusses the numerous tactics that have been identified in the literature to date. A wide variety of steels and alloys are currently being investigated as substrate materials, according to current research. A number of approaches and

[a]kharatdipak18@gmail.com; [b]mpnawathe1968@gmail.com; [c]crpatil333@rediffmail.com

methodologies, as well as contemporary techniques, standards, and process parameters, were discovered with the purpose of increasing their cutting performance. Among these discoveries were: It was discovered that a number of research have attempted to maximise the most efficient process parameters of WEDM technologies, and that these approaches had been applied in several of those studies. WEDM (wire electrical discharge machining) technology, which was initially employed more than 30 years ago, has made significant strides forward in that period. After developing the optical-line follower system in 1974, D.H. Dulebohn was the first to use it to autonomously govern the geometry of the components that would be fabricated using the WEDM process in 1975. After a better understanding of the technology and its possibilities was gained by individuals working in the business, its popularity skyrocketed by 1975. Table 46.1 contains an overview of some of the most important research studies currently being conducted in the area; this is a full description of research in a variety of categories. The information in the supplied Table 46.1 is divided into various categories, including process features and methodology, work material, electrode material, future opportunities for the research, and the research's results.

Inferences from Literature

Despite the fact that numerous academics have examined the relationship between various input parameters and various output parameters, these efforts should be expanded to include more output parameters as well as more input parameters. With the help of the second order response surface modelling technique, a mathematical model relating these output parameters to seven important variable input parameters, namely, Pulse on time (Ton), Pulse off time Toff, Peak voltage (Pp), Water pressure (Pw), Wire feed (f), Machine speed (Vm), and Cutting speed override (O), has been developed. It is also shown in the literature that mathematical programming approaches such as the method of feasible direction, the Taguchi method, and others had been utilised to tackle the optimisation problem in the wire electric discharge machining process. It is therefore possible to conduct fresh research in the subject after reviewing extensive literature and considering their disadvantages. When using the W-EDM process, a high level of surface roughness is generated, which ultimately leads in a large heat affected zone thickness, which results in a sluggish cutting speed (poor productivity) and the possibility of wire breakage (due to high temperature and high wire tension). Following points summarises the scope and inferences from the literature analysis.

- Only a few research have been conducted on the impact of process parameters on machining characteristics and the heat affected zone.
- Only a small amount of research has been published on the subject of changes in mechanical characteristics and surface integrity of WEDM processed materials.
- There have only been a few research on multi response optimisation using response surface methods.
- Investigations into the influence of process parameters on machining quality characteristics, particularly with regard to premature wire failure, are an extremely significant feature of the WEDM.
- Improper parametric setting has been linked to a variety of wear and failure processes, many of which require further investigation and investigation.
- Due to the significant surface roughness generated by the W-EDM process, which leads in a large heat affected zone thickness, sluggish cutting speeds (low productivity) and wire rupture may occur (due to high temperature and high wire tension).
- Due to the lack of manual or manufacturing rules, cutting parameters are not accessible for the material.
- Very few mathematical equations exist between the above performance measurements and controllable parameters. As a result, the investigation of the effects of process parameters on machining characteristics is required to be studied in order to decrease wire breakage and optimise the process.

Future Scope

On the basis of the literature review, several significant future scopes for research are derived and provided in the following sections.

- According to an earlier researcher, only a limited number of studies have been conducted on the impact of process parameters on the machining characteristics and the heat affected zone; thus, further study is needed to improve the process.
- For obtaining good performance metrics and the adjustable parameters, there are no precise mathematical formulae to be found, that can be explored.
- There are no handbook recommendations or manufacturing guidelines available for defining the parameters to cut the material as a result of this situation.

Table 46.1 Recent research and development in the WEDM process

References	*Process characteristics and methodology*	*Work material*	*Electrode material*	*Future opportunity*	*Outcomes of research*
Manivannan and Kumar, 2016	• Feed rate, current, pulse on time, gap voltage. Overcut, taper angle and circularity are all factors. • ANOVA using the Taguchi method.	SS 304	Brass	• High-end microscopic techniques can be used to examine cutting rate, surface roughness, gap current, and dimensional deviation. The effects on mechanical characteristics	• The feed rate and the current are the most effective process parameters.
Marichamy et al., 2016	• Voltage, current, flushing pressure, and pulse on time were taken into account utilising RSM. • Box-Behnken design to analyse reactions such as tool wear rate and material removal rate for checking machinability of -brass material.	Brass	Copper (Cu)	• Aspect ratio, tool penetration rate, hole over cut. • Wear analysis by SEM and EDS • Microstructural analysis • RSM with multifactor and ANOVA • Nonlinear behavior can be studied.	• It was crucial to consider the amount of material that may be removed. Flushing pressure has little effect on tool life because existing tool wear rates are rising. • TWR is found to have optical wear when the MRR is high.
Meena et al., 2017	• Pulse and frequency on the dot. • Overcutting and metal removal rate. • The grey relational analysis of Taguchi.	Titanium	Tungsten carbide (0.4mm) diameter.	• Mechanical qualities can be improved by the use of process parameters. • Use scanning electron microscopy (SEM) to improve accuracy - Aspect ratio, tool penetration rate, tool wear rate and hole overcut.	• Current is the most influential parameter. • Grey relational analysis effectively utilised for optimisation of parameter.
Chandramouli and Eswaraiah, 2017	• Current peak, duration of pulse on, duration of pulse off, and duration of tool lift. Material removal rate and surface roughness are two factors that influence the removal rate. • Taguchi experimental design, ANOVA, MINITAB 17 software. Minitab 17	Precipitation hardening stainless steel	Copper tungsten	• Material characterisation in depth • Multi response optimisation with another technique and Thermal properties from the point of view cutting speed.	• Material removal rate and surface roughness are strongly influenced by pulse on time and discharge current.
Marichamy et al., 2017	• Flushing pressure is measured in psi (pounds per square inch). • Roughness of the surface (Ra). • RSM, Box-Behnken Design, and ANOVA are all examples of RSM.	Duplex brass	Copper	• Thermal properties and effects can be explored. • Multi response optimisation. • Tool wear with SEM	• For duplex brass, EDM machining parameters have a substantial impact on the surface roughness.

(continues)

Table 46.1 Continued

References	*Process characteristics and methodology*	*Work material*	*Electrode material*	*Future opportunity*	*Outcomes of research*
Khullar et al., 2017	• RSM, ANOVA and Genetic algorithms used to analyse pulse on time, pulse off time, current, flushing mode and material removal rate.	AISI 5160 Steel.	Copper 30mm diameter.	• Material characterisation in depth • Mechanical properties not studied. • Thermal properties • -SEM, EDS and XRD analysis	• The machining qualities as they relate to various flushing techniques. On and off times have the biggest influence over material removal rate.
Kumar et al., 2018	• Proportion of titanium diboride as an input process parameter current, timing, and flushing pressure are all parameters that can be controlled with pulse current. • Output parameters include: material removal rate and tool wear rate.	Monel 400™	Copper-titanium diboride alloy	• Need to focused on Multi response optimisation • Microstructural variations due to cutting can be seen. • SEM and EDS analysis for tool wear	• TiB2, flushing pressure, and pulse current all have an impact on material removal rate, but pulse current is by far the most important factor. • Due of the hardness of the TiB2 percent particle, the wear rate of the tool decreases.
Mahanta et al., 2018	• Input parameters voltage, current, pulse time, pulse off time • Output parameters: rate of material removal, power consumption, and roughness of the surface. • A genetic algorithm known as RSM	Al7075 alloy	Copper (12mm diameter)	• Thermal properties and effects can be explored. • Multi response optimisation. • Tool wear with SEM	• It was revealed that the most important component was the current and pulse rate at the time of the experiment. • It has been found that a genetic algorithm has the lowest error rate. • Fabrication of Al/Al alloys.
Ragavendran et al., 2018	• Current, pulse on time, pulse off time. • Material removal rate, surface roughness. • Box-Behnken design RSM, ANOVA, Genetic algorithm, particle swarm optimisation, sensitivity analysis.	Special steel WP7V	Copper Diameter 13.88mm	• Mechanical properties not studied. • Material characterisation in depth • Multi response optimisation with another technique • Thermal properties from the point of view cutting speed.	• Current has most significant parameters. • The sensitivity of pulse off time on surface roughness unaffected by any change in the input parameter levels.
Faisal and Kumar, 2018	• Pulse current, pulse on time, pulse off time, gap voltage. • Material removal rate, Roughness Parameters. • Evolutionary optimisation technique, particle swarm optimisation, biogeography based optimisation technique.	EN-31	Copper	• Optimisation of process parameters for mechanical properties. • Application of scanning electron microscopy SEM for better accuracy • Aspect ratio, tool penetration rate, tool wear rate and hole over cut.	• Biogeography based optimisation technique has better optimised values were obtained like greater material removal rate and less surface roughness.

Sapkal and Jagtap, 2018	• Pulse on time, discharge voltage, capacitance and electrode rotation speed. • Material removal rate, slide gap width, taper ratio-RSM, CCD, ANOVA,	Ti-6Al-4V alloy	Copper tungsten 0.5mm diameter	• Multi response optimisation with another technique for better mechanical properties. • Material characterisation • Wear analysis and SEM	• Due to electrode rotation speed easily remove debris particles during micro EDM drilling process. • Capacitance are the significant parameters.
Magabe et al., 2019	• Input Parameters: Spark gap voltage, pulse on time, pulse off time and wire feed. • Response parameters: Material removal rate, surface roughness. • ANOVA, Empirical modeling, non-dominated sorting algorithm-II. Regression analysis.	$Ni_{55.8}Ti$	Zinc-coated brass wire with 0.25mm diameter.	• Need to focus on Multi response optimisation. • Optimisation of process parameters for mechanical properties • Application of scanning electron microscopy (SEM) for better accuracy	• Higher material removal rate at high voltage, pulse on time, wire feed rate. • Accurately predict the values of MRR, surface roughness with developed empirical model.
Yadav et al., 2019	• Input Parameters: Tool rotation speed, current, pulse on time, liquid flow rate, and gas pressure. • Output parameters: Material removal rate, surface roughness, hole over cut.-RSM	HSS	Copper	• Multi response optimisation with another technique for better mechanical properties. • Material characterisation • Wear analysis and SEM	• Tools with a high rotational speed have a substantial impact on the results of MRR, SR, and hole overcutting. As far as surface roughness and material removal rate were concerned, peak current was the most important factor.
Paul et al., 2019	• Input Process parameters: Pulse on time, pulse off time, pulse current. • Response parameters-Material removal rate, surface roughness. Hybrid methods used for optimisation.	Inconel 800 (Iron-nickel-Chromium alloy)	Oxygen free high conductivity copper	• Other mechanical properties not studied. • Thermal properties from the point of view cutting speed not investigate • Wear analysis by SEM and EDS	• Optimizing the output responses two methods employed hybrid MOORA-PCA • MOORA shows the improvement of MOORA PCA over MOORA method.
Palanisamy et al., 2020	• Discharge current, pulse on time, and pulse off time are the three input parameters. • Material removal rate, reduced surface roughness, and tool wear rate are some of the response metrics. • **L27 orthogonal array and Grey analysis.**	LM6	Copper (12mm diameter)	• Aspect ratio, tool penetration rate, tool wear rate and hole over cut. • Thermal properties from the point of view cutting speed not investigated.	• Discharge current have higher contribution followed by pulse on time and pulse off time.

(*continues*)

Table 46.1 Continued

References	*Process characteristics and methodology*	*Work material*	*Electrode material*	*Future opportunity*	*Outcomes of research*
Gugulothu, 2020	• Input Parameters: Discharge current, pulse on time, pulse off time, dielectric fluid. • Output parameters: Material removal rate, surface roughness. • Taguchi, ANOVA, MINITAB 17 software	Ti-6Al-4V	Copper	• Multi response optimisation • Optimisation of process parameters for mechanical properties • Application of scanning electron microscopy (SEM) for better accuracy	• An experiment was conducted to test the performance of three different types of dielectric fluids. • There is an increase in the MRR when using deionised water and drinking water, whereas there is a decrease when using both.
Mohamed and Lenin, 2020	• Input process parameter: Pulse on time (T-ON), pulse off time (T-off), current (I). • Response Parameters: Surface roughness. • For optimisation Taguchi Technique	Aluminum Alloy (AA6082-T6)	Molybdenum	• Mechanical properties not studied. • Aspect ratio, tool penetration rate, tool wear rate and hole over cut.	• Parameters which affect surface roughness were pulse on time, pulse off time and last current.
Singh et al., 2020	• Input process parameter: Current, pulse on/ off durations and lift settings. • Output parameters: Aspect ratio, tool penetration rate, tool wear rate and hole over cut.-RSM	Ti-6Al-4V alloy	Pure tungsten	• Wear analysis by SEM and EDS • Microstructural analysis • Other Mechanical properties.	• Micro EDM, transition zone, high aspect ratio EDM, and the final critical zone are all examined while analysing the nonlinear dynamic behaviour of the MRR.
Kumar and Mondal, 2021	• Type of tool, rotational speed of the tool, discharge current, pulse on and pulse off periods. • Rate of material removal, rate of tool wear, and depth. Minitab software, TOPSIS, GRA, and ANOVA in Taguchi L27 orthogonal array	Al-2050	Copper, Tungsten, Copper-tungsten alloy.	• Thermal properties from the point of view cutting speed • Wear analysis by SEM and EDS • Microstructural analysis • RSM with ANOVA • Nonlinear behavior can be studied	• When compared to other materials like tungsten or copper-tungsten, copper tools remove the most material. • Compared to a stationary tool, the tool wear rate is reduced while using an EDM rotary tool.
Adyanadka and Madhavrao, 2021	• Pulse on and pulse off times, as well as current, are all factors in determining the speed of the motor. • Material removal rate and surface roughness are two factors that influence the removal rate. • Matlab, Design Expert 12, Minitab, CCD, RSM, Grey-fuzzy technique.	Al5083 with 0.15 wt% addition of Zr.	Molybdenum wire of diameter 0.18mm	• Wear analysis by SEM and EDS • Microstructural analysis Aspect ratio, tool penetration rate, tool wear rate and hole over cut. • RSM with ANOVA • Nonlinear behavior can be studied	• Motor speed, pulse on time, material removal rate, and surface roughness were the three most important factors in determining the performance of the machine. • Improved prediction accuracy by applying RSM and grey fuzzy approach to reduce errors.

- As a result, the challenge is considered to be an investigation of the influence of process parameters on machining characteristics with the goal of minimizing wire breakage while also optimizing the process.
- Only few research has been done to determine if the mechanical characteristics and surface integrity of WEDM worked material have changed.
- There have only been a few research on multi response optimisation using response surface methods.
- Identifying electrode materials while considering their thermal qualities from the standpoint of cutting speed has only been attempted in a limited number of cases.
- Hence it is imperative to explore the effect of process factors by constructing a mathematical model to predict performance parameters using response surface methods, and to compare the results with the findings.
- The focus should be given to reduce the amount of wire breakage while simultaneously increasing the pace of material removal as well as to optimise the process by adjusting the operational process parameters to the optimal level of performance.

Conclusions

In this paper, the recent studies and development in the WEDM process are discussed. In this study, new studies and developments in the WEDM process, as well as their implications were discussed. The research includes a thorough examination of relevant literature, which is made available in its entirety in the report. It also discusses the numerous tactics that have been identified in the literature to date. A study of the literature has been carried out in order to address the approaches and research in the field of environmental decision making (WEDM). There is now ongoing research into a wide range of steels and alloys as substrate materials, with the goal of enhancing their overall performance. Numerous approaches and procedures, as well as contemporary techniques, standards, and process elements, were uncovered during the investigation. Other than achieving the objectives specified in the presentation, the RSM may be used to accomplish other goals as well.

A significant process concern is improved WEDM performance over a range of cutting parameters, including pulse on time, peak current and wire tension, among other things. Process performance measures such as surface roughness, mean residual roughness, and mean residual roughness are examples of what you can measure. It is possible to use RSM to analyse the influence of process parameters and match the mathematical model of the process; all that is necessary for the planning phase is a complete understanding of the process's physical characteristics. Methods such as fitting a regression model to a process and developing a response plot, as well as optimisation strategies that are based on knowledge of the response curve, are all covered in the literature on process optimisation. Therefore, when RSM optimisation is involved, a thorough understanding of statistics is essential in order to properly evaluate the findings.

References

Adyanadka, V. and Madhavrao, R. (2021). Multi-response optimisation and modelling of WEDM using grey–fuzzy and response surface methodology. Int. J. Modern Manuf. Technol. 13(1):7–14. ISBN 2067.

Chandramouli, S. and Eswaraiah, K. (2017). Optimisation of EDM process parameters in machining of 17-4 PH steel using Taguchi method. Mater. Today: Proc. 4(2):2040–2047.

Faisal, N. and Kumar, K. (2018). Optimisation of machine process parameters in EDM for EN 31 using evolutionary optimisation techniques. Technologies 6(2):54.

Gugulothu, B. (2020). Optimisation of process parameters on EDM of titanium alloy. Mater. Today: Proc. 27:257–262.

Khullar, V. R., Sharma, N., Kishore, S., and Sharma, R. (2017). RSM-and NSGA-II-based multiple performance characteristics optimisation of EDM parameters for AISI 5160. Arab. J. Sci. Eng. 42(5):1917–1928.

Kumar, D. and Mondal, S. (2021). Fuzzy-TOPSIS & Grey-Fuzzy for multi-attribute optimisation of process parameters in EDM for Al-2050 alloy. Solid State Technol. 64(2):4002–4017.

Kumar, P. M., Sivakumar, K., and Jayakumar, N. (2018). Multiobjective optimisation and analysis of copper–titanium diboride electrode in EDM of monel 400™ alloy. Mater. Manuf. Process. 33(13):1429–1437.

Magabe, R., Sharma, N., Gupta, K., and Paulo Davim, J. (2019). Modeling and optimisation of Wire-EDM parameters for machining of Ni55. 8Ti shape memory alloy using hybrid approach of Taguchi and NSGA-II. Int. J. Adv. Manuf. Technol. 102(5):1703–1717.

Mahanta, S., Chandrasekaran, M., Samanta, S., and Arunachalam, R. M. (2018). EDM investigation of Al 7075 alloy reinforced with B4C and fly ash nanoparticles and parametric optimisation for sustainable production. J. Braz. Soc. Mech. Sci. Eng. 40(5):1–17.

Manivannan, R. and Kumar, M. P. (2016). Multi-response optimisation of Micro-EDM process parameters on AISI304 steel using TOPSIS. J. Mech. Sci. Technol. 30(1):137–144. doi:10.1007/s12206-015-1217-4.

Marichamy, S., Saravanan, M., Ravichandran, M., and Stalin, B. (2017). Optimisation of surface roughness for duplex brass alloy in EDM using response surface methodology. Mech. Mech. Eng. 21(1):57–66.

Marichamy, S., Saravanan, M., Ravichandran, M., and Veerappan, G. (2016). Parametric optimisation of EDM process on α–β brass using Taguchi approach. Russ. J. Non-Ferr. Met. 57(6):586–598.

Meena, V. K., Azad, M. S., Singh, S., and Singh, N. (2017). Micro-EDM multiple parameter optimisation for Cp titanium. Int. J. Adv. Manuf. Technol. 89(1):897–904.

Mohamed, M. F. and Lenin, K. (2020). Optimisation of Wire EDM process parameters using Taguchi technique. Mater. Today: Proc. 21:527–530.

Palanisamy, D., Devaraju, A., Manikandan, N., Balasubramanian, K., and Arulkirubakaran, D. (2020). Experimental investigation and optimisation of process parameters in EDM of aluminium metal matrix composites. Mater. Today: Proc. 22:525–530.

Paul, T. R., Saha, A., Majumder, H., Dey, V., and Dutta, P. (2019). Multi-objective optimisation of some correlated process parameters in EDM of Inconel 800 using a hybrid approach. J. Braz. Soc. Mech. Sci. Eng. 41(7):1–11.

Ragavendran, U., Ghadai, R. K., Bhoi, A. K., Ramachandran, M., and Kalita, K. (2018). Sensitivity analysis and optimisation of EDM process parameters. Trans. Can. Soc. Mech. Eng. 43(1):13–25.

Sapkal, S. U. and Jagtap, P. S. (2018). Optimisation of micro EDM drilling process parameters for titanium alloy by rotating electrode. Procedia Manuf. 20:119–126.

Singh, R., Dvivedi, A., and Kumar, P. (2020). EDM of high aspect ratio micro-holes on Ti-6Al-4V alloy by synchronizing energy interactions. Mater. Manuf. Process. 35(11):1188–1203.

Yadav, V. K., Kumar, P., and Dvivedi, A. (2019). Effect of tool rotation in near-dry EDM process on machining characteristics of HSS. Mater. Manuf. Process. 34(7):779–790.

47 Influence of vegetable oil based hybrid nano cutting fluids in titanium alloy machining

Vashisht Kant[a], Khirod Kumar Mahapatro[b], and P. Vamsi Krishna[c]

Department of Mechanical Engineering, National Institute of Technology Warangal, Warangal, India

Abstract

The machining of titanium alloy is performed under hybrid nanofluid-based minimum quantity lubrication at different cutting speeds, feeds, and concentrations (nanoparticle inclusions (NPI)). Sesame oil is used as a coolant and the nanoparticles are carbon-nanotubes and molybdenum disulphide (MoS_2) in ratio of 1 is to 2. The cutting forces, cutting temperatures, micro-hardness, and surface roughness were studied at a constant depth of cut. From the experimental analysis, it was observed that the cutting forces and cutting temperatures were first increased and then decreased whereas hardness and surface finish were increased with an increase in NPI%. Cutting temperatures and surface hardness increased whereas cutting force and surface roughness decreased upon the increase in cutting speed. The cutting temperatures, cutting force, and surface roughness were hiked and the surface hardness decreases upon increasing feed rate.

Keywords: Hybrid nano-cutting fluid, MQL, surface integrity, Ti-6Al-4V.

Introduction

Titanium alloy mainly serves the purpose in missiles, naval ships, aircraft, and spacecraft corporations because of its low weight to strength ratio and superior corrosion resistance (Boyer, 1996). But, the machining of titanium alloy is difficult due to features like its lower thermal conductivity (k), high chemical reactivity, lower young's modulus, and high strength. The lower 'k' value of Ti-6Al-4V leads to heat accumulation in the machining zone which results in surface distortion, dimensional inaccuracy, poor surface finish, high tool wear rate, and ultimately failure at an early stage (Sun et al., 2010). These all adverse effects can be minimised by utilisation of cutting fluids as it removes and transport the heat generated in the deformation zones during machining and reduces friction. But the usage of conventional cutting fluids has a few disadvantages like being highly susceptible to microbial contamination leading to environmental pollution and various diseases related to the respiratory system to the people on the work floor (Benedicto et al., 2017). It is difficult to dispose of the cutting fluid contaminated with chips, and it makes the machining process expensive and non-sustainable. Hence, there is a need for alternatives without affecting the functionality and efficiency of machining. The alternatives of conventional cutting fluids are Cryogenics, Minimum quality lubrication (MQL), Solid Lubricants, Vegetable oils, and Nanofluids.

Arul et al. (2021) investigated the machining attributes on machining of alloys by employing nano additives-MQL technique and found it effective in dissipating heat than flood cooling. Impact of additives like Al2O3, polycrystalline diamond (PCD), graphite and water in MQL oil also has been explored on thermal analysis of end milling of Titanium alloy (Shokrani et al., 2021). Various researchers announced enhanced machinability of titanium alloys by cryogenic cooling and minimum quantity lubrication (MQL) (Osman et al., 2019; Gajran, 2020; Bermingham et al., 2012). But these cooling systems are complex and expensive to set up. (Gupta et al., 2021) used MQL with compressed air and found efficient heat dissipation through forced convection, but the usage of MQL possesses limitations in controlling friction and heat generation, especially while machining low thermal conductivity titanium alloys. Solid lubricants such as graphite and molybdenum disulfide can reduce friction between workpiece and cutting tool even being in solid-phase without the need for any oil or liquid medium. Padmini et al. (2017) probed the effect of green nano cutting fluids on machining performance using the minimum quantity lubrication manner. It was found that adding nanographene in coconut oil and using it as a lubricant reduces the surface roughness, cutting temperatures, tool wear, and cutting forces. Hybrid nano cutting fluid improves the heat transfer rates, thermal conductivity, wear resistance, better lubricating ability, etc. This makes the machining process more efficient. Up till now, limited researchers have looked into hybrid nanofluids (HNFs) in the machining of titanium alloys. Zareie and Akbari (2017) used multi-walled carbon nanotubes and magnesium suspensions in the same

[a]vashisht107@gmail.com; [b]khirodmech@gmail.com; [c]vamsikrishna@nitw.ac.in

proportions in water and ethylene glycol to prepare HNF. It was reported as volume percentage increases the dynamic viscosity increases, and its value decreases with a temperature rise.

Vamsi Krishna et al. (2010) reported a notable decrease in flank wear, surface roughness, and cutting temperature by application of solid lubricants (boric acid) in turning of AISI 1040 steel. Vegetable oils offer better results due to the presence of fatty acids that increases the lubrication property of vegetable oils compared to mineral oils. Shaji and Radhakrishnan (2002) probed the effect of solid lubricants like molybdenum disulfide (MoS2), barium fluoride (BaF2), and calcium fluoride (CaF2) in grinding and concluded that the lower tangential force and better surface finish were obtained by the usage of these lubricants.

In this present work, the cutting temperature, cutting force, surface roughness, and surface hardness were studied by changing the cutting fluid concentrations, speed, and feed by maintaining the depth of cut constant. the main aim of this study of this paper is to minimise the usage of cutting fluids which show a negative effect on the environment and workers' health. The MQL is a cooling technique that reduces the usage of cutting fluids 1000 times compared to conventional strategies. The effectiveness of MQL can be improved by the addition of nano-particles to the coolant while supplying in the machining zone. Along with this, the usage of cutting fluid reduces the harms that occur due to usage of conventional cutting fluids. In this study, the CNT/MoS2 nanoparticles with sesame oil at different concentrations were used and the responses were studied by experimentation.

Experimentation

In this work, Ti-6Al-4V IS is used as a workpiece and its length and diameter are 220 and 55 mm respectively. The machining operation was performed by using the PCLNR2525 M12 tool holder and CCMT 120404 uncoated carbide tool on the late machine. To relieve the stress, the aneling of the workpiece was done at 593°C for two hours. Vegetable oil-based hybrid nanofluids used as cutting fluid which is sesame oil as base fluid with nanoparticle suspensions of CNT/nMoS2 in the hybrid ratio of 1:2 and SDS surfactant that is 15 weights % of nanoparticles. The VBHNCFs were prepared for varying concentrations of 0.5%, 1%, 1.5% and 2% NPI by weight. The fluid was initially stirred manually and later sonication process was carried out for approximately 3 hours to obtain homogeneity and to avoid settlement of nanoparticles. The sonication process was carried out using the ultrasonic processor. The ultrasonic frequency was maintained at 22.57 kHz and Variac was kept at 160V.

The turning operation is performed at various cutting speeds (S) (30, 50, and 70 m/min), feed rates (f) (0.1, 0.18, 0.26 mm/rev), and coolant concentrations (0.5, 1, 1.5, and 2 % NPI) by Maintaining the depth of cut value constant (0.5mm). The flow rate in MQL is kept at 7 ml/min at a pressure of 8 bar. The responses in the machining such that the tangential cutting force (CF), cutting temperature (CT), micro-hardness (SH), and surface finish (SR) were measured by using Kistler Piezoelectric dynamometer (Model 5070), Fluke IR Camera (Model Ti400), Vickers Hardness Testing Machine, and Mar-surf (M400) respectively. To achieve better accuracy, each experiment was performed three times and average values are considered. Figure 47.1 shows the experimental setup. Table 47.1 shows the experimental data.

Cutting Temperature

Variations in cutting temperatures with cutting speed are shown in Figure 47.2 at different concentrations. Cutting temperatures increased with speed because, As cutting speed increases material removal rate and

Figure 47.1 Experimental setup for the turning operation

Table 47.1 Experimental data

Ex. No	*S*	*NPI*	*f*	*CT*	*CF*	*SH*	*SR*
1	30	0.5	0.1	165	267	396	0.7
2	30	0.5	0.18	163	425	370	0.9
3	30	0.5	0.26	206	320	364	0.8
4	30	1	0.1	256	156	407	0.7
5	30	1	0.18	248	324	375	0.8
6	30	1	0.26	216	309	368	0.8
7	30	1.5	0.1	256	132	420	0.6
8	30	1.5	0.18	293	173	388	0.7
9	30	1.5	0.26	310	243	372	0.9
10	30	2	0.1	250	119	434	0.6
11	30	2	0.18	286	138	395	0.7
12	30	2	0.26	410	127	381	0.8
13	50	0.5	0.1	193	204	415	0.7
14	50	0.5	0.18	255	322	397	0.9
15	50	0.5	0.26	91	446	393	0.9
16	50	1	0.1	186	185	429	0.7
17	50	1	0.18	228	235	406	0.8
18	50	1	0.26	273	319	395	0.9
19	50	1.5	0.1	182	159	443	0.7
20	50	1.5	0.18	192	220	407	0.7
21	50	1.5	0.26	359	230	395	0.8
22	50	2	0.1	185	131	457	0.6
23	50	2	0.18	211	203	400	0.7
24	50	2	0.26	268	187	402	0.7
25	70	0.5	0.1	151	214	433	0.7
26	70	0.5	0.18	232	260	396	0.8
27	70	0.5	0.26	287	322	397	0.9
28	70	1	0.1	148	208	440	0.7
29	70	1	0.18	158	230	420	0.7
30	70	1	0.26	306	292	382	0.9
31	70	1.5	0.1	146	194	449	0.6
32	70	1.5	0.18	150	221	431	0.7
33	70	1.5	0.26	214	244	399	0.8
34	70	2	0.1	144	174	461	0.6
35	70	2	0.18	150	217	433	0.7
36	70	2	0.26	174	190	428	0.7

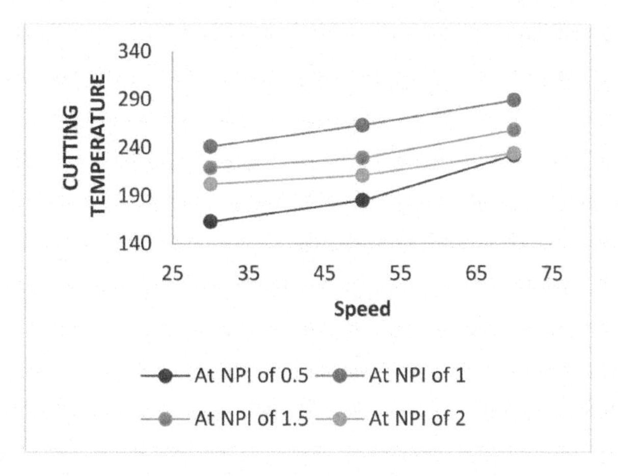

Figure 47.2 Variation of cutting temperature with cutting speed at different NPI at feed = 0.18 mm/rev

strain rate increase in the deformation zones. That causes a large amount of heat generation and generates higher temperature at the interface. It was also noticed that as NPI% is increased from 0.5 to 1 cutting temperatures increase at a faster rate and upon further increasing NPI% the cutting temperatures fall. Both conduction and convection are responsible for the heat transfer process under the application of SBHNF. In the case of 0.5wt%, the heat transfer through convection coefficient dominates the heat transfer through conduction and vice versa when 2wt% SBHNF was used. The cutting temperature got increased at 1wt% NPI SBHNF and then falls with a rise in NPI. This is because there is a trade-off between convective and conductive heat transfer in the case of 1wt% NPI and then with an increase in thermal conductivity, there is the decrease in temperature due to higher conductive heat transfer. Variations in cutting temperatures with feed rate are shown in Figure 47.3 at different concentrations. It was spotted that cutting temperature escalated with feed rate. This was because of the increased section of the chip and higher power consumption due to the higher material removal rate. Equation 1 gives us the regression equation of cutting temperature obtained by Response Surface Methodology (RSM).

$$CT = 124 + 0.62\,S + 229.2\,NPI - 829\,f + 0.0082\,S \times S - 33.2\,NPI \times NPI + 1585\,f \times f - 3.477\,S \times NPI + 6.88\,S \times f + 287\,NPI \times f \quad (1)$$

Cutting Force

Variation of cutting force with cutting speed is depicted in Figure 47.4 at different concentrations. From the figure, it is observed that as cutting speed rises, the cutting force gets reduced with increase in cutting speed. At higher cutting speed the material removal rate is high generates more amount of heat in the cutting zone and increasing the temperature. This high temperature leads to softening of the material and contributed to a reduction in cutting force. From Figure 47.4 it can also observe that cutting forces first increase when NPI wt% is changed from 0.5 to 1 and then they decrease as the further increment is done in the NPI%. When NPI % rises, the viscosity of the fluid also gets increased and that helps in forming a continuous fluid film at the interface of tool and workpiece. As the nanoparticle concentration increases, more we witness

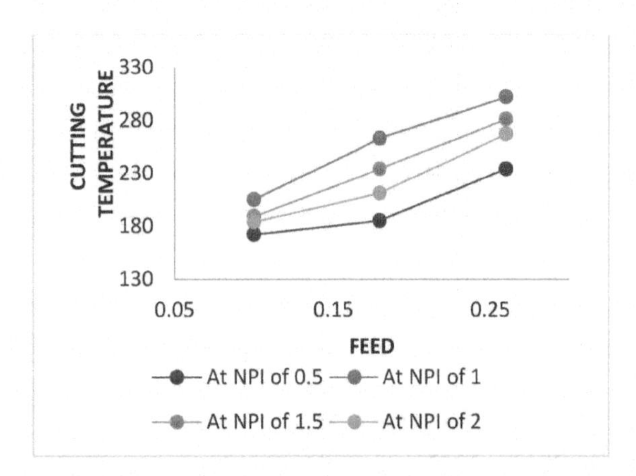

Figure 47.3 Variation of cutting temperature with feed at different NPI at a speed of 50 m/min

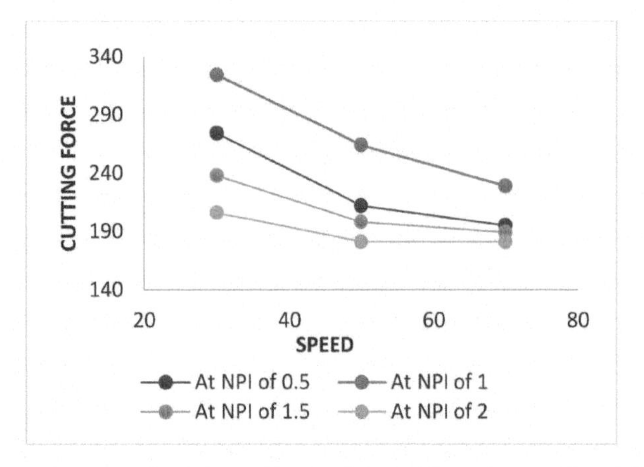

Figure 47.4 Variation of cutting force with cutting speed at different NPI at feed = 0.18 mm/rev

the evolution of this film on the surface of the workpiece and that leads to reduction of cutting force. From Figure 47.5, it is detected that cutting force increased with the increase in feed rate and this is because of a higher chip load per tooth at higher feed rates (Alauddin et al., 1998). Also, at higher feed rates increase in chip thickness was observed that increased the force required for machining due to higher material removal rate. Equation 2 gives us the regression equation of cutting force obtained by Response Surface Methodology (RSM).

$$C\,F = 148 - 0.53\,S - 184.2\,NPI + 2591\,f - 0.0187\,S \times S + 21.6\,NPI \times NPI - 3655\,f \times f + 2.353\,S \times NPI - 2.59\,S \times f - 464\,NPI \times f \qquad (2)$$

Surface Hardness

Variations in Surface hardness with cutting speed are shown in Figure 47.6 at different concentrations. In Figure 47.6 with an increase in cutting speed at all NPI%, micro-hardness of machined surface increases. The quenching rate increased with cutting speed leading to a higher rate of surface hardening at higher speeds. The quenching effect also increases with a growth in the number of nanoparticles. As the NPI% rises, the thermal conductivity of SBHNF increases improving the quenching effect of fluid. As discussed earlier higher quenching effect increases the hardness of the surface. From Figure 47.7 it is seen that surface hardness decreased with the increment of feed rate. This is because at a higher feed rate material removal rate is high and most portion of the cooling load is transferred to the chip. This reduces the cooling effect on the surface and thereby the hardening effect. Due to this low hardening effect lower, surface hardness was observed at a higher feed rate. Equation 3 gives us the regression equation of surface hardness obtained by Response Surface Methodology (RSM).

$$S\,H = 394.1 + 2.145\,S + 18.1\,NPI - 692\,f - 0.01420\,S \times S + 2.75\,NPI \times NPI + 1446\,f \times f + 0.096\,S \times NPI - 0.11\,S \times f - 70.0\,NPI \times f \qquad (3)$$

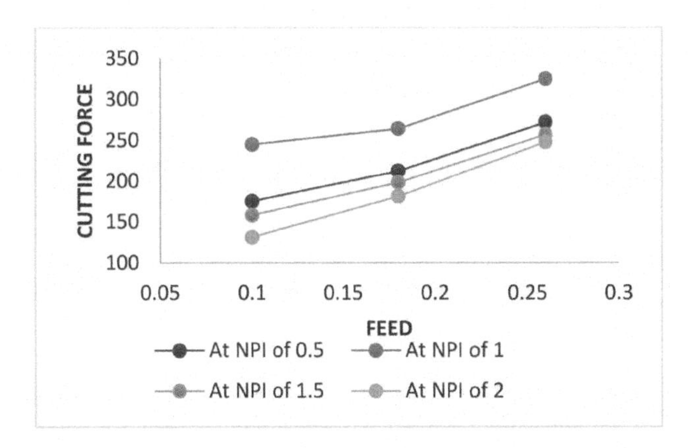

Figure 47.5 Variation of cutting force with feed different NPI at a speed of 50 m/min

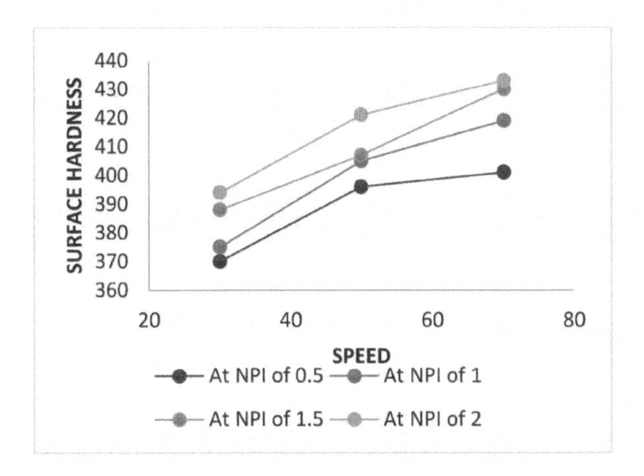

Figure 47.6 Variation of surface hardness with speed at different NPI at feed = 0.18 mm/rev

Surface Roughness

Variations in Surface roughness with cutting speed are exhibited in Figure 47.8 at different concentrations. From Figure 47.8, it was observed that the surface roughness decreased with increasing speed. The tendency of adherence between the tool and workpiece and built-up edge formation were lower at higher speeds. Therefore, the increasing cutting velocity lessens the built-up edge on cutting tool and improved the surface quality (Parhad et al., 2015). Figure 47.9 shows the variation of surface roughness at varying feed rates under MQL mode with speed and depth of cut being constant. It was seen in Figure 47.8 the surface roughness decreased with increased NPI % of sesame oil. This is because the nMoS2 nanoparticles in the form of layered structure offer efficient properties of lubrication and higher viscosity at higher NPI %. An increase in viscosity forms a better film formation that reduced the friction coefficient and improved the surface finish of the machined component. From Figure 47.9 it was detected that the surface

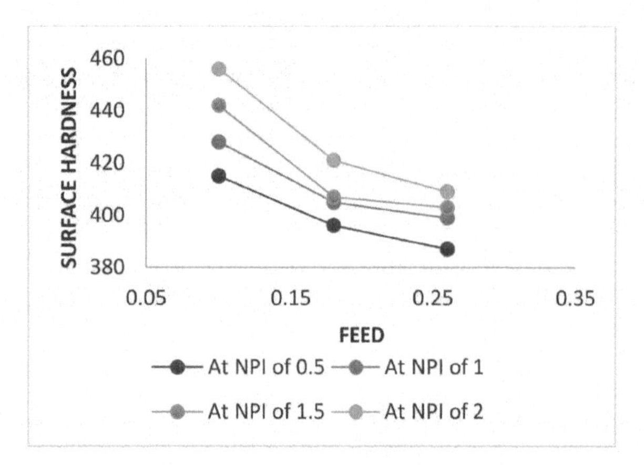

Figure 47.7 Variation of surface hardness with feed feed different NPI at a speed of 50 m/min

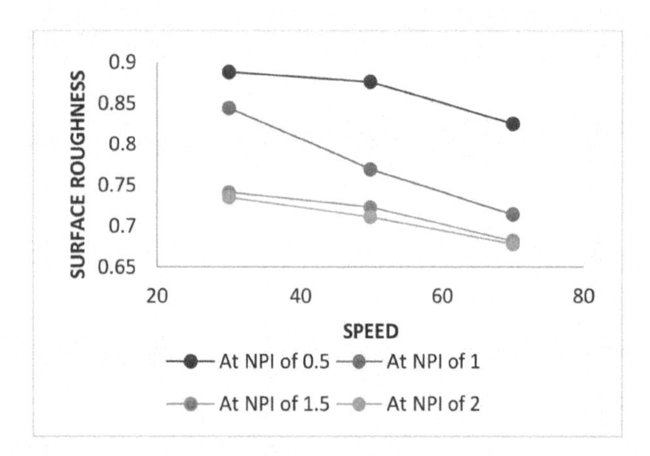

Figure 47.8 Variation of surface roughness with speed at different NPI at feed = 0.18 mm/rev

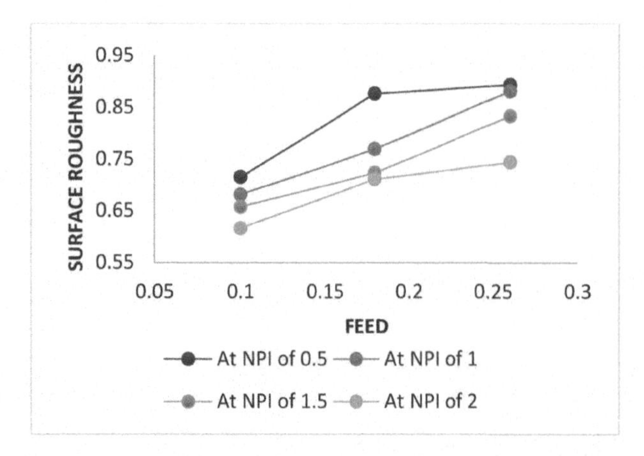

Figure 47.9 Variation of surface roughness with feed at different NPI at a speed of 50 m/min

roughness grew upon increasing feed. As feed increases, MRR thereby leads to higher friction and contact area between cutting tool face and the workpiece were increases. Higher friction leads to higher temperatures, also this temperature may cause less sharp cutting tool and this leads to a poor surface roughness at a higher feed rate.

Equation 4 gives us the regression equation of surface roughness obtained by Response Surface Methodology (RSM).

$$S\,R = 0.546 + 0.00163\,S - 0.0670\,NPI + 2.118\,f - 0.000021\,S \times S - 0.0001\,NPI \times NPI - 3.02\,f \times f - 0.000374\,S \times NPI + 0.00176\,S \times f + 0.026\,NPI \times f \quad (4)$$

Conclusions

Machining of titanium alloy is performed at different speeds, feeds, and NPI% under an MQL hybrid nanofluid environment at a constant depth. The cutting temperature cutting forces, surface roughness, and surface hardness were measured. The Experimental outcome lead to the following conclusions.

- The cutting temperature increased with speed feet and. Whereas it is increased first and then decreased with NPI %. The temperature was increased with NPI % from 0.5 wt. % to 1 wt. % and then decreased 2 wt. %.
- The cutting force decreased with an increase cutting speed and a feed decrease. Whereas it is increased first and then decreased with NPI %. The temperature was increased with NPI % from 0.5 wt. % to 1 wt. % and then decreased 2 wt. %.
- The surface hardness increased with increase in NPI%, cutting speed and it was decreased with increase in feed rate.
- The surface roughness decreased with rise in % NPI, cutting speed and it was increased with increment in feed rate.

References

Alauddin, M., Mazid, M. A., EI Baradi, M. A., and Hashmi, M. S. J. (1998). Cutting forces in the end milling of Inconel 718. J. Mater. Process. Technol. 77:153–159.

Arul, K., Mohanavel, V., Kumar, S. R. Mariduraid, T., Kumar, K. M., and Ravichandran, M. (2021). Investigation of machining attributes on machining of alloys under nano fluid MQL environment: A review. Mater. Today: Proc. https://doi.org/10.1016/j.matpr.2021.11.525.

Benedicto, E., Carou, D., and Rubio, E. M. (2017). Technical, economic and environmental review of the lubrication/cooling systems used in machining processes. Procedia Eng. 184:99–116. https://doi.org/10.1016/j.proeng.2017.04.075.

Bermingham, M. J., Palanisamy, S., Kent, D., and Dargusch, M. S. (2012). A comparison of cryogenic and high pressure emulsion cooling technologies on tool life and chip morphology in Ti-6Al-4V cutting. J. Mater. Process. Technol. 212:752–765. 10.1016/j.jmatprotec. 2011.10.027.

Boyer, R. R. (1996). An overview on the use of titanium in the aerospace industry. Mater. Sci. Eng. A 213:103–114. 10.1016/0921-5093(96)10233-1.

Gajran, K. K. (2020). Assessment of cryo-MQL environment for machining of Ti-6Al-4. J. Manuf. Process. 60:494–450.

Gupta, M. K., Song, Q., and Zhanqiang, L. (2021). Experimental characterisation of the performance of hybrid cryo-lubrication assisted turning of Ti–6Al–4V alloy. Tribol. Int. 153. doi:10.1016/j.triboint.2020.106582.

Osman, K. A., Ünver, H. Ö., and Şeker, U. (2019). Application of minimum quantity lubrication techniques in machining process of titanium alloy for sustainability: A review. Int. J. Adv. Manuf. Technol. 100:2311–2332. 10.1007/s00170-018-2813-0.

Padmini, R., Vamsi Krishna, P., Mahith, S., and Kumar, S. (2017). Influence of green nanocutting fluids on machining performance using minimum quantity lubrication technique. In International conference on nanotechnology; Ideas, Innovations & Intiatives-2017. https://doi.org/10.1016/j.matpr.2019.06.612.

Parhad, P., Likhite, A., Bhatt, J., and Peshwe, D. (2015). The effect of cutting speed and depth of cut on surface roughness during machining of austempered ductile iron. Trans. Indian Inst. Met. 68:99–108.

Shaji, S. and Radhakrishnan, V. (2002). An investigation on surface grinding using graphite as lubricant. Int. J. Mach. Tools Manuf. 42(6):733–740. https://doi.org/10.1016/S0890-6955(01)00158-4.

Shokrani, A., Betts J., and Carnevale, M. (2021). Thermal analysis in MQL end milling operations. University of bath, bath ba2 7ay, United Kingdom. Procedia CIRP 101:358–336.

Sun, S., Brandt, M., and Dargusch, M.S. (2010). Thermally enhanced machining of hard-to-machine materials- a review. Int. J. Mach. Tool Manuf. 50:663–680. 10.1016/j.ijmachtools.2010.04.008.

Vamsi Krishna, P., Srikant, R. R., and Nageswara Rao, D. (2010). Experimental investigation on the performance of nanoboric acid suspensions in SAE-40 and coconut oil during turning of AISI 1040 steel. Int. J. Mach. Tools Manuf. 50(10):911–916. https://doi.org/10.1016/j.ijmachtools.2010.06.001.

Zareie, A. and Akbari, M. (2017). Hybrid nanoparticles effects on rheological behavior of water-eg coolant under different temperatures: An experimental study. J. Mol. Liq. 230:408–414. https://doi.org/10. 1016/j.molliq.2017.01.043.

48 Improvement in student performance using 4QS and machine learning approach

Ruchita A. Kale[a] *and Manoj K. Rawat*[b]

Department of Computer Science & Engineering, SAGE University Indore, Madhya Pradesh, India

Abstract

Teachers must foresee their students' performance tendencies to enhance their teaching abilities. This paper discusses advancements and problems in student performance prediction (SPP) and advances in individualised education. It has developed into a helpful resource for various goals; for instance, a strategic plan may be used to establish a high-quality educational system. This study aims to demonstrate how data mining methods may forecast students' final grades using their past data. This review divides the process of predicting student performance into five stages: data gathering, issue formulation, model development, prediction, and implementation. We performed tests using a data set from the institution and public data established to understand these involved methodologies. The educational dataset, which includes 2500 students, was compiled using an information system representative of a regular university. Finally, existing deficiencies and intriguing future studies are outlined based on the experimental findings from collecting data.

Keywords: Data sets, analysis education, students, prediction.

Introduction

The time has come for those born in this millennium to enrol in universities, and it is necessary to rethink and redefine the way the education system should be changed. The institutes' curriculum and approach are being reconstructed to various questions to alter the educational system. The older system focused on what was taught rather than to whom it was prepared or the teaching method, and this system is getting outdate. Due to technological advancement, the student's learning style has been altered to internet-based learning in the last decade and machine learning (ML) to predict the solution. This has created problems since the educational institutes have not kept up with the changing technology. Intelligent students who are good with technology have already produced new ways to learn through the internet, making their curriculum not compelling enough.

The students connect with the other learners for collaborative learning and hence can teach each other the concepts they do not understand. The curriculum is not effective enough for them (Zhang et al., 2021a). The educators must realise the upcoming technology and be more familiar with improvements in the teaching method with the predictive method to improve the system's performance. When more innovation is made in the curriculum using ML, the students will have more scope to solve problems, and it will become interactive. These new methodologies will keep up with the learners and will change and develop as quickly as possible.

A. Motivation

At the moment, Indian educational establishments face fierce rivalry for students and high-quality education. Laboratory infrastructure is vital for providing a high-quality education in technical and technical programs (Zhang et al., 2021b). The issue with such projects is that laboratory facility architecture is often an afterthought rather than being properly designed and executed holistically. This is particularly true in industrial facilities, where several units are not adequately connected. Numerous laboratories inside these facilities, including CAD labs, CNC labs, and robotics labs, are self-contained. This task describes the existing manufacturing centre of the institution as a case study that identifies its limitations and shortcomings and recommends a general layout based on lean principles that address the weaknesses of the existing facility design (Zhang et al., 2020a). The 4QS and analysis of several factors will be used to improve the curriculum development, reduce student stress, and improve their performance.

Methodology

In this part, it gives an overview of the different models used for the teaching-learning process to improve the overall performance of the system.

---

[a]kaleruchita11@gmail.com; [b]profrawat.sage@gmail.com

A. Models Used for the Teaching-Learning Process

In assessing productive learning methods based on the evolving interests of learners by creating new skills, instructors and clinicians adopt the quotient model of assisted learning (Zhang et al., 2020c). Since competence preparation is used as a coping mechanism for occupational therapists, it supplies a way to understand, organise, and manage learning strategies. The aim is to increase efficiency by manufacturing workplace output components in the target occupation. Figure 48.1 shows the Exiting 4Q model used by the education system.

B. Bloom's Taxonomy Approaches in Higher Education

Bloom's classification is a collection of three hierarchies used to classify pedagogical learning goals into levels of sophistication and specificity. The three lists are learning objectives in the emotional, emotional, and sensory areas (Zhang et al., 2020b). The list of the cognitive regions has become the focus of most formal schooling and is widely used to organise curriculum learning priorities, tests, and assignments. Figure 48.2 shows Bloom taxonomy . Following are the critical categories of bloom taxonomy.

The critical categories are as follows:

- *Knowledge* - recollection of particulars and universals, procedures and techniques, or a remembrance of a pattern, structure, or environment."
- *Comprehension* - Without necessarily linking it to another material or considering its full ramifications, understand what is being conveyed and use the materials or ideas presented.
- *Application* - refers to the use of abstractions and concrete circumstances in particular.
- *Analysis* - is the decomposition of communication into its constituent pieces or parts to articulate the relative hierarchy of concepts and precisely describe the connection between ideas.
- *Synthesis* - necessitates the assembly of pieces and parts to make a complete component.
- *Evaluation* - generates judgments about the utility of materials and procedures for specific uses.

C. Classification of Blooms Taxonomy

To develop this whole field of the learner, some teaching experience should be prepared. Therefore, these three realms are interrelated with each other and thus interdependent too (Fan et al., 2019). Figure 48.3 shows the Flow of bloom's taxonomy, it contain Cognitive Domain, Emotional domain and Psychomotor area.

- *Cognitive Domain*

  Bloom (1956) defined the cognitive domain as 'the acquisition of information and the development of intellectual abilities.' This involves recalling or recognizing facts, procedural patterns, and ideas that develop intellectual talents and skills (Anand, 2019). There are six primary types presented in order of complexity, beginning with the most straightforward action. Consider the categories as a scale of difficulty. The first one must be accomplished before the subsequent one may occur.

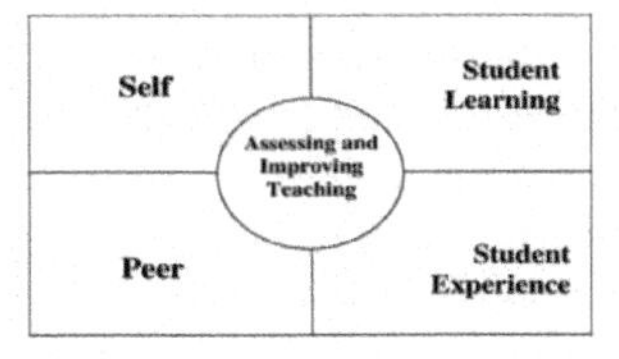

Figure 48.1 Exiting 4Q model used by the education system exiting 4Q model used by the education system

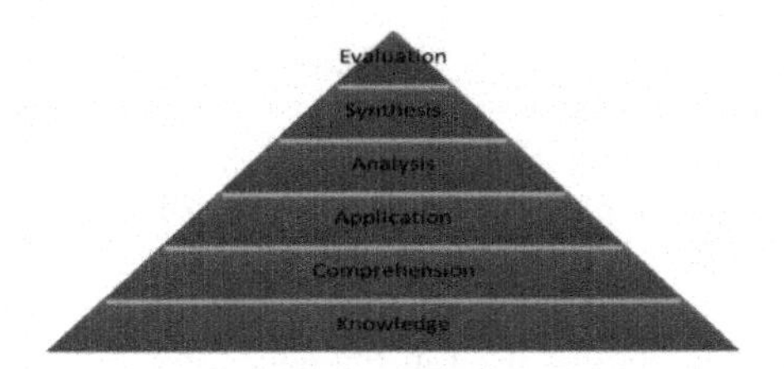

Figure 48.2 Bloom taxonomy

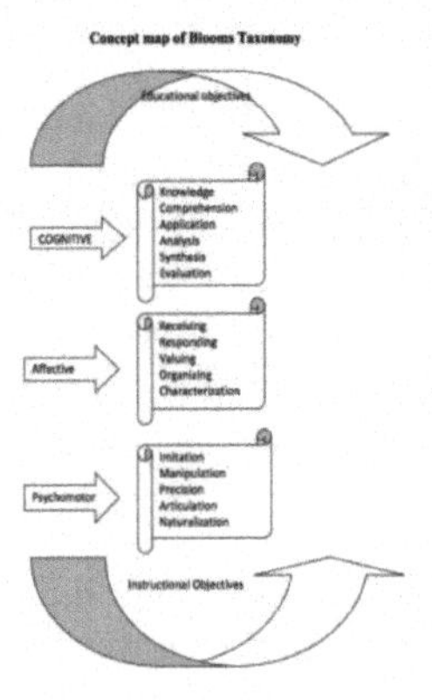

Figure 48.3 Flow of Bloom's taxonomy

- *Emotional domain*
  The emotional domain (Krathwohl, Bloom, and Masia, 1973) contains emotional reactions to objects such as emotions, values, admiration, enthusiasm, motivation, and attitudes.
- *Psychomotor area*
  The Psychomotor Area (Simpson, 1972) includes physical movement, coordination, and the use of exercise. These skills are developed through practice and can be quantified in speed, accuracy, distance, process, or execution strategy.

Use of Machine Learning in the Education System

'ML is defined as a computer technical discipline that uses mathematical without being programmed explicitly for my readers who are new to the word (i.e., constantly increasing performance on a given task. It is usually ignored to suggest that ML will revolutionise education (Juhaňák et al., 2019). Some old school/ college teachers had a tough time getting used to machines that can think and understand. Nevertheless, they would have to come to terms with this new fact eventually.

Following are how ML will revolutionise the field of education-

1. *Increasing efficiency-* By integrating activities like classroom management, scheduling, etc., ML can make educators more effective in terms of artificial intelligence. Teachers should focus on tasks that AI does not do, which involve a human side.
2. *Learning analytics* - When it comes to learning analytics, ML can help teachers understand better the information that the usage of the human brain cannot glean. Systems will carry out in-depth dives into data in this role, sifting through millions of information pieces and drawing comparisons and assumptions that positively influence the learning and teaching process.
3. *Predicative analytics* - When it comes to predictive analytics, ML could draw answers about situations that may occur in the future. For example, 'by using a statistical data set records of middle school students that are more likely to drop out due to academic failure or even their predicted success on a standardised exam, like the ACT or SAT, predictive analytics will notify us.'
4. *Adaptive learning* - ML could be used to remediate failing pupils or challenge gifted individuals in personalised learning. Adaptive learning is a technology-driven or online instructional framework that analyses a pupil's success in real-time and, based on the input, changes teaching methods and the curriculum. Think AI meets committed mentor for math meets customised dedication.
5. *Personalised learning* - ML in personalised learning may give each pupil an individualised educational experience. Personalised learning is an instructional style in which learners direct their learning, go at their speed, and make their own choices on what to learn in certain instances. Ideally, students pick what they are interested in in a classroom using customised instruction, to which the teachers adapt the expectations and curriculum to the needs of the students.
6. *Assessment* - In AI, ML can grade student assignments and reviews more accurately than humans.

Some human interpretation may be needed, but the conclusions would have greater significance and reliability. Using the ML in the education sector, we can predict the student performance on their histological dataset to focus on their future and improve their growth stability.

The summation of this chapter gives a detailed overview of the learning system used in manufacturing industries for improvement and growth. Nowadays, the same method is used in the education system to improve the system's overall structure with curriculum development (Kim et al., 2018). We are using the

4QS proposed scheme to improve the overall performance of learning education with a ML approach to predict the preventive measure we need to consider.

Proposed Work

As a result of the globalisation of the market for commodities, services, and resources on the one hand and the globalisation of issues on the other, education confronts more difficulties than ever.

Another critical issue confronting educational institutions is teacher evaluation. When the public demands more responsibility in education, research indicates that the techniques currently utilised to evaluate teachers are somewhat unexpected (Kushwaha et al., 2019). Teacher assessment is required in light of the study. According to national surveys (NBA), just approximately half of this country's school/college systems employ standardised teacher evaluation processes. Those who use formal programs continue to use inadequate procedures and approaches.

In public schools, the traditional method of evaluating teachers is for the principal to complete a checklist-style form indicating the extent to which a teacher has the stated characteristics and competencies. Typically, following a visitation, the principal terminates the form. In other schools, the assessment is conducted without classroom observation. A conference between the principal and the instructor is omitted to discuss how teachers might improve their teaching techniques. Numerous studies indicate a widespread assumption in many institutions that greater age, experience, and college credit improve performance. The education process is shown in the Figure 48.4.

Evaluations conducted only by the building principal have obvious disadvantages: they are unilateral and subjective, and these processes provide minimal assistance to instructors (Liu et al., 2019). They are consequently viewed negatively by most educators. This shows that difficulties confronting public education today are developing procedures for evaluating teacher performance that is both valid and valuable to the teaching profession.

A committee comprising peers, subordinates, supervisors, and students was offered as a solution to the issue of subjectivity in teacher evaluations.

Today, we reflect and speculate on the evolution of education over the last several decades. Today's most simple definition of "curriculum" is the subjects covered in a college, high school, or university study course. There are distinctions in the course design.

- For instance, a mathematics course taught at one institution may include the same topic, but the instructor may present it differently (Lam et al., 2018).
- nonetheless, the curriculum development process has remained consistent, particularly regarding the key fundamentals. Thus, curricular development must be prioritised.

Figure 48.5 shows the Traditional Approach to Educational Curriculum in education system.

A. Models of Curriculum Development

Then, a successful and effective program will have a range of inputs. Developing such a program must include students, instructors, and administrators. Moreover, to evaluate a teacher's performance with the school's objectives, the measured factors must fulfill Menne's three criteria for the developed framework. Due to the absence of a blueprint for implementation, creating such an evaluation framework is now an intimidating task for local school districts. The difficulty that this work is addressing is developing such a model.

The current curricular models may be classified into two major groups based on their content: process models and product models (Liu et al., 2018).

The model of the product is outcome-oriented. Grades are the primary objective, with a greater focus on the result than on the process of learning. Nonetheless, the process model is more open-ended and

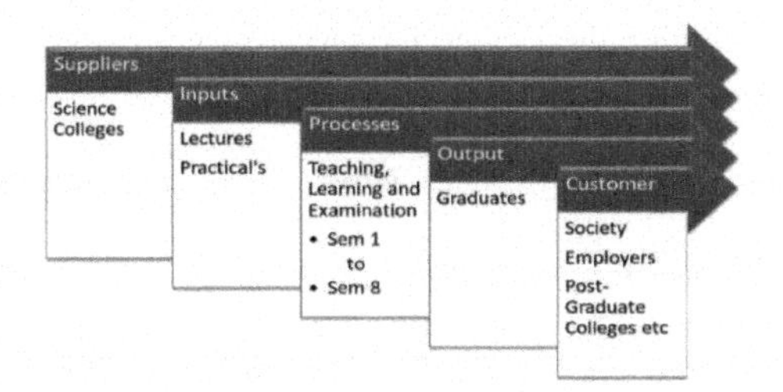

Figure 48.4 Process of the education system

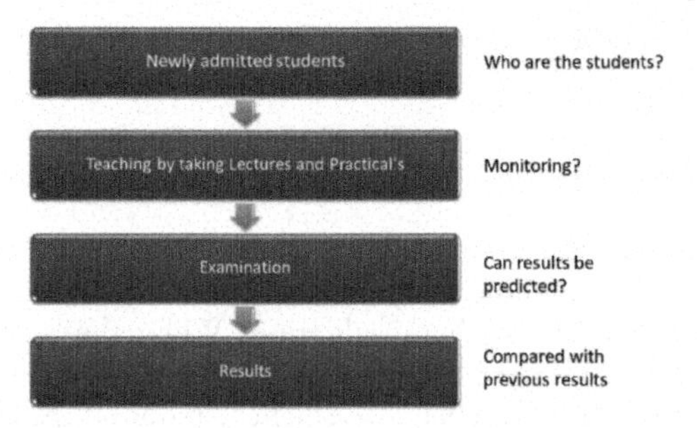

Figure 48.5 Traditional approach to educational curriculum

focuses on learning's progression across time. These two frameworks must be considered while developing programs.

4QS Model Working

Data mining tools, psychology, and educational theories solved student performance prediction challenges. The majority of current research focuses only on classic ML algorithms, oblivious to pre-existing information gained via educational activities. Priors could include: (1) The organisational structure of courses at a school or institution, which affects the goal courses via knowledge linkage across courses. (2) The students' learning curve, illustrating the various weights assigned to objective course expectations. (3) The objective course's requirements, which are critical in predicting the objective course. Additionally, side-information characteristics such as leisure-time learning activities and attitudes toward learning significantly impact course achievement. On the other hand, SPP should prioritise the acquisition of technical norms that may aid in the accomplishment of more actual performance and interpretable findings (Fernandes et al., 2019).

Additionally, several complex aspects affect course grades, such as family, campus life statistics, and learning psychology, which educators have shown affect student academic achievement. This contributes to the model's accuracy but is not a definitive educational conclusion. The following is a fundamental paradigm for enhancing the educational process.

A. 4QS Model for Improving Learning Education

The following study presents how 4QS learn, a technique traditionally used only in education, is tailored to the particularities of the higher education processes and is implemented in a college/university. Teachers and practitioners use the four-quadrant model of helped learning (4QS) to select practical learning approaches based on learners' changing preferences by acquiring new skills.

Following are some quotients that shall be accessed with stress analysis using the 4QS Model:

1. *Intelligence quotient (IQ):* Additionally, intelligence is a predictor of academic achievement. Individual creative behaviours, such as music composition, mathematics, and writing, need a high IQ, as does student performance when confronted with intelligence questions. Figure 48.6 shows the 4QS model used in the education system.

   Before the IQ tests were invented, attempts were made to classify people into intelligence groups by examining their daily life behaviour. Many other forms of behavioural testing are also relevant for validating classifications primarily based on IQ test scores. The type of intelligence by behavioural evaluation in the training room and the category by IQ tests depend on the word 'intelligence' used in the specific case and the mistake in the classification phase in terms of reliability and measurement.
2. *Adversity quotient (AQ):* 'Adverse' refers to a very aggressive or severe sickness. We continue to face all of life's other hardships; it is only through adversity that one learns who is friendly and well-intentioned. The hardship quotient which is comprised of the spiritual quotient (SQ), the emotional quotient (EQ), and the IQ, is a number that indicates an individual's readiness to cope with adversity throughout their life. To endure life's difficulties, it is necessary to build perseverance. Figures 48.7 and 48.8 show the Sample of intelligence quotient (IQ).

   AQ is one of the various success indicators for an individual in life and is often used to gauge the mood, mental stress, persistence, and endurance.

   Emotional Intelligence Quotient (EIQ), Emotional Leadership (EL), Emotional Intelligence (EI), and Emotional Quotient (EQ) are people's abilities to interpret, correctly classify, and mark their own and other emotions, use awareness of emotions to guide thoughts and actions, and monitor and alter their emotions to react to or attain circumstances (Buenano-Fernandez et al., 2019). Figure 48.9 shows the Sample pattern of adversity quotient (AQ).

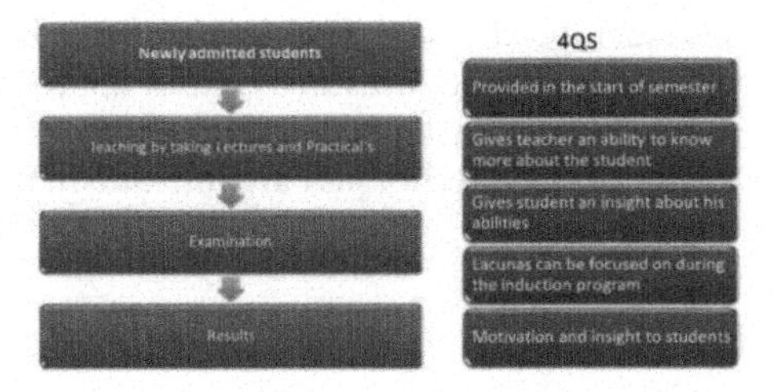

Figure 48.6 4QS model used in the education system

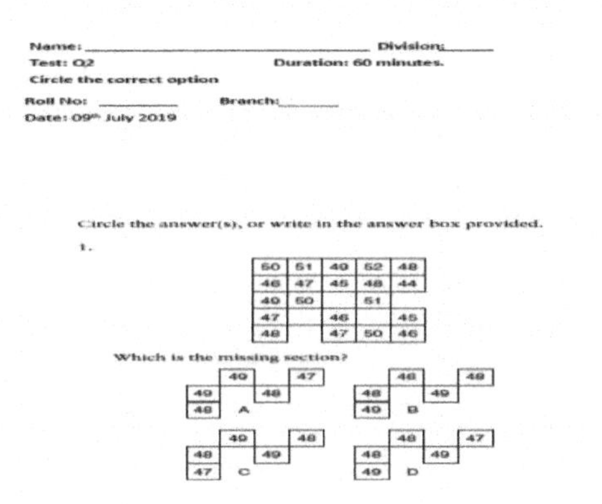

Figure 48.7 Sample of intelligence quotient (IQ)

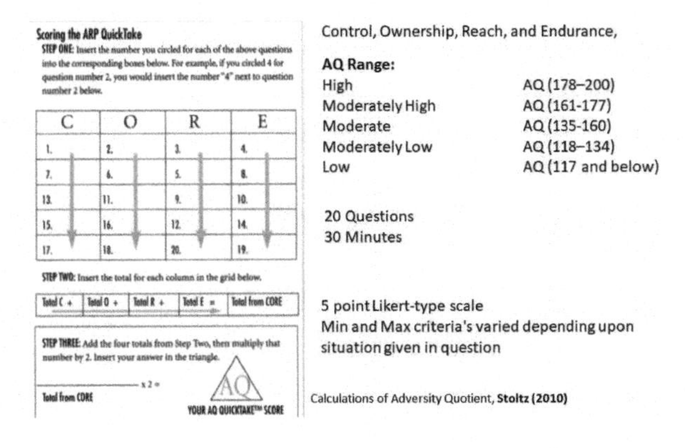

Figure 48.8 Sample pattern of intelligence quotient (IQ)

3. *Emotional quotient:* EQ is a term that refers to your ability to manage your thoughts positively to cope well with life's adversities. Studies say that establishing intimate relationships and entering into group environments is better for people with EQ ratings. Individuals with better emotional intelligence may have a clearer understanding of their psychological state, treating stress efficiently and making them less likely to experience depression. Figure 48.10 shows the Sample pattern of emotional quotient (EQ).

   Following are some questions related to calculating EQ.
4. *Spiritual quotient:* Beyond your mental and psychic capabilities. It is about examining your life and considering what you may contribute to humanity. It is to live modestly and recognise your insignificance in comparison to the immensity of the cosmos. The SQ is a parameter that measures the moral insight of a person; it is as necessary as the IQ and the quotient of emotions EQ. Although IQ looks at logical intelligence, EQ looks at an individual's emotional ability, and an individual's SQ looks at a person's spiritual strength. Spirituality enhances a person's imaginative, receptive, and perceptive ability. You may improve the stability of insight and comprehension with the assistance of trust. A good awareness of SQ motivates people to align their job activities, time with family, and growth inside themselves. Figure 48.11 shows the Sample pattern of spiritual quotient (SQ).
5. *Stress Analysis:* Stress, in day-to-day terms, is an emotion that people face when they are highly loaded and experience difficulties while fulfilling daily demands. Stress affects individual health significantly,

such as cardiovascular disease, attack, stroke, etc. Consequently, stress recognition becomes helpful to control health-related issues generated from stress. Stress can be measured and evaluated on perceptual, conduct, and physiological reactions. A few researchers have proposed different feature extraction and classification techniques. It is based that some of these techniques are intricate in their

1. You suffer a financial setback.
   To what extent can you influence this situation?
   Not at all 1 2 3 4 5 Completely
2. You are overlooked for a promotion.
   To what extent do you feel responsible for improving the situation?
   Not at all 1 2 3 4 5 Completely
3. You are criticized for a big project that you just completed.
   The consequences of this situation will:
   Affect all aspects of my life 1 2 3 4 5 Be limited to this situation
4. You accidentally delete an important email.
   The consequences of this situation will:
   Last forever 1 2 3 4 5 Quickly pass
5. The high-priority project you are working on gets canceled.
   The consequences of this situation will:
   Affect all aspects of my life 1 2 3 4 5 Be limited to this situation
6. Someone you respect ignores your attempt to discuss an important issue.
   To what extent do you feel responsible for improving this situation?
   Not responsible at all 1 2 3 4 5 Completely responsible
7. People respond unfavorably to your latest ideas.
   To what extent can you influence this situation?
   Not at all 1 2 3 4 5 Completely
8. You are unable to take a much-needed vacation.
   The consequences of this situation will:
   Last forever 1 2 3 4 5 Quickly pass

30 Questions 1 Marks each

60 Minutes

Scores between: Rating Range of Marks: 0 to 30

27–30 Very highly exceptional
24–26 High expert
21–23 Expert
19–20 Very high average
17–18 High average
13–16 Middle average
10–12 Low average
6–9 Borderline low
3–5 Low
0–2 Very low

Figure 48.9 Sample pattern of adversity quotient (AQ)

Answer each question or statement by choosing which one of the three alternative responses given is most applicable to you.

1. What is your attitude to beggars in the street?
   A I feel somewhat uncomfortable when I see them and, perhaps, think: *there but for the grace of God go I*
   B I feel sorry for them
   C I give them a wide berth
   Answer ☐
2. Do you tend to laugh a lot?
   A About the same as most people
   B More than most people
   C Less than most people
   Answer ☐
3. Which of the following words best describes you?
   A undaunted
   B unpredictable
   C impartial
   Answer ☐

15 Questions
15 Minutes

Total score 25–30 Excessively emotional
Total score 19–24 Fairly emotional
Total score 13–18 Average
Total score 8–12 Fairly unemotional
Total score below 7 Excessively unemotional

1 point for every 'a' answer,
2 points for every 'b',
and 0 points for every 'c'.

Figure 48.10 Sample pattern of emotional quotient (EQ)

15 Questions
15 Minutes

7 point Likert-type scale
1- Strongly Disagree
7- Strongly Agree

Meaning of studies	(07)
Personal Responsibility	(02)
Positive Connection with other individuals	(01)
Conditions for Community	(01)
Inner Life	(02)
Institute and Individual	(02)

Figure 48.11 Sample pattern of spiritual quotient (SQ)

1. How often do you feel stressed?

( ) always ( ) everyday( ) sometimes ( ) once in a while( ) I dont know ( ) never 2.

2. In what aspect of life you have the most problems?

( ) college ( ) family( ) friends ( ) love life( ) social life/community ( ) I don¶t know( ) others, please specify

____

3. Rate (1-5) the following according to the most difficult problems you have to deal with.

______ college ______ family ______ friends ______ love life ______ community

College Stress

4. Do you think you have more college work than you should have?

( ) yes ( ) no

If yes, what makes most of your college work?

(can answer more than once)( ) assignments ( ) projects( ) tests ( ) extra-curricular activities

5. Do you get bored at college?

( ) yes ( ) no ( ) sometimes ( ) dont know

If yes, why do you get bored? (can answer more than once)

( ) boring lessons ( ) boring teachers ( ) very easy lessons ( ) unexciting environment

Figure 48.12 Sample question of stress analysis

applicability, and they give less precise outcomes in human stress analysis. Figure 48.12 shows the Sample question of stress analysis.

Result and Discussion

The following section gives a details overview of the proposed 4QS model with education and curriculum development on different quotient,

4QS test conduction

Written tests were conducted based on the 4QS model to test the quotients of students at the entry-level questioners were shared with college students.

Data collected includes universities, government-aided colleges, and minority colleges.

Figure 48.14 shows the result analysis of the 4QS model for improving the learning system of education using some predicting questions and some tentative observations on sample students of different classes. Figure 48.13 shows the Test pattern with their details and Figure 48.14 shows the result analysis of the 4QS model for improving the learning system of education using some predicting questions and some tentative observations on sample students of different classes.

Sr. No.	Quotient Tested	Number of Questions	Time Duration
Q1	Adversity Quotient	20	30 mins
Q2	Intelligence Quotient	30	60 mins
Q3	Emotional Quotient	15	15 mins
Q4	Spiritual Quotient	15	15 mins
5	Stress Test	10	10 mins

Figure 48.13 Test pattern with their details

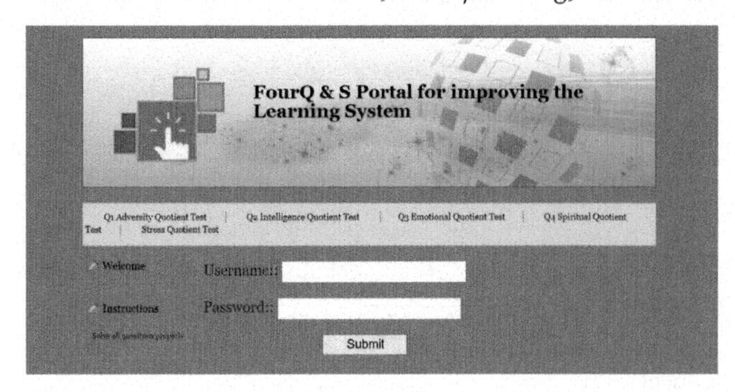

Figure 48.14 4QS portal front page of learning system

The following Figure 48.15 shows AQ test question, Figure 48.16 shows AQ test collected result and Figure 48.17 shows IQ test question of the 4QS Model.

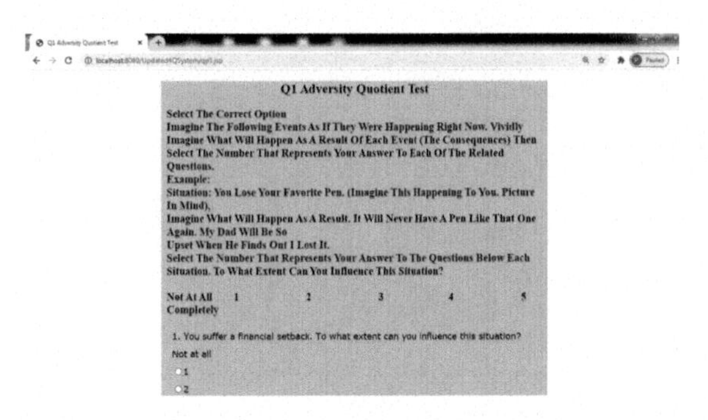

Figure 48.15 AQ test question

Min	92
Max	160
Average	121.828

Rating	Score Range	Sample Count
High	AQ (178–200)	0
Moderately High	AQ (161-177)	0
Moderate	AQ (135-160)	15
Moderately Low	AQ (118–134)	20
Low	AQ (117 and below)	23

Rating	Score Range	C1	C2	C3	C4
High	AQ (178-200)	0	0	0	0
Moderately High	AQ (161-177)	2	1	1	1
Moderate	AQ (135-160)	72	40	43	24
Moderately Low	AQ (118-134)	76	51	31	37
Low	AQ (117 and below)	65	37	24	52

Figure 48.16 AQ test collected result

Q2 Intelligence Quotient Test

1.Circle the answer(s), or write in the answer box provided.

50	51	49	52	48
46	47	45	48	44
49	50		51	
47		46		45
48		47	50	46

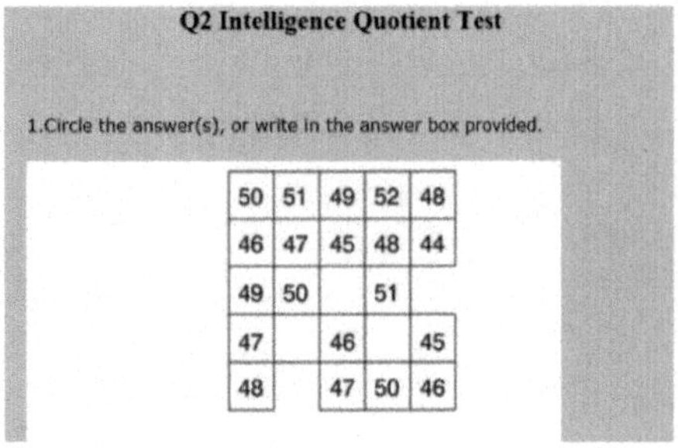

Figure 48.17 IQ test question

The following Figure 48.18 shows the IQ Test collected results of different sample questions with their answer shown below:

Scores between	Rating	C1	C2	C3	C4	Sample
27–30	Very highly exceptional	0	0	0	0	0
24–26	High expert	2	1	0	0	1
21–23	Expert	13	4	1	0	3
19–20	Very high average	49	31	33	5	4
17–18	High average	69	45	29	11	13
13–16	Middle average	45	33	20	29	29
10–12	Low average	30	12	11	43	7
6–9	Borderline low	7	3	5	17	1
3–5	Low	0	0	0	7	0
0–2	Very low	0	0	0	2	0

Figure 48.18 IQ test collected result

The Figure 48.19 shows the implementation of the EQ sample test question and evaluates the performance.

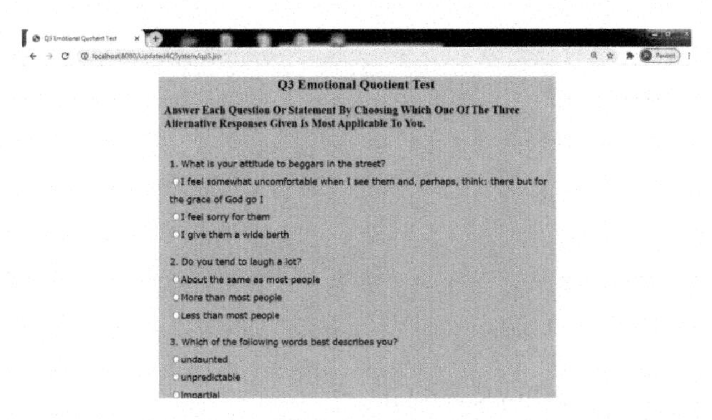

Q3 Emotional Quotient Test
Answer Each Question Or Statement By Choosing Which One Of The Three Alternative Responses Given Is Most Applicable To You.
1. What is your attitude to beggars in the street?
I feel somewhat uncomfortable when I see them and, perhaps, think: there but for the grace of God go I
I feel sorry for them
I give them a wide berth
2. Do you tend to laugh a lot?
About the same as most people
More than most people
Less than most people
3. Which of the following words best describes you?
undaunted
unpredictable
impartial

Figure 48.19 EQ test question

Below, Figures 48.20 and 48.21 shows a detailed overview of student considered for the sample test with their tentative question sets:

Range	Rating	C1	C2	C3	C4	Sample
25–30	Excessively emotional	16	8	3	10	5
19–24	Fairly emotional	116	69	52	51	43
13–18	Average	56	52	43	50	10
8–12	Fairly unemotional	0	0	1	3	0
below 7	Excessively unemotional	0	0	0	0	0

Figure 48.20 EQ test collected result

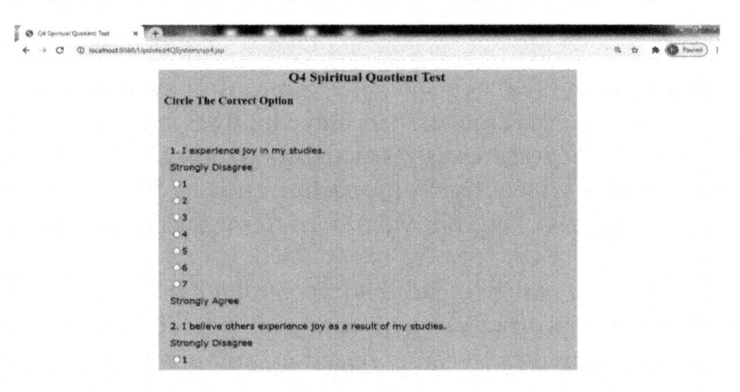

Figure 48.21 SQ test question

The following Figure 48.22 shows the IQ Test collected results of different sample questions with their answer shown below with mean values:

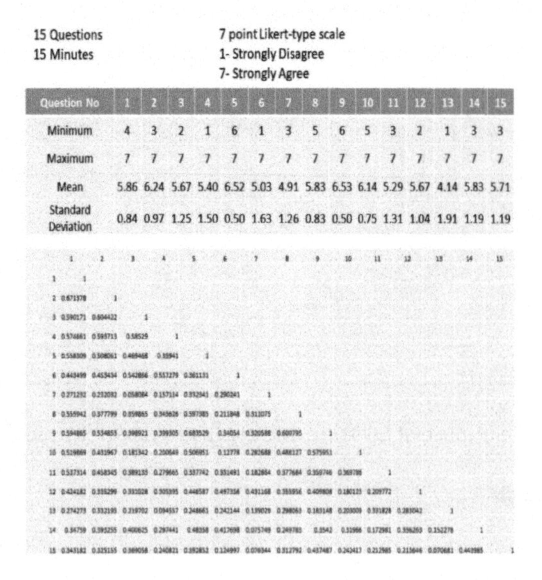

15 Questions
15 Minutes

7 point Likert-type scale
1- Strongly Disagree
7- Strongly Agree

Question No	1	2	3	4	5	6	7	8	9	10	11	12	13	14	15
Minimum	4	3	2	1	6	1	3	5	6	5	3	2	1	3	3
Maximum	7	7	7	7	7	7	7	7	7	7	7	7	7	7	7
Mean	5.86	6.24	5.67	5.40	6.52	5.03	4.91	5.83	6.53	6.14	5.29	5.67	4.14	5.83	5.71
Standard Deviation	0.84	0.97	1.25	1.50	0.50	1.63	1.26	0.83	0.50	0.75	1.31	1.04	1.91	1.19	1.19

	1	2	3	4	5	6	7	8	9	10	11	12	13	14	15
1	1														
2	0.671378	1													
3	0.590171	0.604422	1												
4	0.574661	0.593713	0.58529	1											
5	0.558309	0.508061	0.469468	0.33941	1										
6	0.443499	0.453434	0.542866	0.557279	0.361131	1									
7	0.271232	0.232032	0.058084	0.157114	0.332941	0.290241	1								
8	0.555942	0.377799	0.359865	0.343626	0.597385	0.211848	0.311075	1							
9	0.594865	0.534853	0.398921	0.399305	0.683529	0.34054	0.320588	0.600795	1						
10	0.519869	0.431967	0.181342	0.200649	0.506951	0.12778	0.282688	0.488127	0.575951	1					
11	0.537314	0.458345	0.389135	0.279665	0.337742	0.331491	0.182884	0.377684	0.359746	0.369798	1				
12	0.424182	0.335299	0.331028	0.305395	0.448587	0.497356	0.431168	0.353956	0.409808	0.180123	0.209772	1			
13	0.274273	0.332195	0.219702	0.094557	0.248661	0.242144	0.139029	0.298063	0.183148	0.203009	0.331828	0.283042	1		
14	0.34759	0.395255	0.400625	0.297441	0.48358	0.417698	0.075749	0.249785	0.3542	0.31986	0.172981	0.336263	0.152278	1	
15	0.343182	0.325155	0.369058	0.240821	0.392852	0.124997	0.076344	0.312792	0.437487	0.242417	0.212985	0.213646	0.070681	0.443985	1

Figure 48.22 SQ test collected result

Stress Test

Social Life problems are of more concern
Heavy Mobile users were stressed
Stressed students were easily bullied
Students seek more support from Parents/Mentors related to Social Life.
Figure 48.23 shows the sample stress question for analysing the stress related to questions & answers of students, and class is highly affected by adversity. 40% of students are with low adversity scores.
Most of the students have an IQ level of above average. 13% of the class is still having an IQ below average.
74.13% of students are emotional, and five students are excessively emotional.

Figure 48.23 Stress test collected result

Conclusion

Student performance is vital in educational data mining research since it is a means of evaluating educational outcomes. Additionally, projecting student success may aid learners and educators in improving their respective learning and teaching. However, the present research is restricted to statistical methodologies and educational philosophy and does not emphasise widely used approaches, such as feature learning. Additionally, current research lacks comparative analyses of established procedures using comparable metrics and verified datasets.

With the growth of online education, it is becoming easier to gather large amounts of data to demonstrate the critical nature of SPP. This article summarises prior research on data mining techniques, including data collection, issue formulation, method selection, prediction target selection, and practical implementations. We examined three distinct modes of education, namely online, traditional, and blended learning.

Then, we sorted these papers into several categories to create a study summary. Additionally, we performed assessment tests on two distinct data sets to examine the approaches involved, including a private data set from the university and a public data set. The result demonstrates that (1) Method-learning techniques may achieve a high score on SPP, and (2) SPP could be boosted by feature selection.

Additionally, the assessment argues for the value of prerequisite courses—besides the links between the selected attributes in data sets. Thus, we may use feature selection to identify the most relevant traits associated with a student's performance for improved prediction outcomes. Following the reviews and case studies, we explored the challenges and benefits of present and future research.

We want to expand pedagogical machine-learning methodologies for SPP in the future. On the one hand, this thorough analysis of this emerging field of ML and data mining pushes us to build more widely used approaches for SPP. On the other hand, future research should include past knowledge from education to construct domain-specific machine-learning models. This study can assist the educational system in achieving higher academic achievements while also decreasing educational financing and expense.

References

Anand, M. (2019). Advances in EDM: State of the art. In Software engineering, 193–201. Lviv: Springer.

Buenano-Fernandez, D. and Villegas-CH, W. and Lujan-Mora, S. (2019). The use of tools of data mining to decision making in engineering education—A systematic mapping study. Comput. Appl. Eng. Edu. 27(3):744–758.

Fan, Y., Liu, Y., Chen, H., and Ma, J. (2019). Data mining-based design and implementation of college physical education performance management and analysis system. Int. J. Emerg. Technol. Learn. 14(06):87–97.

Fernandes, E., Holanda, M., Victorino, M., Borges, V., Carvalho, R., and Van Erven, G. (2019). Educational data mining: Predictive analysis of academic performance of public-school students in the capital of Brazil. J. Bus. Res. 94:335–343.

Juhaňák, L., Zounek, J., and Rohlíková, L. (2019). Using process mining to analyse students' quiz-taking behavior patterns in a learning management system. Comput. Hum. Behav. 92:496–506. DOI: 10.1016/j.chb.2017.12.015.

Kim, B.-H., Vizitei, E., and Ganapathi, V. (2018). Gritnet: Student performance prediction with deep learning. arXiv preprint arXiv:1804.07405.

Krathwohl, D. R., Bloom, B. S., and Masia, B. B. (1973). Taxonomy of Educational Objectives, the Classification of Educational Goals. Handbook II: Affective Domain. New York: David McKay Co, Inc.

Kushwaha, R. C., Singhal, A., and Swain, S. (2019). Learning pattern analysis: A case study of moodle learning management system. In Recent trends in communication, computing, and electronics, 471–479, Langkawi: Springer.

Lam, C. M., To, S. M., and Chan, W. C. H. (2018). Learning pattern of social work students: A longitudinal study. Soc. Work Educ. 37:49–65. doi: 10.1080/02615479.2017.1365831.

Liu, Q., Tong, S., Liu, C., Zhao, H., Chen, E., Ma, H., and Wang, S. (2019). Exploiting cognitive structure for adaptive learning. In Proceedings of the 25th ACM SIGKDD international conference on knowledge discovery & data mining, (pp. 627–635), Anchorage, AK.

Liu, Q., Wu, R., Chen, E., Xu, G., Su, Y., and Chen, Z. (2018). Fuzzy cognitive diagnosis for modelling examinee performance. ACM Trans. Intell. Syst. Technol. 9:1–26. DOI: 10.1145/3168361.

Simpson, E. J. (1972). Ofra Walter, Rinat Ezra, Communicating Different and Higher across the Praxis of Bloom's Taxonomy While Shifting toward Health at Every Size (HAES), The Classification of Educational Objectives in the Psychomotor Domain. Gryphon House, Washington.

Zhang, Y., Dai, H., Yun, Y., Liu, S., Lan, A., and Shang, X. (2020a). Meta-knowledge dictionary learning on 1-bit response data for student knowledge diagnosis. Knowl. Based Syst. 205:106290. DOI: 10.1145/3448139.3448184.

Zhang, Y., He, X., Tian, Z., Jeong, J. J., Lei, Y., Wang, T., Zeng, Q., Jani, A. B., Curran, W. J., and Yang, X. (2020b). Multi-needle detection in 3D ultrasound images using unsupervised order-graph regularised sparse dictionary learning. IEEE Trans. Med. Imaging 39:2302–2315. doi: 10.1016/j.knosys.2020.106290.

Zhang, Y., Yun, Y., Dai, H., Cui, J., and Shang, X. (2020c). Graphs regularised robust matrix factorisation and its application on student grade prediction. Appl. Sci. 10:1755. DOI: 10.3390/app10051755.

Zhang, Y., An, R., Cui, J., and Shang, X. (2021a). Undergraduate grade prediction in Chinese higher education using convolutional neural networks. In LAK21: 11th. International learning analytics and knowledge conference, (pp. 462–468).

Zhang, Y., Lei, Y., Lin, M., Curran, W., Liu, T., and Yang, X. (2021b). Region of interest discovery using discriminative concrete autoencoder for covid-19 lung ct images. In Medical imaging 2021: Computer-Aided diagnosis, (Vol. 11597) (International Society for Optics and Photonics), 115970U.

49 Tribological behaviour of Nickel treated with Silicon Nitride prepared through powder metallurgy route

Rajneesh Gedam[a] *and Nitin Dubey*[b]

Department of Mechanical Engineering Madhyanchal Professional University, Bhopal, India

Abstract

The current study presents the tribological behaviour of nickel metal powder reinforced with silicon nitride nano particles through powder metallurgy route (PM route). Cylindrical pre-forms with Si3N4 weight percentage of 0%, 5%, 10%, 15% respectively are prepared at a compaction pressure of 130 KN using a die and punch assembly. Sintering temperature taken as 510^{0}C, 540^{0}C and 570^{0}C using muffle furnace. The sliding wear test on pin-on-disk tribometer is carried out against an EN-8 steel counter face under dry ambient conditions with an applied normal load of 10N, sliding distance 4000 m and 1200 seconds. SEM images are taken to study the surface morphology of the samples. Results revealed an augmented wear resistance due to addition of second phase particle as silicon nitride in the nickel. SEM images showed uniform distribution of the silicon nitride particles in the composite matrix.

Keywords: Hardness, Nickel, Silicon Nitride (Si3N4), wear properties, sintering temperature.

Introduction

Powder metallurgy (PM) is a method that produces metallic forms from metallic powders. The metal or alloy is solid at the start of the PM route and stays solid at the conclusion. Powder metallurgy is becoming increasingly essential in the fabrication sector. Powder metallurgy enables us to work with materials such as refractory materials that are difficult to process, such as sintered carbides (Pradhan et al., 2017; Prakash et al., 2018; Ghorbani et al., 2018; Aoki and Ohtake, 2004). Powder metallurgy is the process of combining tiny powdered materials, compacting them into a desired shape or frame inside a form, and then heating the compacted powder in a controlled environment, known as sintering, to stimulate the arrangement of powder particles to shape the necessary component. Composites are lightweight materials with high mechanical qualities (Singh and Harimkar, 2012; Torralba, et al., 2003). Epoxies and polyester are common matrix materials. Graphite, glass, boron, and other materials are commonly used as reinforcing fibres. Metal matrix and ceramic composite materials are the focus of new developments. Cutting tools constructed of silicon carbide reinforced alumina are being developed, with considerably enhanced tool life (Salem et al., 2017; Mishra and Srivastava, 2017; Kaczmar et al., 2000). A composite material is made up of several different components. Compound materials are incorporated into composites to take use of their properties, resulting in a better version of the material (Van Trinh et al., 2018; Kang et al., 2018; Zhanga et al., 2018). It is found from the different research works that, several researchers have used composites over the substrate to enhance the tribological characteristics of the substrate material but few have used powder metallurgy process to review the surface characteristics. Powder metallurgy route can prove to be highly efficient in improving the wear resistance of the material.

Literature Review

The impact of the aluminium lattice, SiC sizes, and SiC volume on composite microstructure formation and mechanical characteristics were examined by Salem (2017). They used a ball milling machine and powder metallurgy to make volume pieces out of MMCs made of AlSiC in various diameters. After 120 hours of treatment, Al and Al-SiC composites of varying volume fractions were compacted and sintered at 125 MPa and 450°C (Salem, 2017). The produced composite samples' thermal conductivity, electrical resistivity, and micro hardness analyses were then measured. He concludes that as the size of Al-SiC particles shrinks and the volume of Al-SiC expands, the characteristics of MMCs deteriorate. Micro hardness, on the other hand, was raised when the volume fraction of small Al-SiC particles was increased.

Pradhan et al. (2017) studied the wear and friction behaviour of Al-SiC MMC in three different environments: dry, aqueous, and alkaline. As the stress rises, the pin on disc apparatus, in which the specimen

[a]gedamrajnish901@gmail.com; [b]nddubey246@gmail.com

slides beneath an alumina disc at varying speeds, wears out. The alkaline medium appears to have the most wear, followed by aqueous and dry environment.

SEM and EDS are employed for micro structural and wear analysis, and he concludes that in dry mediums, adhesive and abrasive wear are prevalent, whereas mechanical and corrosive wear are prominent in aqueous and alkaline mediums.

Experimentation

A. Die and Punch Preparation

To prepare composite material samples, a well-defined combination of powder pre-forms a die and punch was made by turning the die steel on a lathe and then tempering the punch for high load applications. A pin holder with a capacity ranging from 6 mm to 12 mm is used in the pin on disc machine. Because the powder pre-forms must have enough contact area with the disc, a pin diameter of 10 mm was chosen.

Results & Discussion

A. Effect of 510°C sintering temperature on wear

Figure 49.1 depicts the variance in wear over time. It was discovered that at 510°C sintering temperature, a sample of 0% Si_3N_4 (Silicon Nitride) shows the highest wear rates, followed by 5% Si_3N_4, 10% Si_3N_4, and 15% Si_3N_4 with the least wear visible in the SEM. This experiment uses a 10N applied load and a 4000 metre sliding distance as the test distance.

B. Effect of 540°C sintering temperature on wear

Figure 49.2 depicts the progression of wear over time. It was discovered that at a sintering temperature of 540°C, a sample of 0% Si_3N_4 (Silicon Nitride) shows the highest wear rates, followed by 5% Si_3N_4, 10% Si_3N_4, and 15% Si_3N_4 with the least wear visible in the SEM. This experiment is carried out with a 10N applied load and a sliding distance of 4000 metres. It avoided noisy sound as soon as the specimen rubbed on the wear tester disc.

C. Effect of 570°C sintering temperature on wear

Figure 3 depicts the progression of wear over time. It was discovered that at a sintering temperature of 570°C, a sample of 0% Si_3N_4 (Silicon Nitride) shows the highest wear rates, followed by 5% Si_3N_4, 10%

Table 49.1 Sample preparation

Silicon nitride% at T1 = 510°C	Silicon nitride% at T2 = 540°C	Silicon nitride% at T3 = 570°C
0% Si_3N_4	0% Si_3N_4	0% Si_3N_4
5% Si_3N_4	5% Si_3N_4	5% Si_3N_4
10% Si_3N_4	10% Si_3N_4	10% Si_3N_4
15% Si_3N_4	15% Si_3N_4	15% Si_3N_4

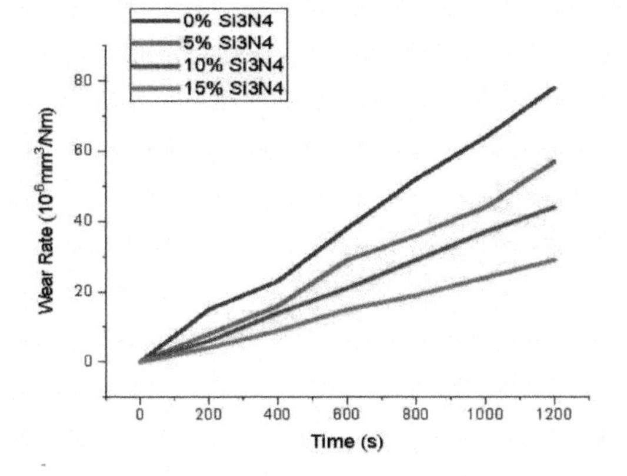

Figure 49.1 Average wear (in micron) with silicon nitride (Si_3N_4) at 510°C sintered specimen

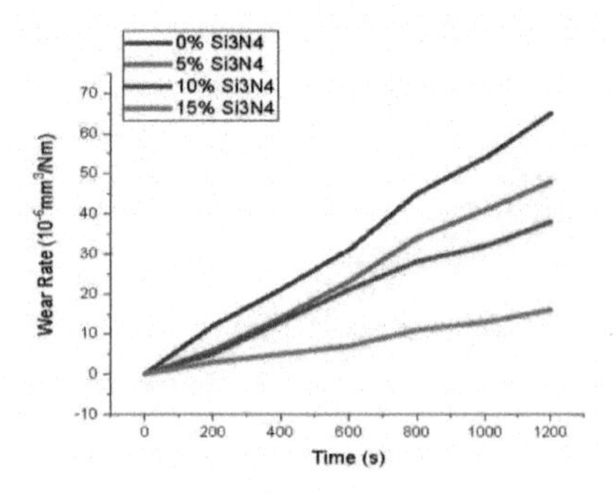

Figure 49.2 Average wear (in micron) with silicon nitride (Si_3N_4) at 540°C sintered specimen

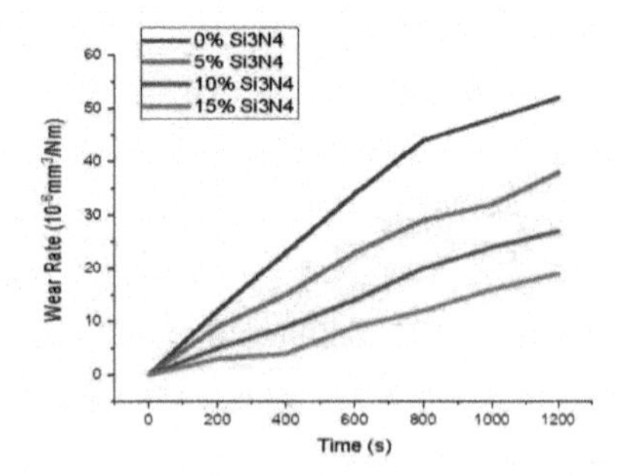

Figure 49.3 Average wear (in micron) with silicon nitride (Si_3N_4) at 570°C sintered specimen

Si_3N_4, and 15% Si_3N_4 with the least wear visible in the SEM. This experiment is carried out with a 10N applied load and a sliding distance of 4000 metres. It avoided noisy sound as soon as the specimen rubbed on the wear tester disc.

Microstructure and Morphology of Composite through SEM

Microstructure of Ni- Silicon Nitride composites (a) Ni-5 percent Silicon Nitride, (b) Ni-10 percent Silicon Nitride, and (c) Ni-15 percent Silicon Nitride composition of Nickel matrix and Silicon Nitride reinforcement of Silicon Nitride are clearly visible in SEM images, according to scanning electron microscope (SEM) analysis. In Ni-5 percent Silicon

Nitride composites, micro cracks and porous microstructure were observed, however porous structure and micro fractures were reduced in Ni-10 percent Silicon Nitride composites. The black region represents the Nickel matrix, whereas the white colour represents the Silicon Nitride particle dispersion. The silicon nitride particles are evenly scattered. Ni-15 percent Silicon Nitride matrix has no porosity when compared to Ni-5 percent Silicon Nitride and Ni-10 percent Silicon Nitride matrix.

For 5% Si_3N_4 Composite

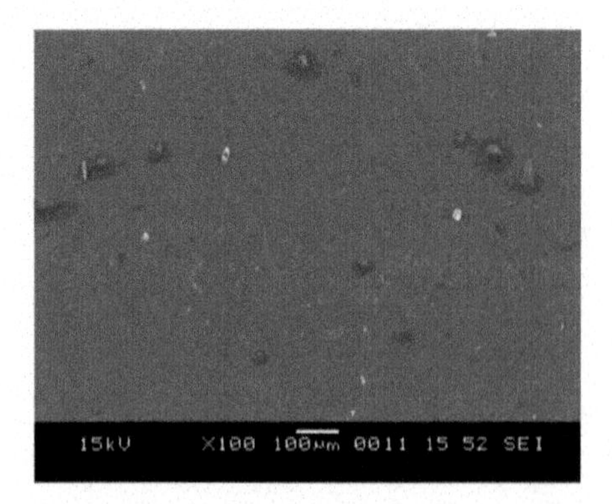

Figure 49.4 5% Si_3N_v composite

Figure 49.4 shows the Microstructure of Ni-Si_3N_4 composites (5% Si_3N_4) which show that silicon nitride particles are distributed over the surface of substrate but are lesser in quantity as visible.

For 10% Si_3N_4

Figure 49.5 10% Si_3N_4 composite

Figure 49.5 shows the Microstructure of Ni-Si_3N4 composites (10% Si_3N_4) which shows that there's equal and continuous distribution of reinforcement material i.e.Si_3N_4 over the base metal nickel. We can observe that the scattering of Si3N4 is more than that of 5% Si3N4 samples.

For 15% Si_3N_4

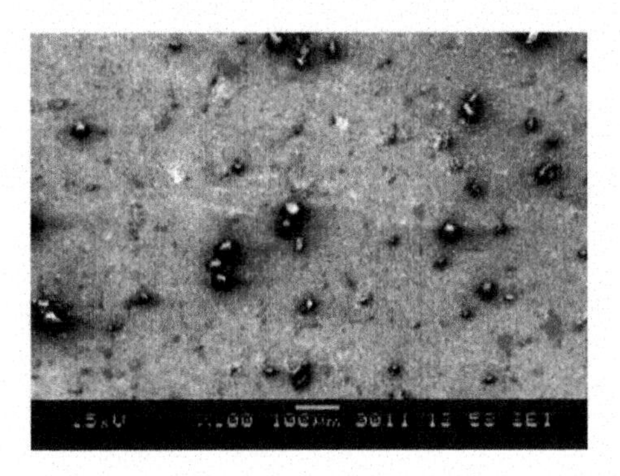

Figure 49.6 15% Si_3N_4 composite

Figure 49.6 shows the Microstructure of Ni- Si3N4 composites (15% Si_3N_4) which shows that there's equal and continuous distribution of reinforcement material i.e. Si_3N_4 over the base metal Nickel.

Conclusion

In this work, wear of Ni-Si_3N_4 composite is investigated at different sintering temperature and percentage incorporation of silicon nitride particles. The results demonstrate that wear is highly dependent on Si3N4 concentration.

1. When compared to 510°C and 540°C, a specimen at 570°C confirms minimal wear.
2. In comparison to 0 percent, 5 percent, and 10 percent Si_3N_4, a specimen at 15% Si_3N_4 confirms minimal wear.
3. Micro cracks and porous microstructure were seen in the Ni-5 percent silicon nitride, however the tiny fractures and a porous structure were reduced in the Ni-10 percent silicon nitride composite.
4. The black region represents the Nickel matrix, whereas the white colour indicates the Silicon Nitride particle dispersion. The silicon nitride particles are evenly scattered.
5. Ni-15 percent Silicon Nitride matrix has no porosity compared to Ni-5 percent Silicon Nitride and Ni-10 percent Silicon Nitride matrix.

Novelty of the Work

The present research includes the wear analysis of nickel incorporated with silicon nitride using the powder metallurgy route. In machine elements, tribological characterisation is of utmost importance to enhance the wear resistance of the elements. The work presented in this paper represents the enhancement of the wear resistance by the use of ceramics and the powder metallurgy process.

References

Aoki, Y. and Ohtake, N. (2004). Tribological properties of segment-structured diamond-like carbon films. Tribol. Int. 37(1112):941–947. doi: 10.1016/j.triboint.2004.07.011.

Ghorbani, A., Sheibani, S., and Ataie, A. (2018). Microstructure and mechanical properties of consolidated Cu-Cr-CNT nanocomposite prepared via powder metallurgy. J. Alloys Compd. 732:818–827. doi: 10.1016/j.jallcom.2017.10.282.

Kaczmar, J. W., Pietrzak, K. and Wlosiński, W. (2000). Production and application of metal matrix composite materials. J. Mater. Process. Technol. 106(1–3):58–67. doi: 10.1016/S0924-0136(00)00639-7.

Kang, B. Lee, J., Ryu, H. J., and Hong, S. H. (2018). Ultra-high strength WNbMoTaV high-entropy alloys with fine grain structure fabricated by powder metallurgical process. Mater. Sci. Eng. A. 712:616–624. doi: 10.1016/j.msea.2017.12.021.

Mishra, A. K. and Srivastava, R. K. (2017). Wear Behaviour of Al-6061/SiC Metal Matrix Composites. J. Inst. Eng. Ser. C. 98(2):97–103.

Pradhan, S., Ghosh, S., Barman, T. K., and Sahoo, P. (2017). Tribological Behavior of Al-SiC Metal Matrix Composite Under Dry, Aqueous and Alkaline Medium. Silicon. 9(6):923–931. doi: 10.1007/s12633-016-9504-y.

Prakash, K. S., Gopal, P. M.,Anburose, D., and Kavimani, V. (2018). Mechanical, corrosion and wear characteristics of powder metallurgy processed Ti-6Al-4V/B4C metal matrix composites.Ain Shams Eng. J. 9(4):1489–1496.doi: 10.1016/j.asej.2016.11.003.

Salem, M. A., El-Batanony, I. G., Ghanem, M., and ElAal, M. I. A. (2017). Effect of the Matrix and Reinforcement Sizes on the Microstructure, the Physical and Mechanical Properties of Al-SiC Composites. J. Eng. Mater. Technol. Trans. ASME. 139(1):1–7. doi: 10.1115/1.4034959.

Singh, A. and Harimkar, S. P. (2012). Laser surface engineering of magnesium alloys: A review. Jom.64(6):716–733. doi: 10.1007/s11837- 012-0340-2.

Torralba, J. M., Da Costa, C. E., and Velasco, F. (2003). P/M aluminum matrix composites: An overview. *J*. Mater. Process. Technol.133(1–2):203–206. doi: 10.1016/S0924- 0136(02)00234-0.

Van Trinh, P., Lee, J., Minh, P. N., Phuong, D. D., and Hong, S. H. (2018). Effect of oxidation of SiC particles on mechanical properties and wear behavior of SiCp/Al6061 composites. J. Alloys Compd. 769:282–292. doi: 10.1016/j.jallcom.2018.07.355.

Zhanga, C., Gao, Q., Lv, P., Cai, J., Peng, C. T., Jin, Y., and Guana, Q. (2018). Surface modification of Cu- W powder metallurgical alloy induced by high- current pulsed electron beam. Powder Technol. 325:340–346. doi: 10.1016/j.powtec.2017.11.037.

50 Drop test finite element analysis of different grades of ASTM A500 for structural lifting frame

Ashish Kumar Shrivastava[1,a], *Ashish Kumar Sinha*[1,b], *Abhishek Choubey*[1,c], *Neha Choubey*[2,d], *Manish Billore*[1,e], *and Juber Hussain Qureshi*[3,f]

[1]Department of Mechanical Engineering, Sagar Institute of Science, Technology & Research (SISTec-R), Bhopal, India

[2]School of Applied Sciences & Languges, VIT Bhopal University, Bhopal, India

[3]Deptartment of Mechanical Engineering, IPS Group of College, Gwalior, India

Abstract

A primary concern of a novel design is to create positive impact on the performance of a new product. International standards, which specify mechanical requirements and evaluate yield stress and deformation properties of products, must be followed during the testing method. Engineers may now design and test new products in virtual environments because to rapid advances in numerical analysis and simulation techniques, quicker processing power, and more memory capacity. These computer based simulations, which are based on finite element analysis (FEA), provide significant information for 3-D planning and manufacturing new items, as well as perfecting existing ones. This strategy has shown to be extremely beneficial to manufacturers, as it allows them to achieve more production at a lower cost per unit, as well as produce engineering components that are simple to build and make the most efficient use of their materials. A drop test is typically used to determine the component's resistance to free fall. This paper focused on the design and (dynamic) analysis of a structural lifting frame in accordance with Det Norske Veritas (DNV) 2.7-1. In this paper, FEA utilized to numerically model drop testing. The simulation setup is developed in according to the requirements of the DNV 2.7-1 standard, and the findings are examined and studied. Based on the outcome, a recommendation for safe handling is suggested. In the worst- situation, the drop test orientation is set to normal to gravity and for 2-D and 3-D modeling, defines drop test height as 2' from lowest point/corner with assign gravity to 3-D model as 386.22 in/sec^2; plane 26 denotes rigid floor and in this analysis obtain the Von misses Stress 28108 Psi and allowable stress value is 46000 Psi with 1.6 factor of safety. The result of displacements analysis at long beam, short beam and corner beam are 0.138 inch and allowable displacements are 0.574, 0.333, 0.428 inches with factor of safety 4.2, 2.4, 3.1 respectively.

Keywords: Crash frame; DNV-2.7-1; dop simulation; FEA, Solidwork.

Introduction

A free fall of a component is referred to as a drop test. One of the functional tests required for DNV certification of a lifting frame is the drop test. As a result, design teams can evaluate practically any vessel drop test without incurring the costs of production and machine time. Impact testing must be carried out numerically in order to reduce design time, improve mechanical performance, and reduce product development costs. The modelling of an impact test for a new product using finite element analysis is the subject to this investigative study. During the impact test, simulation was used to investigate the stress and displacements distributions (Zarei and Kröger, 2007). As a result, explicit finite element systems are increasingly being used to estimate the performance of new product designs rather than physical tests. Frame constructed and certified in accordance with this Standard for Instrument should be strong enough to withstand the normal forces experienced in offshore operations and not completely fail if it is subjected to additional exceptional loads. These frames are commonly used to protect the mechanical machines from any physical damage when they are transferred from one location to another or lifted by crane to be placed on a boat or the ground. Impact causes enormous stresses and strains on a frame, which can eventually cause damage to the equipment inside, resulting in a waste of time and money. Because offshore operations are so expensive, any transportation or equipment failure can result in a significant loss for the organisation. As a result, the crash frame's impact carrying capacity should be determined before it is really tested in order to build it in a more effective way to avoid damage during handling operation (Figures 50.2–50.6). In this work ,we can be determine the strain at the corner, stresses and displacement by dropping by using FEA method. It's very advantages full technology to save the manufacturing cost and tested without the destroy the product. It also reduces the cost and time of physical drop testing (Figure 50.1, (Hu et al., 2009).

[a]ashish.shri.nri@gmail.com; [b]as9610@gmail.com; [c]abhishekchby2@gmail.com; [d]Nehachby2@gmail.com; [e]manishbillore@gmail.com; [f]juber.nri@gmail.com

Materials and Method Selection

Due to significant structural welding, ASTM A500 Grade B was chosen. Carbon steel, cold formed welded, and seamless structural tube in both round and shaped shapes are covered by the ASTM A500 standard specification. ASTM A500 is a standard specification for cold-formed welded and seamless carbon steel structural tubing in round, square, and rectangular shapes, published by the American Society for Testing and Materials (ASTM). In the United States, it is frequently specified for hollow structural components. ASTM A501, which is a hot-formed counterpart of A500, is another similar standard.

Method: Produce a FEA methodology to perform drop test simulation as per parameters considered in standard DNV2.7-1. These parameters are listed below.

The frame must be lowered or dumped on to a concrete or other solid structural plant floor. The frame must be angled so that the bottom of each of the lowermost side and end rails attached to the smallest corner makes a 5°angle with the bottom.

1. The highest height difference between the highest and lowest point of the underside of the frame corners must not exceed 400 mm.
2. The frame must drop freely for at least 5 cm after being released, giving it a speed of at least 1 m/s upon impact.
3. Acceptance criteria: No severe and unending damage will be tolerated. Weld cracks and mild deformities can be rectified.

Boundary condition detail

1. Payload + lifting frame weight = max gross weight (MGW).
2. To replicate internal load for the lifting frame, a payload of thick plate is placed to the lifting frame.
3. A plate with a length of 133.40 inches, a width of 77.20 inches, and a height of 9 inches was designed to reflect a uniform payload across the whole lifting frame.

Figure 50.1 Physical drop test

Table 50.1 Mechanical properties of different grades of ASTM A500 (Galehdari and Khodarahmi, 2016)

Properties	Grade A	Grade B	Grade C
Minimum Yield Strength, Psi	39,000 (269MPa)	46,000 (317MPa)	50,000 (345MPa)
Minimum Tensile Strength, Psi	45,000 (310MPa)	58,000 (400MPa)	62,000 (427MPa)
Elongation at Break, %	25	23	21

Results and Discussion

Define Contacts for 3-D model

Assign the Bonded for welding contact set to all structural Beams in the lifting frame. In solidworks 2016 simulation, bonded entities act as if they were welded, and the computer merges coincident nodes at the interface for a compliant mesh with created nodes and elements.

Due to huge size of model, solidworks' automated mesher chooses a solid type mesh. In solidworks, standard meshing turns on the Voronoi-Delaunay meshing technique for further meshing processes. The software assigns 10 nodes to each solid element when the high quality mesh option is selected: four corner nodes and one node in the middle of each edge (a total of six mid-side nodes). The Jacobian 4 point approach is chosen for analysis because of its high meshing quality. The automatic mesher generates parabolic tetrahedral solid components parabolic elements in high grade mesh (Li et al., 2011).

Following steps are followed for meshing the frame.

1. Import the geometry into the program.
2. Disfeaturing the geometry by deleting small holes, fillets, and little extended surfaces, among other features.

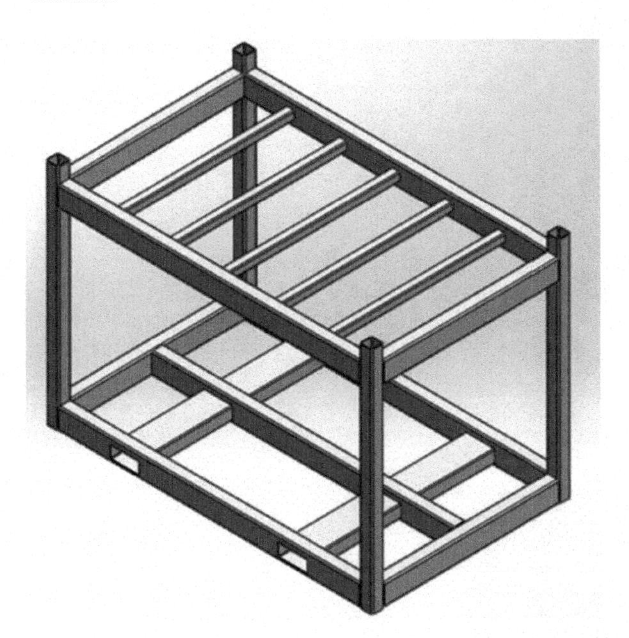

Figure 50.2 CAD model of crash frame

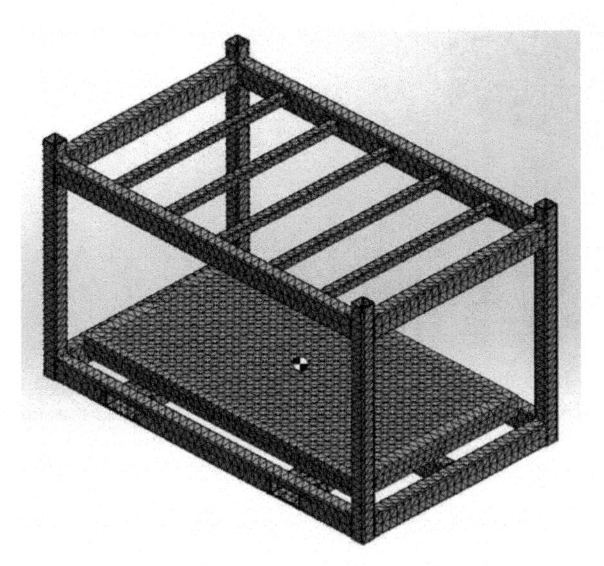

Figure 50.3 Mesh model of crash frame

3. At the welding point, nodes are merged.
4. To acquire an accurate simulation result, a high-quality mesh is chosen.
5. Parabolic tetrahedral solid elements are generated using a high-quality mesh.

Resultant Stresses Plot

The simulation results reveal that no element/geometry exceeds the material's minimum yield strength, indicating that there is no plastic deformation in the frame. Maximum stress is visible on the lowermost corner of the beam, which is the first to contact the unyielding floor. Since it hits the unyielding floor, the maximum stress is seen at the expected place.

Result summary tables

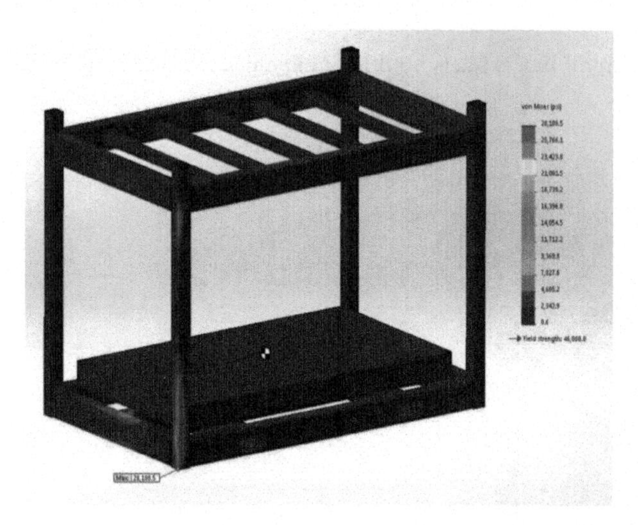

Figure 50.4 FEA model of crash frame stress results

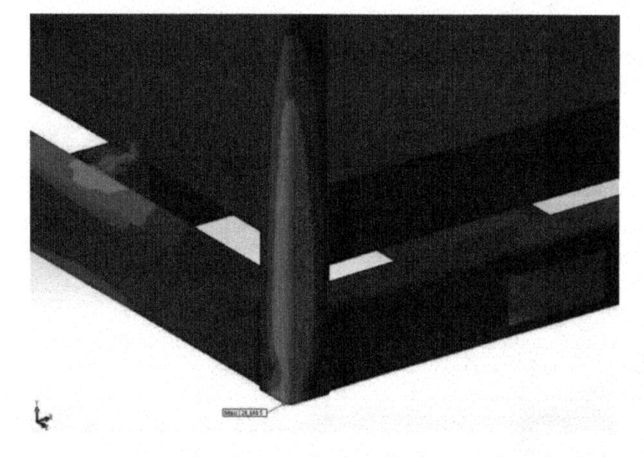

Figure 50.5 FEA Model of crash frame zoom in stress results

Resultant Displacement Plot

Maximum deflection or displacement is noticed on the longer beam of the base frame from the maximum tension inside the frame. If the factor of safety (FOS) is greater than 1.5, the lifting frame has passed the Drop test stress criteria. DNV has established an acceptable deflection limit for each individual beam, as shown in the equation below.

$$y = Ln/250$$

where, y = Allowable deflection.

Ln = Total length of the beam, rail or post.

Corner and base structural beams, which experience the largest bending in drop tests, are key parts of design. The maximum deflection in the longest and most stressed beam is less than the permissible limit. The displacement criterion are applied to the lifting frame (Di Palma et al., 2019).

Resultant Strain Plot

The DNV2.7-1 standard does not indicate how much of a strain is considered substantial. To be recognised as a major unending distortion by bare eyes, a zone of plastic distortion must be quite large. As a consequence, a credible criterion is developed to better interpret the results, as shown below. When the spread of a plastic strain exceeds 10 in length, it is deemed substantial. Figure 50.8 shows the maximum plastic strain plot. The maximum plastic strain recorded is less than the permissible limit of 5; hence the model does not shatter. This plastic strain is distributed across a very small area, as illustrated in the larger view in Figures 50.7 and 50.8.

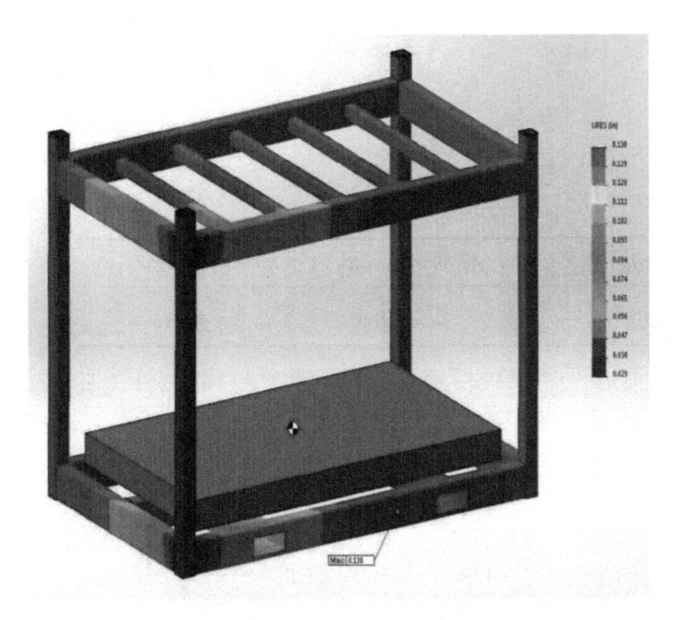

Figure 50.6 FEA model of crash frame displacement results

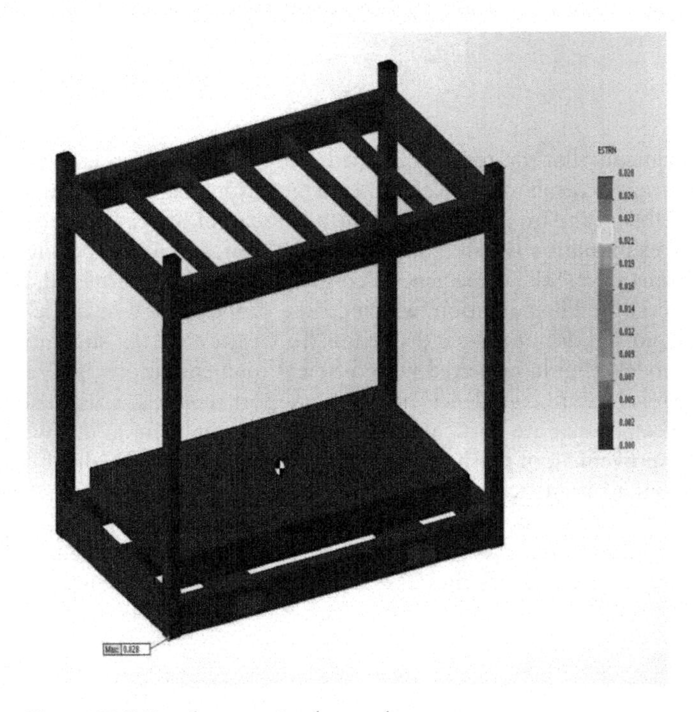

Figure 50.7 Resultant strain plot results

Figure 50.8 Resultant strain plot zoom in results

Table 50.2 von Mises stress summary

Sr.No	*Types*	*Analysis value (Psi)*	*Allowable value (Psi)*	*FOS*
1	Von-Mises stress	28,108	46,000	1.6

Table 50.3 Displacement summary

Sr.	*Beam*	*Analysis value (in)*	*Allowable Value (in)*	*FOS*
1	Long Beam	0.138	0.574	4.2
2	Short Beam	0.138	0.333	2.4
3	Corner Beam	0.138	0.428	3.1

The above tables show the results of drop test analysis, Table 50.2 shows the stress results and Table 50.3 describe the displacements values of analysis.

Conclusion

The It can be interpreted from the experiments that the results obtained using finite element analysis are always variable. Respectable chance variation in result must be determined by existing based on simulation complexity and failure inflexibility. The DNV standard does not provide a value of plastic strain that may be used as a comparison point; the DNV examination is based on visual inspection, and minor plastic distortions are difficult to detect with the human eye. Still, some general correlations can be determined, such as the absence of cracks or rupture strain in the FE simulation, which can be supplemented in the testing findings. In the FEA simulation, there is no significant plastic distortion. To ensure that the simulation's modelling technique is correct, several models should be tested with different load conditions. The simulation results provide the modelling approach more confidence, and they support the mesh size, material models, element selection, and contact sketches that are particular to the problem statement in this work.

This drop test simulation aids in the improvement of crash frame design. Free fall from a height of 5 cm is used in the simulation, which takes about 15 hours to complete. The velocity of the frame can be determined before it strikes the ground, which reduces the time it takes to simulate the velocity.

According to the simulation results, the lifting frame meets all criteria for drop test allowable limits per DNV2.7-1, thereby passing the drop test and ensuring that the design will withstand an accidental drop.

References

Di Palma, L., Di Caprio, F., Chiariello, A., Ignarra, M., Russo, S., Riccio, S., De Luca, A., and Caputo, F. (2019). Vertical Drop Test of Composite Fuselage Section of a Regional Aircraft. ARC Int. Diary. https://doi.org/10.2514/1.J058517

Faidallah, R. S. A., Morad, M. M., Wasfy, K. I., El-Sharnouby, M., Kesba, H., El-Tahan, A. M., El-Saadonyf, M. T., Awnya, A. (2021). Utilizing biomass energy for improving summer squash greenhouse productivity during the winter season: Utilizing biomass energy for improving summer squash greenhouse productivity. Saudi J. Biol. Sci. 29(2):822–830. doi: 10.1016/j.sjbs.2021.10.025.

Galehdari S. A. and Khodarahmi, H. (2016). Design and analysis of a graded honeycomb shock absorber for a helicopter seat during a crash condition. Int. J. Crashworthiness. 21(3):231–241. doi: 10.1080/13588265.2016.1165440.

Hu, D. Y., Yang, J. L., and Hu, M. H. (2009). Full-scale vertical drop test and numerical simulation of a crashworthy helicopter seat/occupant system. Int. J. Crashworthiness. 14(6):565–583. doi: 10.1080/13588260902896433.

Jawad, M., Schoop, R., Suter, A., Klein, P., Eccles, R. (2013). Perfil de eficacia y seguridad de Echinacea purpurea en la prevención de episodios de resfriado común: Estudio clínico aleatorizado, doble ciego y controlado con placebo. Rev. Fitoter. 13(2):125–135. doi: 10.1002/jsfa.

Li, M., Deng, Z., Liu, R., and Guo, H. (2011). Crashworthiness design optimisation of metal honeycomb energy absorber used in lunar lander. Int. J. Crashworthiness. 16(4):411–419. doi:10.1080/13588265.2011.596677.

Zarei, H. R. and Kröger, M. (2007). Crashworthiness optimization of empty and filled aluminum crash boxes. Int. J. Crashworthiness. 12(3):255–264. doi: 10.1080/13588260701441159.

51 Verilog implementation of AES 256 algorithm

Ankita Tijare, Prajwal Yelne[a]*, Ankit Mindewar, and Khushboo Borgaonkar*

Department of Electronics Engineering, Yeshwantrao Chavan College of Engineering, Nagpur, India

Abstract

In the developing world of the digital media most of the instructions are shared via electronic media. Thus there is a need for protecting those data from malicious attacks. Here cryptography comes into existence. It is a method of protecting information and communications using codes, so information can be accessed by those for whom it is intended. Now to encrypt these data we use Advanced Encryption Standard (AES) for securing information. It is generally of 128 bits with varying key lengths. Both encipher and decipher are done with the AES method. In the modern world most of the web browsers uses AES to protect the connection with websites. The simulation of the AES algorithm of the highest key size i.e. (256 bits) is done using ModelSim Intel FPGA starter edition 10.5b. The work discusses the functional simulation of the both encryption as well as decryption process of AES algorithm.

Keywords: Advanced encryption standard, data encryption standard (DES), look up tables.

Introduction

From the first attempts of cryptography which were done several thousand years ago through the enigma code that was used in World War-2 and now modern crypto algorithms such as SHA-3, Argon 2, cryptography has developed by leaps and bounds (Knudsen and Robshaw, 2011).

Due to limitations for data encryption standard due to its low size the National Institute of Standards and Technology came up with a ope process to select a new block of cipher. Thus Rijndael was used because of its security, performance, efficiency and ease in implementation (Surabhi and Meenu, 2021).

The Advanced Encryption Standard (AES) is a symmetric block cipher which encrypts information in blocks of 128 bits. The symmetric cipher implies that the cipher is using the same key for encryption as well as decryption. AES encipher and decipher can be done using 128 bit, 192 bit and 256 bit key (Barrera et al., 2023).

Even with fastest computer we have, it will take many years to solve AES. As a matter of fact cracking AES-256 would take 10^{22} times longer than the universe existed (universe is currently 1.38×10^{10} years old).

The styling of any digital circuit measurement Verilog has two design flows. The first step is to import our Verilog files into the simulation application (Surabhi and Meenu, 2021). Then these simulation tool simulates in code for a particular behaviour of the hardware circuit for input that we describe during testbench as a result our simulation is comparatively quick. Once we are assured that our style is correct, we tend to use them as a hardware synthesis tool to present to a low level gate netlist (Knudsen and Robshaw, 2011). The netlist is then matched to resources on the device we want to use, such as a Field Programmable Grid Array (FPGA), resulting in a working digital circuit.

Advanced Encryption Standard (AES) Algorithm

1. Inputs and Outputs

The AES technique we utilise includes 128 bits of data with values ranging from 0 to 255. These sequences will be called blocks, and the amount of bits will be called lengths. The encryption key utilised has a 128, 192, and 256 bit sequence. The bits in question will be numbered from zero to less than one the length of the sequence. The index of a bit is the number I assigned to it, which varies based on the block and key lengths (Jindal et at., 2020).

2. Algorithm Specification

AES algorithm is sensitive to the length of the input and output blocks and therefore the state is around 128 bits. The input plain text is converted into a state matrix of 4 × 4, which has four words each of 32 bits.

[a]ankita.tijare@gmail.com

128 bits, 192 bits and 256bits are different key size in AES algorithm. According to the AES algorithmic rule. Number of Key (Nk) = 4, 6, or 8 determines the length of the key, this shows how many words are in the encryption key, the size of the word is 32-bit. The key size of the number of rounds determines the number of rounds the algorithm will conduct.

3. Cipher

After N-1 operations of same nature on state matrix which is applying some permutation and combination, the Nth round is little different from first N-1 rounds.

The key used is composed of a one dimensional array of four bits by key expansion routine. The sub-sections are described below which consists of Substitution of bytes subsequent shifting of rows, mixing of columns and adding round key (Sunil, 2020).

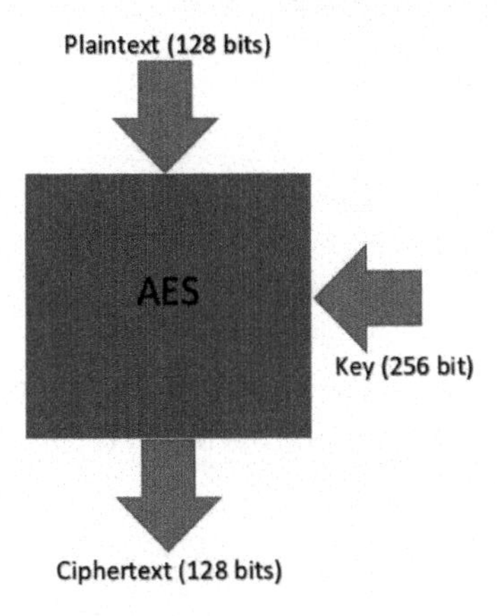

Figure 51.1 AES conceptual schema

Algorithm	*Key length (bits)*	*Block size (bits)*	*No of round*
AES-128	128	128	10
AES-192	192	128	12
AES-256	256	128	14

Figure 51.2 Number of rounds

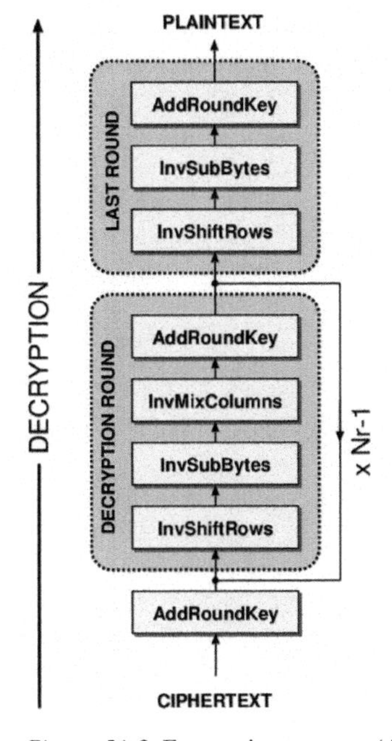

Figure 51.3 Encryption process (Abd Zaid, 2019)

3.1.1. Sub-Bytes Transformation

This transformation is a substitution which is of non-linear type, which is done by using an 8-bit s-box. This makes the algorithm resistant to attacks that are based on simple algebraic properties.

3.1.2. Shift-Rows Transformation

The Shifting of rows is done in a circular nature. First row remains unaltered i.e. shifted by 0 bits towards left. Second row is shifted 8 bits toward left, third row is shifted 16 bits left, and so on.

3.1.3. Mix-Column Transformation

The mix column and the shift row add diffusion to the cipher. In this step each column is processed separately. An effective matrix multiplication in Galois field (2^8) using the polynomial $m(x) = x^8 + x^4 + x^3 + x + 1$.

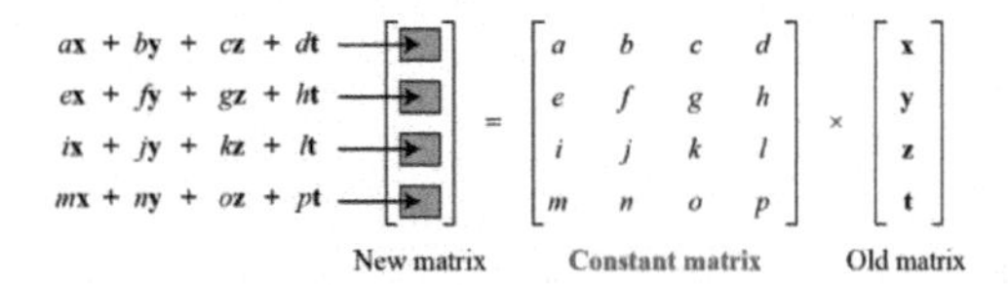

3.1.4. Add-Round-Key Transformation

When adding round key a spherical secret key XORed with the state array one column at a time. Each Spherical Key is made up of N words which is taken from the key schedule. Thus the operation becomes a simple XOR matrix addition.

Inverse Cipher

Include same four step in inverse way.

Key Expansion

The AES algorithm initialises the cypher key as K0 and executes a key enlargement operation to build a key schedule. The key enlargement results in a total of 112 words. The rule requires a large number of words, and each of the round requires a 8 word long round key. Thus, the expansion algorithm generates each round key from previous key by using three functions viz. RotWord, SubWord, and XOR with Rcon.

RotWord performs a linear cyclic vertical rotation to the last column of the initial key.

Next word is passed through the SubWord function which is a substitutes values from a non-linear 16 × 16 substitution box which adds more permutations to the algorithm.

The Rcon function XORs the output of the SubWord with a pre-defined round constant which is different for every round. Finally the output of Rcon is XORed with the first word of the initial key to get ninth word. This whole function is called a Galois function.

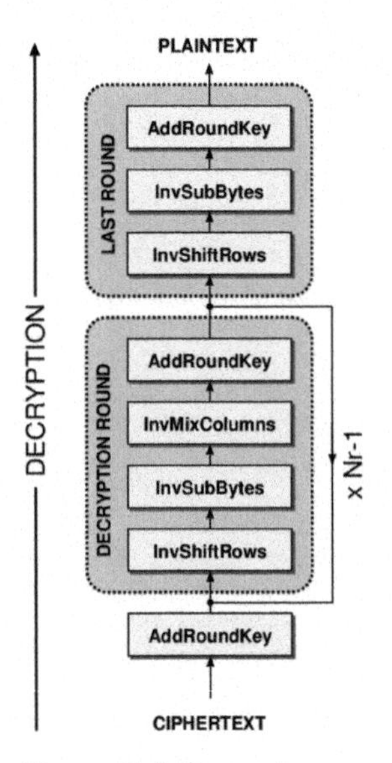

Figure 51.4 Decryption process (Abd Zaid, 2019)

To calculate the rest of the columns we use the Nk+1 word and bitwise add it with the same word of the previous round key. Thus to calculate W10 = W9 ⊕ W2 and so on.

It's worth noting that, the 256 bit key length version of key expansion is different from 192 bit and 256 bit versions. For 256 bit encryption we have 14 rounds and for each round we need eight words long round key, thus the key expansion generates 112 words in total.

Encryption Result

```
VSIM 3> run -all
#

datain=00000000000000000000000000000000,

dataout=46f2fb342d6f0ab477476fc501242c5f
```

In the encryption process we provide 128 bit (32 characters) of plain text in this case is

Plain Text = 00000000000000000000000000000000. With the plain text we also provide a 256 bit encryption key = c47b0294dbbbee0fec4757f22ffeee3587ca4730c3d33b691df38bab076bc558.

The algorithm gives us the output cypher text after encryption process.

Cypher text = 46f2fb342d6f0ab477476fc501242c5f

Decryption Result

In the decryption process we provide 128 bit (32 characters) of cypher text in this case which is

Cypher Text =46f2fb342d6f0ab477476fc501242c5f.

```
VSIM 3> run -all
#

dataout=00000000000000000000000000000000

datain=46f2fb342d6f0ab477476fc501242c5f
```

With the plain text we also provide the same 256 bit key as AES is a symmetric cypher. Encryption key = c47b0294dbbbee0fec4757f22ffeee3587ca4730c3d33b691df38bab076bc558.

The algorithm gives us the output plain text after decryption process.

Plain text=00000000000000000000000000000000

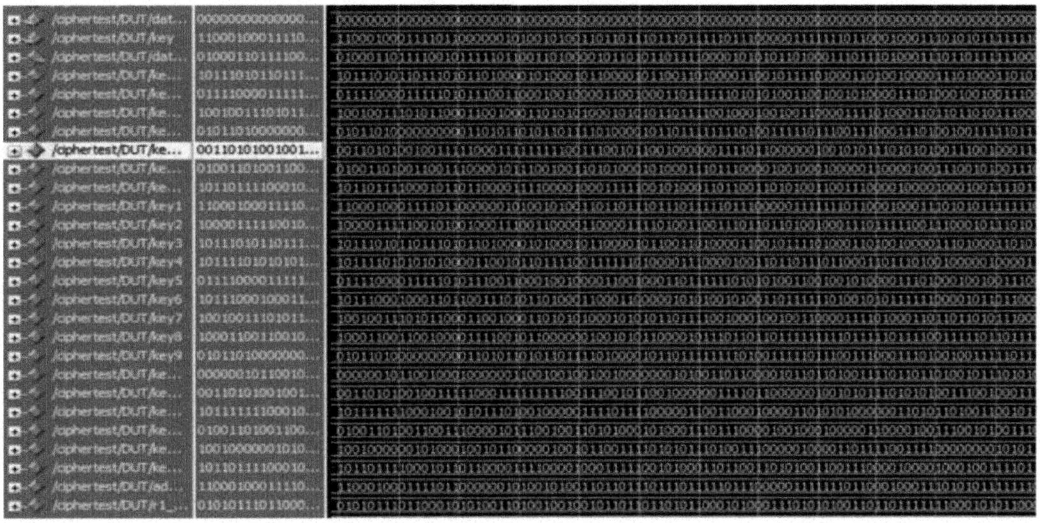

Figure 51.5 Simulation of encryption process

Conclusion and Future Scope

To summarise, this project implements a design using hardware description language Verilog to support key length of 256 bit AES encryption and decryption circuit. We were successfully able to simulate the AES 256 algorithm using ModelSim (Quartus prime 17.1). The encryption and decryption plain text are in accordance. The implemented algorithm has the highest key size which provides high level of security to information. It improves the flexibility. The algorithm can be further optimised in near future to consume less LUTS and slices to make the functional implementation more synthesisable, to do this, we need more analysis of how to reuse. The resource or how to improve the resource utilisation to achieve the better design.

References

Abd Zaid, M. M. 2019. Modification Advanced Encryption Standard for Design Lightweight Algorithms. researchgate.net. doi:10.31642/JoKMC/2018/060104

Barrera, A., Cheng, C. –W., and Kumar, S. 2023. A Fast Implementation of the Rijndael Substitution Box for Cryptographic AES. 3rd International Conference on Data Intelligence and Security (ICDIS). doi: 10.1109/ICDIS50059.2020.00009.

https://www.ibm.com/topics/data-security

Jindal, P., Kaushik, A., and Kumar, K. 2020. Design and Implementation of Advanced Encryption Standard Algorithm on 7th Series Field Programmable Gate Array. 7th International Conference on Smart Structures and Systems (ICSSS). doi: 10.1109/ICSSS49621.2020.9202114.

Knudsen, L. R. and Robshaw, M. J. B. 2011. The Block Cipher Companion (Information Security and Cryptography). Berlin, Heidelberg: Springer.

Sunil, J., Suhas, S. H. Sumanth, B K., and Santhameena, S. 2020. Implementation of AES Algorithm on FPGA and on software. IEEE International Conference for Innovation in Technology (INOCON). doi: 10.1109/INOCON50539.2020.9298347.

Surabhi, P. R. and Meenu, T. V. 2021. Advanced 256-Bit Aes Encyption With Plain Text Partitioning. International Conference on Advances in Computing and Communications (ICACC). doi: 10.1109/ICACC-202152719.2021.9708158.

Soumya, V. H., Neelagar, M. B., and Kumaraswamy, K. V. 2018. Designing of AES Algorithm using Verilog. 4th International Conference for Convergence in Technology (I2CT). doi: 10.1109/I2CT42659.2018.9058322.

52 Wrapper-based feature selection on ransomware detection using machine learning

Rushikesh A. Pujari[a] *and Pravin S. Revankar*[b]

Department of Computer Enginerring and IT, College of Engineering Pune, Pune, India

Abstract

Nowadays, in the world of digitalisation, everyone has an android device. As digitalisation is on the rise to become pervasive, the numbers of cybercrimes are also growing. Malware is one of the biggest threats in new era of smartphones connected to the Internet. Ransomware is a kind of program that threatens to harm you by preventing you from accessing your data. Then the attacker demands a ransom to get data access back. A recent study shows that ransomware can target any kind of device connected to internet. So it is important to detect ransomware attacks. In this study, different machine learning (ML) techniques are used to detect ransomware attacks based on network-traffic features. For this experiment all different ransomware families are combined and relabelled as ransomware, making it a 2-class classification problem, and trained the model. According to results, compared to other supervised ML classifiers, the Random Forest classifier with wrapper-based feature selection technique has achieved the highest weighted accuracy of 87.54%. The suggested strategy is tested and validated using the RandomForest classifier on the CICAndMal2017 and CIC-InvesAndMal2019 datasets, respectively. On these datasets, the proposed system outperforms on seven features from the original independent feature set.

Keywords: Machine learning, malware, network-traffic features, ransomware, wrapper-based feature selection.

Introduction

Mobile phones are said to be one of the necessities as they allow us to keep in touch with the world. Mobile phones provide many vital features like storing personal information, connecting to emergency services like hospitals, and many more. They are becoming an essential part of human life, so bad members of society try various methods to control them. Android operating system and IOS operating system are the two central operating systems widely used. According to reports, Android held the top spot in the global mobile market in January 2022, followed by iOS-based devices (González-Pérezet al., 2022). Due to the widespread usage of Android smartphones, attackers primarily target those smartphones. In the past years, malware attacks on mobile devices increased by nearly double (Humayun et al., 2021).

Ransomware is a form of malicious code that allows an attacker to take control of a victim's machine. It encrypts all data till the victim pays a ransom to the attacker. Mobile ransomware attacks may involve the following actions.

a) Encrypting data and demanding money to decrypt for the affected data.
b) Lock the phone screen so that the user can not use the phone's functions until the user pays the ransom.
c) Steal personal information and threaten to disclose it to the public if the user does not pay the ransom.

There are two types of ransomware on Android devices: Crypto Ransomware: this ransomware encrypts all user data and locks and unlocks it using cryptogenic keys. The attackers then demand a ransom in exchange for the cryptogenic keys. And Locking Ransomware: this sort of ransomware encrypts the phone and prevents it from being used. It displays a message throughout the programs, which remains visible until the user pays the ransom (Scalas et al., 2019). According to Venkatesan (2020), Android malware takes advantage of the platform's SYSTEM ALERT WINDOW feature. This feature allows the Android phone to display notifications that require the user's attention and cannot be dismissed. In Android OS version 8.0 and later, Google included the 'Kill switch' to address this issue. Over three million apps are available on Google's app store. There are, however, numerous third-party app developers. Attackers employ their-party software to carry out their attacks.

[a]pujarira20.comp@coep.ac.in; [b]psr.comp@coep.ac.in

Researchers use Dynamic-based and static-based methods to construct ransomware detection systems. ML uses both methods to improve ransomware detection: The dynamic-based method studies the behaviour in a controlled environment, and the Static-based method evaluates ransomware without executing it (Abuthawabeh and Khaled, 2019).

Recently, supervised ML techniques for identifying ransomware have gotten a lot of attention. The usefulness of supervised learning methods for ransomware detection is examined in this research. Various supervised ML algorithms for ransomware detection are used in the proposed system.

This paper's contributions are summarised as follows:

1. The Wrapper-based feature selection methodology (forward selection) is proposed in this research as a feature selection method.
2. The suggested feature selection technique is validated on the InvesAndMal2019 dataset and tested on CICAndMal2017.
3. Using retrieved features and minimal model construction time, the proposed feature selection technique outperforms the RF classifier on the CICAndMal2017 and InvesAndMal2019 datasets.

This paper is split into six sections. Section II contains some related research on ransomware detection using ML. The dataset's overview is presented in section 3. Section 4 contains the Proposed Methodology. Section 5 evaluation results are presented, section 6 concludes the paper.

Litrature Review

Recent studies show that ransomware is one of the dominant malicious software that encrypts all the system files using cryptogenic keys and demands ransom for decryption of these files. Therefore it is essential to study ransomware attacks. Sgandurra et al. (2016), proposed a ML model called EldeRan. Authors estimated their approach on a dataset consisting of samples of 582 ransomware attacks and 942 benign. A filter-based feature selection and for detection, they used logistic regression. Their methodology is quite good at detecting ransomware. The only drawback is that they only used a small dataset.

Chen et al. (2017) used a dynamic API calls flow graph for ransomware detection. They have used various ML techniques such as support machine vector, Random forest, Naive Bays. Their methodology is effective on their dataset consisting of 168 samples. The only flaw is that they didn't use a standard dataset. Takeuchi et al. (2018), used support vector machines to detect ransomware. Their methodology uses ransomware API calls. They believe that for detecting unknown ransomware support machine vectors are suitable. They employed a dataset with just 588 cases, with 276 ransomware samples. This is the sole constraint.

There is no standard dataset for malware detection, so researchers created their dataset and used different approaches to detect ransomware. Lashkari et al. (2017) created the CICAndMal2017 dataset, which is a conventional dataset. It tries to limit the drawbacks of the dataset that the researchers previously used. They have collected malicious samples from various trusted resources such as VirusTotal Yang et al. (2019), and also they have collected benign applications from the year 2015 to 2017. In the end, they have installed all the applications on real smartphone devices and connected them to a single hotspot to collect network traffic.

Noorbehbahani and Rasouli (2019), used the above standard dataset to detect ransomware. They have used correlation-based feature selection techniques, and various supervised ML algorithms got an accuracy of 83.37. The only drawback of this experiment is that they have not utilised the entire dataset.

Alzahrani et al. (2019), created a behaviour-based ransomware detection called RanDetector. Before applying specific supervised ML models to identify the application, this detection method explores various information connected to ransomware operations.

Table 52.1 Comparision with other state of art techniques

Work	*Feature set*	*ML approach*	*Accuracy*	*Dataset size*
Sgandurra et al. (2016)	Filter based feature selection	SVM, NB	99.20	582-G 982-R
Chen et al. (2017)	Dynamic API calls	Simple Logistic	98.20	85-G 83-R
Takeuchi et al. (2018)	API calls	SVM	97.28	312-G 276-R
Noorbehbahani and Rasouli (2019)	Network Traffic feature	RF	83.37	409761-G 698414-R
Proposed model	Network Traffic feature	RF	87.54	8039-G 10724-R

Then to collect the network traffic feature, a hotspot of the computer is used to capture PCAP files using software such as TCMDUMP. Some advanced malware use time delay for escaping dynamic analysis. To do so, they recorded network traffic in three different scenarios: After installation of the application, 15 minutes before restarting, and 15 minutes after rebooting after installation. They have offered datasets in two formats: PCAP and CSV files.

This paper uses CSV files; CSV files are generated using software CICFlowMeter. The dataset contains the following malware categories:

1. Adware: Adware stands for advertising-supported software; it shows unwanted advertisements on the device. This software aims to get some clicks.
2. Ransomware: Ransomware first encrypts all the data or locks all the applications on the device then asks for a ransom to decrypt it.
3. Scareware: Scareware is the kind of software that manipulates victims to purchase or install something unwanted into the phone.
4. SMS malware: Cybercriminals create malware to target the victim's phone to make unauthorised phone calls or send a text message without the victim's consent.

Each category has different families. For this paper, we have taken the ransomware family. In the dataset, there are 84 independent features out of the four features that are irrelevant, namely Flow_id, Source IP, destination Ip and Timestamp, which are creating unnecessary noise and increasing computational cost. The Fwd Header length feature is coming twice, so one is removed. So, the final dataset contains 79 independent features.

Proposed Methodology

In this Analysis process, different feature selection techniques and supervised machine learning. Firstly, the dataset is prepared as shown in the following Figure 52.1. As there are ten families in the ransomware category. All the small datasets are merged to form a single dataset and relabelled as ransomware. The final dataset has two output values making it a 2-class classification problem, namely ransomware and benign. In the final dataset instances of ransomware are 10724 and instances of benign are 8039. The analysis process includes four stages, as shown in Figure 52.2.

The first stage is Pre-processing; in this step, noise, missing values, NaN values are handled. These all Pre-processing tasks are performed on Jupyter Notebook. With this dataset's help, it is ready for feature selection.

In the second stage, various feature selection techniques are used. ML offers a variety of feature selection strategies, including wrapper, filter, embedded, and hybrid-based methods. The main objective of the feature selection phase is to select the most valuable features. Here the system uses wrapper-based Forward selection techniques. The feature selection process depends on a ML algorithm that we want to fit in. It follows a greedy approach by evaluating every best subset combination. The best seven features are selected in this phase, as shown in Table 52.1.

In the third stage, ML classifiers are used to train the final dataset. Finally, the accuracy of classifiers is measured using the 10 Fold cross-validation approach using following formula.

$$ACCURACY = \frac{TP + TN}{TP + FP + TN + FN} \tag{1}$$

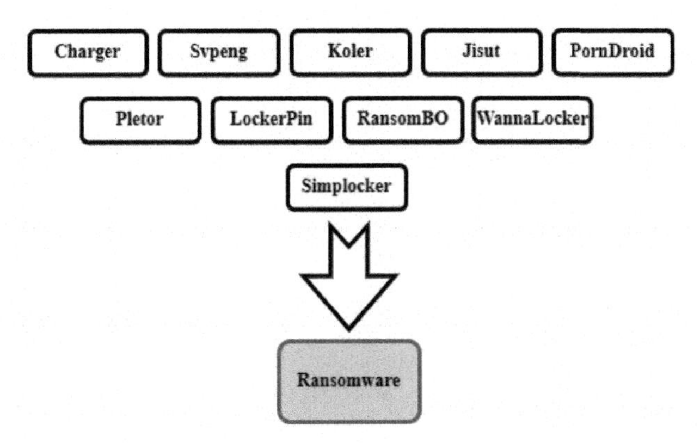

Figure 52.1 Dataset preparation

Here FP represents False Positive, FN represents False Negative, TP represents True Positive, TN represents True Negative. Figure 52.4 shows the evaluation process of the 10 fold cross-validation method. When 10F-CV is applied, the dataset divides into ten segments. Classifiers were trained on nine parts in every iteration and testing is performed on the remaining part. This process is carried out ten times and then it gives average accuracy.

Evaluation

The suggested system specified in Section 3 is successfully implemented on a 28.0 GB Random access memory system with an Xeon(R) CPU E3-1271 v3 @ 3.60GHz using Waikato Information Research Environment (Weka 3.8.5) (Hall et al., 2009). For pre-processing the python library, sci-kit-learn is used. The presented system is evaluated using the CICAndMal2017 dataset. This dataset has a total of 84 network-flow features, excluding the label.

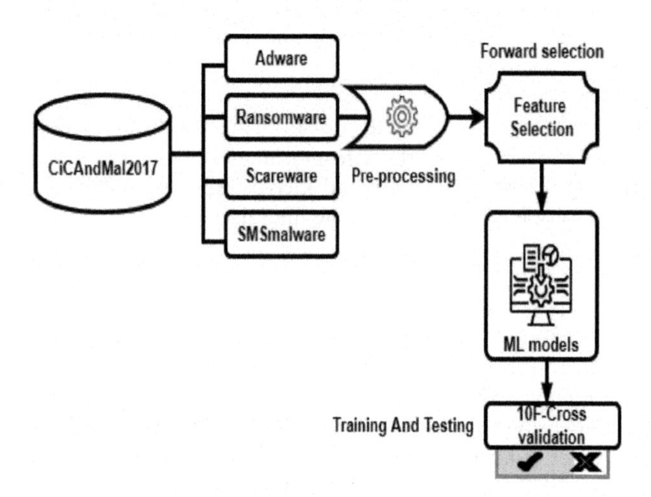

Figure 52.2 Analysis process

Table 52.2 Best 7 features

SourcePort
DestinationPort
Protocol
BwdPacketLengthMin
MinPacketLength
FIN Flag Count
Min_seg_size_forward

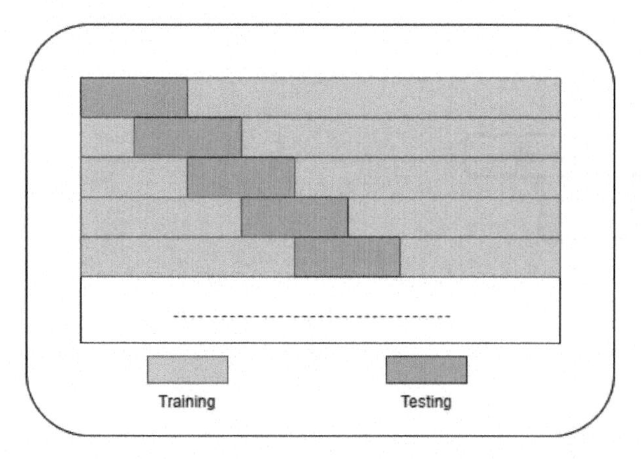

Figure 52.3 10F-cross-validation

Redundant Features such as Source IP, Destination IP, Timestamps, and Flow Id which create noise are removed. Then for the final dataset, various ML classifiers are used such as DecisionTree, RandomForest, KNN-3, KNN-4, KNN-5, ADaBoost ,GB and the total dataset accuracy is shown in Figure 52.4. Figure 52.4 shows Random Forest gives the highest accuracy, followed by the decision tree.

For feature selection, wrapper-based techniques are used. Wrapper-based feature selections are very time-consuming. But they give the best result. The obtained result with 10Fold cross-validation and Random forest classifier is shown in Table 52.1.

Forward selection is performed on the CICAndMal2017 dataset in this experiment, and the same approach is validated on the CICInvesAndMal2019 [14] dataset. Results are shown in Table 52.3. Graphical representation of Table 52.3 is shown in Figure 52.5.

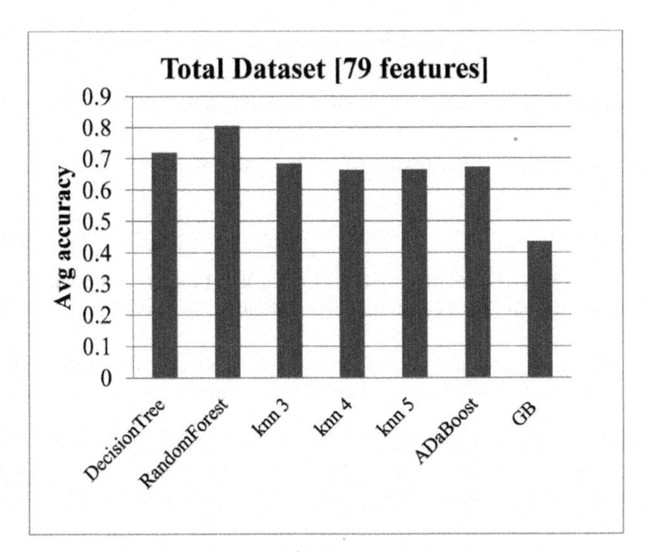

Figure 52.4 Avg. accuracy using train-test split for total dataset for CICAndMal2017

Table 52.3 Best seven features using RF

CICAndMal2017	*Feature selected*	*Evaluation used*	*Accuracy score (%)*
Total Dataset	79	Train-test split	79.31
F. Noorbehbahani[10]	Not mentioned.	10F-cross-validation	83.370
Proposed model	7	10F-cross-validation	87.53

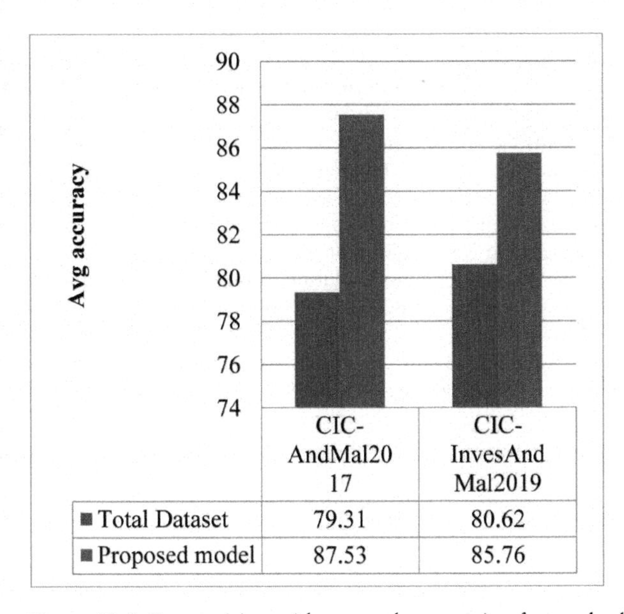

Figure 52.5 Comparision with recent dataset using forward selection +RF

Table 52.4 Validation Result

Dataset	*Total dataset [79 features](%)*	*Proposed Model [7 Features] (%)*
CICAndmal2017	79.31	87.53
CICInvesAndMal2019	80.62	85.76

Conclusion

In the proposed model, the presented system gives an accuracy of 87.53% using the best subset of seven features. Thus the proposed model reached the main objectives of the research paper. Detecting ransomware using all the features is a very time-consuming task, so it is essential to use feature reduction techniques. The proposed system uses the wrapper-based feature selection technique called forward selection is used on the dataset CICAndMal2017. Same method is validated on the dataset provided by CIC called CICInvesAndMAl2019. It also achieved high accuracy of 85.76%.Other feature selection strategies and classification methods will be examined for other datasets of ransomware for future studies.

References

Abuthawabeh, M. K. A. and Khaled, W. M. 2019. Android malware detection and categorization based on conversation-level network traffic features. In 2019 International Arab Conference on Information Technology (ACIT). (pp. 42–47). IEEE.

Alzahrani, A., Alshahrani, H., Alshehri, A., Fu, H. 2019. An intelligent behavior-based ransomware detection system for android platform. In 2019 First IEEE International Conference on Trust, Privacy and Security in Intelligent Systems and Applications (TPS-ISA). (pp. 28–35). IEEE.

Chen, Z.-G., Kang, H.-S., Yin, S.-N., Kim, S. R. 2017. Automatic ransomware detection and analysis based on dynamic API calls flow graph. In Proceedings of the International Conference on Research in Adaptive and Convergent Systems. (pp. 196–201).

González-Pérez, A., Matey-Sanz, M., Granell, C., and Casteleyn., S. (2022). Using mobile devices as scientific measurement instruments: Reliable android task scheduling. Pervasive Mob Comput. 81:101550.

Hall, M., Frank, E., Holmes, G., Pfahringer, B., Reutemann, P., and Witten, I. H. 2009. The WEKA data mining software: an update. ACM SIGKDD explorations newsletter. 11(1):1018.

Humayun, M., Jhanjhi, N. Z., Alsayatc, A., and Ponnusamyd, V. (2021). Internet of things and ransomware: Evolution, mitigation and prevention. Egyp. Informatics J. 22(1):105–117.

Lashkari, A. H., Kadir, A. F. A., Gonzalez, H., Mbah, K. F., Ghorbani, A. A. 2017. Towards a network-based framework for android malware detection and characterization. In 2017 15th Annual conference on privacy, security and trust (PST). (pp. 2332–3309). IEEE.

Noorbehbahani, F., and Rasouli, F. 2019. Analysis of machine learning techniques for ransomware detection. In 2019 16th International ISC (Iranian Society of Cryptology) Conference on Information Security and Cryptology (ISCISC). (pp. 128–133). IEEE.

Scalas, M., Maiorca, D., Mercaldo, F., and Aaron, C. (2019). On theeffectiveness of system API-related information for Android ransomware detection. Comput. Secur. 86:168–182.

Sgandurra, D., Muñoz-González, L., Mohsen, R., and Lupu, E. C. (2016). Automated dynamic analysis of ransomware: Benefits, limitations and use for detection. arXiv preprint arXiv.1609.03020.

Taheri, L., Kadir,A. F. A, and Lashkari, A. H. 2019. Extensible android malware detection and family classification using network-flows and API-calls. In 2019 International Carnahan Conference on Security Technology (ICCST). (pp. 18). IEEE..

Takeuchi, Y., Sakai, K., and Fukumoto, S. 2018. Detecting ransomware using support vector machines. In Proceedings of the 47th International Conference on Parallel Processing Companion. (pp. 16).

Venkatesan, D. 2020. Sophisticated new Android malware marks the latest evolution of mobile ransomware. https://www.microsoft.com/security/blog/2020/10/08/sophisticated-new-android-malware-marks-the-latest-evolution-of-mobile-ransomware/(accessed Feburary 1, 2022).

Yang, P. P. L., Song, L., and Wang, G..2019. Opening the blackbox of virustotal: Analyzing online phishing scan engines. In Proceedings of the Internet Measurement Conference. (pp. 478–485).

53 Combined effect of split injection and EGR on performance and emission of small diesel engine

Dond Dipak Kisan[a] *and Gulhane Nitin Parashram*[b]
Mechanical Department, Veermata Jijabai Technological Institute, Mumbai, India

Abstract

Today, increasing economic power output with a considerable simultaneous reduction in emissions from the diesel engine is the primary concern in the automobile industry. The present study investigates the impact of fuel injection pressure and EGR on the performance and emission attributes of the small diesel engine. The trials were conducted on the small diesel engine test setup for which the conventional mechanical fuel supply system was replaced with an uncomplicated version of the electronic direct injection system. The parameters considered for this electronic injection system study were injection pressure, split fuel injection timing, and quantity, EGR. Trials were carried out for different combinations of split injection and EGR conditions. From obtained results and their analysis, comparisons were made based on the performance and emission aspects of the engine. The analysis shows that split injection and EGR techniques reduce 23% nitrogen oxides (NOx) and 20% hydrocarbon (HC) emissions. But, at the same, there was a slight increase in carbon monoxide (CO) and smoke percentage in engine exhaust. The engine's thermal efficiency and specific fuel consumption show a 10% improvement for trials with split injection techniques. Overall, extensive improvements were observed in the engine's measurable characteristics for split injection and EGR trials.

Keywords: Emission characteristics, exhaust gas recirculation, performance characteristics, split injection.

Introduction

Today the world is facing a problem of finite fossil fuel reservoir capacity and pollution caused by them. Most of the cities of developing countries are getting polluted, and emission from diesel engines is one of the causes of the same. Presently, the government has stringent emission norms to overcome the pollution problem. Hence, this becomes a major challenge for engine manufacturing companies and researchers to go for less polluting engines. Small diesel engines are used for most stationary and mobile applications in the industry and agricultural fields. But till today, small diesel engines use a conventional fuel supply system to deliver an appropriate fuel quantity at the end of the compression stroke. Such traditional injection systems do not have accuracy over the parameters related to combustions such as delay span, injection pressure, injection duration, injection rate, prompt to fuel wastage, and environmental pollution. Earlier, large diesel engines have found significant improvement in efficiency with much lower emissions by using an electronic injection system. The electronic injection system has the adjustability to change the fuel injection policy. Injecting the fuel multiple times at different injection timing is one of the capabilities of such an electronic injection system. Researchers try to retrofit such type of affordable fuel supply system for the low-capacity diesel engine.

In an earlier study, researchers worked out combustion parameters to maximize thermal efficiency and lower the rate of emissions for diesel engines. Parameters like fuel, injection pressure, injection timing, % EGR, compression ratio, and splitting the fuel supply quantity were considered during his study. For the last two decays, authors have used biodiesel as fuel to overcome the problem of fuel crises (Reang et al., 2019; Reang et al., 2020; Dey et al., 2020). Also, many authors have observed the effect of exhaust gas recirculation (EGR) on a conventional small diesel engine's measurable attributes (Thangaraja and Kannan, 2016; Anantha, et al., 2017; Mossa et al., 2019). Authors had used diesel and biodiesel blends as fuel to overcome the problem of fuel crises as well emissions. His research shows a reduction in nitrogen oxides (NOx) percentage due to a drop in the temperature of exhaust gas (EGT). At the same time observed, a small increment in fuel consumption. Now, the researchers want to implement new sensors-assisted injection techniques to improve performance and reduce pollution from small diesel engines. Agrawal et al. (2015) and carpenter et al. (2016) made the modification to the engine setup by replacing the conventional injection system with an existing cheap common rail diesel injection (CRDI) system. They had performed the experiments on the modified setup for different injection pressure (IP) and injection timing (IT). The result analysis shows quite

[a]dpkdnd@gmail.com; [b]npgulhane@me.vjti.ac.in

better results compared to the conventional setup. Dividing the quantity of fuel injected during the combustion phase and injecting it at two different appropriate injection timing was an effective way to minimize the cylinder temperature and emission from diesel engines ((Molina et al., 2021; Jafarmadar, 2013; Anand, 2014; How et al., 2018; Park et al., 2008; Khatamnezhad et al., 2011). Significant improvement in the emission attributes like a large reduction in soot formation from the diesel engine was achieved by Molina et al. (2021) using such type of multiple injection technology. Further, many authors observed the impact of multiple injection techniques on the engine's behaviour and tried to find out the best possible injection timing. The behaviour of combustion, performance and emission attributes under the influence of split injection technology for the direct as well as an indirect diesel engine was studied by Jafarmadar (2013). The author extends his study to get the optimum combination of combustion system parameters to achieve lower NOx and soot concentration without affecting the engine's thermal efficiency. Jain et al. (2017) described the concept of combustion in which the fuel combustion had been shifted towards an increasingly premixed combustion phase. His research observed that successive rises in fuel injection pressure somewhat away from the top dead centre give higher combustion efficiency. Mobasheri (2017) examined the impact of fuel sprays pattern and multiple fuel supply concepts on the diesel engine. Observed results give lower NOx and soot formation at 105° angle. Anand (2014) investigated CRDI single-cylinder diesel engine to reduce the NOx and soot formation. The author employed combined split injection technology and high IP technique during his study. The combustion analysis shows an enhancement in the mixing of air and fuel with a simultaneous increase in the combustion temperature. Such combustion gives lower smoke percentage level. How et al. (2018) analysed the effect of the split injection strategy on the engine's measurable characteristics, which was fuelled with a biodiesel blend. The injected fuel was divided into two injection pulses with a fixed duration of 15° crank angle (CA) between two injections. The results show a significant decrease in the NOx and soot formation at the engine exhaust. A similar type of work was done by Park et al. (2018) and Khandal et al. (2020). They examined the impact of multiple-time fuel supply at different injection timing strategies on single-cylinder diesel engines. Edara et al. (2019) studied the impact of a high-pressure split injection concept along with EGR on the combustion characteristics of a small-sized single-cylinder engine. Khatamnezhad et al. (2011) performed the simulation to forecast the collective impact of multiple injections and EGR technology for a direct injection diesel engine. The result analysis shows that the retarded injection timing gives higher efficiency at 10% EGR rate. Bedar et al. (2017) have done an experimental study on a small medium-duty diesel engine to meet the current emission norms. The engine was fuelled with diesel blended with biodiesel. The value and quantity of parameters such as IP and EGR were varied and variation effects on combustion and emission characteristics were observed. Results show a better performance with a simultaneous reduction in emission for a 10% EGR rate. The remarkable reduction in the soot formation in the diesel engine combustion chamber was observed through the post-injection concept by Oconnor et al. (2021). According to the authors, changes that occurred in the combustion phase were mainly due to post injections.

From the studied literature, less work was observed on the simultaneous impact of multiple injections considering different fuel injection pressure at low EGR rates for small-sized diesel engines. By viewing this gap of work as a motivation for further study, the current paper considered the electronic fuel supply system parameters like IP, timing, and quantity of fuel supply at a low EGR rate. The work shows the analysis and impact of considered parameters on the engine's performance and emission attributes. The main objective of the present study was to obtain optimum parameters for the fuel system to achieve better performance and the most negligible exhaust emission for small-sized diesel engines. Hence, the current research will find helpful for effectively designing sensors-assisted fuel supply systems with optimum parameters.

Experimental Methodology

A 3.5 kW CRDI single-cylinder diesel engine test setup was used for the experimental work. A rotameter was used to measure flow rate and thermocouples were used to measure the temperature of the water that circulated around the engine cylinder jacket. The engine was loaded with the help of an eddy current dynamometer. An arrangement was made to distribute cooled exhaust gas in the predetermined quantity during the suction stroke of the engine cycle. An electronic control unit (ECU) was connected with a CRDI injection system of the engine, which controls the injection system parameters for the engine. The value of injection pressure, the quantity of fuel injection and their injection timing were set through software, which was further connected to the ECU of the injection system. Table 53.1 gives the details about the engine along with the retrofitted CRDI system. The pictorial view of the engine setup with replaced CRDI and connected EGR technology is shown in Figure 53.1. The emitted gases out the engine cylinder-like CO, NOx, and HC were measured by using an (AVL, 444) exhaust gas emission analyzer. The smoke opacity value was obtained through a smoke meter (AVL, 437).

Table 53.1 Engine and retrofitted CRDI system specifications

Engine specifications	
Engine type	Kirloskar
Cylinders	Single (01)
Bore	80 mm
Stroke	110 mm
CR	12 to 18
Rated speed	1500 rpm
Power	5 HP
CRDI system parameters	
Injector	Bosch
Nozzle diameter	0.215 mm
Number of holes	7
Fuel IP	300-1600 bar
High-pressure system	CRDI BOSCH
Data acquisition device	NI-USB -6210
Software operated ECU	Nira (developed)

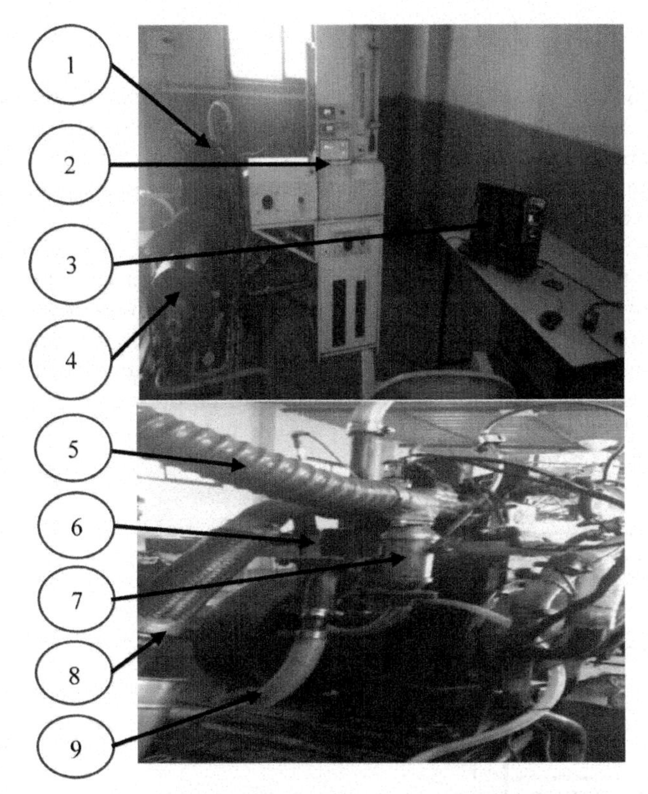

1 – Single cylinder engine
2 – Engine control panel
3 – ECU control software
4 – Dynamometer
5 – Air Supply pipe
6 – EGR control knob
7 – EGR Unit
8 – Exhaust gas pipe
9 – Water circulation for EGR

Figure 53.1 Pictorial view of the experimental setup

The selection of combustion parameters like injection timing and injection quantity as well EGR value were taken from our previous simulation study (Dond et al., 2017) and literature survey. Experiments were carried out with split injection and EGR techniques as per the below combinations.

1. Without (w/o) split injection and EGR
2. Without (w/o) split injection and with EGR
3. With split injection and without (w/o) EGR
4. With split injection and with EGR

For each experiment, the value of IP varied from 350 bar with an interval of 50 up to 600 bar. With the help attached EGR system on the test setup, cooled exhaust gas was supplied by 10% (wt to wt basis) in the engine cylinder along with atmospheric air during suction stroke. The injected fuel quantity was split into two called main and pilot injection. The primary injection, called pilot, was started at 30° CA bTDC with 30 % amount of fuel supplied in this injection span. The secondary injection was called the main injection. The main injection was started at 13° CA bTDC, where left 70% mass of fuel was supplied in this injection interval. For without split injection trials, the pilot injection was absent. The main fuel injection was started at 23° bTDC. All fuel got inserted inside the combustion chamber during this main injection interval. All the combinations experiments were conducted by keeping a 10 kg load on the engine. The engine was set to run with a constant 1500 rpm. The mineral diesel was used to run the engine setup. Each trial was repeated three times, and the mean value was considered to avoid errors in the measurement and analysis of results. The obtained results were explained in detail from the plotted graphs regarding performance and emissions attributes.

Result and Discussion

The result section has been divided into three parts combustion, performance and emission characteristics.

Combustion characteristics

The air-fuel combustion kinematics is important from the point of view of performance and emission attributes. Figure 53.2 gives the variation of cylinder pressure and net heat release rate for different IPs. It was observed that an increase in injection pressure gives a successive rise in cylinder pressure. Before fuel injection, the pressure inside the combustion cylinder was nearly 40 to 50 bar. Hence, an increase in the IP causes a significant pressure difference, which results in fuel getting injected with higher velocity. Such higher velocity fuel penetrates the fuel-air mixture and makes a homogeneous mixture. Also, injected pilot fuel makes the required rise in pressure and temperature inside the cylinder for the main injection period combustion dynamics. Due to all these, many self-sustain flames produced during the main injection combustion period propagated through the entire fuel-air mixture. Hence, an increase in IP helps to increase combustion efficiency. An increase in combustion efficiency gives a rise in net heat release rate and cylinder pressure. Further rise in-cylinder pressure and heat release rate was responsible for producing mechanical work output. There was an approximate 10 to 15% rise in the cylinder pressure and net heat release rate value when compared between the considered injection pressure limit of 300 and 600 bar.

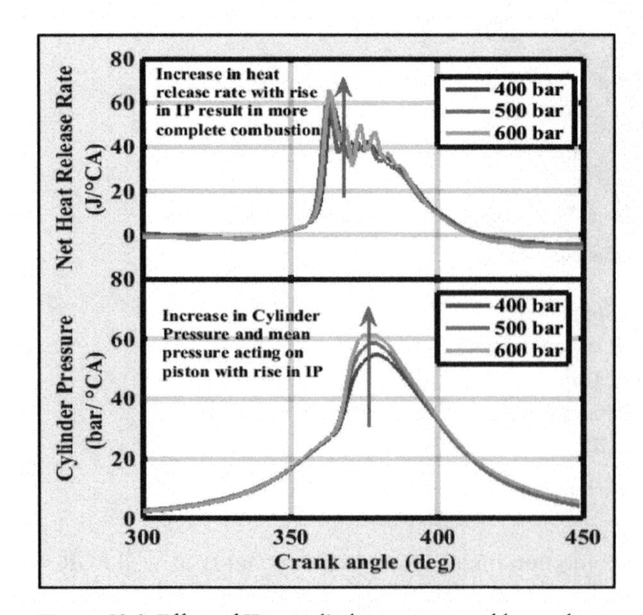

Figure 53.2 Effect of IP on cylinder pressure and heat release rate

Performance characteristics

a) *Brake thermal efficiency (BTE):* The percentage of effective conversion of heat energy released after combustion of fuel inside the engine cylinder into a crankshaft work called BTE. Figure 53.3 gives the effect of rise in IP on BTE. Split injection technology gives higher BTE for all sets of IP. But, the added EGR during suction stoke shows a slight decrement in the BTE and the effect of EGR was found more remarkable with the rise in the fuel IP. The experimental trials conducted with both split injection and EGR technique show higher BTE than those performed with the only EGR system. The progressive rise in IP shows a gradual increase in BTE with a somewhat constant rate for all the sets of experiments. The fall in BTE value observed for the trials taken with EGR technology was recovered when both split injection and EGR techniques were applied. The retrofitted CRDI injection system consists of an electronic injector. The nozzle of the used solenoid injector has more number of equispaced holes with a diameter smaller in size compared to a conventional injector. Such a type of nozzle geometry helps to convert fuel droplets into finer particles at higher IP. These finer particles of fuels lead to a homogenous air-fuel mixture and decrease the physical delay period during the combustion of the fuel. The high value of fuel IP increases the injection rate and reduces the span of fuel required for injecting the necessary fuel quantity inside the combustion chamber. All these factors were in favor of rising mean pressure acting on the piston. Such a higher value mean pressure and less mechanical friction result in up the BTE.

The addition of gases that come out through the cylinder during the exhaust stroke of the cycle with the atmospheric air during the suction stroke of the cycle makes a diluted air-fuel mixture. A diluted combustion mixture decreases the mean value of effective pressure developed inside the cylinder, which was responsible for crankshaft work output. Due to all the above-mentioned factors, less mechanical work and BTE were obtained for added EGR trials. In split injection trials, the fuel supplied during the pilot interval timing raises the pressure and temperature inside the cylinder. The increased value of these two parameters decreases the ignition delay span for the main injection timing of fuel. Due to this reduced delay period, decreased the uncontrolled combustion phase and most of the fuel was burned in a controlled phase for the main injection. All these factors related to split injection were responsible for getting higher combustion efficiency and increased thermal efficiency. Overall 10% rise in the BTE compared to the conventional injection system of the small diesel engine by using split and EGR system on the engine.

b) *Brake specific fuel consumption (BSFC):* The mass of fuel utilised by the engine to generate one kW power at the engine crankshaft during one hour is called brake specific fuel consumption. The BTE and BSFC were opposite in nature. The change in BSFC for successive changes in IP at different considered combinations of experiments was shown in Figure 53.4. The figure observed that a gradual increase in fuel IP gives a constant drop in the BSFC value at all different combinations of trials. The rise in fuel IP causes a much deeper penetration of fuel inside compressed air compared to that with conventional injectors, which inject fuel at 280 bar. Due to such penetration, fuel gets proper mix at all parts of the compressed air, such as near the injector and near the cylinder wall surface, forming a homogeneous air-fuel mixture. Such a homogenous mixture gives higher combustion efficiency with lowering BSFC value. Also, a lower value of BSFC was observed for trials that were taken only with split injection technology. The splitting quantity of percentage fuel burning during the main and pilot injection gets sufficient time and amount to mix with hot air. Hence, dividing the quantity of fuel and sending those at the right time before TDC might be the reason for the lower BSFC. Also, supplying a small amount of fuel during the pilot period raises the sufficient temperature inside the cylinder after

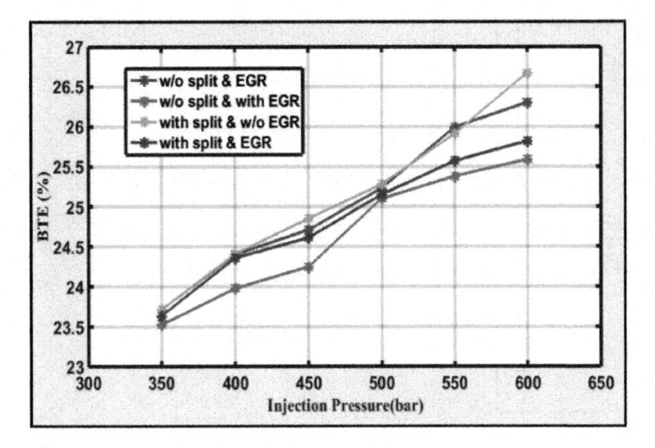

Figure 53.3 Effect on BTE for a different IP

burning. This lowers the span of the delay period for the fuel injected at the main interval or after the pilot injection. The reduced delay period helps to burn the fuel completely and lowers the fuel quantity left for the afterburning phase. But, the added amount of exhaust gas makes the fuel-air mixture so lean to burn entirely and decreases the combustion efficiency. Hence, added EGR was responsible for the rise in BSFC value.

The overall rise in IP gives a significant improvement in the performance level of the engine. The lower value of BSFC was observed at 600 bar IP. The BSFC value was found to be nearly the same for the trials considering split injection technology and without split injection and EGR.

c) *Exhaust gas temperature:* The considerable rise in exhaust gas temperature after the combustion of fuel at the end of the power stroke of the engine cycle is not desirable. Such a high exhaust gas temperature gives to the increased possibility of NOx formation, dissociation of gases, and load on the coolant and lubrication system of the engine. The exhaust gas temperature represents the temperature rise inside the cylinder. In order to keep the cylinder temperature within range and avoid detonation, the maximum IP was kept up to 600 bar. Figure 53.5 gives the variation of EGT with respect to IP for the different combinations of considered trials. The addition of 10% EGR during suction stroke does not show any remarkable reduction in EGT. For the case, without split and EGR trials, it was inferred that exhaust gas temperature was high and increased with the increase in fuel IP. In the case of without split injection trial, the injection timing was set at 23° CA bTDC with total fuel mass getting injected. Due to such advanced injection timing, the span of the delay period increased. The collected fuel during the delay span gets burned immediately and results in a high EGT. The exhaust gas temperature was elevated at 600 bar IP for all the trials. The added EGR act as a barrier to the chain of combustion. Due to this combustion efficiency of the fuel was get decreased and released less heat. Hence, the trials with EGR show a lower value of exhaust gas temperature at all IP. Multiple injection trials also show much lower exhaust gas temperature than all other IP trials. The heat release rate, cylinder pressure, and exhaust temperature were kept at a limit by injecting a low percentage of fuel during the pilot period and maximum during the main period while keeping the main injection near the top

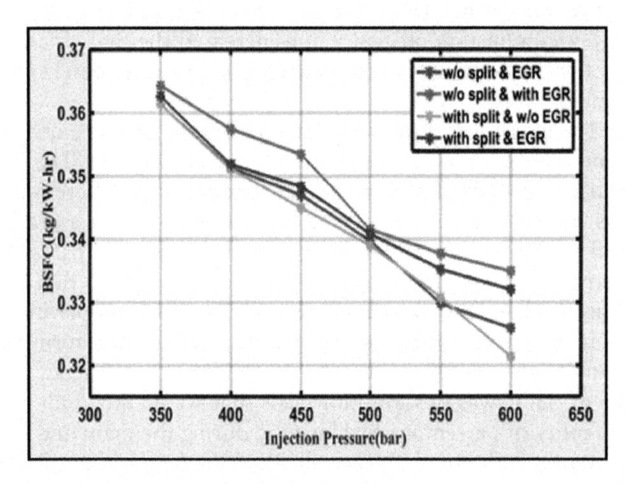

Figure 53.4 Effect on BSFC for a different IP

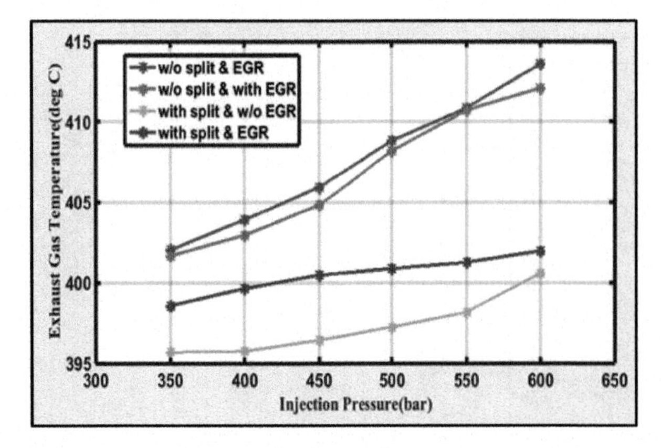

Figure 53.5 Effect on EGT for a different IP

dead centre. Such shifting of combustion towards TDC makes main period combustion continue after TDC. Such delayed combustion causes a drop in the cylinder temperature and EGT. Combined split and EGR also show a lower value of EGT at all IP, but somewhat higher than the split injection.

Emission characteristics

a) *NOx emissions:* Rise in temperature after combustion inside the engine cylinder is the main cause of the NOx emission from the engines. Figure 53.6 shows the variation of mass emission of NOx for different IPs for considered trials on the engine.

A gradual increment in IP shows a successive increase in NOx's mass emission. It was observed that NOx emission was less for the combination of split injection and EGR technology. Split injection without EGR trial shows lower NOx emission, but it was not as low as above mentioned trial. Without split injection and EGR trial shows comparably more NOx emission. The longer delay period was generally observed because of the lower value of temperature and pressure of compressed air. The same has happened in the case of without split and EGR trial. Also, less time was required to send fuel during the pilot and main injection interval when the IP was set at a high value. The total delay period is the summation of physical delay and chemical delay. The physical delay period mainly depends on the injection system parameters like IP and IT. The advanced fuel injection, along with high IP, results in an increased span of the delay period. Hence, accumulated fuel quantity during the delay period was more, which further enlarged the uncontrolled combustion phase with comparably more instantaneous fuel burn during the same. Such a fuel-burning gives a sharp rise in heat release rate for the uncontrolled combustion phase while lowering the heat release rate span. All the things mentioned above were responsible for the rise in-cylinder temperature, which further increases the chances of NOx formation. At the same time, a higher value of IP gives more deep penetration of fuel droplets inside compressed air. This provides proper mixing of fuel with air and nearly complete fuel combustion. Hence, a gradual increase in fuel IP also caused a rise in pressure and temperature inside the cylinder after combustion. Ultimately gradual rise in fuel injection pressure gives a steep rise in NOx emission. The added EGR into fresh incoming air makes the mixture somewhat heterogeneous. Also, exhaust gases absorb the temperature inside the cylinder. Due to this, there was a drop in temperature for trials with EGR techniques. Hence, the addition of EGR lowers the rate of NOx emission. Also, split injection of fuel relatively decreases the temperature after combustion and hence the NOx formation. Therefore, the combined split injection and EGR considerably decrease the NOx emission.

b) *HC emission:* Lower oxygen percentage in the combustion chamber at the time of chemical combustion reaction and heterogenous air-fuel mixture result in some fuel left as an unburn and that goes outside the cylinder during expansion stroke called hydrocarbon emission. Variations of HC emission in PPM at different IP for considered trials was shown in Figure 53.7. It was observed that EGR and split injection combined technology significantly lower the rate of HC formation at high IP. The most negligible value of hydrocarbon emission was observed at 600 bar injection pressure for all combinations of considered trials. Lower IP and without split and EGR trial at advanced IT gives more HC emission. Fuel droplet sizes were reduced mainly because of two reasons. First was a rise in the value injected fuel pressure and second was the used electronic injector. The used electronic injector has more number holes as well decreased size of holes compared to the conventional injector. The decreased nozzle

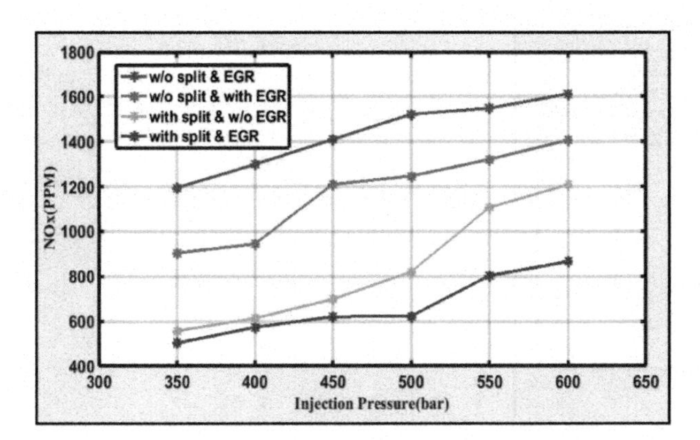

Figure 53.6 Effect on NOx emission for a different IP

hole diameter results in a reduction in the droplet size. Also, increased equispaced holes on the nozzles give proportionate mixing with the air inside the combustion cylinder and quickly evaporates.

The added EGR into the fresh incoming air during the cycle's suction stroke increases the air's average temperature. At the same time, EGR increases the oxygen density of air. Due to all the above reasons, EGR trials also show lower HC emissions. In the case of split injection technology, a proper injection interval gap between the pilot and main injection period gives sufficient time for appropriate mixing of the air-fuel mixture. This was the reason for the observed decrease in HC. The figure shows that both split fuel injection and EGR technology pronto decrease HC emission at higher IP.

c) *CO emission:* The incomplete oxidation of carbon particles present inside fuel results in a CO emission. The percentage CO emission for the different combinations of considered split injection and EGR techniques at different IPs is shown in Figure 53.8. It was observed that both EGR and multiple injection technology affect the formation of CO at different IPs. The split injection and EGR concepts show relatively higher CO emissions at lower IP. The drop in CO emission was observed with increased IP and this drop rate of CO emission was relatively high for split injection trials. The trials only with EGR show a lower value of CO percentage compared to split fuel injection technology. The added EGR into incoming fresh air raises the level of oxygen density. An increased percentage of oxygen leads to the complete oxidation of carbon particles. But, at the same time, the added EGR also diluted the air-fuel mixture. Such a heterogeneous air-fuel mixture inside the engine cylinder decreases carbon particles' oxidation rate. Hence, the trials conducted not considering EGR technology show compared lower CO emissions than the other three cases. The split injection technology shows a lower CO emission only at higher IP. Hence, to decrease CO emission, the considered value of pilot and main injection timing and the fuel injection quantity must be taken at higher fuel IP. Proper selection of IP and injection timing was very much important for CO emission. Fuel IP decides the droplet size of the fuel and fuel injection quantity, whereas injection timing affects the air-fuel mixture formation and a gap between pilot and main combustion. All the factors mentioned above affect CO emission.

d) *Smoke opacity:* Figure 53.9 shows % smoke opacity at different IP for a different combination of split injection and EGR technology trials. The figure clearly shows the least smoke emission for the trials performed without considering splitting fuel injection as well EGR technology. There was no

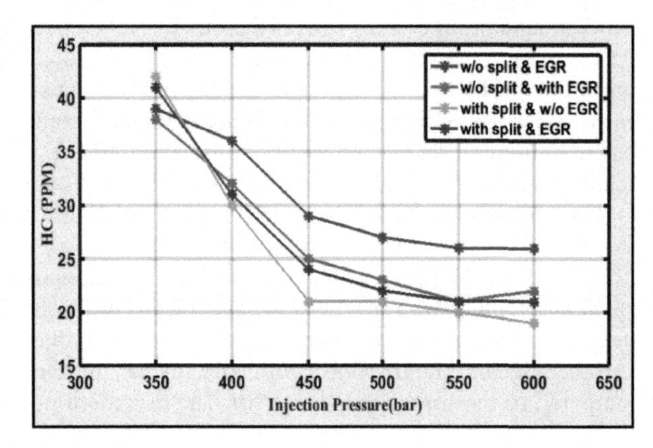

Figure 53.7 Effect on HC emission for a different IP

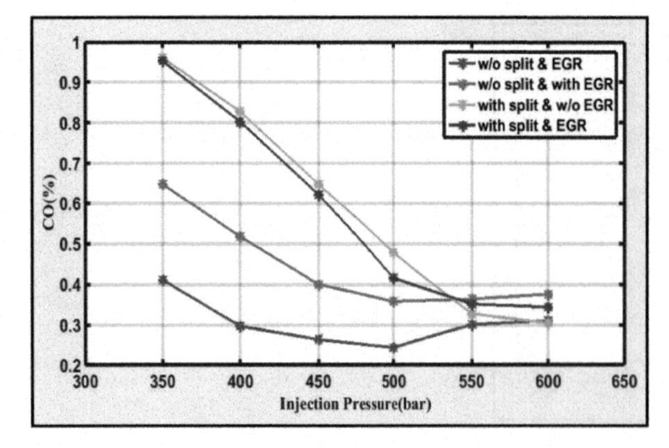

Figure 53.8 Effect on CO emission for a different IP

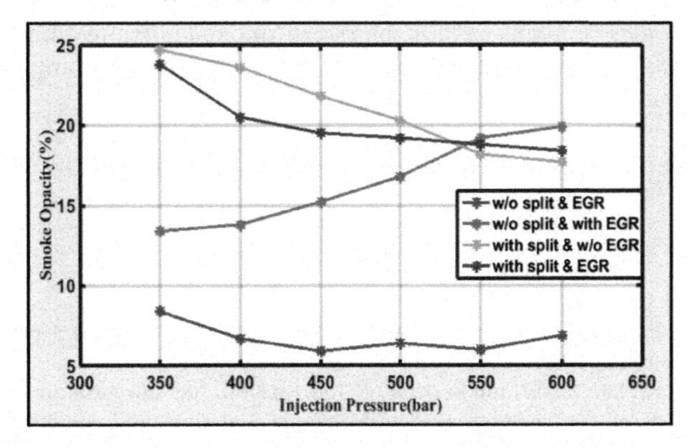

Figure 53.9 Effect on smoke opacity for a different IP

remarkable rise in the smoke emission observed for the without multi injection and EGR trials with the increment of IP from 350 bar to 600 bar. The split injection trial shows higher smoke emissions compared to the remaining three cases, majorly at low IP. The injected fuel quantity was divided into two parts during split injection trials: pilot and main injection. The gap between these two injections in terms of CA was important from the point of view of proper completion of combustion. This gap was found to change with the IP. The high IP increases the gap while low IP decreases this span. There was a short time available for proper fuel-air mixing for the trials that had fewer injection intervals gap between the mentioned two injection periods at lower IP. Therefore, a rich and lean mixture zone was created inside the combustion cylinder. Such a heterogeneous mixture zone breaks the fuel combustion chain and results in the smoky exhaust. The smoke opacity value decreases with an increase in IP for trials with multiple injections. The high IP shorter the span of fuel injection and, at the same time, increases penetration inside high-density air. Also, a small turbulence effect has been generated due to the rise in IP that results in effective fuel-air mixing. All the factors as mentioned above were suitable to decrease the soot formation inside the engine cylinder. The trials with EGR show lower smoke emission for the successive rise in IP up to 500 bar IP than those with split injection. A split and EGR technology combination trial shows a similar path to that of a split injection trial.

Conclusion

The present study investigates the effect of fuel IP along with multiple injections and EGR techniques on the small size diesel engine's measurable combustion, performance and emission attributes. From the obtained results and their analysis following interpretations were made.

1) Multiple injections and EGR techniques greatly influence the engine's characteristics. The combustion characteristics show that there was a 12 to 15%% rise in the cylinder pressure and net heat release rate value for the 600 bar IP.
2) Split injection technology improves BTE and fuel economy. The split injection technology shows a 12% rise in BTE, while it was dropped by 2% due to added EGR during suction stroke. Overall 10% rise was observed in BTE and a similar percentage of the drop was seen in the specific fuel consumption value at 600 bar IP for the combined multi injection and EGR trial.
3) A significant reduction in EGT, nearly about 5-8% drop, was observed for the split fuel injection trials. There was a slight increment in EGT with an increment in IP from 350 to 600 bar for all the trials. Finally, the use of these two technologies decreases the thermal impact on the engine.
4) The EGR technology shows a 14% drop, while split fuel technology gives a nearly 32% drop in NOx emission at all IP compared to the trials conducted within the absence of these two techniques. Overall, a remarkable drop in NOx emission was observed for combined split and EGR trials.
5) A drop of 22% in the HC emission was observed for EGR and multi-injection trials compared to without any of these two technologies at 600 bar IP. Split fuel injection technology shows a 28% drop in HC emission than all other considered combinations trials at all IP expect less than 400 bar.
6) A considerable drop, measured 2030% in the CO emission, was noticed between the without split and EGR trials and with split and EGR trials with an increment in IP from 350 to 600 bar. Overall, slightly higher CO emission was observed for multi injection and EGR trials.
7) There was a 40% rise in the smoke emission observed for the EGR trials with an increment in IP. At the same time, a 28% drop in smoke level was observed for the split fuel technology trials with the

rise in IP. Hence, combining both technologies compensates for the rise in smoke emissions by EGR technology. But, overall, higher smoke emission was observed for both these technologies compared to the trials conducted in the absence of these two technologies.

Overall, combined split injection and EGR concepts are beneficial to enhance the adequate performance of the small-sized diesel engine.

References

Agarwal, A. K., Gupta, P., and Dhar, A. (2015). Combustion, performance and emissions characteristics of a newly developed CRDI single cylinder diesel engine. Sadhana. 40(6):1937–1954.

Agarwal, A. K., Dhar, A., Gupta, J. P., Kim, W., Choi, K., Lee, C., S., and Park, S. (2015). Effect of fuel injection pressure and injection timing of Karanja biodiesel blends on fuel spray, engine performance, emissions and combustion characteristics. Energy Conver. Manag. 91:302–314.

Anand, R. 2018. Simultaneous control of oxides of nitrogen and soot in CRDI diesel engine using split injection and cool EGR fueled with waste frying oil biodiesel and its blends. In Air pollution and control. (pp. 11–44). Singapore: Springer.

Bedar, P., Lamani, V. T., Ayodhya, A. S., and Kumar, G. N. (2017). Combined effect of exhaust gas recirculation (EGR) and fuel injection pressure on CRDI engine operating with Jatropha curcas biodiesel blends. J. Eng. Sci. Technol. 12(10):2628–2639.

Carpenter, A. L., Mayo, R. E., Wagner, J. G., and Yelvington, P.. E. (2016). High-pressure electronic fuel injection for small-displacement single-cylinder diesel engines. J. Eng. Gas Turb. Power. 138(10).

Dey, S., Reang, N. R., Majumder, A., Deb, M., and Das, P. K. (2020). A hybrid ANN-Fuzzy approach for optimization of engine operating parameters of a CI engine fueled with diesel-palm biodiesel-ethanol blend. Energy. 202:117813.

Dond, D. K., Nitin P. Gulhane, and C. L. 2017. Dhamejani. Mathematical Modelling & MATLAB Simulation of Diesel Engine. In International Conference on Advances in Thermal Systems, Materials and Design Engineering (ATSMDE2017).

Edara, G., Murthy, Murthy, Y. V. V. S., Nayar, J., Ramesh, M., and Srinivas, P. (2019). Combustion analysis of modified light duty diesel engine under high pressure split injections with cooled EGR. Eng. Sci. Technol. Int. J. 22(3):966–978.

How, H. G., Haji, H. M., Kalam, M. A., and Teoh, Y. H. (2018). Influence of injection timing and split injection strategies on performance, emissions, and combustion characteristics of diesel engine fueled with biodiesel blended fuels. Fuel. 213:106–114.

Jafarmadar, S. (2013). The effect of split injection on the combustion and emissions in DI and IDI diesel engines. Diesel Engine: Combus. Emiss. Condit. Monitor. 1.

Jain, A., Singh, A. P., and Agarwal, A. K. (2017). Effect of fuel injection parameters on combustion stability and emissions of a mineral diesel fueled partially premixed charge compression ignition (PCCI) engine. Appl. Eenergy. 190:658669.

Khandal, S. V., Tatagar, Y., and Badruddin, I. R. (2020). A Study on Performance of Common Rail Direct Injection Engine with Multiple-Injection Strategies. Arabian J. Sci. Eng. 45(2):623–630.

Khatamnezhad, H., Shahram, K., Jafarmadar, S., and Nemati, A. (2011). Incorporation of EGR and split injection for reduction of nox and soot emissions in DI diesel engines. Therm. Sci. 15(2):409–427.

Lakshmipathi, A. R., Rajakumar, S., Deepanraj, B., and Paradeshi, L. (2017). Study on performance and emission characteristics of a single cylinder diesel engine using exhaust gas recirculation. Therm. Sci. 21(2):435–441.

Mobasheri, R.. (2017). Influence of narrow fuel spray angle and split injection strategies on combustion efficiency and engine performance in a common rail direct injection diesel engine. Int. J. Spray Combus. Dyn. 9(1):71–81.

Molina, S., Antonio, G., Javier, M. –S., and Villalta, D. (2021). Effects of fuel injection parameters on premixed charge compression ignition combustion and emission characteristics in a medium-duty compression ignition diesel engine. Int. J. Eng. Res. 22(2):443–455.

Mossa, M. A. A., Hairuddin, A. A., Nuraini, A. A., Zulkiple, J., and Tobib, H. M. (2019). Effects of hot exhaust gas recirculation (EGR) on the emission and performance of a single-cylinder diesel engine. Int. J. Automot. Mech. Eng. 16(2):6660–6674.

O'Connor, J. and Musculus, M. (2013). Post injections for soot reduction in diesel engines: a review of current understanding. SAE Int. J. Eng. 6(1):400–421.

Park, S., Kim, H. J., and Lee, J. -T. (2018). Effects of various split injection strategies on combustion and emissions characteristics in a single-cylinder diesel engine. Appl. Therm. Eng. 140:422–431.

Reang, N. M., Dey, S., Barma, J. D., and Deb, M. (2019). Effect of linseed methyl ester and diethyl ether on the performance–emission analysis of a CI engine based on Taguchi-Fuzzy optimisation. Int. J. Ambient Energy. 115.

Reang, N. M., Dey, S., Debbarma,B., Deb, M., and Debbarma, J. (2020). Experimental investigation on combustion, performance and emission analysis of 4-stroke single cylinder diesel engine fuelled with neem methyl ester-rice wine alcohol-diesel blend. Fuel. 271:117602.

Thangaraja, J. and C. Kannan. (2016). Effect of exhaust gas recirculation on advanced diesel combustion and alternate fuels-A review. Appl. Energy. 180:169–184.

The authors are thankful to receive funds from VJTI, Mumbai, under R & D project expenses.

54 An intuitive and structured detection system for facial mask using YOLO and VGG-19 to limit COVID-19

Diviya M.[1,a], *Hemapriya N.*[2,b], *Charulatha B.S.*[3,c], *and Subramanian M.*[4,d]

[1]Department of Computer Science and Engineering, Rajalakshmi Engineering College, Chennai, India

[2]Department of Information Technology, St. Joseph's College of Engineering, Chennai, India

[3]Department of Information Technology, Rajalakshmi Engineering College, India

[4]Department of Mechanical Engineering, St. Joseph's College of Engineering, Chennai, India

Abstract

In the current pandemic that we're all facing, a mask has become an important accessory that we need to wear whenever we step out, in order to protect ourselves and those around us. Masks are one of the few COVID-19 preventative measures available, and they play an important role in protecting people's health from respiratory illnesses. All the countries that have beat the pandemic and returned to normalcy attribute their success to people wearing masks. Good surveillance measures that ensure that the public is wearing masks at all times can help us beat this pandemic. We present a method for developing object detection models that can both identify the location of masked and unmasked faces in an image. Due to the potentially disastrous repercussions of false positives for masks, these detectors will be noise-resistant and allow as little room as possible for erroneous positives. Along with this, our approach is also designed with the ability to deploy in mind - utilizing low parameter models to improve real-time inference potentially on edge devices. We intend to suggest two architectures based on YOLO object detection and VGG19. In the YOLO model, the proposed model architecture categorize images of people wearing face masks and no facial mask which has an accuracy of 98.7% for previously unseen test data and an accuracy of 99.8% in the VGG19 model based on the Haar-cascade algorithm. For many countries throughout the world, our research will definitely serve as beneficial tool in reducing the widespread contagious disease

Keywords: Computer vision, Haar-cascade, object detection, VGG19, YOLO.

Introduction

Corona illness has indeed impacted the world severely in 2019. In public places it is must to wear masks so that people can protect themselves. In December of this year, the first corona virus-affected patient was encountered. Since then, COVID-19 virus has been proclaimed as a pandemic that has spread worldwide. A substantial count of the human population is affected largely by the virus. Nearly 34.3 crore affected cases were been declared worldwide as of this writing, with 55.7 lakhs loss of lives. This number is steadily increasing. Corona virus is characterised by an increase in body temperature, a sore throat, a dry cough, tiredness, diarrhoea, and a loss of taste and smell, according to the World Health Organization (WHO) (COVID-19 Dashboard, 2021). As a result, majority of human population are much concerned about the health, and government regard the health of human to be of supreme importance.

Thankfully, studies have shown that surgical face masks can help prevent the pandemic (Eikenberry et al., 2020). Currently, the WHO recommends that those who face with respiratory symptoms or people who care for infected patients and those who have symptoms wear face masks. As a result, mask detection has become an essential computer vision problem for assisting the global civilisation, yet research on this topic is scarce. Face mask detection is the process of determining if a person is wearing a mask or not and where their face is located. This issue is closely associated to the traditional object detection, which is used in identifying several types of things, and face detection, which may be used for identifying specific item, such as a face. Object and face identification holds an extreme area of applications, which includes autonomous driving, education, also spying (Florian et al., 2015; Yu and Chen, 2020).

Furthermore, social isolation is becoming increasingly important. Because to COVID-19, the world's fastest-growing pandemic, we've had to change our ways. Nearly more than 40 million people throughout the globe have been infected with COVID-19., killing over 230 K people, in accordance to the World Health Organization (WHO). So far, the virus has infected people in 213 nations. In medical applications, deep learning algorithms are widely used (Islam et al., 2020). Deep learning architectures is employed in ascertain mask on face. Moreover, a smart city is described as an urban area with a large number of data

[a]diviya.m@rajalakshmi.edu.in; [b]hemuhema2000@gmail.com; [c]charulatha.bs@rajalakshmi.edu.in; [d]subramanianm@stjosephs.ac.in

collecting edge devices. The information is further utilized to carry out a variety of operationalities across the city. Our approach is also designed towards deploying these systems within existing infrastructure, and ensuring low-level optimisation of ML models that allow us to do so.

Handcrafted feature extractors are commonly used in traditional object detectors named histogram of oriented gradients (HOG), scale-invariant feature transform (SIFT), and so on. Viola Jones detector uses Haar feature with integral image method, whereas others use different feature extractors (O'Mahony, 2019). In the field of current object identification, deep learning-based object detectors have lately shown excellent performance and are swiftly gaining attention. Without the use of previous information, deep learning allows neural networks adjust features completely without the need for feature extractors. It is possible that deep learning-based object detectors have a single stage or two stages. It is possible to utilise a neural network to recognise items by seeing just once, such as the single shot detector (SSD) (YOLO). Dual-stage detectors, on the other hand, use two networks, so called the region-based convolutional neural network (R-CNN) (Lin et al., 2017) and the faster R-CNN (Ren, 2015), to conduct a coarse-to-fine detection. Face detection, like general object detection, uses a similar architecture but add up enhanced face-related characteristics, such as facial landmarks in Retina Face (Deng et al., 2020), to increase accuracy. Face mask detection, on the other hand, is the subject of just a small amount of research. Mingjie et al. (2020) suggests the retinal mask detection model, which is combined alongside of cross-class object removal approach. The model makes use of single stage detector that is composed of feature pyramid network to achieve greater precision and recall when compared to the baseline outcome. The authors suggested the use of transfer learning, which is a familiar deep learning approach, to reduce the scarcity of datasets (Das et al., 2020). Whether the face is clearly visible, this approach can tell if it's covered up by a mask. Part of its surveillance responsibilities include the ability to detect a moving face or mask. The sequential CNN model is used to explore optimum parameter values for accurately identifying the existence of masks without over-fitting. Neural network with deep convolutions (Khan et al., 2020) which comes under neural network family shows an outstanding performance in the domain of open computer vision and image processing. Deep CNN involves in learning multiple features of the data. When the available amount of data is large and hardware supports it accelerates the exploration in CNN.

The PASCAL visual object classes (VOC) challenges 1, which ran from 2005 to 2012 (Everingham et al., 2010), were the most famous competition in the realm of early computer vision. Image segregation, object detection, semantic segmentation, and action detection are all included in PASCAL VOC. The best models of Pascal-VOC versions for object detection are VOC07 and VOC12, with the former including 5,000 tr. pictures and 20,000 annotated objects and the latter containing eleven thousand photos and 25,000 annotated objects. The twenty classes of objects which were annotated in these two datasets includes person, Animal class such as bird, cat, dog, and so on. The vehicle class includes like airplane, bicycle, train; and Indoor objects. In the present years, the VOC has steadily gone out of favour, as much large datasets such as ILSVRC and MS-COCO have been provided, and now it has become as a test-bed for new detectors. The Image Net large scale visual recognition challenge (ILSVRC) (Russakovsky et al.,2015) has indeed advanced generic object detection technology. It includes an Image Net-based detection task. There are 200 visual object classes in ILSVRC detection dataset.

Methodology

Delving further into the nuances of Object Detector systems of the day, they've progressed heavily in terms of speed, scalability, and robustness. In order to conduct our object detection tasks, we used the traditional object identification techniques viz YOLO (Bochkovskiy et al., 2020) as well as the classic VGG19 architecture (Simonyan and Zisserman, 2014).

Model 1:YOLO

A model called You Only Look Once (YOLO) is often used in deep learning research. YOLO allied to the family of One-Stage Detectors which is also referred to as the one-shot detection that means the image is looked onto only once. From a more technical standpoint, it works as follows: It is basically a sliding window and yet another classification approach, where the image is looked at and classified for each window. The image is looked in two steps in a region proposal network where the first step is to identify regions which might have objects followed by specifying it. The current unofficial implementation is named YOLOv5. YOLOv5 claims to provide cutting-edge precision while processing at a high frame rate. On the Tesla V100, it obtains a 43.5% AP accuracy for the MS COCO with a 65 FPS inference speed. High precision is no longer the only holy grail in object detection. In the edge devices, we want the model to run smoothly. It's also important to process video input in real time with low-cost technology. That is another major aspect that is covered in this project through the use of YOLOv5 which was designed purely with deploy ability in mind. Our model is evaluated also based on the same metrics as the original YOLOv4 paper follows, which are Mean Average Precision and Recall parameters.

The dataset that would be used is from a combination of many sources, which have been hand-labelled. The dataset would contain annotations for masked and unmasked faces. The dataset contains almost 700 images. After applying various image augmentation techniques, there were around 1000 samples for training and 200 images for testing in the dataset. The dataset had been annotated using the MS-COCO format. It is one of the most efficient and challenging object detection dataset made available in the recent times (Lin et al. 2014). Since 2015, yearly competitions based on the MS-COCO dataset were conducted. There are fewer categories of objects when compared to ILSVRC, but there are more instances present. The most significant advancement of MS-COCO over VOC and ILSVRC is that, in addition to the annotations of the bounding box, every object is yet again labelled using per-instance segmentation to help attain accurate localization. Furthermore, compared to VOC and ILSVRC, it contains more small items with an area of less than 1% of the image and more densely located objects. These aspects bring MSCOCO's object distribution closer to that of the live scenarios. MS-COCO, like Image Net at the time, has become the supreme benchmark standard for the object detection field. On a considerably wider scale than MS-COCO, the open image detection (OID) competition took place in 2018 (Krasin et al., 2017). Item identification and visual link detection are the most important Open Images activities. All annotation was translated to YOLO txt format (Redmon et al., 2016) the model used was YOLO V5. All of the image's annotations and numeric representations of labels are included in this format, as are the labels' human-readable strings. It is possible to deal with the picture annotations even after they have been resized or expanded since they have been normalised to lie in the range of 0 to 1. As a consequence of the DarkNet framework's implementations of different models containing YOLO, it has garnered a lot of interest.

The images acquired by the CCTV footages need to be pre-processed. The image is modified to greyscale in pre-processing step since the RGB coloured image contains more redundant data that isn't needed for detection of masks in human faces. Every pixel of a RGB image was saved with a count of 24 bits. Whereas the greyscale images are stored 8 bits per pixel and supplied enough information for categorisation. The images were then melded into a (64 × 64) shape to keep the input uniform for the architecture. In order to learn the algorithm more rapidly and capture crucial information in pictures, the pixel values of the images are normalised, and these values are in the range of 0 to 1. The model is on the basis of CNN, that is extremely helpful in recognizing patterns in images. Multiple dense neural networks utilize the features generated by CNN for categorization purposes. In this work, we are building upon past methods of face mask detection by introducing the traditional object detection models with custom FPN backbone in producing hopped-up outcomes. In order to identify the maximum number of anchor boxes, the proposed technique uses the mean IoU to rally deployment capabilities and performance.

The proposed model used a dataset created from real-time pictures of masked and unmasked people that was carefully annotated to train and assess our detector in a supervised state. Furthermore, for Adam optimiser training, performance parameters such as accuracy, recall, and mean precision rates were considered. We use the YOLO V5 (Girshick et al., 2014) model, which is an unofficial version of YOLO. As an image's pixels are transformed into bounding box coordinates and class probabilities, object detection in YOLO is often reframed as an individual regression issue. Multiple bounding boxes and class probabilities are predicted simultaneously by a single neural network. In order to get the most accurate results, YOLO trains on the whole collection of photographs. The suggested approach provides a number of benefits over more standard methods of object detection.

The input image is taken into consideration as an entire entity when predictions are made. YOLO visualises the complete image during the test and train process thus it obtains the textual content along with their visual properties. The techniques such as region-based retrieval and sliding window do not follow this approach as a greater context cannot be viewed though. Fast R-CNN is one of the object detection techniques, which does not accurately figure out background patches in an image as objects. YOLO does not perform much background errors like fast R-CNN. Thus, YOLO analyses and identifies generalized representation of objects. YOLO surpasses top supreme recognition algorithms such as DPM and R-CNN by making a consideration of high margin when performing training on real images and testing as in Figure 54.1. Because YOLO is so adaptable, it is less likely to break down when applied to new domains or unexpected inputs.

The proposed network inputs certain characteristics from the input image and predicts the bounding box from many different classes at once as shown in Figure 54.2. Thus, the network takes the whole image into consideration for an iteration including the objects in the images. Mean precision is maintained and the end-to-end transition and real time implementation is possible due to the highly flexible architecture of YOLO. The suggested approach divides a picture into a S × S grid as input data. If the centre of the object is found to be placed inside a grid cell, then the particular grid is in charge of detecting the object.

The confidence ratings are projected for each grid cell in the bounding boxes, which are labelled B. The model's conviction that the box contains an entity of object, as well as the correctness, are represented by the confidence values. Pr(Object) * IOU is a possible definition. To get the final score, use IOU (intersection over union) of the predicted and ground truth values if an item is not present in the grid. Otherwise, confidence ratings must be set to 0. All of the predictions inside each bounding box may be summarised in five different ways. The first two points show where the box's centroid is in reference to the grid cell's boundaries. When creating a picture, all of its dimensions are taken into account. Last but not least, the confidence prediction is a promissory note between the projected box and any actual ground truth box.

Model 2 : VGG19

The VGG19 model is a VGG model variant with 19 layers in total. Other VGG variants include VGG11, VGG16, and many others, but the main advantage of VGG19 is that it has a FLOP count of 19.6 billion. Traditional convolutional neural networks were intended to be improved by AlexNet, which was launched in 2012. VGG is a descendant of AlexNet, but it was given life by a separate Oxford group known as the Visual Geometry Group, and it was christened VGG as a tribute to that group. It props up on and polish up some of its forefathers' ideas, such as the use of deep convolutional layers to improve accuracy. Using the Kaggle face mask dataset, it is possible to build a model that recognizes human faces wearing masks, faces without masks, and incorrectly masked faces. There are 853 images in the PASCAL VOC format in this collection, along with the bounding boxes, that belong to three classes. The three classes include mask on, no mask, and incorrectly worn mask.

As an ideal object identification method, Paul Viola and Michael Jones developed the Haar Cascade classifier in their 2001 publication 'Rapid Object Recognition by Means of Boosted Cascades of Simple Features'. A cascade function is developed using a large number of positive and negative pictures in this machine learning technique. Based on the training, it is used to identify items in other photographs. That is why they are so massive.

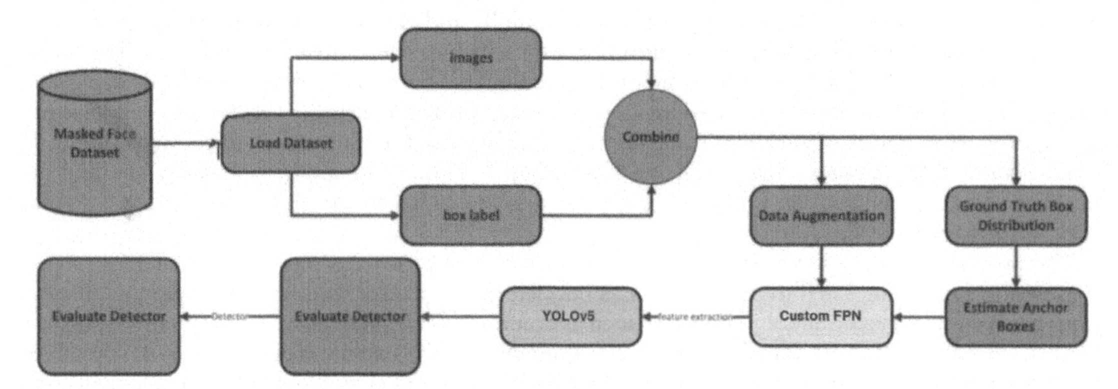

Figure 54.1 The YOLOV5 model's architecture

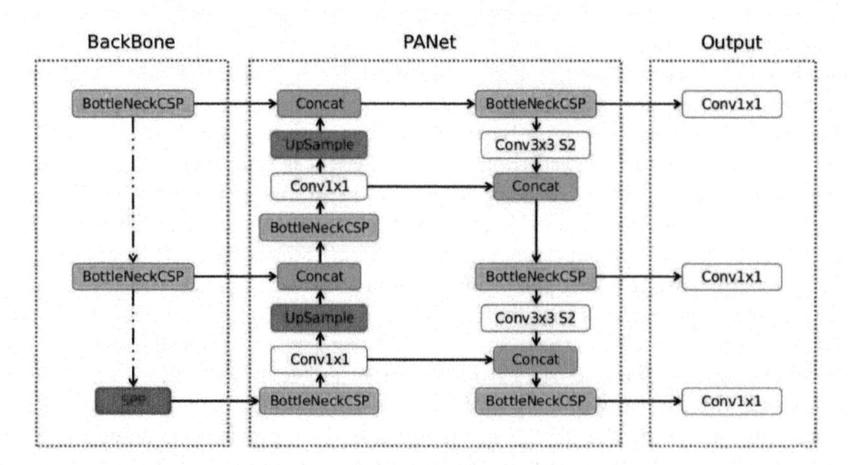

Figure 54.2 Proposed architecture of the YOLO mode

A distinct use case is represented by each xml file, which might include a wide range of information. The feature set for detecting the full body, lower body, eye, frontal-face, and so on is contained in a special xml file. As a result, the xml files are used to pull each feature. As input, the network was given a 224 × 224 RGB image with a fixed size, representing the matrix of shape (224, 224, 3). The only pre-processing is to deduct the mean RGB score from each pixel, which was calculated throughout the full training data set. Spatial padding was used to maintain the image's spatial resolution, and maximum pooling was attained over a 2 × 2 pixel region. Because previous models relied on tan h or sigmoid functions, the Rectified linear unit (ReLu) was created to append non-linearity to rewamp model classification and processing performance, and it proved to be far superior. For the first two levels, which are 4096 pixels in size, there is a 1000-pixel layer at the bottom.

The proposed model makes use of the base model VGG19 architecture as shown in Figure 54.3. All the pre-defined layers are freezed with no changes excepting the last layer. In addition to the existing base model, a dense layer and a flatten layer is added. The sigmoid activation function is used since the probability values are to be predicted. The Adam optimizer function is used which is mainly used for optimizing gradient descent problems. This method is very efficient while working on problems which hold more data with multiple parameters as it is highly effective with minimum memory. It could be portrayed as a hybrid of the typical gradient descent algorithm with momentum along with the RMSP procedure. Save the model in .h5 format and feed in the train data to perform the training operation. Validate the model in the validation set and evaluate the model based on metrics.

The Figure 54.4 depicts the implementation methodology. The images from the dataset are loaded and pre-processed. The image is resized according to the dimensions received by the VGG19 with dimensions 128 × 128. The Keras library performs the required pre-processing using the ImageDataGenerator where horizontal flips and rescaling (255,255,255) are done. The images from the dataset are visualized for better interpretation. Build the model with the fine-tuned layers and input the training data. Run the model and evaluate the metrics. Perform early stopping to avoid over fitting.

Results and Discussion

Model 1: YOLO V5

There are 1000 and 200 images in the training and testing datasets, respectively. Since further training results in regards to the reliability of the developed system, the proposed architecture is made to train for 100 epochs. Table I compares the performance of YOLOv5 models on our dataset. Performance indicators

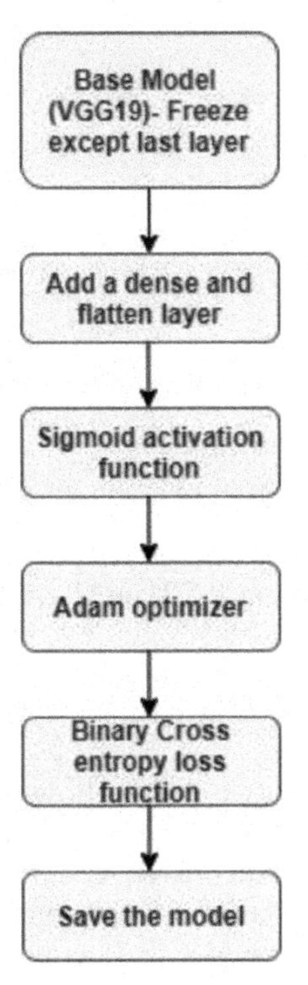

Figure 54.3 Proposed architecture of VGG19 model

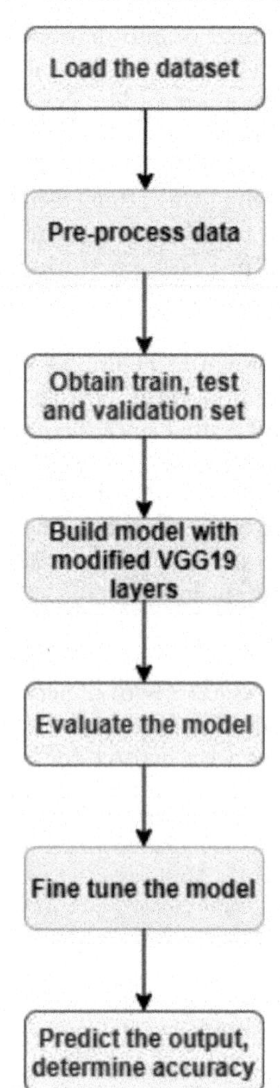

Figure 54.4 Process flow of the implementation for the detection of face mask

Table 54.1 Peformance comparison of YOLO V5 and VGG19 on our dataset

Classifier	*Precision*	*Recall*	*MAP @ 0.5*
YOLOV5S	0.816	0.703	0.732
YOLOV5M	0.851	0.766	0.833
YOLOV5I	0.772	0.633	0.876
YOLOV5X	0.778	0.551	0.701
VGG19	0.98	0.99	Accuracy- 0.99

such as Precision Recall and the average precision are used to evaluate our models' accuracy. There is a clear distinction in performance as we scale to higher parameter models, achieving a 0.701 mAP on the smallest model all the way up to achieving 0.876 on the largest one.

Figure 54.5 represents the inference time taken for the different YOLO models available plotted against the performance tradeoff that occurs. The low parameter YOLO V5 models are much faster for real-time usage, while also outperforming its predecessor models in performance.

(ROC) curve of suggested framework is presented in Figure 54.6. On this graph, we can see how well the classifier does at different levels. According to Equation (1) and Equation n(2), two parameters are displayed in the ROC curve: true positive rate and false positive rate (TPR and FPR). A ROC curve is drawn for each threshold using the given measurements and findings. The binary classifier's performance is measured by the area under the curve (AUC). Assuming all predictions are right, the AUC is 1; assuming all predictions are erroneous, the AUC is zero. With an AUC of 0.845, the suggested classifier is clearly a strong contender.

$$True\,Positive\,Rate = \frac{True\,Positives}{True\,Positives + False\,Negatives} \tag{1}$$

$$False\,Positive\,Rate = \frac{False\,Positives}{True\,Negativee + False\,Positives} \tag{2}$$

MODEL 2: VGG19

According to a work published by Paul Viola and Michael Jones in 2001, Haar feature-based cascade classifiers may be used effectively to identify objects in images. It is basically a machine learning based algorithm wherein cascade function is learned with huge count of true and false images. It is then employed in detecting items in several locations. We intend to crop the faces in the image and use the model developed in the previous phase to identify whether or not the individual faces have a mask. Those who don't use masks are labelled red, while those who do are labelled green. The model outperformed by showing an accuracy of 0.99 when compared with another YOLO model versions as shown in Figure 54.7.

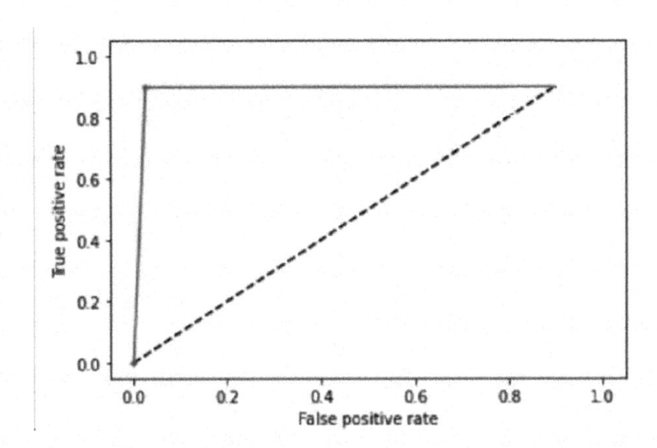

Figure 54.5 Performance comparison on the several variants of YOLO

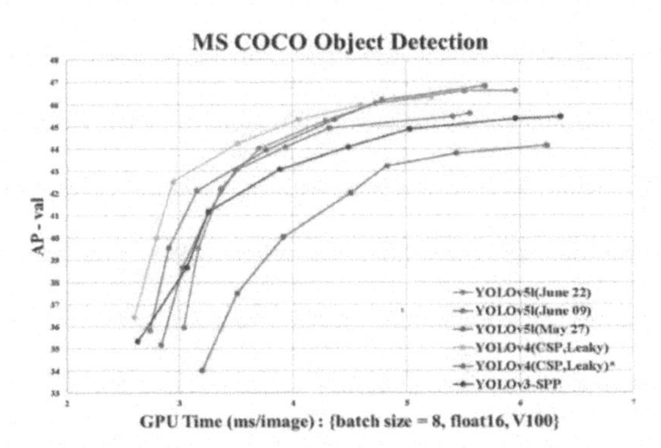

Figure 54.6 ROC curve for YOLO model performance

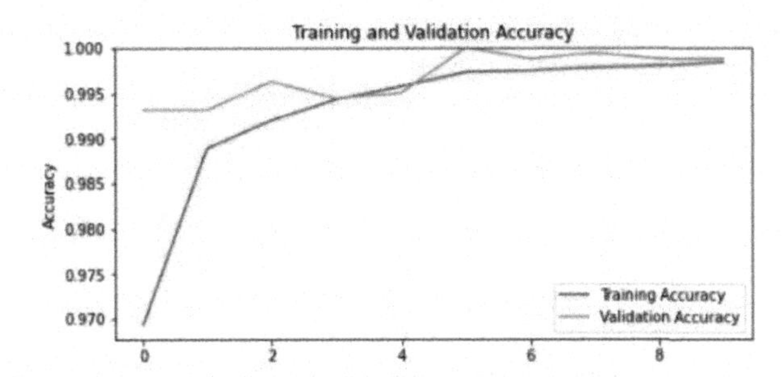

Figure 54.7 Performance of the VGG19 model

Conclusion

Surveillance systems are the need of the hour to curb this pandemic, and ensure conduct in public places. This can also greatly help with tracing back possible contact for those people affected by it. The proposed model strategy for YOLOv5 has been proved to be an effective model for detecting masked faces. Our work outperforms previous work in terms of both performance as well as deploy ability in surveillance systems. We plan to exhaustively study the deploy ability this on edge devices for real-world scenarios, as a note of future study. The model's performance can be further increased by performing hyper parameter tuning. The model's real-world performance can be increased by training the model with bigger datasets and by performing online training. Model deployment is a very important topic in ML research since high parameter models struggle with inference.

References

Bochkovskiy, A., Wang, C. –Y., and Liao, H. Y. -M. 2020. Yolov4: Optimal speed and accuracy of object detection. arXiv preprint arXiv:2004.10934.

COVID-19 Dashboard by the World Health Organization (WHO). Available online: http:// https://covid19.who.int/ (Accessed 9 July 2021).

Das, A., Ansari, M. W., and Basak, R. 2020. Covid-19 face mask detection using TensorFlow, Keras and OpenCV. IEEE 17th India Council International Conference (INDICON).

Deng, J., Guo, J., Ververas, E., Kotsia, I., and Zafeiriou, S. 2020. Retinaface: Single-shot multi-level face localisation in the wild. Proceedings of the IEEE/CVF Conference on Computer Vision and Pattern Recognition.

Eikenberry, S. E, Mancuso, M., Iboi, E., Phan, T., Eikenberry, K., Kuang, Y., Kostelich, E., and Gumel, A. B. 2020. To mask or not to mask: Modeling the potential for face mask use by the general public to curtail the COVID-19 pandemic. Infectious disease modelling. 5:293–308.

Everingham, M., Gool, L. V., Williams, C. K., Winn, J., and Zisserman, A. (2010). The pascal visual object classes (voc) challenge. Int. J. Comput. Vision. 88(2):303–338.

Florian, S., Kalenichenko, D., and Philbin, J. 2015. Facenet: A unified embedding for face recognition and clustering. Proceedings of the IEEE conference on computer vision and pattern recognition.

Girshick, R., Donahue, J., Darrell, T., and Malik, J. 2014. Rich feature hierarchies for accurate object detection and semantic segmentation.r Proceedings of the IEEE conference on computer vision and pattern recognition.

Khan, A., Sohail, A., Zahoora, U., and Qureshi, A. S. (2020). A survey of the recent architectures of deep convolutional neural networks. Artific. Intellig. Rev. 53(8):5455–5516.

Krasin, I., Duerig, T., Alldrin, N., Ferrari, V., Abu-El-Haija, S., Kuznetsova, A., Rom, H., Uijlings, J., Popov, S., and Veit, A. 2017. Openimages: A public dataset for large-scale multi-label and multi-class image classification. 2(3):18. https://github. com/openimages

Islam, M. Z., Islam, M. M., and Asraf, A. 2020. A combined deep CNN-LSTM network for the detection of novel coronavirus (COVID-19) using X-ray images. Informatics in medicine unlocked 20:100412.

Lin, T.-Y., Dollár, P., Girshick, R., He, K., Hariharan, B., and Belongie, S. 2017. Feature pyramid networks for object detection. Proceedings of the IEEE conference on computer vision and pattern recognition.

Lin, T. –Y., Maire, M., Belongie, S., Hays, J., Perona, P., Ramanan, D., Dollár, P., and Zitnick, C. L. 2014. Microsoft coco: Common objects in context. European Conference on Computer Vision.

Mingjie, J., Fan, X., and Yan, H. 2020. Retinamask: A face mask detector. arXiv preprint arXiv:2005.03950.

O'Mahony, N., Campbell, S., Carvalho, A., Harapanahalli, S., Velasco Hernandez, G., Krpalkova, L., Riordan, D., and Walsh, J. 2019. Deep learning vs. traditional computer vision. Science and information conference.

Redmon, J., Divvala, S., Girshick, R., and Farhadi, A. 2016. You only look once: Unified, real-time object detection. Proceedings of the IEEE conference on computer vision and pattern recognition.

Ren, S., He, K., Girshick, R., and Sun, J. 2015. Faster R-CNN: Towards real-time object detection with region proposal networks. Adv Neural Inf Process Syst. 28.

Russakovsky, O., Deng, J., Su, H., Krause, J., Satheesh, Ma, S. S., Huang, Z., Karpathy, A., Khosla, A., and Bernstein, M. 2015. Imagenet large scale visual recognition challenge. Int. J. Comput. Vision. 115(3):211–252.

Simonyan, K. and Zisserman, A. 2014. Very deep convolutional networks for large-scale image recognition. arXiv preprint arXiv:1409.1556.

Yu, H. and Chen, Y. 2020. Autonomous driving with deep learning: A survey of state-of-art technologies. arXiv preprint arXiv:2006.06091.

55 Public acceptance of drones: An approach

M. Junaid Qureshi[1,a], *AkshayKumar V. Kutty*[2,b], *Jayant Giri*[3,c], *Abhiram Dapke*[4,d], *Pallavi Giri*[5,e], *R. B. Chadge*[3,f], *and Neeraj Sunheriya*[3,g]

[1]Yeshwantrao Chavan College of Engineering, Nagpur, India

[2]New York University, USA

[3]Department of Mechanical Engineering, YCCE, Nagpur, India

[4]University of Maryland, USA

[5]Laxminarayan Institute of Technology, Nagpur, India

Abstract

Nowadays, drones are widely used for public, commercial, and scientific applications. Drone technology is getting a boom across the world. However, it is still not clear if the general public accepts the extremely progressive drone technology. This survey was carried out to study and analyse the public acknowledgement of drones applying the Knowledge, Attitude and Practice (KAP) pattern as well as numerical evaluation to reduce the improbability. To have an idea about the general outlook of drones, this KAP model was employed in a new survey report. Although there are several common results with the previous study, we studied the idea of drone acceptance with the subsequent methods. Firstly, by means of knowledge assessments, we found that the community appears to have a moderate perception about the drones. Moreover, we also learned that acceptance intensities concerning drones did considerably vary depending upon the application. Conclusively, various aspects could be accountable for the fluctuating intensities of acceptance throughout the diverse circumstances. The conclusions displayed that drones were not fairly accepted, excluding public security and scientific research purposes. Some general public still considers drones a dangerous technology that bluntly affects their solitude. Furthermore, the community is not knowledgeable of the majority of upcoming drone uses and various present uses. The survey was circulated to the common community (102 fully completed surveys). An analysis was done on this study too. By considering the conclusions, this study suggests that more efforts should be taken to inform the public about drone technology through mass media and scholastic organisations. This could result in developing the repute of drones from predator machineries or privacy stalkers toward a modern concept that benefits humanity.

Keywords: Attitude and practice, drone, drone applications, knowledge, public acceptance, support.

Introduction

Human life has drastically evolved in the past years by the extensive advancements in technologies. Be it the invention of the first light bulb or the discovery of the steam engine by James Watt, they all have contributed to a better and more comfortable human life. Also, no one will argue on the huge influence created by the innovations of transportation technologies on the Homo sapiens - the first aero plane by the Wright brothers or Model-T by Ford. Hardly did anyone think of that one day, we would have a small computer in our hands in the form of a smartphone. Certainly, technologies have altered the way we exist, travel, commune and hang out with our kith and kin, co-workers, acquaintances or even outsiders.

Although drone technologies aren't in mint condition, it has not penetrated our day-to-day lives up until a few years back. Nowadays, the emergence of the utilisations of drone technologies is abundant. Drone technologies have been publicised as a lucrative resolution for security purposes, research, or even cargo delivery. With the hasty development in drone technologies, the interest for drones among the general public, entrepreneurs and even the lawmakers is drastically increasing to apply the technology in a mass of like in-flight photography, surveying, inspection, cargo delivery, rescue operations, and investigation for law implementation (Clothier et al., 2015). Progressively, there is no uncertainty that drones will soon find their path into more areas of application (Boucher, 2015). Without a doubt, drones, if accepted broadly, might have the ability like those machineries stated earlier to influence in what manner our civilisation operates as an organisation. Think of a drone carting your favourite fast food and delivering it to your window on the 10th story of your residence or a drone carrying an investigation process in the super-mart. A lot of us like

[a]junaid.qureshi0911@gmail.com; [b]avk322@nyu.edu; [c]jayantpgiri@gmail.com; [d]abhiramdapke162@gmail.com; [e]pallavijgiri@gmail.com; [f]rbchadge@rediffmail.com; [g]neeraj.sunheriya@gmail.com

viewing sci-fi or superhero shows where drones of different forms and dimensions hover in in the middle and occasionally in and out of constructions, taking off and landing in the vicinity of each other, although are we prepared for these drones to become a vital function of our everyday livings? We are of the same mind as Aydin (2019) stated 'public acceptance of any technology is necessary for realising their benefits fully'. Drone function in communal zones is beyond than a technical topic. To confirm effective execution, communal and psychosomatic aspects of drone functions must be entirely known to improve public acceptance. Is the community prepared for the sci-fi circumstances where an arrangement of drones glides above your head, controlling different types of processes from search and rescue to observation to delivery? How far does the community understand regarding drone machinery? Do the community understand possible advantages of drone technology in their everyday lifetimes? What would be their worries along with troubles? How open are they towards drone technology?

Within this manuscript, the conclusions of review study assessing the community's acquaintance, approaches, and application of drones are exhibited. A comprehensive listing of drone functions was occupied (40 functions) in the study, which assisted the respondents to realise the drones' advantages. Alternatively, future widespread drone application risks were explained to eradicate any unfairness on the opinion scale. The results of this survey could assist experts, scholars, and policymakers decide the subsequent steps in adapting culture to the forthcoming drone era by elucidating the risk and welfares. The subsequent segment presents the research study's background.

A brief history of drone technology

Have you ever thought about how the current trend of drone technology started? The concept of drones started in 1782 in France (Kindervater, 2016) after the Montgolfier brothers deployed unmanned balloons, which was preceded by kites being sailed from a 32-gun frigate to distribute information leaflets in 1806 (Attard, 2017). That is around a couple of centuries earlier than the CIA used the first unmanned killer drone in 2002 in an embattled assassination in Afghanistan (Kindervater, 2016). In 1849, the public intent of the Montgolfiers got a melodramatic twist when the Austrians loaded a couple of unmanned balloons with weapons, invented by Franz von Uchatius, during an undisclosed battle in opposition Venice. A registration for patent for a hovering machinery that can contain explosives was enrolled in 1862. A few months later, one more hot-air balloon carted a container along with a scheduling apparatus that triggered the bomb out was registered in New York. This turned out to be a crucial instant in the past of UAVs for scouting, inspection and directed killings (Attard, 2017) the succeeding episode in the antiquity of drones comprised gyroscopic technology. After being launched from a catapult, a light structure's projectile was examined from where it took off around 50 miles having a 300 lbs bomb. Effortlessly, this revolutionary happening trembled the scientists, and the US Army straightaway appointed the Dayton-Wright, Airplane Company to construct the Kettering Aerial Torpedo – the "Bug". It was invented by Charles F. Kettering–by which it got its name Kettering. Triggered from a quad-wheeled figurine that moved along a moveable path, its structure of internal pre-set controls alleviated and directed it in the direction of the target. It was never productively used in combat, although the US spent $275,000 on the advancement (Attard, 2017). The United States set the revolution in technology by developing advanced armoured drones deployed in the World War II. The US additionally deployed a low elevation hovering drone in over 8000 battles occurrences in the War against Vietnam (1964) called the 'lightning bug'. The cold war became considerable when the CIA positioned some drones above North Korea, China as well as Cuba for inspection and investigation (Central Intelligence Agency, 2005).

Nowadays, drone machinery is not constrained to soldierly works. There are numerous industrial and civilian functions. A machinery instigated primarily from martial intentions is evolving into our world swiftly (like the GPS technology in our day-to-day livings). The situation is crucial to explore the communal view and familiarity on the fresh tasks of the drone machinery for making implementation plans. To lessen the hazards and amplify the advantages of this changeover into the widespread usage of drones shortly, community awareness, approaches, and application of drones need to be documented. Civic acknowledgement of any knowledge is essential for understanding their advantages wholly.

Background

As claimed by the Federal Aviation Administration, United States Department of Transportation's Aerospace Report of 20212024, the upcoming development capability of the drone marketplace for both type, hobbyists and commercial consumers are relatively high (FAA Aerospace Forecast). Looking at the data, there were 626,000 hobbyists registered as of December 2016, and an approximation of about 1.1 million drones goes to the amateurs. F.A.A. predicted an aggregate hobbyist collection of 3,550,000 drones before 2021 in a lower situation while 4,470,000 drones in a best-case situation. Alternatively, F.A.A. estimated four hundred twenty-two thousand industrial drones by 2021 in a lower situation while 1.6 million commercial drones in a best-case situation. Meanwhile, industrial users are obligated to own a remote pilot license allotted by F.A.A. These license predictions could become a helpful sign of the prospect of the drone marketplace. As per the current data, there are about 2.9 k certified remote pilots. Bearing in mind these statistics, there is an instant necessity to recognise a civic belief in drones and a demand for efficient skill acceptance approaches to handle the forthcoming drone modernisation. We have to know what must be done to upsurge communal acceptance of general drone practice in the subsequent groupings as provided by F.A.A.: aerial photography (34%), utility inspection (26%), emergency management (8%), real estate (26%), agriculture (21%), insurance (5%), construction and industrial. F.A.A. also claimed that because of the applications of drones in numerous classifications given above, the addition of the proportions surpassed 100%. A glitch identified here is that in-flight cinematography must be employed for all given classes. The specified 34% rate permits doubts about the correctness of this list (FAA Aerospace Forecast). Furthermore, F.A.A. recorded just the existing functions of drones but overlooked the future purposes. The following section lists prevailing and latent future applications of drones.

Application of drones

As of 2022, the drone size ranges from a purview of approximately 60 meters to Nano-drones as small as 1.5 inches wide. In spite of the load limitations (mean load of 4 lbs as quoted by Austin, 2010), the extensive dimension range of drones permits various uses for the humanity. Table 55.1 exhibits a broad record of present and possible functions of drones. The appliances in modelling or preliminary analysis phases were considered a future purpose. It should be Declared that when in future these functions would originate is out of the extent of this research. The drone's uses noted in Table 55.1 present that the technology is developing into numerous business and civil programs in the time ahead. Presently, the community is supposed to be on an awareness-building phase in the context of drone usages.

(Herron et al, 2014). Yankelovich (1991) described this phase as the community becoming conscious of a developing issue and technology and still collecting data and creating imprints. At this phase, public view can change and be open to disagreements as people comprehend the queries and trade-offs concerned (Yankelovich, 1991). In over-all, assessing these cases by arithmetical methods to decrease ambiguity is the primary the primary purpose of this study. Earlier studies made use of the public's awareness, approaches, and application of drones as well. In the subsequent sections, these findings are exhibited.

Public knowledge of drones

In a survey study, Reddy and DeLaurentis (2016) discovered that 93% of subjects from the typical community had learned about drones, maximum of whom which came from mainstream news media and movies (Reddy and DeLaurentis, 2016). Alternatively, investors understood drones from journalism or individual experience frequently (Reddy and DeLaurentis, 2016). They also studied how knowledgeable respondents believed themselves regarding drones. The knowledge was also advanced by Tam (2011) by inquiring subjects about how informed they are about drones. However, asking the respondents how well-versed they think they are about any modern tech possibly will not truly expose how knowledgeable they stand as actual knowledge and practical knowledge may perhaps not be interrelated (Mostafavi et al., 2014). A series of 5 questions in the form of true-false quiz was allocated to the respondents throughout the study of Reddy and DeLaurentis (2016) to experiment the awareness. The outcomes primarily displayed that the over-all community was uninformed of the technical boundaries and the past of drones (Reddy and DeLaurentis, 2016). This series of five problems did not comprise any function-based knowledge. In different terms, the public's knowledge concerning the Existings and future drone applications was neglected. The acceptance of the technology might be affected by the communal knowledge on the applications of drones. Eyerman discovered that the people had a moderately low degree of knowledge, with quite less than 44% stating that they comprehended just slightly or nothing at all about UAV functions (Eyerman et al, 2013).

There is a necessity to create a dependable and acceptable study instrument that determines definite knowledge instead of observed knowledge of drones. Furthermore, it should be broadened to gain knowledge concerning the broad applications of drones. It is not feasible to develop plans to efficiently cope up the drone upheaval without recognising how familiar the public is with drones. The next segment explains the writing on the view of drones.

Table 55.1 Drone utilisation/applications

	Sample References	*Existing or Future Applications? (As of February 2022)*
Military applications	Military.com, Drones. (2017).	Existings
Recording personal/family events	Alex, C. (2016).	Existings
Search and rescue	Sardrones.org (2017)	Existings
Recording sports events	Sahil, 2016; Pekler, L., 2016)	Existings
Monitoring wildfire and forest fires	(Cohen, J., 2007; Twidwell et al., 2016)	Existings
Transport and deliver cargo	(Hochstenbach et al., 2015; D'Andrea, 2014)	Future
Track suspected criminals or terrorists	Fox5dc.com. (2017).	Existings
Disaster early recognition and disaster aid	(Perks et al., 2016; Al-Rawabdeh et al., 2016)	Existings
Control drug trafficking	(Thompson and Mazzetti, 2011)	Existings
Construction inspecting	(Siebert and Teizer, 2014)	Existings
Control illegal immigration (border control)	(Lee, 2014; Engineering.com.)	Existings
Food Delivery (e.g., pizza drones)	(Reid, 2016)	Future
Building Firefighting	(Rosa, J., 2017; Kutner, 2017)	Existings
Monitoring crop health and growth	(Cohen, 2007; Twidwell et al., 2016)	Existings
Drone racing	(CNBC.com., 2017).	Existings
Emergency response (first aid)	(Bravo, et al., 2016; Kristensen et al., 2017)	Existings
Meteorology measurements	(Jarvis, B., 2014)	Existings
Monitoring nuclear plants for nuclear spills	flyability.com	Future
Archeological surveys	Fernandez-Lozano and Gutierrez-Alonso, 2016; Chiabrando et al, 2011).	Existings
Pesticide spraying	(Atherton, 2016; Rieder, 2014)	Existings
Examining the influences of global warming (e.g., observing icebergs)	(Schroth, 2017)	Future
Monitoring air pollution	(Alvear et al., 2017; Scentroid, DR1000, 2017)	Existings
Home security systems	(McFarland, 2016).	Future
Traffic patrol	(Lozano, 2017)	Existings
Reforestation (planting trees)	[45, 46]	Future
Herding cattle	[51]	Existings
Supplying connectivity via wireless signals	[68, 69]	Future
Passenger transportation	[38, 39]	Future
Disease spread control	[63, 64]	Future
Photogrammetry	[22, 23]	Existings
Thermal examining for discovering damaged insulation as well as air and water leakages	[30]	Existings
Inspecting highways and bridges	[31, 32]	Existings
Insurance claims	[43, 44]	Existings
Treatment of agricultural fields	[47]	Existings
Tracing poaching (prohibited trade of flora and fauna and ecological resources)	[55, 56]	Existings
Timely recognition of oil leaks and pipeline destructions or breakdowns	[59, 60]	Existings
Supplying flotation gear (e.g., life jackets) to the sufferers to assist lifesavers	[61, 62]	Existings
Underwater tasks to observe ocean ecosystems	[70]	Existings
Railway infrastructure monitoring	[72]	Existings
Surveying wild animal ecosystems	[52, 53]	
Archeological surveys	(Fernandez-Lozano and Gutierrez-Alonso; Chiabrando et al., 2011)	Existings

Public practice of drones

Earlier findings were directed upon the knowledge or Research Questions opinion, but no research was observed that highlighted how the drones were utilised by the community. Reddy and DeLaurantis (2016) had a segment for observing the communal use of drones, though the queries they enquired were truly calculating the opinion fairly than performance (Reddy and DeLaurentis, 2016). This report examines the functions community and businesses utilise drones for. The subsequent segment showcases the precise aims of this paper.

Objectives of the study

In the same way as above-mentioned, the community is at a realisation developing phase concerning the drones, yet collecting info and creating impacts, liable to change and open to disagreements trying to comprehend the queries and trade-offs concerned (Yankelovich, D., 1991). It is vital for any developing technology at this phase to observe the knowledge, opinion, and practice (Herron et al., 2014). Inspecting these three factors could aid us progress alteration management approaches for efficient technology implementation. Nevertheless, there is no data on how far the civic understands drones, how they comprehend the profits and hazards of drones, and the connections amongst familiarity, awareness, and the use of drones. Up to now, the studies have explored knowledge, perception, and practice distinctly. This survey discovers the relations between awareness, approaches, and application of drones, however it is likewise valid to every new evolving machinery. This survey report is a developed attempt by means of the KAP model (Knowledge, Attitudes, and Practice Model), enthused by Reddy and DeLaurentis's (2016) study, by mutating their study systematically, considering other related studies in this field as well as the literature review. This conveys what public understand about particular entities, how they experience, and how they operate (Kaliyaperumal, 2004). The KAP model can be applied for any topic concerning. It was initiated from medicinal study; however also been implemented in other areas (Reddy and DeLaurentis, 2016). As an example, Mostafavi et al. employed this technique to research community opinions of advanced backing for infrastructure practices (Mostafavi et al., 2014). Gadzekpoo et al. (2018) employed the KAPi technique for climate transformation coverage among iGhanaian correspondents. The technique is applied in the building researchi area too. As an example, Ibrahim and Belayutham (2019) applied the KAP technique for examining OSHA in the construction sector and organisation, whereas Gohi and Chuai utilised the KAP technique for construction security Goh and Chua (2016). Aydin et al. (2019) utilised it to comprehend how considerably construction companies understand regarding drones, how they remark them, and for what functions they operated them. KAP was also used in farming study by Norkaewi et al. (2010) to examine by means of PPE for agriculturalists.

Table 55.2 exhibits the various research questions created on the KAP model. Some questions did not necessitate arithmetical theory analysis.

Methodology

A review model was created by following the KAP model. The final form of the survey was released on February 15th and concluded on February 25th, 2022. Google forms were used to design the survey as well as for data compilation. The survey was dispersed among some qualified workers and academic students from various regions countrywide to confirm random experimenting. On the whole, 102 respondents finished the survey individually. The survey comprised three main sections: Knowledge, attitudes, and practice. Respondents were primarily requested to confirm the respondents' awareness of the forty drone uses recorded in Table 55.1. Respondents have to tick the application they have heard or have been aware of. The second part was a test comprising 12 True/False problems. This quiz was a reference of the one by

Table 55.2 Research questions created on the KAP model

Category	*Research question*
Knowledge	Research questions 1: '*How do the general public learn about drones? What are the sources of information? Are these sources scientific or solely mass media-based?*'
	Research questions 2: '*When someone buys a drone, does he/she try to learn more about drones? Do the general public have sufficient knowledge regarding drones, their history, technology, and the related laws and regulations?*'
Attitude	Research question 3: '*Do the general public knows future applications of drones as much as on current applications?*'
Practice	Research question 4: '*For what applications do you utilise drones?*'

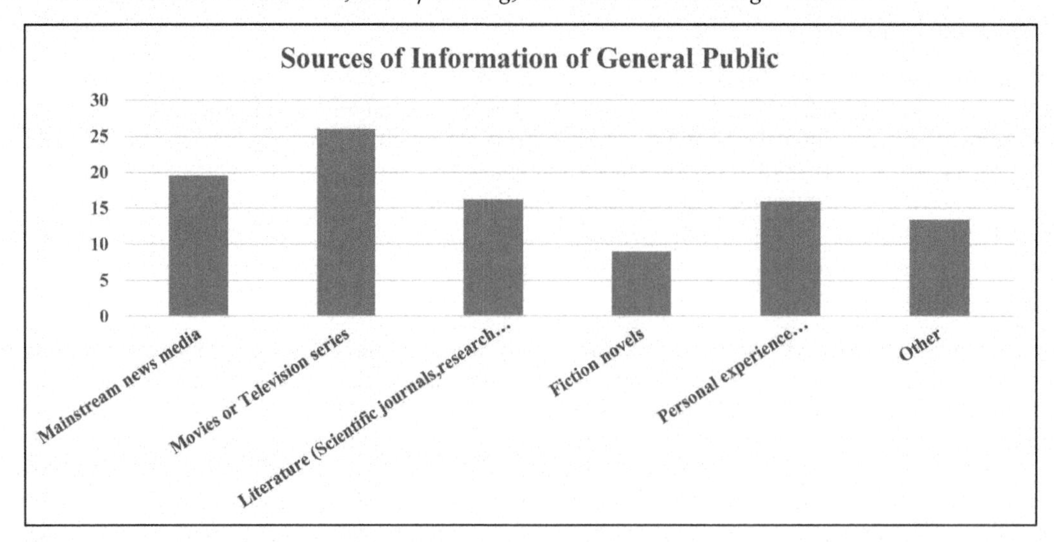

Figure 55.1 Bar chart for over-all communal information sources

Reddy and DeLaurentis (2016) and was developed by multiplied questions and broader subject. The test comprised the subsequent themes: 'past of drones, technical composition of drones, and drone laws as well as regulations.' The participants were told to choose the "I am not sure" alternative if they didn't recognise the response distinctly. Also, the respondents were requested to not make use of any search engines to search the solution. The third section of the acquaintance segment queried the sources of information where respondents learned about drones. The last part of the practice section asked the general public where they tend to use drones.

Results

The research questions given on Table 55.2 were analysed successively. The sequence in the survey arrangement will be pursued for denoting the results: Knowledge, attitudes, and practice.

Knowledge

The broad community studied drones largely from movies and conventional news media by Reddy and DeLaurentis (2016), whereas investors studied mainly from journalism and personal experience (Reddy and DeLaurentis's, 2016). Understanding drones from shows and typical news media could influence acceptance because military works are primarily highlighted in movies and mass media. Afore inspecting the over-all public's familiarity with any technology, I think their data traces should be studied. Therefore, the following research query is focused on further studying the knowledge source of the overall community in India for drones. The precise intention is to imitate Reddy and DeLaurentis's (2016) survey to understand if something has transformed from the time when they directed the study for knowledge resources.

'Research Question 1: How do the general public learn about drones? What are the sources of information? Are these sources scientific or solely mass media-based?'

Survey analysis exposed that the broad community discovered drones from shows, television series, and typical news broadcasting. Each participant was allowed to select more than one choice for the sources of information on drones. According to the data, 19.5% of people voted for mainstream news and& media, 26.0% voted for movies and television series, 16.2% voted for Literature (Scientific journals, research papers, magazines, etc.), 9.0% voted for fiction novels,15.9% voted for personal experience classroom, workshops, etc.) Furthermore, 13.4% voted for

The bold designates the right solution to the question. Other sources. Figure 55.1 shows this data in the form of a bar chart.

"Research Question 2: When somebody buys a drone, does he/she try to learn more about drones? Do the general public have sufficient knowledge regarding drones, their history, technology, and the related laws and regulations?"

There were 12 problems in the quiz, as shown in Table 55.3. Table 55.3 shows the community's quiz answer synopses for each query. The respondents must choose between 'True,' 'false,' and 'Not Sure.' The subjects were asked to choose the 'Not Sure' alternative, except they had accurate information.

Table 55.3 General community answers to the knowledge test (questions amended from Reddy and DeLaurentis, 2016)

#	*Question*	*True*	*False*	*Not sure*	*Correct rate*
1	The first drone was invented in the 1990si	26.5%	**30.4%**	43.1%	**30.4%**
2	The Federal Aviation Administration (FAA) now requires every drone operator to register before flying drones weighing over 0.55 lbs for commercial purposes.	**52%**	6.9%	41.1%	**52%**
3	You will be subject to civil or criminal penalties if you meet the criteria to register your commercial drone but do not register, including civil penalties up to $27,500 or criminal fees up to $250,000.	**36.3%**	9.8%	53.9%	**36.3%**
4	You will be subject to civil or criminal penalties if you meet the criteria to register your commercial drone but do not register, including civil penalties up to $27,500 or criminal fees up to $250,000.	50%	**12.7%**	37.3%	**12.7%**
5	Majority of the drones today are capable of operating autonomously without any human controller.	40.2%	**39.2%**	20.6%	**39.2%**
6	All drones are equipped with cameras.	57.8%	**30.4%**	11.8%	**30.4%**
7	All drones are equipped with GPS.	66.7%	**21.6%**	11.8%	**21.6%**
8	A commercialidrone pilot must be at least 16 years old.	**40.2%**	17.6%	42.2%	**40.2%**
9	It is legal to fly a commercial drone out of the line of sight of the pilot as long as first-person-view cameras are utilised.	37.3%	**23.5%**	39.2%	**23.5%**
10	The maximum ground speed allowed for commercial drones is 80 mph.	28.4%	**13.7%**	57.8%	**13.7%**
11	The maximum altitude allowed for commercial drones is 400 feet.	**31.4%**	9.8%	58.8%	**31.4%**
12	Under iFAA irules, ioperating a idrone ifrom a imoving ivehicle is iprohibited unless the operationi is iover a isparsely ipopulated iarea.	**39.2%**	7.8%	52.9%	**39.2%**

Out of the 12 given questions, 42.16% of the questions were marked as 'True,' 18.70% were marked as 'False' while a great amount of about 35.783% was marked as Not sure.' The results showed a mean correct rate of the solutions of about 30.883%.

Attitudes

Primarily, a complete study of possible aspects that might influence the approaches was done. A continuation study objective examines the variance among public mindfulness of present against upcoming drone utilisations.

'Research question 3: Do the general public knows future applications of drones as much as on current applications?'

Figure 55.2 exhibits the difference of familiarity versus unfamiliarity for each respondent. According to the survey data, 30.3% public were aware of the drone application of recording sports and family events, 30.0% knew the application of drones in military services, while 25.6% of people had an idea about the drone's application in search and rescue. However, some of the applications about which the least number of respondents knew were herding cattle (4.0%), meteorology measurements (7.2%), and disease spread control (7.2%).

Practice

As per the survey data, some of the respondents use drones for hobby at an individual stage (Figure 55.3). The business category mostly comprised professionals who lead wedding photography, while the public security group involved those who operate their drones for investigation and rescue missions. Whereas the scientific grouping had people operating drones for study-related uses. Some subjects fell into more than a few categories.

'Research Question 4: For what applications do you utilise drones?'

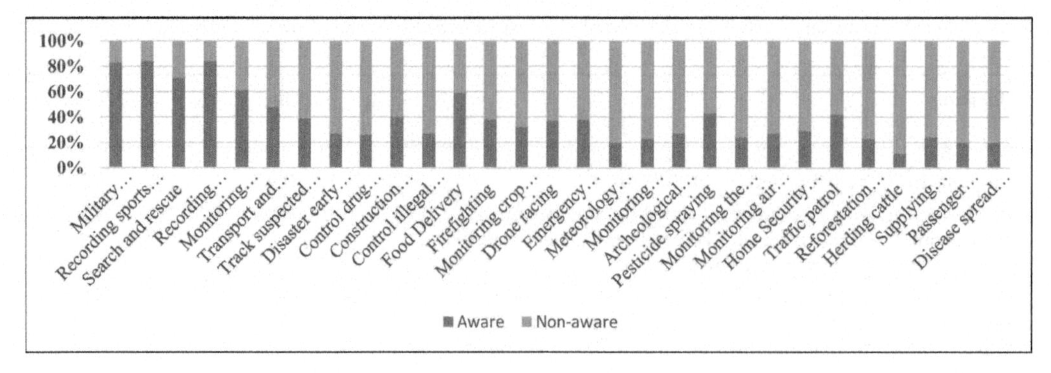

Figure 55.2 Public awareness of drone applications

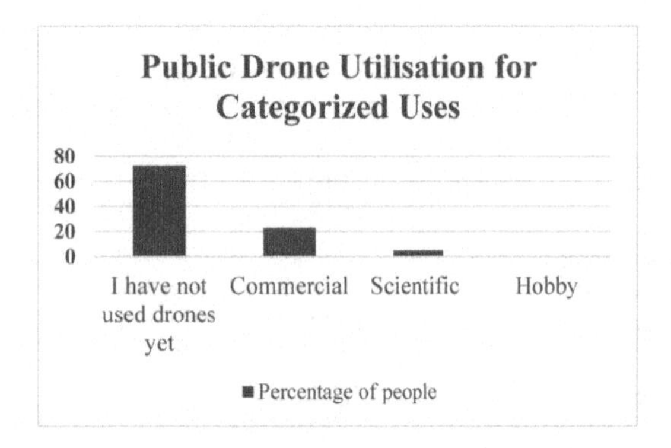

Figure 55.3 Bar chart for public drone utilisation in percentages

Out of the five given categories, 72.3% of people chose the option 'I have not used drones yet', 22.6% use drones for commercial purposes, 4.5% utilise drones for scientific purposes, while only 0.6% people use drones for hobby purposes.

Discussions and Implications

Drones are evolving swiftly, and their functions are developing beyond public learning. There are huge numbers of studies that highlight enhancements on the requirements of drones for instance flight time, freight capability, collision prevention, etc., and competencies like the interaction of groups of drones, AI, and virtual reality enhancements. Though, a satisfactory recognition of policy, ecological, moral, and social inferences of drones loiter far behind their science and technology. In this paper, a non-technical viewpoint on drone acceptance was given withi a complete discussion at drones' knowledge, attitudes, and practice. Also, a social perspective about what the public thinks about drones was tried to understand. The data obtained from this paper would help scholars and researchers understand the public view on drones and their related applications.

Conclusion

This study surveyed the general public's knowledge, attitudes, and practice of drones. We think the utmost significant conclusion is that the community is still at the learning and adopting stage towards drones, and most of the people out there are keen to learn more and more about drones. As in the survey, although respondents were displayed 40 different uses of drones, most people only knew about recording events and military applications. As of the current scenario, drones are largely understood as dangerous machineries for assassination, invading privacy, or models for public. Studies examining the menaces of drones and performing qualitative and quantitative hazards breakdown may be a helpful preliminary stage in the direction of progressing efficient risk moderation and response plans. Also, a detailed report with present remote pilots could be significant for realising the dangers and how to decrease the likelihoods and effects.

Drone modernisation is now taking place. We are certain that the scope of the machinery will be centred not merely on the expansions in technology of drones but particularly centred on the communal acceptance of drones. This might be attained by improved interaction of the public's advantages and hazard extenuation approaches.

References

Clothier, R. A., Greer, D. A, Greer, D. G., and Mehta, A. M. (2015). Risk perception and the public acceptance of drones. Risk Anal. 35(6):1167–1183.

Boucher, P. (2-15). Domesticating the drone: the demilitarisation of unmanned aircraft for civil markets. Sci. Eng. Ethics. 21(6):1393–1412.

Aydin, B. (2019). Public acceptance of drones: knowledge, attitudes, and practice, Technol. Soc. 59:101180.

Kindervater, K. H. (2016). The emergence of lethal surveillance: watching and killing in the history of drone technology, Secur. Dialogue. 47(3):223–238.

Attard, D. 2017. History of Drones. https://www.dronesbuy. net/history-of-drones/,

Central Intelligence Agency. (2005). The CIA World Fact book. New York (N Y): Skyhorse Publishing, Inc.

FAA Aerospace Forecast, Unmanned aircraft systems, https://www.faa.gov/data_ research/aviation/aerospace_forecasts/media/Unmanned_Aircraft_Systems.pdf

Austin, R. 2010. Unmanned aircraft systems: UAVs design, development and deployment. In: ed. I. Moir, A. Seabridge, R. Langton. 28. Aerospace Series. New York (N Y): John Wiley & Sons.

Herron, K. G., Smith, H. C. J., and Silva, C. L. 2014. US Public Perspectives on Privacy, Security, and Unmanned Aircraft Systems. Oklahoma: Technical Report University of Oklahoma.

Yankelovich, D. 1991. Coming to Public Judgment: Making Democracy Work in a Complex World. New York (N Y): Syracuse University Press.

Sahil, P. 2016. Eye in the Sky: Fox Sports Is Bringing Drones to Sporting Events. https://digiday.com/media/eye-sky-fox-sports-bringing-drones-sporting-events/

Flyability Safe Drones for Inaccessible Places. 2017. Inspection of a Nuclear Power Plant. http://www.flyability.com/wp-content/uploads/2017/07/Inspection-of-a-nuclear-power-plant.pdf.

CNBC.com. (2017). K. Song, Drone Racing Is Worth $100,000 in Upcoming Championship (2017). https://www.cnbc.com/2017/06/19/drone-racing-is-worth-100000-in-upcoming-championship.html .

Alex, C. (2016). Wedding Photographer Takes Gorgeous Video Using Drone. https://fstoppers.com/aerial/wedding-photographer-takes-gorgeous-video-using-drone-137941.

Military.com, Drones. (2017). http://www.military.com/equipment/drones

Sardrones.org. (2017). Search with Aerial RC Multirotor, S.W.A.R.M. http://sardrones.org/.

Fox5dc.com. (2017). Stafford Co, Deputies Using Drones to Track Down Suspected Criminals. http://www.fox5dc.com/news/stafford-co-deputies-usingdrones-to-track-down-suspected-criminals .

McFarland, M. (2016). CNN Tech, Forget Your Old Alarm System. This drone will protect your house http://money.cnn.com/2016/11/03/technology/drone-homealarm-system/index.html.

Indiegogo.com. 2017. Hawkeye the Indoor Smart Drone Security Guard. https:// www.indiegogo.com/projects/hawkeye-the-indoor-smart-drone-security-guard-drones#/.

Rosa, J. (2017). New York Post, FDNY Uses Drone to Help Fight Building Fire. http://nypost.com/2017/03/07/fdny-uses-drone-to-help-fight-building-fire/.

Kutner, M. (2017). Newsweek, London Firefighters Used Drone to Battle Grenfell Tower Blaze. http://www.newsweek.com/london-grenfell-tower-fire-departments-drones-626824.

Cohen, J. (2007). Drone spy plane helps fight California fires, J. Sci. 318:727.

Twidwell, D., Allen, C. R., Detweiler, C., Higgins, J., Laney, C., and Elbaum, S. (2016). Smokey comes of age: unmanned aerial systems for fire management, J. Front Ecol. Environ. 14(6):333–339.

Perks, M. T., Russell, A. J., Large, A. R. G. (2016). Technical note: advances in flash flood monitoring using unmanned aerial vehicles (UAVs). Hydrol, Earth Syst. Sci. 20:4005–4015.

Al-Rawabdeh, A., He, F., Moussa, A., El-Sheimy, N., and Habib, A. (2016). Using an unmanned aerial vehicle-based digital imaging system to derive a 3D Point Cloud for landslide scarp recognition. Remote Sens. 8:1–32.

Siebert, S. and Teizer, J. (2014). Mobile 3D mapping for surveying earthwork projects using an unmanned aerial vehicle (UAV) system. Autom. ConStruct. 41:1–14.

Li, M., Nan, L., Smith, N., and Wonka, P. (2016). Reconstructing building mass models from UAV images. Comput. Graph. 54:84–93.

Morgenthal, G., Hallermann, N. (2014). Quality assessment of unmanned aerial vehicle (uav) based visual inspection of structures. Adv. Struct. Eng. 17:289–303.

American Association of State Highway and Transportation Officials, State DOTs Using Drones to Improve Safety, Collect Data & Cut Costs, (2016). https://www.forconstructionpros.com/asphalt/press-release/12187251/american-association-of-state-highway-and-transportation-officials-state-dots-using-drones-to-improve-safety-collect-data-and-cut-costs.

Lee, E. Y. H. (2014). Think Progress, There Have Now Been 10,000 U.S. Drone Flights to Patrol the Southern Border. https://thinkprogress.org/there-have-now-been-10-000-u-s-drone-flights-to-patrol-the-southern-border-9b43c0ac56d3/.

Engineering.com. Autonomous border patrol system could deploy drones on its own, https://www.engineering.com/DesignerEdge/DesignerEdgeArticles/ArticleID/15212/Autonomous-Border-Patrol-System-Could-Deploy-Drones-on-itsOwn.aspx

Thompson, G. and Mazzetti, M.(2011). U.S. Drones Fight Mexican Drug Trade. http:// www.nytimes.com/2011/03/16/world/americas/16drug.html

Hochstenbach, M., Notteboom, C., Theys, B., and Schutter, J. D. (2015). Design and control of an unmanned aerial vehicle for autonomous parcel delivery with transition from vertical take-off to forward flight, Int. J. Micro Air Veh. 7(4):395–405.

D'Andrea, R. (2014). Guest editorial can drones deliver? IEEE Trans. Autom. Sci. Eng. 11:647–648.

Bravo, G. C., Parra, D. M., Mendes, L., and Pereira, A. M. D. J. (2016). First aid drone for outdoor sports activities, Proceedings of the 1st International Conference on Technology and Innovation in Sports, Health and Wellbeing (TISHW 2016), IEEE.

Kristensen, A. S., Ahsan, D., Mehmood, S., and Ahmed, S. (2017). Unmanned aerial system for fast response to medical emergencies due to traffic accidents, World Academy of Science. Eng. Technol. Int. J. Health Med. Eng. 11–11:593–597.

B. Marquand, Meet Your New Claims Inspector: a Drone, (2017) https://www.nerdwallet.com/blog/insurance/drones-home-insurance-claims-inspectors/.

A. Glaser, Insurance Companies Are Preparing Fleets of Drones to Assess the Damage of Harvey, (2017) http://www.slate.com/blogs/future_tense/2017/08/29/insurance_companies_will_fly_drones_after_hurricane_harvey.html.

E.P. Fortes, Seed plant drone for reforestation, Grad. Rev. 2 (2017) 13–26.

L. Fletcher, Reversing deforestation using drones: deforestation, Environ. Manag. 9 (2015) 22.

J. Rasmussen, G. Ntakos, J. Nielsen, J. Svensgaard, R.N. Poulsen, S. Christensen, Are vegetation indices derived from consumer-grade cameras mounted on UAVs sufficiently reliable for assessing experimental plots? Eur. J. Agron. 74 (2016) 75–92.

K.H. Dammer, G. Wartenberg, Sensor-based weed detection and application of variable herbicide rates in real time, Crop Protect. 26 (2007) 270–277.

Atherton, K. D. (2016). This Drone Sprays Pesticides Around Crops. https://www.popsci.com/agri-drone-is-precision-pesticide-machine.

Rieder, R., Pavan, W., Maciel, J. M. C., Fernandes, J. M. C., and Pinho, M. S. (2014). A virtual reality system to monitor and control diseases in strawberry with drones: a project. Int. Congr. Environ. Model. Softw. 3.

Schroth, F. (2017). National Science Foundation Awards Grant for Drone Monitoring of Global Warming. https://dronelife.com/2017/07/10/national-sciencefoundation-awards-grant-for-drone-monitoring-of-global-warming/.

Stuart, M. 2016. Rangers Are Using These Night Drones to Go after Poachers in Africa.http://www.businessinsider.com/anti-poaching-drones-rangers-africa-endangered-wildlife-2016-2.

Corrigan, F. 2017. 8 Top Anti Poaching Drones for Critical Wildlife Protection. https://www.dronezon.com/drones-for-good/wildlife-conservation-protectionusing-anti-poaching-drones-technology/.

Alvear, O., Zema, N. R., Natalizio, E., and Calafate, C. T. (2017). Using UAV-Based systems to monitor air pollution in areas with poor accessibility, J. Adv. Transp. 1–14.

Scentroid, DR1000, Flying Laboratory. (2017). Drone Environmental Monitoring, B. Aydin Technology in Society. 59 (2019) 101180: 13. http://scentroid.com/scentroid-dr1000/.

Cho, J., Lim, G., Biobaku, T., Kim, Parsaei, S. H. (2015). Safety and security management with unmanned aerial vehicle (uav) in oil and gas industry, Procedia Manuf. 3:1343–1349.

Drones at Work, Oil and Gas Drone News. 2017. http://www.dronesatwork.com/category/inspection-drone/oil-gas-drones/.

McSweeney, K. 2016. How Drones Could Save You from Drowning. http://www.zdnet.com/article/how-drones-could-save-you-from-drowning/

Newman, L.H. 2015. Lifeguard Drones Deliver Help to Drowning Victims in 30 Seconds.http://www.slate.com/blogs/future_tense/2015/03/24/lifeguard_drone_ test_in_chile_by_green_solution_and_x_cam.html.

Lab-on-a-drone could improve management of disease outbreaks. http://research.tamu.edu/2016/05/11/lab-on-a-drone-could-improve-management-of-disease-outbreaks/.

A. Hardy, A., Makame, M. Cross, D. Majambere, S. and Msellam, M. (2019). Using low-cost drones to map malaria vector habitats. Parasits Vectors. 10(29):1–13.

Jarvis, B. 2014. Drones Are Helping Meteorologists Decipher Tropical Cyclones.http://www.pbs.org/wgbh/nova/next/earth/drone-meteorology/.

Fernandez-Lozano, J. and Gutierrez-Alonso, G. (2016). Improving archaeological prospection using localized UAVs assisted photogrammetry: an example from the Roman Gold District of the Eria River Valley (NW Spain), J. Archaeol. Sci. Report 5:509–520.

Chiabrando, F., Nex, F., Piatti, D., and Rinaudo, F. (2011). UAV and RPV systems for photo-grammetric surveys in archaeological areas: two tests in the Piedmont region (Italy). J. Archaeol. Sci. 38:697–710.

Lewontin, M. 2016. How Google's SkyBender Drones Will Deliver Internet Access to Remote Areas. https://www.csmonitor.com/Technology/2016/0201/How-Google-s-SkyBender-drones-will-deliver-Internet-access-to-remote-areas.

Goel, V. 2014. A New Facebook Lab Is Intent on Delivering Internet Access by Drone. https://www.nytimes.com/2014/03/28/technology/a-new-facebook-labis-intent-on-delivering-internet-access-by-drone.html?_r=0.

Duarte, M., Oliveira, S. M., and Christensen, A. L. 2014. Hybrid Control for Large Swarms of Aquatic Drones, ALIFE 14: Proceedings of the Fourteenth International Conference on the Synthesis and Simulation of Living Systems. (pp. 785–792). Cambridge: MIT Press

Reid, D. (2016). Domino's Delivers World's First Ever Pizza by Drone. https:// www.cnbc.com/2016/11/16/dominos-has-delivered-the-worlds-first-ever-pizza-by-drone-to-a-new-zealand-couple.html.

Flammini, F., Pragliola, C., and Smarra, G. 2016. Railway Infrastructure Monitoring by Drones, Electrical Systems for Aircraft, Railway, Ship Propulsion and Road Vehicles & International Transportation Electrification Conference (ESARS-ITEC), IEEE Xplore. https://doi.org/10.1109/ESARS-ITEC.2016.7841398.

Pekler, L. (2016). The Future of Drones in Live Sports Coverage and Sports Performance Analysis. https://www.suasnews.com/2016/03/42568/, Accessed date: 1 August 2017.

Reddy, L. and DeLaurentis, B. D. (2016). Opinion survey to reduce uncertainty in public and stakeholder perception of unmanned aircraft, Transportation Research Record, J. Transp. Res. Board. 2600(1):80–93.

Tam, A. 2011. Public Perception of Unmanned Air Vehicles. Aviation Technology Graduate Student Publications. Paper 3.

Yankelovich, D. (1991). Coming to Public Judgment: Making Democracy Work in a Complex World. New York, (N Y): Syracuse University Press.

Herron, K.G., Smith, H. C. J., and Silva, C. L. (2014). US Public Perspectives on Privacy, Security, and Unmanned Aircraft Systems, Technical Report. Norman, Oklahoma: University of Oklahoma.

Eyerman, J., Letterman, C., Pitts, W., Holloway, J., Hinkle, K., Schanzer, D., Ladd, K., Mitchell, S., and Kaydos-Daniels, S. C. 2013. Unmanned Aircraft and the Human Element: Public Perceptions and First Responder Concerns, Institute for Homeland Security Solutions. North Carolina: Research Triangle Park.

Mostafavi, A., Abraham, D., and Vives, A. (2014). Exploratory analysis of public perceptions of innovative financing for infrastructure systems, the US, Transp. Res. A Policy Pract. 70:10–23.

Kaliyaperumal, K. (2004). Guideline for conducting a knowledge, attitude, and practice (KAP) study, Community Ophthalmology. AECS Illum. 4:7–9.

Gadzekpo, A., Tietaah, G. K. M., Segtub, M. (2018). Mediating the climate change message: knowledge, attitudes and practices (KAP) of media practitioners in Ghana. Afr. J. Stud. 39(3):1–23. https://doi.org/10.1080/23743670.2018.146783.

Ibrahim, C. K. I. C. and Belayutham, S. (2019). A Knowledge, Attitude and Practices (KAP) Study on Prevention through Design: a Dynamic Insight into Civil and Structural Engineers in Malaysia, Architectural Engineering and Design Management. https://doi.org/10.1080/17452007.2019.1628001.

Goh, Y. M. and Chua, S. (2016). Knowledge, attitude and practices for design for safety: a study on civil & structural engineers. Accid. Anal. Prev. 93:260–266.

Aydin, B., Yeon, J., Oh, E.I. (2019). Drones in construction sector: knowledge, attitudes, and practice, a pilot survey study, IIE Annual Conference Proceedings, Institute of Industrial and Systems Engineers.

Norkaew, S., Siriwong, W., Siripattanakul, S., Robson, M. (2010). Knowledge, Attitude, and practice (KAP) of using personal protective equipment (PPE) for Chih-Growing farmers in Huarua sub-District, Mueang District, Ubonrachathani province, Thdand. J. Health Res. 24(2):93–100.

56 Challenges and recent advances in automated plant disease recognition and classification using deep learning and machine learning

Vaishali G. Bhujade[1,a], *G. H. Waghmare*[2,b], *and V. K. Sambhe*[3,c]

[1]Department of Computer Engineering VJTI Mumbai, India

[2]Department of Mechanical Engineering YCCE, Nagpur, India

[3]Department of Information Technology VJTI Mumbai, India

Abstract

Automatic crop disease detection and classification pose a variety of challenges in consideration with available tools, and processing techniques and have received considerable attention in the past few decades. In the agriculture industry crop diseases have proven to be responsible for crucial money loss in the whole world. Moreover, it causes a major risk to the food supply. The nation's economy and production are affected by the disease which is found in the agricultural crops, and it causes a major threat to the nation's growth if it is not discovered on time. Within this scenario, automatic disease recognition and classification is a very critical and primary challenge for sustainable farming. Traditional methods were manual which are prone to errors, time-consuming, and costly. In recent years deep learning (DL) along with image processing has garnered tremendous success in a variety of application domains including automatic disease detection. This article is very important for researchers and academicians to review and define several factors influencing the detection process. This study is unique in the following ways:

It provides an in-depth analysis of challenges and solutions for automatic crop disease recognition along with several factors that affect the performance of DL methods.

It also describes the state-of-the-art techniques to overcome these challenges and highlights the gaps in the available literature. It presents different ways to identify crop diseases using machine learning and DL with their pros and cons.

It also identifies major challenges in real-time prediction with their causes and impacts on the performance of methods considered.

Keywords: Deep learning, disease recognition, disease classification, datasets, machine learning, image processing.

Introduction

In India, agriculture is believed as the backbone of the financial system and the farmers could find high-quality crops for their farms. Technical assistance is essential Anyway, for farmers to yield maximum profit and standard manufacture in crop cultivation. In India, 58% of the people depend on agriculture and rural households also; it becomes the fastest-growing developing nation in the world. By the annual report of 20162017 printed by the Department of Agriculture, Cooperation & Farmers Welfare, agriculture provides an important function in India's financial system (Ferdous et al., 2020). Moreover, half of the Indian residents are employed in agriculture and its related activities (as per the 2011 census) also; which gives 17% of the country's Gross Value Added (current price 2015–2016, 2011–2012 series) point (German et. al., 2020; Kaushal et. al., 2020). In the agriculture field, plants are considered a significant source of human livelihood and it fulfils the basic requirements of clothing, food, shelter, and health care. Also, act as a raw material for the medicinal industry (Chen et. al., 2021).

At the present time, the world is known for 374000 species of plants. Because of some biotic and abiotic challenges, some of the species are declared as critically vanished plants. Different kind of pathogens which includes virus, fungi, bacteria, oomycetes, and nematodes attack the plants continuously (Martinelli et. al., 2015). To enter the plants, pathogenic organisms had different kinds of mechanisms which include enzymatic and structural components. Also, it follows various life tactics to enter the plant system (Vishnoi et. al., 2015). The plant's illness decreases the superiority and production of fibre, food, and wood. In the agricultural industry, plant diseases are responsible for crucial money loss the whole worldwide. Moreover, it causes a major risk to the food supply (Kamal et. al., 2019).

[a]vaishali.hardeo@gmail.com; [b]gwaghmare@gmail.com; [c]vksambhe@it.vjti.ac.in

The nation's economy and production are affected by the disease which is found in the agricultural crops, and it causes a major threat to the nation's growth if it is not discovered on time. If it is identified early, the effect of pathogens is minimised with help of using pesticides or their equivalent combat (Ferentinos et al., 2018). To recognise biotic or abiotic factors, different kind of laboratory observations like measurement of pH, soil analysis as well as plant sample analysis is done (Thomas et al., 2018). But this approach is considered a time-consuming and expensive task. Also, pests and climate changes create more effective e challenges in crop yields, and this becomes a serious problem in the food management system. To monitor the plant's condition, quantity as well as quality of the agricultural products, some methods are developed. Furthermore, different frameworks based on Information and Communication Technology (ICT) and artificial intelligence (AI) are utilised by the agriculture field as a support scheme to help the farmers for better farming (Dhingra et al., 2018; Patokar et al., 2018; Pourazar et al., 2019).

Nowadays, automatic identification of plant disease is considered a significant topic that provides benefits to monitoring large fields of crops and routinely recognises an indication of diseases as early as they emerge on plant leaves (Tantalaki et al., 2019). Hence, an automatic, accurate, less expensive and fast method to perform plant disease detection is assumed of great realistic importance. Image recognition is performed by many previous works, and they use specific classifiers to classify the images into diseased or healthy images (Sinwar et al., 2020). Usually, the plant diseases are initially identified by the leaves of the plants because; the disease indications are merely starting to be visible on the leaves. To minimise the monitoring work of large crop farms, automatic diagnosis techniques are needed in which they identify indications of diseases in an early stage on the leaves (Golhani et al., 2018; Nandhini et al., 2018). So, it requires that kind of method to easily identify the disease in plants using images of their leaves to exactly classify them based on the disease of the plant that is suffering from minimum error. In the recent era, deep learning (DL) techniques are extensively applied in several fields of computer vision. Therefore, the research applied this DL technique for disease prediction and achieved a significant advancement in prediction accuracy.

Literature Survey

Some of the recent related works are given below, Bhatia et al. (2020) developed an Extreme Machine Learning (ELM) technique for plant disease forecast using a highly imbalanced dataset. For prediction purposes, this paper used a real-time dataset known as Tomato Powdery Mildew Disease (TPMD) dataset. The imbalanced dataset was resample by different kinds of sampling techniques like Random under Sampling (RUS), Importance Sampling (IMPS), and Random Over Sampling (ROS), and Synthetic Minority Over-sampling Technique (SMOTE) to balance the dataset for the prediction model. Imbalanced TPMD datasets, as well as different balanced datasets acquired from resampling techniques, were utilised in this ELM model. Classification accuracy (CA) and Area under Curve (AUC) parameters were used to analyse the performance of the ELM model. A balanced TPMD dataset performs well rather than an imbalanced dataset.

The quality and quantity of agricultural food production can be affected by plant diseases which cause significant food product reduction in plants. This may lead to severe no grain harvest in plants. So, it requires automatic recognition and analysis of plant diseases extremely in the field of agriculture. Chen et al., (2020) studied a Deep Convolution Neural Network (CNN) was used to recognise plant leaf diseases. Moreover, the Inception module and ImageNet model were utilised for the VGGNet pre-training process. The weights were initialised by using the pre-trained networks on a large ImageNet dataset. The proposed method accomplishes 91.83 % validation accuracy on the public dataset which was greater than existing methods.

Many applications have been developed for the automatic analysis of plant disease based on the accomplishment of DL methods. But these applications are suffered from over fitting problems and, diagnosis of performance is drastically decreased with the use of test datasets from new atmospheres. On the agricultural side, computer vision models and Smartphone penetration create an opportunity for image classification to identify the disease. Cap et al. (2020) developed an efficient data augmentation technique for realistic plant disease analysis. This research developed a leaf GAN method which is a new image-to-image translation system with individual attention mechanisms. The performance of plant disease analysis was performed by this data augmentation tool. Moreover, this tool transforms only a relevant area of the images with different kinds of backgrounds and enhances the flexibility of the training images.

In the smart agriculture field, deep learning, as well as ML approaches were considered a promising solution for fine-grained disease severity classification. Sujatha et al. (2021) studied the performance of the machine as well as DL in plant leaf disease detection. The primary source of human energy production was based on the plants which have nutrias, medicinal, etc. values. In the farming industry, leaf disease identification is considered an important problem in which the plants are affected by diseases at any time during crop farming. In this article, the performance of machine learning (Stochastic Gradient Descent (SGD),

Table 56.1 Machine learning approaches for plant disease identification

Reference	*ML Algorithm*	*Approach*	*Dataset*	*Plant*	*Considered Diseases*
Barbedo et al. (2016)	Maximum Likelihood, Color transformation, and colour histogram	Image processing	Image dataset (Own)	Common Bean Cassava, Citrus, Coconut Tree, Coffee, Corn, Cotton, Grapevines, Passion Fruit, Soybean, Sugarcane, and Wheat	Phaseolus vulgaris L, Manihot esculenta, Citrus sp, Cocos nucifera, Coffea sp, Zea mays, Gossypium hirsutum, Vitis sp, Passiflora edulis, Glycine max, Saccharum sp, and Triticum aestivum
Zhou et al. (2015)	Support Vector Machine	Image processing	Sugar Beet Images (Own)	Sugar beet	Cercospora
Xie et al. (2015)	Multiple-kernel learning (MKL)	Image processing	Data was collected from Wheat, Canola, Soybean, and Corn crop. (Own)	Wheat, Canola, Soybean, and Corn	24 different types of insects (Cifuna locuples, Chilo suppressalis, Sogatella furcifera, etc.)
Yang et al. (2020)	Multiple Linear Regression, Support Vector Regression	Remote sensing	Data collected using Sensors	barley, wheat, and canola	carbon accumulation and meteorological stress indices
Singh et al. (2018)	Extreme Learning Machine Classifier	Remote sensing	AICRP dataset	Potato	Late blight
Raza et al. (2015)	Support vector algorithm with a linear kernel	Visible and thermal imaging	Tomato images (Own)	Tomato	Powdery mildew

Support Vector Machine (SVM)), Random Forest (RF), as well as DL (VGG-16, VGG-19, Inception-v3) are compared by means of citrus plant disease detection. In the case of disease detection, a DL algorithm performs better detection than machine learning techniques.

Crop disease diagnosis plays a significant part to find the disease of plants early to make awareness among farmers. In the agriculture sector, the appearances of numerous crop-related diseases change productivity. To conquer this problem, Khamparia et al. (2020) developed a Deep Convolution Encoder Network for seasonal crops disease prediction and classification. The combination of CNN and auto encoders was combined like a hybrid model to detect the crop leaf disease. This study utilises a 900-image dataset in which 300 images were used for the test set and 600 images were used for the training set. Moreover, this work utilises different convolution filters like 2x2 and 3x3 filters.

Cereals are assumed as the main food source for human consumption so, optimisation of cereal products is considered one of the great challenges in food security. After rice, maize, and wheat, cereal is considered the fourth significant food item in which its production is strongly based on fertilisation treatment. Large economic benefits to the producer are delivered by selecting suitable fertiliser management policies. So, Escalante et al. (2019) developed a method to sustain producers with automatic tools to analyse bare fertilisation management. It receives information from aerial RGB images detained by UAV to simultaneously calculate the nitrogen fertilisation and barely yield. Without increasing cost, the proposed method aims to offer wide-area coverage and a low-cost solution to calculate barely variables. RGB crop field images are captured by the low-cost UAV. CNN is used to extract the features from the images and these features are supplied to predictive models to compute variables of interest.

In plants, chlorophyll content modification is considered a high-quality pointer to disease, environmental, and nutritional stresses on plants.

Table 56.2 Deep learning approaches for plant disease recognition

Reference	*DL Algorithm*	*Approach*	*Dataset*	*Plant*	*Considered Diseases*	*Accuracy*
Jose et al. (2020)	AlexNet, GoogleNet, VGG16, ResNet50 MobileNetV2	Image Processing	Image dataset (Own)	1 coffee	4 rust, leaf miner, Cercospora leaf spot, and brown leaf spot	95.2 (biotic stress classification) 86.51 (severity estimation)
Darwish et al. (2020)	CNN model: VGG16, VGG19 Orthogonal Learning Particle Swarm Optimization (OLPSO)	Image Processing	Public Dataset hosted at Kaggle	1 Maize	Common rust, grey leaf spot, and Northern leaf blight	98.2
Amara et al. (2017)	GoogleNet Architecture (CNN)	Image Processing	Banana leaf Images were collected using a digital camera (Own)	Banana	banana sigatoka, banana speckle	82.88
Zhao et al. (2020)	CNN, Multi-Context Fusion Network	Image Processing	Captured 50000 images with a CCD camera (Own)	19	77	97.5
Fuentes et al. (2017)	R-FCN, Faster R-CNN, and Single Shot Multibox Detector (SSD)	Image Processing	Tomato leaf images are collected by the Camera (Own)	Tomato	Leaf mold, Canker, Gray mold, Low temperature, Plague, Nutritional excess, Whitefly, Miner, and Powdery mildew	83.06

Different kinds of pre-processing methods are developed to minimise the noise from the spectral records to recognise vegetation characteristics such as chlorophyll substances. Also, biochemical properties have been assessed by applying machine learning algorithms; still, integration of machine learning with pre-processing has not been fully estimated. Sonobe et al. (2021) developed five different pre-processing methods with machine learning algorithms to calculate the chlorophyll content in two wasabi cultivars. In general, higher accuracy can be obtained by associating machine learning techniques with pre-processing algorithms.

In oil palm plantations, Basal Stem Rot (BSR) disease caused by Ganoderma boninense is a major disease and it does not control by any effective fungicide. Till now, most researchers applied remote sensing methods for BSR studies. Santoso et al. (2019) developed WorldView-3 to categorise the rigorousness of BSR disease symptoms. A major goal of this article is to estimate rigorousness levels of oil palm tree disease using WorldView-3 imagery with supervised learning methods. For observation, information was gathered from 1923 oil palm trees with different levels of contamination like unhealthy and healthy trees with three levels of symptoms from mild to rigorous. Three different kinds of machine learning methods such as decision tree, random forest, and SVM were applied.

Barbedo et al. (2016) proposed a technique for detecting multiple diseases of plants. The proposed method was based on colour transformation, a pair wise-based classification model, and a colour histogram. The model focuses on the classification of 82 different diseases from 12 different crop species. Zhou et al. (2015) developed a novel algorithm for monitoring the real-time development of Cercospora leaf spot (CLS) in sugar beets using pattern recognition and template matching. The proposed method comprises two frameworks for detection. The first framework was used for tracking leaves from plants by using orientation code matching. For classifying the disease second framework uses a support vector machine (SVM). Post-processing is used to eliminate the noise and increase the precision of the classification process (Zhou et al., 2015).

The authors developed an automatic pest recognition system to recognise crop insects for field crops such as wheat, canola, soybean, and corn. The model used multiple-kernel learning (MKL) techniques in accordance with multiple task sparse representation (MTSR). To enhance the performance of the model the

MTSR has combined several features of insects. The various features such as texture, shape, and colour are extracted from the insect images and these multiple features are combined by MKL. The proposed model was evaluated on 24 different pest species, and the model outperforms the available methods (Xie et al., 2015).

The technique for estimating nationwide yield using carbon accumulation and meteorological stress indices was developed by Yang et al. (2020). The proposed semi-empirical model (crop-SI) comprises the Multiple Linear Regression (MLR) algorithm and Support Vector Regression (SVR) to estimate the yield of numerous crops such as barley, wheat, and canola using remotely sensed data. The Crop-SI combined crop-specific meteorological stress indices (temperature, cold stress, precipitation, etc.) and carbon fixation during the dangerous crop-growth stage. Experiments have shown that the model reduces relative error for every considered crop (Yang et al., 2020).

An efficient model for classifying and estimating the severity of leaf biotic stress for the coffee plant using DL techniques was proposed by Jose G.M. Esgario et al. in 2020. The proposed model consists of CNN architecture with multi-task capabilities for classifying the stress and estimating its severity of it. Five different CNN architectures (AlexNet, GoogleNet, VGG16, MobileNetV2, and ResNet50) have been experimented with, and results showed that ResNet50 outperforms others with a classification accuracy of 95.24%, and severity estimation accuracy of 86.51% (Jose et al., 2020).

An ensemble-based technique for detecting maize leaves diseases using two pre-trained CNNs was developed by Ashraf et al. in 2020. The technique was using pre-trained VGG16 and VGG19 architecture. CNN is sensitive to a variety of hyper parameters; therefore, to select optimal values of hyper parameters, an Orthogonal learning particle swarm optimisation (OLPSO) method was used. The results were compared against two pre-trained CNN models, namely InceptionV3 and Xception. The proposed model outperforms with accuracy = 98.2%, recall = 0.97, F1-score = 0.97, and precision = 0.98 (Ashraf et al., 2020).

Li et al. (2020) proposed a CNN-based model for crop paste recognition. The proposed CNNs based model can recognise ten different pests on crops. The model was experimented on the manually collected dataset by authors by capturing the images by Smartphone and downloading the pictures from search engines, which consists of 5629 images representing ten crop pest species. A GoogleNet model uses Watershed, and the Grab Cut algorithm to remove complicated backgrounds from collected images. The proposed model achieves an improvement in accuracy by 6.22% as compared to the ResNet101 (previous) model. For performance improvement, the data augmentation method was implemented by performing rotation, translation, flipping, mirroring, noise addition, and zooming operations. After data augmentation, the size of the dataset was changed from 5629 to 14475. The training and testing were applied with a 9:1 ratio. The computational complexity of five models, namely ResNet50, ResNet152, VGG16, VGG19, and GoogleNet, were tested. The model gives an average accuracy of 98.91% (Li et al., 2020).

Crop diseases are the major cause of crop reduction in terms of quantity and quality, which leads to an economic loss for farmers and ultimately to the nation. To avoid this early deep learning-based disease detection method was introduced by Amara et al., 2017. This method classifies the banana leaves diseases. The model uses GoogleNet architecture as a Convolution neural network to classify data (Amara et al., 2017).

Singh et al. (2018) gave the method for predicting late blight disease of potatoes using weather parameters such as maximum and minimum temperature, humidity, and rainfall. The proposed method used Extreme Learning Machine Classifier for prediction and evaluated on AICRP dataset, comprising five attributes. The technique identifies the occurrence and severity of the disease. The method experimented with several activation functions, and results showed that the random basis function gives high performance in terms of accuracy (Singh et al., 2018).

Raza et al. (2015) developed an automatic system for detecting tomato plant diseases using stereo and thermal images by combining depth information. The proposed method used a machine learning algorithm that constitutes image acquisition, feature extraction, image registration, depth assessment, and classification phases to develop an automatic disease detection model by combining colour information with depth and thermal data. The method determines the effect of combining two different approaches for detecting tomato leaf disease in the early stages before appearing symptoms. The technique used images acquired from thermal and visible light cameras. To align plant images captured from several cameras, image registration was done. The features extracted from collected images were used to determine depth information. Support Vector Algorithm with a linear kernel was used to delete background information from captured images. The proposed model is evaluated against several traditional methods and provides better classification accuracy of 90% (Raza et al., 2015).

Yushan Zhao et al. (2020) proposed a model for automatically recognising crop diseases by the Multi-Context Fusion Network method deployed in the agricultural Internet of Things in 2020. The model considers 50000 images from in-field observations and extracts visual features using the standard CNN model. The contextual features are extracted from image acquisition sensors. For recognizing crop diseases, the proposed model combines the extracted contextual and visual elements. Experiments consider 19 different crops of three different categories with 77 common crop diseases. Results showed that MCFN achieves 97.5% of recognition accuracy using a deep fusion model (Zhao et al., 2020).

The techniques to detect and classify the pest and diseases of the tomato plant were developed by Fuentes, et al. (2017). This method captures real-time images using a camera with several resolutions. This method combines three different detection algorithms namely Region-based Fully Convolution Network (R-FCN), Faster Region-based Convolution Neural Network (Faster R-CNN), and Single Shot Multibox Detector (SSD). Model evaluation indicates that the proposed method performs well over state-of-the-art methods. It can effectively classify nine different pests and diseases (Fuentes et al., 2017).

Problem identification and motivation

As India is an agricultural country, identification of plant disease is very important to prevent losses in yields. The security of the food is affected by plant disease which is considered an important threat in the agriculture field. The nation's whole economy, population growth as well as agriculture are also affected by this plant disease. Millions of rupees are being spent to protect the crops against several diseases due to the conventional plant disease prediction methods. To avoid this kind of abnormalities, correct and automatic evaluation of plant disease rigorousness is necessary to manage disease management and food safety. So, this inspires the prediction, detection, and management of plant diseases at an initial phase which in turn requires reasons for plant diseases as well as new techniques to identify the diseases. For food security, disease management, and yield loss prediction, automatic and precise evaluation of disease rigorousness is necessary. Due to small interclass variance and large intra-class similarity, plant disease severity classification is much trickier. Another important problem is that the previous techniques can't be able to identify more than one disease in one image or can't be able to recognise multiple occurrences of the same disease in one image. Also, most of the previous segmentation methods can't be able to segment the affected area with correct symptoms and most of the symptoms are not defined correctly which makes it difficult to define the healthy plants and diseased regions of plants. Hence, technologies like machine learning and image processing are very useful to predict plant diseases as well as estimation of disease severity in plants. These motivate us to develop novel techniques to predict multiple plant diseases with a hybrid DL algorithm. So, we have considered a hybrid technique for the prediction of multiple plant leaf diseases in a single leaf and severity estimation using the combination of DL technique with the optimisation algorithm.

From the literature survey done above the following are general observations:

- Traditional methods are too specific: It is identified that most of the available techniques required for precision agriculture are crop-specific, i.e., they have considered only one or two crops. If the technique is for disease identification, then they have considered only one or two crop types with limited varieties. And if the technique is for crop mapping or classification, then it has a maximum of ten crops. Therefore, the generalised model is the need for the current scenario.
- Traditional dataset collection methods are too restricted: Most of the datasets are collected in a specific environment such as particular lighting conditions, angles, etc. Again, most of the datasets are created in a controlled environment. Therefore, we need to generalise a dataset that should consider a variety of crops and diseases and create an outdoor environment.
- Most of the techniques are using image processing to identify and classify diseases. Therefore, we can consider other data acquisition techniques, such as Thermal Imaging or SAR data.
- Earlier methods are either classifying the crops or recognising the disease: We can implement a general model for both tasks.
- Model training: In some systems, the model is trained using samples from specific sites, so the model should be trained using an equal proportion of samples from every considered site.
- Severity estimation: The disease detection techniques can be extended to determine the severity of the disease on a particular scale so that farmers should get insights about giving priority to treating that disease.

Proposed Methodology

Plant disease prediction is the emerging research area in the agriculture field to identify the disease in the plant at an early stage. In this work, convolutional neural network (CNN) is used to predict the disease from the plant. To perform this process, the images are taken in the form of a dataset that contains different kinds of disease-affected images of plants. In the initial stage, the image dataset is pre-processed to execute the further transformations. A pre-processing technique is performed to remove multiplicative speckle noise from the input images. The pre-processed images are then segmented using various segmentation

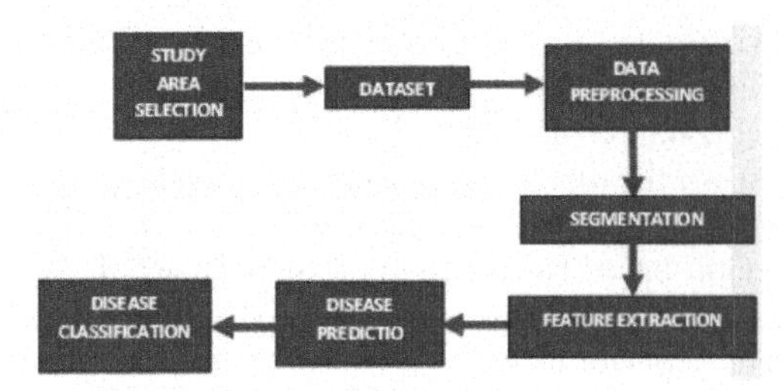

Figure 56.1 Proposed architecture

algorithms. The segmentation method intends to offer image segments based on image features (colour, shape, and texture), which are used in real-world applications. Segmentation in this case can prove effective, by separating a leaf from its background to detect the diseases part of the image. After segmentation, feature extraction is performed to extract useful features like colour, shape, and texture of the plant. In this work, the features are pulled out using different filters. Finally, extracted features are utilised as the input features to the CNN algorithm to correctly predict the multiple diseases from a single plant.

Conclusion

This study efficiently understands available Image Processing, Hyper spectral Imaging, UAV Imaging, Synthetic Aperture Radar Imaging Techniques used for disease detection and classification. The study reveals a survey of sensing systems, Smartphone Sensors, and Mobile Apps needed for accomplishing several tasks in precision agriculture such as disease prediction. The techniques like image processing, machine learning, and DL are used in precision agriculture applications. Some techniques have used passive sensors, whereas others used active sensors for data acquisition.

This study presented the survey of numerous ways for Crop Disease Identification and Biophysical Parameters estimation using ML and DL and identified how these parameters are used in Precision Agriculture applications. The study identified some gaps between the state-of-the-art methods. According to the literature survey done in section II and challenges in the automatic plant disease detection identified in section III, we can conclude that we need to generalise the dataset and model, which should be applicable for most of the precision agriculture applications such as detection and classification of plant diseases and severity estimation.

References

Amara, J., Bouaziz, B., and Algergawy, A. 2017. A deep learning-based approach for banana leafdiseases classification. In Lecture notesin informatics (LNI) (pp. 79e88).

Barbedo, J. G. A. (2016). A review of the main challenges in automatic plant disease identification based onvisible range images. Biosys. Eng. 144:52e60.

Bhatia, A., Chug, A., and Singh, A. P. (2020). Application of extreme learning machine in plant disease prediction for highly imbalanced dataset. Int. J. Stat. Manag. Syst. 23(6):1059–1068.

Cap, Q. H., Uga, H., Kagiwada, S,. and Iyatomi H. (2020). Leafgan: An effective data augmentation method for practical plant disease diagnosis. IEEE Trans. Autom. Sci. Eng. 19(2):1258–1267.

Chen, J., Chen, J., Zhang, D., Sun, Y., and Nanehkaran, Y. A. (2020). Using deep transfer learning for image-based plant disease identification. Comut. Electron. Agric. 173:105393.

Chen. J., Chen, J., Nanehkaran, Y. A., Sun, Y. A. Zhang, D. (2021). Cognitive vision method for the detection of plant disease images. Machine Vision App. 32(1):18. Springer.

Darwish, A., Ezzat, D., and Hassanien, A. E. (2020). An optimized model based on convolutional neural networks and orthogonal learning particle swarm optimization algorithm for plant diseases diagnosis. Swarm Evol. Comput. 52:100616.

Davino, S., Boschetti, M., Davis, C. E., Goulart, L. R., and Martinelli, F., Panno, S., Ruisi, P., Scalenghe, R., Scuderi, G., Villa, P., and Stroppiana, D. (2015). Advanced methods of plant disease detection. A review. Agron. Sustain. Dev. 35(1):125.

Dhingra, G., Kumar, V., and Joshi, H. D. (2018). Study of digital image processing techniques for leaf disease detection and classification. Multimed. Tools. Appl. 77(15):1995120000.

Escalante, H. J., Rodríguez-Sánchez, S., Jiménez-Lizárraga, M., Morales-Reyes, A., De La Calleja, J., and Vazquez, R. (2019). Barley yield and fertilization analysis from UAV imagery: a deep learning approach. Int. J. Remote Sens. 40(7):2493516.

Esgario, J. G. M., Krohling, R. A., and Ventura, J. A. (2020). Deep learning for classification and severity estimation of coffee leaf biotic stress. Comput. Electron. Agric. 169:105–162.

Ferdous, Z., Datta, A., Hasan, A. K., Sarker, A., and Zulfiqar, F. A. (2020). Potential, and challenges of organic agriculture in Bangladesh: a review. J. Crop. Imp. 124. Taylor and Francis.

Ferentinos, K. P. (2018). Deep learning models for plant disease detection and diagnosis. Comput. Electron. Agric. 145:311–318.

Fuentes, A., Yoon, S., Kim, S. and Park, D. (2017). A robust deep-learning-based detector for real-time tomato plant diseases and pest recognition. Sensors. 17(9).

German, L. A., Bonanno, A. M., Foster, L. C., and Cotula, L. (2020). Inclusive business' in agriculture: Evidence from the evolution of agricultural value chains. World Development. 134:105018.

Golhani, K., Balasundram, S. K., Vadamalai, G., and Pradhan, B. (2018). A review of neural networks in plant disease detection using hyperspectral data. Infor. Process. Agri. 5(3):354–371.

Jose G. M. E., Renato A. K., and Jose A. V. (2020). Deep learning for classification and severity estimation of coffee leaf biotic stress, Comput. Electron. Agric. 169:105162.

Kamal, K. C., Yin, Z., Wu, M., and Wu, Z. (2019). Depthwise separable convolution architectures for plant disease classification. Comput. Electron. Agric. 165:104948.

Kaushal, L. A. and Prashar, A. (2020). Agricultural crop residue burning and its environmental impacts and potential causes–case of northwest India. J. Environ. Plan. Manag. 3:121.

Li, Y., Wang, H., Dang, L. M., Sadeghi-Niaraki, A., and Moon, H. 2020. Crop paste recognition in natural scenes using convolution neural networks. Comput. Electron. Agric. 169.

Khamparia, A., Saini, G., Gupta, D., Khanna, A., Tiwari, S., and de Albuquerque, V. H. (2020). Seasonal crops disease prediction and classification using deep convolutional encoder network. Circuits Syst. Signal Process.9(2):818–836.

Nandhini, S. A., Hemalatha, R., Radha, S., and Indumathi, K. (2018). Web enabled plant disease detection system for agricultural applications using WMSN. Wireless Personal Communications. 102(2):725–740.

Patokar, A. M. and Gohokar, V. V. 2020. Precision agriculture system design using wireless sensor network. InInformation and Communication Technology. (pp. 169177). Singapore:Springer.

Pourazar, H., Samadzadegan, F., and Dadrass, J. F. (2019). Aerial multispectral imagery for plant disease detection: Radiometric calibration necessity assessment. Eur. J. Remote. Sens. 2(3):1731.

Raza, S. –E.-A., Prince, G., Clarkson, J. P., and Rajpoot, N. M. (2015). Automatic Detection of Diseased Tomato Plants Using Thermal and Stereo Visible Light Images. PLoS ONE. 10(4): e0123262. doi:10.1371/ journal. Pone.0123262, 2015.

Santoso, H., Tani, H., Wang, X., Prasetyo, A. E., and Sonobe, R. (2019). Classifying the severity of basal stem rot disease in oil palm plantations using WorldView-3 imagery and machine learning algorithms. Int. J. Remote Sens. 40(19):7624–7646.

Singh, B. K., Singh, R. P., Bisen, T., and Kharayat, S. (2018). Disease Manifestation Prediction from Weather Data Using Extreme Learning Machine. 2018 3rd International Conference On Internet of Things: Smart Innovation and Usages (IoT-SIU), 1–16, doi: 10.1109/IoT-SIU.2018.8519908.

Sinwar, D., Dhaka, V. S., Sharma, M. K., and Rani, G. 2020. AI-Based Yield Prediction and Smart Irrigation. Internet of Things and Analytics for Agriculture. (pp. 155180). Singapore: Springer.

Sonobe, R., Yamashita, H., Mihara, H., Morita, A., and Ikka, T. (2021). Hyperspectral reflectance sensing for quantifying leaf chlorophyll content in wasabi leaves using spectral pre-processing techniques and machine learning algorithms Int. J. Remote Sens. 42(4):1311–1329.

Sujatha, R., Chatterjee, J. M., Jhanjhi, N. Z., and Brohi, S. N. (2021). Performance of deep learning vs machine learning in plant leaf disease detection. Microprocess. Microsyst. 80:103615.

Tantalaki, N., Souravlas, S., and Roumeliotis, M. (2019). Data-Driven Decision Making in Precision Agriculture: The Rise of Big Data in Agricultural Systems. J. Agric. Food Inf. 20(4):344–380.

Thomas, S., Kuska, M. T., Bohnenkamp, D., Brugger, A., Alisaac, E., Wahabzada, M., Behmann, J., and Mahlein, A. K. (2018). Benefits of hyperspectral imaging for plant disease detection and plant protection: a technical perspective. J. Plant Diseas. Protec. 125(1):5–20.

Vishnoi, V. K., Kumar, K., and Kumar, B. (2020). Plant disease detection using computational intelligence and image processing. J. Plant Diseas. Protec. 135.

Xie, C., Zhang, J., Li, R., Li, J., Hong, P., Xia, J., and Chen, P. (2015). Automatic classification for field crop insects via multiple-task sparse representation and multiple-kernel learning. Comput. Electron. Agric. 119:123–132.

Yang, C., Randall, J. D., Tim, R. M., François, W., Gonzalo, M., Noboru, O., Alireza, H., Kavina, D., and Lawesd, R. A. (2020). Nationwide crop yield estimation based on photosynthesis and meteorological stress indices. Agri. Forest Meteorol. 284:107872.

Zhao, Y., Leu, L., Xie, C., Wang, R., Wang, F., Bu, Y., and Zhang, S. (2020). An effective automatic system deployed in agricultural Internet of Things using multi-context Fusion Network towards crop disease recognition in the wild. Appl. Soft Comput. J. 89:106–128. https://doi.org/10.1016/j.asoc.2020.106128.

Zhou, R., Kaneko, S., Tanaka, F., Kayamori, M. and Shimizu, M. (2015). Image-based field monitoring of Cercospora leaf spot in sugar beet by robust template matching and pattern recognition. Comput. Electron. Agric. 116: 65–79.

57 Development of a new piped irrigation network for micro irrigation using cropwat and epanet model

Pooja Somani[1,a], *Avinash Garudkar*[2,b], *and Shrikant Charhate*[1,c]

[1]Department of Civil Engineering Amity University Mumbai, India

[2]Faculty of Engineering WALMI Auranagabad, India

Abstract

Conventionally irrigation water is supplied from the dam to the command areas through the Canal Distribution Network (CDN) but inherently it is associated with significant evaporation, seepage and percolation losses. Piped Irrigation Network (PIN) can be feasible alternative for CDN if implemented precisely. It reduces canal conveyance losses substantially and also improves water use efficiency. In this paper an attempt has been made to design cost effective PIN for micro irrigation and applied on initial minors of Pawale irrigation project in Thane district of Maharashtra India. EPANET computer program is used as design tool, with required minimum pressure head at demand nodes and velocity as a constraint to obtain optimal size of pipes in the network. Only commercially available pipe sizes are considered in this study. Discharge requirement at each demand node is worked out by using CROPWAT model considering heterogeneity of command area. The total cost of the network is the summation of cost of each pipe in the network. The network with minimum cost is selected for the design. It is observed that coupling of CROPWAT model and EPANET model have potential to optimise piped irrigation network with micro irrigation in heterogeneous command area.

Keywords: CROPWAT, EPANET, Pawale project, Piped Irrigation Network.

Introduction

Ever increasing population, growth in industrial and economic activities are leading to increase in water demand for agriculture and other purposes *viz.* domestic, industrial and hydropower. As population is increasing in India, it has also witnessed drastic fall in per capita availability of water. The reported per capita water availability in the year 1951 was 5177 cumec, which reduced to 1567 cumec in year 2011 and is near to 1000 cumec in March 2019 (Central Water Commission, 2017; World Bank, 2019).

The competition for water to agriculture from other sectors and increasing food demands of ever growing population, has put relentless pressure on existing water resources for optimal and efficient use of available water for irrigation. As a result, modernisation of existing irrigation system is pivotal so as to use the available water effectively and optimally. It has been also reported that efficient and optimal use of water by improving the irrigation management is one of the feasible alternative to development of new water resources projects.

Conventionally water is supplied to the command areas from the dams through the Canal Distribution Network (CDN) but inherently it is associated with evaporation, seepage and percolation losses. This is major drawback of canal irrigation systems. With the most effective lining and efficient design of canal, efficiency could be achieved as 70% making the overall project efficiency with sprinkler irrigation as 47.25% and overall project efficiency with drip irrigation as 56.7%. Piped Irrigation network can be feasible alternative for canal network, if implemented precisely, can reduce canal conveyance losses substantially (Government of Maharashtra, 2017).

It is observed that the overall project efficiency obtained using the Piped Irrigation Network (PIN) with sprinkler irrigation as 68% and with drip irrigation as 81.23% (Central Water Commission, 2017). This indicates that at least 20% more efficient use of water could be achieved using piped irrigation network. Patil et al. (2018) compared the piped irrigation network with conventional gravity irrigation network and concluded that overall efficiency of the project increased by 13% with PIN. The saved water could be used for irrigating additional land in the command area.

In PIN water is supplied under pressure to the irrigable command area at required pressure and velocity. Unlike open channels, pressurised pipes also facilitate conveyance of water against normal slope in case of hills or ridges. Providing water on non-uniform slopes is also possible in case of PIN. Since crops may be

[a]poojadsomani@gmail.com; [b]as_garudkar@rediffmail.com; [c]sbcharate@yahoo.co.in

grown on the fields above the pipes, buried pipe provides the most direct path from the water source to the fields. Unlike CDN, the command area to be irrigated and the source of supply in terms of elevation and slope doesn't regulate the planning and layout of PIN.

The extent of seepage losses can be very high in CDN, as even in small conventional earthen canal (200m) the conveyance efficiency was just 75.07 % mainly due to seepage losses (Srivastava et al., 2010). This reduces the overall project efficiency to 41 to 48 % for open canal, which actually further reaches to 20 to 35 %, making the need of PIN imperative (Satpute et al., 2012). The use of a pumped irrigation system in the form of buried pipes, rather than open channels may significantly reduce water demand (Bentum et al., 1995; Smout, 1999).

Various computer software are available for pipe network design for urban water supply and PIN. EPANET software is used in the design of the water distribution system(Kumar et al., 2015; Halagalimath et al., 2016). Arora and Jaiswal (2013), proposed methodology for using EPANET with Genetic Algorithm to work out optimal cost of irrigation network. PIN was designed for the Left Bank Canal of Pench Irrigation Project at distributary level using EPANET 2.0 for pipe diameter and compared to critical route approach and optimisation using liner programming in India (Gajghate and Mirajkar, 2019). Sharu and Razak (2020) used EPANET software for understanding operation of drip irrigation system. They carried out study at Laman Sayur, Malaysia Agro Exposition Park Serdang (MAEPS), to investigate the performance of drip irrigation system.

Gajghate et al. (2021) used the Steiner idea to reduce the length of pipes, thus resulting into reduction in overall cost of PIN. As velocity is more in PIN than open channel, transit time is less resulting into lower conveyance losses. Therefore, the delivery of the water is more flexible in terms of duration and frequency of irrigation, as compared to CDN.

The PIN coupled with Micro Irrigation System (MIS) further improves the efficiency. Depending on the crop type, piped irrigation network may be used more effectively in combination with micro irrigation such as sprinklers or drippers, according to (Rao, 2019). Importance of micro irrigation for optimal water productivity was demonstrated by Sivanappan (1994).

In the present study, developed methodology is applied to initial two minors on Pawale irrigation project in Maharashtra. The CROPWAT 8.0 model is used to calculate the crop water requirement considering the heterogeneity of command area. Further EPANET 2.0 is used to design PIN taking into account flexible constraint. CROPWAT has been used by many researchers for various objectives. Surendran et al. (2015) used CROPWAT 8.0 model for computation of crop water requirements of major crops in various agro-ecological zones of Palakkad district of Kerala and compared the same with available water resources of the district.

Mehanuddin et al. (2018) and Ewaid et al. (2019) have emphasised the necessity and efficacy of the CROPWAT model determining crop water requirement required for effective irrigation management.

The input diameters of commercially available pipes are considered so as to meet pressure and velocity requirements of the system. Cost of each possible network is calculated and the cheapest one is selected as the best one to be used in the design. The total cost of the network is the sum of the costs associated with each pipe.

Materials and Methods

Study Area

The developed methodology is applied to Minor 1 and Minor 2 of Pawale Irrigation Project, the location coordinates of the project are 19°17'45" N and 73°25'45" E in Thane district of Maharashtra. This is new irrigation project in which dam is constructed, but considering shortage of water and losses in canal network, it is proposed to design PIN for drip irrigation system, so as to save water. Earlier the project was designed for CDN. Salient features of Pawale Minor Irrigation (MI) Project with CDN are presented in table 57.1.

Methodology

The layout of PIN consist of rising main from the source to initial two minors of the project as shown in Figure 57.1.

The network consists of 63 pipes with 63 nodes inclusive of 35 outlet nodes and one reservoir. The main pipe starts from junction 2 to junction 17 and it has 8 outlets. At junction 17 main pipe branches into minor1 and minor 2. Minor 1 has nodes starting from 18 to 31, out of which 8 are outlet nodes. Minor 2 has nodes starting from 32 to 64, out of which there are 19 outlet nodes.

Table 57.1 Salient Features of Pawale Irrigation Project

S.N.	*Description*	*Details*	
1	Name of project	Pawale MI project	
2	Catchment area	2.087 Sqkm	
3	Average rainfall	246.08 cm	
4	Gross storage capacity	3167.48 Tcum	
5	Live storage of project	3123.017 Tcum	
6	Dead storage of project	44.463 Tcum	
7	Canal	LBC	RBC
	Length of the canal	5.04 km	2.88 km
	Discharge carrying capacity	0.173 cumecs	0.31 cumecs
	Area under (irrigable) Command	90 ha	162.0 ha
8	Gross Command Area	424 ha	
9	Culturable command area	339 ha	
10	Irrigable command area	252 ha	
11	Cropping pattern	Kharif and Rabi	
12	Kharif	65 %	
	i. Paddy	40 %	
	ii. Pulses	10 %	
	iii. Vegetables	15 %	
13	Rabi	100 %	
	i. Paddy	30 %	
	ii. Groundnut	15 %	
	iii. Vegetable	20 %	
	iv. Pulses	25 %	
	v. Milk production grass		

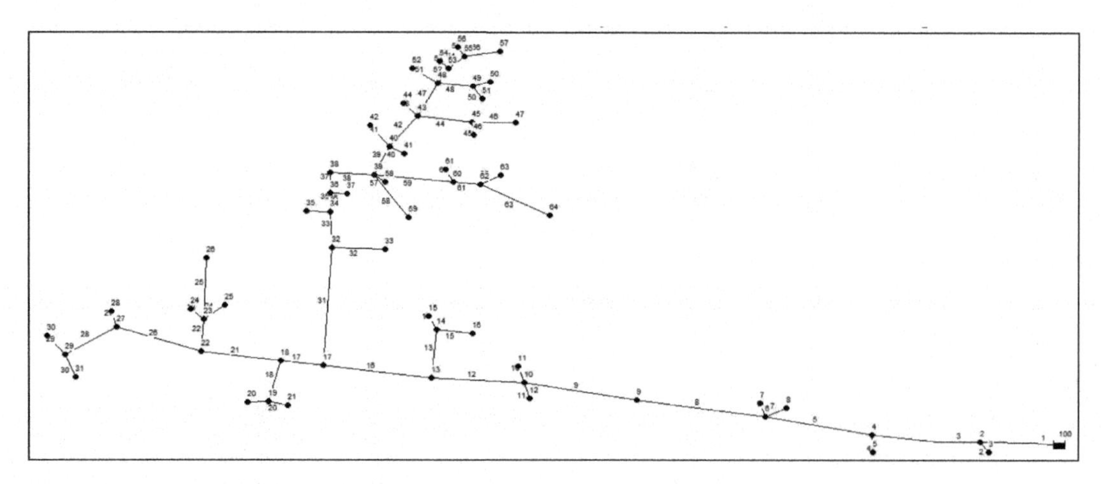

Figure 57.1 The network layout for the study area

Any Irrigation project involves high initial cost of construction of dam and canal network. Even for PIN material cost is major component of total cost. The available head should also be viable for the system's excellent performance. At the same time, the cost of the system must be kept to a minimum. In this case, with the pumping system, the available head is 103.0 m at the source.

EPANET 2.0 was used to construct the pipe networks for two minors of Pawale irrigation project. The pipe status is taken open for this study, as there are no valves. The Input parameters of EPANET 2.0 are elevation of node, demand at outlet nodes, length of the pipe and diameter of pipe. Hazen Willaims equation is used for the calculation of head loss. Also the pipe material is selected and its roughness coefficient is considered. The PVC pipes are used for pipe sizes of 75 mm diameter to 140 mm diameter (with roughness coefficient value of 150) and HDPE pipes are used for sizes having diameters greater than 140 mm (with roughness value of 140) so as to reduce the total cost of network. Flow chart of study is presented in Figure 57.2.

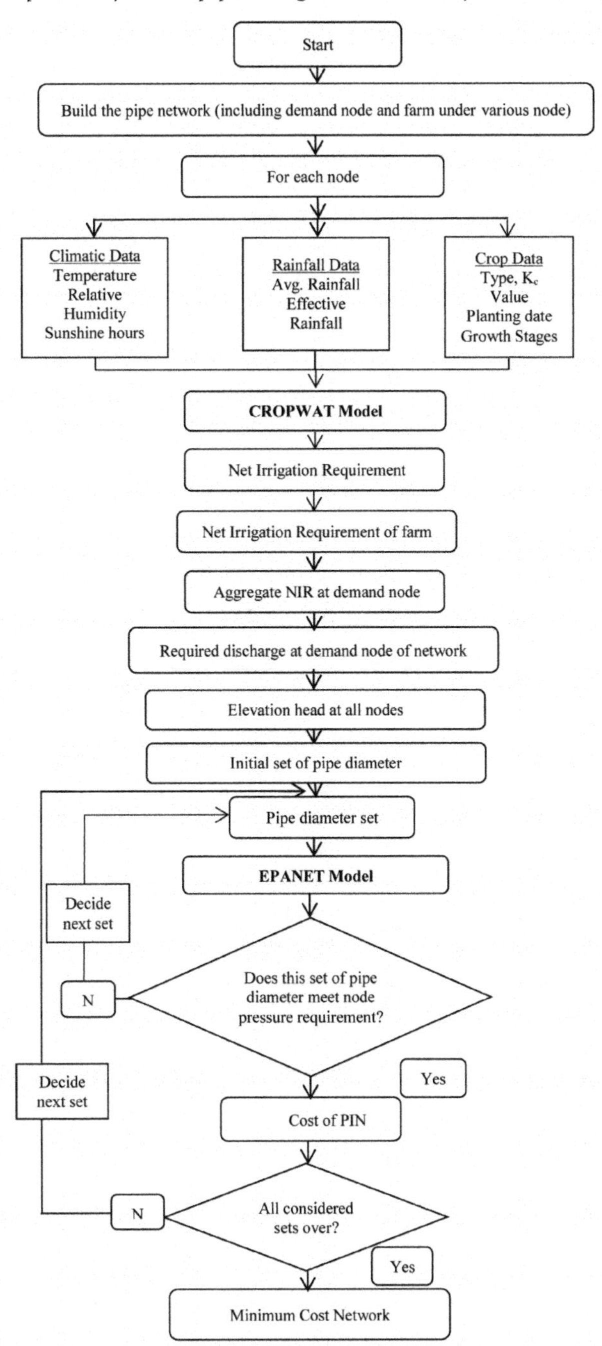

Figure 57.2 Flow chart of study

CROPWAT 8.0 Model

Crop water requirements and irrigation requirements are calculated using the CROPWAT software tool, for each crop considering heterogeneity of command area. Food and Agricultural Organization (FAO), Department of Land and Agricultural Organization, Rome has developed this model. Based on soil, climatic and crop data, CROPWAT 8.0 for Windows is a computer application for calculating crop water requirement and irrigation demands. CROPWAT model is coupled with EPANET model for deign of PIN.

EPANET Model

The Environmental Protection Agency in the United States has developed the EPANET software. Water distribution networks with unlimited pipes may be analysed using EPANET, a widely used tool for this purpose. The data set used in EPANET includes elevation at nodes, roughness parameters for pipe and diameters of pipe. This leads to output in terms of pressure at every node, velocity and head loss in each pipe, concentration of quality parameters at different time. The results obtained may be used for making various analyses of distribution system. The basic assumptions used in EPANET about flow analysis are namely: Incompressible flow, turbulent flow, closed pipe e.g. contaminant injections are modelled as mass/time and full pipe. For constant reservoir levels/tank levels, and water needs, EPANET's hydraulic simulation model estimates junction heads and link flows over a period of time. It is possible to update reservoir levels and junction demands using the present flow solution across time steps, while tank levels are updated according to their set time patterns. Heads and flows at a given time may be calculated by solving the flow conservation equation at each junction and then the head loss relationship at each link in the network. Nonlinear equations must be solved using an iterative method known as hydraulically balancing of the network. To do this, EPANET uses the Gradient Algorithm.

The pipe diameters are selected for this study such that minimum of 20 m pressure head is maintained at every outlet, so as to provide micro irrigation at the tail end. The selected set of diameter also maintains the limiting velocity in the pipe as per standards laid by Central Water Commission (2017).

Results and Discussion

Required discharge considering physiographic and climatic aspects at each outlet node is calculated by CROPWAT model. The details of irrigated areas and discharge requirement at each outlet node and pipe lengths in the network are given in Table 57.2. Cumulative area at each junction is also presented in this table.

Table 57.2 Details of area and discharge at outlet node and pipe lengths in the network

Node/Junction No.	*Demand (Discharge) (LPS)*	*Area/ Cum. area (ha)*	*Pipe (link) No.*	*Pipe Length (m)*
Junc 2	0	251.48	Pipe 1	257
Node 3	6.51	6.08	Pipe 2	16
Junc 4	0	245.4	Pipe 3	493
Node 5	5.59	5.22	Pipe 4	59
Junc 6	0	240.18	Pipe 5	416
Node 7	4.15	3.88	Pipe 6	7
Node 8	7.91	7.39	Pipe 7	7
Junc 9	0	228.91	Pipe 8	511
Junc 10	0	228.91	Pipe 9	510
Node 11	4.7	4.39	Pipe 10	10
Node 12	8.7	8.13	Pipe 11	14
Junc 13	0	216.39	Pipe 12	244
Junc 14	0	9.53	Pipe 13	146
Node 15	5.89	5.5	Pipe 14	65
Node 16	4.31	4.03	Pipe 15	15
Junc 17	0	206.86	Pipe 16	276
Junc 18	0	34.7	Pipe 17	72
Junc 19	0	8.76	Pipe 18	65
Node 20	5.89	5.48	Pipe 19	12
Node 21	3.51	3.28	Pipe 20	7
Junc 22	0	25.94	Pipe 21	182
Junc 23	0	14.47	Pipe 22	192
Node 24	3.63	3.39	Pipe 23	21
Node 25	6.38	5.96	Pipe 24	181
Node 26	5.48	5.12	Pipe 25	20
Junc 27	0	11.47	Pipe 26	216
Node 28	3.94	3.68	Pipe 27	7
Junc 29	0	7.79	Pipe 28	170
Node 30	4.57	4.27	Pipe 29	65

(Continues)

Table 57.2 Continued

Node/Junction No.	*Demand (Discharge) (LPS)*	*Area/ Cum. area (ha)*	*Pipe (link) No*	*Pipe Length (m)*
Node 31	3.77	3.52	Pipe 30	73
Junc 32	0	172.16	Pipe 31	1099
Node 33	5.09	4.76	Pipe 32	367
Junc 34	0	167.4	Pipe 33	375
Node 35	8.82	8.24	Pipe 34	97
Junc 36	0	159.16	Pipe 35	187
Node 37	4.31	4.03	Pipe 36	23
Junc 38	0	155.13	Pipe 37	142
Junc 39	0	88.84	Pipe 38	246
Junc 40	0	71.56	Pipe 39	271
Node 41	5.56	5.2	Pipe 40	16
Node 42	5.13	4.79	Pipe 41	77
Junc 43	0	54.13	Pipe 42	288
Node 44	5.29	4.94	Pipe 43	58
Junc 45	0	12.62	Pipe 44	280
Node 46	6.02	5.63	Pipe 45	19
Node 47	7.48	6.99	Pipe 46	184
Junc 48	0	36.57	Pipe 47	480
Junc 49	0	13.94	Pipe 48	79
Node 50	4.79	4.48	Pipe 49	28
Node 51	10.12	9.46	Pipe 50	29
Node 52	8.85	8.27	Pipe 51	175
Junc 53	0	14.36	Pipe 52	115
Node 54	6.09	5.69	Pipe 53	28
Junc 55	0	8.67	Pipe 54	137
Node 56	5.79	5.41	Pipe 55	93
Node 57	3.49	3.26	Pipe 56	155
Node 58	4.7	4.39	Pipe 57	10
Node 59	3.26	3.05	Pipe 58	202
Junc 60	0	17.28	Pipe 59	184
Node 61	4.56	4.26	Pipe 60	15
Junc 62	0	13.02	Pipe 61	52
Node 63	7.98	7.46	Pipe 62	35
Node 64	5.95	5.56	Pipe 63	488

As discussed in earlier section, to begin with initial set of pipes are considered in EPANET with the trial and error approach amongst available pipe diameters in the market and only those available pipes are considered for subsequent iterations as set of pipes in the network. Total 100 iterations are carried out with EPANET model. Only those networks fulfilling minimum pressure criteria of 20m pressure at outlet node and also range of velocity in the pipeline between 0.6 to 3 m/s as per Central Water Commission (CWC) guidelines were selected as feasible set of pipes for network. Cost of each feasible network was worked out and the network having minimum cost is proposed as optimal network for design. The details of various pipes and their diameters for which total cost is minimum (design pipe network) is presented in Table 57.3. It is observed that max pipe diameter is 500mm and min pipe diameter is 75 mm in the network. Total thirteen type of pipes are used in the network.

The variation of pressure head (m) with respect to junction node is graphically presented in Figure 57.3. It is observed that the pressure ranges from 20.01 m at Node 3 to 45.64 m at Junction 38 on Minor 2.

The variation of pressure head with respect to outlet node is presented in Figure 57.4. As this network is designed for micro irrigation in the entire command area, all outlet nodes are having pressure head more than 20 m. It is observed that available pressure is increasing for the nodes which are away from reservoir.

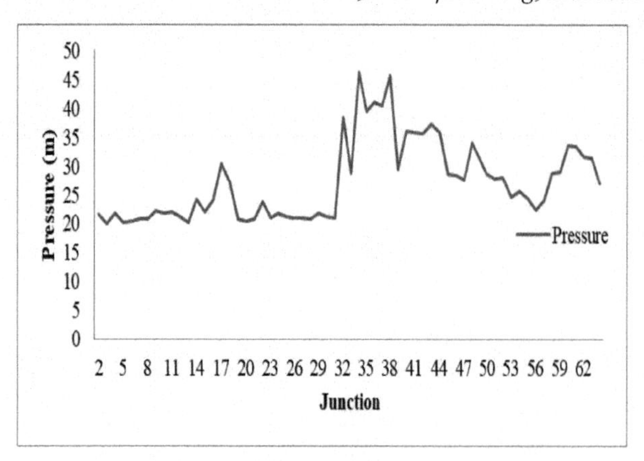

Figure 57.3 Pressure variation at junction node in the pipe irrigation network

Table 57.3 Optimal pipe network design

Link ID	*Diameter (mm)*	*Link ID*	*Diameter (mm)*
Pipe 1	500	Pipe 33	280
Pipe 2	110	Pipe 34	75
Pipe 3	500	Pipe 35	280
Pipe 4	110	Pipe 36	75
Pipe 5	500	Pipe 37	280
Pipe 6	75	Pipe 38	280
Pipe 7	110	Pipe 39	250
Pipe 8	450	Pipe 40	75
Pipe 9	450	Pipe 41	75
Pipe 10	75	Pipe 42	225
Pipe 11	90	Pipe 43	75
Pipe 12	400	Pipe 44	125
Pipe 13	90	Pipe 45	75
Pipe 14	75	Pipe 46	110
Pipe 15	75	Pipe 47	200
Pipe 16	400	Pipe 48	125
Pipe 17	250	Pipe 49	75
Pipe 18	140	Pipe 50	110
Pipe 19	110	Pipe 51	90
Pipe 20	75	Pipe 52	180
Pipe 21	225	Pipe 53	110
Pipe 22	120	Pipe 54	140
Pipe 23	75	Pipe 55	75
Pipe 24	110	Pipe 56	75
Pipe 25	90	Pipe 57	75
Pipe 26	140	Pipe 58	75
Pipe 27	90	Pipe 59	110
Pipe 28	125	Pipe 60	75
Pipe 29	90	Pipe 61	110
Pipe 30	90	Pipe 62	75
Pipe 31	280	Pipe 63	110
Pipe 32	75		

The minimum pressure at the nearest outlet node (Node 3) from the reservoir is 20.01 m. The pressure for all the nodes in Minor 2 is comparatively higher than that at Minor 1. At all the outlet nodes on Minor 1, pressure ranges between 20.48 m to 21.88 m, whereas, for Minor 2, pressure for outlet node ranges from 22.32 m to 40.37 m.

Central Water Commission New Delhi has recommended that variation in the velocity of Piped irrigation network should be between 0.6 and 3 m/s hence this constraint has been maintained in the study [1].

It has been observed that velocity in all the pipes in the network ranges between 0.6 m/s to 2 m/s. It was 0.6 m/s for pipe no 4 to 2 m/s for pipe no 34. In nine pipes velocity is more than 1.5 m/s. Variation in the velocities in all the pipes in the Piped Irrigation Network is shown in Figure 57.5.

Estimation of Cost for the Designed Network

As compared to a CDN, the initial investment cost of PIN is more. However, its feasibility is justified by considering the cost of land acquisition, maintenance and considering more conveyance losses in CDN. The initial cost for the pipe for the optimum set of diameter is calculated and given in table 57.4. Total 100

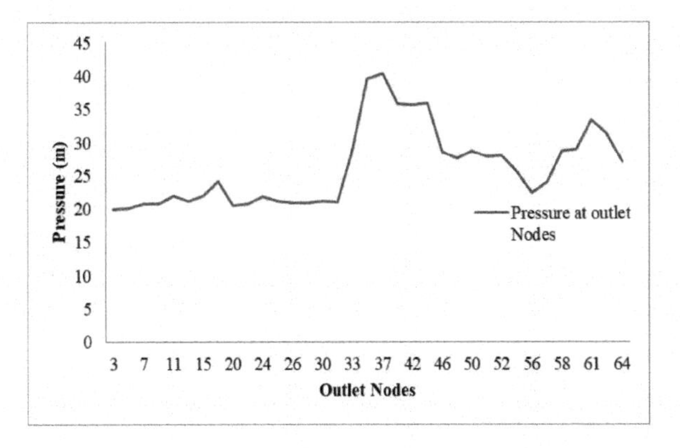

Figure 57.4 Variations in pressure head at piped irrigation network outlet nodes

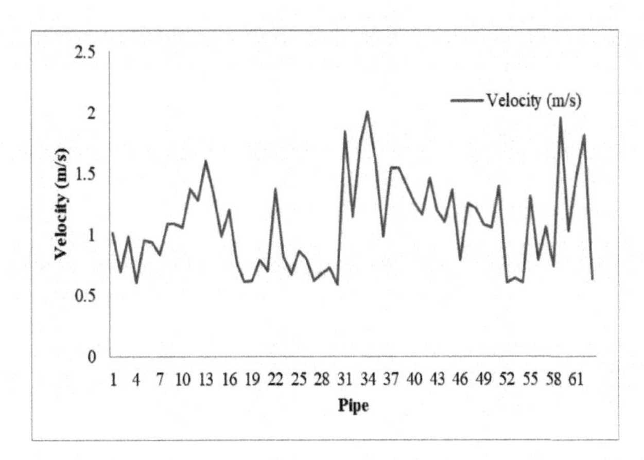

Figure 57.5 Variation of velocities in the pipe network

Table 57.4 Cost of optimal design network

Diameter (mm)	*Total Length (m)*	*Rate /meter (INR)*	*Cost (INR)*
500	1166	4186.2	4881109.2
450	900	3406.3	3065670
400	800	2624.9	2099920
280	2049	1341.45	2748631.05
250	343	1071.65	367575.95
225	450	904.65	407092.5
200	480	699.1	335568
180	180	611.85	110133
140	418	293.65	122745.7
125	642	239.65	153855.3
110	1240	187.4	232376
90	500	142.2	71100
75	1320	105.9	139788
	Total Cost		14735564.7 Say 14735565

iterations were carried out for various set of pipe diameter out of which only 21 networks found to be satisfying constraints. The details of cost estimates by considering common schedule of rates of Maharashtra Jeevan Pradhikaran (CSR-MJP 2019-2020) are given in Table 57.4 (Maharashtra Jeevan Pradhikaran, 2019). The optimal cost of the network was found to be INR 14735565. Thus the average cost of the network is INR 1405 per meter for the micro irrigation system considering Piped Irrigation Network.

Conclusions

Following conclusions can be drawn from the present study

1) Efficiency of irrigation system is substantially improved with pipe irrigation system coupled with micro irrigation.
2) In properly executed PIN, though the net irrigation requirement remains the same, irrigation water demand is reduced due to improvement in water use efficiency.
3) Coupling CROPWAT model with EPANET model has potential to optimise irrigation network for micro irrigation system considering entire command area which is complex and heterogeneous.

Acknowledgement

The authors are thankful to the Pawale Irrigation project authorities at Shahapur, Thane for providing required data.

References

Arora, S. and Jaiswal, A. K. (2013). Optimal Cost of Irrigation Network Design using Epanet. Int. J. Comput. Appl. 68(21):41–45.

Bentum, R. V., Smout, I. K., and Ci, X. Z. (1995). Use of pipelines to improve surface irrigation in Hebei Province, China. J. Irrig. Drainage Eng. 121(6):405–410. doi: 10.1061/(ASCE) 0733-9437(1995)121:6(405)

Central Water Commission 2017. Guidelines for planning & Design of Piped Irrigation Network. Ministry of Water Resources, River Development & Ganga Rejuvenation, New Delhi.

Ewaid, S. H., Abed, S. A., and Al-Ansari, N. (2019). Crop Water Requirements and Irrigation Schedules for Some Major Crops in Southern Iraq, Water. 11:756. doi:10.3390/w11040756, www.mdpi.com/journal/water

Gajghate, P. W. and Mirajkar, A. (2019). Irrigation Pipe Network Planning at Tertiary Level: An Indian Case Study. KSCE J. Civil Eng. 322–335.

Gajghate, P. W., Mirajkar, A., Shaikh, U., Bokde, N. D., and Yaseen, Z. M. 2021. Optimization of Layout and Pipe Sizes for Irrigation Pipe Distribution Network Using Steiner Point Concept. Hindawi Math. Probl. Eng. https://doi.org/10.1155/2021/6657459

Government of Maharashtra. 2017. Guidelines for Pipe Distribution Network (PDN). Water Resources Department, Government Resolution (02/02/2017)

Halagalimath, S., Vijaykumar, H., and Patil, S. N. (2016). Hydraulic modeling of water supply network using EPANET. Int. J. Eng. Res. Technol. 3(3):10221027

Kumar, A., Kumar, K., Bharanidharan, B., Matial, N., Dey, E., Singh, M., Thakur, V. Sharma, S., and Malhotra, N. (2015). Design of water distribution system using epanet. Int. J. Adv. Res. 3(9):789–812.

Maharashtra Jeevan Pradhikaran, 2019. Schedule of Rates. Government of Maharashtra, India

Mehanuddin, H., Nikhitha, G. R., Prapthishree, K. S., Praveen, L. B., Manasa, H. G. (2018). Study on Water Requirement of Selected Crops and Irrigation Scheduling Using CROPWAT 8.0. Int. J. Innov. Res. Technol. Sci. Eng. 7(4):3431–3436.

Patil, S., Talegaonkar, S. D., and Nimbalkar, P.T. (2018). Hydraulic design of Pipe Distribution Network for Irrigation Project. Int. J. Civil Eng. Technol. 9(7):1109–1116

Rao, P. D. 2019. Challenges in planning design and implementation of pipe irrigation network at command level. International Commission on Irrigation and Drainage (ICID) New Delhi, 9th IMIC Conference.

Satpute, M. M., Khandve, P. V., and Gulhane, M. L. 2012. Pipe distribution network for irrigation-an alternative to flow irrigation. Proceedings of 99th Indian science congress. Bhubaneshwar, India. 106–114.

World Bank. 2019. Helping India Manage its Complex Water Resources. https://www.worldbank.org/en/news/feature/2019/03/22/helping-india-manage-its-complex-water-resources (accessed July 25, 2021)

Sharu, E. H. and Ab Razak, M. S. 2020. Hydraulic Performance and Modelling of Pressurised Drip Irrigation System Water. 12, 2295. doi:10.3390/ w12082295, www.mdpi.com/journal/water

Sivanappan, R. K. 1994. Prospects of micro-irrigation in India. Irrig. Drain. 8:4958.

Smout, I. K. (1999). Use of low-pressure pipe systems for greater efficiency. Agric. Water Manag. 40:107–110. doi: 10.1016/S0378-3774(98)00089-4

Srivastava, R., Mohanty, S., Singandhuppe, R., Mohanty., R., Behera, M., and Ray, L.. 2010. Feasibility Evaluation of Pressurised Irrigation in Canal Commands. Water Resour. Manag. 24(12):3017–3032.

Surendran, U., Sushanth, C. M. Mammen, G., and Joseph, E. J. (2015). Modelling the crop water requirement using FAO-CROPWAT and assessment of water resources for sustainable water resource management: A case study in Palakkad district of humid tropical Kerala, India. Aquatic Procedia., 4:1211–1219

58 Application of value stream mapping for effective implementation of the lean system in the small scale industry

Ajay Bonde[a], Shrikant Jachak[b], Akshad Soman[c], Aniruddha Jagirdar[d], Aadarsh Pandit[e], and Satvik Koli[f]

Mechanical Department, Yeshwantrao Chavan College of Engineering (YCCE), Nagpur, India

Abstract

This paper predominantly focuses on the implementation of value stream mapping (VSM) to make a small-scale industry 'lean'. VSM advancements in current manufacturing methods are put forward for small-scale industries. It involves developing a present status guide of the business and surveying something very similar for the identification of waste and bottlenecks in the framework. In light of the suggested arrangements, a future state map is proposed. TIMWOODS, seven waste of lean assembling arrangements are proposed to remove recognised waste. Findings will be beneficial for MSME's. A future state map is also created and analysed for various aspects like TAKT times, process lead time, etc. to achieve better efficiency and overall productivity. Results show an increase of 8% in overall efficiency. Results demonstrate the execution of the lean system which can be applied in different MSMEs.

Keywords: Lean, value stream mapping, TAKT Time, MSMEs, waste assessment questionnaire, waste relation matrix.

Introduction

The ultimate goal of lean manufacturing is to satisfy consumer demands and simultaneously take account of quality by practicing the most efficient methods to manufacture them. 'Lean' was first introduced by John Krafick in 1988 and first implemented by Toyota industries. In this study, we are focusing on the production process to make it more efficient as well to reduce the rejection rate. An efficient manufacturing process can be built by minimising wastes and non-value-added activities in various operations and processes. Usually, lean manufacturing is executed in widely established corporations. In this study, an effort has been made to minimize cycle times and maximise productivity by adopting lean manufacturing concepts through value stream mapping (VSM) in small-scale industries.

Aggregating all value-adding and non-value-adding activities that are necessary for the manufacturing of a particular part or product from raw materials. As indicated by Rother and Shook (1999) have portrayed VSM as a tool that assists in visualization of the flow of material and information as a product makes its way through VSM. The absolute initial phase in VSM is to draw a present status map. After examining the present status map and precluding all the wastes then draw a future state map. VSM has been viewed as to a greater degree a subjective interaction since it has the progression of data connected with assembling and creation as (a) material stream and (b) data stream. The main advantage of VSM is that it is the only tool that includes both (a) material and flow process along with creating a relationship between both of them. The process of drawing the current state map is to scan the wastes and processes that are holding back the high efficiency. Appropriate practical choices are proposed for removing the waste in light of which the future state map is drawn and simulated. The outcomes are then contrasted with the outcomes acquired from the present status guide. Then attempt to build the general effectiveness by lessening the process duration lead time (TAKT time). These outcomes will help Micro little and small scale businesses to know the power and convenience of Lean Manufacturing. The main objective of this case study and research is to develop a framework that will serve the best Micro small and small-scale industries within its limits. It will help to analyse as well evaluate the results after applying the framework in the industry(s).

Literature review and Research gap

Marvel and Standridge (2009) suggested that very few organizations can attain considerable improvements in the system by applying lean. The reason for unattainability is a lack of complete understanding of the

[a]ajaysbonde04@gmail.com; [b]852@ycce.in; [c]akshdsoman@gmail.com; [d]aniruddhajagirdar720@gmail.com; [e]anshulkanhaiyyapandit@gmail.com; [f]kolisatvik25@gmail.com

lean concept. The concept and tools are grossly mistaken due to managerial practices. To effectively implement lean, especially in micro, small and medium enterprises (MSMEs), (Sherif et al., 2013) paper hints at a detailed framework, considering success factors during implementation. Crute et al. (2003) pointed out the need for a broader understanding of industry prior to the implementation of lean. The vast expenses of lean systems need industry-specific implementation as below. For example, Anand and Kodali (2010) highlighted the importance of the development of the lean framework; Smeds (1994) intended towards managing change in a lean organization; Powell et al. (2013) favours enterprise resource planning (ERP) based framework and Shah and Ward (2003) linked successful implementation to age, size and integration of plant. Mackelprang and Nair (2010) found out that overall performance is directly related to lean implementation.

Over the last few years, many organisations across the world are implementing Lean Manufacturing to achieve a competitive advantage over others. For example, Dunstan et al. (2006) studied lean applications in the mining domain. On implementation of lean, the safety measures were reduced by 67%. The manpower cost was also reduced by 3.41.8%, approximately 55% of this section, thereby saving $2 million (Australian) (Gurumurthy and Kodali, 2011).). The lean system can be implemented with unskilled workers and sparing investments within the industry (Hodge et al., 2011). George L. valued specific lean systems in the textile industry. It was achieved by implementing 5S in one industry. This was done by applying 5S in one industry and VSM in other industry. Based on the Interviews conducted, barriers and advantages of applying lean manufacturing were identified, 5S principle is widely accepted because of its ease. Interviews revealed failure on functional fronts of VSM with the majority of industries (Hodge et al., 2011).

Efforts on reducing waste in the rubber industry were attempted using the Waste Relation Matrix (WRM), and Waste Assessment Questionnaire (WAQ) (Amrina and Andryan, 2019). Overall efficiency is the direct indicator of lean implementation (Amrina and Andryan, (2019).

The present work focuses on the strict implementation of cycle time. The prevailing practices are found to be quite casual on this front. In a small scale industry quite often it is so.

The unskilled workers are not interested in the implementation of VSM. This hampers productivity. The reason may be a periodic improvement in the management of cycle time. Such industries find it difficult to reduce wastage of time, in its totality. The phased implementation is a better solution because the understanding of workers goes on increasing as they become more and more conversant.

The concept of VSM quite often reduces the level of satisfaction for whom it is implemented. It is felt that even if we can work out, the total saving of time and implementation becomes a difficulty. It is of course the attitude of the management. 'How to go about implementation?' is the everlasting question. The present-day research produces meager references on this issue. Our work is an attempt to find a solution inside this gap. The plenty of solutions widely admired is under an idle situation. Whereas physical reality should be honoured. The drastic changes in existing systems repose resistance, thereby stalling the complete implementation. This work is the Via-media to update framework definition and its implementers step by step. Such a chronological approach should exhibit asymptotic vicinity to our target.

VSM Methodology

VSM is a graphical and illustrative tool that includes information about the processes in a particular production industry. The current and future state map provides information like total cycle time, Time taken on each workstation, Equipment reliability, Number of workers at each workstation, TAKT time, Work In Progress (WIP) and Quantities of product Level of inventory. VSM is a manually assessing process in which data has to be collected manually by observing and understanding every process carefully. Simultaneously record the processing time. VSM should begin with appropriate planning and intensive arrangement.

Table 58.1 Major case studies on VSM

S.No	*Key Contributors*	*Type of Industry*
1	Mariusz Salwin, Ilona Jacyna –Golda, Michal Banka (2021)	Steel pipe industry.
2	Korakot yuvamitra, Jim Lee, Kanjicai Dong (2016)	Rope manufacturing industry.
3	Dorota sSadnicka, R M Ratnayake (2017)	Aircraft maintenance industry.
4	Palak P. Seth, Vivek Deshpande, Hiren R. Kardani (2014)	Automotive industry.
5	K. Mallikharjuna Rao, Geeta S. Lathkar (2021)	Manufacturing industry.
6	Yonathan Tesfaye, Deepak Panghal (2017)	Garment manufacturing.
7	Miguel malek maalouf,Magdalena Zaduminska (2019)	Food processing industry.

Firstly, we have to identify the processes and product family which include products that undergo similar activities. We have to determine such activities as well processes that are similar and can be manufactured simultaneously in order to increase the productivity.

A current state map must be ready with the help of collected information. This enables manufacturers to see the entire flow of work which can be studied in detail, particularly to find the bottlenecks and interruptions in the complete process. Management can take further action to rectify them. Various types of waste such as overproduction, waiting time, and transport time are identified. Actions are suggested to minimise it and a future state map is proposed. This future state is then stimulated on software and the output from each stimulation is recorded. After studying different aspects of production, the most suitable and favourable proposed plan will be implemented.

Case study

Company background

The industry where the case study is carried out is located in MIDC, Nagpur, Maharashtra. It is a manufacturer and producer of mild steel components, CR steel, stainless steel, etc. The company employs 44 workers and operates in 1920 sq. m. The products manufactured by the XYZ industries are split bush, clutch, eccentric shaft, pin, spacer, small bearing plate, etc. XYZ industry has a working shift of 9 hours a day as one shift and they work 6 days a week. It also consists of an hour-long lunch break.

Process information

A spacer was chosen for the case study. The raw material required to manufacture Spacer is CR5 steel. The process starts with cutting the CR5 rolls into the desired length which weighs 0.680.69 kg. These are then heated in the induction furnace at 950°C. It goes through the hammering process followed by pressing and punching. It undergoes two CNC operations which are turning and chamfering after which drilling is performed. It is then stored after the final inspection.

Current state map

The current state map is a present overview of shop floor workflow; it is generally represented by symbols. The above flow chart helps us to draw a current state map. The current state map starts from the consumer end. The consumer has a net requirement of 20,000 pieces during three months. Our Industry has signed

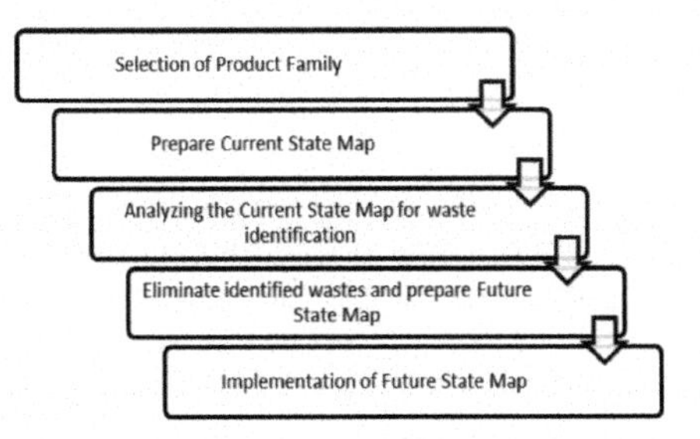

Figure 58.1 Conventional VSM approach

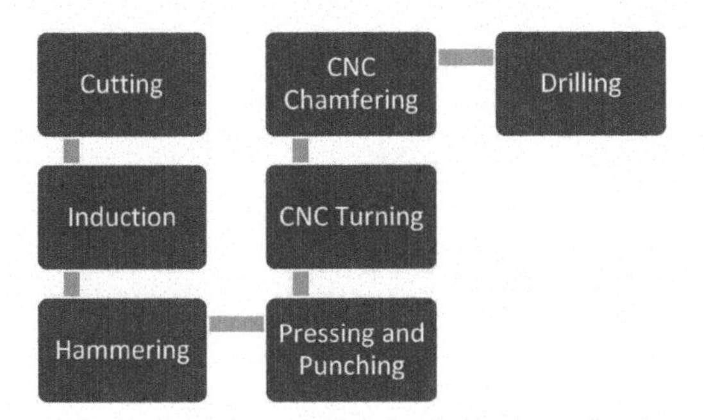

Figure 58.2 Process flow chart

a contract (MoU) with the supplier of raw material for the industry. According to which they buy a fixed amount of raw material for discounted rates and they receive orders through tenders, electronics.

Process Parameters

- Cycle time: Cycle time is the time expected to follow through with a particular job or activity from beginning to end.
- Inventory: The stock essentially alludes to the stock or capacity of merchandise and materials held by an organisation which utilised for offering to acquire benefits.
- Lead time: It is how much all-out time is expected to finish a whole assembling process from the beginning to its conclusion.
- TAKT time: TAKT Time is the amount of time needed in which a particular job must be made to satisfy the client's or consumer's need. It is very well may be determined by separating the networking time accessible to client interest.
- Changeover time: How much time that is necessary to stack and empty the part that is being fabricated and restart the cycle once more.
- Value-added activities: In the assembling system, value-added exercises are the arrangement of cycles or activities that increases the value of the completed item according to the buyer's point of view.
- Non-Value-added activities: In assembling, non-esteem added exercises are viewed as waste as they don't enhance the coproduce quality and accordingly expand shortcomings.

Analysis of Current State map

The complete working time accessible = $[(9 \times 60) - 60] = 480$ minutes.

At the present status, the organisation can make 200 pieces per day.

$$\{TAKT\ time\}_{current} = \frac{Available\ working\ time}{Purchaser\ demand}$$

$$= 480/200 = 2.4 \text{ minutes}$$

Purchaser requests 20000 pieces in 90 days

Complete time accessible = $480 \times 30 \times 3 = 43200$ minutes

$\{TAKT\ time\}_{current} = 480/222 = 2.16$ minutes

The above calculation shows that the relationship between actual and calculated TAKT is not equal.

Future state map analysis

To remove the above-mentioned issues in the present state, different aspects were looked after to find the optimum solution. The concerning issues in the present status are; higher cycle and lead times of assembling processes as well as higher dismissal rates. The ensuing changes were recommended in the present status map:

Firstly, the rejection rate was higher due to the dents and scratches caused during the storage of the final product. Previously, there was no systematic arrangement for storage which led to surface damage and a

Figure 58.3 Sample job-spacer (finished)

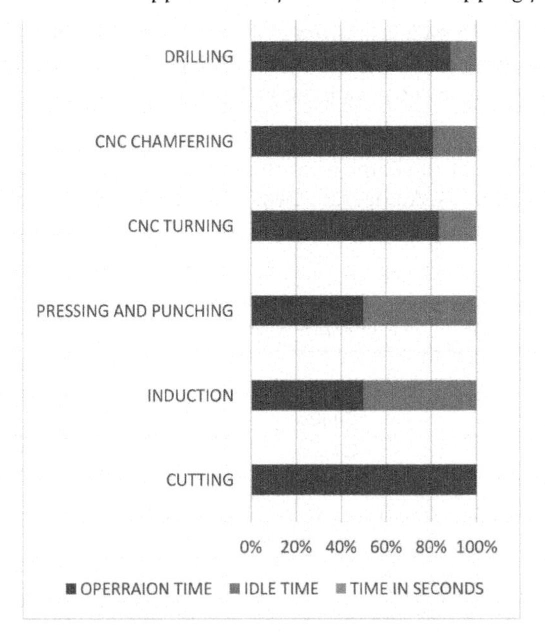

Figure 58.4 Process time chart (current state map)

Figure 58.5 Sample job-spacer (unfinished)

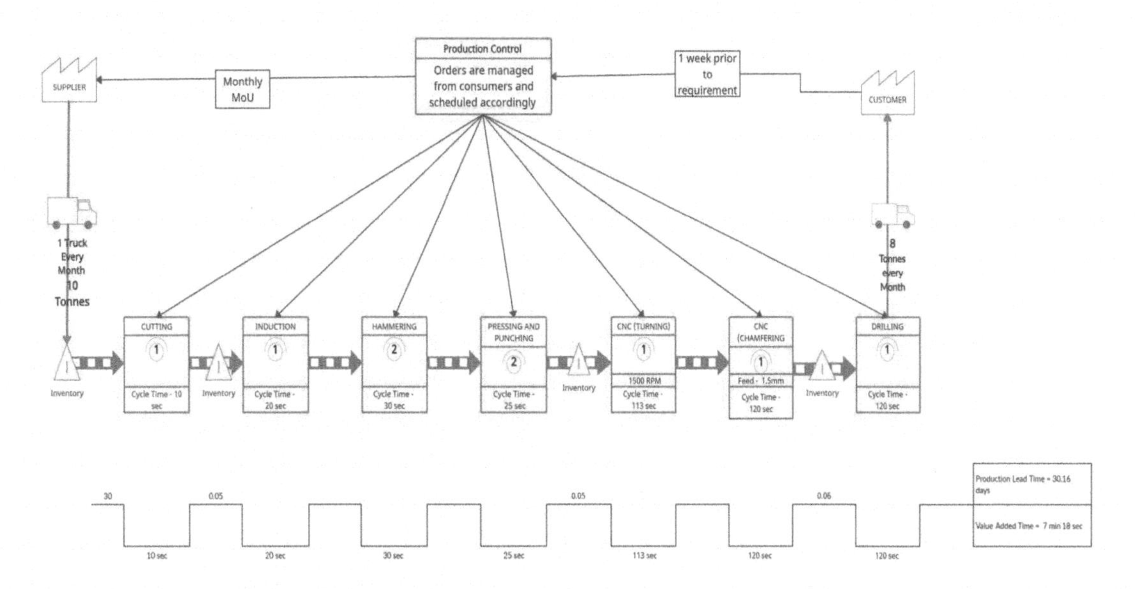

Figure 58.6 Current state map XYZ industries, MIDC, Nagpur

poor surface finish. This issue was resolved by providing a systematic storage arrangement. Furthermore, there were 100 pieces in a lot which was also one of the reasons for the surface damage of the product. The lot size was then reduced to 50 which also assisted in easier packaging and transportation, whereas simultaneously it also reduces the rejection rate.

- Also, analysing the current operations, it was found that the chips from the job adhered to the chuck were causing improper surface finish and uneven thickness. An operator was assigned for removing the chips by pressurized air, which helped considerably.

Table 58.2 Cycle time for various operations

Process	*Operators*	*Current State Map*	*Future State Map*
Cutting	1	10 sec	10 sec
Induction	1	20 sec	20 sec
Hammering	2	25 sec	25 sec
CNC 1 (Turning)	1	1 min 53 sec	1 min
CNC 2 (Chamfering)	1	2 min	1 min 51 sec
Drilling	1	120 sec	120 sec

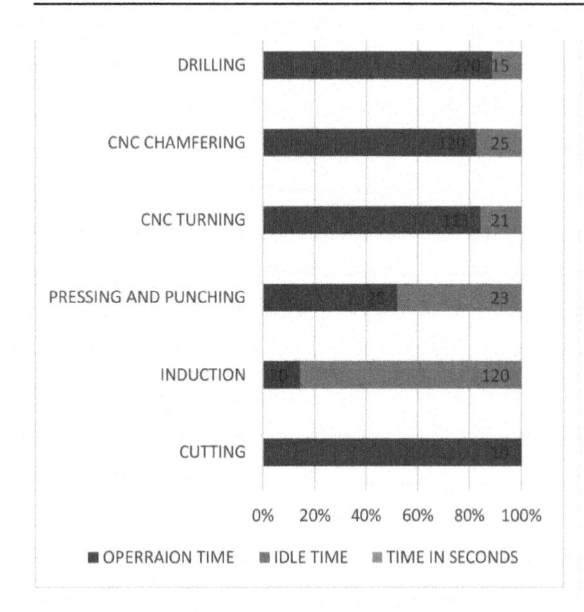

Figure 58.7 Process time chart (future state map)

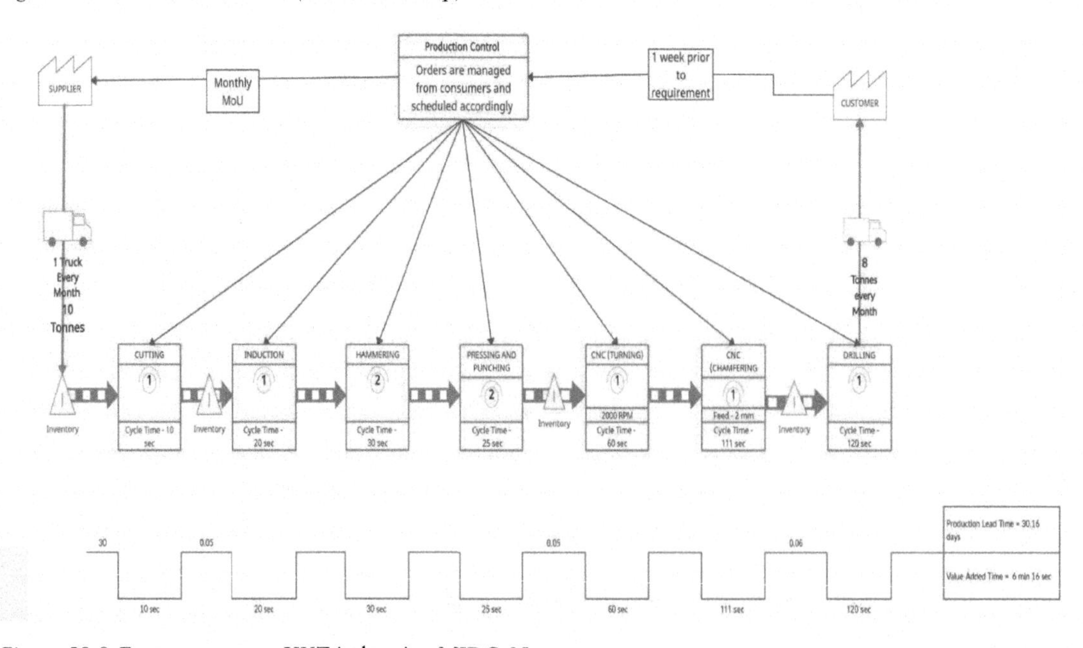

Figure 58.8 Future state map XYZ industries, MIDC, Nagpur

Conclusion

The paper indicated a framework to execute lean manufacturing in the small-scale industry using VSM. Different wastes and shortcomings in the value stream were distinguished by the current state map of the organization. The various wastes that were identified were higher idle times, higher cycle times, improper storage arrangements, and lack of proper scheduling. Preparation of the current state map which entailed targeting and eliminating the root causes of inefficiencies. Non-value-added activities were terminated using the VSM tool and a future state map was developed. The pace of the overall process was considerably increased due to reduced cycle time of manufacturing processes. The reduction in TAKT time was observed as 10%. Also, the total operational time has lessened from 7 minutes 18 seconds to 6 minutes 16 seconds. To minimise the load on workers as well as machines, work order scheduling was performed. It will also be helpful to meet accurate consumer demands.

References

Amrina, E. and Andryan, R. 2019. Assessing Wastes in Rubber Production Using Lean Manufacturing: A Case Study. Doi: 10.1109/IEA.2019.8714925

Anand, G. and Kodali, R. (2010). Analysis of lean manufacturing frameworks. J. Adv. Manufac. Sys. 9(1):1–30. doi: 10.1142/S0219686710001776

Anand, G. and Rambabu, K. (2011). Design of lean manufacturing systems using value stream mapping with simulation. J. Manuf. Technol. Manag. 22(4):444–473. doi:10.1108/17410381111126409

Crute, V., Ward, Y., Brown, S., and Graves, A. (2003). Implementing Lean in aerospace – Challenging the assumptions and understanding the challenges. Technovation. 23:917–928. doi:10.1016/s0166-4972(03)00081-6

Deshkar, A., Kamle, S., Giri, J., and Korde, V. (2018). Design and evaluation of a Lean Manufacturing framework using Value Stream Mapping (VSM) for a plastic bag manufacturing unit. Mater. Today: Proc. 5(2):7668–7677. doi:10.1016/j.matpr.2017.11.442.

Gurumurthy, A. and Kodali, R. (2011). Design of lean manufacturing systems using value stream mapping with simulation A case study. J. Manufact. Technol. Manag. 22(4):444–473. doi: 10.1108/17410381111126409

Hines P. and Rich N. (1997). The seven value stream mapping tools. Int. J. Oper. Prod. Manag. 17(1):4664.

Hodge, G. L., Ross, K. G., Joines, J. A., and Thoney, K. (2011). Adapting lean manufacturing principles to the textile industry. Prod. Plan. Control: Manage. Operat. 22(3):237–247.

Khandelwal, G., Yadav, V., Jain, A., and Jain, R. (2016). Application of VSM approach in Indian SME: a case study. Jaipur: Malviya National Institute of Technology.

Mackelprang, A. W. and A. Nair. (2010). Relationship Between Just-in-Time Manufacturing Practices and Performance: A Meta-Analytic Investigation. J. Operat. Manag. 28(4):283–302. doi:10.1016/j.jom.2009.10.002

Mariusz Salwin, Ilona Jacyna-Golda (2021). Using value stream mapping to eliminate waste: A case study of a steel pipe manufacturer. engineer. MDPI. 14(12 June):1–19.

Powell, D., Alfnes, E., Strandhagen, J. O., and Dreyer, H. (2013). The concurrent application of lean production and ERP: Towards an ERP-based lean implementation process. Comput. Indus. 64:324–335. doi: 10.1016/j.compind.2012.12.002.

Rother, M. and Shook, J. 999. Learning to See Value Stream Mapping to Add Value and Eliminate MUDA. Cambridge: Lean Enterprise Institute.

Serrano, I. Ochoa, C., Castro, R. D. (2008). Evaluation of value stream mapping in manufacturing system redesign. Int. J. Produc. Res. 46(16):4409–4430. doi:10.1080/00207540601182302

Seth, D. and Gupta, V. (2005). Application of value stream mapping for lean operations and cycle time reduction: an Indian case study. 16(1):44–59. doi:10.1080/09537280512331325281

Shah, R. and Ward, P. (2003). Lean manufacturing: Context, practice bundles, and performance. J. Operat. Manag. 21:129–149. doi: 10.1016/s0272-6963(02)00108-0.

Sherif, M., Jantanee, D., and Hassan, S. (2013). A framework for lean manufacturing implementation. Produc. Manufact. Res. 1(1):44–64. doi: 10.1080/21693277.2013.862159

Smeds, R. (1994). Managing Change towards Lean Enterprises. Int. J. Operat. Produc. Manag. 14(3):66–82. doi: 10.1108/01443579410058531

Wan Ibrahim, W. M. K. B., Rahman, M. A., and Bakar, M. R. B. A. (2017). Implementing Lean Manufacturing in Malaysian Small and Medium Startup Pharmaceutical Company. IOP Conf. Ser.: Mater. Sci. Eng. 184:012–016

59 Fluidised bed dryer for agricultural products: An approch

Vivek M. Korde[a], Jayant P. Giri[b], Narendra J. Giradkar[c], and Rajkumar B. Chadge[d]

Department of Mechanical Engineering, Yeshwantrao Chavan College of Engineering, Nagpur, India

Abstract

Fluidised bed dryers are widely used for drying of moist particulate solid products of different shape, size, and compositions. The drying in fluidised bed is very complex phenomenon because of simultaneous mass and heat transport process which is many a times associated with chemical and phase change. The analysis of such process is a challenging task. The literature is full of mathematical models and empirical correlations explaining such process with experimental validation. It is found that empirical correlations may exhibit inconsistency and are incapable to stretch out the large scale of variables and the conditions met. The mathematical models are based on some assumptions and are unable to represent the entire drying process precisely. Thus, they are product or system specific and not generalised. This limits the universality of design and operations of the fluidized bed dryer and further leads to the major upscaling problems. The concepts such as volumetric heat transfer coefficient and mechanistic models seems helpful. The combined computational fluid dynamics and discrete element method approach and artificial neural networks are becoming progressively popular in predicting characteristics of drying for agricultural products in fluidised bed. Some of the problems confronted by fluidised bed dryer can be tackled with aid of external means and latest and innovative technologies. The fluidised bed dryer can be coupled with hybrid drying techniques to enhance the system efficiency and quality of product. Finally, some conclusions are drawn considering the above limitations on which the studies should be focused on in future

Keywords: Agricultural product, drying, fluidised bed dryer, heat transfer.

Introduction

The process of drying of agricultural products (such as grains, vegetables, fruits and leaves) has vital importance as it is required for the benefit of storage, transportation, preservation, and quality improvement. A proper drying technology should be implemented to enhance the agricultural returns in gratitude of the tough effort the farmer has dedicated while cultivating the crops (Chua and Chou, 2003).

The drying process is very complex and diversified as it involves the simultaneous and steady/unsteady heat, mass and momentum transfer associated with/without phase and chemical change. Because of this the perception of drying phenomenon at microscopic level is even now rudimentary (Mujumdar and Huang, 2007). Various drying processes are available in which fluidised bed dryer is most preferred one because of several advantages over others.

The major advantages are large particle-gas contact area, good mixing of solid, very high rates of mass and heat transfer between the gas and particles, easy particle transport and control, smaller drying time and flow area, low maintenance and capital cost, and suitable for large scale applications (Chua and Chou, 2003; Mujumdar, 2006; Daud, 2008; Kunii and Levenspiel, 1991; Gupta and Sathiyamoorthy, 1998).

There are certain limitations also which includes attrition, agglomeration erosion, and poor fluidisation for some products also higher pressure drops (Daud, 2008; Kunii and Levenspiel, 1991; Gupta and Sathiyamoorthy, 1998; Yang, 2003).

A typical FBD is obtained by admitting a hot gas through the base of a bed of particulate matter. The bed remains packed (fixed) at low velocities and the particulate matter remain on the gas distributor plate maintaining contact with each other (Rovero et al., 2012). On raising velocity of the gas, the pressure drop increases across the bed. The bed expands and gets fluidised at a particular velocity, when the load of particles on unit area is equal to the pressure drop, known as minimum fluidisation velocity, u_{mf} (INABA, 2007; Shilton and Niranjan, 1993; Cocco et al., 2014).

Various regimes (Figure 59.1) of fluidisation (bubbling, sludging, turbulent and transport) are observed on gradually increasing the gas velocity (Mujumdar, 2006; Gupta and Sathiyamoorthy, 1998; Grace et al., 2020; Coulson and Richardson, 2019).

[a]vmkorde@gmail.com; [b]jayantpgiri@gmail.com; [c]njgiradkar@gmail.com; [d]rbchadge@gmail.com

The minimum fluidisation velocity, u_{mf}, can be calculated from correlations or obtained experimentally (Gupta and Sathiyamoorthy, 1998; Cocco et al., 2014). Once fluidised, the bed behaves like fluid (Kunii and Levenspiel, 1991; Gupta and Sathiyamoorthy, 1998; Sivakumar et al. (2016) has presented a detailed review on experimental and theoretical studies of FBD for various agricultural products, such as carrot, tea leaves, apple, soybean, paddy, wheat and other products, with effect of different operating parameters. The comparative survey of several driers mentioning the advantages, limitations, and future scope of developments in tabular format is also presented by them.

Drying Kinetics

The drying rate is function of the material characteristic and conditions of fluidisation. Drying period can be divided as constant rate period and falling rate period (first and second) separated by critical moisture content (Yang, 2003).

The line AB, in Figure 59.2, characterises the constant rate of drying period on the drying rate curve in which the free moisture at the particle surface evaporates and the moisture removal rate from the wet exterior is almost same as the internal moisture transfer to the external surface of the particle (Kaur et al., 2015; Srinivasakannan, and Balasubramanian, 2008). The constant rate of drying ceases at critical moisture content, denoted by point B.

Majority of the agricultural products have very small constant rate drying period and long curved falling rate period of drying and hence are dried in the later one (Gupta, and Sathiyamoorthy, 1998).

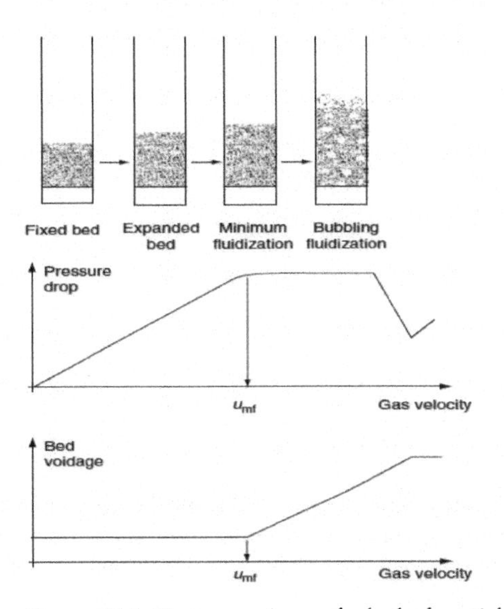

Figure 59.1 Various regimes of a bed of particles at different gas velocities (Mujumdar, 2006)

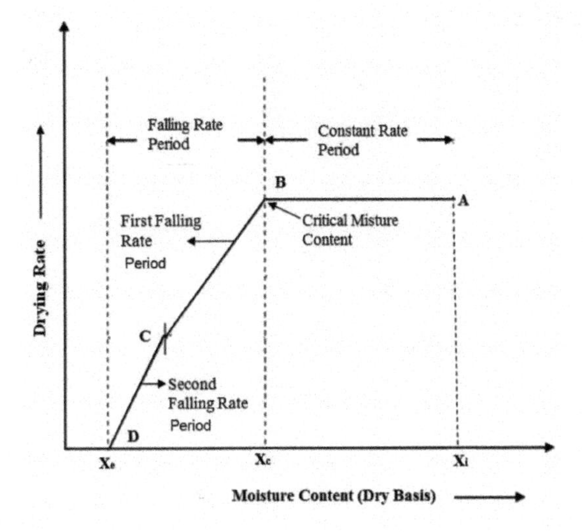

Figure 59.2 Drying curve: Constant rate period and falling rate period (Kaur et al., 2015)

In the first phase of falling rate (line BC) the surface moisture removal rate is more than internal moisture diffusion rate. At point C the surface moisture film is completely evaporated. In the second falling rate (line CD) persists till equilibrium moisture content is reached, which is corresponding to the humidity and temperature to which it is exposed, by sub surface moisture evaporation (Gupta and Sathiyamoorthy, 1998; Yang, 2003; Kaur et al., 2015; Srinivasakannan, and Balasubramanian, 2008).

Gas-Particle Heat Transfer:

In FBD the rate of drying is function of the rate of moisture transfer from the particle to the drying gas. In this drying process the convection is the dominant mode whereas the radiation is neglected as the dryer generally operates below 500°C temperature (Mujumdar, 2006; Grace et al., 2020; Kaur et al., 2015).

The coefficient of heat transfer for convection between interstitial gas and particles is generally low and in the range of 01 to 100 W/m^2-K in bubbling FB, but the interfacial surface area for convection is very large which results in high heat transfer rates in the system (Yang, 2003).. Thus, the isothermal condition in the bed is achieved very quickly within small distance from gas distributing plate (Mujumdar, 2006; Gupta and Sathiyamoorthy, 1998; Yang, 2003; Grace et al., 2020). It is found in the literature that lot of investigators have determined the heat transfer coefficient experimentally for such condition in FBD and presented the correlation for evaluation of the same (Gupta and Sathiyamoorthy, 1998; Ciesielczyk et al., 1997; Ciesielezyk, 2007).

Kunii and Levenspiel (1991) studied selected data of twenty-two investigators (Figure 59.3) for similar conditions of FBD and presented two correlations based on Nusselt, Reynolds and Prandtl number:

$$Nu_p = \left(\frac{h_p d_p}{k_g}\right) \tag{1}$$

and $$Re_p = \left(\frac{d_p u_g \rho_g}{\mu_g}\right) \tag{2}$$

The representing correlations are:

$$Nu_p = 0.028\left(\mathrm{Re}_p\right)^{1.4}\left(\mathrm{Pr}_g\right)^{0.33}\ 0.1 \le Re_p \le 50 \tag{3}$$

$$Nu_p = 1.01\left(\mathrm{Re}_p\right)^{0.48}\left(\mathrm{Pr}_g\right)^{0.33}\ 50 \le Re_p \le 10^4 \tag{4}$$

Empirical correlations developed for the evaluation of heat transfer coefficient should be applied with substantial caution as they may be incapable to deal with the wide range of variables and conditions confronted (Chena, et al., 2005).

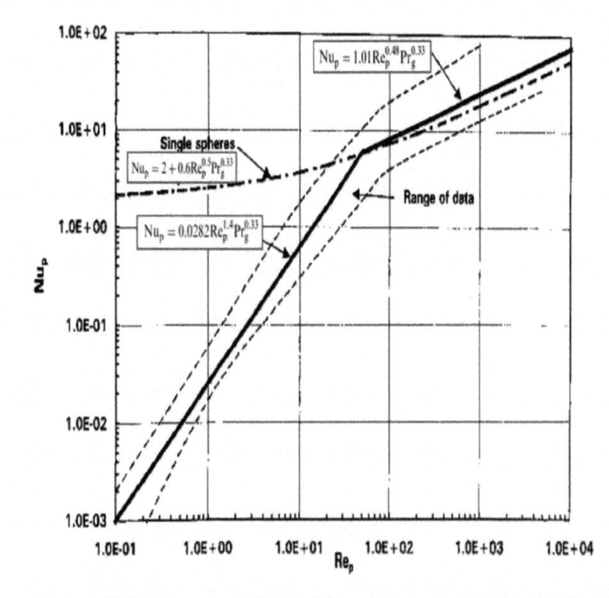

Figure 59.3 Nusselt vs Reynolds numbers for fluidised beds ($Pr_g = 0.7$) (Yang, 2003)

Volumetric Heat Transfer Coefficient and Modified Nusselt Number:

Even though the literature has lot of correlations Nu=f(Re, Pr), for determining the coefficient of heat transfer between the solid particle and the drying medium of gas in the FBD (Gupta and Sathiyamoorthy, 1998; Ciesielczyk et al.,1997; Ciesielezyk, 2007), a general inconsistency is observed in them (Daud et al., 2008; Chena et al., 2005; Poós and Szabó, 2017; Brodkey et al., 1991).

These correlations are generally established on the experimental data for the turbulent gas flow over a fixed sphere, whereas the condition in the bubbling FBD may be different. Further there may be inaccuracy in determining the surface contact (Daud, 2008).

Poós and Szabó (2017) calculated volumetric heat transfer coefficient as a product of heat transfer coefficient and the volumetric interfacial surface area defined as:

$$\alpha = \frac{6(1-\varepsilon)\phi}{d_p} \tag{5}$$

Further the Nusselt Number was modified as

$$Nu' = \frac{h\alpha d_p}{k_g} \tag{6}$$

They have performed the experiments on barley, millet and sorghum and used the available data after reworking for the volumetric heat transfer coefficient to obtain the relation for fluidised bed drying as

$$Nu' = 7.2 \times 10^{-4} Re^{1.68} \tag{7}$$

The relation is applicable for the range of Reynolds number from 17 to 1183 in constant rate drying period.

The revised data with modified Nusselt number exhibits good fit than original data (Figure 59.4). The use of modified Nusselt number is justified as it handles real geometry, size and particle numbers and thus can address more accurately the scaling problem of FBD (Poós and Szabó, 2017; ALVAREZ and SHENE, 2007).

Effect of Operating Parameters on FBD

The major parameters that monitor the rate at which agricultural materials are dried are bed height, gas temperature, particle size, relative humidity, and gas velocity (Kumaresan and Viruthagiri, 2006).

For agricultural materials, with the increment in height of bed, the rate of drying is observed to decrease because of resistance to drying inside the material (Yang, 2003).

The increment in velocity improves the rate of drying in constant drying period while removing the surface moisture but in falling rate period it has little effect due to the internal resistance to moisture transfer by the agricultural material. Thus, the velocity should not be much higher but should be sufficient for required fluidisation (Yang, 2003; Cil and Topuz1, 2010).

The temperature of drying medium has the highest impact on the drying kinetics of particles. Thus, it is found that the increment in gas temperature improves the complex moisture diffusivities and increases the drying rate (Mujumdar, 2006; Cil and Topuz1, 2010).

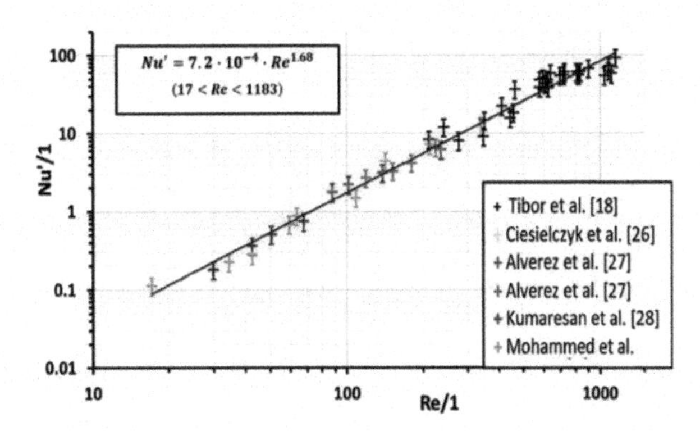

Figure 59.4 Nu' vs Re correlation of literature cited (Poós and Szabó, 2017)

Effect of particle size cannot be predicted appropriately. Geldart (1972) has classified the particles in four different groups (A, B, C and D) based on the difference in density and size of particle for predicting fluidisation behaviour. Generally, with the group B particles the drying time increases with size, but the group A particles show little effect as they are finer and are smoothly fluidised (Mujumdar, 2006; Yang, 2003).

Mathematical Models of FBD

Numerous mathematical models of FBD kinetics are suggested for agricultural products.

The mathematical models indicate the drying characteristics of the specific agricultural product which helps to select drying method and design/sizing a proper drying technique to obtain a better quality of product. These developed models are generally based on various assumptions. Various models are discussed by Mujumdar (2006).

In the diffusion model, it is assumed that the internal moisture diffusion completely controls the drying of single particles in FBD. In empirical model the drying process with three distinct mechanisms is divided into three separate drying periods.

The kinetic model assumes that the process of drying includes both the drying rate periods in which the falling period is linear with constant feed condition and surface contact area. The single-phase model assumes the FBD as a continuum and the mass, momentum and energy balances are considered over it. Two phase model assumes the FBD of two distinct phases, i.e. dilute phase made of bubbles and dense phase which is a phase of emulsion and remains stagnant at condition of minimum fluidisation (Mujumdar, 2006).

Simple exponential time decay models are used for modelling the agricultural products with curvy falling rate period of drying such as Page model, Newton model, two-term exponential model, Henderson and Pabis model and approximate diffusion model. These are completely empirical in nature (Srinivasakannan and Balasubramanian, 2008).

The FBD is a heterogeneous system in turbulent condition with a complex transport phenomenon in varying conditions of fluidisation and material properties. Therefore, it is very difficult to get a reliable model to describe FBD with high accuracy. Thus, the design and operation procedure of FBD are still based on pilot scale testing coupled with some empirical experience (Daud, 2008; Cil and Topuz1, 2010).

Although many research papers are available on FBD, challenges of the diversity and universality of calculation methods applied in designing the process is yet unsolved (Cil and Topuz1, 2010; INABA, H. 2007; Ciesielczyk, 2005). The basic reasons are:

- The intricacy of the process of drying in turbulent diverse systems, along with simultaneous mass, momentum, and energy transport,
- the absence of satisfactory mathematical models defining the system enduring varying quality and hydrodynamic type of fluidisation,
- diverse and time varying physicochemical, mechanical, and other properties of the agricultural materials to be dried.

Spouted Bed Dryer (SBD)

The spouted bed dryer SBD are found to be more appropriate to deal with the agricultural product which are too coarse to be fluidised (Chua and Chou, 2003).

The SBD are available in at least 30 different variants as per requirement (Mujumdar, 2006).

The SBD has several advantages such as high drying rates, shorter drying time, lower drying temperature, reduced pressure drops, relatively lower gas flowrates, handling coarser particles, absence of dead zones, wide range of operating conditions, can use inert particles, viscous paste, and slurries (Chua and Chou, 2003; Rovero, et al., 2012).

The spouted bed can easily be obtained simply by replacing the gas distribution plate at the bottom of FBD by an orifice located at the center or at the apex of central cone on the plate (Rovero et al., 2012; Shilton. and Niranjan, 1993; Brennan, 2006). The gas enters through this orifice into the bed and while travelling upwards creates a circular cavity which opens as a 'spout'. At high flowrates of gas, a fountain of particles in freeboard is generated which returns the particles back to the bed surface. These particles then travel back downwards through the peripheral annulus around the spout and again is recaptured in the spout travelling upwards (Grace et al., 2020).

On comparing the FBD and SBD it is found that drying rates are higher in FBD whereas, specific energy consumption are less in SBD (Mohideen et al., 2012). Concerns about the difficulty in scaling up SBD are reported (Rovero et al., 2020; Jittanit et al., 2013).

Recent Advances in FBD

The fluidizing quality can be improved for the 'difficult to fluidised particles' with Mechanically Assisted Fluidization, Vibrated Fluidized Bed, Agitated Fluidized Bed, Centrifugal and Rotating Fluidized Bed which are discussed in detail by Daud (2008). The Geldart D-Type particles can be fluidised with outstanding ability in swirling fluidised beds (Mohideen et al., 2012).

Several developing technologies are listed and selected innovative drying technologies such as Superheated Stream Drying (SSD), Multi-Stage Drying, Refractance Window Drying, Nanomaterials Drying, Two-Stage Horizontal Spray Dryer are discussed in detail by Mujumdar and Huang (2007).

The latest advances and use of advanced computational tools like CFD for optimisation of the FBD design and operation to improve the thermal efficiency was reviewed along with review of some agricultural products by Haron et al. (2017). CFD–DEM approach is yet a state-of-art technique of simulation (INABA, H. 2007; Hou et al., 2015).

The applicability of artificial neural networks (ANN) in predicting the drying characteristics of agricultural product in FBD was investigated by using feed forward ANN by Topuz (2010). The predictions obtained from the optimum NN were in good agreement with the investigational data.

The extremely latest development in drying technology of agricultural materials such as new methods, new products and modelling and optimisation techniques has been presented by Valarmathi et al. (2017).

Sivakumar et al. (2016) has presented a detailed review on modified FBD with Hybrid drying techniques. The FBD assisted with Microwave, FIR, Ultrasonic transducer, immersed heater bed, Heat pump and Solar were discussed in detail. Utilising the mentioned technologies, the efficiency of system and quality of product can be enhanced with possible heat recovery.

Aghbashlo et al. (2014) has given a review of measurement systems such as infrared moisture sensor, electrical capacitance tomography, near infrared spectroscopy, acoustic emission, focused beam reflectance measurement, microwave resonance technology, optical imaging techniques, electrical capacitance tomography, Raman spectroscopy, spatial filter velocimetry, triboelectric probes and some innovative techniques of control of FBDs.

Research Gap

1) A generalised process to design a fluidised bed dryer is not available because of challenges associated with process of drying and the technology of fluidisation.
2) No reliable mathematical model is available to describe all the four mechanisms properly in the drying kinetics due to the complex nature of the drying process.
3) The particle characteristics, according to Geldart's classification may need some modifications, due to specific geometry for agricultural particles.
4) A uniform particle distributor cannot be expected, in some cases, due to maldistribution of inlet gas in a fluidised bed dryer.
5) The best operating conditions is always a question to reduce the drying duration, consumption of energy and size of dryer in a fluidised bed dryer.
6) Investigations to improve the savings in energy and to minimise the pollution effect are required by employing hybrid and combined FBD techniques of drying.
7) Presently, monitoring techniques for FBD are not well established, hence economical, low-priced, multiuse monitoring technique specially applicable to FBDs should be developed.

Conclusion

Extensive study of drying of agricultural product in FBD has been projected in the available literature. The design of the FBDs is distinctive and mostly product specific. The following conclusions are made considering the developments of FBD from the study:

1) Most of the literature is applicable to the dry particles, hence while applying it to the FBD modifications are expected as the existence of moisture with particle, changes the fluidisation quality at large.
2) Appropriate drying process should be selected by understanding the drying principles. Improper application of mass and heat transfer rates may result in over or under drying of the agricultural material.
3) The fluidisation of Geldart Type-C particles may be enhanced with assistance of external implies such as vibrators, agitators, rotators, and centrifuging in novel FBDs. The theoretical understanding should be improved to develop better models.

4) Concept of the volumetric heat transfer coefficient which is introduced for fluidised bed drying to remove the uncertainties of the determination of the heat transfer surface between the drying medium and the particulates may be utilised for scaling up of fluidised bed dryers.
5) The mechanistic models which are established on surface renewal concept may be utilised while designing and scale-up of the systems with both bubbling dense beds and fast circulating fluidised beds.
6) More focus is required to provide the sound solution for scaling up the spouting units for the preference between larger or multiple units.
7) The swirling fluidised bed exhibited outstanding ability in fluidising Geldart Type-D particles as compared to the usual fluidized beds.
8) The combined computational fluid dynamics and discrete element method approach has a decent ability in defining heat transfer in fluidised beds at a particle scale, although further developments are necessary.
9) ANN is becoming progressively popular in predicting characteristics of drying for agricultural products in the fluidised bed, because of its intrinsic capacity to quantify complex nonlinear interactions automatically.

References

Aghbashlo, M., Sotudeh-Gharebagh, R., Zarghami, R., Mujumdar, A. S., and Mostoufi, N. (2014). Measurement Techniques to Monitor and Control Fluidization Quality in Fluidized Bed Dryers: A Review. Drying., Technol Int. J. 32(9):1005–1051. doi: 10.1080/07373937.2014.899250.

ALVAREZ, P. I. and SHENE, C. (2007). Experimental determination of volumetric heat transfer coefficient in a rotary dryer. Dry. Technol. An://dx.doi.org/10.1080/07373939408962189

Brennan, J. G. 2006. Food Processing Handbook. Weinheim: Wiley-VCH.

Brodkey, R. S., Kim, D. S., and Sidner, W. (1991). Fluid to particle heat transfer in a fluidized bed and to single particles. Int. J. Heat Mass Transf. 34(9):2327–2337.

Ciesielczyk, W., Stojiljković, M., Ilić, G., Radojković, N., and Vukić, M. (1997). Experimental Study on Drying Kinetics of Solid Particles in Fluidized Bed. Sci. J. Facta Univ. Ser.: Mech. Eng. 1(4):469–478.

Ciesielczyk, W. 2005. Batch Drying Kinetics in a Two-Zone Bubbling Fluidized Bed. Drying Technol. 23:1613–1640.

Ciesielezyk, W. 2007. Analogy of heat and mass transfer during constant rate period in fluidized bed drying. dry. Technol. 12(2): 217–230.

Cil, B. and Topuz1, A. (2010). Fluidized bed drying of corn, bean and chickpea. J. Food Proc. Eng. 33:1079–1096. doi: 10.1111/j.1745-4530.2008.00327

Chena, J. C., Graceb, J. R., and Golriz, M. R. (2005). Heat transfer in fluidized beds: design methods. Powder Technol. 150:123–132.

Chua, K. J. and Chou, S. K. (2003). Low-cost drying methods for developing Countries. Trends Food Sci. Technol. 14:519–528. doi:10.1016/j.tifs.2003.07.003

Cocco, R., Reddy Karry, S. B., and Knowlton, T. 2014. Introduction to Fluidization. American Institute of Chemical Engineers (AIChE). CEP Magazine. https://www.aiche.org/publications/cep.

Coulson and Richardson's, 2019. Fluidization. Coulson and Richardson's Chemical Engineering. (2a, pp: 449–554). https://doi.org/10.1016/B978-0-08-101098-3.00010-X

Daud, W. R. W. (2008). Fluidized Bed Dryers — Recent Advances. Adv. Powder Technol. 19:403–418.

Geldart, D. (1972). Types of Gas Fluidization. Powder Technol. 7:285–292. Netherlands: Elsevier Sequoia SA.

Grace, J. R., Bi, X., and Ellis, N. 2020. Essentials of Fluidization Technology. Boschstraße, Weinheim: Wiley-VCH.

Gupta, C. K., and Sathiyamoorthy, D. 1998. Fluid Bed Technology in Materials Processing. Boca Raton: CRC Press. https://doi.org/10.1201/9780367802301.

Haron, N. S., Zakaria, J. H., and Mohideen Batcha, M. F. 2017. Recent advances in fluidized bed drying. IOP Conf. Ser.: Mater. Sci. Eng. 243:012–038

Hou, Q., Gan, J., Zhou, Z., and Yu, A. (2015). Particle scale study of heat transfer in packed and fluidized beds. Adv. Chem. Eng. 46:193–243. http://dx.doi.org/10.1016/bs.ache.2015.10.006

INABA, H. 2007. Heat and Mass Transfer Analysis of Fluidized Bed Grain Drying. (41:pp. 5262). Okayama: Okayama University.

Jittanit, W., Srzednicki, G., and Driscoll, R. H. (2013). Comparison Between Fluidized Bed and Spouted Bed Drying for Seeds. Drying Technol. 31(1):52–56. http://dx.doi.org/10.1080/07373937.2012.714827

Kaur, B. P., Sharanagat, V. S., and Nema, P. K. 2015. Technologies for Foods: Fundamentals & Applications. Fundamentals of Drying. (pp 122). New Delhi: India Publishing Agency.

Kumaresan, R. and Viruthagiri, T. (2006). Simultanious Heat and Mass Transfer studies in drying ammonium chloride in a batch fluidized bed dryer. Indian J. Chem. Technol. 13(5):440–447.

Kunii, D. and Levenspiel, D. 1991. Fluidization Engineering. Boston, MA: ButterworthHeinemann.

Mohideen, M. F., Seri, S. M., and Raghavan, V. R. (2012). Fluidization of Geldart Type-D Particles in a Swirling Fluidized Bed. Appl. Mech. Mater. 110116:3720–3727. doi:10.4028/www.scientific.net/AMM.110-116.3720

Mujumdar, A. S. 2006. Handbook of Industrial Drying. Boca Raton: Taylor & Francis Group, LLC.

Mujumdar, A. S. and Huang, L. X. (2007). Global R&D Needs in Drying. Dry. Technol. 25:647–658.

Poós, T. and Szabó, V. 2017. Volumetric Heat Transfer Coefficient in Fluidized Bed Dryers. Chem. Eng. Technol. www.cet-journal.com, 10.1002/ceat.201700038, https://doi.org/10.1002/ceat.201700038

Poósa, T. and Szabóa, V. (2017). Application of mathematical models using volumetric transfer coefficients in fluidized bed dryers. Sci. Direct Energy Proce. 112:374–381.

Rovero, G., Curti, M., and Cavaglià, G. 2012. Optimization of Spouted Bed Scale-Up by Square-Based Multiple Unit Design. Advances in Chemical Engineering. doi:10.5772/33395

Shilton, N. C. and Niranjan, K. (1993). Fluidization and Its Applications to Food Processing. Food Struct. 12(2):8. https://digitalcommons.usu.edu/foodmicrostructure/vol12/iss2/8.

Sivakumar, R., Saravanan, R., Perumal, A. E., and Iniyan, S. (2016). Fluidized bed drying of some agro products – A review. Renew. Sustain. Energy Rev. 61:280–301. http://dx.doi.org/10.1016/j.rser.2016.04.014.

Srinivasakannan, C. and Balasubramanian, N. 2008. An Analysis on Modeling of Fluidized Bed Drying of Granular Material. Adv. Powder Technol. 19:73–82. doi:10.1163/156855208X291774

Topuz, A. (2010). Predicting moisture content of agricultural products using artificial neural networks. Adv. Eng. Softw. 41(3):464–470. doi:10.1016/j.advengsoft.2009.10.003

Valarmathi, T. N., Sekar, S., Purushothaman, M., Sekar, S. D., Reddy, M. R. S., Reddy, K., and Reddy, N. K. (2017). Recent developments in drying of food products. IOP Conf. Ser.: Mater. Sci. Eng. 197:012–037. doi:10.1088/1757-899X/197/1/012037

Yang, W. -C. 2003. Handbook of Fluidization and Fluid-Particle Systems. Boca Raton: CRC Press.

60 A multi-source fused model for stock prediction using technical and linguistic analysis via ensemble learning

Charanjeet Dadiyala[a] *and Asha Ambhaikar*[b]
Kalinga University Naya Raipur, India

Abstract

Stock Market Prediction is one of the most challenging areas of research due to its complex, uncertain and dynamic nature. The overall market movement and behaviour is governed by various internal and external influential factors and hence simpler linear models using single source input may not be suitable for stock prediction. To reduce the risk and improve the prediction process, a well-defined, more impactful model which can incorporate multiple inputs and non-linear stock behaviour is required. Here, we are extending our previous research by designing and implementing an improved fused model having a CNN designed for stock value pattern analysis and a sentiment analysis engine for stock related tweet and local news analysis. The key finding in the results obtained is that the fused model generates good quality predictions on comparison with the previously designed model. The novelty in our designed model is the idea of using multi-source inputs such as historical stock data, tweets data, news article data etc., and their processing using ensemble learning with various machine learning methods of sentiment analysis and stock value prediction using the leading technical indicators. For more clarity, we have tested and compared the model results with actual values which observed decent accuracy. The results from the fused model are future stock values and the direction of the stock.

Keywords: Ensemble learning, linguistic analysis, machine learning, multi-source, stock prediction, technical analysis.

Introduction

Stock market prediction is one of the most challenging areas of research due to its complex, uncertain and dynamic nature (Gidofalvi and Elkan, 2001; Mostafa et al., 1996). Traditional analysis of stock prediction using historical stock data and leading technical indicators has been one of the most significant methods, but since the exponential growth and innovation in the various machine learning (ML) methods and algorithms, researchers have shown immense interest in implementing the same (Diakoulakis et al, 2002; Vui et al., 2002) . Also, it has been observed that with time, several external factors influence the prediction, such as social media tweets, news articles published in the financial portals etc (Zhang, 2013; Mittal and Goel, 2012).

Hence, here we are proposing and implementing a new predictive model which works on the fused method as several inputs are considered for the prediction and the outputs are fused together to generate the final prediction. Here, we are focusing on the incorporation of news article data, tweets data and their sentiment analysis using ML methods which is then fused with the traditional technical analysis of historical stock data which is based on leading indicators and their calculations.

Here, in this paper, we have used the concept of ensemble learning which basically combines multiple predictions from multiple methods. The concept of ensemble learning uses various ML algorithms in order to generate better predictive output. All the ML algorithms which are combined in ensemble learning can be used initially, but the performance and accuracy of the overall system gets improved when used in ensemble learning mode as compared to the individual method implementation.

There are three main types of ensemble learning, namely bagging ensembles, boosting ensembles and stacking ensembles. Bagging ensembles are based on the usage of the high variance training models on different samples of the training dataset. Examples are Random Forest, Random Subspaces and Bagged Decision Trees etc. Boosting ensembles are based on adding models in linear mode to correct the prediction of the prior models. Examples are AdaBoost, Gradient Boosting Machine, and Stochastic Gradient Boosting etc. Stacking ensembles are based on the learning of combining the best predictions from various models. Examples are voting, weighted averaging, blending etc.

We have dedicated the following sections as mentioned here. Section II covers the related work in the area of stock prediction using ML techniques. Section III is dedicated towards the detailed explanation of architecture and the overall working of the proposed model. Section IV covers the important ML algorithms and

[a]charanjeet1506@gmail.com; [b]dr.asha.ambhailar@gmail.com

methodologies implemented used for prediction, section V covers the experimental results and their graphical representation and section VI covers the final conclusion of the paper including the related future work.

Related Work

A systematic review on various stock prediction methods in last decade has been covered in the paper by Dadiyala and Ambhaikar (2020). A new predictive pattern based model is designed and established where 25 leading technical indicators are used for the technical analysis of the stock using Open, Close, High, Low, Volume of stock are used to create new variable is discussed in (Dadiyala and Ambhaikar, 2021).

As mentioned in paper by Yuan et al. (2020) that the nature of stock market is very uncertain and complex where linear model assumption may be unreasonable and there is a need for a much improved non-linear. In paper by Zhang et al. (2018), multiple inputs were considered for the stock prediction as there are several other factors that affects the movement of the market. In this paper, they have used qualitative data, social media data and web news data for the prediction, making the input as multi-sourced. Whereas in paper by Bouktif et al. (2020), the authors focused on the concept of the predictability of stock market movement direction using an enhanced method of sentiments analysis. In paper by Wen et al. (2019), the authors implemented the idea of using CNN which is combined with two fully connected layers to capture the spatial correlations between historical and current trends and achieve an improvement in the performance. A new prediction model using CNN, LSTM and leading indicators with the aim of more accurate method for feature extraction and eventually better stock prediction is designed, implemented and discussed by Wu et al. (2021). A sentiment computing model based on a convolutional neural network is established to extract and quantify the emotional features is discussed by Zhang et al. (2020). A model to predict stock close values using historical data and proposes that including news data may add more accuracy to the predictive model is designed and implemented and discussed by Vijh et al. (2020).

Overall observation after the detailed literature review was that the previous models were designed and implemented using the either linear models or if non-linear, they restricted their model with one key design algorithm. Moreover, we identified that multiple methodologies combined with multiple input sources can lead to a better and dynamic model design. Hence, keeping all these ideas in mind, we designed a non-linear predictive model with the assumption of integrating multiple source inputs such as historical stock data, tweet data, news article data using various ML prediction algorithms and many sentiment analysis methods.

Working Model

Our proposed model is based on the idea of considering multiple inputs to be used for stock prediction (Figure 60.1). Every input is processed in their own suitable way and their intermediate results are then fused together to produce the final output which gives the future stock values and the direction of the stock.

Methods

There are several ML algorithms and methods implemented for the construction of our proposed model. Methods for fetching the data, sentiment analysis of the fetched data, calculating social media sentiment scores, prediction of stock price values, etc., are covered in this section.

Following Figure 60.2 summarises block-wise methods implemented and various user-defined methods are also systematically mentioned and explained in the following section.

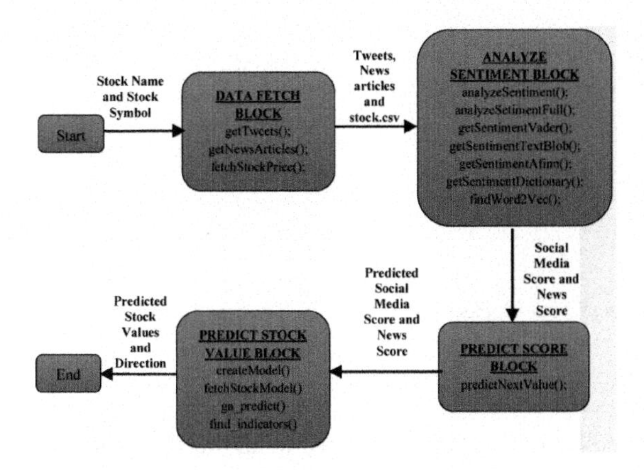

Figure 60.1 System architecture of the proposed fused model for stock prediction

Table 60.1 System blocks and their significant details

Sr. No.	*Block name*	*Input*	*Output*	*Explanation*
1.	Data Fetch block	Stock Name and Stock Symbol	Tweets, News Articles and a csv file with historical stock data	As the user enters the stock name and stock symbol, 100 tweets get fetched using the stock name as the hashtag with the help of Twitter API. With the reference of search query received, relevant news articles get fetched from leading financial portals using FeedBurner. Along with these social media data, historical stock data from 'Yahoo!Finance' website of the said stock (search query in the form of stock name and stock symbol) also gets generated and stored in a csv file named 'stock.csv'.
2.	Analyze Sentiment block	Tweets, News Articles and a csv file with historical stock data	Social Media Score and News Score	All the tweets and news articles fetched from the previous block are fed into this block which is designed for sentiment analysis of the tweets and news, resulting in the output of their scores. Here, we have used a total five methods of sentiment analysis, where 4 standard methods are namely, Vader, TextBlob, Afinn and Dictionary & 1 user-defined method is, namely MainWord2Vec which is an adaptation of Word2Vec. Each method results in their individual sentiment analysis score which is then fused into one final score. So, in the output we get one final score for tweets data namely 'social_media_score' and another final score for news articles data namely 'news_score'.
3.	Predict Score block	Social Media Score and News Score	Predicted Social Media Score & News Score	Here, using a simple linear regression method, new future values of social media score and news score are calculated using the social media score and news score of the data fetched. This block works on the simple idea of how a dependent variable predicts the independent variable.
4.	Predict Stock Value Block	Predicted Social Media Score and News Score	Predicted Future Stock Values and the direction	Here, the social media score and news score which get calculated in the previous block is considered for the final prediction purpose, which is combined with the stock prediction method using leading technical indicators. Here, the model of technical analysis has been extended from the previous paper and infused with the score values of social media and news articles predicted in the previous block. The final output of this block is the future stock values along with the movement of the direction

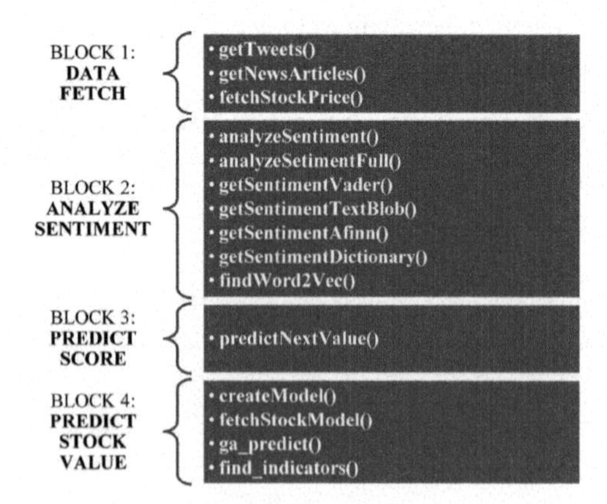

Figure 60.2 List of methods implemented in various system blocks

Design of Data Fetch Block

Data Fetch Block is primarily responsible for data collection and it is dedicated to all the algorithms implemented to fetch significant input data, such as Twitter data, news article data and historical stock data at the run time

This block has three methods, namely, getTweets(), getNewsArticles() and fetchStockPrice(). Here, getTweets() method is used to fetch latest tweets using the search query via Twitter API. Another method, getNewsArticle() is implemented to fetch and collect latest News Articles using the search query. Whereas, fetchStockPrice() method is mainly responsible for fetching the stock price values of user-defined stock at run time.

getTweets() method pseudocode

This method fetches social media tweets using Twitter API taking search query as the input.

Method	**: getTweets (search_Query, totalTweets, show=True)**
Input	: Search query (search_Query), total number of tweets (totalTweets)
Output	: tweet data from Twitter (tweet_data)

Begin

1. Import tweepy
2. Set consumer key, consumer secret, access token and access token secret to access Twitter API
3. Fetch the tweets using the hashtags for the given search query using Cursor() function
4. Return tweet_data

End

getNewsArticles() method pseudocode

This method fetches news articles using Feedburner taking search query as the input.

Method	**: getNewsArticles (searchQuery,show=True)**
Input	: Search query (search_Query)
Output	: Data from news articles and blogs (out_articles)

Begin

1. Import feedparser, re
2. Create feed links (feed_links) using FeedBurner of the prominent news blog articles and financial websites such as NDTV Profit, NDTV News, Gadgets460, NDTV Trending News etc.
3. Fetch news articles using parse() function of feedparser and append in out_articles.
4. Return articles data from news blogs (out_articles)

End

fetchStockPrice() method pseudocode

This method fetches historical stock data from Yahoo! Finance using stock symbol as the input. It fetches and stores the data in a csv file named 'stock.csv'.

Method	**: fetchStockPrice (stockSymbol)**
Input	: Symbol of the stock (stockSymbol)
Output	: Downloaded stock data

Begin

1. Import yfinance
2. Fetch stock data from a start date and an end date using the stock symbol and store it in a variable named 'data'.
3. Return data

End

Design of Analyse Sentiment block

This sub-section is dedicated to cover all the algorithms implemented towards en-semble learning of various ML sentiment analysis methods of all the fetched social media tweet data and news article data to get their sentiment score.

analyzeSentiment() method pseudocode

This method uses two sentiment analysis methods namely, getSentimentVader() and getSentimentAfinn(). Their individual sentiment score gets combined and appended in the python list.

Method	**: analyzeSentiment(statement)**
Input	: Text statement (statement)
Output	: Mean Sentiment Score of the statement (final_val)

Begin

1. Calculate sentiment score 1 using getSentimentVader() method.
2. Calculate sentiment score 2 using getSentimentAfinn() method.
3. Final value of score get calculated using conditional statement of:
   a. If (score1>=0) : score1= 'Positive', else score1='Negative'
   b. If (score2>=0) : score2= 'Positive', else score2='Negative'
   c. These score1 and score2 values get appended in the final python list named score[].
4. Store and return final calculated score as 'final_val'

End

analyzeSentimentFull() method pseudocode

This method uses five sentiment analysis methods namely, getSentimentVader(), getSentimentTextBlob(), getSentimentAfinn(), getSentimentDictionary() and findWord2Vec (). Their individual sentiment score gets combined and appended in the python list.

Method	**: analyzeSentimentFull(statement)**
Input	: Text statement (statement)
Output	: Sentiment Score of the statement (final_val)

Begin

1. Calculate sentiment score 1 using getSentimentVader() method.
2. Calculate sentiment score 2 using getSentimentTextBlob() method.
3. Calculate sentiment score 3 using getSentimentAfinn() method.
4. Calculate sentiment score 4 using getSentimentDictionary() method.
5. Calculate sentiment score 5 using findWord2Vec () method.
6. Calculate the final value of score using conditional statement of:
   a. If (score1>=0) : score1= 'Positive', else score1='Negative'
   b. If (score2>=0) : score2= 'Positive', else score2='Negative'
   c. If (score3>=0) : score3= 'Positive', else score3='Negative'
   d. If (score4>=0) : score4= 'Positive', else score4='Negative'
   e. If (score5>=0) : score5= 'Positive', else score5='Negative'
   f. Append the score values from score1 to score6 in the python list named score[].
7. Store and return final calculated score as 'final_val'

End

getSentimentVader() method pseudocode

This method uses sentiment intensity analyser to calculate the polarity of the sentences and to finally find the mean sentiment score of the input sentences.

Method	**: getSentimentVader(sentences)**
Input	: Text statement (sentences)
Output	: Mean Sentiment Score of the statement (avg_sentiment)

Begin

1. Tokenize the sentence using ntlk.sent_tokenize() method.
2. Select SentimentIntensityAnalyzer() as the analyzer
3. For every token in the sentence, loop:
   a. Calculate the strength of the token using polarity_scores() method
   b. Append the score in the output variable "sentiment_per_sentence".
   c. Calculate the mean of the output variable "sentiment_per_sentence" and store it in the "avg_sentiment" variable.
4. Return "avg_sentiment" value.

End

getSentimentTextBlob() method pseudocode

This method uses TextBlob() to calculate the sentiment score of the input sentences.

Method	: **getSentimentTextBlob(sentence)**
Input	: Text statement (sentences)
Output	: Sentiment Score of the statement (score)

Begin

1. Clean the sentence using clean_sentence() method
2. Calculate the score using TextBlob() method
3. Return the calculated score

End

getSentimentAfinn() method pseudocode

This method uses Afinn() to calculate the sentiment score of the input sentences.

Method	: **getSentimentAfinn(sentence)**
Input	: Text statement (sentences)
Output	: Sentiment Score of the statement (score)

Begin

1. Calculate the score using the Afinn() method
2. Return the calculated score

End

getSentimentDictionary() method pseudocode

This method uses Decision Tree Classifier to predict the sentiment score of the input sentences using a dictionary. Here, a sentiment word dictionary is adopted, which contains 8246 words and their polarity (positive or negative) is mentioned. It also covers whether the word is noun, adjective or pronoun.

Method	: **getSentimentDictionary(sentence)**
Input	: Text statement (sentences)
Output	: Predicted Sentiment Score of the sentence (predict_score)

Begin

1. Apply the CountVectorizer() method on the fetched tweet data
2. Predict the sentiment score by using a dictionary & apply Decision Tree Classifier
3. Store the output in variable 'predict_score'
4. Return predict_score.

End

findWord2Vec() method pseudocode

This method uses the findWord2Vec() from the open source library of genism to predict the sentiment score of the input sentences.

Method	**: findWord2Vec(sentences)**
Input	: Text statement (sentences)
Output	: Predicted Sentiment Score of the sentence (predict_score)

Begin

1. Calculate the sentiment score using the Word2Vec model of the open source library of gensim and store in the output variable 'predict_score'
2. Return predict_score.

End

Design of Predict Score Block

This sub-section is dedicated to cover all the algorithms implemented towards predicting the future sentiment scores of social media and news articles and to get their sentiment score.

predictNextValue() method pseudocode

This method uses Linear Regression to predict the sentiment score of tweet scores and news article scores.

Method	**: predictNextValue(data_vals)**
Input	: Social Media Score and News Score (Tweet_scores, article_scores)
Output	: Predicted Social Media Score and News Score (predict_social, predict_news)

Begin

1. Calculate the future social media score and news score using Linear Regression algorithm using LinearRegression() method.
2. Return predicted value

End

Design of Predict Stock Value Block

This sub-section is dedicated to cover all the significant algorithms implemented towards predicting the future stock values and the direction of the stock using all the processed inputs such as tweet data, news article data, tweet sentiment scores, news article sentiment scores, historical stock data and calculated leading technical indicators, including a newly designed CNN model used for prediction purpose.

createModel() method pseudocode

This method creates a newly designed CNN model with user-defined parameters as mentioned below. This model is then trained for the prediction and it is optimized using Adam optimizer. Cross entropy loss function and accuracy metrics is used for compiling the model.

Begin

1. Design a CNN model architecture of:
    a. Convolutional layers (convolution1D)
    b. Max Pooling layers (MaxPool1D of pool_size=2)
    c. Dropout (frequency rate=0.2)
2. The above step is repeated 4 times to add multiple layers to the architecture. (Nodes=16, 32, 32, 256 and kernel size = 5, 3, 3, 3)
3. Add two deeply connected dense layer with its preceding layer (nodes=64 & activation=relu), followed by the output layer (filter=nclass, activation=softmax)
4. Compile the model with three parameters: optimizer=Adam, loss= cross entropy and metrics = accuracy.
5. Train the model
6. Predict the output using the trained model

End

Architecture of CNN Model

This section covers the overall architecture of the designed CNN model which is summarized in Table 60.2 and it is further depicted in Figure 60.3.

Table 60.2 Designing of the CNN model

Step	*Explanation*
Input	The model is firstly instantiated using the input shape, which specifies the shape of the incoming data which is a 1D array of values. Here, inp=Input(shape=(187,1)) is used.
Creating Input layer	Here, an input layer is added to the model using the Convolution1D method and relu as the activation function.
Creating Hidden layer	Here, several hidden layers are added to the model using the Convolution1D method and relu as the activation function
Creating Output layer	Finally, the output layer was added to the model using the Convolution1D method and softmax as the activation function
Summarizing the model	The model was summarized using model.summary(); and the output is shown in table 60.2. This function returns the detailed textual information about the shape of layers and number of weights in every layer.
Compiling the model	Here, we have specified a loss function of Cross Entropy, which will be used for the model to understand how well it is performing during the training. We have specified metrics as 'accuracy' which calculates the accuracy of the model using simple calculation of dividing the number of accurately predicted records by the total number of records.
Optimiser	Here, Adam Optimizer is used as it is capable of handling noisy and sparse data. It is also touted as one of the best and most popular optimisation methods
Global Max Pooling	It is a pooling operation that selects the maximum element from the region of the feature map covered by the filter. The output of the MaxPoling is a feature map containing the most prominent features of the previous feature map. MaxPool is used with the pool size of 2.
Dropout	It is a technique used to prevent our model from over fitting. It is basically one of the popular regularisation techniques. Here, dropout is used with the frequency rate of 0.1.

The following figure (Figure 60.3) depicts the basic architecture of the designed CNN Model.

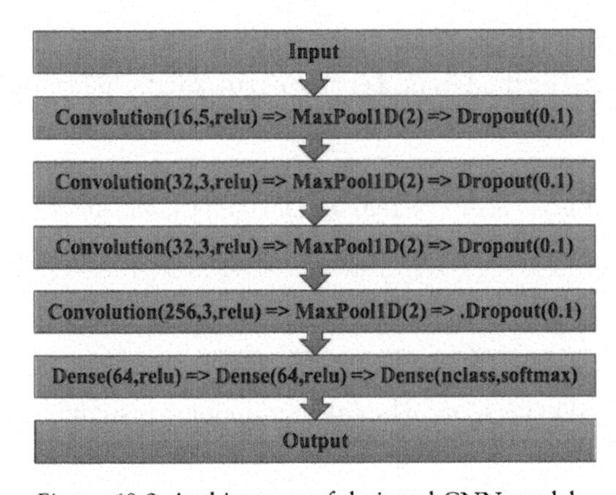

Figure 60.3 Architecture of designed CNN model

The below table (Table 60.3) shows the overall textual summary of the model created, such as number of layers, weights in each layer, activation functions used etc..

fetchStockModel() method pseudocode

Method	*:* fetchStockModel(csv_file, num_values, direction)
Input	*:* A csv file of stock data, total number of prediction and the direction
Output	*:* Future Stock Values

Table 60.3 Summary of the CNN model

Layer (type)	*Output Shape*	*Param #*
input_1 (InputLayer)	[(None, 187, 1)]	0
conv1d (Conv1D)	(None, 183, 16)	96
conv1d_1 (Conv1D)	(None, 179, 16)	1296
max_pooling1d (MaxPooling1D)	(None, 89, 16)	0
dropout (Dropout)	(None, 89, 16)	0
conv1d_2 (Conv1D)	(None, 87, 32)	1568
conv1d_3 (Conv1D)	(None, 85, 32)	3104
max_pooling1d_1 (MaxPooling1D)	(None, 42, 32)	0
dropout_1 (Dropout)	(None, 42, 32)	0
conv1d_4 (Conv1D)	(None, 40, 32)	3104
conv1d_5 (Conv1D)	(None, 38, 32)	3104
max_pooling1d_2 (MaxPooling1D)	(None, 19, 32)	0
dropout_2 (Dropout)	(None, 19, 32)	0
conv1d_6 (Conv1D)	(None, 17, 256)	24832
conv1d_7 (Conv1D)	(None, 15, 256)	196864
global_max_pooling1d (Global)	(None, 256)	0
dropout_3 (Dropout)	(None, 256)	0
dense_1 (Dense)	(None, 64)	16448
dense_2 (Dense)	(None, 64)	4160
dense_3 (Dense)	(None, 5)	325

Total params: 254,901
Trainable params: 254,901
Non-trainable params: 0

Begin

1. Generate the leading technical indicators calculations using ***findIndicators()*** method and store it in the '***indicators***' variable.
2. Create a new CNN model using ***createModel()*** method.
3. Predict the future stock values using ***ga_predict()*** method using a csv file of stock data, indicators and direction values.

End

(Note: ga_predict() and findIndicators() methods are published and explained in the paper by Dadiyala and Ambhaikar, (2021).)

Calculation of the Stock direction

The concept of generating the direction of the future predicted stock values is based on the sentiment score of two values, predicted social media score and predicted news score. The calculation logic is summarised in the table below: (Table 60.4)

Table 60.4 Summary of stock direction calculation

Condition No.	*Predict_social (ps)*	*Predict_news (pn)*	*Direction*	*Label*
	Score			
1.	ps > 1	pn > 1	2	Positive
2.	ps > 1	pn = 0	1	Positive
3.	ps = 0	pn > 1	1	Positive
4.	ps > 0	pn > 0	1	Positive
5.	ps = 0	pn < 0	-1	Negative
6.	ps < 0	pn = 0	-1	Negative
7.	ps < 0	pn < 0	-2	Negative

Experimental Outputs and Discussion

For the following experiments, we have used the following values as mentioned:

1. Total tweets = 100
2. Value predicted = close
3. Dataset d= 01/12/2020 - 01/12/2021
4. Total number of news articles fetched depends on the availability of news articles using the search query

Experiment No.1

Stock Name = Maruti Suzuki, Stock Symbol = MARUTI.NS: Here, we observed that the total number of tweets fetched were 100, and the total number of news articles fetched was 12. The tweet data and news data were analysed using ensemble learning of sentiment analysis and the output is plotted and shown in (b) and (d) respectively. Now, these two individual sentiment scores were used to further calculate the combined score, shown in (e). Now, this combined score is used along with the technical analysis and final values were predicted and shown in (f). The final five predicted values are plotted in (g) and the final output generated by the model is shown in (h). Here, we observed that our model predicted the future values with decent accuracy and performed well with the tweet data, news data and historical stock data. The results and graphs are depicted in Table 60.4.

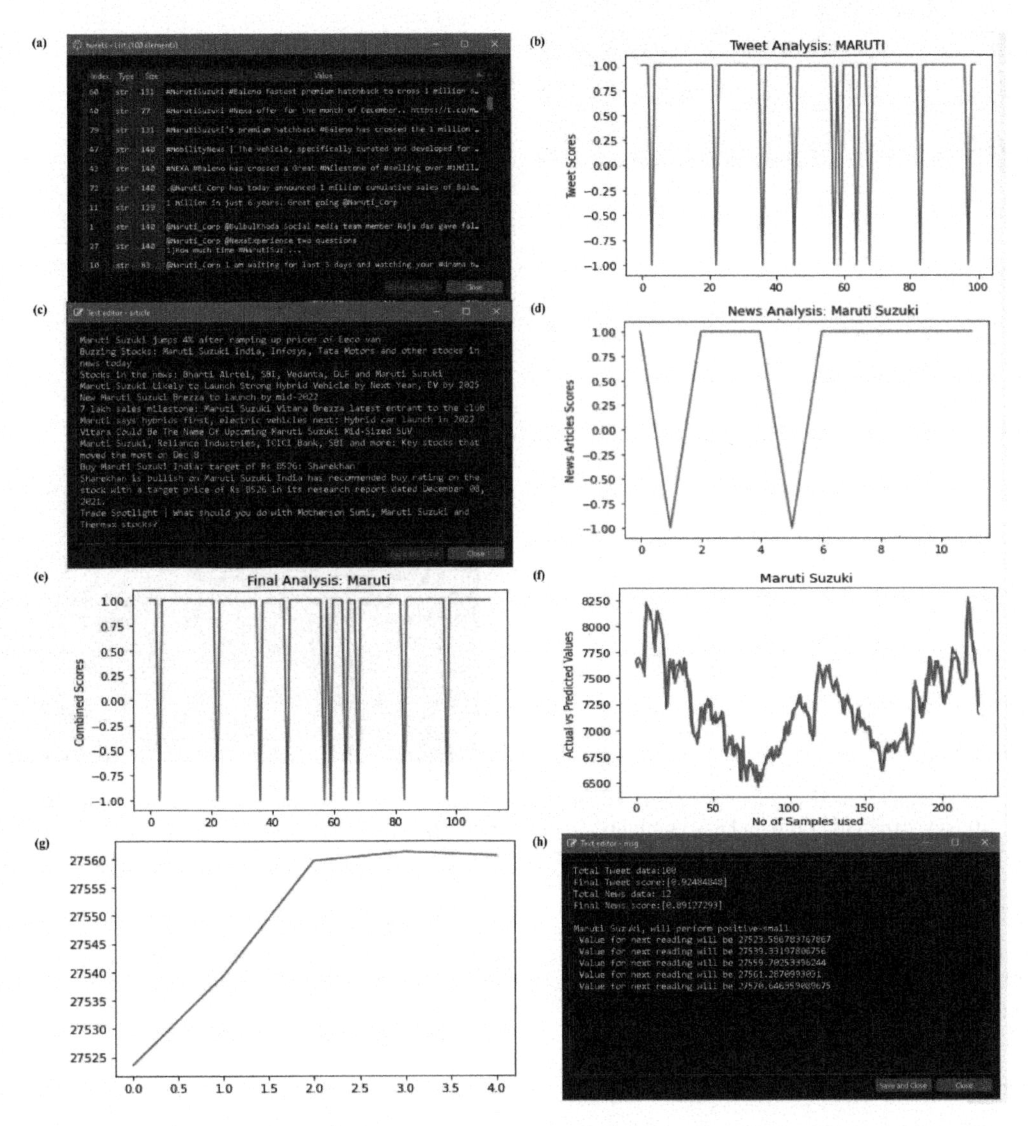

Figure 60.4 Experimental results case 1: (a) Tweets Fetched Screen (b) Tweet analysis graph (c) News fetched screen (d) News analysis graph (e) Final combined analysis graph (f) Actual vs predicted values graph (g) Direction prediction graph with predicted values (h) Final output screen

Experiment No. 2

Stock Name = Tesla, Stock Symbol=TSLA: Here, we observed that the total number of tweets fetched were 100, and the total number of news articles fetched was 23. The results are shown in the table below (Figure 60.5)

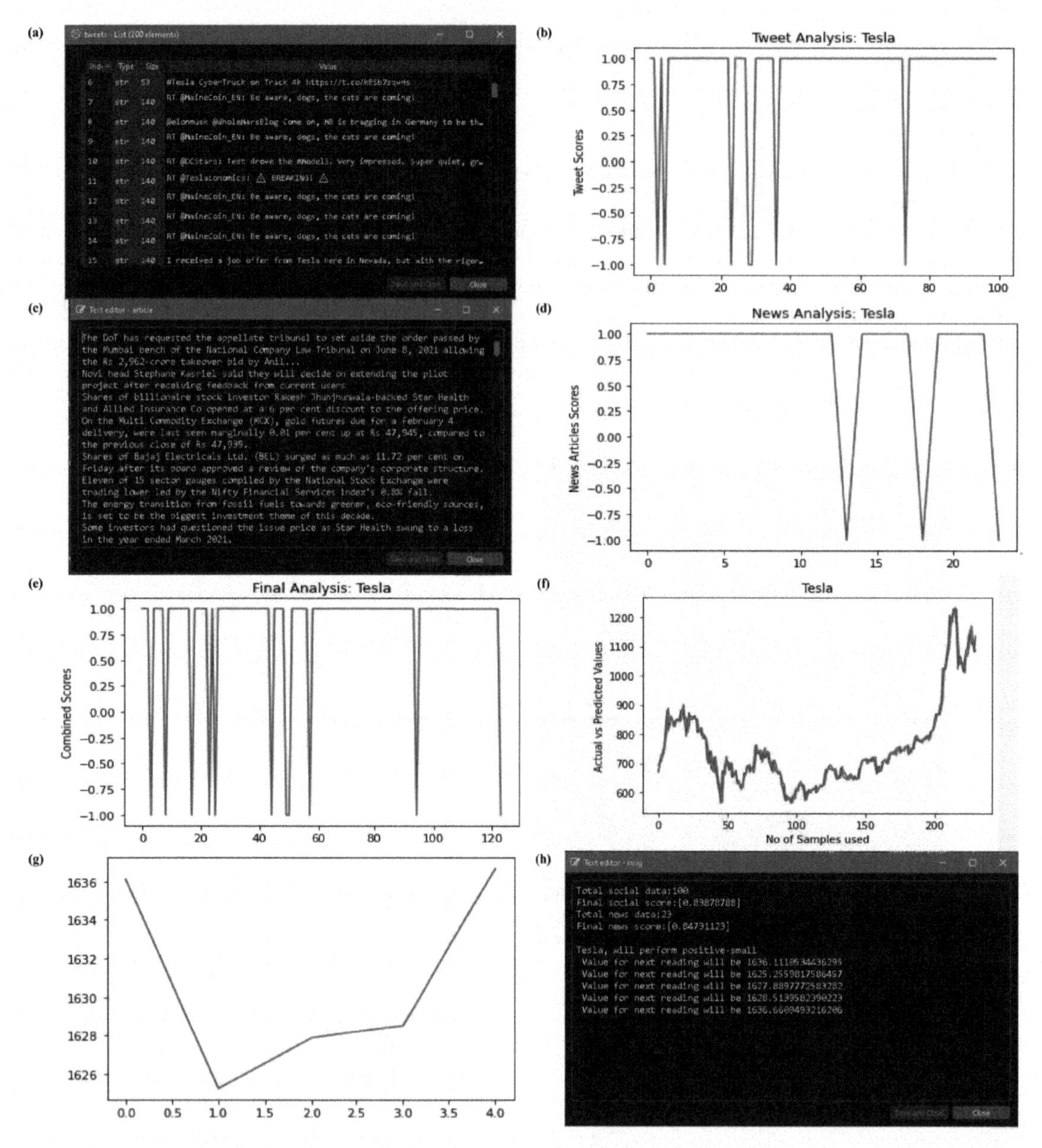

Figure 60.5 Experimental results case 2: (a) Tweets Fetched Screen (b) Tweet analysis graph (c) News fetched screen (d) News analysis graph (e) Final combined analysis graph (f) Actual vs predicted values graph (g) Direction prediction graph with predicted values (h) Final output screen

Experiment No. 3

Stock Name= Amazon, Stock Symbol=AMZN: Here, we observed that the total number of tweets fetched were 200, and the total number of news articles fetched was 9. The results are shown in the table below (Figure 60.6)

Result and Discussion

For the results evaluation, we have compared actual stock values with the predicted values given by the model. Here, simple error has been calculated and then later the accuracy is expressed in percentage. While testing the model, we have witnessed minimum and maximum accuracy of 79.9868154% and 85.8412%

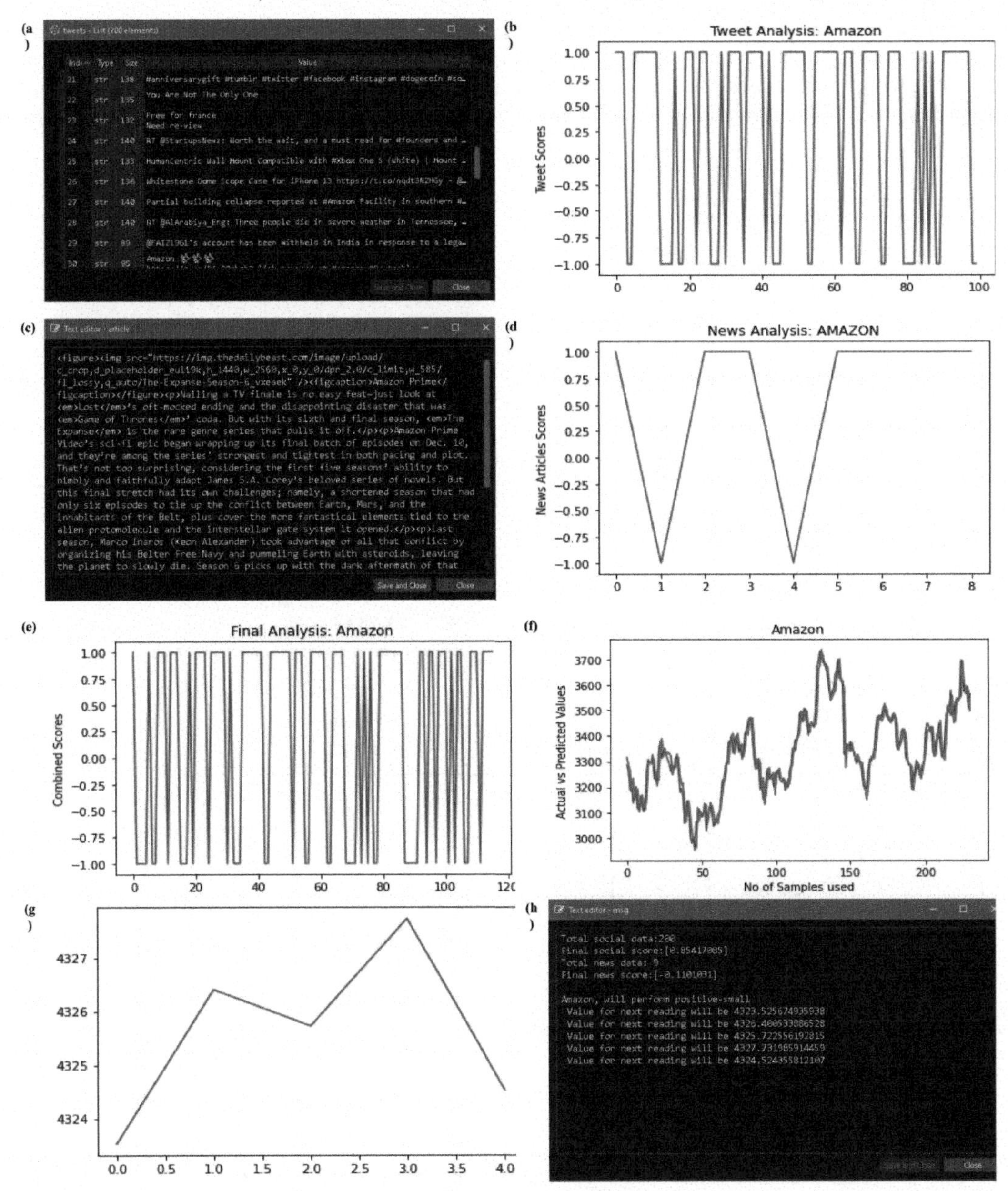

Figure 60.6 Experimental results case 3: (a) Tweets Fetched Screen (b) Tweet analysis graph (c) News fetched screen (d) News analysis graph (e) Final combined analysis graph (f) Actual vs predicted values graph (g) Direction prediction graph with predicted values (h) Final output screen

respectively, and hence those three experimental cases have been included in the paper. Also, we can conclude that our model performs with an average accuracy of 82.056%. The detailed description of the evaluation are covered in the following Tables 60.5–60.7. The following table shows the stock prediction average accuracy of 85.8412% which is related to Experiment no 1 of Maruti Suzuki. Its results and relevant output graphs are shown in Table 60.5 and Figure 60.7.

Table 60.4 Accuracy calculation of experiment No 1

Sr no	*Actual values*	*Predicted values*	*Calculated accuracy*
1	27521.371	27523.586	84.785
2	27535.021	27539.331	85.69
3	27557.120	27559.705	87.416
4	27559.141	27561.287	87.854
5	27574.107	27570.646	83.461

The following table shows the stock prediction average accuracy of 80.338588% which is related to Experiment no 2 of Tesla. Its results and relevant output graphs are shown in Table 60.6 and Figure 60.8.

The following table shows the stock prediction average accuracy of 79.9868154% which is related to Experiment no 3 of Amazon. Its results and relevant output graphs are shown in Table 60.7 and Figure 60.9.

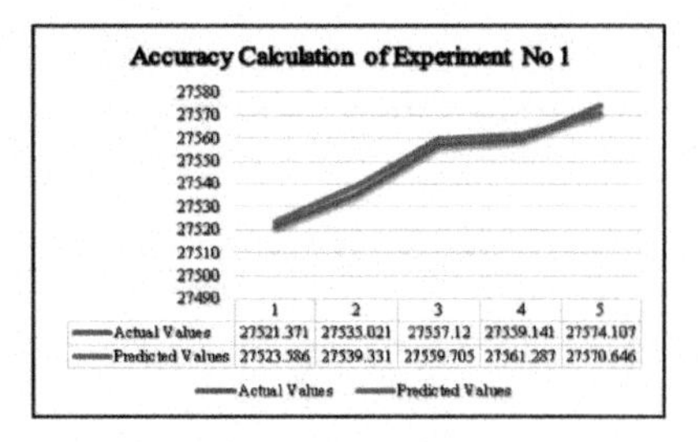

Figure 60.7 Exp 1: Actual vs predicted values graph plot

Table 60.5 Accuracy calculation of experiment No 2

Sr no	*Actual values*	*Predicted values*	*Calculated accuracy*
1	1613.36295	1636.11105	77.2519
2	1609.86457	1625.25598	84.60859
3	1617.83282	1627.88977	89.94305
4	1602.90223	1628.51395	74.38828
5	1612.16206	1636.66094	75.50112

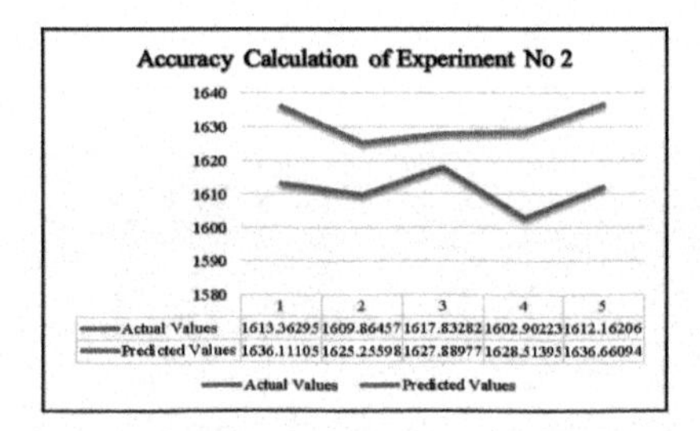

Figure 60.8 Exp 2: Actual vs predicted values graph plot

Table 60.6 Accuracy calculation of experiment No 3

Sr no	*Actual values*	*Predicted values*	*Calculated accuracy*
1	4301.839539	4323.525674	78.313865
2	4305.825688	4326.400533	79.425155
3	4306.518296	4325.722556	80.79574
4	4308.954439	4327.731985	81.222454
5	4304.701218	4324.52435	80.176863

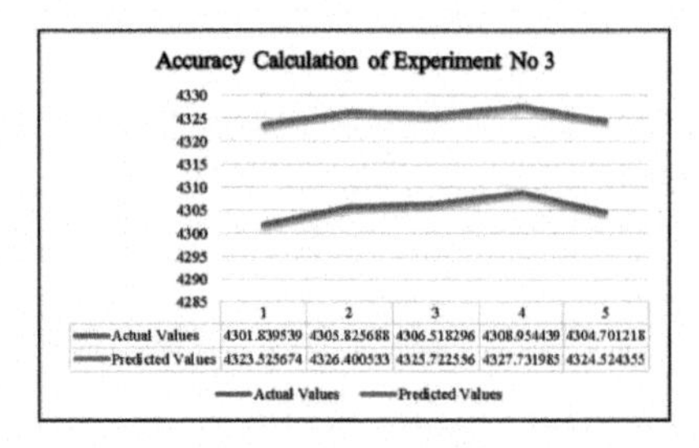

Figure 60.9 Exp 3: Actual vs predicted values graph plot

Conclusion and Future Scope

The purpose of this research was to predict future stock values and the market movements using ML techniques. Three leading companies namely, Tesla, Amazon and Maruti Suzuki were selected for the overall testing and training of the model. The dataset was generated for a year having historical stock data along with 25 leading technical indicators. The input also includes 100 tweets fetched using Twitter API and several news articles fetched from leading financial news portals using FeedBurner.

Many ML algorithms were implemented as the predictors such as CNN, Genetic Algorithm and few regression methods. Total 5 sentiment analysis methods were used for the processing of the tweets and news articles, namely Vader, TextBlob, Afinn, and Dictionary & Word2Vec as the part of ensemble learning.

Our experimental results depicted in the Tables 60.5–60.7 and Figures 60.7–60.9, clearly showed that there was a significant accuracy and improvement in the overall prediction of the stock values. The results observed good accuracy after the comparison with actual values and the generated values.

The idea of comparison of actual vs predicted values are kept with the concept of training and testing the model as per the standard method. We have first trained the model with training data, then tested the model with testing data and finally tested the model using real-time input values. The comparison is actually to show that the model is generating quality prediction within the range of 7980% accuracy, which made us conclude that the design is looks decent and promising. In near future, this accuracy can be improved by re-tuning the model by incorporating many hyperparameters, adding new elements and significant commodities and global input values such as Global Currency Rate, Global News, Commodity Impact, Global Stock Impact, etc. can be incorporated and a more refined model can be developed.

References

Dadiyala, C. and Ambhaikar, A. 2021. Technical Analysis of Pattern Based Stock Prediction Model Using Machine Learning. International Conference on Innovative Computing, Intelligent Communication and Smart Electrical Systems (ICSES), 2021, (pp. 19). doi: 10.1109/ICSES52305.2021.9633961.

Dadiyala, C. and Ambhaikar, A. (2020). Analysis and Survey of Stock Market Prediction Techniques over the last decade.

Wu, J.MT., Li, Z., Herencsar, N., Vo, B., and Lin, J., C-W. (2021). A graph-based CNN-LSTM stock price prediction algorithm with leading indicators. Multimed. Sys. https://doi.org/10.1007/s00530-021-00758-w (2021) [Accessed on 20th December 2020].

Yuan, X., Yuan, J., Jiang T., and Ain, Q. U. (2020). Integrated Long-Term Stock Selection Models Based on Feature Selection and Machine Learning Algorithms for China Stock Market. IEEE Access. 8:22672–22685. doi: 10.1109/ACCESS.2020.2969293. (2020)

Zhang, W., Tao, K. X., Li, J. –F., Zhu, Y. -C., and Li, J. (2020). Modeling and Prediction of Stock Price with Convolutional Neural Network Based on Blockchain Interactive Information. Wirel. Commun. Mob. Comput. 2020:10. https://doi.org/10.1155/2020/6686181

Vijh, M., Chandola, D., Tikkiwal, V., and Kumar, A. (2020). Stock Closing Price Prediction using Machine Learning Techniques. Procedia Comput. Sci. 167:599–606. doi: 10.1016/j.procs.2020.03.326.

Bouktif, S., Fiaz, A., Awad, M. (2020). Augmented Textual Features-Based Stock Market Prediction. IEEE Access. 8:40269–40282. doi: 10.1109/ACCESS.2020.2976725.

Wen, M., Li, P., Zhang, L., and Chen, Y. (2019). Stock Market Trend Prediction Using High-Order Information of Time Series. IEEE Access. 7:28299–28308. doi: 10.1109/ACCESS.2019.2901842.

Zhang, X., Qu, S., Huang, J., Fang, B., and Yu, P. (2018). Stock Market Prediction via Multi-Source Multiple Instance Learning. IEEE Access. 6:50720–50728. doi: 10.1109/ACCESS.2018.2869735.

Bin Yang, Wei Zhang, and Haifeng Wang (2019), "Stock Market Forecasting Using Restricted Gene Expression Programming", Computational Intelligence and Neuroscience, Volume 2019, Article ID 7198962, 14 pages, https://doi.org/10.1155/2019/7198962

Shah, Dev, Campbell Wesley, and Zulkernine Farhana(2018) " A Comparative Study of LSTM and DNN for Stock Market Forecasting", Paper presented at the 2018 IEEE International Conference on Big Data (Big Data), Seattle, WA, USA, December 10–13

Zhang, Jing, Shicheng Cui, Yan Xu, Qianmu Li, and Tao Li. (2018). "A novel data-driven stock price trend prediction system". Expert Systems with Applications 97: 60–69

Hiransha, M., E. A. Gopalakrishnan, Vijay Krishna Menon, and Soman Kp. (2018) "NSE stock market prediction using deep-learning models". Procedia Computer Science 132: 1351–62

Arévalo, Rubén, Jorge García, Francisco Guijarro, and Alfred Peris. (2017). "A dynamic trading rule based on filtered flag pattern recognition for stock market price forecasting". Expert Systems with Applications Expert Systems with Applications 67: 126–39

Pagolu, Venkata Sasank, Kamal Nayan Reddy, Ganapati Panda, and Babita Majhi. (2016). "Sentiment Analysis of Twitter Data for Predicting Stock Market Movements". Paper presented at the 2016 International Conference on Signal Processing, Communication, Power and Embedded System (SCOPES), Paralakhemundi, India, October 3–5.

Dey, Shubharthi, Yash Kumar, Snehanshu Saha, and Suryoday Basak. (2016). "Forecasting to Classification: Predicting the Direction of Stock Market Price Using Xtreme Gradient Boosting"

Peachavanish, Ratchata (2016) "Stock selection and trading based on cluster analysis of trend and momentum indicators". Paper presented at the International Multi Conference of Engineers and Computer Scientists, Hong Kong, China, March 16–18.

Rather, Akhter Mohiuddin, Arun Agarwal, and V. N. Sastry. (2015). "Recurrent Neural Network and a Hybrid Model for Prediction of Stock Returns". Expert Systems with Applications 42: 3234–41.

Ding, Xiao, Yue Zhang, Ting Liu, and Junwen Duan. (2015). "Deep Learning for Event-Driven Stock Prediction". Paper presented at the 24th International Conference on Artificial Intelligence (IJCAI), Buenos Aires, Argentina, July 25–31

Cakra, Yahya Eru, and Bayu Distiawan Trisedya. (2015). "Stock Price Prediction Using Linear Regression Based on Sentiment Analysis" Paper presented at the 2015 International Conference on Advanced Computer Science and Information Systems (ICACSIS), Depok, Indonesia, October 10–11.

Box, George E. P., Gwilym M. Jenkins, Gregory C. Reinsel, and Greta M. Ljung. (2015). "Time Series Analysis: Forecasting and Control" Hoboken: John Wiley & Sons

Ballings, Michel, Dirk Van den Poel, Nathalie Hespeels, and Ruben Gryp. (2015). "Evaluating multiple classifiers for stock price direction prediction". Expert Systems with Applications 42: 7046–56

Kalyanaraman, Vaanchitha, Sarah Kazi, Rohan Tondulkar, and Sangeeta Oswal. (2014). "Sentiment Analysis on News Articles for Stocks". Paper presented at the 2014 8th Asia Modelling Symposium (AMS), Taipei, Taiwan, September 23–25

Lee, Heeyoung, Mihai Surdeanu, Bill MacCartney, and Dan Jurafsky. (2014)."On the Importance of Text Analysis for Stock Price Prediction". Paper presented at the 9th International Conference on Language Resources and Evaluation, LREC 2014, Reykjavik, Iceland, May 26–31

Porshnev, Alexander & Redkin, Ilya & Shevchenko, Alexey. (2013). "Machine Learning in Prediction of Stock Market Indicators Based on Historical Data and Data from Twitter Sentiment Analysis". 440-444. 10.1109/ ICDMW.2013.111.

Zhang, L. (2013). Sentiment Analysis on Twitter with Stock Price and Significant Keyword Correlation. Austin: The University of Texas at Austin.

Mittal, A. and Goel, A. (2012). Stock Prediction Using Twitter Sentiment Analysis. Stanford, California: Standford University.

Bernal, Armando, Sam Fok, and Rohit Pidaparthi (2012) "Financial Market Time Series Prediction with Recurrent Neural Networks".

J. Bollen, H. Mao and X. Zeng (2011) "Twitter mood predicts the stock market", Journal of Computational Science, 2(1), pp.1–8, 2011

Tiwari, Shweta, Rekha Pandit, and Vineet Richhariya. (2010). "Predicting Future Trends in Stock Market by Decision Tree Rough-Set Based Hybrid System with Hhmm" International Journal of Electronics and Computer Science Engineering 1: 1578–87.

D. Harper (2012) "Forces that Move Stock Prices" [Online] Available at: http://www.investopedia.com/ articles/ basics/ 04/ 100804. asp [Accessed on 27th October 2016]

Y.Chen and J.Xie (2008) "Online Consumer Review: Word of Mouth as a New Element of Marketing Communication Mix", Management Science, March 2008, vol 54, no 3, 477–491, 2008.

S.Das and M.Chen, (2007) "Yahoo! for Amazon: Sentiment extraction from small talk on the web", Management Science, 53(9): 1375-1388, 2007

Fu, Tak-chung, Fu-lai Chung, Robert Luk, and Chak-man Ng (2005) "Preventing Meaningless Stock Time Series Pattern Discovery by Changing Perceptually Important Point Detection" Paper presented at the International Conference on Fuzzy Systems and Knowledge Discovery, Changsha, China, August 27–29.

Thissen, U., Van Brakel, R., De Weijer, A. P., Melssen, W. J., & Buydens, L. M. C. (2003). "Using support vector machines for time series prediction", Chemo metrics and intelligent laboratory systems, 69 (1), 35-49.

Schölkopf, B. and Smola, A. J. (2002). Learning with kernels: support vector machines, regularization, optimization, and beyond. MIT press.

Diakoulakis, I. E., Koulouriotis, D. E., and Emiris, D. M. (2002). A Review of Stock Market Prediction Using Computational Methods. In: Kontoghiorghes E.J., Rustem B., Siokos S. (eds) Computational Methods in Decision-Making, Economics and Finance. Applied Optimization, vol 74. Springer, Boston, MA

Vui, C. S., Soon, G. K., On, C. K., Alfred, R., and Anthony, P. (2002). A review of stock market prediction with Artificial neural network (ANN). 2013 IEEE International Conference on Control System, Computing and Engineering, Mindeb., 2013, pp. 477–482. doi: 10.1109/ICCSCE.2013.6720012

Gidofalvi, G. and Elkan, C. (2001). Using news articles to predict stock price movements. Department of Computer Science and Engineering. San Diego: University of California.

Mostafa, A., Yaser S., and Atiya, A. F. (1996). Introduction to financial forecasting". Applied Intelligence 6: 205–13.

Boser, B. E., Guyon, I. M., &Vapnik, V. N. (1992)." A training algorithm for optimal margin classifiers". In Proceedings of the fifth annual workshop on Computational learning theory (pp. 144–152). ACM

61 Simultaneous electrodeposition and electrochemical machining process planning methodology

Vimal Kumar Deshmukh[a], *Sanju Verma*[b], *Nihal Pratik Das*[c], *Mridul Singh Rajput*[d], *and H.K. Narang*[e]

Department of Mechanical Engineering, National Institute of Technology Raipur, Raipur, India

Abstract

Nanotechnology is an area attracting the attention of many research and industrial branches. Nanomaterials have several advantages over bulk materials such as the huge surface-to-volume ratio and very high porosity properties but it is very difficult to control atoms at the nano or extreme micro-level. This paper presents a newly developed process planning to produce nano–microscale features with average deposition structure in the range of 200 nm to 600 nm. A five-stage process plan has been developed. These stages are namely i) sample preparation ii) jet electrodeposition iii) post-deposition treatment iv) jet electrochemical machining v) repetition of the step (ii) to step (iv) till getting desired thickness and result with the post-machining process. The deposited structure has been characterised by field emission scanning electron microscopy images. The FESEM result shows that the deposited feature had a globule structure and size in the range of 200nm to 600nm. The result confirmed that presented process planning has significant potential to fabricate nano-micro scale features with minimal human interference as changing of reference tool is only human interference. The proposed process is rapid and has good accuracy.

Keywords: Copper, electrodeposition, electrochemical machining, micro, nanomaterial, process planning.

Introduction

As the new electronic devices are getting shorter and advance, the new process planning is required to stay in competitive market and become sustainable. The fabrication of thin micro feature requires a process that has control at molecular level. Jet electrochemical deposition and jet electrochemical machining are such two processes that have competitive gain over control on transfer of atoms on micro-scale. The manufacturers have been using electroplating for coating desired material to prevent oxidation, abrasion for the last few decades (Sadana and Nageswar, 1984). The wide application also includes decorative metallic mirror reflection, lubrication and enhancing electrical conductivity. The integration of jet with the electrodeposition process, keeping the scientific principle unaltered, facilitates the manufacturers to build selective deposited features on micro as well Nano level (Rajput et al., 2014). In the era of nanoelectronics and system, having control over atomic size and structure provide an edge to the manufacturing industries to survive in a competitive market. In the past few decades, advanced manufacturing techniques such as jet electrodeposition (Aouaj et al., 2015), additive manufacturing (Blakey-Milner et al., 2021), rapid prototyping etc. have been evolved which can produce micro or Nano-size particles (Dover et al., 1996). Few literatures had been reported the use of ultrasonic energy (Pandiyarajan et al., 2021), chemically reactive agents, acidic and basic solution, laser energy (DeSilva et al., 2004), pulsing of power supply in different electrodeposition methods individually. Simultaneous use of deposition and machining of parts also had been utilized for rapid prototyping by developing a five-axis translation stage (Kim et al., 2002). In Cu–Al_2O_3 composite coating preparation using jet electrodeposition it is seen that increase in Al2O3 nanoparticles in the electrolyte was able to refine the composite coating grains and enhance the deposit inner-structure as well as compactness, which considerably strengthens the composite coating's mechanical property (Fan et al., 2019). In copper coating by pulsed jet electrodeposition decrease in duty cycle significantly improved the coating surface morphology and microstructure (Fan et al., 2020). It had been observed from the literature that there is rarely any research available that reported simultaneous use of jet electrodeposition, electrochemical machining process, ultrasonic energy and pulsed electrodeposition to obtain Nano- micro features. This study is to propose a novel approach to build micro scale features using repetitive jet electrodeposition and electrochemical machining process. A setup has been prepared for jet electrodeposition of Cu by ultrasonic-assisted jet electrodeposition with the help of G-code for XYZ translator and then jet

[a]deshmukh.vimal1920@gmail.com; [b]sanjuverma.sv.sv@gmail.com; [c]nihalpratik2@gmail.com; [d]msrajput.me@nitrr.ac.in; [e]hnarang.me@nitrr.ac.in

electrochemical machining of the deposited particle by M-code for electrochemical machining. The comprehensive process planning method with electrodeposition and electrochemical machining is the prime subject of presented study.

The kinematics of jet electrodeposition

The metallic ion deposition rate follows Faraday's law. Faraday's law states that the extent of occurring the electrochemical process is a direct function or proportional to the electric charge passed. The material deposition rate is a function of current density (J, in mA-cm^{-2}). Each metal has a material density (ρ, in g-cm^{-3}), electrochemical equivalent constant (Z) also called proportionality constant which relates deposition to current density. The linear growth rate and mass deposition rate can be computed by the below relationship.

$$\text{Linear growth rate} = 3.6 \times 10^4 \times \frac{Z \times CE \times J}{\rho} \quad (1)$$

$$\text{Mass flow rate } (\dot{m}) = \frac{IM}{FZ} \quad (2)$$

where CE represents current efficiency, CE has a value nearer to 100 for copper electrodeposition with copper densities (J) up to 50 mA-cm^{-2} but unlike Au and Ni, CE doesn't change with current density (Hui et al., 2019). The electrochemical equivalent constant Z can be calculated as

$$Z = \frac{M}{nF} \quad (3)$$

Where F represents Faraday's constant, 96485 C-mol^{-1}. n represents the number of electrons participating in process M represents the molar weight of metal. The maximum limit of material deposition rate is governed by mass diffusion or transportation and species of metal ions on the cathode surface. I represent the current in mA and A represents the exposed area in cm^2 for electroplating and can be calculated by

$$I = JA \text{ (mA)}$$

In HSSJED, the process is controlled by potential instead of the current, it becomes beneficial in case of continuously changing plated areas. The conventional model given by John Alfred Valentine Butler and Max Volmer predicts the electrode current density (j) by Erdey-Grúz–Volmer equation also known as Butler-Volmer equation.

$$J = J_0 \left[e^{\left(\frac{F\alpha\eta}{RT}\right)} - e^{\frac{(1-\alpha)F\eta}{RT}} \right] \quad (4)$$

Cathode process term   Anodic process term

Where J represents the current density of the electrode and J_0 is the exchange current density and represents the transfer coefficient. In Nernst Equation, the is called activation potential which is a difference of electrode potential (E) and equilibrium potential (Eeq). Mathematically it is

$$\eta = E - Eeq$$

Methodology

The proposed methodology is shown in Figure 61.1. The foremost step was to prepare a sample on which the fabrication of the micro feature would process will take place. The methodology involved five major steps i) sample preparation ii) jet electrodeposition iii) post-deposition treatment iv) jet electrochemical machining v) repeat the step (ii) to step (iv) till getting desired thickness and result.

A. Sample preparation

When it comes to electrodeposition, electronegativity and electrical conductivity are critical properties of materials. Instead of electro-positivity, electronegativity is used as a measurement technique for elements. Because high electropositive material is more likely to be used on the anode to form anions, high electronegative material should be used on the cathode. However, conductivity is also critical while performing electrodeposition. The electronegativity of nonmetallic elements is higher. In the periodic table, electronegativity increases from left to right in groups. However, due to rising atomic radii in periods, it decreases from top to bottom. Because of the electron affinity and the order in which metals are replaced in the periodic table, the cathode and anode materials cannot be chosen at random. Some metals, such as silver, cannot be electrodeposited directly onto other metals, such as iron. The silver (Ag) covering will peel away in thin layers. Instead, copper is electroplated on the iron first, and then a layer of silver is electrodeposited on top (Hui, et al., 2020). Adhesiveness, atomic size difference electron affinity, atomic bonding, and intermolecular interactions are some of the other key parameters. The characteristics such as ample availability, non-hazardous in nature, malleable and good electrical properties are desirable in electrodeposition. Copper has all these characteristics which make it one of the most suitable elements for electrodeposition. For the present study, copper is selected as desired material and copper is the substrate material. The wide application and low cost make the copper suitable for most engineering and electronic application. A copper thin sheet of 0.2 mm thickness has been taken for sample preparation. The one side of the sheet was stuck on a non-conductive base while keeping the other side open for the electrodeposition process.

B. Designing of product

The process had been initiated with the designing of the product part features in CAD software. The dimensions of the part feature were designed in the range of XYZ translator.

Figure 61.2 shows the part design of the standard rectangular shape. The features which require micromachining operations e.g., holes and slots have been extracted. The part file would be the initial input for the proposed system. The CAD system can also generate program g-codes and coordinates for the 3-axis linear translator which was attached to the nozzle assembly.

C. UAJE deposition

The cathodes used were copper plates this is a work piece in which $Cu^{2}+$ will be deposited and the anode is also a copper plate both anode and cathode are submerged in an electrolyte. The electrolyte was prepared by copper sulphate pentahydrate ($CuSO_4 \cdot 5H_2O$) and sulfuric acid ($H2SO_4$). The composition of copper

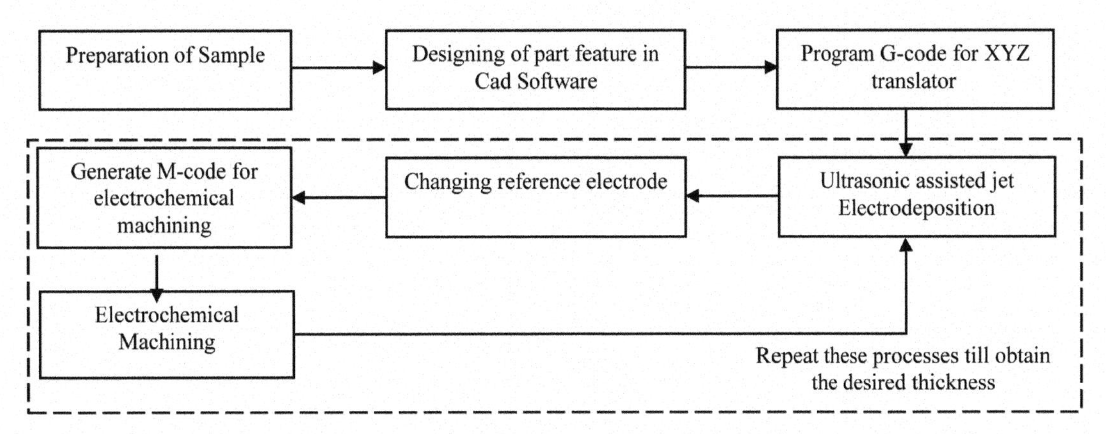

Figure 61.1 Proposed methodology to fabricate micro features using simultaneous electrodeposition and electro-machining

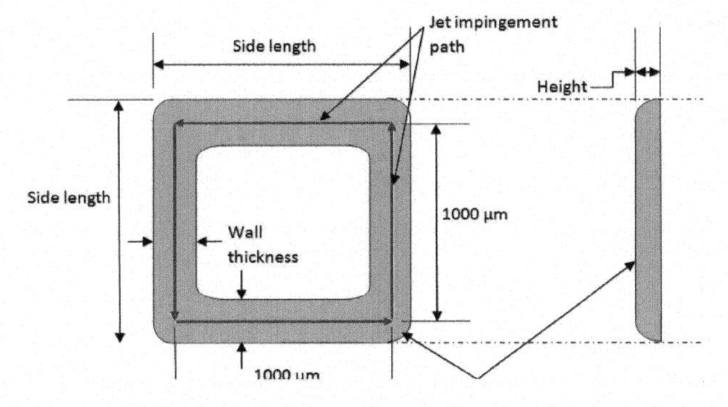

Figure 61.2 Designing of the geometric feature of part

sulphate pentahydrate is 9 gm/l to 22 gm/l and 1M of sulfuric acid is taken for all experiments. Copper was selected as a-deposited material because copper has very good thermal and electrical conductivity. It has wide use in MEMS and other microsystems as well as it can easily be electrodeposited. Jet electrodeposition and the jet electrochemical machining process are done by prepared process planning.

The nozzle of UAJE is attached to the XYZ translator and hence electrodeposits the agitated metallic ions atom by atom at geometrically defined coordinates on the substrate sample plate. The proposed setup of the experiment has been shown in Figure 61.3. The nozzle is connected to an anode (positive electrode) and the substrate material is connected to the cathode of the power source. The fixed based Cartesian coordinate system was used for the proposed setup. The electrolyte and electrode together form an electrolytic cell and when the power sources are attached the electrodeposition takes place. The integration of ultrasonic sources results in the uniform deposition of metal ions. Both anode and cathode come in contact through copper sulphate penta hydrate solution ($CuSO_4.5H_2O$). 1M concentrated H_2SO_4 was also added to the electrolyte solution. Literature reported that acidic electrolyte solutions deposit bright and shiny metal ions. The UAJE deposited the copper ions on the thin metallic aluminium lamina. The height of the deposition ranges from 200 μm to 260 μm. due to electrolysis, the agitated copper ions move from anode to electrolyte and from electrolyte to cathode due to the natural phenomenon of attraction of opposite charges.

The half equations of the electrodeposition are as follows

At anode $Cu \rightarrow Cu^{2+} + 2e^-$

At cathode $Cu^{2+} + 2e^- \rightarrow Cu$

A DC power source was used with constant voltage ranging from 3V to 15V. The electrode gap was set to 3 mm similarly, the concentration of electrolyte was 15 gm/litre. Table 61.1 consists of the experimental conditions.

D. Electrochemical machining operation

The electrodeposited laminated sheet was then placed for the electrochemical machining by changing the polarity and the placing required shape at the cathode as a reference electrode. Another needle-shaped

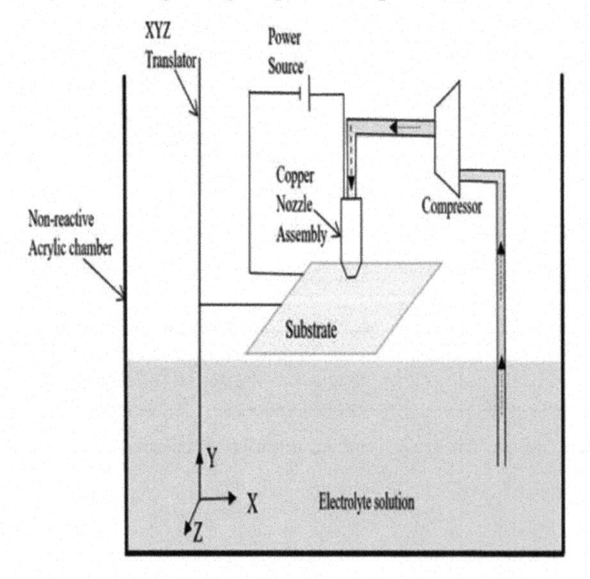

Figure 61.3 The experimental setup for UAJE deposition

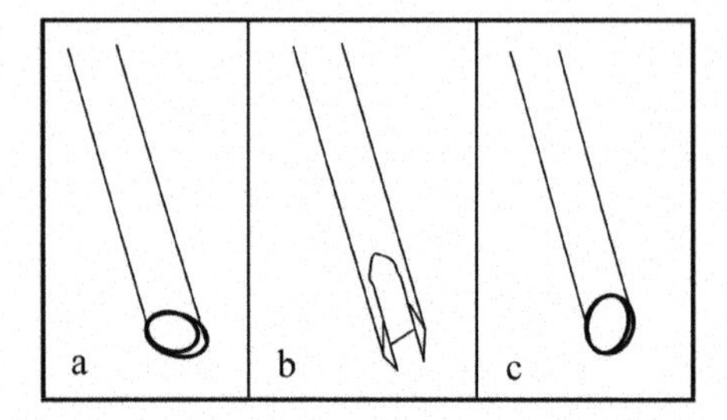

Figure 61.4 Schematic diagram of nozzles (a) off-center single point (b) symmetrical twin element nozzle (c) standard nozzle

Table 61.1 Experimental input and obtained material deposition rate with DC power source

Sr. No.	*Voltage(v)*	*Concentration of electrolyte gm/liter*	*Electrode gap(cm)*	*Ultrasonic effect*	*Deposition rate mg/min*		
					1	*2*	*3*
1	10	15	3	0.8	0.33	0.21	0.21
2	15	15	3	0.8	0.25	0.35	0.44
3	5	10	1.5	Off	0.309	0.48	
4	5	10	2.5	Off	0.233	0.24	
5	10	10	2.5	Off	0.86	1	
6	15	10	2.5	Off	1.73	1.93	

copper rod was used as a reference electrode for electrochemical machining operation. The metal ions move from the thin electrodeposited sheet to the cathode due to localised selective electrochemical machining; the required shape formation takes place at the aluminium substrate. The XYZ controller used the G-code generated to move along the desired path changing the nozzle polarity changes the process. The size of the electrode affects significantly the deposition since a thick cylindrical electrode was used for deposition while a thin needle-shaped electrode was used for the machining process. The repetition of this process has been done to make a complete part. The required number of thin metal deposition depends on the height obtained in a single pass of the designed part. The thickness varies for different scenarios. There is no explicit rule to compute the number of passes. Human reasoning and logic work better in most cases. Post machining operation can be performed after the successful deposition has been completed (Kozak etal., 1996). The literature reported that the 22.5° chamfered nozzle had the highest precision with a deeper cut whereas the symmetrical twin element nozzle led to a wider cut at the same operating condition. The researchers tested the chamfered at different angles e.g. 67.5° and 45° and concluded that the profile parameter significantly affected the surface characteristics (Morsali et al., 2017).

Post machining treatment

The post-machining operation involves the top surface grinding, holes and cuts. The post-machining operation was performed using the self-developed micro milling machine. The required M-codes were generated using Cero parametric manufacturing module.

Open-source Candle software was used to operate the micro milling machine. A high speed 775 spindle motor was used for developing a micro CNC machining setup. The spindle speed was set to 3000 rpm. The global position of the geometric axis was aligned with the local axis by considering the offset in m-code. Figure 64.5 shows a self-developed CNC micromachining setup.

Result

This article proposes electrodeposition and electrochemical machining simultaneously by given process planning shown in Figure 61.1 this process gives more accuracy and time-saving approach because the

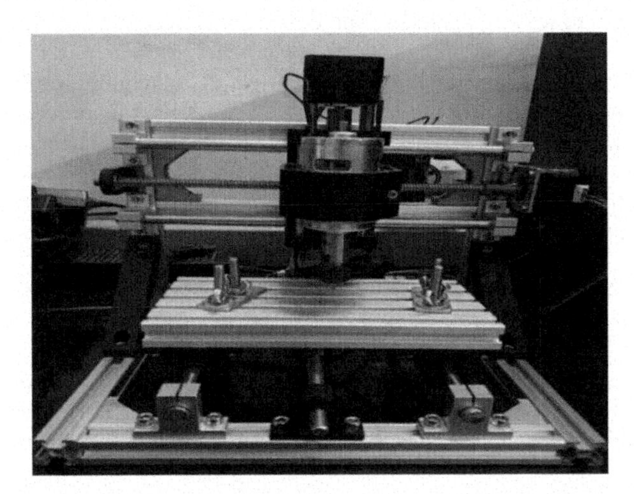

Figure 61.5 Self-developed micro-CNC machining setup

(a)

Voltage - 3vpp
Frequency – 50Hz
Electrode gap – 2.5cm
MDR – 0.00001872gm/min
Ultrasonic effect – 0.75 duty ratio
(a)

Voltage - 5vpp
Frequency – 50Hz
Electrode gap – 2.5cm
MDR – 0.00365gm/min
Ultrasonic effect – off
(b)

Figure 61.6

process is fully automatic only reference tool changing is manhandling process. The Autodesk ArtCAM software was used to generate G-code and M-code. The samples made by this process are characterized by field emission scanning electron microscopy, 'Make&Model: CARL ZEISS UHR FESEM MODEL GEMINI SEM 500 KMAT' shown in Figure 61.6. In this process, the atomic deposition was controlled at the Nano level. The figure shows the structures are at nano size means the process made for jet electrodeposition and jet electrochemical machining simultaneously are capable, accurate and time-saving approaches to other methods. Figure 61.6(a) shows the Field Emission Scanning Electron Microscopy (FESEM) images which is uniformly distributed of copper has the structure of Nano size at 200 nm at 3 Vpp and Figure 61.6(b) shows the structure of Nano size at 400 nm at 5 Vpp which is follows the electrochemical machined oriented path. From both Figure and Table 61.1 it is clear that increase in voltage increases grain size of deposited material and also increases the material deposition rate. Lower voltage gives the fine grain size, good adhesiveness and improved mechanical properties. The particle size find by FESEM are shown in Figure 61.6 are in nano structure then this process is capable to control at nano level.

Discussion and Conclusion

The presented methodology provides a novel approach for the rapid development of thin metallic features. The simultaneous use of ultrasonic-assisted jet electrochemical deposition and electrochemical machining facilitates to fabrication of thin micro features rapidly. The proposed method mitigates the manual intervention in processes. Changing the tool for the electrodeposition and machining process is the only human effort. The surface obtained by the proposed process had a thin porous structure however the density was greater than conventional electrodeposition due to jet impingement. The dimensional accuracy for deposition can be controlled by the nozzle travel speed, applied voltage, electrode gap etc. machine parameter involves spindle speed, feed rate, depth of cut etc. the chamfered nozzle also affect the machining process significantly. Using repeated depositions and machining without any set-up modifications, this procedure completes all manufacturing phases. The lead time and cost of machining are greatly decreased. The potential application for parts prepared by the proposed method involves cooling beds for electronic processors and large electrical conducting areas. The cooling beds used thin copper foils impregnated in phase-change liquids which absorb heat faster than the conventional cooling fan. The proposed method is issuable to manufacture onboard cooling foils.

References

Aouaj, M. A., Diaz, R., El Moursli, F. C., Tiburcio-Silver, A., and Abd-Lefdil, M. (2015). AgInSe2 thin films prepared by electrodeposition process. Int. J. Mater. Sci. Appl. 4(1):35.

Blakey-Milner, B., Gradle, P., Snedden, G., Brook, M., Pitol, J., Lopez, E., Berto, F. (2021). Metal additive manufacturing in aerospace: A review. Mater. Des. 209:110008.

Bari, G. A. D. (2000). Electrodeposition of nickel. Mod. Electroplat. 5:79–114.

DeSilva, A. K. M., Pajak, P. T., Harrison, D. K., and McGeough, J. A. (2004). Modelling and experimental investigation of laser assisted jet electrochemical machining,. CIRP Ann. 53(1):179–182.

Dover, S. J., Rennie, A. E. W., and Bennett, G. R. (1996). Rapid Prototyping using Electrodeposition of Copper. 8:191–198.

Fan, H., Zhao, Y., Wang, S., and Guo, H. (2019). Improvement of microstructures and properties of copper-aluminium oxide coating by pulse jet electrodeposition. Mater. Res. Express. 6(11):115090

Hui, F., Zhao, Y., Wang, S., and Guo, H. (2019). Effect of jet electrodeposition conditions on microstructure and mechanical properties of Cu–Al2O3 composite coatings. Int. J. Adv. Manuf. 105(11):45094516.

Hui, F., Zhao, Y., Jiang, J., Wang, S., Shan, W., and Li, Z. (2020). Effect of the pulse duty cycle on the microstructure and properties of a jet electrodeposited nanocrystalline copper coating. Mater. Trans. 61(4):795800.

Karapinar, D., Creissen, C. E., Rivera de la Cruz, J. G., Schreiber, M. W., and Fontecave, M. (2021) Electrochemical CO2 reduction to ethanol with copper-based catalysts. ACS Energy Lett. 6(2):694–706.

Kim, J., Cho, K. S., Hwang, J. C., Iurascu, C. C., and Park, F. C. (2002). Eclipse-RP: a new RP machine based on repeated deposition and machining. Proc. Inst. Mech. Eng. Part K J. Multi-body Dyn. 216(1):13–20.

Kozak, J., Rajurkar, K. P., and Balkrishna, R. (1996). Study of electrochemical jet machining process. J. Manuf. Sci. Eng. 118(4):490498.

Masanori, K., Katoh, R., and Mori, Y. (1998). Rapid prototyping by selective electrodeposition using electrolyte jet. CIRP Annals. 47(1):161164.

Morsali, S., Daryadel, S., Zhou, Z., Behroozfar, A., Baniasadi1, M., Moreno, S., Qian, D., and Jolandan, M. M. (2017). Multi-physics simulation of metal printing at micro/nanoscale using meniscus-confined electrodeposition: Effect of nozzle speed and diameter. J. Appl. Phys. 121(21):214305.

Pandiyarajan, S., Hsiao, P. J., Liao, A. H., Ganesan, M., Chuang, H. C., and Manickaraj, S. S. M., (2021). Influence of ultrasonic combined supercritical-CO2 electrodeposition process on copper film fabrication: Electrochemical evaluation. Ultrason. Sonochem. 74:105555.

Rajput, M. S., Pandey, P. M., and Jha, S. (2014). Experimental investigations into ultrasonic-assisted jet electrodeposition process. Proc. Inst. Mech. Eng. B. J. Eng. Manuf. 228(5):682694.

Rajput, M. S., Pandey, P. M., and Jha, S. (2015). Modelling of high speed selective jet electrodeposition process. J. Manuf. Process. 17:98–107. doi: 10.1016/j.jmapro.2014.07.012.

Sadana, Y. N. and Nageswar, S. (1984). Electrodeposition of Copper on Copper in the Presence of Dithiothreitol. J. Appl. Electroc Hem. 14(4):489–494.

Ya, L., Zhu, J., Cheng, H., Li, G., Cho, H., Jiang, M., Gao, Q., and Zhang, X. (2021). Developments of advanced electrospinning techniques: A critical review. Adv. Mater. Technol. 6(11):2100410.

62 Aerodynamics and combustion of a realistic annular gas turbine combustor: A simulation study

Nguyen Ha Hiep and Nguyen Quoc Quan

Faculty of Vehicle and Energy Engineering, Le Quy Don Technical University, Hanoi, Vietnam

Abstract

The combustor is a critical component of gas turbine engines. It must efficiently burn fuel/air mixtures, cool combustor walls, dilute the gas mixing entering the turbine, and minimize overall pressure loss. This article discusses numerical simulations of the aerodynamic and combustion processes occurring inside an actual annular gas turbine combustor. A high-accuracy MicroScribe-GX2 3-D digitiser successfully scans a three-dimensional combustor model. The simulations were performed by Star-CCM+ software using RANS and the Steady Laminar Flamelet approach. The results show the flow structure, total pressure loss, temperature field, and gas components in the combustor under nominal conditions with cold and hot tests. In the cold test, the output velocity is lower than in the hot test; the total pressure loss is around 0.05. Approximately 20% of the flow passes through the swirlers for combustion, 30% for cooling the liner walls, and 50% through the dilution zone. The combustible gas has an average temperature of 1120 K and a maximum speed of 126 m/s entering the turbine.

Keywords: Aerodynamic, annular combustor, combustion simulation, gas turbine, steady laminar flamelet, Star-CCM+.

Introduction

The critical role of the gas turbine combustor is to burn the fuel sprayed from the injector with considerable amounts of air provided by the compressor, resulting in a mixture of burning gases at a temperature adequate for the turbine operation. Combustion must occur within a limited volume with minimal pressure loss and the highest amount of heat released (Rolls-Royle Plc. 2015).

The amount of fuel injected is determined by the gas temperature increase by operating regimes. However, the maximum turbine inlet temperature is limited to between 8501700°C, depending on the turbine blade materials. The compressed air temperature increases to 200550°C, and after burning in the combustor, the temperature can be raised to 6501150°C. The combustor must maintain a stable and highly efficient combustion process in all operating modes that require different turbine inlet temperatures depending on the engine thrust or power.

Organising the combustion process in the combustor is a complex task that must meet several requirements, including the following (Lefebvre, 1999): a balance of high burn-out coefficient and minimal total pressure loss, stable combustion, the ability to operate in a wide range of air excess coefficient depending on operating modes, a uniform gas temperature field at the turbine inlet, stable starting under all conditions, low toxic emissions, etc.

There are three approaches to combustor operation study: theoretical, experimental, and simulation. Theoretical investigation employs simplified assumptions and empirical coefficients acquired from several experiments. When the combustor configuration or operating regimes are changed, it is necessary to re-select the experimental coefficients, which commonly results in inaccuracies. Meanwhile, experimental study on the combustor is excessively costly and time-consuming, with difficulty determining the temperature field uniformity at the combustor outlet.

When computer power accompanies the widely used computational fluid dynamics (CFD) simulation, it is possible to obtain adequate results for realistic gas turbine combustors (Popescu et al., 201; Vilag et al., 2019; Konle et al., 2018; Murthy et al., 2018). Additionally, the modelling approach allows the demonstration of the whole parameter field (pressure, temperature, velocity,...) and the distribution of the combustion product fractions inside the combustor, which may not be established experimentally. Aerodynamic and combustion simulations are often carried out using either the Reynolds-Averaged Navier-Stokes equations (RANS, URANS) or Large Eddy Simulations (LES). The LES approach gives more accurate results for turbulent flows but it has less practical uses for real combustors due to the complexity of the liquid fuel spraying, swirling, atomizing, multiphase phenomena, and turbulence combustion model, with the expense of tenfold the computing power (You et al., 2008; Constantinescu et al., 2003, Kim et al., 1999). The RANS approach produces adequate results comparable to experiments evaluating aerodynamic (Popescu et al., 2010; Vilag et al., 2019; Calabria et al., 2015; Poinsot, 2017; Priyant and Selwyn, 2016; Serbin and Burunsuz, 2020) or combustion (Konle et al., 2018; Murthy et al., 2018) characteristics.

The COMOTI Institute's research (Popescu et al., 2010) requires 17 thermocouples (34 locations monitored) to determine the gas temperatures at the combustor outlet. The data are consistent compared to the numerical results obtained using Ansys CFX software. They indicate how combustion occurs in the annular combustor of the TV2-117 engine (Figure 62.1) when various fuels are employed (gasoline, methane, and biogas). The mean circumferential temperature value for the TV2-117 engine was obtained using five thermocouples installed in front of the turbine injector.

The paper (Konle et al., 2018) presented the RANS simulation results of an actual TF50 engine combustor with a multi-physics solver based on the open-source CFD package OpenFOAM to evaluate different re-design options. The study successfully compared thermal paint results with predicted combustor wall temperatures using from 32 CPUs up to 1024 CPUs for 5000 step simulation. Murthy et al. (2018) studied a full-scale model of an actual can-combustor using experimental and RANS numerical techniques by Ansys Fluent software. The results showed that the 45° vane angle swirler was a better recirculation mass compared to the 33° swirler and a lesser pressure loss compared to the 60° swirler.

This article aims to present the simulation results performed using Siemens Star-CCM+ software to better understand the aerodynamics and combustion process in a realistic annular combustor of the TV3-117 gas turbine engine. The engine is a 12-stage axial compressor, an annular combustor, a two-stage compressor turbine, and a two-stage power turbine. The combustor has 12 injectors (with 12 swirlers), up from eight injectors in the previous TV2-117 generation in the research (Popescu et al., 2010; Vilag et al., 2019). Aerodynamic simulations (cold tests) were used to verify that the input parameters (velocity, flow, and compressor pressure ratio) were correct and acceptable for the combustion simulation. The parameters of the engine and combustor at the nominal regime are listed in Table 62.1.

Aerodynamic and combustion models of gas turbine combustor simulation

The gas turbine combustor must meet several requirements: high combustion efficiency, low total pressure loss, high thermal intensity, stable operation in all engine operating regimes, temperature uniformity, and low toxic emissions. Nowadays, the fundamental goal of combustor design is to increase the turbine inlet temperature while minimising overall pressure loss and poisonous emissions.

In modern combustors, the overall total pressure loss is approximately 0.04–0.06. It is categorised as friction loss, turbulent flow loss (local resistance, flow mixing, recirculation,...), and heat loss due to combustion. The first two categories of losses are hydraulic losses, which characterize the combustor performance

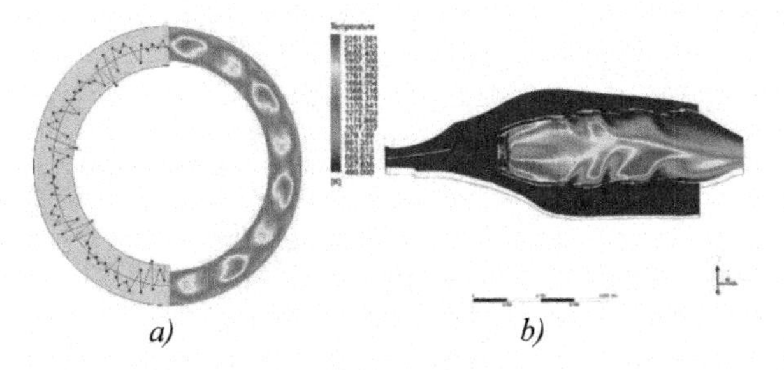

Figure 62.1 Experiments and simulations on the annular combustor TV2-117: a) Temperature uniformity at the combustor outlet; b)The gas temperature along the axial cross-session ([Popescu et al., 2010; Vilag et al., 2019)

Table 62.1 Parameters of TV3-117 engine and combustor at the nominal regime

Parameters	*Value*	*Parameters*	*Value*
Air axial velocity after the compressor, m/s	112	Maximum outlet compressor temperature, K	608
Air axial velocity in front of the liner, m/s	45	Maximum turbine inlet temperature, K	1228
Air axial velocity after the combustor, m/s	158	Mass flow rate, kg/s	8.85
Compressor rotor speed	99%	Pressure after compressor, atm	9.45

almost independently of the combustion process, which can be evaluated under experimental conditions on a test platform without burning (Lefebvre, 1999). Additionally, the combustion efficiency may approach 0.99 (0.98 for the combustor in this study), directly relating to the engine efficiency. Consequently, no boundary conditions for combustion simulation (hot test) are required when just aerodynamic efficiency is considered on the cold test. When numerical investigating the temperature field, the combustion process, and the exhaust gas (hot test), it is required to generate more specific conditions, such as the content and model of the multiphase fuel spray, the reaction mechanisms, and combustion models.

Each of the above issues would need a unique model setup technique, but there is a common requirement that the model is adequate for the physical nature and computing resources. As a result, as seen in Table 62.2, a difference will take place between aerodynamics modelling and combustion simulation.

Additionally, while modelling conjugate heat transfer difficulties, it is critical to employ a Finite Element Method to determine the material properties of the liner walls with cooling channels and holes. This issue requires a more advanced simulation model computer resources (Konle et al., 2018) and is thus inside the future study.

Building a gas turbine combustor simulation model

The combustor geometry model

The MicroScribe GX2 device was used to scan the TV3-117 engine combustor using the 3-D dot technique with an accuracy of 0.23 mm, trimmed to 1/12 sector including the fuel pipe to the injector, as shown in Figure 62.2.

Table 62.2 Combustion and aerodynamic simulation model comparisons

No	*Simulation setup steps*	*Aerodynamic simulation (Cold test)*	*Combustion simulation (Hot test)*
1	Building a 3D combustor model	• Partial sector model of the combustor. • There is no need for a fuel channel. • Ignore or simplify small details.	• The model must include all relevant details, including the injector. • A fuel channel is necessary to generate a fuel/air mixture. • Ignore or simplify fixed details.
2	Constructing a model of finite volume meshing	• High mesh density near liner holes, recirculation zones, and swirlers	• High mesh density in the liner holes, the combustion zone, and the mixing zone with the secondary flows.
3	Establishing mathematical models	• Steady model (RANS) • *k-ω SST* turbulent model • Model of single-component air at the inlet with viscosity that changes with temperature, density changes according to the ideal gas law	• Steady model (Steady Laminar Flamelet). • *k-ω SST* turbulent model. • Multi-component gas. • Spray model • Flamelet model and chemical reaction mechanisms. • Fuel and liquid droplet vaporization modelling (multiphase). • NOx and soot formation models.
4	Setup the boundary conditions	• Boundary conditions according to the realistic combustor working conditions. • Using compressor calculation results or experimental parameters, input velocity field, and turbulence characteristics. • Hydraulic characteristic, i.e., the loss is proportional to inlet air velocity or flow rate.	• Boundary conditions according to the realistic combustor working conditions. • Using compressor calculation results or experimental parameters, input velocity field, and turbulence characteristics. • Requiring to simulate the structure of the air-fuel mixture flow in the combustor; once steady regimes, the temperature will be higher than the mixture ignition temperature.
5	Establish a method for resolving problems	• Second-order discretisation scheme • The number of solve steps: until the combustor inlet and outlet pressures are stabilized.	• Second-order discretisation scheme. • Number of solve steps: until the temperature value is stable at the combustor outlet, flow balance, steady NOx emissions.
6	Analysng and calculating results	• Total pressure, temperature, mass flow rate...	• Total pressure, temperature, mass flow rate, emission distribution

Because the conjugate heat transfer between the combustible gas and the combustor walls is neglected, the liner cooling channels and injector cooling holes (Figure 62.3) are eliminated to avoid meshing errors.

Indeed, the mechanism occurring within the combustor is not perfectly symmetrical, particularly in areas near injectors. Using a Steady Laminar Flamelet model and neglecting ignition, the results demonstrate that the numerical results will repeat cyclically when the reaction flow is steady. Under more ideal conditions, a 1/4 combustor sector (three injectors) would produce the same comprehensive results as an annulus model. However, choosing a 1/12 combustor in this study is reasonable since it eliminates cooling and heat transfer into the liner and simplifies tiny details. The model of 1/12 sector with one injector utilizes less than 3 million polyhedral elements, providing quick convergence with small computer resources (16Gb RAM).

Simulation model

The 1/12 combustor model was built using Siemens NX software, then meshed and simulated in Star-CCM+. The model surfaces are named by the boundary conditions: inlet, outlet, periodic 1, periodic 2, and walls. Star-CCM+ software can mesh with polyhedral elements and automatically handle errors with small and complex details, especially in areas near the liner. The base size of the grid is 1 mm, with the minor elements equal to 2.5% base, boundary layers with an increasing ratio of 1.3.

The mesh model has about 2.6 million polyhedral elements. In Figure 62.4, statistical numbers of elements with cell quality corresponding to a 0–1 scale show that only a few areas have a low quality (cell quality <0.2) with the total number.

Several grid independence tests varying the base size of the grid from 0.5 to 5 mm found a value of 1 mm to be appropriate, giving converging results while saving computational resources. The tetrahedral meshing tests failed to obtain a satisfactory mesh quality.

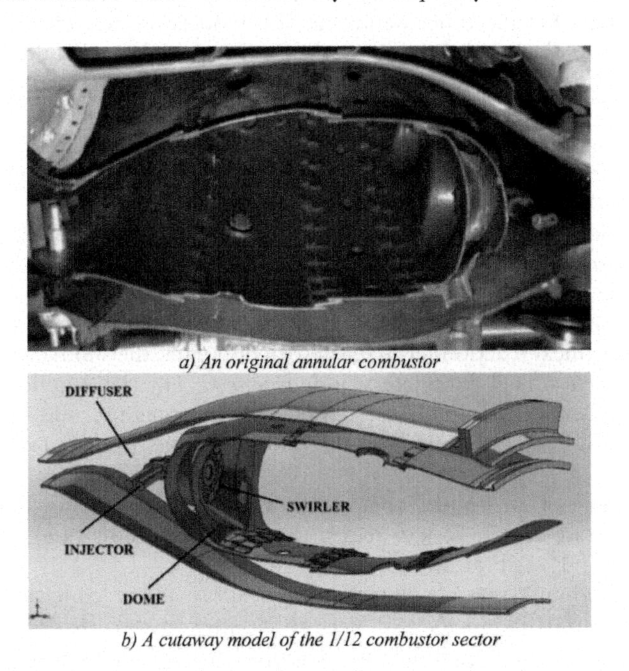

Figure 62.2 The 3-D combustor model

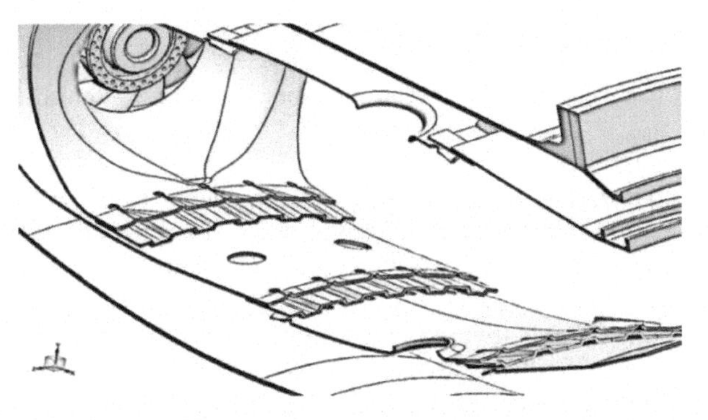

Figure 62.3 Corrugated ribs maintain the temperature of the liner and injector cooling channels

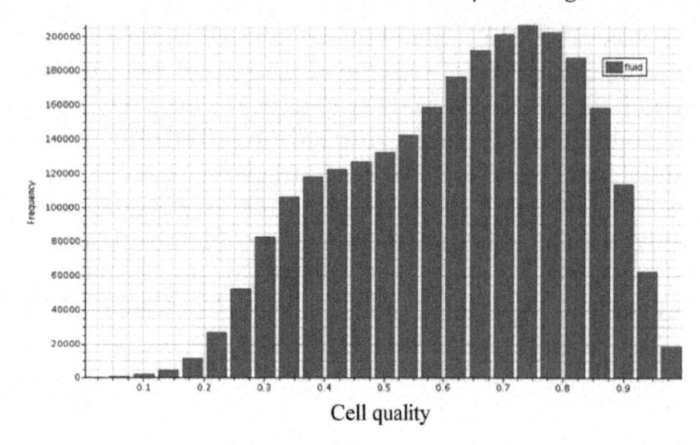

Figure 62.4 Statistical chart of the cell quality on a scale from low to high (0–1)

Boundary conditions

For the cold test, the following parameters are used: pressure 9.45 atm, temperature 610 K, inlet airflow velocity 110 m/s, compressor rotor speed at the nominal regime. The cold test results provide information on the combustor aerodynamic characteristics and serve as a basis for establishing the number of steps required to initialise the combustion simulation. At the 1000th simulation step, begin increasing the gas temperature and activating the flamelet simulation parameters, NOx, and soot emission models. Stable results can be reached up to the 5000th step.

The steady laminar flamelet (SLF) is appropriate for steady combustion, such as in gas turbine combustors that operate in various operating modes. On the other hand, the SLF models are not ideal for combustors with continually fluctuating intensity, changing ambient conditions, or re-ignition. While the combustor flow remains *k-ω SST* turbulent model, the flamelet propagation model is *quasi-steady*; the flame changes shape in response to the flow while maintaining its laminar structure. This model is still widely used in industries due to fewer computing resources while obtaining acceptable temperatures, combustion products, and emissions. Unsteady Turbulent Flamelet models, commonly employed with the LES approach (Murthy et al., 2018; You et al., 2008), need more computational capacity than this study.

The SLF method initialises a variable-resolution SLF table; the more table accuracy, the longer the simulation time. The solver uses the scalar values provided in the SLF table during computation. This approach is substantially faster than dealing with a whole set of chemical processes at every given moment.

The SLF table was prepared using Chemkin software in the public resource (Lu, 2022) and then imported into Star-CCM+ software in the format of a chemical reaction mechanism: reaction data, thermodynamic characteristics, and phase transportation properties. Fuel parameters: *n-dodecane* $C_{12}H_{26}$ (Jet-A1) liquid kerosine, reaction mechanism consists of 31 chemical components, saturated vapor pressure 600 kPa, vaporization temperature 850 K, temperature 320 K, and mass flow rate 0.010926 kg/s (the air excess coefficient is roughly 4.5).

To accomplish rapid convergence of difference of inlet/outlet air mass flow rate, the combustor inlet airflow was set to cold test, and the mass flow rate of 0.7375 kg/s (equal to 1/12 of total flow rate) was replaced by the velocity. A turbulence intensity of 3% is typically used at the inlet.

Centrifugal injector settings are set as indicated in [9]: nozzle diameter 5 mm, spray hollow cone shape (10° inner angle, 30° outer angle), mass flow rate 0.010936 kg/s, 100 fuel particle streams, particle droplet diameter 30 μm, and particle velocity 5 m/s. At the beginning of the calculation, a particle droplet distribution is initialized using a Lagrangian multiphase model.

Results and Discussions

The simulation model of the 1/12 combustor sector, using the SLF method, is suited for steady combustion. The numerical analysis shows that:

- It is necessary to simulate aerodynamics (cold test, (Figure 62.5*b*) to get convergence of the flow, velocity, and pressure parameters (to achieve an equilibrium of inlet and outlet flows), thereby determining the time to activate the NOx and soot emission models, as well as the number of resolution steps in the combustion model (hot test, Figure 62.5*c*).
- By decelerating the gas flow in the diffuser from 110 m/s to 4045 m/s in front of the dome, the velocity field demonstrates that the combustor is aerodynamically effective (Figure 62.5*a*). The airflow entering into combustion zone generates a recirculation zone with a low velocity (less than 25 m/s) to atomize

and mix the fuel. The secondary airflow enters the dilution zone via the liner holes, contributes cooling air, and accelerates to a maximum speed of 126 m/s into the turbine (Figure 62.5c).

- In the cold test (Figure 62.5c), the output velocity is lower than in the hot test, but the flow structure remains the same. In both the hot and cold tests, the total pressure loss in the combustor is around 0.05. This demonstrates that while cold simulation can assess the completeness of the combustor aerodynamics and the turbulent flow structure, the velocity and temperature fields must be defined during the hot test.
- As liquid fuel exits the centrifugal injector, it is atomized and rapidly mixes and vaporizes due to the primary air. Around 25% of the air passes through the swirlers and primary holes to mix and burn the fuel, while the secondary flow cools the hot gas in the dilution zone before entering the turbine (Figure 62.5). This result is compatible with the TV3-117 experimental data: approximately 20% of the flow passes through the swirlers for combustion, 30% for cooling the liner walls, and 50% passes through the dilution zone; in the case of TV2-117, around 25–30% of the flow passes through the swirlers, with the rest entering the liner for mixing and cooling. It is noted that these ratios will vary depending on the operating regimes.
- The gas temperature may approach 2200 K in the combustion zone. When cooling air flows through the liner channels at a low mass flow rate, it surrounds the liner and protects it from burning. The temperature field surrounding the combustor is not uniform in circumferential or radial directions (Figure 62.6). As illustrated in Figure 62.1, this result is consistent with experimental and numerical data (Ansys CFX) for the COMOTI Institute's previous-generation engine combustor TV2-117 (Lefebvre, 1999). The average temperature of combustible gas entering the turbine is 1120 K (Table 62.1), close to the engine data.

Once the fuel is burn-out, the output contains little residual hydrocarbons, soot, and NOx emissions (Figure 62.7).

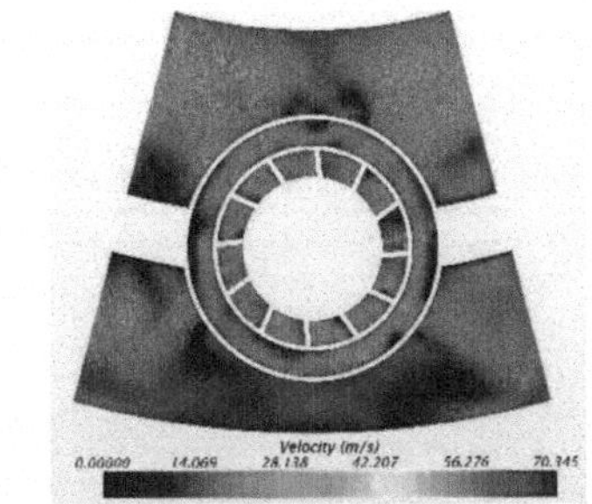

a) The region passing through the swirlers

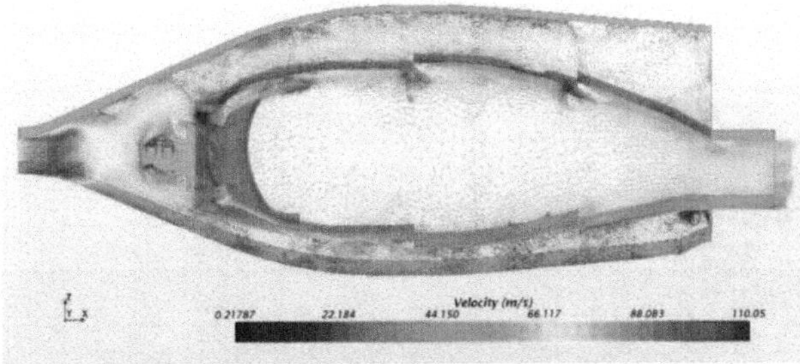

b) the axial cross-section through the injector center (cold test)

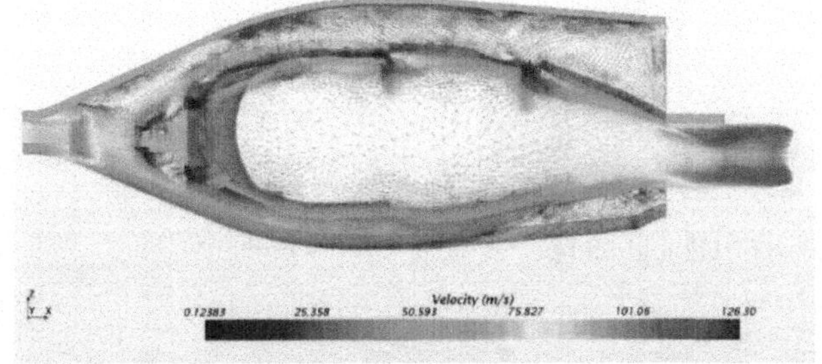

c) the axial cross-section through the injector (hot test)

Figure 62.5 The velocity field in the following regions

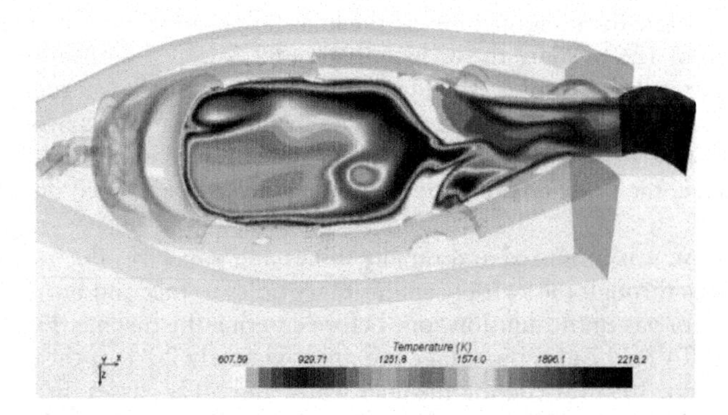

Figure 62.6 The temperature at the periodic surface

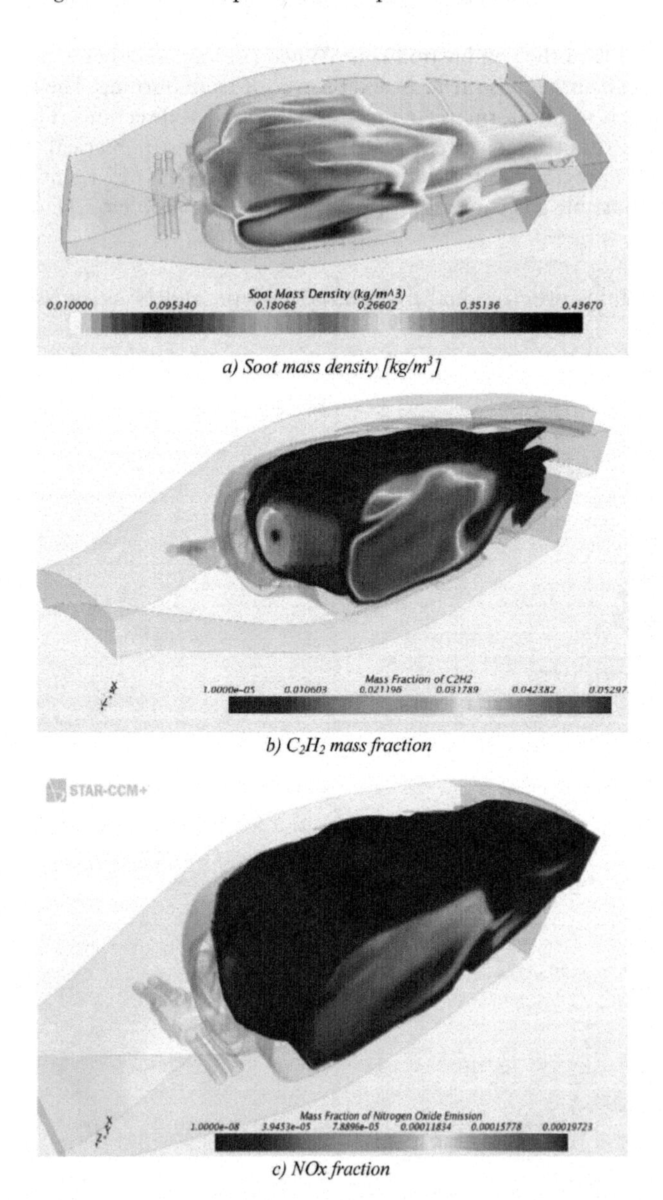

Figure 62.7 Soot mass density and mass fraction C2H2 and NOx

Conclusion

The aerodynamic and combustion characteristics in the realistic annular combustor of the gas turbine engine were simulated using Star-CCM+ software. Combined with a Steady Laminar Flamelet Model, the RANS approach is well suited for simulating gas turbine combustors, obtaining good results with minimal computing resources. The simulation results demonstrate the pressure, temperature, velocity, mixing gas contents, and gas exhaust distributions.

The RANS method demonstrates the appropriate reliability, allowing us to use the unsteady and LES approaches to get more accurate results. It is also possible to change the geometrical models by applying 1/6, 1/4 sector, or full annulus combustor.

References

Calabria, R., Chiarielloa, F., Massolia, P., and Realea, F. (2015). Numerical study of a micro gas turbine fed by liquid fuels: potentialities and critical issues. Energy Procedia. 81:1131–1142.

Constantinescu, G., Mahesh, K., Apte, S., Iaccarino, G., Ham, F. and Moin, P. (2003). A New Paradigm for Simulation of Turbulent Combustion in Realistic Gas Turbine Combustors Using LES. Proceedings of ASME Turbo Expo. 259–272.

Kim, W. W., Menon, S., and Mongia, H. C. (1999). Large-Eddy Simulation of a Gas Turbine Combustor Flow. Combust. Sci. Technol. 143(16):25–62. doi: 10.1080/00102209908924192

Konle, M., de Guillebon, L., and Beebe, C. (2018). Multi-Physics Simulations With OpenFOAM in the Re-Design of a Commercial Combustor. Heat Transfer. 5C: doi:10.1115/gt2018-76578

Lefebvre, A. H. 1999. Gas Turbine Combustion. Philadelphia: Taylor & Francis.

Lu, T. (2022). Combustion, Chemical Kinetics, and Computational Fluid Dynamics. Chemkin data. http://spark.engr.uconn.edu/mechs/C12-lumped.zip (accessed February 22, 2022).

Murthy, M., Bhadkamkar, N., Penumarti, A., Prabbu, S. V., and Sreedhara, S. (2018). Numerical investigation of swirl flow using different swirlers in a real-life gas turbine combustor. J. Flow Vis. Image Process. 25:91–117. doi:10.1615/jflowvisimageproc.201.

Poinsot T. 2017. Prediction and control of combustion instabilities in real engines. Proc. Combust. Ins. 36(1):1–28. doi:10.1016/j.proci.2016.05.007

Popescu, J. A., Vilag, V. A., Petcu, R., Silivestru, V., and Stanciu, V. (2010). Researches concerning kerosene-to-landfill gas conversion for an aero-derivative gas turbine. 639–648. doi: 10.1115/GT2010-23436

Priyant, M. C. and Selwyn, A. 2016. Design and analysis of annular combustor of a low bypass turbofan engine in a jet trainer aircraft. Propulsion Power Res. 5(2):97–107. doi: 10.1016/j.jppr.2016.04.001.

Serbin, S. and Burunsuz, K. (2020). Numerical study of the parameters of a gas turbine combustion chamber with steam injection operating on distillate fuel. Int. J. Turbo Jet Eng. doi: 10.1515/tjj-2020-0029

Rolls-Royle Plc. 2015. The jet engines, 5th edition. Derby, England: Wiley.

Vilag, V., Vilag, J. A., Carlanescu, R., Mangra, A., and Florea, F. (2019). Comput. Fluid Dyn. Simulations. Appl. Gas Turb. Combust. Simul. doi: 10.5772/intechopen.89759

You, D., Ham, F., and P. Moin. 2008. Large-eddy simulation analysis of turbulent combustion in a gas turbine engine combustor. Annual Research Briefs, 219–230, California: Center for Turbulence Research.

63 Comparative study of employing PSO-ANN and BP-ANN to model delamination factor in CFRP/Ti6Al4V drilling

Aakash Ghosh, Aryan Sharma, Navriti Gupta, and Ranganath Muttana Singari[a]

Department of Mechanical Engineering, Delhi Technological University, Delhi, India

Abstract

This paper aims to present a comparison of two optimization methods of improving an artificial neural network (ANN) architecture employed for modelling and predicting delamination factors during the drilling of carbon fibre titanium metal matrix (CFRP/Ti6Al4V). The use of CFRP/Ti6Al4V lies predominantly in the aerospace and the high performance automotive industry, both of which have evolved and are still evolving with the adoption of newer technologies. Neural networks singularly can perform an adequate modelling however with the growing requirement of accuracy and the need for intricate detailing requires the deployment of a much more accurate model. Back-propagation ANN is one of the most commonly used ANN architectures which have known to provide extremely accurate results. For further improvement, metaheuristic optimisation can be applied onto the base ANN to improve the performance of the ANN by decreasing the time to convergence. The paper presents a particle swarm optimization (PSO)-ANN developed on MATLAB to model the delamination. The PSO-ANN is seen to produce an R-value of 0.98 whilst converging to an error of less than 0.01mm with a limited dataset.

Keywords: Aerospace, ANN Automotive, CFRP, Meta-Heuristic, optimisation, PSO.

Introduction

CFRP and CFRP-metal matrices

Carbon fibre has morphed into one of the most essential structural materials of the current industry. The beneficial mechanical properties carbon fibre reinforced polymers (CFRP) possesses make it useful in a plethora of applications. However, the benefits are qualified by the difficulties present in its manufacturing. Furthermore, due to the requirement of stronger composites, the composite-metal matrices have come into the industrial spotlight. CFRP-Metal matrices are an amalgam of the CFRP and a strengthening alloy such as Aluminium or Titanium. CFRP-Metal matrices are used in primarily in aerospace structures due to their impressive strength-to-weight ratio. One prime example of such a material is CFRP/Ti6Al4V which is abundantly used in the fuselage frame of aeroplanes.

Machining defects and delamination

On a limiting note, these materials have been proven to be extremely difficult to machine. Lacking in a conventional metal property, these materials induce extremely high temperatures on the cutting tools under conventional machining methods. Several studies describe the occurrence of a machining fault, termed 'delamination'. This is a surface defect caused in CFRP and its different variants due to the CFRP layers essentially peeling off one another. The most common reason for delamination in CFRP is seen in drilling. A non-dimensional ratio is used to assess the severeness of the delamination, it is known as the 'delamination factor', it is mathematically defined as the ratio between the maximum diameter of the produced hole and the nominal hole diameter as shown in the equation below:

$$F_D = \frac{D_M}{D_0} \tag{1}$$

F_D represents the delamination factor, D_M represents the maximum hole diameter and lastly, D_0 signifies the nominal hole diameter. The effects of delamination on CFRP parts can be negligible, however, over long

[a]ranganath@dce.ac.in

periods of time, it can cause sudden failure of CFRP parts. In the aerospace industry, drilling is recorded to be one of the major machining processes as parts are joined using rivets, and it becomes of utmost importance to minimise structural errors as well as any residual stresses and defects under the extreme physical and thermal loads subjected to aeroplanes.

Literature Review of Past Studies

A study by Hussein et al. (2019) explores the use of vibrational assisted drilling of CFRP-titanium alloys, and discusses the drawbacks of conventional drilling. It is seen that the conventional methods tend to produce parts of low surface integrity and the drilled holes show delamination.

A study carried out by Prasad and Ghodke (2015) established relationships between the material properties of CFRP, machining parameters of Abrasive Waterjet (AWJ) cutting and delamination factor, kerf width and surface roughness of CFRP. It was found that 45° fibre orientation had the highest delamination factor as compared to 60° and 90° fibre orientations.

A study by Chakraborty (2005) noted that a back-propagation based ANN model can be trained to predict the location, shape and size of the delamination in fibre-reinforced composites using the provided natural frequencies. However, it was concluded that real-life data sets would be required to further improve the model.

Krishnamoorthy et al. (2011) conducted an experimental study based on CFRP drilling using a CNC machine. They used a back propagation model trained with gradient descent method to predict the delamination values during drilling of CFRP. Their results showed a maximum error of 0.81% and a post regression analysis of the ANN data shows a linear regression relationship between predicted and actual values of delamination during drilling of CFRP.

Qin et al. (2014) analysed delamination during helical milling of CFRP using a back-propagation ANN model. They assessed the importance of various parameters in order to improve the obtained hole quality. The interaction effects of the various parameters revealed that a high spindle speed with suitable feed rate and screw pitch resulted in lesser delamination. It was recommended that a cutter compensation system be used during helical milling of CFRP in order to prevent the tool wear that occurs at high spindle speeds. The ANN model was found to be suitable for non-linear mapping on account of R values that suggest good correlation (0.98354 for training and 0.95057 for testing). Moreover, it was established that the delamination factor is reduced in helical milling on account of reduced thrust force as opposed to high-speed drilling.

Seo and Lee (1999) investigated damage detection in CFRP laminates by measuring the change in electrical resistance as a criterion for fatigue related damage such as deterioration of stiffness and residual strength. The change in electrical resistance was measured using electrodes that were attached to the surface of the CFRP laminates. The researchers developed a neural network after obtaining the experimental values. The input node was the change in electrical resistance and the output node was the degradation of stiffness or fatigue life. The results showed a good agreement between the experimental values of degradation of stiffness or fatigue life and the predicted values. However, in the final stage, both – the stiffness and the electrical resistance changed in an abrupt manner.

Watkins et al. (2002) conducted an experimental investigation to determine the delamination locations and size in the laminated glass/epoxy cantilever beams by simultaneous measurement of data using piezoelectric sensors, classical beam theory and simulations using FEM. Subsequently, they used EFPI optical sensors to establish the first five modal frequencies of laminated glass/epoxy cantilever beams. The obtained

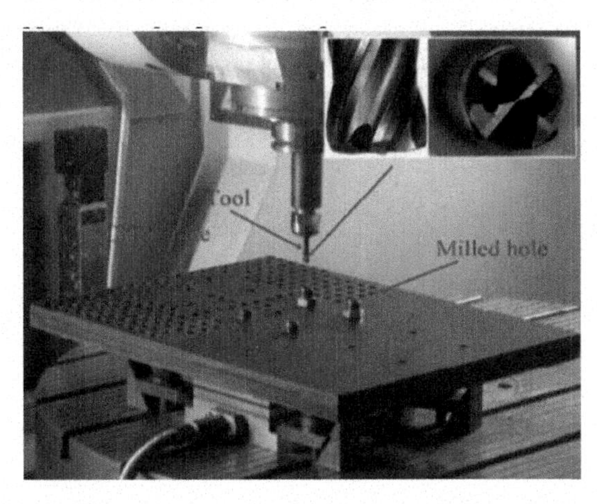

Figure 63.1 Experimental set up used by Qin et al (2014)

frequencies were used in a feed forward back propagation neural network model in order to predict the location and size of delamination failures in the laminated glass/epoxy cantilever beams. The neural network was trained on MATLAB using the classical beam theory. The model predicted the delamination failure sizes with error ranging from 0.9 to 10.7% and an average error rate of 5.9%. The model also predicted the delamination failure locations with error ranging from 0 to 13.9% and an average error rate of 4.7%. The trend in the observations reveals that the neural network over-predicted delamination failure sizes and locations. In this regard, the authors suggest that more accurate training data can be used to improve the results.

Soepangkat et al. (2019) established a back propagation neural network (BPNN) with particle swarm optimization (PSO) with the aim of predicting and optimizing various parameters pertaining to CFRP drilling. They employed a full factorial designed experiment methodology in order to obtain the data for forming the BPNN. Their results indicate that the BPNN had an average error rate of less than 5% and the PSO was found to be effective because the relative error rate between predicted values and experimental values was found to be less than 5%.

Artificial Neural Network

Back propagation artificial neural networks (BPNN) are networks in which the layered hierarchy of neurons consist of high degree of communication amongst one another.

The figure above depicts the flow of information through the neurons. However, it must be noted that the error values are communicated backwards through the layers, whereas the activation only flows from the input to the output layer. Each node of the layers can take multiple inputs from outside world or their preceding layer of neurons, however, they output a single value. BPNNs are most often used for supervised learning, where the example inputs and the correct outputs are known (Wythoff, 1993). These neural networks can be further improved by employing more neurons, however, adding more neurons can lead to more complexity in the program as well as cause over-fitting of the data. In order to optimise these networks, a significant number of studies have used metaheuristic optimisations which develop a neural network with a hidden layer with the optimum features.

Methodology

Karnik et al. (2008) analysed delamination in high-speed drilling of CFRP by means of an artificial neural network (ANN). The study used CFRP laminates that were made using the hand lay-up technique. The laminates were composed of 50% epoxy matrix by weight and 50% plain woven carbon fibres by weight with the fibres oriented in a 0°/90° configuration. The delamination was analysed on the basis of – spindle speed, point angle and feed rate as input parameters. The researchers used 36 sets of experimental observations conducted as per full factorial design (FFD) on a machining centre with 11 kW spindle power and a maximum spindle speed of 10,000 rpm to form the basis of training and testing the ANN model. Their results indicated that the maximum absolute error rate between experimental and predicted values of delamination was found to be 1.46% for the training dataset and 12.5% for the testing dataset. A key takeaway from the research was that for minimizing delamination, high spindle speed, low feed rate and low point angle can be used. The aim of this study is to utilise MATLAB as a tool for preparing two distinctive neural network models, namely BPNN and PSO-ANN and compare their applicability in multi-variable predictive modelling of delamination in CFRP. PSO technique is used for optimising the ANN architecture. PSO-ANN can improve the characteristics of the ANN such that it uses the optimum neurons to model a data set. This serves as the key differentiator between a traditional back-propagation based model and a PSO model.

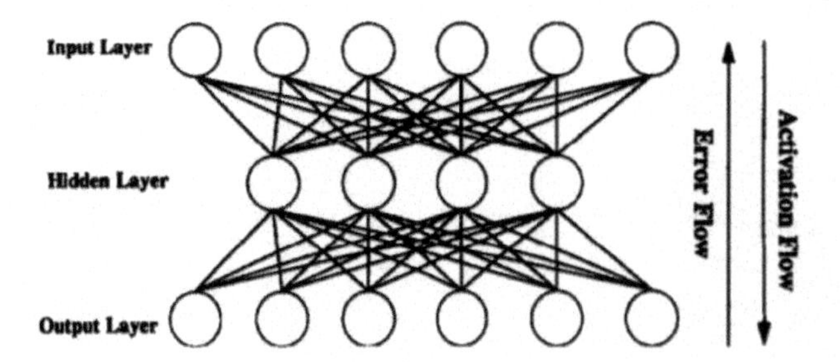

Figure 63.2 Visualisation of information exchange in neurons (Wythoff, 1993)

Modelling and Optimisation Frameworks

The core of this study is based on the modelling frameworks which have been used. The study utilises two models of ANN; namely, BPNN and PSO-ANN). Both of the optimisations are performed on MATLAB (Matrix laboratories).

The BPNN acts as a comparative base in order to realize the effects of integrating a metaheuristic optimization in the neural network framework. Particle Swarm Optimization is a stochastic optimization methodology where the algorithm takes a random approach in guessing the initial values and fits, develops and improves the model as the optimisation proceeds. As aforementioned, PSO has been successfully used in several manufacturing scenarios and in the improvement of predictive modelling. Alam (2016) presents a PSO-ANN developed on MATLAB which can be further developed and utilised in order to suit this study's dataset type and restrictions. The figure below depicts a general flowchart of the PSO-ANN process employed.

Results and Discussions

The experimental dataset is used for training both the BPNN and the PSO-ANN architectures and are then tested with the test dataset. Training of a model is essential to compare the two architectures. It must be recognised that the two models deviate only marginally, but in precision machining, extreme accuracy is required.

The PSO optimisation occurs over 1750 iterations to produce a neural network with two layers, where the hidden layer consists of eight neurons utilising a Tan-Sigmoid transfer function. The output layer uses a linear transfer function. The PSO is able to optimise such that the model fits the training data without any under-fitting or over-fitting. A preliminary NN architecture is fed as a starting point for the optimisation to occur in which the number of neurons in the hidden layer is added as 10, and through PSO the hidden layer is optimised to give the least MSE.

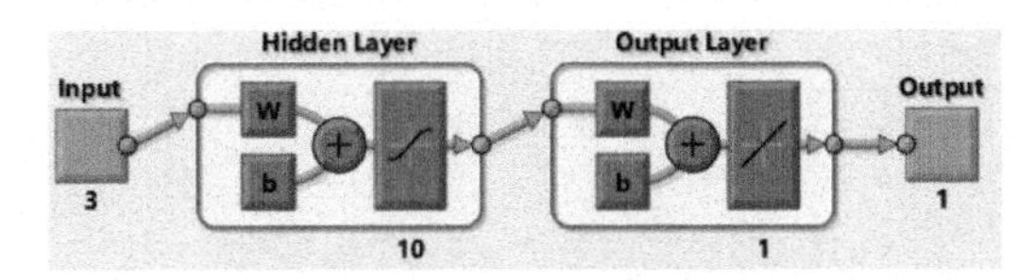

Figure 63.3 Flow diagram of BPNN

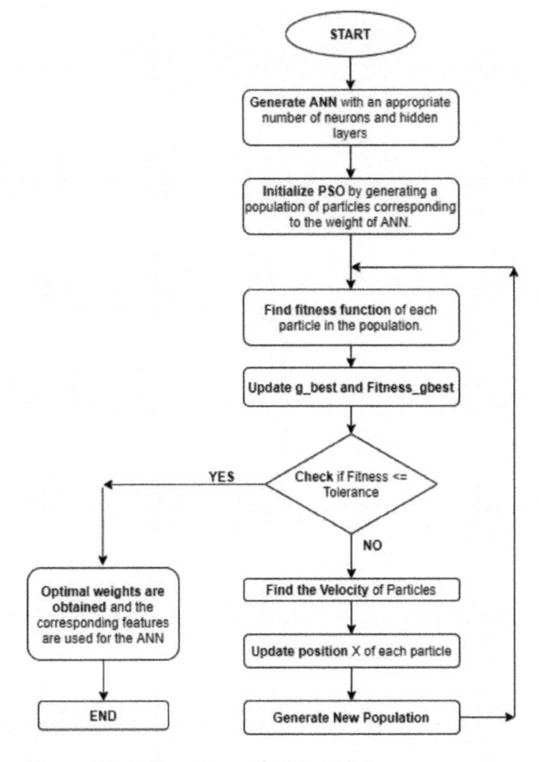

Figure 63.4 Flowchart of PSO-ANN

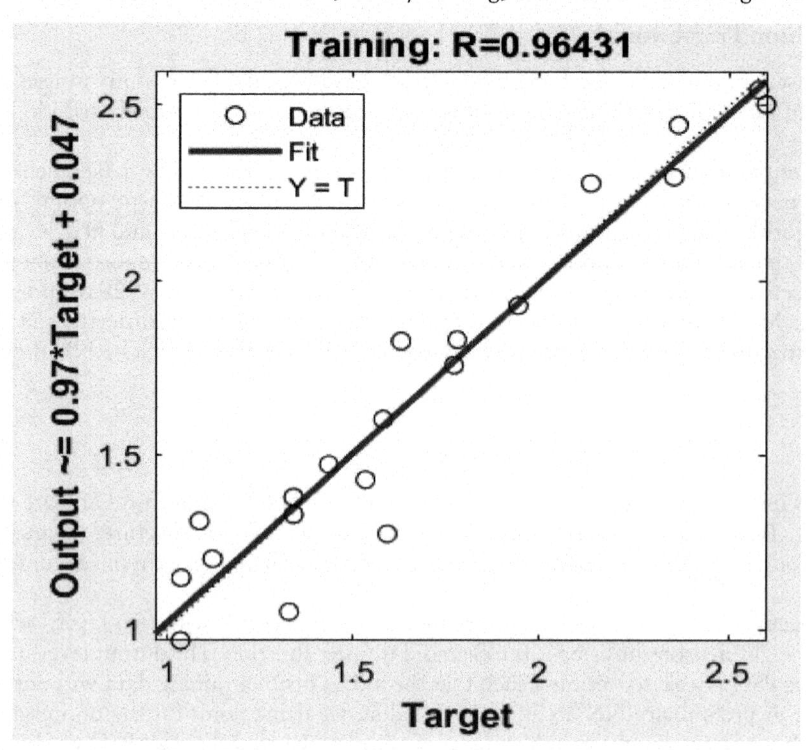

Figure 63.5 BPNN training

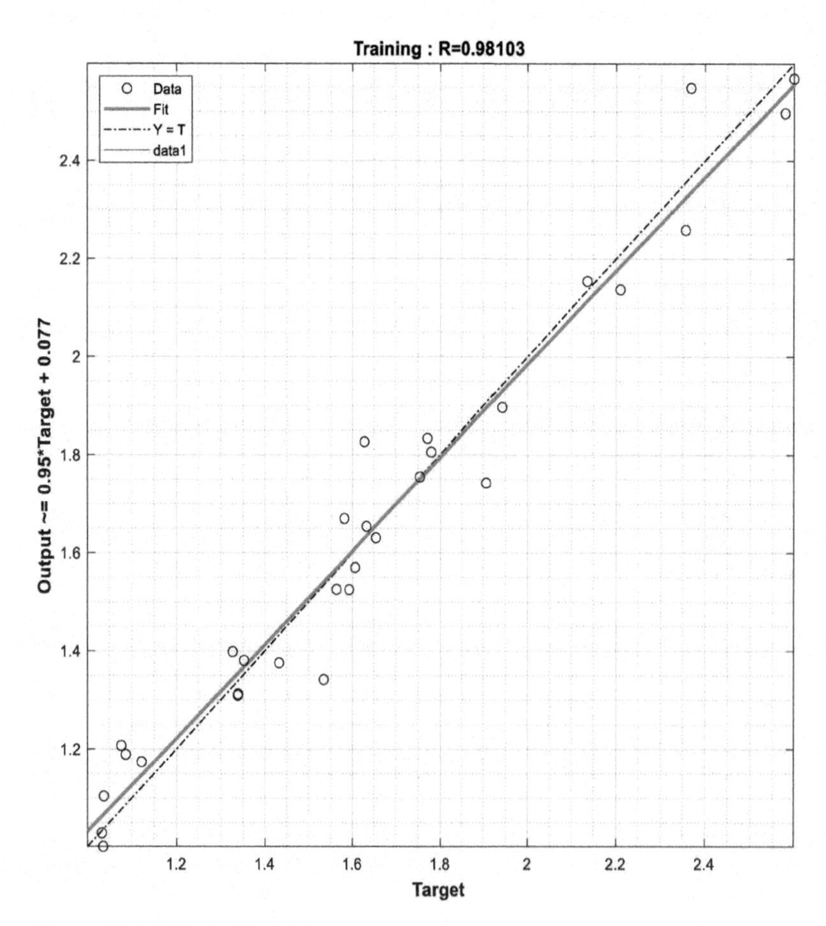

Figure 63.6 PSO-ANN training

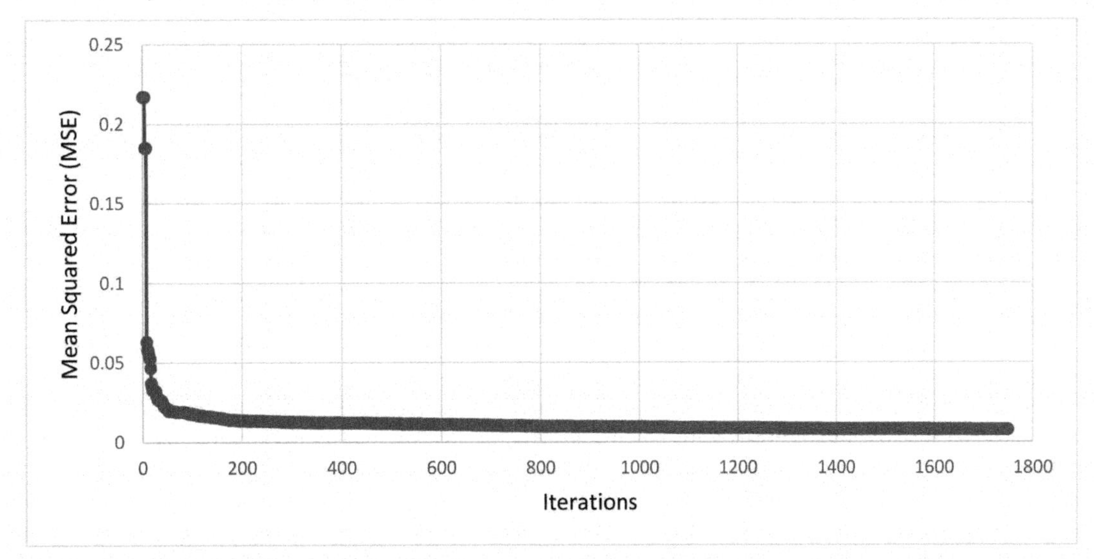

Figure 63.7 Error convergence of PSO-ANN

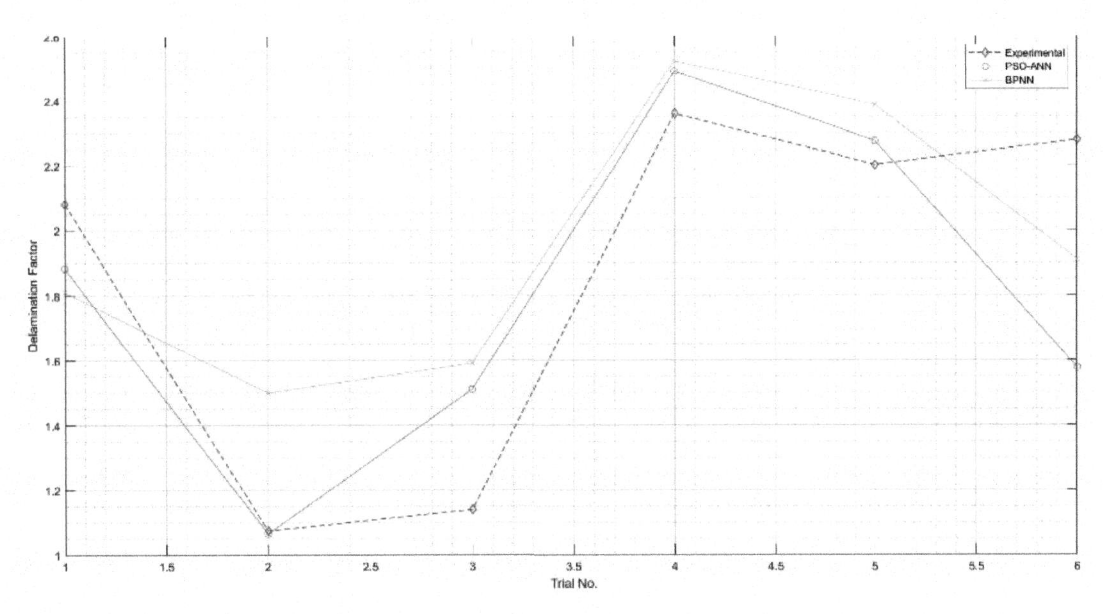

Figure 63.8 Testing data comparison

The graphs above show how the model's predictions fit to the given data for the BPNN and the PSO-ANN. The corresponding R-values offer a good metric to compare how well the two models fit. Given that the training dataset consisted only of 27 data points, the PSO-ANN is able to develop a model with a greater R-value. Furthermore, the merits of PSO-ANN can be further identified in the Mean Squared Error which converges to a value of 0.0076. The figure below shows the convergence of the error values with increasing optimising iterations, it is seen that the major optimisation occurs in the first few iterations as the drop in MSE is extremely large, however, the graph then proceeds to display an asymptotic nature as it levels around 0.0076.

Lastly, Figure 63.8 presents a comparison between the experimental, PSO-ANN and the BPNN values. It is evident that the PSO-ANN better fit the experimental data, however, both models fail to predict the last testing value. This may be a reflection on the lack of training values used to develop the models. However, the accuracy of the PSO-ANN is evidently greater under the first five data sets.

Future Prospects

In agriculture, it can be noticed that weeds (the unwanted plants) often occupy sizeable portions of a field and can survive through adverse conditions and even grow in such conditions. The weeds are versatile and can adapt to change in an expeditious manner. This forms the basis of invasive weed theory and in turn, the invasive weed optimisation (IWO) method. This method of optimisation was found to be highly suitable for solving non-linear, non-convex problems and has been successfully deployed in the fields of bioinformatics, image processing and other forms of industrial engineering. Thus, this method can be used in order to further optimise the conditions for drilling (Kumar et al., 2020).

The suggested models can be further improved by using a larger dataset for both – training and testing. This would enable addition of further information and can be useful for improving the best fit of the results.

It has been observed that complex NNs can be very expensive to optimise and that in some cases, the cost of collecting data is significantly increased. In order to tackle this situation, a Bayesian optimization can be implemented in order to fully use the information from experimental data as it is a highly data efficient method. A Bayesian optimization is useful in such a scenario because it can detect the best points without having to undergo a significant number of function optimisations. As a result of its data efficiency, the Bayesian optimization method finds applications across a number of fields such as A/B testing, recommender systems and hyperparameter tuning. Specific to machining, it has been successfully implemented in optimisation of tool wear during turning operation and for setting up autonomous turning operations (Shahriari et al., 2016).

Conclusion

The study is successful in producing two comparative neural network models which predict the delamination behaviour given the three process parameters. The main points of the study can be briefly organized as the following:

- From the results, it is evident that the PSO-ANN architecture performs better overall with the limited training dataset. While the PSO-ANN provides an R-value of 0.98103, the BPNN concludes with an R-value of 0.96431. Another major consideration can be laid on the error convergence, in which the MSE of PSO-ANN falls to 0.0076.
- From the testing dataset, it can be seen that the PSO-ANN model predicts much more soundly however, fails to predict the 6th data set. The BPNN also fails in predicting the 6th dataset but the margin of error is relatively lower.
- Another major comparison can be built from the time complexity of the two models, while the BPNN is much quicker in building the network, the PSO-ANN takes longer as a result of the in-built optimisation.
- The suggested neural network models indicate that delamination during machining CFRP composite panels can be predicted with a great degree of accuracy and this can also be supported by applications of such networks in turning, milling and other conventional machining operations. However, due to the improved accuracy, the suggested model can be altered in order to meet the specific needs of micromachining of certain materials wherein a high level of accuracy is desirous.

References

Alam, M. N. 2016. Codes in MATLAB for Training Artificial Neural Network using Particle Swarm Optimization. https://doi.org/10.13140/RG.2.1.2579.3524.

Chakraborty, D. (2005). Artificial neural network based delamination prediction in laminated composites. Mater. Des. 26(1):1–7. doi:10.1016/j.matdes.2004.04.008.

Hussein, R., A. Elbestawi, S. M., and Attia, M. 2019. An investigation into tool wear and hole quality During low-frequency vibration-assisted drilling of CFRP/Ti6al4v Stack. J. Manuf. Mater. Process. 3(3): 63.

Karnik, S., Gaitonde, V., Rubio, J., Correia, A., Abrão, A., and Davim, J. (2008). Delamination analysis in high speed drilling of carbon fiber reinforced plastics (CFRP) using artificial neural network model. Mater. Desig. 29(9):17681776.

Krishnamoorthy, A., Boopathy, S., Palanikumar, K. (2011). Delamination prediction in drilling of CFRP composites using artificial neural network. J. Eng. Sci. Technol. 6:191–203.

Kumar, D., Gandhi, B. G. R., and Bhattacharjya, R. K. 2020. Nature-Inspired Methods for Metaheuristics Optimization. Introduction to invasive weed optimization method. nature-inspired methods for metaheuristics optimization, pp. 203–214. Cham: Springer International Publishing.

Prasad, D. U. and R. Ghodke, R. 2015. Investigations of delamination in gfrp material cutting using abrasive waterjet machining. Fourth international conference on advances in mechanical, aeronautical and production techniques - MAPT 2015. 69.

Qin, X., Wang, B., Wang, G., Li, H., Jiang, Y., and Zhang, X. (2014). Delamination analysis of the helical milling of carbon fiber-reinforced plastics by using the artificial neural network model. J. Mech. Sci. Technol. 28(2):713719.

Seo, D. and Lee, J. (1995). Effect of embedded optical fiber sensors on transverse crack spacing of smart composite structures. Compos. Struct. 32(14):5158.

Shahriari, B., Swersky, K., Wang, Z., Adams, R., and de Freitas, N. (2016). Taking the human out of the loop: a review of bayesian optimization. Proceedings of the IEEE 104. 1:148175.

Soepangkat, B., Norcahyo, R., Effendi, M., and Pramujati, B. (2020). Multi-response optimization of carbon fiber reinforced polymer (CFRP) drilling using back propagation neural network-particle swarm optimization (BPNN-PSO). Eng. Sci. Technol. Int. J. 23(3):700713.

Watkins, S., Sanders, G., Akhavan, F., and Chandrashekhara, K. (2002). Modal analysis using fiber optic sensors and neural networks for prediction of composite beam delamination. Smart Mater. Struct. 11(4):489495.

Wythoff, B. 1993. Backpropagation neural networks: A tutorial. Chemometric. Intell. Lab. Syst. 18:115155. https://doi.org/10.1016/0169-7439(93)80052-J

Annexure

RPM	Feed	Angle	Target
4000	1000	85	1.038
4000	3000	85	1.534
4000	6000	85	1.5925
8000	1000	85	1.0365
8000	3000	85	1.0765
8000	6000	85	1.327
8000	9000	85	1.633
40000	1000	85	1.0335
40000	6000	85	1.3385
40000	9000	85	1.1215
4000	1000	115	1.581
4000	3000	115	1.943
4000	6000	115	2.136
4000	9000	115	2.578
8000	3000	115	1.7715
8000	6000	115	2.2095
40000	1000	115	1.339
40000	3000	115	1.3525
40000	6000	115	1.563
40000	9000	115	1.6065
4000	1000	130	1.907
4000	3000	130	1.6285
4000	9000	130	2.598
8000	1000	130	1.655
8000	3000	130	1.78
8000	6000	130	2.3565
8000	9000	130	2.3685
40000	1000	130	1.0865
40000	3000	130	1.4325
40000	9000	130	1.7545

64 Use of machine learning and financial risk profiling for sentiment analysis

Sheetal Thomas[1,a], *Mridula Goel*[2,b], *Parul Verma*[3,c], *and Gunjan Chhablani*[4,d]

[1]Indukaka Ipcowala Institute of Management (IIIM), Charotar University of Science and Technology (CHARUSAT), Anand, India

[2]Department of Economics and Finance, BITS Pilani K K Birla Goa Campus, Goa, India

[3]Department of Electrical and Electronics Engineering, BITS Pilani K K Birla Goa Campus, Goa, India

[4]Department of Computer Science and Information Systems (CSIS), BITS Pilani K K Birla Goa Campus, Goa, India

Abstract

Investor sentiments play a key role in planning of financial portfolio. Sentiment analysis using analytics and machine learning is applied to online expressions (blogs, tweets, etc.) to understand investors emotions in relation to their financial choices. This may however result in inaccurate capturing of sentiments and choices. This paper applies sentiment analysis on primary data gathered as handwritten text about financial choices that an individual would make if they win a lottery. For the purpose of risk profiling, original labels for risk averse, risk neutral and risk seekers were developed by the authors based on literature review. These labels help to train the supervised learning models. Textual Analysis and Machine Learning are applied to understand financial polarities by examining individuals preferred choices to allocate their lottery winnings into different asset classes. The F1-scores across all models show that DistilBERT provides best results from sentiment analysis, while long short-term memory networks (LSTM) are best for risk profiling with F1 scores of 0.501 on test dataset. The findings of the study show that there are no sudden or drastic changes in either investor sentiment or risk attitude on winning a lottery.

Keywords: Sentiment Analysis, Financial Preference, Risk Attitude, Lottery.

Introduction

Modern behavioural finance recognises both sentimental investors and rational investors (Byrne et al., 2008). In recent years, studies have focused on the relationship between sentiments and decision making (Seo et al., 2007). Baker and Wurgler (2007) in their study of investor sentiments suggest that extraordinary sentiments can influence stock market prices.

Computerised methods such as artificial intelligence (AI) and machine learning (ML) are being increasingly employed to analyse investor sentiments (Gladstone et al., 2019; Sohangir et al., 2018; Thomas et al., 2020). It has increased in popularity in the last decade. Introduced in 2003, 'sentiment analysis' is defined as 'determining the subjectivity polarity (positive or negative) and polarity strength (strongly positive, mildly positive, weakly positive etc.) of a given review text; in other words—determining the opinion of the writer' (Nasukawa and Yi, 2003). It is widely applied to datasets such as financial news, blogs, transcripts and podcasts (López-Cabarcos et al., 2020). This study develops a machine learning based application to bring out the sentiments towards windfall gains from a lottery. It classifies resultant investment choices by individuals into risk labels to train the supervised machine learning models.

The data for this study was collected as handwritten text on financial choices that an individual would make if he wins a lottery of INR one crore (US $135,000 approx.). Sentiment analysis requires a set of data to be labelled for effective classification. In carrying out the sentiment analysis existing auto-labels were used (Loughran and Mcdonald, 2011). However available labels were able to capture financial choices to a limited extent. Accordingly, we have developed own set of labels to categorise investors into three risk profiles using literature review and handwritten text.

By employing textual analysis and machine learning to understand financial polarities, the paper identifies how individuals chose to allocate windfall gains from a lottery. As winning a lottery causes a burst of emotions and opens investment choices, the following question was posed to gather responses for sentiment analysis: 'If you won a lottery of INR one crore tomorrow, how would you spend it?'.

[a]sheetalthomas.mba@charusat.ac.in; [b]parulverma0997@gmail.com; [c]mridula@goa.bits-pilani.ac.in; [d]chhablani.gunjan@gmail.com

Recommending investment products based on investors risk preferences helps in developing trust and customer satisfaction (Batra and Kumar, 2018). The use of handwritten text to gather sentiments can help to develop better recommendations for portfolio management.

Research studies have found that measuring investor sentiment is possible, and information gathered through sentiment analysis is observed to be frequently reflected in stock markets performance (Baker and Wurgler, 2007). People tend to provide information about their sentiments by divulging financial plans and preferences through various cues. These sentiments can relate to their level of confidence in a financial product, intention to invest in a particular asset or just their overall future plans (Moons et al., 2013).

The language used by the individual reveals their psychological makeup. The content analysis of expressions can help researchers to identify the underlying sentiment and use the same to predict attitudes, personality and behaviour (Holtgraves et al., 2014). Such assessments of language involve both subjective and objective ratings. The subjective ratings look for words, tones and semantic meaning that can reflect, for example, the risk propensity of an individual (Kjell et al., 2019). On the other hand, objective assessments use a developed dictionary of words to map specific emotions; like sadness is described by use of negatively valenced words and happiness is described by positively valenced words (Kauschke et al., 2019).

In this paper, subjective ratings have been used. This allows to capture the right sentiment underlying the expressions of financial terms. Use of text to analyse financial behaviour of the individual allows to gain important information regarding individual psychology and emotional distress. The text expressed in the event of winning a lottery is used to understand the sentiments of investors, and their risk preferences.

To analyse sentiments, it is important that we have a fairly good sentiment lexicon (Feldman, 2013). Harvard Psychosociological Dictionary is one such dictionary that is available for sentiment analysis. Another dictionary that is available in specific context to analyse sentiments behind financial news and reports is prepared by Loughran and Mcdonald (2011). This work uses the Loughran and Mcdonald dictionary for sentiment labelling and carry out a manual literature review-based labelling for risk profiling.

The work by Wang et al. (2013) emphasises that individual risk preferences are also based on financial sentiments (Wang et al. 2013). One method to assess the financial sentiment is to process financial text inputs with critical linguistic features such as content semantics (Keith and Stent, 2019) or investor sentiment (Malandri et al., 2018).

However, sentiment analysis of financial data is constrained by non-availability of large-scale training data and the difficulty in labelling of the texts, which requires experts from the domain. As a result, the performance of models for financial sentiment analysis is usually much lower than using the same models for application to the general domain (which is referred to as a problem of domain adaptation).

Another study shows that available sentiment analysis labels usually do not perform well in the financial domain (Xing et al., 2021); to overcome this limitation the researchers have created self – labelled (Table 64.2) library to asses individual risk profile and preferred financial product. Different language models were experimented with to identify the one which provides better accuracy rate.

The Bidirectional Encoder Representations from Transformers (BERT) based language model is adapted for financial domain analysis, namely Fin BERT. The research studies using Fin BERT models have shown that it can provide better outputs with smaller training data sets and outperform widely used machine learning methods (Araci, 2019). Given that the dataset is based on primary data, FinBERT models were found to be more suitable.

This paper experiments with supervised machine learning models to identify those with best accuracy. It also provides decision tree plots for sentiment analysis and risk profiling. The next section explains the theoretical framework of the study. It describes various steps of performing sentiment analysis. The concluding section gives the results obtained from the study and discusses its application.

Research Methodology

Data curation

Primary data was collected from 402 participants, these were employees of universities, organisations and self-employed. The questionnaire asked for demographic details and information related to the financial preferences of the individuals. 351 duly filled questionnaires were received out of which 321 clean texts could be extracted and these were used to define the financial sentiments using machine learning. The responses to the open-ended question, 'If you won a lottery of INR one crore tomorrow, how would you spend it?' were highly specific and were used as input into the application to understand sentiments and develop labels for risk profiling. Few samples of statements gathered as responses are shown in Figure 64.1 (flowchart) under heading samples of statements from collected data.

Data description

As most research in the area of sentiment analysis and behavioural preferences is done using textual data, text content from the questionnaire has been extracted. The data also included investment choices across various financial products. These were used to assess whether investor sentiments associated with windfall gains are different from normal financial preferences.

Data processing

Step 1: The handwritten dataset was manually converted to a digital comma separated values (CSV) format and numeric codes were mapped to different responses in Likert score, while the short text answers to the open-ended question were typed as is. This data then had to go through multiple iterations of cleaning manipulation to make it suitable for ML models.

Following are the iterations of data processing that were followed on the raw CSV dataset. The goal of data processing was to create a labelled dataset to extract important words, having financial vocabulary and allow for input-independent ML processing.

Iteration 1—All punctuation marks and special symbols such as brackets, carat, apostrophe and hyphen were removed while full stops and commas were retained.

Iteration 2—Numeric listing of the statements provided by the respondents was removed as it was not contributing to the text examples. Grammatical errors were corrected manually and commas were removed.

Step 2: Sentiment analysis requires a set of data to be labelled for effective classification. The three labels used were – 'risk averse', 'risk neutral', and 'risk seeker'. The polarity strength was also labelled using the Loughran McDonald dictionary-LnM (Loughran and Mcdonald, 2016; Loughran and Mcdonald, 2011). The data was annotated using words from LnM dictionary shown in Figure 64.1, as risk profiling using collected statements. These labels have then been used for training the supervised learning models (Table 64.1). To classify risk preferences of individuals based on the financial products in which they have invested, an extensive literature review was performed. Table 64.2 provides literature review to support labelling of risk preferences and preferred financial products.

Step 3: A few data points were lost as the responses to the open-ended question were incomplete, a total of 11 such questionnaires were dropped. Thus, the final CSV dataset included complete data from 310 respondents. Word clouds, F1 scores, and decision tree plots were generated. Word Clouds are generated using the word cloud python library, where the frequency of words was provided. The words are separated in a sentence; punctuations in the following square-bracketed list [<,>, (,), ', removed from the text, and the frequency is calculated, and stop words are removed.

a) Supervised machine learning for sentiment and risk profile analysis

This kind of learning involves labelled data. The sentiment labels and the risk profile labels created/annotated were used in the data pre-processing step. The text is used as is, after converting it to lowercase.

The tokenisation procedure of text depends on the model used. The various algorithms generate a function that maps inputs to desired outputs—positive, negative, neutral in the case of sentiments, and risk seeker, risk averse, risk neutral in case of risk profiles.

First and foremost, baselines for the tasks were prepared and compared with other models to understand if they brought any improvement. For both the tasks, two baselines were built for a standard of comparison. In the first, all labels are neutral and in the second all labels are randomly assigned. The following models were tested to find best accuracy.

Transformer based models: Transformers were introduced in 2017 and revolutionised the natural language processing (NLP) field. In this study a popular model of BERT is being used. For training or pre-training these models, Hugging Face transformers library was used (Wolf et al., 2020). BERT is a pre-trained language model, and contains deep bidirectional representations of text, based on its vocabulary

Table 64.1 Example of labels provided to words for sentiment analysis

Sentiments	*Labels*
Positive	Assured, Charitable, Dream, Good, Stable, Valuable, Win, Will
Negative	Burden, Disclose, Poor, Wasting, Close
Uncertainty	Almost, Exposure, May, Might, Pending, Probably, Risk

Source: Compiled from words listed in primary dataset based on Loughran McDonald dictionary

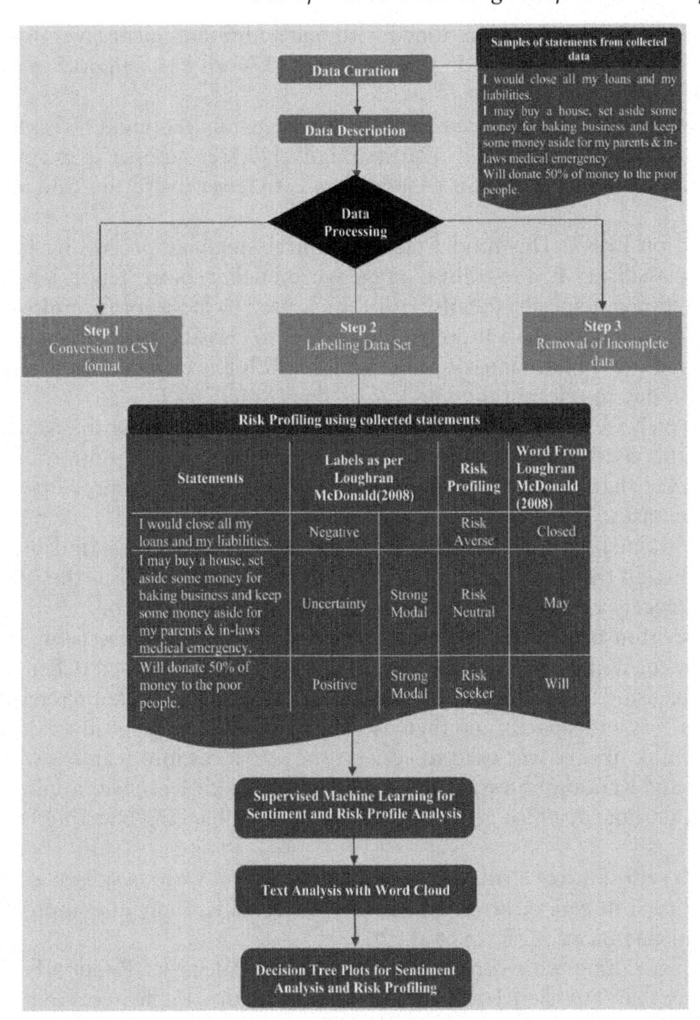

Figure 64.1 Flowchart of research methodology

Table 64.2 Individual risk profile and preferred financial product

Author and year	*Title*	*Source*	*Summary*		
			Risk prefe-rence	*Key factors*	*Product*
Nadeem et al. (2020)	How Investors Attitudes Shape Stock Market Participation in the Presence of Financial Self-Efficacy	Frontiers in Psychology	Risk seekers	Investment intentions of higher returns	Stocks
Akhtar and Das (2019)	Predictors of investment intention in Indian stock markets: Extending the theory of planned behaviour	International Journal of Bank Marketing	Risk Seekers	Financial self-efficacy	Mutual funds and stocks
Hoang et al. (2018)	The seasonality of gold prices in China: Does the risk-aversion level matter?	Accounting & Finance	Risk Aversion	Long term investments and short positions in bullion market	Investment in gold
Muendler (2008)	Risk-neutral investors do not acquire information	Finance Research Letters	Risk Neutral	Requires less or no information regarding asset risk exposure. Would generally hold diversified portfolio	Property, stocks and mutual funds

Source: Compiled from literature review

of about 30,000 tokens. The pre-trained BERT model is fine-tuned with one additional linear layer on top to create a simple classification model. The tokenization is done with standard Word Piece algorithm used with BERT systems (Devlin et al., 2018).

Distil BERT is a version of BERT which has been reduced by 40% in size, while retaining 97% of its language understanding capabilities. DistilBERT is used as it requires relatively lesser number of examples than other deep models. With fewer number of parameters, it generally leads to lesser overfitting and more generalisation (Sanh et al., 2019).

FinBERT is a language model based on BERT. This model tackles natural language processing tasks, specifically financial news sentiment classification. Pre-trained language models require fewer labelled examples and can be further trained on domain-specific details. FinBERT is used to incorporate multi-step training on financial data, thereby using a domain-relevant pre-trained model to classify collected financial text. Even with a smaller training set and only fine-tuning of the model, FinBERT is expected to perform better than BERT for the purported task due to relevant domain knowledge (Araci 2019).

BERT PT is pre-trained BERT, on which we pretrain the study data for five epochs using the Masked Language Modelling, which is a self-supervised learning approach. By doing this, a two-step pre-training is induced. Generally, it has been observed that pre-training on relevant domain tends to improve performance of the fine-tuning or downstream tasks.

BERT PT-TR is a BERT model pre-trained on only the text from the study training set (and hence, excluding the test set). Generally, training on the test set can induce bias in the model, and thus the scores reported may not be general enough. Hence, experiment was conducted with this variant also.

Tree based models are variants of decision trees-based systems. For RandomForest and Decision Tree, the scikit-learn library in python was used. While for XGBoost, the XGBoost library was used (Chen and Guestrin 2016). The tokenization is done using a custom method which takes in all unicode characters, and replaces punctuations and special characters with spaces, and then uses the SpaCy library English tokenizer for splitting. Tf-Idf vectorizer from the nltk library was used to convert the sentences into features which can be fed to the models. For XGBoost and RandomForest, TruncatedSVD from scikit-learn was applied in order to reduce the sparsity of the data produced, which generally helps with such bagging/boosting based models.

Random Forest is a combination of a series of tree structure classifiers trained on various subsets of the data. Random Forest has been widely used in classification and prediction. This is a bagging approach, where a 'bag' of trees is used to perform classification (Reza et al., 2016).

Decision Trees are simple tree-based algorithms, which are used for very small-scale tasks. RandomForest and XGBoost often perform much better than Decision Trees, they also help visualize the features in a tree format, which helps to understand what the trees might be internally focusing on (Quinlan, 1986).

Sequence models are neural networks which are used in a recursive fashion to learn on a time-based or text-based sequence. For this, the PyTorch library, and two bidirectional LSTM layers for classification are used. LSTM is a recurrent neural network (RNN) architecture that is well-suited to classify, process and predict textual data time series given time lags of unknown lengths/duration. It was one of the state-of-the-art text processing neural systems before transformers/BERT models arrived (Hochreiter and Schmidhuber, 1997).

Results and Discussion

Text analysis results with word cloud

The study has used a word cloud to visually interpret the text from the responses received from the open-ended question – 'If you won a lottery of 1 crore (INR) tomorrow, how would you spend it?'. It is useful in quickly gaining insight into the most prominent items in a given text, by visualizing the word frequency in the text as a weighted list. It was found that there were a number of words that were often repeated across all answers indicating that an aggregated analysis of these important common words could give us deep insights into the general investor sentiment across our population of study.

The word cloud (Figure 64.2) is based on financial terms where frequencies of financial terms are clubbed into super-categories in order to better understand financial products/instruments mentioned in the answers. It can be seen that the highest frequency is for words such as 'investment' 'property', 'fixed deposits', 'charity', 'gold', 'bank', 'mutual funds' and so on. The frequency of words 'expense', 'spend', 'buy', describe that individuals want to spend the money from the lottery.

The study also includes a comparative analysis of the results obtained by experimenting with different machine learning (ML) models like XGBoost, LSTM, BERT, RandomForest, Baseline (Neutral), Baseline (Random), BERT-PT, FinBERT, BERT-PT-TR, Decision Tree on the final 310 clean datasets to find out which model is the best fit for the purpose of sentiment analysis. The data is split into two parts, namely, the training data and the testing data using an 80–20 split.

During the process of labelling, it was found that financial sentiment analysis is a challenging task due to the specific language and lack of pre-existing labelled data in the domain. General-purpose models are not effective enough because of the language used in a financial context.

Sentiment modelling

For sentiment analysis, supervised machine learning models were used to analyse the data. The section below provides results of the same.

For supervised classification, linear projection on the sentence embeddings, which is the [CLS] token embedding from the BERT models were used. For LSTM, we use the last hidden state in a similar way. For the tree-based models, the features, as explained in Section 3, are used. Very interesting results, from the macro F1 scores were observed. It was expected that all the models would tend to predict all labels as neutral, as most of the labels in the data are neutral classes (~82%). Hence, they would perform either as good as the neutral baseline, or marginally better than that.

However, it was seen that while bagging/boosting methods—XGBoost, RandomForest are expected to perform better, the Decision Tree, despite being a simpler algorithm, performs much better than its tree-based counterparts and achieves a score of 0.739 (Table 64.3).

The BERT model overfits due to its large number of parameters, achieving a train score of 1, but a test score of only 0.53. Pre-training helps with the learning, thereby improving the score slightly in both

Figure 64.2 Financial product preferences

Table 64.3 Train and test F1 scores for sentiment analysis and financial risk profiling

Model	*Sentiment Analysis*		*Risk Profiling*	
	Train Data (F1 Score)	*Test Data (F1 Score)*	*Train Data (F1 Score)*	*Test Data (F1 Score)*
XGBoost	0.595	0.302	1	0.391
LSTM	0.907	0.403	1	0.501
BERT	1	0.535	0.183	0.186
RandomForest	0.301	0.302	1	0.441
Baseline (Neutral)	-	0.302	-	0.186
Baseline (Random)	-	0.18	-	0.359
BERT-PT	1	0.693	1	0.344
FinBERT	1	0.725	0.885	0.311
BERT-PT-TR	0.723	0.574	0.874	0.332
DecisionTree	0.874	0.739	1	0.414
DistilBERT	0.909	0.771	1	0.291

BERT-PT (0.693) and BERT-PT-TR (0.574). DistilBERT, being a light-weight version of BERT, achieves a test score of 0.771, and still is not as bad as BERT on overfitting, achieving only 0.909 train score. FinBERT model is able to use some of the financial context learned during its training and is therefore, able to perform much better than its BERT counterparts, with the exception of DistilBERT.

The LSTM model falls somewhere in the middle, with the test score of 0.403, meaning that deep learning language models outperform sequence models by a good margin on this task.

Overall, DistilBERT achieves the best performance of all, followed by Decision Trees, followed by FinBERT. The fact that Decision Trees perform as well as these pre-trained deep learning models shows that simple ML algorithms can also perform very well with automated sentiment-labelling. Based on these results future researchers can compare these models on human-annotated sentiment data to understand whether the simple models still perform as well as the pre-trained deep learning models.

While the best decision plots had different depths, the Decision Trees with depth = 5 were plotted to help understand the top few words which the Decision Tree looks at before classification.

Decision tree plots for sentiment analysis and risk profiling

From the Decision Tree below (Figure 64.3), it is observed that the top few most important words, according to the Gini metric (default) are – ‘good’, ‘poor’, ‘win’, ‘dream’, ‘help’ and ‘happy’. The decision rules are mentioned on the top of each parent node. It can be seen that ‘good’ is the most important term as it is on the root node. If the Tf-Idf (term frequency-inverse document frequency) score of ‘good’ is greater than 0.052, it is classified as positive sentiment, otherwise, the rest of the decision rules are checked. In the next step, if the Tf-Idf score of ‘poor’ is greater than 0.112, then the sentiment is ‘negative’, and so on. The leaf nodes are the final classes which are assigned to the examples, based on the decision. Such a plot provides us insights into what the Decision Tree, or other tree-based algorithms, might look at for classification tasks.

It was found that the leaf nodes are pure, i.e. they have examples of only one class present in them, which might mean that model is able to learn the train dataset very well.

For risk profiling, a similar supervised analysis was performed using the risk profile labels during training. Poor results were observed on the test set, while the models often tend to overfit on the training data, getting near 1 macro F1 scores.

The BERT-PT-TR, BERT-PT, FinBERT, and DistilBERT models achieve relatively less test score than the random baseline, but the BERT model fails to get a score even above the neutral baseline (0.186), which is much lesser than the random baseline (0.359).

This can be because the initial starting point of the BERT model is not as good as other BERT-based models, which sometimes affects deep learning systems. The inability of other BERT-based models to cross the random baseline threshold suggests that the problem is learnable, but these models do not have enough data for learning the relationship in this particular task. The same can also be inferred as these models achieve near 1.0 macro F1 train score, which is a clear case of overfitting.

All the tree-based models cross the random baseline, and all of them get very close scores. RandomForest takes over Decision Tree, followed by XGBoost, which could mean that using a bagging-based classifier is better than boosting, as a voting method is used in bagging which proves to be a more effective form of ensemble than boosting. These models outperform the BERT based models. However, these models also overfit on the train data, achieving 1.0 train scores, meaning that there is further scope of improvement by

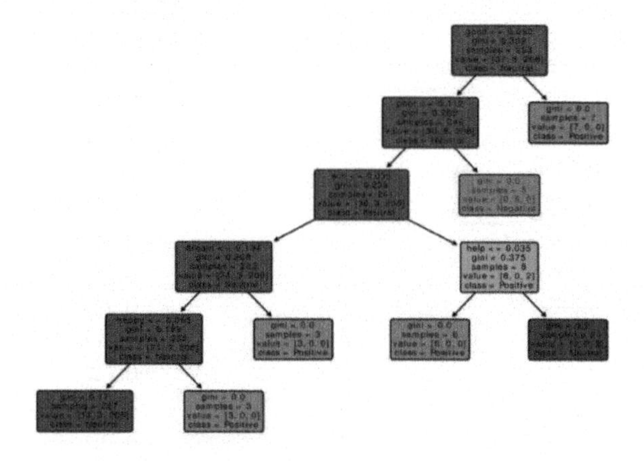

Figure 64.3 Decision tree plots for sentiment analysis

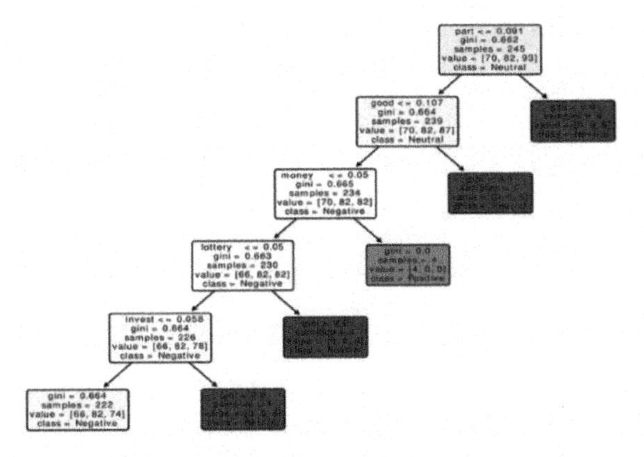

Figure 64.4 Decision tree plots for risk profiling

improving or augmenting the dataset. However, the fact that models achieve the score above the random baseline shows that this is a learnable problem.

The LSTM model outperforms these models, which is somewhat intuitive based on the trend. The neural model is able to identify relationships more complex than the tree-based systems, which only work with decision rules. However, it has a significantly lesser number of parameters, which somewhat reduces the overfitting factor, and helps it reach a test score of 0.501, despite achieving a train score of 1.0.

Overall, in this case, there is a clear order of the category of models: Sequence Model (LSTM) > Tree-based models (RandomForest > DecisionTree > XGBoost) > BERT-based models (BERT-PT > BERT-PT-TR > FinBERT > DistilBERT > BERT).

From the Decision Tree, five most important words, according to the Gini metric (default) are – 'part', 'good', 'money', 'lottery', and 'invest' (Figure 64.4). The decision rules are mentioned on the top of each parent node. The leaf nodes are the final classes that are assigned to the examples, based on the decision. Here, 'positive', 'negative', and 'neutral' stand for 'risk seeker', 'risk averse', and 'risk neutral', respectively. The left arrow is for the case when the given condition is true, and the right is for when the condition is false. For example, at the second node, if the Tf-Idf score of 'good' is greater than 0.107 for a given sentence, then it is classified as 'risk neutral', otherwise, further rule checking is done. This provides very good insights into what the model might be looking at during classification. Note that the best Decision Tree model used all the available features and generated a very large tree, but for ease of visualisation, the tree which uses five features is presented.

Conclusion

The paper uses a unique dataset of individual responses to the open-ended question – 'If you won a lottery of INR one crore tomorrow, how would you spend it?' collected through primary survey. It develops labels for financial data and sentiment analysis. These labels were used to develop a machine learning based application to bring out the sentiments towards windfall gains from a lottery. Such applications can be used to understand sentiments in various financial events.

It was observed that neutral sentiment prevails even after winning the lottery, followed by positive and negative sentiments. This showed that there is no sudden euphoria due to lottery winnings.

By applying different models, it was observed that high F1 score can be achieved on the testing data; thus, demonstrating that it is a learnable problem. It was found that sequence model (LSTM) and tree-based models (Random Forest) gave the best F1 scores for risk profiling. It is also noteworthy that pre-trained deep learning models and Decision Trees perform well showing that simple ML algorithms can also provide good accuracy with automated sentiment-labelling.

The long-term objective of this study is to develop a software that integrates financial sentiment analysis and risk tolerance classification for different financial events, including those from increase in earnings.

References

Ackert, L., Church, B. K., and Deaves, R. (2003). Emotion and financial markets. Economic Rev. - Federal Reserve Bank of Atlanta. 88(2):33–41. http://search.proquest.com/docview/200407218?accountid=79789.

Araci, D. T. 2019. FinBERT: Financial Sentiment Analysis with Pre-Trained Language Models. *ArXiv*. https://doi.org/10.48550/arXiv.1908.10063

Baker, M. and Wurgler, J. (2007). Investor Sentiment in the Stock Market. J. Econ. Perspective. 21(2):129–51. doi:10.1257/jep.21.2.129.

Batra, K. and Kumar, V. (2018). Indian individual investor behaviour: a model based study to meet sustainable and inclusive growth. World Rev. Entrep. Manag. Sustain. Dev. 14(6):705–716. doi:10.1504/WREMSD.2018.097698.

Byrne, A., Brooks, M., and Gifford, B. (2008). Behavioral Finance: Theories and Evidence. Behav. Financ.: Theor. Evidence.1979:1–26. doi:10.2470/rflr.v3.n1.1.

Chen, T. and Guestrin, C. 2016. XGBoost: {A} Scalable Tree Boosting System. *CoRR* abs/1603.0. http://arxiv.org/abs/1603.02754.

Devlin, J., Chang, M. W., Lee, K., and Toutanova, K. (2018). BERT: Pre-Training of deep bidirectional transformers for language understanding. *CoRR* abs/1810.0. http://arxiv.org/abs/1810.04805.

Feldman, R. (2013). Techniques and Applications for Sentiment Analysis. Commun. ACM. 56:82–89. doi:10.1145/2436256.2436274.

Gladstone, J. J., Matz, M. C., and Lemaire, A. (2019). Can Psychological Traits Be Inferred From Spending? Evidence From Transaction Data. Psychol. Sci. 30(7):1087–96. doi:10.1177/0956797619849435.

Hochreiter, S. and Schmidhuber, J. (1997). Long Short-Term Memory. Neur. Comput. 9(8):1735–80. doi:10.1162/neco.1997.9.8.1735.

Holtgraves, T. M., Ireland, M. E., and Mehl, M. R. 2014. Natural Language Use as a Marker of Personality. The Oxford Handbook of Language and Social Psychology. no. May. doi:10.1093/oxfordhb/9780199838639.013.034.

Kauschke, C., Bahn, D., Vesker, M., and Schwarzer, G. (2019). Review: The role of emotional valence for the processing of facial and verbal stimuli - positivity or negativity bias? Front. Psychol. 1–15. doi:10.3389/fpsyg.2019.01654.

Keith, K. A. and Stent, A. (2019). Modeling Financial Analysts' Decision Making via the Pragmatics and Semantics of Earnings Calls. *ArXiv*, no. 2015: 493–503.

Kjell, O. N. E., Kjell, K., Garcia, D., and Sikström, S. (2019). Semantic Measures: Using Natural Language Processing to Measure, Differentiate, and Describe Psychological Constructs. Psychol. Methods. 24(1):92–115. doi:10.1037/met0000191.

López-Cabarcos, M. Á., Pérez-Pico, A. M., Vázquez-Rodríguez, P., and López-Pérez, M. L. (2020). Investor Sentiment in the Theoretical Field of Behavioural Finance. Econ. Res. Ekonomska Istrazivanja. 33(1):2101–2119. doi:10.1080/1331677X.2018.1559748.

Loughran, T. and Mcdonald, B. (2011). When Is a Liability Not a Liability ? Textual Analysis, Dictionaries, and 10-Ks Journal of Finance, Forthcoming. J. Financ. 66(1):35–65. http://papers.ssrn.com/sol3/papers.cfm?abstract_id=1331573.

Loughran, T. and Bill Mcdonald. (2016). Textual Analysis in Accounting and Finance: A Survey. J. Account. Res. 54(4):1187–1230. doi:10.1111/1475-679X.12123.

Malandri, Lorenzo, Frank Z. Xing, Carlotta Orsenigo, Carlo Vercellis, and Erik Cambria. 2018. Public Mood–Driven Asset Allocation: The Importance of Financial Sentiment in Portfolio Management. Cognitive Comput. 10(6):1167–76. doi:10.1007/s12559-018-9609-2.

Moons, W. G., Spoor, J. R., Kalomiris, A. E., and Rizk, M. K. (2013). Certainty Broadcasts Risk Preferences: Verbal and Nonverbal Cues to Risk-Taking. J. Nonverbal B.ehav. 37(2):79–89. doi:10.1007/s10919-013-0146-0.

Seo, M. G. and Barrett, L. F. (2007). Being Emotional During Decision Making- Good or Bad? An Empirical Investigation. Acad. Manage. J. 50(4):923–40.

Nasukawa, T. and Yi, J. (2003). Sentiment Analysis: Capturing Favorability Using Natural Language Processing. In Proceedings of the 2nd International Conference on Knowledge Capture, pp. 70–77. New York (N Y):Association for Computing Machinery. doi:10.1145/945645.945658.

Quinlan, J. R. (1986). Induction of Decision Trees. Mach. Learn. 1(1):81–106. doi:10.1007/bf00116251.

Reza, M., Miri, S., and Javidan, R. (2016). A Hybrid Data Mining Approach for Intrusion Detection on Imbalanced NSL-KDD Dataset. Int. J. Adv. Comp. Sci. Appl. 7(6):1–33. doi:10.14569/ijacsa.2016.070603.

Victor, S., Debut, L., Chaumond, J., and Wolf, T. (2019). DistilBERT, a Distilled Version of {BERT:} Smaller, Faster, Cheaper and Lighter. *CoRR* abs/1910.0. http://arxiv.org/abs/1910.01108.

Sohangir, S., Wang, D., Pomeranets, A., and Khoshgoftaar, T. M. (2018). Big Data: Deep Learning for Financial Sentiment Analysis. J. Big Data. 5(1). doi: 10.1186/s40537-017-0111-6.

Thomas, S., Goel, M., and Agrawal, D. (2020). A framework for analyzing financial behavior using machine learning classification of personality through handwriting analysis. J. Behav. Exper. Financ. 26. doi:10.1016/j.jbef.2020.100315.

Wang, C. J., Tsai, M. F., Liu, T., and Chang, C. T. (2013). Financial Sentiment Analysis for Risk Prediction. Proceedings of the Sixth International Joint Conference on Natural Language Processing. 802–808. http://www.aclweb.org/anthology/I13-1097.

Wartiovaara, M. (2011). Rationality, REMM, and Individual Value Creation. J. Bus. Eth. 98(4):641–48. doi:10.1007/s10551-010-0643-6.

Wolf, T., Debut, L., Sanh, V., Chaumond, J., Delangue, C., Moi, A., Cistac, P. (2020). Transformers: State-of-the-Art Natural Language Processing. In Proceedings of the 2020 Conference on Empirical Methods in Natural Language Processing: System Demonstrations. 38–45. doi:10.18653/v1/2020.emnlp-demos.6.

Xing, F. Z., Malandri, L., Zhang, Y., and Cambria, E. (2021). Financial Sentiment Analysis: An Investigation into Common Mistakes and Silver Bullets. Proceedings of the 28th International Conference on Computational Linguistics. 978–987. doi:10.18653/v1/2020.coling-main.85.

65 Optimisation of airfoil shape based on lift and efficiency using genetic algorithms with PARSEC method

Vinayak H. Khatawate[a], *Raj Anadkat*[b], *and Tanmay Parekh*[c]

Department of Mechanical Engineering, Dwarkadas J. Sanghvi College of Engineering, Mumbai, India

Abstract

Selecting an optimal airfoil is a vital part of the aircraft design process. One can optimize an existing airfoil using traditional interpolation methods to suit their requirements, but the process takes up a lot of time and computational resources. An airfoil geometry is conventionally defined using many coordinates. The method described in the research paper uses a PARSEC method to parametrise and approximate the shape of the airfoil using eleven parameters by denoting it as a shape function and implementing a genetic algorithm to optimize the airfoil under suitable constraints. A MATLAB script is written to estimate the airfoil's aerodynamic characteristics using the vortex panel method and then find the ideal shape using a genetic algorithm. The fitness function is the coefficient of lift (Cl) and the script changes the shape of the geometry until the solver converges to the maximum value of Cl within the defined constraints. The improvement in the Cl and efficiency of optimized NACA 0012 is observed to be 79.34 % and 90.60 %. This method is thus effective to find optimal airfoil in the preliminary design stage.

Keywords: Airfoil shape Optimisation, genetic algorithm, PARSEC method.

Introduction

In aircraft design, the aerodynamic characteristics of a wing play a pivotal role in determining the performance of an aircraft for its desired operation. An important part of designing the aircraft is to design its wing such that it can provide sufficient lift and the required aerodynamic performance. The first step of wing design is to select an appropriate airfoil. It may be possible that an optimised airfoil does not produce much improvement on the overall aerodynamic characteristics of a wing due to the three-dimensional characteristics of airflow, airfoil optimisation usually results in better-designed wings that impact the overall aerodynamic characteristics such as drag reduction and improved wing efficiency.

To optimise an airfoil to enhance its aerodynamic performance, interpolation is a widely used technique. However, because of the large number of coordinates needed to define an airfoil shape, the process is cumbersome and thus a need arises for optimisation using parametrization arises. The method described in this paper uses the PARSEC method developed by Ulaganathan and Balu, (2009) and Sobieczky (1998 defines the airfoil geometry using minimal parameters and further optimises the airfoil using genetic algorithms (Shahrokhi and Jahangirian, 2011; Goldberg, 1989) making it a low-cost aerodynamic computation method based on potential flow (Erickson, 1991) to simplify the preliminary stage of optimisation. Once the optimal result is found, high-fidelity flow simulations can then be used to refine the results. A NACA 0012 airfoil has been optimised in this paper due to its relatively simpler shape and better performance at subsonic flows. A MATLAB script is implemented to couple the PARSEC parameters with predefined bounds with the Genetic algorithm and panel method to obtain an optimised airfoil.

Reported methods for airfoil shape optimization do not account for the efficiency of the new airfoil evolved while optimizing the lift. The method described in this paper takes into account the coefficient of drag and hence the efficiency of the new airfoil set within constraints to optimise the coefficient of lift.

PARSEC Method

PARSEC is a popular and efficient method to parametrise an airfoil. The shape of an airfoil is defined by eleven parameters with them being the upper and lower curvature, upper crest location, lower crest location, leading edge radius, wedge angle of the trailing edge, coordinate of trailing edge, and direction and thickness as denoted in Figure 65.1. In this method, the airfoil shape is described by linear combination of shape functions (Sobieczky, 1998).

[a]vinayak.khatawate@djsce.ac.in; [b]vrajanadkat@gmail.com; [c]tparekh54@gmail.com

These parameters can be used to control the curvature on both the upper and lower surfaces, with their placement influencing the strength and formation of shock waves.

Panel Method

Panel methods, as opposed to high fidelity computational fluid dynamics, solve the potential flow equations without using extensive computational resources and offer precise results, making this method suited for this optimisation needing repeated simulations. It is a fast approach for calculating flows with negligible viscosity and compressibility effects. The method requires the airfoil surface to be discretised into linear panels and assumes that each panel has a constant source strength but a unique value. It is also assumed that the strength of vortices is constant and equal across each panel. The condition that the net vorticity of the flow must be met such that the flow leaving the trailing edge is smooth (Erickson, 1991).

The velocity field's curl is assumed to be zero. The panel approach is depicted in this diagram. The numbering system starts from the trailing edge of the lower surface and moves forward to the leading-edge surface and then to the back of the upper surface of the trailing edge. N-1 panels are defined using N points as shown in Figure 65.2.

Around 200 panels are used in this optimization problem to acquire the pressure distribution around the airfoil. The Boundary condition for flow tangency is applied at the halfway of each panel. Once the tangential velocity at each panel's midpoint is determined, the pressure coefficient at each panel's midpoint can be calculated (Erickson, 1991).

Genetic Algorithm

Genetic algorithms search for optimal solutions based on Darwin's theory of evolution (Goldberg, 1989). They start with an initial population comprised of several candidate solutions called parents (chromosomes). These parents give rise to a new set of offspring for the succeeding generation. While the operators used to evolve these parents are completely random, the genetic algorithm is not. The algorithm runs each chromosome through a fitness function, those with the highest fitness in the population are chosen from the mating pool to reproduce according to a method defined in the reproduction function. The process is repeated, and a mutation operator is introduced to keep diversity in the pool so that the optimised results do not fall into a local minima/maximum. This evaluation and reproduction are repeated such that a given number of iterations have passed, and then the algorithm is terminated. At termination, the algorithm outputs the members (solutions to the optimisation problem) of the population according to the fitness function formed (Mukesh, 2014). The algorithm can be constrained by adding penalties to the fitness function.

Genetic Algorithms are divided into the following phases:

Population initialisation

The procedure begins with initializing the population in the mating pool. Every individual is a set of PARSEC coordinates that lie within the constraints defined. Genes are a set of parameters that define an individual. A chromosome is made up of genes strung together like a string shown in Figure 65.3

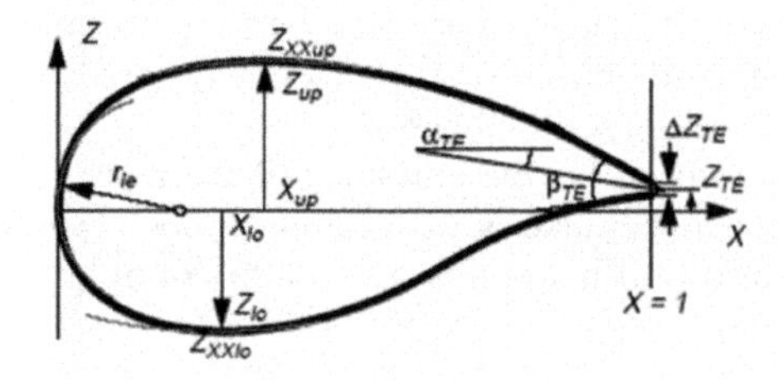

Figure 65.1 PARSEC parameters (Ulaganathan and Balu, 2009)

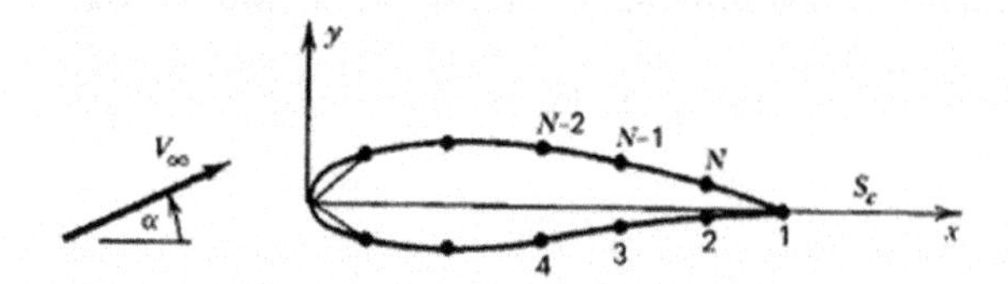

Figure 65.2 Representation of panel geometry

The genes of an individual are represented by a string of 0 and 1 in the genetic algorithm. These binary numbers are used to generate a value between 0 and 1. If the value of a bit lies between 0 to 0.5, the number 0 is set and if it lies between 0.5 to 1, a value of 1 is set. A parameter called population size can be used to define the size of the initial population.

Parent selection

The parent selection process selects the fittest individuals (in this case, those with the highest coefficient of lift) and allows them to pass their genes down to the next generation. These are the numbers that, when compared to other solutions, initially offer the optimisation function a superior value. A selection criterion is set to define a tournament among these individuals. Individuals with the highest fitness scores will be designated tournament winners and shall be selected for reproduction. To ensure that the results do not converge to a local maximum, diversity in every generation is sustained by not always selecting the best individuals and not always excluding the worst individuals. This ensures that the entire domain of the optimisation problem is explored.

Crossover

One of the most important stages in this algorithm is crossover. It is also defined as recombination. Two strings from the mating pool are picked at random to cross over and create superior offspring. This ensures that the algorithm leads to larger values of the lift coefficient, thereby reaching the maxima within the given constraints. The crossover phase is depicted in Figure 65.4.

Mutation

The best solutions to the problem are created using an operator called mutation. It is a minor and random variation in the chromosomes which ensures and preserves population diversity. It is done with a lower probability to maintain convergence since if the probability is set high, the algorithm reduces itself to a random search. Mutation is a component of the genetic algorithm that ensures the search domain is explored. The mutation operator is employed to ensure population diversity and prevent early convergence of the problem. Figure 65.5 depicts the mutation operator where a one ipped with a 0 in the offspring.

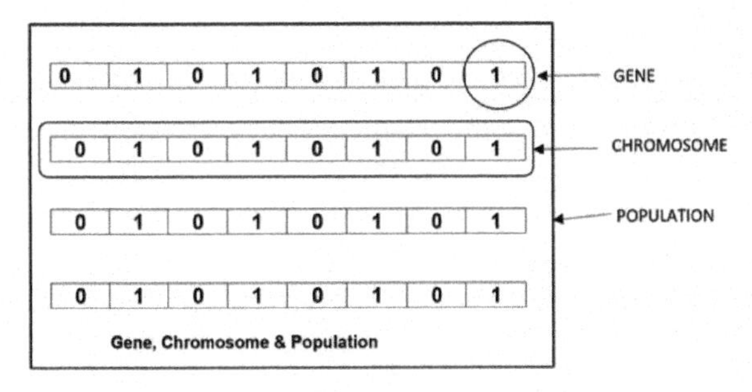

Figure 65.3 Representation of gene, chromosome, and population

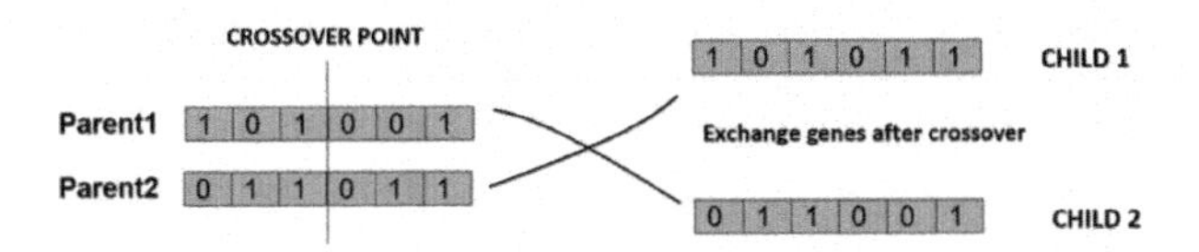

Figure 65.4 Depiction of crossover operation

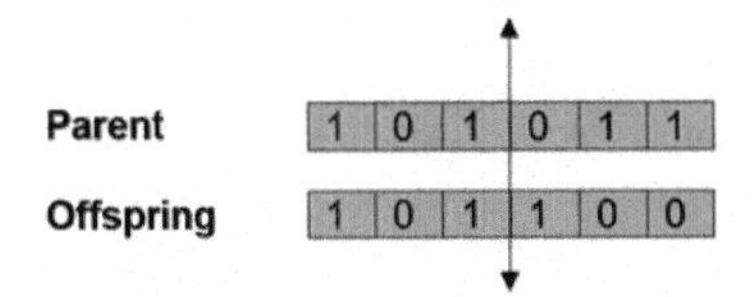

Figure 65.5 Depiction of mutation operation

Assessment of fitness

To find if an individual is capable of competing in the mating pool, a fitness function is used to measure its fitness. A score known as the fitness score is assigned to every individual. The fitness score determines whether an individual will be chosen for reproduction in the pool. For genetic optimisation with constraints, if an individual does not meet the constraints of the optimisation problem, a penalty is applied to its fitness. This renders an individual incapable of survival. As a result, when the program terminates, the offspring are the solutions that adhere to the restrictions and boundaries. In this case, one can define a constraint such as the minimum thickness of the airfoil, and maximum camber.

Termination

If no population of offspring considerably different from the previous generation (convergence) is formed, the process is ended. After that, the genetic algorithm is said to have produced a set of solutions to the problem (Goldberg, 1989).

Optimisation of NACA0012 airfoil

The goal of NACA 0012 shape optimisation is to maximise the airfoil's coefficient of lift. To ensure that the assumptions are valid throughout the optimisation process, a flow constraint is required, as the panel technique is only useful for low-velocity flows.

Various structural constraints are used to prevent unrealistic form outputs. The PARSEC parameters determine the form of the airfoil as shown in Table 65.1. The upper and lower range of values of each PARSEC parameter is fed into the Genetic algorithm. The best set of PARSEC parameters may be found in every generation created using the genetic algorithm. PARSEC parameterisation is used to create the appropriate airfoil profile. The airfoil is next tested for subsonic and incompressible flow at a 5-degree AOA. The panel method is used to assess the pressure distribution throughout the airfoil's surface. The trapezoidal rule is used to compute the coefficient of lift and drag after the pressure distribution is known. This new lift coefficient is compared to the previous one.

Table 65.1 Upper and lower bounds of PARSEC parameters

PARSEC parameters	*Airfoil values*	*Upper range*	*Lower range*
Radius of leading edge, RLE	0.0155	−0.05	0.02
Abscissa of upper crest, XUP	0.2966	0.60	0.023
Ordinate of upper crest, ZUP	0.0600	0.75	0.032
Curvature of upper crest, ZXXUP	−0.4415	0	0.372
Abscissa of lower crest, XLO	0.2966	0	0.078
Ordinate of lower crest, ZLO	−0.0605	0	0.08
Curvature of lower crest, ZXXLO	0.4530	0	−0.63
Ordinate of Trailing edge, ZTE	0	−4.55	−0.65
Thickness of trailing edge, ΔZTE	0.0013	−4.9	0.15
Direction of trailing edge, □TE	0	15	0.187
Wedge angle of trailing edge, βE	7.3600	15.1	−0.02

Optimisation objectives and constraints

Table 65.2 Objective and constraints of optimisation

Angle of attack	5°
Geometric constraints	The maximum thickness of the airfoil should be lower than 10% chord length Minimum thickness should be greater than 1% chord length Optimized cl/cd greater than 40%
Objective	Maximise the lift coefficient
Termination condition	Cl value converged and no change after 1000 iterations

Results

At a 5° angle of attack, the results of the improved PARSEC settings demonstrate an increase in lift coefficient and efficiency. The plot of the original NACA 0012 airfoil and the improved airfoil acquired after the algorithm is ended are shown in Figure 65.7.

The script also exports the coordinates of the optimised airfoil and converts them into the Selig format Dat file used to save airfoil coordinates.

The plot of the coefficient of lift against the angle of attack derived from the code is shown in Figure 65.6. PARSEC parameters and their optimised coefficients are shown in Table 65.3.

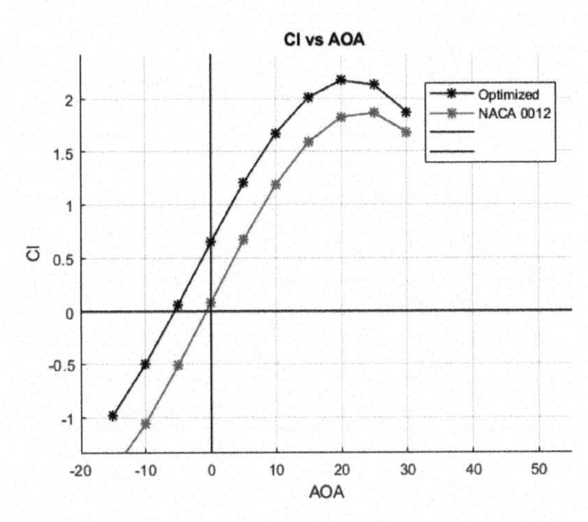

Figure 65.6 The plot of Cl vs AOA

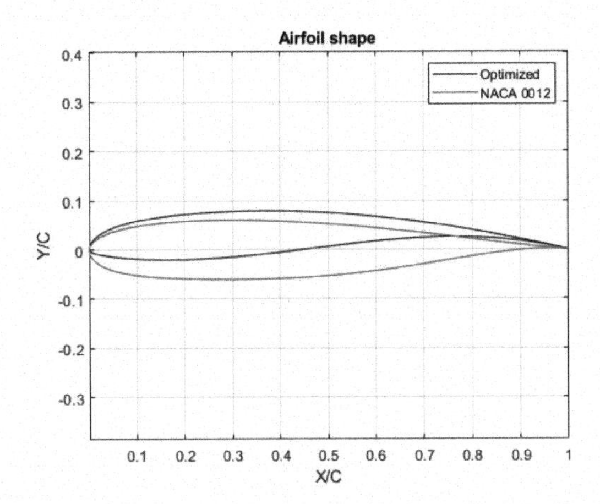

Figure 65.7 Optimised airfoil vs NACA0012

Table 65.3 Original vs optimised PARSEC parameters

Parameters	*NACA 0012 Values*	*Optimized Airfoil*
Leading-edge radius, RLE	0.0155	0.0219
Upper crest abscissa, XUP	0.2966	0.3673
Upper crest ordinate, ZUP	0.0600	0.0764
Upper crest curvature, ZXXUP	–0.4515	–0.6215
Lower crest abscissa, XLO	0.2966	0.18160
Lower crest ordinate, ZLO	–0.0605	–0.0208
Lower crest curvature, ZXXLO	0.4530	0.7193
Trailing edge ordinate, ZTE	0	0
Trailing edge thickness, ΔZTE	0.0013	0
Trailing edge direction, □TE	0	–4.8412
Trailing edge wedge angle, βE	7.3600	15.0576

Alidation of Results using XFLR5

The original NACA 0012 and the optimised airfoil coordinates were exported into a Selig format dat file and run through XFLR5 simulations using Xfoil Direct analysis. shows the XFLR5 simulation results of the coefficient of lift vs angle of attack. The resulting airfoil has a Cl of 1.103 at 5° AOA while the original airfoil has a Cl of 0.615. The improvement in the Cl is observed to be 79.34% with the maximum Cl being 1.514 vs 0.931 of NACA0012 at 10°.

Figure 65.9 shows the simulation results of the efficiency vs angle of attack for both the airfoils. The optimised airfoil has an efficiency of 68.94 at 5° AOA while the original airfoil has an efficiency of 36.17. The improvement in the η is observed to be 90.60 % being the maximum efficiency at 5° in the range of −3° to 12° angle of attack.

Tables 65.4 and 65.5 show the values of the coefficient of lift and efficiency vs the angle of attacks ranging from 3° to 12°.

Conclusion

Using the methods described in this paper, the aerodynamic performance of an airfoil can be improved by combining evolutionary methods such as genetic algorithms with panel methods to save computational resources, reduce design costs, and save computation time. The improvement in the coefficient of lift is observed to be 79.34% and the improvement in the η is observed to be 90.60%. It may be concluded that the aerodynamic shape optimisation approach employed in this study is reliable enough to be applied in the early stages of aircraft design.

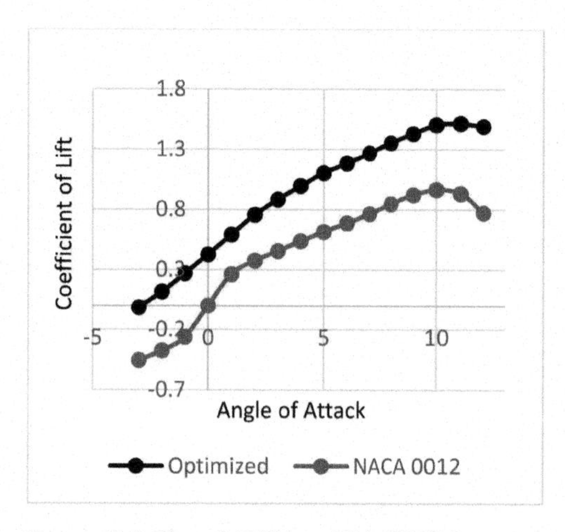

Figure 65.8 Plot of Cl/Cd vs AOA XFLR5

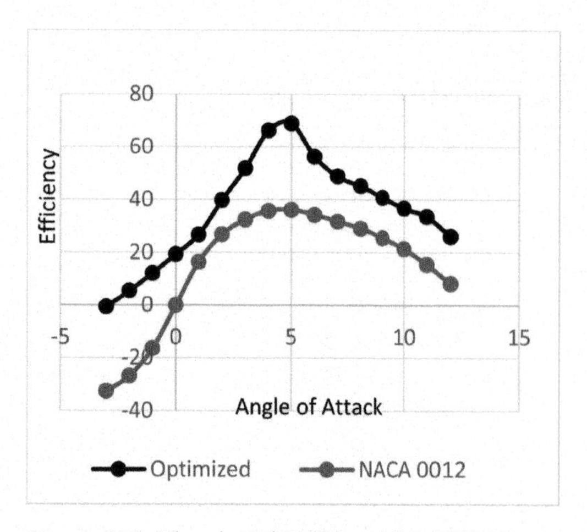

Figure 65.9 The plot of Cl/Cd vs AOA XFLR5

Table 65.4 Values of Cl, efficiency vs AOA

AOA	*NACA0012*	*Optimized*
-3	-0.456	-0.016
-2	-0.375	0.114
-1	-0.262	0.268
0	0	0.425
1	0.262	0.588
2	0.375	0.755
3	0.455	0.881
4	0.537	0.994
5	0.615	1.103
6	0.686	1.181
7	0.763	1.268
8	0.845	1.354
9	0.92	1.429
10	0.963	1.505
11	0.931	1.514
12	0.771	1.49

Table 65.5 Values of Cl, efficiency vs AOA

AOA	*NACA0012*	*Optimized*
-3	-32.57	-0.667
-2	-26.79	5.4286
-1	-16.38	12.18
0	0	19.318
1	16.375	26.727
2	26.786	39.737
3	32.5	51.824
4	35.8	66.267
5	36.176	68.938
6	34.3	56.2381
7	31.792	48.769
8	29.138	45.133
9	25.5556	40.8286
10	21.4	36.707
11	15.517	33.644
12	8.20213	26.1404

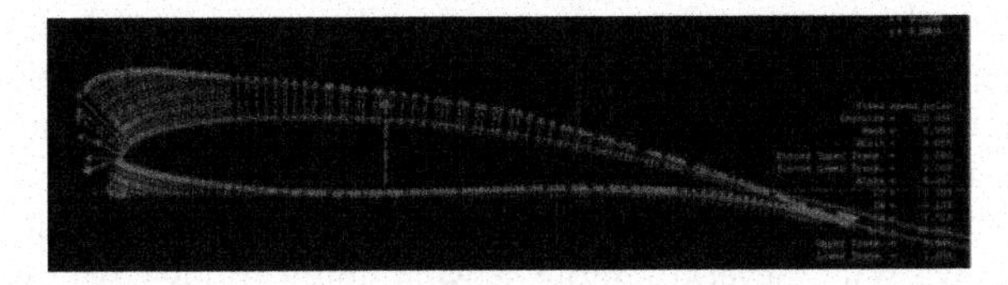

Figure 65.10 Simulation of pressure distribution and displacement thickness

References

Erickson, L. (1991). Panel methods: An introduction. California: Ames Research Center Moffett Field.

Goldberg, D. E. (1989). Genetic Algorithm in Search, Optimization and Machine Learning. Boston, MA: Addison-Wesley.

Mukesh, R., Lingadurai, K., and Selvakumar U. (2014). Airfoil shape optimization using non-traditional optimization technique and its validation. J. King Saud University. Eng. Sci. 26(2):191–197.

Shahrokhi, A. and Jahangirian, A. (2011). An Efficient Aerodynamic Optimization Method using a Genetic Algorithm and a Surrogate Model. Proceedings of the 16th Australasian Fluid Mechanics Conference.

Sobieczky, H. (1998). Parametric airfoils and wings. Notes on Numerical Fluid Mechanics. 68: 71–88, Vieweg,

Ulaganathan, S. and Balu, R. (2009). Optimum Hierarchical Bezier Parameterisation of Arbitrary Curves and Surfaces.

66 Comparative analysis of texture segmentation using RP, PCA and Live wire for Arial images

Pradnya A. Maturkar[1,a] *and Mahendra A. Gaikwad*[2,b]

[1]Electronics Engineering RGCER, Nagpur, India

[2]Information Technology, GHRCOE, Nagpur, India

Abstract

In this paper, comparative analysis of texture segmentation using random projection, principal component analysis (PCA) and live wire for Arial image processing have been done. Random projection (RP) can best be fitted for optimisation of texture feature in comparison with PCA. Segmentation has done by live wire algorithms. Grey level co-occurrence integrated algorithm was used to extract texture features from Arial images. We divided these texture features into two types of feature using K-mean algorithms: stronger features and weaker features. To compare the influence of methods on texture segmentation, the weaker discriminative ability was optimised using PCA and random projection, and segmentation was done using live wire. The results were compared using different datasets. In comparison to previous methods, we have found that RP with live wire produces better results.

Keywords: GMTD, PCA_GMTD, principal component analysis, RP_GMTD, RP_livewire.

Introduction Texture Analysis

Texture analysis (Gonzalez Book; Wu et al., 2012) is particularly challenging because of complexity of the texture pattern. We can easily execute texture segmentation that distinguishes the many textures present in an image using only the knowledge of prominent texture features. Texture feature extraction methods classified into three major types i.e. statistical, structural and spectral. In this paper we have to concentrate on texture segmentation to achieve this we have to find out texture feature of the given images. Here we used GLCIA (Clausi and Zhao, 2003) for texture feature extraction. Total 10 statistics calculated. There are DIS, CON, IDM, INV, COR, UNI, ENT, MAX, VAR and REC (Wu et al., 2012; Clausi and Zhao, 2003). Then extracted features we divided into two by using K mean algorithms i.e. stronger features and weaker feature. We optimised the weaker features by using principal component analysis (PCA) and segmentation done by using GMTD (Wu et al., 2012). To improve the result we optimised the weaker features by using random projection (RP) (Liu and Fieguth, 2012) and segmentation done by using GMTD (Wu et al., 2012). Then we compare the result by using both optimisation techniques (Maturkar and Gaikwad, 2018b). To speed up the segmentation we used live wire algorithm then we compare all result by using different dataset. For monitoring power lines airborne inspection technology used to achieve higher efficiency, accuracy as well as economy.

To see if there is a component missing from the insulator, we must section it. Aerial images recorded by helicopters, on the other hand, frequently show congested backgrounds like grassland, farming, power lines, and tiny rivers. Power lines and tiny rivers have same intensities in these environments. Therefore insulators segmentation has become significantly more complex. The research work's objectives are to use PCA and RP for optimisation and then live wire methods for texture segmentation. During our analysis of the literature, we discovered that only a few people are working on texture segmentation of insulator photos with complicated backdrops. Texture properties distinguish the entire string of insulators as a whole. The texture-based functionality is critical for a variety of applications.

This research compares texture segmentation techniques for aerial image processing, such as Random Projection, Principal Component Analysis, and Live Wire. Several remote sensing applications have used texture-based feature (Clausi, 2000; (Author name missing, 2011); Chowdhury et al., 2011; Ursani et al., 2012). In general, the texture segmentation is determined by both the performance of texture features extraction, optimisation and segmentation.

To speed the process we not only concentrate on segmentation techniques but also concentrate on feature extraction and optimisation also. Till date live wire algorithms mostly used for segmentation of medical image processing. In this paper we used live wire for aerial image processing to improve the result.

[a]pradnya.mathurkar@rgcer.edu.in; [b]mahendra.gaikwad@raisoni.net

Our research work is application based research work where we concentrate on hybrid concept for texture segmentation. Reference paper Wu et al. (2012) used the concept of optimisation and segmentation by Active Counter Model (Zhang et al., 2010; Mishra et al., 2011; Xianghua and Majid, 2008; Lianantonakis and Petillot, 2007; Sagiv et al., 2006; Chan and Vese, 2001) and got a result. Live wire Maturkar and Gaikwad (2017a), Urschler (2000) has achieved good segmentation result over Active Counter Model and new eye sight given to researcher for improving segmentation result. Hence our research paper more concentrates on live wire for segmentation instead of Active counter model. Total research work carried out in different aspect i.e. Introduction II. Features extraction using GLCIA and Clustering by K-mean III. Comparative study of texture segmentation by random projection, PCAand live wire IV.Comparative result for various textures for Brodatz dataset. V. Conclusion.

Feature Extraction Using GLCIA and Clustering by k-mean

GLCIA was used to extract the texture feature of Arial pictures. Using k-means algorithms, extracted features were clustered into two categories. Using the GLCHDH data structure, we calculated DIS, CON, IDM, and INV using GLCIA. Both the GLCHDH and GLCHSH are necessary for COR. GLCHS is used for UNI, ENT, MAX, VAR, and REC. We calculated a total of ten features.

Flow chart of research work
Algorithms for research work:
Step 1: Using GLCIA, extract the feature map.
Step 2: using K means, divide feature maps into two groups.
Step 3: Use PCA to optimise feature maps.
Step 4: Use RP to optimise feature maps.
Step 5: For segmentation, use GMTD algorithms and optimised with PCA
Step 6: For segmentation, use GMTD algorithms and optimised with RP and optimised with RP
Step 6: Use a different database to compare the results

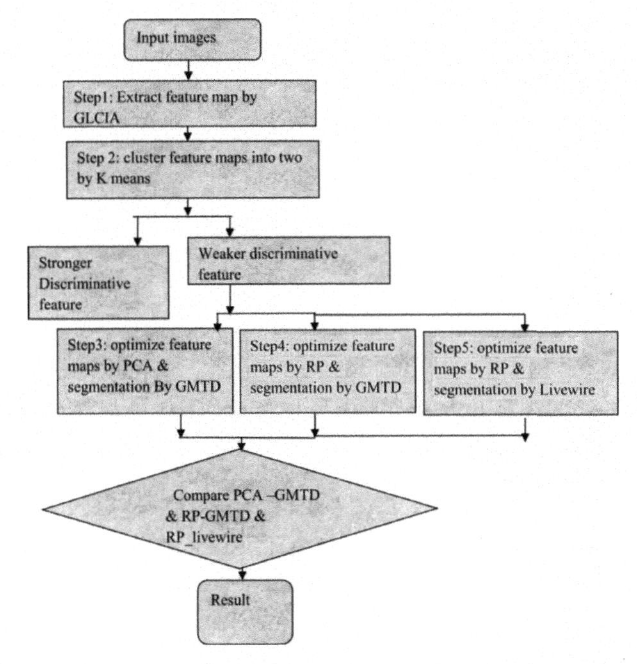

Figure 66.1 Algorithms for research work: Flow chart of research work

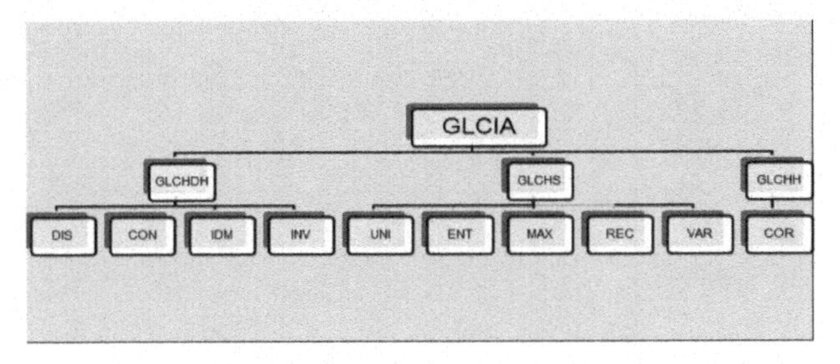

Figure 66.2 GLCIA is used to extract a feature map

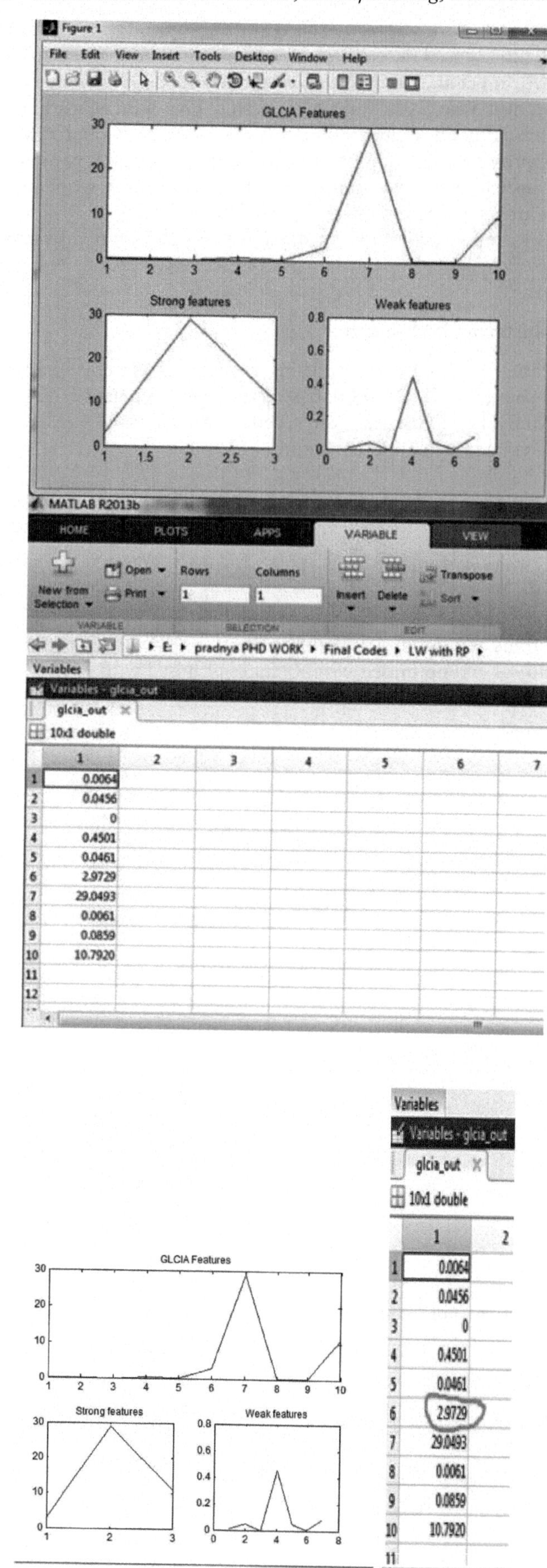

Figure 66.3 Output of GLCIA and K-mean

Comparative Study of Texture Segmentation by Random Projection, PCA and Live Wire

Following clustering, texture features with weaker feature are optimised by PCA and segmented by GMTD, followed by random projection, GMTD segmentation, and finally Live wire techniques. Using the Brodatz dataset, compare the results PCA GMTD (Wu et al., 2012; Clausi, 2000; Pesaresi and Gerhardinger, 2011; Chowdhury et al., 2011; Ursani et al., 2012; Haralick et al., 1973; Li et al., 2008; Maturkar and Gaikwad, 2017; Zhang et al., 2010; Mishra et al., 2011; Xianghua and Majid, 2008) and RP GMTD (Lianantonakis and Petillot, 2007; Sagiv et al., 2006; Chan and Vese, 2001; Maturkar and Gaikwad, 2017; Dasgupta, 2000; Fern and Brodley, 2003; Cande's and Tao, 2005; Dasgupta, 2000; Fern and Brodley, 2003; Cande's and Tao, 2006; Donoho, 2006; Biau et al., 2008; Baraniuk et al., 2008; Achlioptas, 2001; Goel et al., 2005) and RP _livewire (Maturkar and Gaikwad, 2018b; Maturkar and Gaikwad, 2018a; Maturkar and Gaikwad, 2017b; Maturkar and Gaikwad, 2014). The average value of the data set 'S' has been determined. Subtract S from the mean value. From these values a new matrix is obtained. Calculate the covariance, the Eigen values are derived from the covariance matrices, where [V1, V2... VN] are the covariance matrices. Finally, Eigenvectors for the covariance matrix C are calculated. A linear combination of Eigen vectors can be represented as any vector S or. Only the greatest Eigen values are used to create the lower dimensional data collection. Principal components are the components in lower-dimensional space. If the data set is regularly distributed, these principle components must be independent. The scaling of the original variables is often sensitive to PCA.

The following are the steps for using random projections to reduce the dimensionality of data: Assume that by using data set and considering each data point being a dimensional vector such that is a subset of and need by reducing the data to a dimensional space so that is a subset of. By Arranging the data in a matrix form , where is the number of data points and is the dimensionality of the data create a RP matrix R* by using the randn() function in MATLAB. Multiplying the RP matrix with the original data and project the data down into a RP RP space. Here its simple matrix multiplication which guarantees for distance preservation.

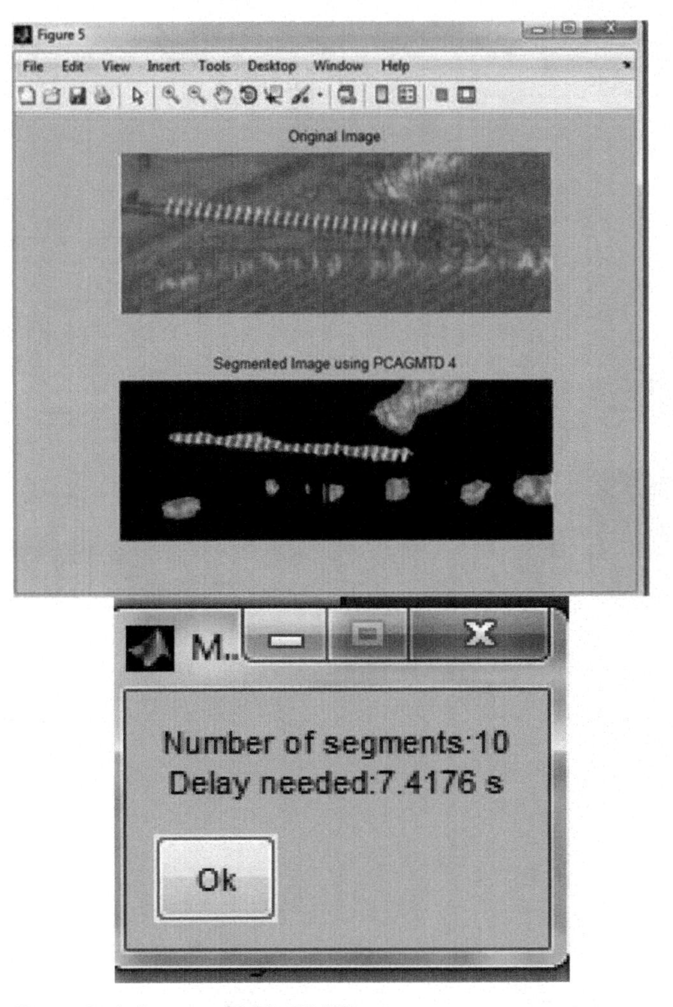

Figure 66.4 Output of PCA_GMTD

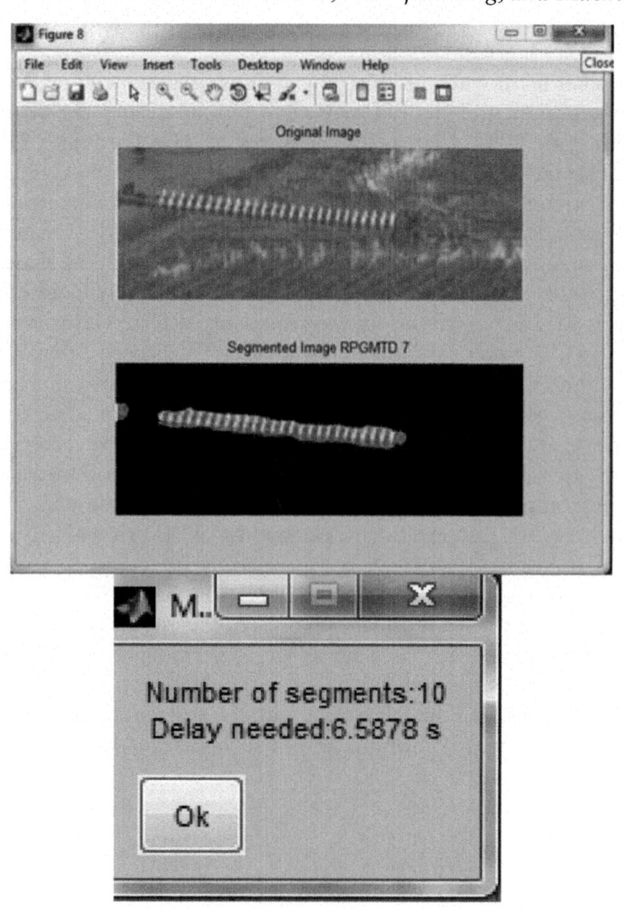

Figure 66.5 Output of RP_GMTD

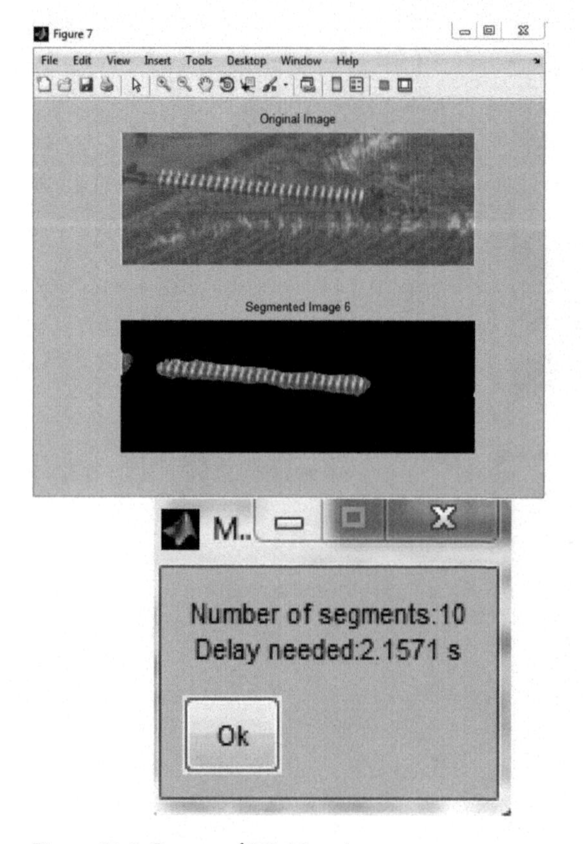

Figure 66.6 Output of RP_Livewire

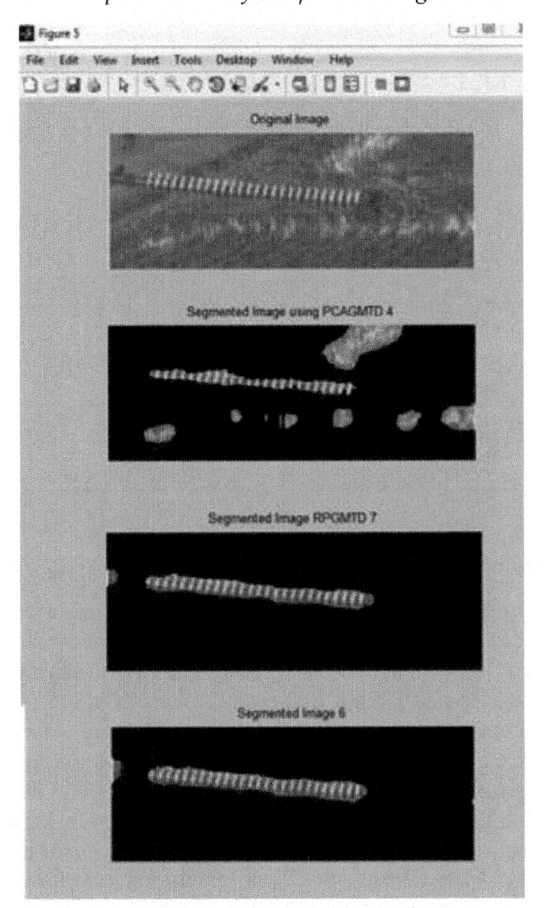

Figure 66.7 Comparison output of PCA GMTD, RP GMTD, and RP Live wire

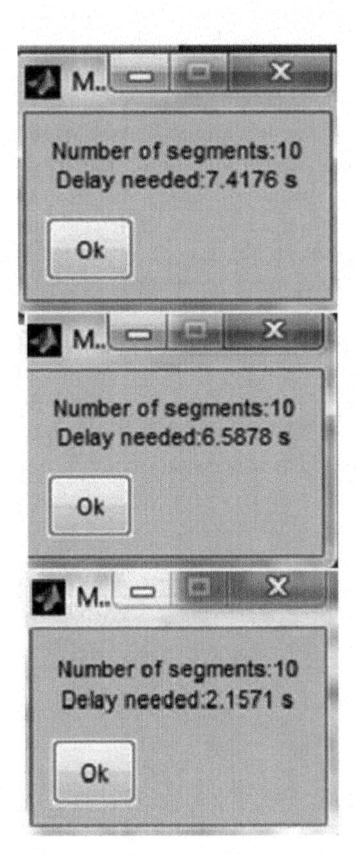

Figure 66.8 Comparison delay output of PCA GMTD, RP GMTD, and RP Live wire

Table 66.1 CPU response time (Ms) comparison between PCA GMTD, RP GMTD, and RP Livewire

Image	*Size*	*Delay PCA_GMTD (s)*	*Delay RP_GMTD (s)*	*Delay Research RP_Livewire (s)*
1.	213*91	1.373	1.360	0.458
2.	195*316	1.746	1.709	0.462
3.	354*879	7.324	5.558	0.5502
4.	151*198	4.380	1.411	0.473
5.	353*327	9.659	7.361	0.4947
6.	398*217	5.279	4.779	0.483
7.	367*228	31.892	17.366	0.500
8.	558*523	42.135	39.0293	0.539
9.	205*393	8.227	7.051	0.50925
10.	224*417	11.625	5.937	0.4694
11.	169*463	11.211	7.965	

Comparative Result for Various Textures for Brodatz Dataset

Initially from Figure 66.1 we have consider images from Brodatz database or insulator images (Maturkar and Gaikwad, 2018b; Maturkar and Gaikwad, 2018a; Maturkar and Gaikwad, 2017; Maturkar and Gaikwad, 2014). Using GLCIA we have found total ten features shown in Figure 66.2. Considering variance (VAR) as a reference we have divided these feature into two types i.e. strong features (Figure 66.3). These weak features are optimised with PCA and segmentation done using GMTD Maturkar and Gaikwad (2018b) and we get output in Figure 66.4. These weak features are again optimised with RP and segmentation done using GMTD and we get output Figure 66.5, compare the result PCA GMTD, RP GMTD and RP Livewire in Figure 66.6. Comparison of three method related to delay output is given in Figures 66.7 and 66.8. Finally we compare all the three methods and compares the delays of various methods by considering total 30 images for various image types like single texture, single object, complex texture and object of interest and other in Table 66.2. RP produces more segments for simple textures; hence it should be utilised when the textures are simple enough. When compared to PCA GMTD, RP GMTD produces the same number of segments for single object textures, hence both methods are capable of delivering decent results for single objects, but RP GMTD has a shorter delay. When comparing the number of segments produced by RP GMTD and PCA GMTD for various textures, both approaches are equally capable of generating decent results, but RP GMTD has a shorter latency. When there is a single object segmentation difficulty, RP delivers a better number of segments for simple textures, hence RP should be employed. The RP Livewire algorithm creates the region of interest that has the unique texture at the output in simple texture scenarios, and may thus be regarded a decent segmentation technique. The RP Livewire method produces astounding results for single texture objects. The RP Livewire method is used to detect complex textures and extract crucial spots at the output. The RP Livewire algorithm can also recognise single objects with high accuracy. By segmenting off the proper region of interest from the input images, the single object detection output displays promising results.

From these we conclude that RP live wire if best fitted for texture segmentation. Also we calculated CPU Response time (ms) with different size images shown in Table 66.1.

From these comparisons, we can find out that segmentation results with different aspect of optimisation and segmentation are more aesthetically pleasant than the alternatives. We also compared the delay by using all three method and same for precision of the proposed system to existing conventional approaches. The results are as follows:

Table 66.2 Compares the delays of various methods

Image Type	*images*	*Delay PCA_ GMTD (s)*	*Delay RP_ GMTD (s)*	*Delay research RP_ Livewire (s)*
Single_ Texture	30	3.871	2.665	2.247
Single _object	30	3.913	2.854	2.258
Complex_Texture	30	4.895	3.374	2.618
Object of interest and other	30	3.227	2.514	1.952

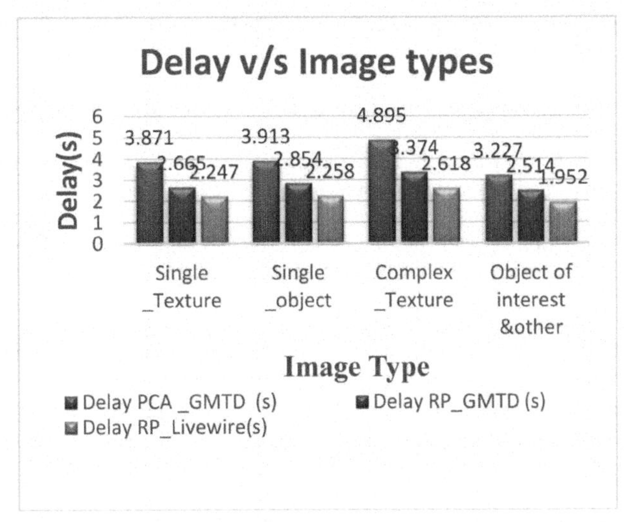

Figure 66.9 Graph delay v/s image types

Conclusion

In this paper comparison we have studied of random projection (RP) and principal component analysis (PCA) for texture optimisation. Comparative analysis has done to identify the computational efficacy of algorithms for texture optimisation. It has been observed that RP is found to be faster than PCA. RP is found to be best suited for highly randomised Gaussian data for texture optimisation. Segmentation by using GMTD (Lianantonakis and Petillot, 2007; Sagiv et al., 2006; Chan and Vese, 2001; Maturkar and Gaikwad, 2017; Fern and Brodley, 2003; Cande's and Tao, 2005; Dasgupta, 2000) for complex images has been done and GLCIA have been adapted to speed up the computational feature. Optimising weaker texture feature of complex images using RP is found to be computationally significant and less expensive in comparison with PCA. When compared to other algorithms, the RP live wire approach performs well in terms of different texture segmentation, single object, and single texture segmentation. The delay reduction is good when compared to other textures, but there is also improvement in other scenarios, which is a good proportion, therefore the provided approach can be used for real-time image processing.

References

Achlioptas, D. (2001). Database-friendly random projections. In Proc. 20th ACM symp. principles of database systems (pp. 274–281).

Baraniuk, R. G., Davenport, M., DeVore, R. A. and Wakin, M. (2008). A simple proof of the restricted isometry property for random matrices. Constr. Approx. 28(3):253–263.

Biau, G., Devroye, L. and Lugosi, G. (2008). On the performance of clustering in hilbert spaces. IEEE Trans. Inf. Theory, 54(2):781–790.

Cande's, E. J. and Tao, T. (2005). Decoding by linear programming. IEEE Trans. Inf. Theory 51(12):4203–4215.

Cande`s, E. J. and Tao, T. (2006). Near-optimal signal recovery from random projections: Universal encoding strategies. IEEE Trans. Inf. Theory 52(12):5406–5425.

Chan, T. and Vese, L. (2001). Active contours without edges. IEEE Trans. Image Process. 10(2):266–277.

Chowdhury, P. R., Deshmukh, B., Goswami, A. K. and Prasad, S. S. (2011). Neural network based dunal landform mapping from multispectral images using texture features. IEEE J. Sel. Topics. Appl. Earth Observ. Remote Sens. (JSTARS) 4(1):171–184.

Clausi, D. A. (2000). Comparison and fusion of co-occurrence, Gabor and MRF texture features for classification of SAR sea-ice imagery. Atmos.-Ocean 39(3):183–194.

Clausi, D. A. and Zhao, Y. (2003). Grey level co-occurrence integrated algorithm (GLCIA): A superior computational method to rapidly determine co-occurrence probability texture features. Comput. Geosci. 29(7):837–850.

Dasgupta, S. (2000). Experiments with random projections. In Proc. 16th Conf. uncertainty in artificial intelligence, (pp. 143–151).

Donoho, D. L. (2006). Compressed sensing. IEEE Trans. Inf. Theory 52(4):1289–1306.

Fern, X. Z. and Brodley, C. E. (2003). Random projection for high dimensional data clustering: A cluster ensemble approach. In Proc. 20th Int'l conf. machine learning.

Goel, N., Bebis, G. and Nefian, A. (2005). Face recognition experiments with random projection. Proc. SPIE 5779.

Gonzalez Book. Gonzalez Book Digital Image Processing, (3rd ed).

Minimisation of region-scalable fitting energy for image segmentation. IEEE Trans. Image Process. 17(10):1940–1949.

Lianantonakis, M. and Petillot, Y. R. (2007). Sidescan sonar segmentation using texture descriptors and active contours. IEEE J. Ocean. Eng. 32(3):744–752.

Liu, L. and Fieguth, P. W. (2012). Member, IEEE Texture classification from random features. IEEE Trans. Pattern Anal. Mach. Intell. 34(3).

Maturkar, P. A. and Gaikwad, M. A. (2017a). RP-live wire algorithms. In Proceeding of by IEEE explorer digital library international conference on power, control, signals and instrumentation engineering (ICPCSI-2017), (pp. 74–78) (dt. 21st & 22nd Sept. 2017).

Maturkar, P. and Gaikwad, M. A. (2014). Texture extraction & segmentation of aerial images of insulator by RP-Live wire algorithms for rural areas. In International journal (IJERT), IC-QUEST 2014, BDCOE, Engg. (19 April 2014).

Maturkar, P. and Gaikwad, M. A. (2017b). RP live wire algorithms. In International conference on power, control, signals and instrumentation engineering (ICPCSI-2017) IEEE with catalog CFP17M84-PRJ: 978-1-5386-0813-5, Chennai, Tamil Nadu, India (during 21th & 22th September 2017), DOI: 10.1109/ICPCSI.2017.8391859.

Maturkar, P. and Gaikwad, M. A. (2018a). Texture feature optimisation of complex images using RP&PCA and segmentation by GMTD. In International conference on recent innovation in electrical, electronics and communication systems (ICRIEECE), 2018 IEEE, 2018 with catalog CFP18P98-PRJ:978-5386-5994-6, Bhubaneswar, Udisa, India (during July 27th &28th 2018).

Maturkar, P. and Gaikwad, M. A. (2018b). Texture extraction & segmentation of complex images by RP –livewire algorithms. In International conference on recent innovation in electrical, electronics and communication systems (ICRIEECE), 2018 IEEE, 2018 with catalog CFP18P98-PRJ:978-5386-5994-6, Bhubaneswar, Udisa, India (during July 27th & 28th 2018).

Mishra, A., Fieguth, P. and Clausi, D. A. (2011). Decoupled active contour (DAC) for boundary detection. IEEE Trans. Pattern Anal. Mach. Intell. 33(2):310–324.

Pesaresi, M. and Gerhardinger, A. (2011). Improved textural built-up presence index for automatic recognition of human settlements in arid regions with scattered vegetation. IEEE J. Sel. Topics. Appl. Earth Observ. Remote Sens. (JSTARS) 4(1):16–26.

Sagiv, C., Sochen, N. and Zeevi, Y. Y. (2006). Integrated active contours for texture segmentation. IEEE Trans. Image Process.

Ursani, A. A., Kpalma, K., Lelong, C. C. D. and Ronsin, J. (2012). Fusion of textural and spectral information for tree crop and other agricultural cover mapping with very-high resolution satellite images. IEEEJ. Sel. Topics. Appl. Earth Observ. Remote Sens. (JSTARS) 5(1):225–235.

Urschler, M. (2000). Image-based verification of parametric models in heart-ventricle volumetry. Graz.

Wu, Q. G., An, J. and Lin, B. (2012). A texture segmentation based on PCA & global minimisation active contour model for aerial insulator images. In IEEE journal of selected topics in applied earth observations and remote sensing accepted.

Xianghua, X. and Majid, M. (2008). MAC: Magneto static active contour model. IEEE Trans. Pattern Anal. Mach. Intell. 30(4):632–646.

Zhang, K. H., Song, H. H. and Zhang, L. (2010) Active contours drivenby local image fitting energy. Pattern Recogn., 43(4):1199–1206.

67 Study of surface wettability to enhance the nucleate pool boiling heat transfer coefficient: A review

Niloy Laskar[a], *Anil S. Katarkar*[b], *Biswajit Majumder*[c], *Abhik Majumder*[d], *and Swapan Bhaumik*[e]

Department of Mechanical Engineering, National Institute of Technology Agartala, Tripura, India

Abstract

This article aims to review and study systematically the consequence of surface wettability on pool boiling heat transfer (PBHT). A significant property of PBHT is surface wettability, which impacts the effectiveness of heat-removal appliances. Although extensive studies have been carried out in recent decades, improving PBHT through changes in surface wettability remains an extremely complex task. Many studies in this literature have been proposed for improving heat transfer coefficient (HTC) and for delaying the onset of critical heat flux (CHF) in the pool boiling by augmenting nucleation rate and the bubble detachment rate. Many of them, however, modify the chemical structure of a solid surface or a liquid simultaneously with the surface topography. As a result, it is difficult to distinguish the consequences of each, as well as it is also hard to tell whether their interactions have cross-effects. However, pool boiling studies reveal that the property of heating surface varies widely according to its influence on wettability. In the current study, it is observed that PBHT on wettable surfaces varied from hydrophilic to hydrophobic with change of nucleation rate. While hydrophobic surfaces encourage nucleation, hydrophilic surfaces enhance CHF. As part of the present review, the effects of surface wettability on the HTC and CHF have been discussed considering the fundamentals of bubble dynamics and nucleation in PBHT. The information obtained from this is then used for understanding a variety of approaches that are developed for enhancing PBHT. Observations of the reported studies indicate that wettability changes could have completely different effects on boiling behaviour and this study lacks information about the concept of HTC and CHF, which needs to be clarified.

Keywords: Heat transfer coefficient, Hydrophilic, Hydrophobic, Pool boiling, Surface Wettability.

Introduction

As the industrialisation and miniaturisation of various electronic appliances grow in popularity, they require high cooling efficiency for a better and more efficient performance. Heat can be removed from a system using phase change heat transfer method such as pool boiling heat transfer (PBHT). In PBHT, heat can be transferred from the heating surface to the working fluid within a small temperature difference, allowing the surface to cool efficiently and rapidly. Nowadays, PBHT is considered one of the most favourable methods of heat transfer, which is commonly used in several industries, including refrigeration and air conditioning units, nuclear plants, electronics cooling, and seawater desalination. There have been extensive researches over the last decade to improve PBHT by the enhancement of surface wettability (Kim et al., 2016; Gong and Cheng, 2015a; Zhang et al., 2017).

Surface wettability is a naturally occurring phenomenon often visible in the environment. Bringing liquid into contact with a solid surface will reduce the surface tension. This is because liquid spread naturally across the surface and become wettable. Surface modification is considered as significant method to enhance PBHT by modifying the surface wettability and morphology. Surface wettability largely associates with the dynamics of the bubble at the interface of the solid and liquid. On a boiling surface, both the contact angle and contact diameter are relevant parameters for predicting bubble departure rate, bubble diameter, and nucleation site density. The applicable parameter for predicting the nucleation site density, departure diameter, and bubble departure rate, on the boiling surface are the contact angle and the contact diameter (Gong and Cheng, 2015b; Bourdon et al., 2015; Kumar et al., 2017). In many studies, surface wettability was enhanced by modifying the surface topography, for instance by roughening the surface or by creating cavities. Additionally, it is possible to alter surface wettability by altering the working fluid, with nanofluids and surfactants, which basically consist of nanoparticles suspended in fluid. The topography as well as

[a]niloylaskar3@gmail.com; [b]anil.katarkar@gmail.com; [c]bmajumdertit@gmail.com; [d]onlyabhik@gmail.com; [e]drsbhaumik@gmail.com

chemistry of a wide range of surfaces have been altered by various authors that it is now very difficult to distinguish the effects.

The potential for achieving high heat transfer coefficient (HTC) values for cooling purposes has been extensively investigated in recent studies. Thus, any cooling system must need to operate within the nucleate boiling region, usually where HTC enhances with the surface temperature. The thermodynamic efficiency of a system is significantly decreased for high superheat temperatures, because of transition of nucleate boiling regime to film boiling regime. The maximum heat flux in the regime of nucleate pool boiling is termed as critical heat flux (CHF). According to first law of thermodynamics, the goal of CHF optimisation is to maximise HTC. The task appears to be complex since the rate of generation of bubbles, release rate, and the size of bubbles determine the increase of HTC and the delay of CHF. In order to improve PBHT, it is necessary to consider the variables that influence wettability dynamics. It is difficult to quantify many of these characteristics, which are influenced by other factors such as manufacturing processes, contamination on surfaces, or oxide film formation.

Fundamental of Surface Wettability

Technology now makes it possible to detect the test surface characteristics based on wettability, which could also be determined by measuring contact angles (CA). To analyse the surface wettability, it is essential to identify the testing liquid contact angles with the surface. Surface free energy (SFE) represents the amount of work required to increase the surface area of a phase. The SFE differs greatly depending on the nature of functional groups available on the surface. The spread of liquids on liquid or solid surfaces is a common occurrence. Microscopy defines spreading as the process of minimizing the total SFE. CA between two surfaces is created by the tangent angle between the liquid and solid surfaces, which is represented by θ. In other word, CA is the intersection angle between the solid-liquid interface and the liquid-gas interface. Figure 67.1 Illustrates various types of wetting surfaces based on CA. A high CA signifies a low SFE or chemical affinity, which is also defined as low degree of wetting. Whereas, a low CA indicates a high degree of wetting as well as a high SFE or chemical affinity. A comparison of the effects of different CA on SFE is presented in Table 67.1.

Figure 67.2 illustrates the various forces acting on the solid-liquid interface. When the liquid droplet is not spreading, all the forces will be at equilibrium. This principle is accountable for the observed CA. Consequently, CA indicates how much surface force is exerted between a liquid and a surface and hence determines the surface wettability in terms of hydrophilicity and hydrophobicity.

Table 67.1 Effects of different contact angles on the SFE

CA	*Degree of Wetting*	*Nature*	*SFE*	*Effect*
$\theta < 5°$	Perfect Wetting	Super-hydrophilic	Increases	Liquid droplets widely spread out.
$5° \leq \theta < 90°$	High Wettability	Hydrophilic	Increases	Liquid droplets spread out.
$90° \leq \theta \leq 150°$	Low wettability	Hydrophobic	Decreases	Liquid droplets beads-up
$\theta \geq 150°$	Perfectly non-wetting	Super-hydrophobic	Decreases	Liquid droplets repelled

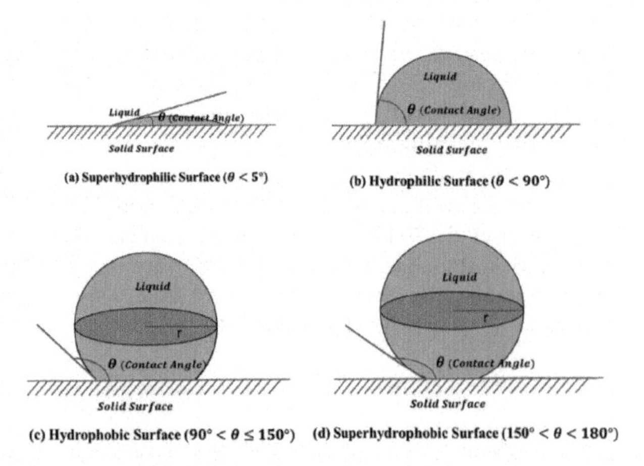

Figure 67.1 Various types of wetting surface based on CA

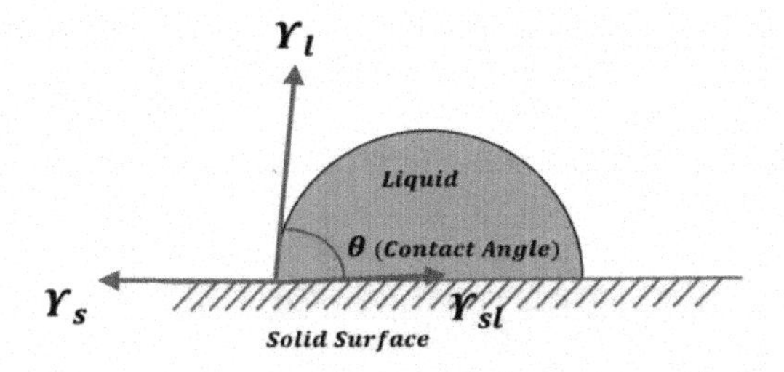

Figure 67.2 Various forces acting on the solid-liquid interface

Hydrophobic surfaces usually require less SFE for inducing an early onset of nucleate boiling than that of hydrophilic surface. Though the bubbles show a tendency to adhere on the surface for a very long time before kicking off from the exterior. This occurs because surface tension prevents bubbles from leaving the surface at the contact point of three-phase. While, the hydrophilic surface helps in facilitating the bubble take-off and suppressing the dry out process owing to the feature of rapid rewetting. For this particular reason, hydrophilic surfaces most often outperform the hydrophobic surfaces underneath high-heat-flux conditions.

Therefore, a heterogeneous surface of wettability has been a reasonable solution to overcome the interchangeability between the virtues of hydrophilic and that of hydrophobic surfaces. The usual way to construct a heterogeneous surface is by depositing a dot patterned hydrophobic mater on a hydrophilic surface. Non-wetting regions excite the initial nucleation of the bubbles at very low heat flux, on the other hand wetting regions spread around prevents the growing bubbles from forming the vapour cover at high heat flux. Heterogeneous wettability surface is shown to improve the HTC as well as the CHF.

Surface Wettability and Surface Topography

Surface wettability played a major role in PBHT, but researchers were unable to perform precise measurement to measure CA and prepare surfaces using topographic and chemical methods. When a liquid boils over a solid heating surface, the contact line between the three phases shifts, expanding and contracting as the phases change. A balance exists between viscous, interfacial, and gravity forces to control its motion. Practically, wettability refers to how long the surface remains wet after a liquid is applied. There are a few methods for quantifying equilibrium contact angles, but the most common is the equilibrium contact angle of a sessile droplet on a flat rigid surface. A relationship between SFE of the solid (Υ_s), CA (θ), the liquid-solid interfacial tension (Υ_{sl}) and the surface tension of the liquid (Υ_l) can be found in Young's equation (Young, 1832), as in (1):

$$\cos\theta = \frac{\Upsilon_s - \Upsilon_{sl}}{\Upsilon_l} \tag{1}$$

When a liquid drops on a surface, if it has a very strong attraction to the solid surface (i.e., water on a highly hydrophilic solid), the liquid will spread out completely on the surface, and CA will tend to zero. Solids with a lower hydrophilic index will have a CA up to 90°. Water droplets typically exhibit CA ranging from 0° to 90° on highly hydrophilic surfaces.

Solid surfaces that are hydrophobic will have CA greater than 90°. Materials having low SFE, such as fluorinated surfaces, have water CA of up to 120° on highly hydrophobic surfaces. Although some surfaces with high surface roughness may have water CA of over 150° commonly referred to as superhydrophobic surfaces (SHS) while few surface may have water CA less than 10° commonly referred to as super-hydrophilic surfaces.

Surface tension of a liquid surface can be determined in different ways. The sessile drop method, however, is favoured because it produces excellent results when used with pure liquid and with surfaces that is prone to aging. CA can be defined by a variety of parameters being considered while measuring it.

Apparent CA or Static CA (θ_{ap}): Macroscopically, it is an equilibrium CA applied to a chemically heterogeneous or rough surface. This angle corresponds to the tangent between the solid-liquid interface and apparent surface of the solid. In fact, θ_{ap} is an equilibrium value of geometric CA (θ). There may be a great deal of difference between θ_{ap} and θ. The measuring parameters such as boundary surface, volume will not be altered forcefully once the liquid droplet is centered on the surface.

Most stable CA (θ_{ms}): It is the apparent CA which is related to a wetting surface having the least possible Gibbs free energy. Equations obtained from Wenzel, or Cassie-Baxter model can provide the value of this CA under the suitable circumstances.

Advancing CA *(θ_a):* It is the maximum apparent CA for a particular wetting surface. According to the Neumann approach, it is usually measured as the drop expands and can be accomplished by increasing the droplet volume.

Receding CA *(θ_r):* It is the minimum apparent CA for a particular wetting surface. It is usually measured as the drop shrinks and can be accomplished by decreasing the droplet volume.

Advancing CA is greater than receding CA, and the difference between the two is acknowledged as CA hysteresis. When a drop falls on a rough surface, it will tend to spread all over the surface. Once the liquid droplet proceeds along the surface, it gets contaminated, and as a result the liquid surface tension vary. In general, advancing CA may be affected by less polar regions, while receding CA may be affected by polar regions and this leads to hysteresis.

Thermal-fluid science takes advantage of the predetermined properties of the working fluid, in order to manipulate solid-gas and solid-liquid interfacial energies. Accordingly, apparent wettability depends on the topography and surface chemistry of the solid (Carey, 2002). Surface chemical properties are implied by Young's contact angle, while surface properties are implied by roughness factor (r) and solid fraction (ϕ) as shown in Figure 67.3. With these parameters, either Wenzel (2002) or Cassie Baxter models (Cassie and Baxter, 1944) can be used to describe the apparent CA of heterogeneous hydrophilic surfaces. The Wenzel model defines the homogeneous surface, as in (2) whereas Cassie-Baxter model, as in (3) deals with heterogeneous wetting surface as seen in Figure 67.4.

$$\cos\theta_{ap} = r\cos\theta \tag{2}$$

$$\cos\theta_{ap} = \phi(\cos\theta - 1) + 1 \tag{3}$$

Generally, hydrophilicity is preferred for heat transfer enhancements in certain engineering appliances, such as refrigeration, air conditioning, electronics cooling, and so on. With better wettability, critical heat flux (CHF) in pool boiling is enhanced, the activation energy for condensate nucleation is lowered, liquid drainage is improved and high thermal conductivity is achieved. Thus, wettability manipulation has become a research topic in recent years. HTC is made more efficient using nano porous structures by introducing topographical variations for enhanced surface wetting. Although, surface wettability can be manipulated using nanoscale technology for small area treatment (Chu et al., 2013; Ahn et al., 2011). Surfaces with super-hydrophilic properties of TiO_2 can be mass produced with UV light and plasma treatment, but they are susceptible to contamination, limiting their longevity in use (Takata et al., 2005; Kim et al., 2002). As a result, it is important to determine whether topographical manipulation is effective in achieving

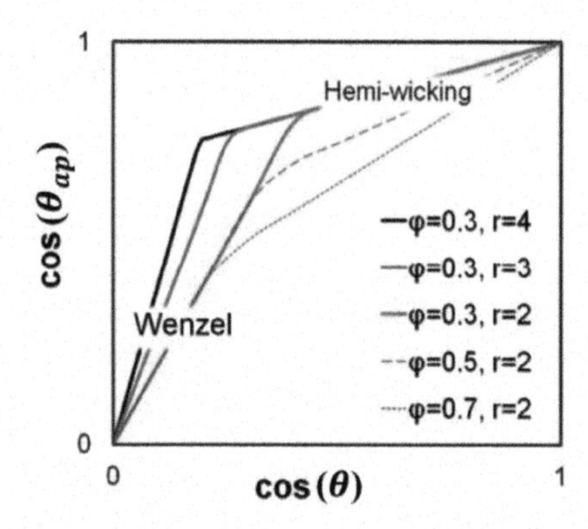

Figure 67.3 Variation of apparent CA (θ_{ap}) with the Young's contact angle (θ) for various ϕ and r on a heterogeneous surface (Zhang and Jacobi, 2016)

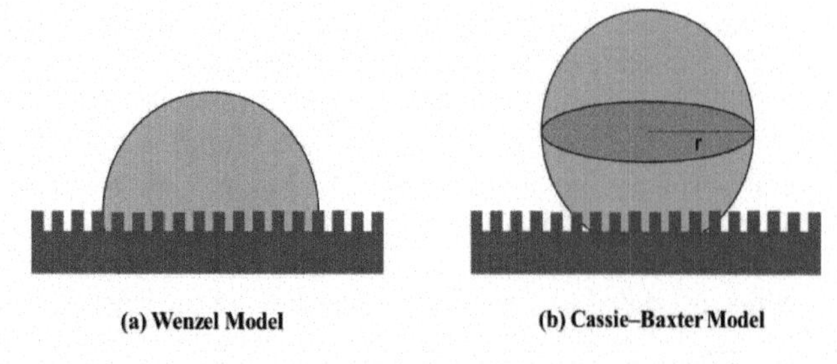

Figure 67.4 Various wetting model (a) Wenzel model (b) Cassie-baxter model

superhydrophilicity on a scalable scale. Table 67.2 illustrates the effect of wettability on nucleation, boiling onset, and bubble dynamics.

Approaches to Modify Surface Wettability

PBHT efficiency varies significantly with the wettability of the heating surface. Hence surface wettability can be enhanced either by modifying the physio-chemical properties of working fluid or modifying the surface using surfactants. The change in one property can influence the change in another, which is evident in many cases. The use of nanoparticles changes the fluid properties, results in a change in the particle settling conditions at the surface. The surface topology can also differ depending on the wetting properties, as was seen with super-hydrophobic surfaces that need particular topographical properties.

The hydrophobic surfaces are favoured by early nucleation of bubbles, resulting in an efficient PBHT with low heat flux. However, the bubbles will detach slowly from the hydrophobic surface at higher heat fluxes owing to the high affinity of the bubbles to the surface, resulting in poor PBHT performance. In order to overcome the major drawbacks of homogeneous wettable surfaces, a biphilic (heterogeneous) wettable surface was developed. The presence of heterogeneous wettable surfaces can result in early nucleation of bubble and delayed surfaces drying, which improves PBHT performance throughout all boiling regimes (Katarkar et al., 2021c; Majumder et al., 2022; Katarkar et al., 2022). Hydrophobic surfaces typically feature large HTC values at low superheat; however, this value remains constant for higher superheats as shown in Figure 67.5. HTC is higher on the hydrophilic surface than on the hydrophobic surface when the superheat is higher. Figure 67.5 illustrates that super-hydrophilic surfaces display the maximum HTC of all the surfaces

Two kinds of biphilic surfaces were developed by (Betz et al. 2011, 2010) creating hydrophilic dots on a hydrophilic surface and hydrophobic dots on a hydrophilic surface. Hydrophobic dots on a hydrophilic surface yielded better PBHT performance, and HTC achieved a 100% improvement. Jo et al. (2011) compared PBHT performance of biphilic surfaces with hydrophobic and hydrophilic wetting surfaces. A

Table 67.2 Effect of wettability on nucleation, boiling onset and bubble dynamics

Mechanism	*Effect of Surface Wettability*
Nucleation	Nucleation is more difficult with high wettability because of the reduced volume of entrapped vapour. Depending on the surrounding fluid temperature, wetting fluid in conical cavities can yield stable or unstable nuclei. Nucleus formation is stable under low wetting conditions (Pinni et al., 2021)
Onset of nucleate boiling (ONB)	Gibbs free energy-based calculations indicate that the energy barrier for boiling is minimum for reduced wettability. The theoretical trends described for nucleation suggest that low wetting conditions are achieved by lower wall superheats
Bubble dynamics	The lack of interfacial tension prevents bubbles from detaching from the surface under low wetting conditions. So, low wetting conditions usually lead to formation of larger bubbles having lesser departure frequencies. In good wetting conditions, bubbles begin to coalesce down, generating a vapor blanket, which reduces wall superheats, so CHF values also reduces.(Katarkar et al., 2021a, 2021b)

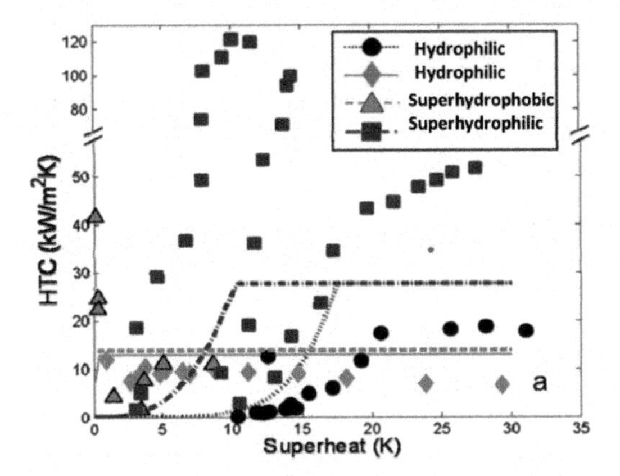

Figure 67.5 HTC v/s superheat curve for various wetting surface (Betz et al., 2013)

biphilic wetting surface generated almost the same CHF values as the hydrophilic wetting surfaces and higher HTC values than the hydrophobic wetting surfaces. The use of new techniques, like MOCVD and PECVD, enabled researchers to prepare a experimental data set to investigate the consequences of surface wettability, without significantly altering the surface topography. A common method involves coating the surface with either hydrophilic or hydrophobic materials. Ujereh et al. (2007) described that coating of carbon nanotubes on copper and silicon substrates enhanced both PBHT and HTC, and reduced the ONB superheat. Heterogeneous wettable surfaces can be created using a wide range of techniques, including screen printing, photolithography, and laser jet printing (Jo et al., 2011; Choi et al., 2016; Huang and Leu, 2013). Further, increasing the concentration of nanofluid and extending boiling time enhances surface wettability, while roughness has complicated impact. An analysis of goniometry and microscopy reveals the mechanisms of wettability modification. Based on the results of the study (Zhang and Jacobi, 2016), it was observed in Al_2O_3 nanoparticle aqueous suspensions, that surface wettability of aluminium surfaces is enhanced by deposition of nanoparticles and development of hydroxides throughout boiling. An overview of various studies that investigated surface wettability modifications to enhance PBHT is shown in Table 67.3.

With degassed water, pool boiling experiments were carried out on stainless steel substrates with varying chemical properties and topographies, according to (Malavasi et al., 2015). Different kinds of boiling curves were observed for hydrophilic surfaces and SHS surfaces as shown in Figure 67.6. The boiling heat flux of SHS was initiated at a lower superheat compared to hydrophilic surfaces, and its behaviour exhibited a quasi-Leidenfrost regime, implying that the classical boiling curve has been significantly modified. Further, the wettability of the SHS with different roughness had a dominant role when the contact angle exceeded a certain value with respect to the nucleation temperatures on the SHS. There is no information provided in this study regarding boiling curves up to the CHF, so further research is necessary.

Table 67.3 An overview of various studies that examined the modification of surface wettability to enhance PBHT

Reference	*Solution*	*Effect of wettability on PBHT*
Wen and Wang (2002)	DI water and Acetone solution having 95% sodium dodecyl sulphate (SDS), octadecyl amine and Triton X-100	With a lower superheat, the same maximum heat flux was obtained for water. Though, using the SDS solution yields the best results
Hetsroni et al. (2001)	Water-soluble cationic surfactant Habon G.	For lower concentrations of surfactant, HTC increases because of the decrease in surface tension, whereas for higher concentrations, HTC decreases because of the increase in kinematic viscosity
Wasekar and Manglik, 1999)	SDS and sodium lauryl sulphate (SLS)	In comparison to pure water, HTC can increase by up to 65%, because bubble dynamics are affected by surface tension. Lower concentrations of surfactant result in lower surface tension. The lowest SDS concentration yielded the highest HTC
Wu et al. (1998)	Aqueous solutions with Aerosol-22, DTMAC, and Triton X-100.	The dynamic surface tension and HTCs cannot be correlated. The boiling incipience cannot be explained solely by surface tension.
Wu et al. (1999)	Aqueous solutions of SDS	The decrease of surface tension leads to a decrease of superheat
Gerardi et al. (2011)	Si nanoparticle	Almost 100% increase in CHF with respect to water, although the HTC drops by 50%
Harish et al. (2011)	Water-based electro-stabilised Al_2O_3 nanofluids	Compared to pure water, CHF increases by up to 120%.
Kim and Kim (2009)	Al_2O_3, TiO_2, and SiO_2 nanoparticle	By coating the heater surface with nanoparticles, CHF can be enhanced.
Kim et al. (2007)	Water soluble nanoparticles of zirconia, alumina, and silica	PBHT performance is enhanced when nanoparticles are deposited on the surface causing a decrease in wettability and increase in nucleation sites
Bang and Chang (2005)	Al_2O_3 in water	Nanofluids have a higher CHF with a lower concentration, but their HTC is always lower than water.
Vassallo et al. (2004)	0.5% Si nanoparticles by volume	Compared to pure water, the CHF can be up to three times higher
Das et al. (2003)	Water–Al_2O_3 nanofluids	With particle concentrations, the wall superheat increases due to particle accumulation on the heated surface

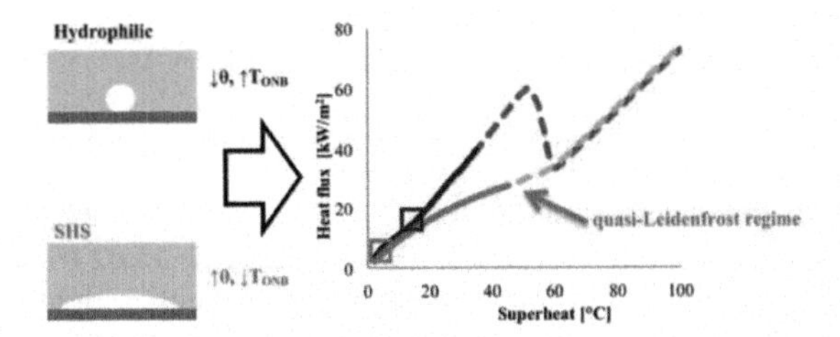

Figure 67.6 Flow diagram for bubble formation on hydrophilic surface and SHS surfaces, with the boiling curve (Malavasi et al., 2015)

Conclusion

This paper discusses the influence of PBHT due to wettability. There have been a number of experimental and theoretical studies on the enhancement of PBHT, which involve changing the properties of the liquid or the surface. Surface nano-coatings and nanofluids are promising strategies to improve HTC and CHF. However further research is necessary to optimise them. Optimisation of these strategies requires an understanding of how these strategies interact with PBHT process. Therefore, fundamental research is crucial to this process. In addition, more specific research is still needed on the morphology and composition of coatings.

After careful investigation of the various approaches to enhance PBHT, that were reviewed in this paper, it is evident that they all act by altering the boundary conditions at the three-phase interface. PBHT affects wettability during the entire boiling procedure, beginning with a nucleation, and continuing with a nucleate boiling and adding a CHF. HTC of hydrophobic surfaces is more at a lower superheat compared to that of hydrophilic surfaces, though CHF of hydrophobic surfaces is very low. Whereas, HTC of hydrophilic surfaces is higher at a higher superheat compared to that of hydrophobic surfaces. This suggests that the wettability at which HTC optimises is related to the surface superheat. Hydrophobic surfaces prevent bubbles from detaching from their surfaces and coalescing with bubbles created at neighbouring sites. With an increase in the surface wettability on hydrophilic surfaces, the bubble departure frequency decreases while the bubble departure diameter increases. A good wettability led to a decrease in density of nucleation sites and a decrease in bubble departure frequency. Hence, wettability effects indicate that non-wettable surfaces provide higher HTC, because they provide more nucleation sites. The wettability effect on PBHT is however not completely understood, since the nucleate boiling is a mixed phenomenon involving wettability, morphology, and roughness. The final section summarises the main findings relative to wettability on PBHT and suggests future research topics.

References

Ahn, H. S., Jo, H. J., Kang, S. H., and Kim, M. H. (2011). Effect of liquid spreading due to nano/microstructures on the critical heat flux during pool boiling. Appl. Phys. Lett. 98(7):98–101. https://doi.org/10.1063/1.3555430.

Bang, I. C. and Chang, S. H. (2005). Boiling heat transfer performance and phenomena of al2o3–water nanofluids from a plain surface in a pool. Int. J. Heat Mass Transf. 48(12):2407–2419. https://doi.org/10.1016/J.IJHEATMASSTRANSFER.2004.12.047.

Betz, A. R., Jenkins, J. R., Kim, C.-J., and Attinger, D. (2011). Significant boiling enhancement with surfaces combining superhydrophilic and superhydrophobic patterns. In 2011 IEEE 24th international conference on micro electro mechanical systems, (pp. 1193–1196). https://doi.org/10.1109/MEMSYS.2011.5734645.

Betz, A. R., Jenkins, J., Kim, C. J., and Attinger, D. (2013). Boiling heat transfer on superhydrophilic, superhydrophobic, and superbiphilic surfaces. Int. J. Heat Mass Transf. 57(2):733–41. https://doi.org/10.1016/j.ijheatmasstransfer.2012.10.080.

Betz, A. R., Xu, J., Qiu, H., and Attinger, D. (2010). Do surfaces with mixed hydrophilic and hydrophobic areas enhance pool boiling? Appl. Phys. Lett. 97(14):1–4. https://doi.org/10.1063/1.3485057.

Bourdon, B., Bertrand, E., Di Marco, P., Marengo, M., Rioboo, R., and De Coninck, J. (2015). Wettability influence on the onset temperature of pool boiling: Experimental evidence onto ultra-smooth surfaces. Adv. Colloid Interface Sci. 221:34–40. https://doi.org/10.1016/j.cis.2015.04.004.

Carey, V. P. (2002). Molecular dynamics simulations and liquid-vapor phase-change phenomena. Microsc. Thermophys. Eng. 6(1):1–2. https://doi.org/10.1080/108939502753428194.

Cassie, A. B. D. and Baxter, S. (1944). Wettability of porous surfaces. Trans. Faraday Soc. 40(0):546–551. https://doi.org/10.1039/TF9444000546.

Choi, C.-H., David, M., Gao, Z., Chang, A., Allen, M., Wang, H., and Chang, C.-H. (2016). Large-scale generation of patterned bubble arrays on printed bi-functional boiling surfaces. Sci. Rep. 6(1):23760. https://doi.org/10.1038/srep23760.

Chu, K. H., Joung, Y. S., Enright, R., Buie, C. R., and Wang, E. N. (2013). Hierarchically structured surfaces for boiling critical heat flux enhancement. Appl. Phys. Lett. 102(15). https://doi.org/10.1063/1.4801811.

Das, S. K., Putra, N., and Roetzel, W. (2003). Pool boiling characteristics of nano-fluids. Int. J. Heat Mass Transf. 46(5):851–862. https://doi.org/10.1016/S0017-9310(02)00348-4.

Gerardi, C., Buongiorno, J., Hu, L., and McKrell, T. (2011). Infrared thermometry study of nanofluid pool boiling phenomena. Nanoscale Res. Lett. 6(1):232. https://doi.org/10.1186/1556-276X-6-232.

Gong, S. and Cheng, P. (2015a). Lattice boltzmann simulations for surface wettability effects in saturated pool boiling heat transfer. Int. J. Heat Mass Transf. 85:635–646. https://doi.org/10.1016/j.ijheatmasstransfer.2015.02.008.

Gong, S. and Cheng, P. (2015b). Numerical simulation of pool boiling heat transfer on smooth surfaces with mixed wettability by lattice boltzmann method. Int. J. Heat Mass Transf. 80:206–216. https://doi.org/10.1016/j.ijheatmasstransfer.2014.08.092.

Harish, G., Emlin, V., and Sajith, V. (2011). Effect of surface particle interactions during pool boiling of nanofluids. Int. J. Therm. Sci. 50(12):2318–2327. https://doi.org/10.1016/J.IJTHERMALSCI.2011.06.019.

Hetsroni, G., Zakin, J. L., Lin, Z., Mosyak, A., Pancallo, E. A., and Rozenblit, R. (2001). The effect of surfactants on bubble growth, wall thermal patterns and heat transfer in pool boiling. Int. J. Heat Mass Transf. 44(2):485–497. https://doi.org/10.1016/S0017-9310(00)00099-5.

Huang, D. J. and Leu, T. S. (2013). Fabrication of high wettability gradient on copper substrate. Appl. Surf. Sci. 280(September):25–32. https://doi.org/10.1016/J.APSUSC.2013.04.065.

Jo, H. J., Ahn, H. S., Kang, S., and Kim, M. H. (2011). A study of nucleate boiling heat transfer on hydrophilic, hydrophobic and heterogeneous wetting surfaces. Int. J. Heat Mass Transf. 54(25–26):5643–5652. https://doi.org/10.1016/J.IJHEATMASSTRANSFER.2011.06.001.

Katarkar, A. S., Pingale, A. D., Belgamwar, S. U., and Bhaumik, S. (2021a). Experimental investigation of pool boiling heat transfer performance of refrigerant R-134a on differently roughened copper surfaces. Mater. Today: Proc. 47(January):3269–3275. https://doi.org/10.1016/J.MATPR.2021.06.452.

Katarkar, A. S., Pingale, A. D., Belgamwar, S. U., and Bhaumik, S. (2021b). Effect of GNPs concentration on the pool boiling performance of R-134a on Cu-GNPs nanocomposite coatings prepared by a two-step electrodeposition method. Int. J. Thermophys. 42(8):124. https://doi.org/10.1007/s10765-021-02876-z.

Katarkar, A. S., Pingale, A. D., Belgamwar, S. U., and Bhaumik, S. (2021c). Experimental study of pool boiling enhancement using a two-step electrodeposited Cu–GNPs nanocomposite porous surface with R-134a. J. Heat Transf. 143(12). https://doi.org/10.1115/1.4052116.

Katarkar, A. S., Pingale, A. D., Belgamwar, S. U., and Bhaumik, S. (2022). Fabrication of Cu@G composite coatings and their pool boiling performance with R-134a and R-1234yf. Adv. Mater. Process. Technol. 0(0):1–13. https://doi.org/10.1080/2374068X.2022.2033046.

Kim, G.-r., Lee, H., and Webb, R. L. (2002). Plasma hydrophilic surface treatment for dehumidifying heat exchangers. Exp. Therm. Fluid Sci. 27(1):1–10. https://doi.org/10.1016/S0894-1777(02)00219-4.

Kim, H. and Kim, M. (2009). Experimental study of the characteristics and mechanism of pool boiling chf enhancement using nanofluids. Heat Mass Transf. 45(7):991–998. https://doi.org/10.1007/s00231-007-0318-8.

Kim, J., Jun, S., Laksnarain, R., and You, S. M. (2016). Effect of surface roughness on pool boiling heat transfer at a heated surface having moderate wettability. Int. J. Heat Mass Transf. 101:992–1002. https://doi.org/10.1016/j.ijheatmasstransfer.2016.05.067.

Kim, S. J., Bang, I. C., Buongiorno, J., and Hu, L. W. (2007). Surface wettability change during pool boiling of nanofluids and its effect on critical heat flux. Int. J. Heat Mass Transf. 50(19–20):4105–4116. https://doi.org/10.1016/J.IJHEATMASSTRANSFER.2007.02.002.

Kumar, C. S. S., Chang, Y. W., and Chen, P. H. (2017). Effect of heterogeneous wettable structures on pool boiling performance of cylindrical copper surfaces. Appl. Therm. Eng. 127:1184–1193. https://doi.org/10.1016/j.applthermaleng.2017.08.069.

Majumder, B., Pingale, A. D., Katarkar, A. S., Belgamwar, S. U., and Bhaumik, S. (2022). Enhancement of pool boiling heat transfer performance of R-134a on microporous Al@GNPs composite coatings. Int. J. Thermophys. 43(4):49. https://doi.org/10.1007/s10765-022-02973-7.

Malavasi, I., Bourdon, B., Di Marco, P., de Coninck, J., and Marengo, M. (2015). Appearance of a low superheat 'quasi-leidenfrost' regime for boiling on superhydrophobic surfaces. Int. Commun. Heat Mass Transf. 63(April): 1–7. https://doi.org/10.1016/J.ICHEATMASSTRANSFER.2015.01.012.

Pinni, K. S., Katarkar, A. S., and Bhaumik, S. (2021). A review on the heat transfer characteristics of nanomaterials suspended with refrigerants in refrigeration systems. Mater. Today: Proc. 44(January):1331–1335. https://doi.org/10.1016/J.MATPR.2020.11.389.

Takata, Y., Hidaka, S., Cao, J. M., Nakamura, T., Yamamoto, H., Masuda, M., and Ito, T. (2005). Effect of surface wettability on boiling and evaporation. Energy 30(2–4):209–220. https://doi.org/10.1016/J.ENERGY.2004.05.004.

Ujereh, S., Fisher, T., and Mudawar, I. (2007). Effects of carbon nanotube arrays on nucleate pool boiling. Int. J. Heat Mass Transf. 50(19–20):4023–4038. https://doi.org/10.1016/J.IJHEATMASSTRANSFER.2007.01.030.

Vassallo, P., Kumar, R., and D'Amico, S. (2004). Pool boiling heat transfer experiments in silica–water nano-fluids. Int. J. Heat Mass Transf. 47(2):407–411. https://doi.org/10.1016/S0017-9310(03)00361-2.

Wasekar, V. M., and Manglik, R. M. (1999). A review of enhanced heat transfer in nucleate pool boiling of aqueous surfactant and polymeric solutions. J. Enhanc. Heat Transf. 6(2-4). (10.1615/JEnhHeatTransf.v6.i2-4.70):135–50.

Wen, D. S. and Wang, B. X. (2002). Effects of surface wettability on nucleate pool boiling heat transfer for surfactant solutions. Int. J. Heat Mass Transf. 45(8):1739–1747. https://doi.org/10.1016/S0017-9310(01)00251-4.

Wenzel, R. N. (2002). Resistance of solid surfaces to wetting by water. Ind. Eng. Chem. 28(8):988–994. https://doi.org/10.1021/ie50320a024.

Wu, W. T., Yang, Y. M., and Maa, J. R. (1998). Nucleate pool boiling enhancement by means of surfactant additives. Exp. Therm. Fluid Sci. 18(3):195–209. https://doi.org/10.1016/S0894-1777(98)10034-1.

Wu, W. T., Yang, Y. M., and Maa, J. R. (1999). Technical note pool boiling incipience and vapor bubble growth dynamics in surfactant solutions. Int. J. Heat Mass Transf. 42(13):2483–2488. https://doi.org/10.1016/S0017-9310(98)00335-4.

Young, T. (1832). An essay on the cohesion of fluids. Abstracts of the Papers Printed in the Philosophical Transactions of the Royal Society of London 1: 171–72. https://doi.org/10.1098/rspl.1800.0095.

Zhang, B. J., Ganguly, R., Kim, K. J., and Lee, C. Y. (2017). Control of pool boiling heat transfer through photo-induced wettability change of titania nanotube arrayed surface. Int. Commun. Heat Mass Transf. 81:124–130. https://doi.org/10.1016/j.icheatmasstransfer.2016.12.007.

Zhang, F., and Jacobi, A. M. (2016). Aluminum surface wettability changes by pool boiling of nanofluids. Colloids Surf. A: Physicochem. Eng. Asp. 506:438–444. https://doi.org/10.1016/j.colsurfa.2016.07.026.

68 Application of PLM in industry 4.0 environment

Yashwanth Jerripothula[a] *and Girish Bhiogade*[b]

Department of Mechanical Engineering Vignan's Institute of Information Technology (JNTUK), Visakhapatnam, India

Abstract

Product lifecycle management (PLM) is one of the emerging sectors in Industry 4.0PLM enables enterprise-wide collaboration for everyone involved in the product lifecycle, from design and supply chain to manufacturing and quality control. PLM helps in better data management. Our goal is to apply the concept of PLM to the manufacturing industry. It is assumed that the manufacturing industry produces three products: a ladder frame chassis, an all-terrain vehicle roll cage (ATV), and an additional frontal protection halo (AFP). By using Siemens Teamcenter 11 PLM software, we have created organisation, project, workflows, schedules, and integrated computer-aided design/computer-aided manufacturing/computer-aided engineering (CAD/CAM/CAE) data with Teamcenter to manage all of the data and processes at each stage of a product or service lifecycle across the industry. PLM acts as a central source of organising the data for collaboration in product data management and configurable new product development process workflows. Better products can get to market quicker by implementing PLM as everyone works from a single central source. This PLM concept can be applied to any large-scale, real-time manufacturing industry by following a similar process.

Keywords: AFP halo, ATV roll cage, ladder frame chassis, product lifecycle management, teamcenter 11.

Introduction

Product lifecycle management

Product lifecycle management (PLM) is a management system of a company that ensures the better output of the product. It checks the product from its raw material to the disposal of the product. It not only takes care of the product but also its portfolios.

PLM is known about the product design data and for further reference in case of failure or less efficiency outputs and to use data for further planning of resource management. A high-end tool like CAD/CAM/CAE software's made feasible in visualising the product's design, simulations, and validation. Where all the data from these tools can be handled with ease using the PLM platform. data from these tools can be handled with ease using the PLM platform.

Siemens teamcenter

Siemens Teamcenter software is an adaptive, modern PLM system that connects people and processes in different departments. It enables us to know our complete product BOM and provides secured access to real-time data to all users. It connects people, processes, and information and easily manages the product changes. It provides information to downstream applications.

Materials and Methods

A. Teamcenter organisation

The Database Administrator has the access to create the User ID, Password, Group, and Role.

- User ID- Each user in Teamcenter has a user ID. It is created by the administrator of your organisation application. You must provide a valid account ID to interact with Teamcenter. I used an admin login to create a total of six new users in my organisation.
- Password-Passwords are created by administrators in the Organisation application.
- Group- The organised collection of users who share data are groups. User accounts can belong to multiple groups and must be assigned to the default group. We created an 'engineering' group and made it the default group for the six new users of the organisation.

[a]yashwanth4100@gmail.com; [b]gebhiogade@gmail.com

- Role- A role is the function or responsibility of the user. The same roles are typically found in many groups (Teamcenter 10.1, 2013a). We created three roles namely, 'design engineer', 'CAE analyst' and 'production engineer', and to each group, we assigned two users so that the work can be distributed among the users in a group.

B. Teamcenter project

References Projects are used to represent and control access to data objects that may be accessible to multiple organisations, such as project groups, development groups, suppliers, and customers. Projects are represented as nodes in a tree (Teamcenter 10.1, 2013a).

To create a project, we need administrator (DBA) login access to the Teamcenter, so we created three projects, which are named as design, analysis, production project groups with administrator access and two added users for each project. The user who created the project becomes the 'project administrator'. The user with privileges to create and administer projects is the project administrator. Users in the project administrator role can modify projects, delete projects, add team members to projects, assign privileges to team members and remove team members from projects. And for the two users in each group, we have assigned one user as 'team administrator and the other user as 'privileged team member' of that project. The user with privileges to modify project information is the team administrator. These privileges apply to the project metadata, not to the data assigned to projects. Users in the team administrator role have access to add or remove team members to projects in which the team administrator is also a member. A privileged team member is a project team member with privileges to assign or remove objects from their projects (Teamcenter 10.1, 2013b; Figure 68.1).

C. Designing in NX

For the implementation of the concept of PLM to the industry, we have assumed an industry that manufactures three different products. Those products are,

1) Ladder frame chassis
2) ATV (all-terrain vehicle) roll cage
3) AFP (additional frontal protection) halo

The designing of these three products is done using Siemens NX 11 modelling software. These three products are designed by following some standard regulations and rulebooks.

The design, analysis, and manufacturing parameters for the three products of the industry are shown in the Table 68.1.

Design of ladder frame chassis

The ladder frame chassis is the backbone of a vehicle that integrates subsystems and absorbs the flexural and torque moments associated with powertrain components. They also receive random dynamic stimuli that occur in off-road conditions. The structure of the frame has a decisive effect on the dynamic performance, and its strength determines the fatigue resistance of the vehicle (Sithik et al., 2014). The box

Name	Status
Engineering Work	
Engineering Work.CAE Engineer	
de01 (de01)	Team Administrator
de02 (de02)	Privileged
Engineering Work.Design Engineer	
jayaram (jayaram)	Privileged
pavan (pavan)	Team Administrator
Engineering Work.Production Engineer	
dheeraj (dheeraj)	Privileged
kundan (kundan)	Team Administrator

Figure 68.1 Status of each user in the project

cross-section side rails provide good flexure resistance and the cross-section increases the torsional rigidity of the frame (Chuaymung et al., 2015).

The ladder frame chassis is designed according to India's central motor vehicles rules (CMVR). According to the 93rd rule of Chapter 5 'Maintenance of Construction Equipment and Vehicles', CMVR states that 'the overall width of the vehicle should not be more than 2.6 meters when measured between the vertical planes of the vehicle axles at right angles' and 'vehicles without trailers'. The total length of the vehicle must not exceed 6.5 meters and must have no more than two axles (Figure 68.2). This is applicable only for the vehicles which do not come under transport. The vehicle dealing in this industry is a MUV (Central Motor Vehicles Rules, 1989). According to Society of Indian Automobile Manufacturers (SIAM), based on the overall length, MUV falls under the 'B2' category. The segment of vehicles that come under the 'B2' category should have a length greater than 4000 mm. As per the design, the dimensions are, the length is 4530 mm and breadth is 1550 mm, hence, it satisfies the CMVR and SIAM rules (Muthyala, 2019).

Design of ATV roll cage:

The ATV roll cage is designed by following BAJA SAE 2021 rulebook Baja Saeindia Rulebook (2021), material selection for ATV roll cage is one of the important factor that has a significant impact on vehicle

Table 68.1 CAD/CAM/CAE parameters

S. No.	*Parameters*	*Ladder Frame Chassis*	*ATV Roll Cage*	*AFP Halo*
		DESIGN		
1	Standards followed	CMVR & SIAM	BAJA SAE 2021 Rulebook	2021 formula-1 technical regulations, FIA STANDARD 8869-2018
2	Modelling Software	Siemens NX 11.0	Siemens NX 11.0	Siemens NX 11.0
3	Material Used	ASTM A710 Steel	AISI1065 Steel	Grade 5 (Ti6Al4V)-Titanium Alloy
4	Cross-sectional Dimensions	Box section (100mm*50m m*6mm) & C - section (100mm*50m m*5mm)	Tubular section (outer dia. 25.4mm) & (Inner dia. 18.4mm)	Tubular section (outer diameter 50mm) & (Inner dia. 46mm)
		ANALYSIS		
5	Analysis	Modal, Torsional, Static Structural	Modal, Torsional, Impact, Roll Over	Quasi-static test, Modal
6	Solver used	NX Nastran	NX Nastran	NX Nastran
7	Solution Type	Sol 101 linear statics & Sol 103 Real Eigenvalues	Sol 101 linear statics & Sol 103 Real Eigenvalues	Sol 101 linear statics & Sol 103 Real Eigenvalues
8	Element	1D (Box¬section & C- section)	1D (Tubular section)	3D Tetrahedral
		MANUFACTURING		
9	Operations performed	Cavity milling, cavity milling finish, rest milling		Cavity milling, rest milling, and hole milling

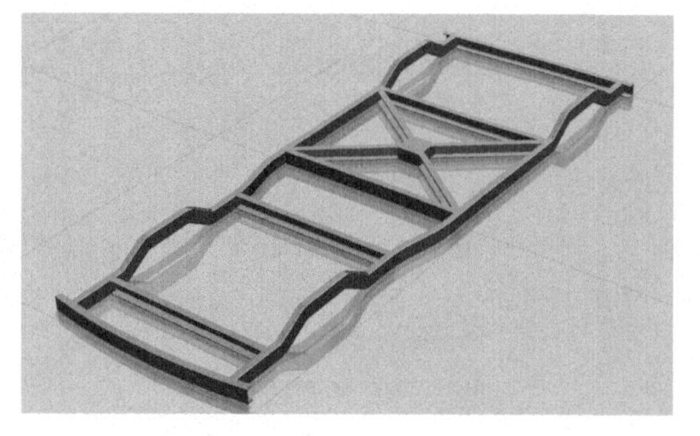

Figure 68.2 CAD model of ladder frame chassis

safety, reliability, and performance. As per the SAE BAJA rulebook (Baja Saeindia Rulebook, 2021), the carbon content of the steel should be at least 0.18%. The material chosen to study the ATV roll cage is AISI 1065 steel, as it offers higher yield strength and greater stiffness (Krishna et al., 2019). The cross-section of the ATV roll cage is a tubular cross-section with outer diameter 25.4 mm. and inner diameter 18.4 mm (Figure 68.3).

Design of AFP halo:

The aditional frontal protection (AFP) Halo is a rigidly attached mechanical structure at three mountings around the driver cockpit. halo is a 'head restraint safety device' used in motorsport to protect the driver in the event of an accident or collision The AFP Halo is designed by following the 2021 formula one technical regulations FIA- (2021) and FIA STANDARD 8869-2018 for single-seater additional frontal protection – halo (FIA Standard, 2018). All the three products are designed in Siemens NX 11, CAD software and are integrated to Siemens Teamcenter 11 PLM software as an item, the design phases are represented as item revisions in Teamcenter software. The CAD File is also saved in .jt (Jupiter Tessellations) format to visualise in the Teamcenter interface directly with no need of opening NX software separately(Figure 68.5).

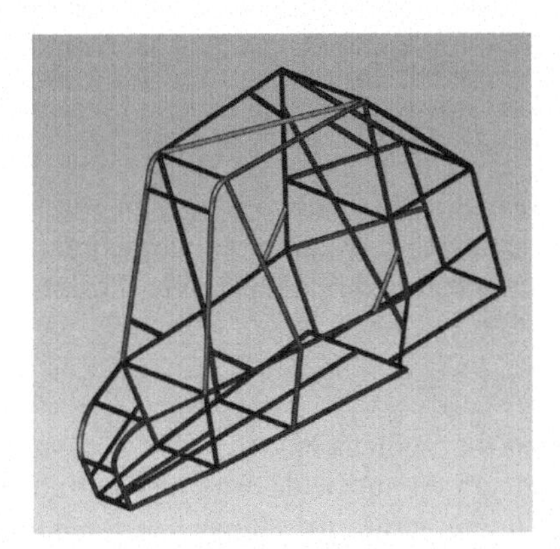

Figure 68.3 CAD model of ATV roll-cage

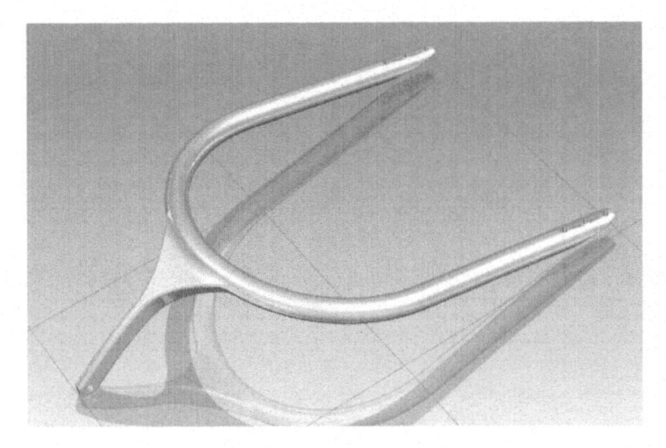

Figure 68.4 CAD model of AFP-Halo

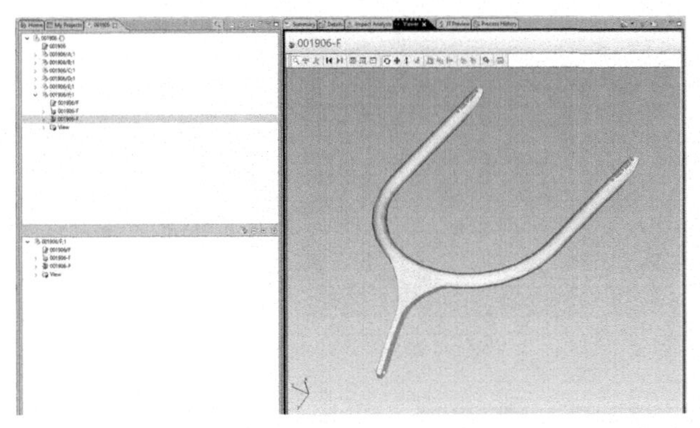

Figure 68.5 3-D model visualisation in Teamcenter

The CAD files from NX 11 are integrated to Teamcenter with the UG Master Dataset. They are also saved in '.jt' format to visualise the model geometry in the Teamcenter viewer directly (Figure 68.5).

Analysis in Simcenter 11.0:

The analysis of these three products is done using Simcenter 11 simulation software. The analysis for these three products is carried out with the help of some standard regulations and journals. The static analysis is carried out by subjecting the structure to constraints and known loads to determine the displacements, stresses, and strains that will serve as a parameter to check the resistance of the construction material used and to evaluate changes in the geometry (Rodrigues et al., 2015). In Simcenter software, after the analysis is being carried out, two files are created. They are,

a) Creating FEM file (.fem)-In Simcenter 11.0 a new file with a .fem extension is created to move into a pre/post environment. In any software, the pre-processing navigator contains, geometric details and Mesh details.
b) Creating SIM file (.sim)- After the geometric details, mesh details and material details are given in the .fem file now we have created a new file with a .sim file extension. In this file, we have access to give boundary conditions such as constraints, loads, and simulation object types.

Analysis integration with teamcenter:

The analysis files (. fem and. sim) from NX 11 are integrated to Teamcenter with two items sim and fem file for each analysis. The CAE files are saved in '.jt' format to visualise the meshing and boundary conditions in the Teamcenter viewer directly with no need of opening Teamcenter integrated NX separately (Figure 68.6).

Manufacturing in NX

The AFP Halo and ladder frame chassis were manufactured using the Siemens NX 11 CAM software. We can program any job using NX software. NX CAM provides integrated and wide-ranging capabilities of NC programming in a single system. A 3-axis mill was used to manufacture the chassis for the halo and ladder frame (Figure 68.7).

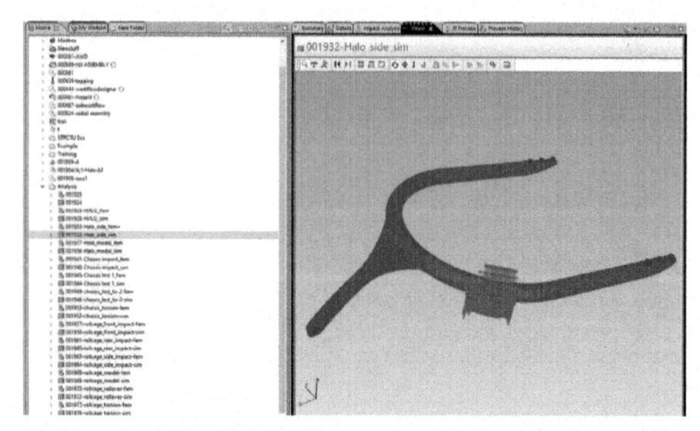

Figure 68.6 Simulation file preview in Teamcenter viewer

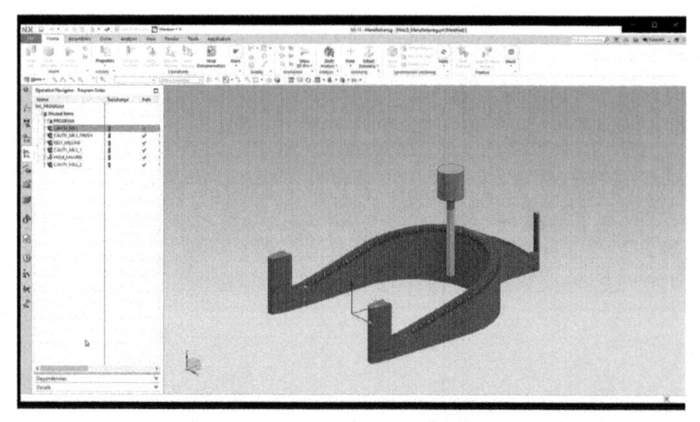

Figure 68.7 Cavity milling operation on a block to generate Halo

Manufacturing integration with teamcenter:

The product structure also incorporates all information and documents related to the product (e.g., CAD models, NC programs, cost analysis, maintenance procedures, overhaul plans) (Schuh et al., 2008). The CAM files (which consist of NC programs) from NX 11 are integrated to Teamcenter with the CAM Master dataset. While integrating, the geometric part file associated with the CAM file is also assigned to the Teamcenter. The CAM files are saved in '.jt' format to visualise the model geometry in the Teamcenter viewer directly with no need of opening NX separately.

D. Workflow

We have created three workflow templates for three products with Administrator access, the workflow templates are assigned to design items and various users are assigned to perform the tasks, do tasks, review tasks and acknowledge tasks. Each user based upon his role performs the task or makes a decision. If the decision-making task is approved or acknowledged, the process flows to the next task, if the decision-making task is rejected or not acknowledged, the process returns to the previous task, representing that the task needs to be performed again (Teamcenter 10.1, 2013a) (Figure 68.8).

E. Schedule manager

For the Industry, we have designed a schedule for each product of the industry. We have created schedules for the three products of the industry starting from designing from scratch till the end of the sales task. In a schedule, we have included design, analysis, manufacturing, post-processing, quality control, inventory, and sales tasks. The analysis, manufacturing, and post-processing tasks are called summary tasks which consist of sub-tasks. We have assigned each task to users depending upon their role, and the workload distribution among the users is also assigned. Task dependencies are given to tasks to give the relation between the tasks. The status of the task can be tracked in a schedule and the user can update the status and determine whether the task is in progress, not started, needs attention to complete, completed, late, or abandoned.

A program view (Figure 68.9) in the name of 'all schedules' is created and all the three schedules for the three products are added to the newly created program view. All the schedules can be opened, viewed and tasks can be tracked in a single window without the need for opening and tracking individual schedules separately (Teamcenter 10.1, 2012).

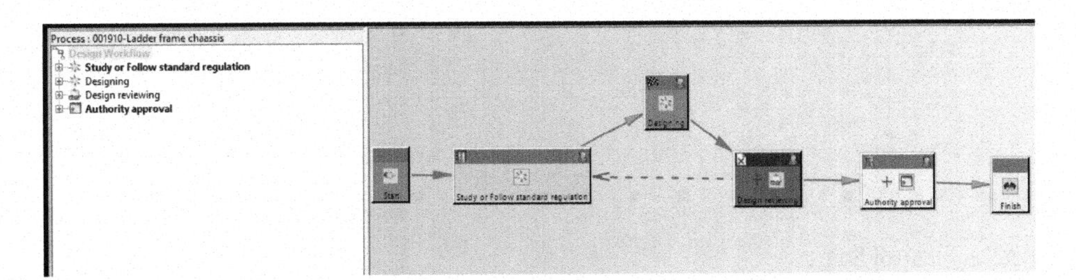

Figure 68.8 Workflow process of ladder frame chassis

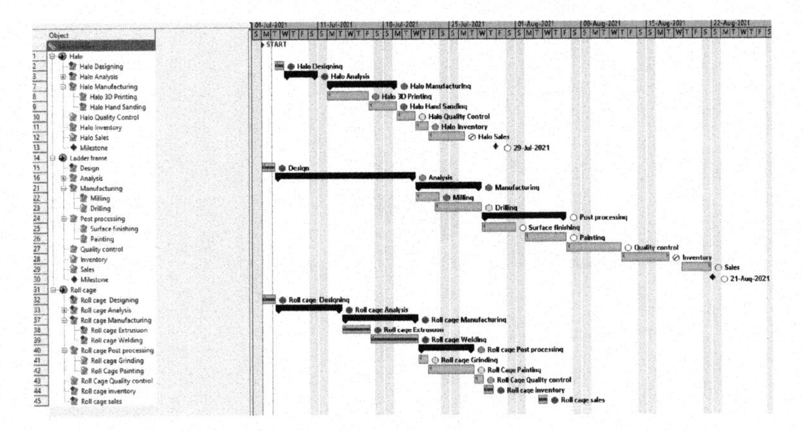

Figure 68.9 Program view of all schedules

F. Access management

The users (employees) in an industry can share their data across the industry through Teamcenter. While sharing the data, each user of Teamcenter has the ability to control the access to the files shared by the user. For example, the user from the design department can share the design data, bill of materials, production drawings, etc. with the production department, and while sharing the data, read access is sufficient for the production department (Figure 68.10). There is not necessary to give them write access, modify, share access, etc. to the production department. This is done to prevent the data from being over-written, modified, erased, etc. while sharing the data (Teamcenter 10.1, 2013a; 2013b).

Results and Discussion

Analysis results of AFP Halo

The maximum deformation obtained in the Quasi-static test 1 (frontal impact) is 9.789 mm. The FIA STANDARD 8869-2018 states that, there should be no structural failure to any part of the structure. Deflection should be less than 17.5 mm when the load on the structure reaches 125 kN (FIA Standard, 2018). Hence the deformation of Halo under Quasi-static test-1 is within the limit (Figure 68.11).

The maximum deformation obtained in the Quasi-static test-2 (side impact) is 36.64 mm (Figure 68.12). According to FIA STANDARD 8869-2018, the maximum load must be at least 125 kN. There should be no structural failure up to 125 kN. Deflection must not exceed 45 mm at a maximum of 125 kN (FIA Standard, 2018). Hence the deformation of Halo under Quasi-static test 2 is within the limit.

Manufacturing results

In NX CAM 11, after all the machining operations are performed on the AFP-Halo, we can see the amount of excess material present for the given halo geometric part file. The maximum excess material thickness is 10.7631 mm. Similarly, for the ladder frame chassis, the maximum excess material thickness is 125 mm after performing contour milling operations. The remaining material can be removed in other machining and post-processing operations (Figure 68.13 and 68.14).

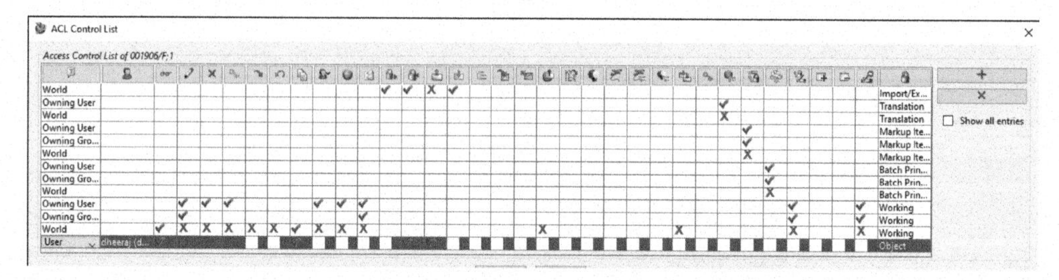

Figure 68.10 Access control list

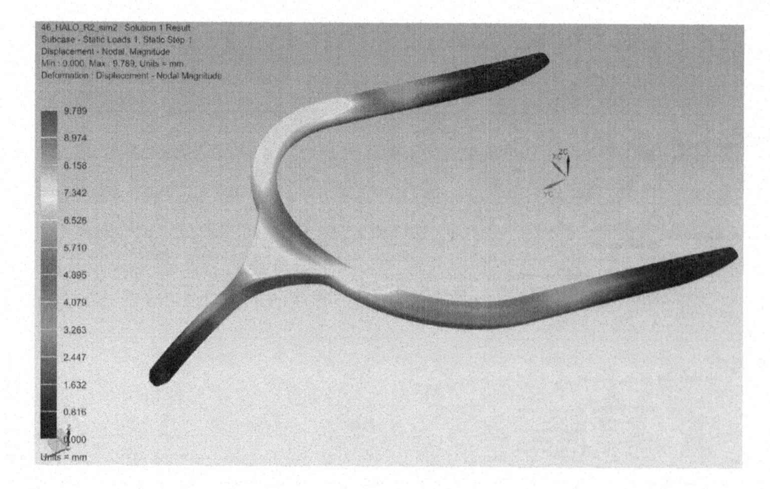

Figure 68.11 Displacement magnitude in AFP Halo in Quasi-static test-1

Result of implementing PLM

The data losses between the departments in an organisation have been reduced as updated information is visible always to all the members of that organisation. The previous data can be retrieved any number of times even after several updates of a product which makes the developer of a product easier. The entire workflow of each part of the product can be tracked, and project managers receive messages or notifications so they can be alerted if someone in the value chain rejects or accepts for further processes. The schedule manager lets the user know whether a task is running at a scheduled time or not.

The access manager makes the information of a user more conservative and prevents the data from being modified or deleted by another user which decreases the data losses. Thin client makes it easy to track the information of the project for clients. The problem can be identified easily in which group it has occurred and which user did that mistake as all the information related to the item can be tracked and seen by the

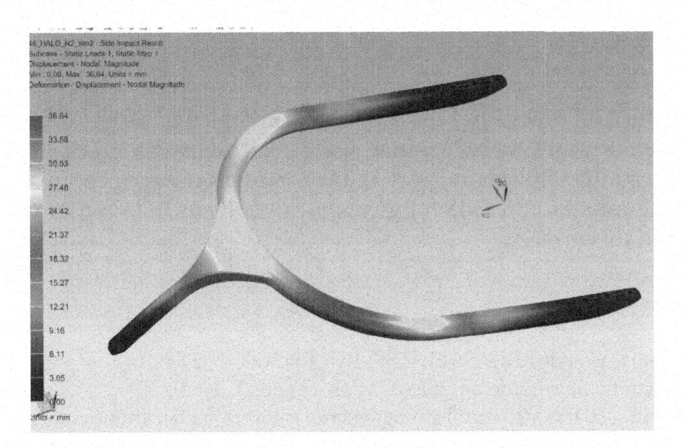

Figure 68.12 Displacement magnitude in AFP Halo in Quasi-static test-2

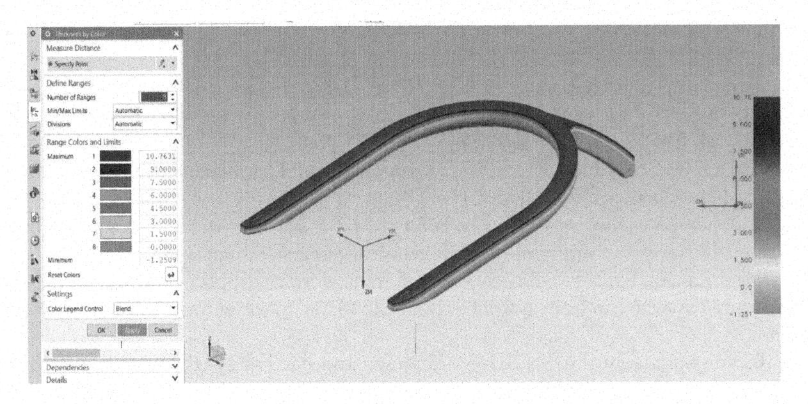

Figure 68.13 Excess material thickness in AFP Halo

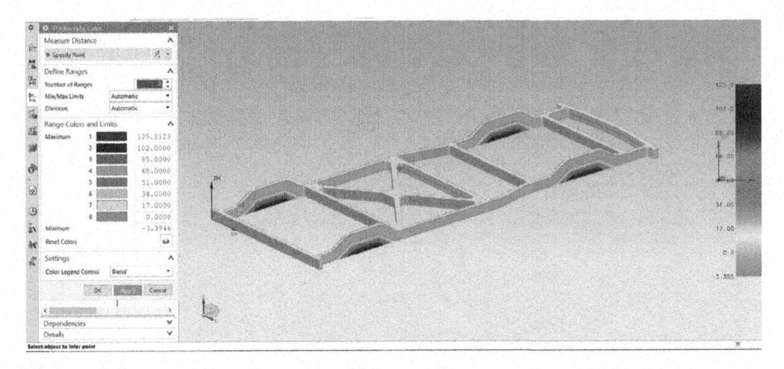

Figure 68.14 Excess material thickness in ladder frame chassis

project administrator. For sales, the project members use the Visualisation Mock-Up tool to make customers easily visualise the product.

Conclusions

The implementation of the concept of PLM helps in providing the centralisation of information. With everybody working from a single central source, the data is always updated which reduces the data losses between the departments and enables the teams to work efficiently, thereby accelerating the process of development and reducing time delay. All these parameters help in reducing the chance of errors, help in securing the data, and increase the revenue of the industry. Hence by using Teamcenter PLM software, we have managed all information and processes at every stage of our industry's product or service lifecycle. And this concept of PLM can be applied to any large-scale manufacturing industry in real-time by following a similar procedure.

Acknowledgement

We thank our trainer Mr. I. Krishna Kanth for all his help and contribution in completing this project. We express our gratitude to Cdr. Gopi Krishna Sivvam, Chief Operating Officer of Centre of Excellence in Maritime and Shipbuilding (CEMS), Visakhapatnam for his support and constant encouragement in completing the project. We sincerely thank all the trainers at CEMS for giving me their heart full support in all stages of the project work and completion of this project.

References

Baja Saeindia Rulebook (2021). https://bajasaeindia.org/pdf/BAJA-SAEINDIA-RULEBOOK-2021.pdf.

Central Motor Vehicles Rules (1989). https://morth.nic.in/sites/default/files/CMVR-chapter1_1.pdf.

Chuaymung, C., Benyajati, C. and Olarnrithinun, S. (2015). Structural strength simulations of ladder frame chassis for light agriculture truck. SAE Technical Paper 2015-01-0090. https://www.sae.org/publications/technical-papers/content/2015-01-0090/.

2021 FORMULA 1 TECHNICAL REGULATIONS (2020, December 16). https://www.fia.com/sites/default/files/2021_formula_1_technical_regulations_-_iss_7_-_2020-12-16.pdf.

FIA Standard (2018). Single-seater additional frontal protection – halo. https://www.fia.com/sites/default/files/fia_standard_8869-2018_afp_v1.2.pdf.

Krishna, S., Shetye, A. M., Mallapur, P., and Jeethray, S. K. (2019). Design and analysis of chassis for SAE BAJA vehicle. IOSR J. Eng. (IOSR JEN), ISSN (e): 2250-3021. https://www.researchgate.net/publication/335110382_Design_and_Analysis_of_Chassis_for_SAE_BAJA_Vehicle.

Muthyala, M. (2019). Design and crash analysis of ladder chassis. Karlskrona, Sweden: Blekinge Institute of Technology. https://www.diva-portal.org/smash/get/diva2:1337405/FULLTEXT02.

Rodrigues, A., Gertz, L., Cervieri, A., Poncio, A., Oliveira, A. B., and Pereira, M. S. (2015). Static and dynamic analysis of a chassis of a prototype car. https://www.sae.org/publications/technical-papers/content/2015-36-0353/.

Schuh, G., Rozenfeld, H., Assmus, D, and Zancul, E. (2008). Process-oriented framework to support PLM implementation. Comput. Ind. 59(2):210–218. https://www.researchgate.net/publication/222431367_Process_oriented_framework_to_support_PLM_implementation.

Sithik, M., Vallurupalli, R., Lin, B. B., and Sudalaimuthu, S. (2014). Simplified approach of chassis frame optimisation for durability performance. https://www.sae.org/publications/technical-papers/content/2014-01-0399/.

Teamcenter 10.1 (2012, August). Teamcenter schedule manager, student guide, MT25700.

Teamcenter 10.1 (2013a). Business modeler IDE guide. https://support.industrysoftware.automation.siemens.com/docs/teamcenter/10.1/PDF/en_US/tdocExt/pdf/business_modeler_ide_guide.pdf.

Teamcenter 10.1 (2013b). Using Teamcenter Student Guide, MT25150.

69 Factor affecting the thermal conductivity of nanofluids – A Review

Dharmender Singh Saini[a] *and S. P. S. Matharu*[b]

Mechanical Engineering Department, National Institute of Technology, Raipur, India

Abstract

When a base fluid is mixed with nanoparticles, the resultant nanofluid possess better TC resulting in improved heat transfer rate. Several researchers reported increase in thermal conductivity of nanofluids when low concentration of nanoparticles (NPs) is mixed with base fluid. The findings of several studies showed that NP size, shape, material, volume fraction, temperature of base liquid and stability of suspension have an impact on nanofluid thermal conductivity. The thermal conductivity of the nanofluid increases as the particle size decreases for the same concentration. Moreover, findings also show increase in thermal conductivity of the nanofluid with increase in temperature. Present work is a review of the various experimental work done by different researchers. The nanofluids are not stable for longer period of time if are not in use. A novel work can be done taking into account the settlement of nanoparticles of nanofluid with respect to time. This work may help the researchers doing experimental work with nanofluids in future.

Keywords: Nanoparticles, nanofluid, thermal conductivity, shape, size, volume fraction.

Introduction

For designing energy efficient systems, the thermo-physical parameters of the conventional fluid governs the heat transfer rate. Nanofluids offer improved heat transfer characteristics, because high TC nanoparticles are suspended in conventional fluids. Choi and Eastman (1995) published their notable work of producing a colloidal combination of metallic NPs and conventional fluids, resulting in nanofluids, a novel class of heat transfer fluid with improved thermal performance. Though the fundamental concept of suspending solid particles in a liquid to increase TC may be traced back to a research done by Maxwell in 1873 (Maxwell, 1873). However, due to specific limitations imposed by micro-sized metallic particles, such as fast sedimentation, erosion, clogging, and huge pressure drop, the system remained ineffective. Nanofluids emerged as a new generation of heat transfer fluids when the advent of particle and interface science brought new options to create nanometer size ultra-fine NP (Kotia et al., 2018; Paul et al., 2011; Saini and Matharu, 2020). However, at low concentrations, these tiny particles have a large surface by volume proportion, resulting in high thermal conductivity, suspension stability, and an increased solid-particle inter-phase leading to improved heat exchange. Nanofluids are also suitable for a variety of applications, including transportation, space technology, microelectronics, solar, nuclear, and refrigeration, as well as biomedical process, due to superior optical features, low sedimentation and clogging, low pressure drop, and little mechanical damage (Ambreen and Kim, 2020).

Nanofluids are classified according to the type of nanoparticles (NPs) used (carbon, metal oxide, metal, rare earth metals, nano-composite, etc.) and the base fluid (Water, EG, synthetic oil, mineral oil, and bio-oil) (Kotia et al., 2018). Nano-fluids are widely used in various heat transfer medium with their high surface by volume proportion and good optical, physical, thermal, and mechanical properties (Li et al., 2021). One-step or two-step method can be used to make nanofluids. In one-step method, NP Synthesis and dispersion in a fluid are done at the same time, whereas in two-step method, synthesised NPs are mixed in base fluid with mechanical/magnetic stirring, ultrasonication, probe-sonication, etc. (Li et al., 2021). The one-step approach is less common for mass production since it requires vacuum environment, which slows down nanofluid production and increases production costs. Although the two-step procedure is very common than one-step approach due to ease of its production of nanofluid, leading to decreased manufacturing costs (Chang et al., 2007).

Recent research also authenticates that NPs can be used to increase TC (Ambreen and Kim, 2020). Song et al. (2017) calculated the TC of Ag/water nano-fluid for the volume fraction of 0.02 to 0.2% of irregular size Ag particle. Pryazhnikov et al. (2017) studied the affect of d_p on k_{nf} of water based nanofluid having SiO_2, ZrO_2, TiO_2 and Al_2O_3 NP of different size in 2% volume fraction. Maheshwary et al. (2017) showed an experimental finding which revealed the impact of particle ϕ, d_p and shape on the k_{nf} of the

[a]dharmender.saini13@gmail.com; [b]spsm.mech@nitrr.ac.in

TiO_2/water nano-fluid. Cubical NPs had the maximum k_{nf}, then rod and spherical NPs. Hemmat et al. (Esfe et al. 2017) measured the affect of d_p, T, and on the k_{nf} of the MgO/water nano-fluid. The experimental research work gave a relationship for k_{nf} based on the examined parameters (Esfe et al., 2017; Agarwal et al., 2017; Chopkar et al., 2008; Yang et al., 2011; Das et al., 2017; Maheshwary et al., 2017; Iqbal et al., 2017; Ceylan et al., 2006; Mahian et al., 2017). The researchers showed a linear variation in TC with NP volume percent (Barbés et al., 2014; Esfe et al., 2015; Maheshwary et al., 2017). Some researchers, showed a non-linear increase in TC with NP volume fraction (Sundar et al., 2013; Wang et al., 1999; Chopkar et al., 2008; Paul et al., 2011; Ambreen and Kim, 2020).

The researcher worked on almost all types of nanoparticles with different sizes and concentrations. Most of the authors mentioned about the stability of the nanofluids but in the literature review for the present research work, the settlement of nanoparticles in nanofluid with time could not be seen. Future work can be concentrated to evaluate the settlement of nanoparticles in nanofluids with time. Authors are working on this field to fill this research gap. More stable is nanofluid, wider will be its applications.

The impacts of different factors that influence the TC of nanofluids are reviewed and discussed in this work. This article includes the effects of NP size, shape, material, NP volume fraction, temperature, base liquid and stability.

Factor Affecting the TC of Nanofluids

TC of nanofluids is an important thermophysical parameter that determines heat transfer efficiency. NP size, shape, concentration, TC of NP, base fluid temperature & its thermal conductivity, stabilisation techniques, influence overall TC of nanofluid (Tawfik, 2017). The individual or combined effect of these parameters must be considered in order to understand the variation in effective TC (k_{nf}) of nanofluid. This section gives the general trends in effective TC of nanofluids realised on of NP size (d_p), NP volume % (ϕ), operating temperature (T), nanoparticle TC (k_{np}), base fluid TC (k_{bf}), NP shape, stability, and measuring techniques (Figure 69.1).

A. Influence of diameter of nanoparticle (d_p) on k_{nf}

Nanoparticle size (d_p) is the most important characteristic that differentiates nanofluids from micro-sized suspensions. This section has a brief discussion of nanofluids dependencies on size of particle. Researchers discovered that a larger surface by volume propotion of NPs results in increased TC of nanofluid, which may be attained with lower particle sizes (Ambreen and Kim, 2020). Al_2O_3 and CuO NPs, as well as deionised water, vacuum pump fluid, motor oil, and ethylene glycol base fluid, were first investigated by Wang et al. (1999). TC was examined with respect to particle diameter and volume fraction of NPs. The result shows that, TC increases as particle size decreases at a certain volume percent of NPs. Chon et al. (2005), studied the TC of Al_2O_3/water nanofluids as a function of NP size (11 nm, 45 nm and 150 nm) for a temperature range 21 to 71. According to investigation, researchers found that the Brownian motion of NPs is an important mechanism of TC for rising temperature and decrease in NP sizes. Kim et al. (2007) confirmed an inverse relationship of TC particle mean diameter of NPs (ZnO and TiO_2) dispersed in water and ethyl glycol in their study. This relationship was linear for different volume fractions. The TC further increases with additional mixing of small sized NPs. For various NP shapes (spherical, square, and needle) and size (50–250 nm), Liu et al. (2006) investigated the k_{nf} of Cu/water nanofluids at 0.2% ϕ. The k_{nf} of nanofluids showed an inverse relationship with the d_p of similarly shaped particles. The k_{nf} enhanced with decreasing the particle size (25–100 nm) for Cu/DO nanofluid (Colangelo et al., 2012). As particle

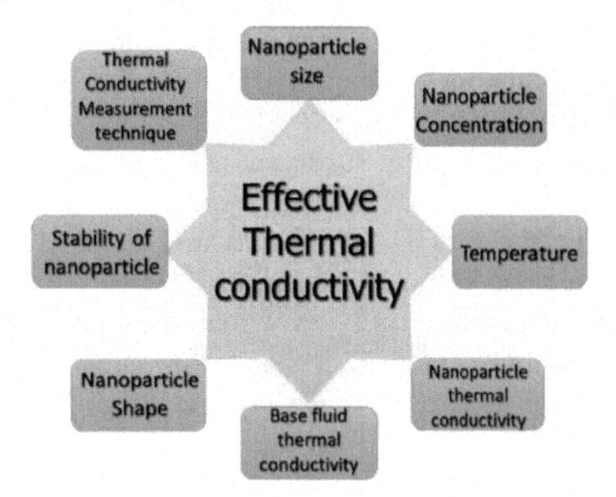

Figure 69.1 Factor affecting the effective thermal conductive of nanofluids

size increases in a given volume of nanofluid, the number of NPs in the sample decreases. Therefore, TC increases when particle size decreases due to the increased surface-to-volume ratio for small sized particles compared to larger size particles.

B. Impact of concentration(ϕ) on k_{nf}

One of the most important researched area in nanofluid studies is the effect of ϕ on k_{nf}. According to several studies, k_{nf} reaches its maximum value at an optimum value of ϕ, further adding more nano-particles decreases, the value of k_{nf} of nanofluids. Moreover, some researchers have discovered a linear correlation with both TC and nanoparticle volume fraction (Barbés et al., 2014; Esfe et al., 2015), (Maheshwary et al., 2017), while others have discovered a non-linear correlation (Sundar et al., 2013; Wang et al., 1999; Chopkar et al., 2008; Paul et al., 2011; Ambreen and Kim, 2020). Maheshwary et al. (2017) used NPs of irregular shape with the same APS to test the TC of TiO_2/water nanofluid depends on NP volume %. The TC increases as particle volume fraction is increased. NP concentration has a greater impact on TC rather than particle size and shape. A non-linear correlation between NP volume fraction and nanofluid TC has been reported in several investigations, which may be due to increased agglomeration at high NP volume fraction. (Agarwal et al., 2017) used KD2 Pro instrument to measure the thermal conductivities of Al_2O_3/water and Al_2O_3/EG nanofluids. The TC of both nanofluids improved as the NP volume fraction is increased. Research work done by Sundar et al. (2013) reflected a more substantial increase in TC with higher NP concentration. The TC of Fe/water nanofluid was determined by Esfe et al. (2015) was a function of NP vol. %. TC increased approximately linearly as the volume fraction of NPs increases. Similarly, a linear relationship of SiO_2/water, Al_2O_3/water, CuO/EG and CuO/water and ZrO_2/water nanofluid was studied by (Iqbal et al., 2017; Das et al., 2017; Barbés et al., 2014). According to the experimental data stated above, the TC of nanofluids increases with NP volume %. Due to the degree of agglomeration, the rate of growth varies for different nanofluids.

C. Effect of temperature (T°C) on k_{nf}

Numerous researchers concluded, that for a given temperature, the TC of the nanofluids were higher than those of the basic fluids. TC may vary with temperature due to Brownian diffusion and Thermophoresis, both of which are directly proportional to the temperature of the nanofluid. Nanofluids are ideal for high-temperature operation for heat exchangers, nuclear reactors, and solar collectors (Ceylan et al., 2006; Al-Shamani et al., 2014). Sundar et al. (2013) evaluated the TC of Fe_3O_4 NPs in EG/water-based nanofluids by weight ratios of 20:80, 40:60, and 60:40 and observed that TC of prepared nanofluid increases with raising temperature. The studied TC of Al_2O_3/water, Al_2O_3/EG, TiO_2/water, Fe_2O_3/water, ZnO/water, CuO/engine oil, CuO/hydraulic oil, CuO/EG and, CuO/water is a function of temperature (Agarwal et al., 2016; Patel et al., 2010; Maheshwary et al., 2017; Sundar, et al., 2013; Kotia et al., 2017; Kole and Dey, 2013). The result was drawn from the literature that rise in thermal conductivities due to Brownian motion at high temperatures. Brownian motion is created by haphazard collisions of molecules of a liquid, which results in random movement of NPs within the base fluid, resulting in increase in thermal conductivity. Increased microscopic motion produced by electrostatic, Brownian, and van der Waals, interactions was also having a role in the rise of thermal conductivity.

D. Influence of nanoparticle TC (k_{np}) on k_{nf}

Investigators conducted studies to examine the impact of NP material on TC of nano-fluids. Results indicated that NP material is a significant influencing parameter for nanofluid TC (Simpson et al. 2019). Various studies have proposed that parameters other than NP thermal conductivity, like size and volume fraction of NP, have a larger impact on nanofluid TC (Pang et al., 2012; Simpson et al., 2019; Kotia et al., 2018; Paul et al., 2011). Table 69.1 show the TC of various NPs. Using the THW approach, Yoo et al. (2007) investigated the TCs of WO_3/EG, Fe/EG, Al_2O_3/water, and TiO_2/water nano-fluids. TiO_2/water nanofluid had a higher TC than Al_2O_3/water nanofluid, despite the reality that Al2O3 has a higher intrinsic TC than TiO_2. This was thought to be due to the TiO_2(25 nm) NPs having 47.92% lower size than Al_2O_3(48 nm) NPs size, resulting in a higher surface by volume proportion in the TiO_2/water nano-fluid. The TC of Fe/EG was greater than WO_3/EG nanofluid. Because the Fe (10 nm) NPs have 73.68% lower size than the WO_3(38 nm) NPs size. It is seen that the TC of suspended NPs has little effect as compared to, the surface area by volume proportion which has a major influence on the TC of the nano-fluid. It was also suggested that metallic NP-containing nanofluids had higher thermal conductivities than ceramic NP-containing nanofluids. Similarly, Xie et al. (2010) evaluated the TCies of nanofluids made up of various oxide NPs dispersed in Ethylene Glycol, including SiO_2, Al_2O_3, TiO_2, MgO, and ZnO. The internal TCies of the NPs were MgO,

Al_2O_3, ZnO, SiO_2, and TiO_2, in the order of highest to lowest. Al_2O_3/EG, MgO/EG, ZnO/EG, SiO_2/EG and TiO_2/EG, nanofluids had the greatest thermal conductivities, followed by Al_2O_3/EG, TiO_2/EG, ZnO/EG. These findings further support the hypothesis that variables other than the inherent TC of NPs have a greater impact on the TC of nanofluids. However, experimental evidence has showed mixed outcomes in several circumstances, with nanofluid TC increasing as NP thermal conductivities increased.

E. Effcect of base fluid TC (kbf) on knf

The previous literature studies show that the TC of synthesised fluid directly proportional to the TC of conventional/base fluid. Table 69.2 shows the base fluid TC with temperature used by various researcher. Table 69.2 also shows that the TC of the base fluid has a major influence on the improvement of nanofluid TC. Chopkar et al. (2007) used EG and water base fluid to disperse Al_2Cu and Ag_2Al. According to the experimental data, TC of water-based nanofluids were greater than those of EG-based nanofluids. CuO NPs suspended in a various base fluid, like DW, EG, and engine oil. Because distilled water has a greater TC than the other two base fluids, it is examined that DW-based nanofluids had larger TC (Agarwal et al., 2016; Barbés et al., 2014). The findings confirmed those of Agarwal et al. (2016) and Barbés et al. (2014), with water-based nanofluids having higher thermal conductivities than EG-based nanofluids for Al_2O_3 and CuO NPs at a given volume %. The thermal conductivities of Al_2O_3/Pump oil and Al_2O_3/engine oil

Table 69.1 Nanoparticles and its thermal conductivities

Reference	*Nanoparticles*	*TC (W/m-K)*
Ramezanizadeh and Nazari (2019)	Ag	429
Patel et al. (2010)	Al	204
Madan and Bhowmick (2021)	Al_2O_3	36
Patel et al. (2010)	Cu	383
Liu et al. (2011)	CuO	33
Feng et al. (2012)	TiO_2	5.6
Park and Kim (2014)	Graphene	3000
Xie et al. (2010)	ZnO	13
	MgO	48.4
Simpson et al. (2019)	Fe_2O_3	7
Iqbal et al. (2017)	SiO_2	1.4
Iqbal et al. (2017)	ZrO_2	2.2

Table 69.2 Thermal conductivity of most commonly used base fluid

Reference	*Base Fluid*	*Temperature (°C)*	*TC (W/m-K)*
(Agarwal et al., 2016)	Water	20	0.629
		30	0.640
		40	0.6582
		50	0.6721
	Ethylene Glycol (EG)	20	0.2441
		30	0.256
		40	0.2481
		50	0.249
	Engine Oil	20	0.140
		30	0.141
		40	0.142
		50	0.1431
(Patel et al., 2010)	Transformer Oil	20	0.112
		30	0.111
		40	0.109
		50	0.107

nanofluids were smaller than those of Al_2O_3/EG nanofluid. The Al_2O_3/engine oil nanofluid seems to have higher TC than the pump oil–based nanofluid. The TC of a nanofluid is said to rise as the TC of the base liquid increases (Wang et al., 1999).

F. Effcect of Shape on knf

NPs are made in a variety of shape like spiral, tube, rod, oval, hexagonal, cubic, triangular and spherical. The TC of nanofluid is influenced by the surface by volume proportion of NPs. The surface area-to-volume ratio of NPs is influenced by its shape as well as their size. Xie et al. (2002) conducted the first organised examination into the influence of SiC shape change in water. The k_{nf} of nanofluids with diameters of 600nm (cylindrical) and 26nm (spherical) were investigated. The importance of the complicated interaction of NPs with base fluids in determining TC increases. The adverse influence of heat flow resistance at the solid-liquid boundary reduces the TC rise expected by the Hamilton–Crosser equation in nano-fluids carry non-circular particles (Timofeeva et al., 2009). Murshed et al. (2005) studied the impacts of particle shape on nanofluid TC (cylindrical NPs 10nm x 40nm, spherical NPs d_p = 15 nm) and found that rod-shaped NPs had higher TC than spherical NPs for all concentration of NPs. Maheshwary et al. (2017) used cubic (87.21 nm), rod (ϕ92.47 nm × 8.27 nm), and spherical (d_p = 23 nm) shaped NPs to explore the TC of TiO_2/water nanofluid. The TC of nanofluids containing cubic NPs was found to be high, while that of nanofluids containing spherical NPs was found to be the relatively low.

It's been proposed that NP shape, instead of NP size, has a more effect on TC. Basics on the literature outcomes, it can be concluded that NP with a larger surface area have a higher nanofluid thermal conductivity. TC was considered to be influenced by shapes of NPs, because of the variation in surface area. Spherical NPs provide a number of advantages over alternative forms of nanofluids, including improved stability and reduced clogging.

G. Stability of nanofluid

If a nanofluid maintains constant suspension for a longer period of time, it means that the nanofluid has good stability. The commonly used methods to measure the nanofluids stability are zeta-potential (Z_p), pH and sedimentation photograph. Surfactant addition is thought to be the straightforward and cost-effective method to increase nano-fluid stability (Wang et al., 2021). However, this surfactant can affect the thermo-physical properties of nanofluid. Das et al. (2016), used SDS, OA, acetic acid (AA) and CTAB as surfactant with water to prepared TiO_2/water nanofluid. The CTAB and AA surfactant had the best stability, resulting in a more uniform dispersion of NPs throughout the base liquid. Further, Das et al. (2017), prepared Al_2O_3/water nanofluid with SDBS, CTAB and SDS surfactant. The SDS surfactant had the best stability compare to other two surfactant. The Z_p of a nanofluid can be used to evaluate the stability of the nano-fluid. Z_p is an electrostatic potential created at the solid-liquid boundary in retort to the comparative mobility of solid particles and fluid (Yang et al., 2011). As per Vandsburger (2009), if the Z_p is approximately ±30 mV, nanofluids should be moderately stable. The stability of nanofluids should be high if the Z_p is approximately ±45 mV. Z_p greater than ±60 mV indicates a very good stability of nanofluid. pH values of nanofluid should be kept below the isoelectric point (IEP) in order to generate stable nanofluids, according to Otterstedt and Brandreth (Otterstedt and Brandreth, 2013). Sedimentation photograph is a cheap but time taking method of assessing stability. In this, the nanofluid is filled in a container and kept at constant temperature and humidity, photos are taken at regular intervals. Stability can be determined by comparing the new images with the previous images taken (Wang et al., 2021).

Detail summary of nanofluid is given in Table 69.3.

Conclusion

This work aims to determine the impact of NP material, size, shape, concentration, and surfactant on the TC of nanofluids. Effect of base fluid has also been taken into consideration. For various base fluid TC of the working fluid is the most essential factor that determines heat transfer rate. Nanofluid TC can be improved by mixing NPs into the base medium, for better heat transfer rate. The review shows that the effective TC is a function of NP size and concentration. At the same concentration, as the particle size reduces, the TC value also rises. Since small particles have a higher surface by volume proportion than larger ones. Literature has shown that the rise in TC is due to Brownian motion at higher temperatures, so nanofluid TC increases as nanofluid temperature increases. This evidence has also shown that nanofluid TC increases as nanoparticle TC increases. It is concluded that the shape of NPs , rather than their size, has a large impact on thermal conductivity. According to previous research, the TC of synthesised liquid is

Table 69.3 Summary of thermal conductivity of nanofluids

Reference	*Nanoparticles*	*Base fluid*	*Size (nm)*	*Concentration*	*Temp (°C)*	*TCR (%)*
Ambreen and Kim (2020)	Al_2O_3	Ethylene glycol	28	8 vol %	RT	40
	CuO	Ethylene glycol	23	14.5 vol %	RT	54
Kim et al. (2007)	ZnO	Ethylene glycol	30	3 vol %	RT	21
	TiO_2	Distilled water	10	3 vol %	RT	11.4
Liu et al. (2006)	Cu (spherical and square)	water	50–100 nm,	0.1 vol %	RT	23.8
Barbés et al. (2014)	CuO	Ethylene glycol	23–37 nm	3 vol %	RT	13
Esfe et al. (2015)	CuO	Ethylene glycol-water (4060%)	-	2 vol %	50°C	15.5
Sundar et al. (2013)	Fe_3O_4	Ethylene Glycol-Water (2080%)	≈13 nm	2.0 vol %	RT	46
Iqbal et al. (2017)	ZrO_2	Deionised water	25 nm	1 vol %	RT	8.5
Yoo et al. (2007)	TiO_2	Deionised water	25 nm	1 vol %	RT	14.4
	Al_2O_3	Deionised water	48 nm	1 vol %	RT	4%
	Fe	Deionised water	10 nm	0.3 vol %	RT	16.5
	WO_3	Deionised water	38 nm	0.3 vol %	RT	13.8
Xie et al. (2010)	MgO	Ethylene glycol	20 nm	5.0 vol %	RT	40.6
	SiO_2	Ethylene glycol	20 nm	5.0 vol %	RT	25.3

RT Room Temperature (30°C)

directly related to the TC of base fluid. The effect in long term stability with different surfactants was studied. It is observed that at an optimum concentration of surfactant, the influence on the TC of nanofluid was insignificant but ensure good stability. Future work should emphasise on settlement of NPs with respect to time and its effect on TC of nanofluid.

References

Agarwal, R., Verma, K., Agrawal, N. K., Duchaniya, R. K., and Singh, R. (2016). Synthesis, characterisation, thermal conductivity and sensitivity of cuo nanofluids. Appl. Therm. Eng. 102:1024–1036.

Agarwal, R., Verma, K., Agrawal, N. K., and Singh, R. (2017). Sensitivity of thermal conductivity for Al2O3 nanofluids. Exp. Therm. Fluid Sci. 80:19–26.

Al-Shamani, A. N., Yazdi, M. H., Alghoul, M. A. Abed, A. M., Ruslan, M. H., Mat, S., and Sopian, K. (2014). Nanofluids for improved efficiency in cooling solar collectors–a review. Renew. Sustain. Energy Rev. 38:348–367.

Ambreen, T. and Kim, M. H. (2020). Influence of particle size on the effective thermal conductivity of nanofluids: A critical review. Appl. Energy. 264:114684.

Barbés, B., Páramo, R., Blanco, E., and Casanova, C. (2014). Thermal conductivity and specific heat capacity measurements of CuO nanofluids. J. Therm. Anal. Calorim. 115(2):1883–1891.

Ceylan, A., Jastrzembski, K., and Shah, S. I. (2006). Enhanced solubility Ag-Cu nanoparticles and their thermal transport properties. Metall. Mater. Trans. A 37(7):2033–38.

Chang, H., Jwo, C. S., Fan, P. S., and Pai, S. H. (2007). Process optimisation and material properties for nanofluid manufacturing. Int. J. Adv. Manuf. Technol. 34(3):300–306.

Choi, S. U. S. and Eastman, J. A. (1995). Enhancing thermal conductivity of fluids with nanoparticles. IL (United States): Argonne National Lab.

Chon, C. H., Kihm, K. D., Lee, S. P., and Choi, S. U. S. (2005). Empirical correlation finding the role of temperature and particle size for nanofluid (Al 2 O 3) thermal conductivity enhancement. Appl. Phys. Lett. 87(15):53107.

Chopkar, M., Kumar, S., Bhandari, D. R., Das, P. K., and Manna, I. (2007). Development and characterisation of al2cu and ag2al nanoparticle dispersed water and ethylene glycol based nanofluid. Mater.Sci. Eng.: B 139(2–3):141–148.

Chopkar, M., Sudarshan, S., Das, P. K., and Manna, I. (2008). Effect of particle size on thermal conductivity of nanofluid. Metall. Mater. Trans. 39(7):1535–1542.

Colangelo, G., Favale, E., de Risi, A., and Laforgia, D. (2012). Results of experimental investigations on the heat conductivity of nanofluids based on diathermic oil for high temperature applications. Appl.Energy 97: 828–833.

Das, P. K., Islam, N. Santra, A. K., and Ganguly, R. (2017). Experimental investigation of thermophysical properties of Al2O3–water nanofluid: Role of surfactants. J. Mol. Liq. 237:304–12.

Das, P. K., Mallik, A. K., Ganguly, R., and Santra, A. K. (2016). Synthesis and characterisation of TiO2–water nanofluids with different surfactants. Int. Commun. Heat Mass Transf. 75:341–348.

Esfe, M. H., Rostamian, H., Shabani-Samghabadi, A., and Arani, A. A. A. (2017). Application of three-level general factorial design approach for thermal conductivity of MgO/Water Nanofluids. Appl. Therm. Eng. 127:1194–1199.

Esfe, M. H., Saedodin, S., Akbari, M., Karimipour, A., Afrand, M., Wongwises, S., Safaei, M. R., and Dahari, M. (2015). Experimental investigation and development of new correlations for thermal conductivity of CuO/EG–Water Nanofluid. Int. Commun. Heat Mass Transf. 65:47–51.

Feng, X., Huang, X., and Wang, X. (2012). Thermal conductivity and secondary porosity of single anatase TiO2 nanowire. Nanotechnology 23(18):185701.

Iqbal, S. M., Raj, C. S., Michael, J. J., and Irfan, A. M. (2017). A comparative investigation of Al2o3/H2o, SIO 2/H 2 O and Zro 2/H 2 O nanofluid for heat transfer applications. Dig. J. Nanomater. 12(2).

Kim, S. H., Choi, S. R., and Kim, D. (2007). Thermal conductivity of metal-oxide nanofluids: Particle size dependence and effect of laser irradiation. J. Heat Transfer. 129(3):298307. https://doi.org/10.1115/1.2427071

Kole, M. and Dey, T. K. (2013). Enhanced thermophysical properties of copper nanoparticles dispersed in gear oil. Appl. Therm. Eng. 56(1–2):45–53.

Kotia, A., Haldar, A., Kumar, R., Deval, P., and Ghosh, S. K. (2017). Effect of copper oxide nanoparticles on thermophysical properties of hydraulic oil-based nanolubricants. J. Braz. Soc. Mech. Sci. Eng. 39(1):259–66.

Kotia, A., Rajkhowa, P., Rao, G. S., and Subrata Kumar Ghosh, S. K. (2018). Thermophysical and tribological properties of nanolubricants: A review. Heat Mass Transf. 54(11):3493–3508.

Li, J., Zhang, X., Xu, B., and Yuan, M. (2021). Nanofluid research and applications: A review. Int. Commun. Heat Mass Transf. 127:105543.

Liu, M.-S., Lin, M. C.-C., Tsai, C. Y., and Wang, C.-C. (2006). Enhancement of thermal conductivity with Cu for nanofluids using chemical reduction method. Int. J. Heat Mass Transf. 49(17–18):3028–33.

Liu, M. S., Lin, M. C. C. and Wang, C. C. (2011). Enhancements of thermal conductivities with Cu, CuO, and carbon nanotube nanofluids and application of MWNT/Water nanofluid on a water chiller system. Nanoscale Res. Lett. 6(1):1–13.

Madan, R. and Bhowmick, S. (2021). A numerical solution to thermo - mechanical behavior of temperature dependent rotating functionally graded annulus disks. In Aircraft engineering and aerospace technology. Bradford, UK: Emerald Publishing Limited.

Maheshwary, P. B., Handa, C. C., and Nemade, K. R. (2017). A comprehensive study of effect of concentration, particle size and particle shape on thermal conductivity of titania/water based nanofluid. Appl. Therm. Eng. 119:79–88.

Mahian, O., Kianifar, A., Heris, S. Z., Wen, D., Sahin, A. Z., and Wongwises, S. (2017). Nanofluids effects on the evaporation rate in a solar still equipped with a heat exchanger. Nano Energy 36:134–155.

Maxwell, J. C. (1873). A treatise on electricity and magnetism, 1. Oxford: Clarendon Press.

Murshed, S. M. S., Leong, K. C., and Yang, C. (2005). Enhanced thermal conductivity of TiO_2—water based nanofluids. Int. J. Therm. Sci. 44(4):367–73.

Otterstedt, J.-E. and Brandreth, D. A. (2013). Small particles technology. Springer Science & Business Media. New York (N Y): Springer.

Pang, C., Jung, J.-Y., Lee, J. W, and Kang, Y. T. (2012). Thermal conductivity measurement of methanol-based nanofluids with Al2O3 and SiO2 nanoparticles. Int. J. Heat Mass Transf. 55(21–22):5597–5602.

Park, S. S. and Kim, N. J. (2014). Influence of the oxidation treatment and the average particle diameter of graphene for thermal conductivity enhancement. J. Ind. Eng. Chem. 20(4):1911–1915.

Patel, H. E., Sundararajan, T., and Das, S. K. (2010). An experimental investigation into the thermal conductivity enhancement in oxide and metallic nanofluids. J. Nanoparticle Res. 12(3):1015–1031.

Paul, G., Philip, J., Raj, B., Das, P. K., and Manna, I. (2011). Synthesis, characterisation, and thermal property measurement of nano-Al95Zn05 dispersed nanofluid prepared by a two-step process. Int. J. Heat Mass Transf. 54(15–16):3783–3788.

Pryazhnikov, M. I., Minakov, A. V., Rudyak, V. Y., and Guzei, D. V. (2017). Thermal conductivity measurements of nanofluids. Int. J. Heat Mass Transf. 104:1275–1282. doi:https://doi.org/10.1016/j.ijheatmasstransfer.2016.09.080.

Ramezanizadeh, M. and Nazari, M. A. (2019). Modeling thermal conductivity of Ag/water nanofluid by applying a mathematical correlation and artificial neural network. Int. J. Low-Carbon Technol. 14(4):468–474.

Saini, D. S. and Matharu, S. P. S. (2020). Synthesis of zirconium Di-oxide powder from micro size to nano-size using high energy ball mill. In AIP conference proceedings, 2273:40003. AIP Publishing LLC.

Simpson, S., Schelfhout, A., Golden, C., and Vafaei, S. (2019). Nanofluid thermal conductivity and effective parameters. Appl. Sci. 9(1):87.

Song, D., Yang, Y. and Jing, D. (2017). Insight into the contribution of rotating brownian motion of nonspherical particle to the thermal conductivity enhancement of nanofluid. Int. J. Heat Mass Transf. 112:61–71.

Sundar, L. S., Singh, M. K., and Sousa, A. C. M. (2013). Thermal conductivity of ethylene glycol and water mixture based Fe3O4 nanofluid. Int. Commun. Heat Mass Transf. 49:17–24.

Tawfik, M. M. (2017). Experimental studies of nanofluid thermal conductivity enhancement and applications: A review. Renew. Sustain. Energy Rev. 75:1239–1253.

Timofeeva, E. V., Routbort, J. L., and Singh, D. (2009). Particle shape effects on thermophysical properties of alumina nanofluids. J. Appl. Phys. 106(1):14304.

Vandsburger, L. (2009). Synthesis and covalent surface modification of carbon nanotubes for preparation of stabilised nanofluid suspensions. Phd diss., McGill University.

Wang, J., Li, G., Li, T., Zeng, M., and Sundén, B. (2021). Effect of various surfactants on stability and thermophysical properties of nanofluids. J. Therm. Anal. Calorim. 143(6):4057–4070.

Wang, X., Xu, X., and Choi, S. U. S. (1999). Thermal conductivity of nanoparticle-fluid mixture. J. Thermophys. Heat Transf. 13(4):474–480.

Xie, H.-Q., Wang, J.-C., Xi, T.-G., and Liu, Y. (2002). Thermal conductivity of suspensions containing nanosised SiC particles. Int. J. Thermophys. 23(2):571–580.

Xie, H., Yu, W. and Chen, W. (2010). MgO nanofluids: Higher thermal conductivity and lower viscosity among ethylene glycol-based nanofluids containing oxide nanoparticles. J. Exp. Nanosci. 5(5):463–472.

Yang, L., Du, K., Zhang, X. S., and Cheng, B. (2011). Preparation and stability of Al2O3 nano-particle suspension of ammonia–water solution. Appl. Therm. Eng. 31(17–18):3643–3647.

Yoo, D.-H., Hong, K. S., and Yang, H. S. (2007). Study of thermal conductivity of nanofluids for the application of heat transfer fluids. Thermochim. Acta 455(1–2):66–69.

70 Finite element analysis of alloy based hip implant under static loading conditions

Mahendra Singh[a], *Dharmender Singh Saini*[b], *S. P. S. Matharu*[c], *and Mobassir Salim*[d]

Mechanical Engineering Department, National Institute of Technology, Raipur, India

Abstract

Recent developments in the field of prosthetics suggest looking for a variety of probable materials and their comparison with existing materials. Metals, ceramics, polymers, and composites are commonly used biomaterials for this purpose. Mostly metal hip implants are usually preferred to repair femur bone fracture because of their better mechanical properties. Presently stainless steel (SS), Co-Cr and Titanium (Ti) based alloys are mostly used for the commercially available bone implants. Studies show that release of Cr (Co–Cr alloys), Nb, V and Ni (Titanium-based) ions may cause harmful tissue reaction if concentration limit of these elements are exceeded than the safe limit. The material selected should be economical, compatible and should be abundant. Magnesium (Mg) and its alloys differ from other biomaterials owing to its compatible mechanical and physical properties with the human bones. Magnesium alloy has been taken as the human hip implant model for the present research work. This study presents a numerical analysis of a partial hip arthroplasty replacement implant using finite element analysis for three biomaterials namely Titanium based Ti-6Al-4V, Magnesium based Mg–Gd–Zn–Zr–Mn and Mg–4.0Zn–0.2Ca. The results show the Magnesium based alloys to be economical but not much durable as compared to Titanium based alloy.

Keywords: Finite element analysis, Human hip implant, magnesium alloy, titanium alloy.

Introduction

The implants currently in use are still not suitable for bone because whatever is being used is either only biomaterial or only biocompatible. But both the quality should remain in the perfect implants. Many developments in the field of biomaterials have been made in recent decades, with materials intended for biomedical purposes evolving across three generations, namely first generation (bio-inert materials), second generation (bioactive and biodegradable materials), and third generation (bioactive and biodegradable materials, the materials designed to stimulate specific responses at the molecular level). Metals, ceramics, polymers and composites are the four types of biomaterials which are in use. Ceramics, such as calcium phosphates, are commonly used as a coating medium in implants as they are non-toxic, biocompatible, and osteo conductive (Tahmasebifar et al., 2016). However, they have poor mechanical properties (lower stress) and a high rate of corrosion in acidic environments, limiting their use as a bone implant in large load bearing areas (Tahmasebifar et al., 2016). The biomaterial selected should be biocompatible and has mechanical properties that are compatible with bone (Mordike and Ebert, 2001).

Polymeric biomaterials are commonly used for bone tissue applications because they can be shaped into complex shapes and have easily modifiable surface properties. Frictional stress can also help to avoid metal-to-metal contact. Additionally, during sterilisation, the chemical and mechanical properties of polymers can be altered to some extent. However, the use of polymers as bio metal is limited. Furthermore, certain toxic additives used in the synthesis of polymers, such as plasticizers, ant oxidisers, and stabilisers, can damage tissue when percolated into body fluid (Mordike and Ebert, 2001). Metal implants are often utilised to heal bone fractures due to their better mechanical properties. (Mordike and Ebert, 2001). Commercially available bone implants are often made of titanium (Ti) alloys, cobalt-chromium (Co-Cr), and stainless steel (SS). Metals were required for long-duration, load-bearing implants because they have high mechanical strength and ductility, along with corrosion resistance. Furthermore, metal inserts with complex structures will be produced by various manufacturing techniques, including casting, machining, and powder metallurgy (Bram et al., 2013). Wear and corrosion have a major effect on the biocompatible materials of metallic implants. Inflammation can be caused by harmful metal ions generated by corrosion and wear in metallic implants, cell apoptosis, and other tissue reactions (Wu et al., 2013; Biesiekierski et al., 2012). By exceeding the % of these elements in body fluid or tissue, the release of V, Nb, Ti (Titanium-based), and Cr (Co–Cr alloys) ions has been confirmed to cause detrimental tissue reaction (Wu et al., 2013;

[a]msk.gsk21@gmail.com; [b]dharmender.saini13@gmail.com; [c]spsm.mech@nitrr.ac.in; [d]mobassirsalim@gmail.com

Biesiekierski et al., 2012). For example, Ni is a highly cytotoxic, genotoxic, carcinogenic, and mutagenic element. However, as the demand for structural materials in the temporary support implant applications has grown, materials that provide short-term structural support and can be reabsorbed into the body after healing have become increasingly popular. This eliminates the need for re-surgery for that support structure (Madan and Bhowmick, 2020).

The compatibility of magnesium-based alloy has withdrawn researchers' attention for prosthetic as well as internal uses despite its non-biodegradability. Mg alloy-based material being used for hip replacement inserts. Mg and Mg-base alloys are distinct from other biomaterials due to their physical and mechanical characteristic that are almost identical to one used in human hip bone. Also, densities and elastic modulus of Mg-based alloys are relatively similar to human bone, which eliminates elastic misalignment between bone and implants (Chen et al., 2014; Dorozhkin, 2014). The density of human bone and magnesium alloy falls in ranges of 1800–1900 kg/m^3 and 1750–1800 kg/m^3, respectively. Mg is also found naturally inside the composition of human skeletons, and it is one of the essential metals for metabolism, as well as Mg absorbed by body fluid. Commonly, Mg alloys contain aluminium (Al) which is well known as neuro-toxic and its higher concentration causes several neurological disorders such as senile dementia, dementia and Alzheimer's diseases. As a result, scientists have recently focused on the development of biologically safe Mg alloys including non-toxic metals such as Mn, Zn, Zr and Ca. The poor corrosion resistance of Mg-based implants reduces reliability and life span and always has been a challenge for researchers and manufacturers. Hip implants carry maximum load among all other human implants because it transfers upper body weight to the ground. Hence it is essential to understand the complete performance behaviour of Mg-based hip implant. Further to compare the various materials for hip implant such as titanium (Ti) alloys, cobalt-chromium (Co-Cr), and stainless steel (SS) are investigated. Due to the high density of implant compared to human bone, only load employed by the metal is considered. Due to, the stress shielding in the bone increases bone loss significantly. Stress shielding possess a major problem with any implant and pre-surgery is required after some time. Hence the role of hip implant material seems to be very important in the field of biomedical engineering. In this present paper, the use of Mg-based alloy as a biomaterial and bioabsorbable/biodegradable implant has been investigated under static conditions to prevent stress shielding. Results having good understanding with the S-N curves for material used for hip implants (Figure 70.1). The S-N value obtained from the experiments is also validated with ANSYS. It is anticipated that the reported results may be helpful for the researchers/manufacturer to produce implants with less stress shielding.

A. Artificial hip implant failure

Failure of an artificial hip implant following surgery may occur in a limited period before the implant's life span. Fracture and damage can happen for a variety of reasons, such as being in a car accident, falling, or

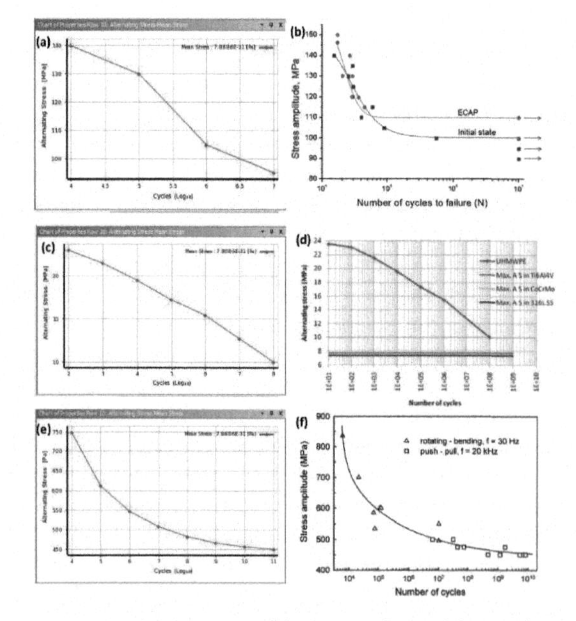

Figure 70.1 S-N curve of biomaterials used for analysis (a) S-N curve of UHMWPE (ANSYS) (b) S-N curve of UHMWPE (Nayak et al., 2020), (c) S-N curve of Ti6Al4V (ANSYS), (d) S-N curve of Ti6Al4V (Wahed et al., 2020), (e) S-N curve of Mg alloy (ANSYS) and (f) S-N curve of Mg alloy (Sezer et al., 2018)

for no apparent reason. Scratches will occur on the surface of the hip prosthesis during general handling, and surgery, increasing the tension at those position and providing a venue for crack propagation. Figure 70.2 depicts the potential for a poorly built artificial hip implant to fail (Čolić et al., 2017). Finite element analysis (FEA) will be used to reduce clinical prosthesis failures and numerically evaluate the design, as it is the most commonly used method for obtaining performance data for new prosthesis materials. In this present work, Mg alloy-based hip implant is analysed numerically under static loading conditions.

Research Gap:- the research work on the materials of present paper had been done by majority of researchers. Still the requirement of artificial hip joint is in large proportion as compare to other artificial implant. Thus, the research work on artificial hip joint with existing materials as well as new materials will continue till the shape and material of the hip joint become economical and durable so that everyone can afford it. In future work, change of shape for the hip joint with the same materials may lead to the affordable outcome.

Numerical Methodology

Geometry: A 3-D Parametric CAD model of Hip prosthetic designed in ANSYS space claim design modular and then import in ANSYS workbench for analysis. Hip implant consists of four parts which are femoral stem (a), femoral head (b), Acetabular lining (c), and Acetabular cup (d) 3-D models are shown in Figure 70.3 and all dimensions in mm.

Meshing: Meshing is a key stage in analysing the model; for appropriate conclusions, a fine mesh is required. Making fine mesh for the entire model its takes a long time, thus turn the entire model into a submodel to save time (Figure 70.4). The element size used for this analysis is 3 mm. Number of node count 25191 and element count 13800 used in present study. A grid independence study was carried out using various mesh sizes. The sensitivity of the current results for the same conditions is verified by considering the different mesh sizes mentioned in Table 70.1. Working with mess size less than 3 mm, the von Mises stress value is less than the current result values. Working with mess size more than 3 mm, then the stress, value is more than the current result values. Therefore, grid 4, 3 mm cell size is selected to conduct all numerical simulations.

Boundary Condition: The model was built with suitable boundary parameters, such as fix bottom surface of implant i.e. stems part along all DoF, and the load was applied in the right direction related to the top of the hip prosthesis's femoral head

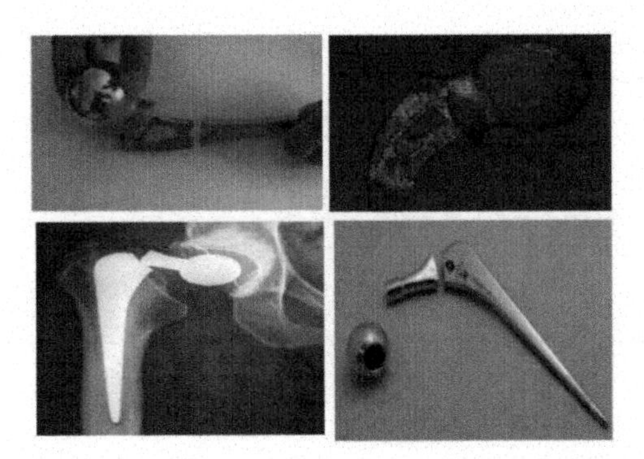

Figure 70.2 Artificial hip prosthesis failure (Sezer et al., 2018) and (Valet et al., 2014)

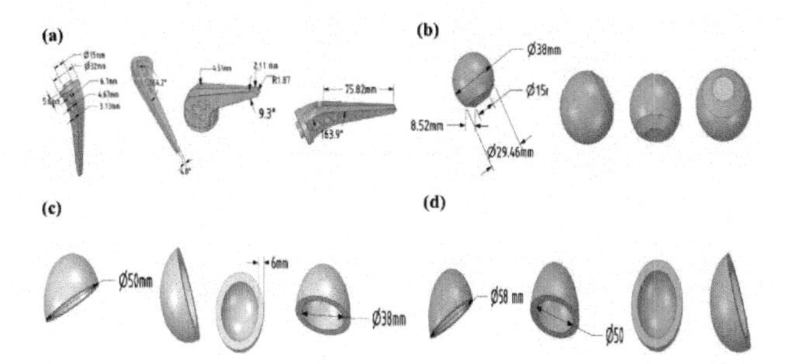

Figure 70.3 3-D model of parts of hip implants (a) Femoral stem (b) Femoral head (c) Acetabular lining (d) Acetabular cup

A. Finite element analysis of hip implant

A total or partially hip replacement is a surgical procedure that involves the removal of elements of the hip joint and the replacement of those parts with artificial prostheses. This work intended to look into the mechanical properties of Mg alloy-based hip prosthetics. In comparison to titanium-based alloys, biomaterials based on titanium, specifically Ti-6Al-4V, are the commonly used alloy for joint prostheses and are listed in the ASTM standard for biomaterials (Nayak et al., 2020; Wahed et al., 2020). The FEA was done with ANSYS commercial tools. Among other biomaterials, magnesium alloy has excellent biocompatibility and biodegradability. Table 70.2 lists the mechanical properties of all Mg-based biomaterial materials.

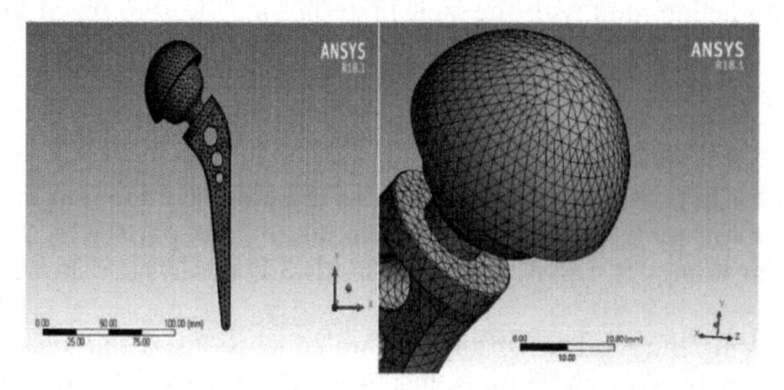

Figure 70.4 Unstructured mesh, nodes and element details

Table 70.1 Size grids used for grid independence

Type	*Element ccount*	*Node count*	*Element size*	*Maximum Von Mises stress(Ti-6Al-4V) at 2490 N*
Grid 1	19253	36580	1 mm	169.21 MPa
Grid 2	17618	33474	1.5 mm	175.44 MPa
Grid 3	15261	28995	2.5 mm	181.65 MPa
Grid 4	13800	25191	3 mm	187.82 MPa
Grid 5	11253	21380	3.5 mm	190.12 MPa

Table 70.2 Mechanical properties of powder processed Mg-based alloys (Sezer et al., 2018)

Mg alloy	*Tensile strength (MPa)*	*Yield strength (Mpa)*	*Elongation (%)*	*Youngs Modulus (Gpa)*	*Compression strength (Mpa)*	*Bending strength (Mpa)*
Mg-4.0Zn-0.2Ca (extruded)	297	240	21.3	45		
Mg-Zn-Y-Nd (hot-extruded)	316	183	I S.6			
Mg-Zn-Y-Nd (CECed)	303	185	30.2			
Mg-1.5Y-l.2Zn-0.44Zr (hot-extruded)	236	178	28		471	501
Mg-3Sn-0.5Mn	240	150	23			
Mg-3Al-4Zn-0.2Ca	198		10.3	44.1		347
Mg-2Zn-0.5Ca-M n (heat-treated)	205		15.7			
Mg-S.3Zn-0.6Ca + l.OCe/La (extruded)	202					
Mg-I Mn-2Zn-l.5Nd (ex truded)	> 285		> 14		> 395	
Mg-Ccl -Zn-Zr-1\10 (extruded)	341	315	21.3			
Mg-Gd-Nd-Zn-Zr (extruded)	267	217				
Mg-Zn-Y-Nd (ECAP)	239	96	30.1			
Mg-Zn-Y-Nd (CEC)	280	4	29.4			
Mg-Zn-Y-Nd (extruded)	242	170	20.9			

B. Material used for hip implant

The key three components of all hip prosthetics are the acetabular cup, acetabular lining, femoral head, and stem lower section. For the stem Ti6Al4V, two materials were considered: Mg–4.0Zn–0.2Ca (extruded) and Mg–Gd –Zn–Zr–Mn (extruded). Biomaterials based on magnesium are a new type of biodegradable material. The FEA study shows that a magnesium-based hip implant can be used in the future with better alloying materials. Table 70.3 lists the biocompatible materials used in this study of the hip implant, along with their physical properties.

C. Loads on hip prosthetic

The mechanical behaviour of the hip prosthetic portion was investigated numerically using 3-dimensional models at force 2.5 to 6.3 kN (Puertolas et al., 2004). To ensure a better design and protection for the prosthetic, thorough analyses of various loads operating on the hip prosthetic must be performed. According to previous research, static FEM studies are usually carried out with tonnes of human body weight magnitude (Puertolas et al., 2004). Still, the effects of sudden body movement loads and body weight will increase by up to 1020% on the prosthesis, and in some situations, the prosthesis can crack or fail due to body exhaustion exercise. To study the discrepancy between the results anticipated by routine implant testing and real-world loads, the prosthesis must be evaluated under static loads corresponding to the body weight, as well as under the highest actual load that is expected to occur during the cycle. The stress analysis on a hip prosthesis is carried out in this paper as the patient performs various activities, such as climbing, slow walking, up and downstairs, and tripping. These loads were calculated using Bergman et al. experimenter's, results on a patient who weighed 860 N and was 25 years old. Under these conditions, it was feasible to determine that the stage in which the patient leans on only one leg, as described in the walking phase analysis, would continue long enough to allow static calculations to estimate issues. The static load used in the numerical analysis was based on a 90 kg individual, and the loads in Table 70.4 were calculated accordingly. Figure 70.5 depicts the physiological condition that leads to the implant being loaded asymmetrically.

D. FEA Simulation and Calculation

Mg as a biomaterial hip implant has been compared to a model made of Ti-6Al-4V-alloy, Mg–4.0Zn–0.2Ca (extruded), and Mg–Gd –Zn–Zr–Mn (extruded). Table 70.3 shows the results of an FEA study using ANSYS to simulate various loadings on an artificial hip implant. This research will also see if mg alloy is appropriate for hip implants. Three-dimensional static idealisations and constitutive models for biomaterial elastic-plastic behaviour are explored in this section of the computational study. The von Mises elastic-plastic typical material model was used in this computational analysis. Since those two Mg

Table 70.3 Properties of material used for analysis (Sezer et al., 2018 and Wahed et al., 2020)

Part	*Material*	*Density (Kg/m^3)*	*Young's Modulus (Pa)*	*Poisson's Ratio*	*Yield strength (Pa)*	*Ultimate strength (Pa)*
Acetabular cup	CoCrMo	8300	2.30E+11	0.3	6.12E+08	9.70E+08
Acetabular lining	UHMWPE	930	6.90E+08	0.29	2.10E+07	4.80E+07
	Ti6Al4V	4430	1.15E+11	0.342	8.80E+08	9.50E+08
Femoral head and Stem	Mg–4.0Zn–0.2Ca (extruded)	1800	45E+9	0.35	240 E+6	297E+6
	Mg–Gd –Zn–Zr–Mn (extruded)	1810	45E+9	0.35	315E+6	341E+6

Table 70.4 Loading of hip joint for different problems (Sezer et al., 2018)

Activity	*Maximum load (% of body weight)*	*Maximum force in joint (N)*
Problem involving slow walking on a flat surface	282	2490
Problem involving climbing upstairs	356	3143
Problem involving climbing downstairs	387	3417
Problem involving tripping	720	6358

alloy-based materials have very good mechanical properties between the Mg alloy category of materials, FEA was used to investigate the mechanical behaviour of Mg alloys with good mechanical properties as biomaterials and compare with titanium-based biomaterials. This study tries to analyse when Mg alloy is used for hip implants.

Results and Discussion of Numerical Simulation

Analyses of 3-D failure and permissible stress were done using von Mises principles; in addition to 3-D deformations, strain areas were used to calculate a stress-based FoS based on implant loads. Shown in Figures 70.6 to 70.17 is a approximate graphic depiction of implant deformation, strain value, and stress value, under the effect of loads of 2500, 3143 N, 3417 N, and 6358 N. Deformation, von Mises strain and von Mises stresses on the stems, caused by the static analysis, are shown in Figures 70.6 to 70.17 respectively for titanium and Mg alloy based hip implant. The graphical representation of FE analysis of titanium-based hip stem is shown in Figures 70.6 to 70.9. The graphical representation of FE analysis of Gd –Zn–Zr–Mn-based hip stem is shown in Figures 70.10 to 70.13. The graphical representation of FE analysis of Mg–4.0Zn–0.2Ca based hip stem is shown in Figures 70.14 to 70.17. The summary of static analysis can be observed from Tables 70.5 to 70.7. Von Mises stresses were evaluated to approximation

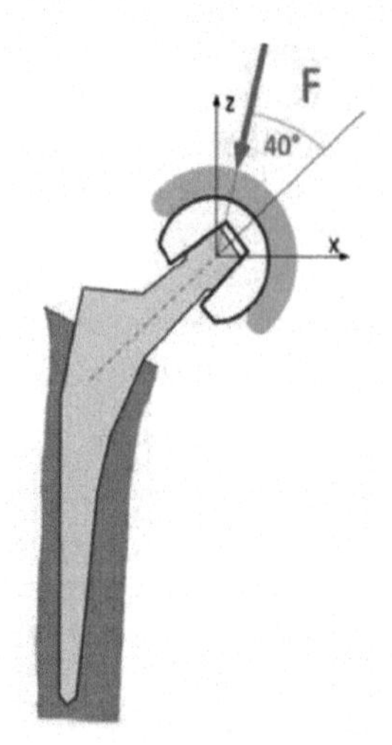

Figure 70.5 Asymmetric loading of a hip arthroplasty (Valet et al., 2014)

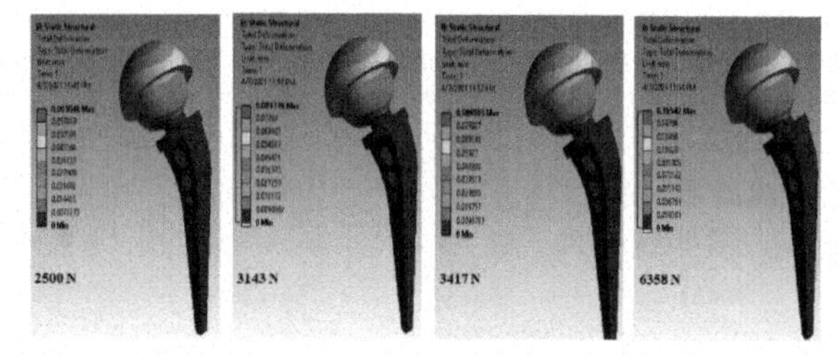

Figure 70.6 Total prosthesis displacement under the effects of maximum load (Ti-6Al-4V)

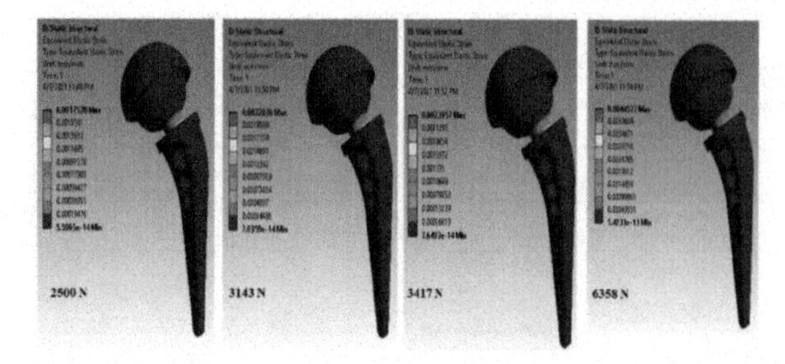

Figure 70.7 von Mises strain in the prosthetic (Ti-6Al-4V) at 2500N, 3143N, 3417N and 6358N

the possibility of prosthesis failure under the effect of different loading conditions which can arise during different physical work by humans shown in Table 70.7 for all materials under different loading conditions. Outcomes of computational analyses of artificial 3-D hip models are shown from Tables 70.5 to 70.7.

Total deformation of the artificial prosthetic for all material used for the analysis is less than 1 mm because a study shows during walking bone bend less than 1 mm so that deformation is not a source of concern shown in Table 70.5. Hip replacement for all material used here is safe during walking implant come its original shape if may get deformed. Maximum von Mises strain values under the effects of different loading conditions are shown in Table 70.6. Here strain values are very less for all biomaterials. Studied

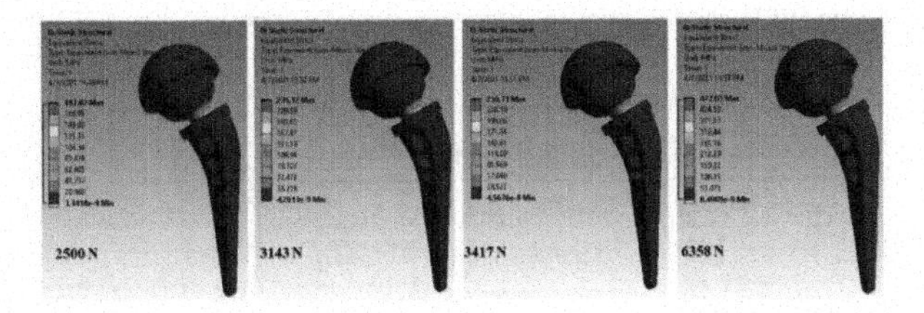

Figure 70.8 von Mises stresses in the prosthetic (Ti-6Al-4V) at 2500N, 3143N, 3417N and 6358N

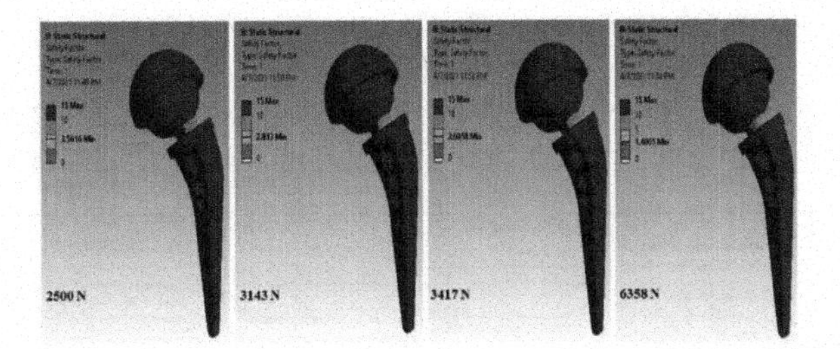

Figure 70.9 Factor of safety value under static load in the prosthetic (Ti-6Al-4V) at 2500N, 3143N, 3417N, and 6358N

Table 70.5 Total prosthesis displacement under the effects of different load condition

Activity	*Maximum force in joint (N)*	*Deformation (mm) (Ti-6Al-4V)*	*Deformation (mm) (Mg–Gd –Zn–Zr–Mn)*	*Deformation (mm) (Mg–4.0Zn–0.2Ca)*
Problem involving slow walking on a flat surface	2490	0.065	0.176	0.172
Problem involving climbing upstairs	3143	0.081	0.221	0.216
Problem involving climbing downstairs	3417	0.088	0.241	0.235
Problem involving tripping	6358	0.165	0.451	0.438

Table 70.6 Maximum von Mises strain values under the effects of different load condition

Activity	*Maximum force in joint (N)*	*Maximum von Mises strain (mm/mm) (Ti-6Al-4V)*	*Maximum von Mises stain (mm/mm) (Mg–Gd –Zn–Zr–Mn)*	*Maximum von Mises strain (mm/mm) (Mg–4.0Zn–0.2Ca)*
Problem involving slow walking on a flat surface	2490	0.0017	0.0047	0.0046
Problem involving climbing upstairs	3143	0.0022	0.0059	0.0058
Problem involving climbing downstairs	3417	0.0023	0.0064	0.0063
Problem involving tripping	6358	0.0044	0.0121	0.0118

von Mises stresses, presented in Table 70.7, are much lower than yield stresses of Ti-6Al-4V(860 MPa), steady under the highest simulated load magnitudes. But Von Mises stresses for the among best of all available Mg-based biomaterial like Mg–Gd –Zn–Zr–Mn and Mg–4.0Zn–0.2Ca larger than the yield strength of 315 MPa and 240 MPa respectively. From Figures 70.9, 70.13, and 70.17, it can be easily seen that the part which has the most minimum factor of safety value is the stem part that needs improvement. Only for titanium even the maximum load the value of the factor of safety is more than 1 everywhere in prosthetic. But for Mg based material the value of facto of safety is less than 1 in many cases. From Figure 70.18, it can be seen that maximum friction zone area is in between femoral head and acetabular cup. Stress analysis gives the idea that Mg-based biomaterial can be used for climbing upstairs, and climbing downstairs and slow walking on a flat surface, but still has to increase mechanical properties in Mg-based biomaterial for tripping activity because this is failing in tripping. Mg-based material needs more reinforcement and

Table 70.7 Maximum von Mises stress values under the effects of different load condition

Activity	*Maximum force in joint (N)*	*Maximum Von Mises stress (MPa) (Ti-6Al-4V)*	*Maximum Von Mises stress (MPa) (Mg–Gd –Zn–Zr–Mn)*	*Maximum Von Mises stress (MPa) (Mg–4.0Zn–0.2Ca)*
Problem involving slow walking on a flat surface	2490	187.82	191.36	186.95
Problem involving climbing upstairs	3143	236.12	239.96	235.03
Problem involving climbing downstairs	3417	256.71	261.01	255.52
Problem involving tripping	6358	477.65	486.81	475.6

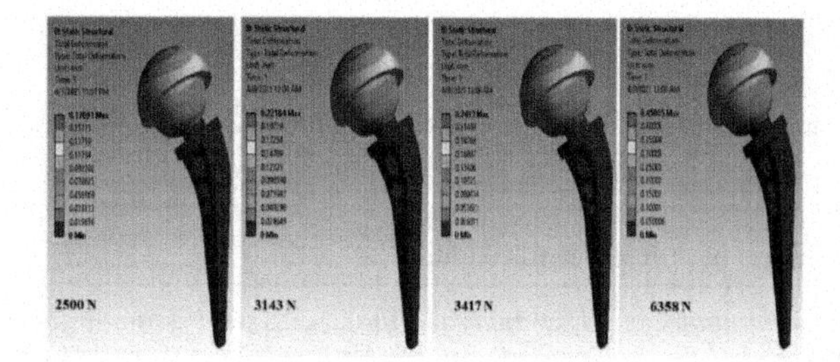

Figure 70.10 Total prosthesis displacement under the effects of maximum load (Mg–Gd –Zn–Zr–Mn) at 2500N, 3143N, 3417N and 6358N

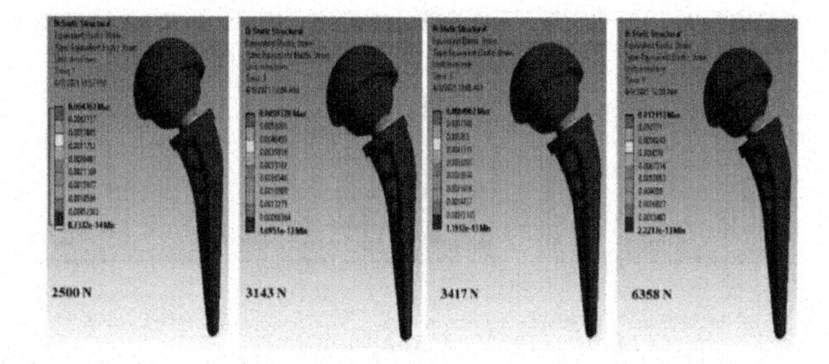

Figure 70.11 von Mises strain in the prosthetic (Mg–Gd –Zn–Zr–Mn) at 2500N, 3143N, 3417N and 6358N

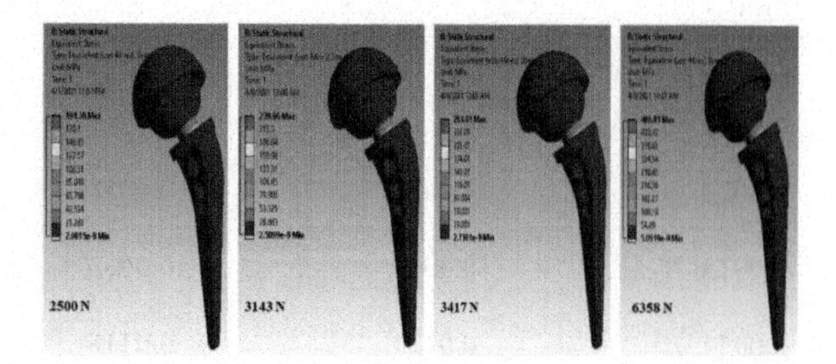

Figure 70.12 von Mises stresses in the prosthetic (Mg–Gd –Zn–Zr–Mn) at 2500N, 3143N, 3417N and 6358N

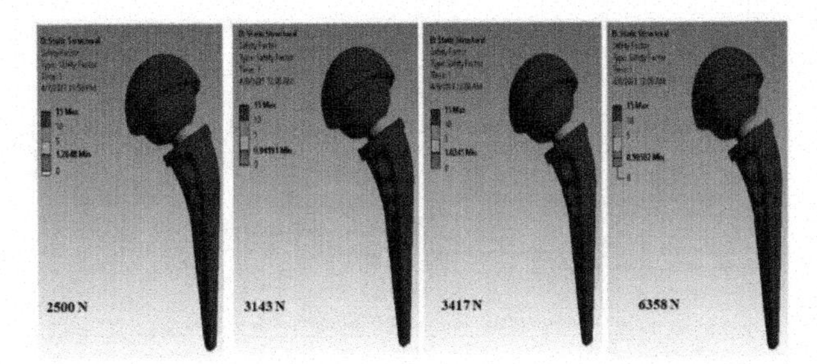

Figure 70.13 Factor of safety value under static load in the prosthetic (Mg–Gd –Zn–Zr–Mn) at 2500N, 3143N, 3417N and 6358N

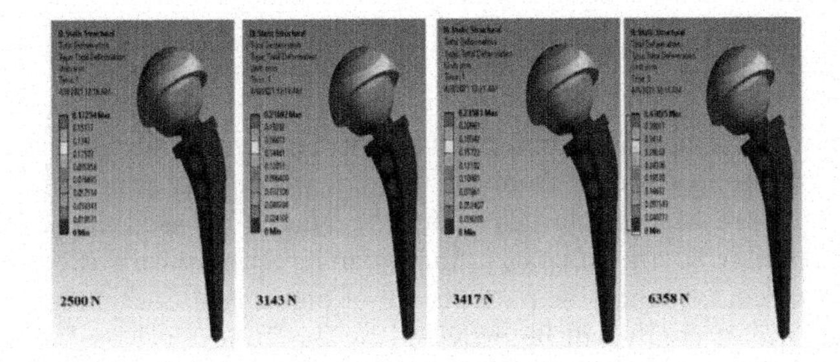

Figure 70.14 Total prosthesis displacement under the effects of maximum load (Mg–4.0Zn–0.2Ca) at 2500N, 3143N, 3417N and 6358N

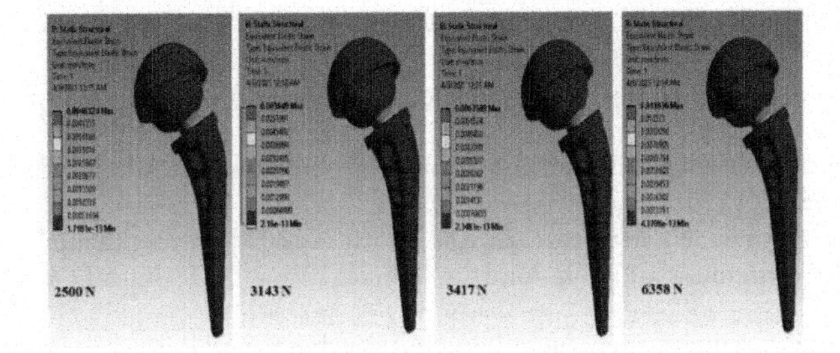

Figure 70.15 von Mises strain in the prosthetic (Mg–4.0Zn–0.2Ca) at 2500N, 3143N, 3417N and 6358N

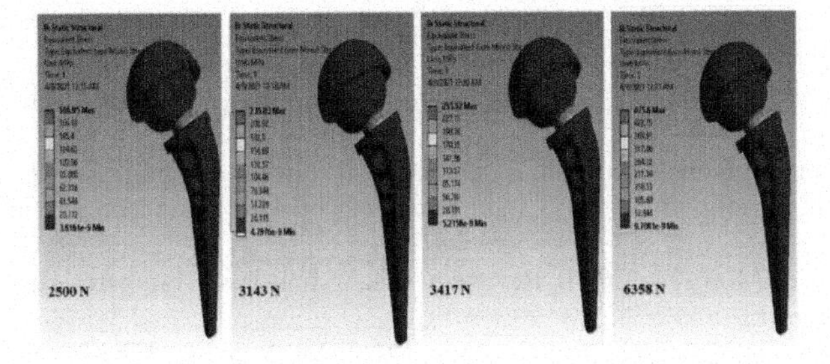

Figure 70.16 von Mises stresses in the prosthetic (Mg–4.0Zn–0.2Ca) at 2500N, 3143N, 3417N and 6358N

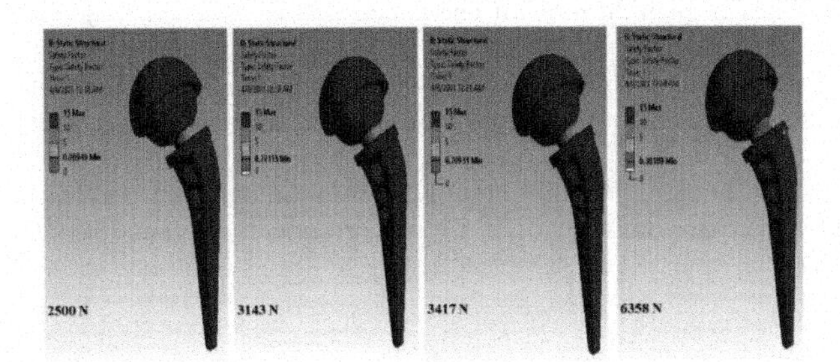

Figure 70.17 Factor of safety value under static load in the prosthetic (Mg–4.0Zn–0.2Ca) at 2500N, 3143N, 3417N and 6358N

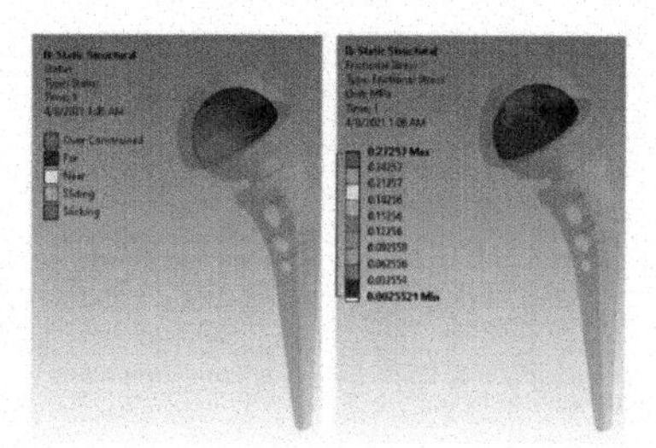

Figure 70.18 Contact status and frictional stresses at the contact zone

coating to increase its mechanical property for extreme use in fatigue. It is clear from this numerical solution that if the implant fails, it can only be in the stem part which is happen in a real situation.

Conclusion

Based on the analysis of results, although the titanium-based biomaterial used for hip implant sustain much higher load, the Mg alloy can be used for moderate loads. Also from the numerical analysis, although the deflections and the related strains are more in both Mg based alloys than the Titanium based alloy, the resulting stresses on all the three materials are similar. So, the Mg based Alloys can be economically used for Hip implants when not much durability is required, i.e. when the patients are of higher age. If the implant is required for young patients, the durability and strength will be provided by the Tungsten based alloy considered for analysis.

References

Biesiekierski, A., Wang, J., Gepreel, M. A. H., and Wen, C. (2012). A new look at biomedical ti-based shape memory alloys. Acta Biomater. 8(5):1661–1669.

Bram, M., Ebel, T., Wolff, M., Barbosa, A. P. C., and Tuncer, N. (2013). Applications of powder metallurgy in biomaterials. Mater. Sci. 520–554.

Chen, Y., Xu, Z., Smith, C., and Sankar, J. (2014). Recent advances on the development of magnesium alloys for biodegradable implants. Acta Biomater. 10(11):4561–4573.

Čolić, K., Sedmak, A., Legweel, K., Milošević, M., Mitrović, N., Mišković, Ž., and Hloch, S. (2017). Experimental and numerical research of mechanical behaviour of titanium alloy hip implant. Tehnički Vjesnik–Technical Gazette 24(3):709–713.

Dorozhkin, S. V. (2014). Calcium orthophosphate coatings on magnesium and its biodegradable alloys. Acta Biomater.10(7)(Elsevier):2919–2934.

Madan, R. and Bhowmick, S. (2020). A review on application of fgm fabricated using solid-state processes. Adv. Mater. Process. Technol. 6(3):608–619.

Mordike, B. L. and Ebert, T. (2001). Magnesium: Properties—Applications—Potential. Mater. Sci. Eng.: A. 302(1):37–45.

Nayak, S., Dhondapure, P., Singh, A. K., Prasad, M. J. N. V., and Narasimhan, K. (2020). Assessment of constitutive models to predict high temperature flow behaviour of Ti-6Al-4V preform. Adv. Mater. Process. Technol. 6(2):244–258.

Puertolas, J., Urries, I., Medel, F., Madre, M., Leiva, K., Gomez-Barrena, E., and Ríos, R. (2004). Fatigue behavior of electron-beam irradiated UHMWPE. In Transactions of the 50th annual meeting of the orthopaedic research society. 211.

Sezer, N., Evis, Z., Kayhan, S. M., Tahmasebifar, A., and Koç, M. (2018). Review of magnesium-based biomaterials and their applications. J. Magnes. Alloy. 6 (1):23–43.

Tahmasebifar, A., Kayhan, S. M., Evis, Z., Tezcaner, A., Cinici, H., and Koc, M. (2016). Mechanical, electrochemical and biocompatibility evaluation of AZ91D Magnesium Alloy as a Biomaterial. J. Alloy. Compd. 687:906–919.

Valet, S., Weisse, B., Kuebler, J., Zimmermann, M., Affolter, C., and Terrasi, G. P. (2014). Are asymmetric metal markings on the cone surface of ceramic femoral heads an indication of entrapped debris? Biomed. Eng. Online 13(1):1–22.

Wahed, M. A., Gupta, A. K., Gadi, V. S. R., Singh, S. K., and Kotkunde, N. (2020). Parameter optimisation in v-bending process at elevated temperatures to minimise spring back in Ti-6Al-4V alloy. Adv. Mater. Process. Technol. 6(2):350–364.

Wu, S., Liu, X., Yeung, K. W. K., Guo, H., Li, P., Hu, T., Chung, C. Y., and Chu, P. K. (2013). Surface nano-architectures and their effects on the mechanical properties and corrosion behavior of ti-based orthopedic implants. Surf. Coat. Technol. 233:13–26.

71 Localisation of brain tumor from brain MRI images using deep learning

Rakhi Wajgi[1,a], *Jitendra Tembhurne*[2,b], *Hemant Pendharkar*[3,c], *and Onkar Deshpande*[1,d]

[1]Computer Technology Department Yeshwantrao Chavan College of Engineering, Nagpur, India

[2]Computer Science Engineering Department Indian Institute of Information Technology, Nagpur, India

[3]Mathematics and Statistics Department College of Arts and Statistics, University of South Florida, St. Petersburg, US

Abstract

Prevalence of metastatic brain tumour is rising over past decade. A total of 40% of all cancers spread to brain. Exact cause of brain tumour is difficult to predict and major population affected by it are children and adults under the age of 40. Manual detection of tumour size, shape and its location in brain MRI images is very tedious, costly and time-consuming process. It needs experienced oncologists to correctly localise tumour. As deep learning has created a remarkable contribution in healthcare domain, we propose to implement two pretrained models and one modified model for localisation of tumour using segmentation in brain MRI images. In this paper, modified deep learning models U-Net and Standard Autoencoder are used for the extraction of image patterns from the BraTS 2020 dataset of 369 graescale images. This dataset of brain MRI images are labelled as T1, T2, T1CE, T2FLAIR and ground truth images. Dice coefficient is used to evaluate the comparison results. Existing U-Net model yields a training accuracy of 99.37%, validation accuracy of 99.07%, dice similarity co-efficient of 0.3393 and validation dice co-efficient of 0.3017. The modified U-Net resulted in a training accuracy of 99.4%, validation accuracy of 99.51%, dice similarity co-efficient of 0.6388, validation dice co-efficient of 0.6526. Furthermore, the Autoencoder revealed an accuracy of 98.48% in training phase, validation accuracy of the model is 98.92%, dice similarity co-efficient of 0.3629 and validation dice co-efficient of 0.3601.

Keywords: Autoencoder, brain tumour, deep learning, dice co-efficient, U-net.

Introduction

Brain tumours are abnormally developing tissues inside the brain that originate from unregulated cell multiplication. These tissues have no physiological function. Tumours induce abnormal neurological symptoms by increasing the size and pressure of the brain, as well as causing oedema. The primary purpose of computerised brain tumour segmentation is to collect clinical data on the presence, type, category, and location of the tumour inside the brain. The information collected through clinical imaging, such as MRI, use as the major source of input, can guide any future interventions, resulting in proper tumour segmentation and treatment (Abd-Ellah et al., 2019). MRI illustrates the layout and architecture of human tissues in detail and is a standard object of reference. DCE- The use of MRI as a clinical tool to determine the vascular supply of various malignancies is common (Nalepa et al., 2020). For the diagnosis of brain tumours, techniques of tumour detection, segmentation, and classification are used. Inside MRI scans, segmentation techniques are employed to localise and isolate various tumour tissues (Liu et al., 2014). Many researchers have contributed significantly to the field of brain tumour diagnostics during the last several decades. Main contribution proposed by them are in the form of methods for segmenting and classifying tumours. The degree of user monitoring and calculation simplicity have been important factors in the clinical adoption of diagnosis systems. However, practical applications are still limited, and despite a significant amount of study, clinicians and neurologists still rely on manual tumour projection, owing to a lack of collaboration between doctors and researchers (Abd-Ellah et al., 2019). Because of its potential to self-learn and generalise, MRI segmentation based on deep learning has gained a lot of interest. Convolutional neural network is one of the most preferred deep learning architectures, and it has excelled in a number of tasks, including detecting alcoholism, atypical breasts, and multiple sclerosis etc. The primary advantage of CNN over traditional methodologies is that it automatically learns critical feature representations of photographs without the requirement for prior knowledge or human interaction. It is a rapidly evolving technology, which is predominantly used in brain segmentation. Deep learning enables several levels of representation

[a]wajgi.rakhi@gmail.com; [b]jitendra.tembhure@gmail.com; [c]hemantpendharkar@gmail.com; [d]onkard09.od@gmail.com

and abstraction, allowing for more information about MRI images and their attributes that is to be collected. Because it aids in non-invasive diagnosis methods, medical imaging is an important aspect of today's healthcare system. It consists of two parts: image creation, reconstruction, image processing, and analysis (Naser and Deen, 2020). Depending on the imaging modality, there are numerous sorts of biomedical images. Clinical images, X-ray imaging, CT, MRI, US, OCT, and Microscopic images are some of the most commonly utilised biomedical imaging modalities (Haque and Neubert, 2020).

Literature Review

In the past, using transfer learning with VGG16's pretrained convolution-base, a DL model based on CNN, i.e. U-Net, is employed for simultaneous tumour segmentation, identification, and grading of low grade glioma (LGG) on MRI images (Naser and Deen, 2020). The contrast improvement phase, as indicated in Abd-Ellah et al. (2019) assists in the formation of a better saliency map, which subsequently accurately separates the tumour region. TNalepa et al. (2020) presented a fully automated deep learning-powered system (Sens-AI DCE) for DCE-MRI analysis of brain tumour patients that uses BraTS data to analyse whole T2FLAIR scans in real time. This method proved to be extremely adaptable, and it has the potential to extract new quantitative DCE-MRI features in the future. DL techniques have enabled unprecedented performance increases in a wide range of biological applications, from automated CT scan processing to skin lesions segmentation, according to the findings in (Haque and Neubert, 2020). After extensive investigation, it was discovered that combining CRF with FCNN with DeepMedic or ensemble is more effective in meeting the segmentation requirements of brain tumour (Wadhwa et al., 2019). RPNet, uses a 3-D fully convolutional architecture to partition brain tissue from 3-D MR images and is effective at processing 3-D MRI images. The recursive residual block RRB), pyramid pooling module (PPM), and deep supervision (DS) method all contribute to improved network performance (Wang et al., 2019). Author used LeNet architecture to achieve classification of AD data with an accuracy of 96.86% which was trained and tested with huge numbers of images (Sarraf and Tofighi, 2016). It allowed researchers to perform feature selection and classification with a unique architecture. Huang et al. (2019) suggested an image segmentation strategy that combines the FCM clustering algorithm with a novel rough set theory and exhibits advantages in both simulated and clinical pictures, with great stability and high accuracy. In two steps, a technique based on segmentation of brain tumours and classification of several brain modalities: 1) The SbDL method for tumour segmentation was validated using the DS rate; 2) the fusion of deep learning and DRLBP features was presented in Haque and Neubert (2020), which was then optimised using the PSO algorithm. In Hu et al. (2019), a novel M2D CNN model has been described that uses three multichannel 2-D CNN networks in tandem to maintain voxel properties of 3-D brain pictures while also integrating three-dimensional information using a fully connected neural network. A multi-path, multi-modal convolutional neural network system for recognising lesions in brain MRI images, as suggested in (Xue et al., 2020). This method is fully automatic and outputs lesions without human intervention. Kamnitsas et al. (2017) proposes an 11-layer deep, 3-D CNN architecture for autonomous brain lesion segmentation that outperforms the state-of-the-art models on difficult data.

Materials and Methods

BraTS 2020 challenge dataset is used here for brain tumour localisation. Challenge data is used for any type of research on brain tumour. The data is downloaded from CBICA's Image Processing Portal (CIBCA, 2020).

Ground truth annotations are included in training data, but not in validation data. The only challenge available to the participants is the testing data. The training data for each individual contains the four structural modalities, ground truth segmentation labels and supporting survival information, age, and resection status, whereas the validation data contains four modalities as mentioned in (Multimodal Brain Tumor Segmentation Challenge, 2020). All BraTS multimodal scans are available as NIfTI files (.nii.gz) which contains a)T2 Fluid Attenuated Inversion Recovery (T2-FLAIR) b) native (T1), c) post-contrast T1-weighted (T1Gd), and d) T2-weighted (T2). They are acquired with variety of clinical protocols and various scanners from multiple (n = 19) data contributors. Figure 71.1 indicates the ground truth images of various parts. All of the imaging datasets are manually segmented using the same annotation technique by one to four evaluators, and their annotations are being evaluated and approved by experienced radiologists. The GD-enhancing tumour, peritumoral oedema, and necrotic and non-enhancing tumour core are among the annotations.

In the clinical domain, MRI garnered increasing attention for brain tumour diagnosis. In today's clinical practice, pictures from various MRI sequences are used to diagnose and delineate tumour compartments. T1-weighted MRI (T1w), T1-weighted MRI with contrast enhancement (T1wc), and T2-weighted MRI are

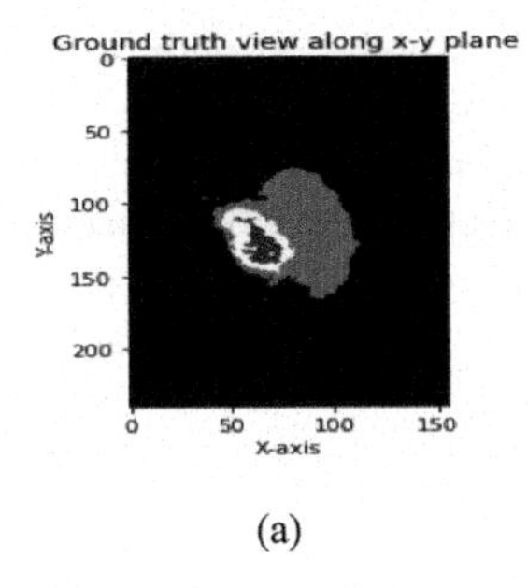

(a)

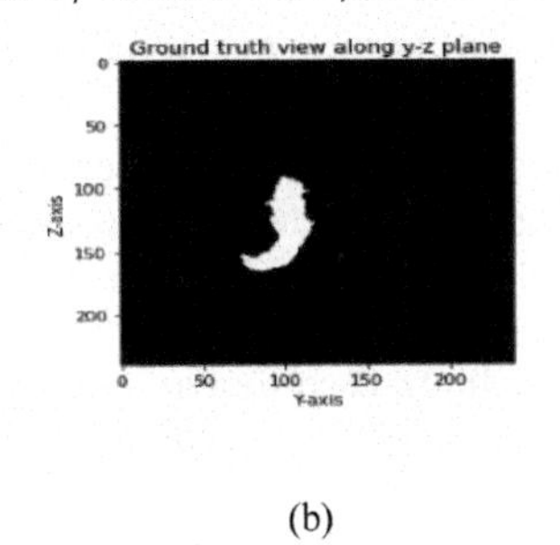

(b)

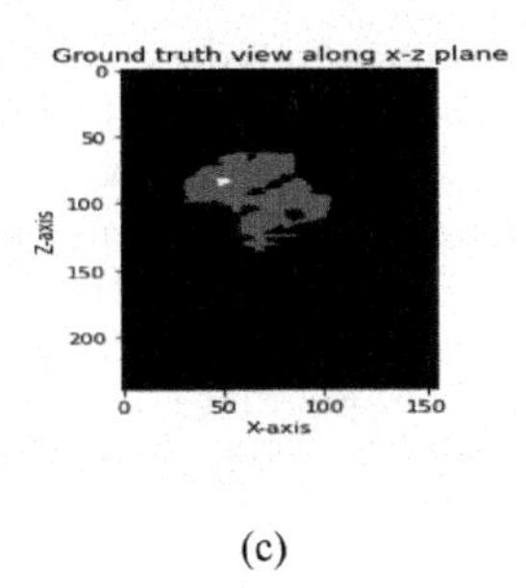

(c)

Figure 71.1 Ground truth images of BraTS dataset

among the sequence pictures (T2w). Fluid attenuated inversion recovery (FLAIR) and T1w has been the most often used sequence picture for analysis of brain tumour shape because it enables easy annotation of healthy tissues. Furthermore, because the blood-brain barrier in the proliferative brain tumour region is disrupted, T1-weighted contrast-enhanced sequence pictures is brightened the brain tumour borders. This helps in improving the contrast of images. The necrotic core and active cell area are easily recognised in these sequence images. The oedema zone in T2w can appear brighter than other MRI sequence images. FLAIR is recognised as a highly effective sequence picture for separating the oedema region from the CSF since the signal of water molecules are muted during the imaging procedure (Keras.Conv2D Class, 2020). Due to computational limitations, we have taken two modalities out of the available four for each patient, and trained CNN models on those. The modalities used by us for training purpose are FLAIR and T1ce. For every patient, each image has dimensions 240 × 240 × 155. On resizing the images to 128 × 128, we take 100 slices of each image, making the image dimensions 128 × 128 × 100. We tried to pick 100 slices from the centre of these 3-D images to ensure all necessary data from the images are covered. In order to reduce data variance, the min-max normalising approach is utilised, which rescales the range of features to a range of (0, 1). In order to train a deep learning model, there is an absolute need to annotate the available data effectively by specialists.

In BraTS dataset, the expert board-certified neuro-radiologists provide the ground truth. These MRI images were scanned by multiple institutions (n = 19), keeping in mind the clinical protocols. The dataset contains data of LGG and high-grade gliomas (HGG) patients. LGG: This type of tumour is classified as Grade 2 tumour making them the slowest growing type of glioma in adults. HGG: This type of tumour is classified as Grade 1 tumor making them the fastest growing type of glioma in adults. Ground truth is in the form of 3-D image. The 3-D ground truth image comprises of voxels of four classes having values 0, 1, 2 and 4 where we use '3' in place of 4 for uniformity during implementation. Components of ground truth with integer value for each voxel:

0: background: It is the image background.
1: necrotic and non-enhancing tumour: Non-enhancing tumours are low grade diffuse gliomas. They typically involve the cortex which is thickened with increased T2 signal.
2: peritumoral oedema: Brain swelling is another name for edema. It's a life-threatening condition in which the brain fills with fluid.
3: GD enhancing tumour: It shows the newly developing tumour.

A. Models used

We have implemented three deep learning models namely U-Net Ronneberger et al. (2015), modified U-net and Autoencoder model. The U-Net architecture, which is derived from fully convolutional neural network, is shown in Figure 71.2 and modified U-Net is shown in Figure 71.3. U-Net, which integrates low-level and high-level information, has been widely employed in medical picture segmentation. There are two pieces to the architecture: a down sampling path and an up-sampling path. There are four phases in the down sampling path. Each stage has two 3 × 3 convolution layers, as well as batch normalisation (BN) and a rectified linear unit (ReLU). Batch normalisation Brownlee (2019) minimises the volume by which unseen unit values change about (covariance shift), and thus is effective at avoiding overfitting. A 2 × 2 max pooling MaxPooling2D Layer (2020) with strides of 2 is attached at the end of each stage for down sampling. After each level, the number of feature channels is doubled. There are four phases in the up-sampling process. Each stage begins with a 2 × 2 kernel up-sampling layer with 2 strides, followed by a convolution layer with BN and ReLU. Following that are two 3 × 3 convolution layers, BN, and ReLU. Concatenating layers connect stages in the down-sampling path with stages in the up-sampling path that have the same resolution to reduce the impact of information loss. In the last layer, a 1 × 1 convolution with 'softmax' activation is used to produce the output. Each convolutional layer is preceded by an up-Sampling layer of size 2 × 2. The number of features is doubled after two consecutive convolutional layers and the size of Image is doubled after each up-sampling layer. The first four convolutional layers in decoding part are of size 3 × 3. In the last layer, a 1 × 1 convolution with 'softmax' activation is applied to produce the required output image. The only difference between U-Net and modified U-Net model is that after down sampling path, dropout layers Dropout Layer (2020) are used to avoid overfitting of model which generated better results. For comparison purpose, default Autoencoder model is also implemented.

Results and Discussion

Implementation of U-Net, Modified U-Net and Autoencoder architectures has led to a deeper insight into the utilisation of Deep Learning Methods using CNN in the field of biomedical imaging, and has yielded a training accuracy of 99.37%, validation accuracy of 99.07%, dice similarity coefficient of 0.3393 and validation dice coefficient of 0.3017 in U-Net architecture. We also achieved training accuracy of 99.4%, validation accuracy 99.51%, dice similarity coefficient of 0.6388 and validation dice similarity coefficient of 0.6526 in Modified U-Net. For Autoencoder we have achieved a training accuracy of 98.48%, validation accuracy 98.92%, dice similarity coefficient of 0.3629 and validation dice similarity coefficient of 0.3601. Realisation of the same has been attained using keras in tensorflow on the GPU provided by Google colab. We have worked on BRATS 2020 dataset of 369 images, with 350 images in the training set and 19 images in the validation set. All three architectures are trained through 40 epochs, where U-Net model took a total of 144 minutes with each epoch running for 3.6 minutes, Modified U-Net model ran for a total of 152 minutes with each epoch taking 3.8 minutes and the Autoencoder took a total of 101 minutes with each epoch taking 2.53 minutes.

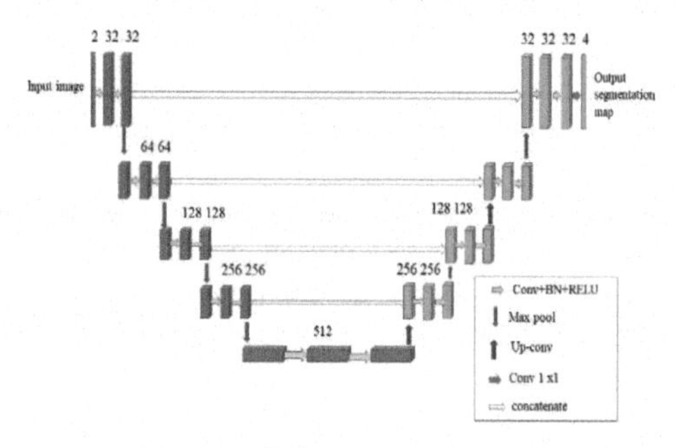

Figure 71.2 U-net model

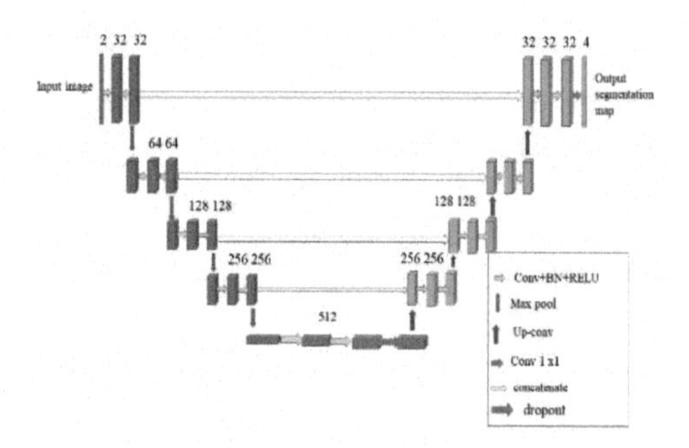

Figure 71.3 Modified U-net model

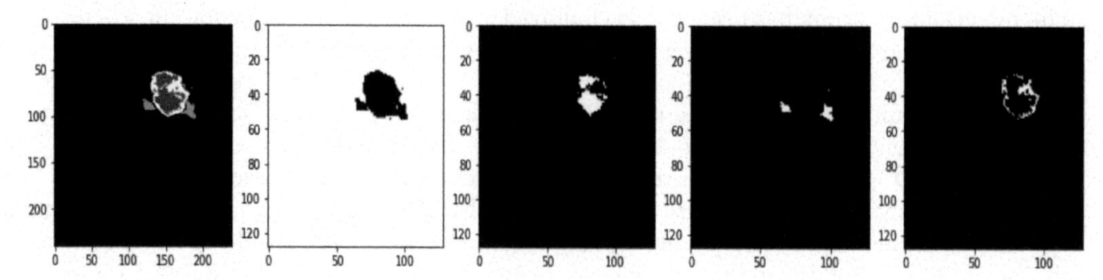

Figure 71.4 Expected result of U-net model

The images depicted in Figure 71.4 are the expected results after applying U-Net model. It comprises of Ground truth, Whole tumour (which consists of tumour core, oedema and enhancing tumour), oedema, tumour core, enhancing tumour. Figure 71.5 shows prediction done by U-Net for various tumour parts in terms of segments that has helped us accomplish our motive of segmentation.

Training and testing accuracy of U-Net is shown in Figures 71.6, 71.7 and 71.8 show expected and predicted results of modified U-Net model with accuracy of training and testing is shown in Figures 71.9, 71.10 and 71.11 show expected and predicted results of Autoencoder model. Figure 71.12 shows training and testing accuracy achieved by Autoencoder model. Values of performance metrics for all three models are shown in Tables 71.1, 71.2 and 71.3. Comparative analysis of all three models is shown in Table 71.4. It is observed from Table 71.4 that, modified U-Net slightly performs better than other two. Similarity between expected and predicted which is measured using dice co-efficient is 0.6388 which is better for modified U-Net model whereas it is lowest for existing U-Net model.

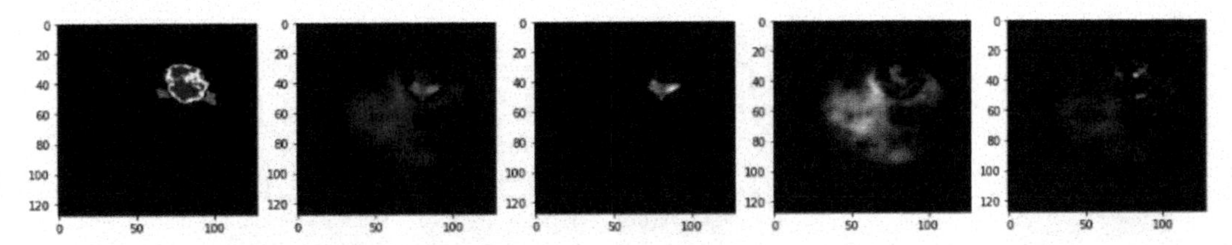

Figure 71.5 U-net prediction

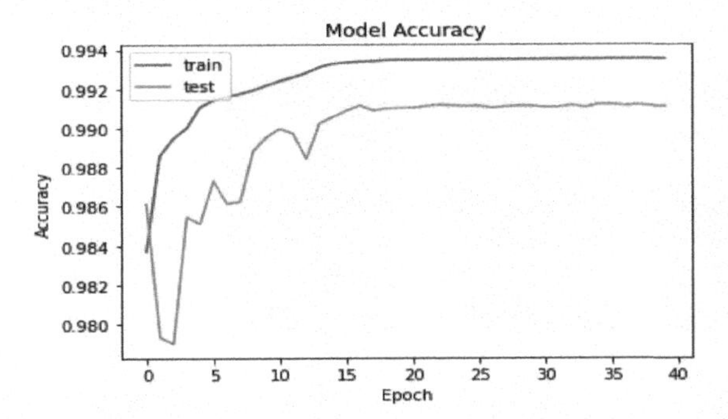

Figure 71.6 U-net accuracy

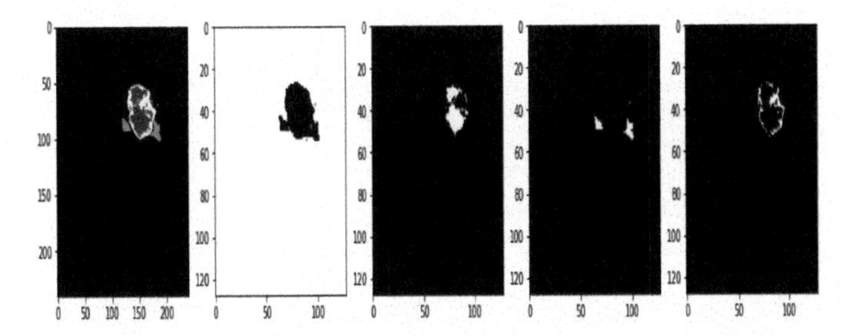

Figure 71.7 Modified U-net expected results

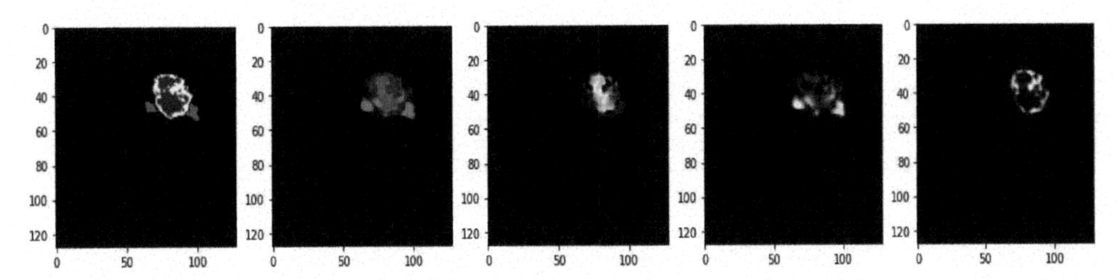

Figure 71.8 Modified U-net predictions

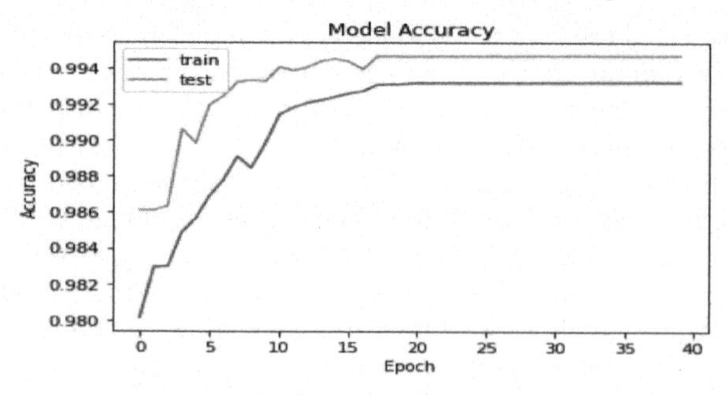

Figure 71.9 Modified U-net model accuracy

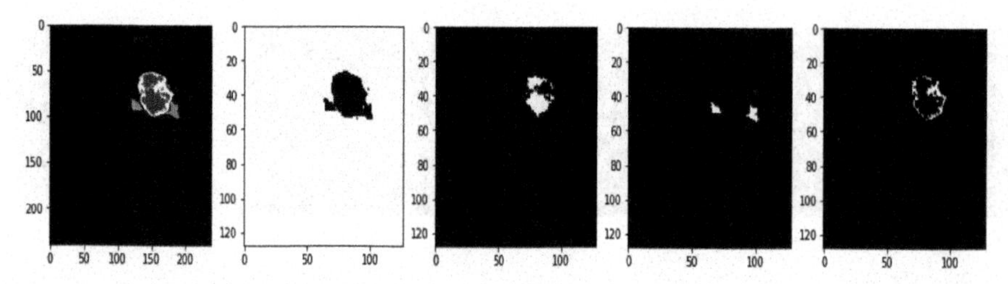

Figure 71.10 Autoencoder expected results

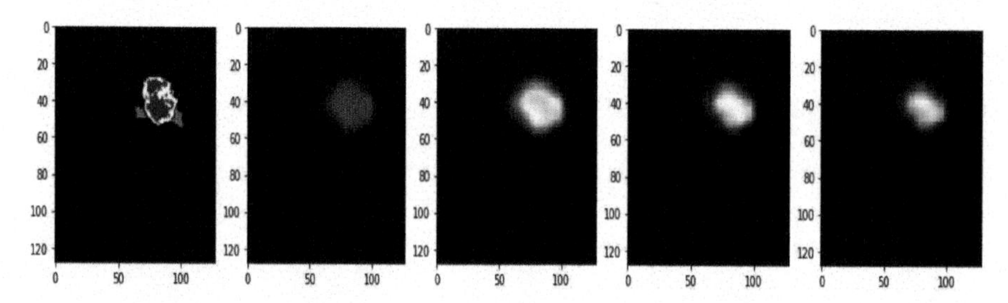

Figure 71.11 Encoder predictions

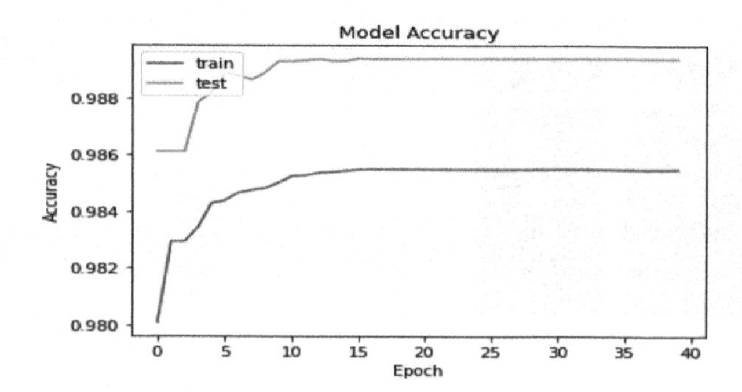

Figure 71.12 Autoencoder model accuracy

Table 71.1 U-Net results

Model Parameters	*Training dataset*	*Validation dataset*
Accuracy	0.9934	0.991
Loss	0.0187	0.0462
Dice	0.4442	0.3639

Table 71.2 Modified U-Net results

Model parameters	*Training dataset*	*Validation dataset*
Accuracy	0.9934	0.9947
Loss	0.0183	0.0157
Dice	0.5990	0.6192

Table 71.3 Autoencoder results

Model parameters	*Training dataset*	*Validation dataset*
Accuracy	0.9864	0.9894
Loss	0.0382	0.0318
Dice	0.3712	0.3746

Table 71.4 Comparison of existing model with proposed models

Model Name	*Training Accuracy*	*Validation Accuracy*	*Dice Co-efficient*	*Validation Dice Co-efficient*
U-Net	0.9937	0.9907	0.3393	0.3017
Modified U-Net	0.9943	0.9951	0.6388	0.6526
Autoencoder	0.9848	0.9892	0.3629	0.3601

Conclusion

This research provides a review of two Deep Learning based models for the purpose of image segmentation and localisation on brain MRI images. We evaluated the similarity amongst these two models based upon dice coefficient and calculated the losses incurred in each. Prior to these approaches, we tried implementing an exact replica of a pre-existing U-Net architecture, but later realised the modified U-Net which has been presented with a few minor changes, i.e., removing the Batch Normalisation segment and adding a Dropout Layer at the beginning of the up-sampling path to avoid overfitting of the U-Net model. We initially tried implementing the Autoencoder by doubling the number of features which yielded vague results, and hence we improvised by halving the number of features that resulted in a better accuracy. Due to limitation of high end infrastructure, only two modalities are considered but accuracy of all models can be improved using Google colab architecture in future.

References

Abd-Ellah, M. K., Awad, A. I., Khalaf, A. A. M., and Hamed, H. F. A. (2019). A review on brain tumor diagnosis from MRI images: Practical implications, key achievements, and lessons learned. Magn. Reson. Imaging 61:300–318.

Brownlee, J. (2019). How to accelerate learning of deep neural networks with batch normalisation, deep learning performance. Jan. (accessed 2021, April 5). https://machinelearningmastery.com/how-to-accelerate-learning-of-deep-neural-networks-with-batch-normalization/

CBICA (2020, December 3). Multimodal brain tumor segmentation challenge (2020): Registration/data request. https://www.med.upenn.edu/cbica/brats2020/registration.html

CIBCA (2020, November 24). CIBCA image processing portal. https://ipp.cbica.upenn.edu/

Dropout Layer (2020, March 12). https://keras.io/api/layers/regularization_layers/dropout/#:~:text=Dropout%20class&text=The%20Dropout%20layer%20randomly%20sets,over%20all%20inputs%20is%20unchanged.

Haque, I. R. I., and Neubert, J. (2020). Deep learning approaches to biomedical image segmentation. Inform. Med. Unlocked 18:100297.

Hu, J., Kuang, Y., Liao, B., Cao, L., Dong, S., and Li, P. (2019). A multichannel 2D convolutional neural network model for task-evoked fMRI data classification. Comput. Intell. Neurosci.

Huang, H., Meng, F., Zhou, S., Jiang, F., and Manogaran, G. (2019). Brain image segmentation based on FCM clustering algorithm and rough set. IEEE Access 7:12386–12396.

Kamnitsas, K., Ledig, C., Newcombe, V. F. J., Simpson, J. P., Kane, A. D., Menon, D. K., Rueckert, D., and Glocker, B. (2017). Efficient multi-scale 3D CNN with fully connected CRF for accurate brain lesion segmentation. Med. Image Anal. 36:61–78.

Keras.Conv2D Class (2020, January 12). Keras.Conv2D Class. May 18, 2020. https://www.geeksforgeeks.org/keras-conv2d-class/#:~:text=Keras%20Conv2D%20is%20a%202D,produce%20a%20tensor%20of%20outputs.

Liu, J., Li, M., Wang, J., Wu, F., Liu, T., and Pan, Y. (2014). A survey of MRI-based brain tumor segmentation methods. Tsinghua Sci. Technol. 19(6):578–595.

MaxPooling2D Layer (2020, February 6). https://keras.io/api/layers/pooling_layers/max_pooling2d/#:~:text=MaxPooling2D%20class&text=Max%20pooling%20operation%20for%202D,by%20strides%20in%20each%20dimension

Nalepa, J., Lorenzo, P. R., Marcinkiewicz, M., Bobek-Billewicz, B., Wawrzyniak, P., Walczak, M., and Kawulok, M. (2020). Fully-automated deep learning-powered system for DCE-MRI analysis of brain tumors. Artif. Intell. Med. 102:101769.

Naser, M. A. and Deen, M. J. (2020). Brain tumor segmentation and grading of lower-grade glioma using deep learning in MRI images. Comput. Biol. Med. 121:103758.

Ronneberger, O., Fischer, P., and Brox, T. (2015). U-net: Convolutional networks for biomedical image segmentation. In International conference on medical image computing and computer-assisted intervention, (pp. 234–241). Cham: Springer.

Sarraf, S. and Tofighi, G. (2016). Classification of Alzheimer's disease using fMRI data and deep learning convolutional neural networks. 2016. arXiv preprint arXiv:1603.08631.

Wadhwa, A., Bhardwaj, A., and Verma, V. S. (2019). A review on brain tumor segmentation of MRI images. Magn. Reson. Imaging. 61:247–259.

Wang, L., Xie, C., and Zeng, N. (2019). RP-Net: A 3D convolutional neural network for brain segmentation from magnetic resonance imaging. IEEE Access 7:39670–39679.

Xue, Y., Farhat, F. G., Boukrina, O., Barrett, A. M., Binder, J. R., Roshan, U. W., and Graves, W. W. (2020). A multi-path 2.5 dimensional convolutional neural network system for segmenting stroke lesions in brain MRI images. NeuroImage: Clin. 25:102118.

72 A novel approach of word sense disambiguation for marathi language using machine learning

Ujwalla Gawande[1,a], *Swati Kale*[2,b], *and Chetana Thaokar*[c]

[1]Department of Information Technology, Yeshwantrao Chavan College of Engineering (YCCE), Nagpur, India

[2]Department of Information Technology, Shri Ramdeobaba College of Engineering and Management, Katol Road, Lonand, Gittikhadan, Nagpur, Maharashtra, India

Abstract

Semantic understanding is the key issue in natural language processing (NLP) systems. A single word with multiple interpretations is a common feature of NLP. Linguistic and structural ambiguity makes speech or written text open to multiple possible interpretations. Humans use the references provided by the world and language communities to solve the task of disambiguation. However, the machine does not have this capability. Machine translation, question answering systems, and information extraction and retrieval are some of the applications of NLP, and their accuracy depends on the accuracy achieved by the word sense disambiguation (WSD) system. The most challenging problem with WSD is to identify, which sense or meaning of a word is used in a particular sentence. There are other problems with predicting the correct word meaning, such as errors in this process and more time consuming. In this paper, we propose a WSD technique using Word's multiple features-based approach for Marathi language. It is one of the morphologically rich languages in India. In this technique, a multi-class classifier is developed that will predict the exact sense of ambiguous words in each context. We conducted all the experiments on the benchmark WordNet dataset developed at Indian Institute of Technology Bombay (IIT Bombay). Experimental results show significant performance improvements in the area of NLP.

Keywords: Disambiguation, machine learning, multi-class classifier, parser, swagger, Wordnet.

Introduction

NLP is an extensive field of machine learning (ML), artificial intelligence (AI), and computational linguistics. It comprises the handling of human language, rather than computer interpreting human understandable language. Natural language has useful information that can be extracted from ubiquitous unstructured text data. We can use the information for many purposes, such as text proofreading, spell correction, automatic chat boat, text translator, question answering application.

Currently, Google translator, search engine understands the different language input using text translation; Google assistant uses voice commands; Cortana, Alexa, such application uses the NLP tools. Human language has a high degree of context-sensitivity and diversity, which can lead to many ambiguities in sentences. Such ambiguous sentences can cause problems when interpreting human language sentences. However, humans' intrinsic capability to interpret natural language based on background, it is a tough task for the machine to identify the correct meaning of the word in a sentence. In the literature, ambiguous word identification, and prediction methods are available, but do not address the issues. Still, it is one of the critical and challenging tasks in NLP. One important parameter is context, while predicting the meaning of a word in a sentence. Again, it is mandatory to investigate the WSD in NLP for Indian languages such as Marathi.

To conduct experiments, we need an effective method and data set. In Marathi language. However, the major problem is the unavailability of datasets and efficient approaches for Indian languages. Therefore, we rely on unsupervised learning to understand and predict the word's meaning. It is difficult for machine learning algorithms to process the original Marathi language. The main applications of WSD are user feedback and comment analysis, and content recommendation systems. However, for raw input text, we need to handle each word in a sentence. Each method has its own limitations. First, the count-based approach gives a special identifier for each word in a sentence. but cannot describe the semantic aspects, background, and organisation sequence of the words, and therefore drives in sparse vectors. These low-dimensional cases can contribute to poor results or models.

[a]ujwallgawande@yahoo.co.in; [b]s12_kale@yahoo.com; [c]thaokarcb@rknec.edu

Types of ambiguity

We relate ambiguity to words that have multiple meanings and interpretations in a sentence. After studying various techniques available in the literature, we divided them into five types of methods, as shown in Figure 72.1. We describe each of the text's ambiguities briefly:

Semantic ambiguity:

When there is more than one way of understanding in the setting, semantic ambiguity appears in the sentence, although it does not contain vocabulary or is potentially semantically ambiguous (Dayal, 2004). For example, consider the following two sentences:

Example: 'The dog is chasing the cat and other animals" and "The dog has been pet for more than thousands of years'

First, 'The dog' similar to animal dog, and Next, context is species.

Lexical ambiguity

When the importance of a word exceeds one, lexical ambiguity occurs in the sentence (Zelta, 2014). For example, consider the phrase 'Features of a bat'. Here in a sentence Bat semantic will be a 'vertebrate, the flying creature' and game context it is 'playing instrument'. Context or application change the meaning of the phrase.

1Syntactic ambiguity

In any expression, syntactic ambiguity can exist at any time, because of the requirements for words in a sentence, a sentence has at least two or more different meanings (Navigli 2009). For example, consider the phrase '*Man saw the kids with the binoculars in playground*'. It is ambiguous statement because man saw the kids with binoculars and the kids saw him with binoculars.

Pragmatic ambiguity

This type of ambiguity comes in scenarios 'different meaning for one sentence'. It can be the ambiguity in the conversation, the ambiguity in the presupposition, or the ambiguity of reference (Navigli, 2009). For example, consider the phrase '*I like you too*'. There are many explanations for this phrase, such as first interpretation can be I like you means just as you like me. Second, I like you means just like everyone else.

Part of speech ambiguity

Specifically, in a sentence, when the ambiguous word object, specific action, singular and plural words (Navigli, 2009). For example,

consider a phrase, '*The green leaves of banana tree*' and '*Mumbai express train leaves Nagpur station.*' In first statement, it is name in plant context. In second sentence it is a verb. In the next section we discuss the importance of Marathi language in the India.

Word sense ambiguity

Accurate meaning of different word in a sentence with different scenarios and the context is defined as WSD. The below example demonstrates the WSD problem.

Example: Sentence-1: **मराठी पाठात तीन अंकआहे**. (Marathi pathat tin aank aahet)

Sentence-2: **नाटकात चार अंक आहे** (Natkat char aank aahet)

In above sentences **अंक** (aanka) is ambiguous word. First, meaning for ambiguous word in a sentence is decimal number means countable and second, it is part, like classification or series. WSD system is unable

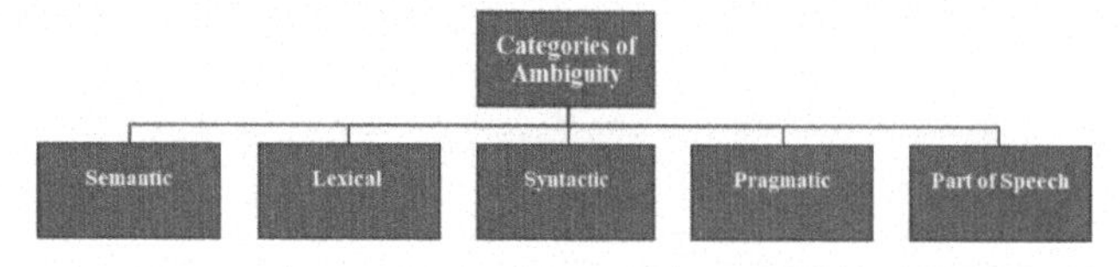

Figure 72.1 Classification of text ambiguity

to pick the correct sense without any additional information based on the context. The success of WSD system relies on two very important resources 1) Sense repository and 2) Sense annotated corpus. Therefore, a WSD system can be easily developed for a language if it has a sense repository like wordnet and considerable amount of sense-annotated corpus in that language. Ample research has been conducted on other languages because of availability of resources. Lack of resource availability acts as a major bottleneck in WSD for Indian languages like Marathi, Hindi, Bahasa, Malay, Punjabi, etc. (Bhingardive and Bhattacharyya, 2010). WSD is a vital task because of following reasons (Roberto and Mirella, 2010).

1. Dataset creation requires lot of time and costly.
2. Requirement of context information while processing a complex ambiguous sentence.

The paper structure is as follows. Section 2, discuss most recent contribution of WSD system in NLP domain. Section 3 describe the proposed approach of WSD. Section 4 describe the experimental results and discussion on results with respect to different scenarios. Paper ends with conclusion and future research direction in this domain.

Literature Survey

Lesk (1986) explored the Lesk algorithm to unravel the WSD problem in phrases. This eventuality has been the initial first machine-understandable language forms a dictionary-based technique. We've completely based this system on the cross - overlapping of the word's meaning within the definition for a specific phrase during a context where it's used or written. Basically, this Lesk algorithm works on selecting a really short phrase from the context which comprises an ambiguous phrase. Singh and Siddiqui (2015) have discussed three algorithms based on corpus statistics. The algorithms have shown excellent results and the authors have remarked that if a sense-annotated corpus is available, then they can pre-compute the conditional probability of the co-occurring words. Laishram et al. (2014) made a primary attempt at building Word Sense Disambiguation system using a supervised method based on decision trees in Manipuri language. Conventional positional and context-based features suggested capturing the correct sense of the ambiguous word, which uses a classification and regression tree (CART) based algorithm to train the classifier. Gopal and Haroon (2016) performed an experiment which used two corpora via ambiguous corpus and sense corpus along with the Naive Bayes algorithm to clarify the Malayali ambiguous words. Naive Bayes' classifier used for locating the contingent probability of various meanings of a word, due to the meaning of ambiguous words in the context, the best probability is selected later. Dhopavkar et al. (2015) analysed how rule-based algorithms can effectively perform Marathi word meaning disambiguation The summary of techniques with the accuracy of every method described within the Tables 72.1 and 72.2. The system achieved an accuracy of about 75% and it clarifies nouns, adjectives and verbs at word level ambiguity.

Table 72.1 Summary of WSD system available in literature

Algorithm	*Language*	*Performance*	*Year*	*References*
Knowledge-based approach	Malayalam	81.3%	2010	Haroon (2010)
Modified Lesk's algorithm	Panjabi	75%	2011	Kumar and Khanna (2011)
Graph based unsupervised method	Bengali	60%	2013	Das and Sarkar (2013)
Decision list	Kannada	Satisfactory	2013	Parameswarappa et al. (2013)
Genetic algorithm	Hindi	91.64%	2013	Kumari and Singh (2013)
Decision tree method	Manipuri	71.75%	2014	Anand et al. (2014)
Support vector machine	Tamil	91.6%	2014	Pal et al. (2015)
Naive Bayes classification	Bengali	8085%	2015	Sankar et al. (2016)
Context similarity unsupervised approach	Malayalam	72%	2016	Vaishnav (2017)
Genetic algorithm	Gujarati	Satisfactory	2017	Vaishnav (2017)
Lesk algorithm	Kannada	Satisfactory	2017	Shashank and Kallimani (2017)
Naive Bayes method	Punjabi	8189% for both model	2018	Pal and Kumar (2018)
Naive Bayes approach	Assamese	91.11%	2019	Borah et al. (2019)
Knowledge-based Approach	Gujarati	Satisfactory	2019	Vaishnav and Sajja (2019)
Deep learning neural Network	Punjabi	9197%	2020	Pal and Kumar (2020)

Table 72.2 Comparative analysis of WSD techniques for different Indian language

Sr. N.	*Approach*	*Algorithm*	*Advantage*	*Disadvantage*
1	Knowledge-based approaches	Lesk algorithm	Fast and less computational complex	Data size is huge for training.
		Semantic similarity	Semantic aspects covered while predicting meaning	Similar context different meaning cause issue understanding semantics.
		Heuristic method	Testing is easy.	Data set is huge and prior knowledge is required.
		Preferences for selection	Manual human intervention is not required.	Word relationship confusion cause issues.
2	Supervised approaches	Conventional Neural Network based approach	Less knowledge base work efficiently.	Hardware dependencies and GPU requirement.
		Conventional Naive Bayes approaches	Simple, easy to implement.	Less data not work efficiently.
		Conventional Decision tree approaches	Easy to understand and implement.	Maintenance cost is more.
		Conventional Decision list approaches	Accuracy is more compare to other method.	Limited dataset cause issue.
		Example-based Method	Reliable system.	Less Performance in complex scenarios.
		Support Vector Machine	No Over-fitting problem.	No transparency
3	Unsupervised approaches	Context clustering or sentence word clustering	No prior experience is required.	Efficient feature selection method is required.
		Co-occurrence graph	Robust features can be generated by assembling small features	Bookmarks cannot be included

Proosed Methodology

The proposed approach focus is to design a novel word's multiple features-based approach for finding the exact sense of words from Marathi language sentences. The proposed system architecture shown in Figure 72.2. It is divided into five stages as 1) Pre-processing, 2) Disambiguation word prediction, 3) Feature extraction, 4) Feature selection, 5) Training input data using supervised approach, 6) Testing and validation. Each of this step describe in brief as follows:

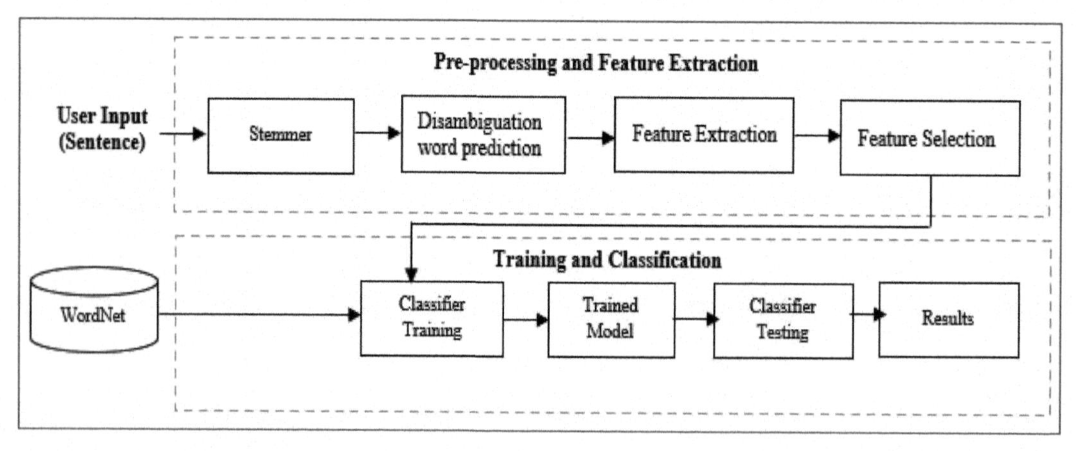

Figure 72.2 System architecture

Pre-processing

The initial process of proposing the model is pre-processing, which is executed based on the input Marathi sentences. In the pre-processing removing stop words, root search, and tokenisation are performed.

Disambiguation word prediction

After completing the pre-processing step, the sentence enters the disambiguation word prediction stage. Here, if the word contains more than one POS, INDEX, ontology ID'S then, that word is considered as ambiguous word. These stages generate the desired output i.e., getting root word, removing stop word and prediction of the ambiguous word.

Feature extraction

In feature extraction, every word in the sentence is converted into a feature. Each word in a sentence represented by the Eq. 1.

$$S = \sum_{i=1}^{n} \{w_1, w_2, w_3, w_4, \ldots. w_n\} \tag{1}$$

Here, sentence represented by '*S*' and words represented by '*w*'. The value of index '*i*' change from *1* to '*n*', to read all the words in the input sentence. Each word represented by different characteristics such as (Noun, Part of Speech, Adverb, Verb, Ontological values, position of word, the input word is ambiguous or not, etc.).

Feature selection

In feature selection, unnecessary features are removed. Only important and distinct features are selected based on the significance of the features in identification of disambiguation. First, according to the noun, part-of-speech and ontology representation of the word, each feature value is converted into a relevant decimal value. Nominal values for features are represented in the following Table 72.3.

Similarly, the nominal ontology values are ranges from 0 to 227, where 0 is assigned to none. Examples of ontologies values with its respective decimal values represented in the Table 72.4.

Next, in the indexing process, the node stores the feature value of the word. Each node individual index values are assigned and sentence 'S'. The ambiguous word represented by the index value 0. The negative number's index range means the left side of ambiguous nodes and the right side represents the ambiguous

Table 72.3 The POS features to decimal value conversion

Sr. No.	*Different types of POS in a sentence*	*Nominal*
1	If the word is None, it is represented by	0
2	If the word is Noun, it is represented by	1
3	If the word is Adjective, it is represented by	2
4	If the word is Verb, it is represented by	3
5	If the word is Adverb, it is represented by	4
6	If the word is Pronoun, it is represented by	5
7	If the word is Helping Verb, it is represented by	6

Table 72.4 Examples of ontology nominal values representation

Representation of Ontology	*Representation of nominal range of values*
पौराणकिव्यक्तरिखा(Mythological Character in word)	173
वशिषनाम(Proper Noun word)	3
मानवनरि्मति(Artifact in a word)	27
प्रत्यक्षातीलठकिाण(Physical Place)	37
मानवनरि्मति(Artifact word)	27
समूह(Group word)	149
गुणवेशषि्ट्य(Quality word)	45
जाणीव* (Perception word)	46
समूह(Group word)	149
काम(Action word)	52
बदलवाचक(Change)	124

node illustrated in the Figure 72.3. We use equations to calculate the characteristics of each node represented in Equation 2.

$$TNF = \sum_{n=1}^{N} NFc \tag{2}$$

Here, *NFc* each node feature values, and *TNF* shows the total features values in the node.

The example sentence "राम मंदरित जात होता" feature values are represented in Table 72.6. The feature vector is created by assigning nominal values as illustrated in following Table 72.7.

The rules for constructing the feature vector are as follows:

- The ambiguous word feature values are represented at Node 0.
- Node index is set to 0, if no word existed.
- The fixed set of features are extracted from each sentence.
- Negative indexed values → left adjacent of ambiguous word in a sentence.
- Positive indexed values → Right adjacent of ambiguous word in a sentence.
- Feature vector size → Nc values.
- Each Node store different ONTOLOGY, POS values.

Finally, the ambiguous words encoded. In encoding, *A* each ambiguous word represented by unique values. The ambiguous word dataset created. Each decimal value is used for unique representation of an ambiguous word. The label encoding method illustrated by Eq. 3.

$$Lu = \sum_{n=1}^{N} Au \times k + Sid \tag{3}$$

Here, *Lu* represents labelled unique code. *k* is constant and *Sid* represents a sense id. *Au* is the word in a sentence.

Classifier training

In classification, a supervised learning method used to create training data sets. The training dataset includes thirty ambiguous words, five hundred Marathi sentences. Training dataset contains one sense of ambiguous word in every sentence. The feature value labelled of sentence is represented in following Table 72.5.

Left side adjacent nodes			*Ambiguous node*	*Right side adjacent nodes*		
N-3	N-2	N-1	N0	N1	N2	N3
←		→ *TNF*				

Figure 72.3 Indexing process of word to node structure

Table 72.5 Multi-class classifier input feature representation

Sentence#	*Feature vector*	*Label code*
S1	Fv1	Lu1
S2	Fv2	Lu2
S3	Fv3	Lu3
S4	Fv4	Lu4
S5	Fv5	Lu5
S6	Fv6	Lu6
S7	Fv7	Lu7
S8	Fv8	Lu8
S9	Fv9	Lu9
S10	Fv10	Lu10
S11	Fv11	Lu11
S12	Fv12	Lu12
S13	Fv13	Lu13
S14	Fv14	Lu14
Sn	Fvn	Lun

Table 72.6 The feature extraction process for the Marathi language sentence

Word	*Root*	*Suffix*	*Replacer*	*Ontology*	*POS*
राम	राम			पौराणकिव्यक्तरिखा(Mythological Character) -173	Noun-1
राम	राम			वशिेषनाम(Proper Noun) -3	Noun-1
मंदरित	मंदरि	ित		मानवनरि्मति(Artifact) -27	Noun-1
मंदरित	मंदरि	ित		प्रत्यक्षातीलठकिाण(Physical Place) -37	Noun-1
मंदरित	मंदरि	ित		मानवनरि्मति(Artifact) -27	Noun-1
मंदरित	मंदरि	ित		पौराणकिव्यक्तरिखा(Mythological Character) -173	Noun-1
जात	जात			समूह(Group) -149	Noun-1
जात	जात			गुणवैशषि्ट्य(Quality) -45	Noun-1
जात	जात			जाणीव* (Perception) -46	Noun-1
जात	जात			समूह(Group) -149	Noun-1
जात	जात			समूह(Group) -149	Noun-1
जात	जा	त		काम(Action) -52	Noun-1
जात	जाणे	त	णे	अवस्थावाचक(Verb of State) -143	Verb-3
जात	जाणे	त	णे	क्रयिावाचक(Verb of Action) -123	Verb-3
जात	जाणे	त	णे	क्रयिावाचक(Verb of Action) -123	Verb-3
जात	जाणे	त	णे	काम(Action) -52	Verb-3
जात	जाणे	त	णे	घडणे(Verb of Occur) -186	Verb-3
जात	जाणे	त	णे	अवस्थावाचक(Verb of State) -143	Verb-3
जात	जाणे	त	णे	हालचालवाचक(Motion) -138	Verb-3
जात	जाणे	त	णे	कार्यसूचक* (Act)	Verb-3
जात	जाणे	त	णे	अवस्थावाचक(Verb of State) -143	Verb-3
जात	जाणे	त	णे	अवस्थावाचक(Verb of State) -143	Verb-3
जात	जाणे	त	णे	हालचालवाचक(Motion) -138	Verb-3
जात	जाणे	त	णे	घडणे(Verb of Occur) -186	Verb-3
जात	जाणे	त	णे	अवस्थावाचक(Verb of State) -143	Verb-3
जात	जाणे	त	णे	घडणे(Verb of Occur) -186	Verb-3
जात	जाणे	त	णे	बदलवाचक(Change) -124	Verb-3
जात	जाणे	त	णे	घडणे(Verb of Occur) -186	Verb-3
जात	जाणे	त	णे	क्रयिावाचक(Verb of Action) -123	Verb-3
जात	जाणे	त	णे	अवस्थावाचक(Verb of State) -143	Verb-3
जात	जाणे	त	णे	सहायकक्रयिापद(Helping Verb) -227	Helping Verb-6
जात	जाणे	त	णे	सहायकक्रयिापद(Helping Verb) -227	Helping Verb-6
जात	जाणे	त	णे	सहायकक्रयिापद(Helping Verb) -227	Helping Verb-6
होता	होणे	ता	णे	बदलवाचक(Change) -124	Verb-3
होता	होणे	ता	णे	घडणे(Verb of Occur) -186	Verb-3
होता	होणे	ता	णे	अवस्थावाचक(Verb of State) -143	Verb-3
होता	होणे	ता	णे	घडणे(Verb of Occur) -186	Verb-3
होता	होणे	ता	णे	घटनादर्शक*(Event) -35	Verb-3
होता	होणे	ता	णे	अवस्थावाचक(Verb of State) -143	Verb-3
होता	होणे	ता	णे	घडणे(Verb of Occur) -186	Verb-3
होता	होणे	ता	णे	अवस्थावाचक(Verb of State) -143	Verb-3
होता	होणे	ता	णे	अवस्थावाचक(Verb of State) -143	Verb-3
होता	होणे	ता	णे	सहायकक्रयिापद(Helping Verb) -227	Helping Verb-6

Table 72.7 Represents the feature vector of the sentence

N-3		*N-2*		*N-1*		*N0*		*N1*		*N2*		*N3*		*Au*
0 0	0 0	1 173	1 3	1 27	1 37	1 149	1 45	3 124	3 186	0 0	0 0	0 0	0 0	10

Classifier testing

Next, we carried out the word tag decoding process. The classifier output is the index_id of the nominals label, which is not similar as sense id of the ambiguous word in a sentence. The label decoding and computation of sense id is represented in Eq. 4.

$$AmbigiousWordId = \frac{LabelNominal - senseId}{maxAmbigiousCount \times k} \tag{4}$$

Here, the *LabelNominal* are the nominals in the sentence. The *LabelNominal* computation represented in Eq. 5. The *senseId* computation represented in Eq. 5.

$$LabelNormal = \sum_{i=1}^{n} Nominals_{predictedIndex} \tag{5}$$

$$senseId = LabelNominal\ MOD\ k \tag{6}$$

Next, we have described the algorithm to predict the correct sense of ambiguous word.

Algorithm: To detect and predict accurate meaning of ambiguous word (Aw) of sentence.

Input: *ambiguosWordSentence*, the ambiguous word sentence
Output: *sentMean*, the meaning of word. *predictedAmbiguousWord*, the ambiguous term

1. Read the input sentence: *ambiguosWordSentence*
2. Extract the features of each sentence.
3. Initialise the multi-class classifier.
4. Predict the appropriate label using multi-class classifier model: *predictedLabel*
5. Decode predicted result by label decoding.
6. [Label decoding process].
7. *sentMean* $\rightarrow$ *predictedLabel* mod *k*
   predictedAmbiguousWord $\rightarrow$ (*predictedLabel* – *sentMean*) / (max*AmbCount* × *k*
8. Display the meaning of sentence meaning and ambiguous words.
9. End

Experimental Results

We have performed the experiments on different sentence of Marathi language of WordNet dataset. It was developed by Dr. Pushpak Bhattacharya with his team at IIT, Bombay. Marathi WorldNet is organised as a semantic network of large electronic databases. Paradigmatic relations such as synonymy, hyponymy, antonymy and entailment etc. are used to construct it (Roberto and Mirella, 2010).

A. Implementation and evaluation metrics

The proposed method is implemented using JAVA language, Nvidia GPU, *i5* Intel having 8 GB RAM. In addition, results are represented using performance evaluation metrics such as accuracy, recall, and precision. The recall calculated using Equation 7.

$$Recall(r) = \frac{TP}{FN + TP} \tag{6}$$

Were,
TP: When input sentence is ambiguous, and word meaning correctly identified.
FN: When input sentence is not ambiguous, and system predict ambiguous.

The precision evaluation is computed using Equation 8.

$$Precision(p) = \frac{TP}{FN + ND + TP} \tag{8}$$

Were,
ND – Not detected.

F-measure of a test's accuracy represented by Equation 9.

$$F - measure = \frac{2 \times P \times R}{P + R} \tag{9}$$

Were,
P – Precision
R – Recall

B. Proposed method results and discussion

Hence, the Marathi WSD is implemented by using several attribute selection methods. The proposed method uses the Marathi word network dataset. The proposed approach tested on 350 Marathi language sentences and for 30 ambiguous words.

Tables 72.8 and 72.9 illustrates the performance of the proposed system using evaluation metrics such as side neighbours, feature nodes, average prediction testing and training time, recall, precision, and F1-Measure Accuracy.

The feature/node and side neighbours' comparison with precision/recall/F1-measure shown in the Figures 72.4 and 72.5 with respect to N gram model and filters.

Figure 72.4 Recall precision and accuracy comparison with respect to N gram model

Figure 72.5 Recall precision and accuracy comparison with respect to filters

The proposed approach gives the recall of 84.21% and precision of 81%. The accuracy of proposed method is 83.3%. Next, we have compared multiple pair attribute selection results with the other methods available in the literature.

Conclusions and Future Research Direction

In this paper, we propose an efficient method for predicting the correct meaning of a sentence and identifying whether the sentence is ambiguous. First, we extract the sentence features in the POS and ontology noun pairs. Next, the feature selection method applies to the extracted features to reduce the dimensionality of the features. At last, these compact features are trained and tested using a multi-class classifier.

Table 72.8 Performance evaluation of multiple pair attribute method (proposed method)

Sr.	*Side Negh*	*Features/ Node #*	*Test set prediction Time (secs)*	*Avg. prediction time/Sen. (secs)*	*Recall (%)*	*Precision (%)*	*F1- Measure (Accuracy) (%)*
1	1	2	3.329	0.010	84.21	83.39	83.8
2	1	4	3.315	0.010	78.95	78.18	78.56
3	1	6	3.418	0.011	82.24	81.43	81.83
4	1	8	3.522	0.011	81.91	81.11	81.51
5	1	10	3.483	0.011	81.58	80.78	81.18
6	1	12	3.883	0.012	81.91	81.11	81.51
7	2	2	3.513	0.011	81.58	80.78	81.18
8	2	4	3.507	0.011	78.95	78.18	78.56
9	2	6	3.694	0.012	81.91	81.11	81.51
10	2	8	3.779	0.012	82.57	81.76	82.16
11	2	10	3.901	0.012	83.22	82.41	82.82
12	2	12	4.061	0.013	83.22	82.41	82.82
13	3	2	3.444	0.011	81.58	80.78	81.18
14	3	4	3.613	0.011	77.3	76.55	76.92
15	3	6	3.748	0.012	80.59	79.8	80.2
16	3	8	3.995	0.013	82.57	81.76	82.16
17	3	10	4.112	0.013	83.88	83.06	83.47
18	3	12	4.26	0.013	83.55	82.74	83.14
19	4	2	3.496	0.011	81.91	81.11	81.51
20	4	4	3.669	0.011	77.3	76.55	76.92
21	4	6	3.942	0.012	78.29	77.52	77.91
22	4	8	4.165	0.013	80.92	80.13	80.52
23	4	10	4.29	0.013	82.57	81.76	82.16
24	4	12	4.749	0.015	82.89	82.08	82.49

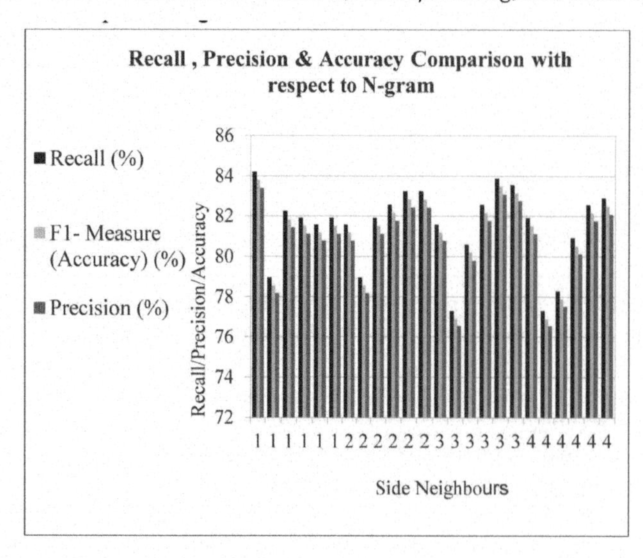

Figure 72.4 Recall precision and accuracy comparison with respect to N gram model

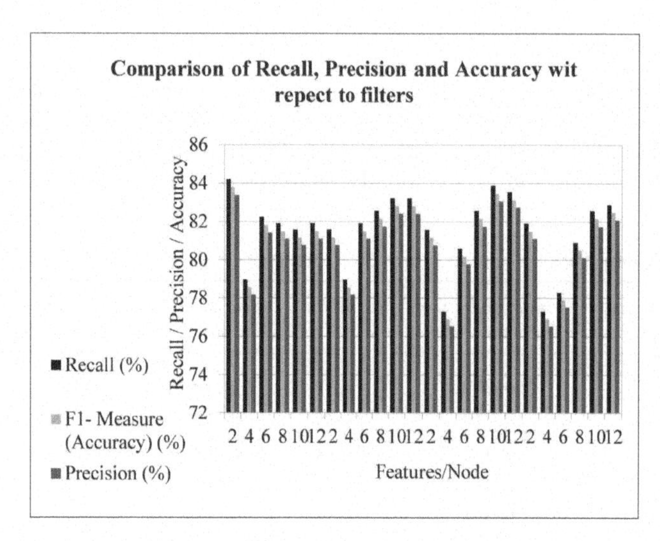

Figure 72.5 Recall precision and accuracy comparison with respect to filters

Table 72.9 Multiple pair attribute selection proposed method comparison with the state-of-the art approaches

Author	*Approach*	*Language*	*Accuracy*
P. Sharma and N. Joshi (2019)	WSD using LESK algorithm	Hindi	71.40%
G. Dhopavkar, M. Kshirsagar, and L. Malik	WSD using Rule based method	Marathi	80%
Namrata Kharate	Lesk algorithm	Marathi	70%
A. Athaiya, D. Modi, and G. Pareek	WSD using Genetic algorithm	Hindi	Only for noun 85%
P. Sharma and N. Joshi	WSD using Knowledge based method	Hindi	71.4%
Proposed approach	Multiple attribute pair selection	Marathi	83.80%

We conducted the experiment on the Marathi WordNet dataset. Our proposed approach outperformed state-of-the-art approaches. In the future, the proposed approach could be useful in different applications, such as Marathi language, semantic understanding, language translation, news and advertisement domain. Again, this approach can be helpful in real-time information extraction applications. It increased the accuracy of Marathi language WSD by increasing the size of training samples and perception inventory.

References

Anand, M., Rajendran, S., and Soman, P. (2014). Tamil word sense disambiguation using support vector machines with rich features. Int. J. Appl. Eng. Res. 9(20):7609–7620.

Bhingardive, S. and Bhattacharyya, P. (2010). IndoWordNet. In Seventh international conference on language resources and evaluation (LREC'10), mediterranean conference centre, (May 17–23), Valletta, Malta.

Borah, P., Talukdar, G., and Baruah, A. (2019). WSD for assamese language. In International journal of advances in intelligent systems and computing. Singapore: Springer.

Das, A. and Sarkar, S. (2013).. Word sense disambiguation in bengali applied to bengali-hindi machine translation. In IEEE 10th international conference on natural language processing (ICON), (pp. 1–10). Noida, India.

Dayal, V. (2004). The universal force of free choice any linguistic variation yearbook. Linguistic Variation Yearbook. 4(1):540. doi: https://doi.org/10.1075/livy.4.02day

Dhopavkar, G., Kshirsagar, M., and Malik, L. (2015). Application of rule based approach to word sense disambiguation of marathi language text. In IEEE sponsored 2nd international conference on innovations in information embedded and communication systems (ICIIECS), (pp. 1–10), Coimbatore: India.

Gopal, S. and Haroon, R. P. (2016). Malayalam word sense disambiguation using Naïve Bayes classifier, 2016 International Conference on Advances in Human Machine Interaction (HMI), 2016, pp. 1–4, doi: 10.1109/HMI.2016.7449181.

Haroon, P. (2010). Malayalam word sense disambiguation. In 2010 IEEE international conference on computational intelligence and computing research, (pp. 1–4).

Kale, S. and Gawande, U. (2020). A novel approach for detection of ambiguity for marathi sentence and development of stemmer. Int. J. Adv. Sci. Technol. 29(5):10184–10192.

Kumar, R. and Khanna, R. (2011). Natural language engineering: the study of word sense disambiguation in punjabi. Int. J. Eng. Sci. 230–238.

Kumari, S. and Singh, P. (2013). Optimised word sense disambiguation in Hindi using genetic algorithm. Int. J. Res. Comput. Commun. Technol. 2(7):445–449.

Laishram, R., Ghosh, K., Nongmeikapam, K., and Bandyopadhyay, S. (2014). A decision tree-based word sense disambiguation system in manipuri language, Adv. Comput. Int. J. (ACIJ) 5(4):17–22.

Lesk, M. (1986). Automatic sense disambiguation using machine readable dictionaries. In ACM International Conference SIGDOC, (pp. 1–10).

Lkshmi, S. and Haroon, R. (2016). Malayalam word sense disambiguation using naïve bayes classifier. In International conference on advances in human machine interaction (HMI), (pp. 1503–1507).

Navigli, R. (2009). Word sense disambiguation: A survey. ACM Comput. Surv. (CSUR), 1–10.

Pal, V. and Kumar, P. (2018). Naive bayes classifier for word sense disambiguation of punjabi language. Malays. J. Comput. Sci. 313:1–20.

Pal, S. and Kumar, P. (2020). Word sense disambiguation for Punjabi language using deep learning techniques. Neural Comput. Appl. 32(8):2963–2973.

Pal, R., Saha, D., Naskar, S., and Dash, S. (2015). Word sense disambiguation in Bengali: a lemmatised system increases the accuracy of the result. In IEEE International Conference on Recent Trends in Information Systems, (pp. 342–346).

Parameswarappa, S., Narayana, N., and Yarowsky, D. (2013). Kannada word sense disambiguation using decision list. Int. J. Emerg. Trends and Technol. Comput. Sci. 2(3):272–278.

Pooja, S., and Nisheeth, J. (2019). Design and development of a knowledge based approach for word sense disambiguation by using WordNet for Hindi. Int. J. Innov. Technol. Explor. Eng. (IJITEE) 8(3), ISSN: 2278-3075.

Roberto, N. and Mirella, L. (2010). An experimental study of graph connectivity for unsupervised word sense disambiguation. IEEE Trans. Pattern Anal. Mach. Intell. (TPAMI), 32(1):1–25.

Sankar, S., Raj, R., and Jayan, V. (2016). Unsupervised approach to word sense disambiguation in Malayalam. Procedia Technol. 24(1):1507–1513.

Shashank, S. and Kallimani, S. (2017). Word sense disambiguation of polysemy words in kannada language. In IEEE international conference on advances in computing, communications and informatics, (pp. 641–644).

Singh, S. and Siddiqui, T. (2015). Utilizing corpus statistics for hindi word sense disambiguation. Int. Arab J. Inf. Technol. (IAJIT) 12(6A):755–763.

Vaishnav, B. (2017). Gujarati word sense disambiguation using genetic algorithm. Int. J. Recent Innov. Trends Comput. Commun. 5(6):635–639.

Vaishnav, B. and Sajja, S. (2019). Knowledge-based approach for word sense disambiguation using genetic algorithm for Gujarati. In Information and communication technology for intelligent systems, (pp. 485–494), Singapore: Springer.

Zelta, N. 2014. Sennet, Adam, "Ambiguity", The Stanford Encyclopedia of Philosophy (Fall 2021 Edition), Edward N. Zalta (ed.), URL = <https://plato.stanford.edu/archives/fall2021/entries/ambiguity/>.

73 Development of methodology and process of Soil stabilisation using fly ash and plastic waste

J. M. Raut[a], S. R. Khandeshwar[b], P.B.Pande[c], Nikita Ingole[b], Radhika Shrawankar[d], and Urvesh Borkar[f]

Civil department, YCCE, Nagpur, India

Abstract

In this work, it is tried to use fly ash and plastic waste for stabilisation of soil. As both the material becomes great problem to environment. If we use this material for construction purpose then it will be win-win situation. Also, some regions contain enough expansive soil which has high shrinkage properties thereby indirectly producing a high percentage of montmorillonite, this creates conditions unsuitable for any construction so it is required to stabilise the soil pre-establishment to avoid differential settlement which causes major damage in buildings and causes road accidents. Lime is a popular chemical stabilisation method because it reduces montmorillonite activity and compaction is popular mechanical stabilisation method. In this article, we have used materials fly ash and plastic waste that are eco-friendly, inexpensive and readily available. In this we have performed different laboratory test and tried to develop methodology of use of fly ash and plastic waste simultaneously.

Keywords: Black cotton soil, plastic waste and Fly ash.

Introduction

Stabilisation of soil is the process of improving and stabilizing the technical features of the soil. It's utilised to improve shear resistance and reduce unqualified soil qualities like permeability and consolidation. This technology is mostly used in the construction of highways and airports. Compaction and pre-consolidation are commonly employed to upgrade soils that are already in good condition. Soil stabilisation, on the other hand, extends beyond increasing the consumption of soft soil and minimizing the cost of soft soil renewal. Chemical modification of the soil material itself, in addition to studies on soil mass interaction, is an important part of this process. Ground stabilisation is sometimes used to make inner city and suburban streets more permeable.

Fly ash is a by-product of power plant combustion of glowing coal that must be buried. However, for the sake of sustainable construction, many governments have encouraged the reuse of certain forms of trash.

Because plastic is not a crucial substance, it cannot degrade. The strategies used to encourage the recycling of plastic garbage have an impact on the environment. The organism is harmed by plastic leaching due to the acidic soil environment. As a result, new ways for removing plastics are required. Plastics have a variety of characteristics, including strength, brittleness, and corrosion resistance. Chemical resistance, insect resistance, abrasion resistance, insulating qualities, and heat resistance We can use plastic trash to strengthen the soil as a treatment for plastic waste. Plastic is used as a soil stabiliser in this novel technology to reduce pollution and improve soil qualities.

There is some researcher who done their contribution in this regards like Anas A (2011) and Arora K.R. has presented how soil bearing capacity can be increased due to addition of waste material. The methodology for determination of bearing capacity adopted from the research material published by Raut, J., khadeshwar, S., Kumar, T., Suryketan, Nagle, R. has been given valuable information about use of plastic strips.

Materials

Soil sample

Soil for experimentation collected from the area around our college campus known as Zilpi, Nagpur and their engineering properties and strength characteristics were analysed. Mostly, expansive soil which locally known as black cotton soil is taken for experimentation. Black Cotton soils are inorganic clays with a compressibility range of medium to high, and they are an important soil category in India. The clay in Black Cotton soil

[a]jmrv100@gmail.com; [b]khandeshwar333@yahoo.com; [c]prashantbpande21@gmail.com; [d]nikitapingole10@gmail.com; [e]shrawankarradhika204@gmail.com; [f]urveshborkar@gmail.com

is largely montmorillonite in structure and colour, and it is black or blackish grey in appearance. The Black Cotton soil has presented a problem to geotechnical and roadway engineers due to its strong swelling and shrinkage characteristics. It is one of India's largest soil deposits, covering roughly 3,00,500 square km. The sieve analysis, liquid limit, plastic limit and specific gravity of soil tests were used to characterise the soil sample and shown Table 73.1 and the apparatus for sieve analysis shown in Figure 73.1..

Fly ash

Fly ash has a low cementitious value on its own, but in the presence of moisture, it interacts chemically and forms a cementitious compound. The strength and compressibility of the soil are improved by the cementitious compound that is generated. It is a thin powder which is grey in colour. It is made primarily of silica that is produced when finely ground coal is burned in a boiler to generate electricity. In the Katol Power Plant in Nagpur, fly ash was collected from locally available cement manufacturers. The properties of fly ash collected are given in Table 73.2.

Plastic waste

Plastic is non-biodegradable waste. As a result, disposing of plastic waste endangering the environment has become serious difficulty. Therefore, using plastic waste as a stabiliser is less hazardous, economical and good for soil embankments as well as help in reduced air pollution. Increased use of plastics in everyday consumer applications has resulted in an ever-increasing fraction of plastic materials that were utilised for a brief time and then discarded ending up in municipal solid trash. As a result, there is a rising need to develop alternate uses for recycled plastic bag waste in order to extend the life of the plastic material and so save the environment from degradation. The properties of plastic waste collected are given in Table 73.3.

Plastic is widely available and plastic possess quality like easily mould in any shape, good insulator, do not rust, etc will help to improve the result.

Plastic was obtained from waste plastic cover (milk and curd packets, plastic bottles, chocolate wrapper, polythene bags) after proper cleaning.

Table 73.1 Properties of soil

Sr. No.	*Properties*	*Value*
1.	Specific gravity	2.62
2.	Liquid Limit (%)	29.5
3.	Plastic Limit (%)	20.2

Table 73.2 Properties of fly ash

S.No.	*Minerals*	*Percentage*
1.	SiO2	48.38
2.	Fe2O3	7.27
3.	Al_2O_3	27.62
4.	CaO	10.54
5.	MgO	2.59
6.	SO_3	3.17
7.	P_2O_5	0.22

Table 73.3 Properties of plastic wastes

Sr. No.	*Description*	*Value*
1	Colour	Clear
2	Unit Weight	0.92
3	Compressive Strength	Poor
4	Ultimate tensile strength	57
5	Modulus of Elasticity	115-455
6	Water absorption (%)	0.02

Methods

Sieve analysis

A set of sieves was used to sieve the dirt. Sieve cloth is usually constructed of spun brass, phosphor bronze, or stainless steel. Sieves are classified by the size of the square opening in mm or microns, according to IS: 14981970. There are sieves ranging in size from 80mm to 75microns available.

Liquid limit

The water concentration at which soil transitions from liquid to plastic is known as the liquid limit. The compression index, which is utilised in settlement analysis, can be determined using the liquid limit test. The clay is virtually liquid at the liquid limit, yet it has a low shearing strength. At that level, the shearing strength is the smallest value that can be measured in the laboratory. The clay minerals present in the soil determine the liquid limit. The bigger the amount of absorbed water and, as a result, the higher the liquid limit, the stronger the surface charge and the thinner the particle.

Plastic limit

The water concentration below which the soil ceases to behave as a plastic substance is known as the plastic limit. When wrapped into a 3mm diameter thread of soil, it begins to crumble. At this water content, the soil loses its flexibility and becomes semi-solid.

Direct shear test

This test is performed to determine cohesion and angle of internal friction of soil sample.

UCS test

Undetermined compression test is a performed test from which we can get unconfirmed compressive strength (UCS) for any rock sample. Saturated and cohesive soils are primarily considered for this test and are generally not recommended. Since the material does not support the lateral containment of the soil. UCS refers to or represents the actual maximum uniaxial compression force that the specimen can withstand without deformation (no failure).

Standard proctor compaction test

Compaction tests are used to figure out how much compaction and water there is on the field. The test establishes a relationship between the amount of water in the sample and its dry density. Using the test relationships, the water content at which the maximum dry density is obtained is computed.

Bearing capacity of soil

The soil's bearing carrying capacity is influenced by soil properties like cohesion, in-situ density soil, voids ratio and angle of internal friction. Because compacted & dense soil has more unit weight correspondingly

Figure 73.1 Sieve shaker

more bearing capacity. If the loading is moderate and bearing capacity is sufficient to take load then shallow foundation is used. If but if the soil strata is weak and intensity of loading is high then deep foundation like pile, well or pier foundation is preferred. So, for construction of any civil engineering structure bearing capacity plays important role. Instead of changing design of foundation we can change bearing capacity of soil. And if we use waste material like fly ash and plastic waste for improving bearing capacity then there is win- win situation. As we are improving bearing capacity as well as saving the environment. We have used following instruments for to determine bearing capacity, which consists of the following:

Frame

It consists of a steel mould that meets IS 1730: 1989 requirements (Steel plates, Sheets, Strips and flats for structural and general engineering purposes Dimensions).

Loading frame and mechanism

The loading mechanism consists of a cylindrical loading cell with an S-shaped censor at the top that detects soil loading. This loading cell is supported by a 30 cm × 30 cm × 5 cm concrete bearing plate with an M25 grade. This loading system. Stainless steel cylinders are used. The S-beam loading cell is named after its shape. The name 'S-Beam load cell' comes from the fact that it is shaped like a S. It has a S shape to it. Under tension or compression, S-Beam load cells can generate an output. Tank levels, hoppers, and truck scales are all examples of applications. A load cell is a sort of transducer that converts force into a measurable electrical output. Although there are many different types of load cells, strain gauge load cells are the most

Figure 73.2 Normal soil specimen undergoing UCS test

Figure 73.3 Framed setup

common. This is the most prevalent type. Except in a few laboratories where precision is required, although mechanical balances are still in use, strain gauge load cells have taken over the weighing market/industry.

Display unit

A display unit is a visual output device that displays information. It provides information on the loading and deflection that occurs as a result of the loading applied to the soil. Wires linked to a beam and a deflection gauge make the connections.

The load is sensed by the gadget when a load is manually applied to a loading cell, which converts it to a digital signal and displays it on the digital indicator's display unit.

Manually loading rod

For hand loading, stainless steel rod of around Fe500 grade is used. Three to four rods, each with a diameter of about 10 mm, are utilised at a time to apply load to the soil. As a result, stainless steels are used in situations where both steel strength and corrosion resistance are required. In lowering acids, um additions improve corrosion resistance and protect chloride solutions from pitting attack.

Figure 73.4 Loading mechanism

Figure 73.5 Digital deflection meter

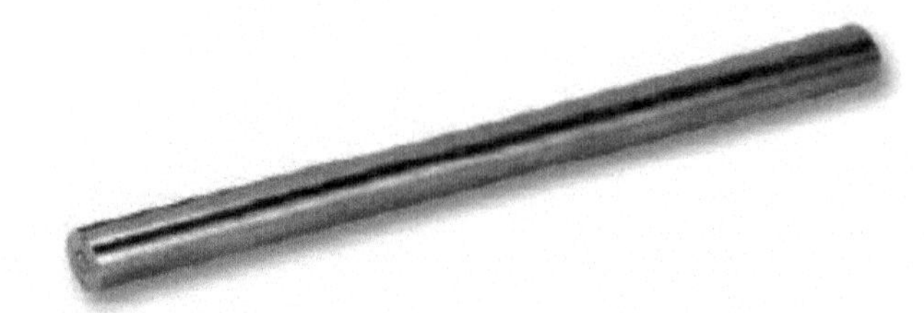

Figure 73.6 Steel rod

L.V.D.T.:

This linear variable differential transformer (L.V.D.T.) is used to determine how much soil settles as a result of loads. As the load on the soil grows, the settlement of the soil increases as well, and this deflection must be measured using L.V.D.T. Internally, it has a spring-like construction, and a casing is placed over the spring. The censor is also used to determine whether or not a settlement has been reached. A wire from the L.V.D.T. is connected to the display unit, which displays the soil settlement reading. The maximum deflection that can be measured is between 0 and 50 mm.

Steel plate

For transferring the weight from the loading cell to the soil, a flat surface is necessary, which can be achieved with a steel plate or a concrete raft. Steel plate was employed in our project.

Result and Discussion

Following test are performed to find out basic properties of soil

Sieve analysis

Sieve analysis is performed in the laboratory and following observation is obtained. It is observed that soil is well graded and clay particle are predominant. Results found in the sieve analysis test are presented in Table 73.4.

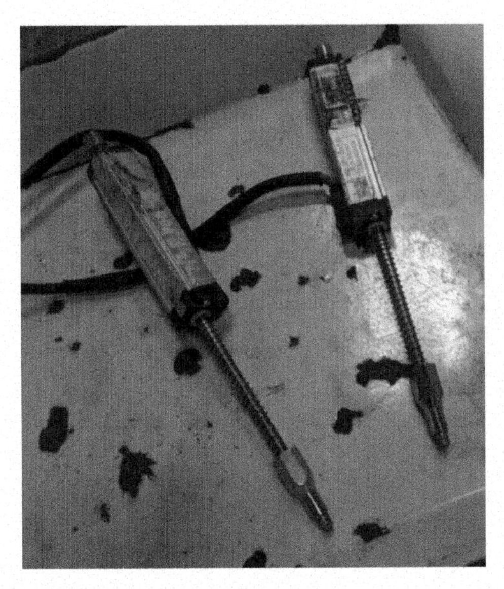

Figure 73.7 L.V.D.T. instrument

Table 73.4 Sieve analysis table

Sieve size (mm)	*Mass retained (gms)*	*Soil mass preserved on Cumulative basis in (gms)*	*% of soil preserved in total*	*Percentage of finer*
4.75	219	21.9	21.9	78.1
2.0	40.2	40.2	62.1	37.9
0.85	282.5	28.25	90.35	9.65
0.425	76.5	7.65	98	2
0.15	11.5	1.15	99.15	0.85
0.075	3.0	0.3	99.45	0.55
PAN	5.5	0.55	100	0

Liquid limit

The liquid limit is calculated of soil having different fly ash content. Figure 73.8 is drawn with liquid limit verses fly ash percentage. As fly ash percentage increases then liquid limit decreases. Results found in the liquid limit test are presented in Table 73.5 and the apparatus shown in Figure 73.8. The graph for liquid limit is shown in Figure 73.7.

Plastic limit

The Plastic limit is calculated of soil having different fly ash content. Figure 73.10 is drawn with Plastic limit verses fly ash percentage. As fly ash percentage increases then Plastic limit decreases. Results found in the Plastic limit test are presented in Table 73.6.

Direct shear test

The direct shear test is conducted on three different plane soil sample with 25% water content and results are presented in Table 73.7. Also same test is conducted on the different combinations of soil sample with fly ash and Plastic waste. The percentage of combinations are specified in Table 73.12. The sample calculations are presented in Table 73.8.

UCS test

The UCS (unconfined compression) test is conducted on soil samples and results are presented in Tables 73.9 and 73.10 and the test apparatus is shown in Figure 73.2 The samples are shown in Figures 73.8 and 73.9.

Standard proctor compaction test

The Standard proctor test is conducted on soil samples and results are presented in Table 73.11.

Table 73.5 Liquid limit table

No. of blows	25
Container no.	82
Wt of container Wo	24.28
Weight of container+ wet soil W1	40.14
Weight of container+Dry soil W2	36.43
W.C. %	30.52

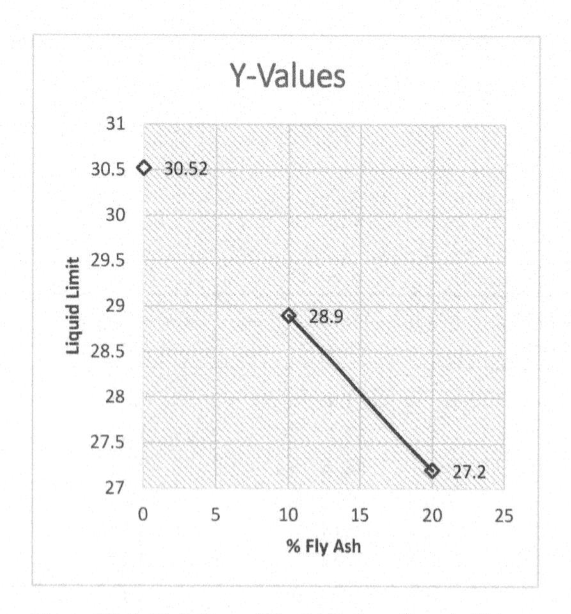

Figure 73.8 Variation of liquid limit vs fly ash (%)

Table 73.6 Plastic limit table

Container no.	3
Weight of container Wo	26.10
Weight of container +Wet soil W1	32.29
Weight of container+ Dry soil W2	31.20
W.C. %	21.37

Figure 73.9 Plastic limit soil thread

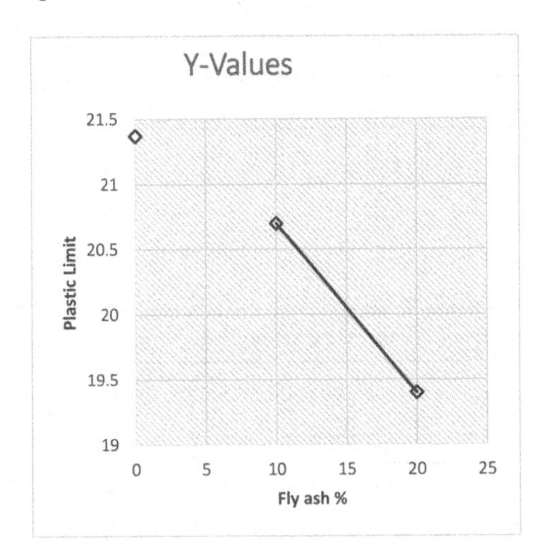

Figure 73.10 Variation of liquid limit vs fly ash (%)

Table 73.7 Normal soil + 25% WC

Load on hanger (W) Kg	2.832 Kg	4.248 Kg	5.664 Kg
Normal load on soil sample (N) Kg (W + W1) × 5 + W2	18.655 Kg	25.735 Kg	32.815 Kg
Normal stress	0.518 Kg/cm^2	0.715 Kg/cm^2	0.912 Kg/cm^2
Proving ring division at failure (D)	8 × 5 = 40	8 × 6 = 48	8 × 7 = 56
Shear force at filure	8.8 Kg	10.56 Kg	12.32 Kg
Shear resistance at failure	0.244 Kg/cm^2	0.293 Kg/cm^2	0.342 Kg/cm^2
Angle of shearing resistance	25.222°	22.283°	20.556°

Table 73.8 Normal soil + fly ash + 25% WC

Load on hanger (W) Kg	2.832 Kg	4.248 Kg	5.664 Kg
Normal load on soil sample (N) Kg (W + W1) × 5 + W2	18.655 Kg	25.735 Kg	32.815 Kg
Normal sress (	0.518 Kg/cm^2	0.715 Kg/cm^2	0.912 Kg/cm^2
Proving ring division at failure (D)	8 × 11 = 88	8 × 13 = 104	8 × 15 = 120
Shear force at Failure	19.36 Kg	22.88 Kg	26.4 Kg
Shear resistance at Failure	0.537 Kg/cm^2	0.635 Kg/cm^2	0.733 Kg/cm^2
Angle of shearing resistance	46.031°	41.608°	38.789°

Bearing capacity of soil

In the loading frame we have filled soil in box of size 0.5 × 0.5 × 0.5 in three layers. Each layer is compacted to 60 number of blows required to achieve maximum dry density at optimum moisture content. Then bearing capacity of soil is calculated in the loading frame as per standard procedure. The test is reputed for the different combinations of fly ash and plastic waste. The different combinations are mentioned in Table 73.12 and the bearing capacity of each combination is calculated. It is observed that maximum bearing capacity is achieved for flay ash 24% and plastic waste of 8%. The different apparatus used in determination of bearing capacity as shown in Figures 73.3–73.7. There may be some more perpetual combinations to achieve the same.

Conclusion

a. The inclusion of fly ash and plastic to expensive soil increases bearing capacity of soil. The combination of 24% of fly ash and 8% of plastic waste gives maximum bearing capacity.
b. The Atterberg limit is slightly influenced by the incorporation of fly ash (which is freely available). The liquid limit is reduced by roughly 25% when 15% Fly ash is added. According to our observations, when 30% fFly ash is applied, the liquid limit reduces by roughly 46%.
c. At 25% water content, the compressive strength increases by 22.88%.
d. Based on the aforementioned, waste material, and test results, it can be determined that raising soil shear strength while decreasing permeability improves soil stability. Compressive strength has also increased as a result of the use of Fly ash and plastic.

Table 73.9 For WC 15%

Sr. No.	*Description of material*	*Unconfined compressive strength (KN/m²)*
1	Normal soil	142
2	Mix soil with 30% Ash + 1% plastic	156

Table 73.10 For WC 25%

Sr. No.	*Description of material*	*Unconfined compressive strength (KN/m²)*
1	Normal Soil	155
2	Mix soil with 30% ash + 1 % Plastic	189

Figure 73.11 UCS specimen of normal soil

Figure 73.12 UCS specimen of mix soil

Table 73.11 Improvement in MDD and OMC

Plastic dimension (mm)	*Percent added*	*MDD (KN/m3)*	*OMC (%)*
0	0	12.81	42
5 × 7.5	2	12.62	28.7
10 × 15	2	11.49	33.8
15 × 20	2	12.01	34.2

S.No.	*Observations*	*Specimen 1*	*Specimen 2*	*Specimen 3*
1.	Mass of empty core cutter (W1)	4036 gm	4036 gm	4036 gm
2.	Mass of core cutter + soil (W2)	5348 gm	5849 gm	5091 gm
3.	Density $\gamma = \frac{(W2 - W1)}{v}$	1.312	1.813	1.055
4.	Dry density $\gamma d = \frac{\gamma}{1 - w}$	1.098	1.279	1.246

Table 73.12 Bearing capacity of soil

Sample No	1	2	3	4	5	6	8	9	10	11	12	13	14	15	16
% of fly ash by weight	0	3	6	9	12	15	18	21	24	27	30	33	36	39	42
% of plastic waste by weight	0	1	2	3	4	5	6	7	8	9	10	11	12	13	14
Bearing Capacity kN/m^2	49.8	58.1	66.2	72.1	81.9	85.3	91.3	95.6	102.3	99.1	97	95.3	94.1	93.2	91.1

e. It's also worth noting that when the amount of Fly ash in the mixture grew, the ideal moisture content increased as well.
f. Smaller strip size and content have seen a significant increase in UCS.

References

Anas, A. (2011). Soil Stabilisation using raw plastic bottles. In Proceedings of Indian Geotechnical Conference, (pp. 15–17).

Arora K. R. (2004). Soil Mechanics and foundation engineering. Delhi, India: Standard Publishers Distributors.

ASTM D698-12 (2012). Standard test methods for laboratory compaction of soil standard effort. West Conshohocken, PA: ASTM International.

Chamberlin, K. S. (2014). Stabilisation of Soil by using plastic wastes. Int. J. Emerg. Trends Eng. Dev. (IJETED) 204–218.

Indian Standard, 2720 (Part 2) –(1973), for moisture content

Indian Standard: 2720 (Part 7) – (1980), for light compaction.

Indian Standard: 2720 (Part 16) – (1987) (Re-affirmed 2002), for UCS.

Kumar, T., Suryaketan P., Hameed, S., and Maity, J. (2018). Behavior of soil by mixing of plastic strips. Int. Res. J. Eng. Technol. 5(5).

Madavi, S. D., Patel, D. and Burike, M. (2017). Soil stabilisation using plastic waste. Open J. Civil Eng. 10(1).

Manuel, M. and Joseph, S. (2014). Stability analysis of kuttanad clay reinforced with PET bottle fibers. Int. J. Eng. Res. Technol. 3(11).

Nagle, R. and Jain, R. (2014). Comparative study of UCS of soil, reinforced with natural waste plastic material. Int. J. Edu. Sci. Res. (IJESR) 4(6):304–308.

Peddaiah, S., Burman, A. and Sreedeep, S. (2018). Experimental study on effect of waste plastic bottle strips in soil improvement. Geotech. Geol. Eng. 36(5):2907–2920.

Raut, J., Khadeshwar, S., Bajad, S. and Kadu, M. (2014). Simplified design method for piled raft foundations. Advances in soil dynamics and foundation engineering, eds. R. Y. Linag, J. Qian, and J. Tao, (pp. 462–471). ASCE.

Sai, M. and Srinivas, V. (2019). Soil stabilisation by using plastic waste granules materials. J. Comput. Eng. (IOSR-JCE). 21(4):42–51.

STM D4318-17 (2017). Standard test methods for liquid limit, plastic limit, and plasticity index of soils. West Conshohocken, PA: ASTM International.

74 Generic recommendation system based on web usage mining concept

Arnav Ekapure[a] *and Aarti Karande*[b]

Department of Information Technology, Sardar Patel Institute of Technology, Mumbai, India

Abstract

There exist different types of recommendation systems depending on the domain of application of the recommendation system. This is due to the fact that there is no general recommendation system that can work effectively in multiple domains. A generic recommendation system can face the cost of production and maintenance. The proposed generic recommendation system is based on web usage mining. Web usage mining techniques are used to discover patterns in web data. The proposed system is more effective than traditional recommendation systems such as collaborative filtering and content-based filtering. The recommendation system is generic thus cost of production and maintenance will be effectively low. The system is lacking a proper recommendation model which can recommend the pages based on already formed association rules in real-time.

Keywords: Generic recommendation system, hd_dbscan algorithm, improved apriori algorithm, web usage mining.

Introduction

The proposed recommendation system is built to recommend web pages effectively and efficiently to the user. Web usage mining has been used effectively as an approach to overcome the traditional deficiency of traditional approaches such as collaborative filtering and content-based filtering.

The remaining paper is organised as follows: Section 2 covers the work related to web usage mining, clustering and existing recommendation system. Section 3 explains the proposed generic recommendation system in detail. Section 4 describes the case study. Section 5 concludes with a conclusion and future work.

Literature Survey

The study focuses on the personalisation of web-based information services using a web usage mining-based recommendation system. The recommendation systems work on the basis of content-based filtering and collaborative filtering. The paper focuses on the issues of Content-based filtering as the difficulty of analyzing the content of web pages and arriving at semantic similarities. The Paper also states that collaborative filtering faces sparsity and scalability problems. Web usage mining can be used to extract knowledge from the web, rather than retrieving information. Moreover, web usage mining aims at discovering interesting patterns of use, by analyzing web usage data (Pierrakos et al., 2003).

Various techniques of web usage mining are surveyed in this paper. Web usage mining comprises data preprocessing, knowledge extraction from data and analysis of the extracted results. Preprocessing involves cleaning, session identification, user identification and path completion. In knowledge extraction, data mining techniques such as association rule mining techniques, clustering techniques, or classification techniques can be applied. In the next step knowledge extraction can be done with the help of the tools that transform the information into knowledge. Various techniques for analyzing the web usage data and are statistical analysis, association rules, clustering, classification, sequential patterns, and dependency modeling (Jain et al., 2017).

Preprocessing constitutes 80% of the mining process. An effective method of preprocessing for web usage mining has been proposed. The method used for user identification is a referrer-based method. This method identifies the unique user on the basis of the user's IP address, user agent and users operating system. Session identification is carried out by dividing the page access of every user into a different sessions. The method used for session identification is a combination of timeout mechanism and maximal forward difference mechanism (Reddy et al., 2014).

The study proposes an effective method for web usage mining. In preprocessing phase, path completion is one of the critical steps. The reasons for path incompletion are mainly, local cache, agent cache, post technique, proxy servers and browser "back" button. Due to these reasons, some entries in log files will be

[a]ekapurearnav@gmail.com; [b]aartimkarande@spit.ac.in

missing although the user had visited them. The proposed solution to this problem can be summarised as, if the page requested by the user is not directly linked with the previous page, then the referrer URL can be checked, and then it can be assumed that the user browsed back with the back button, with the help of cached session of the pages (Rao and Arora, 2017).

The study is focused on finding user patterns in users' web access patterns using web usage mining. It also highlights the drawbacks of association rule mining. The major drawback of association rule mining is that the numbers of rules generated are very high and not every rule is relevant. This limitation can be overcome with the help of clustering. Clustering reduces the input data set, therefore the number of rules is reduced and the probability of the rule being meaningful increases. While creating association rules time spent on the web page is not considered. To identify the interest level of the user, time spent on web pages is an important factor (Langhnoja et al., 2013).

The paper discusses various clustering algorithms and compares them. It mentions some expectations from clustering techniques so that the association rule mining can work efficiently. Those expectations are as follows clustering technique should be scalable, it should be able to work with the high dimensional dataset, it should be able to find clusters of random shapes, it should not be sensitive to outliers, it should not be sensitive to organisation of input records and obtained results from clustering should be understandable and functional (AlZoubi, 2019).

The paper compares various improved variations of the DBSCAN algorithm. Paper puts forward drawbacks of the standard DBSCAN algorithm. The drawbacks are as follows a) For executing the algorithm it requires user input. b) It is not capable of determining clusters in varying density datasets (Rehman et al., 2014).

HD_DBSCAN algorithm is proposed to overcome the issue of finding meaningful clusters in varying density datasets in a database. DBSCAN requires two input parameters epsilon-distance (Eps) and minimum points(MinPts). With the single value of the Eps parameter, it is not possible to discover meaningful clusters in varying density datasets. The proposed algorithm finds multiple epsilon distances to discover clusters in varying density datasets. According to the results the accuracy, as well as scalability, of HD_DBSCAN is much more compared to the traditional DBSCAN algorithm (Jahirabadkar and Kulkarni, 2014).

A recommendation system is proposed by combining collaborative filtering and pattern discovery algorithms. For finding navigational patterns an improved Apriori algorithm has been used. The traditional Apriori algorithm is modified by changing the minimum support value. In this algorithm, candidate itemsets are generated by considering the minimum support value as well as the time duration spent on each web page by the user. Improved Apriori performs well in terms of precision, applicability and hit ratio compared to the standard Apriori algorithm (Suguna and Sharmila, 2013).

HD_DBSCAN algorithm

As a solution to the problem of finding clusters in varying density datasets, the HD_DBSCAN algorithm was proposed. With the help of HD_DBSCAN epsilon-distance parameter can be determined adaptively and efficiently. The getEpsRadius algorithm helps to find multiple epsilon distances which leads to appropriate clustering in high dimensional data and in varying density datasets. The HD_DBSCAN algorithm finds begins with finding multiple epsilon distances for each dimension in the dataset using the getEpsRadius algorithm. The output of getEpsRadius is a set of epsilon distances in each dimension. Then density-based clustering is applied to form clusters using the derived epsilon distances. Figure 74.1 shows the HD_DBSCAN algorithm (Jahirabadkar and Kulkarni, 2014).

```
Algorithm HD_DBSCAN (𝒟ℬ, ε, μ, k)    // Finds clusters in a dimension using various ε-distances
cluster_id =0; E^S=Φ                  // E^S - Set of all ε-distances in all dimensions
n = size(𝒟ℬ);
d = no. of relevant dimensions;
for i = 1, d do
   E^S[i] = getEpsRadius(i);
end for;

For j = 1, | E^S[j]| Do      // For all ε-distances in each dimension
For k = 1, n Do
   p = 𝒟ℬ.get(j);
   If p.class = UNCLUSTERED Then
      If CoreObject(p, 𝒟ℬ, Cluster_id, E[k], μ) Then
         Cluster_id = Cluster_id + 1;
      End if;
   End if;
End for;
End For;
End; {HD_DBSCAN}
```

Figure 74.1 Algorithm HD_DBSCAN (Jahirabadkar and Kulkarni, 2014)

Apriori algorithm

Apriori Algorithm is used to find frequent itemset in a dataset. It can be categorised into sequential pattern mining algorithms.

Let X = $\{i_1, i_2, i_3, \ldots, i_n\}$ represents the set of all products/events concerned in this problem.

Let K = $\{i_1, i_2, i_3, \ldots, i_m\}$denote a set of transactions, with each transaction consisting of a collection of items. Any subset of the item base X can be referred to as itemset. The association rule, which can be represented as X => Y, denotes a relationship between two sets of elements, X and Y. If the fraction of transactions in K that contain both itemset X and Y reaches a specific threshold, called the support threshold, an association rule X => Y is considered a rule of interest. The percentage of transactions that include itemset Y among transactions that contain itemset X defines the confidence for the association rule X => Y (Yuan, 2017).

Proposed system

The proposed system is a new efficient web page recommendation system based on web usage mining. Refer to Figure 74.2 for the system overview. The proposed recommendation system follows the below-mentioned steps.

1. Data preparation.
2. Clustering the weblog files.
3. Pattern analysis and discovering an associative pattern.
4. Web page recommendation.

Data preparation

Data preparation includes the following steps: data collection, data cleaning, session identification, user identification, path completion, time spent calculation, and session duration calculation. The procedures involved in data preparation are depicted in Figure 74.3.

Data collection:

The server automatically generates the logs which record the user's activities and events on the website. The server-side log file has a substantial amount of information. Therefore, in this system, server-side log files are used. The server log file can be obtained from the organisation for which the recommendation system is to be developed. It is not accessible to normal internet users due to security concerns.

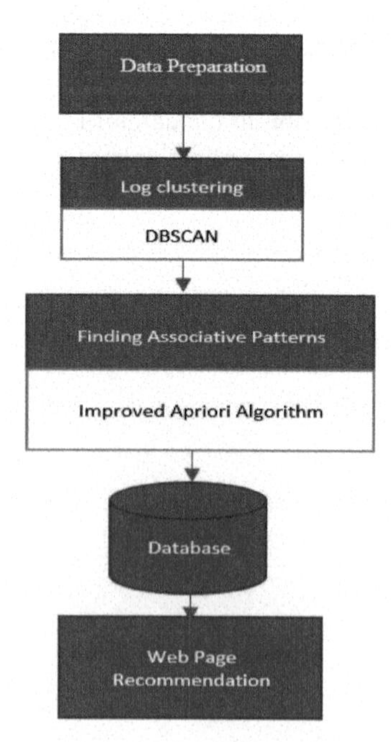

Figure 74.2 Generic recommendation system based on web usage mining

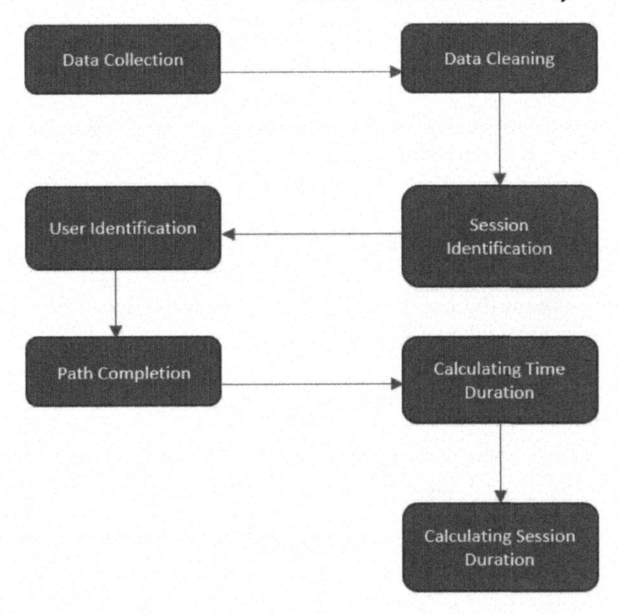

Figure 74.3 Data preparation process

Data cleaning:

The main objective of this process is to remove irrelevant data and outliers. The result of data cleaning is a data file that has the following columns: date, time, client IP, URL access, referrer URL, Operating system and user agent.

User identification:

This process is identifying each unique user accessing the website. The user identification is based on two fields client IP and user agent. The user is assumed to be a unique user if client IP and user-agent fields are combinedly unique. A new column named user_id is added to our data file. Each of the unique is assigned with the user_id.

Session identification:

This process is identifying each different user session accessing the website. All the records accessed by a unique user can be filtered out and grouped together on the basis of user_id. Also, it is necessary to distinguish between the different sessions of the same user. A new column is added to the data file called session_id. Every new session will be provided with a unique session_id. following algorithm is followed for the identification of a new session. The following algorithm is used for session identification

Algorithm Name: SessionIdentification
If (User is new)
    Then (It is a new session)
If (referrer URL field is NULL)
    Then (It is a new session)
If (The time between page requests exceeds 30 minute)
    Then (It is a new session)

All the sessions of the users are identified and they are grouped together on the basis of session_id in the data file.

Path completion:

Proxy servers, cache, and back button usage results in missing log entries. Therefore, it is a tedious task to find the actual navigation path of the user in a particular session. So to create a path that is close to the original path we use the following algorithm:

Algorithm Name: PathCompletion
If (The requested page is unlinked to the last page visited by the user.)
Then

(/*find the page from which the request has been made to the requested page*/
If(the page is present in the referrer field)
Then
(The user has backed using the back button, bringing up the pages present in the cache till a new page is requested. All the records of cached versions of pages are added above the record of the requested page)
Else
(Assumed to be jumped from the last page to the requested page.)
)

As an output of this stage, the data file will contain the log records arranged in the order of navigation path close to the path followed by the user in the particular session.

Time spent on web page:

A new column is added to the data file called Time_spent which refers to the time spent on the webpage by a user. Comparing the time field of the current record and the next record time duration spent on the particular page can be calculated by the following formula.

If there are n records then, time spent for the ith record can be calculated as,

Time_spent on ith record = Time of (i+1)th record - Time of ith record

The data file is now ready for the session duration calculation step.

Session duration:

A new column is added to the data file called session_duration which refers to the time duration of a complete session. Session_duration can be calculated by comparing the timestamp of first record and the last record of the particular session.

If there are n records in a particular session then, the session duration of the ith session can be calculated as,

session_duration of ith session = (time of nth record in ith session) – (time of 1st record in ith session)

The above formula is applied to each session. Then the data file is sent for the clustering step.

Clustering

Clustering is used to group users with similar browsing behaviour. Post Data preparation step the file will contain following five fields: user_ID, URL, Time_Spent, session_ID and session_duration.
where,

user_ID = user_id = the unique user id,

URL = web address of web page,

Time_Spent = time spent on web page

session_ID = session_id = the unique session id

session_duration = session duration of the user

HD_DBSCAN algorithm is used for clustering. As a input it takes DB and μ.

DB = Database containing our preprocessed log file.

No. of dimensions = 5

μ = (No. of dimensions) × 2 = 5 × 2 = 10

Now, it checks for the quality of dimensions. If the clustering quality of dimension is greater than 0.4, then dimension is considered as a relevant dimension and considered for further process. If the clustering quality of the dimension is less than 0.4, then that dimension is considered irrelevant. Only relevant dimensions are allowed to proceed further.

For each dimension, the getEpsRadius algorithm is applied with the same input parameters DB and μ. The getEpsRadius returns the set of epsilon distances for the particular dimension. E^s is a set of all epsilon distances in all the dimensions. All returned values from the getEpsRadius algorithm are stored at E^s. Now, clusters are formed for each epsilon distance in E^s for every record which is left unclustered. The clustered data is sent to the next step for pattern discovery.

Discovering associative pattern

On each cluster improved Apriori algorithm will be applied. The association rules have the basic fundamental of support and confidence.

Support: It tells how popular a particular item set is, measured by the proportion in which it seems.

The improved Apriori algorithm is modified in support value. This algorithm takes into consideration time spent on each web page to be greater than or equal to the minimum specified time along with the minimum support which is a threshold value. Support for each rule is calculated by using (1).

$$support(A->B) = \frac{\begin{array}{l} Number\ of\ tuples\ containing\ both\ A\ and\ B \\ and\ time\ duration\ spent\ on\ A\ and\ B \\ greater\ than\ or\ equal\ to\ the\ specified\ time \end{array}}{Total\ number\ of\ tuples} \quad (1)$$

Confidence: value that represents the reliability of that rule. This also has minimum confidence which is the threshold value. Confidence for each rule is calculated by using (2).

$$confidence(A->B) = \frac{Number\ of\ tuples\ containing\ both\ A\ and\ B}{Number\ of\ tuples\ containing\ A} \quad (2)$$

Rules formed are stored in the database with their support and confidence values.

Web page recommendation

Whenever a user enters, the sequence navigation path of that user is compared with the associative rule stored in the database and if the rule exists for the navigation pattern then the next web page can be predicted and recommended.

Case study

This case study is done on log files of vlab of Sardar Patel Institute of Technology, which is an online course providing platform. The case study follows the following steps: Data preparation, log clustering, forming association rules and web page recommendations.

Data preperation

The data preparation for the case study as per the following steps.

Data collection:

The server log file is collected from vlab of Sardar Patel Institute of Technology, which is an online course provider website. The records present in the log file be 100.

Data cleaning:

In data cleaning process, all records except the status code between 200 to 299 are removed. The File names with extension .jpg, JPEG, CSS, .gif are removed. Also, the records requesting robots.txt are removed. The result of data cleaning is a data file that contains 30 records. These records have the following fields date, time, client IP, URL access, referrer URL and user agent.

User identification:

The user identification is based on two fields client IP and user agent. In the log files, client IP and user-agent are checked and if the combination of these two is unique then the user is assigned a unique user_ID. Four unique combinations of client IP and user-agent are found. This implies that there are four unique users in the case study.

Session identification:

All the records accessed by a unique user can be filtered out and grouped together on the basis of user_id. In the case study, there are four groups. The SessionIdentification algorithm is applied for distinguishing the different sessions of the user. Sixth sessions are identified. Second sessions for first user and first session each for the rest of the users. Each unique session is provided with a unique session_id. As an output, the data file will contain all the records of the users grouped together on the basis of session_id in the data file.

Path completion:

Isn one of the sessions of 1st user, a few of the pages are unlinked to the last page visited by the user. Then the Path_Completion algorithm is applied to all such records to identify the missing records and fill them in at the right place.

This step is executed on all 6 sessions. As an output of this stage, the data file will contain all the session records ordered close to the navigation path followed by the user.

Time spent on web page:

An additional field Time_spent is created to store the time spent on each web page in the data file. For all of the 30 records, time spent on each page is calculated using the formula explained in section 3.1.6. The data file is taken into account for the following stage, which is the session duration calculation.

Session duration:

An additional field session duration is created to store the session duration in the data file. For all six sessions, the time duration of the session is calculated using the formula mentioned in section 3.1.7. The data file is taken into account for the following stage, which is log clustering.

Log clustering

HD_DBSCAN algorithm is applied for clustering with parameters $\mu = 10$ and DB = database containing the datafile. The algorithm will return a data file with the clusters. The derived clusters are then passed for pattern discovery.

Results and Analysis

Clustered data is further used to identify associative patterns. An improved Apriori algorithm is used to determine the rules. For example, if a user spent 3 minute on the 'cyber security and ethical hacking' course page and then 2 minute on 'cryptography' then it is very likely that the same person will spend 3 minute on the 'ethical hacking using Python' page. The following transaction is written in the form [Transaction, Web Page, Time Spent].

[T, Cyber Security and Ethical Hacking, 3] ^ [T, Cryptography, 2] => [T, Ethical Hacking using Pytho]n, 3]

(support = 4%, confidence=80%, minimum time spent = 1 min).

A total of 80% of transactions that contains cyber security and cryptography also contain ethical hacking using Python. 4% of all transactions contain all of the three items. While calculating support time spent on the web page is also checked. 3, 2, 3 are time spent on cyber security, cryptography and ethical hacking using Python respectively. All of the time spent is greater than 1 minute so the rule will get accepted depending on the minimum support value and minimum confidence value. All the six rules formed along with the support value and confidence value will be stored in the database.

When a new user enters, the sequence navigation path of that user is compared with the associative rule. If any of the rules matches the sequence navigation pattern of the user then the page is recommended.

Conclusion and Future Work

The proposed generic recommendation system is solely based on the knowledge extracted from users' navigation patterns and time spent on the web pages. The generic recommendation system is based on the web usage mining concept. HD_DBSCAN clustering technique is used to determine multiple epsilon-distance parameters and to identify clusters in varying density datasets effectively. For pattern identification improved Apriori algorithm was used by modifying the traditional Qpriori algorithm in its support value. The proposed system is cost-effective and efficient compared to traditional content-based and collaborative filtering techniques. The generic recommendation system is easy to implement and can be used as a general recommendation system in multiple domains such as e-commerce, media, insurance, telecom, transport, etc. The system is lacking in a recommendation model that can recommend web pages based on already stored association rules in real-time.

Acknowledgement

Sardar Patel Institute of Technology vlab team for giving their log files. We are grateful for their support. https://vlab.spit.ac.in/ci/

References

AlZoubi, W. A. (2019). A survey of clustering algorithms in association rules mining. Int. J. Comput. Sci. Inf. Technol. 11(2):17–25. doi:10.5121/ijcsit.2019.11202.

Jahirabadkar, S. and Kulkarni, P. (2014). Algorithm to determine ε-distance parameter in density-based clustering. Expert Syst. Appl. 41(6):2939–2946. doi:10.1016/j.eswa.2013.10.025.

Jain, S., Rawat, R., and Bhandari, B. (2017). [IEEE 2017 International conference on emerging trends in computing and communication technologies (ICETCCT) - Dehradun, India (2017.11.17-2017.11.18)] 2017 International conference on emerging trends in computing and communication technologies (ICETCCT) - A survey paper on techniques and applications of web usage mining, (pp. 1–6). doi:10.1109/ICETCCT.2017.8280343.

Langhnoja, S. G., Barot, M. P., and Mehta, D. B. (2013). Web usage mining using association rule mining on clustered data for pattern discovery. Int. J. Data Min. Technique Appl. http://iirpublications.com. ISSN: 2278–2419.

Pierrakos, D., Paliouras, G., Papatheodorou, C., and Spyropoulos, C. D. (2003). Web usage mining as a tool for personalisation: A survey. 13(4):311–372. doi:10.1023/a:1026238916441.

Rao, R. S. and Arora, J. (2017). A survey on methods used in web usage mining. Int. Res. J. Eng. Technol. (IRJET), 4(5): e-ISSN 2395 -0056. p-ISSN 2395-0072.

Reddy, K. S., Varma, G. P. S., and Reddy, M. K. (2014). An effective preprocessing method for web usage mining. Int. J. Comput. Theory Eng. 6(5).

Rehman, S. U., Asghar, S., Fong, S. and Sarasvady, S. (2014). [IEEE 2014 Fifth international conference on the applications of digital information and web technologies (ICADIWT) - Bangalore, India (2014.2.17–2014.2.19)] The Fifth international conference on the applications of digital information and web technologies (ICADIWT 2014) - DBSCAN: Past, present and future. , (pp. 232–238). doi:10.1109/ICADIWT.2014.6814687.

Suguna, R. and Sharmila, D. (2013). An efficient web recommendation system using collaborative filtering and pattern discovery algorithms. Int. J. Comput. Appl. 70:37–44. 10.5120/11945-7755.

Yuan, X. (2017). AIP conference proceedings [Author(s) 11TH Asian Conference On Chemical Sensors: (ACCS2015) - Penang, Malaysia (16–18 November 2015)] - An improved apriori algorithm for mining association rules. 1808,080005. doi:10.1063/1.4977361.

75 Fabrication of hydrophobic textured surface using reverse micro EDM (RMEDM)

Sagar R. Dharmadhikari[1,a], *Sachin A. Mastud*[2,c], *Shital V. Bharate*[2,d], *Aditi K. Dhale*[2,e], *and Saurav R. Yadav*[3,f]

[1]Department of Production Engineering, Veermata Jijabai Technological Institute, Mumbai, India

[2]Department of Mechanical Engineering, Veermata Jijabai Technological Institute, Mumbai, India

[3]Department of Mechanical Engineering, Institute, Mumbai, India

Abstract

Development of functional surfaces with a capability of self-cleaning and water repellency (hydrophobicity) has the prime importance in advance engineering application. Fabrication of such functional surfaces requires micro texture on surfaces. These textures can be machined via micromachining processes. In this paper, the experimental study has been performed to fabricate textured surfaces using Reverse Micro EDM process on Brass workpiece of 10 mm diameter. The effect of Voltage, Capacitance and Feed rate on Machining time has been studied using the Taguchi L9 Orthogonal Array. The surface topography of fabricated micro textures are analysed using SEM followed by contact angle measurement using Digi drop method. The parametric study shows the Voltage is a prime factor affecting the machining time. The maximum contact achieved is about 137.8°.

Keywords: Contact angle, micro EDM, textured surfaces.

Introduction

In nature many examples can be found exhibiting the state of super hydrophobicity such as Lotus Leave, Colocasia, esculanta, dahlia, legs of strider, butterfly wings, etc. (Barthlott and Neinhuis, 1997; Bhushan and Jung, 2006; Gao and Jiang, 2004; Sun et al. n.d). The microstructural studies of the plants reveal the surface characteristics which shows the protruded hierarchical micro and nano pillars with coating wax. The recent years development of arrayed micro textures on the metallic surface to exhibit the self-cleaning and water repellency properties has gain the importance. Surface texture enabled many industrial applications like water repellant textiles, reduction in hydrodynamic friction, and water repellant capacity of solar panel, photovoltaic cell, etc. have been developed. Fabrication of pattern of micro dimples or micro pillars alters the surface energy. Altering the surface properties highly influencing the wettability of the surfaces. Inspiring from the nature, developing the hydrophobic surface artificially is a newer research trend. Many researchers have tried to create the hydrophobic surface artificially using various manufacturing techniques. These functional surfaces can be prepared by additive or subtractive type of manufacturing viz. applying the coating on the surfaces or removing the material from the surface to create the micro pattern on the surface.

The angle created by a liquid molecule with a solid and air is called as Contact angle. The wettability of the surface can be defined by Wenzel state or Cassie-Baxter state of the liquid drop. Zhenyu et al. (2016) had fabricated the concave and convex surface using micro milling process on PMMA and Ti6Al4V and found improved contact angle as compared to smooth surface. Zhang et al. (2016) had studied the effect of micro pillars on the wettability using VOF method in fluent. The study reveals that the pillar height is a critical parameter which governs the wettability of the surface. Roy et al. (2020) had developed mathematical model and validated it by fabricating the array of micro pillars with different geometrical parameters using DXRL, Micro wire EDM and wire wound method on brass substrate. The study reveals that the micro pillar size with 30 μm exhibits superhydrophobic surface and the ratio of pillar size to spacing will largely affect the contact angle and surface with contact angle of 153° has been successfully achieved. Nikam et al. (2021) fabricated the array of micro dimples of various cross section with variable dimple density on HSS disk using micro laser texturing. They have found that micro dimples with 10% dimple density and dimple area of 0.01 mm to exhibit maximum contact angle of 112.6°

Singh et al. (2018) had experimentally investigated the effect of various micro pattern fabricated on polymers and metallic surfaces using laser texturing with varying ariel density on coefficient of friction and wettability. The study reveals that micro pillars of height 30 μm and ariel density 640 pillars/mm^2 gives low coefficient of friction and high contact angle of 140°. Apart from these, numerous researchers tried to

[a]sdharmadhikari.88@gmail.com; [b]samastud@me.vjti.ac.in; [c]svbharate_b19@me.vjti.ac.in; [d]akdhale_b19@me.vjti.ac.in; [e]sryadav_b19@me.vjti.ac.in

mimic lotus leaves hierarchical texturing to improve the surface roughness and transiting the surfaces from hydrophilic to superhydrophobic. Bhushan et al. (2008) studied the hysteresis, static contact angle and tile angle by fabricating micro, nano and hierarchical textures using self-assembly of alkanes and epoxy resin. Study reveals the formation of air pockets due to hierarchical textures greatly influences the hydrophobicity of the surface. Fu et al. (2020) using laser marking treatment process prepared hierarchical structures in the form of Micro strip protrusion and grooves on silicon steel surface to study surface characteristics and wettability. Finding of the study reveals that the prepared surface has the capability of retaining in super-hydrophobic for longer duration with hardness. Prasad et al. (2021) had developed mathematical model to find the effect of hierarchical textures on the wettability. The model then validated experimentally by preparing the samples using diamond turning method. They concluded that the creation of minor pillars on the surface of major pillars (Hierarchical structures) results in increased in Contact angle.

Literature has been revealed that the preparation of patterned surfaces can be achieved via different micro manufacturing process and texturing of surface can be used to derive useful functionality from surface. In this study, the arrayed micro features were fabricated using the RMEDM process. RMEDM a process is a variant of MEDM and capable of creating any complex shapes on conductive metals with higher accuracy. Also, it is more economical than the other process like LIGA and wire electrical discharge machining (WEDM).

Experimental setup and method

In present study the attempt is made to fabricate the array of micro pillars with size 200 µm and height of 150 µm on brass rod of 10 mm diameter using RMEDM (reverse micro EDM) and process parameters effect on machining time is analysed. The work piece of 10 mm dia and 20 mm in length were cut from bulk copper rod using CNC machine and then surface is ground and polished to make it flat. To conduct experiments, Synergy Nano systems Hyper – 15 (table top) micro EDM setup was used as shown in Figure 75.1. Experiments were designed and performed using Taguchi L9 Orthogonal Array. During Micro EDM machining, the spark energy Equation 1 is a critical parameter which governs the material erosion and stability of the process. Another influencing parameter which plays an important role for smooth machining is the evacuation of debris from spark gap. While machining during the RMEDM the effective debris evacuation is a challenging phenomenon as the spark gap lies between few microns range.

$$E = \frac{1}{2} C V^2 \quad (1)$$

Where V – Voltage (volts)
C – Capacitance (pF)

As shown in Figure 75.2, workpiece is mounted on collate and tool is placed at table which fed downwards during machining. During the downward movement of workpiece towards the tool electrode reaching the minimum inter electrode gap initiates the sparks. The initiation of sparks generates the craters on the surface of workpiece which removes the material and to wipe it out from the spark gap is takes place by flushing the dielectric. The movement of tool electrode through the cavities which removes the material and extruded portion were fabricated. For machining the voltage (V), capacitance (C) and feedrate (F) are considered as a control parameter. To evaluate the effect of process parameters on Machining time the depth of machining kept constant to 150 µm and average time for machining is evaluated with recording the machining time for every 50 µm intervals. Table 75.1 shows the Taguchi L9 orthogonal array with

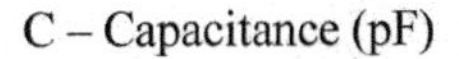
C – Capacitance (pF)

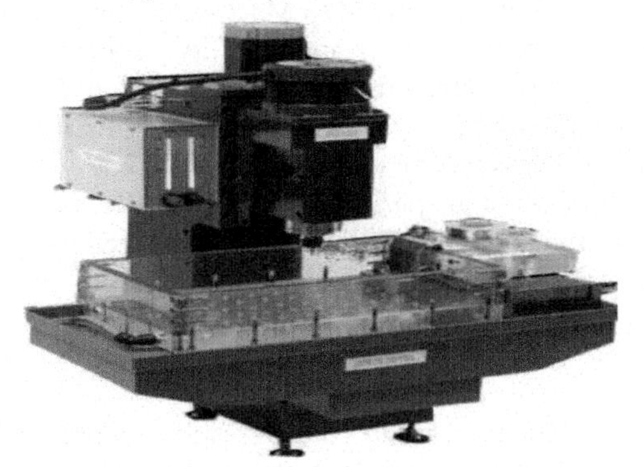

Figure 75.1 Actual photograph of (a) RMEDM setup

Figure 75.2 Electrode positions while machining

Table 75.1 L9 orthogonal array experimental parameters

Details of work piece and tool material			
Work piece: Brass rod of 10 mm diameter			
Tool: - SS316 wire mesh of 200 µm			
Processing Parameters and their levels			
Parameter	*Level – 1*	*Level – 2*	*Level – 3*
Voltage (V)	70	90	110
Capacitance (pF)	100	1000	10000
Feed rate (µm/s)	5	10	15
Dielectric: - EDM oil			
Response parameter: Machining time and hydrophobicity (contact angle)			

setting of parameters used and objectives. The statistical design was carryout and analysed using Minitab 17 statistical tool and the analysis of variance (ANOVA) is performed to evaluate the influencing parameters during machining. For surface characterisation the SEM images were captured. Then the fabricated arrayed micro structures were analysed using Digi drop method and evaluated the contact angle (CA) for the texture surface.

Results and Discussion

Experimental investigation evidenced the effectiveness of proposed method for fabrication of arrayed features for creation of hydrophobic surface. Since the target of the study is to produce arrayed features on larger surface the aspect ratio is not a concern of the study. The experiments were carried out as per L9 orthogonal array and ANOVA is performed to evaluate the critical factor of machining.

Analysis of machining time

Time required for transforming the bulk rod into finish array of structure called as Machining Time. During the machining short circuiting phenomenon is affecting the machining time and stability of the process. In, general the feedrate controls the idol time of machining i.e non-machining time. Statistical analysis to identify the crucial process parameters having effect on machining time was carried out using Analysis of Variance (ANOVA) and Analysis of Means (AOM). Results of ANOVA are listed in Table 75.2.

From the machining time analysis, it is found that, with increase in spark energy time gets reduce. The ANOVA results show that the voltage is a significant factor affecting the machining time followed by capacitance. The Figures 75.4 and 75.5 shows the main effect and interaction plot for machining time.

Effect of voltage (V)

From ANOVA and main effect plot it was observed that with the increase in voltage results in continuous reduction in machining time. At the lower value of 70 V interelectrode gap is very short which obstruct the

Table 75.2 L9 ANOVA for machining time

Factors	*DF*	*Adj SS*	*Adj MS*	*F*	*P*
Voltage	2	31.85	15.92	110.41	0.009*
Capacitance	2	5.38	2.69	18.68	0.051
Feed rate	2	0.13	0.065	0.46	0.687
Error	2	0.28	0.144		
Total	8	37.66			

*Indicates statistically significant factor at 95% confidence level

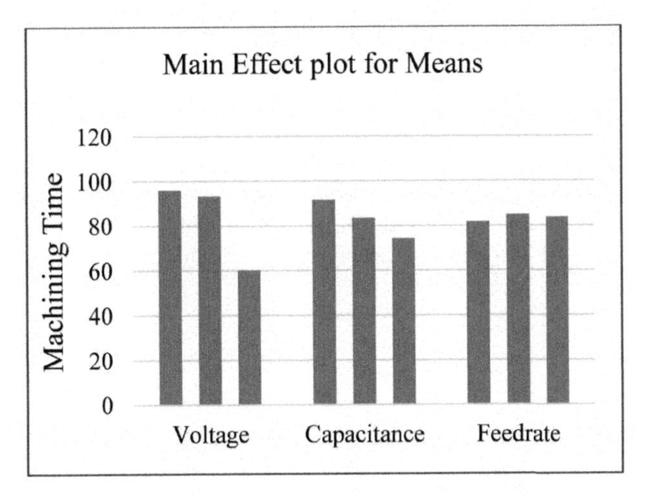

Figure 75.3 Main effect plot of voltage, capacitance and feedrate

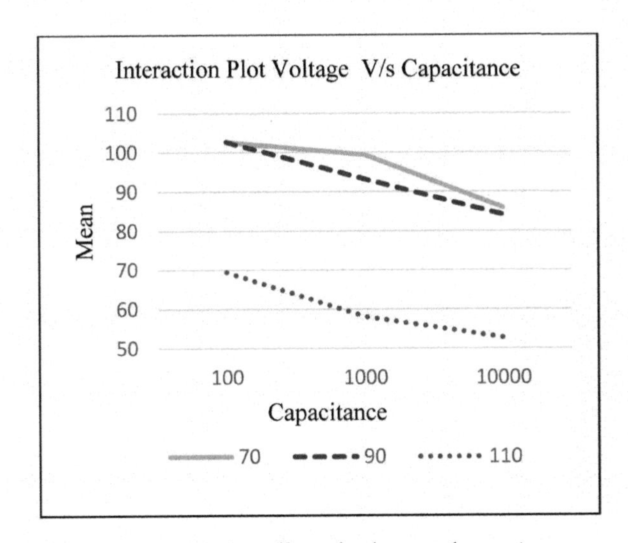

Figure 75.4 Interaction effect of voltage and capacitance

flow of debris and increasing the machining time. At the lower voltages frequent short circuiting occurs which retracted workpiece away from the tool electrode. The inefficient debris flow from IEG results in resolidifying of debris to the machine features was also observed. At 70 V the spark energy produce is not sufficient for expelling the debris resulting the undesired arcing between the tool electrode and workpiece during retraction of workpiece hindering the quality and accuracy of desired surface fabrication. The literature cited also depicted that higher the values of spark energy lower the machining rate with better surface quality due to efficient removal of debris from the gap. Similar trend was observed during machining. At 110 V the time required for machining was observed between 1 to 1.3 hours whereas at 70 V it was 2 to 3 hours. ANOVA and mean effect plot depict that the Voltage is most significant factor as compare to other factors.

Effect of capacitance (C)

It is seen from Figures 75.3 and 75.4 that the capacitance is contributing more after voltage. It was also observed that the machining time is gradually decreasing with the increase in capacitance from 100 pF to 10000 pF. As the micro EDM process is RC circuit base machining the capacitance controls the charging and discharging time. From interaction plot it was observed that the with increase in voltage as well as capacitance from 70 V and100 pF to 110 V and 10000 pF resulting in drastic reduction in machining time.

Effect of feedrate (F)

During the lower energy settings of voltage and capacitance retraction of tool electrode is common phenomenon. These non-machining movement of tool is known as idol time of machine which can be control by feedrate. ANOVA results shows that the effect of feedrate is limited to less than 1% clearly defining the almost negligible effect of feedrate on machining and surface quality of workpiece.

Analysis of surface morphology

After performing the ANOVA to check and study the surface morphology of machine surface scanning electron microscopic images were taken upto (2000 x) magnification level for tip and root of the machine surface. It was observed that the surface created is very rough with lot of molten material restick to the surfaces of micro rods. Also, the debris particles were stuck inside the cavity and resolidified (Figure 75.6).

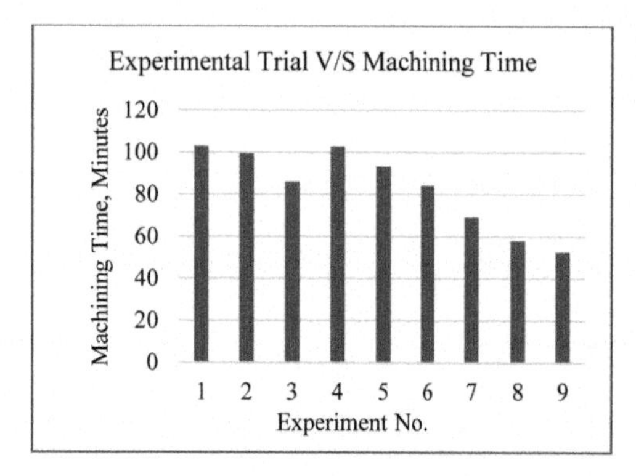

Figure 75.5 Machining time v/s experiment

Figure 75.6a SEM of tool electrode

Figure 75.6b Morphology of fabricated rods at 75 x magnification

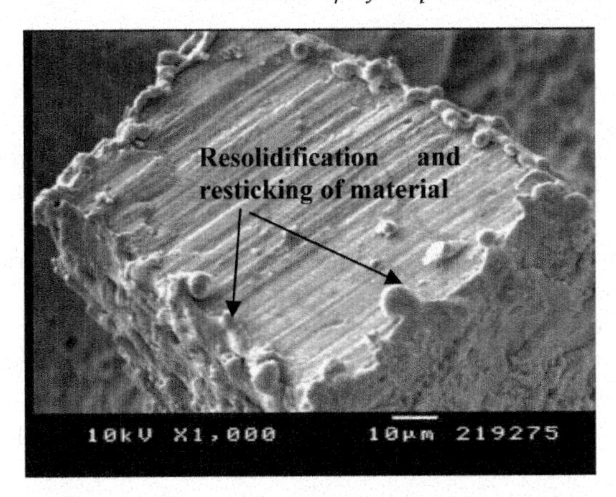

Figure 75.6c Morphology of fabricated single rod at 1000 x magnification

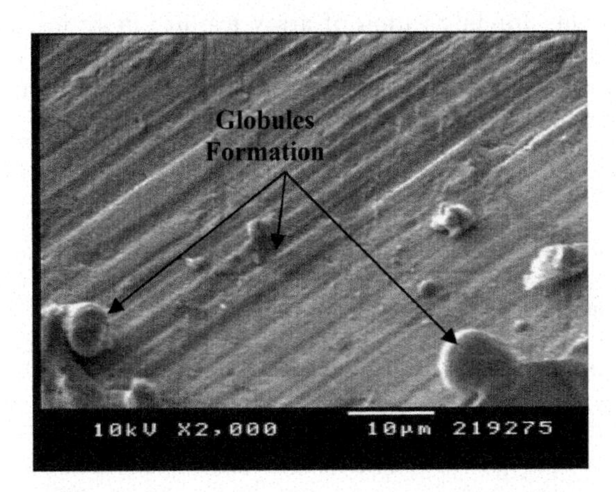

Figure 75.6d Morphology of fabricated single rod at 1000 x magnification

Analysis of hydrophobicity

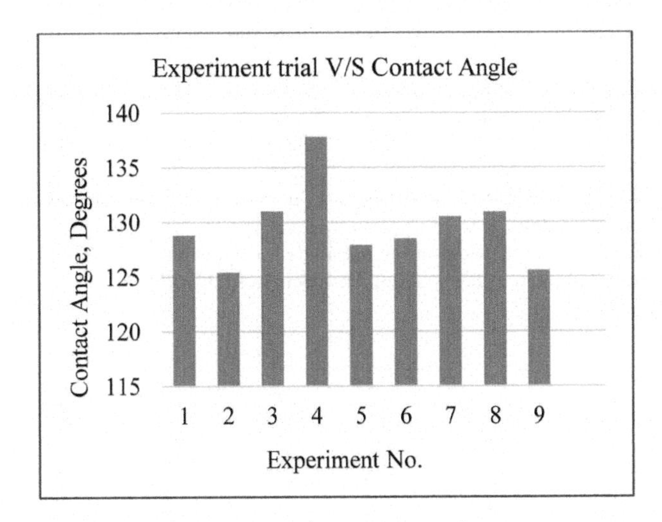

Figure 75.7 Experiment trial v/s contact angle

After studying the surface morphology of the samples then observed for checking the contact angle of the surface textures. Digi drop method was employed to evaluate the effect of process parameters on water contact angle. From Figure 75.7 it was observed that maximum contact angle achieved was 137.8°. It can be seen from Figure 75.7 the contact angle varies between 125.4° to 137.8°.

Conclusion

In this paper details experimental investigation is carried out to study the effect of process parameters on machining time. Surface texture is successfully created on larger surface of 10 mm diameter workpiece. Also, the fabricated samples are then investigated using SEM to study the surface morphology followed by testing hydrophobicity of the texture surface. The following observations have been drawn from the study.

1. Bulk rod of 10 mm diameter had successfully machine to form arrayed micro structures with average height of 150 μm height using RMEDM process.
2. Voltage is a most influential at all the depth of machining structures followed by capacitance.
3. SEM images have shown the surface created with globules formation and sticking of debris particle to the machined features for lower parameter settings of 70 V and 100 pF capacitance.
4. Textured surfaces have exhibited a hydrophobicity with a contact angle of 137 degree with a drop of water.

Future Scope

Though the RMEDM process has shown its capability for fabrication of array features more insights required for fabrication of irregular shapes using the said techniques. Also, the development of effective flushing techniques can be improved the process capability. The feasibility of machining can be further evaluated to developed the micro textures on cylindrical surface.

References

Barthlott, W. and Neinhuis, C. (1997). Purity of the sacred lotus, or escape from contamination in biological surfaces. Planta 202(1):1–8.

Bhushan, B. and Jung, Y. C. (2006). Micro- and nanoscale characterisation of hydrophobic and hydrophilic leaf surfaces. Nanotechnology 17(11):2758–2772.

Bhushan, B., Koch, K. and Jung, Y. C. (2008). Biomimetic hierarchical structure for self-cleaning. Appl. Phys. Lett. 93:093101.

Fu, J., Tang, M. and Zhang, Q. (2020). Simple fabrication of hierarchical micro / nanostructure superhydrophobic surface with stable and superior anticorrosion silicon steel via laser marking treatment. J. Wuhan. Univ. Technol. Mater. Sci. Ed. 35:411–417.

Gao, X. and Jiang, L. (2004). Water-repellent legs of water striders. Biophys. Nat. 432(7013):36–36.

Nikam, M., Roy, T. and Mastud, S. (2021). Wettability analysis of hydrophobic micro-dimpled HSS Surfaces. J. Inst. Eng.

Prasad, K. K., Roy, T., Goud, M. M., Karar, V., and Mishra, V. (2021). Diamond turned hierarchically textured surface for inducing water repellency: Analytical model and experimental investigations. Int. J. Mech. Sci. 193:106140.

Roy, T., Sabharwal, T. P., Kumar, M., Ranjan, P. and Balasubramanian, R. (2020). Mathematical modelling of superhydrophobic surfaces for determining the correlation between water contact angle and geometrical parameters. Precis. Eng. 61:55–64.

Singh, A., Patel, D. S., Ramkumar, J. and Balani, K. (2018). Single step laser surface texturing for enhancing contact angle and tribological properties. Int. J. Adv. Manuf. Technol.

Sun, M., Watson, G. S., Zheng, Y., Watson, J. A. and Liang, A. (2009). Wetting properties on nanostructured surfaces of cicada wings. J. Exp. Biol. 212(19):3148–3155.

Zhang, W., Zhang, R. R., Jiang, C.G. and Wu, C. W. (2016). Effect of pillar height on the wettability of micro textured surface: Volume-of-Fluid simulations. Int. J. Adhes. Adhes.

Zhenyu, S., Zhanqiang, L., Hao, S. and Xianzhi, Z. (2016). Prediction of contact angle for hydrophobic surface fabricated with micro-machining based on minimum Gibbs free energy. Appl. Surf. Sci. 364:597–603.

76 A comparative analysis of different methods for predicting speech in a noisy signal

Keerti Kulkarni[a] *and P. A. Vijaya*

Department of ECE, BNM Institute of Technology, Bangalore, India

Abstract

A noisy speech signal is very difficult to decipher. The noise can be removed or the signal component can be enhanced using different speech processing algorithms. Alternatively, Neural Networks can also be used to predict the signal in a noisy background. A comparative analysis of the signal processing algorithms and the Multi-layer Perceptron model for the speech prediction in the presence of noise is shown in this work. The value of the signal in a noisy environment is predicted using traditional signal processing algorithms such as Least Mean Square Algorithm and the Wiener Algorithm. Their performance is compared to the results obtained with a three-layer neural network. The performance metrics used are Signal to Noise Ratio (SNR) and the Perceptual Evaluation of Speech Quality (PESQ). It is found that the Multi-layer Perceptron model provides a better SNR (19.13 dB) and PESQ score (2.9) compared to the Least Mean Square filter (SNR = 12.31 dB and PESQ = 2.2) and the Wiener filter (SNR = 9.02 dB, PESQ = 2.3). At the same time the computational time taken for the 3-layer Neural Networks (20 seconds) is more than the time taken by both the LMS (5 seconds) and the Wiener filter (9 seconds). It is concluded that the Multilayer Perceptron model performs better than the other two algorithms for the given speech signal corrupted by noise.

Keywords: Least mean square algorithm, multi-layer Perceptron, wiener filter.

Introduction

Noise is an unwanted signal which adds itself to the signal of interest and distorts it. Clarity and the intelligibility of the speech signal can be improved by removing this noise. Many a times, if the noise is very high it becomes very difficult to extract the speech signal components from the noisy environment. A majority of the noise in the speech signal is added in the communication channel. Some noise is also because of the electronic equipment used. Predicting the presence of the signal in a noisy environment can be done either by suppressing the noise or by enhancing the signal component. Adding artificially generated components to the noisy signal so that the noise is cancelled out, is one regularly used process. The speech enhancement techniques can either be based on statistics of the signal (e.g., least means square (LMS) filter and the Wiener filter). These algorithms have been traditional used for speech processing. The problem with this approach is that these algorithms assume a particular probabilistic model and also the fact that the spectral components are highly uncorrelated. But the fact is, the spectral components are correlated. They also require an estimate of the mean and variance of the clear speech signal. This is where the data driven approaches like neural networks (NN) come into picture. Neural networks do not make any assumptions about the underlying structure of the input data. They extract the features from the data itself and predict the future sample depending on the feature extracted. Due to the very random nature and a broad spectrum of the noise that gets itself added to the signal, it is difficult to pinpoint the algorithm that is optimal. The literature surveyed offers a variety of solutions to the problem, some involving algorithms explicitly for speech processing and some others using the generalised machine learning approach. Though all the results are satisfactory, a comparison between the specific speech processing algorithms and the generalised machine learning algorithms has not been done. This work aims to bridge this particular gap by comparing the traditional speech processing algorithms with the machine learning ones.

Literature Survey

LMS has been traditionally used for the estimation of the signal in the noisy environment (Washi et al., 2006; Gupta et al., 2015). The performance is evaluated on both the male and female speech signals

[a]keerti_p_kulkarni@yahoo.com

corrupted by white noise and pink noise. LMS filters have also been used for channel estimation (Gui and Adach, 2013) and beam forming (Jalal et al., 2020). Extracting the speech in the presence of wind using different algorithms like LMS, NLMS and KLMS has been done by (Murugan and Natarajan, 2012). The variation in the performance of the LMS with respect to the step size is analysed by Thunga and Muthu (2020). The performance of the wiener filter with recursive noise estimation is done by Upadhyaya and Jaiswal (2016). Using the wavelets within the wiener filtering improves the filtering performance as shown (Smital et al., 2013). Neural network classifiers is used for the recognition of speech in the presence of noise Hadjahmadi et al. (2008) and it is found to perform better than the traditional methods. Deep learning methods have been used to enhance the speech quality by environmental noise removal (Park et al., 2020). The comparison between the deep learning and denoising autoencoders has been done by the researchers Nossier et al. (2020). Constellation diagrams manifest themselves differently in different SNRs and hence can be used for the estimation of SNR. Xie et al. (2020) have used constellations diagrams based on deep learning for the estimation of SNR. A combination of CNN and LSTM can also be used to predict the SNR in the LTE and 5G systems (Ngo et al., 2020). A comparison between the traditional filtering algorithms and the filtering based on neural networks is done by (Rajini et al., 2019). An adaptive ADALINE NN is designed in MATLAB. Along with the algorithms and methodologies themselves, due importance has to be given to the evaluation techniques. The evaluation of the performance of the algorithms can be subjective or objective. Human listening tests carried out by a panel of listeners can be used for the subjective analysis. The results of such analysis were reported by Hu and Loizou(2006) but it is accurate and reliable under the most stringent conditions. It is also costly and time consuming. Hence there was the need for objective quality measurements (Quackenbush et al., 1988).Several speech quality objective measures like segmental SNR (segSNR) (Hansen and Pellom, 1998) and perceptual evaluation of speech quality (PESQ) which is the ITU-T standard for objective analysis of the speech quality Rix et al. (2001), have been implemented and tabulated. Higher the value of the PESQ score, better is the speech quality.

Methodology

The methodology adopted for implementing the proposed work is shown in Figure 76.1. The signal considered here is a clip of a speech signal captured from a telephonic recording. The modern phone provides enough noise cancellations, hence there is absolutely no noise in the speech signal used. Hence, random noise is added to the signal so that it can be experimented upon. Three different algorithms are then used, least mean square (LMS) Filter, Wiener filter and a multi-layer perceptron (MLP) which is a thre-layer 'vanilla' neural network. The evaluation metric for comparing the performance is the signal to noise ratio (SNR) and PESQ. The time required for executing the algorithm is also an important parameter considered here for the comparison.

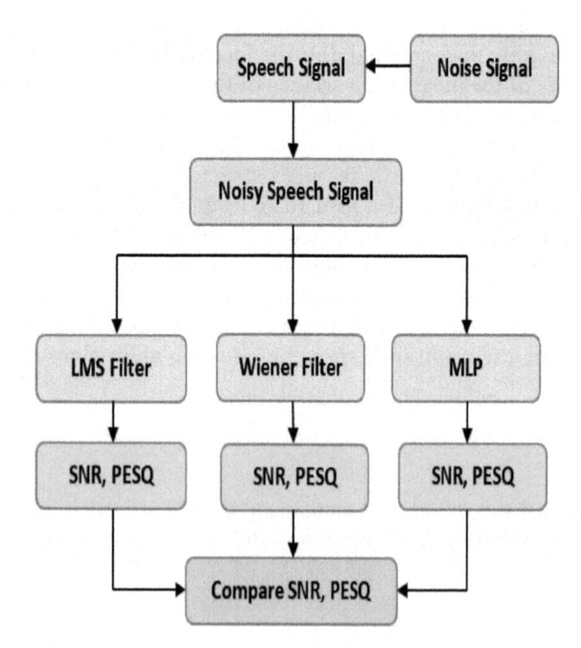

Figure 76.1 Speech signal and noisy speech signal

A. Least mean square filter

One of the most commonly used Adaptive filters is the LMS filter. The coefficients of the adaptive filters regulate themselves depending on the incoming signal. The transfer function of the LMS can be varied by changing the parameters. LMS employs the stochastic gradient descent algorithm for estimating the parameter values. As shown in Figure 76.2, the filter coefficients are directly proportional to the MSE between the desired signal and the original signal. After every pass, the coefficients are updated until they reach their optimal value or till the minimum error reaches a certain threshold point. The step size plays an important role in the convergence of the algorithm. Smaller step size increases the convergence time of the algorithm, whereas bigger step size decreases the convergence time and increases the dependency of the weights on the gradient, which in turn increases the error. The implementation of the LMS is done in five steps as indicated below.

Step 1. Choose the initial filter coefficients (ω), Δ as the difference between the coefficients after successive iterations and the convergence factor. Step size is chosen as the convergence factor. This is done to control the rate of convergence. Mathematically,

$$\omega(0) = 0 \tag{1}$$

$$0 < \mu < \frac{2}{\sigma_{max}} \tag{2}$$

Where μ is the step size that has to be tuned and σ_{max} is the maximum allowed step size.

Step 2. Filter the noisy signal using the above coefficients and the threshold parameters using the following Equation 3.

$$y(k) = \omega^T(k)x(k) \tag{3}$$

Where $x(k)$ is the input noisy signal and $y(k)$ is the filtered output. $\omega^T(k)$ is the transposed matrix of the filter coefficients.

Step 3. Compute the error *err*, by comparing the present output with the previous output as indicated by Equation 4. In the very first iteration, the past output is considered as zero.

$$err(k) = y_i(k) - y_{i-1}(k) \tag{4}$$

Step 4. Update the filter coefficients to be used for the next iteration using the error value as indicated in Equation 4.

$$\omega(k + 1) = \omega(k) + \mu.err(k).x(k) \tag{5}$$

Step 5. If the filter coefficients do not change in the next iteration, the difference between the coefficients is < Δ, then the iteration stops and the filter coefficients are considered final.

B. Wiener filter

Wiener filter minimises the mean square error (MSE). For the implementation of a Wiener filter, it is important to know the spectral properties of both the speech signal and the noisy signal (Jingfang, 2011). Also, this filter assumes the input to be stationary. The block diagram for the implementation of the Wiener filter

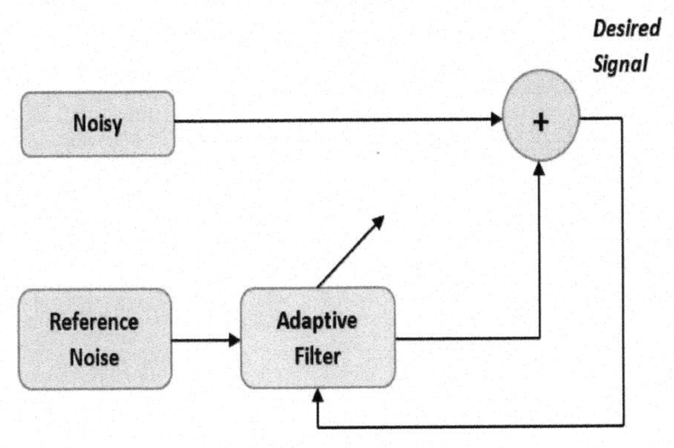

Figure 76.2 Block diagram of LMS filter

is given in Figure 76.3. The disadvantage of this method is that noise reduction comes at a cost of speech signal distortion. Hence, this method is used in cases where this tradeoff is acceptable. The implementation is on similar lines as that of an LMS, but the error function is minimised here. Mathematically, the MSE error (MMSE) criterion is given by Equation 6.

$$J_{MMSE}(\omega) = E\left(\text{err}_k^2\right) \tag{6}$$

Where $P_v(w)$ is the PSD of noise.

C. Multi-layer perceptron (MLP)

The methodology for the speech prediction in the presence of noise using MLP is shown in Figure 76.4. A short time fast Fourier transform (FFT) is calculated for the clean speech signal and the noisy speech signal. This is done to convert the original speech signal (.wav format) to the frequency domain, which is easier for graphical representation. The magnitude part is then fed to the MLP, which predicts the next sample value. This information combined with the phase value is then used to calculate the inverse short time FFT, the output of which is the denoised speech. MLP consists of an input layer, a hidden layer and an output layer (Ghaemmaghami et al., 2009) as shown in Figure 76.5. The activation function used is ReLU with 100 epochs. Literature shows that compared to the other activation functions, the ReLU is computationally

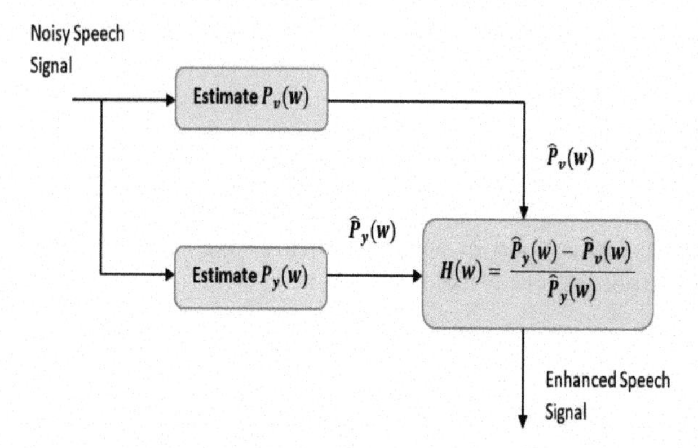

Figure 76.3 Block diagram of wiener filter

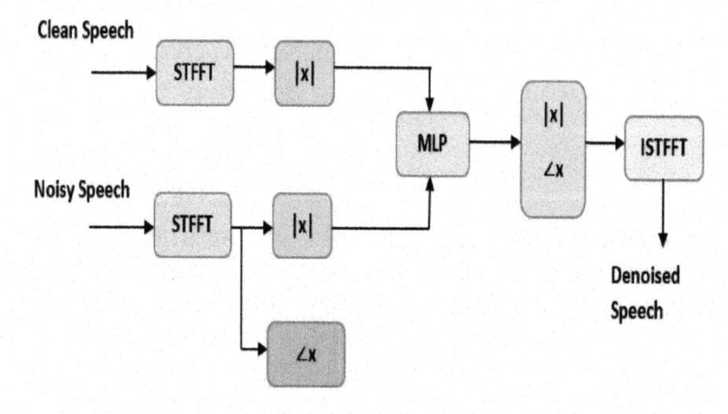

Figure 76.4 Methodology based on MLP

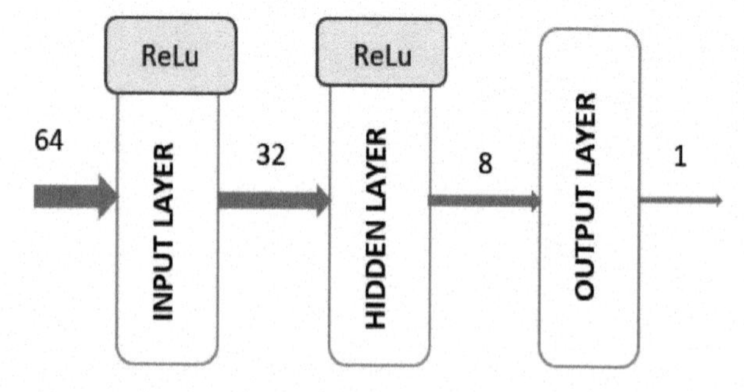

Figure 76.5 Architecture of the MLP

efficient. In this work, a trial and error has been done using the ReLU and the Sigmoid functions and it is found that the results converge faster with the ReLU activation. It is not possible to calculate the perfect weights, hence an error parameter is introduced to get a set of weights with the least differences between successive iterations. Mean square logarithmic error is used as the error metric and the input is the first 64 samples of the speech signal. The output is the 65th sample which is predicted by the MLP at the output layer. After the prediction of the 65th sample, the window slides so that the sample values from 2 to 65 are then used for the prediction of the 66th sample. This process continues till all the samples in the frame are predicted.

Results and Discussion

Figure 76.6 shows the waveform of the first 100 samples of the clip of a speech signal. The blue line denotes the actual signal, to which noise is added randomly. The red waveform indicates the noisy signal. This noisy speech signal then acts as an input to all the three algorithms.

Figure 76.7 shows the prediction of the speech signal using the LMS filter. It can be seen from Figures 76.6 and 76.7 that the prediction closely follows the original signal. Figure 76.8 shows the prediction using the Wiener filter. It can be seen from the figure that the predicted signal does not follow the original signal closely. Figure 76.9 shows the prediction using the MLP implementation of the neural network. The loss function of the MLP training with the number of epochs is shown in Figure 76.10.

For the comparative analysis, the error obtained using the different algorithms is plotted in Figures 76.11 and 76.12. Figure 76.11 shows the prediction error of LMS filter and that obtained using the NN. Similarly, Figure 76.12 shows the error of Wiener filter and NN implementation. All the three errors have not been shown in the same graph, since the LMS error and the Wiener error becomes indecipherable. It can be seen from both the figures that the NN implementation provides the least error.

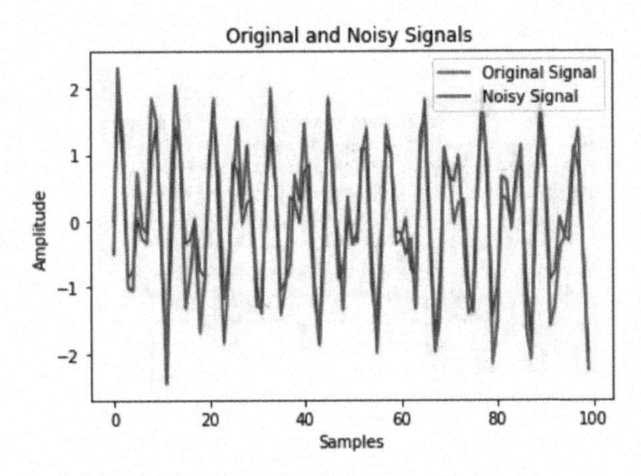

Figure 76.6 Speech signal and noisy speech signal

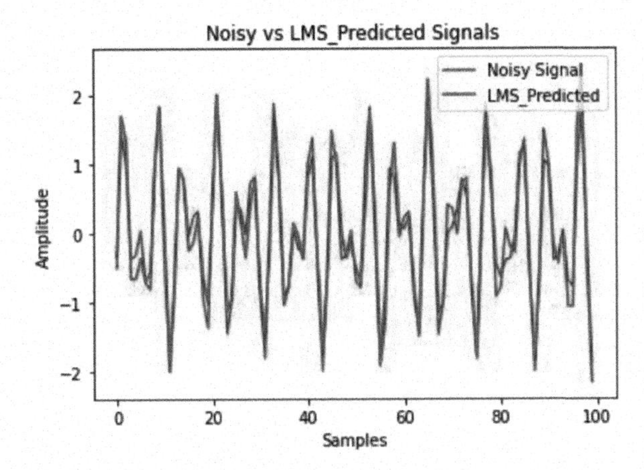

Figure 76.7 Prediction using the LMS filter

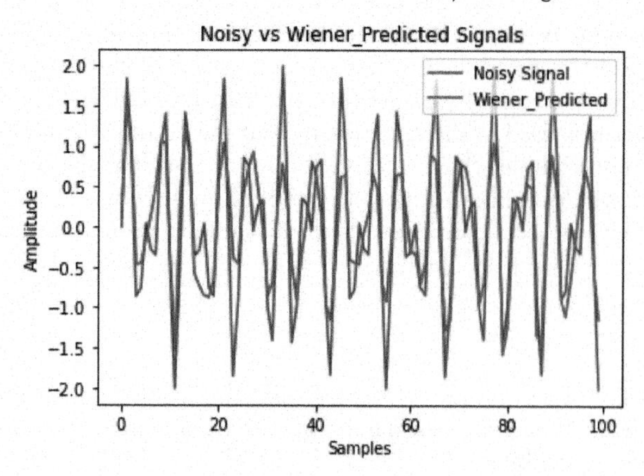

Figure 76.8 Prediction using wiener filter

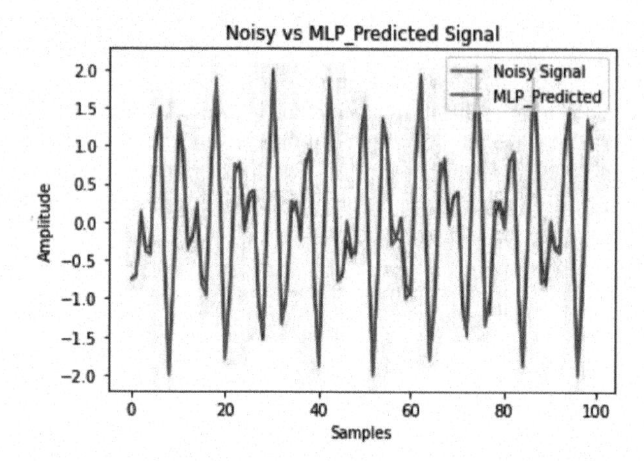

Figure 76.9 Prediction using MLP

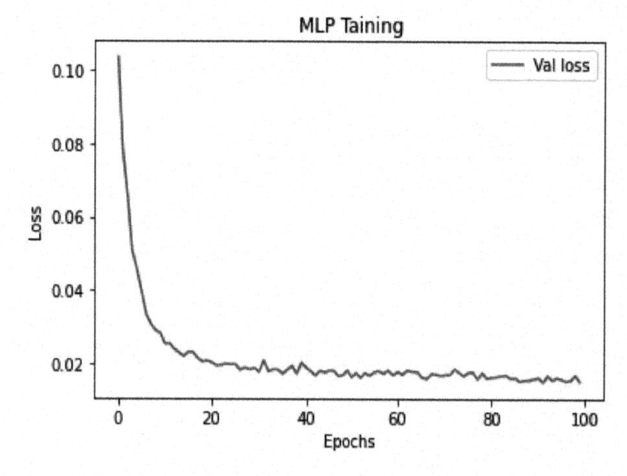

Figure 76.10 Training the MLP

Table 76.1 shows the comparative analysis of the three methods employed here.

The results indicate that the best SNR is obtained with an MLP is the best amongst the three but the time taken for the execution is the maximum. The PESQ score is also better for the MLP implementation.

Conclusion and Future Work

The recent work in the field of speech processing has seen a positive trend towards the use of machine learning approaches for speech predictions in the presence of noise. Whilst this is encouraging, the benefits

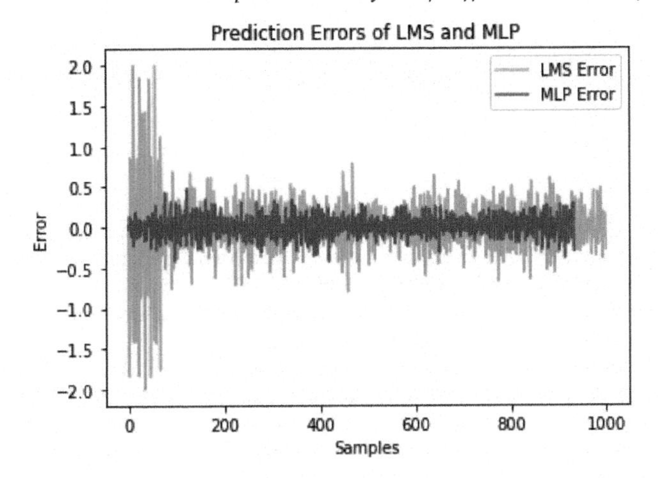

Figure 76.11 LMS error vs MLP error

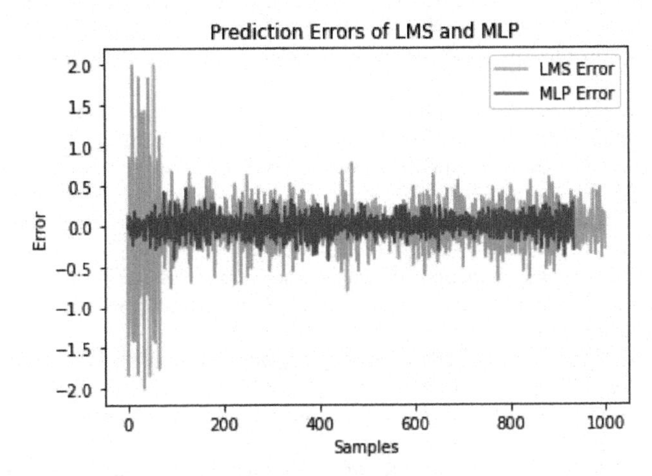

Figure 76.12 Wiener error vs MLP error

Table 76.1 Comparative analysis

Parameters	*LMS*	*Wiener*	*MLP*
SNR (dB)	12.31	9.02	19.13
PESQ	2.2	2.3	2.9
Time (s)	5	9	20

of traditional speech processing algorithms cannot be ignored. In this work, a comparative analysis of the performance of signal processing algorithms and the MLP has been done. The speech clip taken is of 10ms to which noise has been added and then LMS filter, Wiener filter and MLP implementation of Neural Networks are used to predict the signal value in the presence of noise. The performance measure is the SNR, PESQ and the time taken to execute the code. It is found that the performance of the three-layer neural network far exceeds the performance of the signal processing filters at the cost of increasing implementation time. The speech signal used here have a very few samples and hence the training time of the neural network is relatively small. In case of real time scenarios, there will be a tradeoff between the obtained SNR, PESQ and the time complexity. The neural networks are general purpose, they can be trained to predict any type of data, whereas the signal processing filters are specific to the domain and hence the runtime is better than the neural network.

In order to fully analyse the relative pros and cons of the algorithms, it is important to consider a real time noisy speech signal, like the one generated from the engine rooms or the cockpits of fighter planes flying at higher altitudes. The experiments conducted on these signals may give a better insight into the performance of these algorithms.

Acknowledgment

The authors are very grateful to the management of BNM Institute of Technology for providing the necessary infrastructure to carry out the research work.

References

Ghaemmaghami, M. P., Sameti, H., Razzazi F, BabaAli, B., and Dabbaghchian, S. (2009). Robust speech recognition using MLP neural network in log-spectral domain. In 2009 IEEE International symposium on signal processing and information technology (ISSPIT), (pp. 467–472). doi: 10.1109/ISSPIT.2009.5407513.

Gui, G. and Adachi, F. (2013). Improved least mean square algorithm with application to adaptive sparse channel estimation. J. Wirel. Commun. Netw. 204. doi: 10.1186/1687-1499-2013-204.

Gupta, P., Patidar, M., and Nema, P. (2015). Performance analysis of speech enhancement using LMS, NLMS and UNANR algorithms. In International conference on computer, communication and control (IC4), (pp. 1–5). doi: 10.1109/IC4.2015.7375561.

Hadjahmadi, A. H., Homayounpour, M. M., and Ahadi, S. M. (2008). A neural network based local SNR estimation for estimating spectral masks. In IEEE 2008 International symposium on telecommunications (IST), (pp. 608–613). doi:10.1109/istel.2008.4651373.

Hansen, J. and Pellom, B. (1998). An effective quality evaluation protocol for speech enhancement algorithms. In Proc. Int. Conf. Spoken Lang. Process.7:2819–2822.

Hu, Y. and Loizou, P. (2006). Subjective comparison of speech enhancement algorithms. In Proc. IEEE Int. Conf. Acoust. Speech Signal Process. 1:153–156.

Jalal, B., Yang, X., Liu, Q., Long, T., and Sarkar, T. K. (2020). Fast and robust variable-step-size LMS algorithm for adaptive beamforming. IEEE Antennas Wirel. Propag. Lett. 19(7):1206–1210. doi: 10.1109/LAWP.2020.2995244.

Jingfang, W. (2011). Noisy speech in real time iterative wiener filter. In 2011 International conference on mechatronic science, electric engineering and computer. (MEC), (pp. 2102–2105).

Murugan, S. S. and Natarajan, V. (2012). SNR and MSE analysis of KLMS algorithm for underwater acoustic communications. J. Mar. Eng. Technol. 11(3):3–8. doi: 10.1080/20464177.2012.11020267.

Ngo, T., Kelley, B., and Rad, P. (2020). Deep learning based prediction of signal-to-noise ratio (SNR) for lte and 5g systems. In 2020 8th. International conference on wireless networks and mobile communications (WINCOM), (pp. 1–6). doi: 10.1109/WINCOM50532.2020.9272470.

Nossier, S. A., Wall, J., Moniri, M., Glackin, C., and Cannings, N. (2020). An experimental analysis of deep learning architectures for supervised speech enhancement. Electronics. 10(1):17. doi: 10.3390/electronics10010017.

Park, G., Cho, W., Kim, K.-S., and Lee, S. (2020). Speech enhancement for hearing aids with deep learning on environmental noises. Appl. Sci. 10(17):6077. doi: 10.3390/app10176077.

Quackenbush, S., Barnwell, T., and Clements, M. (1988). Objective measures of speech quality. Englewood Cliffs, NJ: Prentice-Hall.

Rajini, G. K., Harikrishnan, V., Jasmin Pemeena Priyadarisini, M, and Balaji, S. (2019). A research on different filtering techniques and neural networks methods for denoising speech. Int. J. Innov. Technol. Explor. Eng. 8(9S2):503–511. DOI: 10.35940/ijitee.I1107.0789S219.

Rix, A. W., Beerends, J. G., Hollier, M. P. and Hekstra, A. P. (2001). Perceptual evaluation of speech quality (PESQ)-a new method for speech quality assessment of telephone networks and codecs. In 2001 IEEE international conference on acoustics, speech, and signal processing proceedings. 2:749–752. doi: 10.1109/ICASSP.2001.941023.

Smital, L., Vítek, M., Kozumplík, J., and Provazník, I. (2013). Adaptive wavelet wiener filtering of ECG signals. IEEE Trans. Biomed. Eng. 60(2):437–445. doi: 10.1109/TBME.2012.2228482.

Thunga, S. S. and Muthu, R. K. (2020). Adaptive noise cancellation using improved LMS algorithm. In Soft computing for problem solving. advances in intelligent systems and computing, ed. K. Das, J. Bansal, K. Deep, A. Nagar, P. Pathipooranam, and R. Naidu, 1048. Singapore: Springer. doi: 10.1007/978-981-15-0035-0_77.

Upadhyay, N. and Jaiswal, R. K. (2016). Single channel speech enhancement: using wiener filtering with recursive noise estimation. Procedia Comput. Sci. 84:22–30. doi: 10.1016/j.procs.2016.04.061.

Washi, T., Kawamura A. and Iiguni, Y. (2006). Sinusoidal noise reduction method using leaky lms algorithm. Int. Symp. Intell. Signal Process. Commun. 303–306. doi: 10.1109/ISPACS.2006.364892.

Xie, X., Peng, S. and Yang, X. (2020). Deep learning-based signal-to-noise ratio estimation using constellation diagrams. Mob. Inf. Syst. 2020:1–9. doi: 10.1155/2020/8840340.

77 Deep learning model for the brain tumour detection and classification

Supriya Thombre[1,a], *Mohini Mehare*[1,b], *Durvesh Manusmare*[1,c], *Riya Kharwade*[1,d], *Mandar Shende*[1,e], *and Vaishali D. Tendolkar*[2,f]

[1]Department of Computer Technology, Yeshwantrao Chavan College of Engineering, Nagpur, Maharashtra, India

[2]Mental Health Nursing, Datta Meghe College of Nursing, Nagpur, Maharashtra, India

Abstract

Cancer is the second leading cause of mortality, after heart disease. The increased demand for automated, dependable, quick, and effective imaging has drawn attention to the subject of medical imaging. Computer-assisted technologies are critical in the diagnosis of brain tumours. These devices make it easier for doctors to identify tumours. Traditional approach errors are minimised. The goal of this study is to use MRI scans to diagnose brain cancer. A deep learning network of CNN models is used in the diagnostic technique. In the first stage, an MRI image of the brain is taken and processed to eliminate noise and change it. The second step in the second stage is to determine the type of malignancy. The conclusions included the algorithm and technique used to handle certain study issues, as well as their pros and drawbacks. The results of this study show that the created image processing algorithm is smart enough to recognise and defining brain tumours in MRI images on its own.

Keywords: Brain tumour detection, computer-aided diagnosis, convolutional neural networks, deep learning, image analysis, image classification, image processing, medical imaging, MRI images.

Introduction

A brain tumour is regarded as one of the most aggressive diseases in both children and adults. Brain tumours account for 85–90% of all primary Central Nervous System (CNS) tumours. Around 11,700 people are diagnosed with brain tumours each year. The 5-year survival rate for people with a cancerous brain or CNS tumour is around 34% for men and 36% for women. Brain tumours are classified as benign, malignant, pituitary, and so on. To improve patients' life expectancy, proper treatment planning and accurate diagnostics should be implemented. Magnetic Resonance Imaging is the most effective technique for detecting brain tumours (MRI). The human body's smallest aberrations can also be identified by Magnetic Resonance Imaging (MRI) images. They can detect the unusual growth and proliferation of brain tumours.

Brain tumours can be identified with various symptoms, such as convulsions, mood changes, hearing loss, vision loss, and muscle movement. Tumours in the primary stage can be easily removed, but in the second stage, there is a risk of a tumour growing back again. This problem mainly occurs due to inaccuracy in detecting the exact location of the tumour.

Automated classification approaches based on Machine Learning (ML) and Artificial Intelligence (AI) has consistently outperformed manual categorisation. As a result, we propose using Deep Learning Algorithms using Convolution Neural Networks to identify and classify brain tumours (CNN). The project's goal is to improve the quality and reliability of real-world MRI data by utilising AI and ML domain expertise. Furthermore, certain treatment ideas will be provided by facilitating access to the programme via the cloud via mobile applications, online browser platforms, and so on.

Literature Review

Marium Malik, Muhammad Arfan Jaffar, Muhammad Raza Naqvi proposed a performance validation algorithm where approximate 170-180 various MRI images are examined. The neoplasm was successfully extracted by converting the image into a grey-scale image as MRI images are in RGB.

The mean filter is useful for the removal of noise formed because of multiple aspects, like environment and sometimes equipment. For enhancing the quality of an image, contrast adjustment is used. The detection of uninterrupted boundaries of an image is useful for extracting information. Image segmentation

[a]supriyathombre@gmail.com; [b]mohinimehare89@gmail.com; [c]durveshmanusmare@gmail.com; [d]riyakharwade161100@gmail.com; [e]mandarshende0808@gmail.com; [f]vaishali.ten@gmail.com

using global thresholding is used for detection of tumours from MRI images. Improved results, reduced execution time, and multiclass classification opportunities for different types of tumours can be done by using an advanced deep learning approach (Malik et al., 2021).

Prewitt edge detection filter and morphological dilation operation are used for finding brain lumps. As the CT scan images have detailing properties, it has been used as an input to the system. The noise produced due to an incorrect detail in an image has been trimmed by using adaptive histogram equalisation. Prewitt edge detection is used to categorise the object boundaries of an image. By adding pixels to the object boundaries, the dilation is applied for extracting tumour regions with an excessive precision rate. AHE is used for the segmentation of CT scan images by enhancing contrast. To acquire improved accuracy, this step is very essential (Soni and Rai, 2021).

The tumour is identified by applying the Naive classifier method. It classifies the tumour into normal and abnormal tumours. Divyamary.D, Gopika.S, Pradeeba.S uses both simulated and real-time images. For eliminating the noises, the pre-processing method like a morphological operation has been used and the features are extracted (Divyamary et al., 2020).

The automatic diagnosing tumour in the brain using Magnetic Resonance (MR) images were proposed by M. Usman Akram; Anam Usman. For detecting the growth of tumours in the brain, MR scans are useful. A brain tumour is detected and segmented in three phases using this approach. The first stage involves acquiring an MR picture of the brain and then pre-processing it to eliminate noise and sharpen it. The brain tumour is segmented from the sharpened image to the second phase after applying the global threshold segmentation. The morphological operations are then post-processed on the segmented image and, for removing any remaining tumours; the tumour masking is carried out in the third stage (Akram and Usman, 2011).

A tumour cell is a cell that grows wildly and consistently in the absence of normal factors. Every year, the major cause of death is brain tumours in humans. In the United States, around half of all brain tumour patients die from primary brain tumours each year. Electronic modalities can be used for diagnosing brain tumours. The most popular electronic modality for diagnosing brain tumours is MRI. T. M. Shahriar Sazzad; K. M. Tanzibul Ahmmed; Misbah Ul Hoque; Mahmuda Rahman proposed an automated approach for brain tumour detection, based on grey-scale MRI images. They have presented an automated system involving initial enhancement for reducing grey-scale colour variations. For improving segmentation, the unnecessary noise was removed by filter operation. The threshold-based OTSU segmentation was used as images are grey-scale instead of colour segmentation. The proposed strategy maintains the accuracy while retaining an acceptable accuracy rate for pathology experts, outperforming the existing available alternatives (Sazzad et al., 2019).

A brain tumour is a type of sickness in which blood clots damage the brain. An MRI picture can be used to see a brain tumour in greater detail. Because of their similar colour, it's difficult to tell the difference between brain tumours and normal tissue. A precise analysis of a brain tumour is required. Segmentation is a solution for analysing brain tumours. For distinguishing the brain tumour tissue from other tissues, like fat, oedema, normal brain tissue, and the cerebrospinal fluid, the medium filtering is used for the MRI images. The segmentation of the tumour requires a thresholding method iterated untill the maximum area is obtained. For marking the inside and outside parts of the brain, the watershed approach is used. For cleaning the skull, the cropping method is used. The areas of the brain, like tumours and tissues, are compared with the segmentation results. The average inaccuracy in the tumour area calculation obtained by this approach is 10% (Wulandari et al., 2019).

A brain tumour is a collection of tissue that has been prepared through the gradual accumulation of irregular cells. It occurs when cells in the brain produce aberrant formations. It has recently become a leading cause of death for many people. Because brain tumours are among the most dangerous of all cancers, prompt detection and treatment are required to preserve life. Because of the growth of tumour cells, detecting these cells is a difficult task. It is critical to compare the treatment of a brain tumour with that of an MRI. Normal, benign, and malignant brain tumours are the three categories. R. Lavanya Devi; M. Macha Kowsalya proposed a neural network approach that will be used to classify the benign, malignant, and normal stages of a brain tumour. For extracting features, the Gray Level Co-Occurrence Matrix was used (GLCM). The Principal Component Analysis (PCA) approach is useful for picture recognition and compression, as well as to reduce the data's dimensionality. A probabilistic neural network is used to automatically classify the stage of a brain tumour (PNN). The brain tumour spread zone is detected as well as segmented using the K-means clustering technique. The spread region has numerous faulty cells. PNN is a quick technique that also has a high level of classification accuracy (Devi et al., 2017).

MRI brain tumour segmentation has been a popular study topic in the medical imaging sector. Detection of brain tumour is particularly beneficial for defining the exact size and location of a tumour. Mahesh Kurnar; Aman Sinha; Nutan V. Bansode proposed a K-means clustering technique for tumour diagnosis based on segmentation and morphological operations. For determining the excised tumour portion's area, the morphological operator is used by extracting the tumour from the pre-processed MRI scanned image after applying K-means clustering (Kurnar et al., 2019).

The segmentation of anatomical parts of the brain is difficult in medical image processing. A strategy for segmenting brain tumours has been devised and verified using MRI data. During the Pre-processing and enhancement stage, the medical image is converted into a standard formatted image. If the appropriate parameters are adjusted correctly, this approach can segment a tumour. For tumour identification, after a manual segmentation method, the 2D & 3D visualisation for surgical planning, better brain tumour shape approximation, and assessing tumour is carried out by the potential use of MRI data (Angel Viji and Jayakumari, 2011).

Pre-processing involves the addition of an MRI brain picture and filtering. Noise is present in MRI imaging; as this noise is not critical, an average filter is employed to smooth the image generated by many filters. The smoothed image is then sent on to the next step. For smoothing photographs, the average filter is a simple and straightforward method (Çinar and Yildirim, 2020).

The noise in the image can decrease the region's ability to generate large areas or can lead to failure. If you encounter noisy images, pre-processing the image by using medium filters is usually convenient. Median filters have classical median filter's strength and edge preservation capabilities. Some fundamental improvements in the image and noise reduction methods are applied during pre-processing. Besides this, different traditions have also been used to identify edges and to segment. These steps are essentially intended to retrieve the image and image superiority to ensure that the tumour is more easily identified. By converting to the grey-scale image, the noise is reduced. The grey-scale image will then be passed to the filter. Using a high-pass filter to filter an image, the weighted average of the surrounding pixels replaces each image's pixel. The weights are determined by the values of the filter and by the size of the filter used, and by the number of surrounding pixels. To improve image quality, the grey image and the filtered image are then combined. For reducing the salt and pepper noise and preserving borders of the image, the median filter is used (Ismael et al., 2019).

Skull stripping is one of the main stages in the processing of biomedical imaging. The heart & lungs analysis doesn't require stripping, but brain image analysis requires it. Before any other picture processing, skull stripping is done. This technique is used to remove all brain tissue other than the brain tissue. All the tissues like skin, fat & skull were removed by this method. Different skull peeling processes exist. For removing the skull tissue, skull stripping is used. The threshold value of normal brain & skull tissue was determined by this method (Haque and Neubert, 2020).

Harish Chetty, Monit Shah, Samarth Kabaria, Saurav Verma, proposed two techniques for brain tumour detection using MRI images. For the detection process, the region growing technique has been verified as favourable. Out of the two techniques, one was focussed on Watershed Segmentation and the other on Symmetric analysis. These two techniques were useful for the detection of brain tumours (Chetty et al., 2017).

For minimising the non-brain tissue contents, each training set needs to be pre-processed. Then, four types of features are extracted from the pre-generated images. The main reasons for MRI intensity being inhomogeneous are the eddy currents caused by gradient field switching and the radio-frequency field non-uniformity in an imaging system. The intensity distribution is affected by many imaging parameters and devices. An automatic method is called Non – parametric. Non -uniform intensity Normalisation (N3) is used to adjust the images for intensity inhomogeneity initially. The non-brain tissues, like the skull and eyes, are removed automatically by using the Brain Extraction Tool (BET). Subject's different MRI modalities are co-registered by using a mutual knowledge-based approach like multi-modality multi-resolution registration (Pedoia et al., 2012).

Texture, intensity, shape deformation, and symmetry are recovered from MRI using Gabor filters. For the segmentation of tumours in MRI, intensity is the basic characteristic. Due to changes in the intensity of tumour and overlapping with the normal tissue's intensity, the tumour detection is not precise and reliable (Ghanavati et al., 2008).

To avoid image forgery in MRI images, the forgery detection method is more demanding. This method is used to identify the authenticity of an image. Copy-move forgery detection technique can be used to identify image forgery (Tembe and Thombre, 2017).

Implementation and Methodology

The goal is to divide the MRIs into three categories, making it a three-class challenge. The Neural Network data will be in .jpeg format.The data was obtained via Kaggle. It had two folders labelled Testing and Training. Both folders had three sub-folders: Meningioma Tumour, No Tumour, and Pituitary Tumour. There were 2044 photographs in the Training folder and 294 images in the Testing folder.

Figure 77.1 represents the flowchart for the approach.

A. Pre – Processing

The mistakenly inserted filthy data is edited or removed if not required. The images were divided into 7:3 ratios for training and testing phases.

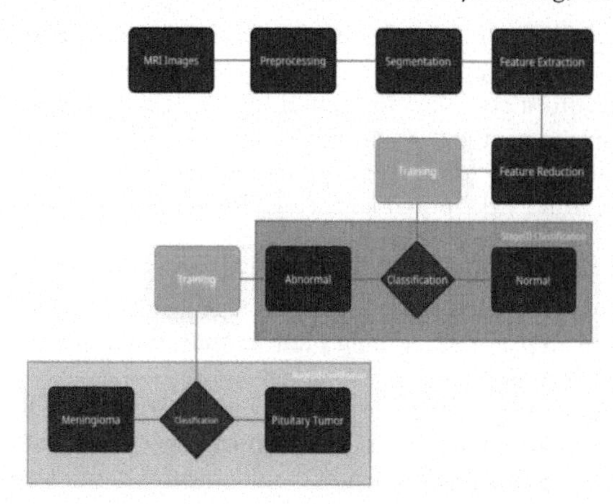

Figure 77.1 Flowchart depicting process of detection of tumour

B. Image Cropping

A dark background is found in the MRI around the brain's core. This black background does not provide useful tumour information and is wasteful if given input into neural networks. It would therefore be good to cut the pictures around the primary contour.

The 'cv2' library cv2.findContours() is used for recognizing the contours. Each image in the dataset goes through this procedure. cv2.findContours() sometimes recognise incorrect contours and may not crop the images correctly. Before moving on to the 'augmentation' phase, such images should be manually removed.

C. Data Augmentation

The amount of collected data was insignificant and can lead to the under-fitting of the model. To increase the amount of data availability, data augmentation is used. This technique uses changes in exposure, rotations, flips, etc. for creating similar images.

For determining the performance of augmentation, several arguments were passed to ImageDataGenerator, which is a Keras.preprocessing.image' library function. Every image in the data set is processed by the augmentation, which reduces the probability of underfitting.

D. Model Building

The cleaned, pre-processed, and augmented data were added to a neural network. The two different techniques, like Artificial Neural Networks (ANN) and Convolutional Neural Networks (CNN) are compared for identifying the most accurate models.

E. Artificial Neural Network

ANN is the computation system inspired by the biological neural network constituting the animal brain. ANN can be simply called a Neural Network.

ANN is the multi-layer fully connected layer structure. The input layer takes each pixel of the inputted image, calculations are carried out using multiple hidden layers, and the output layer produces multiple probabilities. Networks computations and weights are increased by increasing the number of hidden layers.

F. Convolutional Neural Network

CNN is normally used in computer vision. There are various hidden layers, such as pooling, fully connected, normalisation and convolutional layers. For feature extraction, CNN is very helpful, as it can train the features from all the classes. The image's significant elements are retrieved using the pooling layer. For improving picture recognition, the convolutional and pooling layers are beneficial.

The many layers of CNN, such as the pooling layer, aid in extracting the key information from the picture. As a result, integrating convolution layers with pooling layers on top of a basic neural network improves picture recognition.The goal of learning filters thus becomes the training of CNNs. Pooling layers take a block of input data and simply pass on the highest value. Because pooling layers lower the amount of the output and require no additional parameters to learn, they are frequently employed to manage the size of the network and keep the system below a computational limit. A non-linear function is provided by an activation function.

As a result, we create a few CNN models with dense layers of 0,1,2 and layer sizes of 32,64,128 as well as a number of convolution layers of 1,2,3. A combination of these parameters is used to construct the models. As a result, we construct a total of 21 models. The model '3-conv-128-nodes-1-dense' has the maximum accuracy of 91%.

We now know that a model with three convolution layers and one dense layer of 128 nodes has the potential to outperform the prior ANN model. As a result, we are presently fine-tuning the hyperparameters of this model to improve accuracy.

Results

The implemented model is used to identify the existence of tumour tissue in the brain using MRI images of the brain. CNN recognises and classifies the tumour if it exists in the brain (Figure 77.5).

The model trained was successfully able to detect 117 images out of 120 randomly selected images which were tested. This gave an accuracy of 97.5% (Table 77.1). Figure 77.3 shows the Model detecting pituitary tumour successfully.

Pituitary tumours are abnormal growths in the pituitary gland. Some pituitary tumours cause an overabundance of hormones that govern vital bodily systems. Certain pituitary tumours might lead your pituitary gland to generate fewer hormones. The vast majority of pituitary tumours are noncancerous (benign) growths (adenomas).

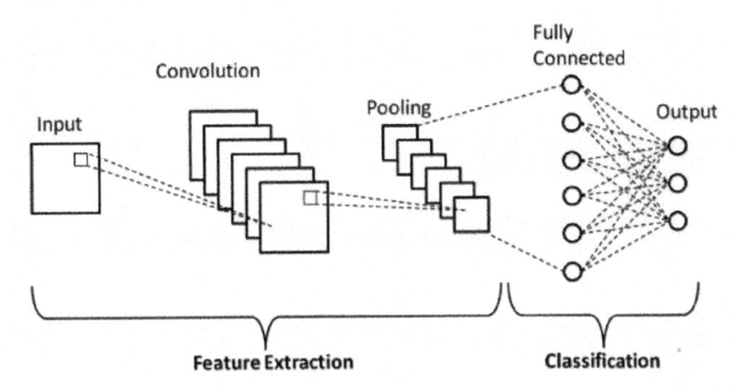

Figure 77.2 Typical CNN architecture

Table 77.1 Number of training and testing sets used

Training Set()	
No. of Normal Images Used	*No. of Abnormal Images Used*
395	1649
Testing Set()	
No. of Normal Images	*No. of Abnormal Images Used*
105	189

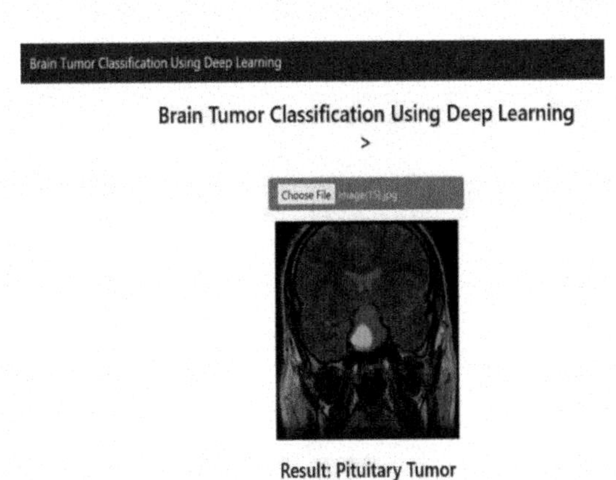

Figure 77.3 Model detecting pituitary tumour successfully

If the model is unable to detect any tumour in the MRI image, then it is labelled as No tumour. The result is shown is Figure 77.4.

The meninges are the membranes that cover the brain and spinal cord. The meningioma tumour emerges from the meninges. Meningioma is often seen as a lobular and axial mass with closed borders. Figure 77.5 shows the Model detecting meningioma tumour successfully.

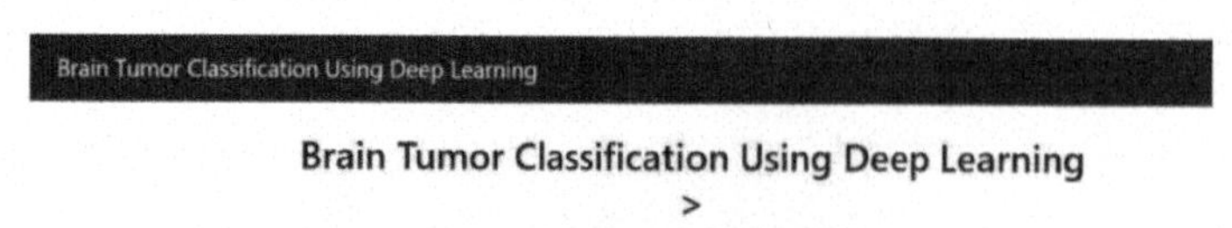

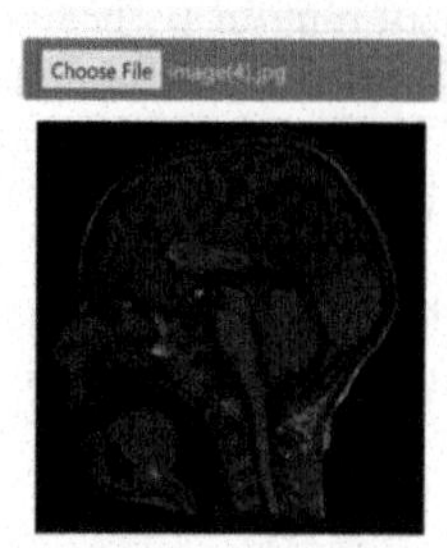

Figure 77.4 Model detecting no tumour successfully

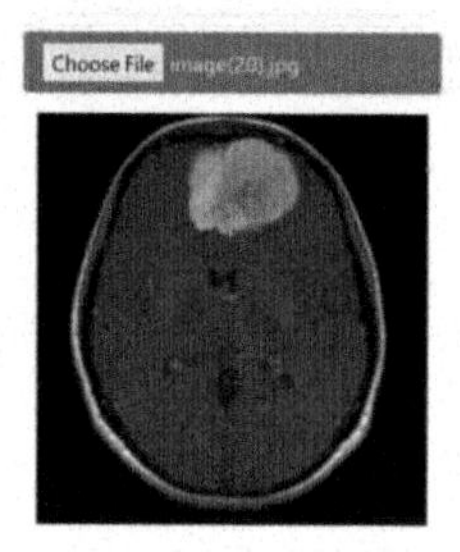

Figure 77.5 Model detecting meningioma tumour successfully

Comparative Analysis

	Our Method	*Brain Tumor Detection from MRI Images using Naive Classifier (Divyamary et al., 2020)*	*Development of Automated Brain Tumor Identification Using MRI Images (Sazzad et al., 2019)*	*Computer Aided System for Brain Tumor Detection and Segmentation (Akram and Usman, 2011)*	*CT Scan Based Brain Tumor Recognition and Extraction using Prewitt and Morphological Dilation (Soni and Rai, 2021)*
Size of Dataset (Number of images)	2338	100	106	100	52
Technologies/ Methods Used	Magnetic Resonance Imaging, Deep Learning	Naive Bayes classifier	OTSU, K-means Clustering	Global Threshold Based Segmentation	Prewitt and Morphological Dilation
Accuracy	97.5%	84%	90%	97%	94.23%
Number of Types of tumors detected	3	1	1	1	2

Conclusion

Various strategies for detecting and segmenting brain tumours from MRI images are presented. It can be shown that numerous approaches are used to detect brain tumours from MRI scans and that in the future, different automatic systems will attain greater accuracy and efficiency. The primary aim for brain tumour detection is to provide the medical diagnosis and to extract the useful and accurate information from the images having negligible error. The model is developed for providing the early diagnosis by identifying the presence of tumour in the brain's MRI images, which is useful for the further treatment.

References

Akram, M. U. and Usman, A. (2011). Computer aided system for brain tumour detection and segmentation. International Conference on Computer Networks and Information Technology, pp. 299–302, doi: 10.1109/ICCNIT.2011.6020885.

Angel Viji, K. S. and Jayakumari, J. (2011). Automatic detection of brain tumour based on magnetic resonance image using CAD System with watershed segmentation. 2011 International Conference on Signal Processing, Communication, Computing and Networking Technologies, 2011, pp. 145–150, doi: 10.1109/ICSCCN.2011.6024532.

Chetty, H., Shah, M., Kabaria, S. and Verma, S. (2017). A survey on brain tumour extraction approach from MRI images using image processing. 2nd International Conference for Convergence in Technology, IEEE.

Çinar, A. and Yildirim, M. (2020). Detection of tumours on brain MRI images using the hybrid convolutional neural network architecture. Medical Hypotheses, Elsevier.

Devi, R. L., Kousalya, M. M., Nivethitha, J. and Kumar, A. N. (2017). Brain tumour classification and segmentation in MRI images using PNN. IEEE International Conference on Electrical, Instrumentation and Communication Engineering (ICEICE).

Divyamary, D., Gopika, S. and Pradeeba, S. (2020). Brain tumour detection from MRI Images using Naïve Classifier. In 6th International conference on advanced computing & communication systems. (ICACCS), IEEE.

Ghanavati, S., Li, J., Liu, T., Babyn, P. S., Doda, W. and Lampropoulos, G. (2008). Automatic brain tumour detection in Magnetic Resonance Images. In IEEE international symposium on biomedical imaging (ISBI).

Haque, I. R. I. and Neubert, J. (2020). Deep learning approaches to biomedical image segmentation. Artificial Intelligence in Medicine, Volume 102, Elsevier.

Ismael, S. A. A., Mohammed, A. and Hefny, H. (2019). An enhanced deep learning approach for brain cancer mri images classification using residual networks. Elsevier.

Kurnar, M., Sinha, A. and Bansode, N. V. (2019). Detection of brain tumour in MRI Images by applying segmentation and area calculation method using SCILAB. 2018 IEEE Xplore.

Malik, M., Jaffar, M. A. and Naqvi, M. R. (2021). Comparison of brain tumor detection in MRI images using straight-forward image processing techniques and deep learning techniques, 2021 IEEE.

Pedoia, V., Binaghi, E., Balbi, S., De Benedictis, A., Monti, E. and Minotto, R. (2012). Glial brain tumour detection by using symmetry analysis. In Proc. SPIE, Med. Imaging 2012 Image Process, (vol. 8314).

Roy, S. and Bandyopadhyay, S. K. (2012). Detection and quantification of brain tumour from MRI of Brain and its Symmetric Analysis. Int. J. Inf. Commun. Tech. Res. 2(6).

Sazzad, T. M. S., Ahmmed, K. M. T., Hoque, M. U. and Rahman, M. (2019). Development of automated brain tumour identification using MRI images. 2019 IEEE.

Soni, A. and Rai, A. (2021). CT scan based brain tumour recognition and extraction using prewitt and morphological dilation 2021 International Conference on Computer Communication and Informatics (ICCCI), 2021 IEEE.

Tembe, A. U. and Thombre, S. S. (2017). Survey of copy-paste forgery detection in digital image forensic In International conference on innovative mechanisms for industry applications (ICIMIA 2017) (IEEE).

Wulandari, A., Sigit, R. and Bachtiar, M. M. (2019). Brain tumour segmentation to calculate percentage tumour using MRI. International Electronics Symposium on Knowledge Creation and Intelligent Computing (IES-KCIC), 2018 IEEE.

78 An experimental study of primitive prediction of crop prices using machine learning

Smita Kapse[1,a], *Devendra Raut*[2,b], *Pushkar Kukde*[1,c], *Shrunkhala Fulzele*[1,d], *Sakshi Meshram*[1,e], *Prathamesh Bhalerao*[1,f], *and Shital Telrandhe*[3,g]

[1]Computer Technology Department, Yeshwantrao Chavan College of Engineering, Nagpur, India

[2]Civil Engineering Department, Yeshwantrao Chavan, College of Engineering, Nagpur, India

[3]Jawaharlal Nehru Medical College, Wardha, India

Abstract

The presence of intermediaries in agricultural finance is a major drawback in maximising the profit of farmers. Due to a lack of financial knowledge, the intermediaries often trick the farmers into selling their crops for extraordinarily low prices, while further selling them for much higher prices. The aforementioned work aims to eradicate the presence of these intermediaries by analysing crop trends using machine learning models and presenting them in an interactive way so that the farmer may foretell the selling price of the crop. This model makes use of the XGBoost (eXtreme Gradient Boosting) algorithm of machine learning since it fits the price to quantity graph plots adequately. The results are plotted using graphic libraries like Matplotlib. The proposed model will help the farmers in judging the approximate crop price for the crop they have grown. It also helps the farmers to decide which crop will be more in-demand and beneficial to them and take middlemen out of the picture which gives more money to their crop.

Keywords: Agriculture, crop price prediction, data analytics, machine learning.

Introduction

Agriculture in India is a prime source of income for 58% of the population. Today, in farm production India ranks second worldwide (Jharna et al., 2017). Crop production is dependent on several factors like climate, soil, temperature, rainfall, irrigation, harvesting, pesticides, and other factors. As of 2018, agriculture contributed 1718% to the country's GDP (Thayakaran et al., 2019). Although India is self-sufficient in food staples, it has way lower productivity of its farms. Crop yields vary significantly between Indian states and so do their prices. With such a large population of the country having the primitive occupation of agriculture, the profits they should be earning are far less. Due to uncertain climatic conditions sometimes, the situation gets worse and farmers have to go through a great loss. The scenario has led to the loss of so many lives of the farmers. A total of 13,754 farmer suicides in 2012 according to the National Crime Records Bureau of India report (Rohitha et al., 2020). During that period, increase in financial debt; various reasons for debt come from the increasing cost of cultivation on farms because of rise in cost of chemical fertilisers and seeds, increase in rate of crop failures, scarcity of water, unstable profits, and trade liberalisation. In 2014 and 2015, farmers experienced a drought for consecutive years and along with a lack of remunerative minimum support prices, it has aggravated the farmers' plight, leading to a fall in real incomes and a rise in debt levels (Cai et al., 2017).

The financial debt also increases due to the interference from the third party in the marketing of the prices. The farmers have to give extra money from their pockets to these third parties. There are several intermediaries between the farmer and the market. The farmers are exploited by these very intermediaries. A farmer gets a menial consumer rupee even when the goods sold are his own produce (Thayakaran et al., 2019). The presence of intermediaries must be wholly removed. Industry-level farmers do not suffer from a large percentage of the profits, but the small farmers do. The small farmers cultivate crops according to their experiences and have difficulty selling their crops at reasonable prices. It is not that uncommon for a farmer to not know about the market's scene. Hence, he cannot decide the crop that will benefit him the most and ends up cultivating crops that are undesirable for that season or year (Rohitha et al., 2020). This paper includes the methodology for analysing crop prices that will help the farmer get information on the crop trends and prices that will help them to cultivate crops as per trend and keep the profit to themselves

[a]smitarkapse@gmail.com; [b]coe@ycce.edu; [c]pushkarkukde@gmail.com; [d]shrunkhalafulzele93@gmail.com; [e]meshramsakshi 99@gmail.com; [f]prathameshbhalerao4@gma; [g]shital.telrandhe88@gmail.com

than pay to the intermediaries. It helps the farmer's not only make a good profit for themselves but also make the country's economy better. The goal is to help the farmers achieve maximum yields with minimum investment by primitive predictions of crop trends.

Primitive price prediction

From the last decade many researchers are contributing to design prediction model for various application domain. In Wankhade et al. (2020), Twitter opinion data is used to predict the election results prior to its declaration. In Yashavanth et al. (2017), Vector Autoregressive model was used to predict the prices of coffee seeds. In Mahabadi et al. (2018), rice potential production is predicted using artificial neural network model, study shows that it depends upon soil pH value, electrical conductivity, and percentage of organic matters. In Sagar et al. (2015), price prediction is done using neural networks Neuroph framework. In Zhang et al. (2018), price prediction is done using Fuzzy Information Granulation and MEA-SVM. This model gives higher prediction rate and quicker calculation speed. In Kalbarczyk et al. (2016) issue of precision agriculture was handle using data driven approach, soil moisture is chosen as key factor for demonstrating the application. This paper emphasises the use of Machine Learning Techniques for primitive prediction and analysis of crop price trends. Predicting the price of a crop mostly depends upon the types of crops planted. Prediction can also be used to minimise losses when some unfavourable climatic conditions occur. For price prediction, the XGBoost algorithm is used. XGBoost algorithm is most commonly used nowadays. It is designed for achieving speed and performance gradient boosted decision trees (Descamps, 2017). It has the advantages of better model performance and faster execution speed. Initially, the datasets are driven from varied sources, filtered, and are then given as input to the XGBoost algorithm. The Algorithm provides us with the desired predictions that are expected from the system (Ng, 2019).

Subsequent sections are arranged as under: Section 2 gives overview of the proposed methodology used in the model of crop price trend analysis. Section 3 discusses experimental results obtained and discussion on it. In last section 4 conclusion of the proposed method is given and future scope is suggested.

Proposed Methodology

The major motivation behind this project is to enable the farmer to make some primitive decisions whilst selling his/her crop (Vohra et al.,2019). To achieve this through machine learning, suitable authentic data was required. So in this research it was taken from the agmarket.gov.in government-hosted website. From the data received it was observed that the prices in a particular state were determined by the district market, thus in this work district-wise classification is performed. Since any model's outcome would be vastly determined by the amount of data, a weekly status of the price for certain crops was acquired. Various factors like global financial markets, famines, global warming, etc. affect the crop price trends. The timeline from 2012 to 2019 was suitable to generate a sensible price for the crop due to the gradual nature of the above factors.

The input parameters that we considered were temperature and rainfall. Other factors like soil quality and humidity were ignored due to their relative nature and volatile differences from region to region in a single district. The weekly timeline of reports also made it easier to analyse exceptional natural cases like excessive rainfall or the lack of it. To decide the regression model, we followed the logical approach applied by the intermediaries and the farmers themselves. Factors like rainfall and temperatures affect the decisions of the farmer to plant a certain crop, which in turn affects the market value of the crop. Thus each input parameter has ramifications on the market value of the crop. A model was required which would consider changes in parameters and would shift its values accordingly, Decision trees were a good first choice since they fit the scenario well but due to their high variance attribute, the results could be highly erroneous. Gradient Boosting machines, on the other hand, provide a continuous decision-making approach and also use weak learners to make complicated decisions. The current trend of the Gradient Boosting model suggests that the XGBoost algorithm is currently the most used and one of the most advanced amongst machine learning models. And this model best fits our requirement so in this proposed work we used the XGBoost algorithm.

The XGBoost algorithm

The XGBoost algorithm stands for the eXtreme Gradient Boosting algorithm. It uses decision trees integrated with various optimisation techniques like tree-pruning, handling of missing values, and regularisation. It utilises 'out-of-core' computing to handle big data frames and smart cache storage to store gradient statistics. It also implements a parallelised threading approach to further optimise and improve runtime. Unlike traditional Machine Learning algorithms, XGBoost does not require data normalisation. Cross-compatibility over platforms offers further extensibility and flexibility to implement the algorithm. Integrated implementation is also available in the scikit-learn library.

The highly sophisticated algorithm is an optimised version of the Gradient Boosted decision tree algorithm. The mathematical computation process of XGBoost is explained as follows (Guest_blog, 2018)

For a 2D training set consisting of *n* inputs, $\{(x_i, y_i)\}_{i=1}^{n}$, a loss function $L(y, F(x))$, and number of iterations is M

Step 1: $F_0(x) = \arg_\gamma \min \sum^{n} L(yi, \gamma)$
Step 2: Repeat step 2 for! m = 1 to M
2.1 Compute pseudo-residuals,

$$\gamma_{im} = -\left[\frac{\delta L(yi, F(x_i))}{\delta F(x_i)}\right]_{F(x)=F_{m-1}(x)} \quad \text{for } i = 1 \text{ to } n$$

2.2 Train weak learners using $\{(x_i, \gamma_{im})\}_{i=1}^{n}$
2.3 Compute multiplier

$$\gamma_m = \arg_\gamma \min \sum_{i=1}^{n} (Lyi,\ F_{m-1}(x_i) + \gamma h_m(x_i))$$

2.4 Model updation $F_m(x) = F_{m-1}(x) + \gamma_m h_m(x)$
Step 3: Output $F_M(x)$
Following hyper parameters of XGBoost is use:

- objective = 'reg:squarederror'
- colsample_bytree = 0.8
- learning_rate = 0.1
- aax_depth = 10
- reg_alpha = 10
- n_estimators = 1200

The description of the above hyper parameters is as under:

- objective:
  Specify the learning task and the corresponding learning objective
  reg: squared error: regression with squared loss.
  Mean squared error is usually used to estimate the performance of the regression model.

  $$MSE = \frac{1}{n}\sum_{i=1}^{n}(y_i - \tilde{y}_i)^2$$

- colsample_bytree:
  colsample_bytree is the fraction of features (randomly selected) that will be used to train each tree. The colsample_bytree specifies the percentage of features or columns that will be used for building each tree. Since the dataset contains few columns, a value between 0.8–1 will be beneficial
- learning_rate: It is usually in between 0.1 to 0.01. To get better performance at expense of time, the learning rate is usually inversely proportional to the number of trees.
- max_depth:
  Determine the maximum depth of the tree. Since a higher value may cause the model to over fit, the value of 10 was chosen due to the relatively smaller size of the dataset while not over fitting the model.
- reg_alpha:
  It is the L1 regularisation term on weights.
- n_estimators:
  It determines the number of rounds the model will work for. Usually optimal for the value of 10^2 for large datasets and 10^3 for small datasets.

XGBoost algorithm for price prediction

XGBoost is a relatively modern technique to solve regression problems. Most of the work done in similar prediction papers uses either decision tree algorithms, Support Vector Machine (Rachana et al., 2019), or K-Nearest Neighbour (Singh et al., 2017). With the fluctuating nature of crop price values in current scenarios, which are altered by multiple factors, the XGBoost is predicted to perform a bit better than the above techniques due to its optimisation and implementation (Rohit et al., 2020; Sahu, 2019).

Akin to the standard machine learning model preparation process, the dataset will first be segregated for testing and training. Selected data will then be split into input and output patterns. The model will be trained using the training data before being used to make predictions. The price of every commodity will be the output pattern while all the other remaining factors will be considered as input patterns.

The algorithm is widely used in stock market prediction due to its faster processing time. The learner used for this research work will either be the gblinear booster, which learns only from linear relations, or the DART (Dropouts meet Multiple Additive Regression Trees), which drops trees in order to tackle over-fitting. To compute a prediction, this algorithm sums predictions of all its trees. In XGBoost the number of trees used is controlled by n_estimators argument and by default it is 100. An individual tree is not a great predictor on its own, but by aggregating across all trees, this algorithm will provide a robust estimate of many cases. XGBoost can be anticipated to outperform classic regression algorithms like KNN due to the discrete nature of values of crop prices in the market.

XGBoost algorithm flowchart

Following Figure 78.1 depicts the flowchart of the XGBoost algorithm. First, we load the data from the final cleaned dataset. After the data is loaded, it is split into labels and data to construct a Data Matrix (DMatrix) because DMatrix is a special dataset used by the XGBoost Algorithm for performance optimisation. The loaded data is split into training and testing datasets. Here, we've split the data as 95% for training and 5% for testing. In the next step, the XGBRegressor is initialised by passing appropriate hyper parameter values. The XGBRegressor is then trained on the aforementioned split training data to calculate the Root Mean Square Error (RMSE) Value which gives an estimate of how well the model is predicting the testing data. The same training data, structured as a DMatrix, is used to train the model n times, and this process is called kfold cross- validation as the procedure is repeated k times. The purpose of this is to

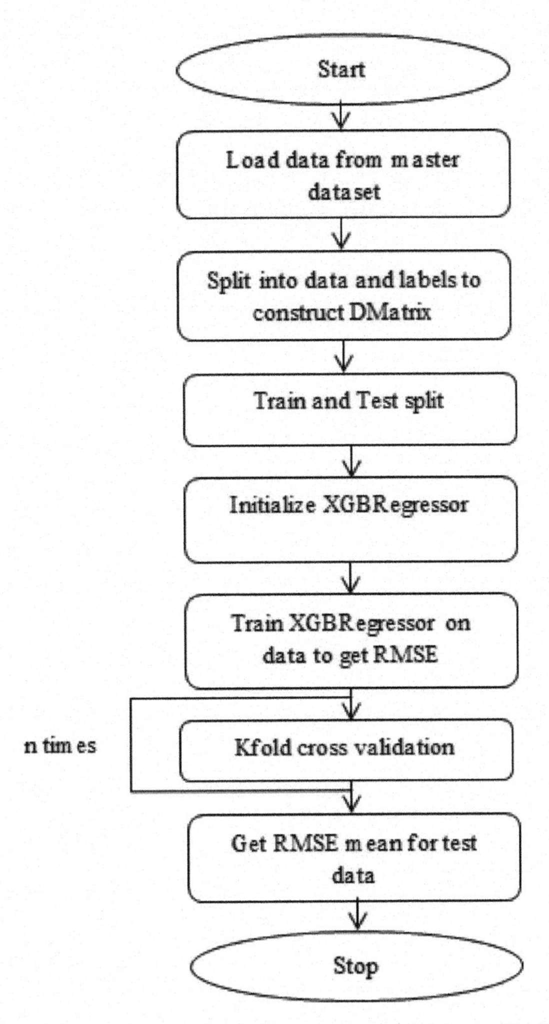

Figure 78.1 Flowchart of XGBoost algorithm used in the proposed system

improve the accuracy of the given model. In the last step, the Root Mean Square Error Value is calculated for the testing data.

Flowchart of crop price prediction model

The flow with which the model will proceed is as shown in the following Figure 78.2. Data accrual will be done in three parts. The crop price dataset will be acquired from agmarket.gov.in. The temperature and rainfall dataset will be acquired from dateandtime.com. The data acquisition process is fully automated with the help of python scripts which extensively use the selenium library to automate website clicks. A master dataset is compiled from the above datasets which will be the source of all training and testing data. The regression outputs generated from the model will be displayed as not only values but also in terms of graphs and charts as well.

Experimental Results and Discussion

Data visualisation

A graphical way of representing information and data is done by a data visualisation tool. In the proposed system, data visualisation techniques in Python is used for graphical representation of the information. To visualise the data, we have used the Jupyter Notebook as a tool. Visual elements such as pie charts, bar graphs, and double bar graph have been used. We have used the Matplotlib library for plotting graphs in Python along with pandas and NumPy libraries for visualising our data. The visual elements are created by the data collected from three websites needed for three important factors, i.e., crop prices, temperature, and rainfall. The website used to collect the crop prices is agmarket.gov.in, it was indiawris.gov.in for rainfall, and power.larc.nasa.gov for temperature. The data is collected for the year 2016–2019 for all three factors. 95% of the total number of rows in the dataset were used to train the model, while the remaining 5% were used to test the model's prediction.

Each data set is then organised in a way that will suit the purpose of our visualisation. The entire data visualisation process is carried on by dividing the data into particularly three seasons namely rabi, kharif, and zaid. For performing experiments and predicting crop prices we consider the data for the Nagpur district of Maharashtra, India.

Figure 78.3 demonstrates a pie chart for actual rainfall in Nagpur District for the year 2019. The Rainfall dataset is used. The actual rainfall (mm) column is considered. The row count is 48. The parameters used are seasons of agriculture in India i.e. kharif, rabi, and zaid. The visual describes the distribution of rainfall (%) within a year with regard to seasons.

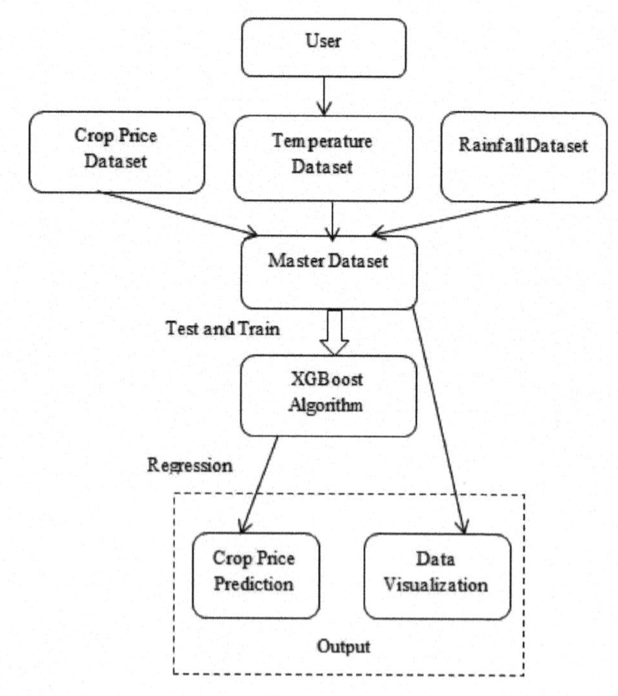

Figure 78.2 Flowchart of crop price prediction model

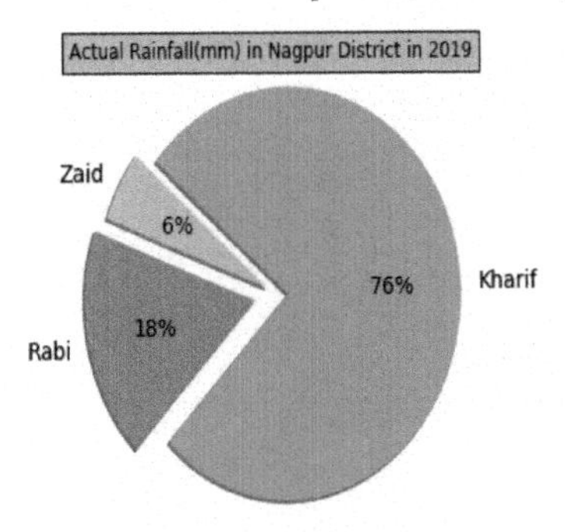

Figure 78.3 Distribution of rainfall within a year with respect to seasons

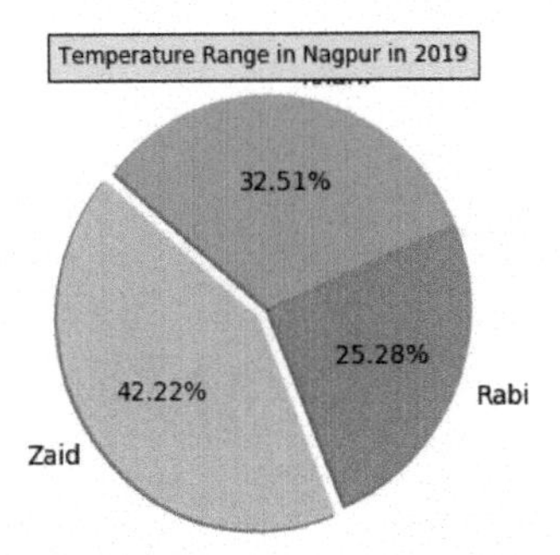

Figure 78.4 Distribution of temperature (%) within a year with respect to seasons

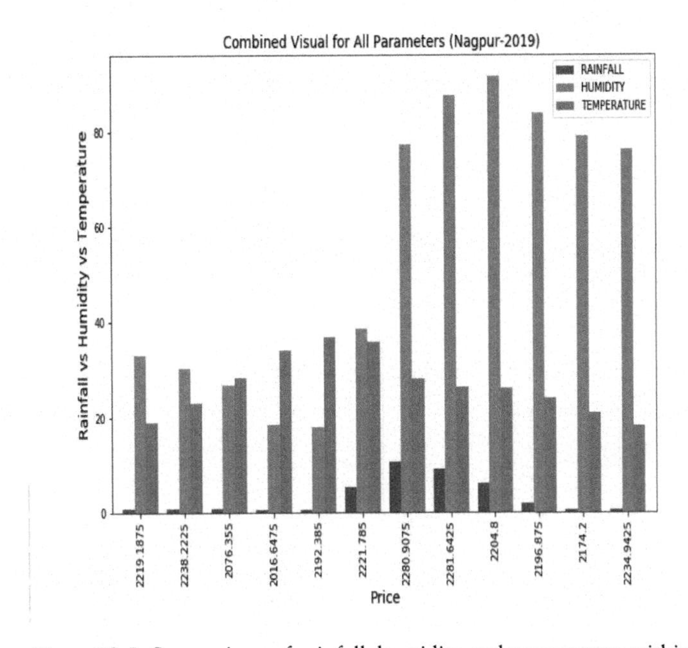

Figure 78.5 Comparison of rainfall, humidity and temperature within a year with respect to price

Figure 78.4 depicts a pie chart for the temperature range of Nagpur District for the year 2019. The temperature dataset is used here. For plotting graph from the dataset temperature_range column is considered. The row count is 48. The parameters used are seasons of agriculture in India i.e. kharif, rabi, and zaid. The visual describes the distribution of temperature (%) within a year with respect to seasons.

Figure 78.5 depicts a bar graph for the combined dataset of temperature, rainfall, and humidity vs crop price for Nagpur district. Rainfall, temperature and humidity are considered as parameters are on Y-axis while the Price parameter is considered on X-axis. The crop considered here is wheat.

Consider an instance from the above bar graph. The temperature range is 28.04 (°C), rainfall is 10.04 (mm) and humidity is 77.21. So, depending on specified parameters the price rate of a specific (here we consider wheat) crop is Rs. 2280.9075 per quintal.

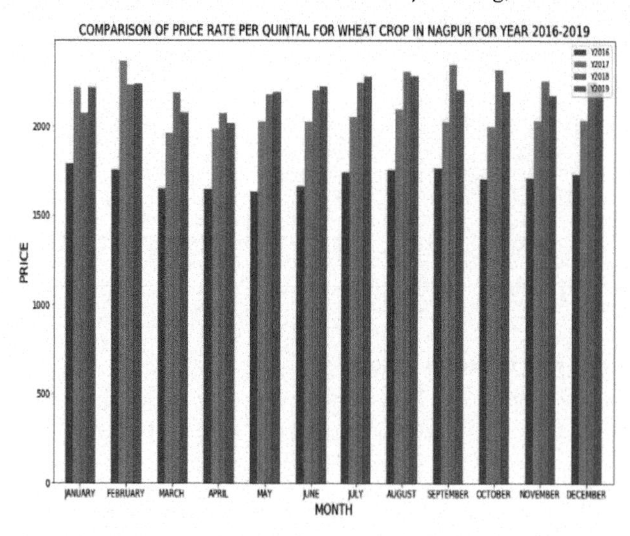

Figure 78.6 Comparison of prices of the wheat crop from the year 2016 to 2019

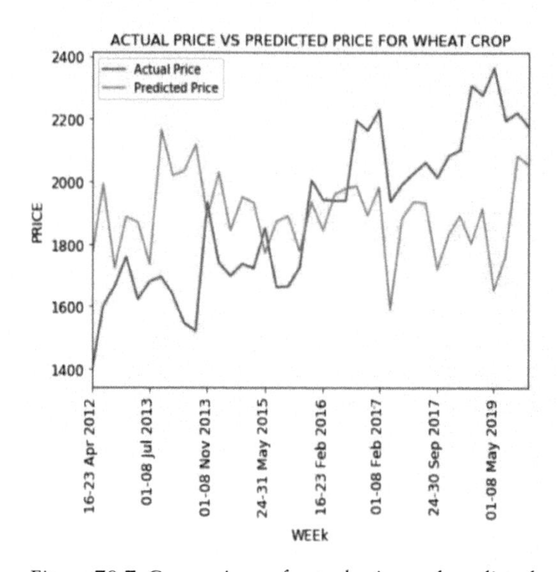

Figure 78.7 Comparison of actual price and predicted price for wheat crop in Nagpur district

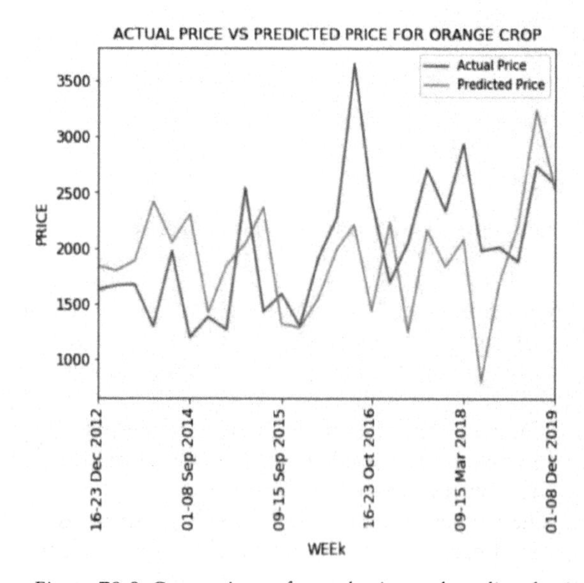

Figure 78.8 Comparison of actual price and predicted price for orange crop in Nagpur district

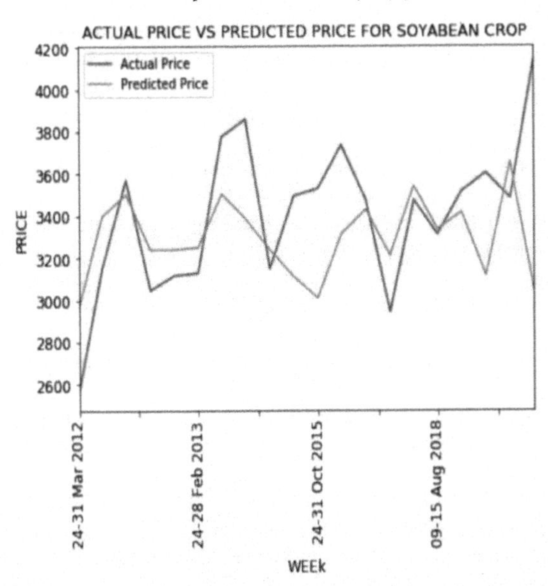

Figure 78.9 Comparison of actual vs predicted price for soyabean crop in Nagpur district

Figure 78.6 shows a bar graph that gives us a comparison of prices of the wheat crop from the year 2016 to 2019. The months are plotted on the X-axis whereas prices on the Y-axis.

From the above graph, we can infer facts such as the price of wheat crop in 2017 was found to be the highest in February. Also, wheat was the cheapest in May of 2016. The year 2017 and 2019 has referred to the same range of prices in January. We can also infer the growth of wheat prices with the growing years. Let's take an instance for October. In 2018, the recorded price rate is 2315.9275 Rs. which is the highest compared to others.

Figures 78.7–78.9 show the graph, which gives us the visual of actual values and the predicted values of the Wheat, Orange, and Soyabean crop for the different years. The graph shows Maximum accuracy with the actual value to the predicted values.

Conclusion and Future Work

A model for a crop-price trend prediction system is proposed in this paper. This model will be built on data from a government website called agmarket.com. The database is specific to Maharashtra for all weeks, from the year 2016 to the year 2019, for the wheat crop. To test our proposed training model on this data, a machine learning algorithm XGBoost algorithm is used. Prediction is done based on the previous year's database. The experimental results show that the proposed model will give the price prediction of the crop, like in which month the price goes highest or lowest, it can also give prediction regarding weather conditions, etc. Based on this inference, farmers can take the decision for the cultivation and selling of crops. From the proposed model price of the crop can be inferred with 80% efficiency. In this way proposed model helps the farmers to decide which crop will be more in-demand and beneficial to them for cultivation, and take middlemen out of the picture.

The above working model can further be improved and made more effective by increasing the batch size, as well as testing the models on larger datasets. The proposed system can be extended to include other factors like soil, and previous crop cycles subject to proper data acquisition.

References

Cai, Y., Moore, K., and AdamPellegrini, A, Elhaddad, A., Lessel, J., Townsend, C., Lessel, J., Solak, H., and Semret, N. (2017). Crop yield predictions - high resolution statistical model for intra-season forecasts applied to corn in the US. https://www.researchgate.net/publication/316278341_Crop_yield_predictions_-_high_resolution_statistical_model_for_intra-season_forecasts_applied_to_corn_in_the_US

Descamps, B. (2017). Regression prediction intervals with XGBOOST. Published in Towards Data Science. https://towardsdatascience.com/regression-prediction-intervals-with-xgboost-428e0a018b. (accessed April 25, 2017).

Guest_blog. (2018, September 6). An End-to-End Guide to Understand the Math behind XGBoost. https://www.analyticsvidhya.com/blog/2018/09/an-end-to-end-guide-to-understand-the-math-behind-xgboost/

https://ieeexplore.ieee.org/stamp/stamp.jsp?tp=&arnumber=7501673

Jharna, M., Eyappa, S. N., and Ankalaki, S., (2017). Analysis of agriculture data using data mining techniques: application of big data. J. Big Data. 4:20:115. https://journalofbigdata.springeropen.com/track/pdf/10.1186/s40537-017-0077-4.pdf

Kalbarczyk, H. Z. and Iyer, R. K. (2016). A Data Driven Approach to Soil Moisture Collection and Prediction. IEEE International Conference on Smart Computing.

Mahabadi, N. Y. (2018). Use of the Intelligent Models to Predict the Rice Potential Production. Int. Acad. J. Innov. Res. 5(2).

Ng, Y. (2019). Forecasting Stock Prices using XGBoost. https://towardsdatascience.com/forecasting-stock-prices-using-xgboost-a-detailed-walk-through-7817c1ff536a (accessed October 26, 2019)

Rachana, P. S., Rashmi, G., and Shravani, D., Shruthi, N., and Kausar, S. R. (2019). Crop Price Forecasting System Using Supervised Machine Learning Algorithms. International Research J. Eng. Sci. Technol. (4):8054807 https://www.irjet.net/archives/V6/i4/IRJET-V6I41037.pdf

Rohit, R. and Baranidharan, B. (2020). Crop Yield Prediction Using XGBoost Boost Algorithm. International J. Recent Technol. Eng. 8(5):35163520.

Rohitha, R., Vishnu, R., Kishore, A., and Chakkarawarthi, D. (2020). Crop Price Prediction and Forecasting System using Supervised Machine Learning Algorithms. International J. Adv. Res. Comput. Commun. Eng. 2729. https://ijarcce.com/wp- content/uploads/2020/03/IJARCCE.2020.9306.pdf

Sagar, P., Uttam, P., and Sidnal, N. (2015). Prediction of Future Market Price for Agricultural Commodities. International J. Syst. Softw. 1017. http://www.publishingindia.com/ijsse/70/prediction-of-future-market-price-for-agricultural-commodities/394/2804/ https://www.ijrte.org/wp-content/uploads/papers/v8i5/D9547118419.pdf

Sahu, H. (2019). Stock Prediction with XGBoost: A Technical Indicators' approach. https://medium.com/@hsahu/stock-prediction-with- xgboost-a-technical-indicators-approach-5f7e5940e9e3 (accessed Jan 28, 2019).

Singh, V. and Abid, S., (2017). Analysis of soil and prediction of crop yield (Rice) using Machine Learning approach. Int. J. Adv. Res. Comput. Sci. Softw. Eng. 5(8):1254–1259. http://www.ijarcs.info/index.php/Ijarcs/article/view/3830/3516

Thayakaran, S., Suganya, S., and Puvipavan, P., Manogarathash, M. P., Gamage, A., and Kasthurirathna, D. (2019). Agro-Genius: Crop Prediction using Machine Learning. International J Innov. Sci. Res. Technol. 4(10):243249 https://ijisrt.com/assets/upload/files/IJISRT19OCT1880.pdf.pdf

Vohra, A., Pandey, N., and Khatri, S. K. (2019). Decision Making Support System for Prediction of Prices in Agricultural Commodity. Amity International Conference on Artificial Intelligence. 345348. https://ieeexplore.ieee.org/document/8701273

Wankhade, A., Warkade, D., Shirke, S., Patole, S., Kapse, S., and Ardhapurkar, S. (2020). Social Intelligence Monitoring Based Election Prediction Using Sentimental Analysis in R for Twitter. A J. Comp. Theory. 10571065. http://www.jctjournal.com/gallery/132-feb2020.pdf

Yashavanth, B. S., Singh, K. N., Paul, R. K., Paul, A. K. (2017). Forecasting prices of coffee seeds using Vector Autoregressive Time Series Model. Indian J. Agric. Sci. 87(6). http://epubs.icar.org.in/ejournal/index.php/IJAgS/article/view/70960

Zhang, Y. and Sanggyun,N. (2018). A Novel Agricultural Commodity Price Forecasting Model Based on Fuzzy Information Granulation and MEA-SVM Model. Hindawi Mathematical Problems in Engineering. 110. https://downloads.hindawi.com/journals/mpe/2018/2540681.pdf

79 Design novel algorithm for sentimental data classification based on hybrid machine learning

Jyoti Srivastava[a] *and Neha Singh*[b]

Department of ITCA, Madan Mohan Malviya University of Technology Gorakhpur, India

Abstract

Natural language processing can be used to identify emotional content in text or spoken utterances in general. One example of this type of processing is sentimental analysis. Online monitoring and listening tools analyse and characterise emotions in various ways, and each has a varied level of performance and accuracy in doing so. Market feedback can be used to provide quantitative service measurement and sentiment analysis of teacher reviews. The qualitative measurement of accuracy, on the other hand, is complex since it requires feature extraction and machine learning methods for example support vector machines (SVM) and Naive Bayes classifier. Correct analysis and interpretation of sensations because sentiment analysis is far more sophisticated, the SVM classifier outperforms the Naive Bayes classifier in terms of accuracy and speed. We will look at how to classify the sentiment of a teacher review dataset in this paper. Create a new technique to identify sentimental data using a fusion-based algorithm.

Keywords: Classification, machine learning, SVM classifier.

Introduction

Analysis of sentiment is a dimension evaluation of the mind, sentiments, feelings and attitudes of human beings. Approaches of sentiment analyses can typically be split into two groups, machine learning and lexical approaches. The first uses machines for the description of emotions of polarity. Such methods usually require a lot of marked training material. Nevertheless, it is a challenge in itself to collect appropriate labelled data. Lexicon related approaches are used to measure the spectrum of sentiments of the comments. They then group the assessments by emotional values in positive or negative categories. Sentiment lexicon is the basis for a lexicon-based system, so it has drawn great attention naturally to how to create a sentiment lexicon (Lochter et al., 2016). The methods of Sentiment generation are dispensed normally into two types: dictionary and corpus. The look at suggests that the domain-specific Sentiment lexicon leads in comparison with a widespread emotion lexicon to more potent sentiment analysis. For instance, a positive word is long when it is used for explaining the standby time of teacher review in the summary that comments on your mobile, but when a printer prints a letter, it is a bad word (Awwad and Ieee, 2016). No efficient way to discover and evaluate domain-dependent Sentiment Lexicons is still accessible despite a great deal of study. There are some shortcomings in the current methods: 1. these methods are applicable only to some areas with the regular use of emoticons; 2. Human-annotated information is needed; 3. the lexicon contains more optimistic or negative terms of Sentiment. Yan and Zhang (2012) propose a hybrid algorithm to produce a cross-domain sentiment word list, used to characterised teacher review sentiment. To establish emotion classification model for teacher review, combine lexicon-based methods and a supervised computer learn methods. Figure 79.1 shows various sentiment analysis (SA) process.

Relater Work

The opinionated text in sentiment analysis is essential in making decisions depends on its analysis. For the system of opinion mining, the opinionated text is gathered and viewed it as the input and at the time of accomplishing this, it is very important to know about the terms related to opinion mining. An objective expression made by the user concerning definite entities, events or objects, and its features is referred as facts (Li et al, 2018). Similarly, a subjective expression that depicts the person's emotion, sentiments and the performance appraisal regarding entities, objects, events and its attributes is referred as opinion. Normally opinions can be conveyed on everything that may be a person, product, particular topic or business industries. This specifies that, the most opinions are conveyed on target entities that are having own elements and characteristics. Therefore, in opinion mining, an item may be cut up into the hierarchical ranges relying

[a]sriv.jyoti1996@gmail.com; [b]nehaps2703@gmail.com

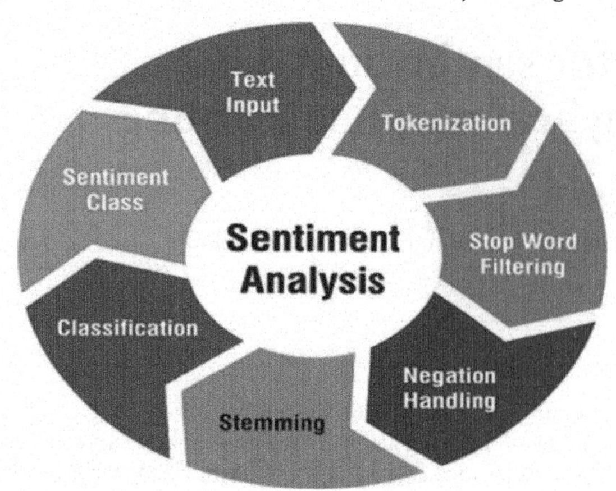

Figure 79.1 SA process

at the element-of relation. Han et al.(2018) then, we suggest a finalised lexical sentiment analysis method utilising a domain-specific sentiment lexicon created by a process of production of domain-specific sentiments. The work is focused on databases which are available to the public. Results suggest that the proposed lexical emotion analysis system utilising field-specific lexicons generated by the proposed approach performs well.

Big Data technologies provide the required and efficient methodology for the collection, storage and processing of large datasets belonging to different variety of data types and samples (Park et al., 2015). Educational data mining is a popular field for knowledge analysis using data mining algorithms in educational fields (Pai et al., 2017). In this paper we used few gaining knowledge of algorithms to assess the school of an educational group efficiently on the premise of the remarks acquired from the scholars. Our theoretical model uses emotional analysis and algorithms to interpret student input to capture the emotions. This model offers the staff of a specific educational institution an objective and efficient way to determine. The staff may be evaluated with this new model and evaluated with different parameters, to be able to help us to enhance the instructional and education standard.

The sentiment analysis use of natural linguistic, computational and text analysis which identifies, retrieves, and studies the opinion, the polarity of sentiments from the text (Cocea and Fallahkhair, 2014). Polarity of emotions is typically negative or positive, although it is often neutral. Earlier studies have shown that an analysis of sentiment is more effective when used in some areas (Awwad and Ieee, 2016). The key subject of sentiment study in the field of education was e-learningwith no work into the input from the classroom (Jiang et al., 2018; Anandarajan et al., 2018). Even if e-learning and classroom education sound identical, the relationships between students and lecturers are different and the lecturer must come by student comments come at same time. Feedback from classroom students is distinct from that of remote students since of different circumstances and problems that student may face. Students studying may have issues such as contact absence. Class input may also be about environments for classes such as hot classrooms. When a model is trained, terms must contribute to the intent of main aim to provide best education must be trained using input from the classroom.

In order to perform sentimental analysis, it proposes a Lexicon-Naive Bayesian (HL-NBC) hybrid system. In addition, a subject classification precedes the Sentiment analysis system (Rodrigues et al., 2019). Tweets are categorised in various categories and insignificant tweets are flushed. The suggested approach for unigram and bi-gram features is contrasted with HL-NBC classifier. Our proposed HL-NBC system can improves sentiment classification and provides 82% accuracy, which is better than other methods, from different approaches. The sentimental research is also completed in less time than conventional approaches and increases the processing time by more than 93%.

The following conclusions can be taken from the study (Li et al., 2018). (1) Digital sentiment analysis focused on deep learning will remain a hot subject. (2) In the context of multimedia sensitivity analysis in particular for visual sense analysis, the use for similar areas of current semanthetic research technologies (reasoning technologies) will be advantageous. (3) Our focus needs to extend current methods for analysis of textual sentiments to a combined study of visual-textual sentiment.

The relation between textual and visual content should also be taken into account. The study of audio sentiments, video sentiments and multimodal emotion study will concentrate on the audio-visual material (not limited to self-timer videos) of social networks. (4) More effort should be devoted. In brief, this thesis performed an exhaustive overview of social multimedia SA literature, recognised important open issues and explored future directions and patterns of research. To those researchers interested in this area, it will be a valuable guide.

The enormous interest among Internet users in social network sites is rising rapidly in comparison to the worldwide increase in technological levels (Anuprakash et al., 2019). Twitter is the social networking site where the updates of users communicate with the comments that are labelled 'tweets.' For this study the program would classify the person's actions as proactive, hostile and proactive-aggressive when evaluating the user's cumulative amount of optimistic, negative, or neutral opinions in all of his/her entries. The proliferation of social network data offers incentives for interpreting criminal-minded interactions, which often presents computational challenges for instructional purposes in making sense of the social media data.

For starters, in the new age online media networks, Twitter, have become exceedingly popular, because the amounts of consumers that utilise them on a regular basis are increasingly growing (Mir et al., 2019). Data distributed across these stages is their most enticing feature, because it is considered to be expedient and informative in financial terms. Another incredibly seductive aspect of these processes is the manner in which consumers will articulate what needs to be next to zero regulation.

Hybrid based mostly approch

It is a mixture of ML and lexicon-based methodology, which is often used in many essential techniques with sentimental lexicon express main position. We have a tendency in this to outline the various terms used in the mining opinion.

Fact: It has really occurred, or is completely the case, in fact. Opinion: Associate degree opinion can be a traditional read or opinion on one issue, not necessarily focused on fact or evidence.

Subjective paragraph: Whether it just reflects one's thoughts, a phrase or document is subjective or mild.

Objective sentence: The corresponding objective degree sentence shows some details and established globe info.

Piece: a private object or device, particularly one which is a part of a display, selection or package.

Review: A review can be a text comprising a series of words for a chosen object and has consumer views. A analysis may either be, or could be arbitrary or impartial.

Established features: Recognised features of square measure default attributes of the attached platform on which users give scores highly.

Sentiment: Sentiment can be a word of polarity signifying the manner in which an idea or viewpoint is conveyed. We have a propensity to use emotion as a viewpoint on one thing in an additional unique context. For e.g., the term 'battery life' inside the expression 'this smartphone has a wonderful battery life' may be wonderful.

Opinion expression: The expression correlated with the degree opinion can be a mixture of head word and adjective. Usually, the head word may be a facet of the nominee and thus the suffix may be a thought reflecting any viewpoint

Polarity of opinion: Polarity of thought or good judgement Orientation reflects the polarity of numerical values shared by the consumer or customer.

- Polarity: Polarity may be an orientation measure in three directions. A perception is either optimistic bad, indifferent in this, whatever.
- Rating: The bulk of reviews platforms use polarity star scores, defined by stars within the spectrum from one to five square score.

A comparison of the variances achieved using different characteristics on the validation collection is shown in Flowchart 79.1.

Proposed algorithm

Several steps are:-

Step 1: Remove all stop words, punctuation, and symbols from the data to make it cleaner. The parts of the data set with a frequency larger than one are considered.

Step 2: The words are repeated in data sets are compared with the words that are available in vocabulary and also in training datasets.

Step 3: Examine the data set gathered from the training data for matching words s-set or its subgroup that is comprised of items greater than one, and the s-subset that is comprised of items greater than one of the repeated data set of the new dataset.

Step 4: The equivalent possibility values of matched data set for every target class are gathered.

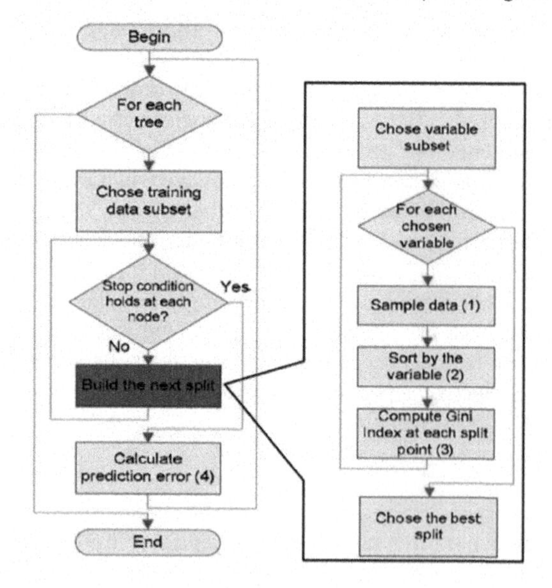

Figure 79.2 Open source language R

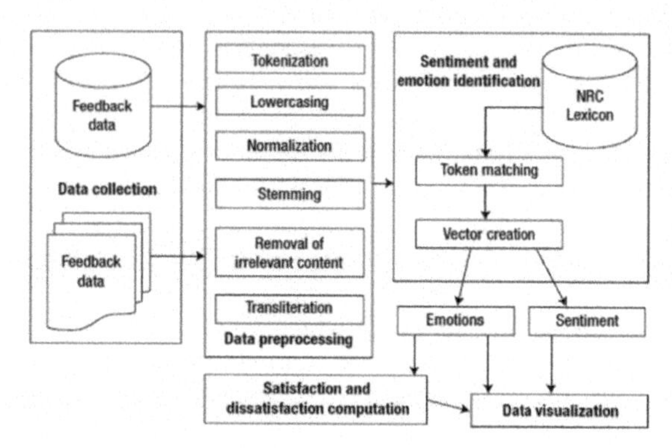

Figure 79.3 Proposed sentiment analysis system architecture

Step 5: The probability should be deliberated.
Step 6: Score algorithm is practiced to compute the range in which the attributes must be lying.
Step 7: The probability class should be evaluated by executing the algorithm of expectation maximisation.
Step 8: The dataset arranged in the class maintaining maximum probability as cleaned dataset.

Open source language R is used by the system to perform pre-processing and sentiment analysis of data as shown in Figure 79.2.

A. Classification

Firstly, let's create seed for generators of random numbers.

B. BOW + Naive Bayes

Bayes: It's fun to see what kind of effects we might get from a basic model like that. The representation of the bag-of-words is binary, so Naive Bayes classifier seems like a good algorithm for beginning the experiments.

C. Additional features

To avoid forcing any other algorithm main model limit of info, try to attach some functionality that would help to distinguish tweets.

A common sensei indicates to the other word such as good marks and may be critical in deciding the mood. Main functions are applied to the model for the data:

Logic behind extra features

We should perceive how (a portion) of the additional highlights separate the informational index. Somei of ithem, i.e number outcry imarks, inumber of pos/neg emojis do ithis truly well. In spite of the great detachment, those highlights once in a while happen just on little subset of the preparation dataset.

More features - word2vec

The general execution of the past classifiers could be improved by show main parameters modifications, anyway there's not ensure on how enormous the addition will be.

On the off chance that the out-of-the-rack techniques didn't performed well, it appears that there's very little in the information itself. The following plan to include more into information model is to utilise word2vec portrayal of a tweet to perform order.

The word2vec permits to change words into vectors of numbers. Those vectors speak to extract includes, that portray the word likenesses and connections (i.e. co-occurrence).

The word2vec is that procedure on the vectors roughly keep the qualities of the words, with the goal that joining (averaging) different words sentences are used for probably going to speak to the general subject of the sentence.

It should be modified to word2vec category in order to use GloVe-prepared layout in the gensim collection. The main distinction between these arrangements is that word2vec content records start with two numbers: documents are to be measurement number.

Sadly, this scale of the material archive is over 1.9 GB and word processors cannot be used to access and change it in a fair period of time, this C # scrap contains a need line (sorry, it's not Python, but I had memory issues with the text encoding in Python).

Using Word2Vec

word2vec API from gensim library: are used for different classed

Extra features from word2vec

Other than the 200 extra highlights from the word2vec portrayal, I had a thought of three additional highlights. On the off chance that word2vec permits to discover similitude between words that implies it can discover likeness to the particular feeling speaking to words. The primary thought was to process similitude of the entire tweet with words from names: positive, negative, unbiased. Since the object was to discover the notion, I imagined that it will be smarter to discover comparability with progressively expressive words, for example, great and terrible. For the unbiased supposition, we utilised word data, since the vast majority of the tweets with impartial conclusion were giving the data.

Logic ibehind word2vec iextra features

So as to appear, why the three additional highlights fabricated utilizing word2vec may help in feeling examination. The graph beneath shows what number of tweets from given class were commanding (had most elevated worth) on closeness to those words. Despite the fact that the words doesn't appear to isolate the assessments themselves, the contrasts between them notwithstanding different parameters, may enable the order to process - i.e. when tweet has most noteworthy incentive on good similarity it's almost certain for it to be characterised to have positive assumption.

Test data classification

Subsequent to discovering best cross-approved parameter for the XG Boost, it's a great opportunity to stack the test information and anticipate supposition for them. Last classifier will be prepared in general preparing set. End score will be uncovered when the Tweets od angry rivalry will end The information will be sent out to CSV record in group containing two segments: Id, Category. That 4000 test tests with obscure circulation of the feeling names. Figures 79.3–79.5 show the required result.

```
make_np_array_XY()
make_np_array_XY()
type(svc)= <class 'sklearn.svm.classes.SVC'>
svc= SVC(C=1.0, cache_size=200, class_weight=None, coef0=0.0,
  decision_function_shape=None, degree=3, gamma='auto', kernel='linear',
  max_iter=-1, probability=False, random_state=None, shrinking=True,
  tol=0.001, verbose=False)
Y_test:
[ 1.  0.  1.  1.  1.  0.  1.  1.  0.  0.  1.  0.  1.  0.  1.  0.  1.  1.
  1.  1.  1.  0.  1.  1.  0.  1.  1.  1.  1.  1.  1.  0.  1.  1.  1.  0.
  1.  0.  0.  1.  1.  0.  1.  1.  0.  1.  1.  1.  1.  1.  1.  1.  1.  1.
  1.  1.  1.  0.  0.  1.  1.  0.  1.  0.  0.  1.  1.  1.  1.  1.  0.  0.
  1.  1.  1.  0.  0.  1.]
Y_predict:
[ 1.  1.  1.  1.  1.  1.  1.  1.  1.  1.  1.  1.  1.  1.  1.  1.  1.  1.
  1.  1.  1.  1.  1.  1.  1.  1.  1.  1.  1.  1.  1.  1.  1.  1.  1.  1.
  1.  1.  1.  1.  1.  1.  1.  1.  1.  1.  1.  1.  1.  1.  1.  1.  1.  1.
  1.  1.  1.  1.  1.  1.  1.  1.  1.  1.  1.  1.  1.  1.  1.  1.  1.  1.
  1.  1.  1.  1.  1.  1.]
Got 54 out of 78
f1 macro = 0.41
f1 micro = 0.69
f1 weighted = 0.57
```

Figure 79.4 It shows training and testing result

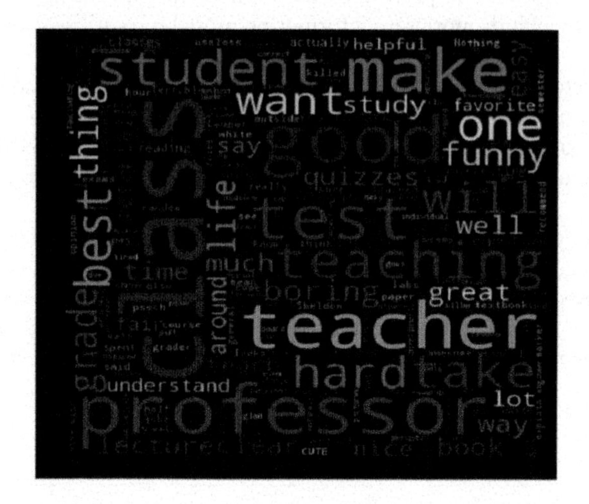

Figure 79.5 It shows result based of the word

```
Collected 562 feature sets
Training set size = 337 reviews
Test set size = 225 reviews
Accuracy on the training set = 0.9406528189910979
Accuracy of the test set = 0.72
Most Informative Features
                  suck = True           0.0 : 1.0    =    8.6 : 1.0
               nothing = True           0.0 : 1.0    =    8.6 : 1.0
              favorite = True           0.0 : 1.0    =    7.2 : 1.0
                 write = True           0.0 : 1.0    =    5.9 : 1.0
                  hate = True           0.0 : 1.0    =    5.9 : 1.0
                saying = True           0.0 : 1.0    =    5.9 : 1.0
               willing = True           1.0 : 0.0    =    5.6 : 1.0
                     . = None           0.0 : 1.0    =    5.1 : 1.0
                always = True           1.0 : 0.0    =    5.0 : 1.0
                   lot = True           1.0 : 0.0    =    4.7 : 1.0
                  told = True           0.0 : 1.0    =    4.6 : 1.0
            ridiculous = True           0.0 : 1.0    =    4.6 : 1.0
               comment = True           0.0 : 1.0    =    4.6 : 1.0
                 sense = True           0.0 : 1.0    =    4.6 : 1.0
                 waste = True           0.0 : 1.0    =    4.6 : 1.0
                   ask = True           1.0 : 0.0    =    4.6 : 1.0
                 every = True           1.0 : 0.0    =    4.4 : 1.0
                 tough = True           1.0 : 0.0    =    4.2 : 1.0
                  help = True           1.0 : 0.0    =    4.0 : 1.0
                office = True           1.0 : 0.0    =    3.9 : 1.0
```

Figure 79.6 shows the Naïve Bayes Classifier with accuracy 80.64516129032258%

Conclusion

Using emoticons as noisy markers to train data is an efficient method to do remote supervised learning. Using this approach, ML algorithms (Naive Bayes, maximum entropy classification) may achieve high precision when classifying the sentiment. This study compares the performance of three classifiers. Model: The model is made up of four parts: collecting data, pre-processing, extracting feature vectors, and categorizing them. While each classifier performs well, it can be stated that a machine-learning approach holds a lot of potential for sentiment classification. As a result, the findings of this investigation will aid us in understanding how individuals perceive various things. It is believed that if all three algorithms have more data, they will be better at classifying emotions. The work can be enhanced by removing Internet slang and including additional machine learning classifiers.

References

Altrabsheh, N., Cocea, M., and Fallahkhair, S. (2014). Sentiment Analysis: Towards a Tool for Analysing Real-Time Students Feedback. 2014 IEEE 26th International Conference on Tools with Artificial Intelligence. doi:10.1109/ictai.2014.70

Anandarajan, M., Hill, C., and Nolan, T. (2018). Learning-Based Sentiment Analysis Using RapidMiner. Adv. Anal. Data Sci. 243–261. doi:10.1007/978-3-319-95663-3_15

Awwad, H. and Ieee, A. A. (2016) Performance comparison of different lexicons for sentiment analysis in arabic. 2016 Third European Network Intelligence Conference (Enic 2016). 127–133. https://doi.org/10.1109/enic.2016.25

Chithra, R. G, Harshitha, G. M, Anuprakash, M. P., and Rakshitha, H. B., (2019). Behavioural Analysis of Tweeter data : A Classification Approach. , Int. J. Eng. Res. 7(8).

Jiang, J., Lu, Y., Yu, M., Li, G., Liu, C., Huang, W., and Zhang, F. (2018). Sentiment Embedded Semantic Space for More Accurate Sentiment Analysis. Lecture Notes in Computer Science. 221–231. doi:10.1007/978-3-319-99247-1_19

Park, S., Lee, W., and Moon, I.C. (2015). Efficient extraction of domain specific sentiment lexicon with active learning. Pattern Recogn. Lett. 56:38–44. https://doi.org/10.1016/j.patrec.2015.01.004

Hamilton, W. L., Clark, K., Leskovec, J., and Dan, J. (2016). Inducing domain-specific sentiment lexicons from unlabeled corpora. Proceedings of the 2016 Conference on Empirical Methods in Natural Language Processing, (pp. 595–605). https://aclanthology.org/D16-1057

Han, H., Zhang, J., Yang, J., Shen, Y., and Zhang, Y. (2018). Generate domain-specific sentiment lexicon for review sentiment analysis. Multimed. Tools. Appl. 77(16):21265–21280. doi:10.1007/s11042-017-5529-5

Krishnaveni, K. S., Pai, R. R., and Iyer, V. (2017). Faculty rating system based on student feedbacks using sentimental analysis. 2017 International Conference on Advances in Computing, Communications and Informatics (ICACCI). doi:10.1109/icacci.2017.8126079.

Li, Z., Fan, Y., Jiang, B., Lei, T., and Liu, W. (2018). A survey on sentiment analysis and opinion mining for social multimedia. Multimedia Tools and Applications. doi:10.1007/s11042-018-6445-z

Lochter, J. V., Zanetti, R. F., Reller, D., and Almeida, T. A., (2016) Short text opinion detection using ensemble of classifiers and semantic indexing. Expert Syst Appl 62:243–249. https://doi.org/10.1016/j. eswa.2016.06.025.

Mir, A. A, Joshi, K., and Oberoi, A. (2019). A Credibility Analysis System for Assessing Information on Twitter, Int. J. Eng. Res. 8(1).

Rodrigues, A. P. and Chiplunkar, N. N. (2019). A new big data approach for topic classification and sentiment analysis of Twitter data. Evolution. Intelligence. 15(9). doi:10.1007/s12065-019-00236-3

Yan, L. and Zhang, Y. (2012). News Sentiment Analysis Based on Cross-Domain Sentiment Word Lists and Content Classifiers. Lecture Notes in Computer Science. 577–588. doi:10.1007/978-3-642-35527-1_48.

80 Design and development of bull operated mulching machine suitable for small scale farmers in India

Sandip S. Khedkar[a], *Abhiroop Sarkar*[b], *Aditya Sankale*[c], *Ameya Kumbhare*[d], *Arpit Mishra*[e], *and Ashish Yesankar*[f]

Department of Mechanical Engineering Yeshwantrao Chavan College of Engineering, Nagpur, Maharashtra, India

Abstract

This research paper analyses a bull-operated integral approach for the mulching process which can help the low-scale farmers of India. For this integration, a safe and simple model was designed in such a way that both the laying and retrieval process can be carried out while taking the previous literature works about mulching into consideration.

The literature survey was done and it was observed that there is a lack of prototypes with an integrated mechanism in the market that can be afforded by low scale farmers. The two key mechanisms this research focuses on are laying and retrieval. The retrieval mechanism is performed using components like duck foot, chain-sprocket mechanism and flexible rod while the laying process uses components like the laying rod, supporting wheels, and soil collector plates. To design the machine, a 3-D CAD model was designed on Fusion 360 after calculating all the design calculations manually and the same was analysed based upon the forces calculated, where the design was found to be safe and the stresses were well under the permissible limit.

Keywords: Agricultural, design, laying, mulching, retrieval, crop-production.

Introduction

Around 38% of the global land surface in the world and more than around 60% of the total land surface in India is occupied for agricultural purposes thus making it one of the highest water consumers around the globe (Sharma and Bhardwaj, 2017). Taking into consideration the increasing water shortage, there is a need for rain-fed cultivation for maintaining the food supply. This is where the process of 'mulching' arises. Mulching provides the best possible solution for increasing the yield and crop quality with very fewer amount inputs compared to the conventional method. Mulching is a process where mulch paper is being laid all over the crops and removed after its proper usage. Mulch paper covered over the soil creates a microclimate condition which makes the conditions favourable for the crop growth and increases its yield. It is being observed that the overall yield in mulching is twice compared with the conventional method. Mulching provides multiple advantages over conventional farming methods in agriculture. Mulching conserves soil moisture, as it is observed there are plenty of abiotic items or components that are accountable for the minimisation of moisture content from the soil of the agricultural field. Mulching minimises the erosion of soil and soil compaction as mulching products and materials preserve the soil from biotic factors like wind and water erosion phenomenon and it reduces the compaction of the agricultural soil which might affect the roots badly while simultaneously minimising the growth and development of crops. Mulching regulates the soil temperature, as mulching packs the above soil surface which is beneficial in controlling or maintaining the temperature of soil which helps for overall crop production and growth. Mulching improves soil fertility as it has been observed that natural mulches provide many advantageous effects on the quality of soil with respect to improving levels of nutrients. There exist several research studies that show the beneficial effects or impacts of mulches on crop production which include germination, seedling's transplantation, and overall effective performance of crop production. Mulching materials are capable of reducing the evaporation losses which conserves the soil moisture content thus causing a reduction in water irrigation requirements. Mulches are found to be beneficial in helping for the nutrition of several beneficial species which decreases the occurrence of some diseases in crop production.

[a]sskhedkar@ycce.edu; [b]abhiroop.org@gmail.com; [c]adityasankale007@gmail.com; [d]kumbhareameya619@gmail.com; [e]arpitmishra040301@gmail.com; [f]ashishyesankar84@gmail.com

Literature Review

This paper mainly focuses on various effects and applications of using mulching on soil and conservation of water through it (Sharma and Bhardwaj, 2017).

This paper discusses the novel construction, combinations, and arrangements of components which would be an optimum cost and easy enough to be operated by an unskilled farmer (Padawal et al., 2017).

The paper discusses various advantages of using mulching and different types of mulching paper which can increase crop productivity (Kader et al., 2017).

This paper is based upon the design and development of a laying mulch machine that is also equipped with a punch hole mechanism and thereby reducing the labour cost (Bhargava et al., 2020).

To avoid possible contamination of plastic mulch in soil, this paper focuses on the winding mechanism which helps in removing the plastic mulch from the soil after the arrival of the harvesting season (Khazimov et al., 2021).

This patent comprises a prototype equipped with a conveyor and a retrieval mechanism necessary for digging out the mulch sheet from the soil (Rocca, 2012).

This paper illustrates a design based on laying mechanism such that the amount of people required in doing the job is less when compared with the traditional approach (Tipayale et al., 2017).

The paper discusses the various side effects of plastic mulch remains and the pollution caused by it. It further focuses on designing a rotary nail tooth mechanism that can be used for the retrieval process (Guo et al., 2019).

The study is based on using draught animals in the agriculture field and how bullock-driven mechanisms prove to be an integral part of our agriculture sector (RuTAG IIT Delhi, 2014).

Even today, there is a need for research on possible retrieval mechanisms. There are many patents available for laying but only a few for retrieval and even those designs are extensive massive structures that the general farmers cannot afford. This leads to a very intricate situation where there are no prototypes available in the market which can do both laying and retrieval of mulch as an integrated mechanism hence this paper focuses on the possible integrated mechanisms affordable to farmers.

Methodology

The mulching machine works as an integration of both the laying and retrieval process. Laying process in mulching refers to the laying of the mulch paper over the bed. Laying is conventionally carried out by labourers by small-scale farmers or the prototypes available as an attachment for the tractors by large-scale farmers. Thus, this makes this process tedious and cost-consuming. Therefore, our design eliminates these circumstances and situations and offers a very easy operation reducing manual work by labourers and cost. Retrieval process refers to the process where the laid mulch by laying procedure will be taken out after the proper interval of time. There exist many different prototypes for this process which act as attachments for tractors. This makes this process uneasy and costly. The methodology under this prototype makes this process easier and more cost-effective. Thus, this integrated approach will be beneficial for both these processes laying and retrieval processes.

Laying

These are the active components of our prototype that will act during our laying processes:

- Mulching roller rod: It is used for holding the mulch roll during the laying process. And In order to attach/detach the mulch roll to it, a bunch of bolts are used which can be tightened and loosened when required to remove the rod for loading the mulch roll.
- Supporting wheels: Often known as press wheels, these wheels are used for pressing the mulch sheet coming from the mulch roller over the soil. It helps in the uniform flow of the laying process.

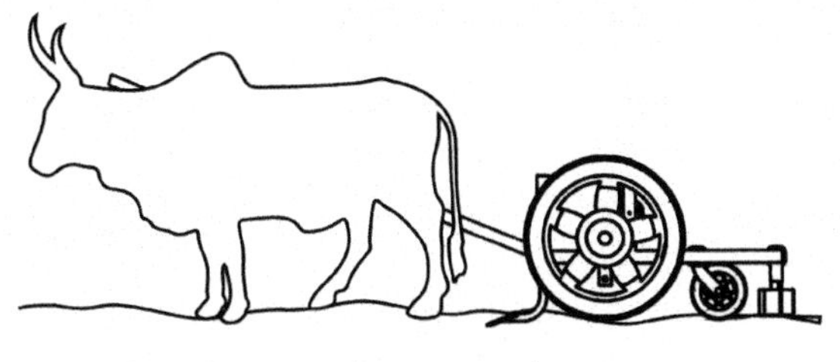

Figure 80.1 Conceptual representation

- Soil collector plates: These are used to gather soil from the field and distribute it over the laid mulch paper so that it will not get loosened and reduce the tension necessary for the mulch laying process to occur.
- Supporting rod: It is used for supporting the mulch paper before it falls on the mulching bed, it helps in providing uniform tension force and avoiding any possible wrinkled paper. The hinge joint mentioned below provides some oscillatory motion such that external factors like vibration will not affect the laying process.

The proposed solution to the tedious laying and retrieving job of mulching paper is to design and fabricate a semi-automated mulching machine which will be an integration of both laying as well as the retrieval mechanisms.

The model represented here helps us in visualising the following mechanism easily.

The mechanism consists of a supporting rod and a mulch roll along with a bunch of supporting wheels. Initially, a few metres of mulch paper are manually laid and fixed at the one end of the mulching bed, and then as the prototype moves forward with one end of mulch being fixed in the field, the mulch rolls out automatically because of the pull/tension force exerted by the fixed end. This rolled out paper is tuck/sink in the soil with the help of supporting wheels and further soil is spread over it using the soil collector plates. This will be continued until we reach the end of the mulch bed. And then the paper is cut out and the machine is prepared for the next mulching bed.

Retrieval

The main components required in the retrieval operation of mulch paper are duck foot blades, flexible rod, supporting rod, mulch wrapper, and chain sprocket mechanism. The mulching machine is designed according to the different widths of the bed. One is required to adjust the width of cultivator blades, the width of supporting wheels, and the width of collector plates according to bed sizes. Adjustments in the width are accomplished by a sliding arrangement. For sliding adjustment, holes are drilled on the mainframe which helps components like duck foot and supporting wheels to change their position accordingly, they can be fixed with the help of nuts and bolts.

After cultivation of the crop, the mulch paper will be deep inside the soil from its end and therefore initially, we must remove the mulch paper from the bed manually and tie it on the mulch wrapper. The mulch paper would stream from a flexible rod to a mulch wrapper via a supporting rod that helps in maintaining proper tension. A flexible rod is fixed at the front end in between duck foot rods which helps in removing the soil particle stuck to mulch paper and helps to loosen the contact with the surface. When the machine moves in a forward direction, the main wheels and its axle start rotating. The axle joint of the main wheel contains a chain sprocket mechanism which in turn rotates the mulch wrapper. The rotation of the mulch wrapper causes the mulch paper to be wrapped along with its rotation. The mulch wrapper/retrieval roller is designed in such a way that it should not slip with the mulch paper at the time of winding. Therefore, small circular rods are attached to the circular disc along its periphery instead of using one solid rod to provide firm gripping at the time of mulch wounding.

The design consists of construction, design calculation, analysis, and CAD models of the mulching machine.

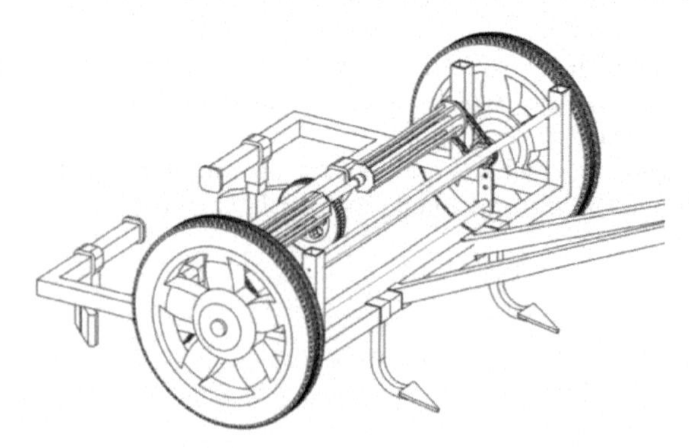

Figure 80.2 CAD model

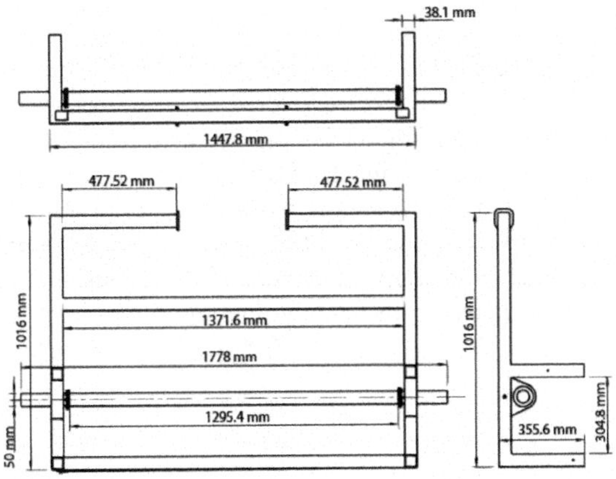

Figure 80.3 Dimensional analysis

Design calculations

Shaft

(Maximum shear stress) theory was used here for the design of Hollow shaft where do = 50 mm and Di = 36 mm.

$$\text{Design torque, } Td = \frac{60 \times P \times K_l}{2 \times \pi \times N} \tag{1}$$

Where $K_l = 2.1$(Design data book Table XI-5) [10]

$$Td = \frac{60 \times (0.7 \times 746) \times 2.1}{2 \times \pi \times 13} = 805.534\ N.m = 805.534 \times 10^3\ N.mm$$

Shear stress,

$$\tau = \frac{16 \times Td}{\pi \times D_o^3}$$

$$\tau = \frac{16 \times 805.534 \times 10^3}{\pi \times D_s^3}$$

Here for the diameter of the shaft, ds were calculated above, $\tau = 52.365$ MPa

From the Design Databook, the value of Syt for material SAE-1045 is 545 MPa and from *TRESCA* theory of Maximum shear stress theory:

$$\text{Abs.}(\tau max) <= Syt/2.N \tag{2}$$

where N is factor of safety

$$\text{Abs.}(\tau max) <= \frac{545}{2 \times 2}$$

Abs.(τmax) <= 136.25 MPa

It can be seen that the above-calculated stress was under the permissible limit. Henceforth, the design is safe for a diameter of Do = 50 mm and Di = 36mm.

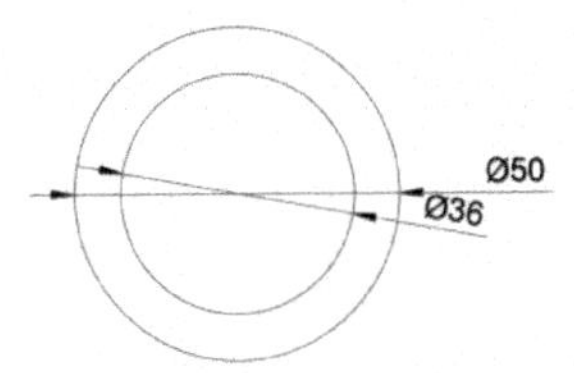

Figure 80.4 Cross-section of shaft

Bull-force analysis

The following assumptions were considered while calculating the load exerted by two bulls in order to pull the machine:

The average height of the bull is considered to be 1.524 m. According to the 'comparative study of bullock driven tractors (RUTAG IIT Delhi, 2014), the horsepower Hp of two bullocks during ploughing, ranges from 0.291.87 Hp. There are many instances when these bullocks can produce more work than a mini tractor. The ground clearance for our prototype is to be 0.254 metres, so the coupling would be mounted at 0.2794 metres. So, the effective height will be the difference between bull height and coupling height and that would be 1.2446 metres. The horizontal distance is assumed to be 3 metres for proper transmission of required power.

Bull height, (Maddock) $Hb = 1.524\ m$

Coupling height, $Hc = 0.2794\ m$

Effective height, $H' = 1.2446\ m$

Horizontal distance $Hd = 3\ m$

Elevation angle $\psi = tan\text{-}1(H'/Hd) = 22.531$ *degrees.*

The length of rods for a plane passing through A' CO

$sin\ \psi = H'/Lb$

$Lb = H' / sin\ \psi = 1.2446 \times sin\ (22.531) = 3.2480\ m.$

Assuming distance between links to be $Ld = 0.45\ m.$

For a plane passing through ABO at an angle ψ from horizontal

$tan\ \lambda = (Lb + Lc)/(0.5 \times Ld) = 15.012$

The angle between coupling rod and frame $\lambda = 86.188$.

Total length of the rod L:

$sin\ \lambda = (Lb + Lc) / L$

$L = (Lb + Lc) / sin\ \lambda = 3.405\ m.$

Effective length $Le = L - Lc' = L - (Lc/sin\ \lambda) = 3.2546\ m.$

Considering Hp of bullock to be 1.08 (avg,),

Power $P = 1.08 \times 746 = 805.68\ W$

Considering speed v to be 1.5 kmph or 0.41 m/s,

Pulling force delivered by bullocks,

$F\ bull = P/v = 805.68/0.41 = 1965.073\ N$

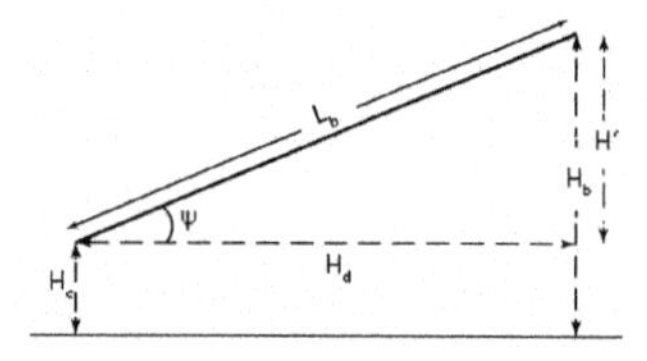

Figure 80.5 Side plane

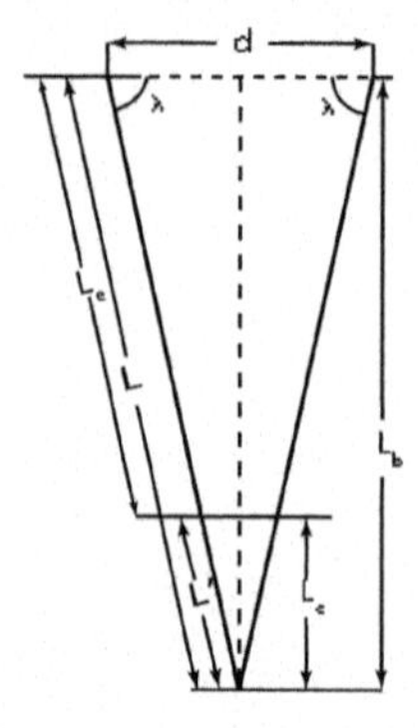

Figure 80.6 Top plane

Force distributed for points A and B are:

F (bull A) = 982.536 N F (bull B) = 982.536 N

Duck foot

Soil cultivator has a curved shape, they can be approximately analysed using the angle of the cutting point at power tip (as effective rake angle). They have symmetry across the x-z plane.

Thus, they can be analysed as Flat plates. By using McKyes (1985) model for development of cutting blade, this is as follows:

$$P = (\rho.g.\ d^2.\ Nf + Cd.\ Nc + q.d.\ Nq + Ca.d.\ Nca).\ w \quad (3)$$

$$P = (1.6 \times 9.81 \times 0.0762^2 \times 4.2 + 18 \times 0.0762 \times 4.4 + 0 \times 0.0762 \times 16 + 12 \times 0.0762 \times 2.1) \times 0.07$$

$$P = 0.583\ KN$$

$$H = P.sin(\alpha + \delta) + Ca.dw.cot(\alpha) = 0.557\ KN$$

Soil collector plates

Taking d as 0.005m and width as 0.2034m here. For soil collector plates, equation (3), similar to duck foot the calculations can be done as:

$$P = (\rho.g.\ d^2.\ Nf + Cd.\ Nc + q.d.\ Nq + Ca.d.\ Nca).\ w$$

$$P = (1.6 \times 9.81 \times 0.025^2 \times 4.2 + 18 \times 0.025 \times 4.4 + 0 \times 0.025 \times 16 + 12 \times 0.025 \times 2.1) \times 0.2034$$

$$P = 0.1065\ KN$$

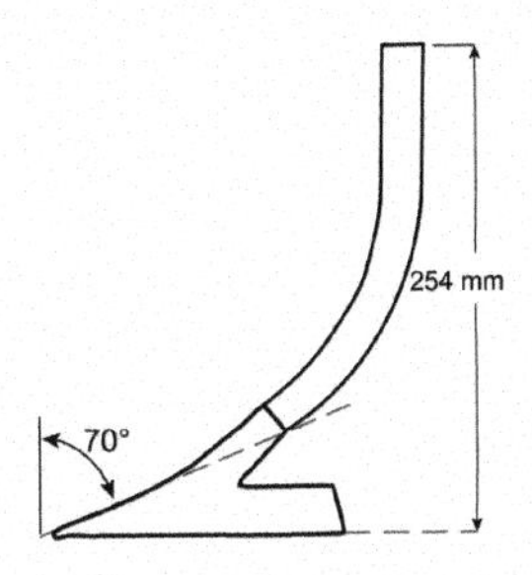

Figure 80.7 Duck foot

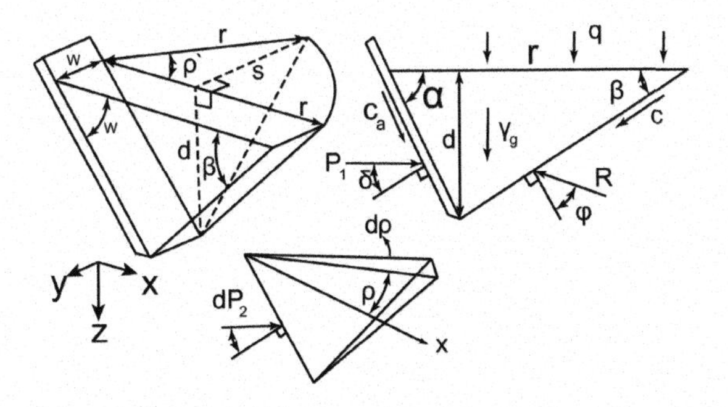

Figure 80.8 The three-dimensional model

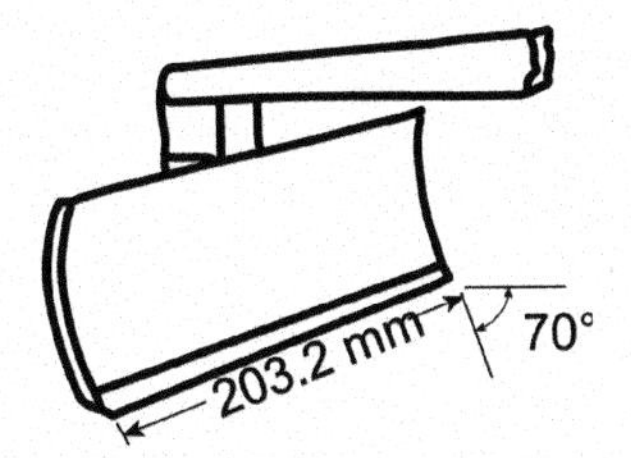

Figure 80.9 Soil collector model

CAD model

Analysis

The first step of analysis was pre-processing, where the average element size of model under study was 5% of model size, using scale mesh size part the element order was 'parabolic' and the curved mesh elements were created where the maximum turn angles on curve was 60°. The maximum adjacent mesh size ratio and maximum aspect ratio were 1.5 and 10 respectively.

There were four mesh refinements steps taken to improve the accuracy of our study where the convergence percentage was set at 10% and the base line accuracy for result was 'Von-Mises stress or maximum distortion strain energy theory'. The contact tolerance was set at 0.1 mm, The solids under the study had a total of 654334 nodes and 362593 elements.

A rigid constraint was applied on the voke with Ux, Uy and Uz fixed. Figure 80.11 describes the various forces acting on the frame with the help of a colour based (dof) description.

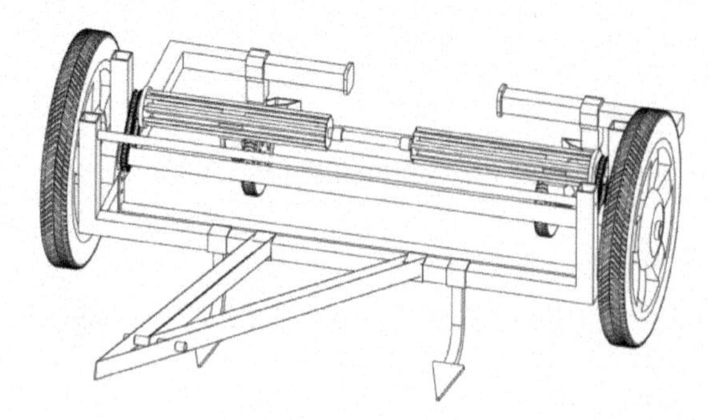

Figure 80.10 Wireframe model

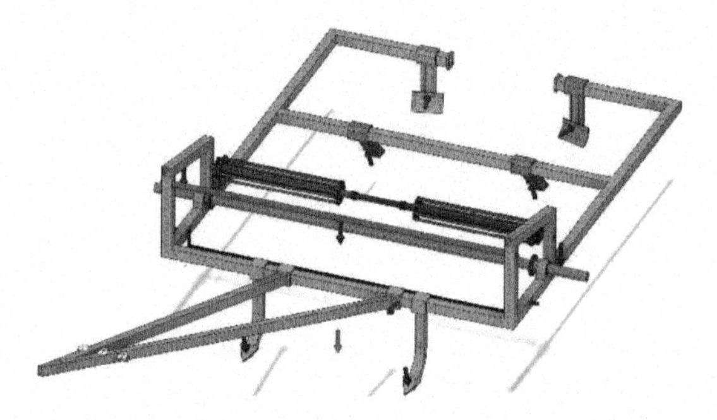

Figure 80.11 Acting forces and boundary conditions

The various forces are described below in the following ways:

Table 80.1 Forces applied

Force	*Component*	*Fx*	*Fy*	*Fz*
Gravity	Body m/s^2	0	–9.807	0
Force 1: F1	Duck foot 1	0 N	310 N	–557 N
Force 2: F2	Duck foot 2	0 N	311 N	–557 N
Force 3: F2	Coupler 1	356.4 N	481.3 N	778.9 N
Force 4: F3	Coupler 2	–267.5N	362.3 N	873.3 N
Force 5: F3	Soil Plates	49.62 N	0 N	–75.09 N
Force 6: F4	Main shaft	0 N	–965 N	0 N
Force 7: F4	Laying Rod	0 N	50 N	N

Results:

Factor of safety: 0 8

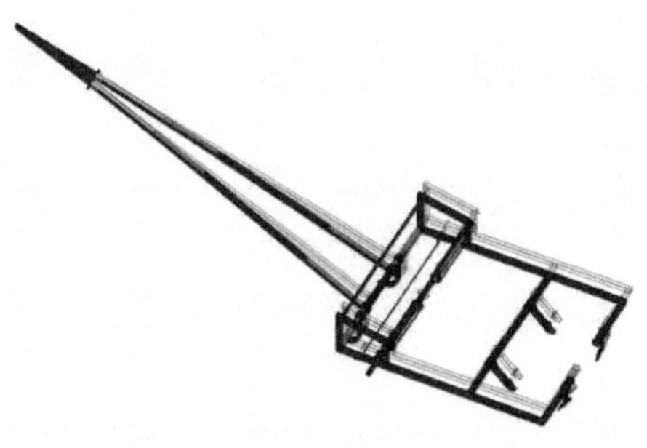

Maximum distortion strain energy theory: [MPa] 0 315

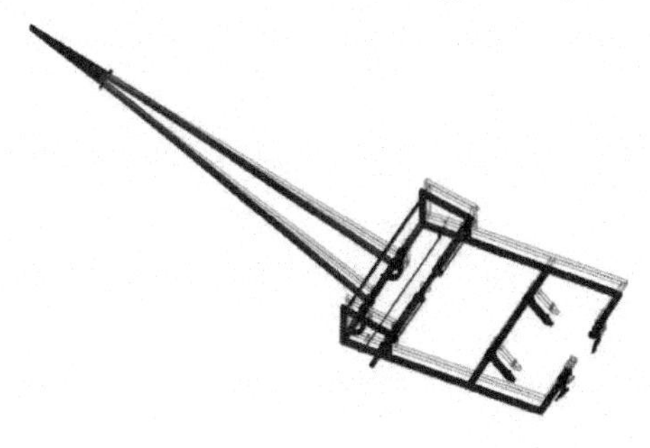

First principal stress: [MPa] −517 3822

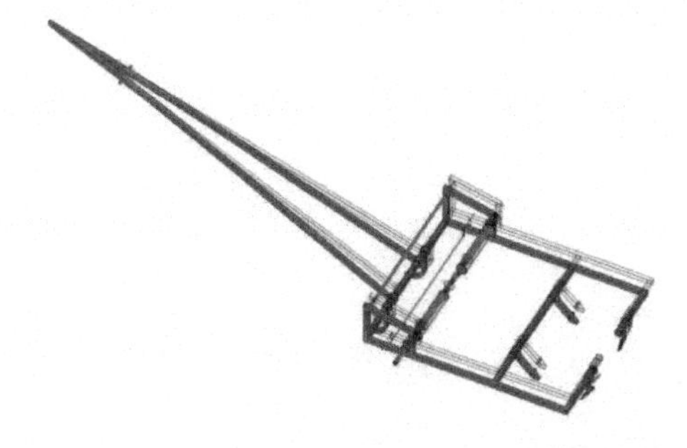

Third principal stress: [MPa] −169.5 945.1

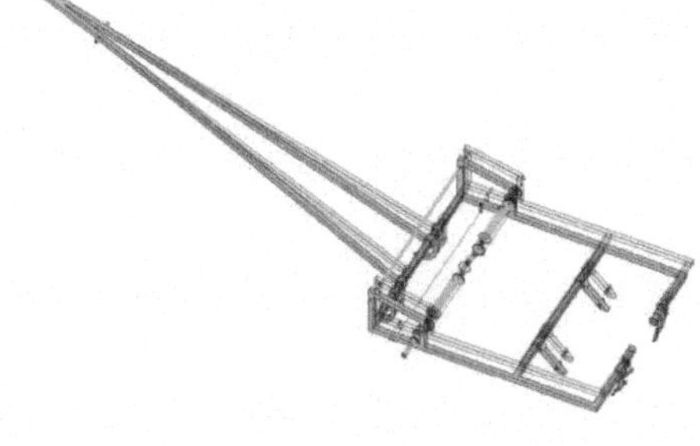

Displacement: [mm] 0 16.26

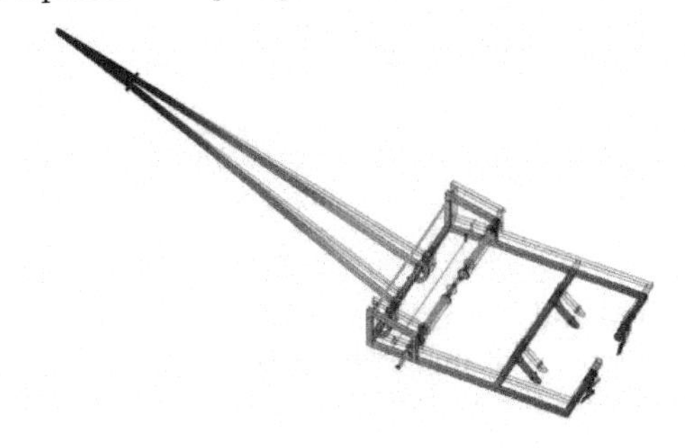

Table 80.2 Result summary

	Safety factor	
Name	*Minimum*	*Maximum*
	Factor of safety	
Per Body	3.7663	15
	Stress	
MDST	5.893 MPa	315 MPa
Normal YY	153 MPa	1893 MPa
Normal XX	–743.1 MPa	48 MPa
Normal ZZ	44 MPa	3759 MPa
Shear YZ	–135.6 MPa	999.3 MPa
Shear XY	103.9 MPa	988.2 MPa
Shear ZX	–94.3 MPa	989.1 MPa
3rd Principal	169.5 MPa	945.1 MPa
1st Principal	–517.2 MPa	–260 MPa
	Displacement	
Total	0 mm	16.26 mm
Y	–6.324 mm	0.02758 mm
X	–0.0881 mm	0.02687 mm
Z	–15.36 mm	0.06221 mm
	Reaction Force	
Total	0 N	9149 N
Y	–6391 N	7773 N
Z	–6211 N	6880 N
	Contact Pressure	
Total	0 MPa	403 MPa
Y	–88.3MPa	1329 MPa
X	–11.1 MPa	241.4 MPa
Z	–106.2 MPa	1315 Mpa
	Strain	
3rd Principal	–0.0184	3.26E–05
1st Principal	–2.29E–05	0.02302
Equivalent	4.72E–10	0.0213
Normal YY	–0.01027	0.00743
Normal XX	–0.001408	0.002618
Normal ZZ	–0.00843	0.01455
Shear YZ	–0.01747	0.01237
Shear XY	–0.01391	0.01223
Shear ZX	–0.01059	0.01225

Results

The analysis was done successfully with the help of Fusion 360 and it was observed that the loads and stresses acting on the prototype when it is under performance are well under the permissible limit and henceforth making this design safe. This result is in accordance with the analytical solution solved (Tables 80.1 and 80.2).

Discussions

Current research based on mulching is majorly limited to laying or soil pollution caused by mulch sheets when not retrieved at the end of the crop season. This led the research paper to focus on a possible bull-operated integrated approach that involves laying and retrieval mechanisms in a single machine. The machine is designed such that it can be afforded by the farmers.

A literature survey was done in order to observe the research gaps and use them in our research after analytical alterations/modifications. After making assumptions based on the practical mulching process into consideration, a 30D-CAD model was designed on fusion 360 (Figure 80.10). The design involved design calculations of duck foot, soil collector plates, frame size, and shaft. In the end, an analysis/simulation was done where the design was found to be safe.

Conclusion

The present research was done in order to invent an integrated mechanism of the mulching machine that can do both laying and retrieval processes. This integration would help the agricultural workers in saving capital and energy, as the manual mulching process requires high labour costs and the process is time-consuming.

The paper included inspirations from many pieces of literature and used them based on the research gaps observed, further research involved designing a 3d cad model based on the design calculations of coupling, duck foot, soil collector plates, and practical applications of mulching on the farm which included the factors like mulch-bed size, soil types. A sliding variable mechanism was also designed to account for different sizes of mulching paper in the market. Analysis/simulation was performed after the design was made on Autodesk Fusion 360 and the results were observed to be safe with stresses being well under the permissible limit and the minimum factor of safety being 3.7633.

References

Bhargava, R. A., Monisha. J. B., Prithvi Bhushan, S., Rukmini, and Tripathi, R. (2020). Design and Fabrication of Mulch Paper Laying Machine. Int. J. Scientif. Eng. Res. 11(6):437–441.

Guo, W., Wang, X., Lu, B., Hu, C., Zhang, P., and Hou, S. 2019. Design and Experiment of Rotary nail tooth type Residual Plastic Mulch Recycling machine. 2019 ASABE Annual International Meeting 1900126. doi: 10.13031/aim.201900126

Kader, M. A., Senge, M. A., Mojid, M. A., and Ito, K. (2017). Recent advances in mulching materials and methods for modifying soil environment. Soil Tillage Res. 168:155–166. Available doi: :10.1016/j.still.2017.01.001.

Khazimov, K. M., Niyazbayev, A. K., Shekerbekova, Z. S., Urymbayeva, A. A., Mukanov, G. A., Bazarbayeva, T. A., Nekrashevich, V. F., and Khazimov, M. Z. (2021). A novel method and device for mulch retriever. J. Water Land Dev. 46(49):85–94. doi: https://doi.org/10.24425/jwld.2021.137100.

McKyes, E. (1985). Soil Cutting and Tillage. Developments in Agricultural Engineering. 7:12–17. https://www.sciencedirect.com/bookseries/developments-in-agricultural-engineering/vol/7/suppl/C.

Padawal, N. T., Mali, R. D., Nandgavakar, S. D., Ramdas, S. V., Sutar, U. P., and Badkar, D. S. (2017). Design and Development of New Mulching Machine for Agriculture. J. Adv. Sci. Technol. 13(1):1–8. doi: https://doi.org/10.29070/JAST.

Rocca, A. R. (2012). Plastic mulch retriever. patent no. Australia: Rocca Manufacturing Pty Ltd

RUTAG IIT Delhi (2014, December, 12). Comparative Study of Bullock Driven Tractors. Delhi: doi: http://rutag.iitd.ac.in/rutag/?q=projects/bullock-driven-tractors.

Sharma, R. and Bhardwaj, S. (2017). Effect of mulching on soil and water conservation -A review. Agricul. Rev. 38(4):311–315.2017. doi: 10.18805/ag.R-1732

Shiwalkar, B. D. (2017). Design Data book for Machine Elements. Maharashtra: Denett & Co.

Maddock, B. Holstein Friesian Cattle. https://www.dimensions.com/element/holstein-friesian-cattle.

Tipayale, A., Salunke, M. S., Thete, S. U., Thete, T. S., and Thete, S. B. (2017). Advance mulching paper laying machine. Int. J. Scientif. Res. Dev. 5(3):217–219. doi: https://doi.org/10.29070/jast.

81 Experimental and numerical analysis of jet impingement heating on cylindrical body

Ashvin Amale[1,a], *Neeraj Sunheriya*[2,b], *R. B. Chadge*[1,c], *Pratik Lande*[1,d], *Jayant Giri*[1,e], *and S. G. Mahakalkar*[1,f]

[1]Department of Mechanical Engineering, YCCE, Nagpur, India

[2]Department of Mechanical Engineering, NIT, Raipur, India

Abstract

Jet impingement is a term mainly used to attain a high heat exchange process in a wider range of applications like commercial and domestic frameworks. Whenever precise and rapid thermal control is required, both cooling and heating air jets can be used. In this paper, the heat transfer predictions of hot air jets impinging on circular cylinders were investigated using computational fluid dynamics (CFD). The distribution of local Nusselt numbers for different Reynolds numbers (8,000–100,000), and variation of distances between the jet and cylindrical body (H/D ratio) from 1 to 8 for different pressure and temperature was determined around the cylinders. The K-ε and K-ω turbulence model is used separately to get better result outside the boundary layer and inside the boundary layer, and the SST turbulence model is used to get combine the application of both the k-w and K-ε model. The validation of CFD simulation with Experiment needs to be performed for its acceptance. The validation of the CFD results is done by experimentation with the help of resistance temperature detectors (RTDs) to find out the temperatures at various points on the cylinder, blower, and heater to get hot air at different velocities. The heat transfer distribution and various important flow characteristics around the cylinders are evaluated and found to be dependent on the H/D ratio with different Reynolds numbers and Pressure. Some deviations in the non-isotropic region are found. Heat transfer increases as the Reynolds number increases. Heat transfer around the cylindrical body is better if the H/D ratio is more than 2 to 4. Though the heat distribution is more the temperature distribution over the surface of the cylinder decreases when H/D increases beyond H/D = 3. Heat transfer increases for a higher Reynolds number.

Keywords: Computational fluid dynamics, cylinder, digital anemometer, experiment, jet impingement, Nusselt number, resistance temperature detector.

Introduction

The heat transfer, fluid flow, and turbulence, from the impinging jets to the cylindrical product may be calculated using computational fluid dynamics (CFD). CFD modelling is an effective technique for investigating complex flow and heat transport. It may be used to explore the influence of various factors on the heat-transfer distribution and flow pattern on the product's surface, resulting in a decrease in process time in the required experimental testing. Heat transmission is quick. Methods can be employed to reduce process time and enhance output rate. Jet impingement is a quick convective heat-transfer technique that consists of focused jets impinging on the product's surface. It can be utilised to accelerate heat operations in the food sector and other industries. Impinging jets have been widely utilised and developed to improve heat and mass transfer in a variety of applications, including cooling gas turbines, drying paper and textiles, and cooling electronic components. The impinging jet normally forms four areas, which are represented in Figure 81.1 and are referred to throughout this project. The free jet zone (where the jet axial velocity is nearly constant to its so-called value) interacts with the atmospheric air, resulting in a chain of macrostructures. Below these two sectors, the jet steadily decelerates until it reaches the stagnation region when nearly all kinetic energy is converted into a static pressure increase. Following contact, the jet flow is redirected along the target surface in the wall jet zone, which includes the boundary layer, recovering much of its ideal velocity before relaxing upon expansion and eventually splitting and recirculating.

The novelty of the work

The most conventional method of heating is practiced everywhere whereas the jet impingement technology is a very efficient heating technology. Therefore, a design engineer must get complete data regarding the Jet Impingement, Heat, and Nusselt number distribution on the surface of the body.

[a]a_amale@rediffmail.com; [b]rbchadge@rediffmail.com; [c]neeraj.sunheriya@gmail.com; [d]patiklande@gmail.com; [e]Jayantpgiri @gmail.com; [f]sachu_gm@yahoo.com

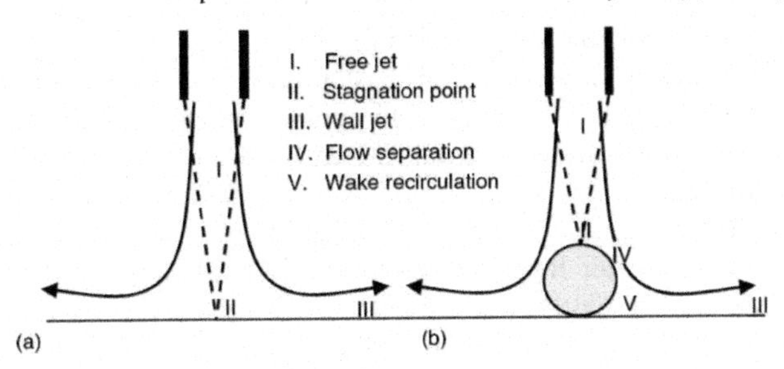

Figure 81.1 Distinguished regions in jet impingement

CFD analysis is a better and faster method for studying the heat and Nusselt number distribution. To study the detailed effects of designing jet impingement, CFD analysis is used. Commercial CFD packages available in the market provide the user with a variety of options for analysis.

There are several turbulence models available in commercial software ANSYS FLUENT 14.0. These turbulence models use a different set of equations to solve the Navier-Stokes equations for a fluid flow. As a result of these differences, the results obtained from using different models are different for the same case, It is necessary for the design engineer who wishes to use CFD tools for analysis of Jet impingement temperature and Nusselt number distribution to have ample knowledge of these turbulence models and their applicability and suitability to the case which is under study.

Objectives of the Work

An experimental study concerning the heat transfer between a cylindrical body and hot air from a single round jet is proposed. The primary objectives of this research can be summarised below.

- To analyse the heat transfer rates along the jet impingement area for the close nozzle to jet spacing over the cylindrical body.
- To monitor the heat transfer at a reference temperature (48°C) for varying Reynolds Numbers.
- To study the effect of potential core jet impingement on local heat transfer rates.
- To study the effect of jet height to cylindrical diameter ratio (H/D ratio).
- To study the effect of jet diameter to cylindrical diameter ratio (d/D ratio).

Literature Review

The heat transfer of a jet impinging on a cylindrical surface is governed by several parameters. It represents a standard thermo-fluids problem with a strong combination between heat transfer and fluid mechanics. Several researchers have performed experimental, numerical, and analytical studies on this problem. The present section aims at highlighting some of the findings that are closely related to the topic of this work.

As explained earlier, jet impingement heat transfer finds its use in a variety of industrial applications.

Olsson et al. (2004; 2007) carried out at domain pressure 0.1 MPa and 2°C, by keeping solid food product at 35°C; they reported that when we increase the Reynolds number the heat transfer also increases around the cylinder. It is found that the heat transfer for a specific Reynolds number does not much dependent on (H/D). Sarkar and Singh (2004) studied the effect of constrictions in a jet for an array of multiple jets, and the effect of the roughness of the impingement surface on the flow characteristics. According to Olsson et al. (2005), the maximum average heat transfer is recorded for cylinders beneath two impinging jets with a shared distance of H/D = 2. Heat transmission is decreased over longer and shorter distances (H/D = 1 and 4). Heat transmission is greatest with the smallest aperture (d/D=1) at a jet distance of H/D = 4. Larger apertures considerably minimise heat transmission. Carmela Dirita et al. (2007) investigated how heat transfer rate in food is affected by conduction. Chougule et al. (2011) investigated the effect of the H/d ratio on a heat sink flat plate using jet impingement. While Gori and Bossi (2000) investigated the impact of impinged free cylinders but made no conclusions on the influence of adjacent restricting walls. Downs and James (1987) analysed the influence of geometric and temperature effects, interference and cross-flow, turbulence levels, incidence, and surface curvature on the parameters of heat transfer from round and slot impinging jets. The experiment was carried out by Lee and Lee (2000) for a better understanding of the effects of nozzle outlet configurations on the enhancement of heat transfer of an axisymmetric impinging orifice jet in the stagnation area. Three different orifice nozzles with the same exit diameter D were evaluated for nozzle-to-plate spacing of H/D. Sagot et al, performed an experimental study of the Gas-to-Wall heat transfer configuration for around air jet impinging on a round flat plate to estimate an average

Nusselt number correlation. Through an enthalpic balancing of the enclosure, the average wall heat transfer coefficient may be determined using simultaneous measurements of mass flow rate and characteristic temperatures (hot jet, cool wall, enclosure exit). Carmela et al concentrated on air impingement cooling of cylindrical meals, which was designed and quantitatively studied during the early step of cooling/chilling operations. The temperature distributions within and on the surface of the food, as well as the accompanying flow field caused by the jet–food interaction, are among the results. Lee et al. investigated the effect of the convex surface curvature on the local heat transfer from an axisymmetric impinging jet. The jet Reynolds numbers range from 11000 to 50000. And the H/D ratio ranges from 2 to 10. Verboven et al. studied the effect of heat and mass transfer coefficients to control food surface temperature and moisture inside the microwave oven using CFD. Rosana studied the comparative effect of drying food products by using superheated steam and hot air under the same conditions.

Experimental Setup

The experimental apparatus consists of Nichrome wire of diameter 1 mm fitted over the length of 1 meter of the end of the pipe. Jet impingement flows through the rounded hole pipe with a diameter of 25 mm. The test cylindrical body with a diameter of 25 mm was used. Resistance temperature detectors (RTDs) were used to visualise the temperature distribution on the target surface. Later analysed the heat transfer between the impinging jet and the heated target surface. Figure 81.2 below shows the general setup of the experiment. The cold air from the blower was circulated through an air filter to a flow regulator. The air from the regulator then goes through an orifice meter of 2.5 cm Throat diameter, which was used for metering the flow. Air then flows through a 2.5 cm diameter Nichrome wired fitted heating pipe and reaches the 25 mm cylindrical steel body. The target surface was heated by hot air flowing through the nozzle attached to the heating pipe and resistance temperature detectors were used to display the temperature achieved over the cylindrical surface through a digital indicator. The heating of the pipe is done by adjusting the voltages through a dimmer stat so that air temperature from the jet issuing from the nozzle is maintained up to 48°C. The jet speed was controlled by a by-pass valve, and the corresponding flow rate is measured with the orifice. The velocity of the air coming out from the nozzle is measured by a digital anemometer. Nichrome wire is wounded on the pipe fitted with the nozzle at the end; winding is spread one meter of span on the pipe. The size of the Nichrome wire is 1 mm in diameter and the length of the wire is 7 m. Current is passed through to produce heat. The digital ammeter is used to measure the current flow through the circuit, the reading obtain from the ammeter is 1.5 Amp.

The Dimmer stat is used to maintain voltage across the circuit, voltage across the circuit is maintained at 90 V. Therefore, the power supplied through the circuit is 135 watts, which is sufficient to heat air at 48°C. The digital anemometer is used to read the air velocity coming out of the nozzle, keeping the anemometer normal to the direction of flow of the air. It read the velocity of air as well as the temperature of air impinges on the targeted surface. The velocity of air is kept at 6 m/s for validation of CFD results with an experimental setup (Figure 81.3).

Resistance temperature detectors are used to measure the temperature over the cylindrical surface by keeping the bulb of RTD's at a different location on the surface varying from 0° to180° (Figure 81.4).

Numerical Analysis

For the solution of complex motions of turbulent flows with minimum required duration and computational resources, a turbulence model is commonly used. Turbulence models that are both robust and accurate are essential components of the ANSYS FLUENT model package. The offered turbulence models have

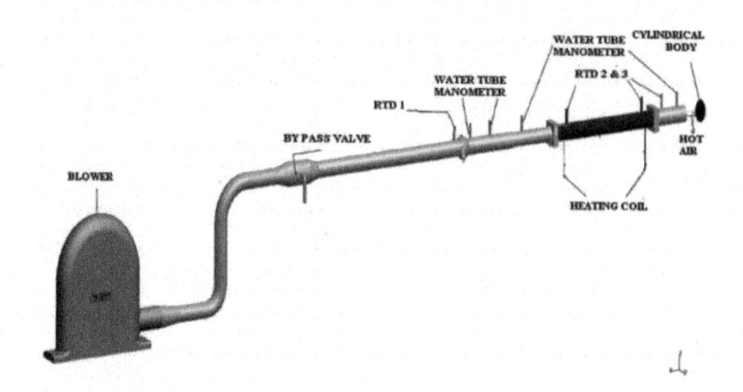

Figure 81.2 General layout of the experimental setup

Figure 81.3 Experimental set-up

Figure 81.4 Digital anemometer

Table 81.1 Range of the instruments used

Instrument	*Range*
Blower	A/C induction motor, RPM = 2800, Amp = 7, Volt = 230, HP = 1
Dimmer stat	Input voltage = 240 V 1Ph A/C Output voltage = 0 to 240 V A/C
Resistance temperature detectors	Temperature = –70°C to 150°C
Digital temperature indicator	Resolution = ± 0.1°C+
Digital anemometer	Resolution = ± 0.1, RPM = 159 – 7512, Temperature = -10°C–50°C
Nichrome wire	Diameter = 1 mm

a wide range of applications and incorporate the impacts of various physical phenomena such as buoyancy and compressibility. The use of extended wall functions and zonal models has been used with special attention to solving difficulties of near-wall accuracy. The heat transfer through porous media under Buoyancy induced flows forced convection, and mixed convection can be represented. The first law of thermodynamics serves as the foundation for the energy equation.

Governing equations

The governing equations used in the present investigation are the RANS as follows:

Continuity Equation:

$$\frac{\delta u}{\delta x} + \frac{\delta v}{\delta y} = 0$$

Momentum Equation in X-direction:

$$\frac{\delta}{\delta t}(\rho u)+\frac{\delta}{\delta x}(\rho uu)+\frac{\delta}{\delta y}(\rho vu)=-\frac{\delta p}{\delta x}+\frac{\delta}{\delta x}\left\{(2\rho(v+v_t))\frac{\delta u}{\delta x}\right\}+\frac{\delta}{\delta y}\left\{\rho(v+v_t)\left(\frac{\delta u}{\delta y}+\frac{\delta v}{\delta x}\right)\right\}$$

Momentum Equation in Y-direction:

$$\frac{\delta}{\delta t}(\rho v)+\frac{\delta}{\delta x}(\rho uv)+\frac{\delta}{\delta y}(\rho vv)=-\frac{\delta p}{\delta y}+\frac{\delta}{\delta x}\left\{\rho(v+v_t)\left(\frac{\delta u}{\delta x}+\frac{\delta v}{\delta y}\right)\right\}+\frac{\delta}{\delta y}\left\{2\rho(v+v_t)\frac{\delta v}{\delta y}\right\}$$

Energy Equation:

$$\frac{\delta}{\delta t}(\rho c_p T)+\frac{\delta}{\delta x}(\rho c_p uT)+\frac{\delta}{\delta y}(\rho c_p vT)=\frac{\delta}{\delta x}\left\{\rho c_p(a+a_t)\frac{\delta T}{\delta x}\right\}+\frac{\delta}{\delta y}\left\{\rho c_p(a+a_t)\frac{\delta T}{\delta y}\right\}$$

Turbulence model for jet impingement

It is critical to select the appropriate turbulence model and wall treatment. Heat transport near solid surfaces necessitates the use of the boundary layer technique.

- **A Transport equations for the standard K-ε model**

$$\frac{\delta}{\delta t}(\rho k)+\frac{\delta}{\delta xi}(\rho k u_i)=\frac{\delta}{\delta xj}\left\{\left(\mu+\frac{\mu t}{\sigma k}\right)\frac{\delta k}{\delta xj}\right\}+G_k+G_b-\rho\acute{\varepsilon}-Y_M+S_K$$

$$\frac{\delta}{\delta t}(\rho\acute{\varepsilon})+\frac{\delta}{\delta xi}(\rho\acute{\varepsilon}\, u_i)=\frac{\delta}{\delta xj}\left\{\left(\mu+\frac{\mu t}{\sigma\acute{\varepsilon}}\right)\frac{\delta\acute{\varepsilon}}{\delta xj}\right\}+C_{1\acute{\varepsilon}}\frac{\acute{\varepsilon}}{k}(G_k+C_{3\acute{\varepsilon}}G_b-C_{2\acute{\varepsilon}}\rho\left(\acute{\varepsilon}^2/k\right)+S_{\acute{\varepsilon}}$$

Where turbulence generation is represented by Gk. Kinetic energy is due to mean velocity gradient, and Gb is turbulence kinetic energy production owing to buoyancy. The contribution of fluctuating dilation incompressible turbulence to the total dissipation rate is represented by YM. C1, C2, and C3 are constants; k and are turbulent Prandtl numbers for K and, Sk, and S are user-determined source terms.

- **B Transport equations for the standard K-ε model**

$$\frac{\delta}{\delta t}(\rho k)+\frac{\delta}{\delta xi}(\rho k u_i)=\frac{\delta}{\delta xj}\left[\Gamma k\frac{\delta k}{\delta xj}\right]+G_k-Y_k+S_k$$

And

$$\frac{\delta}{\delta t}(\rho\omega)+\frac{\delta}{\delta xi}(\rho\omega u_i)=\frac{\delta}{\delta xj}\left[\Gamma_\omega\frac{\delta\omega}{\delta xj}\right]+G_\omega-Y_\omega+S_\omega$$

Gk indicates the creation of turbulent kinetic energy owing to mean velocity gradients in these equations. G denotes the rate of generation of a particular dissipation rate. k and are the effective diffusivities of k and, respectively. Yk and Y indicate the turbulence-induced dissipation of k&, whereas Sk and S are user-defined source terms.

Boundary conditions

Table 81.2 The boundary conditions used in the simulation

Sr. number	*Zone name*	*Type*	*Temperature*	*Velocity*
1	Jet Inlet	Inlet	48° c	6–70 m/s
2	Outlet	Outlet	30° c	-
3	Wall	Wall	30° c	-
4	Jet side	Wall	50° c	-
5	Body	Wall	30° c	-

Geometric Modelling

The entire domain is filled with stale air. CFD software (ICEM meshed and fluent solver) is used to solve the energy equation and 2-D Navier-Stokes equations using the conventional turbulence model. These equations are then integrated with momentum and continuity equations to evaluate the flow fields under thermal and turbulent conditions. The turbulence model utilised is the shear stress transport (SST) K-w model and K- model, both of which have been determined to operate the best among the existing turbulence models for this flow configuration and have also been chosen for their simplicity, computational economy, and general acceptability. It is assumed that the flow is constant, incompressible, and two-dimensional. The effects of buoyancy and radiation heat transmission are ignored, and the fluid's thermo physical parameters such as density, specific heat, and thermal conductivity are believed to remain constant. Figure 81.5 depicts a schematic representation of the physical geometry and computational domain.

Geometric meshing

To implement the boundary condition of undisturbed flow, the effect of different domain side length values was initially monitored (no overall recirculation). Finally, nozzle diameters of 25 mm were chosen along the x-axis. A triangular unstructured grid with around 40,000 cells was used (Figure 81.6). This allows for the resolution of temperature velocity and pressure gradients in the boundary layer caused by airflow direction, heating, and impingement. A grid independence positive check was also done on a grid ranging from 30,000 to 50000 cells and by altering the grid across the body surface from 350 to 1000 cells.

Throughout, a finite-volume segregated solver with second-order unsteady implicit formulation was used, with SIMPLE pressure–velocity coupling and second-order Upwind stencils for all other variables

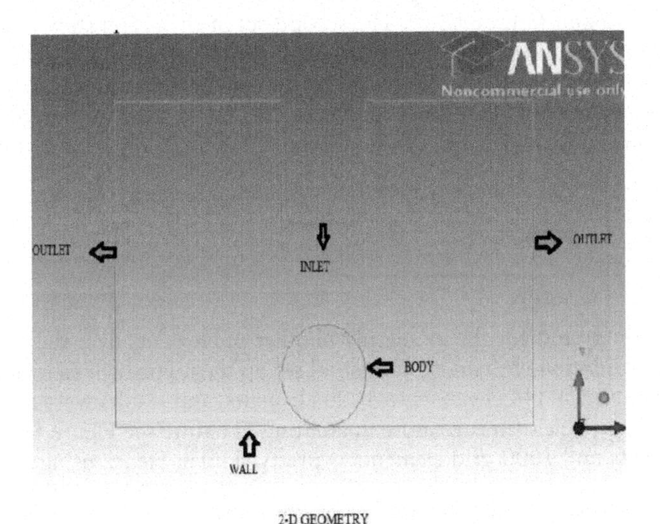

Figure 81.5 2-D physical geometry and the computational domain

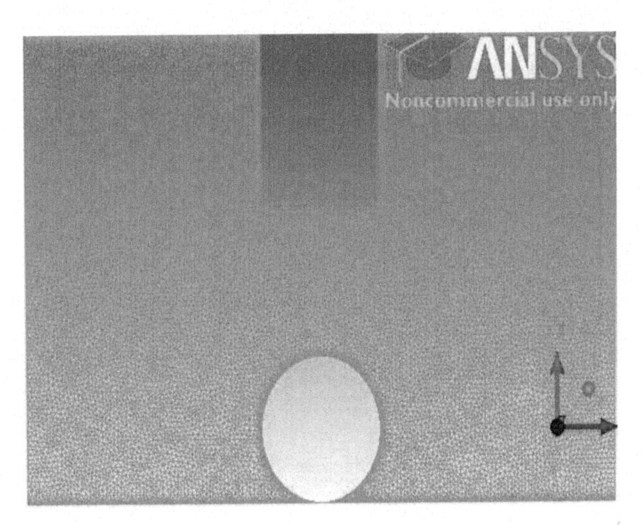

Figure 81.6 Triangular unstructured grid (H/D = 1)

such as momentum, turbulent kinetic energy, and specific dissipation rates (Fluent user's guide, November 2011). The residuals were kept to 1×10^{-4} for all variables since smaller values resulted in longer processing durations on a Pentium (R) with single processor elements (Windows 10 OS, 1.7 GHz, 4 GB RAM), as needed by the velocity gradients at process start-up. The numerical simulations are carried out with the help of the commercial CFD solver ANSYS FLUENT version 14.0. To get good heat transfer estimates, the flow and turbulence fields must be precisely solved.

For all parameters affecting heat transfer, a higher resolution plan is employed. For the pressure, momentum, turbulent kinetic energy, particular dissipation rate, and energy, a higher-order discretisation technique is utilised. The equations for flow, turbulence, and energy have all been solved. The flow was defined at the nozzle's inlet boundary condition in the fluid domain, with a measured velocity of 8 m/s and a static temperature of 421 K. The wall surface was treated with a non-slip coating. In the fluid domain, there is also an opening boundary condition in which the flow establishment is subsonic, the relative pressure is 0 Pa, the operating temperature is 300 K, and the turbulence intensity is 3%.

The no-slip condition was given in the solid domain with an initial temperature of 300 K. The fluctuation of thermal and physical characteristics of air with temperature is ignored to simplify the results.

Results and Discussions

The experimental data is used to calculate the convective heat transfer coefficient for heat transfer between the target steel body and the impingement jet. The convective heat transfer is expressed in non-dimensional form through the Nusselt number. The following section highlights the important results obtained as part of this work. The experimental and CFD results are expressed in different sets of graphs. The first set shows the grid independency of CFD results with an experimental setup for the non-dimensional distance (H/D ratio fixed at 1 and 2) from the jet stagnation point (Figure 81.7). The second set of the graph shows the validations test. The third set of a graph consists of heat transfer variation at various Reynolds numbers and the fourth set of graphs consists of the effect of surface curvature (d/D ratios) on heat distribution. The readings are plotted for five different Reynolds numbers 8000 to 100,000 at 48°C. The effect of different H/D ratios for velocity is shown, and the effect of jet height to cylinder diameter for temperature distribution is also explained in this section.

Grid independence test

Grid independence test for H/D = 1

The grid independence test is used to verify the results for the increasing number of elements over the surface of the body. In this case, H/D = 1, after validation with an experimental set up the number of elements over the surface of the target body is increased from 346 elements to 1000 elements, and results were analysed. It seems that results obtained from 346 grid elements are more fluctuating in nature see Figure 81.9. And solution obtained from 575 grid elements and 1000 grid elements is pretty similar. Hence using 575 grid elements complete analysis is done.

Grid independency test for H/D = 2

The simulations' grid independence was tested, as illustrated in Figure 81.8. In the simulations, the mesh comprises 40,000 cells in total, with 396 to 1100 elements in the boundary layer near the wall. A revised

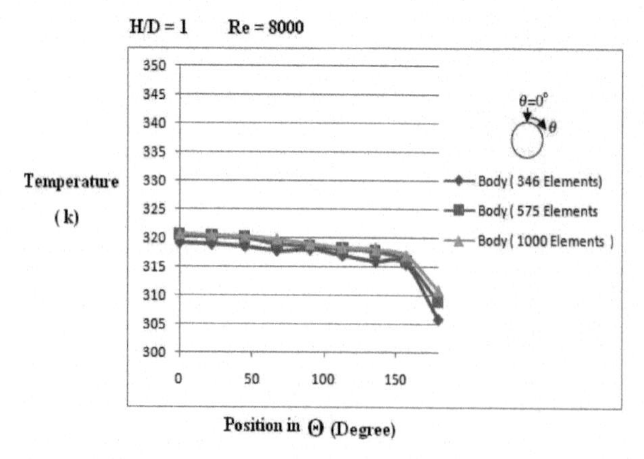

Figure 81.7 Grid independency test of H/D = 1. For a different number of elements on the body

mesh with four times the number of cells (1100 elements) produced the same results as the mesh with 546 elements. Mesh with 396 elements was producing erratic readings when compared to others, hence mesh with 546 elements was chosen as the standard reference to proceed with further reading in all circumstances.

Validations

Validation with CFD & Experimental result for H/D = 1

After validation with an experimental setup for H/D = 1, the surface temperature of the cylinder in both CFD and experimental analysis seems to be closer. The temperature difference in both cases CFD and experimental at the impact point of the jet is around 1% to 8% beyond the separation point. Separation, in this case, occurs at around 120° from the impact point of the jet therefore the error is more for H/D = 1. The minimum temperature difference between CFD and experiment results is 0.8 K at the impact point of the jet and the maximum temperature difference is 3.8 K at bottom of the surface.

Validation with CFD and Experimental result for H/D =2

Validation of H/D = 2 is done at Re = 8000, and the temperature of hot air is kept at 48°C, the results obtained from CFD and experimental setup in this case also seems to be closer. But the heat distribution of hot air over the surface is more compared to H/D = 1, this is because of more height of the jet to the diameter of the target body.

As we increase the height of the jet to the diameter of the target body the separation point increases and heat distribution also increased. The purpose of this work is to achieve a higher temperature on the surface of the cylindrical body instead of getting good heat distribution over the surface, this can be possible by maintaining a lower H/D distance from 1 to 3. The temperature difference in CFD and experimental results is obtained around 0.6% to 6% maximum. The minimum temperature difference between CFD and

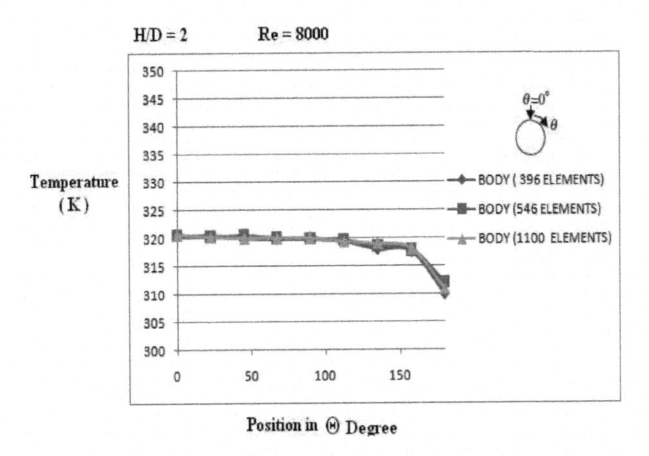

Figure 81.8 Grid independency test with H/D = 2 at various body elements

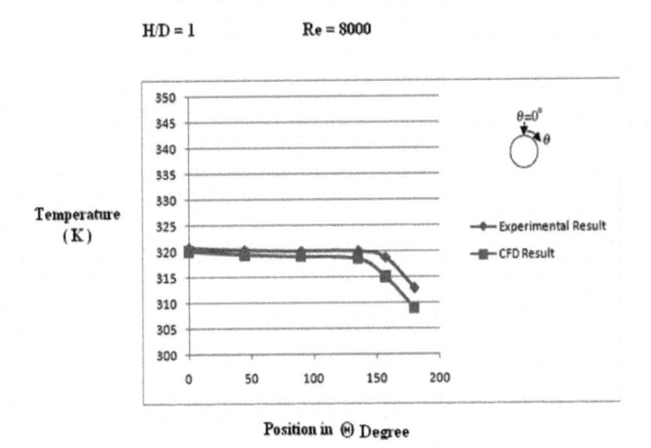

Figure 81.9 Experimental and CFD results comparison for H/D = 1, Re = 8000

experiment results is 0.2 K at the top of the cylindrical surface and the maximum temperature difference is 3.6 K at the bottom of the surface.

Validations has been carried out at fixed temperature and Reynolds number, in this case for H/D = 1, H/D = 2 and H/D = 3. The Reynolds number is taken as 10000 at 48° c, after being validated from the Experimental setup. The further reading and graphs have been taken by varying the Reynolds number from 10000 to 100000 at a fixed temperature in ANSYS software using fluent as a solver. And various graphs were obtained, plotted, and analysed.

Validation with CFD and

The above three graphs show the validation done with an Experimental setup showing close relation between CFD and Experimental results up to H/D = 2, beyond this if we increase the ratio of H/D then temperature achievement is sloped down in both the CFD and experimental cases as shown in graph H/D = 3. But the main intention of this work is to achieve the temperature at various H/D ratios by varying the Reynolds number, thus work is stopped at H/D = 3 ratios, thus various readings and graphs were drawn between these three ratios i.e. (H/D = 1.2 and 3). The minimum temperature difference between CFD and experiment results is 0.2 K at the impact point and the maximum temperature difference is 2 K at the bottom of the cylindrical surface (Figure 81.11).

Comparisons of Experimental results, K-ε and K-ω turbulence Model

The data obtained from experimentation is compared with both the turbulence models used for validation purposes as shown in Figure 81.10. It is clear from the above figure (Figure 81.12) that the result obtained from the K-ε model is more close to experimental data than the K-ω model. Hence thereafter K-ε model is used for further analysis and simulations done in CFD.

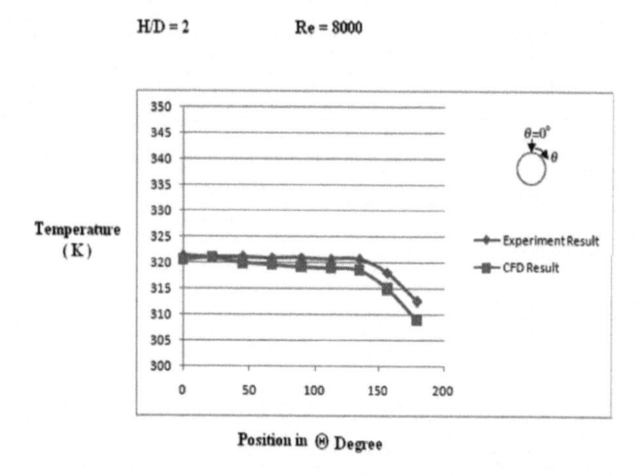

Figure 81.10 Experimental and CFD results comparison for H/D = 2, Re = 8000

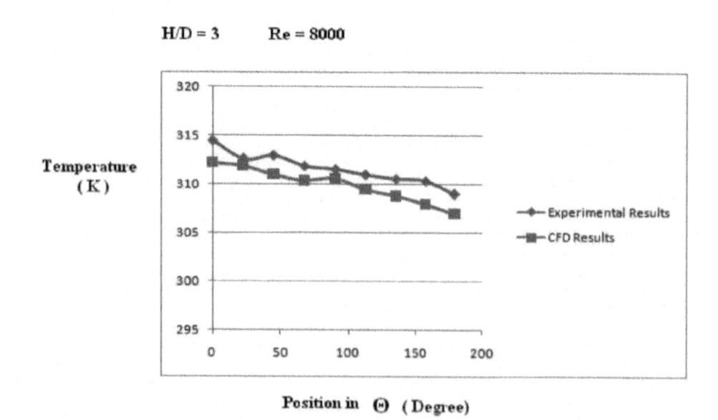

Figure 81.11 Experimental and CFD results comparison for H/D = 3, Re = 8000

As the Reynolds number increases, so does the heat transmission around the cylinder. The heat transfer does not affect significantly (H/D), compared to the literature (Figure 81.14).

Heat transfer variation at various Reynolds number

Nusselt number variation at Re = 10000

Several sources claim that the optimum heat transfer occurs when H/D = 5 to 8 (Gori and Bossi, 2000). Round jets impinge on a cylinder or a convex surface in a wide range of Reynolds numbers are used in these experiments (8000 to 100,000). In general, as the Reynolds number falls, so does the maximal heat transmission. The degree of turbulence in the jet also influences heat transmission; more turbulence intensity equals higher heat transfer.

By using CFD we have obtained the increase in heat transfer around the area of the cylinder with increasing Reynolds number while experimenting with the temperature distribution reduces at a higher H/D ratio. At the lower height of the jet to the diameter of the target body (H/D), the flow separation was higher beyond 90° of the impact of the jet thus the Nusselt number distribution was also low in that region, but when the height of the jet to the diameter of target body increases wake formation region get reduced beyond 135° of jet impingement.

The possible jet core was discovered to be short, measuring around 1 to 2 times the height of the jet to the diameter of the target body. Olsson et al. (2004) and 2005 discovered that the potential core is around four slot jet widths longer. The flow entrainment in the jet is influenced by the degree of confinement. The nozzle in the impingement system shown in Figures 81.12 and 81.13 was built as an unconfined jet, which resulted in flow entrainment of air that differed from that seen in earlier studies and may explain the observed discrepancy. Turbulence intensity was considerable in the shear layers, as predicted. Before colliding with the cylinder, the jet lost kinetic energy to the environment.

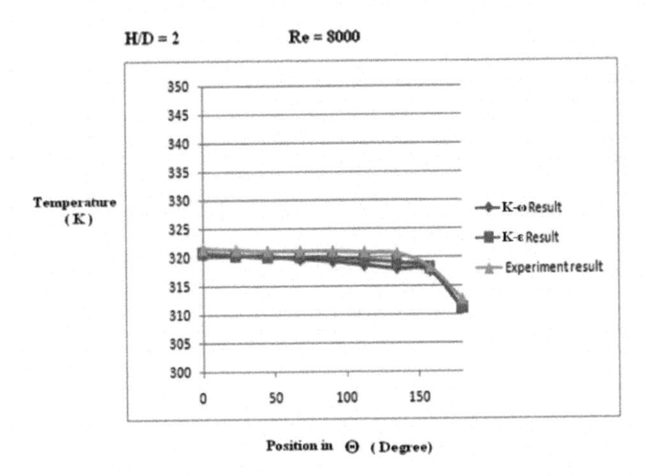

Figure 81.12 Comparisons between experimental data, K-ε and K-ω results

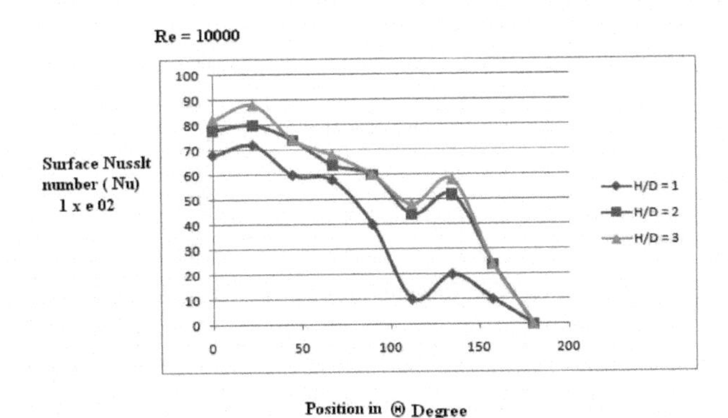

Figure 81.13 Heat transfer variation at Re = 10000 for different H/D ratios

Nusselt number variation at Re = 20000

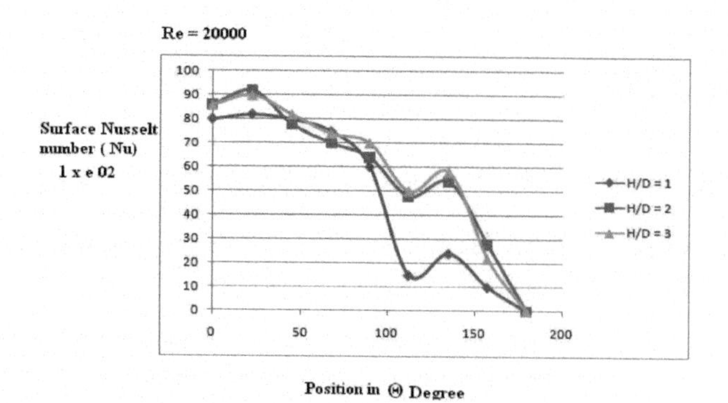

Figure 81.14 Heat transfer variation at Re = 20000 for different H/D ratios

Nusselt number variation at Re = 50000

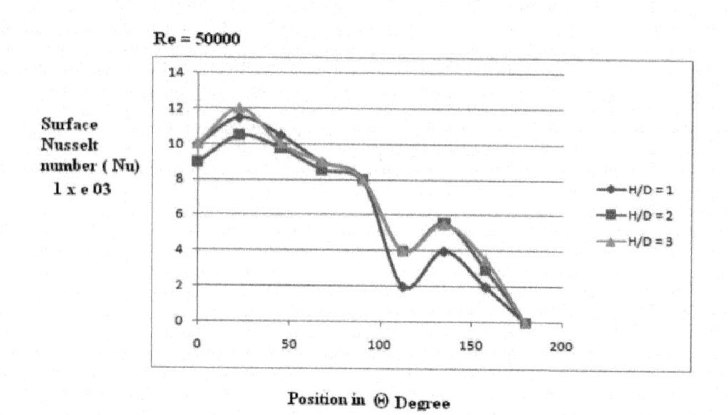

Figure 81.15 Heat transfer variation at Re = 50000 for different H/D ratios

Nusselt number variation at Re = 100000

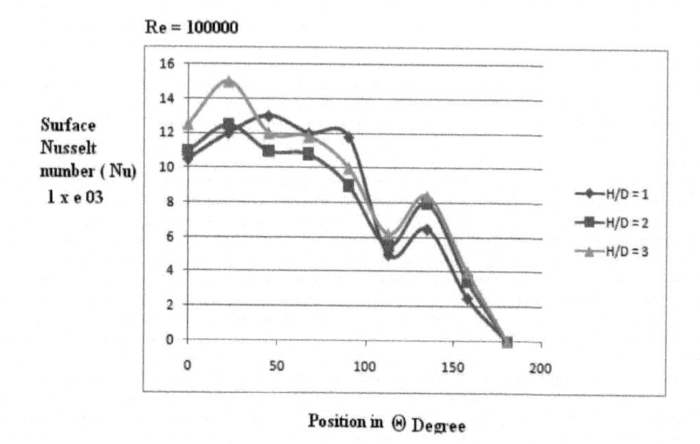

Figure 81.16 Heat transfer variation at Re = 100000, for different H/D ratios

Effect of surface curvature (d/D ratio) on heat distribution

The influence of surface curvature (ratio of jet width to cylinder diameter) was also studied. The surface curvature was discovered to alter the Nusselt number distribution around the cylinder and at the stagnation point. The Nusselt number rises as the surface curvature increases (Figure 81.14–81.17). It should be noted that the Nusselt number and the Reynolds number are predicated on the constant jet diameter (d). The cylinder's diameter (D) is altered while its height (H) remains fixed (D = 25 mm and H = 50 mm). The same Reynolds number implies the same mass flow, and by adjusting the by-pass valve, the velocity of the jet may be increased.

Effect of H/D on heat transfer

Effect of H/D = 1 on heat transfer

The velocity contours in Figures 81.18 and 81.19 depict the spreading flow for H/D = 1, 2 for Re 10000. Lower H/d ratios result in more heat transmission due to a reduction in impingement surface area. Higher H/d ratios result in more momentum exchange between the impinging and surrounding fluids, causing the jet diameter to expand and spread across a larger surface area. When the H/d ratio is low, the same volume of fluid distributes across a smaller surface area, resulting in a greater heat transfer rate.

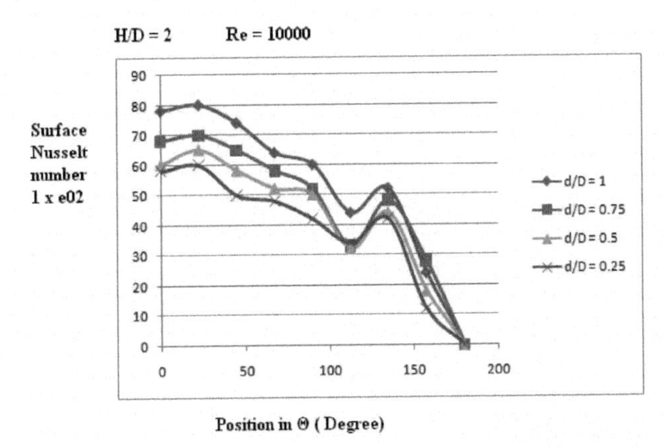

Figure 81.17 Heat transfer variation for different surface curvature (d/D ratio), Re = 10000 and H/D = 2

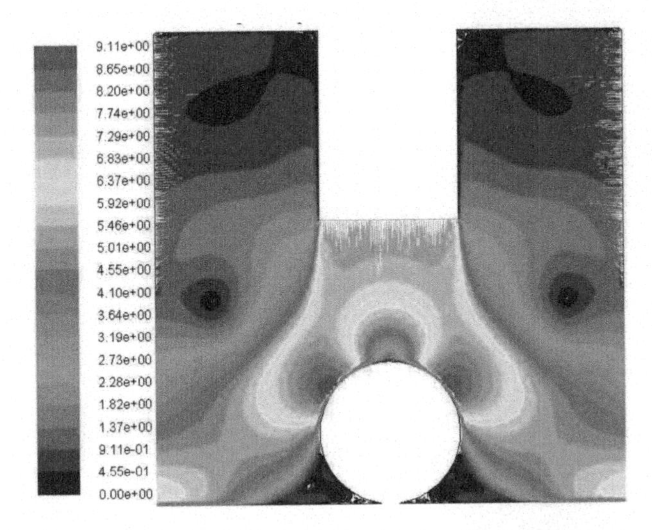

Figure 81.18 Velocity contour for H/D = 1at Re = 10000

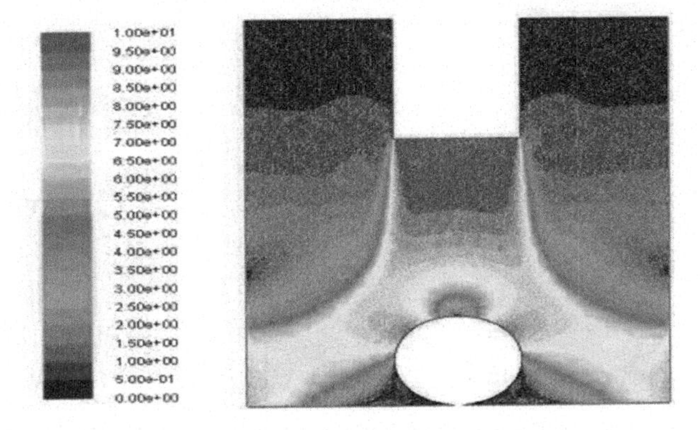

Figure 81.19 Velocity contour for H/D = 2 at Re = 10000

Effect of H/D = 2 on heat transfer

In the event of a larger H/D ratio, fresh air is mixed with the jet before impingement, which is undesirable, and the outside jets deflect away from the intended impingement. At H/D = 2, good results are achieved, and increasing the H/D ratio from 2 to 3 reduces the temperature at Re 8000 from 320 K to 315 K. The following image shows that the velocity profile is not uniform over the surface; near the contact point, velocity is insignificant, but the pressure is maximum, causing the jet to regain velocity on either side of the surface via full conversion of pressure energy into kinetic energy.

Effect of H/D on temperature distribution

Effect of H/D = 1 on temperature distribution

The spread flow is shown by temperature contours in Figure 81.20 at H/D = 1 for Re = 10000, the temperature distribution over the surface is more efficient, around 92% maximum. There is more wake region that reduces the heat distribution over the surface beyond 120° from the impact point. As the H/D ratio increases the heat distribution over the surface increases since the wake region reduces as the height of the jet increases. This causes more contact of hot air with the cylindrical surface, because of the entry of fresh air from surrounding into the path of hot air impinging to cylindrical surface the temperature of hot air reduced significantly and the efficiency of impingement get reduced up to 10% at H/d = 3 (Figures 81.21 and 81.22). Thus analysis of jet impingement beyond the H/D = 3 ratio increases the cost of production.

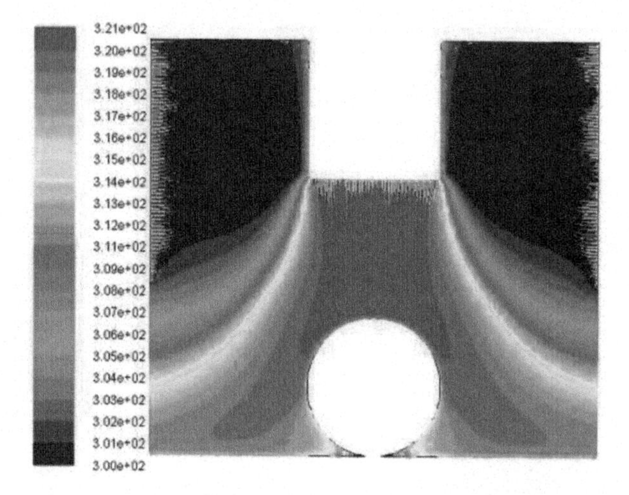

Figure 81.20 Contour of temperature distribution of H/D = 1 at Re = 10000

Effect of H/D = 2 on temperature distribution

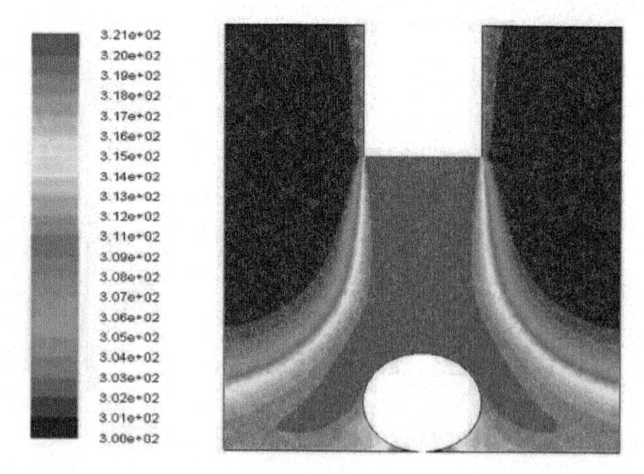

Figure 81.21 Contour of temperature distribution of H/D = 2 at Re = 10000

Effect of H/D = 3 on temperature distribution

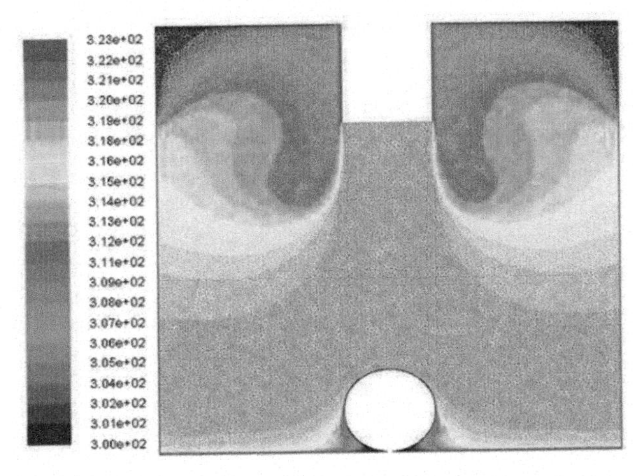

Figure 81.22 Contour of temperature distribution of H/D = 3 at Re = 10000

Conclusions

Computational fluid dynamics is an excellent method for estimating flow and heat transport.

The primary disadvantage is that an appropriate turbulence model must be chosen. For estimating wall-bounded flow and heat transfer, an RSM is frequently recommended. The velocity predicted by CFD modelling is often consistent with experimental results. Turbulence models based on the K-approach (the K- and SST models) that use a low Reynolds-number model near to the surface predict heat transfer better than the K-model in impingement flows.

The two-equation K-, K-, and SST models are common. For estimating wall-bounded flow and heat transfer, an RSM is frequently recommended. The velocity predicted by CFD modelling is often consistent with experimental results. In impingement flows, turbulence models based on the K-approach (the K- and SST models) forecast heat transfer better than the K- model utilising a low Reynolds-number model near to the surface. Validation of the simulation has been carried out by experimental results & using the K-ε turbulence model. On the upper part of the cylinder, the simulations predicted the heat transfer characteristics very well and were acceptable. Some deviations in the non-isotropic region are found. Heat transfer increases as the Reynolds number increases. Heat transfer around the cylindrical body is better if the H/D ratio is more than 2 to 4. Though the heat distribution is more the temperature distribution over the surface of the cylinder decreases when H/D increases beyond H/D = 3. This is due to the entrainment of fresh air from the surrounding into the path of the hot air jet impinging on the body which reduces the temperature of the jet. To get a better result of temperature distribution, the H/D ratio must lie in the range of 1 to 3.

References

Angioletti, M., Nino, E., and Ruocco, G. (2005). CFD turbulent modeling of jet impingement and its Validation by particle image Velocimetry and mass transfer measurement. Int. J. Thermal Sci. 44(4):349–356.

Chougule N. K, Parishwad G. V, Gore P. R, Pagnis S, and Sapali S. N. 2011. CFD Analysis of multi-jet air impingement on flat plate. Proceedings of the World Congress on Engineering.

Dirita, C., De Bonis, M. V., and Ruocco, G. (2007). CFD turbulent modeling of jet impingement and its validation by particle image Velocimetry and mass transfer measurements. J. Food Eng. 81:12–20.

Downs, S. J. and James, E. H. 1987. Jet impingement heat transfer - A literature survey. American Society of Mechanical Engineers (Paper), 87-HT-35. 1–11.

Gori, F.and Bossi, L. (2000). On the cooling effect of an air jet along the surface of a cylinder. Int. Commun. Heat Mass Transf. 27(5):667–676.

Lee, J. and Lee, S. J. (2000). The effect of nozzle configuration on stagnation region heat transfer enhancement of axisymmetric jet impingement. Int. J. Heat Mass Transf. 43:3497–3509.

Olsson, E. E. M., Ahrne, L. M., and Tragardh, A. C. (2004). Heat transfer from a slot air jet impinging on a circular cylinder. J. Food Eng. 63:393–401.

Olsson, E. E. M., Ahrens, L. M., and Tragardh, T. C. 2004. Prediction of optimal heat transfer from slot air jets impinging on cylindrical food products using CFD. In: Proceedings of the 9th International Congress on Engineering and Food (ICEF 9). Montpellier, France.

Olsson, E. E. M., Ahrens, L. M., and Tragardh, A.C. (2005). Flow and heat transfer from multiple slot air jets impinging on circular cylinders. J. Food Eng. 67:273–280.

Olsson, E. E. M. and Tragardh, C. 2007. CFD modeling of jet impingement during heating and cooling of foods. Computational Fluid Dynamics in Food Processing, (pp. 487504). Lund: CRC Press

Sarkar, A. and Singh, R. P. (2004). Air impingement technology for food processing: visualization studies. Swiss Society of Food Science and Technology. 37:873–879.

82 A cost-effective water management framework for reuse of grey water in thermal power plant boilers on usage of biomaterials

Raja K.[1,a], *Rupesh P. L.*[1,b], *Vivek Panyam Muralidharan*[2,c], *S. V. S. Subhash*[1,d], *and K. Pranay Chowdary*[1,e]

[1]Department of Mechanical Engineering, Veltech Rangarajan Dr Sagunthala R&D Institute of Science and Technology, Chennai, India

[2]Department of Chemistry, Veltech Rangarajan Dr Sagunthala R&D Institute of Science and Technology, Chennai, India

Abstract

Water is the main constituent of the thermal power plant to generate electric power from thermal energy extracted from steam. Water used for the conversion of steam in the boiler should be free from pollutants in order to enhance the amount of steam produced from the boiler. The thermal energy extracted depends on the amount of steam produced at high pressure. If water with poor characteristics such as low pH (acidic in nature), high hardness (> 500 ppm), etc. used in the boiler, it may lead to corrosion of tubes in the boiler and scale formation on the tube surface. The current research focuses on the development of a cost-effective framework for effectual water management to produce and use low hardness water in boilers. The present research deals with the investigations on treating the wastewater from nearby educational institutions using the biomaterials such as seeds of Moringa oleifera along with a hydrated form of potassium aluminium sulphate (alum) along with aeration at different time intervals for primary filtration. The experiments have been repeated using these biomaterials at different proportions to extract water with properties matching the boiler feed water. The outcomes of the present study indicate that treated water produced from one litre of grey water through the proposed technique suits the properties of feed water for the boiler which in turn increases the efficiency and decreases the fouling factor.

Keywords: Alum, efficiency, fouling factor, feed water, hardness, Moringa oleifera, pH.

Introduction

Water is the most effecting way of medium through which all essential vitamins and minerals will be carried out along the bodies of living organisms, and it also plays a vital role in easing the work of enzymes in living organisms. Water also helps to maintain body temperatures in plants and animals (Oliveira et al., 2005; Ettouney et al., 2002; Darwish et al., 2003; Khawaji et al., 2008). Grey water is the term described as water segregation from a domestic wastewater collection system that will be reused on site. This water can be collected from a variety of sources such as washing machines, bathroom sinks, showers, bathtubs. It also contains some soap and detergent residues, but it is clean enough for non-potable uses. Many buildings or individual dwellings have systems that capture, treat, and distribute grey water for irrigation or some other non-potable uses. Brackish water is the mixture of freshwater and saltwater which refers to a water source that is somewhat salty but not as salty as seawater (Jorgensen et al., 1998; Parry et al., 1984). The exact amount of salinity will be varying depending on the environmental factors which cannot be precisely defined. The variation of salinity usually measures from 10–32% while the average salinity of freshwater sources is around 0.5%. These water sources are most commonly seen at transitional points of water where fresh water meets seawater (van Nieuwenhuijzen et al., 2000; Harp et al., 1996).

The current technologies employed for the treatment of grey and brackish water are available in huge numbers. But these techniques can purify polluted water in a less effective manner. To achieve water management at a high rate these current trends can be blended. Usage of screens of coarse sides, band screens, and micro strainers can be used for the removal of solid particles which forms a basis for primary filtration. The content of algae can be pulled out on the usage of gravel and fine sand. The elimination of other contaminants and pollutants can be achieved through the usage of activated carbons (Duff and Hodgson, 2001; Gernjak et al., 2003; Esplugas et al., 2007).

[a]rajamech24@gmail.com; [b]rupeshkumar221@gmail.com; [c]anandvivek16@gmail.com; [d]samisubhash@gmail.com; [e]Kspc2012@gmail.com

The steam produced from the boiler decides the efficiency of the boiler. The amount of steam extracted depends on the quality of feed water used. The fouling factor also plays a vital role in the elevation of efficiency. The fouling factor remains to be less if more amount of freshwater is used (Herrera Melián et al., 2002; Han et al., 2013). The amount of feed water reduces after the production of steam due to which the demand for water required for the boiler increases. Due to the reason for water scarcity, the grey water obtained from the educational institutions can be treated using a filtration technique and its properties can be adjusted to the properties of feed water and can be used in boilers thereby decreasing the demand for freshwater.

Al-Qahtani (2008) enumerated the utilisation of wastes inclusive of their peel from varied fruits' layers for the removal of toxic and heavy elements such as ions of zinc, cadmium, and chromium from the grey water. On the usage of the kiwi layer in preliminary experiments, it was observed that particles of diameter 1 mm were retained maximum. It was also proved from the author's research that the heavy elements present in grey water were pulled out on utilisation of layers of kiwi and tangerine than the layer of banana as these fruits exhibit a high ability of biosorption. The delineation of electrochemical techniques for efficient water management was delivered by Igunnu et al. (2014) and they declared that a blend of compounds of inorganic and organic matter was present in processed water accompanying the production of huge by-products. It was also explained that the proffering of pure water along with the recovery of valuable metals from produced water can be achieved through the concepts of photoelectrochemistry. De Koninga et al (2008) deepened a major key element associated with the treatment of wastewater that the selection of efficient technique depends on the quality of polluted water and its reusable entreaty. A system for pasteurization of water using solar energy was developed by Duff and Hodgson (2005). The system proposed by the authors was proved to be a correct choice for the alternative to the present techniques of pasteurization. This system uses the principle of density difference without the working of any moving parts. Gupta et al. (2012) initiated a practical methodology for the treatment of wastewater and reuse it based on the evolution of water quality parameters. The work conducted by Das et al (2014) concentrates on the removal of contaminants of all sizes (macro to nano) through the usage of CNT membranes. The research outcomes indicate that CNT membranes have a high removal rate of salt and pollutants than the conventional ones and also it proves to be a low energy consumption technology with features of self-cleaning and fouling resistance. A detailed survey has been done by Bhatnagar et al. (2015) on the usage of peels of agricultural waste in the field of water treatment. It was found from the survey that these are used as natural adsorbents of contaminants present in the polluted water and they also proved to be low-cost technology. A brief discussion on the current technologies used for water purification through membranes has been carried by Werber et al. (2016) where the limitations of these technologies were also studied in brief for the selection of appropriate technology based on the requirement. It also been observed from the study that the resistance to fouling is enhanced through membrane design modification. The technological development in the field of polluted water treatment with the usage of different lists of adsorbents has been studied by Singh et al. (2018). The mechanism behind the adsorption of impurities by the listed absorbents was also studied by the author to address the advantages and drawbacks of these technologies. A systematic assessment of water treatment technologies for the removal of heavy metals using sustainable approaches has been carried out by Bolisetty et al. (2019). The impact of combining the effects of electromagnetic radiation and technologies of electrocoagulation in the field of water purification has been studied by Hashim et al. (2021). The outcomes of the research indicate that the incorporation of these two combinations results in around more than 90% of impurities removal. A detailed interpretation of the usage of solar energy for water purification has been done by Zhou et al. (2021) in which it has been proved that the usage of solar power is a favourable sustainable technology for efficient water treatment. An exploration of the usage of laccase in the eradication of pollutants of microsize as a biocatalyst has been taken by Zhou et al. (2021). The detailed investigation on the impact of parameters such as the efficiency of pollutant eradication and mortification of enzymes along with carrier adsorption on the usage of laccases in immobilised form. Mahdavi Far et al. (2022) outlined a detailed study on the utilization of membranes made up of nanocomposites along with particles of zeolite in the performance assessment of traditional membranes used in water purification. It was noted from the research outcomes that the membrane's performance has been enhanced with the utilisation of zeolite. A structured assessment on the outline of usage of carbon nano fibres in water purification has been carried out by Wang et al (2022). A brief discussion on the impacts and future strategies on carbon nanofibres incorporation was also done by the author. An outlook on the usage of MOF in the treatment of grey water has been projected by Ihsanullah (2022). MOF is used as a material of absorbent to extract the contaminants such as dyes and heavy metals from polluted water.

In recent years, the usage of biomaterials in the field of water purification has been carried out by several researchers. Moringa oleifera is one of the most commonly used natural absorbents in the purification of contaminated water (Aho and Lagasi, 2012; Bichi, 2013; Beltrán-Heredia et al., 2012; Sánchez-Martín et al., 2012; Yarahmadi et al., 2009; Polepalli and Rao, 2018). All the research carried out using Moringa oleifera concentrates on the complete purification of contaminated water and converting it to drinking water. But the current work concentrates on the usage of contaminated water in micropower plant boilers after a considerable reduction in values of pH and hardness using Moringa oeifera and alum as pollutant absorbents (Aho and Lagasi, 2012; Bichi, 2013; Beltrán-Heredia et al., 2012; Sánchez-Martín et al., 2012; Yarahmadi et al., 2009; Polepalli and Rao, 2018). The present research depicted here also focuses on the development of a micropower plant in which power can be generated using contaminated water collected from educational institutions. The usage of contaminated water may corrode the boiler tubes and leave salt deposits if they have low pH and high hardness. In order to eradicate this issue, efforts have been made in the present work towards a reduction in hardness and enhancement in pH. The data survey about the usage of water (in litre) for cleaning in educational institutions has been explored. The wastewater treatment plant used in these institutions at present was studied in deep to recognize the demerits and the advancements to be made. The major drawback of present wastewater treatment is water management in a cost-effective way. This drawback in the present treatment has been framed as a base for the hypothesis of the current study. The establishment of a low-cost experimental setup for wastewater treatment has been focused on in this research.

Experimentation

Experimental test rig

Greywater of 1 litre was collected from the author's institution from the grey water collection tank and passed for filtration in stage 1. At stage 1, Moringa oleifera at different proportions of weight mixed with the grey water. At stage 2, water collected from stage 1 dissolved with a hydrated form of potassium aluminium sulphate at variable masses. The properties of water collected from stage 1 and stage 2 have been tested simultaneously using a laboratory setup available at Vel Tech University. The clarified water from stage 2 was aerated at different time intervals to distil water to reduce hardness and it was passed to softener where its properties are further changed adjustable to boiler feed water properties. The experimental test rig used for the proposed water treatment was depicted in Figure 82.1. The flow of experimentation was depicted in Figure 82.2.

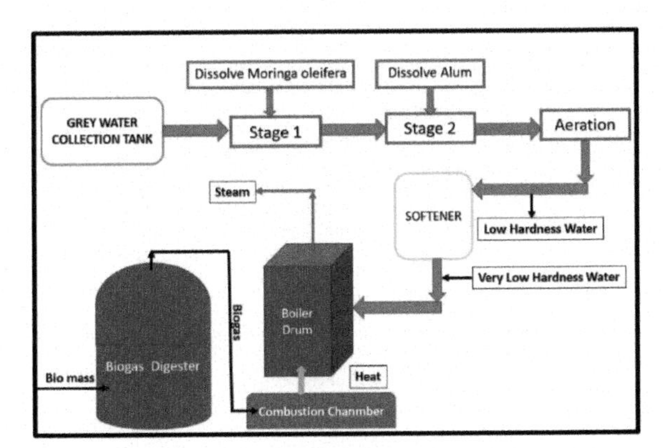

Figure 82.1 Experimental test rig

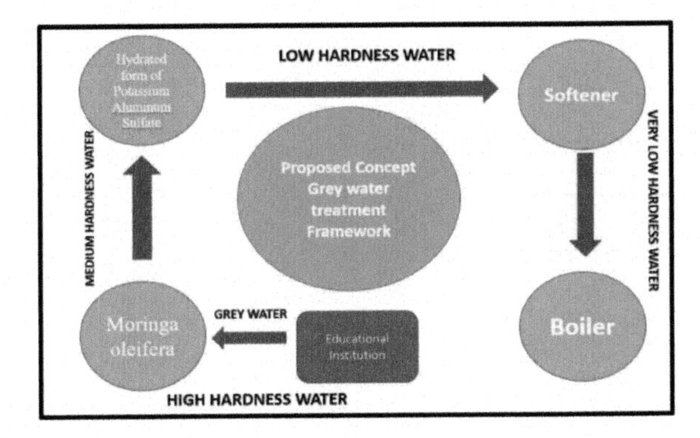

Figure 82.2 Experimentation

The treated water samples at each stage were collected and the values of pH and hardness were evaluated for each sample using the available testing methods in the author's institution. Figure 82.3 shows the images of water samples at each stage. As the grey water passes through different stages, the colour of the water also changes, and finally, the water from the softener becomes colourless.

Experimental tests for pH and hardness

The pH meter shown in figure 82.4 makes used to evaluate the pH value of water samples as shown in Figure 82.3. The presence of Hydroxide alkalinity, carbonate alkalinity, and Bicarbonate alkalinity responsible for hardness in water samples was determined using volumetric analysis as shown in Figure 82.4. The alkalinity of these compounds was evaluated using (1) and (2).

The quality parameters such as pH evaluated using pH meter and hardness are calculated based on the input values of experimentation using the Equations from (1) to (2) given below and the values are listed in Table 82.1.

$$OH^{-}Alka = \frac{[2P - M\} \times N_{HCl}}{V_{water}} \times 50 \times 1000\,ppm \tag{1}$$

$$CO_3^{2-}Alka = \frac{[2M - 2P\} \times N_{HCl}}{V_{water}} \times 50 \times 1000\,ppm \tag{2}$$

where 'M' represents Burette reading of Methyl Orange and 'P' represents Burette reading of phenolphthalein. The above parameters are evaluated for individual samples of untreated and treated water at various stages proposed in the current work.

Figure 82.3 Samples of treated water-stage wise

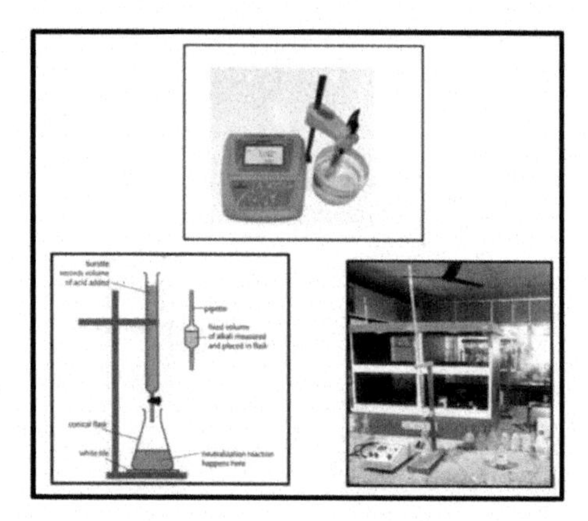

Figure 82.4 Laboratory tests

Table 82.1 Treated water sample-characteristics

Sample	*Properties*	
	pH	*Hardness (ppm)*
Grey water	4.5	1625
Stage 1 water	7	260
Stage 2 water	10.5	85
Aerated water	11	10
Softener water	11.5	2

Results and Discussion

The experimental results derived from the above formulae were depicted as plots in the below figures to expose the variation between the pH and hardness for each water sample in the current study.

The variation of pH value along with the weight of Moringa oleifera was depicted in Figure 82.5. It was evident from Figure 82.5 that the value of pH increases along with an increase in the addition of Moringa oleifera at stage 1. The Greywater at initial was more acidic whose pH was less than 7 and as the Moringa oleifera dissolves in the water the value of pH increases near to 7 and the water becomes neutral

The variation of hardness along with the weight of Moringa oleifera was depicted in Figure 82.6. It was evident from Figure 82.6 that the value of hardness decreases along with an increase in the addition of Moringa oleifera at stage 1. The hardness of grey water at the initial was 1625 ppm and as the Moringa oleifera dissolves in the water the value of hardness decreases and it was 380 ppm.

The variation of pH value along with the weight of alum was depicted in Figure 82.7. It was evident from figure 82.7 that the value of pH increases along with an increase in the addition of alum at stage 2. The value of pH of water from stage 1 was 7 and as the alum gets dissolves in the water the value of pH increases above 7 and the water becomes alkaline.

The variation of hardness along with the weight of the alum was depicted in Figure 82.8. It was evident from figure 82.8 that the value of hardness decreases along with the increase in the addition of alum at stage 2. The hardness of water at stage 1 was 380 ppm and as the alum dissolves in the water the value of hardness decreases and it was 85 ppm.

The experiment has been repeated five times with 45 grams of Moringa Oifera and 16 grams of alum dissolved in polluted water in order to evaluate the uncertainty in the value of pH and hardness. As from the results of the first experiment, the value pH and hardness are optimal only at 35 grams of Moringa

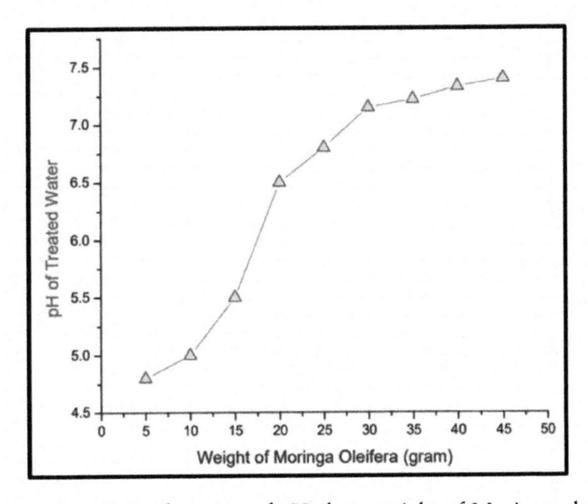

Figure 82.5 Alteration of pH along weight of Moringa oleifera

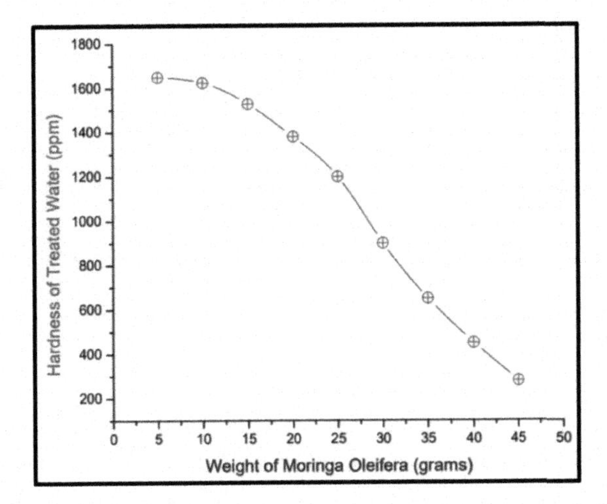

Figure 82.6 Hardness variant with weight of Moringa oleifera

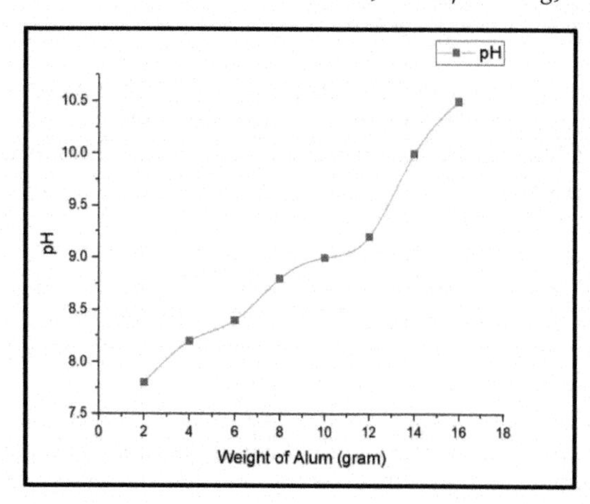

Figure 82.7 Variation of pH with weight of alum

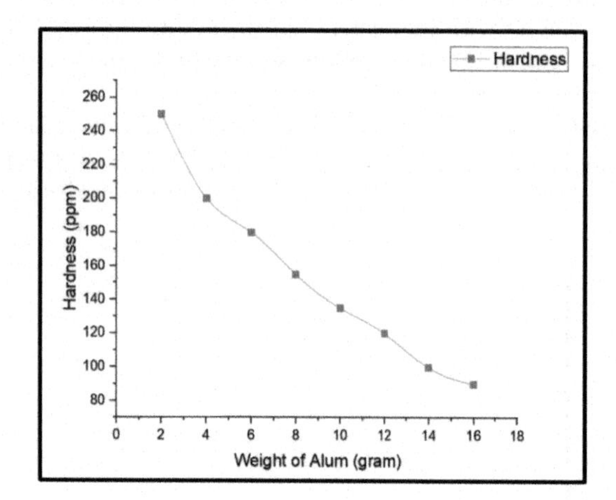

Figure 82.8 Hardness variant with weight of alum

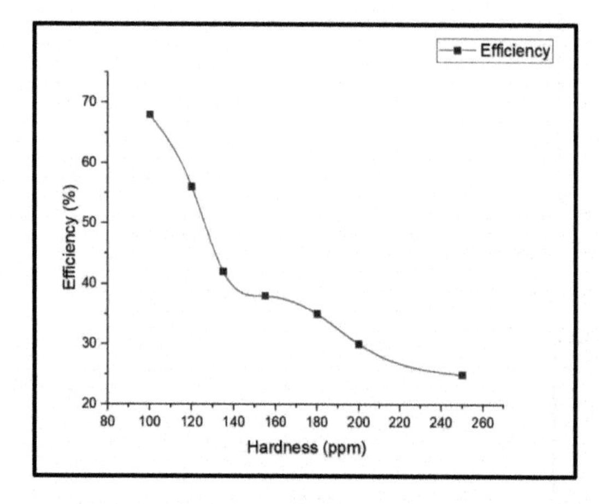

Figure 82.9 Alteration of boiler efficiency along hardness

oleifera and 16 grams of alum. The uncertainty in the value of pH and hardness has been calculated based on the below Equation (3).

$$\sigma = \sqrt{\frac{\Sigma(x_i - \mu)^2}{N}} \tag{3}$$

where σ is the standard deviation; N is the number of readings; x_i is the values of pH and hardness and μ is the mean of measurements. The values of pH and hardness at stage 2 after repeating five times are listed in below Table 82.2.

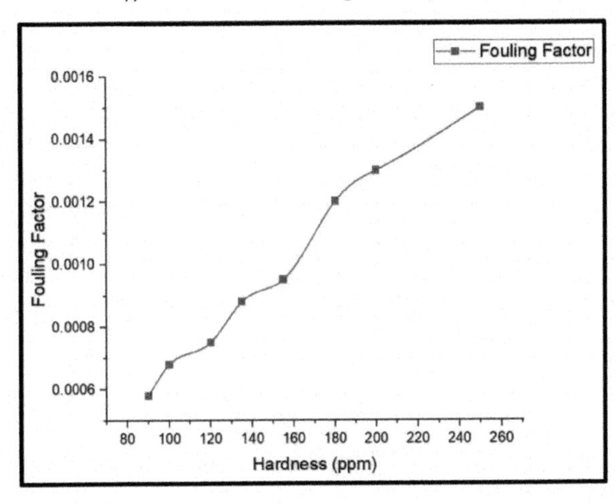

Figure 82.10 Alteration of fouling factor along with hardness

Table 82.2 pH and hardness-repeatability results

No. of Experiments	*Properties*	
	pH	*Hardness (ppm)*
1	10.5	85
2	10.8	90
3	10.45	84
4	9.95	79
5	10.15	78

The uncertainty of pH and hardness found from the data in Table 82.2 using Equation 3 is 10.37±0.294 and 83.2±4.35 respectively.

The boiler efficiency (η_b) is calculated using the below Equation (4).

$$\eta_b = \left[\frac{H-h}{Q}\right] \times 100\% \tag{4}$$

where Q is the calorific value of fuel (kJ/kg); H is the steam enthalpy (kJ/kg); h is the feed water enthalpy (kJ/kg). The enthalpy of steam varies with the hardness present in the feed water. As the hardness decreases, the enthalpy of steam generated increases thereby the boiler efficiency increases.

The value of the fouling factor (R_d) is calculated using the below Equation (5).

$$R_d = 1/U_d - 1/U \tag{5}$$

where U_d is the overall heat transfer coefficient of boiler after fouling and U is the overall heat transfer coefficient of clean boiler.

The overall heat transfer coefficient of a clean boiler is 300 W/m^2K and the value of U_d varies based on the hardness value of feed water supplied to the boiler.

Figure 82.9 represents the change in the efficiency of the boiler along with the value of the hardness of treated water after aeration. The efficiency of the boiler depends on the steam produced from feed water. If the hardness of feed water becomes less, the steam production will be high, and thereby the boiler efficiency increases. It was evident from the plot that at 250 ppm, the efficiency is 25% and at 85 ppm, the efficiency is 68%.

Figure 82.10 represents the change in the fouling factor along with the value of hardness of treated water after aeration. The fouling factor depends on the alkalinity deposits from feed water. If the hardness of feed water becomes less, the alkalinity deposits will be low, and thereby the fouling factor decreases. It was evident from the plot that at 250 ppm, the hardness is 0.00158 and at 85 ppm, the hardness is 0.0005. It was evident from the experimental outcomes that there observed a productivity of high pH and low hardness water of 800 ml from 1 litre of grey water collected using the proposed experimental set up and methodology as shown in Figures 82.1 and 82.2 respectively.

Conclusion

A cost-effective framework for effectual water management to produce and use low hardness water in boilers was developed in the current work. The investigations on treating the wastewater from nearby educational institutions using the biomaterials such as seeds of Moringa oleifera along with the hydrated form of potassium aluminium sulphate (alum) along with aeration at different time intervals for primary filtration were done. The outcome of the repetitive experiments indicates that treated water produced from one litre of grey water using 48 grams of Moringa oleifera along with 12 grams of alum through aeration for 80 minutes has pH value of 10.5 and hardness was 85 ppm. The hardness of treated water further reduced using softener. With the usage of treated feed water in boiler, the efficiency was 68% and fouling factor was 0.0005.

Acknowledgment

This work supported by the Research & Development Centre of Vel Tech Rangarajan Dr Sagunthala R & D Institute of Science and Technology, Avadi, Chennai, India.

References

Aho, I. M. and Lagasi, J. E. (2012). A new water treatment system using Moringa oleifera seed. Am. J. Sci. Ind. Res. 3(6):487–492.

Al-Qahtani, K. M. (2008). Water Purification using different Waste Fruit Cortex for. Bioresourc. Technol. 99:4420–4427.

Beltrán-Heredia, J., Sánchez-Martín, J., Muñoz-Serrano, A., and Peres, J. A. (2012). Towards overcoming TOC increase in wastewater treated with Moringa oleifera seed extract. Chem. Eng. J. 188:40–46.

Bhatnagar, A., Sillanpää, M., and Witek-Krowiak, A. (2015). Agricultural waste peels as versatile biomass for water purification–A review. Chem. Eng. J. 270:244–271.

Bichi, M. H. (2013). A review of the applications of Moringa oleifera seeds extract in water treatment. Civil Environ. Res. 3(8):110.

Bolisetty, S., Peydayesh, M., and Mezzenga, R. (2019). Sustainable technologies for water purification from heavy metals: review and analysis. Chem. Society Rev. 48(2):463–487.

Darwish, M. A., Al Asfour, F., and Al-Najem, N. (2003). Energy consumption in equivalent work by different desalting methods: case study for Kuwait. Desalination. 152:83–92.

De Koning, J., Bixio, D., Karabelas, A., Salgot, M., and Schaefer, A. (2008). Characterisation and assessment of water treatment technologies for reuse. Desalination. 218(13):92–104.

Duff, W. S. and Hodgson, D. (2001). A passive solar water pasteur- ization system without valves. Proceedings of the 2001 Annual Conference of the American Solar Energy Society, April, Washington, DC, American Solar Energy Society, Boulder, Colorado.

Duff, W. S. and Hodgson, D. A. (2005). A simple high efficiency solar water purification system. Solar Energy. 79(1):25–32.

Das, R., Ali, M. E., Abd Hamid, S. B., Ramakrishna, S., and Chowdhury, Z. Z. (2014). Carbon nanotube membranes for water purification: A bright future in water desalination. Desalination. 336:97–109.

Esplugas, S., Bila, D. M., Krause, L. G. T., and Dezotti, M. (2007). Ozonation and advanced oxidation technologies to remove endocrine disrupting chemicals (EDCs) and pharmaceuticals and personal care products (PPCPs) in water effluents. J. Hazard. Mater. 149:631–642.

Ettouney, H. M., El-Dessouky, H. T., Gowin, P. J., and Faibish, R. S. (2002). Evaluating the econom-ics of desalination. Chem Eng Prog. 98:32–9.

Gernjak, W., Krutzler, T., Glaser, A., Malato, S., Caceres, J., Bauer and Ferna´ndez-Alba, R. (2003). Photo-Fenton treatment of water containing natural phenolic pollutants. Chemosphere. 50:71–78.

Gupta, V. K., Ali, I., Saleh, T. A., Nayak, A., and Agarwal, S. (2012). Chemical treatment technologies for waste-water recycling—an overview. RSC Adv. 2(16):6380–6388.

Han, Y., Xu, Z., and Gao, C. (2013). Ultrathin graphene nanofiltration membrane for water purification. Adv. Function. Mater. 23(29):3693–3700.

Harp, J. A., Fayer, R., Pesch, B. A., and Jackson, G. J., (1996). Effect of Pasteurization on Infectivity of Cryptosporidium parvum Oocycsts in Milk and Water. Appl. Environ. Microbiol. 62:2766–2868.

Hashim, K.S., Shaw, A., AlKhaddar, R., Kot, P., and Al-Shamma'a, A. (2021). Water purification from metal ions in the presence of organic matter using electromagnetic radiation-assisted treatment. J. Cleaner Product. 280:124427.

Herrera Melián, J. A., Doña Rodríguez, J. M., Viera Suárez, A., Tello Rendón, E., Valdés do Campo, C., Arana, J., and Pérez Peña, J. (2000). The photocatalytic disinfection of urban waste waters. Chemosphere. 41:323–327.

Igunnu, E.T. and Chen, G. Z. (2014). Produced water treatment technologies. Int. J. Low-carbon Technol. 9(3):157–177.

Ihsanullah, I. (2022). Applications of MOFs as adsorbents in water purification: Progress, challenges and outlook. Current Opin. Environ. Sci. Health.100335.

Jorgensen, A.J., Nohr, K., Sorensen, H., and Boisen, F. (1998). Decontamination of drinking water by direct heating in solar panels. J. Appl. Microbiol. 85:441–447.

Khawaji, A. D., Kutubkhanah, I. K., and Wie, J. (2008). Advances in seawater desalination technologies. Desalination 221:47–69.

Mahdavi Far, R., Van der Bruggen, B., Verliefde, A., and Cornelissen, E. (2022). A review of zeolite materials used in membranes for water purification: History, applications, challenges and future trends. J. Chem. Technol. Biotechnol. 97(3):575–596.

Oliveira, E. P., Santelli, R. E., and Cassella, R. J. (2005). Direct determination of lead in produced waters from petroleum exploration by electrothermal atomic absorption spectrometry X-ray fluorescence using Ir-W permanent modifier combined with hydrofluoric acid. Anal. Chim. Acta. 545:85–91.

Parry, J. V. and Mortimer, P. P. (1984). Heat sensitivity of Hepatitis A virus determined by a simple tissue culture method. J. Med. Virol. 14:277–283.

Polepalli, S. and Rao, C. P. (2018). Drum stick seed powder as smart material for water purification: role of Moringa oleifera coagulant protein-coated copper phosphate nanoflowers for the removal of heavy toxic metal ions and oxidative degradation of dyes from water. ACS Sustain. Chem. Eng. 6(11):15634–15643.

Sánchez-Martín, J., Beltrán-Heredia, J., and Peres, J. A. (2012). Improvement of the flocculation process in water treatment by using Moringa oleifera seeds extract. Braz. J. Chem. Eng. 29(3):495–502.

Singh, N. B., Nagpal, G., and Agrawal, S. (2018). Water purification by using adsorbents: a review. Environmen. Technol. Inn. 11:187–240.

van Nieuwenhuijzen, A. F., Evenblij, H., and van der Graaf, J. H. J. M. (2000). Direct wastewater membrane filtration for advanced particle removal from raw waste- water. Proceeding of 9th Gothenburg Symposium, Istanbul, Turkey, (pp. 2–4).

Wang, J., Zhang, S., Cao, H., Ma, J., Huang, L., Yu, S., Ma, X., Song, G., Qiu, M., and Wang, X. (2022). Water purification and environmental remediation applications of carbonaceous nanofiber-based materials. J. Cleaner Product. 331:130023.

Werber, J. R., Osuji, C. O., and Elimelech, M. (2016). Materials for next-generation desalination and water purification membranes. Nat. Rev. Mater. 1(5):115.

Yarahmadi, M., Hossieni, M., Bina, B., Mahmoudian, M. H., Naimabadie, A., and Shahsavani, A. J. W. A. S. J. (2009). Application of Moringa oleifera seed extract and poly aluminium chloride in water treatment. World Appl. Sci. J. 7(8):962–967.

Zhou, X., Zhao, F., Zhang, P., and Yu, G. (2021). Solar water evaporation toward water purification and beyond. ACS Mater. Lett. 3(8):1112–1129.

Zhou, W., Zhang, W., and Cai, Y. (2021). Laccase immobilization for water purification: A comprehensive review. Chem. Eng. J. 403:126272.

83 Experimental optimisation of crushing parameters in stone crusher using taguchi method

Shubhangi P. Gurway[a] *and Padmanabh A. Gadge*[b]
Department of Mechanical Engineering, G H Raisoni University, Saikheda, India

Abstract

This paper presents the application of Taguchi optimisation process in stone crushers which are the major role players in quality concerns of construction industry. The objective of the present work is to investigate the influence of various crushing parameters on production yield. Crushing parameters namely eccentric speed, feed rate, closed side setting and throw are selected for experimental design approach where L9 orthogonal array was formed and selected for experiments. The effect of selected crushing parameters has been analysed by ANOVA and is optimised by using signal-to-noise (SN) ratio as a statistical method. Analytical results i.e. main effect plots of SN ratios and regression analysis witnessed that closed side setting put a major impact on production yield whereas eccentric speed, throw and feed rate does not have notable influence on the same.

Keywords: ANOVA, orthogonal array, performance optimisation, S/N ratio, Stone Crusher, Taguchi Optimisations.

Introduction

Stone crushers are one of those important part of commination process which convert the large size useless rocks into useful stones finding wide range of application in construction right from home to buildings to transports and other utilities. Depending upon the area of application where the crushed stones demanded, the process of crushing may change (Mali et al., 2015). In general, the crushers usually apply three steps in this size reduction process namely primary, secondary and tertiary crushing.

With the ever increasing technological advancement the demand for various infrastructural projects is rising drastically. And hence most of the countries focus on the effective utilisation of all the resources available to them. Stone crushers are among those useful resources. To become competitive in market most of the infrastructure companies worldwide now increasing their interest towards the quality of crushes aggregate as they are part of whole foundation of any infrastructure. In the developing countries like India due to lack of economic and technical restrictions very less attention is given to these aggregate processing plants which indirectly affect the production yield and quality of crushed rock as well.

The crushing parameters are always act as a major role players in influencing the quality of the crushed aggregates. The crushing parameters right from crusher speed, feed gradation to crusher settings not only affect the specific energy conversion of the crusher but also affect the aggregate quality. Similarly, the parameters might also affect the mechanical properties of finally crushed aggregates to some extent. The flakiness index of the crushed aggregates are sometimes independent of closed side setting but the reduced feed rate may yield a low graded stone (Fladvada and Onnelac, 2020). The reduction in feed particle size distribution of material can reduce the flakiness of the material which is again one of the measure of aggregate quality (Ramos et al., 1994).

The optimisation techniques are generally used to select the best solution among the most available alternatives. In aggregate production units these techniques are mostly used by authors to achieve the subsequent amount of results in their studies. Which are discussed below:

Lee (2012) optimise the compressive crushing by improving the energy efficiency of the crusher. The complex compressive behaviour of four different rocks and two different iron ore were studied and mathematically modelled. In order to theoretically optimise the compressive crushing, genetic algorithm approach was used. It has been observed that the optimality of the crushing is totally depend upon the type of objective function formed because each crushing sequence is depend upon different production situation. It has been observed that the current situation where the crusher works is not effective in achieving optimisation and run with only 30-40% of performance efficiencies. Theoretical studies performed by author and by comparing those with the existing method used in cone crusher it has been observe that the successful implementation of optimisation techniques with good results are totally depend upon the type of crusher.

[a]shubhangi.pbcoe@gmail.com; [b]padmanabh.gadge@ghru.edu.in

Implementation of optimisation results of aggregate and mining application have been performed on prototype and tested on full scale experiment, analysis of which indicate that cone crusher performance can be improved in terms of size reduction ratios and production yield.

Deepak (2010) uses optimisation in design and analysis of jaw plate of single toggle jaw crusher where the author analysed the wear between jaw plate and material particle which is the main reason behind increase in the energy consumption. Kinematic analysis of commercially available swing jaw plate of 0.9m wide with 304mm and 51mm top and bottom opening respectively has been done using finite element analysis. Parametric design package (CATIAP3V5R15) has been used in designing and modelling of the jaw plate and three dimensional model of pitman has been developed. Similarly CATIAP3V5R159 (GENERATIVE STRCTURAL ANALYSIS) programming has been used to perform analysis. The deigned stiffened jaw plate achieves 25% saving in energy.

Numbia et al. (2014) proposed an optimal energy control model in vertical shaft impact crushers which are basically used in tertiary crushing process for fine production aggregates. TOU tariff and other system limitations which may leads to system underperformance has been taken into account for optimal energy control. Various control parameters like conveyor feed flow rate, VSI crusher rotor feed rate and cascade flow rate have been used. Similarly, energy cost as one of the performance index has been selected and a control strategy has been developed. The objective function of the proposed control strategy was quadratic functions with all the constraints being linear. So Matlab 2013 optimisation tool with quadprog function has been utilised. In order to simulate the function of proposed model A 600kW (800 hp) Barmac B9100SE VSI crusher was used with no recirculating material ($\beta = 0$). 2013/2014 Eskom Megaflex for high demand season weekday is used as TOU tariff and Matlab 2013 Optimisation Toolbox with linprog functions has been used to solve the problem.

Wu et al. (2021) proposed an optimisation of crushing chamber of cone crusher which basically consist of mantle and concave which work together to facilitate crushing. Some key structural parameters like length of the parallel zone, eccentric angle, bottom angle of the mantle has been consider to analyse their effect on productivity and product quality. To move forward for achieving the optimisation the amount of blockage layer was analysed and the traditional mathematical model for crusher productivity has been analysed by considering amount of uplift in the chamber. The proposed study has been carried out on C900 cone crusher and comparative analysis has been done with gyratory crusher. The performance was simulated using Discrete Element Method which shows the consistent results with numerical analysis results. Result witnessed better performance of revised mathematical model for crusher productivity by considering the uplift in the blockage layer. The length of the parallel zone and the bottom angle of the mantle are in the range of 140190 mm and 50–60° respectively. It was also observed that bottom angle of mantle put positive impact on productivity but the length of parallel zone negatively affect it. The eccentric angle is within the range of 1.4°–2° and its decrease has a negative effect on the productivity and product quality.

While et al. (2004) optimise the crusher performance and its flow sheet design by proposing a multi objective evolutionary algorithm which is basically belongs to the family of population-based algorithm use to solve the problem in computer. This approach helps to create and evaluate the internal geometry of crusher against multi performance objectives. The author suggested to follow multi objective approach in rock processing as the whole process consist of number of machines focusing on one may not optimise the crushing plant completely.

The case study was conducted in crushing plant where the commination circuit comprised of an input stream, a cone crusher for generating aggregates and one separation screen which control the stone size by rejecting the oversise material. The final size required from the process is –32mm.This single objective of achieving –32mm of final product is done by increasing the speed of shaft and reducing the closed side setting. With the selected control factors, series of experiment have been performed and the algorithm was run for 50 population size and 200 generations for each run. Results depict the increase in capacity by 140% and around 10% in P80 which is representative of 80 % size of aggregate less than k mm. The proposed solution proved to be best for most of the plant capacity and size related issues.

Blessing Ndhlal (2017) have perform the optimisation of crushing plant dealing with production of copper concentrates from oxide, supergene and sulphide ores. The average output of the plant never attain the production of more than 16.2ntoms per hours. During case study it has been observed that discrete and moderate changes in the process put major impact on the crushing plant performance. The process model describing the dynamic operation of an Osborn 57S gyrashere cone crusher has been investigated and number of crushing parameters right from eccentric speed to CSS, federate, crusher power etc have been selected for optimizing sequence where it has been observed that Closed Side Setting is most influential control parameter. Self-tuning controller has been design to determine the real insight of optimisation in the system and Simulink/MATLAB environment has been used to determine the parameter of PID controller. The control model which is basically a function of cavity level and power drawn by the cone crusher is then simulated and tested outcome of which is then validated against the real Mowana crushing process control upgrade.

Hulthén (2010), perform real time optimisation of cone crusher by considering closed site setting and Eccentric speed as one of major parameters for optimisation. For optimal process the author proposed model by following the hypothesis that it's not possible to achieve optimality based on some fixed parameter as the crushing process vary continuously. So because of this reason it is challenging task to maintain speed and CSS as well. To achieve this author apply the frequency convertor to the crusher motor supply which help in continuous adjustment of speed. To continuously monitor flow rate mass flow sensors are used and results are communicated to the operator on his computer. The algorithm has been developed for real time data collection like closed side setting, eccentric speed, flow rate etc. The crushing stages where different speed algorithms have been applied have then come up with the increased performance of 4.2% and 6.9%, which is actually increased by almost 20% when applied in real case.

Lindstedt and Bolander (2015) perform automation and optimisation of gyratory crusher with the objective of productivity improvement. The problem of initial failure of the crusher within vary short span of its installation have been witnessed. In this case controlling of charge level according to different characteristic property of ores have been observed as the important parameter which could be reason of failure. As a solution author proposed the automatic control of the steady state setting of charge level by developing a mathematical model and design different.

Control algorithms which were tested at different level of the analysis.

GAP Analysis

Significant number of research have been carried out in performance optimisation of stone crusher by different researchers which is discussed in Table 83.1.

Most of the published research article focused on the energy efficiency and other part of the process but, the studies related to aggregate production quality and increasing production yield have been underestimated. Similarly optimisation of such process parameters with the objective function of improving the production yield of aggregate have rarely been covered by the researchers. Most of the research work has concentrated on design, analysis and experimentation of the crushers.

Whole aggregate production units are foundation of construction industry but always taken for granted in research concern.

To bridges the gap between these areas a Design of Experiment (DoE) have been planned and ANOVA have been done on each significant parameter to find out the effect of each parameter on the quality and yield of stones.

The main objective of the work is to find out the effect of selected parameters on quality of the product and the production yield through Taguchi which is one of the significant tools for finding put best results with less experimental design cost. Taguchi approach is best suitable tool for finding the optimal values of the process variable. The design of orthogonal array can be defined with reduced number of experiments which can be effectively used in finding out the effect of number of variables in single parameter.

Methodolgy

The adopted methodology is divided into four different phases which is shown in Figure 83.1 and described below:

Phase 1: Planning

In this phase the aggregate production process was observed thoroughly and the operating parameters controllable or uncontrollable will be identified followed by Taguchi Design of Experiment where L orthogonal array are formed by selecting different performance parameters with levels.

Phase 2: Phase two involve the conduction of onsite experiment for determining production yield and cumulative weight fraction according to the predetermined combination of the parameter setting obtained in phase one.

Phase 3: The data collected in phase 2 is then analysed by finding the SN ratio and ANOVA to find out the effects of the experimental factors on the performance of the crusher and thus optimise the parameters.

Phase 4: The data is finally validated to know which operating parameter will positively affect the production yield.

Table 83.1 Summary of research work carried out on crushers

Sr.No	*Type of crusher consider in work*	*Crusher specification*	*Type of Study*	*Summary/remark*	*Publication year*	*Author/Reference*	*Country*
1.	Secondary crusher	Cone crusher Model Svedala H6000	Design+ analysis	The cone crusher performance is analysed by using Bonded Particle Method to study Rock breakage mechanics. DEM is used as a tool for the analysis.	2012	Quist (2012)	Sweden
2.	Primary crusher	Jaw crusher	Expt. + Nume.	FEM analysis can be used as a powerful tool for analysing the failure of the component in jaw crusher.	2013	Rusiński et al. (2013)	Poland
3.	Primary crusher	Jaw crusher	Expt. + Nume.	Regression analysis was done in order to find out the effect of mechanical properties of rock on the specific power utilisation of jaw crusher	2014	Korman et al. (2015)	Croatia
4.	Primary crusher	Gyratory crusher	Analytical+ design	Control system was designed and added to the crusher for controlling the charge level. Simulation was done to monitor the work. Productivity increased.	2015	Lindstedt and Bolander (2015)	Sweden
5.	Primary crusher	Jaw crusher	Design	Lighter weight swing jaw plate of jaw crusher was designed using CatiaV5R15 and Finite element analysis of the same was done by ALGOR V19.	2016	Khalidurfeasif and Qureshi (2016)	India
6.	Primary crusher	Jaw crusher	Design + analysis	Dynamic and kinematic analysis of all parts of the crusher has been done to improve its performance.	2016	Abdulkarim et al. (2016)	Iraq
7.	Secondary crusher	Cone crusher	Expt.	Experimental analysis of cone crusher using RMT-150B rock mechanics testing system	2016	Ma et al. (2016)	China
8.	Secondary crusher	Cone crusher	Expt. + Analysis	Simulation of The laboratory crusher Morgårdshammar B90 has been done using DEM and validate with laboratory experiments. MATLAB image analysis is use to find out the performance parameters.	2016	Johansson et al. (2016)	Sweden
9.	Primary	Single toggle jaw crusher	Analysis	Kinematic analysis was done in order to find the effect of any kind of alteration on the performance of the Single toggle jaw crusher.	2017	Oduori et al. (2016)	Kenya
10.	Primary crusher	Jaw crusher	Analysis	Adaptive neuro-fuzzy interference system (ANFIS) was use to predict the specific power consumption of the jaw crusher with 96% of accuracy when compared with regression model.	2019	Abuhasel (2019)	Saudi Arabia
11.	Primary crusher	Jaw crusher	Design + Analysis	New design of jaw plate involving toothed profile was designed using CATIA software and Finite element analysis was done using ANSYS.	2019	Ashok and Rao (2019)	India
12	Aggregate type	Granite stone crusher	Design + Analysis	The design of stone crusher for crushing particle size between 25-135mm to about 24-20.2mm was done. ANSYS software is used to find out high region stress area as major factor for safe working of proposed equipment.	2020	Tauyanashe et al. (2020)	Zimbabwe
13.	Secondary crusher	cone and horizontal shaft impact (HSI) crushers	Comparative study	Statistical Analysis was used to compare the performance parameters of both crushers. Pearson Correlation Matrix was used to correlate the variables with performance	2020	Köken and Qu (2020)	Turkey, China
14.	Secondary crusher	Inertia Cone crusher	Design	Modelling of real time dynamic model of inertia cone crusher using EDEM software. The designed model is verified through industrial scale experiment on a GYP1200 inertia cone crusher.	2020	Cheng et al. (2020)	China
15.	Secondary crusher	Cone crusher	Analytical	The evaluation of size reduction process was done by using analytical method. Crushability test was conducted using Bonded particle method.	2020	Köken (2020)	

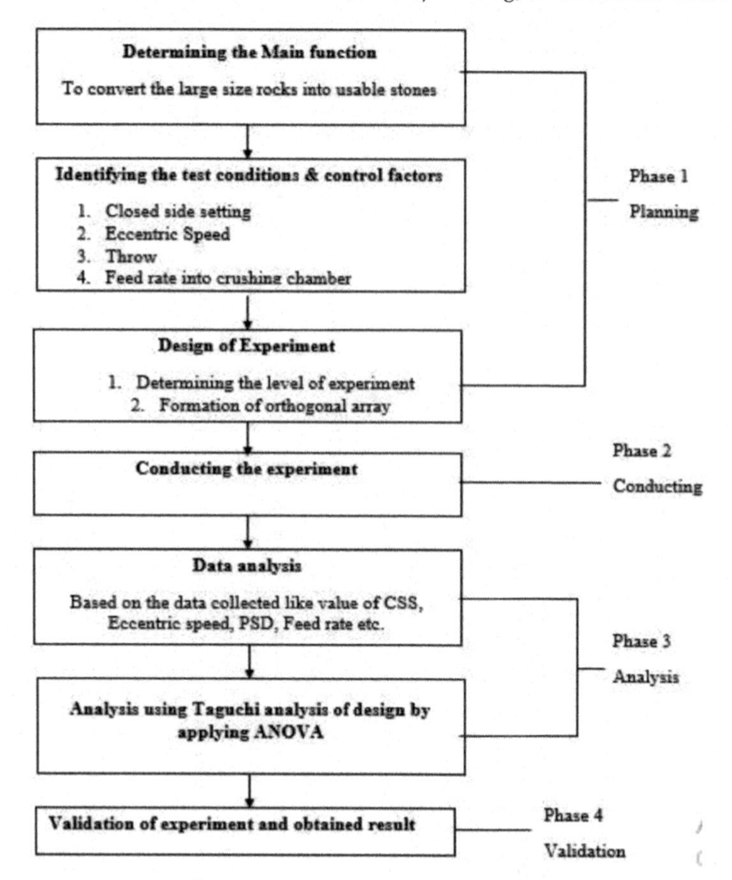

Figure 83.1 Methodology

Experimental design approach

Taguchi Design of Experiment is one of the promising optimisation tool in reducing the number of test required in a design procedure and finding out the optimal solutions with less test trials. Therefore we have adopted Taguchi optimisation to conduct the experiment. An orthogonal array OA, L09 were applied. In the proposed work we have Eccentric speed, Closed Side Setting and Feed rate as process parameters which are controllable factors. The level of parameters are represented in Table 83.2. To achieve the subsequent optimisation result we have selected production yield as a performance parameters for this work. The effect of selected process parameters on the performance parameters was evaluated with onsite experiment and laboratories test. Moreover, ANOVA is used as an analytical tool to show the effects of the experimental factors on the performance of the crusher and thus optimise the parameters (Foster, 2000).

Identifying the main function and its side effect

Main function: Crushing of rocks in variable range stones of size 1 mm, 5 mm, 10 mm, and 20 mm, from basalt rock.

Side effects: variation in product quality and rejection from customers because of higher flakiness index of crushed rocks.

This step also involve the identification of controllable and uncontrollable factors that are directly associated with the objective function. These factors are listed in Table 83.1.

While constructing the matrix it is considered that those factors which actually influence the product quality and Production yield should be taken into considerations while other factors can be considered as noise factors.

Identification of testing conditions and quality characteristics

Performance parameter: Production yield

Table 83.2 Crusher controllable and uncontrollable factors

Sr.No	*Crushing factors*	
	Controllable	*Uncontrollable*
1	Eccentric speed	Environmental condition
2	Feed rate	Vibrations
3	Closed side setting	Operator skills
4	Throw	

Identifying the objective function

For optimisation of any static problem there are three signal to noise ratios which need to be consider for designing of experiment, which are:

1. Smaller The Better
2. Larger The Better
3. Nominal is The Best

Objective functions like cost of production should always be minimised so for that 'Smaller The Better' SN ratio suited best. For objective functions like Quality and production yield should be maximise to improve the performance of the industries so 'Larger The Better' SN ratio fitted best for that. In some of the cases like thickness of surface and other product characteristics should be sufficient enough according to requirement of process for such kind of objective functions 'Nominal the Best' SN ratio should be chosen (Introduction to Taguchi Method, ????).

In our design of experiment we have selected increased production yield as objective function which should be maximised.so we have selected 'Larger The Better' SN ratio.

Objective function:
Larger the better: We have decided to optimise the quality and production yield so for both of the performance parameters, the larger the SN ratio best optimised will be the result. Therefore S/N ratio for the function is used.

The S/N ratio is calculated for each of the parameter combination.

Larger-is-better S/N ratio using base 10 log can be calculated as:

$$S/N = -10 \text{ Log}_{10} (\Sigma (1/Y^2)/n)$$

Where, Y represents the responses for the given factor level combination and n represents the number of responses in the factor level combination.

Identification of control factors and their levels

The factors and their corresponding levels were decided for conducting the experiment based on a data collected during the questionnaire session with the operator and the technicians of the crushing plant. Similarly the standard operating manual has been used for deciding the parameters. The crusher variable and their respective levels are shown in Table 83.3.

Table 83.3 Crusher variable and their levels

Parameters	*Level of performance*			*Performance parameter*
	1	*2*	*3*	
Eccentric speed (rpm)	1440	1440	1420	Production yield
Closed side Setting (mm)	30	32	35	
Feed rate (ton/hr.)	110	120	125	
Throw	36	29	29	

Formation of orthogonal array

The experiment has been designed on MiniTab-14 software three levels and four factors where possibility of two orthogonal array L9 and L27 was obtained as alternatives. Taguchi L09 orthogonal array was selected for initial Design of Experiment, with four variables namely Eccentric speed, closed side setting, feed rate and throw which may have significant effect on the performance and have represented in Table 83.2. Based on the initial investigation and to select the near optimal range Taguchi L09 orthogonal array have been formed which is shown in Table 83.4 (Abuhasel, 2020).

Experimental setup

After formation of orthogonal array various settings have been carried out in the crushing plant with different combination of every parameters. Figure 83.2 shows the onsite experimental setup where the selected parameters were set at different combination values found in designed L9 orthogonal array.

Experimental Result

To find out the significant results in production yield, average cumulative weight was determined from the various testing combinations of the selected parameters which are described in Table 83.5. The Taguchi L09 DOE

Table 83.4 Taguchi standardise L09 orthogonal array for selected factors and level

Eccentric Speed	*Closed Side Setting*	*Feed Rate*	*Throw*
1	1	1	1
1	2	2	2
1	3	3	2
1	1	2	2
1	2	3	1
1	3	1	2
3	1	3	2
3	2	1	2
3	3	2	1

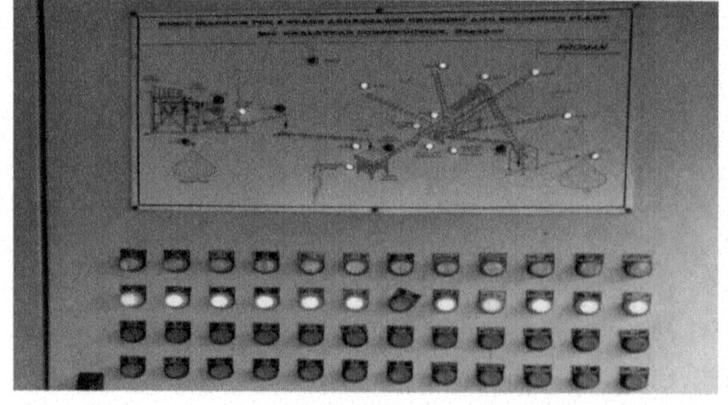

Figure 83.2 Onsite experimentation setup with different parameter settings in crusher

method was used to optimise the decided parameters to achieve higher production rate. As we select larger the better SN ratio the larger value will reflect the better characteristics, as per the equation shown below.

$$S/N = -10 \times \log(\Sigma(1/Y^2)/n)$$

Where n is the number of experiments and y is the measured variable values.

Table 83.4 shows the details of all the experimental factors and the DOE tool applied to optimise the crushing parameters to achieve high production yield. The table also represents the signal-to-noise ratio calculated for each cumulative weight.

The quality of aggregate product used during the construction plays a very important role, ineffective and unsafe aggregate is not acceptable by any of the work. Again the quality of the aggregate produced is vary application wise to comply with this sieve analysis is used which determines the distribution of aggregate particle by size and the tested results are then use to verify whether the design is comply with the desired production requirement or not. In sieve analysis the dry representative of the sample is weighted and the placed in the top of the sieve screen trays which are arranged in descending orders with largest opening at the top and largest at the bottom. The stack is then agitated manually using sieve shaker for a designated period of time and then the operator weights the material on each sieve using one of the two type of method discussed below:

1. Cumulative method
2. Fractional method

Cumulative method: In this method each sieve fraction, beginning with the coarser material in previously used weighing pan and measured and each retained fraction is added.

As each cumulative mass is retained we divide the cumulative mass by total mass and multiply by to calculate the percentage retained. To calculate the cumulative % passing, subtract the cumulative % retained from 100.

Cumulative % retained = (Cumulative mass/total mass) × 100

Cumulative % passing = 100 – (Cumulative % retained)

Fractional method: the content of each sieve fraction is separately weighted and discard the weighted material completely. As the operator weighted the content of each sieve fraction, we divide the mass retained on each sieve by the total mass and multiply by 100 to calculate the percentage of fraction retained. Similarly to calculate the percentage fractional passing, subtract the percent retained from percent of fractional retained.

% Fractional retained = (Fractional mass/total mass) × 100

% Fractional passing = (% Fractional retained) – (% retained on sieve below)

Analysis and Discussion

The test were conducted and SN ratio for each level of experiment have been calculated. The data were plotted as shown in Figures 83.3 and 83.4. Similarly Table 83.6. Shows the responses to various SN ratios

Table 83.5 Measured cumulative weight fraction under different crushing condition with SN ratio

Eccentric speed (rpm)	*Closed side Setting (mm)*	*Feed rate (tonn/hr)*	*Throw (mm)*	*% Cumulative weight Fraction*
1440	30	110	39	7
1440	32	120	29	13
1440	35	125	29	47
1440	30	120	29	8
1440	32	125	39	15
1440	35	110	29	50
1420	30	125	29	5
1420	32	110	29	11
1420	35	120	39	62

for larger the better condition. It shows that among all the four parameters Closed Side Setting is most significant parameters which create the major impact on mean SN ratio followed by feed rate, where eccentric speed and throw put a negligible impact on the cumulative weight fraction. 'Figure 83.3' represent the residuals are linear with normal probability plot and have no identifiable error due to time or data.

A regression method was used to identify the effect of these selected parameters though which it is easy to correlate the measured variable with measured variable. The empirical relationship is formed by mean of regression Equation (1).

In order to analyse the measured data ANOVA at the confidence interval of 95% ($p<0.05$), have been performed which is represented in Table 83.7. And results shows that the cumulative weight fraction is increased with increased closed side setting and have major impact on output production quality and yield whereas rest of the other parameters put a negligible effect on it.

Regression equation

$$\% \text{ Cumulative weight fraction} = 519 - 0.28 \text{ eccentric speed} - 1.96 \text{ closed side setting} - 0.23 \text{ feed rate} + 0.10 \text{ throw} \quad (1)$$

Table 83.6 Response table representing signal to noise ratios larger is better

Level	*Eccentric Speed*	*Closed Side Setting*	*Feed Rate*	*Throw*
1	26.32	22.21	24.04	24.28
2	24.21	34.42	26.41	26.18
3	-	18.11	24.29	
Delta	2.11	16.32	2.38	1.90
Rank	3	1	2	4

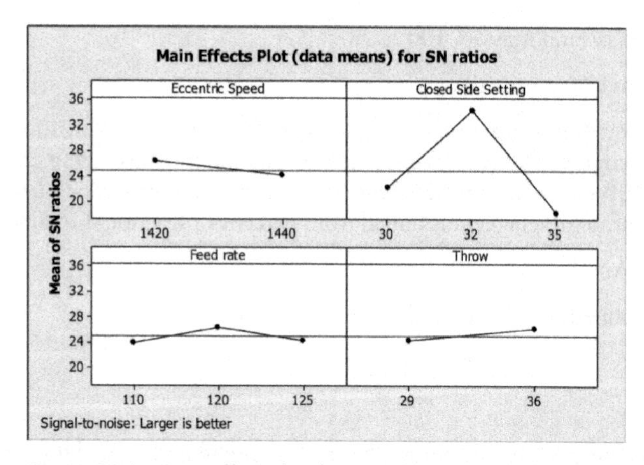

Figure 83.3 Main effect plot for signal to noise ratio larger is better for each individual parameter showing its effect on cumulative weight fraction

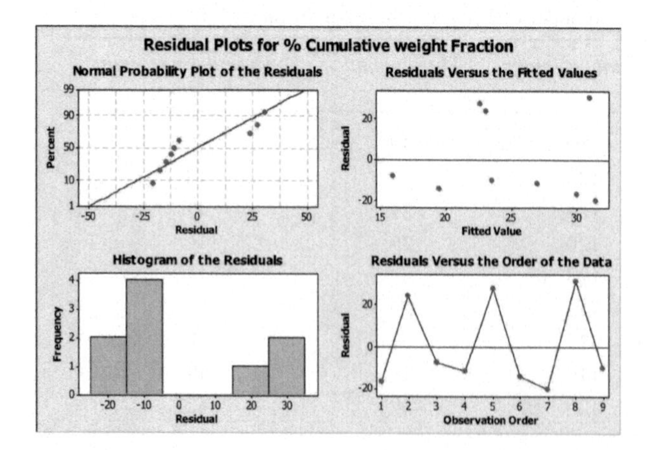

Figure 83.4 Residual plot for cumulative weight fraction %

Table 83.7 ANOVA

Source	*DF*	*SS*	*MS*	*F*	*P*
Regression	4	230.9	57.7	0.07	0.989
Residual error	4	3520.0	880.0	0	0
Total	8	3750.9	0	0	0

Conclusion

In order to improve the production yield, the effect of different process parameters on it has been identified by Taguchi optimisation method. L9 and L27 orthogonal arrays have been come as an option out of which we first selected L9 OA as an initial stage of study and perform the experiment as per the standard OA combinations. With the aid of this analytical and experimental investigation it has been observed that quality of stone and production yield is directly depend on the eccentric speed, closed side setting, feed rate and throw. Analysis of data using ANOVA at 95% confidence level have been performed followed by regression analysis where it was found that the cumulative weight fraction was most influenced by closed side setting. Rest of the three parameters have a negligible effect on cumulative weight fraction.

For future work the effect of the selected operating parameters on the objective function under environmental condition as an uncontrollable noise factor is planned to get real insight of production yield in different seasons.

References

Abdulkarim, J. M., Rahim, D. Y., AL_shamikh, A. J., Ali, N. M., and Tofiq, H. A. H. (2016). Development design for jaw crusher used in cement factories. Int. J. Sci. Eng. Res. 7(6):445–456. ISSN 2229-5518.

Abuhasel, K. A. (2019). A comparative study of regression model and the adaptive neuro-fuzzy conjecture systems for predicting energy consumption for jaw crusher. Appl. Sci. 9(18):3916. https://doi.org/10.3390/app9183916.

Abuhasel, K. A. (2020). Aggregate production optimisation in a stone-crushing plant using the taguchi approach. Proc. Natl. Acad. Sci. India, Sect. A Phys. Sci. doi: 10.1007/s40010-020-00658-0

Ashok, E. and Rao, L. N. V. N. (2019). Design and analysis of jaw plate with various tooth profiles using FEM. Int. J. Basic Appl. Res.

Cheng, J., Ren, T., Zhang, Z., Liu, D., and Jin, X. (2020). A dynamic model of inertia cone crusher using the discrete element method and multi-body dynamics coupling. Minerals 10(10):862. https://doi.org/10.3390/min10100862.

Deepak, B. B. V. L. (2010). Optimum design and analysis of swinging jaw plate of a single toggle jaw crusher. Master diss., National Institute of Technology Rourkela.

Fladvada, M. and Onnelac, T. (2020). Influence of jaw crusher parameters on the quality of primary crushed aggregates. Miner. Eng. 151:106338.

Foster, W. T. (2000). Basic taguchi design of experiments. In National association of industrial technology conference, Pittsburgh, PA.

Hulthén, E. (2010). Real-time optimisation of cone crushers. Ph.D diss., Chalmers University of Technology Göteborg, Sweden.

Johansson, M. Johannes, Q., Magnus, E., Erik, H. (2016). Cone crusher performance evaluation using DEM simulations and laboratory experiments for model validation. In 10th International comminution symposium. Article in Minerals Engineering (September 2016). Available online at www.researchgate.net.

Khalidurfeasif, S. P. and Qureshi, U. (2016). Design and finite element analysis of swing jaw plate of jaw crusher with stiffener. Int. J. Adv. Eng. Res. Dev. 3(9):53–59.

Köken, E. (2020). Evaluation of size reduction process for rock aggregates in cone crusher. Bull. Eng. Geol. Environ. 79:4933–4946. Available online via www.springer.com.

Köken, E. and Qu, J. (2020). Comparison of secondary crushing operation through cone & horizontal shaft impact crushers. In International multidisciplinary scientific geoconference surveying geology and mining ecology management, SGEM. (Vol. 2020 August, Issue 1.1, p. 789796). https://doi.org/10.1016/j.mineng.2010.07.013.

Korman, T., Bedekovic, G., Kujundzic, T. and Kuhinek, D. (2015). Impact of physical and mechanical properties of rocks on energy consumption of jaw crusher. Physicochem. Probl. Miner. Process. 51(2):461–475. http://dx.doi.org/10.5277/ppmp150208.

Lee, E. (2012). Optimisation of compressive crushing. PhD diss., Chalmers University of Technology Goteborg, Sweden.

Lindstedt, S. and Bolander, A. (2015). Automation and optimisation of primary gyratory crusher performance for increased productivity. Master diss., Luleå University of Technology.

Ma, Y., Fan, X., and He, Q. (2016). Prediction of cone crusher performance considering liner wear. Appl. Sci. 6(12):404. https://doi.org/10.3390/app6120404.

Mali, A. V., Morey, N. N. and Khtri, A. P. (2015). Improvement in the efficiency of the stone crusher. Int. J. Sci. Eng. Technol. Res. 5:2265–2271.

Ndhlal, B. (2017). Modelling, simulation and optimisation of a crushing plant. magister of technologiae: Engineering: Electrical. University of South Africa.

Numbia, B. P., Xiaa, X. and Zhangb, J. (2014). Optimal energy control modelling of a vertical shaft impact crushing process. In The 6th international conference on applied energy – ICAE2014. Energy Procedia 61:560–563.

Oduori, M. F., Munyasi, D. M. and Mutuli, S. M. (2016). Analysis of the single toggle jaw crusher kinematics. J. Eng. Desig. Technol. 13(2):213–239. https://doi.org/10.1108/JEDT-01-2013-0001Quist, J. (2012). Cone crusher modelling & simulation. Mater thesis., submitted to Department of product & production development, chalmers university of technology, Goteborg Sweden, (2012). Available online via https://publications.lib.chalmers.se/.

Ramos, M., Smith, M. R. and Kojovic, T. (1994). Aggregate shape-prediction and control during crushing. Quarry Manag. 23–30.

Rusiński, E., Moczko, P., Pietrusiak, D., and Przybyłek, G. (2013). Experimental and numerical studies of jaw crusher supporting structure fatigue failure. J. Mech. Eng. 556–563. https://doi.org/10.5545/sv-jme.2012.940.

Tauyanashe, C., Mushongo, R. N., Chinguwa, S., Sakala, T., and Nyemba, W. R. (2020). Design of a small-scale granite stone crusher. Procedia CIRP. 91:858–863. doi:10.1016/j.procir.2020.03.119

While, L., Barone, L., Hingston, P., Huband, S., Tuppurainen, D. and Bearman, R. (2004). A multi-objective evolutionary algorithm approach for crusher optimisation and flowsheetdesign. Miner. Eng. 17(1112):1063–1074.

Wu, F., Ma, L., Zhao, G. and Wang, Z. (2021). Chamber optimisation for comprehensive improvement of cone crusher productivity and product quality. Hindawi Math. Probl. Eng. 2021(Article ID 5516813):1–13.

84 Automated classification of ovarian tumors using GLCM and Tamura features with Light GBM classifier

Smital Dhanraj Patil[1,a], *Pramod Jagan Deore*[1,b], *and Vaishali Bhagwat Patil*[2,c]

[1]R. C. Patel Institute of Technology, Shirpur, India

[2]RCPET's Institute of Management Research and Development, Shirpur, India

Abstract

Ovarian abnormalities, such as ovarian cysts, tumours, and polycystic ovary (PCO) are one of the most serious concerns for women's health after breast cancer. In females, ovarian cysts must be accurately diagnosed to make informed decisions at the appropriate time. Classification and identification of ovarian masses are quite complicated due to complexity and resolution of image background. So features of such images play a vital role in automatic classification. This paper aims to implement the classification of ovarian abnormalities using GLCM and Tamura texture features. Various texture features like coarseness, contrast, directionality, and roughness are extracted and Light GBM classifier is used to achieve the classification. Performance measures like F1-score, precision, accuracy, and recall are used to analyse the results of classification. The classification accuracy for combination of GLCM and Tamura texture features is found to be 72% for proposed work without segmentation.

Keywords: Classification, GLCM, LightGBM, ovarian masses, tamura texture feature.

Introduction

Ovaries are reproductive glands. These glands produce reproductive cells. Each female has two ovaries, which play a crucial role in a woman's reproductive system. The ovaries are placed on either side of the uterus and generate the hormones oestrogen and progesterone, which are associated with menstruation and pregnancy.

Ovarian mass is a cyst or abnormal tissue development (tumour) that occurs in one or both of the female ovaries and can be benign or malignant.

- Ovarian cyst: Cysts are small sacs in the ovaries that fill with air, body fluids, or other substances.
- Ovarian tumour: Abnormal growth of unwanted tissue is called a tumour.
- Ovarian mass: A non-cancerous or cancerous ovarian mass might be a cyst or a tumour.
- Polycystic ovary (PCO): PCO is a female endocrine condition characterised by cysts in the ovary and caused by a hormone imbalance that affects roughly one out of every ten women. These are follicular cysts that are small in size.

Ovarian abnormalities, such as ovarian cysts, tumours, and polycystic ovary (PCO) are some of the most serious concerns for women's health after breast cancer.

The most noninvasive medical imaging technique for ovarian mass identification is diagnostic ultrasound (US). Ultrasound (US) imaging is one of the most extensively used imaging modalities for diagnosis in medical practice. Ultra sonographers and radiologists use this imaging modality to visually inspect the acquired ultrasound image for the detection of tumours that could be benign or malignant. It's portable, precise, and fully human-friendly. The efficiency of a visual assessment or evaluation is frequently determined by radiologist's skill, as it is the most usual procedure. An incorrect diagnosis could lead to a pointless surgical procedure. As a result, unskilled operators must be given support tools in order to increase their diagnostic accuracy.

As a result, there is a need to extract relevant textural information from raw images.

Hence there is a need to extract meaningful features describing textural properties from raw images.

[a]smitalpatil55@gmail.com; [b]pjdeore@yahoo.com; [c]vaishali.imrd@gmail.com

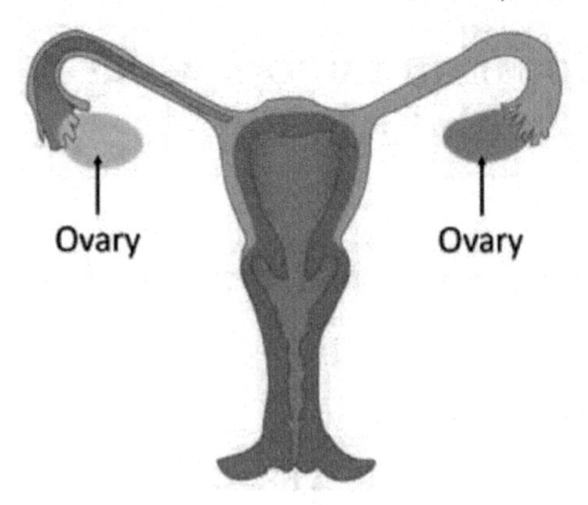

Figure 84.1 Female reproductive system

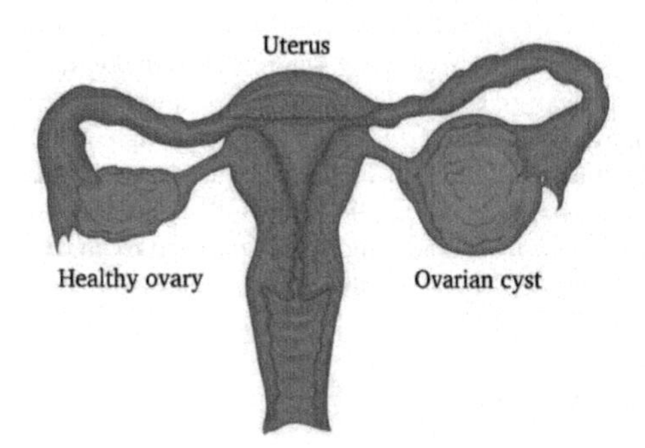

Figure 84.2 Ovarian cyst

Related Work

A combination of Tamura and Gabor method for feature extraction used in Imran (2019) for colour and texture- based image retrieval.

In Karmakar et al. (2017) used Tamura features and suggested a set of kernel descriptors. In this kernel descriptor framework, each pixel plays an equal role in matching two image patches.

Chi et al. (2019) proposed Principal Component Analysis for feature selection. A subset of the original feature vector is obtained by using the framework of the principal components of the feature set, where the features reflect the most typical attributes for the textures in the provided image dataset).

Bagri and Johari (2015) used a fusion of texture and shape features, as well as the texture Gray Level Co-occurrence Matrix and Hu-moments, to extract features.

The Hausdorff distance representing the overall similarity of the medical image feature set, was used by Xiaoming et al (2018). This is used for extraction of the textural features of medical images using a fusion of Tamura texture features and wavelet transform features (Xiaoming et al., 2018). Authors discovered that the proposed method is more accurate than using Euclidean distance.

A detailed survey of regarding extraction of textural features is provided by Anne Humeau-Heurtier (2019a). The texture features introduced by Tamura et al. (1978) correlate to human visual perception. They proposed six basic texture descriptors.

A comparison of three feature extraction approaches that are Histogram of Gradients (HOG), Local Binary Pattern (LBP), and Tamura features was done by R. Beaulah Jeyavathana, Dr. R. Balasubramanian and Mr. Anbarasa Pandian in (Jeyavathana et al., 2017). They found that the combination of these three approaches give better performance in diagnosis of tuberculosis using chest radiographs.

Rufo et al. (2021) used the LightGBM approach to create an accurate model for diabetes diagnosis. The results of the experiments suggest that the produced diabetes dataset is useful in predicting diabetes mellitus.

Comparison of various machine learning approaches for miRNA identification in breast cancer patients, including eXtreme Gradient Boosting (XGBoost), Random Forest (RF), and Light Gradient Boosting Machine (LightGBM) was done in (Wang et al., 2017). The accuracy and logistic loss of each algorithm were compared, and LightGBM was found to perform better in several areas.

For computer aided diagnosis (CAD) systems, the textural features in Ultrasound (US) images of the breast, prostate, thyroid, ovaries, and liver are used for diagnosis of cancer (Faust et al., 2018).

Preetha and Jayanthi (2018) used a hybrid GLCM and GLRLM features from mammograms of breast tissues to reach high accuracy in classifications. Kiruthika et al. (2018) have employed the discrete wavelet transform (dwt) for despeckling. Automatic recognition is done using texture and intensity-based segmentation algorithms and ovaries are classified depending on their morphology.

A wide range of fields and applications make use of Texture analysis that includes texture classification to segmentation to image synthesis and pattern recognition. Different feature extraction methodologies have been introduced, and a thorough survey of texture feature extraction approaches has been done in (Humeau-Heurtier, 2019b).

An artificial neural network (ANN) technique for follicle segmentation using Enhanced Fruit Fly Optimisation (IFFOA) is proposed by Nilofer and Ramkumar in 2021. The statistical GLCM is utilised to introduce the feature extraction algorithm, and the ANN is trained to classify the features.

To characterise the bone X-ray images, Mall et al. (2019) used a variety of machine learning algorithms. For anomaly detection, linear SVM, Logistic Regression, LBF SVM and Decision Tree classifiers are employed.

Extraction of second order statistical texture information for image motion estimation using grey level co-occurrence matrix (GLCM) is very well explained by Mohanaiah et al. (2013).

Methodology

The following figure shows the methodology of the proposed work. The proposed system uses Tamura features for feature extraction and Light GBM classifier for classification of images of benign and malignant ovarian masses.

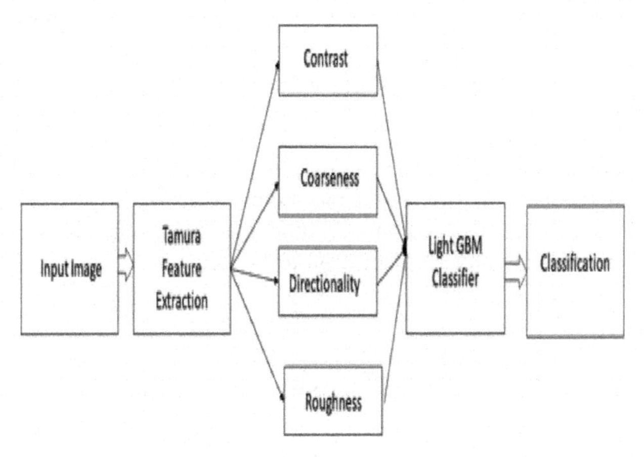

Figure 84.3 Methodology

Tamura Texture Features

The value of 'coarseness-contrast-directionality' (CCD) for each pixel in an image is used to create a Tamura image. This image was created for a combined RGB distribution. Tamura textural features include six features that includes coarseness, regularity, roughness, contrast, line-likeness, and directionality.

The following four features are considered:

Contrast: Contrast gives measure of how much the grey level (q) of the image (I) changes and how skewed their distribution is towards black or white. It can also be defined as the difference between an image's maximum and minimum pixel intensities.

Directionality: The edge strength and directional angle are taken into account while determining directionality. They're calculated using Prewitt's edge detector and derivatives.

Coarseness: Coarseness measures the scale of texture. For a given window size, a texture with fewer texture components is said to be coarser than one with a larger number. The scale and repetition rates of texture correspond to coarseness.

Roughness: It is a combination of coarseness and contrast.

GLCM Features

A key spatial element for distinguishing items or regions of interest in an image is texture. A prominent texture-based feature extraction method is the grey level co-occurrence matrix i.e. GLCM. The GLCM calculates the textural relationship between pixels by working on the image's second-order statistics. For this process, two pixels are commonly used. The frequency of distinct combinations of these pixel brightness values is calculated by the GLCM. It represents the frequency of forming pixel pairs. GLCM attributes of an image are represented as a matrix with the same number of rows and columns as it's grey values.

It works by computing the frequency with which pairs of pixels with particular values appear in an image in a specific spatial relationship, creating a GLCM, and retrieving statistical measurements from it.

The elements of this matrix are determined by the two provided pixels. Based on their surroundings, both pixel pairs can have distinct values. These matrix components comprise second-order statistical probability values depending on the grey value of the rows and columns. The GLCM features used for the proposed work are contrast, correlation, dissimilarity, energy, and entropy etc. These properties are used to construct a GLCM feature matrix that can correctly represent an image with fewer parameters.

Feature extraction is a data pre-processing step that splits and simplifies a big amount of raw data into smaller groups. By extracting all these features using GLCM and Tamura feature extraction method, we can use it for classification.

Light GBM Classifier

The Light Gradient Boosting Machine (LightGBM) is a unique gradient boosting with a few alterations that make it especially useful. It is based on classification trees, however, the leaf splitting is done more successfully at each level. It builds on gradient boosting by including autonomous feature selection and focusing on boosting samples with higher gradients. This can result in significant improvements in training speed and prediction accuracy.

Dataset

The dataset consists of total 187 ultrasonography images of ovarian masses out of which 112 images are benign (non-cancerous) and 75 are malignant (cancerous) (source: osf.io/n9abq) (Martínez-Más, 2019). Using simple random sampling techniques, the images are divided into training and testing sets for measuring the performance of proposed work without applying any segmentation techniques.

Performance Measures

Accuracy: The percentage of correctly classified instances can be used to calculate accuracy.

$$\text{Accuracy (A)} = TP + TN/TP + FP + TN + FN \quad (1)$$

Precision: It can be defined as the ratio of true positives to total positive forecasts. It refers to the accuracy of a model's positive prediction.

$$\text{Precision} = TP/TP + FP \quad (2)$$

F1-Score: The weighted average of Precision and Recall is termed as the F1 Score.

$$\text{F1 Score} = 2 \times (\text{Recall} \times \text{Precision})/(\text{Recall} + \text{Precision}) \quad (3)$$

Recall: Recall indicates the quantity of correct positive predictions produced out of all possible positive predictions.

$$\text{Precision} = TP/TP + FN \quad (4)$$

Here,
TP (True Positive) = Positives forecast correctly
TN (True Negative) = Negatives forecast correctly
FP (False Positives) = Positives forecast incorrectly
FN (False Negatives) = Negatives forecast incorrectly

Result

After extracting GLCM and Tamura features and applying Light GBM classifier, we found that the overall accuracy of the proposed system is 72%, with precision (0.71 and 0.67), recall (0.97 and 0.12) and, F1-score (0.82 and 0.21) respectively for benign and malignant tumours.

Table 84.1 Performance of classification with tamura feature

SN	*Type of ovarian cyst*	*No. of samples*	*Precision*	*Recall*	*F1-Score*
1	Benign	36	0.71	0.97	0.82
2	Malignant	16	0.67	0.12	0.21
3	Overall Accuracy		72%		

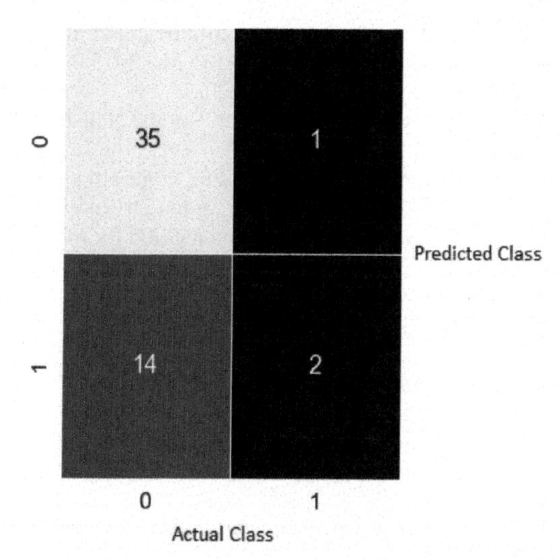

Figure 84.4 Confusion matrix

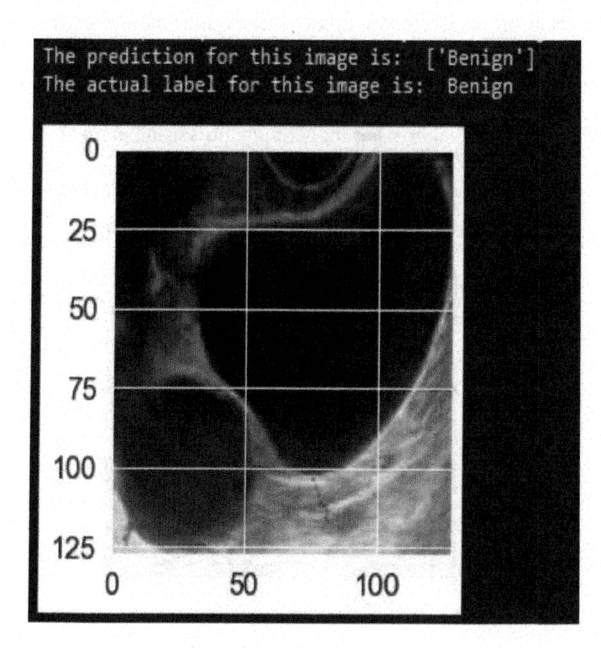

Figure 84.5 Result with prediction

Conclusion

In Medical image processing, classification and identification of ovarian masses are quite complicated due to complexity and resolution of image background. So features of such images play a vital role in automatic classification. In this proposed work, we have used GLCM and Tamura texture features for feature extraction without doing segmentations. The combination of GLCM and Tamura features with Light GBM classifier is used for image classification. Experimental results show that the proposed method gives 72% accuracy. The results can be improved much further by using different segmentation techniques before feature extraction is applied. Also in future we can use hybrid feature extraction techniques with different classifiers to improve accuracy of proposed method.

References

Bagri, N. and Johari, P. K. (2015). A comparative study on feature extraction using texture and shape for content based image retrieval. Int. J. Adv. Sci. Technol. 80:41–52. http://dx.doi.org/10.14257/ijast.2015.80.04.

Faust, O., Acharya, U. R., Meiburger, K. M., Molinari, F., Koh, J. E. W., Yeong, C. H., Kongmebhol, P. and Ng, K. H. (2018). Comparative assessment of texture features for the identification of cancer in ultrasound images: A review. Biocybern. Biomed. Eng. 38(2):275–296. ISSN 0208–5216. https://doi.org/10.1016/j.bbe.2018.01.001.

Humeau-Heurtier, A. (2019a). Texture feature extraction methods: A Survey. IEEE 7:2169–3536.

Humeau-Heurtier, A. (2019b). Texture feature extraction methods: A survey. IEEE Access 7:8975–9000. doi: 10.1109/ACCESS.2018.2890743.

Imran, B. (2019). Content-based image retrieval based on texture and color combinations using tamura texture features and gabor texture methods. Am. J. Neural Netw. Appl. 5(1):23–27. doi: 10.11648/j.ajnna.20190501.14.

Jeyavathana, R. B., Balasubramanian, R. and Pandian, A. (2017). An efficient feature extraction method for tuberculosis detection using chest radiographs. Int. J. Appl. Environ. Sci. 12(2):227–240. ISSN 0973–6077.

Karmakar, P., Teng, S. W., Zhang, D., Liu, Y. and Lu, G. (2017). Improved tamura features for image classification using kernel based descriptors. In 2017 International conference on digital image computing: Techniques and applications (DICTA) (pp. 1–7). doi: 10.1109/DICTA.2017.8227447.

Kiruthika, V., Sathiya, S. and Ramya, M. M. (2018). Automatic texture and intensity based ovarian classification. J. Med. Eng. Technol. 42:604–616.

Mall, P. K., Singh, P. K. and Yadav, D. (2019). GLCM based feature extraction and medical X-RAY image classification using machine learning techniques. In 2019 IEEE conference on information and communication technology, (pp. 1–6). doi: 10.1109/CICT48419.2019.9066263.

Martínez-Más, J. (2019). Evaluation of machine learning methods with fourier transform features for classifying ovarian tumors based on ultrasound images. [Online]. Available: osf.io/n9abq.

Mohanaiah, P., Sathyanarayana, P. and GuruKumar, L. (2013). Image texture feature extraction using GLCM approach. Int. J. Sci. Res. Publ. 3(5). ISSN 2250-3153.

Nilofer, N. S. and Ramkumar, R. (2021). Follicles classification to detect polycystic ovary syndrome using glcm and novel hybrid machine learning. Turk. J. Comput. Math. Edu. 12(7):1062–1073.

Preetha, K. and Jayanthi, S. K. (2018). GLCM and GLRLM based F eature E xtraction technique in mammogram images. Int. J. Eng. Technol. 7(2.21):266–270.

Rufo, D. D., Debelee, T. G., Ibenthal, A. and Negera, W. G. (2021). Diagnosis of diabetes mellitus using gradient boosting machine (LightGBM). Diagnostics 11(9):1714. doi: 10.3390/diagnostics11091714.

Tamura, H., Mori, S. and Yamawaki, T. (1978). Textural features corresponding to visual perception. IEEE Trans. Syst. Man Cybern. 8(6):460–473.

Wang, D., Zhang, Y., and Zhao, Y. (2017). LightGBM: An Effective miRNA classification method in breast cancer patients. ICCBB.

Xiaoming, S., Ning, Z., Haibin, W., Xiaoyang, Y., Xue, W. and Shuang, Y. (2018). Medical image retrieval approach by texture features fusion based on hausdorff distance. Math. Probl. Eng. Hindawi 2018. https://doi.org/10.1155/2018/7308328.